D0078887

SIXTH EDITION

MICROBIOLOGY IN PATIENT CARE

JOSEPHINE A. MORELLO
A.M., Ph.D.—Director of Hospital Laboratories
Co-Director, Clinical Microbiology Laboratories
Professor, Departments of Pathology and Medicine,
The University of Chicago, Chicago, Illinois

HELEN ECKEL MIZER
R.N., A.B., M.S., M. Ed.—Formerly Professor, Department of Nursing,
Western Connecticut State University, Danbury, Connecticut

MARION E. WILSON
M.A., Ph.D., D.L. (Hon.)—Formerly Director of Microbiology,
Public Health Laboratory Services, New York City Department of Health

PAUL A. GRANATO
Ph.D.—Associate Professor, Department of Microbiology and Immunology,
SUNY Health Science Center at Syracuse, Syracuse, New York

Boston, Massachusetts Burr Ridge, Illinois Dubuque, Iowa
Madison, Wisconsin New York, New York San Francisco, California St. Louis, Missouri

WCB/McGraw-Hill

A Division of The ***McGraw-Hill*** *Companies*

MORELLO, MIZER, WILSON, GRANATO: MICROBIOLOGY IN PATIENT CARE

This book is printed on recycled, acid free paper containing 10% postconsumer waste.

1 2 3 4 5 6 7 8 9 0 KGP/KGP 9 0 9 8

ISBN 0-697-25763-0

Publisher: Kevin Kane
Sponsoring editor: Ron Worthington
Developmental editor: Terry Stanton
Marketing manager: Tom Lyon
Project manager: Marilyn Rothenberger
Production supervisor: Deb Donner
Designer: Chris Reese
Cover designer: K. Wayne Harms
Photo research coordinator: John Leland
Art editor: Joyce Watters
Compositor: Graphic World, Inc.
Typeface: Bembo
Printer: Quebecor, Kingsport

Library of Congress Cataloging-in Publication Data

Microbiology in patient care.,—6th ed. / Josephine A. Morello
[et al.]
p. cm.
Includes bibliographical references and index.
ISBN 0-697-25763-0
1. Medical microbiology. 2. Allied health personnel.
I. Morello, Josephine A.
[DNLM]: 1. Microbiology—nurses' instruction. 2. Communicable Diseases—nurses' instruction. QW 4 M62605 1998]
QR46.W749 1998
616'.01—dc21
DNLM/DLC
for Library of Congress

http://www.mhhe.com

CONTENTS

PART ONE
BASIC PRINCIPLES OF MICROBIOLOGY 1

PART TWO
MICROBIAL DISEASES AND THEIR EPIDEMIOLOGY 233

PREFACE

In this, the sixth edition of *Microbiology in Patient Care,* the authors have again attempted to reflect the expanding knowledge in the fields of basic microbiology and infectious diseases. As before, the primary intention of this book is to serve those whose chosen profession requires an intelligent role in management and control of infections in patients. For students entering the allied health professions (e.g., nursing, infection control practitioners, operating room or cardiopulmonary technicians, respiratory therapists, physical therapists, dietitians, dental hygienists, optometric technicians, physicians' assistants, and others), it is important not only to grasp the fundamentals of the science of microbiology, but also to understand its clinical and epidemiologic applications. Thus, the text is designed to provide these students with a basic knowledge of the principles of microbiology and epidemiology without undue emphasis on technical details.

The organization of the material in this edition is similar to that of the previous edition. The three sections of part 1 cover the basic principles of medical microbiology and provide an introduction to epidemiology and infection control. The three sections of part 2 describe important human infectious diseases, organized according to the common route of entry of the microbial agent and the nature of the infecting microorganisms (bacteria, viruses, fungi, parasites). This latter part is designed to be of value during the student's classroom years, and afterward, as a ready reference source. In keeping with current educational trends, outlines, objectives, and key words (with a pronunciation guide) for each chapter have been included to focus and aid student learning. We have again used questions at the end of each chapter to assess the student's ability to apply classroom knowledge of microbiologic principles to patient care situations. A bibliography and recommendations for additional readings complete each chapter. A glossary is also included.

NEW MATERIAL AND FEATURES

Newly recognized aspects of infectious diseases caused by known and previously unknown microorganisms have been added to this revision. The chapter on immunology (chapter 6) incorporates the latest knowledge of this rapidly changing field. Chapter 11 has been completely rewritten to conform to the new Guideline for Isolation Precautions in Hospitals published by the Centers for Disease Control and Prevention. Sections on microorganisms whose pathogenesis was just beginning to be understood when the last edition was prepared have been expanded. New diagrams and illustrations have been added when needed for improved comprehension.

Also new to this edition are icons that correlate text to *Microbes in Motion,* the microbiology tutorial program on CD-ROM, by Gloria Delisle and Lewis Tomalty, from WCB/McGraw-Hill. Throughout the text and in selected figure legends, icons refer students to specific segments of the CD that amplify text discussion or that provide further exploration of a concept. In addition, chapter-opening pages feature a box that directs the student to portions of *Microbes in Motion* that can serve as a supplement to or study aid for that chapter. These tie-ins to the CD-ROM—which includes a number of narrated animations and videos, along with quizzes, a pronouncing glossary, and many other features—provide students with an interesting and helpful tool for learning microbiology, and offer instructors a way to expand the scope of their course. Look for this icon in the text: .

As in previous editions, most chapters of part 2 conclude with a table summarizing important features of the disease entities described therein: agent, entry route(s), incubation and communicable periods, laboratory diagnosis, prevention, therapy, management, and transmission-based precautions. In keeping with the other sections of this book, each of these reflects the latest knowledge or patient care recommendations.

J.A.M.
H.E.M.
M.E.W.
P.A.G.

ACKNOWLEDGEMENTS

The authors gratefully acknowledge those many colleagues who provided advice and guidance during the preparation of this edition. In particular, we appreciate the photographic contributions of Dr. Edward Bottone, Dr. Paul Beaver, and Mr. Gordon Bowie. This time-consuming task could not have been completed without the support of our families and friends whose patience and understanding were essential.

We also acknowledge the role of WCB/McGraw-Hill in encouraging us to prepare this new edition. Liz Sievers, *Editor,* and Terry Stanton, *Developmental Editor,* have provided expert advice and assistance throughout the preparation of this manuscript. Marilyn Rothenberger, *Project Manager,* carefully and with good humor guided the many processes that resulted in this final product. Her vigilant attention, together with that of the production team, K. Wayne Harms, *Designer,* John Leland, *Photo,* and Joyce Watters, *Art,* are responsible for the look and editorial precision of this edition.

We thank the following reviewers, who provided many helpful observations and comments about the manuscript:

LIST OF REVIEWERS

D'Maris Allen	Austin Community College
Ann Brown	CUNY Medgar Evers College
Mary Ann Cardani	University of Michigan-Flint
Helen Foster	Santa Fe Community College
Richard Goodman	University of Southern California
Robert Jochen	Blue Ridge Community College
Philip Loh	University of Hawaii-Manoa

PREVIOUS EDITION

Marlene McCall	Community College of Allegheny County
Leigh W. Callan	Floyd College
Barbara Stephens	Westmoreland County Community College
Janet Dettloff	Wayne County Community College
James O. Murray, SM (AAM)	Academy of Health Sciences Fort Sam Houston, TX
Ellen P. Digan	Manchester Community College
Robert J. Boettcher	Lane Community College
Chin C. Lee	King's College
Pamela B. Fouchè	Walters State Community College
Rudy Locklear	Robeson Community College
Charles F. Denny	University of South Carolina-Sumter
David F. Oetinger	Kentucky Wesleyan College
Barbara A. Prindiville	Northeast Wisconsin Technical College-Green Bay

SIXTH EDITION

MICROBIOLOGY IN PATIENT CARE

BASIC PRINCIPLES OF MICROBIOLOGY

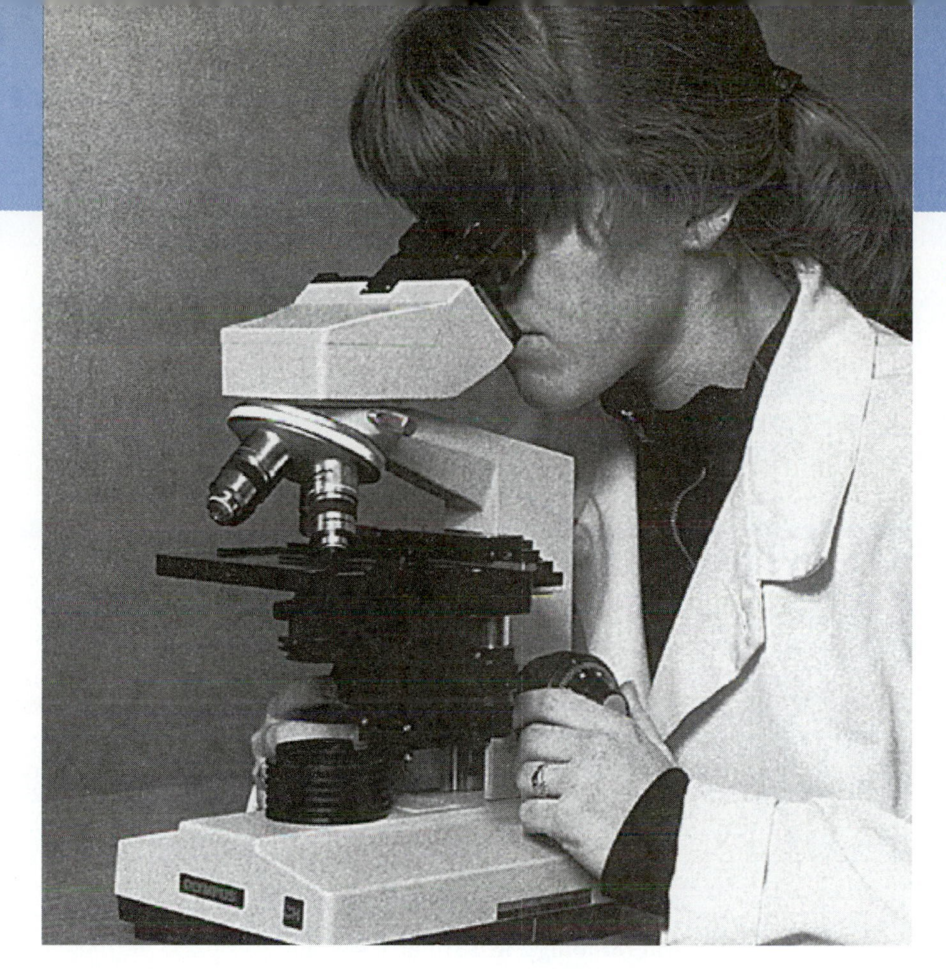

SECTION I

Character of Microorganisms

CHAPTER 1

The Evolution of Microbiology as a Modern Science

OUTLINE

OBJECTIVES

On completing this chapter the student will be able to:

1. discuss the discovery of microorganisms and their recognition as important agents of disease.
2. explain the need for developing adequate tools and methods for studying microorganisms.
3. describe the modern microbiologist's functions in the diagnosis and management of infectious diseases.
4. define the role of the allied health professional in the care and treatment of infectious diseases, and describe the importance of that role in the broader context of "microbiology in patient care."

KEY WORDS

Animalcules (**AN**-i-mal-kewles)
Bacteriology
Biogenesis (bio-**GEN**-eh-sis)
Contagion
Deoxyribonucleic (dee-oxy-**RIBE**-oh-nuke-**LEE**-ic) acid (DNA)
Germ theory of disease
Infection
Koch's postulates (**COKE'S PAHST**-u-lates)
Microbes
Microbial (my-**CRO**-bial) disease
Microbiologist
Microbiology
Microscope
Spontaneous generation

NOTE

In this and all future chapters, a guide to pronunciation of many key words is provided. **BOLD CAPITALIZED** letters indicate the primary emphasis, and underlined lowercase letters indicate secondary emphasis in words of several syllables. When a syllable is not emphasized, it appears in ordinary type.

Introduction

Microbiology is the study of those living organisms that, in most of their forms, are so small that they cannot be seen by the unaided human eye. The magnifying lenses of microscopes are required to make them visible. To understand the importance of this science and the vast "unseen world" it explores, we should know something about its beginnings and development and its emergence as a vital part of medical practice today.

Microbiology is a science that has grown tremendously during the twentieth century. The history of its development parallels, and is related to, the dramatic progress that occurred in the last half of the nineteenth century. The remarkable extent of this progress, in such a relatively brief time span, becomes even more obvious when compared with the developments of any similar time span.

Until the late 1800s, advances in medical knowledge and practice were slow and unexceptional in comparison with later discoveries. The primary reason for this lack of progress was the undiscovered relationship between microorganisms and infectious diseases. Although the **microscope** had been invented in the early part of the seventeenth century, another two hundred years passed before its potential value to medicine was realized. The pace of progress surged forward when the microscope was used to study diseased human tissues and cells, revealing that minute, living organisms, visible only under the microscope, were responsible for some of the most dreadful of human diseases. By the end of the nineteenth century, it was clear that afflictions such as diphtheria, tuberculosis, typhoid fever, cholera, gonorrhea, and many others that had plagued human populations throughout recorded history were caused by organisms that could at last be seen and studied with the proper tools and methods.

Suddenly a new understanding of the mechanisms of many such diseases was at hand, and a new science, later to be called **microbiology** (*micro* = small; *bios* = living; *-ology* = study), began to emerge.

This new science derived its name from its focus on the study of living organisms that are too small to be seen with the human eye. The newly discovered organisms, being so minute, were called **microbes,** or *microorganisms.* Actually, some had been seen through one of the earliest microscopes invented, long before their importance in disease was recognized. Since that time it has become increasingly apparent that the invisible microbial world is enormous in size and in its effect on our lives.

Microorganisms share all parts of the world we live in; they are involved with plants, animals, and human beings in the great biologic cycles of matter and energy that characterize the ongoing physical life of our planet. An immense number of microbial species are free-living in our natural environment, while many others live in continuous association with us on our body surfaces, or on animals and plants. These relationships lead to biologic interplays and interdependencies essential to our welfare, but they may also result in disease if we lose our physiologic balance.

Medical Microbiology

Among the vast numbers of species of microorganisms with which our lives are bound, there are a relative few whose properties and interactions with the human body may lead to the development of disease. These are of concern in the field of medical microbiology where great progress has been made in defining the nature and properties of harmful microorganisms and the disease processes they induce. Such studies have led to the understanding that **infection** usually represents a balanced relationship between microbes and other living things, whereas **microbial** (or *infectious*) **disease** results from a more damaging interplay between microorganisms with injurious properties and the infected body's responses to them. It has also become clear that the properties of a particular microbe that are responsible for harmful effects in one individual or species do not necessarily induce injury in another. This observation confirms that infection is a series of interactions between two living organisms, each exerting opposing forces on the other. The degree or lack of injury to the infected (or *host*) organism reflects its defensive mechanisms as well as the properties of its opponent, the microbe. These concepts provide an important basis for applying microbiologic principles of medicine to patient care.

The management and control of infectious diseases today require the concerted efforts of people trained in many disciplines: chemists, clinical **microbiologists,** epidemiologists, nurses, pathologists, physicians, pharmacists, respiratory therapists, and other health care professionals. Successful treatment of a microbial disease depends on its prompt, accurate diagnosis. This is the physician's area of skill and responsibility using evidence provided by laboratory specialists. The control and prevention of these diseases are also essential because they are so frequently communicable and can involve whole populations in epidemics. Here the epidemiologist's special knowledge of microorganisms and the manner in which they are transferred from person to person (or among animals, and from animals to humans) is applied. The capacity of microorganisms to survive in the natural environment, and their reactions to changing environmental conditions are also of great epidemiologic importance. The health care professional, as a member of this team of specialists, plays an important role in each aspect of the management of infectious disease—in diagnosis, treatment, control, and prevention of active infection. In a very real sense the health care professional's education and experience combine basic elements of each specialty involved in patient care. In particular, nurses must be

- Conversant with diagnostic signs of infectious disease and methods of treatment, particularly the patient's response to antimicrobial drugs.

- Able to participate in making the diagnosis by accurately reporting the patient's symptoms and by ensuring well-timed, careful collection of specimens ordered for laboratory studies to identify the cause of infection.
- Competent in applying appropriate precautionary measures in the bedside care of patients so that communicable diseases are not spread from one to another and new infections are not superimposed upon patients who are ill with other conditions.

Control of active infectious disease requires a working knowledge of the usefulness and applications of physical and chemical agents that suppress or kill microorganisms, as well as familiarity with sources of potentially dangerous microbes, routes by which they spread, and their portals of entry into the human body. *Prevention* of infectious disease depends on adequate control of these sources and transmission routes. Frequently, it also relies on specific measures designed to improve individual resistance to infectious microorganisms. Specific host resistance is referred to as *immunity* and plays a vital role in prevention.

Although health care responsibilities extend across all areas of infectious disease management, the health care professional must above all be an epidemiologist, with a solid knowledge of the nature of infectious diseases and available means for controlling or preventing them. In every field of patient care, whether in hospitals, public health facilities, physicians' offices, or industry, a knowledge of the basic principles of microbiology is essential. The need for such a background of preparation in the health care professional's curriculum is emphasized by the dramatic ease, speed, and frequency of human travel today, which brings distant diseases closer to home, opens up new pathways for communicability of infections, and creates new problems in their control.

HISTORICAL DEVELOPMENTS

In beginning the study of microbiology, we should take a brief look back at the context within which it evolved and the people involved in its development.

EARLY DISCOVERIES AND THE MICROSCOPE

The broad field of biology embraces all sciences that inquire into the nature of life—that is, into the structure, character, and behavior of living forms. These inquiries have been going on consistently since the beginning of recorded history. For centuries botanists, zoologists, chemists, and physiologists have sought to define life in its many forms at every observable level. When the search has been hindered because methods for accurate observation were not available, efforts were made to devise better tools or more refined techniques. In general, these efforts have enlarged our visual and intellectual range; as a result, increasingly minute structural features have been made visible and more subtle details of function have become better understood (figs. 1.1 to 1.8). So it was that the world of microscopic life, long imagined but never before seen, finally became visible late in the seventeenth century when a curious eye looked through an enlarging lens at a suspension of material that contained living microbes, swimming and moving about.

It was a simple beginning, of little apparent consequence at the time. The observer was Antony van Leeuwenhoek, a Dutch merchant, who had a hobbyist's interest in lens grinding and optics. Although he was an amateur, his skill with lenses was extraordinary. He developed an untiring interest in the microscopic world his lenses revealed to him as he examined drops of lake water, pepper infusion, scrapings of plaque from his teeth, fecal suspensions, and other fluids. His lenses could magnify up to about 300 diameters, an enlargement that enabled him to see innumerable "incredibly small" living creatures in these materials. With great objectivity, he recorded his observations of these little **"animalcules,"** as he called them, in a long series of letters written to the Royal Society of London over a period of fifty years. In a letter dated 1683, and another in 1692, he included drawings that clearly depicted forms identifiable as bacteria (see fig. 1.2 and Perspective 1.1).

Since Leeuwenhoek was not a scientist, and indeed had received relatively little schooling, he may have been unaware that the existence of microorganisms had been postulated, but not confirmed, by medical workers concerned with the causes of diseases spread by **contagion.** Nonetheless, the objective nature of his reports to the Royal Society left no doubt about the validity of his observations. His contribution was all the more remarkable in view of his personal simplicity. Certainly he could never have dreamed that he would one day come to be known as the founder of a new science of far-reaching influence—microbiology.

SPONTANEOUS GENERATION VERSUS BIOGENESIS

Microbiology had a tentative beginning but did not become a science in the disciplined sense for nearly two hundred years. Although Leeuwenhoek's animalcules were believable and their existence could be confirmed, they were not taken seriously at first. One reason was that the simple observation of microbial life was difficult to interpret without some means of cultivating microorganisms and studying their properties. Certainly their relationship to the diseases of humans, other animals, or plants was not immediately recognizable and received little attention.

Over the ages there were strong popular beliefs and superstitions as well as astute, educated theories about the cause of contagious diseases, but these ideas had to be subjected to scientific testing before real progress could be made. It has been recognized from ancient times that diseases can spread

FIGURE 1.1 A replica of Antony van Leeuwenhoek's early microscope. His lenses gave him magnifications up to about 300 diameters, revealing amazing microbial forms, which he called animalcules.

Courtesy Carl Zeiss, Inc. Thornwood, NY

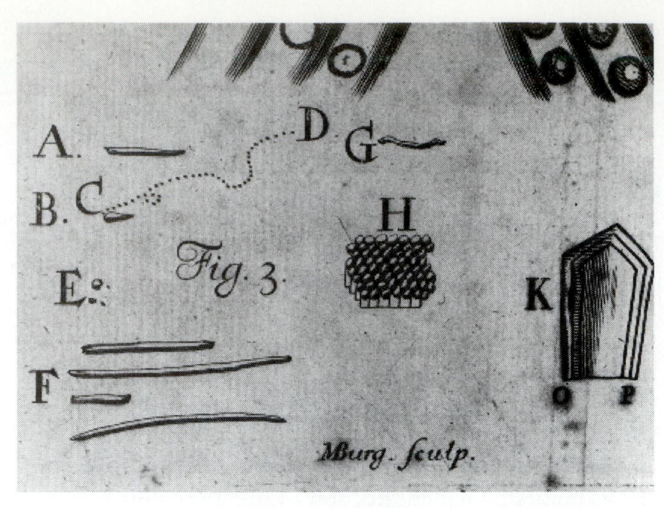

FIGURE 1.2 Some of Leeuwenhoek's "animalcules" as he illustrated them in letters written to the Royal Society in London. These organisms were seen in scrapings from his teeth. **A** through **G** are clearly recognizable as bacterial forms: **A** is a rod-shaped bacterium, or bacillus (pl., bacilli). **B** is a shorter, motile rod, whose path is indicated by the dotted line between **C** and **D. E** depicts small, spherical bacteria, or cocci (sing., coccus). **F** represents some long, filamentous bacilli, and **G** appears to be a spirochete with loose flexible coils of a type commonly found in the healthy mouth. Compare the size and shape of the coccus in **E** (enlarged by Leeuwenhoek for the drawing) with the photomicrographs in figure 1.4 and colorplate 1b. [See Perspective 1.1 figure.]

Courtesy Royal Society, London.

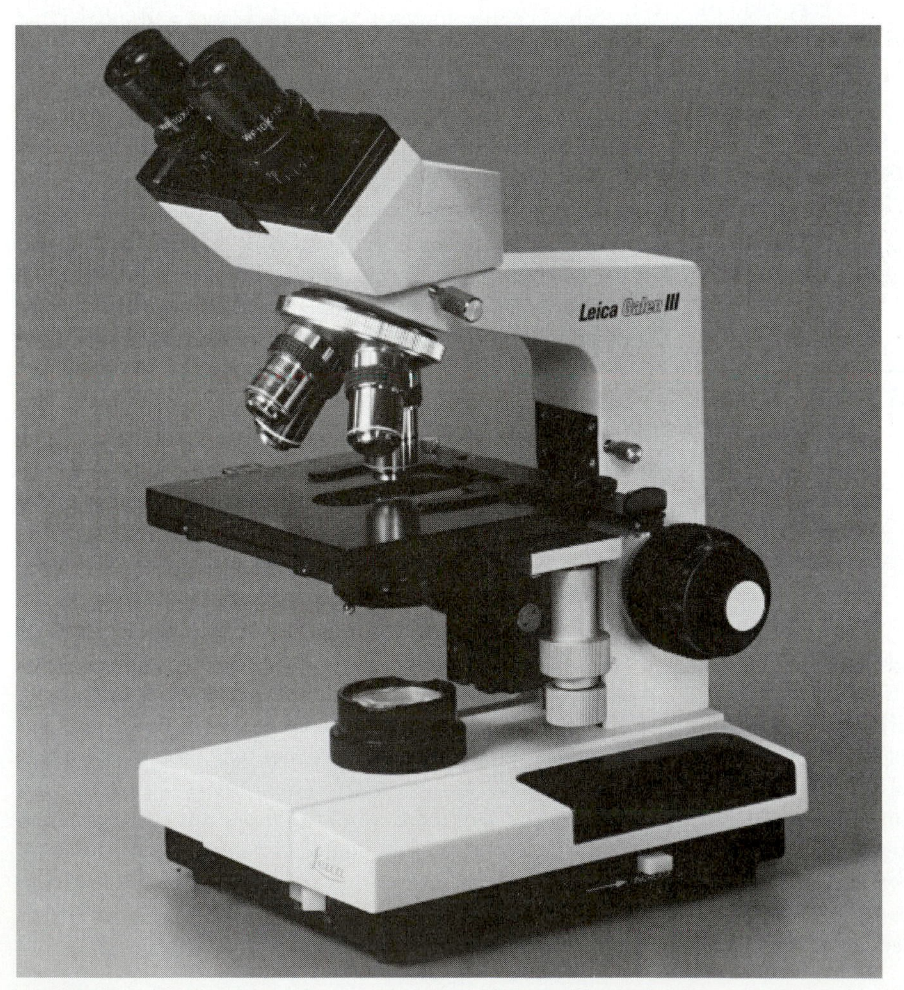

FIGURE 1.3 A binocular compound microscope. This is a precision instrument used by today's medical microbiologist. With the oil immersion objective lens and a 10× ocular, microorganisms can be magnified approximately ×1,000.

Courtesy Leica Inc., Buffalo, NY.

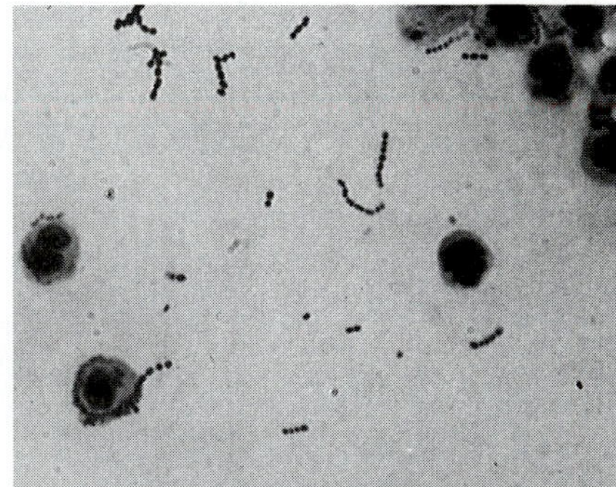

FIGURE 1.4 Spherical-shaped bacteria (cocci) arranged in chains viewed through the compound microscope shown in figure 1.3 at a total lens magnification of ×1,000. Although their shape is distinctive (compare with Leeuwenhoek's drawing, fig. 1.2 E), their size at this magnification is minute in contrast with that seen by electron microscopy (figs. 1.6 and 1.8). The large spherical cells are white blood cells.

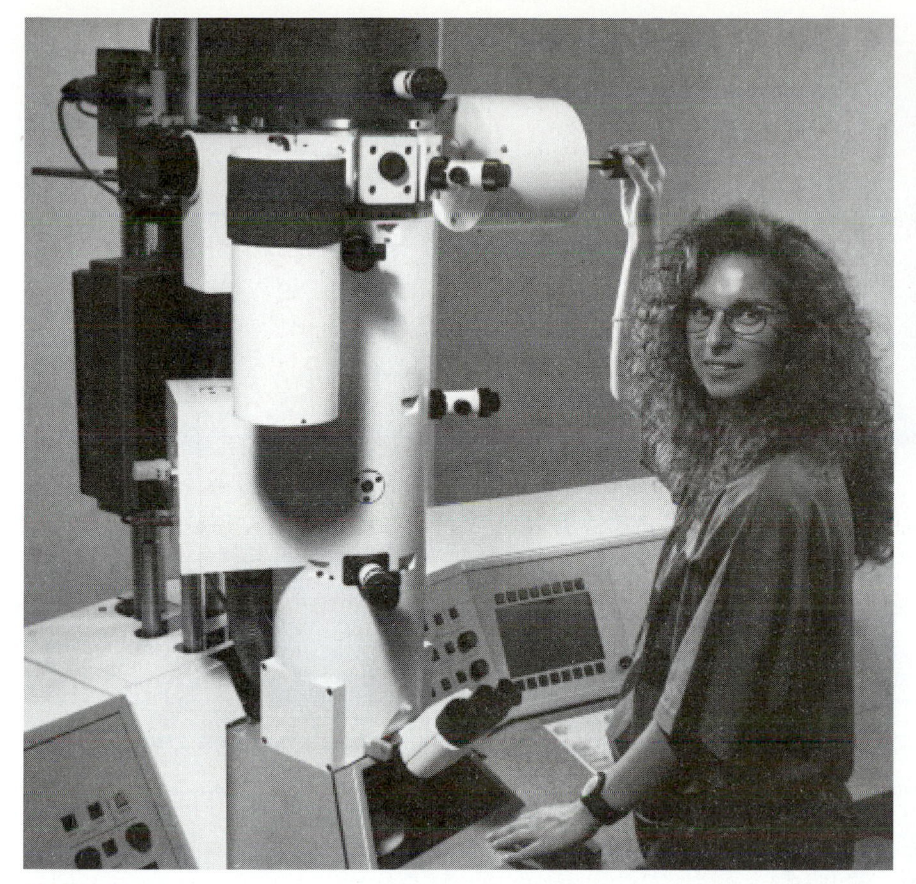

FIGURE 1.5 The electron microscope provides magnifications from ×10,000 to ×200,000 or more depending on the type of equipment used. In this instrument a stream of electrons, rather than light rays, is directed at the object to be examined.

Courtesy Carl Zeiss, Inc. Thornwood, NY.

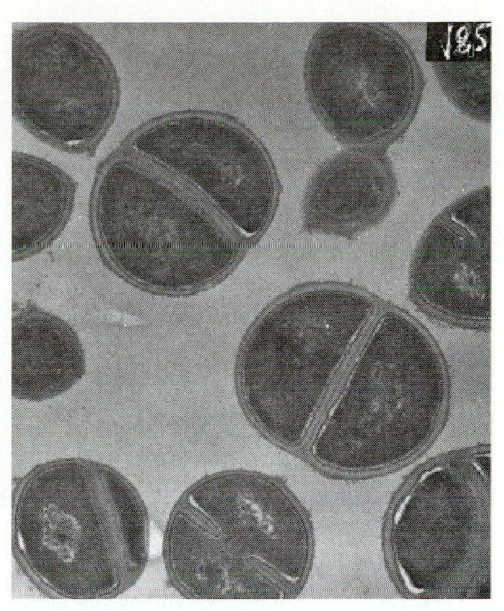

FIGURE 1.6 An electron micrograph of cocci in which cross walls are clearly visible at this magnification (×32,000).

Reproduced from V. Lorian, and B. Atkinson, "Some Effects of Subinhibitory Concentrations of Penicillin on the Structure and Division of Staphylococci," *Antimicrobial Agents and Chemotherapy.* 7:864, 1975. American Society for Microbiology.

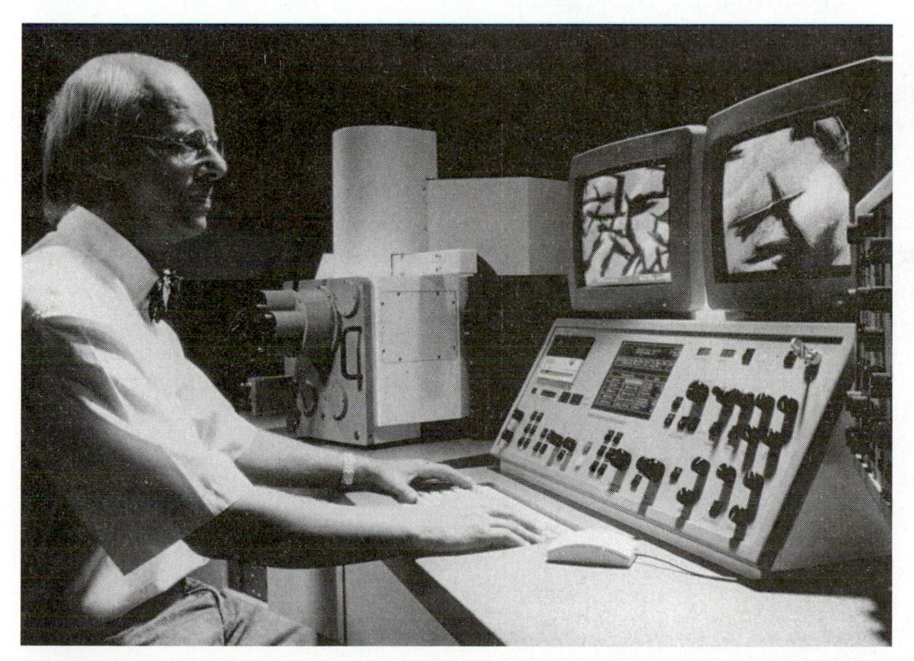

FIGURE 1.7 A scanning electron microscope. A magnification range from ×15 to ×10,000 can be obtained with this instrument. Further enlargements are achieved photographically.

Courtesy Carl Zeiss, Inc. Thornwood, NY.

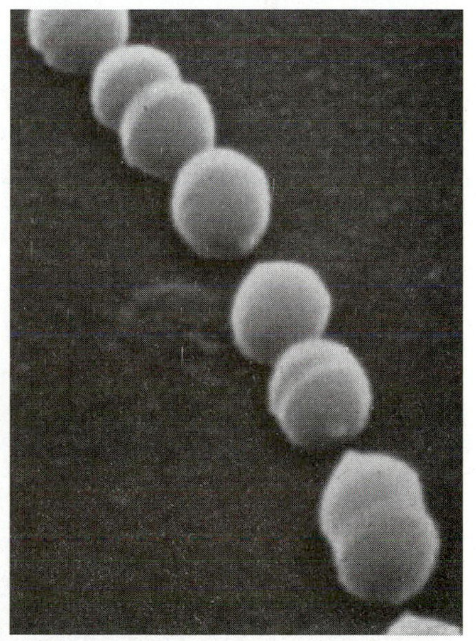

FIGURE 1.8 Spherical bacteria (cocci) as seen with the scanning electron microscope. Although the final magnification in the photograph (×6,500) is about one-fifth of that of figure 1.6, note the three-dimensional quality here. The scanning electron microscope does not permit visualization of the internal structures of microorganisms, but it does provide a new perspective on surface structures, spatial relationships, and dimensions.

William G. Barnes, Ph.D., John D. Arnold, M.D., and Arthur Berger, Department of Pathology, General Hospital and Medical Center and Harry S. Truman Research Laboratory, Kansas City, MO.

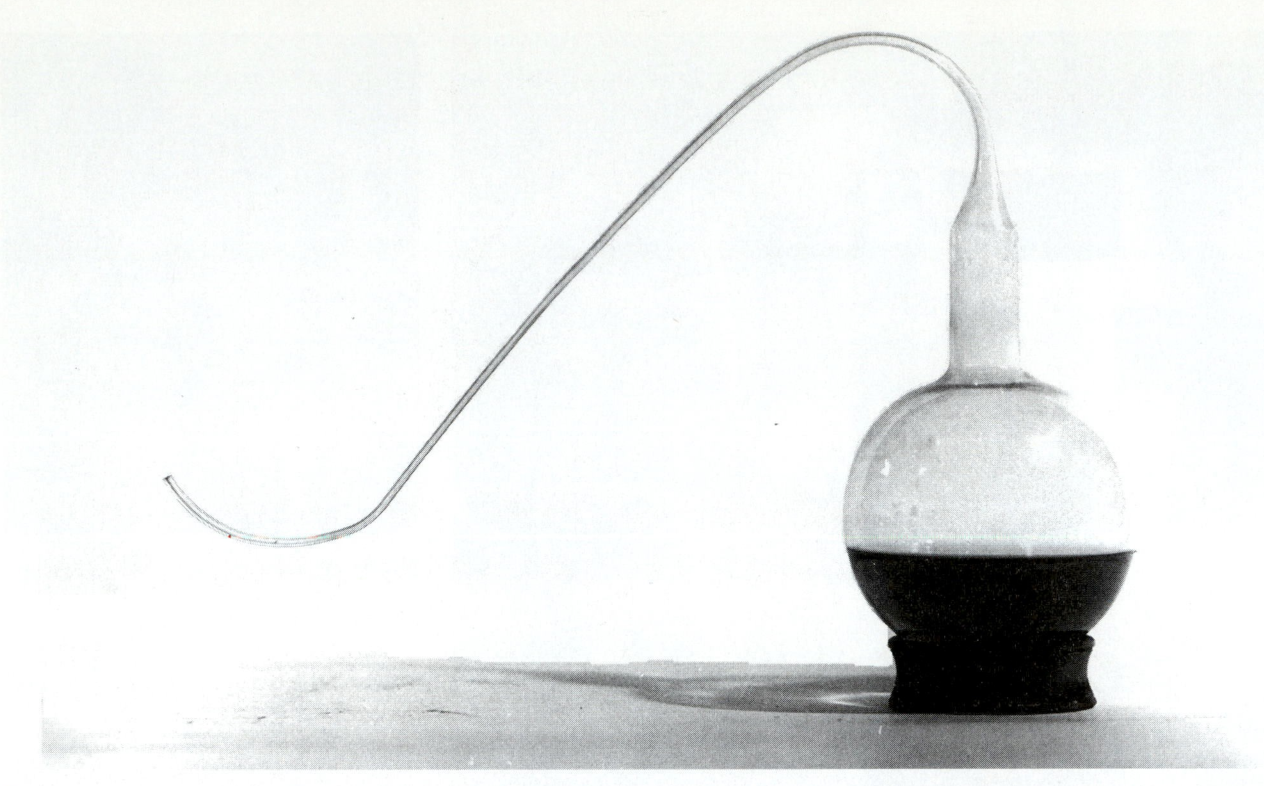

FIGURE 1.9 Louis Pasteur's swan neck flask, which he used in his experiments to disprove spontaneous generation theories. From Conserveteur du Musée Pasteur, Paris France. © Institut Pasteur.
Courtesy Royal Society, London.

from person to person, and consequently many forms of control were devised. These controls were more or less successful according to whether they were based on pure superstition or clear observation. Efforts to explain the nature of such diseases and the dreadful epidemics they so often caused led to some discerning theories in a number of instances. Both Girolamus Fracastorius in the sixteenth century and Athanasius Kircher in the seventeenth wrote with perception and foresight about living but invisible agents that might cause contagions. At the time, however, neither these nor conflicting ideas could be proved.

Through the centuries the question of the origin of life itself was interwoven with the problem of contagion. From the time of the early Greek philosophers the theory that life could be generated from nonliving matter had enjoyed strong support. By 1700, scientific experiments had dispelled belief that animals of visible size arose by **spontaneous generation.** But when microorganisms were discovered, this theory was still applied to them. Thus, until the middle of the nineteenth century, Leeuwenhoek's little animalcules remained in the center of a controversy concerning their origin and were often the subjects of poorly designed experiments that were claimed to prove spontaneous generation. For example, flasks of vegetable broth that clouded with the growth of microbes after several days of exposure to air were thought to provide such proof. Nothing was known about the presence of bacteria in the air or their adherence to dust particles settling from air into open flasks containing nutrient material. Furthermore, nothing was known about the ability of some bacteria to form heat-resistant endospores. As a result, experiments in which bacterial growth could be demonstrated even in flasks of heated broth were similarly open to misinterpretation.

Despite the work of several investigators whose well-controlled experiments led to results that refuted the theory of spontaneous generation, the controversy continued until 1860. In that year, Louis Pasteur, a French biochemist and physicist, conducted a series of classic experiments demonstrating that no growth occurs in flasks of broth protected from contamination by airborne microorganisms. In one type of experiment, he introduced beef bouillon into a long-necked flask and then drew the neck of the flask out into a long, sinuous shape (fig. 1.9). When he boiled the liquid in the flask, vaporization forced the air out of the opening at the end of the long neck. As the liquid cooled, air was pulled back into the flask but had to pass over condensed moisture in the neck that washed it free of dust and other particles, so the bouillon remained clear and free of growth. Control flasks were also heated, but immediately afterward the long necks were broken off. As a result, air pulled into the fluid during cooling was not first washed but delivered dust particles containing viable microorganisms directly into the nutrient fluid, which soon became cloudy with microbial growth.

In another series of experiments, Pasteur used an idea of earlier scientists and introduced cotton plugs into the necks of his flasks. The cotton acted as a filter, effectively

PERSPECTIVE 1.1 *Antony van Leeuwenhoek: The Father of Microbiology*

Antony van Leeuwenhoek (1632–1723) was a Dutch linen-draper, haberdasher, and town official in the city of Delft who in his spare time ground and mounted lenses to make a simple microscope (see fig. 1.1 and Perspective fig. 1.1). This hobby led him to discover an amazing unseen world of living microorganisms (which he called "little animalcules"). He made meticulous notes and drawings of his observations and then wrote letters to the Royal Society of London describing his findings. These letters, written in old-fashioned Dutch, spanned more than fifty years and provide extraordinarily accurate descriptions of the many different microbial forms that Leeuwenhoek was the first to see.

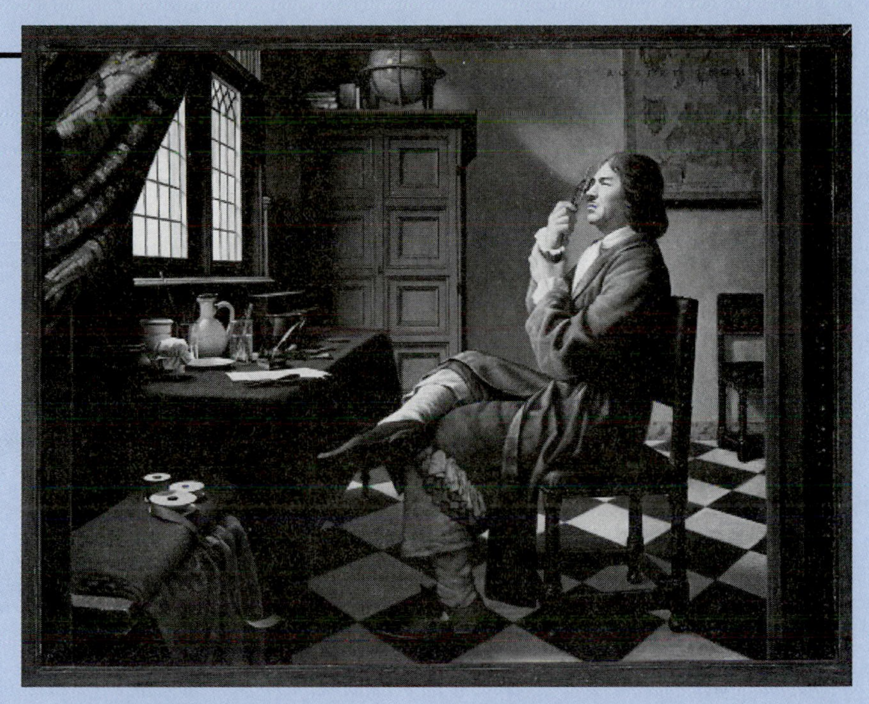

Antony van Leeuwenhoek at work with the microscope he constructed.
Painting by Robert A. Thom, © 1959, Parke-Davis Division of Warner Lambert Co.

The following excerpt from one such letter demonstrates his keen powers of observation. The drawings he sent with it are reproduced in figure 1.2.

Dated September 17, 1683, this particular correspondence contains some "Microscopical Observations about Animals in the Scurf of the Teeth." It was printed in the Philosophical Transactions of the Royal Society of London, Vol. 14, May 20, 1684, No. 159, pages 568–574 (1 plate).

... Tho my teeth are kept usually very clean, nevertheless when I view them in a Magnifying Glass, I find growing between them a little white matter as thick as wetted flower: in this substance tho I do not perceive any motion, I judged there might probably be living Creatures.

I therefore took some of this flower and mixt it either with pure rain water wherein were no Animals; or else with some of my Spittle (having no Air bubbles to cause a motion in it) and then to my great surprize perceived that the aforesaid matter contained very many small living Animals, which moved themselves very extravagantly. the biggest sort had the shape of A. [authors' note: see the drawing in Fig. 1.2] their motion was strong and nimble, and they darted themselves thro the water or spittle, as a Jack or Pike does thro the water. These were generally not many in number. The 2d. sort had the shape of B. these spun about like a Top, and took a course sometimes on one side, as is shown at G. and D. they were more in number than the first. In the 3d. sort I could not well distinguish the Figure, for sometimes it seemed to be an Oval, and other times a Circle. These were so small that they seem'd no bigger than E. and therewithal so swift, that I can compare them to nothing better than a swarm of Flies or Gnats, flying and turning among one another in a small space. Of this sort I believe there might be many thousands in a quantity of water no bigger than a sand, tho the flower were but the 9th part of the water or spittle containing it. Besides these Animals there were a great quantity of streaks or threds of different lengths, but like thickness, lying confusedly together, some bent, and others streight, as at F. These had no motion of life in them, for I well observed them, having formerly seen live Animals in water of the same figure.

Such remarkable descriptions from an observer who had no training or background in science were preserved by the Royal Society, which elected van Leeuwenhoek a Fellow in 1680. Although neither he nor anyone else at the time understood the significance of his animalcules, Antony van Leeuwenhoek is honored today as the father of microbiology because he was the first to provide enduring and indisputable evidence that there is an "unseen world."

Ref: Brock, T., *Milestones in microbiology. American Society for Microbiology, Washington, D.C., 1975.*

straining out dust particles and keeping the nutrient in the flask free of microorganisms from the air. (Cotton plugs are still used occasionally in test tubes or flasks, but screw caps and slip-on metal or plastic closures serve the same purpose and are easier to use.)

The results of Pasteur's experiments led to the conclusion that life at every level must be self-reproducing, rather than spontaneous. This theory of **biogenesis** (i.e., life arises from preexisting life) began to take precedence, and the way was clear for the development and application of scientific

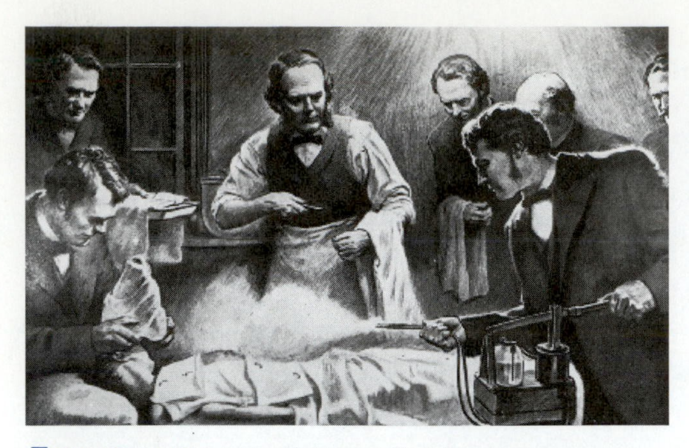

FIGURE 1.10 Lord Joseph Lister, an English surgeon, was so impressed with Louis Pasteur's germ theory of disease that he revolutionized surgical practice by spraying phenolic acid to combat microorganisms in the air and so reduced infections in surgical patients.
Bettman Archives

methods to study the nature of microscopic organisms and the mechanisms of their self-replication.

FIGURE 1.11 Dr. Robert Koch (1843–1910) in his laboratory. At about the same time as Pasteur, and without prior knowledge of Pasteur's work on the germ theory of disease, Koch began to discover organisms responsible for various diseases. Using the principles of Jacob Henle (1840), Koch developed the techniques for applying these principles (which have since become known as Koch's postulates).
Bettman Archives

THE GOLDEN AGE OF BACTERIOLOGY

Pasteur's greatest contributions evolved from his studies showing that certain microorganisms cause spoilage of wines and that still others produce disease in silkworms, both of which were matters of great economic importance to France at that time. The experimental proof that the agents were of microbial origin led Pasteur to use a similar approach in his study of anthrax in sheep and of certain human infections. The successful identification of particular microorganisms as the cause of individual diseases of plants, animals, and human beings then led him to reexamine and reformulate the old **germ theory of disease,** which attributed the cause of infection to microbial agents. He provided sound scientific evidence on this point for the first time, and this proof soon came to the attention of the great English surgeon Joseph Lister, who applied the theory in his surgical practice (fig. 1.10).

Lister speculated that if Pasteur was right about the ubiquitous presence of microorganisms in the air, and if these could fall into a flask of nutrient that supported their growth, then such organisms could also enter open surgical wounds and cause the bloodstream and wound infections that so often led to death after surgery. With this idea in mind, Lister began to use a phenolic solution in the operating room, using it on his hands and instruments and spraying it into the air around his patients. He also applied it to the dressings used for surgical or traumatic wounds. As a result, his surgical patients had a very low fatality rate compared to those of other surgeons who used no precautions to prevent complications because they did not believe them to be "infections."

At about the same time but without knowledge of Pasteur's work, a German physician named Robert Koch began to discover the organisms responsible for various human diseases (fig. 1.11). With his students and associates, Koch developed many of the techniques for cultivation, staining, and animal experimentation essential for the study of microorganisms and the interpretation of their behavior as agents of disease.

A number of years before (in 1840), Jacob Henle, another German scientist, had again outlined the germ theory of disease, stating that its proof would be provided by a series of necessary tests. These tests included (1) recognition of the infectious microbe in every case of human disease but not in healthy persons, (2) laboratory isolation of the organism from the patient and its pure cultivation, free of other microbes, (3) demonstration that it could produce disease when inoculated into healthy laboratory animals, and (4) recovery of the organism in pure laboratory culture from infected animals that had displayed typical disease following their inoculation. Henle had no technical means to prove the validity of these "postulates." It remained for Koch to develop the techniques (including the pure culture technique) for applying these principles (which have since become known as **Koch's postulates**). Using such methods, he and his associates provided, in rapid succession, proof of the microbial origin and nature of such major human diseases as tuberculosis, diphtheria, typhoid fever, cholera, and gonorrhea.

Thus the field of microbiology achieved scientific status. However, the greatest excitement was generated in the area of medical microbiology where advances were so obviously important to human welfare. The contributions of Pasteur, Lister, and Koch established the truth of the germ theory and provided the basis for today's understanding of infectious diseases. The laboratories of Pasteur in France and Koch in Germany attracted students from around the world, who came to learn the methods and ideas of **bacteriology.** In their turn, these students made further contributions that caused the end of the nineteenth century to be called the "golden age of bacteriology."

MICROBIOLOGY TODAY

In the three centuries that have elapsed since Leeuwenhoek's first glimpse of bacteria and protozoa, microbiologic investigations have resulted in the accumulation of a vast body of knowledge. The discovery in the early 1940s of **deoxyribonucleic acid** or **DNA** (see chaps. 2 and 3) and its role in the heredity of bacteria ushered in the era of molecular biology. By studying how bacterial DNA regulates cell functions in all living organisms, scientists have made dramatic advances in agriculture, environmental studies, and medicine. Henle and Koch would have been astounded to know that the microbial nature of certain diseases of previously unknown cause has been discovered by finding specific microbial DNA in the tissues of afflicted persons rather than by growing the microorganism in pure culture or producing the disease in animals. The knowledge gained has contributed to our understanding of the nature of all life and lightened immeasurably the burden of human disease.

The microbial world, as we recognize it today, contains a tremendous variety of organisms that exhibit amazing diversity in size, structure, and activity. Microorganisms of one type or another are found wherever life is possible—in soil, water, and air, and in association with plant, animal, or human life. Many of their vital functions of growth and reproduction, respiration, biosynthesis, and excretion, as well as their responses to environmental stimuli, have the same basic character as those of more structurally complex forms of life. Indeed, the study of microorganisms has served to explain many fundamental biologic mechanisms operative in higher organisms as well. Biologists working in the fields of genetics, biochemistry, and biotechnology have found microorganisms to be useful and profitable tools in their studies of the nature of life processes. Furthermore, it is increasingly obvious that microbial forms of life make an essential contribution to the maintenance of balanced relationships among all living things.

In the field of medical microbiology, emphasis is placed not only on the relatively small group of microbes that adversely affect human health but also on the interplay of factors that affect the ability of humans to maintain a balance in the world they share with so many microbial types. Students in health care professions require more than just familiarity with the microorganisms of medical importance. The fundamental principles of microbiology and of host resistance to infection, as they are currently understood, apply to every phase of patient care. The crisis of infection and the threat of epidemic have been minimized for the modern world by the development of specific measures for their treatment, control, and prevention. But problems continue in many countries where factors such as poverty, poor dissemination of knowledge, and civil unrest prevent the application of these measures. In addition, new, previously unrecognized, or reemergent diseases caused by microorganisms including the human immunodeficiency virus of AIDS, Ebola virus, and hantavirus remind us that we cannot lower our vigilance against the threat of infectious diseases.

It should also be understood that new problems of infection have resulted from many modern medical and surgical treatments. The widespread use of antimicrobial agents has not eliminated infectious microorganisms but indeed has often led to the emergence of new strains that are resistant to these drugs. Prolonging the life of patients with debilitating organic diseases has created a population abnormally susceptible to infection, for whom the risk of infection is life threatening. The development of sophisticated surgical techniques for organ transplantation and for treating cancers and diseases of many other organs demands rigid protection of patients from the infections that may be acquired during medical therapy, surgery, or the difficult period of postoperative recovery.

For the health care professional, an understanding of the means by which patients develop infection forms an important basis for the total approach to every patient's care.

MICROBIOLOGY IN PATIENT CARE

Primary effort is made in this text to provide health care professionals with the most practical and applicable information regarding microbial diseases, including bacterial, rickettsial, viral, and fungal infections and diseases caused by animal parasites.

With regard to the care of the patient ill with an infectious disease the emphasis is twofold. First, a knowledge of the suspected or diagnosed disease process itself, and of its agent, is essential so that its course from onset to outcome can be anticipated to the patient's best advantage. Second, the communicability of the infection is of vital concern. Health care professionals must understand the common sources of infection, as well as pertinent routes of entry to, and transfer from, the body. In this way they can help limit each case of active infection and prevent its spread to other patients in a hospital, to other members of a household, or to those occupied in patient care, whether or not they are in direct contact.

The infectious diseases are presented in part 2 of the text, in an order based primarily on (1) currently available information as to the most probable route of entry of the microbial agent, (2) routes of exit and transmission, or (3) the nature of the organisms involved—that is, fungi, bacteria, viruses, protozoa, or helminths. If students learn to associate the infectious agent of a given disease with its routes of communicability and entry, their techniques as graduates will be oriented toward the medical needs of the patient and the simultaneous protection of others (including themselves) from the possibility of infection transfer.

Before considering individual diseases, however, the student must learn some basic facts about microbial behavior, the fundamental relationships between microbe and host in health or disease, and the general principles of prevention and control as they apply to most infectious diseases. The microbial world is outlined broadly in the following chapter, and the succeeding chapters of part 1 describe the basic principles of microbiology.

Questions

1. Briefly describe Antony van Leeuwenhoek's contribution to microbiology.
2. Describe Louis Pasteur's experiments that ended the "spontaneous generation" controversies.
3. How did Pasteur's work on the "germ theory of disease" influence Dr. Joseph Lister's surgical practice?
4. How did Robert Koch and his associates prove the microbial origins of human diseases such as tuberculosis, diphtheria, cholera, and gonorrhea?
5. How has molecular biology replaced Koch's postulates in the identification of new diseases?
6. How have modern medicine and technology contributed to new microbiological problems for the patient?
7. What information is essential for the health care professional involved in patient care?

Bibliography

Boatman, E. S., M. W. Berns, R. J. Walter, and J. S. Foster. 1987. Today's microscopy. *BioScience.* 37 (6): 384–394.

Brock, T. D. ed. 1975. *Milestones in microbiology.* Washington, D.C.: American Society for Microbiology.

Brock, T. D. 1988. *Robert Koch: A life in medicine and bacteriology.* Madison, Wis.: Scientific Tech Publishers.

Bulloch, W. 1979. *The history of microbiology.* New York: Dover Publications.

Clendening, L., ed. 1960. *Source book of medical history.* New York: Dover Publications.

Dobell, C. 1960. *Antony van Leeuwenhoek and his "little animals."* New York: Dover Publications.

Dubos, R. 1986. *Louis Pasteur: Free lance of science.* Jersey City: DaCapo.

Geison, G. L. 1995. *The private life of Louis Pasteur,* Princeton, N.J.: Princeton University Press.

Krasner, R. I. 1995. Pasteur: High priest of microbiology. *ASM News* 61(11):575–579.

McNeill, W. H. 1974. *Plagues and peoples.* New York: Anchor Press.

Nightingale, F. 1969. *Notes on nursing: What it is, and what it is not.* New York: Dover Publications.

CHAPTER 2

The Microbial World Defined

OUTLINE

Microbes in Motion

This chapter covers the general nature and structure of microorganisms and describes the important differences between bacteria, fungi, protozoa, and animal parasites. The following books and chapters of the Microbes in Motion CD-ROM may be a useful preview or supplemental study aid to your reading: Bacterial Structure and Function: Internal Structures; Cell Membrane; Cell Wall; Bacterial Groups; Gram-Positive Cells; Gram-Negative Cells; External Structures. Microbial Metabolism and Growth: Genetics. Viral Structure and Function: Viral Structure; Viral Replication; Viral Assembly. Parasitic Structure and Function: Protozoan Classification; Cestodes; Nematodes; Trematodes. Fungal Structure and Function: General Eucaryotic Structures; Specific Fungal Structures; Metabolism and Growth; Taxonomy and Nomenclature. Miscellaneous Bacteria in Human Hosts: Chlamydia. Control of Microorganisms: Physical Control; Chemical Control. Antimicrobial Action: Approach to Antimicrobial Study; Cell Wall Inhibitors; Agents Affecting Nucleic Acids. Mechanisms of Antimicrobial Resistance: Transfer and Spread of Resistance.

OBJECTIVES

On completing this chapter the student will be able to:

1. describe the general nature and structure of microorganisms.
2. distinguish procaryotic from eucaryotic cells.
3. describe the methods used to classify living organisms.
4. outline important differences between bacteria, viruses, fungi, protozoa, and animal parasites.
5. discuss the importance of each group of microorganisms in human disease.

KEY WORDS

Ameba (ah-**ME**-bah)
Anticodon
Apicomplexa (ape-ee-com-**PLEX**-ah)
Archaebacteria (ar-**KAY**-back-**TEAR**-e-ah)
Ascomycota (**ASK**-oh-my-**COAT**-ah)
Ascospore (**ASK**-oh-spore)
Ascus (**ASK**-us)
Autotrophic (auto-**TROPH**-ic)
Bacillus (bah-**SILL**-us)
Bacteriophage (back-**TEAR**-ee-oh-**FAHGE**)
Basidiomycota (bah-**SID**-e-oh my-**COAT**-ah)
Basidiospore (bah-**SID**-e-oh-spore)
Basidium (bah-**SID**-e-um)
Binary (**BY**-nary) fission
Blastospore
Capsule
Cell membrane
Cell wall
Cestode (**SESS**-toad)
Chlamydiae (clam-**ID**-ee-ee)
Chlorophyll (**CHLORO**-fill)
Chloroplast (**CHLORO**-plast)
Chromosome
Ciliophora (sill-i-**OFF**-or-ah)
Classification
Coccus (**COCK**-us)
Codon
Conidia (co-**NID**-e-ah)
Conidiophore (co-**NID**-e-oh-for)
Cytoplasm (**SITE**-oh-plaz-em)
Deoxyribonucleic (dee-oxy-**RIBE**-oh-nuke-**LEE**-ic) acid (DNA)
Deuteromycota (**DUE**-toe-roe-my-**COAT**-ah)
Double helix
Ectoplasm (eck-toe-**PLAZ**-em)
Elementary body
Encystment
Endoplasm (**END**-oh-plaz-em)
Endoplasmic reticulum (end-oh-**PLAZ**-mick ree-**TICK**-you-lum)
Endospore
Endotoxin
Eubacteria (**YOU**-back-**TEAR**-ee-ah)
Eucaryotic (**YOU**-carry-ought-ic) cell
Flagella (fla-**GELL**-ah)
Flagellate (fla-gell-**ATE**)
Flatworm
Fluke
Fungi (**FUN**-jie)
Fungi Imperfecti (im-per-**FECT**-eye)
Gene
Genotype (**JEAN**-oh-type)
Golgi (**GOAL**-gee) apparatus
Helminth
Hemoflagellate
Heterotrophic (het-er-oh-**TROPH**-ic)
Lipopolysaccharide (lip-oh-poly-**SACK**-are-ide)
Lysosome (**LYE**-so-sohm)
Messenger RNA (mRNA)
Mitochondria (my-toe-**CON**-dria)
Monera (mo-**NERE**-ah)
Morphology (more-**FOLL**-oh-gee)
Mycoplasmas (my-co-**PLAZ**-mas)
Nemathelminth (**NEEM**-at-hell-minth)
Nomenclature (**NO**-men-**CLAY**-ture)
Nucleic (nuke-**LEE**-ic) acid
Nucleolus (nuke-**LEE**-oh-luss)
Nucleotide (**NUKE**-lee-oh-tide)
Nucleus
Obligate intracellular parasite
Organelle
Peptidoglycan (pep-tid-oh-**GLY**-can)
Phenotype (**FEE**-no-type)
Photosynthesis (photo-**SIN**-the-sis)
Phylum (**FILE**-um)
Pili (**PILL**-eye)
Plasma membrane
Plasmid (**PLAZ**-mid)
Platyhelminth (**PLAT**-ee-hell-minth)
Polymerase (pol-**IM**-er-ase)
Prion (**PREE**-on)
Procaryotic (**PRO**-carry-ought-ic) cell
Proglottid (pro-**GLOTT**-id)
Protista (pro-**TEES**-ta)
Protoplasm (**PRO**-toe-plaz-em)
Protozoa (pro-toe-**ZO**-ah)
Pseudopod (**SUE**-doh-pod)
Reticulate (re-**TICK**-you-late) body
Rhizopoda (rise-oh-**PO**-dah)
Ribonucleic (**RIBE**-oh-nuke-**LEE**-ic) acid (RNA)
Ribosome (**RIBE**-oh-sohm)
Rickettsiae (rick-**ETT**-si-ee)
Roundworm
Spirilla (spy-**RILL**-ah)
Spirochete (spy-roe-**KEET**)
Sporangiospore (spore-**ANG**-e-oh-spore)
Sporangium (spore-**ANG**-e-um)
Sporulate
Strobila (**STROBE**-ill-ah)
Tapeworm
Taxonomy (tax-**ON**-oh-me)
Teichoic (tie-**CO**-ic) acid
Transcription
Transfer RNA (tRNA)
Translation
Trematode (**TREE**-ma-toad)
Trophozoite (troph-oh-**ZO**-ite)
Virion
Viroid
Virus
Zoomastigina (zo-oh-mast-i-**JEAN**-ah)
Zygomycota (**ZY**-go-my-**COAT**-ah)

Introduction: A Brief Survey of the Microbial World

The microorganisms that cause human disease are diverse in biologic nature, size, and complexity of structure. In fact, some of them do not meet the definition of a true microorganism as an entity. When we are dealing with infectious disease, we are concerned not only with the recognition of the organism involved, but also with the nature of the illness and the principles of treatment, prevention, and control. These basic principles are similar for all infectious diseases so any causative organism, whatever its nature, falls within the province of the medical microbiologist and of all others responsible for patient care.

Among the organisms that cause human disease, the largest are the parasitic worms. These invertebrate animals are visible to the naked eye and thus are not strictly microorganisms although their developmental forms (eggs and larvae) are microscopic in size. The smallest of the group—the viruses—are simpler in structure than other microorganisms and cannot replicate outside of living cells. Between these two extremes are the true microorganisms—bacteria, protozoa, and fungi—distinguished from all other organisms by the comparative simplicity of their structural organization. Their similarities and dissimilarities to other living things serve as the basis for their taxonomical grouping.

To begin the study of this varied group, we will consider briefly the nature of all living things and then see how they can be categorized.

The Nature and Kinds of Living Things

Basic Structures

The higher forms of plants and animals are complex organisms composed of many cells that make up differentiated tissues and organs specialized for particular functions. The microorganisms, which will be discussed in detail later in this chapter, are composed of single cells or are simple multicellular organisms. In plants, animals, and microbes the fundamental unit of structure and function is the individual cell. All cells have basic features and activities in common: they are bounded by a cell membrane, and they are composed of a colloidal mixture of proteins, lipids, and **nucleic acids** called **protoplasm** (from the Latin words *proto,* meaning "first," and *plasm,* meaning "molded"). Protoplasm may also be referred to as **endoplasm** (*endo* = within) or **cytoplasm** (*cyto* = cell). In addition, cells contain either an organized **nucleus** held within a nuclear membrane, referred to as eucaryotic (*eu* = true, *caryo* refers to the nucleus, being derived from the Greek word *karyon,* for "nut" or "kernel"), or functional nuclear material lying free, that is, a procaryotic (*pro* = before) nucleus. Other features that distinguish **procaryotic** and **eucaryotic cells** are detailed in table 2.1. These will be described further as we expand our discussion of microorganisms.

Organization of Eucaryotic and Procaryotic Cells

Figure 2.1 is a diagram of the eucaryotic cellular organization typical of fungi, protozoa, and algae, and also of animals and plants. The major feature distinguishing the more complex eucaryotic cell from a procaryotic cell is that eucaryotic cells contain membrane-bound structures, called **organelles,** to perform specialized functions. These organelles include the nucleus, endoplasmic reticulum, Golgi apparatus, mitochondria, lysosomes, and chloroplasts. The nucleus contains **deoxyribonucleic acid** or **DNA**, the component in which the cell's genetic information is encoded. Eucaryotic DNA is organized into multiple **chromosomes** complexed with basic proteins called histones. The nucleus contains at least one **nucleolus,** which is involved in the formation of **ribosomes,** structures that are subsequently used in protein synthesis.

The remaining organelles are found in the cell cytoplasm. The **endoplasmic reticulum,** which extends throughout the cytoplasm, is divided into two types, smooth and rough. The rough endoplasmic reticulum is covered with ribosomes, the sites of protein synthesis. After synthesis, proteins migrate through the membrane-bound channels of the endoplasmic reticulum to the **Golgi apparatus** where they are packaged into vesicles. Some of these vesicles are then transported to the **plasma membrane,** where they release their contents to the outside of the cell. Other vesicles called **lysosomes** contain a wide variety of hydrolytic enzymes and are involved in the degradation of large molecules such as proteins, lipids, and nucleic acids inside the cell. **Mitochondria** are the primary site of energy production within the cell. Mitochondria contain their own DNA and synthetic machinery and are capable of self-replication. **Chloroplasts** also contain DNA and are found in most plants and algae. These organelles contain **chlorophyll** which is important in the process of **photosynthesis.**

All cells are surrounded by a cell or plasma membrane, which is used to transport nutrients into the cell and waste materials out of the cell. The eucaryotic plants and algae also have a cell wall composed of cellulose or chitin, which gives the cell rigidity.

Figure 2.2 shows the organization of a procaryotic cell. As can be seen from this diagram, procaryotic cells are much less complex and, as such, much less specialized than eucaryotic cells. Despite this primitive structure, procaryotic cells can perform all the functions necessary for life. The various structures of a procaryotic cell will be discussed in detail later in this chapter.

Principles of Classification

Classification, or **taxonomy,** is a science in its own right. Its purpose is to unify and summarize existing knowledge of the nature of living forms and to provide biologists with a common language through which this knowledge can be communicated precisely. Classification recognizes both the differences and the similarities of organisms, indicating

TABLE 2.1 A COMPARISON OF PROCARYOTIC AND EUCARYOTIC CELLS

	Procaryotes	Eucaryotes
Nuclear Structures and Events		
Nuclear membrane	Absent	Present
Chromosome structure	Single molecules of double-stranded DNA (i.e., haploid)	One or more pairs of double stranded DNA (i.e., diploid)
Nuclear division	No mitosis	Mitosis
Cytoplasmic Structures and Events		
Ribosome size	70S*	80S
Mitochondria, chloroplasts, Golgi apparatus, endoplasmic reticulum	Always absent	Usually present (chloroplasts in green plants only)
Flagella	Submicroscopic when present	Microscopic when present
Cell wall	Usually present and chemically complex	Chemically simple when present (in fungi and plants)

* S = Svedberg unit, a measurement of mass as determined by sedimentation velocity in an ultracentrifuge.

Fungal Structure and Function
General Eucaryotic Structures: Cell Wall; Cell Membrane; Internal Structures.

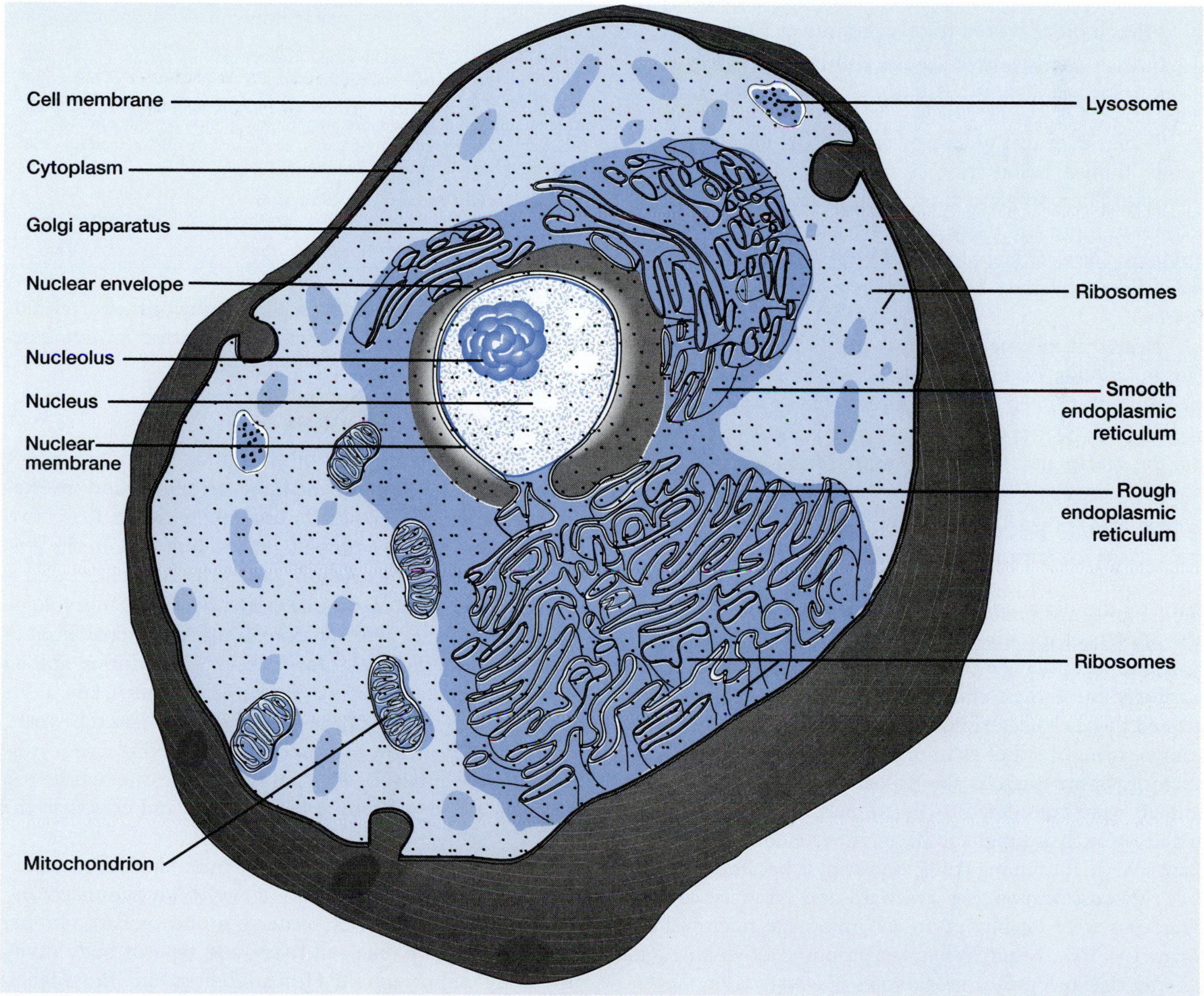

FIGURE 2.1 Diagram of a eucaryotic cell. Note membranous boundaries of the nucleus and other internal structures. Compare the form and structures of this cell with that shown in figure 2.2.

Fungal Structure and Function
General Eucaryotic Structures: Cell Membrane; Internal Structures

possible origins or developmental pathways. It not only describes structure (anatomy) but relates it to function (metabolic, reproductive, or other processes).

Nomenclature, or the system for naming organisms appropriately, depends on the classifying scheme employed. The system in common use today was developed in the eighteenth century by a Swedish botanist named Carl von Linné (translated as Carolus Linnaeus in the common Latin language of nomenclature). The Linnaean system has been greatly modified and enlarged but remains the basis for modern classification. It begins by dividing the living world into kingdoms (the original two were *Plantae* and *Animalia*). Further subdivision classifies related organisms into smaller, more specific, and mutually exclusive groups on the basis of fixed criteria. These groups are listed below:

- The primary division of all plant, animal, and microbial life is a *Kingdom*.
- In each kingdom related classes constitute a *Phylum*.
- In each phylum related orders constitute a *Class*.
- In each class related families constitute an *Order*.
- In each order related tribes constitute a *Family*.
- In each family related species constitute a *Genus*.
- In each genus similar organisms constitute a *species*.

To illustrate this system of naming organisms, we can classify human beings first as members of the kingdom *Animalia*. Then, by describing their structures and characteristics in more and more detail, we can place them in the **phylum** *Chordata* (subphylum *Vertebrata*), class *Mammalia*, order *Primates*, family *Hominidae*, genus *Homo*, and species *sapiens*.

Note that the scientific name finally assigned to an organism consists of the two words designated genus and species, both printed in italics such as *Homo sapiens* for humans. The first letter of the genus name is always capitalized, and the species name is always written in lowercase letters.

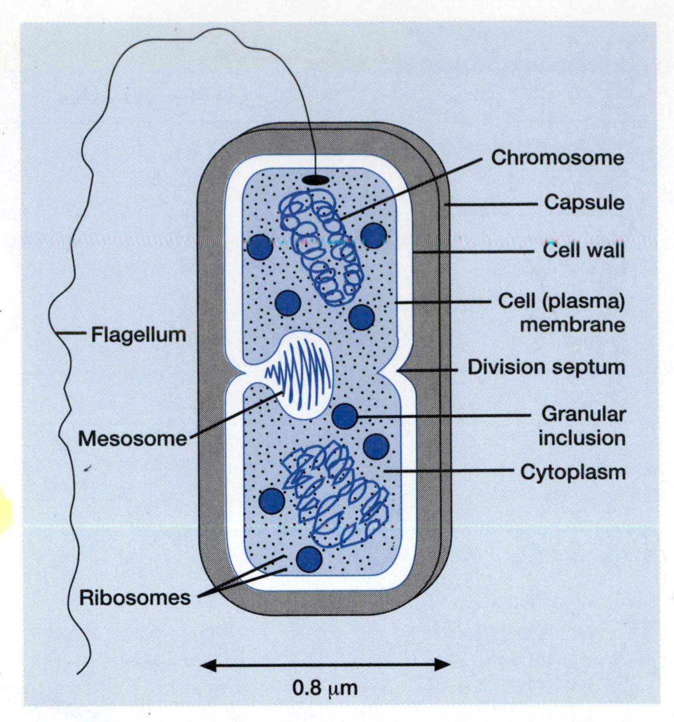

FIGURE 2.2 The generalized structure of a bacterium (procaryotic cell). Note the lack of a nuclear membrane and the absence of mitochondria, Golgi apparatus, and endoplasmic reticulum. The bacterial DNA (chromosome) is believed to be a highly folded circular structure. The mesosome is an irregular invagination of the cell membrane and may play a role in cell respiration and division.

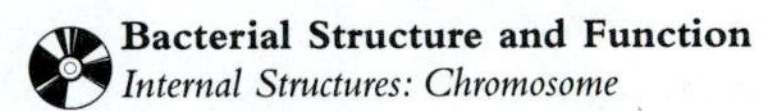
Bacterial Structure and Function
Internal Structures: Chromosome

THE FIVE KINGDOMS OF LIVING THINGS

Prior to the discovery of microorganisms, almost nothing was known of the evolutionary origins of plants and animals, and intermediate forms were not yet recognized. Accordingly, the earliest efforts to categorize microorganisms stressed those characteristics that related them to the lowest known forms of either plant or animal life. The large, motile microorganisms such as the protozoa (*proto* = first; *zoa* = animal) were considered to be animals, and those that did not move, such as fungi and bacteria, were considered to be plants. With continuing study, however, it became apparent that this classification was awkward and arbitrary, so new kingdoms were established to accommodate microbial life forms. In 1969, Robert Whittaker proposed a five-kingdom scheme that recognizes procaryotes (*Monera*) as the ancestors of all living eucaryotes. The four eucaryotic kingdoms are distinguished by their modes of nutrition, developmental patterns, and morphological and phylogenetic relatedness. Table 2.2 shows the current classification for living organisms.

The Kingdom *Animalia*

The first phyletic divisions of the animal kingdom separate it into two large groups containing vertebrate and invertebrate animals. The earliest distinction between these two types appears during embryologic development. In the embryonic vertebrate, a structure called the notochord develops, from which the backbone is later derived. Embryologic evidence, together with anatomic and functional distinctions, provides the basis for further subdivisions among higher vertebrates. External anatomic features taken alone may be misleading, as in the case of the fish and the whale, for instance. Both are anatomically equipped to live in water, but have little else in common, since the whale is a mammal and therefore of a higher order and class than the fish.

Some of the factors that determine animal classifications are cellular differentiation (if any); body symmetry and axis orientation (radial or bilateral, round or flat); number and type of embryonic cell layers; and type of body cavity and body segmentation. Human beings are the highest members of the animal kingdom, in terms of both structure and function. Although the human body develops from a

TABLE 2.2 CLASSIFICATION OF LIVING ORGANISMS INTO FIVE KINGDOMS

Kingdom	Major Characteristics	Important Members
Animalia (Animals)	Eucaryotic; heterotrophic; multicellular; tissue differentiation and specialization; vertebrate or invertebrate	Humans and other mammals, birds, reptiles, insects, mollusks, helminths
Plantae (Plants)	Eucaryotic; autotrophic, photosynthetic; multicellular; tissue differentiation and specialization	Vascular plants, conifers, ferns, worts, mosses
Fungi	Eucaryotic; heterotrophic, nonphotosynthetic; unicellular or multicellular; latter have differentiated structures, e.g., mycelium, hyphae, conidia (spores)	Molds, yeasts, mushrooms, plant smuts and rusts
Protista	Eucaryotic; heterotrophic or photoautotrophic; unicellular or multicellular; motile forms have specialized structures such as flagella or cilia	Protozoa; slime molds; red, brown, and green algae, including seaweeds
*Monera**	Procaryotic; heterotrophic or photoautotrophic; unicellular, some filamentous or mycelial; some have specialized structures such as flagella, pili, endospores	Cyanobacteria (blue-green bacteria) and eu bacteria (true bacteria)

* Sometimes referred to as *Prokaryotae.*

single cell formed by the union of two single cells, the cellular subdivisions that subsequently occur result in a highly complex, multicellular animal. As we pass down through the animal kingdom to the invertebrates, the animals become structurally more simple, yet their differentiated organs remain capable of very complex function.

Among the lowest invertebrates there are a number of animal parasites (the worms or **helminths**) of medical importance. These are described in the last section of this chapter.

The Kingdom *Plantae*

The vast world of plants is divided into several phyla based, as in the animal kingdom, on anatomic structures and developmental and evolutionary relationships. All plants are **autotrophic** ("self-nourishing"), containing chlorophyll that allows them to synthesize essential nutrients from simple inorganic substances and light energy. The two major groups of plants are the bryophytes (worts and mosses) and the tracheophytes (vascular plants, including trees). Plants are of great economic importance to human beings, but do not play a role in infectious disease production.

The Kingdom *Fungi*

Fungi formerly were classified as primitive members of the plant kingdom, having no roots, stems, or leaves. They differ significantly from members of the plant kingdom by not containing chlorophyll. Therefore, they are dependent on complex organic foods derived from other organisms, dead or alive (they are called **heterotrophic,** meaning that their food comes from others).

The four subdivisions of fungi, including those of medical importance, are described later in this chapter.

The Kingdom *Protista*

The kingdom **Protista** (also called *Protoctista*) consists of two eucaryotic groups: unicellular organisms such as algae, which are autotrophic, chlorophyll-containing protists; and the protozoa, which are heterotrophic, single-celled forms that formerly were classified as animals. Seaweeds are a kind of algae, as are the familiar green furry scums seen so often on the surface of still ponds. Most protozoans are free-living and contribute significantly to the breakdown of organic materials in the soil and water. As will be discussed later, a few of them are responsible for producing important human diseases.

The Kingdom *Monera*

The kingdom ***Monera*** includes the **eubacteria** ("true bacteria") and the *cyanobacteria,* previously known as the blue-green algae (or blue-green bacteria) because they contain chlorophyll. These microorganisms have simple nutritional and reproductive processes and a procaryotic nuclear structure that lies free in the cytoplasm.

Among the most studied of human infectious diseases are those caused by eubacteria. The text later reviews these agents and classifies them in more detail, along with the medically significant protozoa, and multicelled parasites and fungi. In part 2 the discussion of each infectious disease includes a description of the causative organism.

THE PROCARYOTES AND VIRUSES

We have seen that eubacteria are procaryotic organisms. Classified with them are the rickettsiae and chlamydiae, two groups of extremely small bacteria, which (unlike other

TABLE 2.3 THE SMALLEST METRIC UNITS

Unit	Symbol	Metric Equivalent	Inch Equivalent
Meter	m		39.37 in.
Centimeter	cm	1/100 m	1/2.5 in.
Millimeter	mm	1/10 cm	1/25.4 in.
Micrometer	μm	1/1000 mm	1/25,400 in.
Nanometer	nm	1/1000 μm	1/25,400,000 in.
Angstrom	Å	1/10 nm	1/254,000,000 in.

bacteria) can live only within the cells of the animals they infect. The viruses also lead a strictly intracellular existence, but they are classified separately because they are structurally very different from procaryotes.

Included with the procaryotes are a group of bacteria known as the **archaebacteria** (*archaios* = ancient). Because they grow in unusual, adverse environmental conditions, for example at extremely high temperatures such as in hot springs; under oxygen-free conditions where they produce methane gas; or in environments with high salt concentrations (e.g., Great Salt Lake), they are thought to have evolved early in the formation of earth. An enzyme derived from one species of archaebacteria, however, has an important role in modern diagnosis of disease. This enzyme is a heat-resistant polymerase that is an essential component of the polymerase chain reaction, as described in chapter 4.

BACTERIAL MORPHOLOGY

The recognition and classification of the many thousands of bacterial species begins with a study of their structure and form, or ***morphology.*** Their size and shape, the patterns they form as they divide and multiply, and their cellular structures provide the first obvious clues to their identification.

Size

The measurements of bacteria and other microorganisms usually are expressed in metric units. Table 2.3 lists these units and relates them to the meter and to the inch (see also fig. 2.14).

Bacteria vary widely in size measuring approximately 0.5 to 1.5 μm (500 to 1,500 nm) in diameter and from 1 to 6 μm (1,000 to 6,000 nm) in length. The exceptions are the rickettsiae and chlamydiae, which are even smaller. The optical tools and stains required to examine them are discussed in chapter 4, together with laboratory methods for cultivating them.

Shape

Each bacterium exhibits one of three basic shapes (colorplate 1a–f). (1) Individual cells may be spherical. These round cells are called **cocci** (sing., **coccus,** a Greek word for "berry") (colorplate 1a–c). (2) Individual cells may be shaped like small sticks. These are called **bacilli** (sing., **bacillus,** a Latin word for "staff" or "rod"), or rods, to avoid confusion with members of the genus *Bacillus* (colorplate 1d). (3) Individual cells may be spiraled in shape, and they are called **spirilla** (sing., **spirillum,** from the Latin word for "coil"). These forms have one to three fixed curves in their rigid bodies.

The **spirochetes** ("animated hairs") are spiraled bacteria that are much longer and more tightly coiled than spirilla. Their bodies are very flexible and they propel themselves by rapid rotation around their long axes as well as by undulatory motions (colorplate 1f). Such complicated movements suggest that these organisms possess more advanced cellular mechanisms than other bacteria, yet they have a procaryotic nuclear structure.

Group Patterns

The most frequent method of reproduction among bacteria is asexual binary fission in which each individual cell divides in half, forming two new cells. As the cells accumulate, distinctive groupings may form that are useful for characterizing bacteria.

Coccal Patterns

Among the cocci, fission may occur at different planes within the round cell. Cocci that divide along one plane only tend to arrange themselves in pairs (*diplo*cocci) (colorplate 1a) or in chains of varying length (*strepto*cocci) (colorplate 1b). When the division occurs alternately along each of two planes, groups of four (tetrads) or cubelike packets of eight are the characteristic result. Haphazard splitting on several planes produces an irregular cluster of cocci, bunched together like grapes (*staphylo*cocci) (colorplate 1c).

Bacillary Patterns

The patterns formed by bacilli are more limited because they divide only across their short axes. They may appear as end-to-end pairs (diplobacilli), or they may line up in chains (streptobacilli). The two new rods formed by the split of a parent cell sometimes bend away from the point of fission and form V, X, or Y patterns. Very short, small rods can resemble cocci (the term *coccobacillary* is often used for this), but they do not form packeted or clustered groups (colorplate 1d).

Spirilla

Like bacilli, the spirilla split on their short axes. When they remain close to each other after division, they may align in short chains (colorplate 1e) or in long, sinuous formations. The spirochetes always appear as individual cells (colorplate 1f), although they may clump and tangle together.

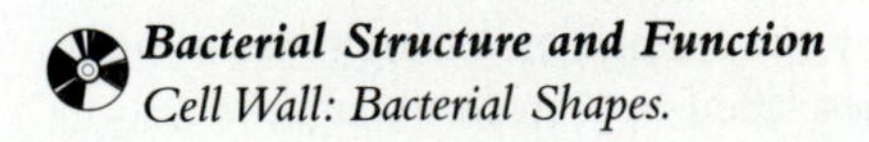
Bacterial Structure and Function
Cell Wall: Bacterial Shapes.

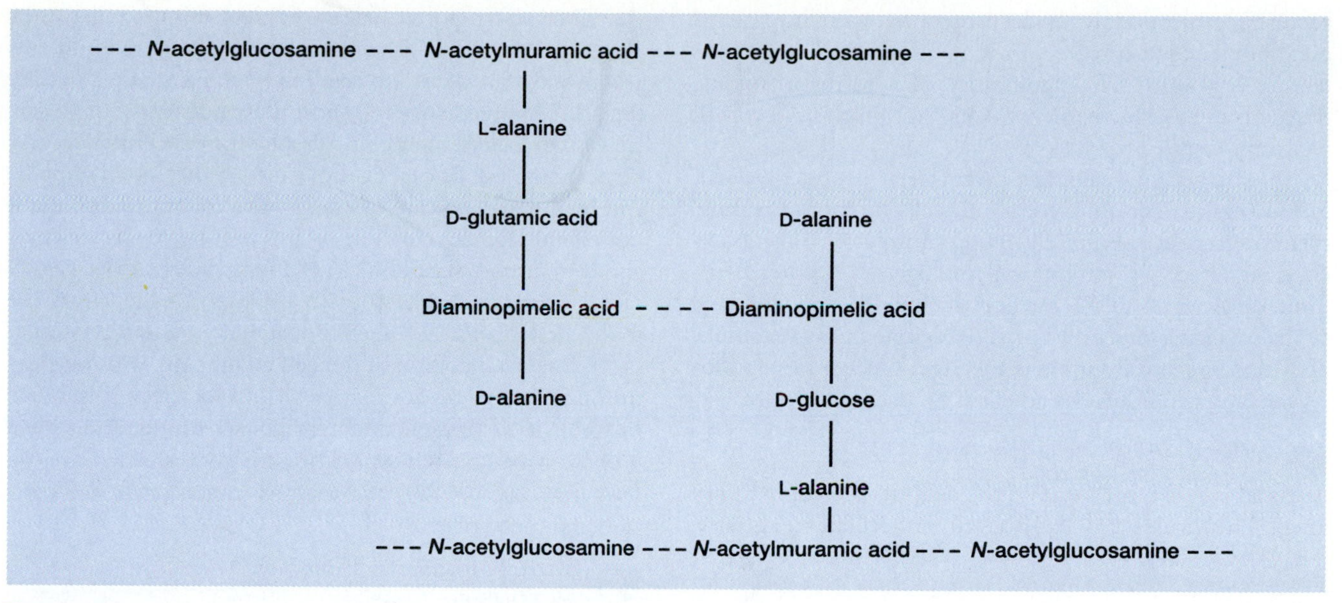

FIGURE 2.3 The procaryotic peptidoglycan structure. Note repeating units of the two sugars, *N*-acetylglucosamine and *N*-acetylmuramic acid linked by small groups of amino acids. Other amino acids are found in some bacteria.

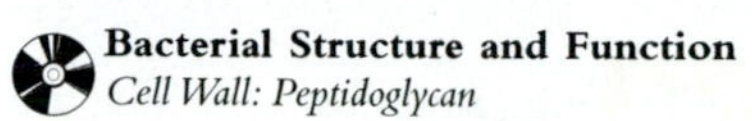
Bacterial Structure and Function
Cell Wall: Peptidoglycan

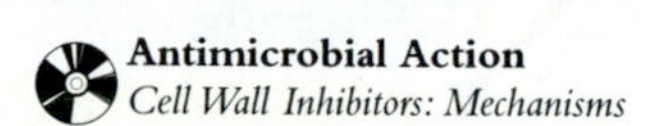
Antimicrobial Action
Cell Wall Inhibitors: Mechanisms

Cell Wall

The bacterial **cell wall** is a chemically complex, multilayered, rigid structure whose major function is to prevent the cell from rupturing because of high osmotic pressure within the cell. The length and cross-linked structure of cell-wall layers give bacteria their typical shape as cocci, bacilli, or spirilla.

Most bacteria can be placed into one of two groups, gram-positive or gram-negative, based upon chemical and physical differences in their cell walls. The differential staining of bacteria by the Gram method (see chap. 4) demonstrates the cell-wall differences.

The inner layer of the cell wall is known as **peptidoglycan**. The gram-positive cell wall contains far more peptidoglycan than does the gram-negative cell wall, but lacks the additional lipid layers found in all gram-negative cells. Gram-positive cell walls may also contain **teichoic acids,** which are composed of repeating units of sugar alcohols (either glycerol or ribitol) that are not found in gram-negative cell walls. These acids may be important in regulating cell-wall synthesis in gram-positive bacteria. The peptidoglycan layer is found in almost all procaryotes. This structure (fig. 2.3) consists of repeating units of two sugars, *N*-acetylmuramic acid (MurNAc) and *N*-acetylglucosamine (GlcNAc). The compound MurNAc is unique to the procaryotic world. In addition to the two sugars, peptidoglycan contains small groups of amino acids (L-alanine, D-glutamic acid, diaminopimelic acid, and D-alanine are shown) that cross-link or attach one peptidoglycan layer with each subsequent layer.

In gram-negative bacteria, the outer layer of the cell wall is **lipopolysaccharide (LPS),** also referred to as **endotoxin** (see chap. 5). When LPS is broken down into its two major components, lipid A and polysaccharide, many of the toxic effects noted in infections caused by gram-negative bacteria are specifically associated with lipid A. The LPS layer also acts as a barrier to prevent passive diffusion of certain substances through the cell wall. Gram-negative bacteria are intrinsically more resistant to antimicrobial agents and less susceptible to ionic detergents than are gram-positive bacteria because of the LPS permeability barrier.

Although most procaryotes are either gram-positive or gram-negative, a few bacteria have a chemically unique cell wall and do not stain with the Gram method. A notable human pathogen in this category is the tubercle bacillus, *Mycobacterium tuberculosis.* Mycobacteria have lipid-rich cell walls that resist penetration of the Gram stain dyes. Instead, these procaryotes are stained using an acid-fast procedure (see chap. 4).

A few procaryotes do not have cell walls and, as might be expected, these bacteria survive only under conditions that compensate for their osmotic fragility. **Mycoplasmas** are bacteria that lack the capacity to produce cell walls under any circumstances but behave like true bacteria in other respects. Mycoplasmas, along with *Legionella pneumophila* (the agent of Legionnaires disease) and influenza A viruses, are leading causes of primary atypical pneumonia in humans (see chaps. 12 and 13).

Bacteria that normally produce a cell wall may lose their ability to do so through mutation, or by exposure to agents like penicillin that inhibit cell-wall synthesis. Bacteria without cell walls are known as L-forms, or cell-wall deficient variants. The term *L-form* commemorates the Lister Institute in London where L-forms were first described. L-forms, which may be produced by any species of bacteria, can multiply under appropriate environmental conditions and may revert back to normal cell-wall containing organisms. The significance of L-forms in human disease is unclear but some workers have proposed a role in recurrent disease.

No structure analogous to bacterial peptidoglycan is found in human or other eucaryotic cells. It is not surprising, therefore, that many clinically useful antimicrobial agents exert their effects by inhibiting peptidoglycan synthesis. Antimicrobial agents in this category include all penicillins and cephalosporins, vancomycin, and bacitracin. For example, penicillin specifically inhibits the cross-linkage of peptidoglycan layers and thus has no effect on eucaryotic cells.

Bacterial Structure and Function
Cell Wall: Cell wall; Peptidoglycan
Gram-Positive Cell: Gram-Positive Cell Wall; Teichoic Acids
Gram-Negative Cell: Cell Wall Exceptions

Capsule

The bacterial cell wall is often surrounded by a viscous polysaccharide or protein coat produced under the direction of the cell's genes and referred to as a slime layer or **capsule** (colorplate 2a). A capsule can have a variety of functions, but its most important one in disease production is as an antiphagocytic structure. Encapsulated bacterial pathogens resist phagocytosis (ingestion) by certain white blood cells (neutrophils, macrophages; see chap. 6), especially in areas such as the pleural and peritoneal cavities. This is most clearly seen in the case of *Streptococcus pneumoniae,* the most common cause of bacterial pneumonia. Studies in non-immune mice have demonstrated that encapsulated *S. pneumoniae* strains resist phagocytosis, while nonencapsulated strains are readily phagocytized. Thus only encapsulated strains of *S. pneumoniae* have disease-causing ability, or virulence.

Bacterial Structure and Function
External Structures: Glycocalyx

Cell Membrane

Lying beneath the thick outer cell wall is the **cell membrane,** which mediates the transport of materials into and out of the cell. Some molecules, such as water and oxygen, can move passively across the cell membrane. The direction of this passive movement (or diffusion) depends on the concentration of the dissolved molecules (solutes) on either side of the membrane. Solutes move from an area of high concentration to an area of low concentration until the concentration on both sides of the membrane is equal. This type of solute transport mechanism is called *passive diffusion.* Some solutes can be transported across the cell membrane by a process called *facilitated diffusion,* in which the solute is transported by a protein carrier molecule located in the cell membrane. These carrier proteins pick up solutes on one side of the cytoplasmic membrane and transport them, chemically unchanged, to the opposite side of the cell membrane.

The most important transport mechanism is called *active transport.* Microorganisms usually live in environments where most nutrients are at a lower concentration outside the cell. Through active transport these nutrients can be carried across the cell membrane despite the concentration gradient by the use of carrier molecules in the cell membrane and by the expenditure of energy. The transported solute is not chemically changed during this process. Molecules may also be transported across the cell membrane by *group translocation,* a process similar to active transport except that the molecule is chemically altered during the transfer process.

Close examination of the cell membrane by using electron microscopy reveals that at various locations it is folded in upon itself (invaginated), forming a rounded sac filled with its own convolutions. This structure is called a *mesosome* (figs. 2.2 and 2.4) and may be important in cell division and metabolism.

Bacterial Structure and Function
Cell Membrane: Transport

Flagella

Many bacteria possess **flagella** (sing., **flagellum**), which are fine threadlike, wavy appendages that provide the cell with the capacity for directional motion, or motility. Originating in the cytoplasm, each flagellum is composed of a single strand of a protein molecule that repeats itself in a spiral pattern, extending through the cell membrane and wall to the exterior. They are extremely fine in diameter (about 25 nm), but usually as long as or longer than the cell itself (fig. 2.5). Because they are so fine, special staining techniques are required to observe them with the light microscope. In general, the cocci are not flagellated and so are nonmotile, but most spirilla and many bacilli are motile, and some possess many flagella. The number of flagella and their position on the cell (either at one or both poles, or on the circumference) may be useful features in the identification of bacterial species.

Pili (Fimbriae)

Some bacteria display numerous filamentous surface appendages called **pili,** or *fimbriae.* These are very fine structures that can be seen only with the special staining techniques and higher magnifications of electron microscopy. They differ from flagella in that they are smaller in diameter, shorter, straighter, and do not provide motility, although, like flagella, they are composed primarily of protein. They appear to provide bacterial cells with the ability to adhere to inert surfaces as well as to other cells. Some bacteria have a specialized "sex" pilus through which nuclear material (DNA) can be transferred to another bacterial cell in a process known as bacterial conjugation (see fig. 3.5). "Sex" pili are also the sites for attachment of certain viruses (bacteriophages) that infect bacterial cells. Viruses will be described in more detail later in the chapter.

Bacterial Structure and Function
External Structures: Pili; Flagella

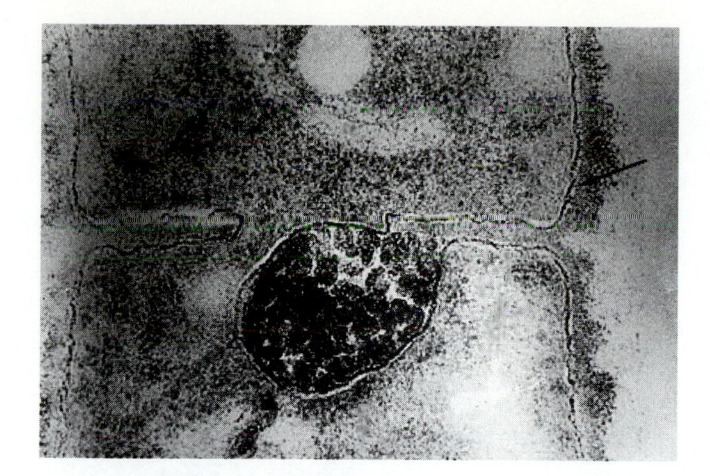

FIGURE 2.4 In this electron micrograph of a portion of a bacterial cell, the cell membrane (arrow) is clearly visible beneath the outer cell wall. Two invaginations of the membrane can be seen, one of which has formed a rounded sac (mesosome) filled with membrane convolutions.

Reproduced from D. J. Ellar, D. G. Lundgren, and R. A. Slepecky, "Fine Structure of *B. Megatherium* During Synchronous Growth." *Journal of Bacteriology* 94:1189, 1967. American Society for Microbiology.

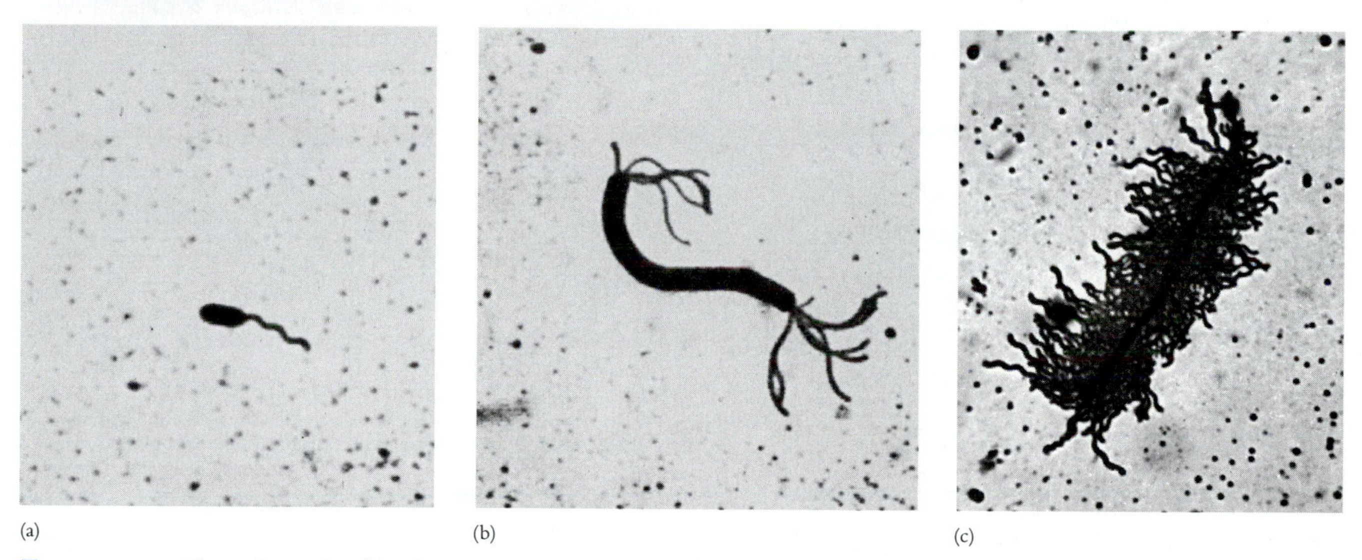

FIGURE 2.5 Photomicrographs of flagellated bacteria (×2,000). (a) *Pseudomonas aeruginosa* has a single (unipolar) flagellum and is said to be monotrichous. (b) *Spirillum serpens* is lophotrichous (tufted), with bipolar flagella. (c) This *Proteus* species is peritrichous, or flagellated around its entire surface.

Photos courtesy Dr. Einar Leifson, *Atlas of Bacterial Flagellation*, 1960, Academic Press.

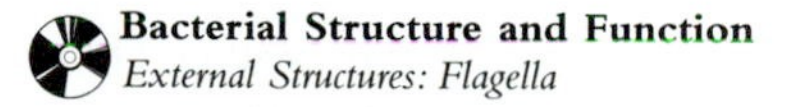

Bacterial Structure and Function
External Structures: Flagella

Internal Structures

The dense cytoplasm of bacterial cells contains deoxyribonucleic acid (DNA), the functional nuclear material, and different forms of **ribonucleic acid (RNA)** in a matrix of protein, carbohydrate, lipid, and other organic substances.

Nuclear Material

Bacteria and other procaryotic cells do not have a true nucleus compartmented within a nuclear membrane. They do, however, possess nuclear material that is supercoiled in the cytoplasm and is visible through electron microscopy as a mass (sometimes two or more) of fibrils (fig. 2.2). These fibrils are composed primarily of DNA, lying free in the cytoplasm.

DNA

In all living organisms DNA is the genetic material. The characteristics exhibited by a microorganism reflect the genetic information coded within its DNA and are expressed through its chemical and physical reactions to a particular environment. Specific nucleotide sequences contain the hereditary information in **genes**, which are carried on a specific structure called the chromosome. The numbers and kinds of genes of any organism are referred to as its **genotype.** The properties of an organism that are expressed at any particular time in its environment are its **phenotype.**

Certain viruses, however, contain not DNA but the chemically similar substance RNA. DNA and RNA collectively are known as the nucleic acids and are composed of smaller subunits, or building blocks, known as **nucleotides.** In turn, nucleotides are made up of three kinds of molecules: a sugar molecule (deoxyribose in DNA), an adjacent phosphate group, and an attached nitrogen base. Alternating sugar and phosphate molecules form the backbone of the DNA molecule with the nitrogen bases pointing inward (fig. 2.6). Four different nitrogen bases are found in DNA,

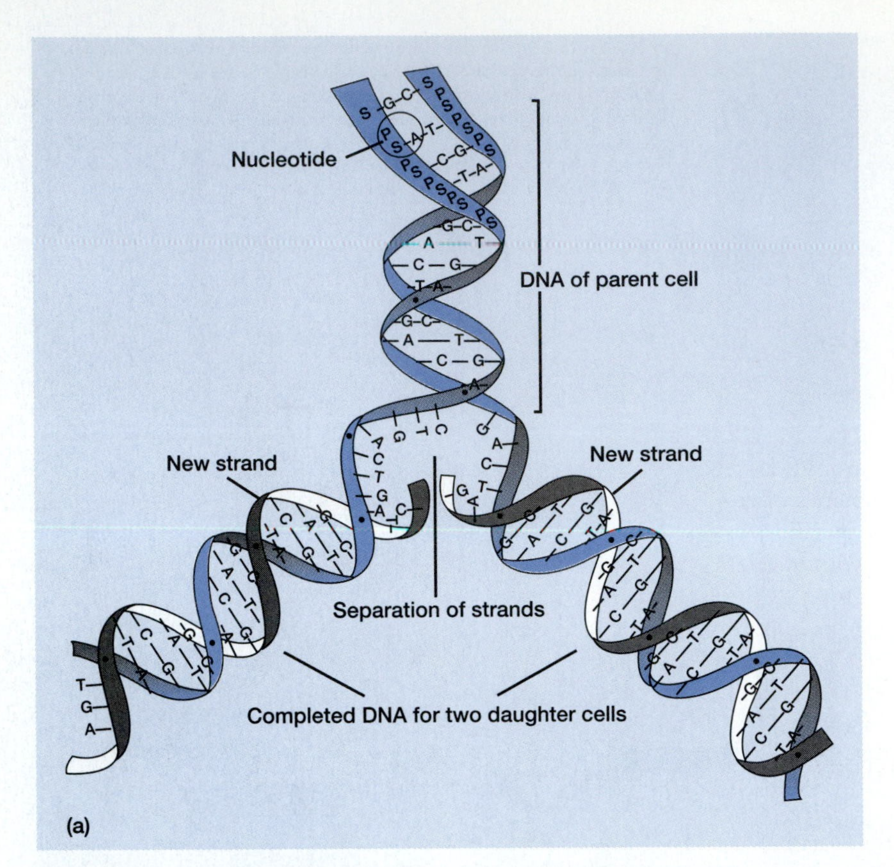

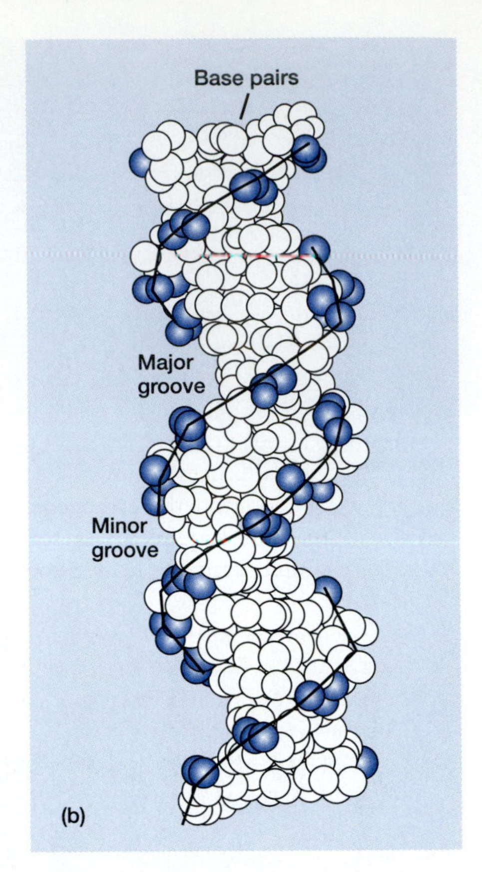

FIGURE 2.6 (a) Diagram showing duplication of a DNA double helix in a dividing bacterial (procaryotic) cell. Nitrogen bases (A = adenine, C = cytosine, G = guanine, T = thymine) link the deoxyribose (S = sugar)-phosphate (P) groups in a repeating pattern (nucleotide) along each DNA strand. The order of nucleotides is unique for each type of bacterium and is duplicated precisely for the progeny of each bacterial species. In the drawing above, the DNA molecules of the daughter cells are identical copies of the DNA of the parent cell. (b) Structural model of the DNA double helix. The major and minor grooves are formed as the two strands twist about each other.

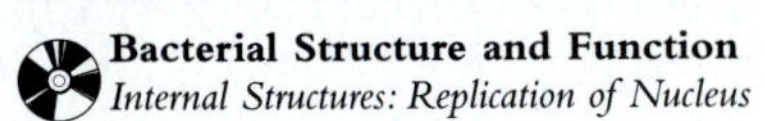
Bacterial Structure and Function
Internal Structures: Replication of Nucleus

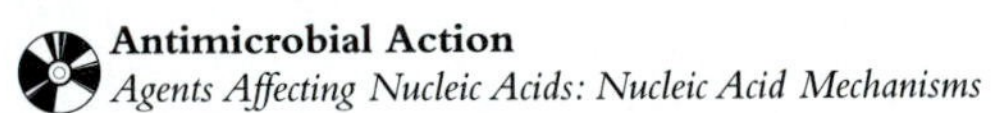
Antimicrobial Action
Agents Affecting Nucleic Acids: Nucleic Acid Mechanisms

two purines (adenine and guanine) and two pyrimidines (thymine and cytosine). Chemically, these bases are flat, ring-shaped molecules. As seen in figure 2.6a, there is specific pairing between the nitrogen bases of the DNA strands, with adenine (A) always pairing with thymine (T) and guanine (G) always pairing with cytosine (C). This complementary double-stranded (**double helix**) structure of DNA was discovered by James Watson, Francis Crick, Rosalind Franklin, and Maurice Wilkins in 1953 and is one of the great scientific discoveries of this century.

As mentioned earlier, bacteria divide by a mechanism known as **binary fission** in which the bacterium separates into two daughter cells, each containing the same components as the parent (see fig. 2.7). Before bacterial cell division can occur, the DNA must replicate and does so by a mechanism that is suggested by the double helical structure of the DNA molecule itself. The two strands separate, and each strand then serves as a template for the synthesis of new strands (fig. 2.6a). Since the adenine of one strand pairs only with a thymine on the opposite strand, and guanine pairs only with cytosine, the new strand is an exact replica of the previous parental strand. The entire process of DNA replication is catalyzed by an enzyme called DNA **polymerase.** The new DNA molecules formed by this enzyme consist of one original (parental) strand and one new strand that twist around each other to form a grooved structure (Fig 2.6b).

Bacterial DNA The chromosome of a bacterium such as *Escherichia coli* consists of a single circular molecule of doubled-stranded DNA (see fig. 2.2) with a molecular weight of approximately 3×10^9. This corresponds to about four million base pairs and is enough DNA to code for (i.e., provide the cell with instructions for) the synthesis of approximately 3,000 proteins. Although no absolute proof is available, it is reasonably safe to assume that the DNA of all bacteria is structurally similar to *E. coli* DNA.

Eucaryotic cells are diploid; their chromosomes occur in pairs. Bacteria are haploid because their chromosome is a single molecule. During the period of logarithmic growth when bacteria are in the process of active cell division, a bacterial cell may contain more than one copy of its circular chromosome. This is not a true diploid state but rather a situation in which DNA replication has occurred faster than cell division. Although we refer to bacterial DNA as a chromosome, it is not structurally analogous to the true chromosomes found in eucaryotes. Unlike true chromosomes, bacterial chromosomal DNA is circular and not complexed with histones (basic proteins). Many bacteria also contain extrachromosomal DNA structures called **plasmids.**

Plasmids Plasmids are important carriers of extra genetic knowledge for processes such as conjugation (see F plasmids, chap. 3) and antimicrobial resistance (R factor). Recently, plasmids have been used as vehicles to insert foreign (eucaryotic) DNA into bacteria, thereby allowing them to manufacture nonbacterial proteins such as human insulin and interferon (a process known as recombinant DNA technology). In addition, the pathogenic capabilities of certain microorganisms require the presence of specific plasmids. For example, strains of *E. coli* associated with traveler's diarrhea not only produce enterotoxins that cause intestinal disease (see chaps. 5 and 16) but also surface factors that allow these strains to attach to and colonize cells in the small intestine. Enterotoxins and surface factors are coded for by plasmids called *ent* (for enterotoxin), which are found only in the enterotoxin-producing strains of *E. coli.*

Structurally, plasmids are small (less than 3% of the total DNA), circular bodies of double-stranded DNA that are found in most bacteria (figs. 2.8, 2.9). Plasmids multiply independently of the major chromosomal body of DNA, are inherited by each daughter cell at the time of cell division, and are not essential for basic cellular functions (metabolism, reproduction). Each cell contains a varying number of copies of a particular plasmid, usually one per chromosome for large plasmids and 10 to 20 per chromosome for small ones. The number of copies is controlled by the plasmid itself, and a bacterium may contain more than one kind of plasmid (compatible plasmids).

In a way, plasmids may be thought of as symbionts (see chap. 5) within their host cells because they supply it with adaptive genetic information. A good example of this is the antimicrobial resistance plasmids that benefit their bacterial host by providing an adaptive trait, the resistance to a variety of antimicrobial agents. Plasmids that possess the necessary genetic information to confer resistance to an antimicrobial agent are often referred to as *resistance transfer factors* or RTFs.

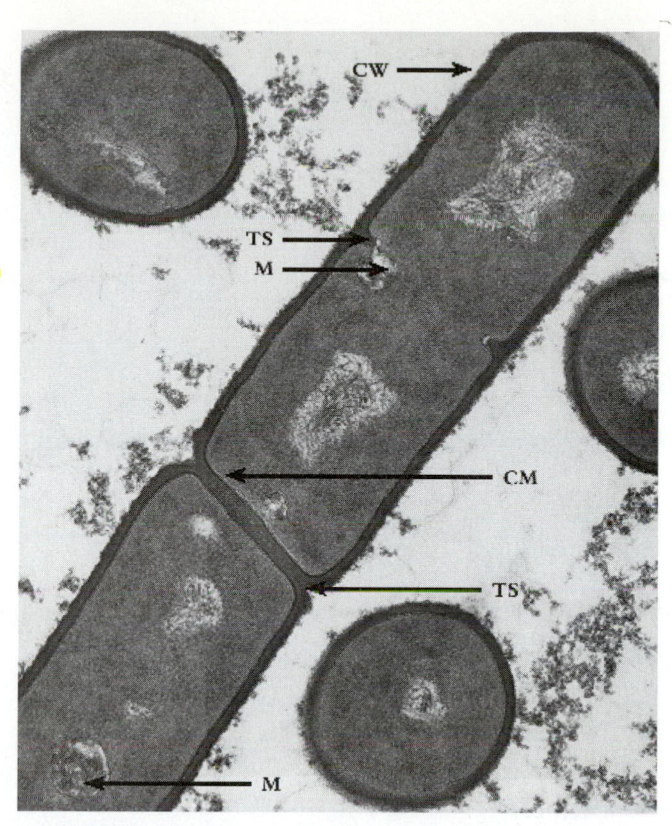

FIGURE 2-7 Electron micrograph of a thin section of a gram-positive bacterium, *Bacillis subtilis.* In this photo one can clearly see the cell wall (CW), cell membrane (CM), mesosomes (M), and the transverse septum or crosswall that is formed when bacteria divide (TS). Note that the cell in the upper half of the photo is beginning to divide again and that a mesosome is developing at the site of the new transverse septum that is forming.

Courtesy L. Caro, Universite de Geneve, Geneve, Switzerland.

RNA

Except for the viruses, which contain either DNA or RNA, microorganisms contain both nucleic acids. There are several structural differences between RNA and DNA. RNA is usually single stranded; the sugar in RNA is ribose; and the DNA pyrimidine base, thymine, is replaced in RNA by the structurally similar base uracil (U). Like DNA, RNA consists of two purines (A and G), two pyrimidines (C and U), a sugar, and a phosphate group.

There are three kinds of RNA that function in protein synthesis: *ribosomal* RNA (rRNA) (a structural component of the ribosomes), *messenger* RNA, and *transfer* RNA. **Messenger RNA (mRNA)** is produced as a complementary copy of a specific DNA nucleotide sequence. The function of mRNA is to obtain and carry genetic information encoded in the DNA (the "message") to the location where proteins are synthesized on the ribosome. The mRNA provides the molecular instructions for the production of the microbial cell's various proteins. **Transfer RNA (tRNA)** is the shortest of the RNA molecules and is used to transfer specific amino acids to the ribosome, the site of protein synthesis. There are at least 20 different types of tRNA.

Ribosomes

Ribosomes, the sites of protein synthesis in all cells, are composed of 67% RNA and 33% protein. The bacterial cytoplasm is dense with ribosomes (33 to 45% of the dry weight), illustrating the importance of protein synthesis to the bacterial cell. Unlike eucaryotic ribosomes, bacterial ribosomes are not associated with the endoplasmic reticulum because it is not found in procaryotes. Instead, ribosomes may cluster at sites along the bacterial cell membrane.

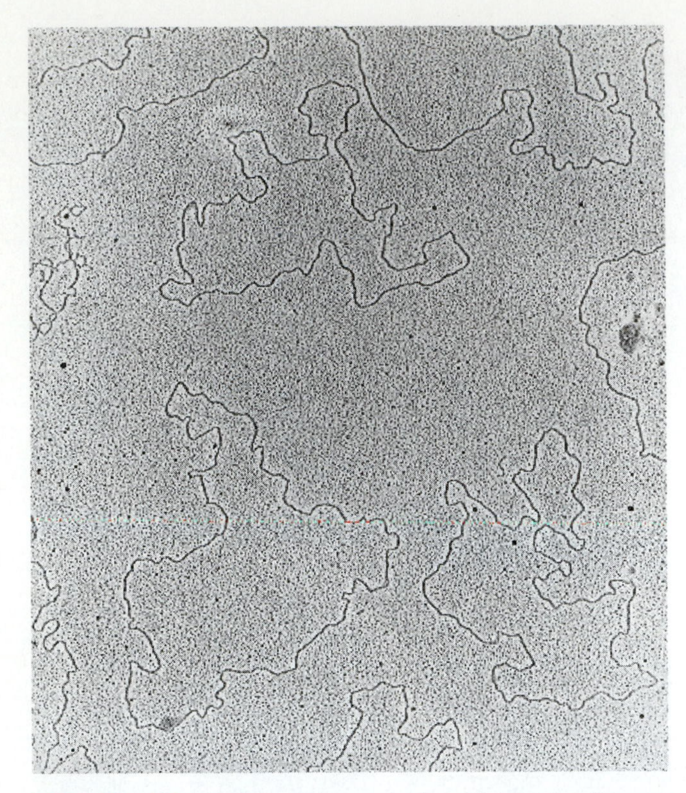

FIGURE 2.8 An electron micrograph of several molecules of bacterial plasmid DNA, each occurring in a closed loop (circular DNA) (×32,430).

Courtesy Dr. Richard Novick, The Public Health Research Institute of the City of New York, New York, NY,

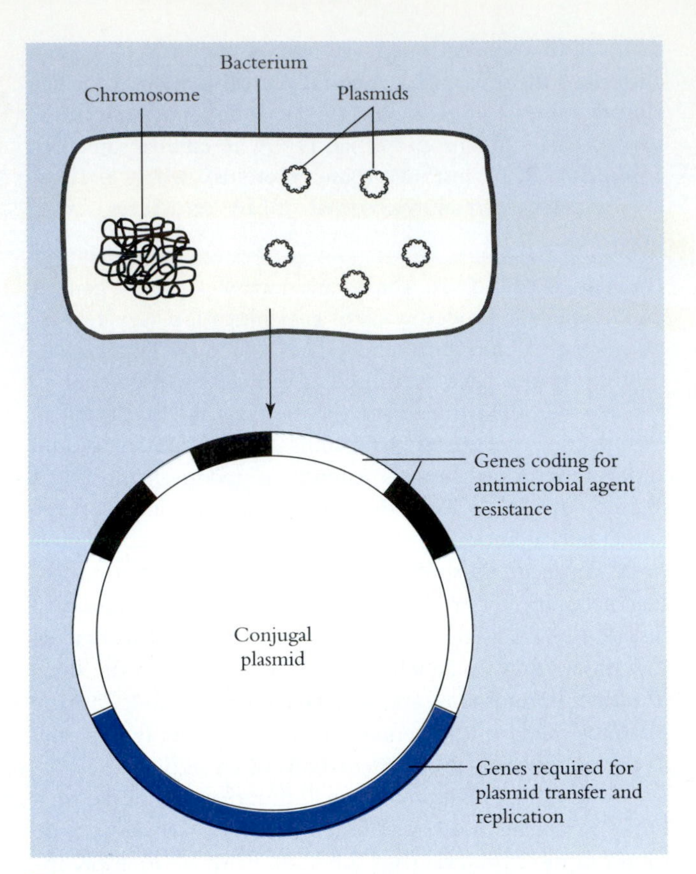

FIGURE 2.9 Many bacteria contain extrachromosomal bodies of CNA know as plasmids. Plasmids provide bacteria with adaptive genetic capability such as antimicrobial resistance.

Mechanisms of Antimicrobial Resistance
Transfer and Spread of Resistance: Transfer/Spread

Bacterial ribosomes are 70S in size whereas eucaryotic cytoplasmic ribosomes are 80S (S = Svedberg unit, a measurement of mass as determined by sedimentation velocity in an ultracentrifuge). Although this 10S difference may seem trivial, it is a basis for the selective toxicity of most antimicrobial agents that act by inhibiting protein synthesis. Erythromycin, chloramphenicol, and gentamicin are examples of clinically useful drugs that inhibit bacterial, but not human, protein synthesis. These agents can bind to 70S (bacterial) but not to 80S (eucaryotic) ribosomes, although certain other agents, such as the tetracyclines, can bind to both. In part, the tetracyclines are clinically useful because they can penetrate many bacterial cells but not human cells.

Proteins and Protein Synthesis The function of DNA in living organisms is to direct the synthesis of all protein matter. Indeed, a molecular definition of the gene is that amount of DNA coding for one complete protein. With a few exceptions, it also appears that the basic way in which DNA codes for proteins is similar in all forms of life from bacteria to plants and animals. Proteins, the most common organic molecules found in living things, are composed of smaller subunits called amino acids. These in turn consist of carbon, hydrogen, oxygen, nitrogen, and small amounts of sulfur. There are about 20 different kinds of amino acids (table 2.4) contained in all proteins. The molar amounts and specific order of the amino acids vary from protein to protein. Small proteins contain only a few amino acids (51 in the case of the hormone insulin), while large proteins consist of over 1,000; the average protein has approximately 400 amino acids.

The individual amino acids making up the protein are joined in chainlike fashion by chemical bonds called peptide bonds. As a protein is being synthesized, it is frequently called a polypeptide chain, referring to the numerous peptide bonds occurring in each protein. It is the unique amino acid sequence that gives the protein a particular shape; the shape determines its biological function. DNA determines a specific amino acid sequence by a genetic coding mechanism and, in turn, the specific amino acid sequence determines the function of a protein.

The importance of proteins to bacteria and all other living things cannot be understated. Many cellular structures (membranes) and organelles (ribosomes) are composed, in part, of proteins; and the metabolism of all living things occurs only because of specific proteins called *enzymes* that function as biologic catalysts. In higher life forms, such as vertebrates, additional biologically important proteins include the antibodies (see chap. 6) and some hormones.

The key to the DNA protein code is in the sequence of the nitrogen bases. A sequence of three nitrogen bases (called a **codon**) is needed to code for each amino acid in a protein. A glance at the genetic code (table 2.5) indicates

TABLE 2.4 THE 20 COMMON AMINO ACIDS FORMING PROTEINS

Amino Acid	Common Abbreviation	Amino Acid	Common Abbreviation
Glycine	Gly	Lysine	Lys
Alanine	Ala	Arginine	Arg
Valine	Val	Asparagine	Asn
Isoleucine	Ile	Glutamine	Gln
Leucine	Leu	Cysteine	Cys
Serine	Ser	Methionine	Met
Threonine	Thr	Tryptophan	Trp
Proline	Pro	Phenylalanine	Phe
Aspartic acid	Asp	Tyrosine	Tyr
Glutamic acid	Glu	Histidine	His

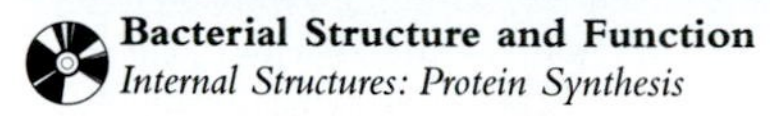

that it is expressed in the language of RNA, not DNA (note U is present but not T), because proteins are synthesized in two steps called **transcription** and **translation** (fig. 2.10). The four bases (A, G, C, U) can combine into 64 codons (4^3 different combinations of three bases), therefore, some amino acids have more than one codon (table 2.5). Under normal circumstances some codons (UAA, UAG, UGA) do not code for any amino acid, but rather specify where the protein ends or stops.

During transcription, an enzyme called *DNA dependent RNA polymerase* makes an mRNA copy of one DNA strand that makes "genetic sense" for the amino acid sequence of a particular protein. The RNA polymerase recognizes and copies this strand in a complementary fashion to make mRNA. As the genetic message in DNA is transcribed into mRNA, T (thymine) in DNA codes for A (adenine) in mRNA, A codes for U (uracil, *not* thymine as in DNA), C (cytosine) codes for G (guanine), and G codes for C. Thus for every protein, molecules of mRNA are transcribed from DNA and contain the specific codons for the amino acid sequence of that protein.

During the translation step of protein synthesis, mRNA, tRNA, and ribosomes interact functionally. The mRNA molecules attach to ribosomes and are translated into proteins. To understand this step, note in figure 2.10 that one end of the mRNA molecule is referred to as the 5′ end and the other as the 3′ end. The starting codon for proteins is AUG, which occurs near the 5′ end of the mRNA. The ribosomes begin the process of translation here and move stepwise down the mRNA molecule toward the 3′ end. Each codon specifies the position of a particular amino acid (table 2.5), and these amino acids are joined to each other, in order, by the peptide bonds. Transfer RNA (tRNA) molecules bring each amino acid to the proper site on the ribosome. Each of the 20 amino acids has a separate tRNA molecule that recognizes the proper codon for its attached amino acid by virtue of the **anticodon.** The anticodon is a series of three nitrogen bases in the tRNA molecule that is complementary in sequence to the three nitrogen bases in a codon. Thus, pairing occurs between the codon and the anticodon (A with U and G with C) ac-

TABLE 2.5 THE GENETIC CODE

First Letter in Codon	Second Letter in Codon: U	C	A	G	Third Letter in Codon
	Phe*	Ser	Tyr	Cys	U
U	Phe	Ser	Tyr	Cys	C
	Leu	Ser	Stop	Stop	A
	Leu	Ser	Stop	Trp	G
	Leu	Pro	His	Arg	U
C	Leu	Pro	His	Arg	C
	Leu	Pro	Gln	Arg	A
	Leu	Pro	Gln	Arg	G
	Ile	Thr	Asn	Ser	U
A	Ile	Thr	Asn	Ser	C
	Ile	Thr	Lys	Arg	A
	Met	Thr	Lys	Arg	G
	Val	Ala	Asp	Gly	U
G	Val	Ala	Asp	Gly	C
	Val	Ala	Glu	Gly	A
	Val	Ala	Glu	Gly	G

*The code for Phe is UUU or UUC; for Tyr, it is UAU or UAC.

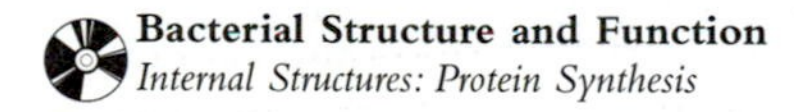

cording to base-pairing rules. Figure 2.10 shows the serine tRNA (anticodon = AGA) in the proper position at the serine codon (UCU) on the mRNA molecule. While still attached to its tRNA molecule, this newest amino acid is joined by a peptide bond to amino acids previously brought in line by their specific tRNA molecules. Thus, the polypeptide chain grows and now contains four amino acids. Once each amino acid is in its proper position, its tRNA molecule is released alone (see lysine tRNA in fig. 2.10). This process continues as the ribosome moves toward the 3′ end of the mRNA molecule until a specific stop signal (UAA, UAG, UGA) is reached, specifying completion of the protein molecule. The protein is then released from the last tRNA molecule and the ribosome, and it is transported to the cell location where the protein functions biologically. Although the starting codon for proteins is AUG, the codon

FIGURE 2.10 Proteins are manufactured in two steps: transcription, and translation. During transcription, an mRNA copy is made complementary to one of the DNA strands. During translation, amino acids are joined by peptide bonds in the proper order because of codon (mRNA) and anticodon (tRNA) recognition.

Bacterial Structure and Function
Internal Structures: Protein Synthesis; Transcription; Translation; Protein Assembly

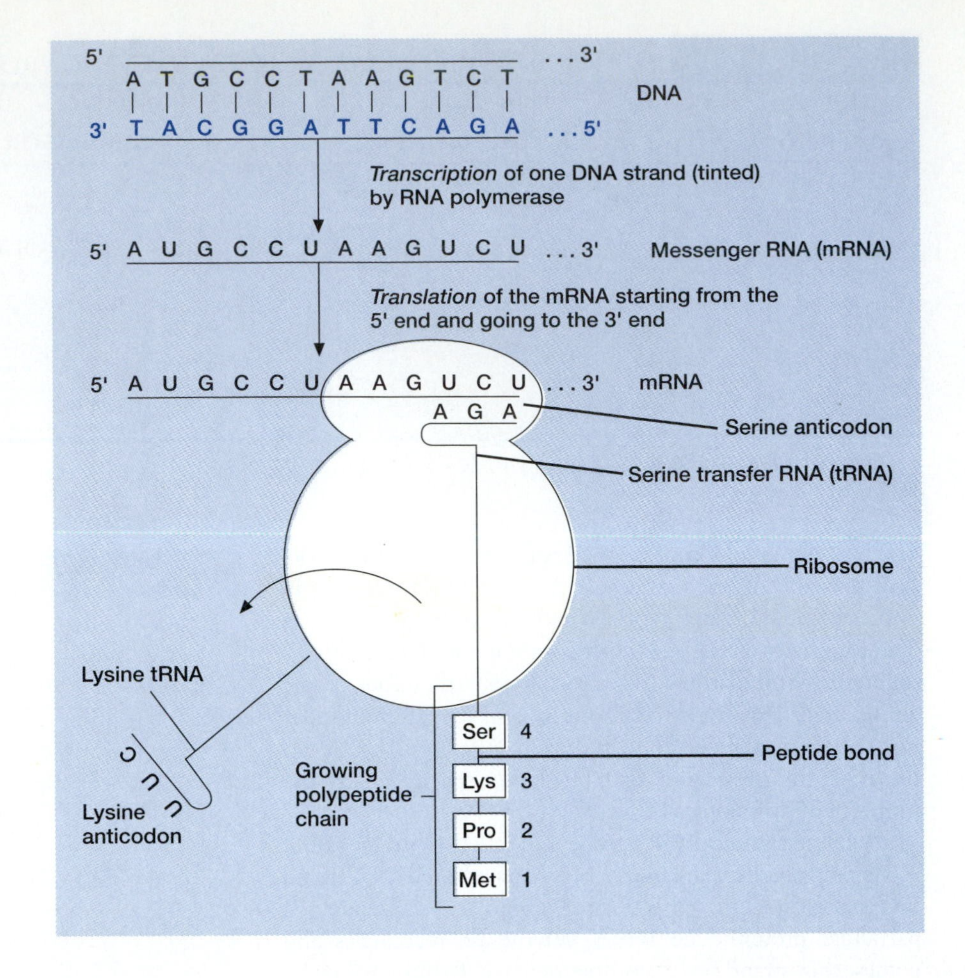

for methionine, fewer than 50% of all proteins begin with this amino acid because a post-translational mechanism exists that cleaves it from the beginning position of most proteins.

Protein synthesis has important implications for the health care professional: (1) Certain proteins are required for a variety of important biological functions in humans (enzymes, hormones, antibodies, structural components), yet others, in bacteria (tetanus, botulinal, and diphtheria exotoxins, staphylococcal enterotoxin), are directly responsible for diseases caused by the organisms that produce them. (2) The synthesis of proteins demonstrates how the genetic material (DNA) functions at the molecular level. (3) Many of the antimicrobial agents in current use exert their antibacterial effects by inhibiting protein synthesis and so it is possible to understand their mode of action.

Bacterial Endospores

When normal bacterial growth ceases because essential nutrients are depleted, a few genera have the capacity to **sporulate,** that is, to form a very resistant, dehydrated structure called an **endospore** (*endo* = within) (fig. 2.11). Medically important bacteria that form endospores are in the genus *Bacillus* (aerobic endospore formers) and the genus *Clostridium* (anaerobic endospore formers). The process by which actively growing, vegetative cells form endospores is shown in figure 2.12. During a stage of replication, the tightly coiled bacterial DNA becomes more dense, forming an axial filament, and one DNA molecule moves to the pole of the bacterial cell. The cell membrane invaginates to form the endospore septum, which moves inward to enclose a complete DNA molecule and give rise to the *forespore.* Successive protective layers of modified peptidoglycan (cortex) and protein coats are added around the forespore. In some cases an exosporium also forms (fig. 2.12). Finally, the vegetative cell dissolves, and the completed endospore, now referred to as the free endospore or spore, is released.

Even though endospores contain DNA, ribosomes, and some enzymes, they do not divide and are metabolically inactive. Essentially, an endospore is a resting cell that, in a favorable environment, can again become a metabolically active, dividing vegetative cell in a process known as *germination.* Unlike vegetative cells, endospores are resistant to the effects of most physical and chemical agents such as heat, antimicrobial agents, and disinfectants. The high resistance of endospores to heat is a truly unique property. While most vegetative cells are killed within minutes in boiling water (100° C), endospores may survive hours of such boiling. It is postulated that the heat resistance of endospores is due to the dehydrated nature of the endospore and to the high amounts of Ca^{2+} and a unique compound, *dipicolinic acid* (DPA), that endospores contain. The DPA and Ca^{2+} may form chemical complexes with proteins in the endospores,

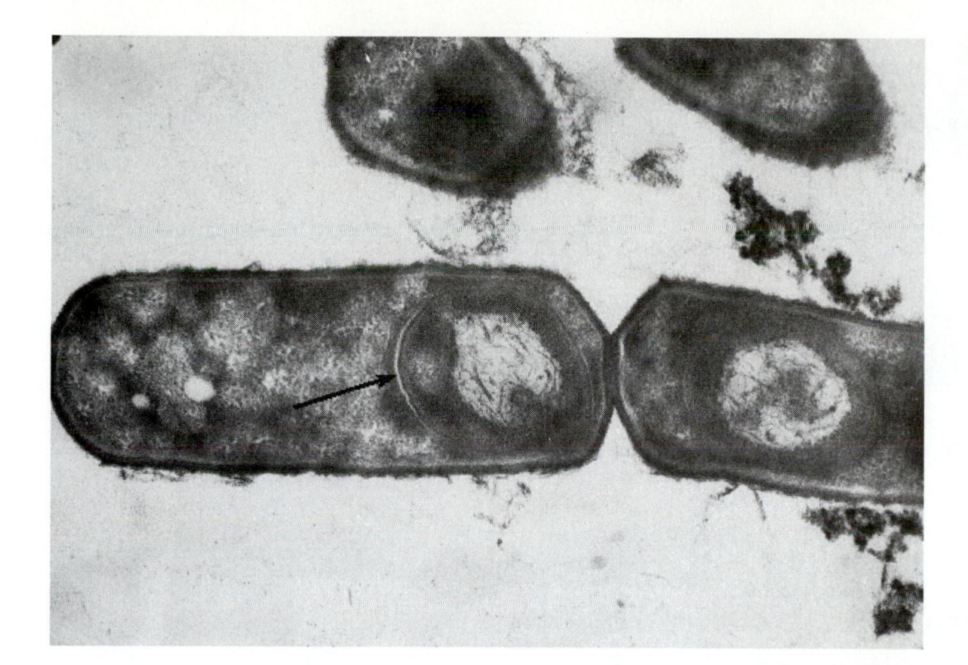

FIGURE 2.11 Two sporulating vegetative bacilli. The arrow marks the spore wall in one bacillus. The cell's nuclear material is preserved within the spore and will form a new vegetative cell on germination (×32,000).

Reproduced from D. J. Ellar and D. G. Lundgren, "Fine structure of sporulation in *Bacillus cereus* grown in a chemically defined medium," *Journal of Bacteriology*, 92: 1759, 1966. American Society for Microbiology.

thereby protecting them from heat denaturation (destruction). Even though they are highly heat resistant, endospores are killed by the standard autoclaving conditions used in the microbiology laboratory and by hospitals to sterilize certain surgical supplies.

Sporulation is important to species of *Bacillus* and *Clostridium* because their natural ecological habitat is the soil, an environment that is frequently unfavorable for the survival of vegetative cells. Endospores can survive for long periods in soil, tolerating temperature extremes and lack of nutrients and water.

Although endospores do not directly cause the diseases such as anthrax, botulism, and tetanus that are associated with pathogenic *Bacillus* or *Clostridium* species, they do play an indirect role in those conditions. Garden vegetables may contain botulinal spores from the soil. If these vegetables are improperly canned, the endospores may germinate into vegetative cells of *Clostridium botulinum,* which produce the botulinal exotoxin. When ingested, the toxin gives rise to the symptoms of botulism (see chap. 16). Although the endospores are highly heat resistant, botulinal exotoxin is totally inactivated by boiling for ten minutes. Boiling canned vegetables before eating makes them safe for consumption, even if they are contaminated with botulinal toxin.

RICKETTSIAE AND CHLAMYDIAE

The **rickettsiae** (named for H. T. Ricketts who discovered them) were once thought to be transitional forms between bacteria and viruses because of their intermediate size and because, like viruses, they are capable of growth only within higher living cells. The **chlamydiae** are smaller still than rickettsiae and were once classified as "large viruses." Both groups are now classified with the bacteria for, although they are much smaller and strictly dependent on an intracellular life, their structures and properties indicate that they are true bacteria.

Rickettsiae

Structurally, rickettsiae display all the features of bacteria except that none of them is flagellated and they do not form endospores. They have a typical gram-negative bacterial cell wall, contain DNA and RNA, and divide like bacteria. They are very narrow, short rods (0.3 × 1 μm), coccobacillary or nearly coccal in shape (figs. 2.13 and 2.14). Their patterns are *pleomorphic* (of variable form); that is, they may occur singly or in pairs, sometimes in short chains, and occasionally in long filaments. Because of their small size they require special stains and the highest magnifications of the light microscope to be seen.

In the laboratory, rickettsiae survive and grow only in animal cells grown in the test tube (cell culture), or in intact animals such as embryos of fertilized chicken eggs, suckling mice, or guinea pigs. They cannot be studied on artificial nutrient media because they lack the metabolites necessary for growth, which must be provided by living cells. Rickettsiae characteristically grow and multiply intracellularly in the cytoplasm surrounding the nucleus (see fig. 2.13).

In nature, rickettsiae occur chiefly as harmless parasites of insects such as lice, ticks, and mites. Sometimes the insect becomes diseased and dies of rickettsial infection, but more often the relationship is not damaging. Furthermore infected insects transmit rickettsiae to their offspring through their eggs. Rickettsiae can be transmitted to humans through the bite of infected blood-sucking insects. When this happens, the human infection is likely to result in serious disease. Rickettsial diseases in humans include typhus fever (not to be confused with typhoid fever), Rocky Mountain spotted fever, rickettsialpox, and Q fever (see chaps. 12 and 22). Ehrlichiae, related organisms in the same

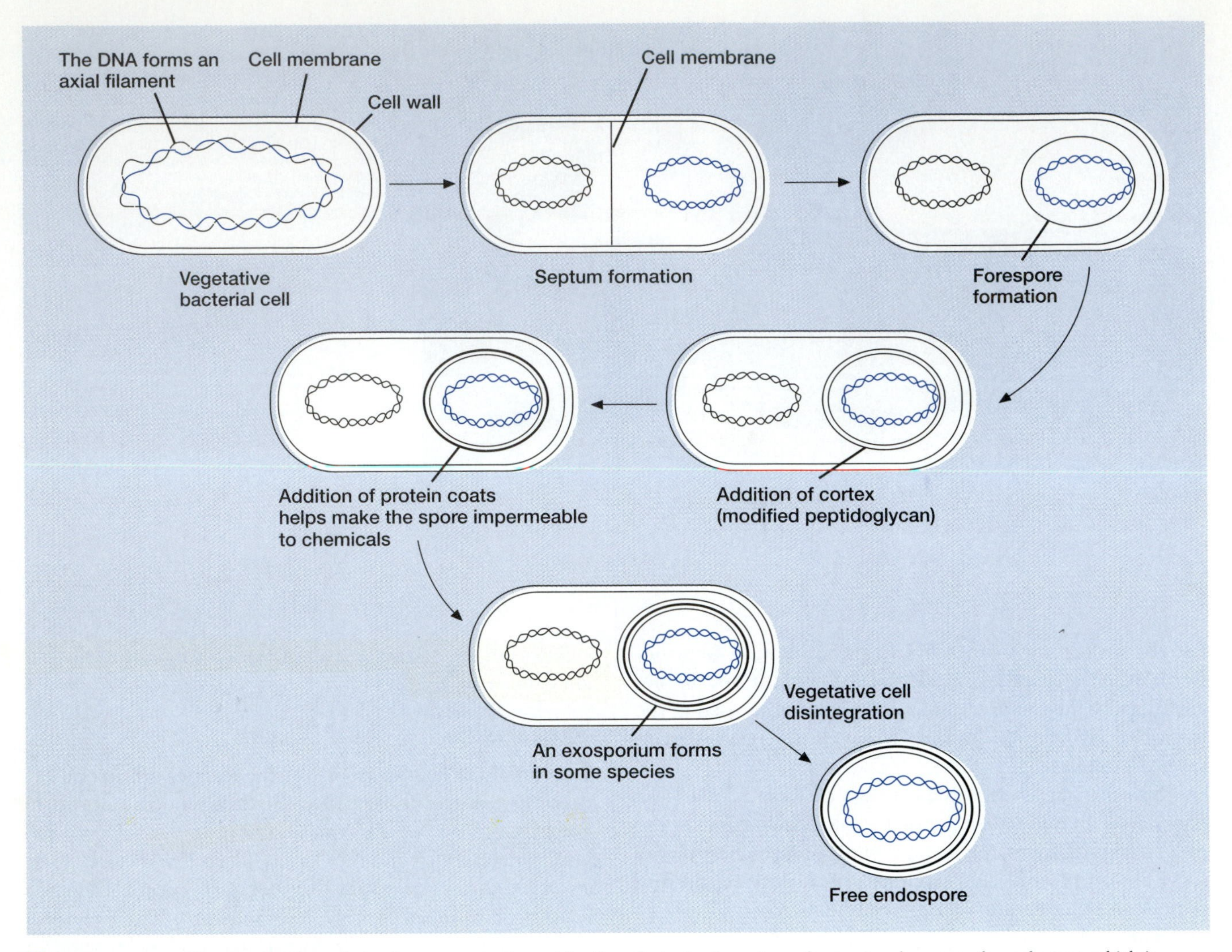

FIGURE 2.12 The process of sporulation by a vegetative bacterial cell results in the formation of a structure known as the endospore, which is resistant to physical (heat) and chemical agents.

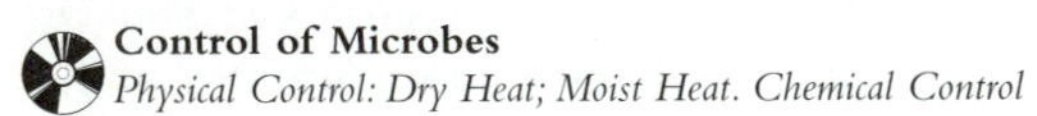

family as rickttsiae, infect human white blood cells to produce a disease known as ehrlichiosis (see chap. 22).

Chlamydiae

The word *chlamydia* means "thick walled." The very small size and intracellular life of these organisms led to their classification originally as viruses, but it is now known that they differ from viruses in several respects. The structure and composition of their cell walls are like those of bacteria, and like bacteria also, they possess both DNA and RNA, multiply by binary fission, contain ribosomes, and are susceptible to many antibacterial agents. Chlamydiae are nonmotile, coccoid organisms, ranging in size from about 0.2 to 1 μm (200 to 1,000 nm). They rely totally on living host cells for energy to carry out their metabolic processes and display a unique developmental cycle within the cells. When a single chlamydial particle, called an **elementary body,** enters such a cell, a membranous vacuole forms around it. The small chlamydia reorganizes into a larger particle called a **reticulate body.** Within the vacuole the reticulate body grows and divides. As division repeats, some reticulate bodies decrease in size within the cell and become typical elementary bodies. When freed from the cell, each of these new small particles is capable of infecting another cell.

Chlamydiae are associated with several human diseases, including trachoma, pneumonia, and psittacosis, and two sexually transmitted diseases, lymphogranuloma venereum and nongonococcal urethritis (see chaps. 12 and 19).

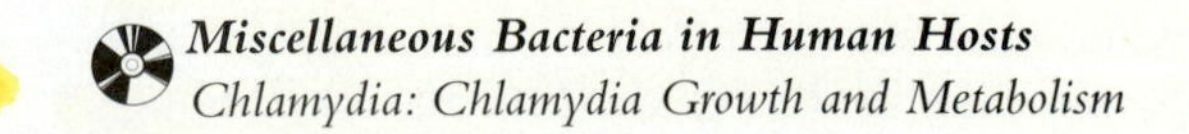

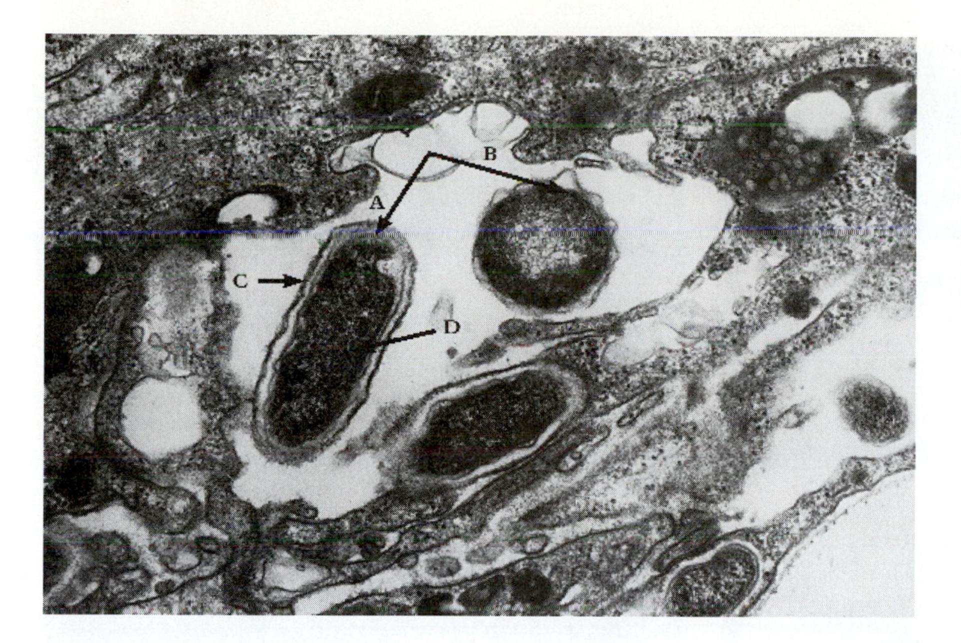

FIGURE 2.13 Electron micrograph of rickettsiae lying intracellularly within tissues of their wood tick vector. Note similarity in shape (bacillary and coccal forms) to other true bacteria (arrows *A* and *B*). The cell walls of these organisms, and their cytoplasmic ribosomes, are clearly visible at ×28, 000 (arrows *C* and *D*).

Courtesy Drs. Lyle P. Brinton and Willy Burgdorfer, Rocky Mountain Laboratory, USPHS, Hamilton, MT.

Miscellaneous Bacteria in Human Hosts
Zoonoses: Rickettsia

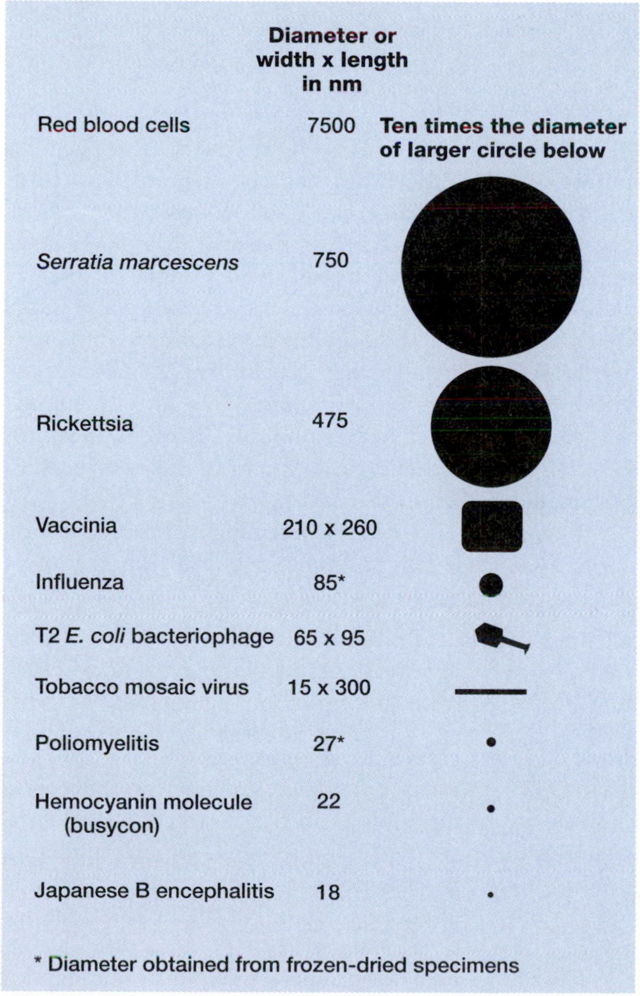

	Diameter or width x length in nm	
Red blood cells	7500	Ten times the diameter of larger circle below
Serratia marcescens	750	
Rickettsia	475	
Vaccinia	210 x 260	
Influenza	85*	
T2 *E. coli* bacteriophage	65 x 95	
Tobacco mosaic virus	15 x 300	
Poliomyelitis	27*	
Hemocyanin molecule (busycon)	22	
Japanese B encephalitis	18	

* Diameter obtained from frozen-dried specimens

FIGURE 2.14 In the diagram above compare the size of a red blood cell with that of various microorganisms.

Reproduced from *The Unseen World*, by copyright permission of *The Rockefeller University*.

VIRUSES

Morphology and Classification

Viruses are submicroscopic particles that reproduce only when they are inside living cells of animals, plants, or bacteria. Viruses that parasitize bacteria are known as **bacteriophages** ("bacteria eaters"), or simply *phages.* The size of viruses ranges between 20 and 300 nm in diameter (1 nm = 10^{-9} m), considerably smaller than most bacteria, and they cannot be seen with the compound light microscope (figs. 2.14 and 2.15a). Yet, some of the largest viruses (e.g., vaccinia virus) are slightly larger than the smallest bacteria (chlamydiae). Since viruses are not cells, they have no cytoplasmic membranes and do not contain cytoplasm.

In its simplest state a virus consists of a protein coat called a *capsid* surrounding a core of nucleic acid, either DNA or RNA. The nucleic acid core and capsid together are referred to as a *nucleocapsid*. The nucleocapsid in some viruses is surrounded by an envelope (enveloped virus), while others have no envelope (naked viruses) (fig. 2.16). Viruses also contain internal proteins that are necessary for viral reproduction within the host cell. For example, the RNA-dependent RNA polymerase of the rabies virus (an RNA-containing virus) is necessary for viral reproduction and is not supplied by the rabies virus host cell. Enveloped viruses also have spikes composed of glycoproteins (carbohydrate-protein complexes) on their surfaces.

The viral nucleocapsid may have one of three shapes: icosahedral, or a crystalline shape (an icosahedron consists of 20 equilateral triangular faces); helical, or an elongated rod shape; or complex, unusual shapes. Examples of these three are adenovirus, tobacco mosaic virus, and bacteriophage T2 (see fig. 2.15b), respectively. In addition, the envelope itself, when present, may impart a shape that is

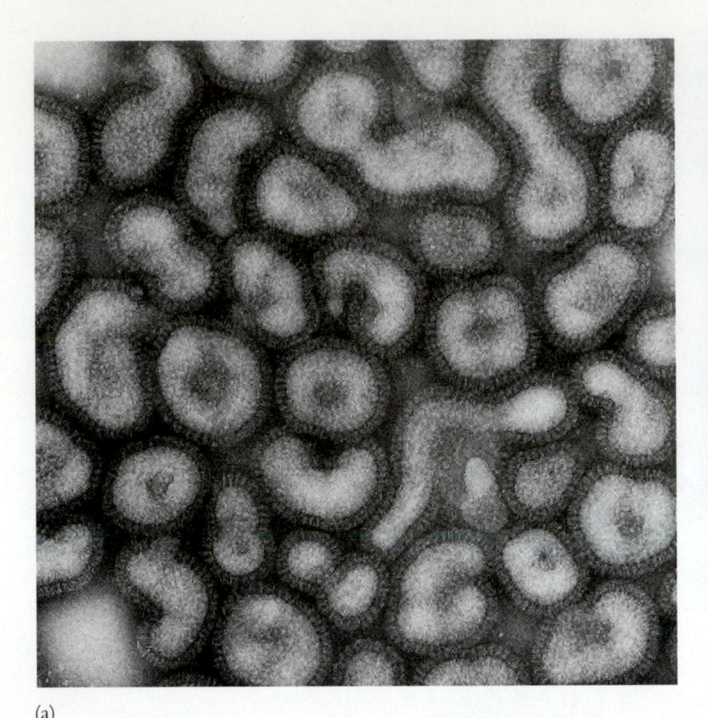

(a)

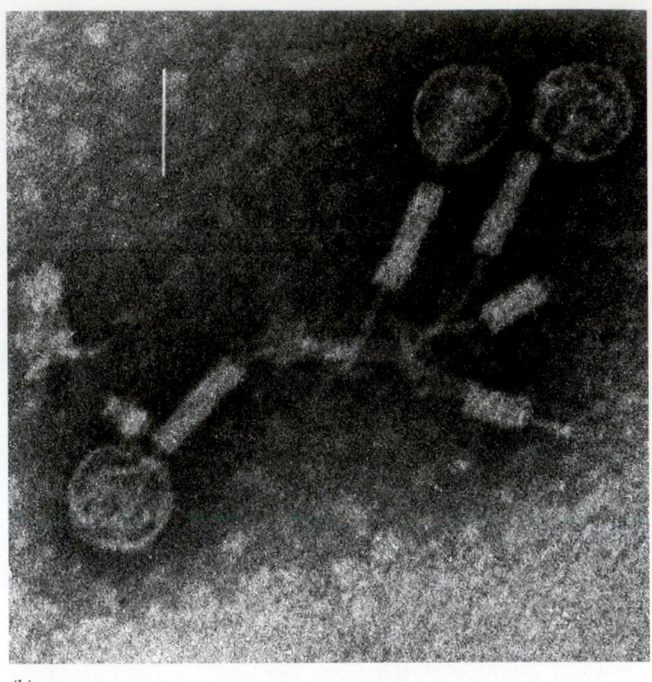

(b)

FIGURE 2.15 (a) Human influenza virus. Note the flexibility of the envelope and projections on the surface. (b) Bacteriophages (×250,000, bar = 60 nm).

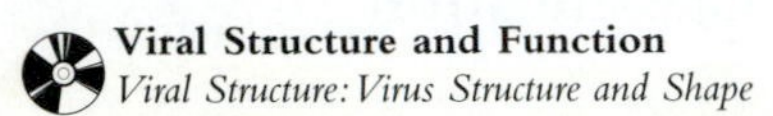

Viral Structure and Function
Viral Structure: Virus Structure and Shape

different from that of the nucleocapsid alone. Figure 2.16d illustrates a virus similar to the influenza virus whose nucleocapsid is helical but whose envelope confers a somewhat spherical shape on the entire virus particle (a complete virus particle is also known as a **virion**). Table 2.6 shows how some of this information is used to classify certain human viruses.

Reproduction

Viruses are **obligate intracellular parasites.** Since they lack organelles such as ribosomes, they cannot manufacture their own proteins and thus cannot generate their own energy. They are totally dependent on the protein-synthesizing and energy-generating machinery of their host cells for reproduction and often cause harm to or the death of these cells in the process.

Animal virus reproduction differs in some respects from bacteriophage replication (see chap. 3). The process also differs between the DNA- and the RNA-containing viruses. Poliovirus reproduction can serve as the prototype for the replication of animal RNA viruses (fig. 2.17). The virus adsorbs (attaches) to specific chemical receptors on the host cell's cytoplasmic membrane, and the entire virion is engulfed into the cytoplasm. If a viral envelope is present, it is removed during this entry process. Inside the host cell, enzymes digest the viral capsid to expose the poliovirus nucleic acid, a single-stranded RNA molecule (called +RNA) that serves as the messenger RNA for producing poliovirus proteins. The host cell's ribosomes translate the viral +RNA into one large protein molecule. This molecule is cleaved by cytoplasmic enzymes into a number of proteins to form new viral capsids and an enzyme called a *replicase.* Replicase is an RNA-dependent RNA polymerase that, in a manner similar to its DNA counterpart, makes copies of the viral +RNA strand. In the host cell cytoplasm, the +RNA copies and the viral capsid proteins are assembled into progeny virions. About 500 virions per infected cell are produced. These new virions are released into surrounding tissue when the host cell lyses (ruptures). The events in animal RNA virus replication typically occur in the host cell's cytoplasm, rather than in the nucleus.

Adenovirus is a DNA-containing animal virus primarily associated with self-limited human illness involving the respiratory tract and the eye (conjunctivitis). Adenovirus reproduction is diagrammed in figure 2.18 and can serve as the prototype for animal DNA virus reproduction. A key difference between DNA and RNA animal virus replication is that the progeny virions produced by almost all DNA viruses are assembled in the host cell's nucleus rather than its cytoplasm. Adenovirus replication causes many harmful effects in the host cell; these effects include halting host DNA and protein synthesis, inhibiting mitosis, producing chromosomal abnormalities, and, ultimately, causing cell disintegration with the release of progeny virions.

The retroviruses (for example, the human immunodeficiency virus, the cause of AIDS; see chap. 19) are RNA viruses but have a complex replication cycle that involves DNA as well as RNA and events that take place in both the cytoplasm and the nucleus (see fig. 2.19). After the virus en-

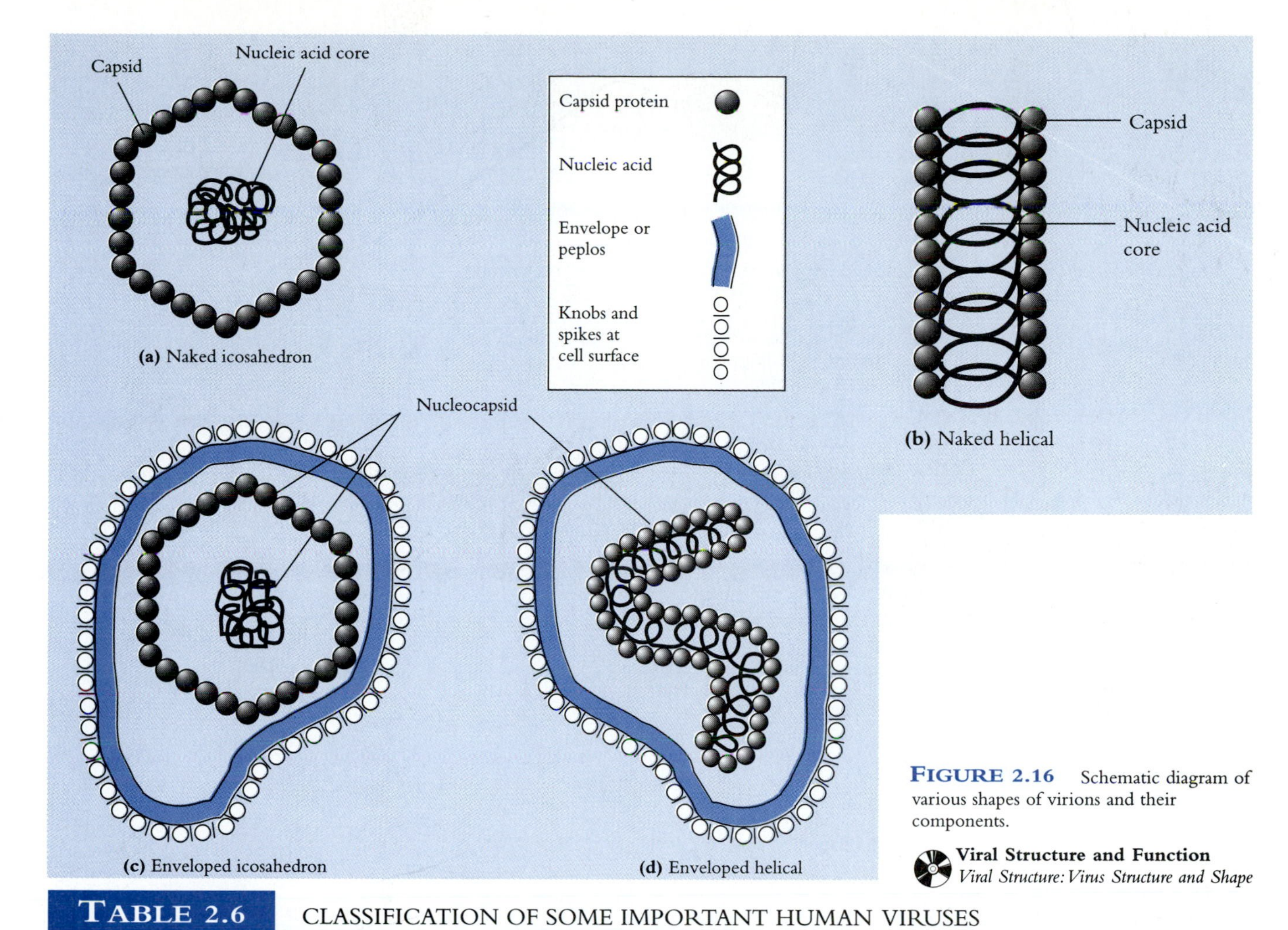

FIGURE 2.16 Schematic diagram of various shapes of virions and their components.

Viral Structure and Function
Viral Structure: Virus Structure and Shape

TABLE 2.6 CLASSIFICATION OF SOME IMPORTANT HUMAN VIRUSES

Virus Group	Human Diseases	Envelope	Capsid Symmetry	Nucleic Acid
Adenovirus Papovavirus	Respiratory infections Warts	Absent	Icosahedral[a]	DNA
Herpesvirus	Chickenpox (varicella) Shingles (zoster) Cold sores (herpes simplex) Infectious mononucleosis	Present		
Poxvirus	Smallpox (variola)	Present	Complex	
Rotavirus Picornavirus	Intestinal infections Polio, intestinal infections, colds	Absent	Double shell Icosahedral	RNA
Togavirus	Encephalitis, yellow fever	Present		
Orthomyxovirus Paramyxovirus Rhabdovirus	Influenza Measles, mumps Rabies	Present	Helical[b]	
Retrovirus	AIDS	Present	Icosahedral	

[a]An icosahedron is a polyhedron having 20 faces (the Greek word *eikosi* = 20; *-hedron* is a crystal or geometric figure).
[b]A helix is a spiral or coiled form.

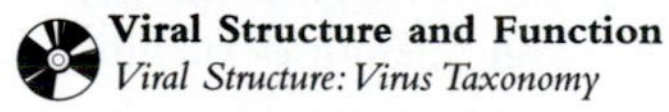
Viral Structure and Function
Viral Structure: Virus Taxonomy

ters the cell, a DNA copy of the +RNA is made by a unique viral enzyme called reverse transcriptase (an RNA-dependent DNA polymerase). The RNA strand is then degraded by a ribonuclease, and a complementary DNA strand is formed, again with the aid of reverse transcriptase. The double-stranded DNA must integrate into the host chromosome as a provirus before transcription and translation of viral proteins can begin. Viral structural proteins and the reverse transcriptase enzyme are synthesized, and assembly of the virus takes place at the host cell cytoplasmic

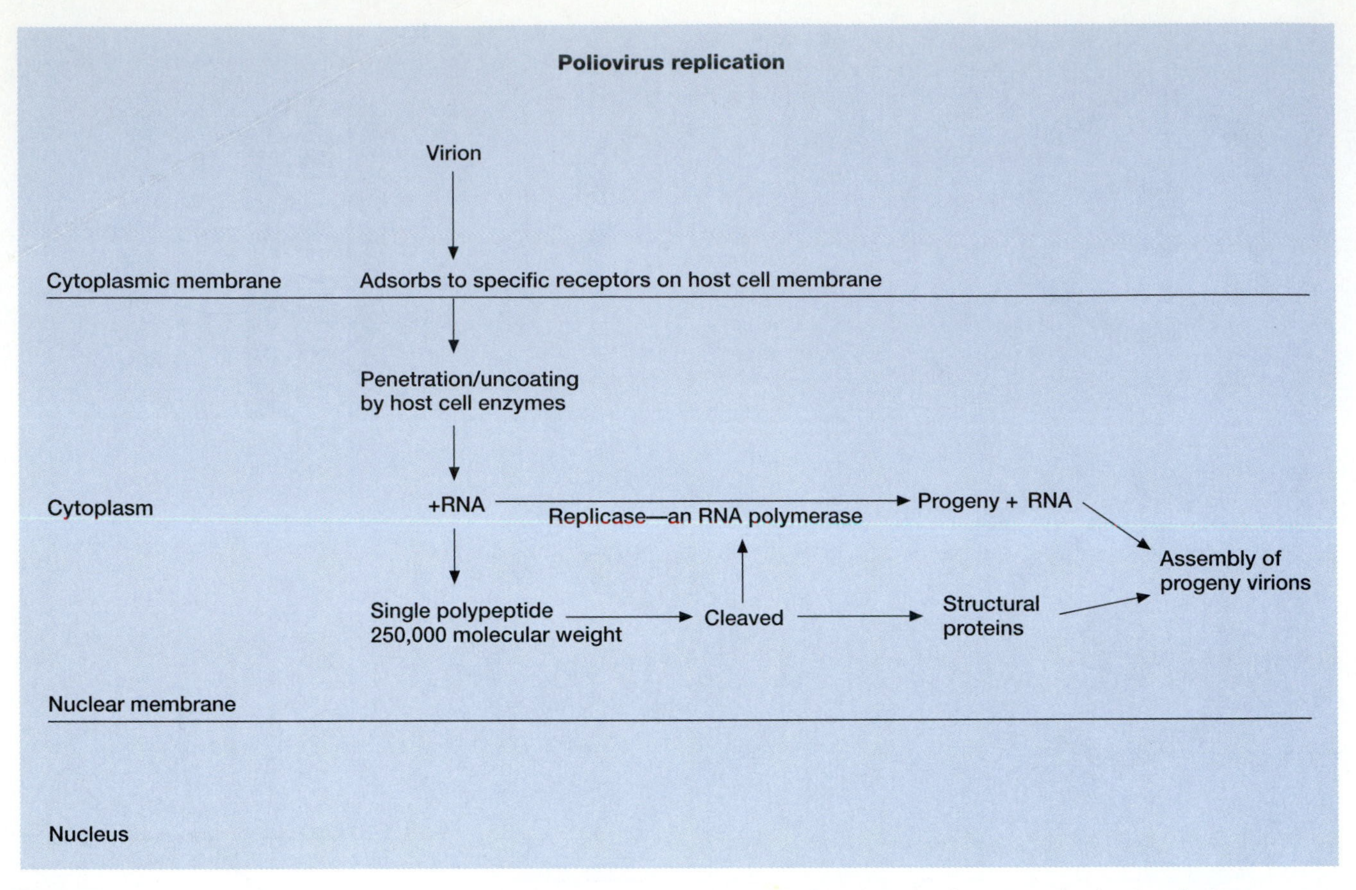

FIGURE 2.17 The sequence of events in the replication of poliovirus, an RNA virus. Note that no nuclear events take place in poliovirus replication.

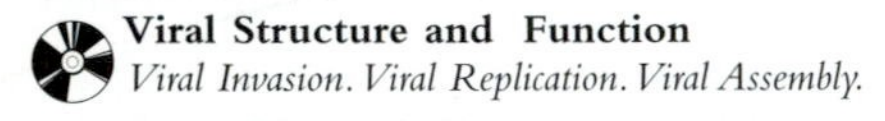
Viral Structure and Function
Viral Invasion. Viral Replication. Viral Assembly.

membrane. In some retroviruses, integration of the provirus into the host cell chromosome can convert the cell into a tumor cell.

Viral reproduction illustrates a major mechanism of viral pathogenesis, that is, destruction of the host cell by the replicating virus. In addition, as we have seen for retroviruses, viral DNA may become stably integrated into the host cell DNA to form a provirus and bring about changes in the host cell that result in either transformation or lysogenic conversion (see bacteriophage reproduction, chap. 3). The provirus hypothesis, postulated by Howard M. Temin, provides a model for how tumor viruses may induce cancer by establishing a lysogenic state in their host cells. In most cases, however, viruses alone do not appear to cause human cancer.

Understanding the replication of retroviruses and their possible role in tumor cell formation was such a great advance in virology that in 1975, Howard Temin and David Baltimore, who independently discovered the reverse transcriptase enzyme, were awarded the Nobel Prize.

Many antimicrobial agents such as penicillins, cephalosporins, tetracyclines, erythromycin, gentamicin, and sulfonamides inhibit bacterial metabolic processes but have no antiviral effects whatsoever because viruses lack all self-metabolic processes. Antimicrobial agents are sometimes prescribed during viral illnesses to prevent secondary bacterial infection (to which persons may become more susceptible) rather than to directly attack the virus.

Only a few antiviral agents are currently available in comparison with antibacterial agents. A better understanding of the mechanism of viral replication has led to the recent development of additional effective drugs, especially for the human immunodeficiency virus (see table 9.8). Examples of available drugs include acyclovir, which is used for treating some serious herpes simplex virus infections and amantadine can be used prophylactically to prevent influenza A virus disease in high-risk groups. Hope for a broad spectrum antiviral agent had been placed in *interferon,* a substance produced by the body itself in response to viral infection. Interferon, which may be thought of as a natural antimicrobial agent, causes biochemical changes in cells that enable them to resist viral attack. Recombinant DNA technology has provided a mechanism for cloning the interferon genes in the bacterium *E. coli.* Large quantities of human interferon have been produced cheaply using this technology. Its use in treating patients, however, has been limited by as yet unsolved problems of toxicity.

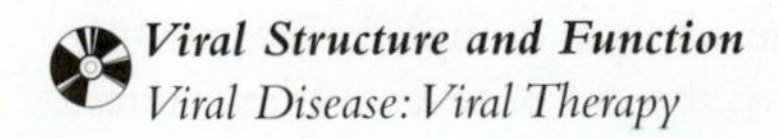
Viral Structure and Function
Viral Disease: Viral Therapy

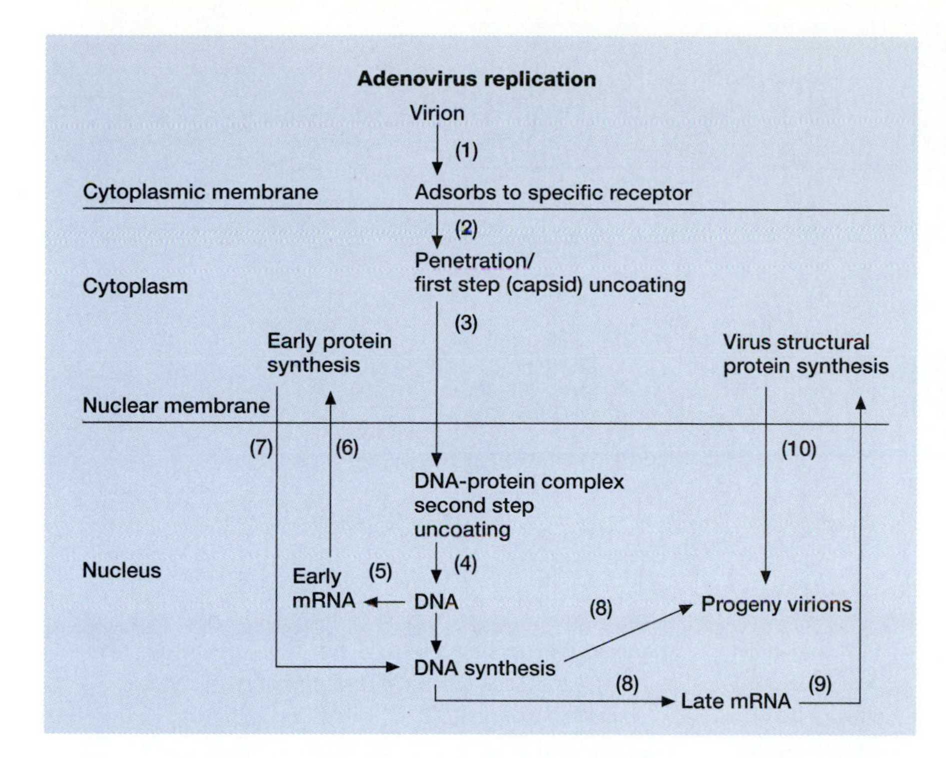

FIGURE 2.18 The sequence of events in the replication of adenovirus, a DNA virus: (1) adsorption by specific receptors on both the virus and the host cell; (2) penetration of the entire virion; (3,4) uncoating by host enzymes and transportation to host nucleus; (5) transcription of viral DNA by host cell enzymes into viral specific messenger RNA; (6,7) early viral proteins manufacture multiple copies of adenovirus DNA; (8,9,10) some of this DNA produces viral structural proteins and some assembles into the progeny virions.

Viral Structure and Function
Viral Invasion. Viral Replication. Viral Assembly.

There is no totally convincing answer to the question of whether viruses are living agents (simple microbes) or nonliving complex chemicals. Since viruses are not composed of cells, they are at odds with the cell theory that states "all living things are composed of cells." However, the cell theory was put forth approximately 60 years before viruses were discovered and almost 100 years before they were seen and their noncellular nature known. Given the proper conditions and appropriate host cell, viruses can direct their own reproduction using the host cell's machinery; this ability to reproduce is characteristic of living things. Viruses seem to occupy an ambiguous position on the fringe of current classifications of living things. Simpler entities since discovered, such as the plasmids and viroids, may replace viruses in that position. Richard P. Novick has suggested that plasmids are subcellular organisms poised on the threshold of life. Thus our concept that all life is cell based may need to be reappraised. Novick suggests that any nucleic acid system controlling its own replication should be considered an organism.

VIROIDS

Viroids, short strands of RNA, lack the protein coat found on RNA viruses, and are even smaller agents of infectious diseases than viruses. (Virus nucleic acid has a molecular weight of 10^6 or more, while the potato spindle-tuber viroid has a molecular weight of only about 10^5, considerably smaller than even the smallest viruses.) To date, viroids have been associated only with certain plant diseases; however, it is likely that they may cause human and animal diseases as well. All known viroids are infectious RNA molecules, but new viroids may be found that are naked DNA molecules, lacking a protein coat.

PRIONS

Certain central nervous system diseases of humans and animals are caused by agents called **prions** (proteinaceous infectious particles) that have unconventional properties and are markedly different from both animal viruses and viroids. Diseases caused by prions include scrapie, a progressively destructive and fatal disease of sheep and goats, and kuru and Creutzfeldt-Jakob syndrome, which are related fatal diseases in humans. Since they require many years to develop, these diseases are referred to as slow infections.

The scrapie agent consists exclusively of a protein referred to as PrP (for *P*rion *P*rotein), and exhibits unusual physical, chemical, and biologic properties. Among these unusual properties are a molecular weight of only 3×10^4, a filamentous form, a size smaller than any known viroid, and a remarkable resistance to radiation and enzymes. There are many theories as to how this protein can cause disease but the answer has so far eluded scientists who study it.

In 1996, the cattle industry in the United Kingdom suffered great economic loss when a variant form of Creutzfeldt-Jakob disease diagnosed in 10 people was said to be related to a degenerative brain disease of cattle known as bovine spongiform encephalopathy or "mad cow disease." The cattle appeared to have acquired this prion disease when, as calves, they were fed meat and bone meal supplements from cattle carcasses. Although the link between the two diseases was not proved conclusively, imports of beef from Britain were banned in most European countries, and

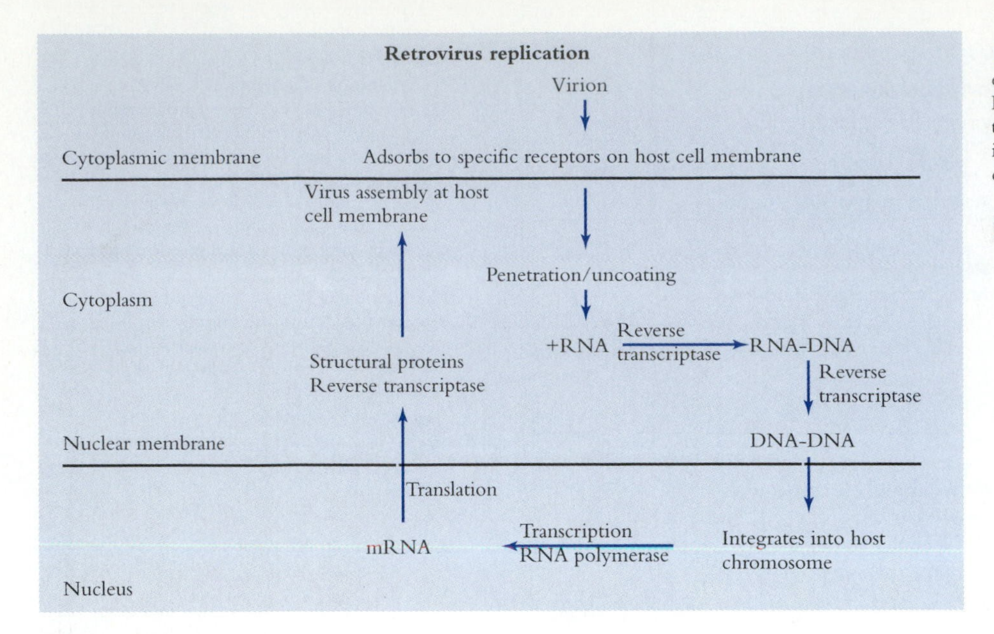

FIGURE 2.19 The sequence of events in the replication of a retrovirus, an RNA virus. Replication is more complex than that of other RNA viruses and involves events that take place in both the cytoplasm and nucleus (see text).

Viral Structure and Function *Viral Invasion. Viral Replication. Viral Assembly.*

beef consumption in the United Knigdom fell drastically. Studies did not detect evidence of this variant form of Creutzfeldt-Jakob disease in the United States. The increased awareness of the disease has stimulated research that may provide insights into the characteristics of this unique agent and the epidemiology of the disease produced.

THE EUCARYOTES

The protozoa and fungi are eucaryotic, higher microorganisms. Their individual cells are considerably more complex and larger than bacterial cells. Both are very widely distributed in nature, and both groups include genera that can produce disease in plants, animals, and humans.

PROTOZOA

The animal kingdom probably began with the single-celled protozoa. Many thousands of species of these simple animal cells exist, exhibiting wide variation in their biologic properties, size, shape, and structure. ***Protozoa*** are the largest of the unicellular microorganisms, but they range greatly in size. Some are just barely visible without the help of a microscope; others approach the size of the largest bacteria and must be magnified nearly 1,000 times to be seen clearly. They are structurally simple but their physiologic activities—food intake and utilization, excretion, locomotion, reproduction—are quite complex.

The protozoa are subdivided into four phyla on the basis of the structural means of locomotion possessed by their mature forms (fig. 2.20). The most important types of protozoa in terms of human disease are the following:

Rhizopoda *are the simple ameboid forms that move by bulging and retracting their protoplasm in any direction.*

Ciliophora *possess cilia, or fine, minute hairs, that cover the outer cell surface. Rapid rhythmic beating of the cilia moves these cells along with efficient speed.*

Zoomastigina *possess long, whiplike flagella projecting from their bodies. The lashing of these threadlike extensions provides rapid motility.*

Apicomplexa *have no external means of locomotion, but some immature forms display an ameboid, gliding motility. This group has a complex reproductive cycle, maturing as spores or sporozoites.*

Each of these groups contains organisms associated with human disease. Some protozoan infections are minor, but others are quite serious or even life threatening (see chap. 18).

Rhizopoda (Amebae)

The cell bodies of **amebae** (from the Greek word *amoibe,* meaning "change") are constantly altering in shape as they glide about.

Morphology

Active "vegetative" amebae are called **trophozoites** (*tropho* = taking nourishment). They have a thin cell membrane, a granular cytoplasm containing ingested food and food vacuoles, and a well-defined nucleus within a demarcated membrane. (They range in diameter from about 10 to 60 μm.) The granular endoplasm is sharply demarcated from the clear, homogeneous **ectoplasm** (*ecto* = without) that separates it from the cell membrane (fig. 2.20a). The external membrane seems to function as the eyes, ears, nose, and fingers of the cell, responding sensitively to the slightest change in the surrounding environment. The cell moves continuously as the membrane extends in a fingerlike projection first in one direction, then in another, and the endoplasm flows into each bulge, then retracts and surges elsewhere in an endless quest. The cell protrusions are called **pseudopods,** or false feet, because of the motion they provide. They also are the means by which food is ingested, reaching out to surround and engulf nutrient particles (unlike bacteria, amebae are not restricted to soluble nutrient).

COLORPLATE ONE

Morphology of Common Pathogenic Bacteria

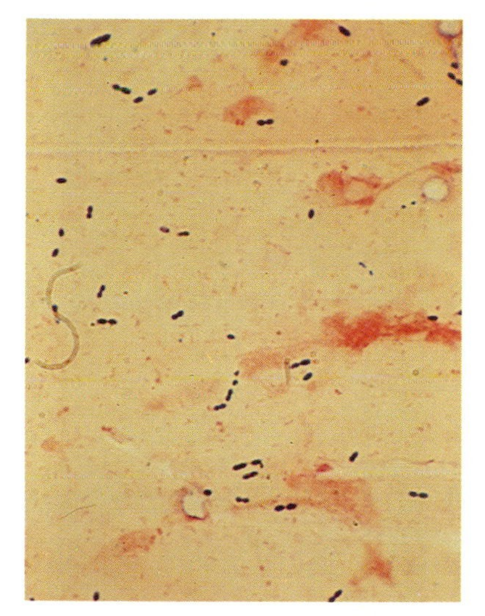

a) Cocci (spheres) in pairs (diplococci).

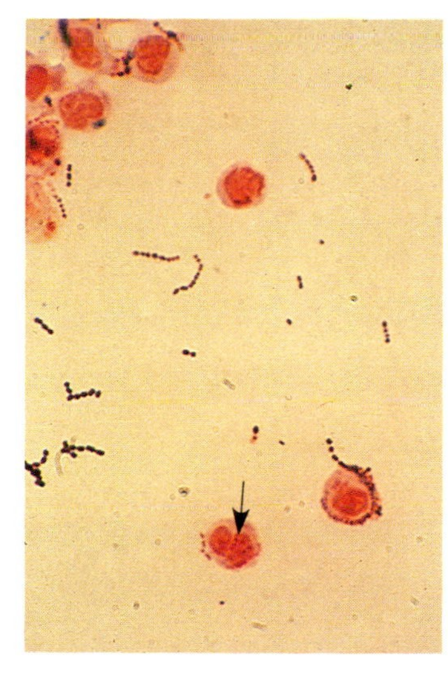

b) Cocci in chains (streptococci).

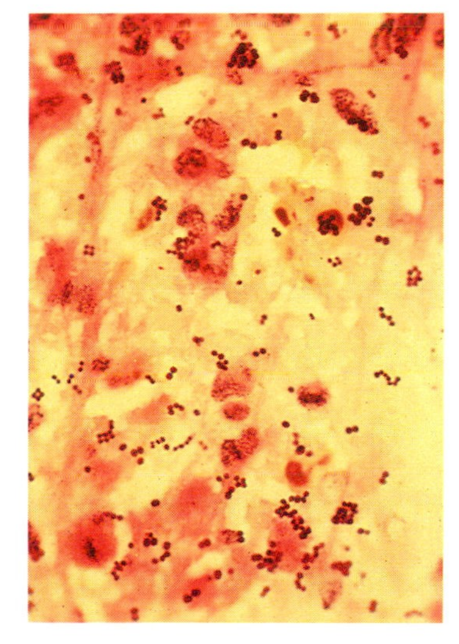

c) Cocci in irregular clusters (staphylococci).

Courtesy Dr. E.J. Bottone, Mt. Sinai Hospital, New York, NY.

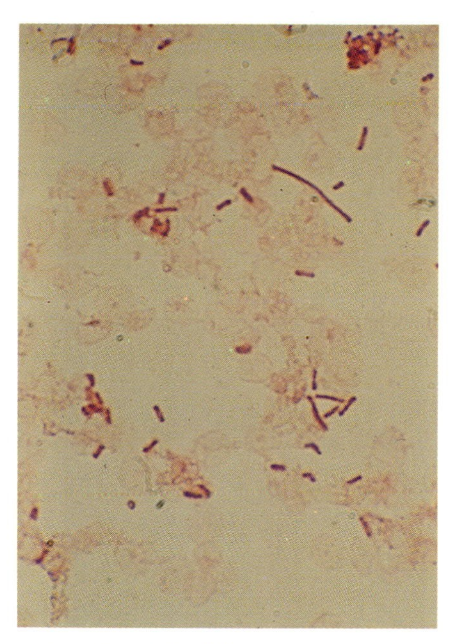

d) Straight rods (bacilli).

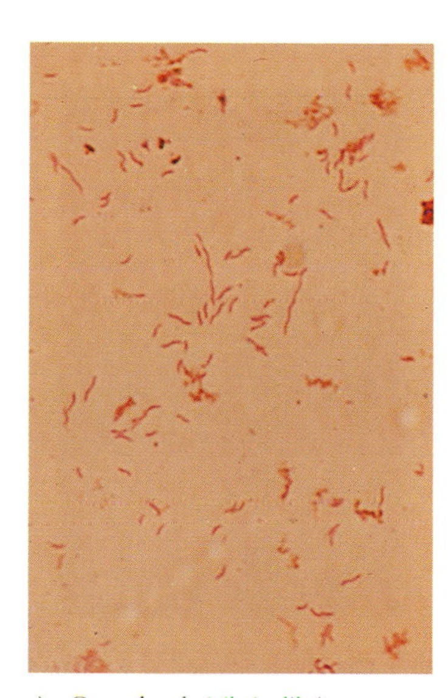

e) Curved rods (vibrio-like).

Courtesy Dr. E.J. Bottone, Mt. Sinai Hospital, New York, NY.

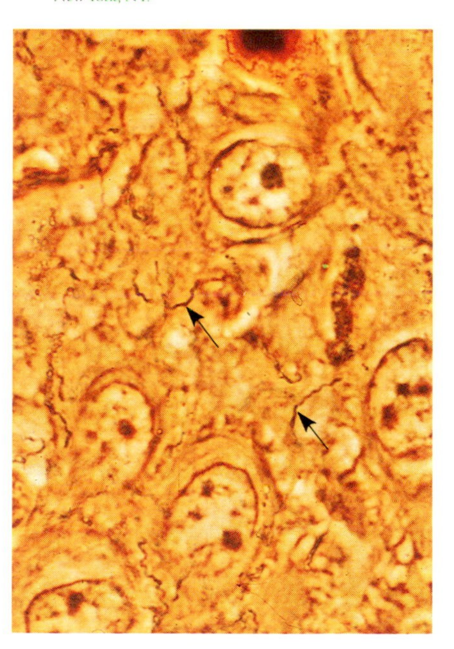

f) Spirochetes (flexible spirals) (arrows).

Courtesy Dr. E.J. Bottone, Mt. Sinai Hospital, New York, NY.

Photomicrographs illustrating morphologic characteristics of some bacteria.

Bacterial Structure and Function

Cell Wall: Bacteria Shapes. Bacteria Groups: Gram Stain Categories

Microscopic Observations of Pathogenic Bacteria Using Special Methods

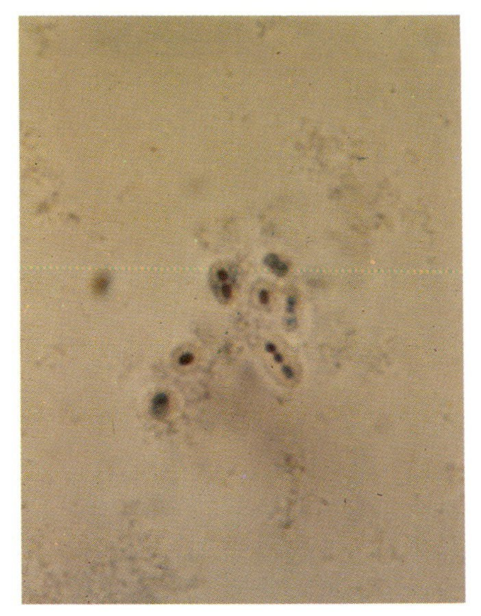

a) A quellung reaction in which pneumococcal cells have been treated with pneumococcal antiserum. The pneumococcal capsule (seen as a halo around the bacteria) appears to swell and is sharply defined.

Courtesy Dr. E.J. Bottone, Mt. Sinai Hospital, New York, NY.

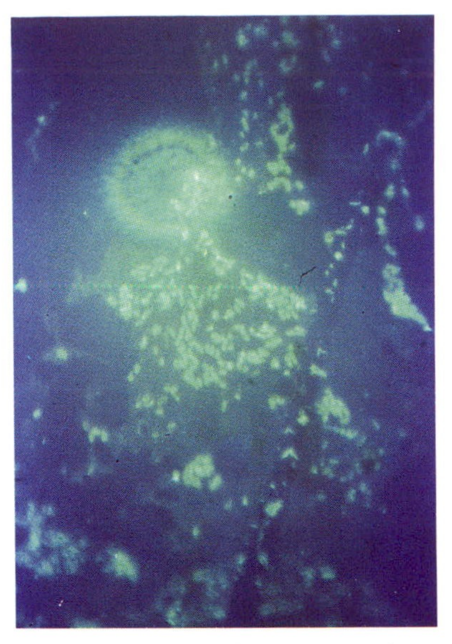

b) Fluorescent antibody preparation of *Legionella pneumophila*. A smear of a patient's tracheal aspirate has been treated with *L. pneumophila* antibody conjugated with a fluorescent dye. The bright fluorescent *Legionella* bacilli stand out against the dark background when viewed microscopically with an ultraviolet source.

Courtesy Dr. E.J. Bottone, Mt. Sinai Hospital, New York, NY.

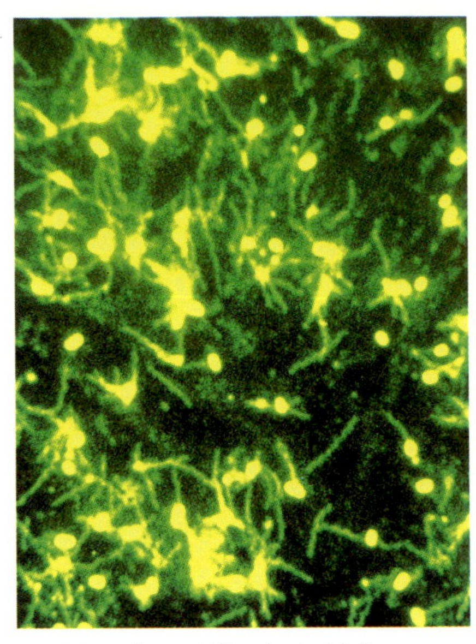

c) *Salmonella enteritidis* stained with fluorescent dye (×500).

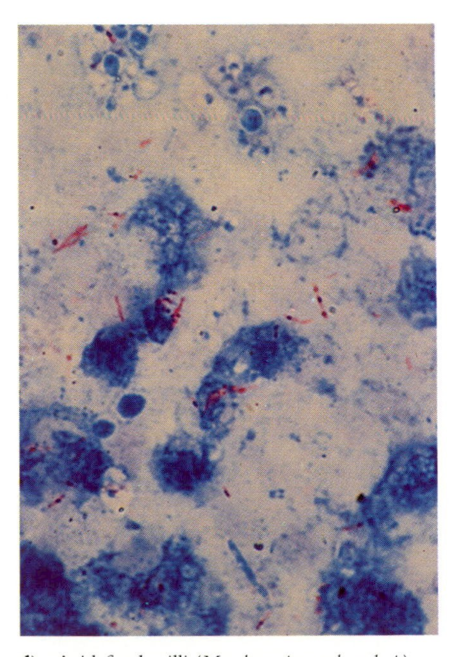

d) Acid-fast bacilli (*Mycobacterium tuberculosis*) in an acid-fast stain of sputum from a tuberculosis patient. The acid-fast bacilli appear as red, beaded rods against a blue background.

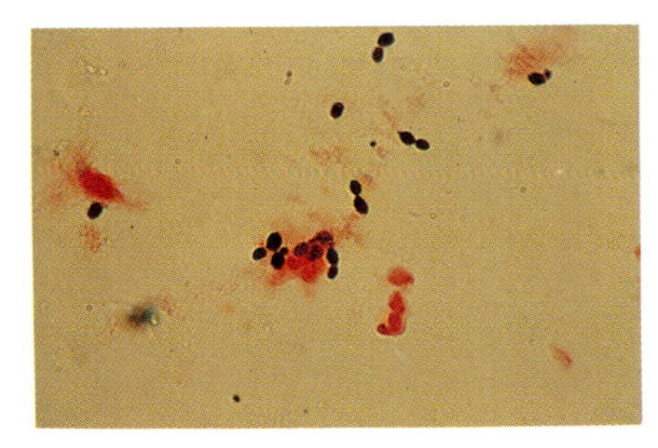

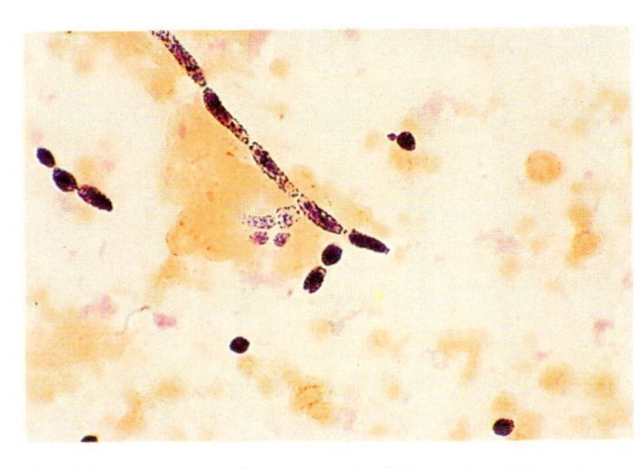

e) Gram stain of yeast cells (*Candida albicans*). (top) Gram-positive budding cells from a sputum smear; (bottom) yeast from a culture of patient's blood. Note that filamentous irregularly staining hyphae have formed.

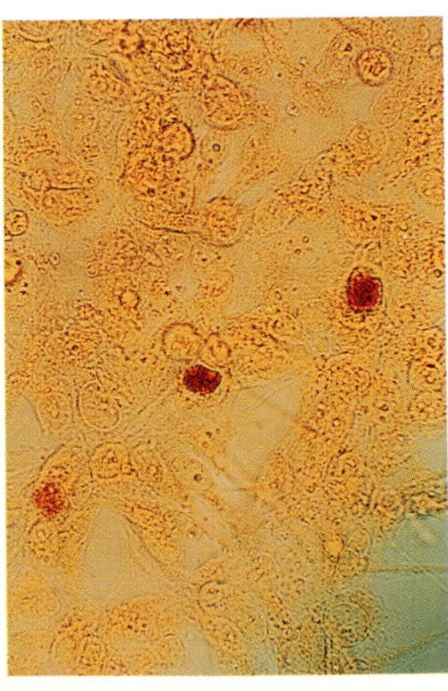

f) Inclusions of *Chlamydia trachomatis* in McCoy cell culture. The glycogen-containing inclusions stain dark brown when the cells are treated with an iodine solution.

Courtesy Dr. E.J. Bottone, Mt. Sinai Hospital, New York, NY.

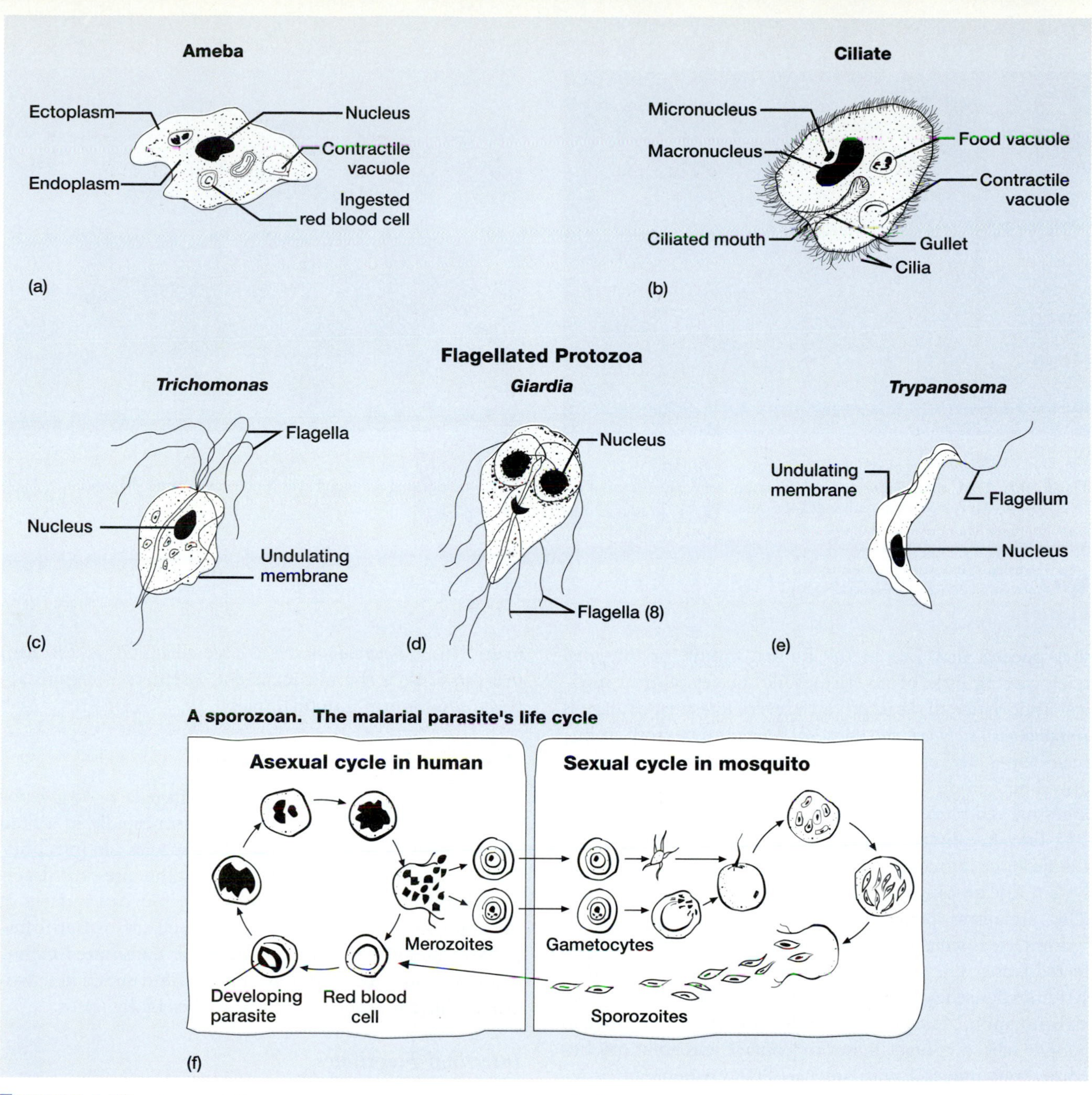

FIGURE 2.20 Diagrams of four types of protozoa. (a) An active ameba. (b) A ciliated protozoan (*Balantidium coli*). (c, d, and e) Three types of flagellated protozoa. (f) Developmental stages of the malarial parasite, a sporozoan (*Plasmodium* species) of the phylum Apicomplexa.

Parasitic Structure and Function
Protozoa Classification: Amoeboid Protozoa; Ciliated Protozoa; Flagellated Protozoa; Sporozoan Protozoa

Once in the cell, the food is taken into vacuoles where it is digested and metabolized, and the waste is excreted through the cell membrane.

Reproduction and Encystment

Amebae reproduce asexually by simple binary fission following mitotic division of the nucleus. Each trophozoite, when it divides, gives rise to two new cells. As part of their life cycle, some amebae are capable of passing into a cyst stage. The trophozoite rounds itself up, the membrane thickens and becomes hyaline (glassy), internal activities slow down, and the cell enters a resting stage. This stage is important in the survival and transmission of some pathogenic amebae because it enables them to survive under adverse environmental conditions. During **encystment** mitotic nuclear divisions occur and several nuclei are produced within one cyst. When excystment occurs, the trophozoite emerges with its multiple nuclei and undergoes fission and nuclear division, eventually forming new uninucleate trophozoites.

Amebic Infection

Free-living species of amebae abound in soil and water throughout the world, but few are parasitic. A number of

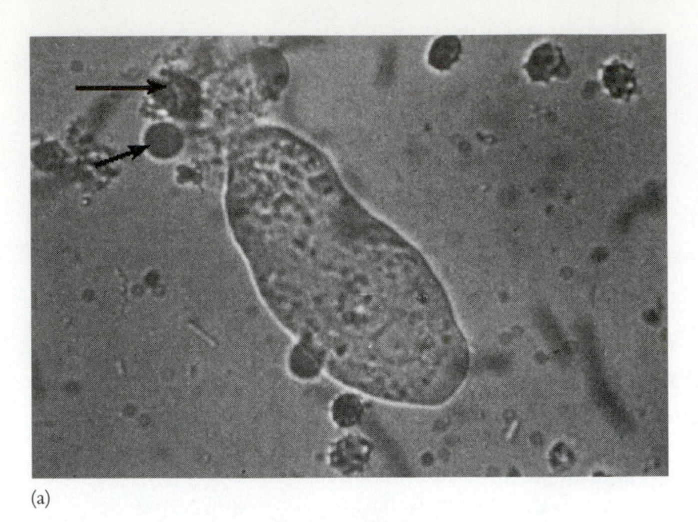

(a)

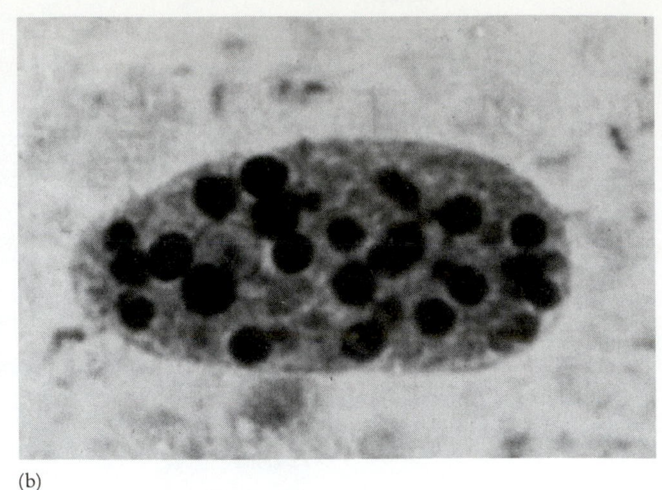

(b)

FIGURE 2.21 (a) *Entamoeba histolytica* trophozoite. The organism is in the process of ingesting red blood cells (see arrows) in this preparation (×2,500). (b) An ameba filled with red blood cells.

(b) Dr. Paul Beaver, Tulane University, New Orleans, LA.

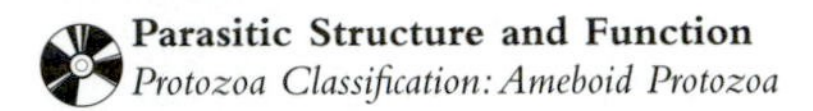

Parasitic Structure and Function
Protozoa Classification: Ameboid Protozoa

such species may live in the human mouth or intestinal tract, causing little or no harm. One species of great medical importance that can produce severe intestinal damage is *Entamoeba histolytica,* the cause of amebic dysentery in human beings. The enzymatic activities of this organism in the bowel may erode the intestinal lining as it nourishes itself by ingesting red blood cells (fig. 2.21a, b).

Two free-living amebae, *Naegleria* and *Acanthamoeba,* can produce human brain or eye infections. Central nervous system disease caused by *Naegleria* is almost invariably fatal. The organisms penetrate through the nasal passages of swimmers, especially divers, who come in contact with infected water. The amebae penetrate to the central nervous system where they multiply rapidly causing tissue injury and inflammation. *Acanthamoeba* produces corneal infections in people whose contact lenses or contact lens solutions become contaminated with amebae. These patients often require corneal transplant to prevent blindness.

Ciliophora (Ciliates)

These organisms have a permanent ovoid shape of fixed dimensions. The largest among them, the paramecia, can be detected with the naked eye, but most are microscopic. Their ciliated surfaces keep them in constant swimming motion in a watery environment. They have more complex structures than amebae, including a mouth and gullet (also lined with cilia that sweep food in), an excretory pore, and two nuclei, one much larger than the other. The smaller or micronucleus is responsible for cell reproduction, and the macronucleus directs the cell's other activities (fig. 2.20b). Reproduction is usually asexual but may also occur by conjugation. Some, but not all, ciliates display an encystment stage; the cysts also contain two nuclei.

Most ciliates are free-living. One species inhabits the intestinal tracts of animals and may, infrequently, infect humans. This is *Balantidium coli,* a large ciliate (70 × 200 μm) that can cause a chronic, recurrent dysentery if it gains access to the human intestinal tract.

Zoomastigina (Flagellates)

All the members of this group possess flagella, or long ectoplasmic threads that propel the organism rapidly in a fluid environment. A number of **flagellate** species can infect human beings. These fall into two groups: the intestinal flagellates that live in the bowel or on other mucosal surfaces of the body; and the blood and tissue flagellates, or **hemoflagellates** (*hemo* = blood). The latter are transmitted to humans by biting insects and live in the soft tissues. These two groups differ morphologically as well as biologically.

Intestinal Flagellates

Morphologically, intestinal flagellates have a fixed, ovoid shape (about 10 × 20 μm), with three to eight flagella that propel them. Species of the genus *Trichomonas* (fig. 2.20c) belong to this group and are characterized by a short, undulating membrane as well as four flagella on the anterior rounded end. They have a single nucleus and occur only as trophozoites (without a cyst stage). *Giardia,* another intestinal flagellate, has two nuclei (note the bespectacled appearance this gives them in fig. 2.20d) and eight flagella, but no additional locomotive membrane. *Giardia* species frequently display a cyst stage with four nuclei.

These two genera are often found in the human intestinal tract where they may live harmlessly. *Giardia* is frequently associated with intestinal disorders when conditions permit it to multiply to very large numbers. One species of *Trichomonas* is commonly found on the genital mucosa of both sexes. If it multiplies excessively there, it can cause troublesome irritations, particularly in women.

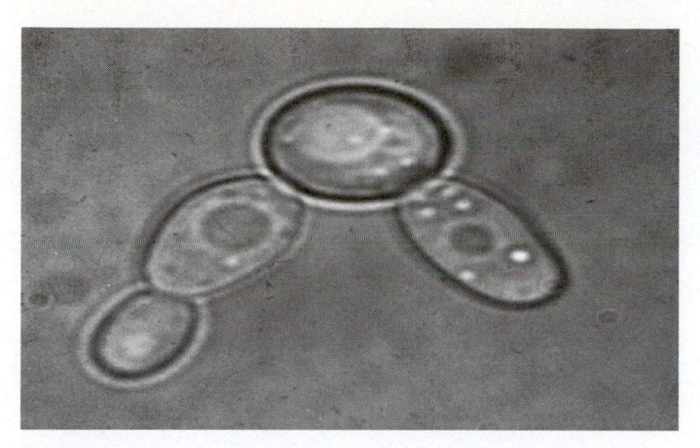

FIGURE 2.22 A photomicrograph of a budding yeast cell (*Saccharomyces* species) (×1,000).
Reproduced from S.S. Schneierson, *Atlas of Diagnostic Microbiology,* 1974, Abbott Laboratories.

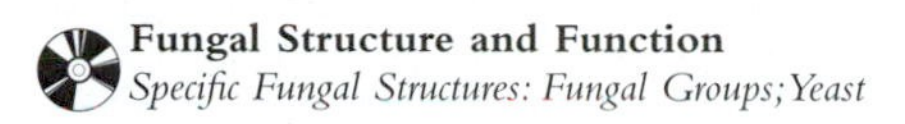

Fungal Structure and Function
Specific Fungal Structures: Fungal Groups; Yeast

Hemoflagellates

Medically, hemoflagellates are the most important flagellates. When biting insects deposit them in the body, they enter the bloodstream and are carried to various sites. They multiply asexually, inducing damage in the soft tissues where they localize (liver, brain, or other organs). The two major genera of this group are *Trypanosoma* and *Leishmania.* The former, when mature, are thinly elongated (2 × 20 μm) with a single flagellum, a long undulating membrane, and a central nucleus (fig. 2.20e). These organisms have a complex cycle of asexual development, some stages occurring in the insect, others in infected humans or animals. *Leishmania* also have several developmental stages, including a flagellated form similar to a trypanosome but lacking an undulating membrane, and a small, nonflagellated oval form that lives intracellularly in humans. African sleeping sickness (trypanosomiasis) and kala-azar (leishmaniasis) are examples of hemoflagellate diseases.

Apicomplexa

The life cycle of **Apicomplexa** involves two means of reproduction. Developmental stages of an asexual type of nuclear division occur in vertebrate animals such as humans, but other transitional forms are produced by sexual union and nuclear division, usually in insects. The cells resulting from such sexual fusion and division are called *sporozoites.* Blood-sucking insects, notably some species of mosquito that prey on vertebrates infected with these organisms, ingest sexual forms of the parasite during the blood meal. Further reproduction in the mosquito leads to the production of sporozoites, which are then injected into the blood of the next animal upon which the mosquito feeds. A reservoir of infection is thus maintained by cyclic transmission of the organism from one animal to another. Malaria is perhaps the best example from among the human diseases caused by Apicomplexa. Malaria is caused by one of several species of the genus *Plasmodium* and is transmitted by the *Anopheles* mosquito (fig. 2.20f).

FUNGI

The fungi are an extremely large and diverse group distributed universally. They range in size and complexity from the simple unicellular yeasts to the multicellular molds and mushrooms.

Fungus Morphology

The body of a multicellular fungus is composed of a mat of long, branching, filamentous tubes containing cytoplasm, in which nuclei are strung out at irregular intervals. In some fungi these tubes are interrupted at numerous points by septations, or cross-walls. The entire mat is referred to as the mycelium (*mykes* is the Greek word for fungus), and the individual filaments that compose it are called *hyphae.* The hyphae form originally from the elongation of a germinating, reproductive cell, microscopic in size, called a *spore* or *conidium.* Some of the yeasts, which are classified with the true fungi, fail to produce a mycelium and merely pinch off daughter cells, or buds, at their germinating tips (fig. 2.22). These yeasts are therefore unicellular.

As more and more mycelium is produced by the growing, branching hyphae, the fungus reaches visible proportions. Its eventual size will depend on the continuing availability of nutrient and on other environmental conditions. Most molds do not grow to a height of more than 0.5 to 1 cm, but they extend laterally along the surfaces of, and downward into, the material supporting them. That portion of the plant extending downward and into the medium is called the *vegetative mycelium,* and that which extends above the surface is the *aerial mycelium.* Usually the reproductive structures are found in the aerial portion, and their presence contributes to the powdery or fuzzy look of the mold.

The fungal kingdom consists of four divisions: *Zygomycota, Ascomycota, Basidiomycota,* and *Deuteromycota* (the *Fungi Imperfecti*).

These divisions are based on both visible and microscopic features of their mycelia, hyphae, and spores and, most important, on their type of sexual reproduction. The *Fungi Imperfecti* are so called because their sexual stages are rare or have not yet been found. Most of the fungi that cause human disease fall in this group, but the others also have some members of medical importance.

Zygomycota

The group ***Zygomycota*** displays a type of sexual reproduction in which a zygote is formed from the fusion of a pair of cells derived from the same or a different fungus (fig. 2.23a). Germination of the zygote leads to production of a new generation. Asexual spores are also formed by the fungi of this subdivision (fig. 2.24a). The mycelium is characteristically nonseptate. It gives rise to specialized hyphae that extend aerially, each developing at its tip a round, saclike structure called a **sporangium.** Within this sac many asexual

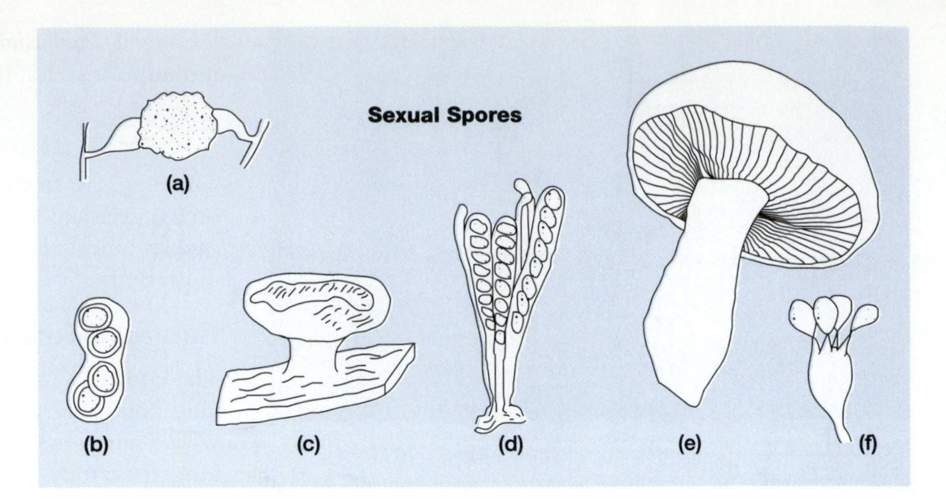

FIGURE 2.23 Some types of sexual spores of fungi. (a) is a zygospore, or zygote, commonly formed by zygomycetes; (b) is an ascus containing four sexual spores of an ascomycete; (c) and (d) represent more complex ascus formation seen in the higher ascomycetes; (e) is a basidiomycete (a mushroom), and (f) shows its basidium bearing basidiospores.

From N.F. Conant, D.T. Smith, and J.L. Calloway, *Manual of Clinical Mycology,* 3d ed. Copyright © 1971 W. B. Saunders Publishers, Philadelphia, Pennsylvania. Reprinted by permission.

Fungal Structure and Function
Specific Fungal Structures: Fungal Groups; Mold. Metabolism and Growth: Fungal Reproduction

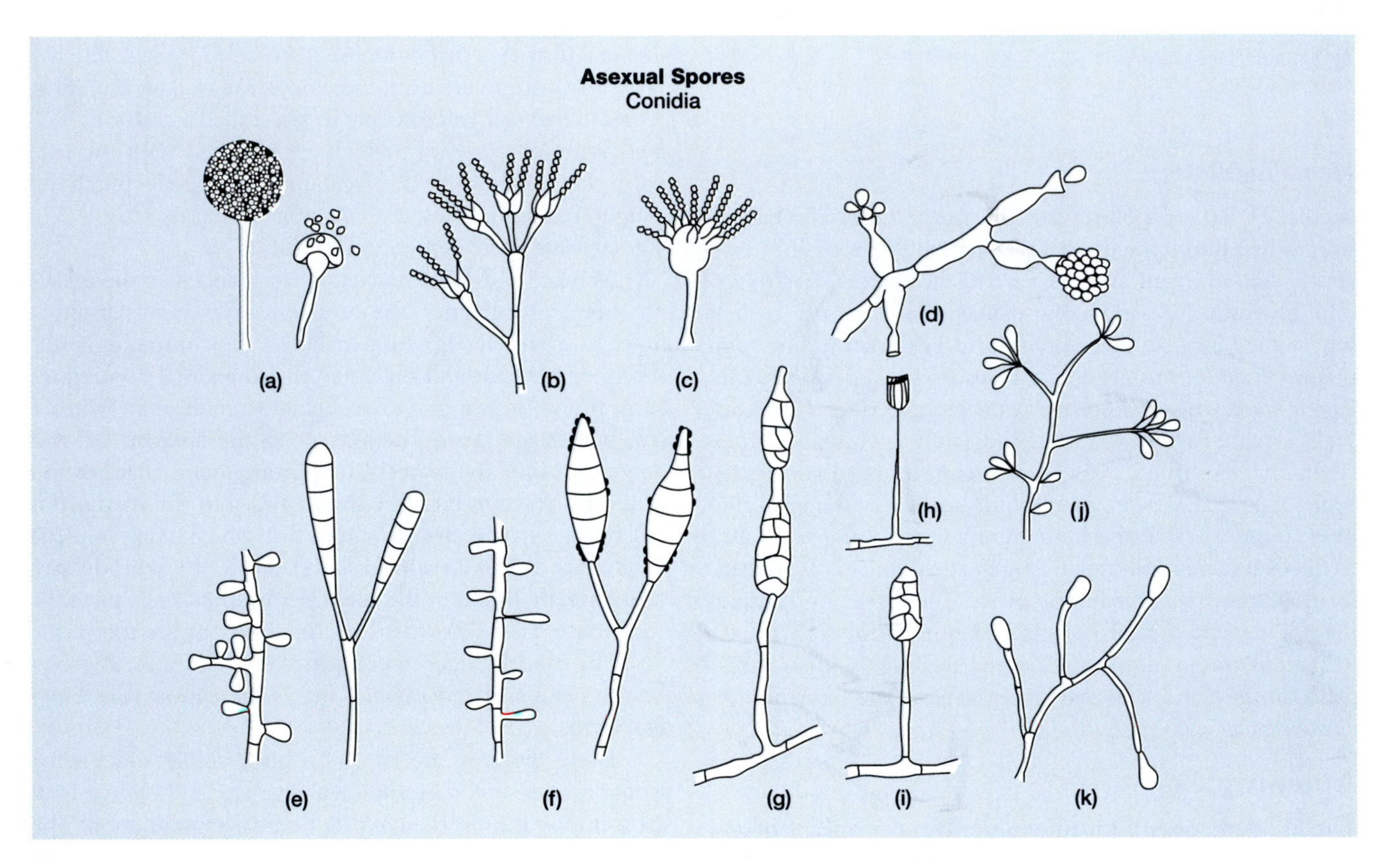

FIGURE 2.24 These drawings illustrate several types of asexual spores (conidia) produced by different fungi. The term "conidia" (sing., conidium) is broadly used to denote asexual spores borne directly on the fungal hypha or on specialized spore-producing structure called conidiophores.

From N. F. Conant, D. T. Smith, and J. L. Calloway, *Manual of Clinical Mycology,* 3rd ed. Copyright © 1971 W. B. Saunders Publishers, Philadelphia, Pennsylvania. Reprinted by permission.

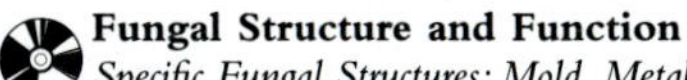

Fungal Structure and Function
Specific Fungal Structures: Mold. Metabolism and Growth: Fungal Reproduction. Taxonomy and Nomenclature: Fungal Morphology

sporangiospores develop, and when these are released by rupture of the sac (fig. 2.24a), each may germinate to produce new hyphae and a new mycelium.

One of the most familiar members of the Zygomycota is the common black bread mold, *Rhizopus nigricans.* Some other members of the genus *Rhizopus* are associated occasionally with human disease, usually through opportunistic circumstances. This is true also of species of *Mucor* and *Absidia,* two other genera of this group (see Zygomycosis, chap. 14).

Ascomycota

Ascomycota is the largest of the fungus subdivisions. It includes the true yeasts as well as some mildews and molds. The characteristic sexual spores of this group always occur in even numbers (usually eight), within a sac called an **as-**

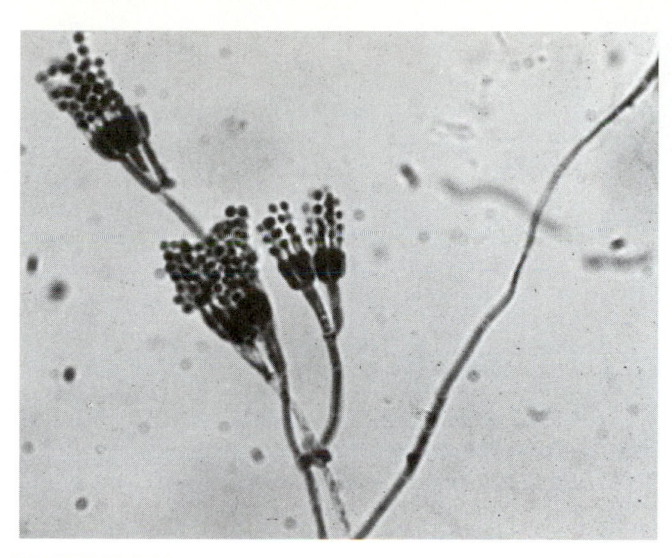

FIGURE 2.25 *Penicillium notatum* growing in a slide culture (×236). Courtesy Dr. John W. Rippon, Pritzker School of Medicine, University of Chicago.

Fungal Structure and Function
Specific Fungal Structures: Fungal Groups; Mold

Antimicrobial Action
Approach to Antimicrobial Study: Origins of Antibiotics

cus, from which they take the name **ascospore** (fig. 2.23b). The asexual spore of the yeasts is a budding cell called a **blastospore** (fig. 2.22). Many mold ascomycetes have specialized hyphae (**conidiophores**) that produce asexual spores called **conidia,** borne in chains or singly as shown for b, c, or k in figure 2.24. Conidia is a general term referring to a kind of asexual spore seen in many molds.

Bakers' yeast, and other yeasts used in brewing beer or in wine making, are well-known examples of ascomycetes. Most yeasts of medical importance are included in the genus *Candida.* These are not classified with the ascomycetes but with the deuteromycetes. Several species of *Candida* are often found living as commensals on human mucosal surfaces among other microorganisms there, but if circumstances permit these yeasts to overgrow, or if they have an opportunity to find their way into deeper tissues, severe infection can result.

Basidiomycota

The familiar mushrooms and toadstools are included in the group ***Basidiomycota,*** together with the rusts and smuts that cause disease in plants. A club-shaped structure called a **basidium** produces the typical sexual spore, a **basidiospore.** In the case of mushrooms, the vegetative mycelium develops underground and pushes upward a compact mass, or button, that develops into the small, commonly seen umbrellalike structure. The basidia form on the gills underneath the umbrella, producing basidiospores that drop to the ground, giving rise to new plants if the right conditions exist (fig. 2.23e, f). With the exception of the yeast *Cryptococcus,* this group contains no members capable of causing infectious disease in humans, but some mushrooms are medically important because they contain a substance that is poisonous if eaten. *Cryptococcus* is associated with serious diseases of the lung, brain, and other vital organs.

Deuteromycota (Fungi Imperfecti)

Deuteromycota is a large and very mixed group of fungi in which a recognizable sexual stage has not always been observed and whose final classification is uncertain. Its members produce many types of asexual spores, which form an important basis for identification of the species, together with other microscopic details and visible appearance (fig. 2.24d–j).

Most of the fungi important in human mycotic diseases are grouped with the **Fungi Imperfecti.** A number of them cause infections involving only superficial tissues, that is, the skin and mucous membranes. "Athlete's foot" and ringworm are examples of such infections. A small group of fungi is also associated with serious systemic diseases, in which various vital organs, including the brain, may be involved. Among these are diseases such as histoplasmosis, blastomycosis, and coccidioidomycosis in which the tissue involvement may become extensive and treatment difficult.

Role in Antimicrobial Agent Production

The role played by some fungi in treatment of many bacterial diseases of humans has grown dramatically in importance ever since the famous ascomycete, *Penicillium* (fig. 2.25), was first demonstrated to produce penicillin, a substance found to be strongly antagonistic to many bacteria. Since that discovery, the search for such antibacterial agents (commonly called *antibiotics* or, more recently, antimicrobial agents), among fungi and other organisms has grown in both scope and intensity. Untold numbers of fungi and fungal products have been studied exhaustively in laboratories around the world in the effort to produce drugs that can be safely used in the treatment of human infectious diseases. Most of these products have had to be discarded either because they are toxic to humans or because they are relatively inefficient in their action on bacteria. However, a handful of clinically useful, relatively safe agents has emerged from the search and has become an inseparable part of the medical approach to infectious disease. In an effort to hasten the discovery of new antimicrobial agents, the pharmaceutical industry has begun to use new methods such as genetic modification of known antibiotic-producing organisms and recombinant DNA technology (see chap. 3) for isolating genes involved with antibiotic production.

ANIMAL PARASITES

A number of parasitic, invertebrate worms, or helminths (the Greek word for worm, derived from the meaning "to roll"), cause human and animal diseases. In their adult forms most worms are large enough to be seen without the help of a magnifying lens. A microscope is typically necessary to diagnose the presence of their immature stages (ova, or eggs, and larvae) in the excretions or tissues of infected patients. Microscopic and other microbiologic techniques are sometimes used in their study, and the basic principles of prevention and control of helminthic diseases are similar to those applicable in true microbial infections.

MORPHOLOGY

The helminths are soft-bodied invertebrate animals whose anatomic structures include differentiated muscle tissue; nervous, excretory, and reproductive systems; and often a simple digestive tract. They range in size from nearly microscopic dimensions (the adult pinworm is just barely visible without a magnifying lens) to a meter or more in length (some adult tapeworms can be measured with a yardstick).

There are two major groups of helminths: the nemathelminths, or roundworms, and the platyhelminths, or flatworms. They differ in anatomic respects as well as in their reproductive methods and life cycles.

Nemathelminths

The **roundworms** or **nemathelminths** (also often called nematodes) are cylindric with bilateral symmetry. The outer skin, or integument, of roundworms is a tough cuticle that may be smooth, rough, or spiny. Their bodies are not segmented, and they are pointed at both ends. A few members of this group, notably the hookworms, have hooks or cutting plates around the mouth for attaching themselves to the intestinal wall of the animal or human whose body they have entered. Roundworms have a complete digestive tract and excretory system, as well as a nervous system. The sexes are usually separate. The larger female produces eggs constantly, as many as 200,000 a day in some cases. After fertilization larval development begins within the egg. When the larvae are hatched, they may go through several subsequent stages before the adult form is reached.

The cyclic development of roundworms from adult to adult, through egg and larval forms, varies greatly among species. In the sections of part 2 where helminthic diseases are discussed (chaps. 18 and 20), the life cycle of each worm is considered in its relation to human infection, as well as to methods of prevention and control.

Among the important nematodes associated with human disease are hookworm and *Ascaris* (fig. 2.26), which infect the intestinal tract; the *trichina* worm, whose larvae become lodged and encysted in muscle tissue; and the *filarial worms,* whose adult and larval forms live both in the blood and in various tissues.

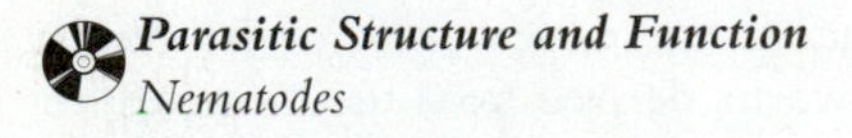
Parasitic Structure and Function
Nematodes

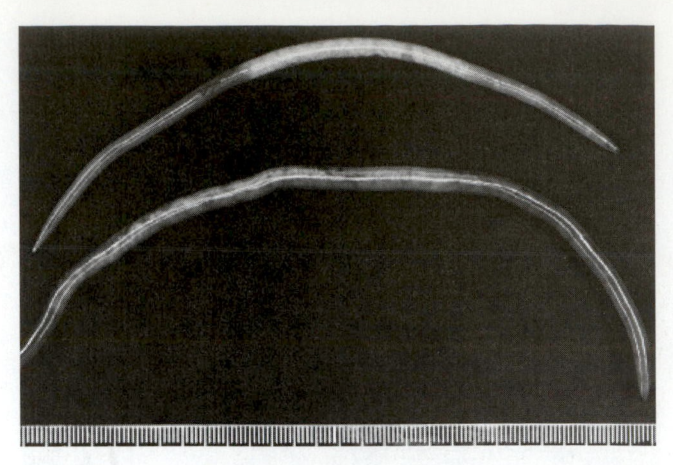

FIGURE 2.26 Adult forms of *Ascaris lumbricoides,* a nematode or roundworm. The ruler measures 15 cm (see also figs. 18.8, 18.9, and colorplate 8a).

Platyhelminths

The **flatworms** or **platyhelminths** also display bilateral symmetry and organized tissues differentiated for special functions. They are oriented dorsoventrally, hence the term *flat.* They vary greatly in shape and size, some being thinly elongated and segmented, others being nonsegmented and thickly ovoid. On the basis of such anatomic differences, the flatworms are subdivided into two major groups: the cestodes, which are tapeworms, and the trematodes, often called flukes.

Cestodes

The bodies of **tapeworms** or **cestodes** are flattened, elongated, and composed of individual segments, each of which contains both male and female sex organs. This combination of sexes in one individual is termed *hermaphroditism.* The series of segments begins with a head, or scolex, equipped with muscular suckers, and sometimes with a circlet of hooks as well, which permit attachment to the inner intestinal wall of the human or animal host. Just below the scolex is an area of growth, called the neck, from which arises a series of maturing segments, or **proglottids,** each containing fully developed sex organs, and nervous and excretory systems, but no digestive tract. At the far end of the series, mature segments become filled with eggs and are said to be *gravid.* The whole length of a tapeworm is called the **strobila,** its full reach varying from three to four proglottids to several hundred depending on the type of worm (fig. 2.27). Eggs are extruded from the segments into the host's intestines, or the gravid proglottids may break off and be passed from the body in the feces, together with free eggs. Sometimes long chains of proglottids are passed, providing startling visible evidence of infection. As long as the scolex and neck remain attached to the bowel wall, however, regeneration of segments continues and the strobila can grow again, even if all proglottids have broken away.

The eggs of tapeworms must pass through stages of larval development before the adult form is reestablished. For

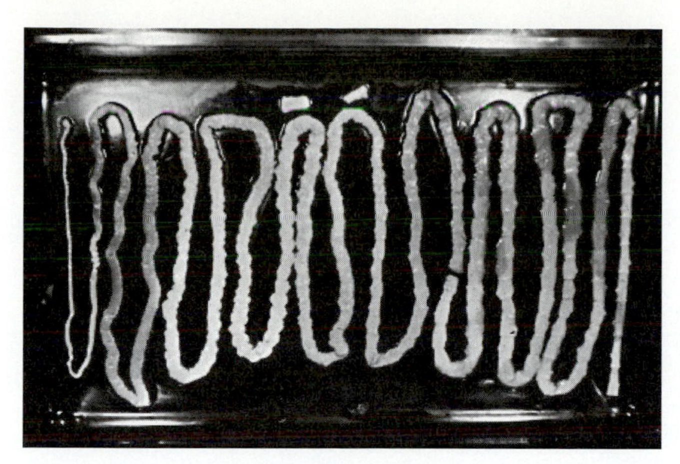

FIGURE 2.27 This tapeworm (*Taenia saginata*) is 285 cm long. The entire strobila (fully developed worm) with scolex (head) was recovered from a case of human infection. This dish measures 27.5 × 40 cm.

the most part these intermediate stages require one or more additional animals in which they can develop. Most of the tapeworms that infect humans live in the intestinal tract in the adult form (e.g., the so-called beef, pork, and fish tapeworms). In one important case, humans may harbor the larval stage of a small tapeworm, *Echinococcus,* whose adult form lives harmlessly in the intestinal tract of dogs. The larvae penetrate human (or other animal) tissue and develop structures called *cysts,* or *hydatids* (fluid-filled sacs). The condition that results is spoken of as hydatid disease (see fig. 18.19).

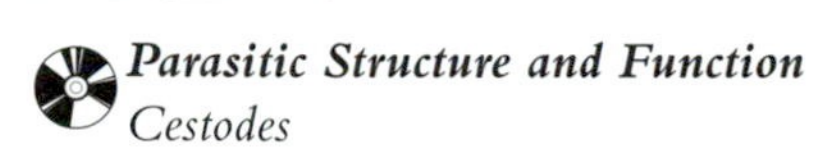

Parasitic Structure and Function
Cestodes

Trematodes

The **trematodes,** or **flukes,** are grouped on the basis of anatomic and functional differences into (1) those that are flattened, thickly leaf-shaped, and hermaphroditic, and (2) those that are elongated, round, and bisexual. None of the flukes is segmented and they have only a rudimentary digestive system. All of them have a tough outer cuticle and muscular suckers for attachment to the host. Humans and other mammalian animals are the hosts for the adult forms, which inhabit and produce eggs at various sites in the body, while consecutive larval stages require one or more aquatic animals to complete their development.

The common names given to flukes refer to those sites in the body where the adult usually lives and produces its eggs. On this basis they are divided into groups known as blood (the schistosomes), intestinal, liver, and lung flukes. Structurally, the blood flukes are elongated, round, and bisexual. The adults are quite small, comparable in size to the smallest roundworms. They live within blood vessels, the female producing her eggs in small capillaries of such organs as the large intestine, liver, or urinary bladder. All the other flukes have a thickly ovoid shape and are hermaphroditic. The intestinal fluke lives in the lumen of the upper part of the small bowel, the liver fluke in bile ducts, and the lung fluke in lung tissue. The eggs produced by flukes at these various sites may often be trapped in tissues where they cannot develop further but stimulate defensive reactions in the host. Many eggs, however, may find an exit route from the body through feces, urine, or sputum. If and when they are deposited in water, the ova hatch into larval forms, which then pass through several generations in aquatic hosts. Humans are infected by ingesting the larvae of infected fish, crustaceans, or water plants (hosts differ for different types of flukes), or by direct penetration of skin by a larval form swimming freely in infected water. Development of the adult then proceeds in the final host, and the production of eggs continues the cycle.

The insects that have great importance in the transmission of many infectious diseases, helminthic or microbial, are described in chapter 8 in the discussion of sources and routes of infection.

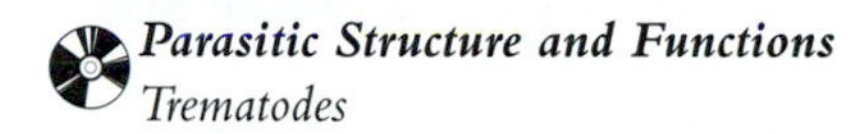

Parasitic Structure and Functions
Trematodes

QUESTIONS

1. How are microorganisms distinguished from all other living things?
2. What is the basis for classification within the plant and animal kingdoms?
3. What features distinguish procaryotic and eucaryotic cells?
4. What is the ancestor of living eucaryotic cells?
5. Describe the basic shapes and group patterns of bacteria.
6. What is the significance of protein synthesis in bacteria?
7. Explain the function of each of the following structures in bacteria. a. cell wall b. cell membrane c. mesosome d. flagella e. endospore f. nucleus g. ribosome h. capsule i. pili j. plasmids
8. How do rickettsiae and chlamydiae differ from other bacteria?
9. Describe the basis for virus classification. What is a virion? How does it function?
10. How does viral replication differ from that of bacteria?
11. How does reproduction in RNA viruses differ from that of DNA viruses?
12. How do retroviruses differ from DNA and RNA viruses?
13. What is the basis for protozoan classification? List one example of a pathogen in each group.
14. What is the basis for classification of fungi? List one example of a pathogen in each group.
15. What is the basis for helminth classification? List one example of a pathogen in each group.

BIBLIOGRAPHY

Barghoorn, E. S. 1971. The oldest fossils. *Sci. Am.* 224 (5): 30–43.

Becker, W. M. 1986. *The world of the cell.* Menlo Park, Calif.: Benjamin/Cummings.

Brock, T. D. 1988. The bacterial nucleus: A history. *Microbiol. Rev.* 52: 397–411.

Brock, T. D., and M. A. Madigan, 1990. 6th ed. *Biology of microorganisms.* New York: Prentice Hall

Cohen, S. N., and J. A. Shapiro. 1980. Transposable genetic elements. *Sci. Am.* 242 (2): 40–49.

Darnell, J. E. 1985. RNA. *Sci. Am.* 253 (4): 68–78.

deDuve, C. 1985. *A guided tour of the living cell.* New York: Scientific American Books.

Diener, T. O. 1987. *Viroids.* New York: Plenum Press.

Dixon, B. 1994. *Power unseen: How microbes rule the world.* San Francisco: W. H. Freeman.

Dubos, R. 1976 *The professor, the institute, and DNA.* New York: Rockefeller University Press.

—1962. *The unseen world.* New York: Rockefeller University Press.

Essex, M., and P. J. Kanki. 1988. The origins of the AIDS virus. *Sci. Am.* 259 (4): 64–67.

Felsenfeld, G. 1985. DNA. *Sci. Am.* 253 (4): 59–67.

Gallo, R. C. 1987. The AIDS virus. *Sci. Am.* 256 (1): 47–56. 1986.

—The first human retrovirus. *Sci. Am.* 255 (6): 88–98.

Gootz, T. D. 1990. Discovery and development of new antimicrobial agents. *Clin. Microbiol. Rev.* 3 (1): 13–31.

Grivell, L. A. 1983. Mitochondrial DNA. *Sci. Am.* 248 (6): 78–89.

Haseltine, W. A., and F. Wong-Staal. 1988. The molecular biology of the AIDS virus. *Sci. Am.* 259 (4): 52–62.

Jahn, T. L., E. C. Boree, and F. F. Jahn. 1978. *How to know the protozoa.* 2d ed. Dubuque, Iowa: Wm. C. Brown.

Lappe, M. 1982. *Germs that won't die: The medical consequences of the misuse of antibiotics.* Garden City, N.Y.: Doubleday.

LeGuenno, B. 1995. Emerging viruses. *Sci. Am.* 273 (4): 56–64.

Margulis, L., and K. Y. Schwartz. 1982. *Five kingdoms: An illustrated guide to the phyla of life on earth.* San Francisco: W. H. Freeman.

Moulder, J. W. 1984. Looking at chlamydiae without looking at their hosts. *ASM News.* 50 (8): 53–62.

Novick, R. P. 1980. Plasmids. *Sci. Am.* 243 (6): 103–127.

Pollack, R. 1994. *Signs of life: The language and meanings of DNA.* Boston: Houghton Mifflin.

Porter, K. R., and J. B. Tucker. 1981. The ground substances of the living cell. *Sci. Am.* 244 (3): 57–67.

Ptashne, M., A. D. Johnson, and C. O. Pabo. 1982. A genetic switch in a bacterial virus: Two regulatory proteins interact with a region of viral DNA. *Sci. Am.* 247 (5): 129–140.

Radetski, P. 1991. *The invisible invaders: Viruses and scientists who pursue them.* Boston: Little Brown.

Scott, A. 1985. *Pirates of the cell: The story of viruses from molecule to microbe.* New York: Basic Blackwell.

Shapiro, J. A. 1988. Bacteria as multicellular organisms. *Sci. Am.* 258 (6): 82–89.

Spector, D. H., and D. Baltimore. 1975. The molecular biology of the poliovirus. *Sci. Am.* 232 (5): 25–30.

Varmus, H. 1988. Retroviruses. *Science* 240: 1427–1435. 1987.

—Reverse transcription. *Sci. Am.* 257 (3): 56–64.

Vidal, G. 1984. The oldest eukaryotic cells. *Sci. Am.* 250 (2): 48–57.

Weaver, R. F., and P. W. Hedrick. 1995. *Genetics.* 2nd ed. Dubuque, Iowa: Wm. C. Brown.

Weber, D. J., and W. A. Rutala. 1988. Bacillus species. *Infect. Control. Hosp. Epidemiol.* 9 (8): 368–373.

Weiss, E. 1982. The biology of rickettsiae. *Annu. Rev. Microbiology* 36: 345–370.

Woese, C. 1981. Archaebacteria. *Sci. Am.* 244 (6): 98–122.

Woese, C. 1987. Bacterial evolution. *Microbiol. Rev.* 51 (2): 21–71.

CHAPTER 3

Microbial Life

OUTLINE

Microbes in Motion

This chapter covers the biochemical and physiologic nature and activities of microorganisms and their economic uses. Also discussed are mechanisms of bacterial growth, survival, reproduction, and mutation. The following books and chapters of the Microbes in Motion CD-ROM may be a useful preview or supplemental study aid to your reading: Bacterial Structure and Function: Cell Membrane; External Structures. Microbial Metabolism and Growth: Growth; Metabolism; Genetics. Viral Structure and Function: Viral Invasion; Viral Replication; Viral Assembly. Fungal Structure and Function: Metabolism and Growth. Mechanisms of Antimicrobial Resistance: Acquisition/Expression of Resistance Genes; Transfer and Spread of Resistance.

OBJECTIVES

On completing this chapter the student will be able to:

1. explain briefly how the biochemical and physiologic nature and activities of microorganisms contribute to continuation of life on earth.
2. discuss some of the economic uses of microorganisms.
3. describe briefly the production and use of antimicrobial agents in the treatment of human and animal disease.
4. cite several genetic mechanisms involved in bacterial growth, survival, and resistance to antimicrobial agents.
5. describe some mechanisms of bacterial reproduction, including mutation and the transfer of genetic material through conjugation, transduction, and transformation.
6. indicate the potentially practical applications of gene splicing.

KEY WORDS

Aerobic
Anaerobic
Antibiosis (anti-by-**OH**-sis)
Antimicrobial (anti-my-**CROW**-bial) agents
Bacterial genetics
Bacterial growth curve
Bacterial transposons
Bacteriophage (back-**TEAR**-ee-oh-**FAHGE**)
Binary (**BY**-nary) fission
Catalyst
Chemosynthetic autotroph (**KEEM**-oh-sin-**THET**-ic **AUTO**-troph)
Chemosynthetic heterotroph (**HET**-er-oh-troph)
Coenzymes (**CO**-**EN**-zimes)
Conjugation
Deamination (dee-am-ee-**NATION**)
Death phase
Denitrification
Dominant gene
Enzyme
Exponential (logarithmic) phase
Extrachromosomal DNA
F plasmid
Facultative anaerobes
Fermentation
Gene splicing
Generation (doubling) time
Genetic recombination
Growth curve
High-energy bond
High frequency of recombination (Hfr)
Holozoic (**WHOLE**-oh-**ZO**-ic)
Infectious drug resistance
Jumping genes
Lag phase
Logarithmic (log-ah-**RITH**-mic) phase
Mutation
Nitrification (**NIGHT**-ree-fee-**KAY**-shun)
Nitrogen cycle
Nitrogen-fixing
Nonsense mutation
Obligate intracellular parasite
Oxidation-reduction reaction
Parasite
Photosynthesis (photo-**SIN**-the-sis)
Photosynthetic autotroph
Photosynthetic heterotroph
Plasmid
Point mutation
Putrefaction (**PEW**-tre-**FACTION**)
R factors
Reading-frame shift
Recessive gene
Recombinant DNA
Respiration
Restriction endonucleases (end-oh-**NUKE**-lee-aces)
Saprophytic (sap-roe-**FIT**-ic)
Stationary phase
Substitution mutation
Substrate
Transduction
Transformation
Transition mutation
Transversion mutation

THE BIOSPHERE

Geologists speak of our planet as displaying three zones: the lithosphere, or outer solid part composed of rock and soil; the hydrosphere, the aqueous envelope of the earth, containing its many bodies of water; and the atmosphere, the gaseous envelope or surrounding mass of air. Biologists speak of another zone, the *biosphere,* meaning that portion of the planet where life exists. The biosphere is maintained through a constant cycle of energy and chemical supplies among earth's diverse living organisms, although overwhelming human activities increasingly threaten these vital exchanges. Microorganisms have always played a profoundly important role, together with plants and animals, in the cyclic maintenance of earth's environment.

PLANTS, ANIMALS, AND MICROBES

Plants and animals maintain and share the planet's life systems by taking out, transforming, and returning essential factors. For example, the metabolic activities of green plants provide atmospheric oxygen for human and animal **respiration.** The green, light-sensitive pigment chlorophyll in plants enables them to use light energy to manufacture carbohydrate from water and atmospheric CO_2, a process known as **photosynthesis** (*photo* = light, *synthesis* = manufacture). Oxygen, a by-product of the reaction, is returned to the atmosphere, whereas the carbohydrate provides energy for other plant processes. A simple formula for the plant photosynthetic reaction is

$$6CO_2 + 6H_2O \xrightarrow[\text{chlorophyll}]{\text{light}} C_6H_{12}O_6 + 6O_2$$

where $C_6H_{12}O_6$ represents the carbohydrate formed. During their respiration animals inhale oxygen from the air, use it in their own vital chemical reactions, and return CO_2 to the atmosphere during exhalation. This cycling of oxygen and CO_2 supports the differing needs of plants and animals, while the light energy, briefly fixed in photosynthesis, is returned into the biosphere as heat. Constant exchange among living things ensures the availability of materials needed for life processes.

The influence of microorganisms in the biosphere is far-reaching. In the soil, some of them live on the roots of plants, providing them with inorganic nitrogen compounds synthesized from atmospheric nitrogen. Plants are unable to use nitrogen as an element and could not survive without the chemical transformations accomplished by their microbial root and soil companions. Microorganisms also decompose dead plants and animals, converting their components into nutrient substances usable by plants. In oceans, rivers, and lakes, microbial activities similarly assure rotations of nutrients essential to higher aquatic life forms. In the bodies of animals, microorganisms living on surface tissues often make important contributions to the host's well-being. For example, microorganisms in the gastrointestinal tract of ruminant animals (e.g., cows or sheep) digest the cellulose of grass and hay to provide nutrition. Also, substances that are manufactured by intestinal organisms can be used by animals in their own growth processes. As we know, microbes can also produce harmful and destructive effects. They cannot only cause disease in higher organisms, but they are responsible for costly deteriorations of food and many materials.

To understand the way microorganisms interact with our world, let us define some of their major nutritional patterns and see how these explain their activities.

NUTRITIONAL PATTERNS

In order to grow and carry on their vital processes, all forms of life must have a source of usable nutrients and energy. The key nutritional element in all living structures is carbon, which must be converted to the complex organic materials of protoplasm. Energy is available directly from sunlight or from the chemical breakdown of organic and inorganic compounds.

Originally, organisms were classified simply as either *autotrophic,* being relatively independent of other forms of life in their nutrition, or *heterotrophic,* deriving their food from organic matter manufactured by other living things (*hetero* = other; *auto* = self; *troph-* implies nourishment). To better characterize nutritional patterns, organisms are classified according to their principal carbon source and the nature of their energy source. Thus **photosynthetic autotrophs** such as plants, algae, and some bacteria use CO_2 from the atmosphere and light energy to manufacture organic building blocks. The group of **chemosynthetic autotrophs,** which consists of only a few genera of bacteria, also uses CO_2 as its principal carbon source, but its energy is obtained from the chemical oxidation of inorganic compounds rather than from light.

The heterotrophs cannot use inorganic carbon directly but must use organic compounds such as carbohydrates, proteins, and lipids (fats) as their major carbon source. A small group of bacteria comprise the **photosynthetic heterotrophs,** which rely on light for their energy source. Human beings, animals, protozoa, fungi, and most bacteria are **chemosynthetic heterotrophs.** They depend on the chemical energy released from a variety of oxidation-reduction reactions of organic compounds. Table 3.1 compares the carbon and energy sources used by each nutritional type of microorganism.

Heterotrophs that live on the dead organic material produced by or derived from other organisms are said to be **saprophytic** (*sapro* = dead or rotten). This term also implies that the organism absorbs organic nutrients in soluble form, whereas a heterotroph capable of ingesting solid food is termed **holozoic** (*hol-* = whole). Among the true microorganisms, amebae and many other protozoa are holozoic. Their membranous walls and specialized structures enable them to ingest particulate food and excrete wastes. Many bacteria and fungi are saprophytic because their rigid cell walls are capable of exchanging only soluble materials with their environment.

TABLE 3.1 COMPARISON OF AUTOTROPHIC AND HETEROTROPHIC ORGANISMS

Nutritional Type	Principal Carbon Source	Energy Source	"Lifestyle"
Photosynthetic autotroph	CO_2	Light	Independent
Chemosynthetic autotroph	CO_2	Oxidation of inorganic compounds	Independent
Photosynthetic heterotroph	Organic compounds	Light	Dependent, saprophytic
Chemosynthetic heterotroph	Organic compounds	Oxidation-reduction of organic compounds	Dependent, saprophytic, or parasitic

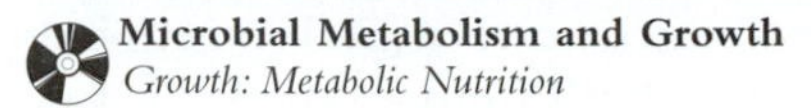

Microbial Metabolism and Growth
Growth: Metabolic Nutrition

Saprophytic microorganisms vary widely in their nutritional needs. Many can survive in a wide range of environments as long as essential nonliving organic nutrients are available. Those that live in the bodies of animals or plants, deriving support at the expense of their hosts, are referred to as **parasites.** Some saprophytes may take up a parasitic life if the opportunity presents itself. Or, they may live independently in the environment as their circumstances or specific nutritional needs require.

When parasitic organisms damage their hosts sufficiently to disrupt normal functions and cause illness, they are said to be *pathogenic* (disease producing). Many of them have rather strict nutritional needs for the organic substances present in their parasitized hosts. Such organisms are said to be *fastidious* because they cannot survive in an environment that does not provide their specific nutritional requirements. They can be cultivated in the laboratory on complex inanimate nutrients, but in nature they are usually restricted to parasitism.

Some microbial pathogens cannot use their hosts' prefabricated nutrients but survive only within living host cells, making them even more demanding parasites. These **obligate intracellular parasites** (e.g., rickettsiae, chlamydiae, and viruses) require specific substances produced within the actively metabolizing cells of their hosts.

MICROBIAL ACTIVITIES

The metabolic processes of all forms of life—plant, animal, and microbial—have certain common vital components. Among these, the most important are the **enzymes.** These organic **catalysts** promote the uptake and use of raw materials, the synthesis of materials needed for cell energy and function, and the degradation of unneeded substances that then become available again in the environment. As catalysts, enzymes initiate and direct chemical reactions without taking a direct part in them. They not only act within cells but may be secreted by cells into the surrounding environment. Here they may break down solid food particles to make them soluble for absorption through cell walls or in other ways make the environment more supportive to the life of the cell.

Enzymes are usually quite limited, that is, *specific,* in the kind of reactions they catalyze. Some are involved only in protein digestion, some break down particular carbohydrates, and others act on fatty substances. With these specific limitations on their activity, a battery of different enzymes is required to carry out the many metabolic processes of both multicellular and unicellular organisms. Many different microbial and higher organisms contain the same enzymes or enzyme systems—a fact that is not surprising when one considers the uniform distribution of **substrates** (substances on which enzymes act specifically) such as carbohydrates, lipids, and proteins. Some general metabolic processes in which enzymes participate are described below.

RESPIRATION AND FERMENTATION

The microorganisms that cause human disease or have important economic impact generate energy through a series of **oxidation-reduction** reactions in the processes of respiration and fermentation. Oxidation-reduction reactions involve the transfer of electrons or hydrogen atoms from one compound to another. The compound that transfers the electrons loses energy and becomes oxidized whereas the compound that accepts the electrons gains energy and becomes reduced. Transfer of electrons is often accomplished by helper substances known as **coenzymes** (because they accompany the enzymes that promote the particular metabolic reaction), which bind temporarily with the electrons.

In the process of respiration, organisms require oxygen to metabolize energy-yielding products. The oxygen is the final acceptor of electrons and becomes reduced to water by the addition of two hydrogen molecules. During metabolism of carbohydrates by the respiratory process, a large amount of energy is produced. This energy is transferred to a chemical bond, referred to as a **high-energy bond,** in the compound adenosine triphosphate (ATP). ATP is an important carrier of chemical energy in all living cells. When the high-energy bond is broken, the energy is released and is available for use in other metabolic processes. In the process of **fermentation,** glucose is broken down to a variety of end products such as lactic acid or ethanol, depending on the enzymes available to the organism. In contrast to respiration, the final hydrogen acceptor is an organic compound rather than oxygen. Although some ATP is formed during fermentation, the process also requires the input of energy, and thus, fermentation is not as energy efficient as is respiration.

Organisms that require oxygen for growth are referred to as **aerobic** and carry out respiratory metabolism. Many organisms that live in soil and water, such as the fungi and a few bacteria that cause human disease, are strict aerobes. For example, it is not surprising that the organism responsible for the pulmonary disease tuberculosis is a strict aerobe and resides primarily in the oxygen-rich lung when it causes disease. Other organisms are strictly **anaerobic** and cannot survive in the presence of oxygen. Their metabolism is primarily of the fermentative type. Unlike aerobes, anaerobes lack enzymes, such as superoxide dismutase, that are needed to break down toxic oxygen products formed during growth of anaerobes in air. Other factors may also be responsible for their extreme oxygen sensitivity but these have not yet been clarified. Most of the harmless bacteria that live in the human intestinal tract and pathogens such as the tetanus and gas gangrene organisms (species of *Clostridium*) are strict anaerobes. In the case of the pathogens, the hardy endospores of the clostridia can lie dormant in soil for many years, but the vegetative bacteria that germinate from the spores cannot survive in oxygen. In the microbiology laboratory, special methods that exclude oxygen must be used to isolate anaerobes from the infectious processes they cause (see chap. 4). A third group of bacteria is referred to as **facultative anaerobes;** that is, they use oxygen for their metabolic processes if it is present, but they also can carry out fermentative metabolism in its absence. Most of the pathogenic bacteria that will be discussed in part 2 of this book belong in this group.

Other Metabolic Processes

In addition to carbohydrates, microorganisms also require nitrogen to carry out their life processes. Nitrogen is usually obtained from amino acids, but as we shall see later in this chapter, some soil bacteria have the ability to use gaseous nitrogen directly from the atmosphere to form ammonia. The ammonia is subsequently assimilated into amino acids, which are the building blocks of proteins. Carbon-containing compounds and energy formed during the processes of respiration and fermentation are often the starting point for amino acid synthesis and for the synthesis of purines and pyrimidines, the subunits of RNA and DNA.

Microorganisms have developed an incredible array of biosynthetic and degradative processes that enable them to survive and multiply under a wide variety of conditions. Most of these microbial processes are of no consequence to human hosts although in some instances they may be of benefit. Only a small number of the biochemical activities of microbial pathogens are related to disease production, as will be described in later chapters.

Some important metabolic activities of fungi and bacteria are reviewed next.

Fungi

The yeasts and molds are heterotrophic saprophytes and are widely distributed in nature. Most of the fungi that cause human disease have a primary habitat in the soil but can become parasitic if given the opportunity. The pathogenic fungi are a very small group compared with the vast numbers of free-living species that do not display parasitism. Many fungi cause plant diseases, however, and some molds are well known for their ability to grow on and destroy nonliving organic substrates, such as those in fabrics and foods. Mildew and bread molds are good examples.

The yeasts and their metabolic processes have been known and put to good use for hundreds of years. With the help of their enzymes, yeast cells use glucose for energy, converting it, under aerobic conditions (in which free molecular oxygen is available), into carbon dioxide and water. If free oxygen is not available (i.e., conditions are anaerobic), these organisms break down the sugar to ethyl alcohol and CO_2. These reactions in simple formulae are

$$\underset{}{C_6H_{12}O_6 + 6O_2} \xrightarrow[\text{Yeast enzymes}]{\text{Aerobic}} \underset{\text{Carbon dioxide}}{6CO_2} + \underset{\text{Water}}{6H_2O} + \text{Energy}$$

$$C_6H_{12}O_6 \xrightarrow[\text{Yeast enzymes}]{\text{Anaerobic}} \underset{\text{Ethyl alcohol}}{2CH_3CH_2OH} + \underset{\text{Carbon dioxide}}{2CO_2} + \text{Energy}$$

Glucose metabolism by yeasts is of value in making bread. In the leavening process, carbon dioxide released from the breakdown of sugar contributes to the rising of the bread. The main energy food for many animal cells, including human, is also glucose, which is oxidized through the activity of enzymes in muscle cells to produce carbon dioxide and water and, most important, energy.

The fermentation process by which yeasts convert glucose to alcohol leads to the production of wines and other alcoholic beverages. In wine making, grapes are the natural source of the carbohydrate, glucose. Strains of the winemaker's yeast also are found in nature growing with other microorganisms on the skins of grapes; this combination is called the "bloom." The bloom produces natural fermentation of the crushed grapes, which gives the wine its characteristic flavor, or individual bouquet (fig. 3.1). In bad years, if the bloom is overgrown by microorganisms that produce an unwanted flavor, the winegrower uses a laboratory-grown culture of yeast to ferment the grapes. This is the method typically used in the production of North American wines. True vintage wines are produced through natural fermentation by the bloom, which is why knowledge of the locale and year of the grape harvest is important to their purchase.

FIGURE 3.1 The flavor of wines can be enhanced by aging in wooden wine casks (foreground) after a yeast culture has been added to ferment the grapes. Modern wineries use the stainless steel tanks for aging large amounts of wine.

Fungal Structure and Function
Metabolism and Growth: Fungal Metabolism; Fungal Growth

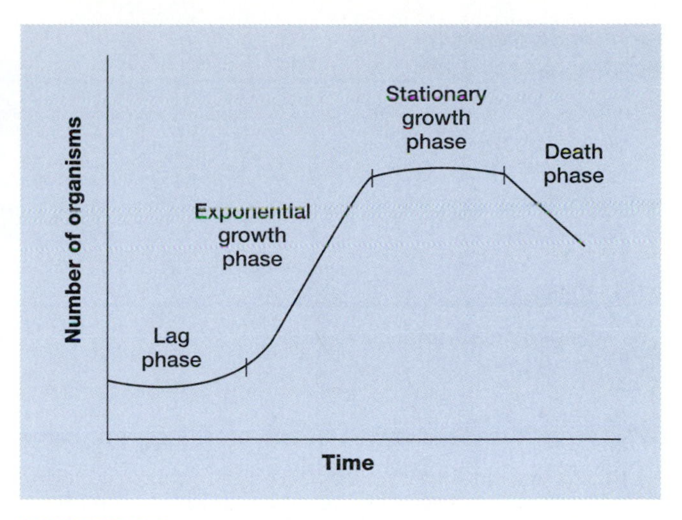

FIGURE 3.2 Bacterial growth cycle in a liquid culture medium.

Microbial Metabolism and Growth
Growth: Growth Curves

BACTERIA

As we have seen, some bacteria are autotrophic, but many are saprophytic heterotrophs. These are widely distributed in soil and water and in the intestinal tracts and on the surface tissues of animals and human beings, thriving for the most part on nonliving organic material. Some of the pathogenic bacteria are obligate parasites. Two important genera, *Rickettsia* and *Chlamydia,* are noted for their strictly intracellular parasitism.

Bacterial Growth

Bacteria multiply by **binary fission,** a process in which a single bacterium divides into two daughter cells. Each daughter cell is identical to its parent cell and can continue to grow and divide. The number of cells increases geometrically in the division process as the population doubles with each succeeding generation. The time required for a single cell or population of cells to double is called the **generation time** or **doubling time.**

The generation time is a genetic characteristic that varies for different microorganisms and is affected by the conditions under which the organism is growing. For example, in an optimal growth medium in the laboratory, *Escherichia coli,* the common intestinal bacillus, divides within 20 to 30 minutes whereas the tuberculosis organism requires 12 to 18 hours to divide. In the human body where conditions are not as favorable as in a culture medium, the doubling times may be longer.

This growth process can only take place when the environmental conditions are appropriate (table 3.2) and the proper chemicals, or nutrients, are present. These nutrients may be organic or inorganic compounds or a combination of both. The basic nutrients needed to support bacterial growth include a carbon source, such as glucose, for energy production and biosynthesis; a nitrogen source, usually in the form of nitrate or ammonia, for the production of amino acids and nucleotides; various minerals such as sulfur and phosphorus; and water, for dissolving the nutrients so that they can be transported across the cytoplasmic membrane into the cell.

When the density of bacterial growth is monitored in a liquid culture medium a typical **growth curve** (fig. 3.2) made up of several important phases is seen: (1) the lag phase; (2) the exponential (or logarithmic) phase; (3) the stationary phase; and (4) the death phase. Each phase is characterized by different physiologic and biochemical activities occurring within the cell.

Lag Phase

When bacteria are introduced into a growth medium, cell numbers do not immediately increase; this is referred to as the **lag phase.** During this time, the cell is adjusting to its new chemical and physical environment by increasing its metabolic activity and producing enzymes that will allow it to grow in the medium. While this increased metabolic activity is taking place the bacterial cells do not divide

TABLE 3.2 ENVIRONMENTAL REQUIREMENTS FOR BACTERIAL GROWTH

pH	Neutrality (pH 7) optimum for many; acidity or alkalinity can be adjusted to meet needs of individual organisms
Water	Essential for all microbial growth
Oxygen	*Aerobes* require atmospheric oxygen; *anaerobes* cannot grow in presence of free oxygen; *facultative* organisms grow under aerobic or anaerobic conditions; *microaerophiles* require that atmospheric oxygen be reduced in total amount
Temperature	Optimum ranges: *psychrophiles,* 5–10° C; *mesophiles,* 35–37° C; *thermophiles,* 50–60° C
Light	Not required for bacterial growth; pigment production may be affected by light; ultraviolet of sunlight may kill bacteria

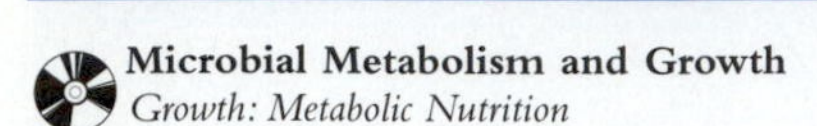

at all or divide only slowly. The line on the growth curve representing the lag phase is nearly horizontal. Cell division does not occur or accelerate until near the end of this phase.

Exponential Phase

The **exponential** (or **logarithmic**) **phase** is an active period in which the bacteria grow and divide at the highest rate possible given their genetic potential, the nature of the medium, and the conditions under which they are grown. Their rate of growth is constant as the organisms divide and double at regular time intervals (i.e., the generation time).

Stationary Phase

As essential nutrients, especially the carbon and nitrogen, that support growth, are depleted and as toxic end products of bacterial metabolism accumulate in the culture medium, the bacterial growth rate declines. This is referred to as the **stationary phase.** Many bacteria in the nutrient-depleted medium die while others continue to grow using nutrients obtained from the dead cells. In the stationary phase the number of cells dying and the number of cells dividing is balanced so the number of living cells in the culture medium remains constant.

Death Phase

The **death phase** is essentially the reverse of the exponential phase in that cells are dying at a geometric rate. Factors responsible for the death phase are the depletion of nutrients and the accumulation of toxic waste materials in the culture medium.

Environmental Activities of Bacteria

The biochemical processes carried out by bacteria are tremendously versatile and enable them to adjust to changing environmental conditions. During their growth and metabolism, bacteria use and convert a wide variety of organic substrates. When the supply of a particular nutrient dwindles or other modifications in their surroundings occur, many bacteria can readily adapt. They do so in part by calling into play alternative enzyme systems, by altering cell-wall or membrane constituents, or, in some species, by forming resistant endospores. Genetic mutation and selection are responsible for some of these mechanisms, which will be further explained in the section on bacterial genetics. The ability of bacterial populations to survive in an altered environment becomes of great practical importance when human infections are treated with **antimicrobial agents** or when physical and chemical agents such as heat, ultraviolet radiation, and disinfectant solutions are used to control bacterial growth. To ensure their continuance certain bacteria may produce enzymes that inactivate antimicrobial agents, change their method for using a particular substrate, start using another kind of available nutrient, or enter into the endospore stage.

In the natural world the metabolic processes of bacteria play an indispensable role in maintaining and nourishing all life on our planet. The decay and **putrefaction** of animal wastes and dead plants and animals are due in large part to the action of bacterial (and fungal) enzymes that break them down to forms they and other organisms can absorb and transform into new compounds for their own cellular structures and functions.

A brief review of the **nitrogen cycle** (fig. 3.3) illustrates this point. Nitrogen gas is a major component of the earth's atmosphere (76% by weight) and the ultimate source of nitrogen compounds in living organisms, most of which cannot use it in the elementary form. Several bacterial species work in concert to convert inorganic nitrogen (N_2) into the substances used by plants to synthesize amino acids and, subsequently, the proteins essential for all life processes (see chap. 2). So-called **nitrogen-fixing** bacteria (present in the soil) first transform atmospheric nitrogen into ammonia (NH_3). In the aerobic process of **nitrification,** other organisms convert ammonia to nitrites (NO_2^-) and nitrates (NO_3^-). Plants in turn convert (or *assimilate*) nitrates into the amino groups (NH_2^+) essential for amino acid and protein synthesis. During **deamination,** bacteria break down the organic protein matter of dead plants and animals, changing the amine groups back into ammonia. In the anaerobic process of **denitrification,** nitrites and nitrates are transformed to gaseous nitrogen (N_2) as the cycle continues.

Economic Uses of Bacterial Activities

The metabolic activities of bacteria have been put to a variety of productive uses. Bacteria, like yeasts, use carbohy-

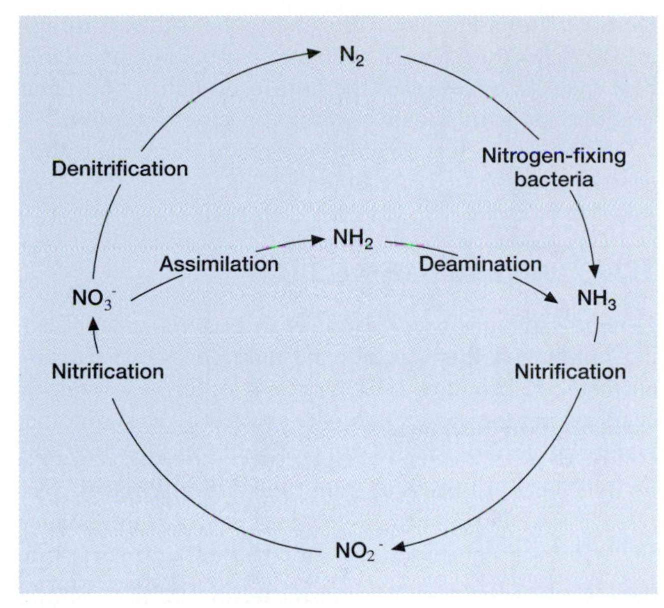

FIGURE 3.3 The nitrogen cycle in nature—bacteria play an important role in each step.

FIGURE 3.4 The great variety of cheeses and wines is made possible by the metabolic activities of bacteria, yeasts, and other microbial organisms.

Courtesy Milk Promotion Services, Inc., South Windsor, CT.

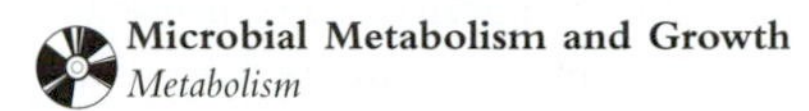
Microbial Metabolism and Growth
Metabolism

Fungal Structure and Function
Metabolism and Growth: Fungal Metabolism; Fungal Growth

drates for food, converting them into acids, alcohols, and gases. The ability to use a particular carbohydrate, or other organic compound, depends on whether or not a bacterial species possesses an enzyme that can act specifically on that substrate to change it chemically. This in turn directs the choice of the organism for the job to be done. When wines, or other alcoholic beverages such as hard cider are made, fermentation of the sugar to alcohol is usually assigned to yeasts. If desired, the alcohol may be further degraded into acetic acid, or vinegar, by introducing a culture of *Acetobacter* bacteria whose oxidative enzymes conduct the transformation, as illustrated by the equation

$$\underset{\text{Ethyl alcohol}}{2CH_3CH_2OH} + \underset{\text{Oxygen}}{O_2} \xrightarrow[\textit{Acetobacter}\text{ enzymes}]{\text{Aerobic}} \underset{\text{Acetic acid}}{2CH_3COOH} + \underset{\text{Water}}{2H_2O}$$

Another group of organisms called lactobacilli characteristically produce lactic acid when they ferment glucose. For this reason lactobacilli are useful to the dairy industry for preparing sour cream, yogurt, and cheese:

$$\underset{\text{Glucose}}{C_6H_{12}O_6} \xrightarrow[\textit{Lactobacillus}\text{ enzymes}]{\text{Anaerobic}} \underset{\text{Lactic acid}}{2CH_3CHOHCOOH}$$

The production of many popular foods is made possible by such bacterial conversions (fig. 3.4). Other practical uses of bacterial activities include the preparation of animal hides for leather manufacture, plant fibers used to make fabrics or rope, tobacco and tea leaves to improve their flavor, and industrial alcohols and acids. Many industries of basic importance to our economy have developed by using the chemical talents of bacteria and other microorganisms to produce materials that are essential to our progress and useful in our pleasures.

Conversely, the bacterial profits of the economic world are balanced against the expenditures of ingenuity, effort, and time required to circumvent or control the less desirable features of bacterial activity. Spoilage of foods and other valuable materials is costly and has stimulated major research and development in the field of food preservation. Methods used to prevent the growth of putrefactive (decay-producing) bacteria in food must not be deleterious to flavor or nutritional content and must not be toxic on ingestion by humans. Natural physical methods of preserving foods include dehydration, cold storage, freezing, and radiation: chemical methods employ the antibacterial effects of salting, pickling in vinegar or in high concentrations of sugar, smoke curing, and so forth. Antimicrobial agents also are used to suppress bacterial growth in some foods, including fish and poultry. This procedure is controversial, however, because it may allow bacteria that carry antimicrobial-agent-resistance genes (see R factors) to survive preferentially and become widespread in the human population.

Bacterial Activities and Human Health

In medicine and public health microbial activities are turned to our greatest advantage in (1) the production of antimicrobial agents and (2) the disposal of sewage and wastes.

Some antimicrobial agents are products of microbial metabolism excreted by the cells into the environment. In the soil, as in the human body where many microorganisms live together, compounds formed by one species may be quite toxic to another. In other cases, they may support the growth of neighboring organisms or have an antagonistic effect referred to as **antibiosis** (against life) (see colorplate 3b). The great names in the discovery and development of antimicrobial agents for medical purposes are Sir Alexander Fleming, Sir Howard Florey, and Ernest Chain, in England, and Selman Waksman in the United States. All are recipients of Nobel Prizes acknowledging the significance of their work in this field. Fleming first observed the antibacterial effects of the mold *Penicillium;* Florey and Chain identified the substance, penicillin, as responsible for these effects and opened the door to its medical application; and Waksman who began extensive investigations in the United States, discovered streptomycin, and stimulated many major research efforts that followed.

Antimicrobial drug production is a major activity of many pharmaceutical companies today. Research has led to the successful development of a number of useful drugs, such as ampicillin, nafcillin, piperacillin, and other penicillin derivatives produced by making chemical changes in the original molecule. Discovering new agents and synthesizing effective modifications of existing ones are serious challenges in the pharmaceutical industry. Many bacteria rapidly develop resistance to new antimicrobial agents (see chap. 9), often because of indiscriminate use by physicians in hospitals and the community.

In sewage and waste disposal, advantage is taken of the digestive activities of putrefactive bacteria. One of the common methods for handling sewage is to direct it slowly over beds of gravel and sand so that solid matter has time to settle out and be acted upon by the various bacteria in these beds. As water seeps through, it is cleaned and purified, pathogenic bacteria are removed and die, and the water is made ready again for consumption. Digestion of the settled material by bacteria converts it into a form that can be collected, dried, and used as fertilizer. The purification of water is discussed in more detail in chapter 10, together with other measures of concern to public health in control of infectious diseases.

The intestinal bacteria of humans and animals also break down food residues by digestive processes. Some of these organisms synthesize specific growth factors called vitamins that they require in their metabolism. This activity provides a ready supply of vitamins for the host, who also requires it for certain growth processes. The production of vitamin K as well as some elements of the vitamin B complex by intestinal bacteria contributes to the host's well-being.

Pathogenic Effects of Bacteria

The biochemical versatility of bacteria is critical to the complex interrelationships between the microbe and its host. The host's ability to resist harmful microbial effects and the organism's ability to survive in spite of this resistance constitute a competition at the cellular level. Host and parasite enzymes may compete for the same substrate, and each organism may produce substances that are antagonistic to the other. In finding alternative metabolic pathways for using nutrients, the bacterial cell is often much more versatile than the cells or tissues of the human or animal body. This versatility provides it with a strong capacity for survival.

The toxic factors directly involved in microbial pathogenicity are discussed in chapter 5.

Bacterial Genetics

Genetic variations occur naturally in bacteria as in all living things, and they can also be induced by experimental methods. In either case, variations occur as a result of changes in the genes on the bacterial chromosome. **Bacterial genetics** is the study of genes and heredity and their effect on the structure and function of bacteria.

Mutations

Transmissible changes in the characteristics of the cell that are due to gene alterations are called **mutations.** They probably occur in bacteria with the same frequency as in higher organisms but are manifest more frequently because bacteria are functionally haploid; therefore, "recessive" mutations are not masked as they may be in diploids. (In diploids a mutation in a gene of one chromosome set usually has no effect on the function of the equivalent gene in the other set of the pair. Thus, the **dominant gene** is one whose effect is expressed; its partner gene, the **recessive gene,** is unexpressed and masked.) Furthermore, the rapid multiplication of bacteria may produce large populations of mutant cells. Mutations are thus common among bacteria, affecting one or more characteristics such as cell-wall formation, pigment or toxin production, enzyme formation, or ability to resist the action of antimicrobial agents.

Mutations may develop as a consequence of natural or spontaneous events, or may be induced by various physical or chemical agents. The result of the mutation depends on the effect that it has on transcription of the DNA code to mRNA and on protein synthesis. For example, the change of one DNA base pair causes the complementary base to be substituted in the mRNA. The resulting codon is read by a different tRNA, and therefore, an amino acid different from the one specified by the original DNA code is placed in the protein molecule that forms. Such a change in the amino acid sequence of the protein can result in an altered or nonfunctional protein.

Changes in a single DNA base are called **point mutations** and can involve either transition or transversion mutations. **Transition mutations** are substitutions at a single site in the DNA where one purine base (adenine or guanine) is changed to the other or one pyrimidine base (thymine or cytosine) is changed to the other. **Transversion mutations** occur when a purine base is replaced by either of the pyrimidine bases or a pyrimidine base is replaced by either of the purine bases.

Several types of point mutations can occur; these include substitution mutations, addition mutations, and deletion mutations. A **substitution mutation** is the insertion of an incorrect base in place of the usual one within a base pair. This type of error, called a missense, results in a codon change that is significant if it results in the coding for a different amino acid. Another possible effect is the creation of a stop signal, which results in the formation of an incomplete protein. This is referred to as a **nonsense mutation.**

Mutations may involve either the deletion or addition of one or more purine or pyrimidine bases. These changes cause misreading of the genetic code and such mutations are called **reading-frame shifts.** Both deletion and addition mutations result from abnormal stretching or twisting of the DNA strands and may cause the synthesis of nonfunctional proteins.

TRANSFER OF GENETIC MATERIAL

Bacteria are known to exchange genetic material between genera and species. This exchange can occur through the transfer of entire chromosomes or their fragments, or through the transfer of **extrachromosomal DNA** molecules called **plasmids** from one bacterium to another. Plasmids usually exist and replicate independent of the host cell's chromosome and genetically code for properties nonessential to the maintenance of bacterial life. However, these DNA molecules may contain genes for characteristics of particular use to some bacteria in certain environments. Some examples of valuable and adaptive features associated with plasmids include antimicrobial resistance, synthesis of metabolic enzymes for unusual substrates, and increased virulence.

Genetic recombination is the process by which new combinations of chromosomal genes are obtained from two different types of cells. On a molecular level, a new chromosome is formed from the DNA contributed by the two parental cells. In bacteria, recombinant chromosomes can form from three different processes—**conjugation, transduction,** and **transformation.**

Conjugation

Conjugation involves the direct transfer of DNA from one bacterial cell (the donor) to another (the recipient). For conjugation to occur, specific interaction between the donor and recipient is required. Donor cells possess sex pili that stabilize the contact between donor and recipient and through which DNA may be transferred (fig. 3.5).

F Plasmids

In order to conjugate, the donor cell must also possess a conjugal plasmid DNA (see chap. 2). The best known and most studied of these is the **F plasmid,** or the fertility factor. F DNA codes for the production of a sex pilus, for the many functions involved in transfer of DNA to the recipient, and for its own replication. It can exist in two physical states within the bacterial cell. In the *autonomous* state, F exists separately from the main chromosomal DNA; the bacterial cell is called F^+. In the *integrated* state, F recombines with the chromosomal DNA and becomes part of the chromosome; the resulting cell is called **Hfr** or **high frequency of recombination** (fig. 3.6). The integration of F into the chromosome to produce an Hfr is a reversible event, that is, F can leave the integrated (Hfr) state and become autonomous (F^+). During the detachment process F occasionally carries some chromosomal genes with it (as in fig. 3.6 where F carries the genes for lactose catabolism and proline synthesis). Autonomous F with one or more chromosomal genes attached is called an F prime factor (F').

Genetic transfer between an F^+ donor and an F^- recipient is shown in figure 3.7. Only plasmid DNA (F, in this example) is transferred from donor to recipient; donor chromosomal DNA is not transferred. The same general events occur during the transfer of all conjugal plasmids and are not unique to F. The important event in F^+ conjugation is that the plasmid replicates as it is transferred from donor to recipient (fig. 3.7b); thus, both donor and recipient contain this plasmid and both are now donor cells (fig. 3.7c). If the transferred plasmid has genes for resistance to one or more antimicrobial agents, it is not difficult to see how one resistant donor cell could convert a large population of susceptible recipient cells to resistance in a short time.

It should be noted that the F plasmid is not a totally stable structure and can be lost quite easily from cells possessing it. Thus, not all cells that conjugate with the F plasmid (and probably other plasmids as well) are permanently converted into donors. The transfer of the autonomous plasmid from donor to recipient is a rapid process, occurring in a matter of minutes; therefore, the entire plasmid is transferred rather than just a portion of it. F' transfer, often called *sexduction,* takes place in a manner similar to that of F^+ transfer and also results in the formation of two donor (F') cells.

Conjugation may not take place in all bacteria. It has been studied most in certain gram-negative bacilli such as *E. coli* and related members of the family *Enterobacteriaceae.* Conjugation systems are also known in certain gram-positive bacteria. For example, *Enterococcus faecalis,* an organism prevalent in the human intestinal tract, transfers a variety of plasmids by conjugation, several of which are associated with antimicrobial resistance.

Transduction

Transduction involves the transfer of DNA from one bacterial cell to another by means of a bacterial virus or **bacteriophage** that carries the DNA from donor to recipient (fig. 3.8). To understand this process, it is necessary to examine how bacteriophages reproduce (fig. 3.9).

Bacteriophage Reproduction

Certain bacteriophages, such as the T phages of *E. coli,* reproduce strictly by a lytic cycle (fig. 3.9, steps 1–9) in which the virus is said to be *virulent.* The T phages adsorb to receptors on the host cell surface, inject their DNA into the cell (fig. 3.9, steps 2–3), take over its metabolic machinery, and force the cell to produce hundreds of new viral

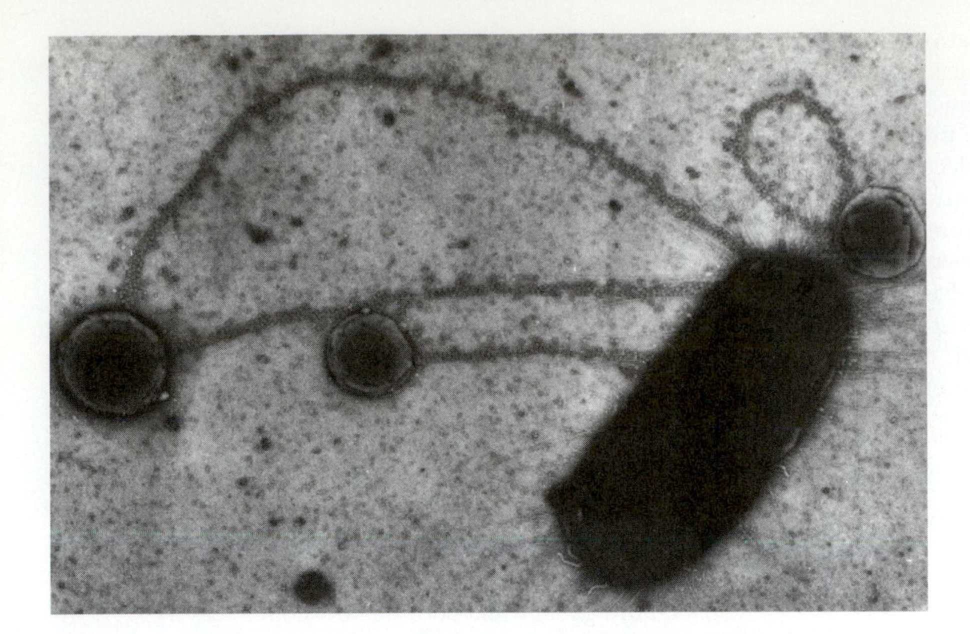

FIGURE 3.5 Sex pili connect an F+ donor cell with three minicells. Plasmid DNA of the larger cell passes through the pili to the small cells.

Courtesy Dr. Roy Curtiss, III, Washington University, St. Louis, MO.

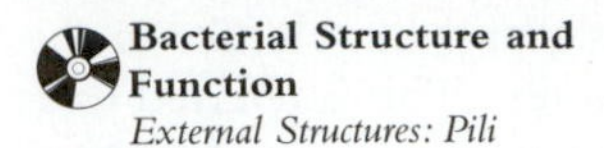

Bacterial Structure and Function
External Structures: Pili

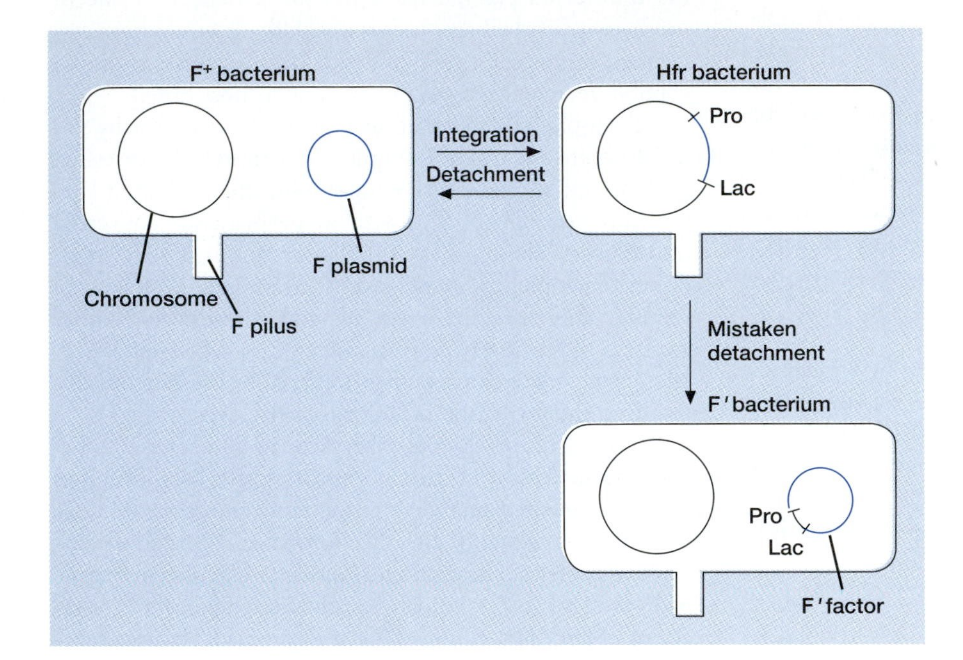

FIGURE 3.6 A schematic representation of the formation of F+, Hfr, and F′ donor cells. Note that the F plasmid can exist in two physical states: separate from the chromosome in F+ cells and integrated into the chromosome in Hfr cells.

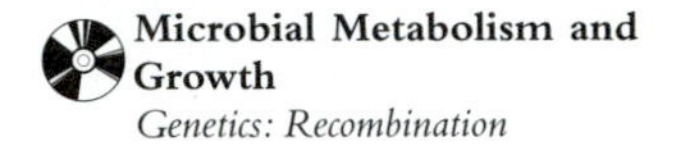

Microbial Metabolism and Growth
Genetics: Recombination

particles, or virions (fig. 3.9, steps 5–8). As a result, the host cell dies, and the virions are released. If suitable new host cells are available, these virions continue the lytic cycle of reproduction.

Other bacteriophages, such as phage lambda (λ), reproduce either by the lytic cycle or the lysogenic cycle (fig. 3.9, steps 1–3; 10–13); such viruses are said to be *temperate.* The unique feature of the lysogenic cycle is that new virions are not produced, but rather the viral DNA becomes integrated into the host cell DNA (fig. 3.9, step 10) and is replicated with it (fig. 3.9, steps 11–13). This viral DNA, now called a *prophage,* is not expressed (translated into viral proteins); therefore, the host cell remains viable. As the host cell DNA replicates prior to cell division, the integrated viral DNA is replicated with it. The prophage, however, can be "awakened" in a process called *induction* (fig. 3.9, step 12). When this occurs, the viral DNA is fully expressed and the lytic cycle commences resulting in virion production and host cell death. Prophage induction may occur spontaneously or can be stimulated by radiation or certain chemical agents.

Host cells containing a prophage may acquire new properties, a process called *lysogenic conversion.* Both the toxin of the diphtheria bacillus (diphtheria exotoxin) and the streptococcal pyrogenic exotoxin (scarlet fever toxin) are produced only by cells that have a prophage. These toxins therefore are products of the prophage rather than the bacterial genome.

Transducing Viruses

Most transducing viruses are temperate. Except for those highly virulent (lytic) viruses (such as T_2) that degrade their

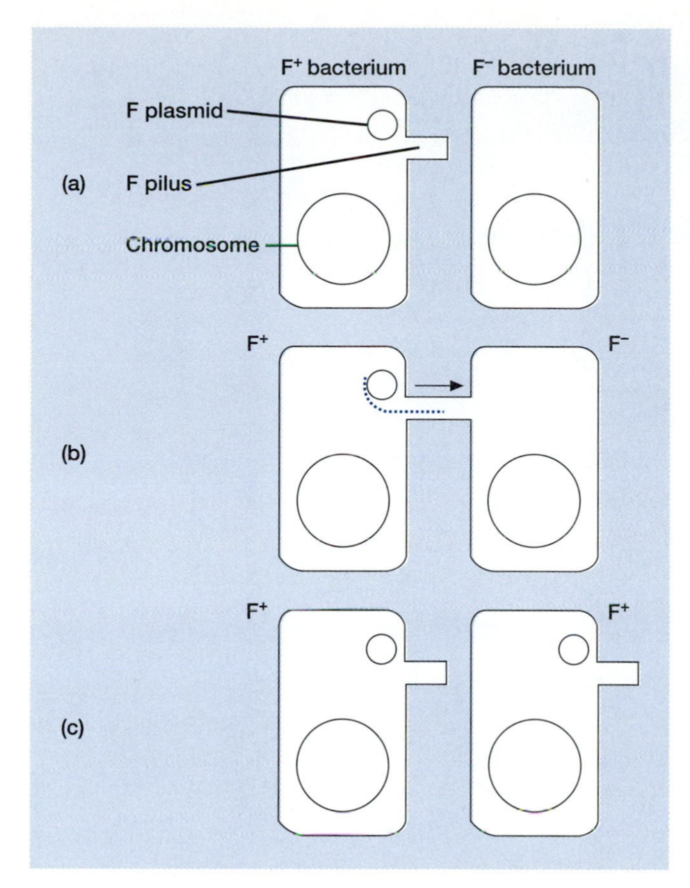

FIGURE 3.7 Conjugation between an F^+ donor and an F^- recipient (a) leads to the formation of two donor F^+ cells (c) because the F plasmid is replicated at the time of its transfer (b).

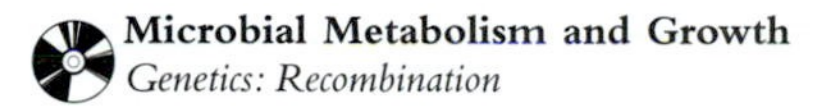

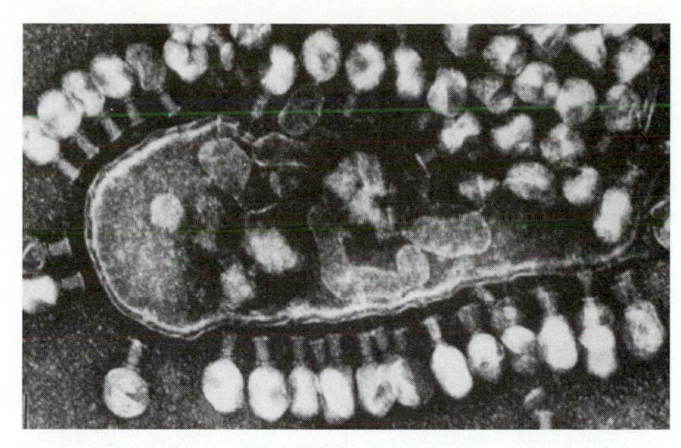

FIGURE 3.8 Electron micrograph of *E. coli* infected with bacteriophages. Each virus possesses a tailpiece and globular head that contains bacteriophage genome. The tailpiece attaches to the bacterial cell and is a conduit for the viral DNA, which passes into the bacterium (×36,500).

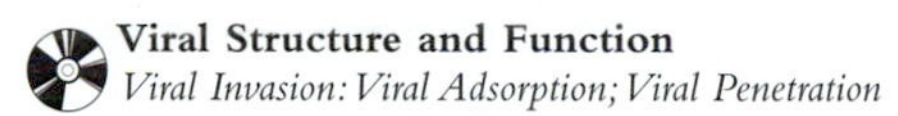

host cell's DNA down to nucleotides, virulent bacteriophages may form transducing viruses also.

Transduction may be either generalized in which case any host genes may be transferred, or specialized, in which case certain genes are transferred preferentially. In generalized transduction, a virus in the lytic cycle "accidentally" packages host cell DNA, rather than viral DNA, into the viral particle (fig. 3.10). Such accidents are rare, and the resulting transducing phage particles are defective because they lack some or all of the necessary viral DNA to carry out viral replication. If this transducing particle infects a bacterial cell, new virions cannot be produced. The bacterial DNA carried by the virus, however, can recombine with the DNA of the new host cell and thus potentially provide it with new properties. Because a virus is very small, only a few host genes can be carried by any transducing particle. As a result, the total amount of genetic information transferred by transduction is small by comparison to that transferred by Hfr's during conjugation.

In specialized transduction a temperate virus integrates between host genes. Phage lambda integrates between the genes for galactose catabolism and biotin synthesis (fig. 3.11). Occasionally, when this prophage is induced, a mistake is made that leads to the formation of defective lambda-transducing particles. Because lambda integrates between the galactose and biotin markers, the transducing particles formed preferentially contain the galactose or biotin markers in comparison to all the other host cell genes.

Transformation

Transformation involves the uptake by a bacterial cell of naked DNA released into the surrounding medium by another bacterial cell. The source of the naked DNA may be natural release by bacteria during death and autolysis, or it may be isolated in the laboratory. In a few bacterial genera, such as *Streptococcus, Bacillus,* and *Haemophilus,* transformation appears to be a major mechanism for gene transfer. Many other bacteria are transformable, however, and this process is probably quite widespread throughout the bacterial world.

Figure 3.12 depicts the uptake of a small linear DNA fragment and its integration into the genome of a recipient cell. Like transduction, transformation involves the transfer of only small amounts of DNA, a few genes at most. Certain *Streptococcus pneumoniae* strains may have acquired increased resistance to penicillin by transformation. These penicillin-resistant organisms were first noted in 1967 and now constitute about 20% of all pneumococcal strains in the United States.

In addition to providing basic genetic information for gene mapping and studies of linkage relationships, the genetic mechanisms of conjugation, transduction, and transformation have important ramifications with respect to drug resistance, toxin production, and other microbial virulence factors.

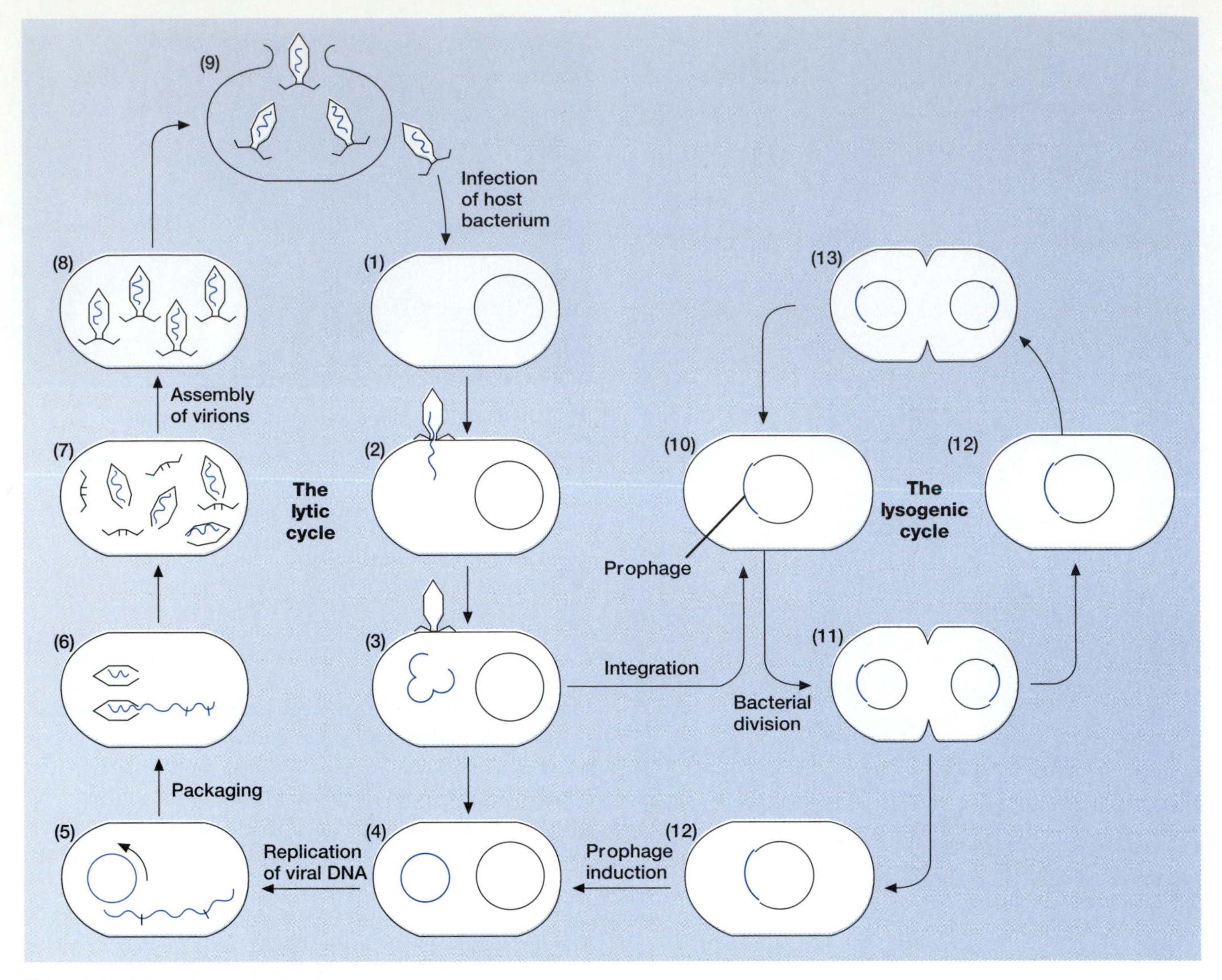

FIGURE 3.9 Bacteriophage replication may occur by the lytic (1-9) or lysogenic (1-3, 10-13) cycle. In the lytic cycle new virions are produced by the host cell under the direction of the viral DNA. In the lysogenic cycle the viral DNA forms a prophage by integrating into the host cell DNA. The prophage is replicated every time the cell divides, but virions are not produced. A prophage may be induced (12, 4-9), leading to the lytic production of virions.

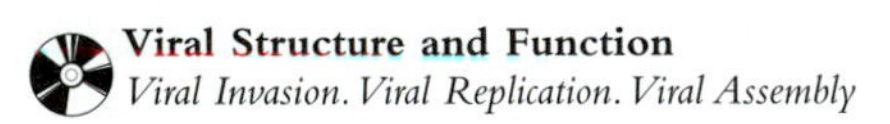

R Factors and Infectious Drug Resistance

R factors were discovered in Japan in 1959 and are now commonly found in clinical isolates of gram-negative enteric bacilli (Enterobacteriaceae) and anaerobic bacilli (*Bacteroides*). These intestinal bacteria often are resistant to three or more currently used antimicrobial agents. Since R factors are conjugative plasmids, they can be transferred from bacterium to bacterium within the family Enterobacteriaceae, a phenomenon known as **infectious drug resistance.** Potentially, bacterial populations may become drug resistant more rapidly by acquiring R factors than by mutation and selection. An important characteristic of R factors is that they can be transferred between different bacterial species (and even some genera) as well as between cells of the same species.

Structurally, R factors are large plasmids consisting of two functionally distinct regions: the *resistance transfer factor* (RTF) region codes for those functions concerned with replication and transfer of the plasmid; and the *resistance determinants* (genes) code for resistance to a variety of antimicrobial agents (fig. 3.13). Resistance determinants are not limited to those shown in figure 3.13 but, at present, might include genes for penicillin, kanamycin, and gentamicin resistance as well. Many resistance determinants are located in mobile transposons (see next section) and may be picked up or lost from R factors as transposons migrate from plasmid to plasmid or plasmid to chromosome.

Commonly now, strains of *Neisseria gonorrhoeae* (the agent of gonorrhea) that inactivate penicillin by producing the enzyme penicillinase have acquired an R factor with the

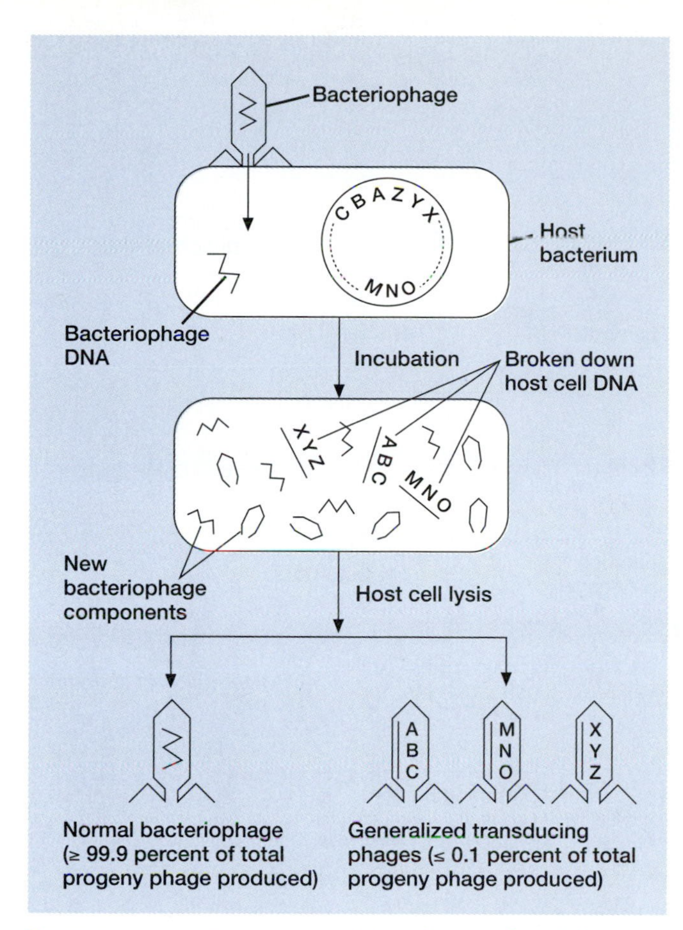

FIGURE 3.10 Generalized transducing phages are formed during viral replication by an accidental event in which host cell DNA is packaged (incorporated) into the viral capsid in place of viral DNA. These transducing phages are defective viruses because they lack essential viral DNA required to carry out viral replication in a host cell.

Microbial Metabolism and Growth
Genetics: Recombination

penicillinase-producing determinant on it. Not all penicillin resistance in bacteria is R-factor mediated, however. Staphylococcal penicillinase, produced by almost all clinical isolates of *Staphylococcus aureus,* is coded for by a gene located in a nonconjugative plasmid unrelated to R factors. The staphylococcal plasmids can be transferred by transduction.

The medical significance of R factors is obvious in that they represent a major mechanism for gram-negative enteric bacteria to acquire drug resistance. Before 1955, infectious drug resistance was rare. By 1965, however, with the development and widespread use of many new antimicrobial agents, most enteric bacteria isolated from hospitalized patients contained R factors coding for resistance to three or more agents. Although R factors existed before the antimicrobial era, these drugs have served as a selective mechanism to increase the population of resistant bacteria. The existence of R factors does not argue against the use of antimicrobial agents but rather strongly argues for their judicious use.

Bacterial Transposons (Jumping Genes)

Bacterial transposons are DNA segments that move within a bacterial cell from one plasmid to another, or from plasmid to chromosome, or from chromosome to plasmid. Unlike the typical process of recombination, this movement, or transposition, does not require DNA homology (relatedness) and greatly enhances the spread of transposon genes between bacteria. Genes coding for antimicrobial resistance, for synthesis of certain toxins, for resistance to toxic mercury compounds, and for certain metabolic processes may be located on transposable genes. Transposition is not limited to the bacterial world; it was first demonstrated by Barbara McClintock in the common corn plant, *Zea mays,* and has recently been shown to occur in animals as well. For her discovery, McClintock was awarded the Nobel prize for Medicine in 1983. Transposition then, in addition to mutation and homologous recombination, is a mechanism for producing biological diversity.

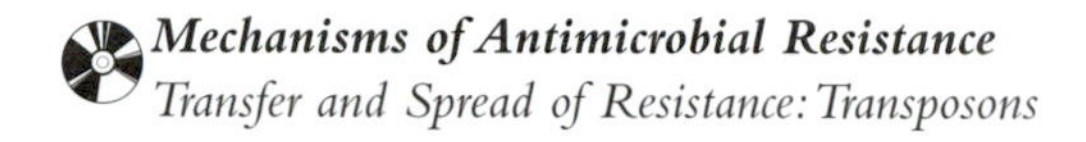
Mechanisms of Antimicrobial Resistance
Transfer and Spread of Resistance: Transposons

Recombinant DNA (Gene Splicing)

Scientists have been altering the genetic makeup of microorganisms for many years. However, it has been only recently that the direct manipulation of DNA has been possible. The deliberate modification of an organism's genetic information by directly changing its nucleic acid genome is often called *genetic engineering* and is accomplished by a collection of methods known as **recombinant DNA** technology, or **gene splicing.** In this technology the DNA or gene responsible for a particular phenotype is identified, isolated, and purified. The purified gene is then fused with other pieces of DNA to form a recombinant DNA molecule. This recombinant DNA molecule is then inserted into an organism that need not even be in the same kingdom as the original gene donor. The recombinant gene is then grown or propagated in the recipient cell, a process referred to as *gene cloning.*

In a remarkable example of recombinant DNA technology, DNA containing the human gene for insulin production was inserted into a bacterial plasmid. This plasmid was then reintroduced into a nonpathogenic *Escherichia coli* strain that now synthesizes a protein identical to human insulin in structure and function. Because the bacteria multiply rapidly and attain high numbers (10^9/ml in culture), large quantities of the human protein can be produced quickly and inexpensively. Insulin produced in this way is now commercially available and used by many persons with diabetes who need it to control their levels of blood glucose. In comparison with the previously available insulin, which was purified from pig pancreas, human insulin exerts its effect more rapidly and avoids potential allergic reactions to the foreign pig protein (see chap. 6)

FIGURE 3.11 The formation of specialized transducing particles of phage lambda is understood by observing the formation of the lambda prophage. In the lysogenic cycle lambda DNA is not expressed but is integrated into the host cell (*E. coli*) DNA between the genes for galactose (gal) and biotin (bio). Lambda DNA is believed to circularize to allow for its integration into the *E. coli* chromosome. Induction of the lambda prophage normally would be the reverse of this process and lead to the lytic production of lambda virions. On rare occasions a mistake is made in the induction process, and the host DNA to the left (gal) or right (bio) of the prophage replaces some phage DNA, leading to the formation of lambda dg (defective, galactose transducing) and lambda db (defective, biotin transducing) specialized transducing particles.

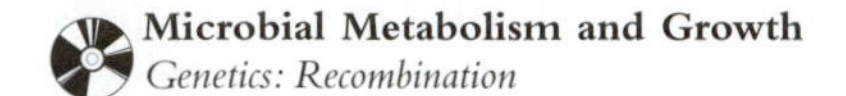
Microbial Metabolism and Growth
Genetics: Recombination

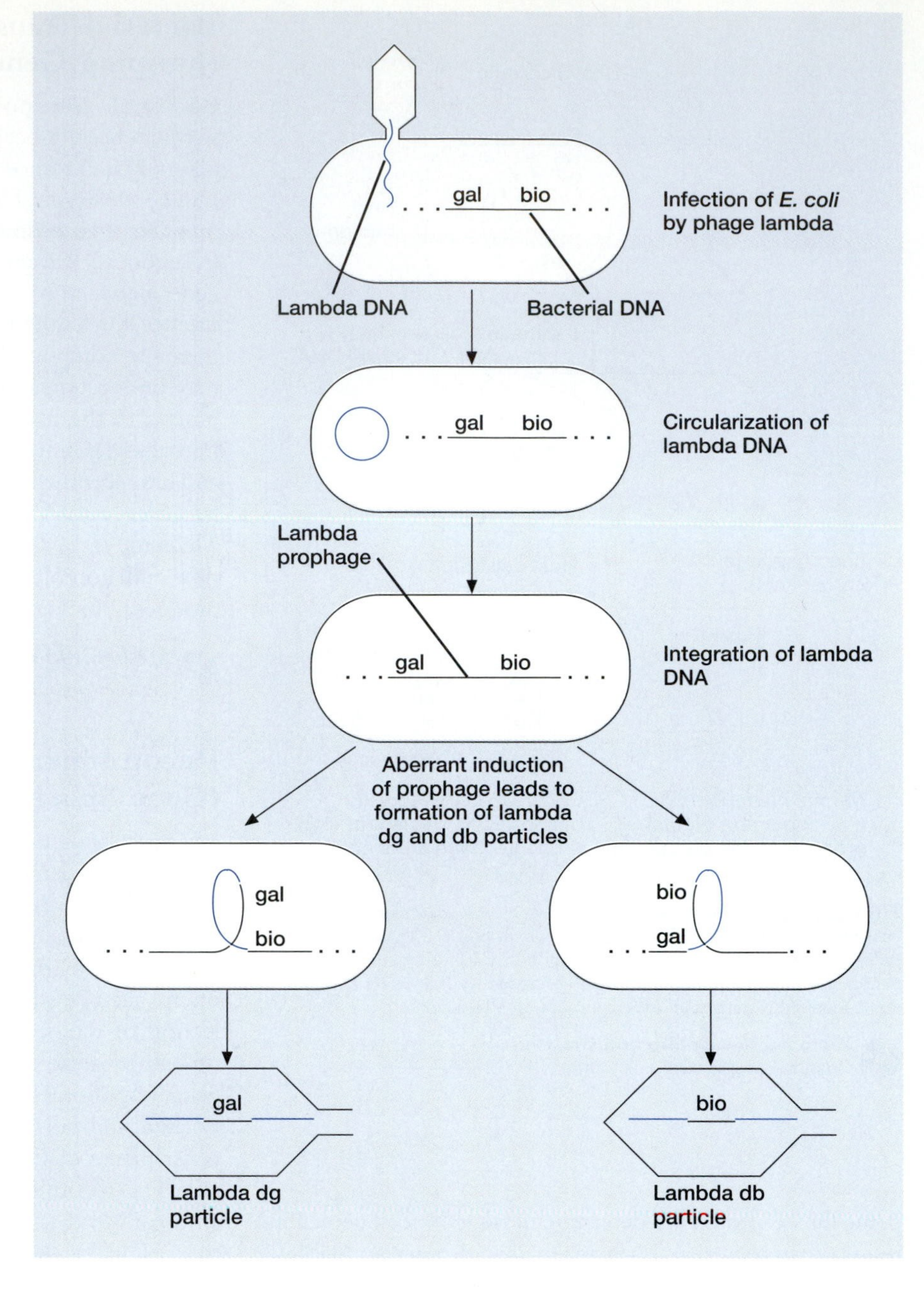

Recombinant DNA provides the technology for mass production by bacteria of useful human proteins. In addition, the bacterial plasmid (known in recombinant DNA terminology as a *vector*) with the inserted foreign DNA amplifies or clones the foreign DNA so that multiple copies of a pure gene are obtained. This procedure provides scientists with the material necessary to study gene structure and regulation in higher forms of life.

Restriction Endonucleases

How has it become possible to isolate bacterial plasmids, cut them open, splice foreign DNA into them, and reintroduce them into a bacterial host? An important technological advance was the discovery of bacterial restriction enzymes or **restriction endonucleases.** Restriction enzymes make cuts in DNA molecules, not at random, but only at specific base pair sequences; they cut the DNA every time such a base pair sequence occurs. More than 200 restriction enzymes with over 70 different specific cleavage sites are known.

In part, bacteria contain restriction enzymes to protect against invasion by viral DNA, which, if not broken down, frequently leads to the lytic production of virions and cell death. To prevent destruction of their own DNA by restriction enzymes, bacteria modify it, sometimes by adding a methyl group to certain nitrogenous bases.

Significance of Recombinant DNA Technology

In addition to insulin, several other medically significant human proteins are now being produced by *E. coli* through re-

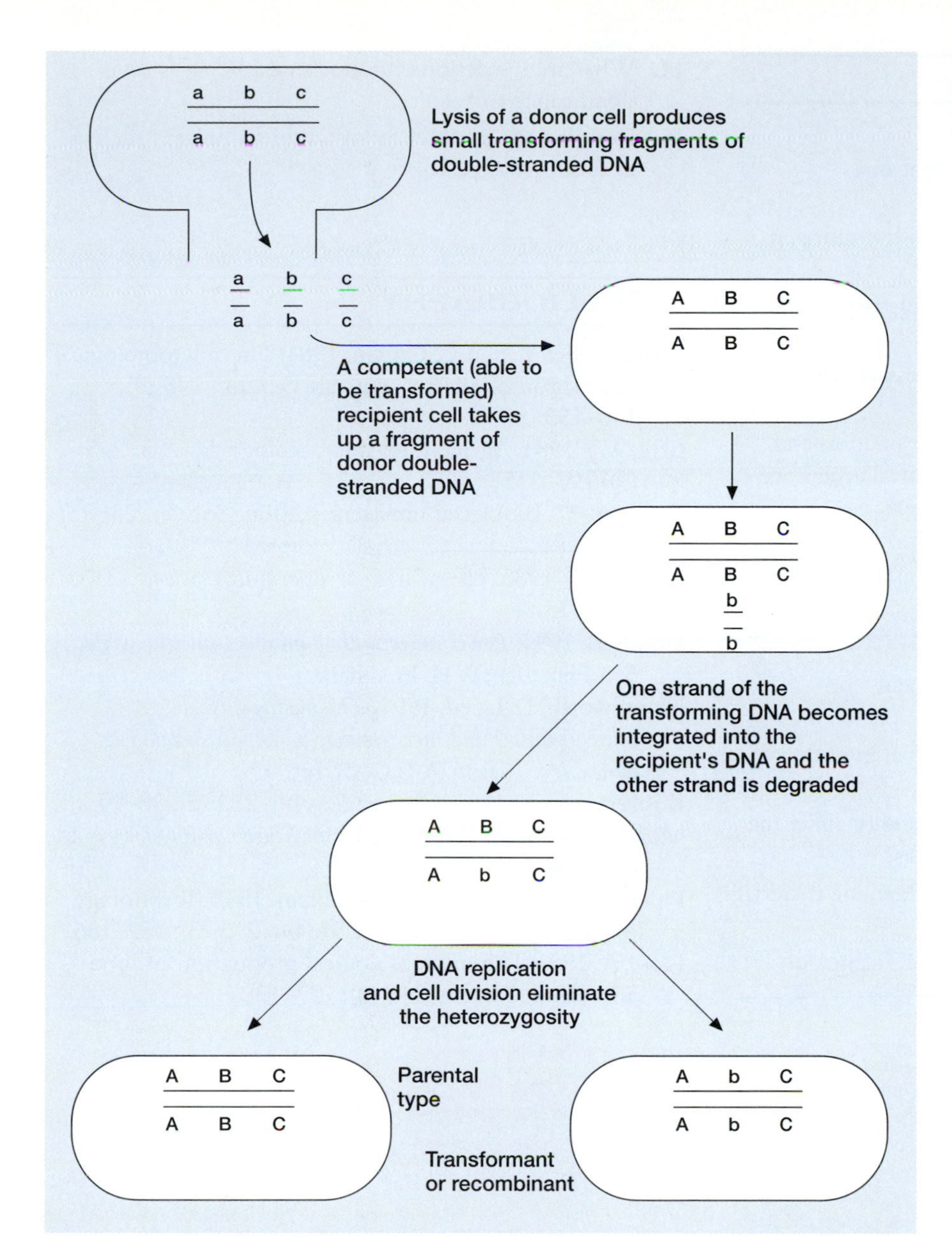

a b c
a b c
Lysis of a donor cell produces small transforming fragments of double-stranded DNA
a b c
a b c
A B C
A B C
A competent (able to be transformed) recipient cell takes up a fragment of donor double-stranded DNA
A B C
A B C
b
b
One strand of the transforming DNA becomes integrated into the recipient's DNA and the other strand is degraded
A B C
A b C
DNA replication and cell division eliminate the heterozygosity
A B C
A B C
Parental type
A b C
A b C
Transformant or recombinant

FIGURE 3.12 A schematic representation of the uptake and integration of transforming DNA by a recipient cell. The letters define small regions (one or two genes) of donor DNA (small letters) and recipient DNA (capital letters). The lower- and uppercase letters may be thought of as alleles or allelic regions.

Microbial Metabolism and Growth
Genetics: Recombination

combinant DNA technology: human growth hormone; the hormone somatostatin; interferon, an antiviral substance that is effective also in treating some forms of cancer; and TPA (tissue plasminogen activator), which dissolves blood clots. The potential for developing other useful proteins is limitless. In 1977, Genentech became the first genetic engineering company to manufacture medically important drugs using recombinant DNA technology. Since then many other biotechnology companies have sprung up and recombinant DNA technology has become a big business. With extensive ongoing efforts, recombinant DNA technology will certainly have a significantly beneficial effect on human life.

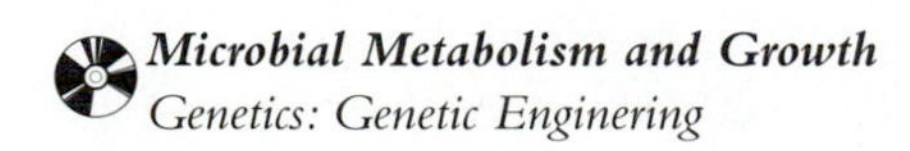
Microbial Metabolism and Growth
Genetics: Genetic Engineering

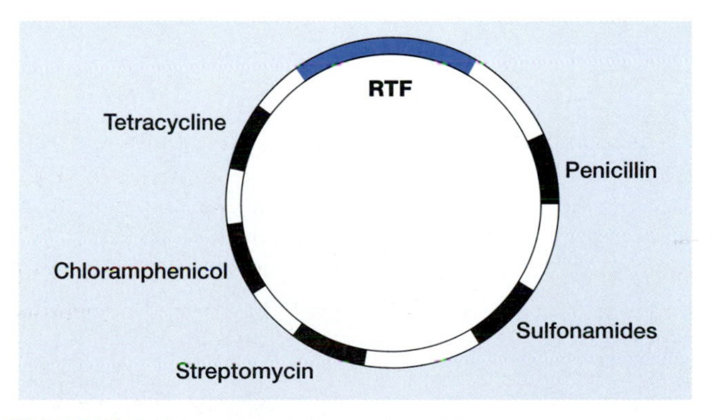

RTF
Tetracycline
Penicillin
Chloramphenicol
Sulfonamides
Streptomycin

FIGURE 3.13 R factors are conjugal plasmids carrying a variety of antimicrobial resistance genes. Genes coding for tetracycline, chloramphenicol, streptomycin, sulfonamides, and penicillin resistance are shown, but the number of resistance genes and the specific antimicrobial agents involved vary from R factor to R factor. The RTF (resistance transfer factor) portion controls replication and transfer of the plasmid.

Mechanisms of Antimicrobial Resistance
Acquisition/Expression of Resistance Genes; Transfer and Spread of Resistance.

QUESTIONS

1. What is the biosphere?
2. How are plants and animals dependent on microorganisms for survival?
3. Briefly describe the role of bacteria in the nitrogen cycle.
4. Compare autotrophic and heterotrophic organisms. List several examples of each.
5. What are enzymes and why are they essential to microbial activity?
6. Compare the chemical processes of respiration and fermentation. What are the final chemical acceptors? How does the energy production differ?
7. How do aerobes and anaerobes differ?
8. List some beneficial and harmful activities of each of the following microorganisms: yeasts, molds, protozoa, bacteria, viruses.
9. What are the environmental requirements for bacterial cultures?
10. Describe the four phases of the bacterial growth curve.
11. How does changing the oxygen availability affect the metabolic breakdown of sugar by yeasts?
12. Of what importance is the transfer of genetic material among bacteria?
13. What are transposons? What is the significance of transposition to infectious disease?
14. What are restriction endonucleases? What is their significance to bacteria?
15. Describe the production of a human protein such as insulin by bacteria.

BIBLIOGRAPHY

Aharonowitz, Y., and G. Cohen. 1981. The microbiological production of pharmaceuticals. *Sci. Am.* 245 (9): 140–153.

Brill, W. J. 1981. Agricultural microbiology. *Sci. Am.* 245 (9): 198–215.

—1977. Biological nitrogen fixation. *Sci. Am.* 236 (3): 68–81.

Brock, T. D. 1985. Life at high temperatures. *Science.* 230: 132–138.

Dixon, B. 1994. *Power unseen: How microbes rule the world.* San Francisco: W.H. Freeman.

Hawksworth, D. L., ed. 1991. *The biodiversity of microorganisms and invertebrates: Its role in sustainable agriculture.* Tucson, AZ: CAB, Int.

Kosikowski, F.V. 1985. Cheese. *Sci. Am.* 252 (5): 88–99.

Moat., A. G., and J. W. Foster. 1988. *Microbial Physiology.* 2d ed. New York: Wiley.

Plucknett, D. L., and D. L. Winkelman. 1995. Technology for sustainable agriculture. *Sci. Am.* 273 (3): 182–186.

Rose, A. 1981. The microbiological production of food and drink. *Sci. Am.* 245 (3): 127–138.

CHAPTER 4

Tools and Techniques in Microbiology

OUTLINE

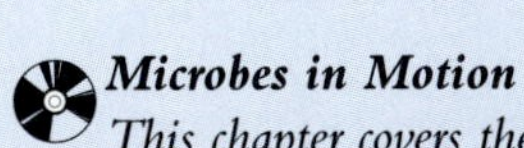

Microbes in Motion

This chapter covers the various tools and techniques used by microbiologists to aid in the diagnosis of infectious diseases and determine the antimicrobial susceptibility of bacteria. The following books and chapters of the Microbes in Motion CD-ROM may be a useful preview or supplemental study aid to your reading: Bacterial Structure and Function: Bacteria Groups. Microbial Metabolism and Growth: Growth. Anaerobic Bacteria: Characteristics. Control of Microbes: Physical Control. Antimicrobial Action: Antibiotic Susceptibility Testing.

OBJECTIVES

On completing this chapter the student will be able to:

1. discuss various tools and methods used by the modern microbiologist to aid the diagnosis of infectious disease.
2. describe the component parts of compound and electron microscopes.
3. state the practical uses of compound versus electron microscopes.
4. list several microscopic techniques and staining methods for studying microorganisms.
5. indicate how staining methods can best be used for direct microscopic examination of clinical specimens.
6. explain the nature, preparation, and uses of at least three types of culture media for growing and identifying pathogenic bacteria or other microorganisms.
7. outline general methods for culturing clinical specimens, interpreting laboratory results, and making final identification of any pathogenic organisms.
8. describe commonly used serologic methods for identifying microorganisms.
9. outline one method commonly used for determining the antimicrobial susceptibility of an isolated pathogenic organism.
10. name some special methods that may be used to refine laboratory diagnoses of infectious diseases.
11. indicate how laboratory computers are useful in performing, reading, and reporting results

KEY WORDS

Acid-fast staining
Aerobic incubation
Amplification (am-pli-fee-**CAY**-shun) techniques
Anaerobic incubation
Antimicrobial (anti-my-**CROW**-bial) susceptibility testing
Asepsis
Autoclave
Bacterial culture
Biological safety cabinet
Compound light microscope
Culture media
Dark-field microscopy
Direct smear techniques
Fluorescence microscopy
Gram stain
Magnification
Nucleic acid probes
pH
Phase contrast microscopy
Polymerase (poll-**IM**-er-ase) chain reaction (PCR)
Prereduced media
Quantitative culture of urine
Radiometric detection methods
Resolution
Scanning electron microscope
Serologic (seer-oh-**LOGIC**) techniques
Specimen culture
Staining methods
Steam sterilization
Sterilization by filtration
Transmission electron microscope
Wet preparations for microscopy

INTRODUCTION

The microbiology laboratory plays an essential role in the diagnosis of infectious diseases. In order to function effectively, however, the laboratory requires clinical information from the patient's physician and, most importantly, an appropriate and well-collected specimen for examination. The health care professional who is well informed and uses correct procedures for specimen collection is invaluable to the laboratory. General guidelines for specimen collection are given in chapter 7, but it should be noted that specific details may vary to some extent in different patient care situations.

To provide reliable results, the laboratory must have tools and techniques for visualizing microorganisms and, when appropriate, for isolating them in culture and identifying them by morphologic and physiologic properties. Some of the basic equipment and techniques of the microbiology laboratory and the general principles of their use are described in this chapter.

TOOLS

MICROSCOPES

There are three types of microscopes available for the study of microorganisms. The **compound light microscope** (figs. 4.1 and 4.2a) is the most practical tool for the diagnostic laboratory, providing sufficient magnification to observe the size, shape, surface structures, and staining properties of all microorganisms except viruses. Adaptations of the compound microscope permit organisms to be viewed by dark-field, phase, or fluorescence microscopy. The **transmission electron microscope** is a more sophisticated instrument that has made it possible to study virus morphology and to visualize the internal structures of all microbial cells. The **scanning electron microscope** has provided dramatic new observations of cellular surfaces and has shed light on the mechanisms by which microorganisms attach to the surface tissues of animals and humans.

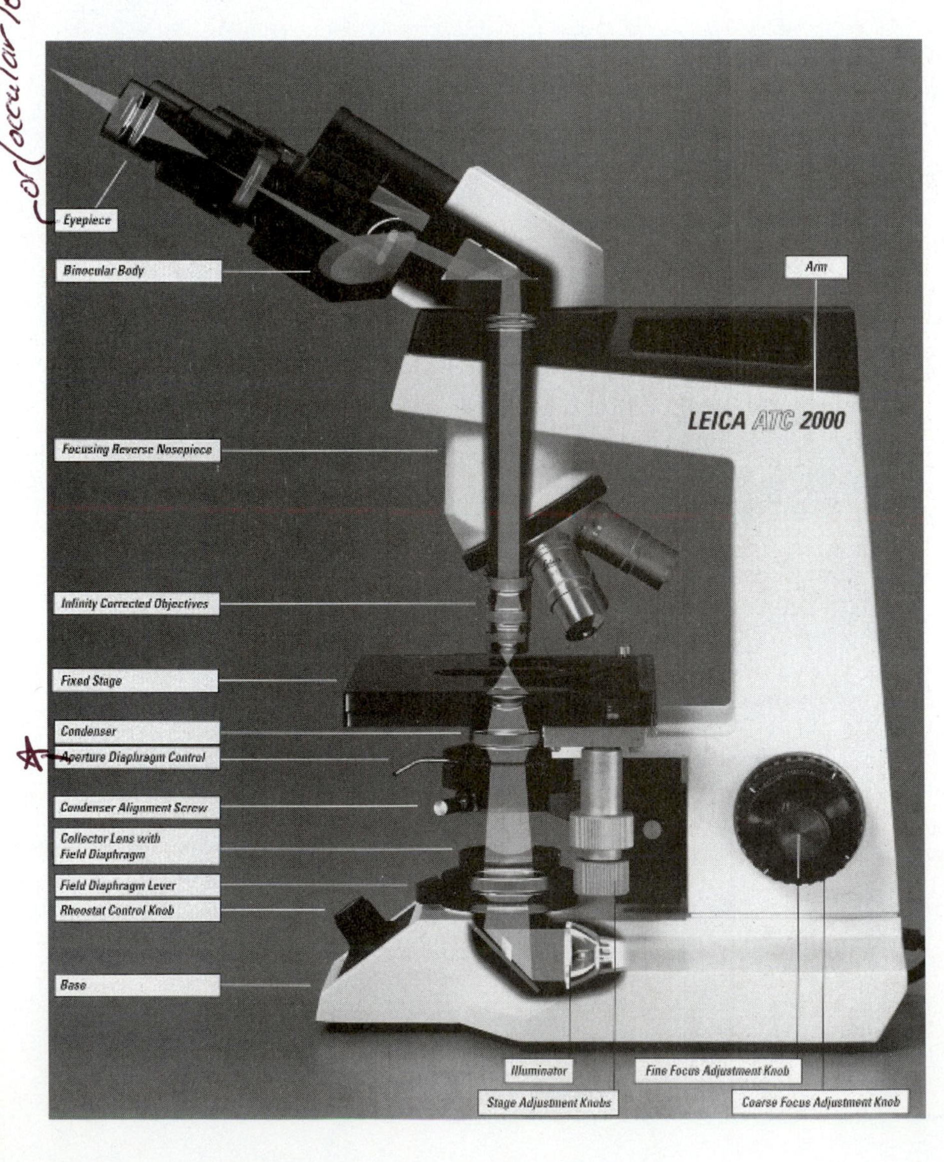

FIGURE 4.1 A compound light microscope with parts labeled for easy identification. The pathway of light is also illustrated.

Courtesy of Leica Inc., Optical Products Division.

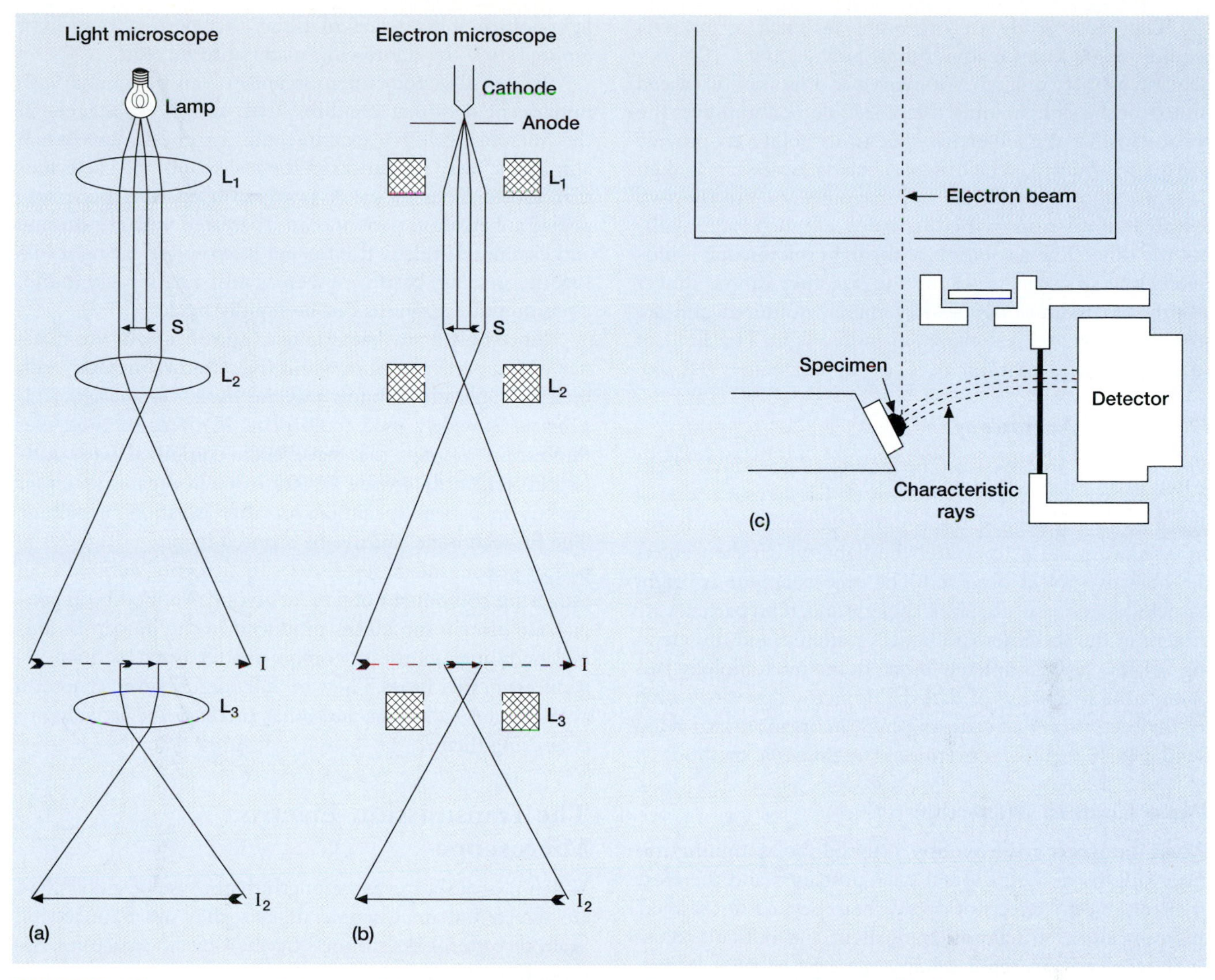

FIGURE 4.2 Comparison of the optics of the compound light microscope (a) with the mode of operation of the RCA electron microscope (b) and the scanning electron microscope (c). In the light microscope, a system of glass lenses (L) focuses light rays on the object (S), recollecting them to form first a primary image (I), then a highly magnified image. In the electron microscope, a beam of electrons, rather than light rays, is focused by a system of magnetic fields (L) that serves the same function as the lenses of the light microscope. In diagrams (a) and (b), L_1 is the condensing lens, L_2 is the objective lens, and L_3 is the projector lens. The scanning electron microscope (c) also focuses an electron beam on the specimen. Rapid back and forth scanning of the beam creates a flying spot of radiation that is collected on and activates a viewing screen to produce an image.

(a) and (b) Dr. Thomas F. Anderson, The Institute of Cancer Research, Fox Chase Cancer Center, Fox Chase, PA. Courtesy Mrs. Wilma Eaton Anderson.

The Compound Light Microscope

The compound light microscope has two sets of magnifying lenses placed at opposite ends of a tube that is mounted on a stand. The uppermost or viewing lens is called the *ocular.* The other lens, nearest the object being magnified, is called the *objective.* Usually there are three objectives attached to a revolving nosepiece. On some microscopes, the revolving nosepiece is raised or lowered for viewing; on others the tube is raised or lowered to bring objects into focus. The individual objectives magnify the object 10, 45, or 100 times (×10, ×45, ×100). This **magnification** is further increased by the ocular, which is usually a ×10 lens. Thus, the combined enlargement provides a total magnification of 100 times, 450 times, or 1,000 times.

The platform that holds the slide being viewed is called a *stage.* Just below the centrally placed opening in the stage is the *substage condenser,* which functions to collect and concentrate the light and to direct it upward through the object on the stage. The object absorbs and bends the light, so that, when it passes onward through the objective and ocular, it is patterned in the image of the object (fig. 4.2a). The condenser is fitted with an iris diaphragm to regulate the amount of light passing into it. Raising or lowering the position of the condenser also changes the amount of light it can collect. Light is reflected from its source into the condenser by a mirror attached at the base of the microscope, or from a light bulb built into the base. In general, the higher the magnification desired, the more intense the light must be, but the degree of illumination needed varies also with the density of the object. For this reason, stained material usually requires more light than unstained preparations.

The close study of organisms as small as bacteria requires 1,000 times magnification provided by a 10× ocular and a 100× objective immersed in a drop of oil placed on top of the slide holding the object. The oil improves the **resolution** of the objective, that is, its ability to provide sharpness of detail, which is particularly necessary at high magnifications. Resolution and magnification are the two essentials of microscopy, neither being of much value without the other. The resolution of the light microscope is limited because, as objects diminish in size, they can no longer absorb and redirect light—they merely scatter it and are seen with less and less clarity in visible light. The limit of resolution of the light microscope is approximately 0.2 *µm*.

Dark-Field Microscopy

The light microscope can be equipped for **dark field microscopy** with a special condenser that focuses a cone of light on the specimen. Objects in the specimen, such as microorganisms and cells, scatter the light, which then enters the objective and is observed. The objects appear as bright particles against a nearly black background. If no particles are present in the specimen, no light is scattered, and the viewing field appears completely black. In the microbiology laboratory the major use of dark-field microscopy is to search for the bacterium that causes syphilis, an organism that is too slender to be seen by conventional microscopic methods.

Phase Contrast Microscopy

Phase contrast microscopy is useful for examining microorganisms in their natural state, unstained and therefore unaltered by the effects of dyes. When they are not stained, microorganisms are almost transparent and difficult to see under the ordinary light microscope. Nevertheless, advantage can be taken of the fact that, when light passes through objects of different density or refractive index, the light waves emerge with altered characteristics (amplitude and frequency); that is, they are "out of phase." By means of a special condenser and objectives, the light microscope can be converted to a phase contrast microscope, through which these variations in light waves become visible. Microorganisms can then be seen to stand out sharply from their surrounding medium. When larger cells such as protozoa or tissue cells are viewed under phase microscopy, the nuclei and other internal structures can be readily distinguished because they vary in density from the cytoplasm and thus affect light waves differently. Also, the exact size and shape of microorganisms can be determined with the phase contrast microscope because they have not been distorted by the effects of staining.

Fluorescence Microscopy

Fluorescence microscopy depends on the principle that fluorescent compounds can absorb the short invisible rays of ultraviolet light and emit longer visible light waves. Equipment for fluorescence microscopy incorporates (in the compound microscope) an ultraviolet rather than a bright light source and a series of filters that permits only light emitted from the fluorescing material to be seen.

In the laboratory microorganisms can be stained with fluorescent dyes that combine with specific substances in the microbial cell. For example, the dye rhodamine has an affinity for waxy substances in the cell wall of the bacterium that causes tuberculosis. A portion of sputum from suspected tuberculosis patients can be treated with rhodamine and examined under a fluorescent microscope. If bright yellow (fluorescing) bacilli are seen against a dark background, a presumptive diagnosis can be rapidly made.

Since only a few bacteria react specifically in this manner with a particular fluorescent dye, another method, with broader applications, known as the *fluorescent antibody (FA) technique* is widely used (colorplate 2b). This method uses fluorescent reagents that have been combined with antibodies to identify a wide variety of microorganisms, either directly in patient specimens or after isolation in culture. The FA technique can also be adapted to aid in diagnosis of past or present infectious diseases by detecting antibodies in the serum component of patient's blood. Antibodies, in general, are protein substances produced in the human or animal body in response to foreign matter (such as microorganisms) in the tissues. They are fully described in chapter 6 as are the FA technique and other methods for detecting or measuring them.

The Transmission Electron Microscope

When rays of shorter wavelength than those of visible light are directed at microscopic objects, they are bounced off again in patterns determined by the object's structures. Although the human eye cannot see short waves, other sensitive surfaces can record them. A photographic plate coated with a responsive material records the impact of short waves, producing a picture that reveals the design of the object from which they were bounced.

Operating on this principle, the electron microscope (figs. 1.5 and 4.2b) sends a stream of short-wave electrons through a vacuum chamber (to avoid their scatter by air) to strike a microscopic object in its path. Instead of absorbing these waves, as a larger object would do with visible light, the object scatters them in a pattern that forms its image. Circular electromagnets attract the scattered electrons still farther outward, thus enlarging the image, which is then focused on a fluorescent screen. Electrons hitting the screen cause visible fluorescence that can be viewed directly. The image can also be photographed and further enlarged, 200,000 times or more depending on the equipment used (fig. 4.2b).

Because the electron microscope has much greater resolving power than the light microscope, structures as small as 0.001 µm (1 nm) can be clearly visualized. Objects to be examined must be extremely thin. Even bacteria, for example, must be sliced into thin sections before their internal structures can be viewed. In one type of preparation, the

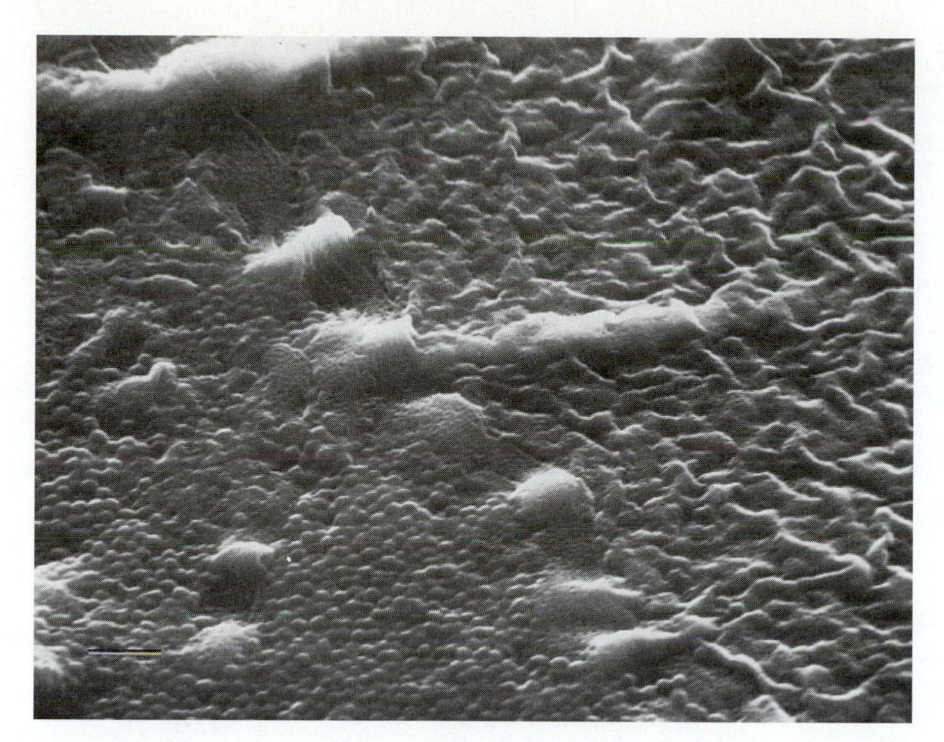

FIGURE 4.3 Scanning electron micrograph of a colony of *Streptococcus pneumoniae* growing on a blood agar plate. Note the three-dimensional aspect of individual cocci comprising the colony in the left half of the photo. In the right half, older cocci in the growth have lysed (dissolved), and this area of the colony has collapsed. The bar in the lower left-hand corner represents 10μm.

From E. L. Springer and I. L. Roth. "Scanning electron microscopy of bacterial colonies." *Canadian Journal of Microbiology* 18: 219, 1972. Photo by Dr. Ivan L. Roth. Reproduced by permission of The National Research Council of Canada.

cells are fixed and dehydrated in an organic solvent, embedded in plastic, and cut on a special instrument (an ultramicrotome) fitted with a diamond knife. Several slices are prepared and treated with special stains such as osmic acid, uranium, lead, or permanganate. Such materials scatter electrons well (thus providing good contrast) because they are composed of atoms of high atomic weight. Thin sections of bacteria prepared in this way reveal fine details of internal cellular structures (see fig. 2.4). Other techniques include shadowing, in which only microbial outlines are seen, and negative staining used especially for examining very small structures such as viruses.

A number of examples of electron micrographs of microorganisms can be found in chapters 1 and 2 and elsewhere in the text. In these photographs, note how the instrument provides details of both the surface and internal morphology of microscopic cells.

The Scanning Electron Microscope

The development of the scanning electron microscope (figs. 1.7 and 4.2c) has opened up another approach to the visualization of microbial surface structures. Unlike the flat effects seen in electron or light photomicrographs, the photographic images obtained with this instrument have a remarkable three-dimensional quality (fig. 4.3). With this microscope, an electron beam is directed at the specimen to be studied and scans back and forth across it. The material to be examined must first be coated with a thin film of a heavy metal such as platinum. The metallic coating scatters the electrons in an image-forming design onto a viewing screen. The latter is activated by the flying spot of electrons and illuminates the subject in a picture essentially like that seen on a television screen (fig. 4.2c). Photographs can then be made of the image appearing on the screen (fig. 1.7).

Magnifications ranging from ×15 to about ×10,000 can be obtained with the scanning electron microscope. However, the instrument can be used only to study the surface structures of an object; internal details cannot be visualized by this method.

A comparison of the features and uses of the three principal types of microscopes used by microbiologists is provided in table 4.1. Note that the compound microscope can be adapted for various purposes, all of which require visible light. Condensers, objectives, oculars, and filters can all be adapted to influence the visual effect and provide the desired type of image. Both electron microscopes, however, produce an image on a fluorescent screen formed by an electron beam rather than a beam of visible light focused on the specimen.

GENERAL EQUIPMENT

The equipment needed to successfully cultivate and identify microorganisms varies according to the type of organism to be studied, the environmental conditions it requires, and the techniques that must be used to demonstrate its properties. Most diagnostic microbiology laboratories, particularly those in small hospitals, cannot afford the full range of equipment, staff, space, and time required to identify every type of microorganism that might be associated with infectious disease. The facilities of specialized laboratories, including those of city or state health departments, can be used when uncommon organisms are encountered or elaborate techniques are required. However, certain basic items of equipment are

TABLE 4.1 COMPARISON OF THREE TYPES OF MICROSCOPES: FEATURES AND USES

Type of Microscope	Limit of Resolution[a]	Magnification Range	Major Components	Commonly Used for Examination of:	Methods for Preparing Specimens
Compound Microscope					
Light	0.2 μm	×100–×1,000[b]	Light condenser; objectives, ocular	All microorganisms except viruses	Simple stains, differential stains, wet mounts
Dark-field	0.2 μm	As above	Special condenser	Spirochete of syphilis	Wet mount
Phase contrast	0.2 μm	As above	Special condenser; objectives	Tissue cells; all microorganisms except viruses, rickettsiae, chlamydiae	Wet mount
Fluorescence	0.2 μm	As above	UV light source; special filters	Many bacteria; serum antibodies	Fluorescing dyes (for TB bacillus); fluorescent reagents combined with antibodies
Transmission Electron Microscope	0.001 μm (1 nm)	×10,000–×1,000,000	Electron beam in vacuum chamber; image on fluorescent screen	All microorganisms, especially viruses, surface and internal structures	Stains of high atomic weight, shadowing, negative staining; slicing technique for bacteria
Scanning Electron Microscope	0.01 μm (10 nm)	×15–×10,000	As above, but electron beam scans in flying spot	Microbial or cellular surface structures and features	Coating with heavy metal

[a]Approximate.
[b]Can be increased to ×2,000 with a ×20 ocular.

essential for the successful operation of every microbiology laboratory, and these will be described here.

Incubators

Most microorganisms have specific temperature requirements for growth and must be cultivated in a constant-temperature incubator. Modern incubators have an insulated interior chamber with a thermostatically controlled heating element to provide constant temperature (fig. 4.4). In addition, large models are often equipped with automatic devices that maintain constant humidity and provide for even circulation of heated air. If the laboratory frequently deals with different types of organisms that vary in their temperature requirements, it must have a separate incubator for each desired temperature range. Most pathogenic bacteria, however, grow best at 35 to 37° C and fungi at 28 to 30° C.

While microorganisms are under incubation, their atmospheric requirements must be met. Under **aerobic incubation,** aerobic organisms must have adequate access to the ambient air (the closed vessels that contain them must not be airtight). Microaerophilic organisms must have lowered oxygen tension and usually also require additional CO_2 in the atmosphere. Modern incubators can meet this requirement through a device that admits a flow of metered CO_2 and air into the chamber. A simple, inexpensive method for providing increased levels of CO_2 involves the use of a "candle jar." Cultures are placed in a wide-mouthed jar with a screw-capped top such as a pickle or mayonnaise jar. A candle is then placed in the jar, the candle is lighted, and the jar is closed by screwing the lid down tightly. The candle burns briefly within the closed jar, using some of the available oxygen from the air and giving off CO_2. The candle stops burning when there is not enough oxygen left to support the oxidation and the jar is then placed in the incubator.

Anaerobic organisms require **anaerobic incubation** conditions in which atmospheric oxygen is removed from their environment. The most common method is to place cultures in an "anaerobic jar" fitted with a vented lid through which air can be evacuated and a hydrogen-containing gas introduced. A "cold" catalyst (palladium-coated alumina pellets) that is active at room temperature is also placed in the jar. The pellets catalyze the union of hydrogen with any free oxygen remaining in the jar to form water.

The final atmosphere in the jar (which is then placed in the incubator) permits growth of obligate anaerobes. Many laboratories now use commercially available foil envelopes containing reagents that generate hydrogen and

FIGURE 4.4 A technologist places a rack of cultures in the incubator that is maintained at 35-37° C (mesophilic range), the proper temperature for optimum growth of many pathogenic bacteria.

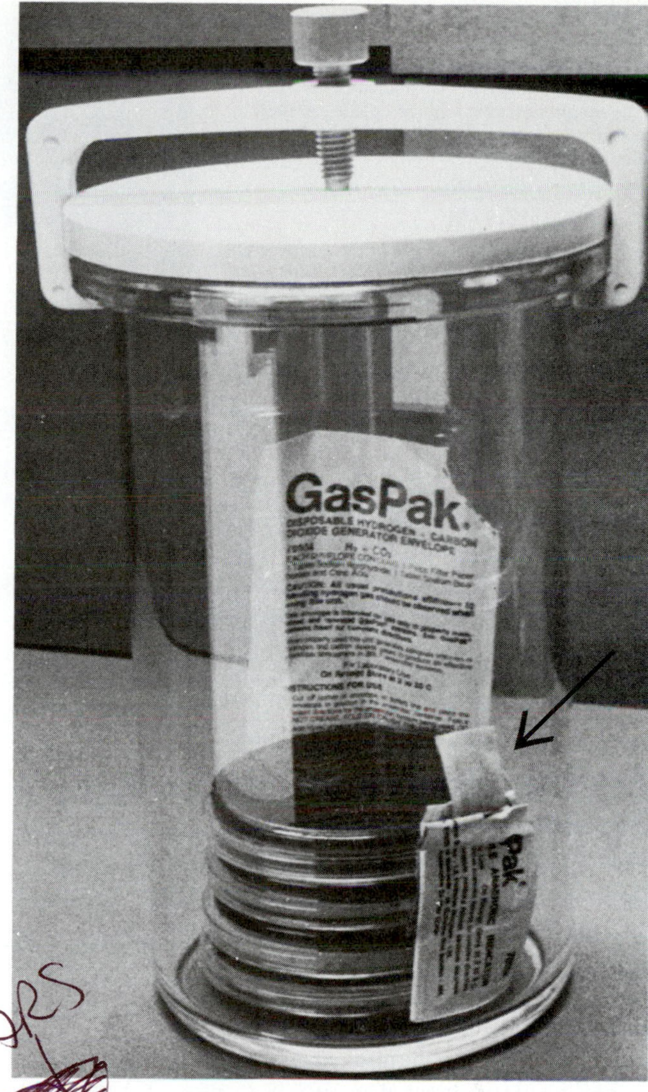

FIGURE 4.5 An anaerobic jar for cultures to be incubated anaerobically. The tight-fitting lid contains a catalyst. The large foil envelope has been opened to receive 10 ml of water, delivered by a pipette. With the lid clamped in place, hydrogen generated from substances in the large envelope combine with oxygen in the jar's atmosphere. This combination is mediated by the catalyst and forms water, which condenses on the sides of the jar. Carbon dioxide is also given off by substances within the large envelope, contributing to the support of growth of fastidious organisms. The smaller envelope has also been opened to expose a pad (arrow) soaked in methylene blue, an indicator used to detect whether oxygen is present. When first exposed, this pad was blue in color. It is now colorless, which indicates that there is no free oxygen left in the jar. The jar now contains an anaerobic atmosphere and can be placed in the incubator.

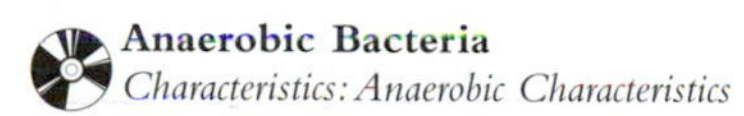
Anaerobic Bacteria
Characteristics: Anaerobic Characteristics

CO_2 when water is added to them. The envelope is placed with the cultures in an anaerobic jar with an unvented lid, water is added to the reagent envelope, and the jar is immediately clamped tightly shut. The released hydrogen combines (through the mediation of the catalyst) with the free oxygen in the jar to form water, and the jar is then incubated. The CO_2 given off helps support the growth of fastidious anaerobes (fig. 4.5).

The efficiency of anaerobic jars or incubators in providing sufficient absence of oxygen must be checked each time they are used. A "redox" dye such as methylene blue is a useful indicator because it is colorless when reduced, under anaerobic conditions, and blue when oxidized in the presence of oxygen. An envelope containing a pad soaked in methylene blue is opened and placed with the cultures in the chamber to be incubated. During incubation, the color of the pad indicates whether or not anaerobic conditions have been achieved and are being maintained.

Media and methods for cultivating anaerobes and other microorganisms are described in a later section of this chapter.

Miscellaneous Equipment

Other major items of equipment for cultivating and identifying microorganisms include refrigerators and freezers for storing perishable supplies such as culture media, blood, serum, antimicrobial agents, or isolated organisms to be saved for special studies. Centrifuges are also needed to concentrate liquid specimens before culture when it seems probable that a large volume of specimen may contain only a few of the suspected microorganisms (e.g., urine

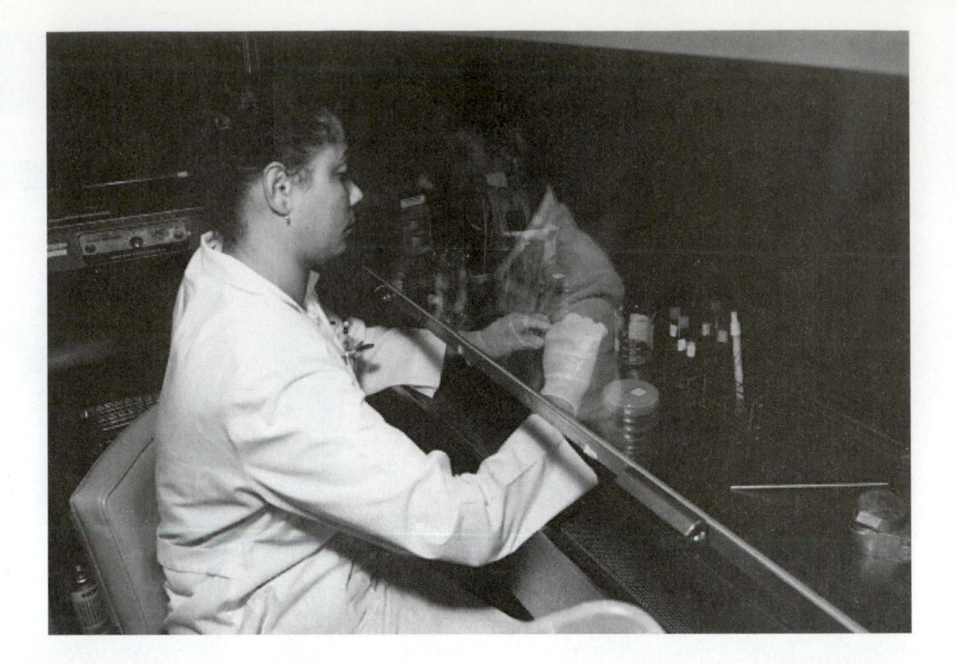

FIGURE 4.6 A biological safety cabinet. The airflow inside the cabinet directs the contaminated air to a filter at the top, preventing outside contamination. The laboratory technologist works through an opening in the hood, wearing disposable gloves for further safety.
Photography by Gordon Bowie.

to be examined for tubercle bacilli). Centrifugation spins the organisms down with the sediment, which is then cultured after the supernatant material is drawn or poured off.

Biological Safety Cabinets

Also referred to as laminar flow hoods, **biological safety cabinets** are used in several areas of the laboratory where risk of personnel infection is high. These include areas where patient specimens are initially received and set up for cultures, and where cultures for tubercle bacilli, fungi, and viruses are processed. Technologists work with their hands through an opening in the front of the cabinet (fig. 4.6). The airflow inside is directed in such a manner that contaminated aerosols cannot exit through the opening but are carried through a High Efficiency Particulate Air (HEPA) filter at the top of the cabinet. Any microorganisms present are trapped in the filter, and clean air is piped into the building exhaust system or vented directly to the outside.

Microbiologic Glassware and Plasticware

Microbiologic glassware is similar to that used in a chemistry laboratory with its assortment of test tubes, flasks, beakers, pipettes, and the like. One of the items specifically designed for microbiologic purposes is the petri dish (or plate). It is named for its originator, Richard Petri, a German bacteriologist, and has been a standard item since 1887 when he first designed it. It is simply a shallow, covered dish for the convenient, safe handling of solidified culture media upon which bacteria and a few other microorganisms can be grown and isolated from each other (colorplates 3a, c).

Petri dishes and most other vessels are available in disposable plastic. Plasticware is widely used in microbiology laboratories. It must be presterilized by the manufacturer for most microbiologic uses.

Sterilizing Equipment

All materials used in the microbiology laboratory, including glass and plasticware, must be scrupulously clean and, in most instances, sterile before use. Cleanliness is essential because even trace amounts of chemical impurities can adversely affect the growth or activities of microorganisms. Sterility is imperative because the presence of contaminating organisms (from the environment) in specimen cultures would interfere with the search for significant pathogens. Before disposable items are discarded or reusable glassware cleaned, specimens or cultures must be sterilized to destroy pathogenic organisms they may contain. Sterilizing equipment is therefore essential to safe laboratory operation.

Sterilization

Sterilization implies the destruction of all forms of life, and sterility means, in our microbiological application, "free of any microorganisms." Sterilization is accomplished by physical or chemical agents, the principles of which are discussed fully in chapter 9. Heat, especially applied as steam under pressure, is a principal agent of microbial destruction, and the most reliable one.

Steam Sterilizers (Autoclaves)

Moist heat, applied as steam under pressure in an **autoclave,** is used in the laboratory for sterilizing culture media, discarded cultures, dirty glassware, and many other items that would be damaged by dry heat (fig. 4.7). Modern autoclaves that provide a drying cycle after sterilization is completed can also be used for sterilizing clean glassware.

The autoclave is an indispensable piece of equipment, not only in the laboratory but elsewhere in the hospital, in operating rooms and central supply areas, for instance,

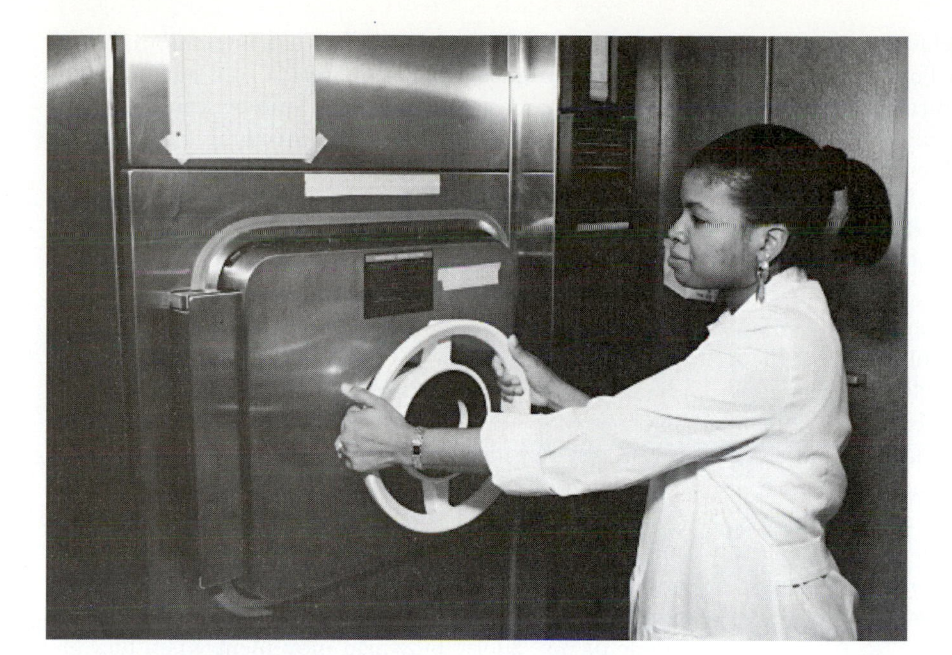

FIGURE 4.7 A laboratory technologist operating an autoclave (or steam sterilizer).
Photography by Gordon Bowie.

Control of Microbes
Physical Control: Moist Heat

where sterilization of items for surgical procedures and other patient uses is of critical importance.

Steam sterilization can be accomplished in 15 to 45 minutes at 121 to 125° C (250 to 256° F) with 15 to 20 lb of steam pressure. The time depends on the size, contents, and distribution of the load, as will be discussed more fully in chapter 9.

The autoclave is satisfactory for sterilizing clean, empty glassware provided damp cotton plugs, paper wraps, or other fabrics used are permitted to dry undisturbed. Microorganisms deposited on wet porous surfaces can contaminate the vessel or material within the wrap by passing along the moist channels between fibers to the interior.

It is important to lay empty tubes, flasks, bottles, and other such vessels horizontally in the autoclave with their plugs or caps loosely in place. In this way the air in them, which is heavier and cooler than steam, can run out and down, allowing steam to cover their interior surfaces.

Vessels containing liquids such as culture media or other solutions can be sterilized in an upright position because these liquids vaporize at autoclave temperatures. The rising vapor displaces the air in the vessel upward and out of the tube or flask while the pressure of the steam prevents any violent bubbling of the liquid. When the sterilizing period is ended, the steam in the chamber is permitted to escape very slowly, so that the liquid can cool below its boiling point while enough pressure remains to prevent its bubbling over.

Safe Disposal of Glassware and Plasticware

All dirty glass or plasticware from the laboratory that has been used for collecting patients' specimens or for any part of the culture procedures must be sterilized before it can be discarded. The autoclave is always used for glassware to ensure that organisms within the depths of liquid materials are rapidly killed and to spare glassware from the damage it would suffer if the contents were baked onto its surfaces with dry heat. Plasticware may be autoclaved or placed in special biosafety bags and sent for incineration.

After sterilization, plastic items can be discarded, but glassware must be scrupulously cleaned and free of chemical impurities that might, even in trace amounts, adversely affect bacterial or viral cultures. Mechanical dishwashers adapted for laboratory glassware are an integral part of the operations of modern laboratories. These machines provide a final rinsing with distilled or deionized water that removes inorganic residues. The glassware is then allowed to drain dry, or it is dried with heat.

Filters

Another type of sterilizing equipment used in microbiology laboratories is the filter. **Sterilization by filtration** is used to remove bacteria or larger microorganisms from fluids or solutions that cannot be heat sterilized. Although filters are made of different porous materials (asbestos, diatomaceous earth, sintered glass, cellulose membranes, collodion) of varying grades of pore size, those most frequently used in diagnostic microbiology are cellulose acetate membranes with pore sizes of 0.22 or 0.45 μm. When solutions are filtered through them, particulate matter, including microorganisms, is retained according to the relative size of particles and pores.

Most viruses and rickettsiae are filterable; that is, they pass through the pores of common filters because of their very small size. In bacteriologic work this is not of practical importance, provided the filtered solution does not contain living cells that could support viral growth and multiplication. Filtration is used therefore to prepare bacteria-free fluids for various purposes. For example, some of the enriching substances in culture media are heat sensitive and cannot be added to basal ingredients before autoclaving. Components such as blood serum, vitamin solutions, antimicrobial agents, or carbohydrates are passed separately through filters of suitable pore size. Using sterile pipettes or syringes, the filtrate is then added to previously sterilized culture medium.

Another use for filtration is to concentrate organisms by passing the specimen through a sterile filter. This method is an alternative to centrifuging liquid clinical specimens that may contain only small numbers of bacteria or fungi. Any organisms present are trapped on the filter, which is then placed in culture medium where the organisms can grow. In use, the sterile filter is mounted on a sterile flask fitted with a side arm. A piece of rubber tubing connects the flask arm to a vacuum, and the solution is pulled through the filter into the flask.

TECHNIQUES

ASEPSIS

The extensive preparation of glassware and other items would be wasted if it were not for the scrupulous care with which microbiologists subsequently handle their work. They must use meticulous technique when handling cultures to ensure the safety of laboratory personnel and to protect the cultures from contamination by stray organisms from the environment. Beginning with sterile media, glass, and plasticware, the microbiologist manipulates every tool and culture so that the microbes under study cannot contaminate hands, clothing, or any part of the working area, or become mixed with extraneous organisms. Every effort is made to maintain **asepsis** (the condition of being sterile or germ-free) in their work. The aseptic or sterile technique is one designed to prevent "sepsis," or infection, and to avoid unwanted microbial growth.

By learning where, when, and under what circumstances microbes may grow and multiply, the microbiologist develops an awareness of their presence (a "seeing eye" for the invisible) and an aseptic technique to circumvent their possible effects. The phrase implies a set of attitudes, as well as of methods, that is fundamental and essential for all who are concerned with the diagnosis and care of patients with infectious diseases and the protection of others from the transfer of infection.

MICROSCOPY

Bacteria and larger organisms can be examined microscopically in the living state, suspended in a fluid, or in dried stained smears in which the microbes are no longer alive.

Wet Preparations

Microorganisms can be studied in the living state to determine such characteristics as motility, shape, size, and group arrangements. When they are in fluid suspensions, or so-called **wet preparations,** the larger forms, such as yeasts, fungi, protozoans, and helminth ova, can be seen with good resolution using ×100 and ×450 magnification. Internal structures can often be identified without staining, but to clarify these structures, a dye, taken up by the cells without distorting them, can be used. Weak solutions of methylene blue or iodine are examples of such stains. Bacteria are too small and transparent to be seen with good resolution in wet preparations even with the oil-immersion lens (×1,000), but their motility as well as their patterns and arrangements such as chaining, clustering, or packeting can be observed.

In wet preparations, all particles suspended in the fluid, including bacteria, exhibit a kind of vibratory motion called *brownian movement.* This motion is rapid but nonprogressive and purposeless. It is caused by the continuous movement of molecules of the liquid, which constantly bombard the suspended particles. All bacteria, whether they are truly motile or not, display brownian movement when they are suspended in a wet preparation. In contrast, the activity of motile organisms is progressive and directional and may have a characteristic mode (e.g., spinning, undulant, dashing, sluggish). Bacterial flagella cannot be seen in wet preparations under the compound light microscope, but when motile protozoans, which are much larger than bacteria, are examined in this way, locomotive structures (flagella, cilia, pseudopods) can be seen in action. Wet preparations are also useful for demonstrating the capsules of microorganisms. Some yeasts, for example, possess large capsules that can be visualized by adding a drop of India ink to the preparation to create a negative stain. The ink provides a dark background against which the capsule appears as a clear, glassy zone surrounding the organism.

Stained Preparations

Dried, stained smears permit better microscopic resolution of bacteria and of internal structures in larger organisms. Stained smears examined with the oil-immersion lens can provide a sharp view of even very small bacteria when they are thinly spread and well separated.

Aniline dyes such as methylene blue, gentian or crystal violet, and carbol fuchsin are basic dyes that react with nucleic acids in the bacterial cell cytoplasm. When they are applied singly as a simple stain, aniline dyes reveal only individual cell morphology. However, when the dyes are added in a specific sequence in combination with other reagents, the result is a differential stain that allows certain types of bacteria to be distinguished from each other. The Gram stain and the Ziehl-Neelson stain (or its Kinyoun modification) are the most commonly used differential stains in diagnostic microbiology.

When material to be studied microscopically has been smeared thinly on the surface of a clean glass slide using the inoculating loop or a sterile swab, it must be allowed to dry completely in air. Drying fixes the material to the slide so that it will not wash off during staining. When dry, the slide is passed quickly through the Bunsen flame to "heat-fix" the material still more firmly to the glass. Excessive heating may distort cells, however, and must be avoided. Alternatively, less distortion occurs if the smear is fixed by placing a few drops of methyl alcohol on the slide and allowing it to air dry. The preparation is now ready for staining by one or more **staining methods.**

Gram Stain

The **Gram stain** was devised by the Danish physician, Christian Gram in 1884 while he was attempting to demonstrate bacteria in lung tissue. To perform this stain, a solution of gentian violet is first applied to the fixed smear. This stains any bacteria present a purple color. Then an iodine solution is added to act as a mordant to "fix" the dye firmly in the cells. The slide is washed with water, and an alcohol decolorizing solution is applied briefly. At this point, only certain types of bacteria, referred to as gram-positive, retain the purple dye. Other bacteria, referred to as gram-negative, lose the purple color; that is, they are decolorized. Unless another *counterstain* (usually the red dye, safranin) is applied to the smear, gram-negative organisms are not visible when viewed microscopically.

The Gram stain reaction is characteristic for each species of bacterium. The exact mechanism is not yet fully understood but is presumed to be related to differences in the composition of the bacterial cell walls. One explanation is that gram-positive bacteria have thick cell walls that become dehydrated by the alcohol decolorizer, trapping the insoluble gentian violet–iodine complex inside. In contrast, the structure of the thin gram-negative wall and its outer membrane permits entry of the decolorizer and the purple dye–iodine complex is readily washed out.

In addition to separating bacteria by their specific staining reaction, the Gram stain also provides clues about their susceptibility and the characteristics of the diseases they produce. For example, most gram-positive bacteria are susceptible to penicillin, whereas gram-negative bacteria are resistant to this drug (see chap. 9). Infection with a gram-negative bacterium may result in a severe condition known as endotoxin shock, which can be rapidly fatal. It should not be surprising to learn that the bacterial cell wall plays an important role in its susceptibility to antimicrobial agents and that endotoxin is a component of the gram-negative outer membrane.

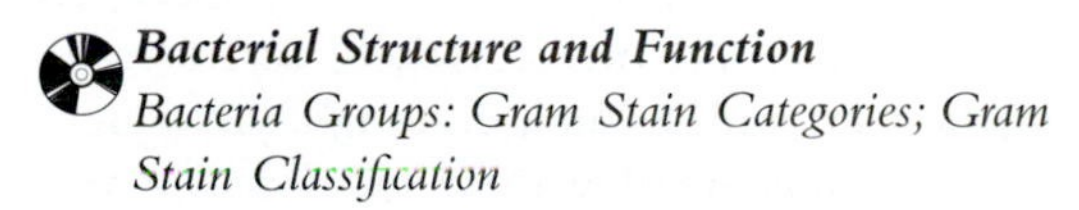
Bacterial Structure and Function
Bacteria Groups: Gram Stain Categories; Gram Stain Classification

Acid-Fast Stain

Like the Gram stain, the Ziehl-Neelson, or **acid-fast stain,** is performed in several steps. The first dye added to the smear is carbol fuchsin. When this dye is combined with a wetting agent (as in the Kinyoun modification of the Ziehl-Neelson stain) or is gently heated to its steaming point on the slide, it can penetrate the characteristic lipid substances in the cell walls of certain bacteria, such as the tubercle bacillus (the agent of tuberculosis). These cells then stain a bright red color, as do other bacteria on the slide. The second step, washing with an acid-alcohol solution, decolorizes organisms that do not have the specific waxy lipid in their cell walls. In the third step the decolorized bacteria are counterstained with methylene blue. Bacteria such as the tubercle bacillus that retain the red color of carbol fuchsin are said to be acid-fast because they resist the action of the acid in the decolorizer. Bacteria that take up the blue counterstain are nonacid-fast. This procedure is useful for directly examining specimens from patients suspected of having tuberculosis and for staining organisms that have grown in culture of these specimens.

Other Stains

Large organisms such as protozoa, the ova and larvae of animal parasites, yeasts, and fungi usually require only simple stains. They are often studied in wet mounts to which a penetrating stain can be added (iodine for parasites, lactophenol cotton blue or the fluorescing calcofluor white for fungi). The fine structures of bacteria (flagella, capsules, spores) require special stains, as do intracellular organisms such as the rickettsiae and chlamydiae, growing in aggregates within parasitized cells. Stains used to visualize organisms within host cells must differentiate the organisms from the host cell structures.

CULTURE TECHNIQUES

This section deals primarily with the cultivation of those pathogenic bacteria that are not obligate intracellular parasites. Rickettsiae, chlamydiae, and viruses require the presence of living cells to grow in culture, and yeasts, molds, and a few cultivable protozoan species each have special requirements. Yet the basic principles for preparation and handling of such cultures are similar to those used in the study of bacterial populations.

The Nature and Preparation of Culture Media

Pathogenic bacteria obtain nourishment from their parasitized host. Therefore, to cultivate bacteria in the laboratory, **bacterial cultures** must contain the required nutrients for growth. A laboratory culture medium can be liquid, referred to as a broth, or can be solid, formed when agar is added to the broth medium.

Composition of Culture Media

The composition of a good **culture medium** (plural, **media**) requires careful adjustment of a number of ingredients and factors. In general, nutritional media must contain products of partial protein breakdown, since many bacteria cannot digest intact protein. Protein components, such as peptides, peptones, and amino acids, are required for synthesis of bacterial cell protoplasm and for cellular structure. These are commonly provided in the form of partially cooked meat or meat extracts. Bacterial syntheses require energy; therefore, carbohydrates such as glucose, lactose, or starch must be provided. Minerals and vitamins may be added in purified form or as components of some natural food. Minerals are often present as their salts in meat digests, while yeast extract is an excellent source of vitamin B_1. Water is always present in large proportion. Other ingredients can be added for special purposes.

A medium can be made *selective* by incorporating one or more ingredients that will inhibit growth of some organisms while encouraging multiplication of others. A *differential* medium can be prepared by including a substance to which different bacteria respond differently, thus making it possible to distinguish them by their growth characteristics in such a medium. The medium may be used as a liquid broth, or agar may be added to solidify it at temperatures below 45° C (113° F).

Media prepared with meat extracts are usually referred to as nutrient agar or broth media, while those prepared with meats that have been partially digested by slow cooking or by addition of enzymes are called infusion broths or agars. Animal blood (usually sheep) may also be added for further enrichment. For this purpose whole blood that has been defibrinated (the fibrin removed to prevent clotting) is used in a proportion of 5 to 10% by volume. For growth of some bacteria, hemoglobin must be liberated from the blood cells by heating the blood gently or adding it to a hot medium. In the process, hemoglobin changes from its customary bright red color to a rich brown, so that an agar medium prepared with heated blood is called chocolate agar. If an enriched medium without the color or turbidity of blood is desired, other materials can be substituted such as human or animal serum, yeast extract, or extra protein components.

An important additive for solidifying liquid media is agar. This complex carbohydrate, which is isolated from red algae, has the unique property of solidifying at 45° C but melting only at temperatures near 100° C. Thus, it is a solid at the usual incubation temperature for bacterial pathogens (35° C). Another important feature of agar is that it is not digested by bacteria and, therefore, does not serve as a nutrient in the medium.

The culture media described here contain mixed and varying amounts of nutrients. Occasionally, for performing special microbial assays or determining specific nutritional requirements, a synthetic medium is required that contains exact quantities of chemically pure compounds. Such media are usually tedious to prepare and are seldom used or needed in the diagnostic microbiology laboratory.

pH Adjustment

Before the medium is complete and ready for use, it is important to ensure the stability of its acid-alkaline balance, or **pH.** (pH is expressed as a mathematical value from 1 to 14 and used to indicate the degree of acidity or alkalinity of a solution.) Most bacteria require a pH near neutrality, that is, neither actively acid nor alkaline. This means that not only must the finished medium be adjusted to neutrality, but it must also be buffered against any major changes in pH that may occur as a result of bacterial growth in the medium. Bacterial growth products accumulate in an artificial medium because they cannot diffuse away as they might in a natural environment. Large shifts in the pH of the medium toward acidity or alkalinity can occur, therefore, and since bacteria are sensitive to such changes, they may die. For this reason it is desirable to incorporate buffers into the medium. Buffers are compounds that react with free acids and alkalis and, in so doing, bind them so that they are no longer active in solution. Continuing buffer action throughout the period of bacterial growth prevents large pH changes in the medium.

Microbiologists often want to be certain that a culture maintains a stable pH, or they may be looking for a change in pH that occurs when organisms use some ingredient in the medium. For these reasons, indicator dyes that assume a different color at various pH values are frequently added to culture media. A shift in pH, as the culture grows, can be detected by a visible change in the color of the medium. For each indicator it is necessary to know the range of colors that occur in response to pH changes. By comparing the color of the test with the standard colors for the indicator at each pH value, an estimate of the degree of acidity or alkalinity developing in the test can be made. Finer measurements can be made with electrical pH meters when it is necessary to be more exact (fig. 4.8).

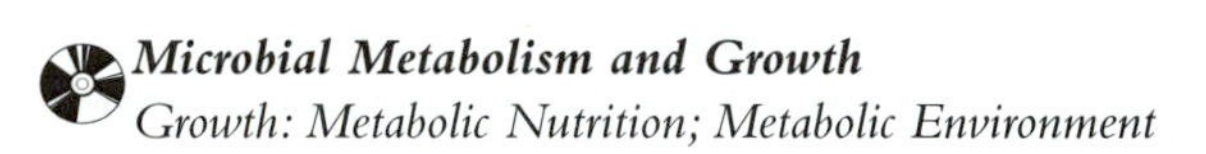

The Final Step

The finished product must of course be free of any extraneous organisms from the general environment. Such organisms would contaminate the medium with their own growth products, compete for nutrient with the species being studied, and make accurate interpretation of results impossible. The final step in the preparation of culture media, therefore, is sterilization, either in the autoclave, as described, or by filtration if heat-sensitive ingredients are involved.

It is important that there be no time delays between the steps taken to prepare the media. Otherwise, microorganisms could begin to grow and, although subsequent sterilization of the media will kill these organisms, it will not restore the changes in nutrient ingredients that they may have produced or remove any impurities they may have added.

Uses of Culture Media

Media can be used in either solid or liquid form, and solid media may be employed in a number of ways.

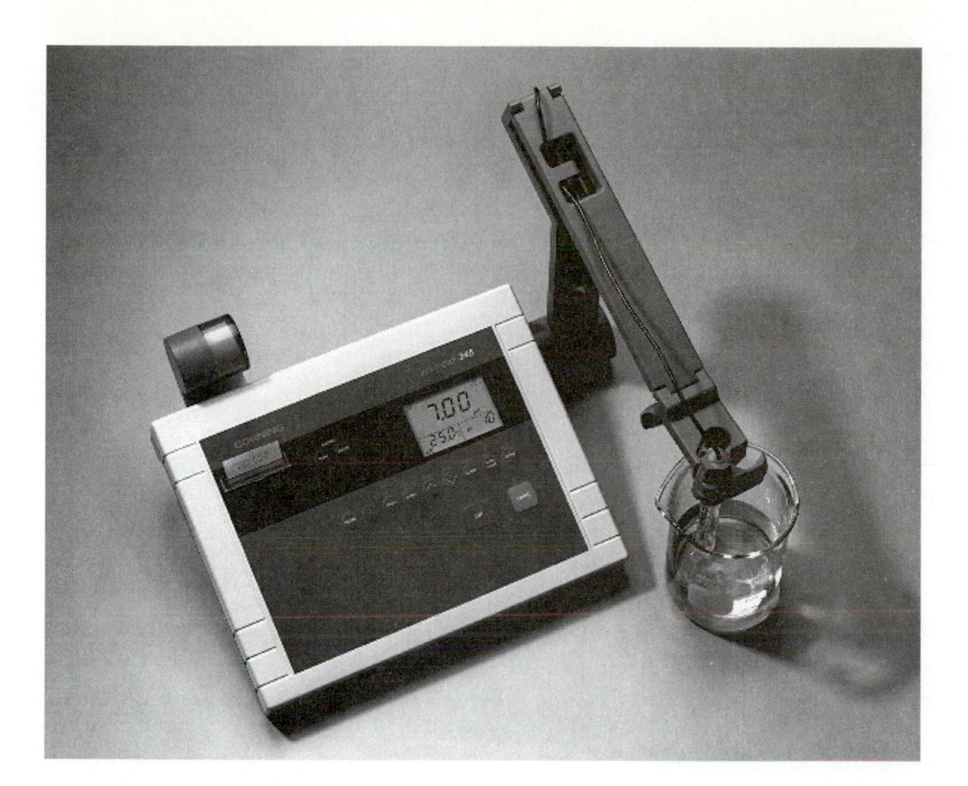

FIGURE 4.8 The pH of the solution being tested appears as a digital readout on this electrical pH meter.

Printed with permission of Corning Incorporated, Corning, NY 14831.

Microbial Metabolism and Growth
Growth: Metabolic Environment

Liquid Media

As mentioned, a medium used in liquid form is usually called a broth. It may be prepared in test tubes in small quantities (5 to 10 ml) or in flasks or bottles. Sometimes a small shallow layer of broth in a large flask is desirable, to provide good oxygenation of the culture. Conversely, to limit atmospheric oxygen, the broth may be dispensed in tall, narrow tubes. For some purposes it may be useful to layer a little broth over part of a solid agar slant so that organisms will have a very wet but solid surface for growth.

Solid Media

Most commonly, agar media are plated by pouring them into petri plates. This is always done after the medium has been sterilized in tubes, flasks, or bottles and has cooled to about 50° C (or has been melted down again after solidification). If desired, the material to be cultured can be placed in the petri dish or added to the cooled agar before it is poured. In this case the organisms grow throughout the depths of the medium as well as on the surface, and the preparation is called a *pour plate.* If the medium is poured before it is inoculated, the material for culture is streaked across its hardened surface with the inoculating loop or a sterile swab. This results in growth of organisms across the surface along the lines drawn by the tool. Such a preparation is called a *streak plate* (colorplate 3a).

Agar medium also may be dispensed into a number of test tubes which, after sterilization, are laid at an angle to cool. When the agar has solidified, the medium will have a sloping surface called a *slant.* Slants are inoculated with bacteria by streaking the surface with the inoculating wire, which may also be plunged, or stabbed, to the bottom of the tube through the solid butt of medium below.

Tubed agar media can also be solidified in an upright position for inoculation by a stab of the wire down through the column. Such a *stab culture* may be used to demonstrate the characteristic way a particular organism grows along the path of the wire. This depends on whether it is motile and can diffuse outward from the stab through the medium as well as on its ability to grow at varying depths where oxygen availability becomes a factor.

Prereduced Media

For work with fastidious anaerobic bacteria, highly reduced, oxygen-free media are required. Liquid media are prepared in the conventional manner; however, once they have been boiled sufficiently to remove oxygen (as determined by the redox indicator, resazurin), anaerobic gas is bubbled through while the broth is distributed into tubes. The residual air in the tubes is replaced with oxygen-free gas, the tubes are closed tightly with rubber stoppers, and finally they are placed in a press that prevents the stoppers from blowing off during autoclave sterilization. This medium, referred to as **prereduced,** *anaerobically sterilized* (PRAS), must be inoculated under a special device that excludes oxygen while the tube is open.

Fortunately, most disease-producing anaerobic bacteria are not extremely affected by small amounts of oxygen. They grow on freshly prepared agar media incubated in an anaerobic jar or in freshly boiled broth containing reducing agents that lower the amount of available oxygen.

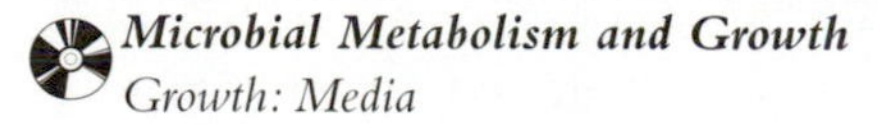

SPECIMEN CULTURE

The work of the clinical microbiologist begins with the arrival of a specimen in the laboratory. The specimen must be accompanied by a request from the patient's physician for a culture. However, even before the specimen reaches the laboratory, the results of that culture will be influenced by the method and timing of its collection and transport. If these are inadequate or incorrect, the results may be also, and, even more unfortunately, this source of error may pass unnoticed so that the results are not subjected to proper interpretation. Specimen collection, often performed at the patient's bedside by a physician or other health care professional rather than by the microbiologist, will be emphasized many times in this text because of its importance to proper diagnosis and patient care.

Direct Smears and Preliminary Information

Time and appropriate action are of the essence in medical microbiology. Specific measures for treating the patient seriously ill with an acute infection may depend on news from the laboratory regarding the nature of microorganisms, if any, being isolated from the patient's specimens and their response in culture to the action of clinically useful antimicrobial agents. Between the time cultures are started and identifications are complete, the laboratory can provide tentative, preliminary information concerning the results to guide those who are caring for the patient. Even incomplete results can be used to make decisions as to whether or not the patient should be treated with specific isolation precautions or certain antimicrobial agents. For example, **direct smears** made immediately from the specimen and stained by the Gram method, may provide important clues to the nature of organisms present (gram-positive or gram-negative; cocci or bacilli; chained, clumped, or paired; intra- or extracellular), and a good smear report may guide the physician to the first choice of treatment. Direct smears may also help the microbiologist determine which culture methods to use.

Initiation of Culture

The nature of the specimen, the kind of information desired from it, and the smear reading all determine the type of **specimen culture** procedures initiated. As we shall see, many areas of the human body are parasitized by microorganisms that cause no harm, but different organisms are found in different areas because environmental conditions vary from one part of the body to another. Thus a particular kind of specimen, such as a throat swab or a sample of feces has its own "normal flora." These organisms will grow out in the media as a *mixed culture,* in which any pathogenic microorganism that may be present must be recognized. The suspected pathogen must then be separated from the others and grown alone, in *pure culture,* so that it can be identified by observing its appearance, properties, and behavior (colorplate 3a).

Not only does the normal flora vary with different specimen types, but the pathogenic organisms that can be expected in different specimens vary also. For example, one would not expect to find the diphtheria bacillus in a fecal sample or the agent of a sexually transmitted disease in a surgical wound culture. On the other hand, certain types of specimens, such as blood, urine, and spinal fluid, are normally sterile in the healthy body and contain microorganisms only by accidental contamination or as a result of an infectious process. Therefore, any organism found in such a specimen may be suspect.

The initiation of culture procedures is thus carefully guided by the specimen, its source, and the diagnostic possibilities. A battery of different media is inoculated with the specimen in order to provide a wide range of bacterial requirements (e.g., enriched for fastidious organisms; selective to reduce competition but still permit the suspected organism to grow; differential to indicate different kinds of organisms growing upon it (colorplate 3c); reduced for anaerobes). Broth as well as agar media are usually inoculated. A larger volume of specimen can be added to the liquid medium, and thus there is a better chance that organisms present in only small numbers will be isolated. The widest possible screening, consistent with practical limitations on time and space, is provided to cover the widest possible range of microbiologic and clinical expectation.

Day-to-Day Readings

Each day after cultures of specimens are started, the microbiologist makes observations of the growth occurring in each of the media inoculated and plans the next step that will narrow identifications to a final decision as quickly as possible. The work is organized in a logical sequence designed to establish a pattern of facts from a number of positive and negative clues, in much the same way as a detective sets out to find evidence and fit it together.

Isolations

Once a specimen arrives at the laboratory, it is checked for proper patient and source identification. With the inoculating loop, a portion of the specimen is streaked across the surface of suitable agar media, and a broth may also be inoculated. When the streaking technique is performed properly, colonies at the terminal end of the streaks are not crowded together but grow individually and apart from each other during incubation. Each isolated colony represents growth originating from a single bacterium and thus is a pure culture. Isolated colonies are picked from the surface with the loop or straight wire and transferred to other media for identification through observations made of their characteristic growth and biochemical reactions. These second sets of cultures are referred to as *subcultures.*

Final Identification

Pure subcultures of isolated organisms are examined after 18 to 24 hours of incubation. If they provide sufficient information, final identifications are made. If not, further subcultures may have to be made to test additional properties. Carbohydrate media are inspected for evidence of degradation provided by the color of the pH-sensitive indicator and, depending on the identification system used, by the presence or absence of gas.

Other types of enzyme activity characteristic for a group or a species can be demonstrated quickly and simply with pure subcultures. Some strains of staphylococci, for example, produce an enzyme called coagulase because it coagulates blood plasma. In the diagnostic microbiology laboratory, this enzyme can be readily assayed by incubating bacterial colonies in a tube with rabbit plasma. If coagulase is produced by the organisms, the plasma will clot, usually within four hours. Coagulase production together with characteristic colonial morphology, pigment production, and microscopic appearance might lead to a report of *Staphylococcus aureus.* This organism is coagulase positive and is known for its ability to cause serious infections and for the ease with which it spreads from one person to another.

Another example of an easily identified enzyme is urease, possessed by certain organisms commonly encountered in stool specimens. As its name implies, this enzyme breaks down urea to ammonia and carbon dioxide. When the organisms are incubated in a urea broth, the pH becomes highly alkaline as a result of ammonia release, which can be observed as a color change of the pH indicator in the broth.

In some instances, it is possible to identify a bacterium correctly by using only a few such tests, but often an extensive biochemical "profile" is needed. Because it is expensive and time-consuming to make and keep a wide variety of culture media on hand, many laboratories now use commercially available, multimedia identification kits. The most commonly used kits contain a wide variety of culture media dried either on disks or in small cups on a plastic strip. These media are rehydrated with an aqueous suspension of the organism to be identified and are incubated for 4 to 24 hours. The reactions are read and interpreted by the technologist who then identifies the organism, usually with the aid of a computer-generated code book. Although not every reaction is needed to identify each microorganism, the simultaneous testing of multiple biochemical reactions ensures that most bacteria will be identified in the shortest time possible, without the need for setting up additional tests.

Within recent years, several automated systems for identifying bacteria have been developed. The test organism is inoculated into a biochemical test panel purchased from the instrument manufacturer. The test panel, which is similar to the multimedia test kits, is placed in the instrument where biochemical reactions that occur are monitored by a light sensor that detects ("reads") changes in the indicators in the test media. When sufficient biochemical activity has occurred (usually within 3 to 16 hours), the instrument transmits the readings to its computer which interprets all results and then automatically identifies the isolate based on the test results.

Even in laboratories that use such systems, however, skilled technologists are needed to perform microscopic and cultural examinations, to determine which organisms need further testing, and to double-check the instruments.

With the availability of such methods, most reports are available within 24 to 48 hours of the time the culture was initiated. Meantime, the microbiologist keeps careful notes and records of tests and results, so that final reports can be made with clear, factual support.

Serologic Identification of Microorganisms

Conclusive identification of microorganisms can often be obtained using **serologic techniques.** These tests rely on the use of reagents containing antibodies, substances that, in the laboratory, can be made to react in a visible way with a specific microorganism. Antibodies are produced in the serum of humans and animals in response to being infected, or immunized with a microorganism or its components; substances introduced into the body that stimulate the production of antibodies are called *antigens* or *immunogens.* Antibodies react only with the immunogen that stimulated their production or one that is closely related chemically. Serologic reactions can confirm the identity of a microorganism when the nature of the antibody in a given serum reagent is known. The nature of immunogens and antibodies is discussed in chapter 6.

ANTIMICROBIAL SUSCEPTIBILITY TESTING

Antimicrobial susceptibility tests can be performed as soon as individual organisms are isolated from the specimen in pure culture. It is not good practice to perform these tests with mixed cultures. In a mixture it may be impossible to determine the response of each organism to a particular antimicrobial agent; the mixed species may compete with each other for growth on the medium, adding to any antagonism exerted by the drug; or one species may be capable of inactivating the antimicrobial agent so that its inhibitory effect on the other bacteria is not apparent. In any of these cases, results are difficult to interpret and may mislead the physician. Therefore, it is usual for the laboratory to plate the specimen on agar media and to test only isolated bacterial growth. An exception to this general rule may be made if the situation is urgent, and if stained smears of the specimen reveal the presence or predominance of only one type of organism. In this case, susceptibility testing can be done directly from the initial specimen. However, it is always best to repeat the test later using a pure culture of each clinically significant organism isolated from the specimen to confirm the first results.

A number of methods exist for testing the antimicrobial susceptibility of isolated organisms. Some are highly quantitative but expensive to perform; others are semiquantitative but more practical and faster.

Disk Agar Diffusion Method

The susceptibility testing technique used most frequently is called the disk agar diffusion method. For reasons of accuracy, only rapidly growing aerobic and facultative bacteria are tested by this technique. Commercially available, small filter paper disks of uniform size and shape are impregnated with antimicrobial agents by the manufacturer. Each disk contains a known concentration of one such agent. The organism to be tested is first "seeded" on the surface of a suitable nutrient agar plate (Mueller-Hinton agar) by streaking it in a way that permits confluent growth across the entire plate. A number of disks containing different antimicrobial agents are then placed on the plate, spaced at regular intervals. The plate is incubated aerobically at 35° C for 18 to 24 hours and examined the following day. During incubation, the drug in each disk diffuses into the agar so that the medium surrounding the disks contains a gradient of drug concentrations. The highest concentration of each drug is in the area nearest to the disk. Meantime, the organism seeded across the plate grows everywhere except in those areas where it encounters a concentration of an antimicrobial agent sufficient to interfere with its growth. When the plate is examined the next day, "zones of inhibition" may be seen around disks that contained drugs inhibitory to the organism; that is, such disks are surrounded by clear zones where no growth has occurred. The zones are obvious because they are rimmed by confluent growth beyond the areas of drug activity where the organism has grown uninhibited (colorplate 3b). Disks containing antimicrobial agents that did not affect the organism at all are surrounded by growth that extends right to their edges.

Several factors influence the results of susceptibility tests and these factors must be standardized each time the test is performed. These include the culture medium, size of the inoculum, concentration of drug in each disk, and measurement of the diameter of each observed zone of inhibition. A method of testing recommended by the National Committee for Clinical Laboratory Standards (NCCLS) specifies the techniques for standardization of each step of the procedure. This method is known either as the NCCLS or modified Bauer-Kirby method (after the originators of the standard disk-diffusion method).

When the NCCLS technique is followed, zones of inhibition around the antimicrobial disks are measured exactly and the zone sizes are compared with those listed in an interpretive table that relates zone sizes to organism drug susceptibility (table 4.2). The results place the organism into one of three categories: resistant (R), intermediate (I), or susceptible (S) to each of the antimicrobial agents tested. When the variables in the disk agar diffusion method are carefully controlled and zone sizes properly interpreted, the reported results provide the physician with an almost quantitative estimate of the organism's response to each antimicrobial agent. For example, a result of susceptible (S) or resistant (R) clearly advises for or against the use of a specific drug, respectively; when an interpretation of I is made, the physician may request further tests, use higher than normal drug doses, or select an alternative antimicro-

TABLE 4.2 ZONE DIAMETER INTERPRETIVE TABLE

		Diameter of Inhibition Zone (mm)		
Antimicrobial Agent	**Disk Concentration**	**R**	**I**	**S**
Ampicillin[a]	10 μg	≤13	14–16	≥17
Carbenicillin[b]	100 μg	≤13	14–16	≥17
Cefoxitin	30 μg	≤14	15–17	≥18
Cephalothin	30 μg	≤14	15–17	≥18
Clindamycin	2 μg	≤14	15–20	≥21
Erythromycin	15 μg	≤13	14–22	≥23
Gentamicin	10 μg	≤12	13–14	≥15
Methicillin[c]	5 μg	≤ 9	10–13	≥14
Penicillin G[c]	10 U	≤28	—	≥29
Penicillin G[d]	10 U	≤14	—	≥15
Sulfonamides	250 or 300 μg	≤12	13–16	≥17
Tetracycline	30 μg	≤14	15–18	≥19
Vancomycin	30 μg	≤ 9	10–11	≥12

Source: Adapted from *Performance Standards for Antimicrobial Disk Susceptibility Tests,* 6th ed. National Committee for Clinical Laboratory Standards NCCLS 1997. The material is constantly being updated, and you should obtain the latest information from NCCLS.

Notes: Zone sizes appropriate only when testing

[a]Gram-negative enteric organisms

[b]*Pseudomonas*

[c]Staphylococci

[d]Enterococci

Antimicrobial Action
Antibiotic Susceptibility Testing: Disk Diffusion Testing

bial agent to which the infecting microorganism is fully susceptible.

The battery of drugs tested is chosen based on a consultation between microbiologists, pharmacists, and physicians taking into account factors such as the clinical usefulness of the antimicrobial agent and the feasibility of testing it by the routine disk method. In keeping with the concern of rising health care costs is the consideration of whether to test expensive drugs when equally effective but less costly alternatives are available.

Dilution Methods

Broth Dilution

A more fully quantitative measurement of drug effect can be obtained by diluting the test antimicrobial agent in broth. Under this method a solution of antimicrobial agent is diluted serially in tubes of Mueller-Hinton broth to achieve varying drug concentrations. Each tube is then inoculated with the organism of interest and incubated. The organism grows only in those tubes containing a concentration of the drug too low to inhibit growth of the organism. The lowest concentration of antimicrobial agent that prevents visible growth in the broth is called the *minimal inhibitory concentration* (*MIC*). Antimicrobial agents can thus be titrated rather exactly to determine the most effective dose required to inhibit the organism in the test tube. From this information one can predict whether a sufficient amount of a drug can be administered to control the patient's infection.

For patients whose natural antimicrobial defenses are not functioning well, it is important to know not just whether the agent inhibits the organism in question but whether it kills the organism. In the broth dilution method this may be determined by subculturing portions of broth from those tubes that have no visible growth onto drug-free agar medium. In many instances, organisms from the tube containing the MIC and from one or two higher concentrations in the series will grow on the subculture, indicating that they have been inhibited but not killed. The lowest concentration of antimicrobial agent that kills as well as inhibits the bacterium, as demonstrated by failure to grow in subculture on a drug-free plate, is called the *minimal bactericidal concentration* (*MBC*). In practice, a 99.9% reduction in the number of organisms inoculated into the tubes containing the antimicrobial agent is considered a satisfactory bactericidal effect. For some organisms and drugs, however, a bactericidal effect is never achieved.

Agar Dilution

The minimal inhibitory concentration of antimicrobial agents may also be determined by the agar dilution method. Varying concentrations of the agent are incorporated into a series of nutrient agar plates (usually Mueller-Hinton agar). The test microorganism is then inoculated onto a portion of each plate, and the plates are incubated overnight at 35° C. Again, the lowest concentration of drug that inhibits growth is the MIC. Unlike the broth method where only one organism can be tested in each set of tube dilutions, up to 32 organisms can be tested on each plate of the series in the agar dilution method. The MBC cannot be determined by this technique, however.

In contrast to the qualitative results of the NCCLS method, dilution methods provide the physician with a quantitative measurement of microorganism susceptibility to antimicrobial agents. The MIC and MBC values are reported in terms of micrograms per milliliter (μg/ml) of the drug; for example, "MIC of methicillin for *Staphylococcus aureus* is 1 μg/ml." This quantitative information is especially important for physicians when they are treating serious infections such as bacterial endocarditis or meningitis, or when highly resistant organisms such as *Pseudomonas* species are causing infection.

Etest

The Etest is a recent advance in antimicrobial susceptibility testing that combines the ease of performing a disk diffusion assay with the ability to obtain a quantitative result, such as with dilution assays. The Etest consists of a plastic-coated strip impregnated with a particular antimicrobial agent and on which is printed a range of antimicrobial agent concentrations. When an Etest strip is placed on an agar plate inoculated with the organism to be tested, a preformed antimicrobial gradient diffuses from the strip into the agar medium during the incubation period. If the organism is susceptible to the particular antimicrobial agent on the strip, an oval-shaped zone (ellipse) of inhibition is produced (fig. 4.9). The MIC is read off the printed scale on the strip at the point where the zone of growth inhibition intersects the strip. Etest strips containing gradients of several different antimicrobial agents can be placed on one agar plate. Although the method is accurate and convenient, it is more expensive to perform than some of the other methods and is usually reserved for testing in specific clinical situations such as when resistant staphylococci or fastidious or anaerobic organisms are isolated.

Microdilution

Preparing dilutions is time consuming and not practical for small hospital laboratories with limited personnel. Commercially prepared broth microdilution panels are now available that are simple to use and give results comparable to the agar and broth macrodilution tests. The basic unit for this technique is a small, plastic tray containing approximately 96 wells. The wells are filled by the manufacturer with a series of concentrations of several antimicrobial agents. In the laboratory the technologist need only inoculate the test organism into each well, using a special device that inoculates all wells simultaneously. After overnight incubation each drug-containing well in the tray is examined for the presence or absence of visible bacterial growth. The MIC and MBC of each antimicrobial agent are determined in the same manner as for broth dilution (fig. 4.10).

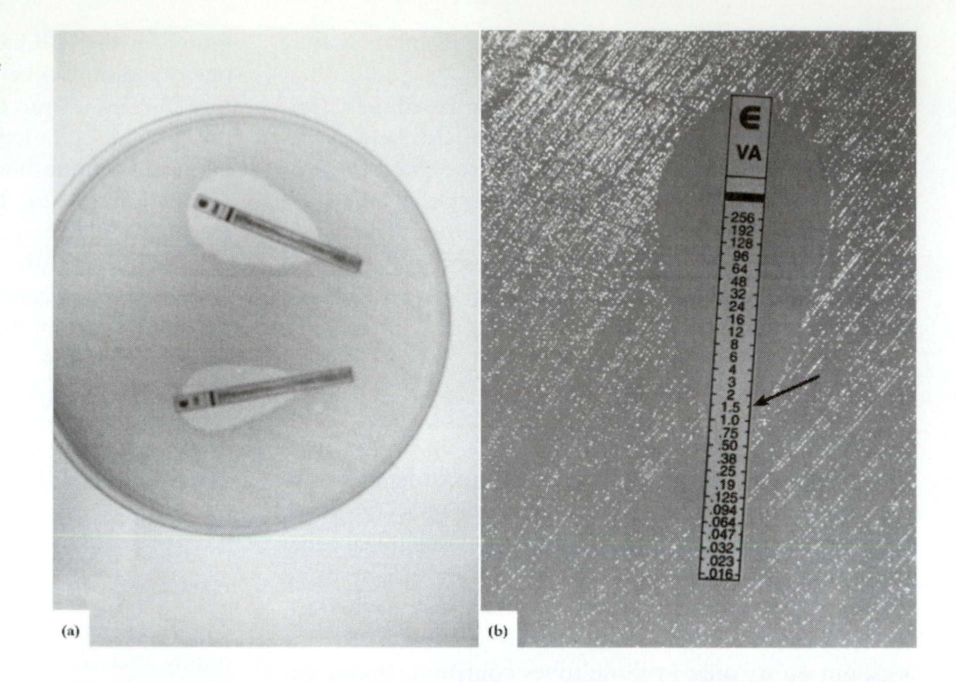

FIGURE 4.9 Etest strips are placed on a culture plate inoculated with the organism to be tested. If the organism is susceptible to the antimicrobial agent on the strip, an oval-shaped zone (ellipse) of inhibition is produced (a). As shown in b, the lowest concentration of the antimicrobial agent vancomycin (VA) that inhibits the microorganism is 1.5 micrograms per milliliter (arrow).

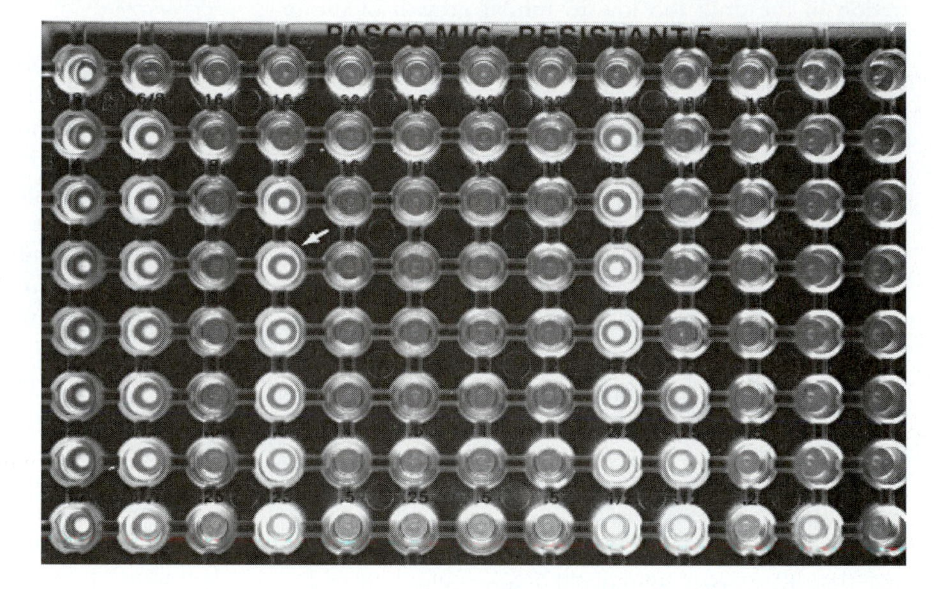

FIGURE 4.10 A microdilution susceptibility test. A series of concentrations of several different antimicrobial agents are contained in the tray wells. If the bacterium inoculated into the tray wells is resistant to the concentration of antimicrobial agent within, the organism grows and settles as a white button (arrow) at the bottom of the well. The clear wells indicate that the organism is inhibited by the concentration of antimicrobial agent inside.

Antimicrobial Action
Antibiotic Susceptibility Testing: Broth Dilution

Automated Methods

Several automated systems are available commercially for performing antimicrobial susceptibility testing (fig. 4.11). Different instruments are used in each system but the basic principles of testing are similar. A suspension of the organism is inoculated into a disposable plastic multiwelled card or tray, each well of which has been filled by the manufacturer with a predetermined amount of antimicrobial agent. A single well that contains no drug serves as the growth control. The inoculated cards or trays are then incubated directly in the instrument. At various time intervals, optical scanners measure the growth turbidity in the control well and the drug-containing wells. Bacteria growing in the control wells cause increasing cloudiness that is detected by the scanner. In the drug-containing wells, where growth inhibition has occurred, the turbidity will be reduced in proportion to the degree of inhibition. A computer in the instrument compares measurements of growth in the control and drug-containing wells, and after an appropriate time interval (usually three to eight hours), calculates S, I, and R values for each antimicrobial agent tested. A printing device then prints out these results on a report form. Modifications of these instruments are available to allow MIC determination and organism identification.

A variety of instrumentation is also available to help technologists read and interpret the susceptibility results of microdilution panels. The most sophisticated instruments optically read individual wells of the microdilution trays, calculate MIC values, and print out a susceptibility report.

The expense of these instruments to laboratories is offset by the reduced labor necessary to perform large num-

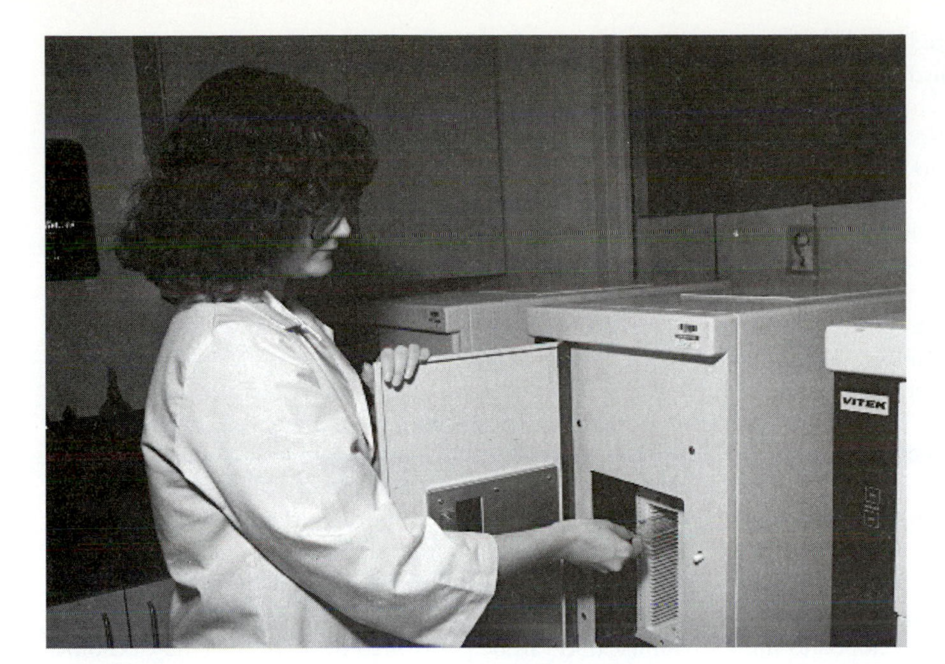

FIGURE 4.11 The laboratory technologist is using an automated instrument (Vitek) to determine antimicrobial susceptibility.
Photography by Gordon Bowie.

bers of daily susceptibility tests and the shortened time required to obtain test results.

Antimicrobial Assays of Patient's Body Fluids

For patients with serious infections the laboratory may need to determine whether concentrations of antimicrobial agents sufficient to inhibit the offending organism are present in blood, urine, or other body fluids. In addition, certain antimicrobial agents are toxic to body tissues at high concentrations, and it is important to measure these levels in the patient to be sure toxic levels are not reached. The amount of drug in body fluids may be estimated by making several dilutions of the fluid, then adding a suspension of the specific pathogen. After the test is incubated overnight, the tubes are examined for the absence of turbidity, which indicates growth inhibition. A general rule applied to levels of antimicrobial agents in the serum of patients with endocarditis is that adequate therapy is achieved when a serum dilution of at least 1:8 inhibits the patient's organism. A variety of other methods, more complex to perform but more accurate, are available. These include a disk diffusion assay and assays that use fluorescent (FA) or enzyme-linked (EIA) antibodies prepared against antimicrobial agents (see chap. 6). These methods are preferable for detecting toxic drug levels.

SPECIAL TECHNIQUES

Quantitative Cultures

It is often important to know not only what organisms are present in a particular specimen, but how many. The mere presence of an organism in the specimen, even a potential pathogen, does not necessarily mean that it is the cause of the patient's disease. It may simply be an organism normally present in the area of the body from which the specimen was derived or a contaminant from the environment. Its occurrence in large numbers, however, may indicate its significance.

This question is of particular interest in the case of urine samples. Specimens of urine for culture are usually voided into a sterile container. Thus, small numbers of organisms normally found on the external urethra may contaminate the specimen. If, however, the sample is collected with a special clean technique and is sent to the laboratory immediately, any organisms present in the urine will not have an opportunity to multiply. A numerical count of the bacteria that grow out in culture provides a better evaluation of their significance than a mere report of their presence. In general, when fewer than 10,000 colonies per milliliter of urine are grown, they are considered to be urethral contaminants rather than urinary tract pathogens.

A **quantitative culture** is prepared by placing a measured quantity of the sample on one or more solid media, counting the number of colonies that grow, and translating the figure arithmetically to express the number of organisms present in 1 ml of sample. It is usually sufficient for the laboratory to report the approximate range of the count rather than an exact figure, such as "less than 10,000 organisms per milliliter" or "more than 10,000 organisms per milliliter." It cannot be overemphasized, however, that the reliability of a urine colony count depends on the adequacy of specimen collection and the speed with which the culture can be initiated. If a time lapse is unavoidable before the culture can be started, the specimen should be refrigerated to prevent bacterial multiplication in the interim.

New Developments

A number of special methods for detecting or identifying microorganisms in as short a time as possible have been developed in recent years. Rapid microbiologic diagnoses allow appropriate treatment to be instituted in much shorter times than were possible with conventional techniques. As a result, hospitalized patients can be discharged much earlier than before with important clinical and economic impact. Before implementing such techniques, however, diagnostic laboratories must consider their accuracy, ease of use, and space and funds available for any equipment required.

Detection of Bacteremia

Microorganisms circulating in the bloodstream (bacteremia) represent a serious threat to the patient and such infections can be rapidly fatal. Standard detection methods require inoculation of several milliliters of blood into a broth medium and at least daily visual examination of the broths for signs of microbial growth. At the end of five to seven days, a portion of the broth is subcultured to a solid agar medium to check for the presence of organisms that do not produce visible signs of growth. This standard method is slow, cumbersome, and very labor intensive. Over the past 25 years, a variety of automated methods have been developed to simplify and speed up this process. The current generation of such methods relies on instruments that detect the presence of CO_2 released as a by-product when microbes metabolize various substrates such as glucose in the culture medium. Depending on the instrument manufacturer, the CO_2 is detected by use of colorimetric or fluorescent sensors or by changes in gas pressure across a membrane as the CO_2 (or other gases) is released. All of these systems monitor the culture broths every 10 to 12 minutes for signs of microbial growth, thus reducing the time to detect most positive cultures from 48 to 72 hours to 6 to 24 hours. Although the instruments are relatively expensive, the cost is more than recovered by savings in technologist time and patient benefits. One such instument is illustrated in figure 4.12.

Nucleic Acid Probes and Amplification Methods

As described in chapter 2, the bacterial chromosome contains genes that are responsible for coordinating all activities in the bacterial cell. The chromosome of each bacterial genus and species contains unique sequences of genes. By taking advantage of this uniqueness, molecular biologists have devised clever methods for identifying microbes in clinical specimens or in culture. The complementary strands of the double helix of the DNA molecule (see fig. 2.6a) separate at high temperature and reform (reanneal) again when the temperature is lowered. When unique segments of single-stranded DNA are tagged with a radioactive or enzyme marker, they will reanneal with any complementary DNA in a reaction mixture and the product can be detected by its radioactive signal or the activity of the enzyme.

In clinical use, specimens or cultures are mixed with the unique, tagged DNA sequences under appropriate conditions and the reaction product is examined for presence of the marker after any unbound, tagged reagent is removed. A positive test is indicated by detection of the radioactive or enzyme signal, which signifies a match between the known DNA added to the specimen and the organism present. Similar tests can be performed to detect specific ribosomal RNA sequences as well. In this case, because more copies of ribosomal RNA than DNA are present in each bacterial cell, the test is more sensitive; that is, it will detect smaller numbers of organisms. These procedures are rapid and can be completed in two to four hours. Therefore, they have been used primarily to detect or identify microorganisms that grow slowly in culture such as the tuberculosis organism (chap. 12) or organisms that require special culture methods such as the chlamydial agent of sexually transmitted disease (chap. 19).

Two important drawbacks to these techniques are that first, you must know for which microorganism you are searching so that the appropriate DNA or RNA reagent is used and second, too few bacteria may be present in a clinical specimen to be detected. To overcome the second problem, a variety of methods referred to as **amplification techniques** have been devised. The most popular of these (for which Dr. Kary Mullis won the Nobel prize) is called

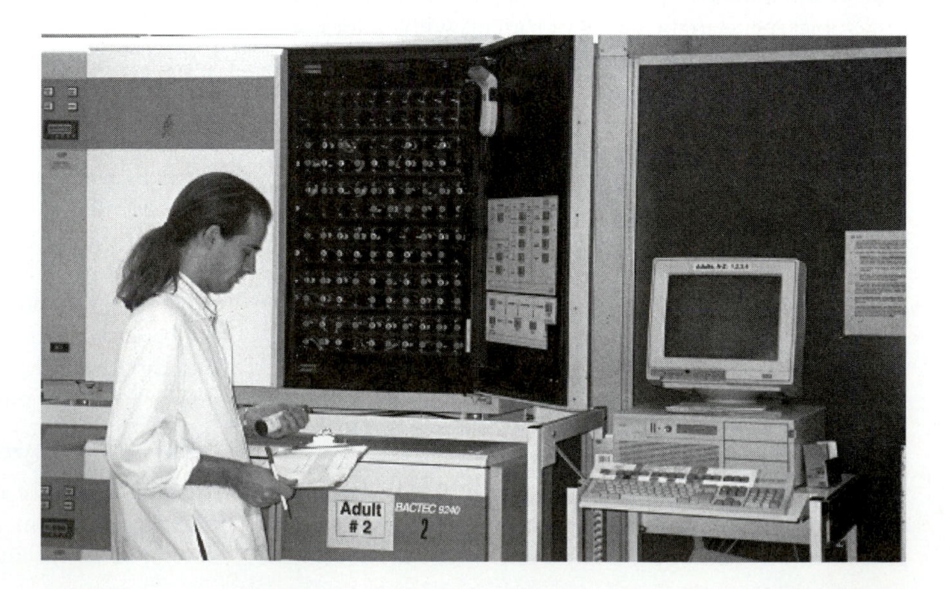

FIGURE 4.12 A technologist using an instrument (BACTEC) that detects microbial growth in blood cultures by a fluorescent sensor. Bottles of broth inoculated with patients' blood are placed in the slots in the machine (top) and monitored for microbial growth every 10 minutes. The computer terminal displays the status of each slot.

Laboratory Observations of Pathogenic Bacteria

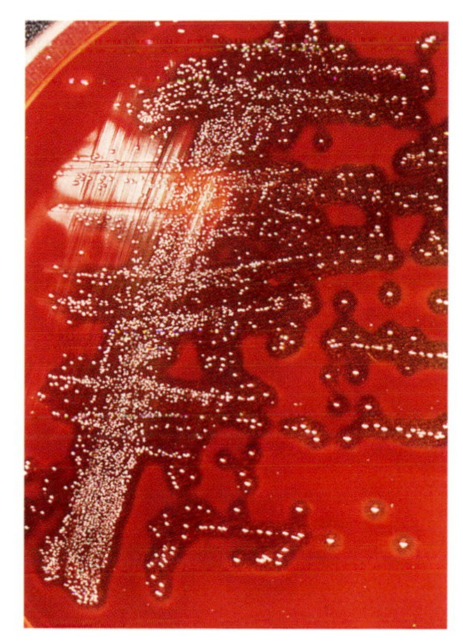

a) *Streptococcus pyogenes* growing on a blood agar plate. The clear areas of red cell lysis surrounding the punctuate colonies (beta hemolysis) are caused by the organism's streptolysin enzyme.

Courtesy Dr. E. J. Bottone, Mt. Sinai Hospital, New York, NY.

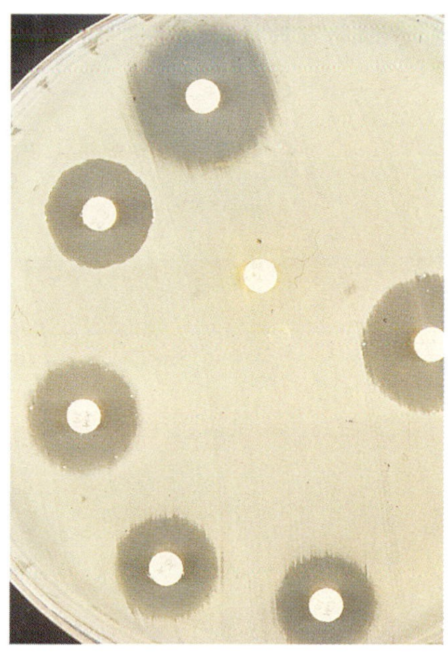

b) A disk diffusion antimicrobial susceptibility test. The clear zones of bacterial growth inhibition around the disks containing antimicrobial agents demonstrate antibiosis.

c) A MacConkey agar plate with *Escherichia coli* (pink, lactose-fermenting colonies) growing on the left-hand side and a *Salmonella* species (colorless, lactose-nonfermenting colonies) on the right.

Courtesy Dr. E. J. Bottone, Mt. Sinai Hospital, New York, NY.

d) Bacterial satellitism (an example of commensalism). Colonies of *Haemophilus influenzae* that require both X factor (hemin) and V factor (a coenzyme, NAD) grow only around *Staphylococcus aureus* colonies on this blood agar plate. The blood provides the needed X factor and the staphylococcus provides the V factor.

e) The CAMP test. When a group B streptococcus (*Streptococcus agalactiae*) is streaked at right angles to a hemolytic *Staphylococcus aureus* (long straight streak down middle of plate), areas of synergistic hemolysis in the shape of a beta-hemolytic arrow are formed.

Courtesy Dr. E. J. Bottone, Mt. Sinai Hospital, New York, NY.

f) Growth of *Mycobacterium tuberculosis* on Lowenstein-Jensen slants. The green tube on the left is uninoculated, showing the normal color of the medium. The tube in the center has the characteristically heaped, dry, and rough growth of *Mycobacterium tuberculosis.* The tube on the right shows growth of *Mycobacterium kansasii,* which produces a yellow pigment when its colonies are exposed to light.

Human White Blood Cells Involved in Defense Mechanisms

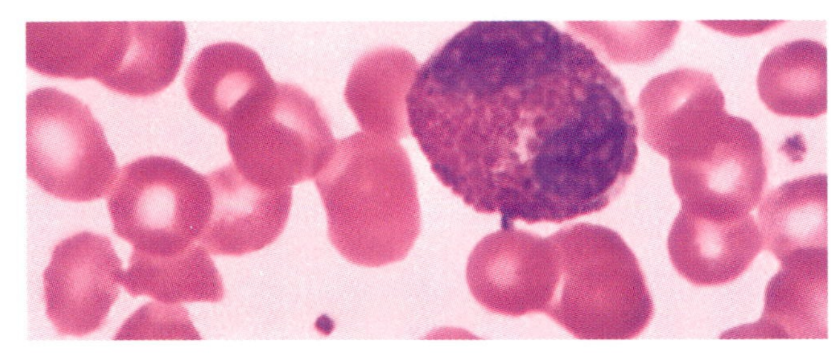

a) Eosinophil (granulocyte).

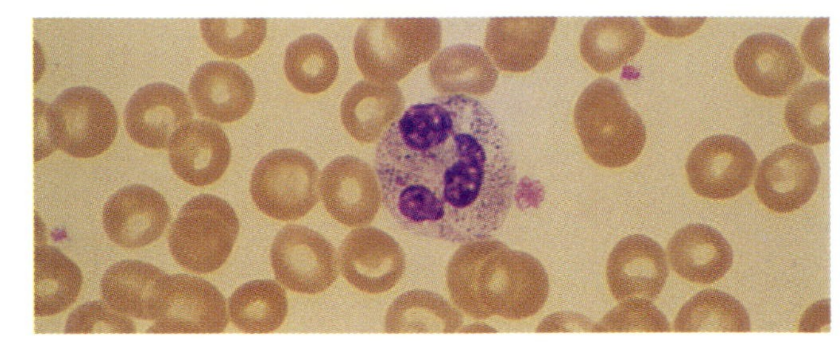

b) Polymorphonuclear neutrophil (granulocyte).

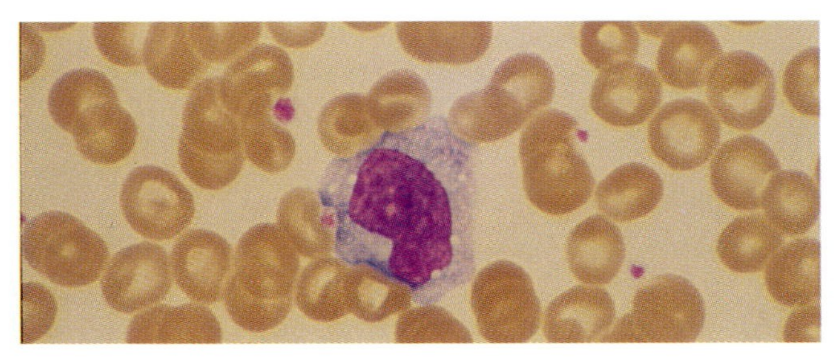

c) Monocyte.

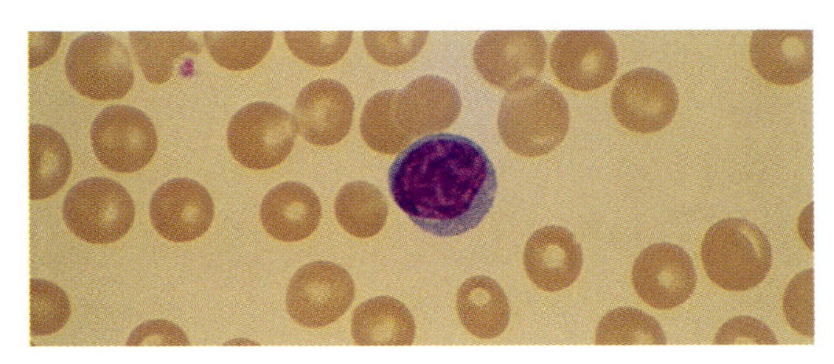

d) Lymphocyte.

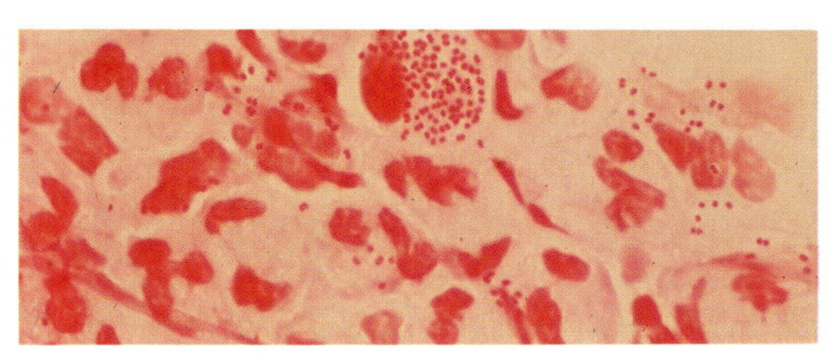

e) Gram-negative, bean-shaped diplococci (*Neisseria gonorrhoeae*) in a gram-stained smear from a male urethral exudate. Many bacteria are seen in a single polymorphonuclear neutrophil; other cells contain only a few bacteria or no diplococci.

Some important cells of the blood. The small spherical cells in photographs (a–d) are red blood cells (×2,000).

(a–d) courtesy of Dr. James W. Vardiman, University of Chicago.

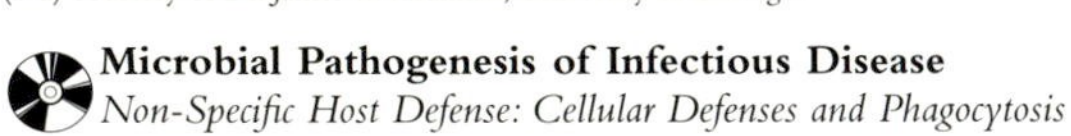

Microbial Pathogenesis of Infectious Disease
Non-Specific Host Defense: Cellular Defenses and Phagocytosis

Reactions Caused by Cellular Immune Responses

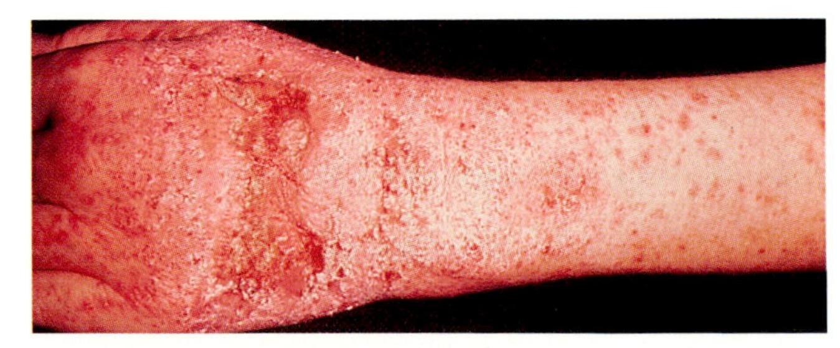

f) Contact dermatitis. This patient used 5-flouorouracil topical cream to heal skin lesions caused by sun damage. After 10 days, she developed contact dermatitis (an inflammatory allergic condition to the cream).

Courtesy of K. Soltani, M. D., University of Chicago.

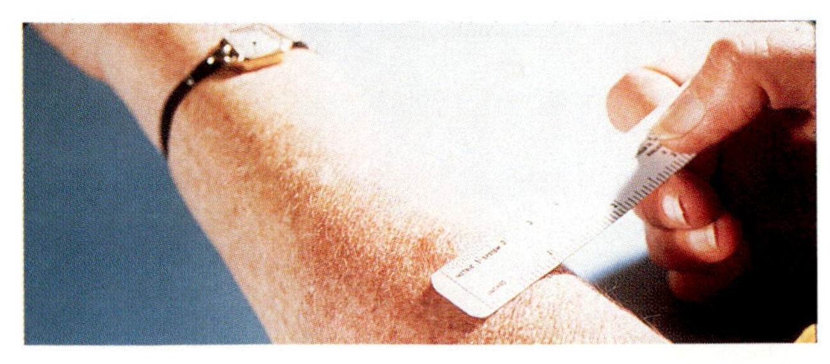

g) A positive Mantoux test for tuberculosis (at 48 hours). The area of the intradermal injection shows induration of at least 10 mm in diameter and can be felt by lightly pressed fingers.

Centers for Disease Control, Atlanta, GA.

the **polymerase chain reaction** or PCR. In this method, the specimen is heated to separate the DNA strands, known bacterial DNA sequences called primers are added to the mixture, and if they match the unknown single-stranded DNA they anneal to it. Because the primers are shorter sequences than the original DNA strands, nucleic acids and a heat-resistant polymerase enzyme (originally isolated from an archaebacterium living in a hot spring in Wyoming) are added to the mixture to complete the formation of the double-stranded DNA. The new double-stranded DNA is again separated, annealed with new primers, and extended with nucleotides in the presence of the polymerase enzyme. This cycling is continued for up to 40 cycles during which the original DNA sequences are increased or amplified a million-fold and, thus, many copies are available for detection. The technique is so powerful that very strict precautions must be taken in the laboratory to prevent introduction of extraneous, contaminating DNA into the reaction mixture. This method is in routine use in many research laboratories and is expected to find its way quickly into diagnostic use.

Use of Computers

Computer systems are widely used throughout clinical laboratories, particularly in large hospital centers. In the diagnostic microbiology laboratory, as in clinical chemistry or hematology, they provide a means for rapidly transmitting laboratory reports to patients' charts and efficiently storing laboratory data (see fig. 7.2). Periodic tabulations of these data can be of great value in monitoring the types of microorganisms being isolated from patients within the entire hospital or in certain units such as surgical, pediatric, or maternity wards. Similarly, a watch can be kept on the antimicrobial susceptibility patterns of various organisms to the antimicrobial agents in common use within the hospital so that any gradual or sudden emergence of resistant strains can be detected promptly. Computerized data can also provide early warning of an outbreak of infection and possible clues to the origins of recurring infections in the hospital or in the surrounding community.

These and many other potential applications of computer systems make them a valuable adjunct in the laboratory approach to patient care.

QUESTIONS

1. Name the parts of a light microscope and give the function of each.
2. How is the total magnification of an object determined? What is resolution?
3. What is the purpose of dark-field microscopy?
4. What is the advantage of phase contrast microscopy?
5. Describe the image seen through a scanning electron microscope.
6. Describe how the electron microscope enlarges the view of an object.
7. Why is a wire inoculating loop often used to transfer microorganisms in the laboratory?
8. Briefly describe the steps in the Gram stain. Why is it called a differential stain?
9. Describe the steps in the acid-fast stain. When is its use indicated?
10. What is the advantage of the fluorescent antibody technique?
11. Why is careful preparation of laboratory glassware and media important? Describe the uses of culture media.
12. What factors involved in specimen collection can influence the results of culturing in the laboratory?
13. Outline the flow of work in the microbiology laboratory in handling a specimen submitted for smear, culture, and susceptibility.
14. Describe the modified Bauer-Kirby (NCCLS) technique. Why is it necessary to perform antimicrobial susceptibility tests on pure cultures of isolated organisms?
15. Briefly describe the advantages of automation in the microbiology laboratory.

BIBLIOGRAPHY

Boatman, E. S., M. W. Berns, R. J. Walter, and J. S. Foster. 1987. Today's microscopy. *BioScience* 37 (6): 384–394.

Bottone, E. J., R. Girolami, and J. M. Stamm, eds. 1984. *Schneierson's atlas of diagnostic microbiology.* 9th ed. No. Chicago: Abbott Laboratories.

Everhart, T. F., and Hayes, T. L. 1972. The scanning electron microscope. *Sci. Am.* 226 (1): 54–60.

Fessia, S., P. Fawcett, C. MacVaugh, and S. Ryan. 1988. *Diagnostic clinical microbiology: A benchtop perspective.* Philadelphia: W. B. Saunders.

Gerhard, P., R. G. E. Murray, W. A. Wood, and N. R. Krieg, eds. 1994. *Methods for general and molecular bacteriology.* Washington, D.C.: American Society for Microbiology.

Howells, M. R., J. Kirz, and D. Sayre. 1991. X-ray microscopes. *Sci. Am.* 264 (2): 88–94.

Kee, J. L. 1991. *Laboratory and diagnostic tests with nursing implications,* 3d ed. East Norwalk, Conn.: Appleton and Lange.

Kirsop, B., and J. J. Snell. 1984. *Maintenance of microorganisms: A manual of laboratory methods.* San Diego: Academic Press.

Morello, J., H. Mizer, and M. Wilson. 1997. 6th ed. *Laboratory manual and workbook in microbiology: Applications to patient care.* Dubuque, Iowa: WCB/McGraw-Hill.

Murray, P. R., E. J. Baron, M. A. Pfaller, F. C. Tenover, and R. H. Yolken. 1995. *Manual of clinical microbiology,* 6th ed. Washington D.C.: ASM Press.

Tainted laboratory solution mimics disruptive outbreak. 1990. *Hosp. Infect. Control* 17 (2): 20–21.

SECTION II

A diagnostic microbiologist at the laboratory bench.
Photography by Gordon Bowie.

Character of Microbial Diseases

CHAPTER 5

Humans and Microbes in Health and Disease

OUTLINE

Microbes in Motion

This chapter describes the important interplay between microorganisms and the host, which determines whether disease will be established. It also covers the various routes by which microbes may enter the body to cause disease and the properties of microorganisms that produce damage to the host. The following books and chapters of the Microbes in Motion CD-ROM may be a useful preview or supplemental study aid to your reading: Bacterial Structure and Function: External Structures. Viral Structure and Function: Viral Invasion; Viral Pathogenesis. Parasitic Structure and Function: Nematodes; Trematodes. Gram-Positive Organisms in Human Hosts: Normal Flora. Gram-Negative Organisms in Human Hosts: Normal Flora. Microbial Pathogenesis of Infectious Diseases: Principles of Pathogenicity; Nonspecific Host Defense; Adherence; Toxins. Concepts of Epidemiology of Infectious Disease: Terminology.

OBJECTIVES

On completing this chapter the student will be able to:

1. state the role of microorganisms in maintaining ecologic balances in nature.
2. define and give examples of the patterns of interdependency among microorganisms and between microorganisms and other living things.
3. describe the normal flora of human skin and mucosa.
4. discuss the interplay between the microbial and host factors involved in infection and infectious disease.
5. explain why knowledge of the means by which an infectious disease is acquired is essential to understanding its epidemiology.
6. cite the four major entry routes of infection to the human body.
7. identify some of the infectious microorganisms most commonly associated with each entry route.
8. describe the types of damage microorganisms cause that enable them to gain a foothold in the body and to cause disease.
9. define pathogenicity, virulence, invasiveness, and toxigenicity.
10. identify remaining questions about how the seesaw balance is maintained between the host and the microbe in health and disease.

KEY WORDS

Commensalism
Ecology
Endotoxins
Entry routes of infection
Exotoxins
Invasiveness
Microbial ecology
Mutualism
Normal flora
Parasitism
Pathogenicity
Resident flora
Toxigenicity
Transient flora
Virulence

Ecology

An outstanding feature of life in earth's biosphere is interdependency. As John Donne said of the human condition "No man is an island unto himself . . . ," so we can say that no living organism can survive without support systems provided by other forms of life. The science of **ecology** is concerned with the biologic interrelationships that exist in the natural world, linking organisms with each other and with their environments.

An understanding of ecologic relationships requires study of the environmental pressures that may influence the coexistence of biologic species or affect the direction of their evolution. As the human species has developed its enormous technologic and cultural capabilities, it has created many new pressures on the natural world. Steadily increasing human populations and their needs have endangered the maintenance of fragile ecologic balances.

Microbial ecology

The ability of any given area to support an animal and plant population depends on the activities of microorganisms in soil. The decaying, putrefactive, and fermentative processes of these microorganisms release vital chemical elements, such as nitrogen, carbon, potassium, and phosphorus, from the organic components of dead material in the area. The released elements are essential to living plants for synthesis and growth, and the plants in turn are a source of food for animals. When animals and plants die, the soil microorganisms again break down their complex organic substances, thus perpetuating the cyclic turnover of matter and energy.

The populations of microorganisms residing together in the soil are themselves dependent on the available food material, the temperature, oxygen concentration, and pH of the soil, and the metabolic products of microbial neighbors. As we have seen, these metabolic wastes are sometimes antagonistic (antibiotic or antimicrobial) and sometimes beneficial to adjacent species. When the food or water content of soil fluctuates, microbes must adjust to the new conditions or be eliminated. As acids, carbon dioxide, and other products of microbial activity accumulate, further adjustments must be made. Some organisms may flourish under the changed conditions, while the activity of others may be diminished. Marked and lasting shifts in **microbial ecology,** the interrelationships between microbes and their environment, may affect the higher life of the area. Because their growth periods are much longer than those of microorganisms, plants and animals may suffer severe and lasting damage if their microscopic chemists, cooks, and purveyors dwindle off or vanish from the scene. Without the saprophytic bacteria, yeasts, and molds of the soil—the "chefs of the underworld"—to transform the substance of dead organisms into elements essential for the living, our world would become cluttered with death, and life would cease altogether.

Ecologic patterns

Four general patterns categorize the biologic interrelationships of living things. These are scaled in terms of their dependency on one another. At one end of the scale there may be *independence* between two different kinds of organisms living side by side but indifferent to each other. The other three patterns involve a close relationship in which there is some type of interdependency between two diverse types of organisms. These associations are described collectively by the word *symbiosis,* meaning literally "a living together." When the dependency is mutual so that each organism derives some benefit, trivial or vital, from the other, the pattern is known as **mutualism.** (In common practice, the terms *symbiosis* and *mutualism* are often used interchangeably.) If only one of the two associates benefits from the relationship, while the other is unaffected by it, the term **commensalism** is used, and the partners are referred to as *commensal* organisms. In the third pattern of association, **parasitism,** one organism derives support from another at the latter's expense. These concepts are illustrated in figure 5.1.

Mutualism

Mutualistic relationships exist among microorganisms living together in the human body as well as between these microorganisms and their host. Bacteria living in the intestinal tract receive their primary support from food materials present there. Some species may benefit each other by digesting or synthesizing substances important to both. In turn, the human host can absorb some of the products synthesized by certain bacterial species and put them to use. An example is the bacterial production of vitamin K in the intestine. This vitamin is essential to humans for the synthesis of a component involved in blood clotting. Humans derive vitamin K from dietary sources, however, and are not ordinarily dependent upon intestinal bacteria for its synthesis.

In chapter 3, the explanation of the interaction between nitrogen-fixing bacteria and certain plants describes a mutualistic relationship (see fig. 3.3). Since the largest supply of fixed nitrogen is of biologic origin, attention has been focused on the nitrogen-fixing bacteria in an effort to clarify and expand their natural role in the chain of nutritional resource production. If methods of biotechnology (see chap. 3) can be used to increase the rate of nitrogen fixation by genetically manipulating organisms, the entire world could benefit by the potential to increase the food supply.

Conversely, the antagonisms exerted by microorganisms upon each other in the natural world are important to plant, animal, and human life as well because they serve to control microbial populations and prevent overgrowths of species that might be harmful in large numbers (fig. 5.2). Particular mutualisms and antagonisms among organisms hold many implications for humans in their effort to control the environment to their advantage. Each of the following are approaches that may have value: the use of antimicro-

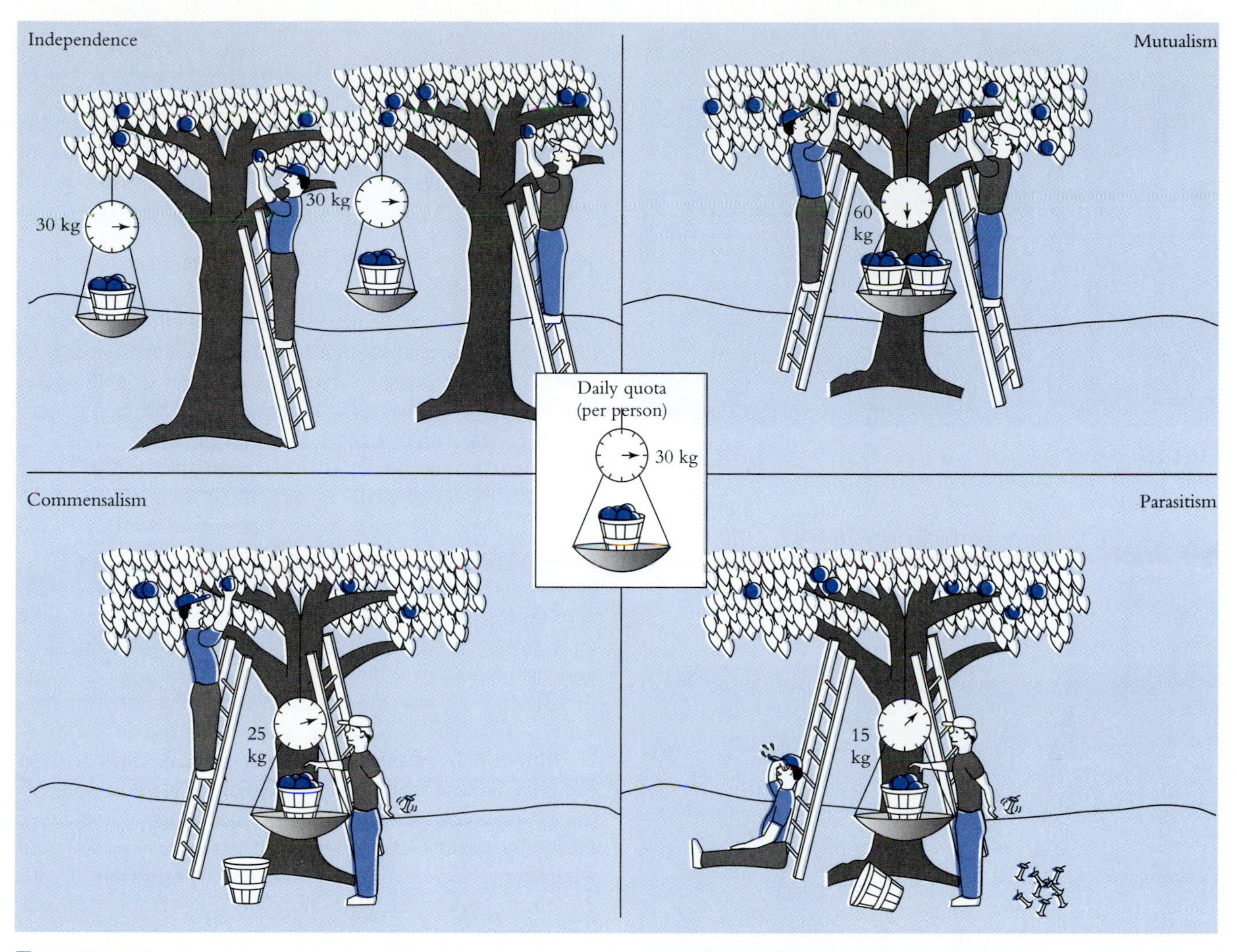

FIGURE 5.1 Biologic interrelationships. Using fruit pickers for an analogy, these diagrams illustrate the following concepts: (a) Independence: each of the two fruit pickers easily meet their daily requirement working independently of each other on different trees (or the same tree). (b) Mutualism: two pickers working together on the same tree can increase their yields by combining efforts to harvest the best areas of the tree. (c) Commensalism: the commensalist enjoys the fruit of the working partner's labor without making a contribution, but also without significantly affecting the partner's output. (d) Parasitism: the parasite reaps all the benefits at the expense of the working partner who is exhausted and cannot maintain the normal quota.

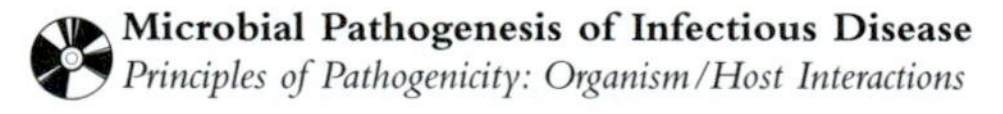
Microbial Pathogenesis of Infectious Disease
Principles of Pathogenicity: Organism/Host Interactions

bial agents or chemical substances to control infectious diseases; the deliberate introduction into the body of one species of organism to supplant or control the growth of others; a specific attack on one member of an interdependent pair to destroy the other indirectly. The interest in the relative merits of these methods to control insect pests as well as microorganisms is ongoing. For example, it has been shown that termites have a mutualistic relationship with certain flagellated protozoans that live in their intestinal tracts (fig. 5.3). When the termite has foraged successfully for a woody meal, the protozoa in its intestine produce enzymes that digest cellulose fibers, thus benefiting the termite. In return the protozoa are housed and fed; neither can get along without the other. If the flagellates are eliminated from this association, the termites die within 10 to 20 days; when the flagellates are reinstated, the insects survive. This vital interaction suggests that it may be possible to control the insect by measures designed to eliminate its mutualist companion.

Commensalism

Commensalism is a common type of association between humans and their microbial populations. Organisms commensal with humans are supported by them but do not affect their welfare one way or the other. Commensalism differs from parasitism in that the microorganism derives support from nonliving organic substances on the host's surfaces without being a source of harmful competition to living cells and tissues of the body. On the other hand, the relationship offers no benefit to the host, who remains indifferent to it.

Such relationships are subject to change since many commensals living harmlessly on the body's surfaces have a

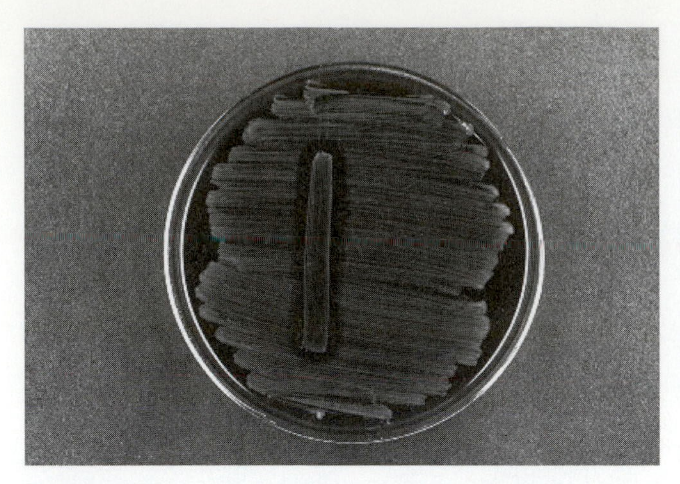

FIGURE 5.2 The inhibition of *Legionella pneumophila* (a potentially harmful human pathogen) by *Staphylococcus saprophyticus* (vertical line).
Courtesy of Dr. Alan Flesher, Ph.D.

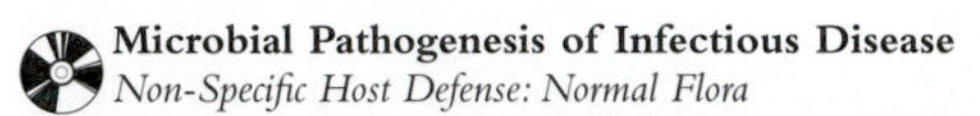

FIGURE 5.3 The termite and its intestinal protozoan partner provide an excellent example of mutualism. A protozoan in the termite's intestinal tract secretes enzymes that digest cellulose eaten by the insect. Each organism would die of starvation without the other.
U.S. Department of Agriculture, Washington, D.C.

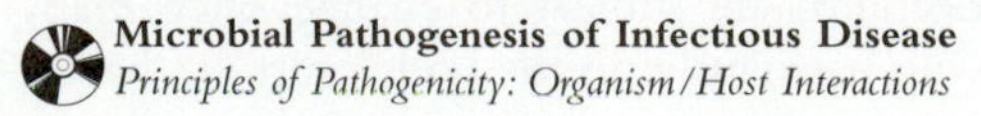

capacity for parasitism. Harmful interactions may occur when host tissue resistance becomes weakened in some way to allow microbial invasion of body tissue. Conversely, parasitic organisms may in some instances adjust to a more independent life. Microorganisms may also participate in commensalistic relationships with each other as shown in colorplate 3d.

Parasitism

Parasitism implies some degree of harmful effect induced in the host by an organism that itself benefits from the association. A parasite actually living in the tissues or cells of the host is called an *endoparasite,* as in the case of the hookworm (fig. 5.4) and other microbial parasites. Large organisms (ticks or lice) that attach themselves to the skin or outer mucosal surfaces of a human or animal host and that take their nourishment from living host tissue or blood, are called *ectoparasites.* The period of attachment may be temporary, as in the case of a biting mosquito, or more lasting, as occurs with ticks or the deeper burrowing mites. The term *infestation* refers to the presence of ectoparasites on a host, as opposed to *infection* by endoparasites.

The nature and extent of the harm suffered by the parasitized host may be slight and unrecognized or seriously life threatening. In other words, parasitism is distinct from disease. Characteristics of both the parasite and host determine whether disease will result from their interaction. From the parasite's point of view, parasitism is successful when the organism strikes a balance in its relationship with the host. The host furnishes required support for the parasite's activities, yet continues to function within reasonably normal limits.

The parasite's requirement for a certain host species or kind of host tissue in which to localize may be specific or broad. Thus, plant parasites generally cannot live in animal tissues, whereas animals may share their parasites to a wide extent. Humans acquire many parasitic worms from the animals with which they associate or whose flesh they eat. As we have seen in chapter 2, some helminths have complex life cycles involving more than one host and require particular tissues for support and development. Also, among viruses and rickettsiae, cell and tissue specificity is often quite marked, whereas bacteria, yeast, and molds are usually more versatile about the host tissues in which they localize.

Certain microorganisms are obligate parasites and require living cells to perform vital functions for them. Other parasites do not require a living partner and can support themselves, if need be, on nonliving organic matter. These are referred to as *facultative parasites.* As we have seen, many parasitic species of bacteria, yeasts, and molds, when taken from the human body on which they have been dependent, can be cultivated in the laboratory on nonviable nutrient media. The microbiologist's purpose is to grow the organisms of interest in pure culture so that each may be studied in isolation without interference from others. There are no pure cultures in nature, however, and relationships that exist

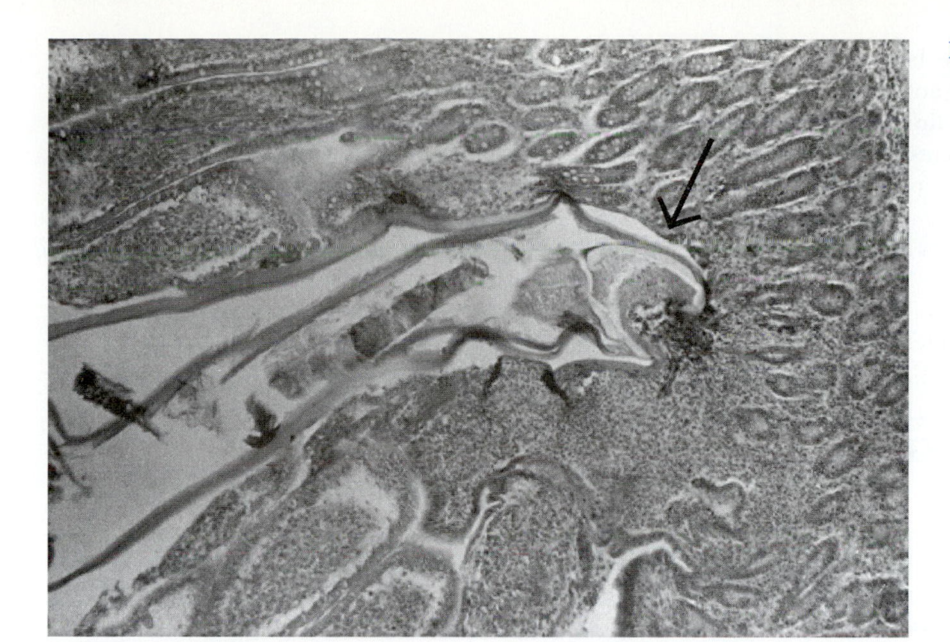

FIGURE 5.4 Photomicrograph of a hookworm (arrow) attached to the intestinal wall of its host.

between living organisms are not static. Between the two ecologic extremes of a free-living, saprophytic existence and obligate parasitism, a gradient of symbiotic relationships reflects the diverse abilities of living things to respond to each other and to changing environmental pressures.

HUMANS AND THEIR MICROBES

THE BODY'S "NORMAL FLORA"

From the moment of birth, humans live in the midst of the microbial world. For example, the average adult human is made up of about 10 trillion cells. Living in and on that average human are about ten times as many microorganisms, or 100 trillion microscopic beings that constitute the **normal flora.** In addition, a constant traffic of microorganisms enters the body in air inhaled, in food eaten, on surfaces of things touched or placed in the mouth. Microorganisms constantly exit the body in excretions from the intestinal tract, mucosal surfaces, and skin. Under normal circumstances this traffic flows only over certain body surfaces such as the skin and mucosal linings of the respiratory, alimentary, and genitourinary tracts. Any daring excursions into deeper tissues beneath these linings are met (often with lethal effects) by the healthy body's defensive mechanisms.

Skin and mucosal surfaces offer many suitable conditions for local growth and multiplication, or colonization, of commensal microorganisms. Dead or dying cells in the outer layers of skin, the secretions of skin and mucosal glands, and nutrients in the intestinal tract offer abundant organic nourishment for these microbial residents that live there on either a temporary or permanent basis. Just as in the world around us, human colonization by microorganisms is influenced by local environmental conditions of different body areas such as types of nutrient, pH, moisture, darkness, oxygen supply, and temperature. Other factors that may be involved in this ecology include individual or racial differences and the geography and sociology of human populations.

RESIDENT AND TRANSIENT FLORA

Commensal organisms living on skin and mucosal surfaces are referred to, collectively, as **resident,** or normal, **flora** (sometimes called *indigenous microbiota*). The nature and variety of these microbial populations are often distinctive for different regions of the body, being influenced by the structure and physiology of surface lining cells. They fluctuate also in response to such host factors as age, activities, or skin type as well as to external environmental influences. However, within these limits there is a fixed and well-defined normal microbial flora distributed over body surfaces, which tends to reestablish itself whenever it is disturbed. Members of the resident flora are capable of parasitism and may become pathogenic if the opportunity arises, but their potential for inducing active or severe disease is small under ordinary circumstances. They are harmless on surface tissues but may be "opportunists" causing disease if introduced into the bloodstream or tissues. Predisposing factors that lower the normal host defenses (accidental or surgical injury, malnutrition, certain drugs given for cancer treatment and after organ transplantation) set the stage for resident flora to cause opportunistic infection.

Since microorganisms are ubiquitous and their contacts with the body constant, it is not difficult for new organisms to deposit on body surfaces. However, most new acquisitions do not establish themselves permanently. They may be washed away easily from external areas, die in competition with resident organisms, fail to survive the body's conditions, or simply be ushered out in the body's excretions.

Organisms that are present on the body temporarily (for a few hours or days) but that do not become firmly entrenched are referred to as **transient flora.** Occasionally, under suitable conditions, some transients may hold their place long enough to become resident commensals or establish themselves in a parasitic relationship with the host. In that case, depending on their properties and on the host's resistance, active disease may ensue.

Table 5.1 summarizes the indigenous microbial flora found in various areas of the human body.

INFECTION AND INFECTIOUS DISEASE

From the point of view of the host, *infection* can be defined as a state in which the host harbors microorganisms that survive and multiply in or on body tissues. Commensalism is included in this definition, and therefore it can be said that the human host lives in a state of infection throughout life. The definition also includes, but does not necessarily imply, infectious disease.

The potential for disease exists when microorganisms live in a parasitic relationship with their host or shift to this type of relationship from a commensalistic or saprophytic one. The extent of harm caused by a parasite may be clinically unrecognizable or perfectly obvious because of symptoms that range from mild to severe. In any case damage depends on the properties and functions of both the parasite and the host. When the damage is sufficient to become clinically apparent, the term *infectious disease* then applies, meaning a noticeable state of abnormality induced by a living organism.

MICROBIAL FACTORS

The manner in which a disease originates and develops is referred to as its *pathogenesis,* from the Greek words for disease (*pathos*) and origin (*genesis*). Accordingly, the word **pathogenicity** denotes the ability of a microorganism to produce infectious disease. The degree of pathogenicity possessed by a parasitic organism is expressed by the term **virulence,** which is a function of two principal attributes; **invasiveness,** the ability to get into host tissues, survive, multiply, and spread; and **toxigenicity,** the ability to produce substances that are toxic to host cells and tissues. Pathogenic microorganisms differ in virulence according to their relative display of such properties.

HOST FACTORS

The severity of an infectious disease and the total degree of damage done indicate the virulence of the pathogenic causative (etiologic) agent as well as the host's capacity for defense against invasion and toxicity. The host's defense mechanisms function to protect the body from entry by microorganisms and to eliminate or control those that succeed in entering. Unbroken skin and mucous membranes offer natural barriers to intrusion and within the body many specific defenses operate. Among these defenses are phagocytosis (the ingestion of foreign particles, including microorganisms, by white blood cells and certain other tissue cells); the inflammatory response (the process by which tissue reacts to injury in order to limit and control it locally); and the specific immune response (the ability to resist a particular infectious agent, or its toxic products). This form of resistance can take place either through the development of immune substances (*antibodies*) that react specifically with the agent or its products or by increased activity of specialized cells of the immune system (*cell-mediated immunity*). The kind of tissue involved in the infection or the type of injury suffered are particularly important to the host. The vital functions of certain areas of the brain or the action of the heart, for example, may be seriously impaired by a small injury, while other organs, such as lungs, liver, or kidneys, may continue to function efficiently even when extensively damaged.

MICROBE AND HOST TOGETHER

In healthy people, most infectious diseases result from the introduction of pathogenic transient microorganisms into the tissues. On normal skin and mucosal surfaces, the transient pathogen must survive the activities and antagonisms of host cells and any commensal organisms already present. At the same time the normal resistance of the host is tested. The infecting pathogen may be exceptionally virulent or may arrive in unusually high numbers; or, an unnatural route of entry may be opened up, as in the case of surgical or accidental trauma to skin or mucosa. When the host's defensive mechanisms are diminished by preexisting circumstances, such as organic disease, fatigue or other stress, certain medical treatments, or the extremes of age, the host's own commensals as well as transient pathogens have a better opportunity to establish themselves within tissues. The possibility for developing active disease is thereby enhanced. In the ensuing struggle for survival, the total interplay of forces between an infecting microbe and the host may lead to a complete victory of one over the other in a relatively short time (acute infection) or to a prolonged relationship (chronic infection) precariously balanced between them. The symptoms displayed by the host vary accordingly from none at all (subclinical infection) to those of a sudden, acute process or a slow, insidious one. Between the extremes there are many variations in the severity as well as the duration of symptoms.

The process of infection begins at the body site where the infectious microbe enters the host. Therefore, for those who are responsible for preventing and controlling infectious diseases the **entry routes of infection** are highly important. These entry routes and the properties and activities of microorganisms that are associated with their pathogenicity and virulence (i.e., with their ability to establish themselves in human tissues and cause damage) are described next.

TABLE 5.1 MICROBIAL FLORA INDIGENOUS TO THE HUMAN BODY

Anatomic Location	Microorganisms	Comments
Skin	Staphylococci (*S. epidermidis, S. aureus*) Streptococci (viridans, nonhemolytic strep) Enterococci Anaerobic cocci (*Peptococcus* and *Peptostreptococcus*) Corynebacteria (diphtheroid bacilli) Propionibacteria (anaerobic diphtheroids) Gram-negative enteric bacilli (*Escherichia* and *Enterobacter* species) Mycobacteria (acid-fast bacilli) Yeasts and fungi	In addition, exposed skin may harbor many transients, which may remain for hours, days, or weeks; transients are readily removed, but reduction in numbers of residents is temporary at best
Conjunctiva	*Staphylococcus epidermidis* Streptococci (viridans and nonhemolytic strep) *Moraxella (Branhamella) catarrhalis* *Neisseria* species Corynebacteria (diphtheroid bacilli) Microaerophilic gram-negative bacilli	Washing action of tears and their content of lysozyme help to control microorganisms
Nose and Throat	Staphylococci (*S. epidermidis, S. aureus*) Streptococci (*viridans,* nonhemolytic strep, and enterococci) *Streptococcus pneumoniae* *Moraxella (Branhamella) catarrhalis* *Neisseria* species Corynebacteria (diphtheroid bacilli) *Haemophilus* species Anaerobes Yeast (*Candida*)	Action of nasal ciliary cells, swallowing, mucous secretion, and lysozyme of saliva help to control microorganisms
Alimentary Tract *Mouth*	Staphylococci (*S. epidermidis, S. aureus*) Streptococci (viridans, nonhemolytic strep, *S. mutans*) Enterococci *Lactobacillus* species *Fusobacterium* species *Bacteroides* species Spirochetes *Actinomyces* species Yeasts	Swallowing action, lysozyme of saliva, and good dental hygiene help to control microorganisms; nutrition important in preventing dental caries
Esophagus	No indigenous flora	Organisms swallowed with food or saliva remain only temporarily
Stomach	No indigenous flora	Gastric acidity is too high for microbial growth
Intestines (adults)	Anaerobic bacilli (species of *Bacteroides, Fusobacterium, Bifidobacterium*) Anaerobic cocci Gram-negative enteric bacilli (species of *Escherichia, Enterobacter, Proteus*) Enterococci Yeasts Protozoa	The intestines contain much larger numbers of microorganisms than any other area of the body; anaerobes predominate 1,000:1
Genital tract	Staphylococci (*S. epidermidis, S. aureus*) Enterococci Nonhemolytic strep—sometimes viridans strep Lactobacilli Gram-negative enteric bacilli (species of *Escherichia, Enterobacter, Proteus*) Anaerobic cocci Anaerobic bacilli (species of *Clostridium, Porphyromonas, Prevotella*) Mycobacteria (acid-fast bacilli) Spirochetes Yeasts Protozoa (especially *Trichomonas* species)	In the female, lysozyme content of cervical secretions and the acidity of the normal adult vagina control the numbers and variety of microorganisms establishing in this area; lactobacilli predominate between puberty and menopause

Note: In the bacteriologic culture of specimens obtained from these areas of the body, the search for pathogenic organisms or for commensals that have assumed a pathogenic role is complicated by the variety of indigenous bacterial species encountered in such specimens.

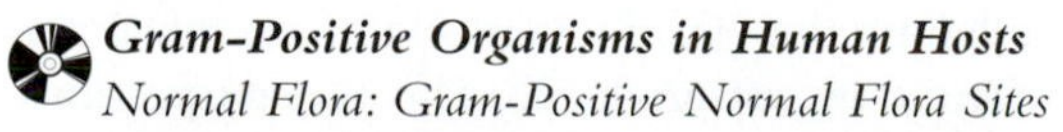

Gram-Positive Organisms in Human Hosts
Normal Flora: Gram-Positive Normal Flora Sites

Gram-Negative Organisms in Human Hosts
Normal Flora: Gram-Negative Normal Flora Sites

Microbial Pathogenesis of Infectious Diseases
Nonspecific Host Defense: Normal Flora

INFECTION ENTRY ROUTES

GENERAL CONSIDERATIONS

Epidemiologic techniques designed to prevent infectious disease or to limit its spread from one person to another are based on a knowledge of the entry, exit, and transmission routes taken by pathogenic microorganisms. The point of entry marks the beginning of infection and also affords one of the last opportunities to prevent it. Once infection is established, the ultimate site(s) where the microbe localizes will determine whether or not it can later exit from the body, through what portal(s) it may leave, and by what route(s) it may be transmitted to another individual.

Reducing the human (or mammalian animal) body to a simple anatomic concept, we can think of it as a cylinder traversed by a central tube, the alimentary canal (fig. 5.5). Two major blind sacs, the respiratory and urogenital tracts, diverge inward, one at the upper end of the canal, the other from the region near the lower opening. The external body surface is covered by skin which, with its relatively impermeable cornified, nonviable, outer layer, provides protection and insulation from the environment as well as sensory communication with it. This protective covering is interrupted only at the locations of the eyes and the openings of the alimentary, respiratory, and urogenital tracts. The external covering of the eyes is a thin transparent layer of viable cells called the conjunctiva. The conjunctiva and surface linings of the inner tracts are covered by mucous secretions that protect them and keep them moist. For this reason they are commonly called *mucous membranes,* or mucosae (sing., mucosa).

Microorganisms may penetrate the body through the skin and inner surface linings, although the skin is less easily entered than the thinner membrane surfaces. There are four major avenues through which microorganisms may enter the body: (1) they may enter with inhaled air into the respiratory tract; (2) they may be ingested with material taken into the mouth and gastrointestinal tract; (3) they may use their pathogenic properties to infect or penetrate the skin or superficial mucous membranes and be limited to these tissues or spread from there to deeper sites; and (4) they may be deposited directly into tissues beneath the skin and mucosae when these barriers are penetrated or traumatized (injured) by some other agent such as an insect or animal bite or a penetrating wound. Entry by avenues 1, 3, and 4, that is routes other than the intestinal tract, is also referred to as the *parenteral route.*

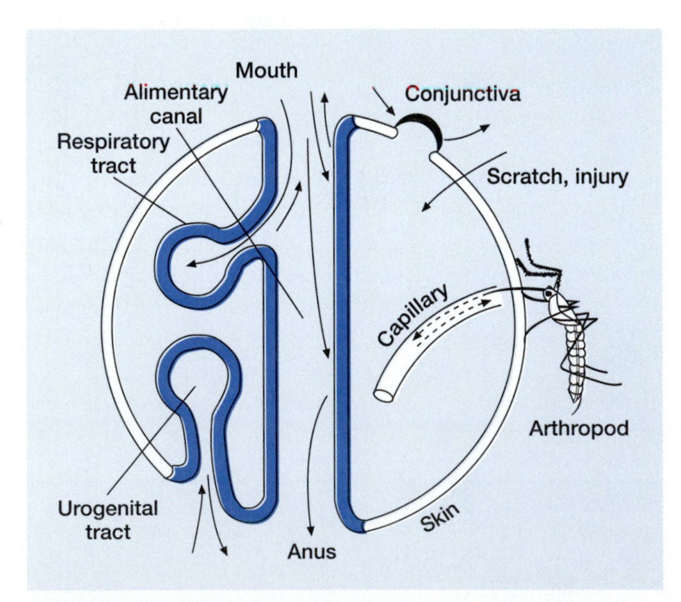

FIGURE 5.5 Bodily surfaces as sites of microbial infection and shedding.

From Cedric A. Mims, *The Pathogenesis of Infectious Diseases,* 1987. By permission of Academic Press Ltd, London.

Establishment of Infection

The entry route does not necessarily determine the site where the infective organism will ultimately establish itself and multiply. The areas where the parasite may gain a foothold are dictated by its own properties, by the biochemical environment afforded by host tissues, and by the host's many defense mechanisms. In order to colonize or penetrate the body's surface tissues, microorganisms must first become attached. This process is called *adherence.*

A number of specific microbial attachment mechanisms have been clarified in recent years, although some are as yet unknown. For example, *Streptococcus mutans,* an organism involved in tooth decay, attaches to tooth enamel through its synthesis of glucans and by the action of a binding site on its cell surface. In the case of poliovirus and some other viruses, viral capsid protein reacts with a specific receptor on the host tissue cell. *Mycoplasma pneumoniae,* the bacterial agent of "atypical" pneumonia, attaches by a "foot" to a specific acid component on the surface of a respiratory epithelial cell. Electron microscopy indicates an apparent role of pili in the attachment of the gonococcus (the agent of gonorrhea) to the surface of urethral epithelial cells (fig. 5.6). Experimental infections of rabbits with cholera indicate that the bacteria (*Vibrio cholerae*) adhere by their surface coats directly to the tips of the brush border of the villi in the intestinal tract (fig. 5.7).

Once a pathogenic microorganism has entered the host, attached to a body surface, and found conditions favorable for multiplication, it may induce damage in one or both of two ways: (1) it may remain localized at or near the site of entry and damage the surrounding tissues mechanically or physiologically, or (2) either the organism itself or its toxic products, or both, may disseminate through the body, producing injury to tissues far removed from the entry point. The parasite itself may have invasive qualities enabling it to spread through the tissues, or it may be carried passively in the bloodstream or lymph and thus be distributed widely. Microbial toxins are also disseminated by way of blood and lymph flow from a local site of microbial colonization where they are produced. The general possibilities for localization along the entry route, or the dissemination therefrom, are discussed for each of the four avenues of entry.

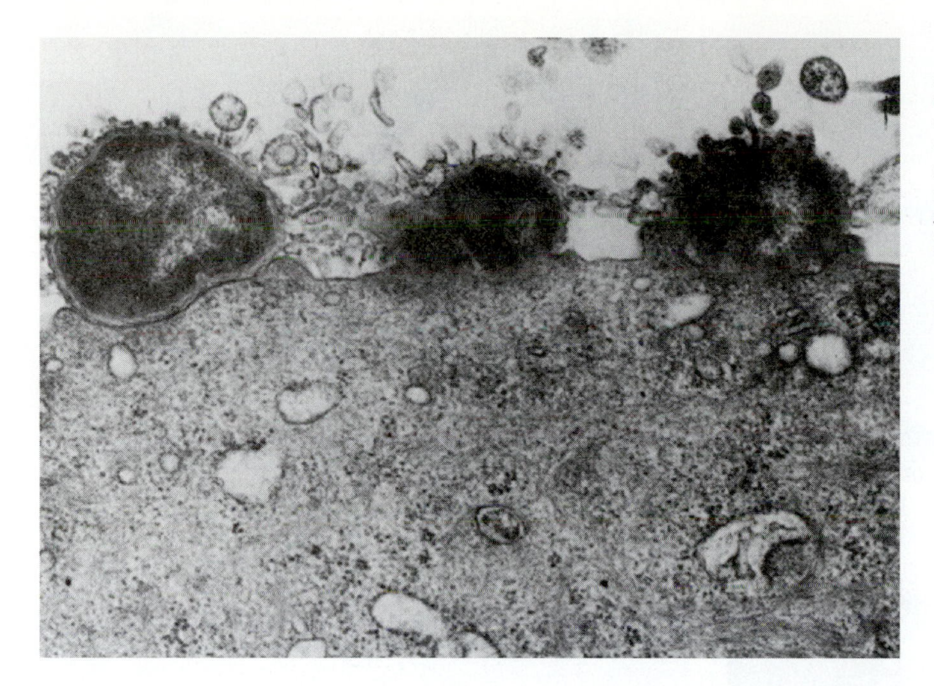

FIGURE 5.6 Electron micrograph showing gonococci that have closely attached by pili to the surface of a urethral epithelial cell. The membrane of the host cell appears to be pushed up around the organisms to form cushionlike structures, as an initial stage of engulfment (×16,000).

Reproduced from M. Ward and P. J. Watt "Adherence of *Neisseria gonorrhoeae* to urethral mucosal cells: an electron microscopic study of human gonorrhea," *Journal of Infectious Diseases* 126:601, 1972. University of Chicago Press.

Bacterial Structure and Function
External Structures: Fimbriae; Pili

Microbial Pathogenesis of Infectious Diseases
Adherence: Adherence Mechanisms; Receptors

RESPIRATORY ENTRY

The respiratory tract offers an easy access route for infectious microorganisms (fig. 5.8) and is the most difficult one to control. Infections acquired by this route have worldwide distribution because of face-to-face contact among people (see section IV).

Microorganisms are inhaled through the nose or mouth in the droplets of moisture sprayed into the air by talking, coughing, or sneezing. People may harbor many commensal or parasitic organisms in their oral, nasal, and pharyngeal secretions. The quantity of droplets reaching the air and the distance to which they scatter depend on the force and frequency with which people talk, cough, or sneeze (see the illustration on p. 235).

Fine droplets lose their moisture quickly by evaporation and linger in the air with their burden of microbes. The larger droplets of mucous spray settle rapidly from the air and are not likely to be inhaled. They may, however, contaminate the hands of individuals from whom they were shed or the hands of passersby. Any infectious microorganisms present may then be transmitted by touching or rubbing to susceptible upper respiratory mucous membranes such as those of the nose or eyes.

When microorganisms enter the nose or mouth, they may lose the competition for nutrient or attachment sites to other organisms already present there, or they may be destroyed by the activities and secretions of host cells. Many survivors of these hazards are caught in and moved along with mucus secreted by cells lining the upper respiratory tract. The ciliated epithelial cells keep the mucus moving and draining into the throat, where it is either swallowed or expectorated, or into the nose, from which it can be blown out. Microorganisms swept into the gastrointestinal tract may be killed by the acidity of gastric secretions or by competition encountered farther along from the intestinal commensals.

The organisms present in inhaled air are thus subjected to filtration, mechanical removal, or rerouting before the air is carried down into the deeper reaches of the respiratory tract. Only a relative few reach the lungs. Those that survive and entrench locally may induce damage in the respiratory tract or elsewhere if they or their toxic products, or both, are disseminated from that site.

Localization in the Respiratory Tract

The most familiar example of an inhalation infection localized in the upper respiratory tract is the common cold. More than 100 viral agents have been associated with this mild disease. When a virus localizes within mucosal cells of the pharynx and nasopharynx, it interferes with their normal functions and results in the characteristic nasal irritation, sore throat, and profuse watery discharge that replaces normal mucus secretion. Bacterial pathogens that commonly localize in the upper respiratory tract include *Staphylococcus aureus,* the beta-hemolytic streptococci, pneumococci, and *Haemophilus influenzae.* Among these, beta-hemolytic streptococci are probably the only significant cause of acute bacterial pharyngitis, although *H. influenzae* can produce a severe epiglottitis in children who are not immunized against it.

Members of the normal intestinal flora, such as enteric bacilli or enterococci, may also find their way to the throat but are usually of little consequence unless aspirated into the lower respiratory tract.

Many of the bacterial species introduced transiently in inhaled air or by contaminated hands may become established as commensals in the normal nose or throat.

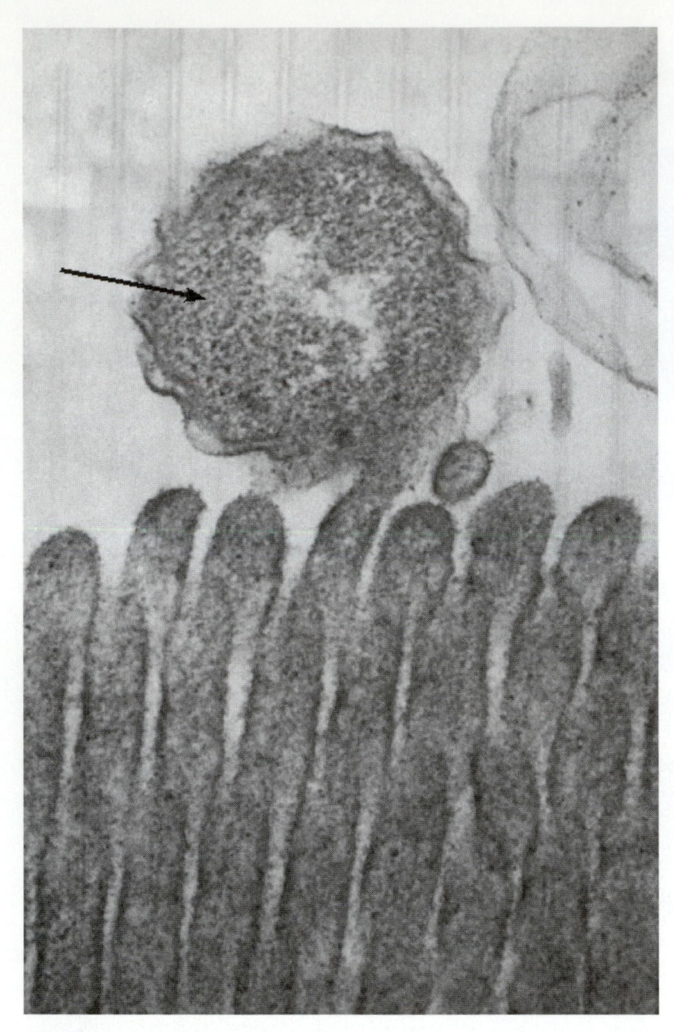

FIGURE 5.7 Electron micrograph showing a cross section of *Vibrio cholerae* (arrow) adhering directly to the edge of the intestinal brush border. The outer membrane of the vibrio is in contact with the surface of a microvillus (×130,000).

Reproduced from E. T. Nelson, J. D. Clemens, and R. A. Finkelstein, "*Vibrio cholerae* adherence and colonization in experimental cholera: electron microscopic studies," *Infection and Immunity* 14:543, 1976. American Society for Microbiology.

Microbial Pathogenesis of Infectious Disease
Adherence: Adherence Mechanisms; Receptors

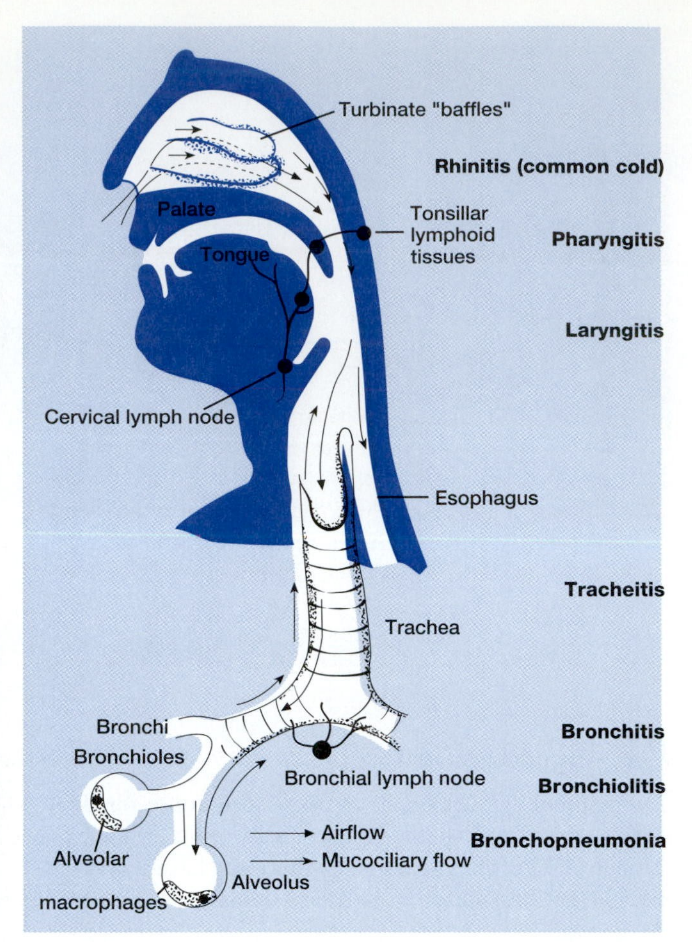

FIGURE 5.8 Principal sites of infection in the respiratory tract.

From Cedric A. Mims, *The Pathogenesis of Infectious Diseases*, 1987. By permission of Academic Press Ltd, London.

However, if the pharyngeal membranes are injured by a virus infection, excessive dryness, or some other factor that interferes with their normal functions and secretion of mucus, commensals may multiply excessively and add to the preexisting damage.

In the lower respiratory tract, infectious diseases acquired by inhalation and aspiration of infectious droplets include influenza and other pneumonias of viral origin; tuberculosis, psittacosis, and the pneumonias caused by bacterial pathogens, such as pneumococci, staphylococci, or the plague bacillus (pneumonic plague); and fungal diseases such as histoplasmosis or coccidioidomycosis. Pulmonary disease may also be caused by members of other major groups of organisms (rickettsiae, protozoa, helminths), but the entry route in these cases is not usually through droplet inhalation.

As air carrying infectious droplets passes from upper to lower respiratory passages, microorganisms are filtered out so that few remain in the air that finally reaches the alveolar spaces of the lungs. Depending on the organism and its environmental requirements, localization may occur at different points along the way (fig. 5.8). A cold virus producing symptoms in the upper respiratory passages may pass downward from its original focal point and infect cells of the bronchial linings, but the influenza virus probably does not find a suitable locale until it reaches the smaller branches of the bronchial tree. The tubercle bacillus, when inhaled, usually reaches the alveolar bed and localizes there, but it may also penetrate lung tissue, reach adjacent lymphoid tissue, and be carried from there in lymph and blood.

Pathogenic cocci may reach the lungs directly, localizing en route in bronchial passages or in the lung beds, and then spread peripherally through adjacent lung tissues. Spores of pathogenic fungi, such as *Histoplasma* and *Coccidioides* species, must reach the depths of the lungs in fair numbers and in finely disseminated particles to establish primary pulmonary disease. Fine dust from the soil, in areas where these organisms are prevalent, carries the infective spores.

Dissemination of Microorganisms or Their Toxins from the Respiratory Tract

Following entry by inhalation, infective organisms can have serious effects in tissues far removed from the respiratory portal. This may happen if (1) the pathogen penetrates or is carried to other parts of the body, or (2) toxigenic substances secreted by the organism multiplying within the respiratory tract are distributed through the body by the blood or lymph.

Penetration and spread of an infectious organism are well exemplified by the meningococcus (*Neisseria meningitidis*), the bacterial agent of a severe systemic infection. This organism commonly finds a respiratory entry and localizes in the nasopharynx. It may reside there as a harmless transient, or it may produce a local infection that is usually inapparent. However, it can reach the bloodstream directly from this site, be circulated through the body, and relocalize in vital areas such as the meninges, the inner membranous coverings of the brain and spinal cord, which then become greatly inflamed. Inflammation of this tissue, whatever the cause, is called *meningitis* (hence the name *meningococcus*) although this organism is not the only infectious agent capable of producing meningeal injury. *N. meningitidis* may also produce multiple lesions in skin, joints, eyes, ears, and lungs.

There are many other systemic infections whose agents find a primary point of entry in the respiratory tract, localize there, and later disseminate to other tissues. The viral agents of poliomyelitis and infectious hepatitis may localize initially in the throat, then in the bowel, and be carried to other vital tissues. Tuberculosis, pulmonary anthrax, and fungal diseases of the lungs are other examples. As in meningococcal infection, disease effects are produced both by the disseminated organism and by tissue reaction to it, with symptoms referable to the location of various foci of infection.

The dissemination of toxigenic substances produced locally by microorganisms in the respiratory tract can also result in systemic disease. When microbial toxins circulate in the blood and damage tissues, the condition is spoken of as a *toxemia,* the word meaning "toxins in the blood." (When organisms themselves are present in the blood, words such as *bacteremia* or *viremia* are used.)

Diphtheria and scarlet fever are well-known examples of diseases originating as local infections in the nose or throat but characterized also by serious effects of toxemia. In both instances the infective organism entrenches near its entry point in the pharynx. An acute sore throat results from the damage each organism causes to the membranes of underlying tissues, and more importantly, each secretes a characteristic exotoxin that is distributed widely by the bloodstream. These and other bacterial exotoxins have distinct properties that affect different cells and tissues in different ways. The diphtheria toxin damages the heart muscle and nerve tissue, as well as other organs, by interfering with protein synthesis in these tissues. The streptococcal toxin associated with scarlet fever affects the blood vessels and epithelial cells of the skin, producing the typical scarlatinal rash (tables 5.2, 5.3).

GASTROINTESTINAL ENTRY

The gastrointestinal tract may offer admission to members of any of the microbial groups. They may enter on anything taken into the mouth, but the most common sources of infection by this route are food, water, milk, and fingers.

Many microorganisms that are swallowed do not survive their journey through the intestinal tract. Some are destroyed in the mouth by the lysozyme activity of saliva; others are killed by the stomach's acidity or by the activity of digestive enzymes in the small intestine. Infectious disease arises when pathogenic microorganisms manage either to localize in the bowel, producing damage at this site; penetrate the intestinal wall and disseminate; or produce substances toxic for human cells from their site in the bowel.

TABLE 5.2 DISTINCTIONS BETWEEN BACTERIAL ENDOTOXINS AND EXOTOXINS

Endotoxins	Exotoxins
1. Components of gram-negative bacterial cell walls	1. Secreted by or released from bacterial cells into surrounding medium (usually by gram-positive bacteria)
2. Complex molecule (lipid, protein, and carbohydrate)	2. Protein in nature
3. Heat stable	3. Generally destroyed by heat
4. Stimulate production of antibodies of questionable protective value (chap. 6)	4. Stimulate production of antibodies (antitoxin) of high protective value (chap. 6)
5. Toxic effects nonspecific with regard to microbial agent; produce similar symptoms (fever, weakness, generalized aches)	5. Toxic effects specific to each exotoxin; clinical symptoms distinctive
6. Endotoxic diseases include typhoid and paratyphoid fever, gonococcal and meningococcal disease	6. Exotoxic diseases include botulism, cholera, tetanus, gas gangrene, scarlet fever, diphtheria (table 5.3)
7. Antitoxins not available	7. Antitoxins available for treatment (botulism, tetanus, diphtheria)

Microbial Pathogenesis of Infectious Diseases
Toxins: Endotoxins; Exotoxins

TABLE 5.3 DISEASES CAUSED BY BACTERIAL EXOTOXINS

Disease	Type of Toxin	Nature and Site of Injury
Bacterial food poisoning		
Staphylococcal enteritis	Enterotoxin	Stimulates nervous response that results in vomiting and intestinal hypermotility
Botulism	Neurotoxin	Interrupts motor nerve impulses; muscle paralysis (e.g., respiratory muscles)
Cholera	Enterotoxin	Stimulates enzyme that causes hypersecretion of water, chloride, bicarbonate from small intestine
Scarlet fever	Pyrogenic exotoxin	Affects epithelial cells of skin and mucosa of upper respiratory tract
Toxic shock syndrome	Pyrogenic exotoxin	Elicits damaging inflammatory response; causes fever, rash, shock
Diphtheria	Exotoxin	Interferes with protein synthesis in many tissues, especially cardiac muscle, nervous tissue, kidney
Tetanus	Neurotoxin	Interrupts motor nerve impulses; muscle spasm (e.g., face, neck, respiratory muscles)
Gas gangrene	Lecithinase	Destroys lecithin in muscle tissue locally, or in red blood cells
	+	
	Collagenase	Destroys collagen protein in muscle tissue locally

Localization in the Gastrointestinal Tract

Bowel infections are most commonly caused by protozoa, helminths, certain species of bacteria, and, in infants, viruses. These organisms localize in either the small or large bowel, the site usually being characteristic for the particular parasite. Mucosal cells may be damaged and eroded, with consequent microbial invasion of deeper tissues. Symptoms often include nausea and vomiting, diarrhea, intestinal spasm with accompanying pain, and, in some cases, sloughing of the mucosal lining with bleeding into the bowel.

The protozoan associated with serious human disease is *Entamoeba histolytica,* the agent of amebic dysentery. Giardiasis, caused by *Giardia lamblia,* has become more common but is less severe than amebiasis. Helminth infections are caused by the common roundworms (pinworm, hookworm, *Ascaris*) and the tapeworms.

Important intestinal diseases of bacterial origin include gastroenteritis induced by strains of *Salmonella* and *Campylobacter;* bacillary dysentery caused by species of *Shigella* (often called shigellosis); and cholera, a severe disease of the small bowel caused by the toxin of a small curved rod, *Vibrio cholerae.*

Fungal infections of the bowel are uncommon and usually transient, the agents being relatively incapable of invading intestinal or other tissue. The yeasts of the genus *Candida,* however, may infect and damage the mucosal surfaces of the mouth and throat as well as the lower bowel.

Dissemination of Microorganisms from the Gastrointestinal Tract

Typhoid and paratyphoid (or enteric) fevers are excellent examples of bacterial diseases whose agents enter the body through the alimentary canal, usually in contaminated water or food, but penetrate the intestinal lining directly. The organisms localize first in the lymphoid tissue of the small bowel wall and disseminate from there by lymph and blood to other organs, notably, the liver, gallbladder, and kidneys. The typical symptoms of typhoid and enteric fevers are produced by this systemic invasion, not by localization in the bowel.

Entamoeba histolytica may also erode its way into the bowel wall, enter the portal circulation, and reach the liver, where it induces abscess formation. From this site it can extend directly through the diaphragm into the lungs or be carried from such sites to establish itself in other areas, including the brain.

In some helminthic infections adult worms living in the intestinal tract produce eggs or larvae capable of penetrating the intestinal wall. They are picked up in the bloodstream and distributed widely. When deposited in human tissues, these larval forms cannot complete their development into mature adults. Some of them die as a result of host reactivity and later become calcified, but others remain alive in the tissues for long periods. Trichinosis is a serious roundworm disease of this type in which larvae invade muscle tissue (fig. 5.9).

The agents of two very important virus diseases also commonly localize in the bowel, as well as in the upper respiratory tract, but produce disease by invading other tissues. These are the poliovirus, which causes its most serious effects when distributed to cells of the central nervous system, and the hepatitis A virus, which invades and damages liver cells. A number of other viruses have been implicated in acute gastroenteritis. These are discussed in chapter 17.

The Toxins of Intestinal Pathogens

Most of the bacterial pathogens that damage the intestinal tract (so-called enteric pathogens) have been assumed to do so through the action of their **endotoxins.** These substances are components of the outer layer of gram-negative

FIGURE 5.9 *Trichinella spiralis* larvae (arrow) encysted in human muscle (×1,000).
Centers for Disease Control, Atlanta, GA.

Parasitic Structure and Function
Nematodes: Tissue Life Cycles

bacterial cell walls that usually are not liberated into the environment until the cell disintegrates (see table 5.2). Endotoxins produce marked physiologic effects on the body including fever, altered numbers of white blood cells, decreased blood pressure, hemorrhage, and even shock and death. Enteritis caused by *Salmonella* and most species of *Shigella* is considered to be endotoxic disease, but the exact relationship of endotoxin to the damaging effects produced is still not known.

In certain intestinal tract diseases the primary damage is caused by bacterial **exotoxins.** These toxins remain localized in the bowel and are not absorbed from the intestine. The vibrios that cause cholera, for example, multiply in the small intestine, producing a type of exotoxin called *enterotoxin*. This toxin binds to small bowel mucosal cells activating an enzyme in the cell membrane that causes enormous hypersecretion of chloride, bicarbonate, and water. The effect is massive diarrhea with acidosis. Some strains of *Escherichia coli,* once regarded as harmless intestinal commensals, also have been found to produce enteric disease. In some instances their production of an enterotoxin (exotoxin) leads to a choleralike intoxication. Other *E. coli* strains cause a *Shigella*like dysentery, and some strains are now known to be one of the primary agents of "traveler's diarrhea."

Ingestion of food contaminated with bacteria or their toxin products is responsible for bacterial food poisoning (also known as foodborne disease). The specific types of this disease and their causative organisms are described in chapter 16.

SKIN AND SURFACE MEMBRANE ROUTES

The skin and external mucosa offer another possible portal of entry for microorganisms, but the variety of pathogenic microbes that find their way by these paths is more limited than in the two routes already discussed. We have seen that skin surfaces have a large commensal flora, varying with the anatomic area, sex, and age. These surfaces are normally impervious to the many microorganisms with which they come in contact, and most of these are easily washed away or mechanically removed. Breaks in the skin or mucosa or some unusual capability of an infecting organism apparently is required before microorganisms can enter and establish themselves (see section VI).

Localization in the Surface Tissues

Local infectious disease of the skin or mucous membranes may be caused by yeasts and other fungi (*Candida* species, ringworm, and "athlete's foot" fungi), by bacteria (notably staphylococci or streptococci), by a few viruses (herpes virus, papillomavirus (warts)) or by the larval forms of helminths (the "ground itch" induced by hookworm larvae, "swimmer's itch" caused by larvae of *Schistosoma*).

Differences in the type of skin lesion produced by various kinds of microorganisms are related in part to the capacities and opportunities of the organism to localize in particular anatomic areas or in certain layers of the epidermis or dermis, and in part to the kind of reaction they elicit from the host. Thus the so-called "superficial" fungi infect only the outermost layers of skin or its appendages. Some characteristically involve the hairy skin (ringworm of the scalp), others the smooth skin surfaces, skin folds (between fingers or toes, axilla, or groin), or nails. Similarly, the herpes simplex virus of the "cold sore" shows a preference for cells of the oral mucocutaneous border. Often the host reaction is a minimal to very superficial infection, but if the organism involves deeper layers of the dermis, more vigorous defenses are brought to bear. Helminthic larvae entering the skin from contaminated soil or water and wriggling about under the epidermis, unable to penetrate further, elicit a good deal of response from host cells whose job it is to deal with foreign intruders. (These are the phagocytes, blood and tissue cells that ingest foreign particulate matter.) Large organisms are not easily inactivated, however, so the reaction around them tends to be diffuse until they die and can be walled off.

Bacterial infections of the skin generally involve structures in the deeper layers of the dermis and elicit greater and more effective host reaction. Bacteria generally cannot penetrate intact skin, but they can enter hair follicles or ducts of sweat and sebaceous glands. The secretions of these glands discourage the establishment of microorganisms because of their acid pH and because they contain antimicrobial substances such as fatty acids and lysozyme. When bacteria successfully establish in these sites because of their invasive properties or some local abnormality in the skin or its functions, the cells of the dermis make every effort to localize the infection and prevent its spread. They are assisted by phagocytes brought to the area by the bloodstream. Thus the typical lesion of a staphylococcal infection of the skin is a "pimple" or "boil," a small abscess circumscribed by a defensive wall of material laid down around it by the host. The pus that forms in an abscess contains many phagocytes, which ingest the bacteria, while the tissue formed around the area keeps them localized. This type of host response is described more fully in chapter 6.

Mucous membrane colonization by specific staphylococcus strains has been associated with the toxic shock syndrome (TSS), a disease primarily observed in young, menstruating women who use tampons (see chap. 19). In almost all instances staphylococci isolated from the patients' genital tracts produce a toxin called toxic shock syndrome toxin, or TSST-1. The organisms themselves do not invade, but rather the toxin disseminates and is responsible for the symptoms, which include high fever, a characteristic rash, and low blood pressure. Eventually, shock and even death can result. The exact relationship between tampons and TSS is not known, but tampon use is thought to provide the appropriate environment for multiplication of the staphylococci and subsequent toxin production. In addition, staphylococci on the hands may contaminate the tampon as it is inserted, thereby initiating the disease.

The conjunctiva is normally kept moist and healthy by the continuous secretion of tears and frequent blinking, which mechanically keep the eyes clear of foreign particles. Infections of the conjunctiva may result if defects in the lacrimal gland or lid disease interferes with these cleansing mechanisms. More frequently, however, minor injuries to the conjunctival surface from foreign bodies rubbed in by contaminated fingers are responsible for conjunctivitis. A common bacterial conjunctivitis, called "pink-eye," is spread by fingers and may travel rapidly through a household or schoolroom. Pink-eye is caused by *Haemophilus aegyptius* (a species closely related to *Haemophilus influenzae*), a fastidious gram-negative rod.

Chlamydiae are responsible for the severe conjunctival disease known as trachoma, also easily transmitted by fingers or contaminated objects. Chlamydial agents grouped in the species *Chlamydia trachomatis* also cause inclusion conjunctivitis and two forms of sexually transmitted disease, lymphogranuloma venereum and so-called nongonococcal urethritis. The genital tract is probably involved not only in sexual transmission of chlamydiae but transferral by hands and to the newborn as well (see chap. 19).

Several viruses are associated with viral conjunctivitis, especially certain adenoviruses that cause epidemic keratoconjunctivitis. In this condition, the conjunctivae become inflamed and the cornea may become temporarily keratinized and opaque.

Dissemination of Infection through or from a Superficial Entry Point

Some of the organisms mentioned previously, particularly streptococci and staphylococci, may produce substances that enable them to spread through the skin, often involving wide areas. This is the case in erysipelas, a streptococcal infection of the skin characterized by a diffuse and rapidly spreading involvement of dermal tissues, with an accumulation of fluid (edema) but little pus formation. In impetigo, an infection caused by staphylococci or streptococci, separately or together, superficial blisters form on the involved skin that break easily and liberate infected fluid to spread to adjacent areas. The invasive quality of such infections affords further risk of extension to deeper tissues. The organisms may enter lymphatic or blood vessels within the lesion and be distributed through the body, inducing severe damage to more vital tissues.

Sexually transmitted diseases constitute the most important and common group of infections acquired by direct contact with microbial agents that enter through superficial mucous membranes. The genital mucosa is the usual route of entry of the causative organism, which is transmitted directly from an infected to an uninfected individual.

Sexually transmitted diseases are caused by several microorganisms. Syphilis, among the most serious of these diseases in its long-range effects, is caused by a spirochete, *Treponema pallidum.* Gonorrhea (caused by a gram-negative diplococcus, *Neisseria gonorrhoeae,* also called the gonococcus) is a sexually transmitted disease that is rampant among all sexually active age groups. Nongonococcal urethritis caused by *Chlamydia trachomatis* (an agent discussed earlier in association with conjunctival diseases) is now the most common cause of sexually transmitted disease and is even more prevalent than gonorrhea. Viral genital infections on the increase today are caused by herpes simplex virus type 2 (and, less commonly, type 1) and by human papillomaviruses. Three somewhat less common, but nonetheless troublesome bacterial diseases are chancroid, or "soft chancre," caused by *Haemophilus ducreyi;* granuloma inguinale, caused by *Calymmatobacterium granulomatis;* and lymphogranuloma venereum, also caused by *C. trachomatis* but clinically different from nongonococcal urethritis. Bacterial organisms known as mycoplasma are also known to be associated with sexually transmitted diseases and have been isolated from "nonspecific" infections of the cervix, urethra, and prostate. Their role as agents of such infections and of certain types of reproductive failures has been suspected but not established beyond doubt. Sexual transmission is also one route of spread of the human immunodeficiency virus, the cause of AIDS. The clinical disease produced, however, does not affect the genital tract directly.

The nature and symptoms of the systemic diseases induced by different sexually transmitted agents vary widely and depend on the tissues involved following dissemination of the organisms from their local entry site. These diseases are discussed in chapter 19.

PARENTERAL ENTRY

Finally, we must consider the possibility of entry of infectious agents directly into the deeper tissues of the body or into the bloodstream by penetration of, or following deep injury to, the skin and mucosal barriers.

Penetration to Deep Tissues

Without significant injury to or infection of the skin or mucosa, penetration may be accomplished in one of four general ways: (1) direct entry of the organism into the blood capillaries or lymphatic vessels of the skin, without localization in the dermis; (2) injection of the organism by a biting or bloodsucking insect; (3) injection of organisms by a contaminated needle, syringe, or solution; and (4) entry of organisms from skin or mucosal surfaces along the pathway of an indwelling intravenous or urinary catheter.

Direct entry of an organism to the parenteral tissues, as an active rather than a passive accomplishment, is relatively rare. Intact skin and external mucous membranes present a strong mechanical barrier to microorganisms, but injured or infected skin may permit access. The health of the skin and surface membranes is of great importance, particularly when the risk of infection is high, as in the handling of infectious material from sick patients.

The best examples of organisms that can actively penetrate human skin are certain helminth larvae, notably the hookworms and *Schistosoma* (the blood flukes). Infective hookworm larvae on the surface of moist soil or *Schistosoma* larvae swimming in water (following their emergence from their intermediate snail hosts) attach to the surface of the skin, wriggle through, and are picked up in the bloodstream from the capillary beds in the lower layers of the skin. From there they find their way to tissues in which development into the adult form may occur.

Injection of an organism by a biting or bloodsucking insect is the access route for a number of human diseases. With the exception of the fungi, certain members of each of the major classes of microorganisms may gain entry in this fashion.

Occasionally, humans are accidently infected by certain species of rickettsiae that are transmitted by the bites of ticks or mites. Ordinarily, these rickettsiae are passed around only among wild rodents and other animals that the insects parasitize. Two diseases acquired in this way are Rocky Mountain spotted fever and the Oriental infection known as tsutsugamushi. Another infection, Lyme disease, is transmitted to humans by the bite of a deer tick. The infection, caused by the spirochete *Borrelia burgdorferi,* was first reported in Lyme, Connecticut, but reports of the disease are increasingly widespread in the United States. Other diseases such as malaria and yellow fever are *primarily human infections,* kept going in the human population by mosquito vectors that transmit them from one person to another (fig. 5.10). Between these two extremes are a number of diseases vectored by insects (mosquitoes, biting flies, lice, ticks, mites), whose microbial agents find suitable reservoirs in several animal species as well as humans.

The microbial agents of diseases transmitted to humans by biting insects include protozoa (such as the plasmodia responsible for malaria, the trypanosomes of "African sleeping sickness," and species of *Leishmania*); roundworms of the group known as filaria; a few bacteria (such as the plague bacillus); the rickettsiae of typhus fever, rickettsialpox, and others mentioned earlier; and a number of viruses (such as yellow fever, dengue fever, and Eastern and Western equine encephalitis viruses). In addition, the relapsing fevers caused by spirochetes of the genus *Borrelia* are vectored from human to human by body lice or from infected animals to humans by ticks.

Many of these diseases occur naturally only in tropical or semitropical areas of the world, primarily because of climatic restrictions on the insect vector. However, geographic boundaries no longer prevent the mixing of human populations as they once did, so the recognition and treatment of these diseases may be important in any area to which people have traveled after becoming infected. Also the prevention and control of such diseases have become even more pressing as people in third-world countries struggle to reach contemporary standards of health.

The injection of organisms by a contaminated needle, syringe, or solution is an ever-present possibility wherever these items are used. Pathogenic, or merely "opportunistic," organisms may be introduced in this way into parenteral tissues or the bloodstream, sometimes with disastrous results. The viruses of acquired immunodeficiency (AIDS) and type B hepatitis (serum hepatitis) are good examples. Because the route of entry is unnatural, however, some organisms may not be able to survive. Also, some microorganisms appear to be totally incapable of pathogenicity and are easily eliminated by the host's defenses. In some instances the host may react to protein antigens of the organism and develop symptoms of hypersensitivity (see chap. 6). Constant use of strict techniques of sterilization and asepsis offers the only insurance against infection by this route.

Entry of organisms from skin or mucosal surfaces along the pathway of an indwelling intravenous or

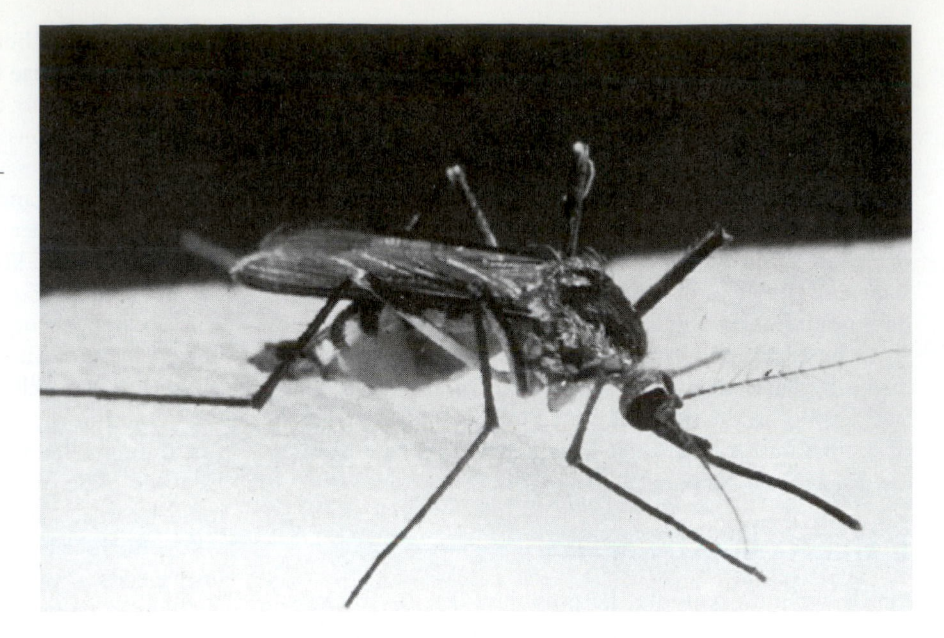

FIGURE 5.10 A mosquito of the genus *Aedes* transmits yellow fever.
Centers for Disease Control, Atlanta, GA.

Concepts of Epidemiology of Infectious Disease
Terminology: Vectors. Horizontal Transmission—Vectorborne

urinary catheter represents an important way in which infections are acquired in hospitals (see chap. 21). These microorganisms are often members of the host's normal resident flora such as staphylococci or gram-negative enteric bacilli and are generally of low disease-producing ability. Their entrance into sites that are normally sterile, however, provides them the opportunity to multiply unchecked, and the infections they produce can be life threatening. Prevention depends on strict attention to aseptic techniques.

Deep Injury to the Skin and Mucosal Barriers

Trauma to the skin or mucous membranes from surgery, accident, or the bite of an animal opens up many possibilities of entry by microorganisms of all types.

During surgery these possibilities are strictly limited or eliminated by sterilizing instruments, linens, and clothing, and by using aseptic techniques in the operating room. Commensals of normally low pathogenicity from the skin or mucosal surfaces may take advantage of the surgical situation, however, and enter to infect the wound or more distant tissues. Bowel surgery or bowel perforation in areas ulcerated by infection or other causes creates the risk of exposing the peritoneal cavity to microorganisms usually confined to the intestinal tract, and peritonitis may result. Similarly, major or minor wounds in the mouth resulting from tooth extraction or other oral surgery may admit normal mouth flora to deeper tissues or to the bloodstream, with consequent risk of systemic infection.

An animal bite introduces microorganisms that may be normal residents of the animal's mouth but that are pathogenic for human tissues. The microorganisms may also have been pathogenic for the animal and may have caused it to bite because it is ill. Among microorganisms that can be transmitted to humans in this way, the one with the most serious potential is the rabies virus. Rabies in humans is almost always fatal, although its development may be prevented if the rabies vaccine is administered following the bite of a rabid animal. Rabies is often associated with dogs, which acquire it from the bites of other animals. Wild animals such as skunks, raccoons, and bats appear to be the reservoir of this disease.

Other accidental wounds, such as lacerations, gunshot wounds, or crushing injuries, may also introduce a variety of microorganisms into deep tissues from the body's surfaces or from the penetrating or crushing objects. The risk of infectious disease is always increased in such instances, not only because the barriers are broken, but because the local tissue defenses are handicapped by the injury. Organisms then have greater opportunity to multiply and to penetrate deeper before the tissue has a chance to heal and defend itself. Normal blood supply to the injured tissue may be cut off or seriously reduced if blood vessels are damaged, thus preventing adequate levels of bloodborne substances that help in tissue defense from reaching the injured area. Such substances would include any antimicrobial agents administered to assist in combating infection. For these reasons, the prompt surgical cleaning of such wounds, with removal of foreign objects, dead tissue, or blood clots, is of foremost importance to prevent infection and to promote healing.

The Toxins of Wound Pathogens

Anaerobic spore-forming bacilli of the genus *Clostridium* produce powerful exotoxins. If they contaminate deep, traumatic wounds that provide appropriate conditions for their growth and toxin production, serious disease results. These organisms are frequent members of the normal intestinal flora of humans and of many animals including dogs, horses, and cattle. Their resistant endospores therefore are widely distributed in soil and dust and may survive there for very long periods of time. It is not surprising that the endospores are frequently present on objects that inflict wounds, such as rusty nails, sharp sticks, animal claws, knives, and bullets. When clostridial endospores enter a surgical or accidental wound, they germinate if conditions permit. The actively growing bacilli then produce toxins that either exert their effect locally (causing gas gangrene, for example) or disseminate to other body areas causing tetanus or, as it is sometimes called, lockjaw (see chap. 23). Although clostridial endospores may be found in the hospital environment in dust and on linens, the source of infection for most surgical patients is organisms from their own intestinal tracts.

DAMAGING MICROBIAL PROPERTIES

Obviously, the production of infectious disease in humans does not depend solely on the opportunity for microorganisms to enter the body. If this were true, illness would be the rule rather than the exception. The biologic resistance of the human host with its many defenses provides a partial explanation for the fact that infection is more common than infectious disease. The other half of the explanation lies in the fact that most microorganisms with which humans are surrounded have few or none of the properties associated with pathogenicity and virulence.

PATHOGENICITY

Following their entry into the body, pathogenic microorganisms display their virulence in terms of two general qualities: invasiveness; or ability to attach, multiply, and spread in the tissues, and toxigenicity, or production of substances injurious to human cells and tissues. The properties that add up to virulence among the microbes pathogenic for humans are not always fully understood, but a number of factors appear to make a definite contribution to the pathogenesis of some infectious diseases.

VIRULENCE

Virulence is commonly measured in terms of the numbers of a particular organism that are required to kill a given species of experimental animal (or its cells in culture) under well-defined experimental conditions. Factors that must be defined in this measurement are related to variations in both the host and the pathogen. The host, or its isolated cells in the culture tube, must be susceptible to the organism. The intact animal must be infected through an entry route that permits the organism to establish itself and induce disease; the route of administration may be intravenous, intramuscular, intraperitoneal, or intracerebral. The numbers of organisms introduced, and the period of time they require to induce disease or death, or both, must also be defined. Virulence is then expressed in quantitative terms for the organism: the minimum lethal dose (MLD), or the smallest number of organisms inducing disease and death of a specified host, or its cells, in a stated time; or, the fifty percent lethal dose (LD_{50}), which allows for variations in host resistance and states the numbers of organisms required to kill 50% of all the infected animals or cell cultures within a certain time.

These terms reflect both the invasiveness and the toxigenicity of a pathogenic organism under controlled conditions and help to explain differences in virulence as they are observed under natural conditions. Thus, the natural occurrence of infectious disease in an individual host may be the result of entry of large numbers of relatively avirulent organisms or small numbers of a virulent one.

INVASIVENESS

Specific explanations for the invasive character of virulent organisms have been offered. In general, they relate to the ability of an organism, once it has gained access to the body of a suitable host, to (1) grow and multiply, (2) protect itself against the defenses of the host, and (3) penetrate and spread throughout the tissues.

Microbial Pathogenesis of Infectious Diseases
Principles of Pathogenicity: Pathogenicity Definitions

Factors Involved in Growth and Multiplication within the Host

For an organism to produce disease, it must use available nutrient within the host's tissues and multiply there. If sufficient vital nutrients are not present, the invading organism cannot grow and therefore appears to be avirulent. The failure to grow may depend on additional factors operating at the entry site such as the presence or absence of oxygen, the prevailing pH or temperature, and competition from host cells or commensal residents for a particular growth factor (such as a vitamin or mineral). In some instances use of available nutrient by the invading organism may deprive host cells of growth factors vital to their health, so tissue injury may result from such starvation whether or not other microbial properties are involved.

Factors That Protect Microorganisms from Host Defenses

Phagocytosis, or ingestion by specialized blood and tissue cells, is one of the important processes by which the body deals with foreign particles or microbes that find their way to these sites. The invasiveness of some bacterial pathogens appears to be related to their ability to resist phagocytosis or to destroy the phagocytes.

The ability to resist the action of phagocytes is due in most cases to the surface components present on the bacterial cells. Many species of bacteria possess capsules of viscous polysaccharide material (e.g., strains of pneumococci) or polypeptides (anthrax bacilli), or they possess surface proteins (e.g., M proteins of certain streptococci). Such components of the microbial cell either prevent their ingestion by phagocytes or, as is the case with the waxy cell wall of the tubercle bacillus, enable them to survive intracellularly.

The ability to kill phagocytes is associated with the production of leukocidin, a substance that is toxic for leukocytes, or white blood cells. Some leukocytes are highly phagocytic. When they enter an injured area from adjacent capillaries, they may be inactivated and killed by leukocidin secreted into the tissue by invading organisms, such as actively growing streptococci and staphylococci.

Phagocytic cells contain many cytoplasmic granules called *lysosomes* (see chap. 6). These are small membranous sacs containing various enzymes that are important in intracellular digestion. One of these enzymes, lysozyme, is capable of lysing some bacteria after they have been engulfed by the phagocyte by breaking down a component of their cell walls. In some instances the tables may be turned by virulent microorganisms taken up by phagocytes. For example, a hemolytic toxin (hemolysin) of some streptococci is not only capable of destroying host red blood cells but may also destroy phagocytes that have engulfed the organisms. Hemolysin release within the phagocyte has an immediate effect on the lysosome granules, causing them to explode and discharge their content of digestive enzymes into the cell's cytoplasm. The "suicidal" result is that the phagocyte's enzymes digest and liquefy its cytoplasm, and the cell dies. Staphylococcal hemolysins can kill phagocytes as well, but the virulence of staphylococci is also related to their production of a nonhemolytic leukocidin that acts on phagocytic cell membranes, breaking down the lysosomal sacs with the same lethal effect.

Factors That Contribute to the Penetration and Spread of Microorganisms

A number of the extracellular products of bacterial growth have been associated with specific kinds of damage to host tissue that permit the offending organism to spread. Several of these products are enzymes for which the substrates are components of host tissue. Following is a list of some of these enzymes:

Collagenase *an enzyme that disintegrates the proteins of collagen, a structural protein of connective tissue.*

Mucinase *an enzyme that digests the mucin produced by many mucosal lining cells.*

Hyaluronidase *an enzyme that breaks down hyaluronic acid, the gellike matrix of connective tissue.*

Lecithinase *an enzyme that destroys lecithin, a component of many human cells including red blood cells.*

Kinase *an activator of an enzyme present in plasma. When activated, the enzyme, called fibrinolysin, can dissolve fibrin clots.*

Hemolysins *substances that can destroy red blood cells. Some hemolysins, such as lecithinase, are enzymes; others are not.*

The production of one or more of such substances by a microorganism multiplying within host tissues causes additional injury, disrupts the normal host cell architecture, interferes with its defensive activities, and promotes the spread of the organism to adjacent cells and tissues. Organisms may enter the blood and lymph stream and be distributed to other parts of the body. The flow of blood into the infected region is often reduced or cut off during the course of such events so that the concentration of both nutrient and defensive substances brought by the blood is further diminished.

Gas gangrene, caused by *Clostridium perfringens,* is a graphic example of tissue destruction resulting from the enzymes of an invading organism. The anaerobic bacilli get their start in traumatized tissue to which the supply of blood and oxygen is low. As they multiply, they produce many exotoxins, including enzymes with collagenase and lecithinase activity. These substances break down healthy tissue on the margins of the injured area, enlarging the damage, releasing more nutrient for the organisms from host cells, and extending the zone of reduced blood supply. Continuing growth of the organisms means continuing production of enzymes and rapid extension of tissue injury.

The spread of streptococci through the tissues may be associated with their production of hyaluronidase. Many strains of streptococci and staphylococci secrete hemolysins (called, appropriately, streptolysins or staphylolysins) as well as kinase (streptokinase, staphylokinase). Kinases are thought to prevent the organisms from being trapped within fibrin clots formed during the body's defensive inflammatory response (see chap. 6). Mucinase is produced by the influenza virus attached to mucosal cells of the respiratory tract. By altering the surface characteristics of mucin, changing it from a viscous to a thin watery substance, mucinase helps to ensure the penetration and attachment of the influenza virus to ciliated epithelial cells (fig. 5.11), thus contributing to its invasiveness.

Toxigenicity

Endotoxins are components of the outer portion of cell walls of gram-negative bacteria that are released into the surrounding area, usually after death of the cell. Chemically, endotoxins are composed of lipid and carbohydrate. The lipid portion is responsible for the physiologic effects of endotoxin (e.g., fever, blood vessel changes that may lead to hemorrhage, shock, death). The carbohydrate portion is the antigen that stimulates the production of circulating antibody against endotoxin. Infections caused by the endotoxin-containing gram-negative bacteria are often associated with a great deal of tissue and blood vessel damage. Although many of the symptoms of these diseases are referable to the release of endotoxin, the exact mechanism for its role in disease production is still under investigation. New studies indicate that many harmful effects of endotoxin are related to a substance known as tumor necrosis factor whose production is stimulated by endotoxin (see chap. 6). Table 5.2 differentiates between endotoxins and exotoxins.

Exotoxins, as we have seen, are substances, usually protein in nature, produced and secreted by bacteria during

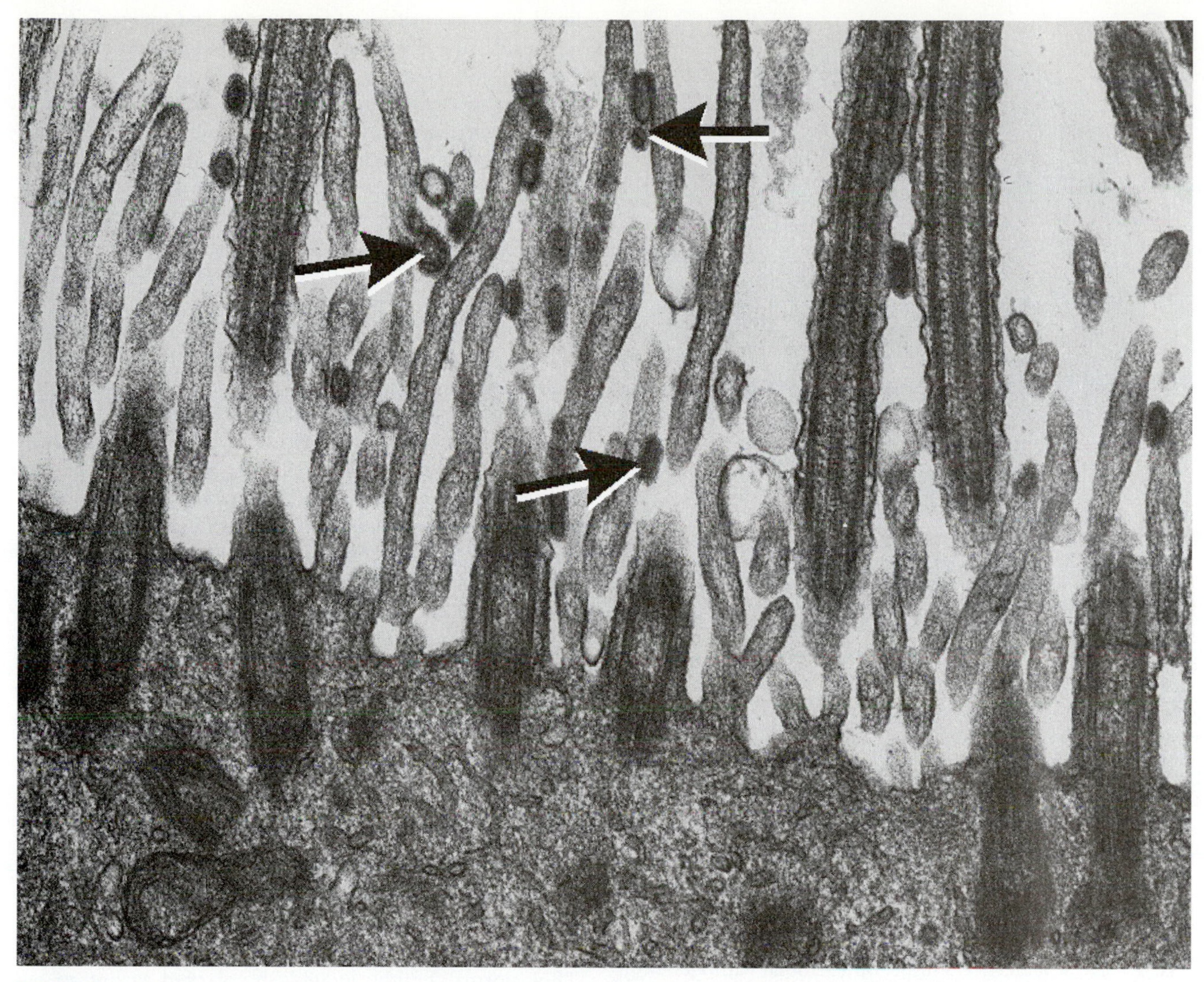

FIGURE 5.11 Electron micrograph showing a portion of a ciliated epithelial cell from an organ culture of guinea pig trachea after incubation with influenza virus for one hour at 4° C. Virus particles (arrows) are attached to cilia and to microvilli.

Reproduced from Cedric A Mims. *The Pathogenesis of Infectious Disease,* 3rd ed., 1987, Academic Press, London. Photograph courtesy of Dr. R. R. Dourmashkin, Clinical Research Centre, Middlesex, England.

Viral Structure and Function
Viral Invasion: Virus Adsorption. Viral Pathogenesis: Host Cell Damage

their growth. Only a few bacterial pathogens secrete exotoxins. These toxins differ from each other chemically as well as in their biologic effects, each producing a specific type of injury to particular cells and tissues (table 5.2). The important toxin-producing bacteria have been mentioned in connection with their usual entry routes into the human body and the manner in which their toxins induce disease. These diseases are summarized briefly in table 5.3.

OTHER FACTORS CONTRIBUTING TO INFECTIOUS DISEASE

In addition to chemical injuries that may be caused by products and secretions of microorganisms, a good deal of mechanical damage may be done. As foreign bodies, their presence in the tissues is irritating and may be mechanically disruptive. A mass of worms in the intestinal tract, for example, may cause intestinal obstruction. The presence of helminth larvae in organs such as liver, bone, or muscle may disrupt normal architecture and interfere with tissue function. They may also block the flow of blood through a vessel and thus diminish the nutrition of adjacent tissues.

Physical effects are often aggravated further by the nature of the host's reaction to a foreign body. The supply of blood to an injured area usually is increased, and both cells and fluid from the blood exude into the tissue, causing swelling. This increases the pressure and further disrupts the functions of the tissue. As the reaction proceeds, fibrinous tissue is laid down around the area (fig. 5.12). When the infection subsides, the damaged region may be filled in with fibrinous scar tissue, then further solidified with deposits of calcium, or it may be completely cleared and returned to normal.

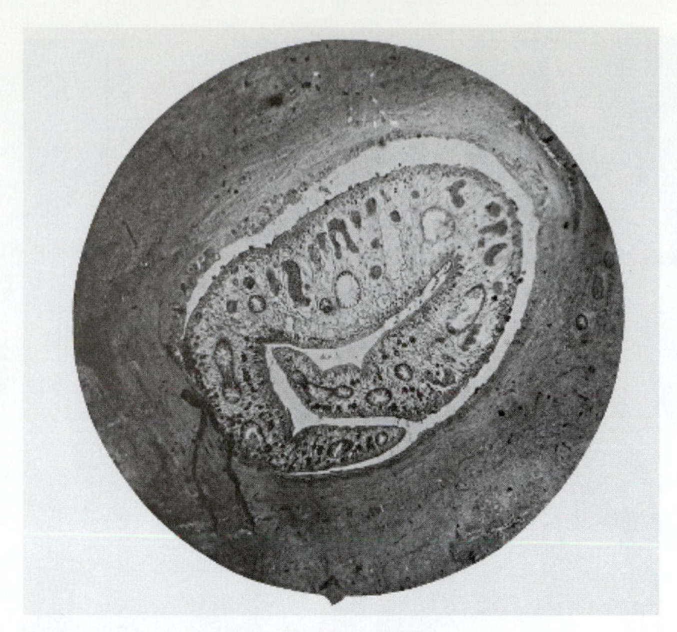

FIGURE 5.12 An adult trematode, or fluke (*Clonorchis sinensis*), established within a bile duct (×10). Its presence there injures the architecture and function of the duct. The host's inflammatory reaction and the fibrosis that occur later also narrow or obstruct the duct.
Courtesy Dr. Kenneth Phifer, Rockville, MD.

Parasitic Structure and Function
Trematodes: Trematodes Life Cycles

Another form of injury that may result from the host's reactions to an invading organism is known as *hypersensitivity.* If the host becomes overreactive to a microorganism or its products, the tissue at the site of reaction is often damaged. Hypersensitivity is developed through the same mechanism that leads to immunity and is discussed in chapter 6 with other host responses to infection. Table 5.4 summarizes the several mechanisms of microbial pathogenesis.

UNANSWERED QUESTIONS CONCERNING MICROBIAL DISEASE

Many answers have yet to be found explaining the pathogenesis of some infectious diseases. Virulence has been defined as a combination of invasiveness and toxigenicity; yet neither of these is synonymous with production of disease. For example, some viruses may widely invade the human body without producing apparent ill effects; some endotoxic bacteria are harmless commensals of the intestinal tract (e.g., strains of *Escherichia coli*), yet their close relatives are correlated with severe diarrheal disease, and some organisms display strict host specificity with regard to virulence (as in the case of the strain of tubercle bacillus that is virulent for humans and the guinea pig but does not produce disease in the rabbit).

The evaluation of an organism as the virulent agent of a given disease was once governed by the conditions or "postulates" originally outlined by Robert Koch in the late 1800s. These conditions require the repeated isolation of the organism from cases of the clinical disease having similar or identical symptoms; its repeated recognition in pure laboratory culture; its ability to produce the characteristic disease in a suitable experimental animal; and its reisolation from the infected animal. These criteria satisfactorily established the microbial agents of many infectious diseases, particularly those caused by bacteria and higher organisms. However, they are difficult to apply to viruses, and they do not explain virulence, as such, or host specificity.

More likely, virulence, like infection, is an expression of the combined properties of the host, or a host species, and the pathogen. Infection is a dynamic state in which two organisms continuously oppose each other. In a simple analogy, it might be said that the relationship between a parasitic microorganism and the host that harbors it is rather like a seesaw, the position of the board depending on the weight each opponent can bring to bear on the other (fig. 5.13). It depends also on the balance point, and whether its placement allows either adversary an advantage over the other. Thus, normal, healthy individuals balanced against an organism of low virulence may easily hold their own, but persons whose resistance has been lowered for some reason might have less success with the same organism. In the same way, virulent microbial species may induce more or less serious disease, depending on the host's ability to control the situation. Their failure to do so at all in a certain type of host may reflect the weight of biochemical and immunological

TABLE 5.4 MECHANISMS OF MICROBIAL PATHOGENESIS

Routes of Entry	Microbial Activities (Pathogenicity and Virulence)
A. Respiratory tract (droplets, droplet nuclei)	A. Attachment and multiplication at entry site
B. Gastrointestinal tract (milk, water, food, fingers, fomites)	1. Mechanical damage
C. Skin and mucosa	2. Metabolic competition
D. Other parenteral entry	3. Chemical damage (toxins)
1. Trauma	4. Allergenic effects (host hypersensitivity)
a. Accidental	B. Invasion and multiplication at local or distant sites; factors 1, 2, 3, and 4, as above
b. Surgical	C. Toxigenic factors
2. Injection	1. Exotoxins
a. Accidental (syringe, needle)	2. Endotoxins
b. Insect vectors	3. Enzymes (e.g., lecithinase, hyaluronidase)

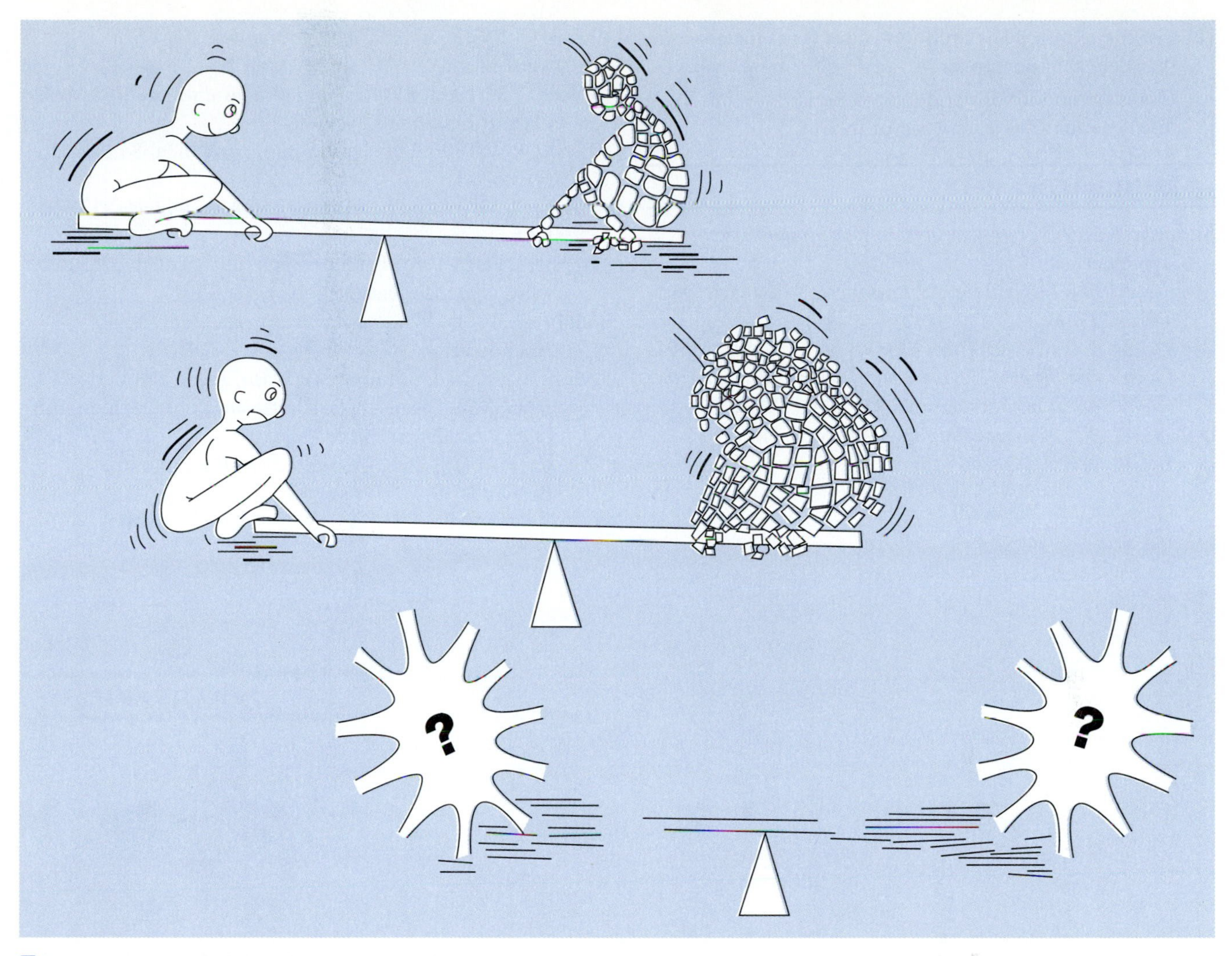

FIGURE 5.13 The balance between the microbe and the host is constantly changing. The virulence of the organism is pitted against the resistance of the host. What factors tip the seesaw in favor of the host? In favor of the organism?

Microbial Pathogenesis of Infectious Disease
Principles of Pathogenicity: Organism/Host Interactions

properties brought to bear by one type of host as compared with another.

We have considered some of the factors that add weight to the parasite's position on the board; now let us turn our attention to human hosts and the power with which they can balance or tip the board in their own favor.

QUESTIONS

1. Define ecology.
2. Define the terms parasitism, commensalism, mutualism, and symbiosis. Give an example of each and explain how it fits the definition.
3. What is the difference between transient and resident flora of the skin? Of what importance is this in surgery?
4. List examples of flora indigenous to the skin, genital tract, upper respiratory tract, lower respiratory tract, and conjunctiva. What is the significance of the normal flora?
5. Define pathogenicity and virulence. How do they influence the severity of an infectious disease?
6. What host factors are involved in resistance to infectious disease?
7. List the routes through which microorganisms gain entry to the human body.
8. Give examples of a disease that may be acquired through each entry route of infection.
9. What determines whether or where a microorganism will ultimately establish itself and multiply in the human body?
10. What are the limitations of Koch's postulate's when attempting to identify the causative organism of a new disease? What method is used today for this purpose?
11. How do pathogenic microorganisms display their virulence following entry into the body?

12. List the extracellular products of bacterial metabolism that help microorganisms to penetrate and spread.
13. Compare endotoxins and exotoxins. List several diseases caused by each type of toxin.

BIBLIOGRAPHY

Andrews, M. 1977. *Life that lives on man.* New York: Taplinger.

Atkins, E. 1984. Fever: The old and the new. *J. Infect. Dis.* 149: 339–347.

Cartwright, F. F. in collaboration with M. D. Biddiss. 1991. *Disease and history: The influence of disease in shaping the great events in history.* New York: Dorset Press.

Dubos, R. 1971. *Mirage of health.* New York: Harper and Row.

Eron, C. 1981. *The virus that ate cannibals.* New York: Macmillan.

Ewald, P. W. 1993. *Evolution of infectious diseases.* New York: Oxford University Press.

Ewald, P. W. 1993. The evolution of virulence. *Sci. Am.* 268 (4): 86–93.

Gladwell, M. 1995. The plague year. *New Republic;* 17–24 July: 38–46.

Isenberg, H. D. 1988. Pathogenicity and virulence: Another view. *Clin. Microbiol. Rev.* 1(1): 40–53.

McNeill, W. H. 1976. *Plagues and peoples.* New York: Anchor Press.

Mims, C. A., N. Dummock, A. Nash, and J. Stephen, eds. 1995. *Mims' pathogenesis of infectious disease,* 4th ed. San Diego: Academic Press.

Zinsser, H. 1975. *Rats, lice, and history.* Boston: Little, Brown.

CHAPTER 6

Host Defense Mechanisms in Microbial Infection

OUTLINE

Microbes in Motion

This chapter covers the natural and specific immune defense mechanisms by which the healthy human resists infection and the consequences of dysfunction of the normal immune system. The following books and chapters of the Microbes in Motion CD-ROM may be a useful preview or supplemental study aid to your reading: Viral Structure and Function: Viral Pathogenesis; Viral Detection. Fungal Structure and Function: Diseases. Microbial Pathogenesis of Infectious Disease: Nonspecific Host Defense.

OBJECTIVES

On completing this chapter the student will be able to:

1. differentiate between the natural and immune defense mechanisms through which the healthy human body resists infection.
2. cite the natural defense mechanisms the body uses to resist invading microorganisms.
3. discuss the nature and mechanisms of specific resistance to infection (immunity).
4. describe humoral and cellular immune responses to infection.
5. explain the dysfunctions of the normal immune system that may lead to conditions such as hypersensitivity or to serious immunodeficiency diseases.

KEY WORDS

Actively acquired immunity
Agglutinins (a-**GLOOT**-in-ins)
Allergic contact dermatitis (ACD)
Anamnestic (an-am-**NEST**-ic) response
Anaphylaxis (an-a-fill-**AX**-sis)
Antibody
Antigen
Antigen-antibody specificity
Antigenic determinants
Atopic (a-**TOP**-ic) allergy
Autoimmune diseases
B cells (lymphocytes)
Basophils (**BASE**-oh-fills)
C-reactive protein
Capsular antigens
Cell-mediated immune response (CMI)
Cellular immunity
Chemotaxis (keem-oh-**TAX**-is)
Clonal selection theory
Combined immunodeficiency
Complement
Cytokines
Cytotoxic T cell
Dendritic cells
Enzyme immunoassay (EIA)
Eosinophils (**EE**-oh-**SIN**-oh-fills)
Erythrocytes (ee-**RITH**-roe-sites)
Exudate (**EX**-u-date)
Fibrinogen (fibe-**RIN**-oh-gen)
Flagellar (flag-**ELL**-are) antigens
Fluorescent antibody
Gamma globulin (**GLOB**-u-lin)
Graft rejection
Granulocytes (**GRAN**-u-lo-sites)
Hapten (**HAP**-ten)
Humoral (**HUME**-or-al) immunity
Hypergammaglobulinemia
Hypersensitive
Hypogammaglobulinemia
Immune complex disease
Immunoblots
Immunodeficiency diseases
Immunodiffusion
Immunogen (**IMM**-u-no-gen)
Immunoglobulins
Inflammatory response
Interferon (inter-**FEAR**-on)
Interleukins (inter-**LOO**-kins
Latex agglutination
Leukocytes (**LEW**-co-sites)
Lymphatic system
Lymphocytes
Lymphokines
Lysozyme (**LICE**-oh-zyme)
Macrophages (macro-**FAY**-ges)
Major histocompatibility complex
Monoclonal antibody
Monocytes
Mononuclear phagocyte system (MNPS)
Mucociliary (mew-ko-**SILL**-i-airy) elevator
Neutrophils (**NEW**-tro-fills)
Nonspecific resistance
Opsonins (**OP**-sone-ins)
Passively acquired immunity
Phagocytes (**FAG**-oh-sites)
Phagocytosis (**FAG**-oh-site-**OH**-sis)
Phagolysosome (**FAY**-go-**LICE**-oh-sohm)
Plasma
Platelets
Polymorphonuclear neutrophil (PMN)
Precipitins (pre-**SIP**-it-ins)
Properdin (**PRO**-per-din)
Purulent (**PURE**-u-lent)
Recombinant DNA vaccines
Red blood cells
Serology
Somatic (sohm-**AT**-ic) antigens
Specific resistance (immunity)
T cells (lymphocytes)
Toxoids
Tumor necrosis factor
Vaccines
White blood cells

In order to survive in this world of antagonistic biologic forces, the human body has developed many defense mechanisms. Some of these are *nonspecifically* directed against any foreign substance that may find its way into the body. Others are *specific* mechanisms of resistance (or immune mechanisms) that have evolved to protect the body against invasion by parasitic microorganisms. Most of these responses contribute to human health and well-being, but others result in self-damage, causing hypersensitivity or autoimmune disease. In addition, deficiencies in the immune system may seriously impair the body's capacity to resist even minor infections.

This chapter will consider the most important factors, both nonspecific and specific, that enable the human host to resist infection and describe some of the disease conditions resulting from disorders of the immune system.

NONSPECIFIC RESISTANCE TO INFECTION

The general resistance offered by the healthy body to most microorganisms is nonspecific in two ways: (1) it is effective against microbes as a group and does not operate against any single organism, and (2) in many instances it operates against nonliving foreign matter as well, inactivating chemical substances that do not necessarily originate in living organisms.

The most important and best-defined factors of **nonspecific resistance** are associated with the protective effects of the skin and mucous membranes and the cells of the mononuclear phagocyte system (MNPS), also known as the reticuloendothelial system (RES).

ENTRANCE BARRIERS

The skin and mucous membranes offer the first barrier to microorganisms coming into contact with the body. If these structures are intact and functional, most microbes are unable to enter into deeper tissues. This is due to the architectural and the physiologic properties of these tissues, as well as to the constant supply of phagocytes and plasma components carried to them by the bloodstream.

Architecture of Skin and Mucous Membranes

The skin is composed of a continuous sheath of closely bridged cells covering all exposed surfaces and supported beneath by dense connective tissue. In addition, the outermost surface of the skin is layered over with keratin, a protein substance that is not only tough and dense but waterproof.

Most of the upper and lower respiratory tract is lined by ciliated epithelial cells. These cells are bathed by mucus secreted from goblet cells and subepithelial glands. Microorganisms or other foreign particles carried into the lower airways are trapped in mucus and swept up to the back of the throat by the beating action of this **mucociliary** escalator. The alveoli of the lungs do not have ciliated cells or mucus, but are lined by phagocytic cells, which engulf any foreign particles that progress that far. In the nose, the nasal hairs as well as the ciliary lining sweep particles to the back of the throat to be swallowed. In the upper part of the nasal cavity, the "baffle plates" formed by the turbinate bones are covered by a ciliated mucosa that acts as a trap for small particles. In the mouth, the saliva provides a mechanical flushing action, and the movements of the tongue and the reflexes of swallowing ensure that most ingested or inhaled particles are carried to the stomach where the acidic pH is lethal to most microorganisms.

Physiology of Superficial Tissue

The physiologic defenses of the skin include its secretions, notably those of the sebaceous and sweat glands, which have an antimicrobial effect. Fatty acids of the sebaceous gland secretions discourage some fungi and bacteria from colonizing on the skin or scalp. The low pH and high salt concentrations of sweat provide a similar protection. **Lysozyme** is an antimicrobial substance present in mucus, tears, saliva, and other body secretions, as well as in phagocytes. This enzyme can break down some bacterial cell walls by splitting the peptidoglycan molecules (see chap. 2).

In the stomach, the acidity of gastric secretions destroys many microorganisms that are swallowed. Farther along the intestinal tract, the action of the body's digestive enzymes, together with microbial competition from commensal flora, keeps the surviving newcomers in check. Normally, most ingested transient microorganisms are either destroyed in the intestinal tract or finally removed with passage of fecal material from the body. When organisms do attach and penetrate, their resulting disease production depends on the extent of microbial multiplication and spread, damage from exo- and endotoxins, and the body's defensive reactions.

In the genitourinary tract, the surface tissues are also covered by a mucous film. The direction of mucus flow, its pH, salt concentration, and lysozyme content, as well as the activity of resident commensals, tend to discourage the localization of transient organisms. The pathogenic organisms of the sexually transmitted diseases, however, have a special opportunity for entry during sexual intercourse with an infected partner and then are usually able to attach to and penetrate the genital mucosa, establishing new infection.

When the skin or mucous membranes are damaged or diseased, their ability to withstand infection or to prevent the penetration of microorganisms to other tissues is compromised. Even minor wounds and abrasions may open up entry routes for microbes. Moisture is important to the integrity of surface structures, healthy skin being relatively dry, mucous membranes normally moist. Skin becomes macerated in the continuing presence of excessive moisture, such as may accumulate under tight, wet dressings. Under these conditions damage to the skin's structures decreases its effective resistance to infection. Conversely, excessive dryness of mucosal surfaces, particularly those of the respiratory tract, also predisposes to infection. Other malfunctions, for example those resulting from the occlusion of glandular ducts to the surface (lacrimal, salivary, sebaceous, or sweat glands) offer still another opportunity for microorganisms to cross the surface barriers.

BLOOD BARRIERS

Blood is a vital and dynamic factor in the body's resistance to infection. Not only does it constantly supply oxygen and nutrient to all tissues, but it carries away waste products of cellular metabolism that would become toxic if they were to accumulate. If circulation is diminished or cut off from any area, tissue injury or death (necrosis) may result, and the involved area becomes more susceptible to the activities of microorganisms.

Blood is comprised of a fluid portion called plasma and of cellular constituents. Both have important functions in responding to foreign substances including microbes and their products.

Plasma

Plasma is a colloidal solution of proteins in water containing carbohydrates, lipids, vitamins, hormones, electrolytes, and dissolved gases—all of which are carried to and exchanged in the tissues nourished by the bloodstream. Antimicrobial agents and other drugs are carried in the plasma as well. The proteins present in largest quantity are albumin, fibrinogen, and globulins. Albumin is responsible for much of the osmotic force of the blood and **fibrinogen** is a vital factor in blood clotting. Globulins occur in three major groups: alpha, beta, and gamma. However, **gamma globulins** are of greatest importance in immunity. During the development of *specific* resistance, gamma globulins known as **immunoglobulins,** or **antibodies,** are synthesized. These circulate in the blood or other fluids to provide humoral immunity (*humor* is a Latin word meaning moisture, used in medieval medicine to denote a body fluid or juice).

Antibodies are produced in specific response to particular foreign substances (called *immunogens* or *antigens*) and become widely distributed in body fluids, including plasma. Infectious microorganisms usually stimulate antibody production so that the concentration of antibodies in plasma characteristically rises during the course of an infection. Other plasma components that play a nonspecific role in infection occur normally, without specific stimulation. Their activities are not directed against any one organism but work against infectious agents in general. These nonspecific substances may participate, however, in the reactions that occur between antibodies and microorganisms, supplementing the antibody effect or bringing it to completion.

One such plasma constituent is **complement,** so called because it makes an important contribution to the effect of antibodies. Complement is a complex of proteins normally present in plasma. Acting alone it has no destructive effect on microbial or other foreign cells. If these cells have stimulated antibody production, however, the antibody combines specifically with them. Complement then also attaches to the combination of foreign cell and antibody and brings about the death and sometimes, lysis, of the cell. Complement activity also requires the presence of calcium and magnesium, both of which are normally present in the blood in ionized form.

Cellular Constituents of Blood

The cellular elements of blood include **erythrocytes,** or **red blood cells,** white blood cells, and platelets (thrombocytes). Each type of cell originates from stem cells that are produced in the bone marrow (colorplate 4a–d).

The blood components that are important in nonspecific resistance are the **white blood cells,** or **leukocytes:** granulocytes, monocytes, and lymphocytes. The **granulocytes** are further classified as neutrophils, eosinophils, and basophils, depending on their staining characteristics. They have multilobed nuclei (they are said to be *polymorphonuclear leukocytes,* although this term, shortened to poly or PMN, is generally reserved for the neutrophils) and contain numerous *lysosomes,* which are "bags" of hydrolytic enzymes and other degradative substances. The lysosomes are important for digesting foreign materials, such as bacteria, engulfed by the PMN. In blood smears stained with the common differential stains, neutrophils have neutral-staining or azure-colored granules; eosinophils have acidophilic, red granules; and basophils have basophilic or dark blue granules. **Neutrophils** are the most numerous of the white blood cells and are most important for their phagocytic activity. **Eosinophils** (colorplate 4a) play a role in allergies involving complicated mechanisms that are beginning to be understood. Their numbers increase in allergic reactions that include those resulting from certain parasitic infections. **Basophils,** like the mast cells of the tissues, produce histamine, a blood vessel dilator released in allergic reactions. They also contain heparin, an anticoagulant that prevents blood clotting.

The **polymorphonuclear neutrophils (PMN)** (colorplate 4b) are the most abundant of the body's phagocytic cells. They are produced continuously in the bone marrow and discharged in large numbers into the circulating blood. There they survive for only a few days, but their loss is balanced by new arrivals from the bone marrow, which reserves a large supply of these cells against emergency or sudden demand. In any tissue with an adequate blood supply, PMN are constantly available to protect against entry of foreign material. The presence of a foreign substance, especially a living microorganism, provokes a chemical stimulus that attracts phagocytes to the scene. This attraction is spoken of as **chemotaxis.** The polymorphonuclear cells have an ameboid type of motility. When attracted by a chemotactic stimulus, they migrate out of the capillaries (a process known as *diapedesis*), make their way to its source, and attempt to eliminate the foreign object by, quite literally, devouring it. (The Greek word for "eating" is *phagein.*) Moving as amebae do, these phagocytes surround and engulf particulate matter, digesting it if possible (fig. 6.1).

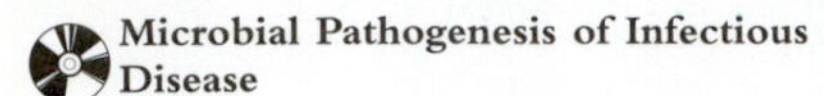

FIGURE 6.1 A human polymorphonuclear neutrophil ingesting a short chain of *Streptococcus pyogenes* (×1,200),

Reproduced from G. Mandell, "Polymorphonuclear neutrophilic phagocyte," *MCV Quarterly* 13:57–59, 1977. Photo by J. Sullivan and G. Mandell.

Microbial Pathogenesis of Infectious Disease
Nonspecific Host Defense: Cellular Defenses and Phagocytosis

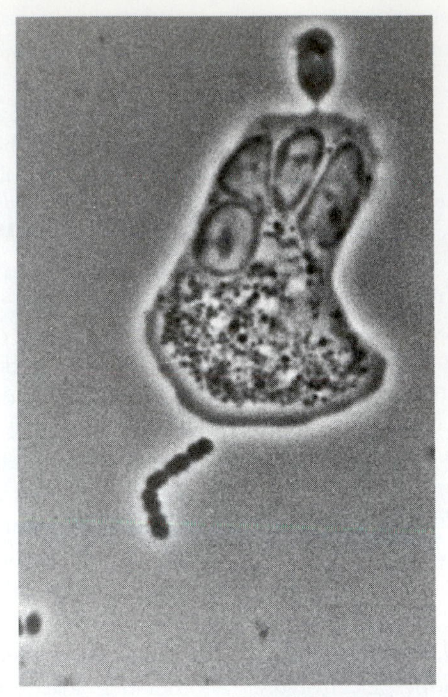

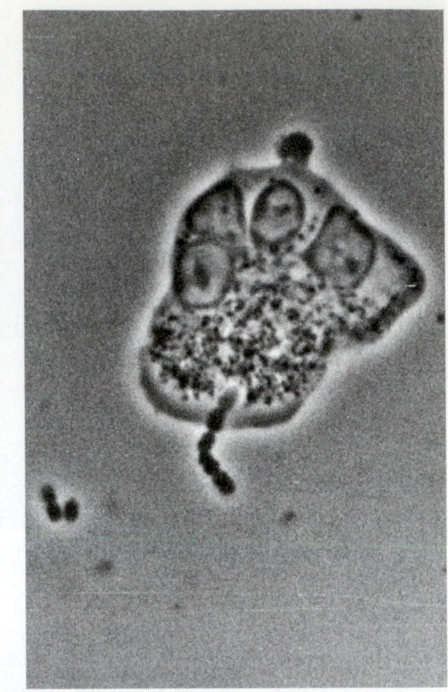

The **monocytes** (colorplate 4c) also arise from stem cells in the bone marrow. These are mononuclear cells, larger than PMN, that are precursors of **macrophages** (also known as mononuclear phagocytes). Monocytes are also chemotactically attracted out of the circulating blood to sites of infection in the tissues where they become enlarged, mature, phagocytic macrophages that secrete substances known as **cytokines** or monokines. Cytokines play an important role in the early events of inflammation (see table 6.2). Macrophages are not so numerous as PMN, nor is there a large reserve of them in the bone marrow. They are widely distributed throughout the body; however, many are in fixed positions within the mononuclear phagocyte system (reticuloendothelial system), as we shall see.

Lymphocytes (colorplate 4d) also occur in the bloodstream, but their major role is in specific immunity and hypersensitivity. Populations of lymphocytes known as B cells produce antibodies that circulate in the fluid portion of the blood and are responsible for the immunity known as humoral immunity (humor is the Latin word for liquid). A second type of lymphocyte, the T cell, produces cytokines called **lymphokines,** which help to kill virus-infected and tumor cells and play a part in cellular (cell-mediated) immunity. Humoral and cell-mediated immune responses will be described in detail in a later section.

Colorplate 4a–d illustrates the appearance of red and white blood cells in a stained smear of circulating blood, and table 6.1 provides a list of some of the constituents of blood with their ranges of normal concentration.

Platelets are small, cell-like structures that have no nucleus. They play an important role in hemostasis (arresting blood flow) by sealing off breaks in blood vessels and helping to facilitate the interaction of clotting factors found in blood. Once a clot is formed, platelets cause the clot to contract, thus preventing the vessel from being obstructed. Like basophils and mast cells, platelets contain substances that play an important role in allergic reactions.

TABLE 6.1 SOME CONSTITUENTS OF CIRCULATING BLOOD IN NORMAL ADULTS

Constituent	Normal Concentration Range
Erythrocytes	Males 4.3–5.9 millions of cells per mm^3 Females 3.5–5.5
Neutrophils	3,000–6,000 cells per mm^3
Lymphocytes	1,500–3,000
Monocytes	280–500
Eosinophils	50–250
Basophils	15–50
Albumin	35–50 g/l
Fibrinogen	2–4
Immunoglobulin G	6–16
Immunoglobulin M	400–3500 mg/l
Total protein	60–83 g/l

LYMPHATIC SYSTEM

Tissues that make up the **lymphatic system** are scattered throughout the body. Lymphoid tissue serves as a major center for antibody production as well as the site where foreign material that enters in lymph (the fluid drained from body tissues into lymphatic capillaries) is removed. Lymph nodes of the upper respiratory tract (fig. 6.2), cervical region, gastrointestinal system, genitourinary tract, and the inguinal area are strategically located to pick up microorganisms that get past the main entry barriers. Here most of them are filtered out by phagocytes that line the numerous

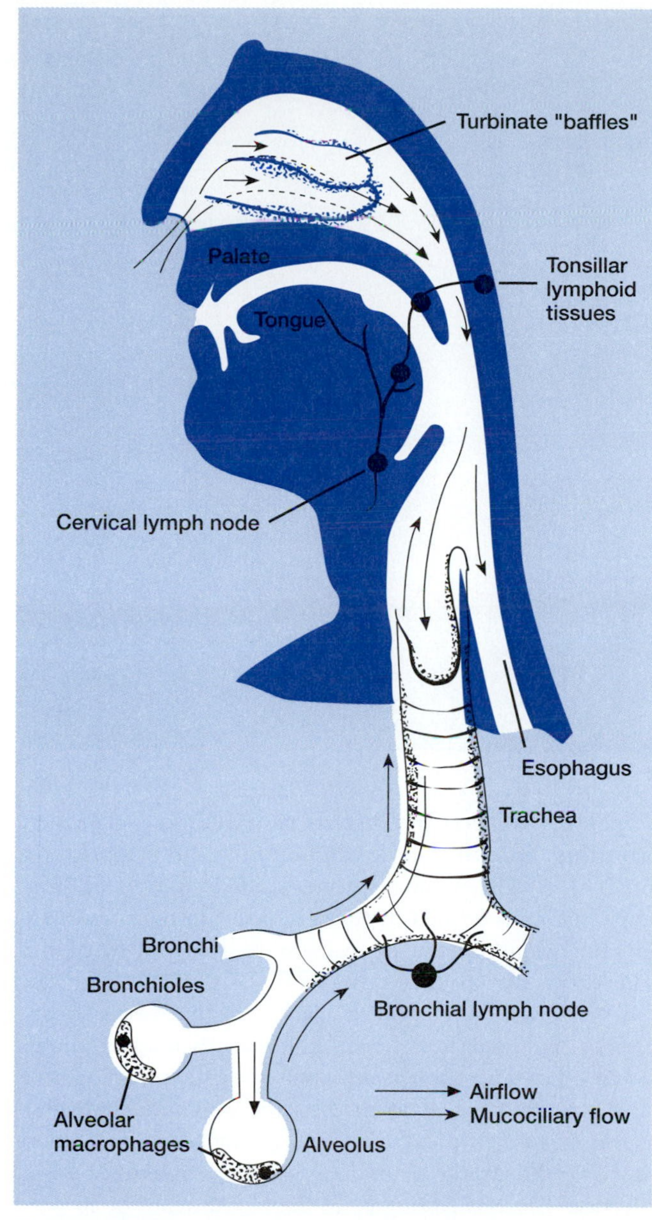

FIGURE 6.2 Mechanisms of nonspecific resistance to respiratory tract infection.

From Cedric A. Mims, *The Pathogenesis of Infectious Diseases,* 3rd ed.1987. By permission of Academic Press Ltd, London.

Microbial Pathogenesis of Infectious Disease
Nonspecific Host Defense: Normal Flora; Mechanical and Chemical Control

branching channels within the node. Some may escape to the efferent lymph vessel only to be filtered out in the next or succeeding nodes. The lymphatic vessels themselves can be penetrated readily by bacteria, but such intruders are quickly drained to a node in the area and filtered there. (Fig. 6.3 shows the structure and circulation of a lymph node.)

Lymphoid tissues may become chronically infected at a major microbial entry point such as the upper respiratory tract. In this way the effective function of tonsils and adenoids may be reduced, and these tissues may then constitute a possible source of infection to other areas through their efferent lymphatics. Under these circumstances, surgery to remove tonsillar and adenoidal tissue is often recommended by physicians after duly considering the consequences of losing any protective value they may afford.

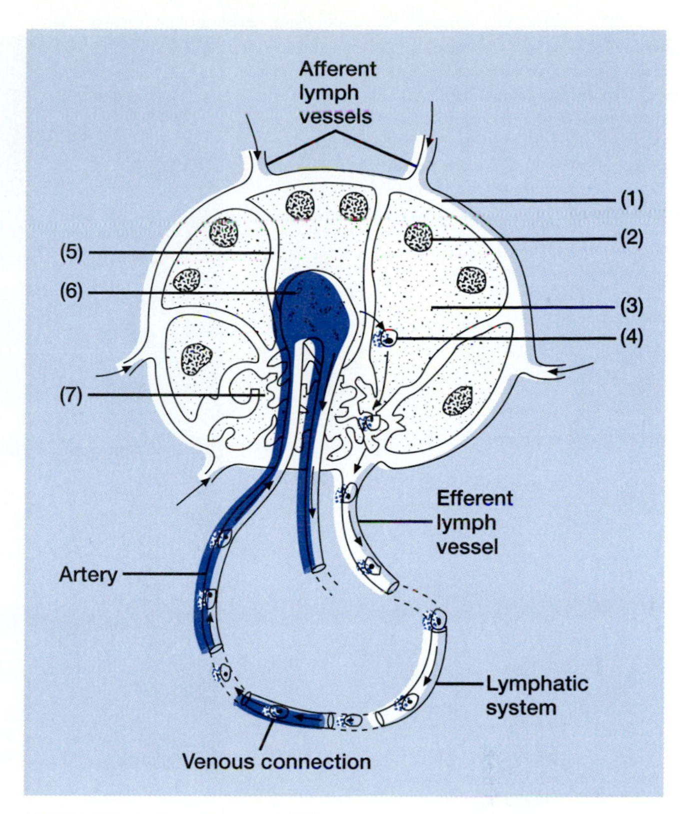

FIGURE 6.3 Diagram of the structure and circulation of a lymph node: (1) Sinus lying beneath the capsule of the node. (2) Nodule, or follicle, of densely packed lymphocytes lying within the superficial cortex of the node. (3) The deep cortex of the lymph node contains a field of lymphocytes, enroute from the blood to the lymph stream. (4) A lymphocyte in passage from the blood vessels to the deep cortex and from there to the efferent lymph vessel. (5) One of a system of lymphatic channels running through the cortex into the medulla. (6) Capillaries and venules of the blood circulation. (7) Medulla of the lymph node, with sinuses and cords of lymphocytes. The lymph flow from all channels empties into the medullary sinuses and collects in the efferent lymph vessel.

MONONUCLEAR PHAGOCYTE SYSTEM (MNPS)

The term **mononuclear phagocyte system,** or **MNPS,** (formerly known as the reticuloendothelial system) is a collective reference to phagocytic cells scattered widely throughout the tissues of the body, including the lymph nodes. Some of these cells are fixed in the reticulum, a network of loose connective, interstitial tissue, and some along the endothelium, the lining of vessels and sinusoids (capillary expansions). Fixed macrophages are abundant in the spleen, bone marrow, lungs, liver, lymph nodes, and central nervous system. In some of these anatomic sites they have particular designations: in the central nervous system they are called *microglia;* in the liver, *Kupffer cells;* and in the lungs, *alveolar macrophages.*

Whatever their location, the principal function of the MNPS cells is phagocytosis and removal of foreign or useless

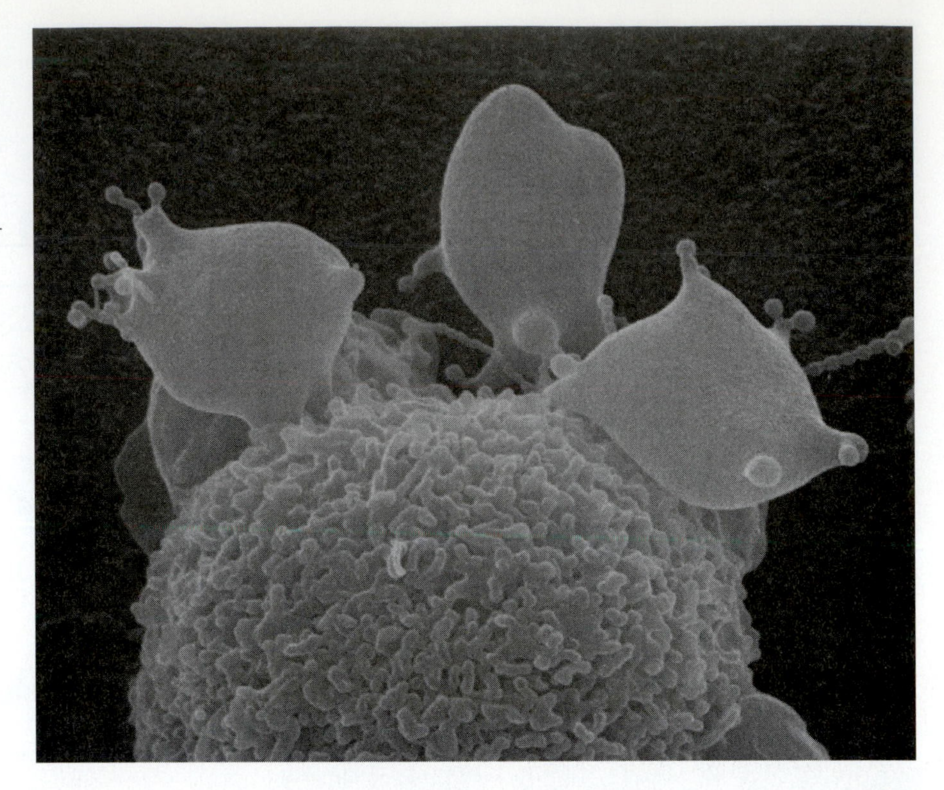

FIGURE 6.4 A scanning electron micrograph of three opsonized erythrocytes in contact with the folds of a macrophage. The cells of the mononuclear phagocyte system clear a local scene of microorganisms and other intruders along with the body's own cellular debris, such as worn out erythrocytes (×8,600).

Reproduced from J. M. Orenstein and E. Shelton, "Membrane phenomena accompanying erythrophagocytosis," *Laboratory Investigation* 36:363, 1977. U.S. and Canadian Division of The International Academy of Pathology, Inc.

particles, living or dead. They have a great capacity for clearing a local scene not only of microorganisms and other unwanted intruders but of the body's own cellular debris, including dead or dying leukocytes, red blood cells (fig. 6.4), and the fragments of other disintegrating cells. When the battle of infection is over, it is the macrophages and the fixed mononuclear cells that clean up the rubble and help to restore order so that tissues can function normally once more. As we shall see, phagocytosis of microbes by MNPS cells is also an essential first step in the immune response subsequently effected by B and T lymphocytes.

PHAGOCYTOSIS

Probably most microbes that manage to penetrate the body's outer barriers and reach the tissues find themselves imprisoned within **phagocytes** (see fig. 6.1).

The process of **phagocytosis,** first described by Metchnikoff, can be divided into stages: contact, ingestion, intracellular destruction and digestion, and exocytosis, the elimination of undigestible waste materials.

Contact

Contact with neutrophils may be a chance encounter, but these PMN respond swiftly by chemotaxis when soluble microbial products are released in the tissues. Attachment of the foreign particle to the phagocytic cell surface and the ingestion that follows is aided greatly by complement, acting together with specific antibodies that may also be circulating in plasma. Antibodies, or immunoglobulins, are molecules that combine specifically with surface constituents of microorganisms, making them easier for the phagocytes to ingest and digest. These antibodies are called **opsonins,** from the Greek word meaning "to prepare food for." The enhancement of phagocytosis by means of antibody coating is termed the *opsonic effect.* Opsonization may also be helpful in the destruction by phagocytes of the body's own worn out cells (fig. 6.4).

When antibody is not available, phagocytosis is facilitated by the presence of fibrin, which serves as a sort of trap to pin down microorganisms. The same effect may operate on the rough surfaces formed by the polymorphonuclear neutrophils when they adhere to the walls of small blood vessels or the alveoli of the lungs, as they sometimes do in infection. Bacteria passing by in the bloodstream or air passageways are caught on irregular edges, and surface phagocytosis then occurs (fig. 6.5).

Ingestion

The ingestion stage of phagocytosis resembles food intake by amebas. Bacteria are ingested as the cytoplasmic membrane of the leukocyte invaginates to encircle the microbes or other particles. As a result, the ingested material becomes enclosed within a membranous vacuole called a *phagosome.*

Intracellular Destruction, Digestion, and Exocytosis

Once the phagosome has formed, the lysosomal granules (see chap. 5) of the neutrophil migrate to this vacuole, fuse their own membranes with its wall, and form a **phagolysosome.** Various enzymes within the lysosomal sac (acid hydrolases, lysozyme, proteases, myeloperoxidases, and other toxic products derived from the increased oxygen

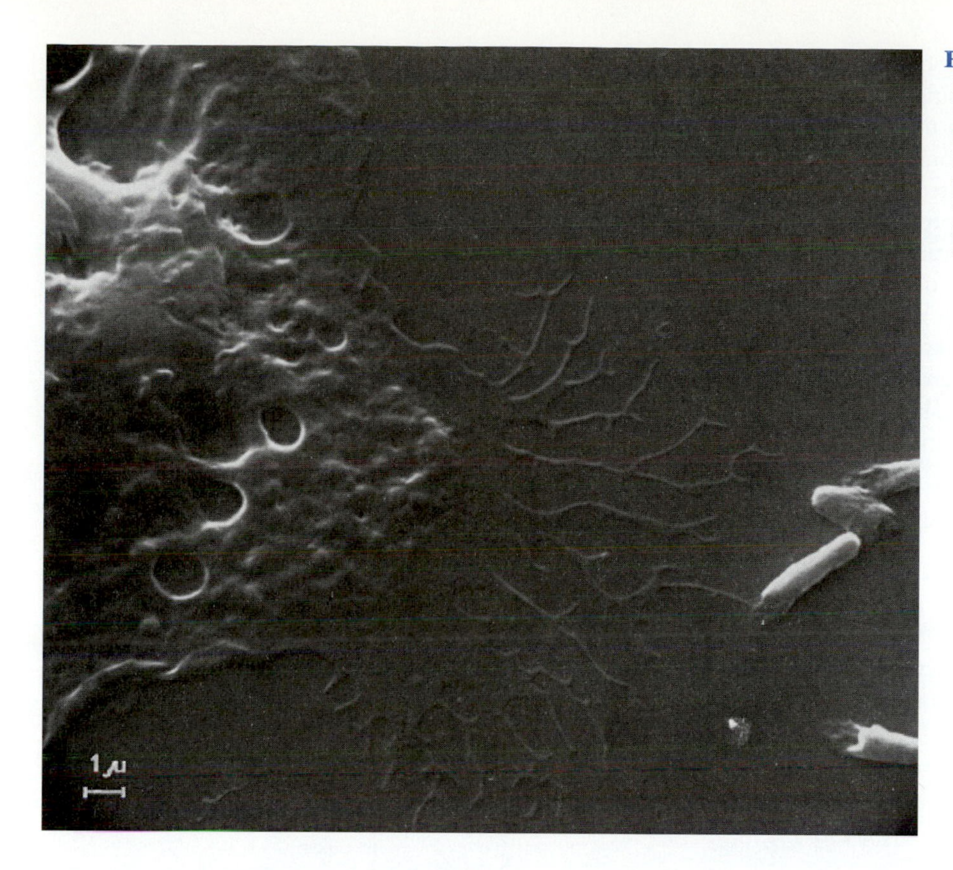

FIGURE 6.5 Scanning electron micrograph of a human polymorphonuclear leukocyte on glass illustrating surface phagocytosis. Note the fernlike nature of the phagocyte's cell membrane as it spreads across the smooth surface approaching the bacilli at the right of the field (×5,000).

Reproduced from A. S. Klainer and I. Geis, *Agents of Bacterial Disease,* 1973. J. B. Lippincott Company, Philadelphia.

consumption that accompanies the initial events in phagocytosis) then proceed to kill the microorganisms and digest the microbial cell contents (fig. 6.6). An important bactericidal reaction is the activity of myeloperoxidase in combination with hydrogen peroxide (formed in the PMN during phagocytosis) and chloride ions. As a result of the reaction, chlorine is incorporated into the bacterial cell wall, disrupting its structure and contributing to cell death. Residue or waste from microbial digestion is then extruded from the phagocyte by the process of exocytosis.

Cytoplasmic granules other than lysosomes contain cationic (positively charged) proteins that also have a killing (bactericidal) action on ingested bacteria. Neutrophils probably possess other mechanisms as well for dealing with ingested microorganisms, but these are not yet fully understood.

Phagocytosis in macrophages is essentially the same as that described for PMN, but there are some differences. Although macrophages display chemotaxis, they respond to different chemical stimuli than those that attract PMN, and they do not migrate through tissues so readily. Their lysosomal enzymes are different; that is, they do not contain the myeloperoxidase enzyme or certain other bactericidal proteins found in PMN granules. These factors account for differences in the distribution of the two types of phagocytes in tissues and explain why some types of organisms killed by PMN may survive and grow within macrophages, and vice versa. Also, macrophages normally contain fewer granules than do neutrophils, but in the immune process they are activated by lymphocytes to produce more. They have a much longer life in the tissues than do PMN and are often in-

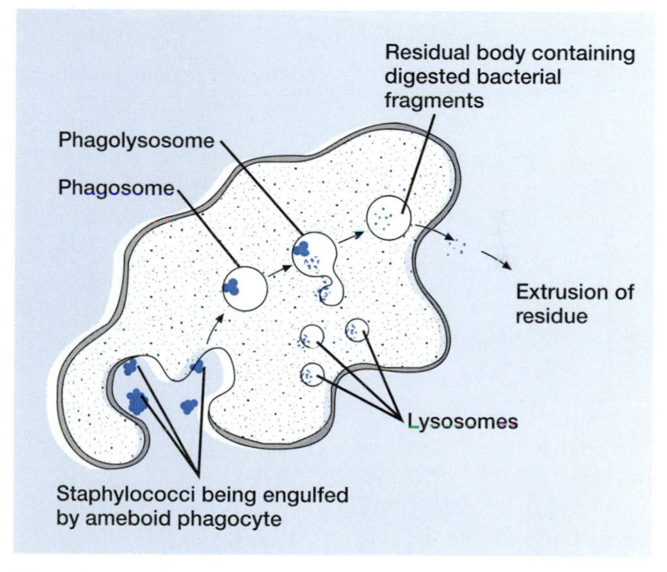

FIGURE 6.6 Diagram showing phagocytosis and digestion of bacteria.

volved, as shown previously, in digesting the remains of dead or dying leukocytes or other body cells (see fig. 6.4).

Pathogenic microorganisms may resist phagocytosis or survive it once taken up. Capsulated organisms, for example, may be difficult for phagocytes to deal with unless the capsule is first coated with capsular antibody. One part of the antibody molecule binds to the organism and another binds to a complementary (receptor) site on the phagocyte surface, thus serving up the opsonized bacteria to the phagocyte.

Streptococcal hemolysins and staphylococcal leukocidins destroy the lysosomal membrane, releasing the PMN digestive enzymes with disastrous effects on the cell's cytoplasm (see chap. 5). Indeed the virulence of some pathogens, such as tubercle bacilli or brucellae, is associated with their ability to multiply within phagocytes. In the process they may kill the leukocytes, but first they may be carried within these cells through the bloodstream to other tissue sites, far from the point at which they were picked up. When such organisms emerge from the cells they have destroyed, they may localize in the new site; if conditions there are conducive to their growth, they may be picked up again by other phagocytes, or they may be met by more specific defenses.

INFLAMMATORY RESPONSE

The body's normal response to infection or tissue injury is called *inflammation*. It constitutes a marshaling of all specific and nonspecific defense mechanisms.

The **inflammatory response** (fig. 6.7) is designed to localize foreign substances at their site of entry, to prevent their penetration into other areas, and to assist in the restoration of injured tissues to normal function.

As the inflammatory response begins at the site of injury, the walls of small blood vessels in the vicinity dilate (under the influence of cytokines and other substances released by injured cells) and become more permeable. This allows plasma, rich in immunoglobulins, fibrinogen and other proteins, and phagocytic cells to pass through more readily. The phagocytes stick to the lining of blood vessels, then pass between the endothelial cells by diapedesis, and migrate to the affected tissues.

Initially, PMN are the most numerous cells present, but in later stages of inflammation, monocytes and tissue macrophages predominate. In certain chronic infections the tissue macrophages fuse together to form giant cells, sharing their many nuclei. Even if they cannot destroy ingested organisms, these fused cells assist in segregating them from adjacent tissue. Segregation proceeds most efficiently through fibrosis, the continuing formation of a fibrous connective tissue wall around the site. Within the active zone, plasma components exert their effects and cellular efforts

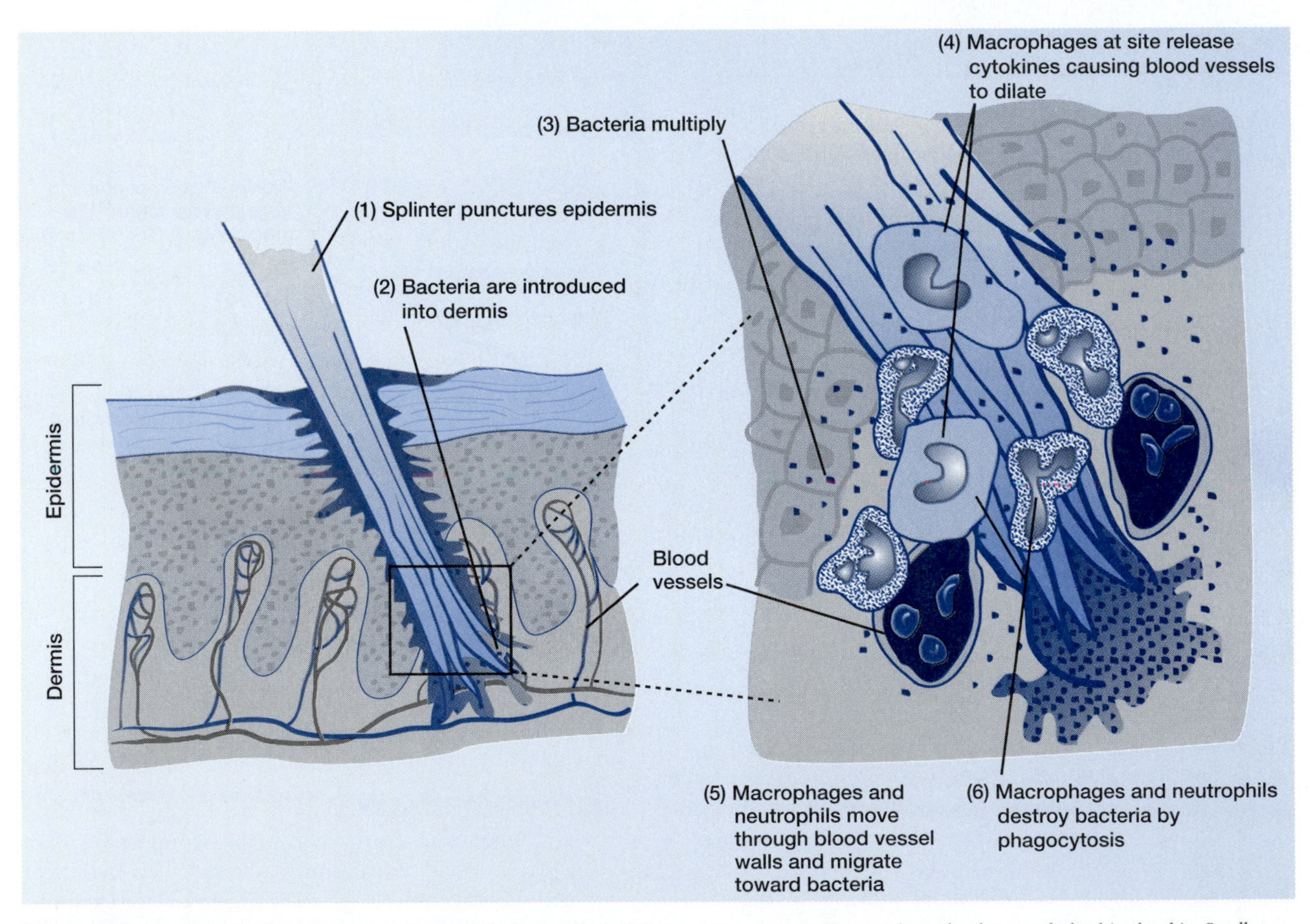

FIGURE 6.7 Injury to human tissues is characteristically met by an inflammatory response. Here a splinter has become lodged in the skin. Small blood vessels in the area become dilated and congested with cells. Ameboid, phagocytic neutrophils, and macrophages migrate from these vessels toward the site of injury. They surround and attempt to ingest both the splinter and the bacteria it has introduced into the subcutaneous tissue. Plasma escaping from the swollen blood vessels causes swelling. Redness of the local skin area results from the dilation of these vessels.

Microbial Pathogenesis of Infectious Disease
Nonspecific Host Defense: Cellular Defenses and Phagocytosis

continue. The outcome of the battle depends on the host's vigor weighed against the nature and extent of the injury. Many microorganisms die under these conditions, but some may remain viable though quiescent for long periods of time.

The collection of fluid and cells that forms within the area of injury is called the inflammatory **exudate.** This material may vary a good deal in consistency and character. When it is thick and yellow with large numbers of leukocytes, it is called a **purulent,** or *suppurative,* exudate (more commonly known as *pus*). A microbial agent that elicits this type of response is said to be pus-producing, or *pyogenic.* When the exudate contains more plasma than cells and is therefore more watery in consistency, it is called a *serous* exudate. If it contains sufficient fibrinogen to form a fibrin clot, it is referred to as a *fibrinous* exudate.

If it occurs superficially in skin or mucous membranes, the inflammatory reaction usually becomes quickly evident; redness, heat, pain, and swelling develop locally as the blood supply to the area increases and as exudate accumulates and presses on the adjacent nerves. Individual tissue cells are compressed as the spaces between them fill up. As the area becomes walled off with fibrinous material and a fringe of connective tissue, it grows more tense and hard to the touch. An abscess forms, reddened on its periphery by excessive blood supply and yellowed at the center where the purulent cellular exudate is heaviest. Eventually it may rupture spontaneously through the surface (suppurate), or it may be opened surgically to drain its contents.

Abscess formation in the deeper tissues and organs of the body follows much the same course, provoking symptoms of tenderness or tissue dysfunction. A deep abscess may rupture and disseminate its contents to adjacent tissues unless it can be drained surgically. However, the inflammatory reaction may resolve without surgical intervention. Most successfully, inflammation clears completely before permanent tissue damage is done; the exudate is absorbed, cellular debris is completely phagocytized, fibrinous deposits are removed, and the area returns to a state of normal function. With less success, the process may result in permanent disfigurement of the involved tissue by its partial or total replacement with fibrous (sometimes calcified) scar tissue, or an intermediate state in which a viable but subdued organism is held in check by chronic inflammation. In the latter case the outcome is determined by the host's ability to react defensively over a long period of time, as well as the capacity of the infectious microorganism to survive the host's strengths and to revive its own when the opportunity presents. This is often the situation in chronic infectious diseases that may remain arrested or become slowly progressive, such as tuberculosis and brucellosis.

The inflammatory reaction is a vital defense, providing early response to, and warning of, infection. It is mediated by the nonspecific and specific circulating antimicrobial factors including neutrophils, monocytes, lymphocytes, macrophages, cytokines, fibrinogen, complement, and antibodies. These factors result in a rising concentration of host metabolic products, a lowered pH, and a rise in local temperature within a walled-off area. All of these are injurious to microbial cells. The price paid by the host depends on the degree of mechanical injury induced locally by exudate and swelling and the amount of chemical damage from toxic products of disintegrating cells and microorganisms. Toxic substances may be carried off in the bloodstream to exert damaging effects on distant tissues. A secondary fall in the number of leukocytes of the peripheral blood is sometimes a consequence of this dissemination.

Another change in the blood commonly occurring as a result of tissue inflammation is an increased erythrocyte sedimentation rate (ESR). This abnormality may be due to changes in the proportions of the plasma proteins. As the gamma globulin fraction of plasma increases with antibody production, the changed ratio of this protein to albumin causes erythrocytes to settle more rapidly.

Other Factors Appearing in Blood

Certain substances increase or arise newly in blood during the inflammatory and immune responses to infection. Several of these are discussed next.

C-Reactive Protein

A globulin called **C-reactive protein** that is normally present in trace amounts increases during the inflammatory process. Its name reflects the fact that it is reactive with (is precipitated by) a component of the pneumococcus known as C carbohydrate. Its presence in the blood is not always related to pneumococcal infection, however; it increases as a result of any noninfectious inflammation and appears during the course of other microbial invasions as well. C-reactive protein has some opsonic properties that, in conjunction with complement, act to kill microorganisms before specific antibody is produced. It also has a modulating effect on the early inflammatory and immune response and is often referred to as an *acute phase reactant.*

Cytokines

Cytokines are a group of soluble molecules that have a variety of physiological activities. They regulate the inflammatory response and stimulate specific cell types to proliferate and differentiate. Cytokines are divided into two types: lymphokines and monokines, produced by lymphocytes and monocytes, respectively. Those synthesized in response to a specific antigenic stimulus are described later (see table 6.4). Those that mediate natural immunity to protect the host against microorganisms are described here (see table 6.2).

Interferon **Interferons** are a family of proteins, certain types of which are manufactured primarily by fibroblasts (connective tissue cells) and mononuclear phagocytes. The most potent inducer of interferon production is viral infection. Cells infected with a virus release interferon, which binds to neighboring cells and causes them to synthesize new proteins that inhibit viral multiplication. Thus, interferon does not act on the infected cell that produces it but

TABLE 6.2 EFFECTS OF CYTOKINES THAT MEDIATE NATURAL IMMUNITY

Cytokine	Cell Source	Cell Target	Activity
Interferon	Mononuclear phagocytes, fibroblasts, other	All	Antiviral, inhibits cell proliferation
		Natural killer cells	Activates natural killer cells
Tumor necrosis factor	Mononuclear phagocytes, T cells	Endothelial cells of blood vessel	Enhances inflamatory response, blood flow
		Neutrophils	Activates neutrophils
		Mononuclear cells	Stimulates cytokine production
		Liver	Induces acute phase reactants
Interleukin 1	Mononuclear phagocytes, other	Endothelial cells of blood vessel	Enhances inflammatory response
Interleukin 6	Mononuclear phagocytes, T cells	Mature B cells	Increases antibody production
		Liver	Induces C-reactive protein
Interleukin 8	Mononuclear phagocytes	Leukocytes	Stimulates chemotaxis
Interleukin 12	Mononuclear phagocytes	Mature B cells	Increases antibody production
		Liver	Induces acute phase reactants

on adjacent cells that are now resistant to viral infection. Interferon also inhibits cell proliferation, activates certain lymphocytes important for killing viruses (called natural killer cells), and increases production of a factor that helps lymphocytes recognize viral antigens.

Because interferon inhibits cell proliferation, it is used in selected conditions as an antitumor agent; for example, in patients with hairy cell leukemia and for Kaposi sarcoma in patients with AIDS (see chap. 19). These therapeutic uses are possible because of recombinant DNA techniques (see chap. 3) that have permitted synthesis of large quantities of biologically active interferon. The source is an *Escherichia coli* strain containing a plasmid that codes for interferon production.

Tumor Necrosis Factor **Tumor necrosis factor** (TNF) is critical in the host response to gram-negative bacteria and perhaps other infectious microorganisms. TNF is synthesized primarily by mononuclear phagocytes in response to contact with the lipopolysaccharide in the outer membrane of the gram-negative bacterial cell wall (endotoxin, see table 5.2). TNF derived its name from the observation that when tumor-bearing animals were treated with endotoxin, TNF appeared in their serum and the tumor cells were killed. The effects of endotoxin (fever, lowered blood pressure, and even shock and death) are now known to be mediated primarily by high concentrations of TNF.

At low concentrations, TNF has a number of critical functions early in the inflammatory process that protect the host against microbial invasion. It increases the diameter and permeability of blood vessels, bringing more blood rich with immunoglobulins, complement, and leukocytes to the site of the injury or invasion. These blood vessel changes are responsible for the redness and heat (increased blood flow) and pain and swelling (accumulation of fluid) characteristic of inflammation. TNF causes the endothelial cell surface of blood vessels to become sticky so that leukocytes initially, and then monocytes and lymphocytes, adhere to blood vessels and accumulate at local sites of inflammation (fig. 6.7). TNF also activates inflammatory leukocytes to kill microbes, stimulates mononuclear phagocytes and other cells to produce cytokines, and has an interferon-like protective effect against viruses.

Interleukins **Interleukins** (IL) may be produced by either monocytes or lymphocytes. The major source of those produced early in infection (primarily IL-1, IL-6, IL-8, and IL-12) is the activated mononuclear phagocyte. IL-1 has a function similar to TNF; that is, it acts on the walls of blood vessels to increase blood flow and brings protective cells and plasma components to the site of inflammation. IL-8 chemically attracts leukocytes (it is a chemotactic factor), and IL-6 and IL-12 activate lymphocytes so that antibody production is increased. IL-6 is also responsible for stimulating the liver to produce the acute phase reactant, C-reactive protein, which as described above, acts to opsonize and kill bacteria. The effects of interleukins produced by *activated* lymphocytes are described in table 6.4.

OTHER FACTORS

The factors in blood just described are important in keeping growth of harmful microorganisms in check until a specific immune response can develop. In some instances, these substances alone can rid the body of the infectious agent and further immunity does not develop. If they fail, however, they have set the stage for the synthesis of immunoglobulins and activation of macrophages and lymphocytes that continue the battle until the host or microbe is successful. These specific processes are described in the subsequent sections of this chapter.

A variety of other factors that are poorly understood may determine the outcome of the interaction of the host and the microorganism. These include age, nutritional status, stress, and genetic factors. As research methods become more sophisticated and immunologists are able to dissect more finely the incredibly complex regulation of the immune system, it is becoming clear that an understanding of genetic control will provide many answers.

SPECIFIC RESISTANCE TO INFECTION

THE IMMUNE SYSTEM: DEVELOPMENT AND FUNCTIONS

Immunity can be defined as a state of specific, relative resistance to infection. The previous sections described *nonspecific* mechanisms by which the host attempts to prevent establishment of infection. **Specific resistance** implies that resistance is directed against a particular type of microorganism. The specificity of resistance is the single most important characteristic of immunity.

In describing the nonspecific mechanisms of resistance to infection, we have spoken continuously of the body's reaction to "foreign" substances or particles. It is the immune system that enables the body to recognize a foreign agent as not being "self" and to take specific action for neutralizing, killing, and eliminating it. This action begins early in the processes of nonspecific resistance and involves some of the same cells and systems, notably lymphocytes and the lymphoid tissues. Sometimes the activity of the immune system leads to tissue damage, as seen in allergic disorders and other hypersensitive states, or autoimmune diseases in which a failure to recognize "self" results in an attack on the body's own cells and tissues. These problems are discussed in the last section of this chapter.

In order to recognize "self," the immune system must be able to distinguish the characteristic proteins or protein-linked components of body cells and tissues from those of an alien substance or agent introduced into the tissues. This distinction provokes certain lymphocytes to produce antibodies specifically directed against the foreign matter, or immunogen. Other lymphocytes become sensitized to the invading immunogen by reacting directly with it, and they also release a number of chemical factors called *lymphokines* designed to destroy it. Thus lymphocytes are responsible for both humoral immunity, mediated by antibodies (immunoglobulins) circulating in the blood, and cellular immunity, induced by intact sensitized lymphocytes and their products.

Before examining the mechanisms of humoral and cellular (cell-mediated) immunity, we will consider the nature of antigens, immunogens, and antibodies, and of the lymphocytes involved in these immune reactions.

Antigens, Immunogens, and Antibodies

Antigens and Immunogens

Immune responses arise as a result of the body's exposure to foreign substances, usually protein macromolecules. A compound that evokes this immune response is referred to as an antigen or immunogen, but there is a functional distinction between these two terms. An **immunogen** is any substance that induces an immune response *and* specifically binds to the resulting antibody. In contrast, an **antigen** is an agent that binds specifically to an antibody but may not evoke the immune response. Thus, all immunogens are antigens, but not all antigens are immunogens.

Although the terms are used interchangeably (and will be throughout this text), the distinction between immunogens and antigens is necessary. For example, low-molecular-weight compounds such as antimicrobial agents are not immunogens because, by themselves, they cannot induce immune responses. If the antimicrobial agent is coupled with a much larger protein molecule, however, the resulting conjugate may induce an immune response directed against the various parts of the conjugate, including the antimicrobial agent. When this type of conjugate is formed, the low-molecular-weight compound is referred to as a **hapten** (from the Greek word *haptein,* meaning "to grasp"), and the larger molecule is referred to as a *carrier.*

The uptake and processing of immunogens by tissue macrophages and specialized, branching **dendritic cells** in lymph nodes appears to be an initial critical step in most immune responses. Simple ingestion of foreign proteins, for example, does not generally result in antibody formation because the proteins are broken down in the digestive tract before their components are absorbed into the tissues.

The chemical structure of an antigen determines the specificity of its interaction with an antibody. Various chemical groups on the antigen molecule are responsible for its immunogenicity; that is, they are its **antigenic determinants,** or *epitopes,* accounting for its effectiveness in stimulating antibody formation and for the specificity of its reactions. Although the basic constituents of all living protoplasm are proteins, carbohydrates, and lipids, each species is chemically and antigenically unique. Proteins, in particular, afford unlimited variety because their constituent amino acids interlock in many distinct combinations (see chap. 2). Thus the unique proteins of one species are foreign to those of another and can provoke an antibody response if introduced into the tissues of the other species. In human beings, this situation occurs most commonly with microbial antigens. Even within the same species, tissue antigens among individuals, referred to as *histocompatibility antigens* (*histo* is Greek for tissue) are distinct. The ABO red blood cell group antigens that determine part of a person's blood type are another example. When blood is transfused or organs are transplanted from one individual to another with different ABO group or histocompatibility antigens, severe hemolytic reactions and tissue rejection, respectively, occur.

Microbial Immunogens and Antigens

Many constituents of microorganisms are protein in nature, are linked to protein, or are large carbohydrate molecules and, therefore, immunogenic. The bacterial cell has been described as a mosaic of antigens in three dimensions. Furthermore, many of the substances produced by microorganisms as metabolic by-products are immunogenic. A given organism may contain or produce many different immunogens that can induce the formation of different antibodies, but antigens are also present. For example, a bacterial cell may include these distinctive immunogens or antigens (all are referred to as antigens for convenience):

Somatic antigens *of the cell body (soma) itself. These are sometimes called O antigens. In some bacteria they are the endotoxins of the cell. Some cellular antigens occur only on the surface of the cell and are called surface antigens.*

Capsular antigens *in species that form a capsule.*

Flagellar antigens *(also called H antigens) in motile species.*

Exotoxins *and other protein substances secreted by some species.*

Antibodies produced in response to such characteristic antigens in the infected patient help to provide diagnostic proof of an infectious disease (see Interactions of Antigens and Antibodies, following).

Antibodies

The definition of an antibody is the converse of that of an antigen. An antibody is a specialized protein substance produced by an animal host in response to a foreign antigen in its tissues, and it is capable of reacting specifically with that antigen.

Antibodies are immunoglobulins (Igs) produced primarily by plasma cells, which are derived from highly specialized lymphocytes (see B Lymphocytes, following) in response to the presence of antigen. They circulate in blood and have a number of different biologic functions.

Structurally immunoglobulins are made up of four polypeptide (multiply linked amino acids) chains arranged in a Y pattern. Each molecule has two chains of heavy molecular weight (H chains) and two of light molecular weight (L chains) linked together by disulfide bonds (fig. 6.8a). The outer ends of each arm of the antibody molecule carry a unique amino acid sequence. These reactive or antigen-combining sites bind the antibody molecule to its antigen and are referred to as the *Fab region* (*F* for fragment, *ab* for antigen-binding). Antibody molecules have two such combining sites; that is, they are bivalent but some types of antibody form aggregrates of two (IgA) or five (IgM) such structures (see later section, Classes of Immunoglobulins). Thus the number of combining sites of different immunoglobulins varies (for example, 10 for IgM, fig. 6.8b). Certain antibodies with only one combining site are called *incomplete* or *blocking antibodies.* The larger the number of combining sites, the more avid and effective is the antibody in its interactions with an antigen.

The remaining portion of the antibody Y is called the *Fc region* because this protein fragment can be readily crystallized (*F*ragment *c*rystallizable). It has an amino acid sequence that is constant and characteristic for its class. In the major classes of antibody (IgG and IgM), it is this Fc portion of the molecule that activates complement and gives it opsonic properties. In the latter case, when the antigen-binding sites have attached to a microbial surface antigen, coating the microorganism, the free end of the antibody molecule mediates linkage of the coated particle to a neutrophil or macrophage, thus promoting phagocytosis.

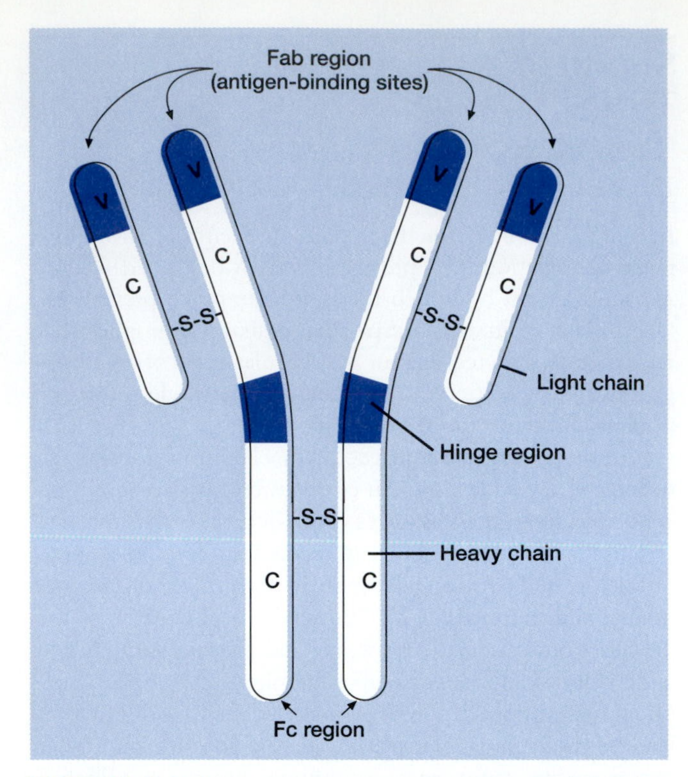

FIGURE 6.8(a) Diagram of the Y structure of an immunoglobulin molecule (IgG). The shaded areas marked V indicate sites where amino acid sequences are variable and specifically reactive for antigen (Fab region). The clear sites marked C represent constant amino acid sequences characteristic for an immunoglobulin class. Disulfide bridges are marked S–S. The "hinge" region provides the molecule with flexibility, permitting it to swing in or out for matching and binding to antigenic sites. The tail of the Y is known as the Fc region.

Antigen-Antibody Specificity

The surfaces of the combining sites of the antibody molecule are reciprocal, or converse, to the structure of the antigen molecule, especially its determinant groupings. Therefore, when antigen and antibody combine, their surfaces dovetail like two pieces of a jigsaw puzzle or a key in a lock (fig. 6.9). **Antigen-antibody specificity** refers to this complementary arrangement of their surface structures, an arrangement that provides the basis for the specificity of immune reactions.

The reciprocal structure of the antigen-antibody combining sites also accounts for the fact that differences in antigens are "recognized" by their respective antibodies; thus only specific combinations occur. The specificity of immune reactions is often exquisitely accurate in detecting even subtle molecular differences in the combining pairs. Such reactions manipulated in vitro (in the test tube) sometimes afford the only means by which protein materials can be shown to differ. The immunity of human or animal hosts to infectious agents is similarly exact; different microbial antigens induce the formation of specifically different antibodies. Conversely, antigenic similarities in closely related microbial species are detectable in overlapping antigen-antibody reactions.

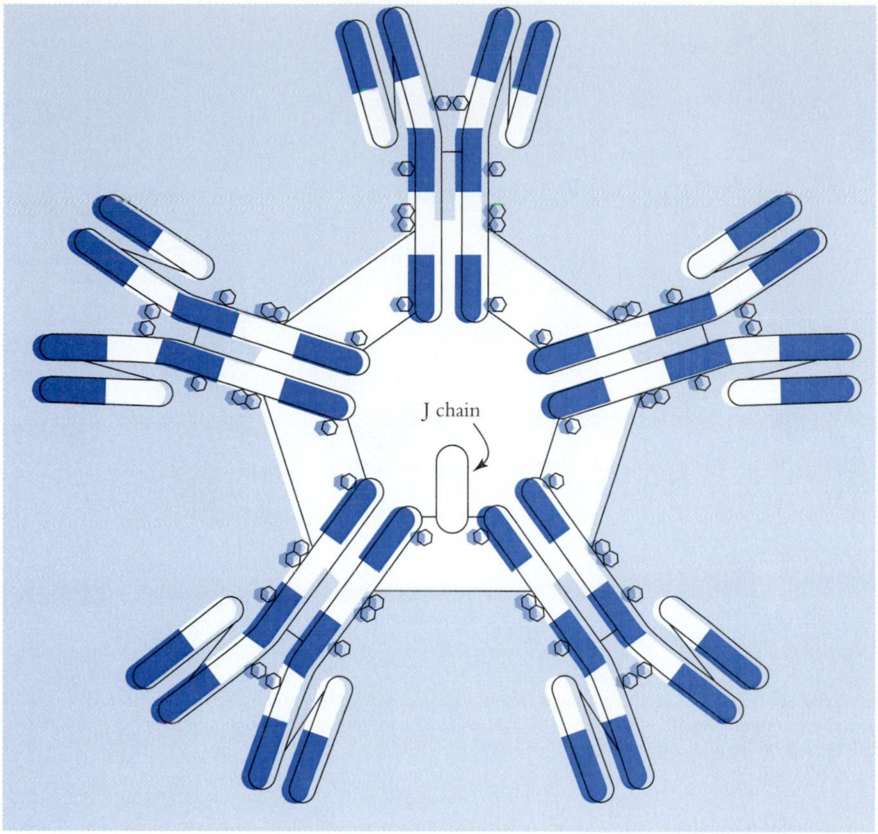

J chain

FIGURE 6.8(b) IgM structure illustrating the pentameric arrangement of the basic immunoglobulin Y-shaped units.

Classes of Immunoglobulins

Five different classes of immunoglobulins are found in plasma or serum. (Serum is the clear portion of blood remaining after clot formation.) They are named and abbreviated according to the types of heavy chains they possess, each designated by a Greek letter: *gamma* = IgG; *mu* = IgM; *alpha* = IgA; *epsilon* = IgE; and *delta* = IgD. Each class possesses different antibody activity and serves specialized functions.

IgM is the major component of the primary antibody reaction and appears first in a given immune response (fig. 6.10). It is a very large molecule with a high molecular weight, its structure consisting of five units joined together in circular fashion by a J (joining) chain (fig. 6.8b). Because of its large size, IgM is unable to diffuse through fine cell membranes and is found almost exclusively in the bloodstream. Although it comprises only about 7% of total plasma Ig, 90% of IgM is present in the vascular system. IgM has more antigen-binding sites than the IgG molecule shown in figure 6.8a, and so is more avid in combining with foreign invaders. It also has a greater capacity for activating complement than do smaller molecules, and thus it is a better opsonin. This powerful antibody is formed early in response to a new infection; therefore IgM is particularly valuable in giving the host a head start in the race to prevent an invading microorganism from gaining a permanent foothold or from spreading. When the secondary immune response gets underway, IgM is gradually replaced by newly formed IgG antibodies. For this reason the detection of IgM for a particular microbial antigen is indicative of either recent or continuing infection. For example, the presence of IgM antibodies for rubella in a pregnant woman with a history of recent rubellalike illness indicates that she did (or does) indeed have rubella, whereas the presence of rubella IgG alone would mean that she is probably not currently infected or has passed the infectious stage of the disease.

IgM antibodies are also the first to be formed if the fetus becomes infected during its uterine development. Antibody formation becomes possible after the fifth or sixth month, and when such formation occurs, the antibodies are of the IgM type. Accordingly, the presence of antimicrobial IgM in the cord blood of a newborn is indicative of intrauterine infection, since these antibodies cannot be transferred across the placenta from the mother's blood.

The natural blood group antibodies (ABO) are of the IgM class. In addition, specific antitoxins directed against the exotoxins of a number of microorganisms, including those that cause diphtheria, tetanus, and botulism, are composed of IgM, together with IgG.

IgG is the principal antibody of the secondary immune response, appearing upon subsequent exposures to the same antigens that initially stimulated IgM production (fig. 6.10). Approximately 75 to 80% of the Igs circulating in the bloodstream are IgG, and this antibody also finds its way

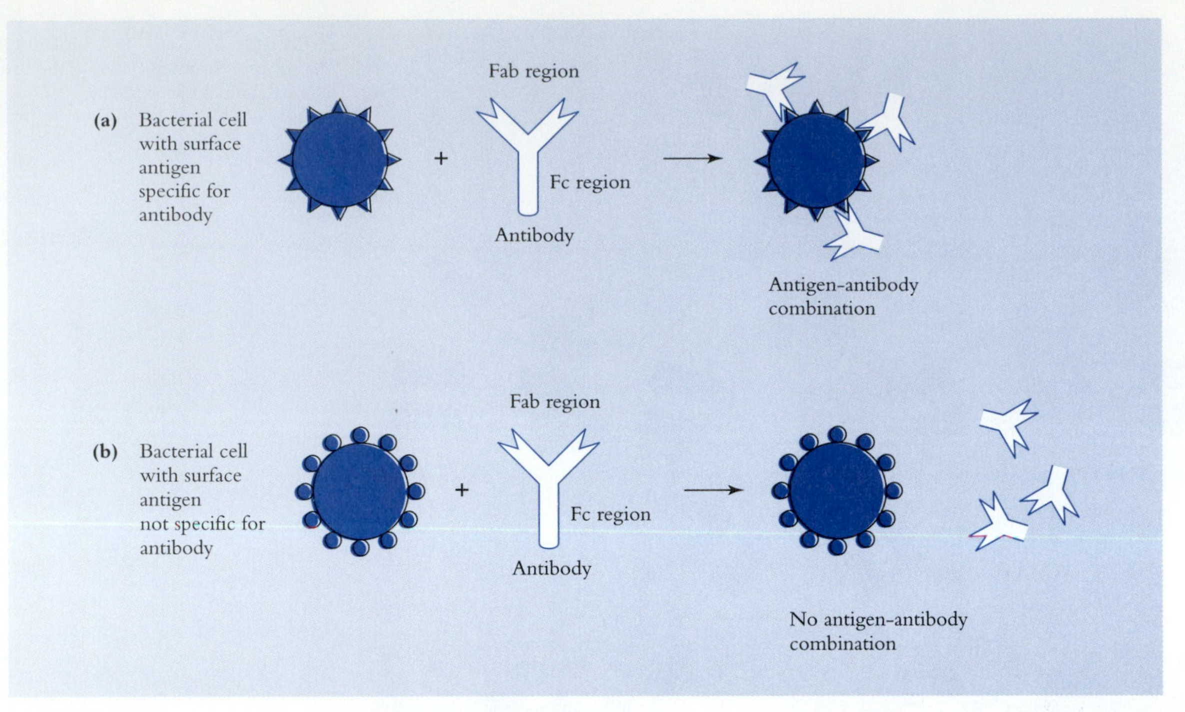

FIGURE 6.9 The specificity of antigen-antibody reactions is explained by the complementary arrangement of their surface structures. In (a), the Fab region of antibody is reciprocal in structure to bacterial surface antigens and, therefore, antigen and antibody combine. In (b), no antigen-antibody reaction takes place because the bacterial surface antigens and Fab region of antibody are not complementary.

into other internal secretions such as spinal, synovial, and peritoneal fluids and lymph. IgG is active against all the major microbial antigens. It is the only Ig class capable of crossing the placental barriers; therefore it serves to protect the fetus and newborn against those infections for which the mother has developed this type of antibody (naturally or through artificial immunization). This form of transferred (or passive) immunity of newborns is only temporary; but by the time it has waned, they are usually capable of producing their own IgG antibodies.

IgA is found to only a slight extent in serum, but secretory IgA (sIgA) is the principal immunoglobulin in external secretions such as those of mucosal surfaces (respiratory, gastrointestinal, urogenital tracts) and tears, saliva, bile, urine, and colostrum. Whereas IgM is a five-unit structure, sIgA consists of two units held together by a J chain and a so-called secretory component. The latter helps transport sIgA into external secretions and may protect it from degradation by gastrointestinal enzymes. These antibodies are important in resisting infections of the surface tissues of the body and increase in amount during mucosal infections. They enter the blood by way of the lymphatic system, so serum IgA levels are raised at the same time. Maternal sIgA in colostrum and milk helps to boost the passive immunity of the newborn, who has already acquired IgG antibodies by way of placental transfer in utero.

IgD is found in only small quantities in serum, and therefore has been difficult to study. Serum levels of this Ig increase during chronic infections, but IgD increases have not been associated with any specific infection. IgD, along with IgM, is found in the cell membrane of human B lymphocytes, where it functions as an antigen receptor in the initial stages of the immune response. IgD is thought to activate the B cell, a process that ultimately leads to immunoglobulin secretion.

IgE antibodies occur in only slight concentrations in serum but play an important role in hypersensitivity of the immediate (anaphylactic) type. This is a damaging role, inducing tissue injury. The IgE molecule has an affinity for attaching itself, by the Fc portion, to certain cells of the body, notably mast cells and basophils. When its exposed antigen-specific combining sites (Fab) bind with an antigen, the combination triggers the release of biologically active substances, including histamine, from the mast cell or basophil. The results are damaging to surrounding tissues and responsible (depending on the degree and site of the reaction) for the symptoms of such conditions as anaphylactic shock, allergic asthma or rhinitis (hay fever), urticaria (hives), and allergic (atopic) dermatitis. Attachment of IgE to mast cells and basophils in the skin and other organs sensitizes them to the antigens that induce these allergic responses. Such antigens are referred to as *allergens.* IgE has been called, variously, *reaginic* antibody, *atopic* antibody (atopy refers to allergy, or, literally, "out of place" reactions), *skin-sensitizing* antibody, or *anaphylactic* antibody. Allergies are discussed in the section of this chapter covering hypersensitivity.

Not all effects of IgE are harmful to the host. Once this immunoglobulin binds to mast cells, chemicals are released that act on blood vessels to increase blood flow and bring antibodies and other inflammatory cells to sites of infection. Still other substances induce contraction of smooth muscles in the intestinal and respiratory tracts resulting in diarrhea

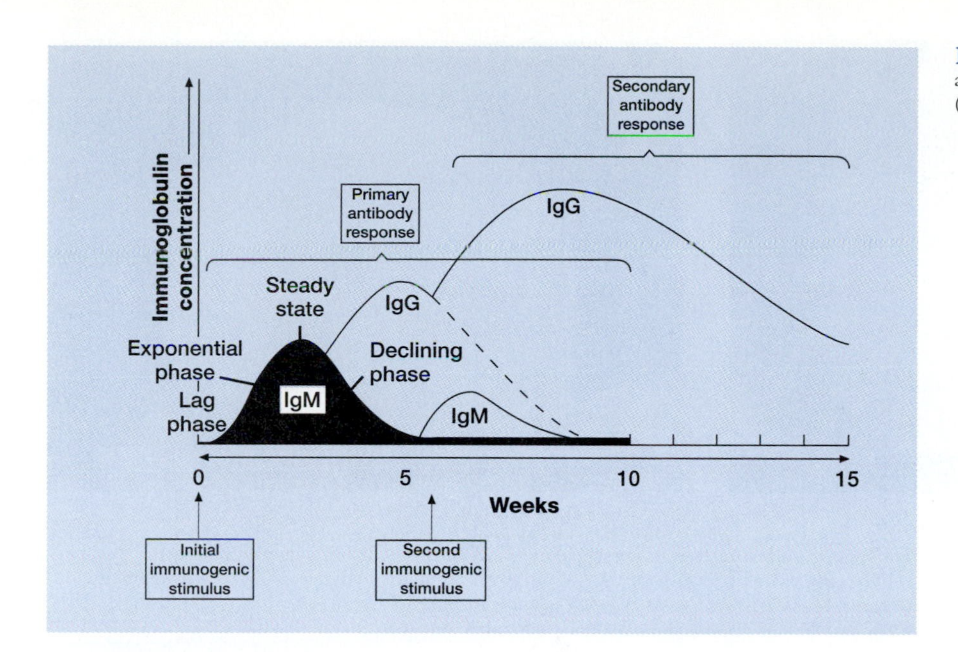

FIGURE 6.10 The primary and secondary antibody responses to an immunogenic (antigenic) stimulus.

TABLE 6.3 IMPORTANT PROPERTIES OF IMMUNOGLOBULINS

Property	IgM	IgG	IgA	IgE	IgD
Heavy chain	Mu	Gamma	Alpha	Epsilon	Delta
Molecular weight	900,000	150,000	160,000–450,000	190,000	180,000
Serum concentration (mg/100 ml)	100	1,100	250	0.01	3
Placental transfer	–	+	–	–	–
Activation of complement	+	+	–	–	–
Beneficial functions	Agglutinating Activates complement Provides antigen receptor on B lymphocytes	Neutralizes toxin, viruses, bacteria Agglutinating Opsonizing Activates complement	Neutralizes toxin, viruses Blocks intestinal entry of bacteria	Brings about changes in vascular permeability Sensitizes mast cells Role in infestations by intestinal pathogens	Provides antigen receptor on B lymphocytes Activates cells to produce immunoglobulins
Harmful effects	Complexes formed with antigen can induce injury, e.g., serum sickness, glomerulonephritis	Same as IgM	?	Immediate hypersensitivity, local or systemic, including anaphylaxis	?

or vomiting to expel intestinal pathogens or increased secretion of mucus and bronchial contractions in the lungs to more effectively deal with respiratory pathogens. In infestations with intestinal worms (see chap. 18) the levels of IgE increase, indicating that it plays an important role in the host's responses.

Some of the important properties of the major classes of immunoglobulins are summarized in table 6.3.

Immune Functions of the Lymphoid System

The body's immune responses to antigenic stimuli are a function of the lymphoid system and its specialized cells. Like other cells of the circulating blood, lymphocytes are derived from stem cells located in the bone marrow.

Stem cells are the undifferentiated precursors of the mature red and white blood cells found normally in the circulating blood. Undifferentiated stem cells proliferate throughout life as a self-renewing reservoir, replenishing the pool of mature cells used up during normal activities. These stem cells are considered *pluripotent* because they can develop into any one of several more differentiated cell lines (see fig. 6.14). Differentiation is regulated by a variety of soluble factors in serum that control the extent and direction of maturation. Once a cell has differentiated in any direction, it is *unipotent;* that is, committed to becoming a cell of a single type only.

The two major pathways of cell differentiation are myeloid and lymphocytic. In myeloid differentiation, stem cells originating in the bone marrow develop into the

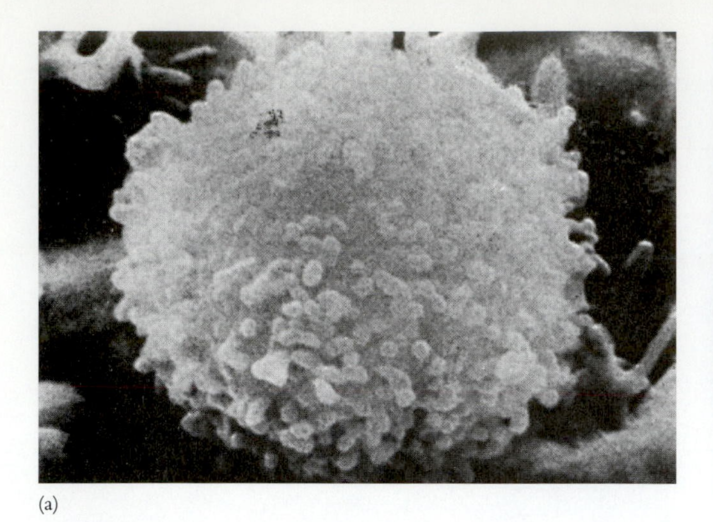
(a)

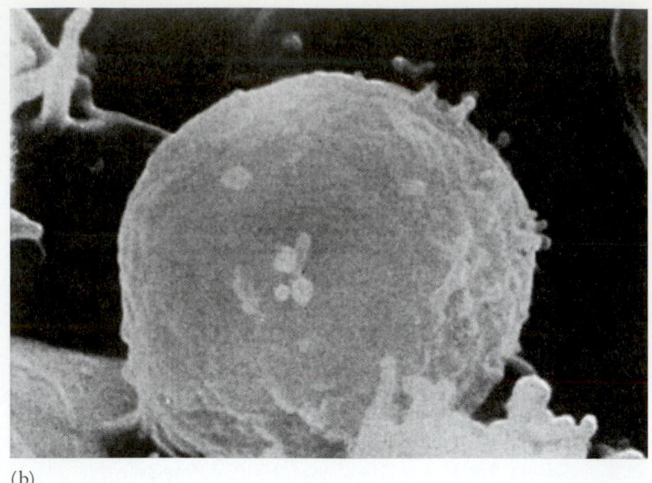
(b)

FIGURE 6.11 The membrane surface of the B lymphocyte (a) is covered with immunoglobulin, which is specific for a distinct antigen (b). T lymphocytes become sensitized by certain antigens and may interact with these antigens in a way protective to the body providing cellular immunity.

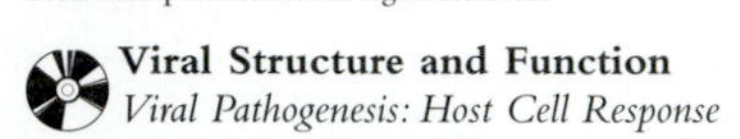
Viral Structure and Function
Viral Pathogenesis: Host Cell Response

precursors of erythrocytes, platelets, eosinophils, basophils, neutrophils, and monocytes. In lymphocytic differentiation two distinct cell types are produced, each involved in the development of different immune responses. One type of population is referred to as **B lymphocytes,** or **B cells,** because in birds (where they were first recognized) the differentiation takes place in a lymphoid organ called the bursa of Fabricius. The avian bursa is associated with the intestinal tract. Its equivalent in human beings and other animals has not been positively identified but is thought to be primarily the bone marrow. The other type of lymphocyte population is comprised of **T lymphocytes,** or **T cells,** so called because their conversion occurs in, and is dependent on the thymus gland. The bone marrow and thymus are considered *primary* lymphoid organs.

The transformation of stem cells into B and T lymphocytes is probably complete in humans in the first few months after birth. These cell populations then migrate to *secondary* lymphoid organs, the lymph nodes and spleen. Eventually, the B cells form the antibodies that provide humoral immunity. T cells become sensitized lymphocytes, responsible primarily for cellular immunity, though they often play a role in humoral immunity as well (fig. 6.11).

Phagocytic cells of the MNPS are also important for B and T cell function. Phagocytized antigens are processed intracellularly and then presented (on the phagocyte surface) to lymphocytes to initiate specific immune responses. Phagocytes are also a major component of the inflammatory response in cellular immune reactions. In addition, dendritic cells in the lymph nodes appear to present viral antigens to T cells.

Clonal Selection Theory

In the past, several theories were proposed to explain how lymphocytes recognize immunogens and antigens and subsequently produce highly specific antibodies. In the 1950s Sir McFarlane Burnet developed the **clonal selection theory,** which has been confirmed by experimental methods. According to this theory, certain small populations, or clones, of lymphocytes bear distinct antibody molecules (receptors) on their cell membranes. Production of this surface antibody is genetically determined and not dependent on prior contact with any antigen. When such lymphocytes encounter an antigen matching up with their surface receptor, the antigen binds there and stimulates the lymphocyte to proliferate and differentiate into antibody-producing cells (as will be described in detail in the next section). The antibody produced then reacts with the same antigen that stimulated its formation. Thus, the human and animal body contains lymphocyte clones with surface antibody receptors (primarily IgM with some IgD) capable of recognizing the wide variety of foreign materials likely to be encountered throughout life.

The essential features of the clonal selection theory may be summarized as follows: (1) B and T lymphocytes of all antigenic (immunogenic) specificities exist before contact with antigen (immunogen). (2) Each lymphocyte carries immunoglobulin or T cell receptor molecules of a single specificity on its surface. (3) Under appropriate conditions, antigens can stimulate lymphocytes to produce progeny with identical antigen specificity. (4) Lymphocytes potentially reactive with "self" antigens are deleted or in some way inactivated to ensure that no immune response to host components develops.

B Lymphocytes

B lymphocytes are the antibody-synthesizing cells that originate from stem cells and that are responsible for humoral immunity. During development, each B lymphocyte acquires a genetically determined surface immunoglobulin specific for a distinct antigen. Upon exposure to its specific antigen, the B lymphocyte activates and differentiates in one of two ways: proliferating cells may differentiate into plasma cells or some may transform back into resting mature B lymphocytes. Plasma cells are the specialized, fully differentiated cells that synthesize and secrete antibody with a single antigen-binding specificity (identical V regions, fig. 6.8). The resting mature B lymphocytes, known as memory B cells, can later be activated to differentiate into plasma cells that secrete antibody at an accelerated rate.

As these antigen-induced processes take place, the phenomenon of *class switching* also occurs. Progeny of B lymphocytes involved in the early IgM response switch over to synthesize IgG, IgA, or IgE molecules under the influence of T cell factors. Thus, a number of cells produce different immunoglobulin types but the progeny of a particular clone all produce antibody with the same antigenic specificity.

B lymphocytes are distinguished from other lymphoid cells by the presence of immunoglobulins on their membrane surfaces. These surface membrane immunoglobulins function as specific antigen receptors and serve to mark or identify different B lymphocyte populations. Most B lymphocytes in the spleen and peripheral blood of mature animals exhibit IgM and IgD, which bind antigen and may independently determine how antigen is recognized as foreign (nonself and, therefore, immunogenic). Antigen binding to surface immunoglobulins may be responsible for activating and differentiating B lymphocytes into antibody-producing plasma cells and memory B cells. Sometimes this cell activity induces a state of tolerance during which no antibody is produced against the immunogen. The plasma cell, once produced, secretes antibodies with the same antigen-binding properties as the antigen receptor molecules on the surface of the parent B lymphocyte. Thus, the process of B lymphocyte selection and activation by antigens lends support to the clonal selection theory.

Kinetics of Antibody Production

Antibodies for a particular immunogenic substance are not detectable in serum until an individual has been exposed to the immunogen. Even then, contact with the immunogen does not invariably lead to an immune response; a number of factors may influence whether an immune response is generated. These factors include (1) the immunogenicity of the antigen; (i.e., its ability to be recognized as foreign by the recipient); (2) its physicochemical nature; (3) the dosage received; (4) the number and frequency of exposures; and (5) the particular animal species involved. Certain characteristics of the actual immune response to an antigen, however, are common to all humoral antibody responses.

The first exposure to a particular antigen is referred to as the *primary event* and the antibody response that ensues is called the *primary response* (see fig. 6.10). The primary response includes several stages. The first is a latent or lag phase in which a variable time, generally one to two weeks, elapses before antibody is detectable in serum. The latent period includes the time needed for T and B cells to make contact with the antigen and for the B lymphocyte to proliferate and differentiate into plasma cells. Plasma cells must also secrete a sufficient quantity of antibody to be detected in serum. If an antibody assay with low sensitivity is used, more antibody will be required for detection and the latent phase will appear longer. During the exponential phase, the concentration of antibody in serum increases exponentially. In the primary response, IgM is the first class of antibody produced and, in some instances, it is the only class of immunoglobulin made. If IgG antibody is produced, it generally appears at about the same time that IgM production stops. The steady state stage is characterized by a balanced production and degradation of antibody and is followed by the declining phase in which the immune response begins to shut down and the concentration of antibody in serum declines rapidly.

Although production of antibody after primary contact with immunogen may cease entirely after several weeks, the immunized individual is left with a "memory" of this antigen exposure in the long-lasting memory B lymphocytes. Later, when the person is exposed again to the same antigen, the secondary immune response develops; here the antibody is produced faster and in higher concentrations than in the primary response. In addition, the response matures such that antigen has a greater affinity for binding to the antibody. Antibody production may also continue for a longer time, and serum levels may persist for months or even years.

The secondary response is often called the **anamnestic response,** from the Greek term *anamnesis,* meaning "recall" or "memory." In anamnestic responses there is little or no lag period and higher concentrations of antibody are generally produced that last longer than those in the primary response. The anamnestic response provides a selective immunologic advantage for people who survive the first contact with an infectious agent and then are reexposed to the same pathogen at a later time. In addition, the anamnestic response serves as the basis for immunization programs and explains the need for and effectiveness of booster shots (see fig. 6.10).

T Lymphocytes

The conversion of lymphoid stem cells into T lymphocytes depends on an intact thymus, at least in the early years of life. Later, T cells find their way to the lymph nodes, spleen, bone marrow, and the circulating blood. In the blood they represent the larger portion (65 to 80%) of the small lymphocytes found there. Their life span is much longer (months or years) than that of B cells.

T lymphocytes participate in many processes; among those processes are the tuberculin skin reaction, rejection or acceptance of certain tissue grafts, the regulation of immunoglobulin production, and defense against certain microorganisms and cancers. T lymphocytes are involved in essentially all immune reactions either as effector (directly attacking) cells or as regulators of humoral and cellular responses.

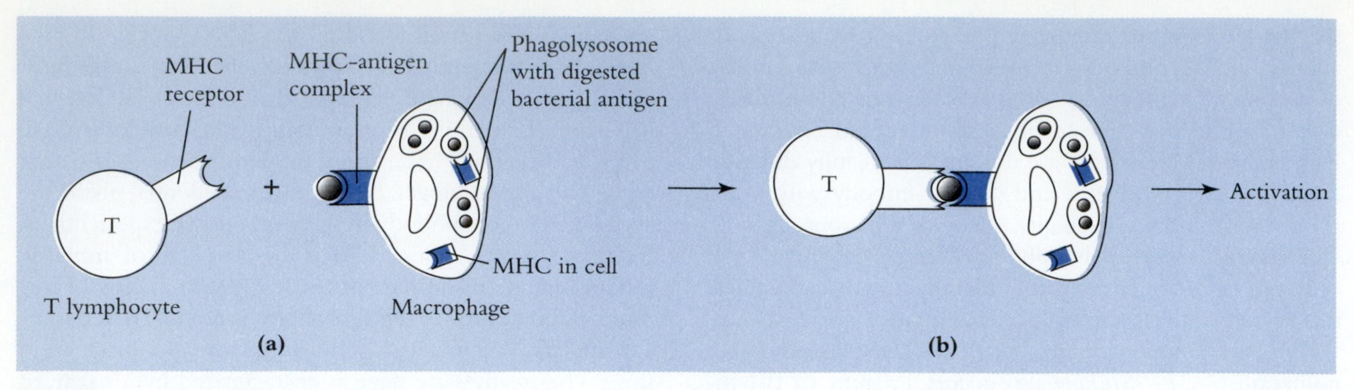

FIGURE 6.12 (a) Bacterial antigen digested in the macrophage is carried to the phagocyte cell surface bound to the major histocompatibility complex (MHC). (b) The MHC receptor on the T lymphocyte combines with the MHC-antigen complex. The T cell becomes activated to divide and produce lymphokines, which in turn activate the macrophage for more efficient bacterial killing.

Thymus-derived T cells do not secrete immunoglobulins. When they react with certain antigens, they become sensitized, or "immunologically committed." They may then interact with those antigens in a way that is protective to the body, providing cellular immunity (often referred to as **cell-mediated immunity,** or **CMI**). These interactions are called into play particularly for defenses against intracellular microorganisms involved in chronic, persistent infections. CMI reactions are also involved in the T cell recognition of tumor proteins produced by cancer cells. As cancer cells multiply in the body, their normal protein antigens become altered and are treated as "nonself" by specialized T cells (called *killer* or **cytotoxic T cells**), which then attempt to destroy them. These cells are also responsible for the rejection of foreign grafts and tissue transplants. CMI mechanisms produce harmful effects in the body as well. T cells induce the changes seen in delayed hypersensitivity, including those of some infectious diseases (e.g., tuberculosis) and of autoimmunity.

Besides helper and killer cells, other T cells act to suppress the immune response in certain instances, and may be responsible for preventing autoimmune disease (immune reactions against "self" antigens). These cells are called the *suppressor T cells.* These cells, however, have also been found to inhibit antitumor responses, thereby enhancing tumor growth. Interactions of T helper and T suppressor cells are important for regulating the immune response.

T Lymphocyte Maturation

In the thymus, T cells develop and differentiate in distinct stages during which they acquire a number of genetically determined T cell surface markers. The surface markers are associated with specific steps in the development and activation of mature T cell subpopulations or subsets.

In mammalian species, a cluster of genes is located on a single chromosome in a region known as the **major histocompatibility complex** (MHC). These genes encode the major histocompatibility (tissue) antigens found on body cells. MHC antigens are recognized by different T cell types and are essential for immune recognition reactions. Most mature T cells cannot recognize free foreign antigens circulating in blood, but respond to antigens only under particular conditions. The foreign antigen and the MHC must both be displayed on a cell surface for a T cell immune response to occur. Two important cells that process foreign antigens and present them to T cells are macrophages, which process bacteria and other particulate antigens, and dendritic cells, which present viral antigens. Thus, the antigen receptor on T cells differs from the antigen receptor on B cells. The B cell receptor is immunoglobulin, which binds free antigen, whereas the T cell receptor binds to a complex composed of antigen and a MHC molecule on an antigen-presenting cell (fig. 6.12).

Once the T cell encounters its specific antigen in the manner just described, it becomes activated and produces a lymphokine known as interleukin 2 (IL-2). As described earlier (see Cytokines), lymphokines are a subset of proteins known as cytokines, which mediate natural immunity or are secreted in response to antigen and affect the behavior of various cell types. IL-2 causes the activated T cell to multiply rapidly, giving rise to thousands of lymphocytes with the same antigen receptor, and to further differentiate so as to carry out (effect) specific functions. These are referred to as effector T cells.

T Cell Subsets

T cells all contain surface antigens that provide the basis for their classification using "cluster of differentiation" (CD antigens) nomenclature. Cells with CD4 and CD8 antigens are important effector cells that are specialized to deal with microbial pathogens. CD8 T cells, also referred to as cytotoxic or killer cells, destroy cells that are infected with viruses. CD4 cells are divided into two groups, inflammatory T cells (T_H1) and helper T cells (T_H2). T_H1 cells activate macrophages, which can then more readily destroy phagocytized intracellular pathogens. Thus, T_H1 cells are important in effecting cell-mediated immunity. On the other hand, T_H2 cells activate B cells to secrete immunoglobins and thus effect the humoral immune response. Cytotoxins and lymphokines produced by CD8 and CD4 cells, respectively, and their effects on the immune response are listed in table 6.4.

TABLE 6.4 CYTOTOXIC EFFECTS AND LYMPHOKINES OF T LYMPHOCYTES

Cell Type	Agent	Activity
CD8 (cytotoxic, killer cell)	Cytotokines	Make holes in cell membranes; enzymes disrupt cell proteins
	Interferon-γ	Activates macrophages, increases MHC activity, inhibits viral replication
CD4		
T_H1	Interleukin-2	Causes proliferation of T cells; activates T_H and B cells
	Interferon-γ	Activates macrophages, increases MHC activity
	Tumor necrosis factor-β	Kills cells, activates macrophages
T_H2	Interleukin-4	Activates B cells, mast cells; increases IgG1 and IgE production; promotes growth and survival of T cells
	Interleukin-5	Promotes IgA synthesis; increases eosinophils
	Interleukin-6	Promotes growth and differentiation of T and B cells, release of acute phase reactants
	Interleukin-10	Stimulates mast cells; inhibits T_H1 cells, increases MHC activity

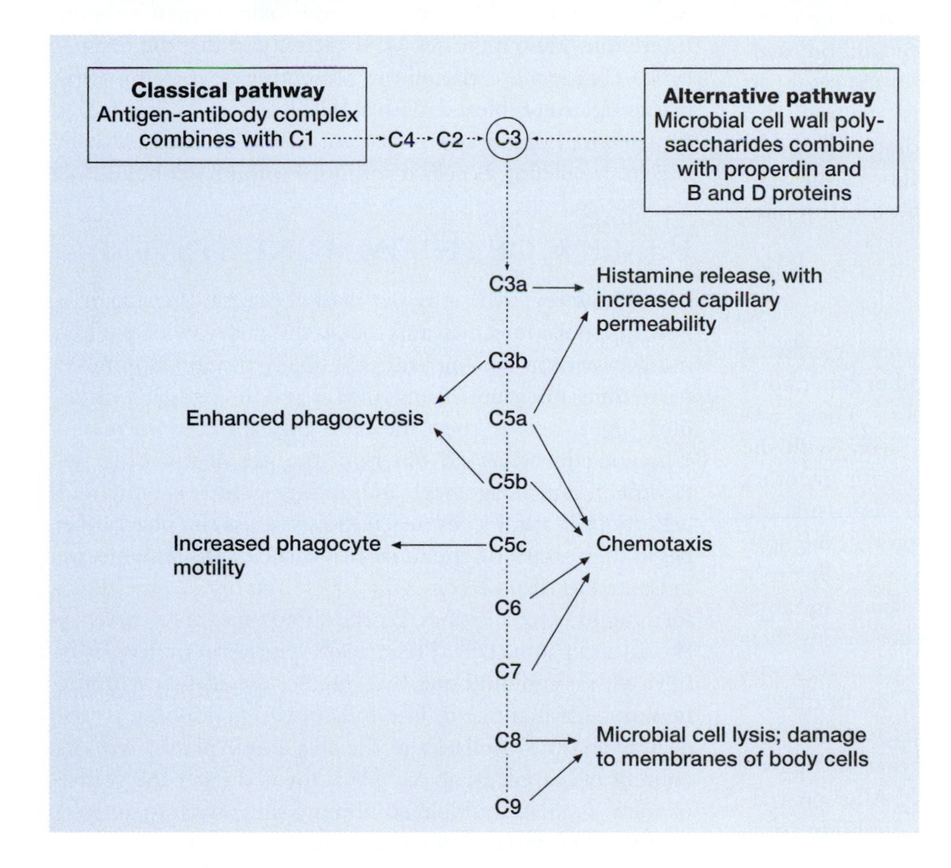

FIGURE 6.13 Role of complement and properdin in the immune system: activation pathways and functions.

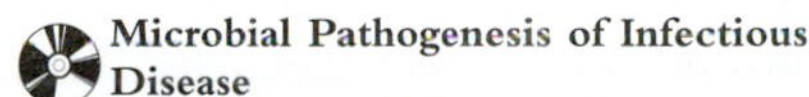
Microbial Pathogenesis of Infectious Disease
Nonspecific Host Defenses: Cellular Defenses and Phagocytosis

CD4 cells are also of special interest and importance in AIDS (see chap. 19). They are a primary target for the human immunodeficiency virus, which attaches to a receptor site on CD4 cells and destroys them. As a result, the immune functions carried out by these cells (table 6.4) are no longer active, and patients with AIDS become infected by many microorganisms that normal hosts are able to resist.

Role of Complement and Properdin in the Immune System

Complement (C) and **properdin** are proteins that occur naturally in the bloodstream and play important roles in the human defense system. Complement consists of a complex series of enzyme proteins numbered C1 through C9, with intermediate fragments split off during activation indicated by a letter (e.g., C3b, C5a). In the classical complement pathway, the first component, C1, is activated after it combines with a complex of antibody bound to antigen. The subsequent order of proteins activated is C4, C2, C3, then sequentially through C9. In the properdin or alternative complement pathway, an antigen-antibody complex is not necessary; instead activation is accomplished by certain polysaccharides (primarily from bacterial and fungal cell walls). The sequence begins with the activation of three distinct proteins—P, B, and D—then C3, and continues on as in the classical complement pathway (C5 through C9).

As the complement cascade (as it is often called) proceeds, particular complement proteins perform a variety of functions related to the tissue inflammatory process (fig. 6.13):

1. C3a and C5a cause histamine release from mast cells and increased capillary permeability.
2. C5a, C5b, C6, and C7 attract neutrophils and macrophages to the site of injury.
3. C5c causes increased motility of leukocytes, but not lymphocytes.
4. C3b and C5b cause enhanced phagocytosis of microorganisms and their antigens; this is one of the most critical complement activities.
5. C8 and C9 result in cytolysis—that is, lysis of microbial cells (primarily gram-negative bacteria)—and also in damage to membranes of other body cells.

Since the properdin pathway does not require the presence of specific antibody, the protective action of the complement system can be initiated early (before antibody synthesis) to help combat microbial intruders. For the most part the complement effects are directed against the cells that initiated their sequential activity. The cascading sequence is controlled by inhibitors and by the instability of the activated components so that they cannot diffuse to produce tissue damage.

Humoral immunity

Humoral immunity is the immune response conferred by specific antibodies in the circulating blood or other body fluids, which interact with foreign antigens. These immunoglobulins are designed to react exclusively with the immunogen that induced their formation or with any other antigen having a nearly identical molecular configuration. Antibodies can act in one of three ways to protect the host: they can neutralize an organism or its toxin by binding to it and thus prevent it from attaching to and damaging target cells; they can opsonize (coat) bacteria, thus permitting their ready ingestion by macrophages; or they can activate complement. When complement is activated, an antibody-complement complex is deposited on the bacterial surface. The result is that the microbial cell lyses or is taken up by a macrophage or both may occur sequentially. After neutralization of organisms or toxins, the antigen-antibody complex is also taken up by macrophages. Both IgM and IgG activate complement but IgG has the capacity also for neutralization and opsonization. IgA can neutralize foreign microbes at body surfaces.

Humoral immunity is especially effective against those microorganisms that are rapidly taken up and destroyed in phagocytes (such as staphylococci and streptococci) and against bacteria that release potent toxins (diphtheria and tetanus bacilli). Humoral immunity can be transferred passively from one individual to another since antibodies are present in cell-free plasma.

Cellular immunity

Cellular immunity, or the cell-mediated immune response (CMI), is a function of the T lymphocytes (thymus-derived cells). As described earlier (see T Lymphocyte Maturation), when a T cell reacts to antigen-MHC complex on an antigen-presenting cell, it divides into clones of identical cells that release chemical mediators called lymphokines (table 6.4).

Lymphokines perform various actions with important end results. For example, interferon-γ produced by CD8 cytotoxic cells inhibits viral replication in infected body cells and also activates macrophages. The interferon-γ produced by CD4 inflammatory cells (T_H1) also plays a critical role in activating macrophages, enabling them to kill intracellular microorganisms that resist degradation in nonactivated cells. When activated, the macrophages fuse their lysosomes more efficiently into phagosomes, thus exposing the ingested bacteria to bactericidal lysosomal enzymes (see earlier section on Phagocytosis) and toxic oxygen radicals. Interferon-γ also increases MHC activity so that the phagocytic cells are more efficient in presenting degraded microbial antigen complexed with MHC to T cells (fig. 6.12). Other lymphokines are important in killing infected cells and in stimulating B cells to produce immunoglobulins.

Review of the immune system

Let us now review briefly the natural course of events in a new microbial infection and follow the dual role of the immune system in dealing with it. Usually the initial number of invading microorganisms is small and the antigenic stimulus slight, but as they multiply, this stimulus increases. Chemotactic signals go out from the site of microbial attachment, and phagocytes and antigen-sensitive lymphoid cells (both B and T cells) reach the scene. As the phagocytes begin their attack on the microbial particles, the antigens are presented to B and T cells, which begin to divide repeatedly, forming individual clones. B cells differentiate into memory B cells and plasma cells. Plasma cells synthesize molecules of IgM with combining sites that interact specifically with the new microbial antigens. The inflammatory response is underway, and the capillaries of the area exude plasma with its content of complement. As IgM is formed (over the course of a few days), its multiple combining sites attach to antigen on microbial surfaces, forming a sticky coat that immobilizes and agglutinates the living organisms. This immune complex activates the enzyme precursor of complement to form C_1 and the resulting cascade of C components. This activity enhances phagocytosis of the immune complexes now being formed more steadily, releases more chemotactic factors, and increases the inflammatory reaction. Complement fixation may also bring about direct cytolysis of the microbial cells.

If the site of infection is on a mucosal surface, of the upper respiratory tract for example, sIgA is formed and combines with the antigen. In this case, however, there is little or no seepage of plasma into the area and, hence, no complement is present. Opsonization and phagocytosis must therefore proceed without complement fixation.

In the meantime, the T cells have cloned, the sensitized T lymphocytes have combined with microbial antigens through their surface receptors, lymphokines are being released, and a crescendo of tissue reactions is proceeding. If all these defenses are successful, the invaders cease to multiply, and they succumb to the host's forces. Without further antigenic stimulus, B and T lymphocytes withdraw, leaving the field to the macrophages to clean up. These lymphocytes will "remember" the occasion, however, and, if called upon again to meet the same organisms, will do so with built-in capacity for more rapid specific responses. Most important, the B lymphocytes begin within a few days of the initial infection to produce the IgG antibodies that replace IgM and may persist in the blood for long periods.

Humoral antibodies are formed and T cells become sensitized against a great variety of microbial components and products so that, in time, the normal person's immunity to infection becomes broader (see Mechanisms of Immunization, following). The strength of the immunity depends on the nature and frequency of antigenic stimuli as well as on the vigor of the immune response. Deficiencies in either the B or T cell systems lead to serious problems.

Figure 6.14 summarizes the mechanisms of immunity, indicating the role of B and T cells, phagocytes, and complement.

INTERACTIONS OF ANTIGENS AND ANTIBODIES

Serology

The interactions of antigens and antibodies are often observable, or can be made so, whether they occur in vivo (in the body) or in vitro (in the test tube). **Serology** is the in vitro study of serum for its antibody content, and *serologic reactions* are the interactions of antigens and antibodies. These reactions are used routinely to identify an antibody or an antigen when one of these reactants is known. The serology laboratory commonly works with the serums of patients suspected of having infectious disease, testing them for the presence of antibodies to known microbial antigens. The microbiology laboratory, conversely, tests unknown microbial isolates with serums of known antibody content to confirm their identity.

Serologic techniques can be performed in a quantitative manner to provide information as to the relative amounts of antigen or antibody present in a given system. This is done by using one of the two reagents in a constant quantity and diluting the other until it fails to produce further reaction. When a patient's serum is being tested with a particular microbial antigen, for example, the serum is diluted in a serial fashion. A row of tubes containing equal amounts of the saline diluent is set up; a measured amount of serum is placed in the first tube and mixed well. Then a measured portion of this dilution is placed in the second tube and mixed, this process being repeated down the line to obtain a graded series of dilutions. The antigen is then added to each tube in a constant volume. When the tubes are examined for evidence of antigen-antibody combination, the last dilution of serum that shows reactivity is reported, using the reciprocal of the dilution figure. This figure is spoken of as the *titer of the serum* and provides an estimate of the level of antibody present in a unit volume. For example, if the last reactive dilution of the serum was 1:100, the titer is 100.

Serologic diagnosis of infectious disease is attempted by demonstrating the presence of specific antibodies in the patient's serum. In this case it is often necessary to test two samples of serum, one obtained from the patient soon after the onset of symptoms, the other taken ten days to two weeks later. The reason for this is that, as we have seen, antibody production increases during the course of active infection, stimulated first as the microbial agent begins to grow and multiply in the body, then rising steadily in response to a continuing antigenic stimulus. A patient in the early acute stage of illness may have no detectable serum antibodies against the causative microorganism, or the serum titer may be low, in the range of 20 or 40. Such a level of antibody is of little significance because it may, in such small amounts, merely reflect previous artificial immunization with the microbial antigen or a titer lingering from some previous attack of the infection. If the test is repeated in a week or two, however, and the titer is found to have increased at least fourfold (increased to 80 or 160 using the example above), this is considered serologic evidence of current active response to the microorganism in question. This type of serologic information is often very valuable, both in diagnosing the present disease and in estimating the past experience of the patient with a given antigen.

Distinctions between IgM and IgG antibodies, made with the aid of appropriate tests, help to determine whether a patient is in an early or still active stage of infection, or whether active infection has subsided. Since IgM is produced in the primary immune response, its presence indicates new or very recent infection. The presence of IgG indicates a secondary immune response, whereas the height of the IgG titer observed at two-week intervals provides clues as to whether or not infection is subsiding. IgG titers usually reach a peak during clinical convalescence and wane gradually thereafter but may persist at low levels for long periods. Chronic, persistent infection may induce elevated levels of both IgM and IgG antibodies.

To study a patient's serum diagnostically for its antibody content, particular laboratory tests are ordered to confirm the clinical or bacteriologic evidence of the patient's infectious disease. These tests are often named in ways that reflect the nature of the antibody or reaction expected to result from the test. The following section describes a number of these laboratory tests.

In Vitro Reactions

In laboratory studies, blood plasma or serum of human and animal origin is used as a source of antibody, and antigens are obtained from a variety of sources including microorganisms. When an antibody combines with its specific antigen, the reaction may have one of several observable results,

FIGURE 6.14 Mechanisms of immunity.

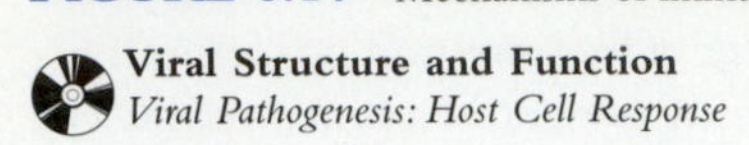
Viral Structure and Function
Viral Pathogenesis: Host Cell Response

Fungal Structure and Function
Diseases: Nonspecific Host Defenses; Immune System

Microbial Pathogenesis of Infectious Disease
Nonspecific Host Defenses: Cellular Defenses and Phagocytosis

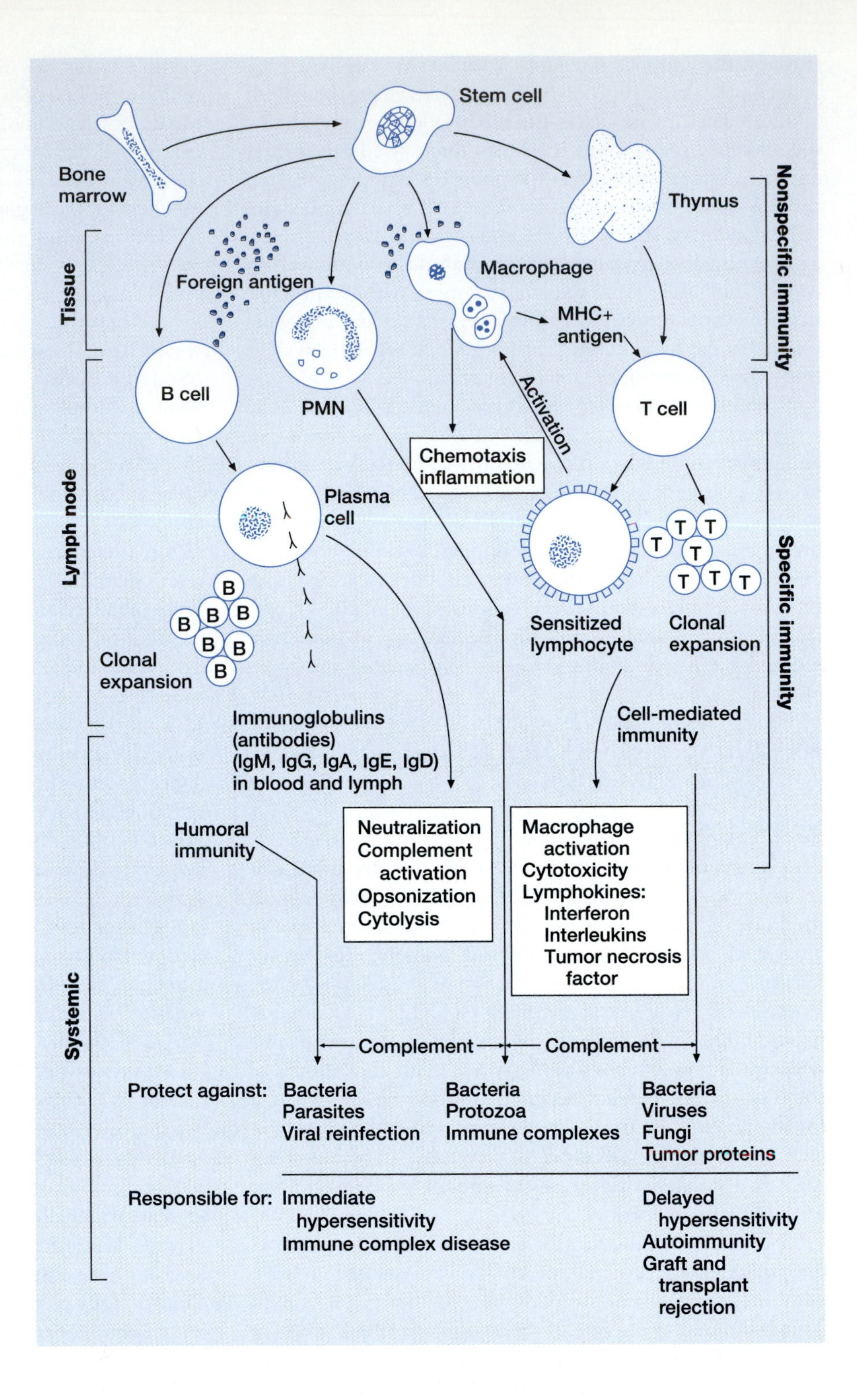

depending on the nature and condition of the antigen as well as the presence of other nonspecific factors. Consequently, antibodies are often described according to the type of reaction that occurs. In the following sections, a few of the more common serologic assays performed in serology or microbiology laboratories are described. Other tests that were in common use, such as the complement-fixation test, are no longer performed. These have been replaced by newer, less complex, and less labor-intensive assays.

Agglutinins

Antibodies that cause a visible clumping of the antigen are called **agglutinins.** This type of reaction occurs if the antigen is on the surface of or adherent to a large particle (e.g., a microbial cell, a red blood cell, a latex particle). Since antibodies have multiple combining sites, each antibody links up with at least two antigen molecules. These antigen molecules may then link with other antigen molecules by attaching to other antibodies. This latticelike structure grows until large clumps form and settle out of suspension. Saline in physiologic concentration must also be present to encourage particles to adhere. The suspending solution becomes clearer as more particles clump and settle out, and the agglutinated masses become visible.

Precipitins

Antibodies that cause antigens to visibly settle out of the solution in which they are dissolved are called **precipitins.** In this case, the antigen in solution is unassociated with any cell or particle. Soluble free antigen combines with antibody, and in the presence of saline the complexes formed become visible as a fine precipitate in the bottom of a test tube.

In another form of precipitation test called *gel diffusion,* or **immunodiffusion,** the reaction takes place in a supporting agar gel.

The test involves making wells in the agar gel and placing an antigen solution in one well and serum antibody in an adjacent one. Each diffuses through the agar. If the antigen and antibody combine specifically where they meet, visible precipitation lines are formed. If they do not combine, no precipitate can be seen.

One application of this technique is made in the Elek plate method of testing *C. diphtheriae* isolates for toxigenicity. The diphtheria toxin is an antigen and will react visibly with its antibody if the two are brought in contact. For this test, a strip of filter paper impregnated with serum containing the specific antibody (antitoxin) is embedded in an agar plate, and an isolated strain of a suspected diphtheria bacillus is streaked across the agar plate, perpendicular to the strip. The diphtheria toxin is an exotoxin (see chap. 5). If toxin is produced by the test strain, it will diffuse into the agar medium from the area of growth along the streak and meet the antibody diffusing from the paper strip. A line of visible precipitation will form at the juncture and the strain will be reported to be pathogenic (toxigenic). Nonpathogenic strains of the diphtheria bacillus do not produce toxin and therefore show no lines of precipitation in this test. A known culture of a toxigenic strain of the diphtheria organism is always included to ensure that the test and reagents are working properly.

Figure 6.15 illustrates the techniques of slide agglutination and gel diffusion, as well as the use of the Elek plate.

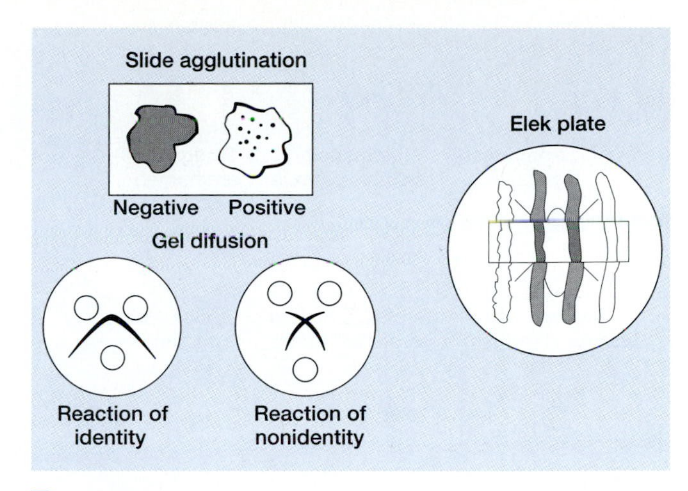

FIGURE 6.15 Diagrammatic examples of slide agglutination, gel diffusion, and the Elek plate techniques used to demonstrate antigen-antibody reactions.

Fluorescent Antibodies

The **fluorescent antibody** technique provides another means of visualizing antigen-antibody interactions. This method involves the preliminary treatment of antiserum with a fluorescent dye that conjugates firmly with the antibody globulin. The antibody is thus "labeled," and when it combines with its specific antigen, the site of combination fluoresces visibly under a microscope fitted with an ultraviolet illuminator. As the ultraviolet rays pass through the specimen, the dye-conjugated antibody fluoresces brightly on the surfaces of cells it has coated. The use of a fluorescein-labeled antibody makes it possible to locate microorganisms quickly and identify them in smears from clinical specimens or in tissue sections. Intracellular viruses can also be visualized by this method.

Conjugation of antiserums with fluorescent dyes is expensive and time-consuming; therefore, it is impractical to label a battery of antiserums, each directed against a different microbe. To solve this problem, an indirect method of fluorescent staining can be used. It not only helps to identify unknown microorganisms but aids in determining the presence of specific antibodies in patients' serums. Human serum is first injected into an animal, which responds by producing antibodies to human globulin (itself an antigen). Serum from the animal is then conjugated with fluorescent dye, providing a labeled antihuman antibody. To detect microbial antigens, the preparation is exposed to unlabeled human serum of known antibody content. If a specific antigen-antibody combination occurs, the microbial cells become coated with human globulin, but the reaction is not visible. Now the fluorescent-labeled antihuman-globulin antibody from the immunized animal is added to the preparation. It combines with the human globulin coating the cells, providing a new outer layer that now fluoresces in ultraviolet light. This indirect method makes it possible to use a number of unlabeled human antiserums to screen smears for the presence of antigens. The smears can then be treated with one labeled antihuman globulin, which identifies those cells that are coated with specific human antibody. These techniques are illustrated diagrammatically in figure 6.16. The effects of immunofluorescent staining of microorganisms can be seen in colorplate 2b.

To determine whether antibodies against a specific antigen are present in a patient's serum, it is allowed to react with that antigen. The preparation is then treated with labeled antihuman globulin and observed for fluorescence. The principle is the same as just described, but in this case the identity of the antigen, rather than the antibody, is known. This test is most widely used for the detection of antibodies against *Treponema pallidum,* the causative agent of syphilis (see chap. 19). In practice, when detecting microbial

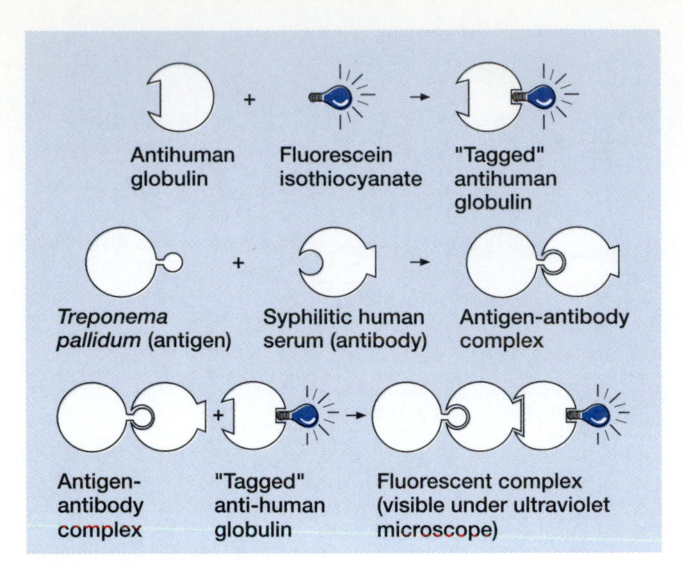

FIGURE 6.16 Diagrammatic illustration of the indirect fluorescent antibody reaction.

Courtesy of Becton Dickinson Microbiology Systems, Cockeysville, MD.

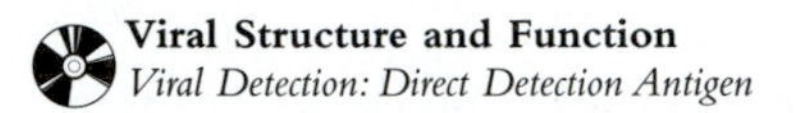
Viral Structure and Function
Viral Detection: Direct Detection Antigen

antigens by fluorescent antibody techniques, specific antiserums from animals other than humans are used because it is more practical to immunize them against a variety of agents. The fluorescent-labeled antiglobulin serum must then be prepared in a second animal species.

Other Labeling Techniques

In recent years a number of techniques have been developed for labeling either antigens or antibodies with markers that can be detected after antigen-antibody combination has taken place. Two of these are described briefly below.

Enzyme immunoassay (EIA), sometimes called enzyme-linked immunoassay (ELISA), is a method used to detect antibody in a patient's serum produced against an infecting microorganism. To perform the test, the antigen in question (usually the suspected microorganism or microbial extract) is immobilized in a plastic tube or microwell. The patient's serum is then added to allow any specific antibody present to combine with the immobilized antigen. A wash step follows to remove unbound antibody; then an antihuman globulin linked to an enzyme is added. The enzymes chosen (usually horseradish peroxidase or alkaline phosphatase) act on substrates that are colorless, initially, but degrade to colored end products because of the enzyme action. The test is "developed" by adding the colorless substrate for the enzyme to the mixture. A colored end product that can be seen visually or detected by an instrument is the indicator of a positive test. This method is used to detect a wide variety of bacterial, viral, and parasitic antibodies.

Immunoblots (also referred to as *Western blots*) can be used to detect antibody in a patient's serum. In this procedure, a known antigen, or a mixture of known antigens, is separated by electrophoresis on a gel. The material is then transferred ("blotted") onto a strip of nitrocellulose to which the antigen(s) bind strongly. Patient serum is applied to the nitrocellulose strip; if specific antibody is present it binds with the antigen. The strips are then washed to remove unbound antibody and, as in EIA tests, treated with an enzyme-linked antihuman immunoglobulin reagent. The enzyme-linked reagent combines with any human antibodies that have attached to their specific antigen on the nitrocellulose strip. The substrate for the enzyme is added and any antigen-antibody complexes formed are revealed by the colored end product of the reaction. Modifications of this technique include the use of radiolabeled or fluorescent dyes instead of an enzyme combined with the antibody. The immunoblot method is widely used to detect antibody to the human immunodeficiency virus (see AIDS, chap. 19).

Antigen Detection Methods

Methods are available for detecting microbial antigens present in patients' specimens or on microorganisms isolated in culture. Two common techniques and their important uses are described below.

In the **latex agglutination** test, antibody is bound to latex particles, which serve as the antibody carrier. The particles can be coated with antibodies produced against a number of different pathogenic microorganisms and serve as convenient reagents for their immunologic identification. Latex agglutination is often used to detect microbial antigens that are in solution, either in culture broth or body fluids. In meningitis caused by *Cryptococcus neoformans,* for example (see chap. 14), the antigenic cryptococcal capsular polysaccharide dissolves in the patient's cerebrospinal fluid and can be detected by mixing a drop of the fluid with latex particles coated by the cryptococcal polysaccharide antibody. The clumping of the latex, which takes only minutes to occur, is a positive test. Latex tests can detect the same soluble antigens as precipitin tests, but they detect much smaller amounts of antigen. Therefore, they are more sensitive because the large particles on which the antibody is coated form visible reactions more readily.

Enzyme immunoassays can be used for antigen as well as antibody detection. In this case, the unknown antigen (often an extract of patient specimen that has been immobilized on a filter paper) is reacted with a known antibody that has been linked to an enzyme. If the antibody is specific for the immobilized antigen, the addition of the substrate for the enzyme produces a colored end product (fig. 6.17). This test is the basis for rapid detection (10 minutes to 4 hours) of many microorganisms or their products directly in patient specimens. Thus the streptococcus of streptococcal pharyngitis can be detected in throat swabs (chap. 12), the microbial toxin causing pseudomembranous colitis in stools (chap. 16), and the respiratory syncytial virus in nasopharyngeal washings (chap. 13).

Monoclonal Antibody

An exciting development in immunology is the use of hybridomas to produce highly specific antibodies. Hybrido-

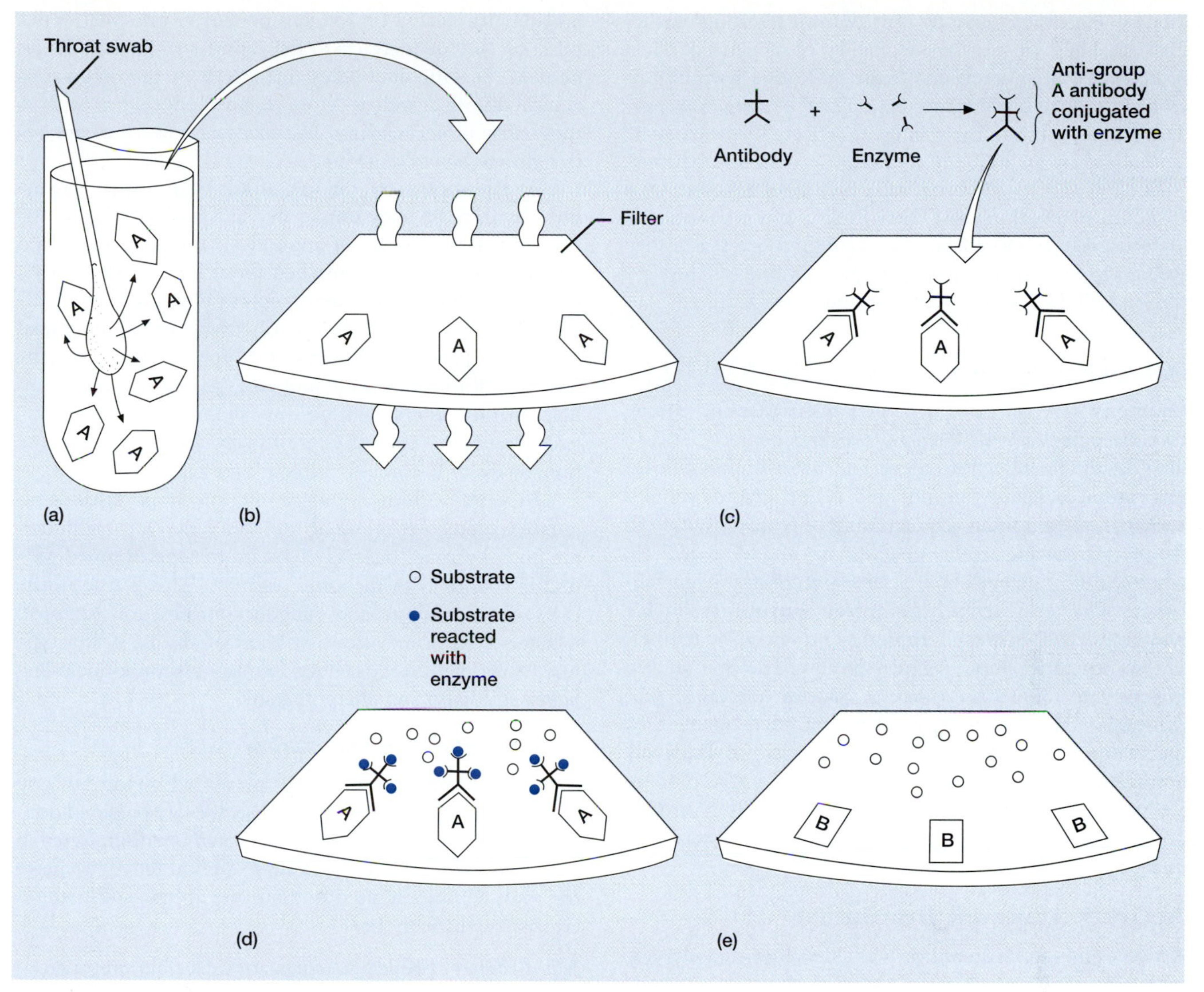

FIGURE 6.17 Diagram of an enzyme immunoassay (EIA or ELISA). (a) A throat swab from a patient with streptococcal pharyngitis is placed in a tube with extraction solution, which extracts the group A antigen. (b) The extraction solution is then passed through a filter where the group A antigen attaches to its surface. (c) After a wash step, an antibody against the group A *Streptococcus,* which is linked to an enzyme, is added to the filter where it attaches to the group A antigen. (d) After another wash, a colorless substrate specific for the enzyme is added and is split to a colored end product when it comes in contact with the antibody-bound enzyme. (e) If an antigen other than group A was present, no antibody would bind, unbound antibody would be washed away, and no color reaction would be seen. This method is used to detect a variety of bacterial, parasitic, and viral antigens.

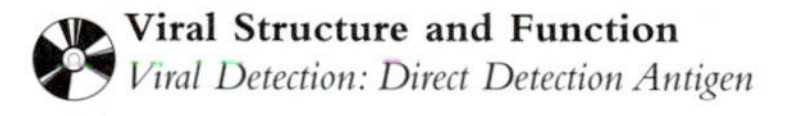
Viral Structure and Function
Viral Detection: Direct Detection Antigen

mas are hybrid cells formed by fusing antibody-producing plasma cells with rapidly multiplying myeloma (bone marrow tumor) cells in vitro. The plasma cells are taken from the spleens of animals that have been immunized with an appropriate organism, or other antigen, to induce the required pure or **monoclonal antibody.** Each hybridoma quickly yields a large population of cells (clones) that, by virtue of its plasma cell component, is secreting the highly specific, single (monoclonal) antibody. At present, these highly purified antibodies are valuable reagents for a variety of diagnostic immunological studies such as those described in this chapter.

In addition, monoclonal antibodies are being tested as highly specific antitumor agents. These antibodies could kill tumor cells, which have foreign antigens on their surfaces, without affecting normal tissue cells. As another example, monoclonal antibodies prepared against the lymphokine IL-2 are being examined as a means to prevent rejection of transplanted organs. Inhibition of IL-2 production would prevent proliferation of T lymphocytes, which are responsible for graft rejection (see later section on Graft Rejection). Such studies are still in the experimental stage.

In Vivo Reactions

Many of the reactions demonstrable in the test tube may occur in the body when antibodies combine with antigens in the tissues. Microorganisms agglutinated or lysed by an antibody with the assistance of complement in vivo are thereby inactivated and prevented from exerting further damage. They are also more readily phagocytized when coated with an antibody. Exotoxins and other harmful microbial products are similarly neutralized by antitoxins produced in response to their antigenic effects. Viruses may be neutralized by an antibody if enough is present in the circulating blood to combine with such agents before they have localized within cells. These in vivo effects of antibodies have given rise to still another term for them; that is, they are said to be *neutralizing,* or *protective,* antibodies. The next section will discuss how these antibodies are acquired.

Mechanisms of Immunization

Immunity may occur as a natural phenomenon, existing even though there has been no previous contact between the host and the infectious agent. The factors responsible for this natural immunity are not well defined. Much more is known, however, about acquired immunity that results from the production and activity of antibodies and sensitized cells when the host defends itself against intruding foreign substances. The term **actively acquired immunity** implies humoral or cell-mediated immunity produced by the host in response to an immunogenic stimulus. The immunogen may be encountered in a natural way (by clinical or subclinical infections) or introduced in an artificial manner (by vaccination) to induce a host immune response. **Passively acquired immunity** means the host did not produce the protective antibodies but received them from another host, either naturally by placental transfer or artificially by injection.

Actively Acquired Immunity

Active immunity acquired through infection may develop whether or not the infection results in clinically recognizable disease or remains subclinical. In either case the immune response continues for as long as immunogenic stimulation persists. Some destruction and removal of humoral antibodies occurs at a slow rate, however, so that there is a maximum peak for antibody levels in the blood. This peak is maintained while the immunogen remains in the body, then slowly declines when it disappears. The decline may be quite rapid in some cases, occurring in a matter of days, but in many instances (e.g., diphtheria and rubella) low levels of antibody are detectable for many weeks, months, or years after the original infection. Cellular immunity may be even more durable. Upon reentry of the immunogen into the body at a later time, there is more rapid and abundant antibody production as well as a more efficient cellular immune response. This phenomenon is referred to as the *anamnestic* (memory) *response.*

For some commonly encountered organisms such as the meningococcus (which is carried in the pharynx of 5% or more of the population), the protective antibody is often found in the bloodstream in the absence of any clinical infection. Antibodies to other bacteria, such as *Haemophilus influenzae,* appear in a similar fashion, but their formation is probably stimulated by antigens present in intestinal strains of *E. coli* that are identical to the capsular antigens of *H. influenzae.* Specific antibodies that occur in the absence of clinical disease as well as "cross-reactive" antibodies (such as those that protect against *H. influenzae* infection) are referred to as *natural antibodies.*

In other instances, the first attack by a virulent organism may find the body unprepared and unable to avoid disease. Initial production of antibody is not immediate, and protective levels are not reached for at least 10 to 14 days, during which time acute symptoms of illness become manifest. In the days before the discovery of antimicrobial agents, this period was often critical for the patient (as, indeed, it still may be), until antibodies joined the defense to help turn the tide, sometimes very dramatically, and convalescence began. Antibodies are usually at their peak at the time recovery becomes complete, then fall away slowly.

In certain chronic, progressive infectious diseases no durable, protective humoral immunity develops, although the body becomes relatively resistant to superimposed, second infections with the same organism. This is true in the case of tuberculosis and fungous diseases, for example. Chances for recovery from, or arrest of, disease of this type rest primarily with the cellular immune response (including hypersensitivity) and chemotherapy.

Active Artificial Immunization

Many infectious diseases can be prevented by immunizing techniques employing **vaccines.** Vaccines are prepared from an infectious agent or its products treated or administered in such a way that they will not cause clinical illness. Some of the most commonly used immunizing agents and methods are discussed briefly next.

Killed Vaccines Killed vaccines are injectable preparations containing microorganisms grown in laboratory culture (or in experimental animals, chick embryos, tissue culture, etc.), separated from the growth medium, and killed by heat or a chemical agent. Formalin and phenol are useful for this purpose. Such vaccines are usually injected subcutaneously, less often by intracutaneous or intramuscular routes. As a rule, multiple injections are required to produce durable protection; for example, two or three spaced over a few weeks' time and one or more "booster" shots (for the anamnestic effect) given at yearly intervals. The Salk poliomyelitis vaccine, and influenza and cellular pertussis vaccines are examples of killed preparations of organisms given in this manner.

Living, Attenuated Vaccines Virulent microorganisms can be treated by a method that eliminates their pathogenic properties without killing them or inactivating their antigens. In this condition, an originally virulent organism is said to be *attenuated.* The most successful methods for attenuation included drying (Pasteur's original rabies vaccine contained dried, attenuated virus [see fig. 23.3]); continued, prolonged passages of the organism through artificial culture media to induce mutations; and cultivation and passage of the organism through hosts other than humans. Occasionally, advantage can be taken of the close relationship of strains of organisms causing disease in humans or an animal host, but not both. The agent of bovine tuberculosis, for example, causes a serious infection in cattle. It is antigenically similar to the human tubercle bacillus and also can induce human disease. However, a strain of the bovine bacillus has been attenuated by many years of culture on artificial media in the laboratory and is used as a vaccine against human tuberculosis (BCG).

Attenuated vaccines provide more lasting immunity than killed vaccines because the organisms multiply in the host for a period of time, providing continued immunogenic stimulus. This is one reason why the attenuated live Sabin polio vaccine largely replaced the killed Salk vaccine although both are now used in sequence (see table 10.1). Another reason is that the polio virus multiplies in the intestinal tract and stimulates mucosal (secretory IgA) as well as humoral antibody production. Thus, the attenuated vaccine provides immunity more nearly like that of the natural infection.

Toxoids **Toxoids** are nontoxic but immunogenic substances derived from toxins. Microbial exotoxins, such as those of the diphtheria and the tetanus bacilli, can be rendered nontoxic, having first been separated from the bacterial cells that produce them. The toxin present in the filtrate of a liquid culture is first treated with heat or formalin to destroy its toxicity, but not its immunogenicity. It may then be precipitated with alum to provide an injectable preparation that is slowly absorbed from the tissues, thus ensuring a prolonged immunogenic effect. Toxoids are usually given intramuscularly according to the schedule shown in table 10.1.

Recombinant DNA Vaccines One important application of recombinant DNA technology is the ability to genetically engineer (bioengineer) microorganisms to produce safe vaccines. In 1986, a genetically engineered hepatitis B vaccine was licensed for human use. The **recombinant DNA vaccine** is produced by inserting a plasmid containing the gene for hepatitis B surface antigen into a yeast cell (see chaps. 3 and 21). The surface antigen is then synthesized by the yeast and is collected for vaccine use by disrupting the yeast cells and purifying the antigen using chemical and physical methods. Previously, hepatitis B vaccine was prepared by inactivating live virus collected from the plasma of infected persons. The immunogenicity of the recombinant vaccine is comparable to that of the previous vaccine but avoids the possibility of live virus surviving the inactivation process. Administering the vaccine routinely has significantly reduced the incidence of this dangerous infection (see table 10.1).

Passively Acquired Immunity

Passive immunity is acquired naturally through placental transfer of maternal antibodies (serum IgG) to the blood of the developing fetus in utero. The congenital immunity acquired by infants depends on the infections their mothers have had or have been immunized against as well as on the level and protective qualities of the antibodies involved. The mother's antibodies (including sIgA) are also contained in her milk, particularly the colostrum, and these reinforce the breast-fed child's immunity. As the maternal antibody declines, children become more susceptible to infection, but by this time (at the age of four to six months) their own IgG antibody mechanism becomes functional. Their exposure to common microorganisms at this stage is likely to lead to durable immunity against them, and they can be given further protection against dangerous pathogens by active immunization with vaccines and toxoids.

Passive Artificial Immunization

Passive immunity acquired artificially can be useful in providing temporary protection to a susceptible individual who has been exposed to a very hazardous infection. The source of antibodies is immune serum from a person or animal that has been actively immunized by infection or by vaccination. (Cellular immunity can be passively transferred only by injecting sensitized lymphocytes.) Most commonly, a "pool" of serum from a number of individuals of the same species is obtained, the gamma globulin content is separated out of the serum, and this purified antibody is administered. The concentration of serum obtained by this method ensures a high level of antibody in a small volume per dose. Pooled gamma globulins of human origin may contain antibodies to a number of common human diseases such as poliomyelitis, infectious hepatitis, measles, and others.

The use of passive antibodies in the prevention of disease constitutes an emergency measure only, since the immunity provided does not persist beyond two to three weeks. Susceptible persons who have been exposed or are at risk of exposure to a disease that might have serious consequences for them are suitable candidates for passive immunization. Such persons might include the very old, the very young, or people receiving immunosuppressive therapy, whose immune and nonspecific resistance mechanisms are inadequate. Passive immunization of all contacts of a serious infection, such as infectious hepatitis, may be advisable if natural or artificial active immunity is not widespread because an effective vaccine is not available or the disease is not common. Passive immunization may prevent a specific disease from developing altogether or lessen the severity and duration of the infection. In the latter case, active, effective immunity usually follows as a result of direct stimulation of antibody production by the immunogen. Whenever effective vaccines or other immunizing agents are available, however, active immunization is preferable because it ensures better and more durable protection. Another reason to use active immunization where possible is that animal serums are foreign proteins for humans, and after they are administered, their immunogenic activity may lead to a hypersensitive state. The "serum sickness" that develops can cause devastating tissue injury. Even if no adverse reaction follows the initial serum dose, injecting serum from the same animal species at another time may produce a violent, immediate reaction with possible fatal consequences.

TABLE 6.5 HOST VERSUS MICROBE

Nonspecific Resistance Factors	Specific Resistance Factors
I. Skin and mucosa	I. The immune system
A. Architecture	A. Humoral immunity
Mucociliary elevator	B lymphocytes (produce immunoglobulins (antibodies) of five classes: IgM, IgG, IgE, IgA, IgD)
Keratin	B. Cellular immunity
B. Physiology	T lymphocytes (produce lymphokines that activate macrophages, stimulate B and T cells, increase antibody production, kill cells)
Lysozyme	II. Acquired immunity
Intestinal enzymes	A. Active
pH	1. Natural (via infection or disease)
Flushing action	2. Artificial (via immunization with vaccines, toxoids)
II. Blood	B. Passive
A. Circulates nutrients, removes wastes	1. Natural (via placenta or colostrum)
B. Complement	2. Artificial
C. Leukocytes	a. Serum antibodies (immunoglobulins)
III. Mononuclear phagocyte system	b. Sensitized lymphocytes
A. Fixed macrophages	
B. Circulating macrophages	
IV Lymphatic system	
A. Filtration	
B. Lymphocytes (specific)	
V. Phagocytosis	
A. Neutrophils	
B. Macrophages	
VI. Inflammatory response	
A. Localization of agent	
B. Biochemical effects	
VII. New products or changes	
A. C-reactive protein	
B. Cytokines	
VIII. Other	
A. Age	
B. Nutritional Status	
C. Stress	
D. Genetic factors	

With respect to the treatment of infectious disease, immune serum is of limited usefulness in this era of antimicrobial therapy. However, it is helpful in the treatment of certain diseases caused by microbial exotoxins. In diphtheria, tetanus, or botulism, the immediate administration of antitoxin is needed to neutralize the toxin before it can attach to tissue cells. Speed is essential because even the antitoxin cannot prevent damage to cells on which the toxin has already bound. In addition, in cases of diphtheria and tetanus, administration of antimicrobial agents is essential to eliminate the organisms from their localized sites of growth and multiplication.

Summary of Resistance Mechanism

We have seen that many of the mechanisms of nonspecific resistance are closely meshed with those of the immune system, and that the dual aspects of specific immunity may overlap. The most outstanding factors that protect the human host from the inroads of infectious disease are summarized in table 6.5.

DISORDERS OF THE IMMUNE SYSTEM

Diseases directly attributable to the functions of the immune system may be related either to hypersensitivity, mediated through humoral or cellular factors, or to immunodeficiencies (i.e., functional failures of either humoral or cellular mechanisms) or both. The dual nature of the immune system is again in evidence as we consider these problems.

HYPERSENSITIVITY

Characteristics

We have seen that the immune system plays a highly protective role in defending the body against the intrusion of foreign antigens, including those of microbial invaders. This same system may become injurious, however, if its responses are excessive or overreactive, that is, **hypersensitive.** The consequent damage to the body may be quite serious and, in some instances, even lethal.

Most of the injury results from interactions of antigens with antibodies or sensitized cells, which trigger the normal immune responses to an excessive degree: increased capillary permeability with edema and inflammation, release of histamine and other biologically active substances, and lymphokine effects. The general nature and symptomatology of these reactions depend on whether they are mediated by antibodies or by sensitized T lymphocytes. The former are responsible for reactions of the immediate type of hypersensitivity, the latter for the delayed type. Immediate hypersensitivity includes the systemic reactions of anaphylaxis and serum sickness, and the more localized tissue responses of allergy (atopy). Delayed hypersensitivity provides the mechanism for the "allergy of infection," allergic contact dermatitis, and some autoimmune diseases.

Immediate Hypersensitivity

The manifestations of immediate hypersensitivity depend on several factors: the nature of the antigen, the frequency and route of antigen contact, and the type of immunoglobulin reacting with it. The initial dose of antigen (there may be several doses, or exposures) to which an individual later becomes sensitive is known as the *sensitizing dose.* A later dose of the same antigen creating tissue damage is called the *eliciting dose,* or the *shocking dose* if the injury is profound. Immediate reactions begin within minutes of contact with the eliciting dose. They may disappear within an hour or so if there is no widespread, residual damage. If the antigen is introduced directly into the tissues by injection or an insect sting the clinical result will be systemic, as in anaphylaxis or serum sickness. When the route of contact is superficial, involving the epithelial tissues of the skin or the mucosal linings (respiratory, conjunctival, intestinal), the symptomatology is more localized at those surface sites, as in asthma, hay fever, or allergic rhinitis and conjunctivitis. These local reactions are referred to as symptoms of allergy (meaning "altered reactivity"), or atopy (out of the way, out of place). They continue while contact with the antigen(s) persists, the symptoms reflecting the frequency of exposure; that is, they may be constant, fluctuating, seasonal, sporadic, and so on.

The antigens of immediate hypersensitivity are usually called *allergens,* especially when referring to localized allergies. They may also be labeled more specifically as injectants, inhalants, ingestants, or contactants. They may be complete immunogens or hapten antigens bound to a large carrier molecule. The best-known allergens of atopy include pollens, dust, animal danders, feathers, and foods. The injectants inducing systemic reactions may be any foreign protein to which the body has been sensitized, but the most common offenders are animal serums (used for passive immunization), drugs (penicillin and some other antimicrobial agents), and bee venom, including that of related stinging insects (e.g., yellow jackets, wasps, and hornets).

The antibodies involved are usually of the IgE class, but IgG (acting together with complement) and sometimes IgM, are implicated in immune complex diseases such as serum sickness. IgE is responsible for anaphylaxis and the common allergies; for this reason it has many names, the most frequently used being reaginic, atopic, skin-sensitizing, or anaphylactic antibody. When it is formed in response to the sensitizing dose of antigen, very little IgE circulates in the blood, for it has a tendency to attach to tissue cells, particularly tissue mast cells and basophils.

Understanding that the degree and site of injury in hypersensitivity will vary with circumstances, we can say that

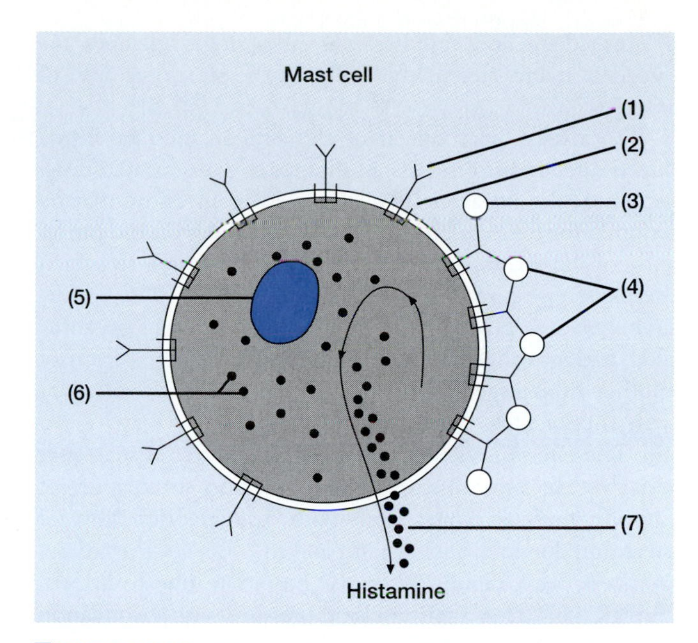

FIGURE 6.18 Diagram of histamine release from mast cell granules in immediate hypersensitivity. (1) IgE antibody attaches to special cell receptor (2). (3) Antigen attaches to IgE antibody. (4) Antigen molecules create cross-linkage of IgE antibodies. (5) Mast cell nucleus. (6) Cell granules containing histamine. The arrow within the cell indicates enzyme activation and energy release following steps (3) and (4), with a subsequent burst of histamine from the cell (7).

the eliciting or shocking dose of antigen sets off the following train of physiologic events:

1. Antigen combines with antibody at the site of antibody attachment to tissue cells.
2. Mast cells and basophils release a number of physiologically active substances, but primarily histamine, from their intracellular granules (fig. 6.18).
3. As a result of histamine release (together with other "mediators," e.g., serotonin from blood platelets and kinins formed from plasma proteins), there is a sudden dilatation of blood vessels, contraction and spasm of smooth muscle (particularly in the bronchioles and small arteries), and increased capillary permeability with edema.

If these effects are massive, there may be a sudden drop in blood pressure, with circulatory collapse and shock. Extensive smooth muscle contraction brings about severe bronchospasm, and edema results in laryngeal constriction. These clinical symptoms are characteristic of anaphylactic shock. Lesser or more localized effects occur in immune complex disease and in atopic allergies.

Anaphylaxis

Anaphylaxis is a term coined to imply an effect opposite to that of prophylaxis, which signifies the prevention of disease. The extreme injury that can be produced by hypersensitive immune mechanisms is anaphylactic shock, which stands in marked contrast to the prophylactic, or protective, effect usually associated with immunity. It occurs as a result of immediate and simultaneous physiologic changes induced at many sites in the body by the shocking dose of antigen.

To achieve such an effect, the antigen must be introduced directly into the tissues in a fairly concentrated dose. Intramuscular, intravenous, or intracardiac injections may be required for experimental animals, depending on the purity and concentration of the antigen and the degree to which the tissues were sensitized after the sensitizing dose of immunogen was given. The allergens involved may be complete antigens such as the foreign proteins of the serum of another animal species, or they may be incomplete antigens such as penicillin.

The effects induced by the shocking dose of such materials in sensitive human beings include stimulation of smooth muscle contraction with sudden difficulty of breathing due to bronchospasm and to edema of the larynx; facial edema; a rapidly increased heart rate due to arterial constrictions, then a fall in blood pressure with vasodilation and shock. A generalized urticarial reaction may occur as well, with giant "hives" developing at the site of injection of the antigen and also over large areas of skin. Intense reactions can result in death within minutes or a few hours.

Patients may recover from an anaphylactic episode if the shocking dose, the degree of their sensitization, the strength of their constitution, or all of these factors permit. Before the administration of potentially shocking agents, patients should be tested for possible sensitization.

Atopic Allergy

Unlike anaphylaxis, which requires injection of the allergen, **atopic allergies** occur naturally through inhaling, or ingesting, or making contact with the antigen. The well-known problems of asthma, hay fever, hives, and other skin rashes exemplify this type of hypersensitivity. In these instances the cellular injury induced by antigen contact with antibody-sensitized tissue is limited in scope. While these diseases are incapacitating, they do not have systemic effects. The antigen-antibody mechanism constitutes the basis of allergic symptoms, but other factors such as emotional stress, hormonal changes, and physical reaction to environmental stimuli may influence the response markedly. The capacity for hypersensitivity probably has a genetic basis, as evidenced by the distribution of allergic disorders in families. Actual sensitization of individuals, however, depends on their contacts with particular immunogens.

Immune Complex Disease

A number of diseases appear to be due to the formation of antigen-antibody combinations (immune complexes) that fix complement. These are called **immune complex diseases.** Antibodies of the IgG or IgM class are primarily involved, since these are the only immunoglobulins that bind complement to any extent (see table 6.3). The antigens are soluble proteins or nucleic acids. Some are derived from viruses or bacteria, but not all of them have been identified.

Serum sickness is a type of immune complex disease that was initially observed in the preantibiotic era. Serum from immunized animals, usually horses, was often given prophylactically to provide immediate passive immunity against a variety of infecting agents. Diphtheria antitoxin prepared from horse serum, for example, is still used to prevent the effects of intoxication in diphtheria patients (who are not actively immune to diphtheria toxin). However, the antitoxin and other protein antigens in horse serum are foreign to human tissues and therefore induce antibody formation. After the initial dose of serum, antihorse antibodies begin to reach a significant level in the blood and tissues in about 8 to 12 days. If the original immunogen (horse serum) has been eliminated from the body by this time, no untoward reactions take place. If sufficient immunogen remains, the antibody reacting with it and complement produce symptoms beginning at the time antibody level is reaching its height. More frequently, serum sickness is provoked by a second dose of serum, from the same animal species, on some later occasion.

The symptoms of serum sickness are similar in many ways to those of anaphylaxis, but less severe. They include sudden fever; hives; edema of the face, hands, and feet; lymphoid reactions with swelling of many nodes; and often an intense reaction at the site of serum injection. Edema of the laryngeal and other upper respiratory tissues can be a severe threat to normal respiration. The illness continues until all the antigen has been inactivated or eliminated.

Penicillin and other drugs may induce reactions similar to serum sickness in hypersensitive individuals, and these are seen more commonly now than serum sickness. For this reason use of a particular antimicrobial agent in treatment of infectious disease may be contraindicated. The patient's history of possible reactions to drugs should be carefully considered before they are administered. Evidence of hypersensitivity can also be obtained in some cases by skin testing.

The injury of immune complex disease results from deposition of small immunoglobulin and antigen complexes in blood vessel walls of various organs (especially kidney glomeruli). This triggers the complement cascade, which induces an inflammatory response mediated primarily by neutrophils. Immunoglobulin E acting on mast cells to release factors that sensitize blood vessel linings probably plays a role in this process as well. Immune complexes are responsible for the glomerulonephritis that may follow streptococcal infection, and glomerulonephritis also occurs in systemic lupus erythematosus, where the immunoglobulins are directed against a number of antigenic tissue substances including DNA. Rheumatoid arthritis is a chronic inflammatory joint disease caused by deposition of immune complexes composed of IgG itself and anti-IgG (rheumatoid factors) on synovial membranes.

Delayed Hypersensitivity

The sensitized T lymphocytes of cell-mediated immunity are responsible for the immunologic injury of the delayed type of hypersensitivity; circulating antibodies are not involved. T cells respond to antigenic stimuli by producing lymphokines, which stimulate an influx of activated macrophages, and by cytotoxic effects (fig. 6.14). These benefits of cellular immunity can be damaging in some circumstances. (Immunity is like a fire: it can warm one's house when restricted to the fireplace or burn it down if it goes out of control.)

The term *delayed* originally arose from the fact that positive skin tests with a specific antigen require about 24 hours to become visible and about 48 to 72 hours to reach their full development, in contrast to the immediate wheal seen in skin reactions involving antibody. Delayed hypersensitivity can be passively transferred by injection of intact sensitized lymphocytes although this practice is seldom carried out.

Delayed hypersensitivity is operative in microbial allergies (allergy of infection) and allergic contact dermatitis. These are distinct from the atopic allergies of immediate hypersensitivity with their constitutional or hereditary basis. CMI is also responsible for the autoimmune diseases and provides the mechanism for foreign graft rejection as well as tumor immunity.

Allergy of Infection

Hypersensitivity is a frequent complication of infections and may sometimes be a major feature of an infectious disease. In many instances the organism involved may be relatively inert, immobilized within a lesion created by the body's inflammatory response, or have few virulent properties; yet the disease it induces is of long duration, persistent, and chronic because of hypersensitization. Infection of the tissues by parasitic worms, fungous infections, viral and bacterial diseases—all may have an allergic component. Any type of tissue may be involved depending on the location of CMI reactions.

Skin tests may be performed to detect microbial allergies. Delayed skin reactions are those that develop in not less than 24 to 48 hours following intracutaneous injection of the antigen. After some hours redness and edema appear, and increasing induration (hardening) of the area develops. Vasodilation and infiltration of the site with plasma and leukocytes account for these events. The cellular infiltrate often becomes quite dense, containing predominantly mononuclear cells. Tissue necrosis may occur in the indurated area when the reaction is severe. As the reaction subsides, redness and swelling disappear within a day or so, but the induration may persist for weeks.

This type of skin test result is typical when microbial antigens are used to demonstrate the hypersensitivity of infection. It provides a diagnostic clue as to the nature of an infectious disease or its stage of development. A positive skin test alone, however, does not necessarily indicate currently active infection, but only the fact that infection (clinical or subclinical) with the agent has occurred at some time prior to the test. It may be of assistance in the diagnosis of a large number of diseases, whether they are of bacterial (tuberculosis), fungal (histoplasmosis, coccidioidomycosis), or helminthic (trichinosis) origin. The antigen employed in most instances is a cell-free extract of the organism, containing one or more of its allergenic constituents (see colorplate 4g).

Allergic Contact Dermatitis (ACD)

Allergic contact dermatitis may be induced by many simple chemical agents. The symptoms of ACD may range from those of a minor skin rash to the appearance of large blisterlike lesions. These vesicles, or *bullae,* as they are called, are surrounded by an area of erythema; they may ooze or become crusted, and they are usually intensely itchy. Scratching not only may lead to infection of the lesions but may distribute the antigen to other areas of the body.

The most common antigens of ACD include such widely different agents as metals (nickel, mercury), soaps, cosmetics, disinfectants, ointments, and drugs (colorplate 4f). Plant substances, particularly the oleoresins of poison ivy, poison oak, and poison sumac are notoriously allergenic. Since many of these substances are not composed of large molecules characteristic of complete immunogens, it is thought that they function as haptens, combining with skin proteins that act as carrier molecules. Such antigenic compounds are especially effective in activating T lymphocytes. Repeated direct contact with the allergen is generally the cause of ACD, but indirect exposure may also induce it. The oily resin of poison ivy, for example, may be carried on clothing or in the smoke from burning plants. Skin testing to determine the causal agent is usually done by a patch test method in which the antigen is taped to the skin, either directly or on a piece of gauze or other inert fabric, and kept in contact for 24 hours. The reaction is read at that time and again after 48 hours.

Corticosteroids are used to treat severe ACD. The condition can be completely cleared only by avoiding contact with the allergen.

AUTOIMMUNE DISEASES

Failure of the immune system to recognize "self" and its activity against the body's own tissue antigens may lead to the development of autoallergies, or **autoimmune diseases.** Not all the factors involved in the pathogenesis of such diseases are fully understood, nor is it always clear whether cellular or humoral immunity is responsible, or both. Possibly an imbalance between them or a defect in one of these mechanisms may contribute to autoimmunity, but certain other mechanisms are also thought to be operative, as follows:

1. Some tissues of the body are relatively isolated from the circulating lymphocytes of the blood and lymph. Thus, under normal circumstances proteins of the crystalline lens of the eye, spermatozoa, thyroid tissue, islet cells of the pancreas, and brain tissue are not encountered by the lymphocytes of the immune system. The introduction of such proteins into the

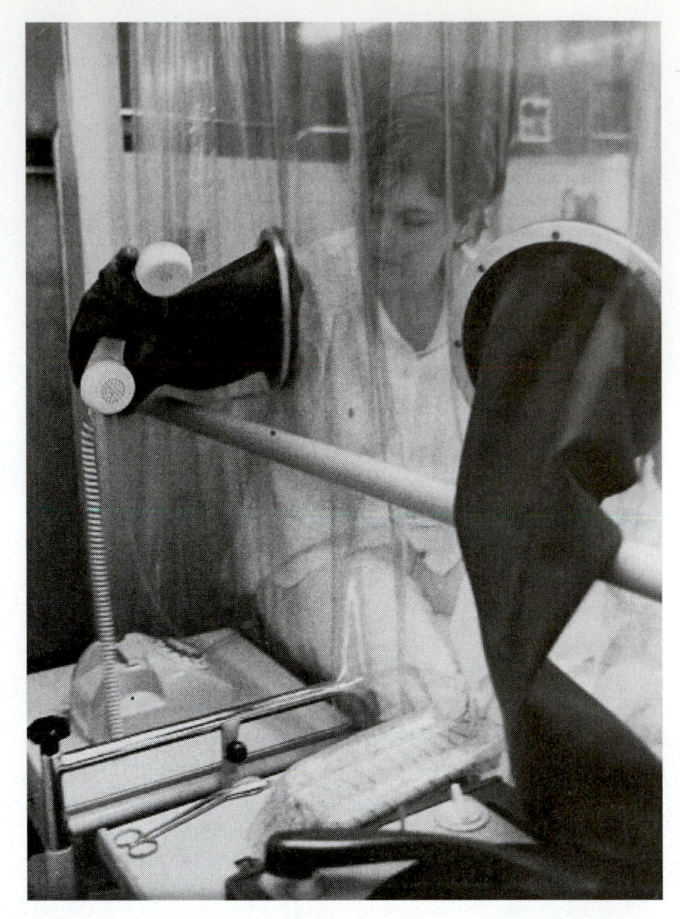

FIGURE 6.19 This patient has received a bone marrow transplant for the correction of a blood disease. She is in a specially designed aseptic room due to the immunosuppression that makes her more susceptible to infections.

World Health Organization. Photo by E. Schwab.

circulating blood through some disease process, such as viral infection or surgical procedure, may result in activation of specific lymphocytes that react to them as foreign antigens. The clinical result may be an allergic endophthalmitis, aspermatogenesis (failure to produce viable sperm), thyroiditis, autoimmune diabetes, or encephalitis, depending on the tissue antigen involved.

2. An antibody formed in response to a specific immunogen may cross-react with a different one if both antigen molecules have one or more constituents in common, "recognizable" by the antibody. Thus, an antibody produced in response to a foreign protein, such as a streptococcal antigen, might also react with a tissue protein of similar molecular structure. This may be the basis of some immunologic characteristics of diseases such as rheumatic fever and glomerulonephritis occurring as sequelae to streptococcal infection.
3. Certain blood disorders, such as autoimmune hemolytic anemia, are caused by the action of antibodies on antigens of the red blood cells. In some cases the sensitization has been attributed to the administration of certain drugs (quinidine, sulfonamides) that may be responsible for altering the antigenic specificity of the host cell.
4. A genetic defect may operate in such a way as to bring about antibody production against some normal human protein. This type of effect appears to have influence in the patient with rheumatoid arthritis whose serum contains an IgM or IgG directed against normal human IgG. In other diseases (lupus erythematosus, polyarteritis nodosa, scleroderma, and others) an autoimmune destruction of the nuclei of certain types of normal human cells takes place.

Graft Rejection

Attempts to replace defective or diseased organs with tissue transplantation are sometimes defeated by the body's own immune mechanisms. **Graft rejection** will occur if the donor's tissues are sufficiently foreign to the recipient to cause them to be attacked and destroyed by T lymphocytes and, to some extent, antibodies. Donors for transplantation must be chosen on the basis of close matching of their major histocompatibility complex tissue antigens with those of the recipient. In humans, these tissue antigens are referred to as *human leukocyte antigens* (HLA), some of which (called class I) are expressed on all tissues and others (class II) expressed primarily on macrophages and lymphocytes. The genetic regulation of class I and class II antigens is highly complex and multiple forms (50 to 100) are found. As a result, it is extremely rare for any two individuals to have identical human leukocyte antigens. Since these are under genetic (i.e., hereditary) control, however, family members often share common, though not identical, tissue antigens and therefore are preferred donors. Procedures that inhibit the immune system (immunosuppressive techniques) are also used to prepare the recipient for a transplant.

The care of immunosuppressed patients is extremely challenging (fig. 6.19), for such patients have little or no resistance to even minor infections. Their recovery, and the success of transplantation, depend heavily not only on the immunocompatibility of the grafted tissue but also on their ability to withstand infectious diseases. Pneumonia and other infections caused by the patient's commensalistic organisms may constitute life-threatening complications until immunosuppressant therapy can be discontinued and the immune system's functions can recover.

Tumor Immunity

Cancer cells differ antigenically from normal cells and are therefore subject to attack by the immune system, particularly CMI. Although T cells can and do destroy cancer cells, tumors or malignancies of the lymphatic system may progress if their growth outstrips the ability of the immune system to keep pace, just as the battle of infection may be lost in the establishment of infectious disease. The

TABLE 6.6 DIFFERENCES BETWEEN IMMEDIATE AND DELAYED TYPES OF HYPERSENSITIVITY

	Immediate Type	Delayed Type
Mediating system	Humoral immunity	Cellular immunity
Factors involved	Primarily IgE antibodies without complement IgM and IgG involved with complement in immune complex diseases	T lymphocytes and their lymphokines
Clinical examples	Anaphylaxis, atopic allergies (asthma, hay fever, hives), immune complex diseases (serum sickness, rheumatoid arthritis, glomerulonephritis)	Microbial allergies, allergic contact dermatitis (chemicals, poison ivy, etc.),
Tissue events	Capillary dilatation and increased permeability, edema, smooth muscle contraction (histamine, serotonin, kinin effects)	Inflammatory response, macrophage activation, cytotoxicity (T cell and lymphokine effects)
Passive transfer	Circulating antibodies transferred passively by serum injection	Transferred passively by injection of lymphoid cells or their extracts

propensity of the cellular immune system to destroy "non-self" tumor cells offers hope that immunotherapy, particularly in patients with immune defects, may be a valuable approach to the treatment of cancer.

Summary of Hypersensitivity

Some of the important differences between immediate and delayed types of hypersensitivity are summarized in table 6.6.

IMMUNODEFICIENCY DISEASES

Disorders of the immune system (**immunodeficiency diseases**) create serious problems in resistance to infection. Immunodeficient persons are highly susceptible to infection, giving almost any microorganism they meet the chance to be a pathogen or a real pathogen the chance to display more virulence. The incidence of cancer is also much greater among such individuals than in the general population.

Immunodeficiencies may be congenital (primary), iatrogenic (induced by immunosuppressive drugs), or the result of malignancies of the lymphatic system. Those that are not congenital are referred to as secondary immunodeficient disorders. We will classify them here on the basis of their origin in B cell abnormalities of function, T cell defects, or a combination of these. Defects of phagocytic cells have also been discovered.

B Cell Abnormalities

Aberrant production of immunoglobulins (gamma globulins) is called a *gammopathy.* The abnormality may appear in the form of an overabundance, as in hypergammaglobulinemia, or a decreased production or increased loss as in hypogammaglobulinemia.

Hypergammaglobulinemias

Each class of immunoglobulins is believed to be produced by a different specific clone of B cells. If one type of clone produces an overabundance of its immunoglobulin, the condition is called *monoclonal gammopathy* (MG). Biclonal or triclonal gammopathies are rare, but *polyclonal gammopathy* (PG), characterized by a general increase in all classes of immunoglobulins by many clones, is more common among patients with **hypergammaglobulinemias.**

PG can develop from a variety of diseases including infections, diseases of the liver, immune complex disorders such as rheumatoid arthritis or systemic lupus erythematosus, and malignancies. MG disorders are called *plasma cell dyscrasias* because they result in an overproduction of one type of immunoglobulin from a particular clone of proliferating plasma cells (mature B cells). Multiple myeloma is a malignant disease of plasma cells associated with MG. Large quantities of the immunoglobulin are filtered through the kidneys in this disease, appearing in the urine where it is called *Bence Jones protein.* Renal disease is a prominent feature of the later stages of multiple myeloma. Because high levels of abnormal immunoglobulins are accompanied by decreased levels of normal gamma globulins, the patient's susceptibility to infection often becomes the final cause of death.

Hypogammaglobulinemias

A decreased synthesis, increased catabolism (destruction), or loss of immunoglobulins through the urinary or intestinal tracts may all lead to **hypogammaglobulinemias.** Complete failure of immunoglobulin production is a congenital defect called *agammaglobulinemia.* Normally, the fetus and newborn infant can synthesize some IgM, acquiring most of their IgG passively from the mother. Normal babies produce their own IgG by the end of the first year of life. Infants with agammaglobulinemia may appear normal until this time or until their maternal IgG has disappeared, but then they become abnormally susceptible to infection. Because they have a normal CMI response to infection, hypersensitivities may develop.

Secondary hypogammaglobulinemias develop in adults and children as a result of bone marrow and lymphoid disorders. Increased catabolism causes myotonic dystrophy, and disorders of the gastrointestinal tract or kidneys may lead to dangerous losses of gamma globulins by these routes.

Low levels of immunoglobulins can be treated by injections of gamma globulin, the frequency depending on the severity and/or permanence of the condition.

T Cell Abnormalities

Congenital absence of the thymus gland leaves little hope for an affected infant to survive long or normally. This condition, called *DiGeorge's syndrome,* leaves the infant open to severe bacterial, viral, or fungous infections, as well as more liable to develop cancer. Such children must be raised in complete protective isolation, with no direct physical contact with others. Although their immunoglobulins are normal, the complete absence of CMI provides ample opportunity for microorganisms to invade disastrously.

Even within their short life spans, thymus-deficient children show an increased incidence of cancer as compared with normal children. These tumors tend to arise in the lymphatic or mononuclear phagocytic systems, often being of the lymphosarcoma or reticulosarcoma type. Studies of malignancies of the lymphoid system have contributed enormously to an understanding of the monitoring role of T lymphocytes in natural cancer prevention. It is when the T cell function is absent or overwhelmed that the incidence of cancer rises.

Combined Immunodeficiency

Congenital absence of bone marrow stem cells results in dual deficiency of B and T cells that is usually fatal within the first year of life. This total immunologic incompetence is a **combined immunodeficiency** called the *Swiss type* of immunodeficiency. Recently, techniques for the transplantation of bone marrow with competent stem cells have had some success in restoring such children to normalcy.

Phagocytic Cell Abnormalities

Abnormalities in phagocytic cells prevent them from carrying out their primary functions: engulfing and destroying bacteria that have been opsonized with antibody and complement; and killing intracellular bacteria, such as the tuberculosis bacterium, responsible for chronic infections. One defect, known as leukocyte adhesion deficiency, prevents adhesion and migration of phagocytes across blood vessel walls. As a result, the cells are unable to get to the sites of multiplying bacteria, ingest opsonized bacteria, and prevent widespread infection. With other defects, such as in chronic granulomatous disease and Chediak-Higashi syndrome, the phagocytes are unable to produce certain oxygen radicals that are toxic for microorganisms and fuse lysosomes properly with phagosomes, respectively. In these cases, because of the defective killing of intracellular bacteria, persistent infections and often death result. In each of these diseases a different critical pathway for microbial destruction is impaired allowing unchecked growth of the organisms.

Questions

1. What are the most important host factors involved in nonspecific resistance? Give examples of the protection conferred.
2. How does the blood act as a barrier to microorganisms?
3. What is the role of the mononuclear phagocyte system (MNPS) in nonspecific resistance to infection?
4. How does phagocytosis differ for macrophages and neutrophils?
5. What is the purpose of the inflammatory response? How is it mediated? What are the cardinal signs of inflammation?
6. List at least four substances that increase in blood during the inflammatory and other nonspecific responses to infection
7. What is specific resistance?
8. Define antigen, immunogen, antibody, and antigen-antibody specificity
9. Describe the structure of an immunoglobulin. How do the five classes of immunoglobulins differ?
10. What is humoral immunity? cellular immunity?
11. What is the function of B cells? T cells? How do they differ?
12. What is the function of the T cells in the CD4 group? Of what significance is this in AIDS?
13. What is acquired immunity? Passive immunity? Give examples of each.
14. What is serum sickness?
15. Describe several antibodies used in laboratory diagnostic testing.
16. Describe the anamnestic response. Why is it important?
17. What is the chief distinction between hypersensitivity and immunity?
18. What is anaphylaxis? atopic allergy?
19. What important information can skin tests provide?
20. What are autoimmune diseases? Describe one.
21. What are immunodeficiency diseases? Describe one.
22. When does graft rejection occur? What methods are used to lessen the possibility of rejection?

Bibliography

Ahmed, R., and D. Gray. 1996. Immunological memory and protective immunity: Understanding their relation. *Science* 272(5258): 54–60.

Bochner, B. S., and L. M. Lichtenstein. 1991. Anaphylaxis. *N. Engl. J. Med.* 234 (25):1785–1790.

Brown, S. 1990. Behind the numbers on the CBC. *RN* 53 (2): 46–48, 50–51.

Bruce, J. L., and S. K. Grove. 1992. Fever, pathology and treatment. *Crit. Care Nurs.* 12 (1): 40–49.

Cerrato, P. 1990. Does diet affect the immune system? *RN* 53 (6): 67–68.

Chapel, T. A., H. D. Riley, G. Russo, and E. Septimus. 1988. Cutaneous signs of infection. *Patient Care* 22 (13): 185–188, 193, 197.

Cohen, I. R. 1988. The self, the world, and autoimmunity. *Sci. Am.* 258 (4): 52–60.

Coleman, R. M., F. M. Lombard, R. E. Sicard, and N. J. Rencricca. 1992. *Fundamental immunology* 2nd ed. Dubuque, Iowa: Wm. C. Brown.

Desowitz, R. S. 1987. *A thorn in the starfish: The immune system and how it works.* New York: Norton.

Edelson, R. L., and J. M. Fink. 1985. The immunologic function of the skin. *Sci. Am.* 256 (6): 46–53.

Englehard, V. H. 1994. How cells present antigens. *Sci. Am.* 271 (2): 54–61.

Fearon, D. T., and R. M. Locksley. 1996. The instructive role of innate immunity in the acquired immune response. *Science* 272(5258): 50–54.

Golde, D. W. 1991. The stem cell. *Sci Am.* 265 (6): 86–93.

Janeway, C. A. 1993. How the immune system recognizes invaders. *Sci. Am.* 269 (1): 72–79.

Janeway, C. A., and P. Travers. 1996. *Immunobiology. The immune system in health and disease.* 2nd. ed. New York: Garland Publishing.

Krier, J. P., and R. F. Mortensen. 1990. *Infection resistance and immunity.* New York: Harper/College.

Kuhn, M. A. 1990. Anaphylaxis versus anaphylactoid reactions: Nursing interventions. *Crit. Care Nurse* 10 (5): 121–136.

Lawrence, J. 1985. Immune system in AIDS. *Sci. Am.* 253 (6): 84–100.

Lichtenstein, L. M. 1993. Allergy and the immune system. *Sci. Am.* 269 (1): 117–124.

Marrack, P., and J. Kappler. 1993. How the immune system recognizes the body. *Sci. Am.* 269 (1): 82–89.

Marrack, P., and J. Kappler. 1986. The T cell and its receptor. *Sci. Am.* 254 (2): 36–45.

Mims, C. A., N. Dummock, A. Nash, and J. Stephen, eds. 1995. *Mims' pathogenesis of infectious disease.* 4th ed. San Diego: Academic Press.

Mitchison, A. 1993. Will we survive? *Sci. Am.* 269 (3): 136–144.

Nossal, G. J. V. 1993. Life, death and the immune system. *Sci. Am.* 269 (3): 52–62.

Patel, R., and C. V. Paya. 1997. Infections in solid-organ transplant recipients. *Clin. Microbiol. Rev.* 10 (1): 86–124.

Paul, W. E. 1993. Infectious diseases and the immune system. *Sci. Am.* 269 (1): 91–97.

Paul, W. E. 1993. *Fundamental immunology.* 3d. ed. New York: Raven Press.

Rose, N. R., and I. R. Mackay. 1992. *The autoimmune diseases.* New York: Academic Press.

Roueche, B. 1991. Antipathies. In *Medical detectives.* New York: Plume.

Sheehan, C. 1990. *Clinical immunology: Principles and laboratory diagnosis.* Philadelphia: Lippincott.

Silverstein, A. M. 1989. *A history of immunology.* New York: Academic Press.

Smith, K. A. 1990. Interleukin-2. *Sci. Am.* 262 (3): 50–54.

Steinman, L. 1993. Autoimmune disease. *Sci. Am.* 269 (1): 107–114.

von Boehmer, H., and P. Kisielow. 1991. How the immune system learns about self. *Sci. Am.* 265 (4): 74–81.

Wade, B. H., and G. L. Mandel. 1983. Polymorphonuclear leukocytes: Dedicated professional phagocytes. *Am. J. Med.* 74(4): 686–693.

Weissman, I. L., and M. D. Cooper. 1993. How the immune system develops. *Sci. Am.* 269 (3): 64–71.

Wenzel, R. P. 1988. Interaction of man and microbes: Implications of the AIDS epidemic for hospital epidemiology. *J. Infect. Control* 16 (5): 214–220.

Wigzell, H. 1993. The immune system as a therapeutic agent. *Sci. Am.* 269 (3): 127–134.

CHAPTER 7

Diagnosis of Infectious Diseases

OUTLINE

OBJECTIVES

On completing this chapter the student will be able to:

1. discuss the roles of the microbiology laboratory and health care professional in ensuring the correct diagnosis of infectious disease.
2. state the differences between acute, chronic, and latent stages of infectious disease.
3. explain the need for appropriate specimens in the correct laboratory diagnosis of each stage of infectious disease.
4. indicate the importance of proper specimen selection, timing, collection, and handling to the work of the microbiology laboratory.
5. describe how the activity of the microbiology laboratory contributes to the diagnosis and treatment of infectious disease.

KEY WORDS

Acute infectious disease
Acute period
Chronic infectious disease
Convalescent period
Fever
Incubation period
Latent infectious disease
Prodromal period
Rash

INTRODUCTION

According to a national health survey, more than 55 million people in the United States experience an acute infectious or parasitic disease each year. This figure represents an incidence of one out of every four people or, looked at another way, more than 200 million days of bed disability and more than 24 million days lost from work. Therefore, the early recognition and treatment of disease is of urgent importance to the patient and to all who have been exposed to infection. Insofar as possible, the family and community contacts of the patient with a communicable disease must be considered and protected from risk. The prompt, accurate diagnosis of infectious disease is the primary responsibility of the physician. If the patient is admitted to a hospital, appropriate safeguards must be provided to prevent cross-infection of other patients and of members of the hospital staff. Since the route and ease of transmission vary with the nature and stage of the infectious disease, diagnostic accuracy may be doubly important. The hospital's health care professionals including laboratory personnel can offer immediate assistance to the physician so that diagnosis and treatment of the patient may proceed without delay, and suitable precautions may be established for the protection of others.

To play their essential roles well, health care professionals involved in patient care must be alert to the symptomatology and communicability of infections. They must also understand the circumstances in which the laboratory's diagnostic role can best be fulfilled. A background of theoretic and practical instruction in the nature of infectious diseases should provide a concept of the rhythms of host-parasite interactions and of the corresponding steps required for diagnosis, therapy, prevention, and control of infection.

In this chapter we will review some of the broad patterns of infection, their clinical recognition, and the part the laboratory can play in diagnosis.

CLINICAL STAGES OF INFECTIOUS DISEASE

Clinical infections are often described as being either *acute* or *chronic,* depending on the severity of host damage produced by the infectious microorganism and the length of the illness. In general, acute infections are characterized by a rising pitch of symptoms that continues for as long as the microorganism or its products are actively inflicting damage. The symptoms subside and disappear either when host defenses become effective or when appropriate antimicrobial treatment is administered. Chronic diseases are produced by parasites that provoke less reaction in the host as they become established. The organisms can survive the host's defenses and maintain some level of damaging activity over a longer period of time. References to diseases as chronic or acute are only generally applied; one type often reverts to the other. Thus, acute disease may subside and persist as chronic infection, or the chronic disease may flare up into an acute episode of infection.

In some instances a latent infection may exist in which the pathogen is present in host tissues but produces no reaction at all. When the host defense mechanisms are lowered for some reason, the organism may reemerge to produce an acute episode of infection.

ACUTE DISEASE

Human responses to **acute infectious diseases** occur in four stages, defined on the basis of the activities of the infectious agent and the symptoms produced. These are the incubation period, the prodromal period, the acute period of active disease, and the convalescent period of recovery.

Incubation Period

At the first encounter between parasite and host, the pathogenic properties of the parasite and the defense mechanisms of the host come into play. The parasite requires adequate living conditions and time to multiply before producing its harmful effects. The host resists this parasitic intrusion by exerting nonspecific and specific mechanisms of resistance. No outward signs of infection are visible during this **incubation period,** when the "incubating" parasite is attempting to establish itself. If the organism survives and multiplies, the symptomless incubation period comes to an end when the parasite's activities begin to damage host cells. The length of this period varies. It may be very short (two or three days) for some microorganisms, or quite long (weeks or months) for others, with an average of one to two weeks.

Prodromal Period

In its Greek origin the word *prodromal* means "running before." In medical use, the **prodromal period** refers to a brief time, usually one or two days, in which early symptoms appear, preceding the development of acute illness. Persons in this stage feel mildly uneasy or indisposed, experiencing so-called *malaise.* They may have a "scratchy" sore throat, some headache, or gastrointestinal discomfort, but often they are unable to point out the source of their problem. This gradual onset of symptoms provides the earliest evidence of the activities of a pathogenic microorganism, or its products, in host tissues. It may reflect the general dissemination of the organism through the body from its portal of entry, or it may coincide with injury developing at the site of localization, for example, in the throat or in the gastrointestinal tract. On the other hand, some acute infections have an abrupt onset, illness developing with little advance warning.

Acute Period

The **acute period** is a time in which the host-parasite interactions reach full intensity, producing symptoms of illness corresponding to the degree and site of injury. The inflammatory response of the host is in full swing, and immune mechanisms come into play. If this is the host's first encounter with the invading organism, protective antibodies are not present during the early days of acute illness. If the pathogen has been encountered previously, the rapid production of a high level of antibodies (anamnestic reaction) may shorten the period of acute symptoms or alleviate the symptoms altogether. Conversely, a preexisting hypersensitivity is likely to contribute to tissue injury and heighten the acute symptoms, although hypersensitivity reactions more commonly occur during the later stages of chronic infection.

Fever

Fever is defined clinically as any temperature above 100.5° F (37.8° C) when measured orally or above 101.5° F (38.4° C) when measured rectally. Fever is perhaps the most common symptom of infectious disease, but it occurs also in persons with a wide variety of malignancies and metabolic illnesses. The substances that induce fever, including most bacteria, fungi, and viruses, are called *pyrogens.* In gram-negative bacteria, pyrogenic activity is associated with the endotoxin portion of the cell wall. Pyrogens interact with certain host cells, primarily neutrophils and monocytes, which in turn release protein substances known as endogenous (or leukocytic) pyrogens. These endogenous pyrogens are thought to act on temperature-regulating centers in the brain, altering their "setting" and resulting in fever.

The pattern of fever seen from day to day in the acute stage of infection may reflect the nature of the disease, its progress, or the manner in which it spreads in the body. In some acute bacterial infections fever may rise to a peak, continue with little fluctuation while infection persists, then subside as the organism is subdued or eliminated. In other instances the patient's temperature may rise and fall markedly every few hours. This pattern is often seen when "showers" of organisms are released periodically into the bloodstream from some localized site in which they are multiplying. The periodicity and height of fever peaks may reflect the frequency and intensity of such releases.

When organisms are present in the bloodstream, the suffix *-emia* (i.e., "in the blood") is attached to a word descriptive of the microbe involved. Thus, we speak of bacteremia, viremia, or, sometimes, just septicemia, meaning sepsis or infection of the blood. Normally, microorganisms are easily removed from the blood, but in fulminating infections they may disseminate rapidly from an original focus to localize in many other sites.

Febrile reactions are also influenced by the age and general condition of the affected person. Children, for example, have the frightening ability to develop sudden high fevers for reasons that may be quite insignificant or extremely serious. Very old people may also display this kind of unstable temperature control. Patients in shock may have no fever or even a depressed body temperature. Fever is accompanied by a rise in pulse and respiration rates, as well as general metabolic activity. It often induces restlessness, irritability, loss of appetite, headache, or diffuse pains. Some pathogenic microorganisms may find the host's increased body temperature and metabolic activities antagonistic to their growth requirements, but fever is not necessarily an effective defense mechanism.

Rash

Rash is another characteristic of the acute phase of certain infectious diseases. In some infections the appearance of a rash and its distribution on the body may suggest the diagnosis, as in the case of scarlet fever, measles, or chickenpox. In various diseases the lesions that appear in the skin, upper respiratory tract, and oral mucosa may result from different processes of infection. For example, in the "pox" diseases, the cells of the skin may be one of several sites where the organism is localized. Alternatively, skin cells may be injured by the effects of a circulating toxin, as in scarlet fever, or they may be involved in a hypersensitive response to microbial antigens, as in certain fungal infections. Petechial lesions (from an Italian word meaning "flea bite") appear in the skin when the walls of its small blood vessels rupture, spilling blood into surrounding tissue. Such rupture may occur if the cells lining the vessel are injured by microbial constituents or actually infected by microorganisms. The latter occurs in systemic meningococcal disease or in rickettsial infections such as Rocky Mountain spotted fever, which are characterized by the appearance of petechial hemorrhages over large areas of the body. The rash of an infection may be fleeting in its duration, or it may appear pronounced and be persistent. When skin cells within a particular area die as a result of infection, the local inflammatory reaction may clear the site of cellular debris, leaving no trace. If the damage is more severe, inflammation may continue to fibrosis and scar formation. This is the case in some severe "pox" diseases, the healed skin lesions remaining as permanent scars, or pocks.

Localized Symptoms

The symptoms of many acute infections are referable to the original or extending sites of localization of the infectious agent. Nausea, vomiting, and diarrhea are frequent hallmarks of gastrointestinal infection, together with the pain of intestinal spasm. The symptoms of different kinds of intestinal infections may be similar; but the time and nature of their onset, and their duration and fluctuations, may furnish important clues as to their possible bacterial, viral, or helminthic etiology.

Respiratory infections may be difficult to distinguish because a cough, sore throat, or catarrhal discharge is common to many. In a few instances the morphology of a lesion may be characteristic or suggestive; a streptococcal sore throat is typically edematous and red, or "beefy"; in diphtheria a tough pseudomembrane may be formed over the pharynx or the larynx; some virus infections produce small, clear blisters, or vesicles, on the reddened wall of the pharynx. Similarly, X ray of the lower respiratory tract may be diagnostically useful by revealing the size, shape, density, and distribution of lesions.

Abscess formation and other manifestations of the inflammatory process, wherever they occur, produce symptoms referable to the local tissue injury. Inflammatory infections of the brain or other central nervous system tissue may be associated with symptoms such as drowsiness, mental confusion, a stiff neck, a spastic or flaccid paralysis of involved muscle groups, or other neurologic signs of motor nerve damage.

Convalescent Period

Survival and recovery from the acute phase of disease depend on the extent and nature of the damage done, the strength of the patient's resistance mechanisms, and the adequacy of the medical support received. When it is possible to offer specific therapy aimed directly at elimination of the infectious microorganism, the acute stage may be shortened, the degree of damage lessened, and the strain on natural host reserves reduced.

The **convalescent period** begins with the decline of fever, often accompanied by a feeling of weakness but great improvement. It is at this point that antibody production has usually reached a peak and, if the level is sufficiently protective, contributes significantly to inactivation of the infecting organism (see chap. 6). The persistence of antibody may then provide the individual with a long-lasting immunity. It should be noted that prompt, successful antimicrobial therapy may not only abort the disease but also lower the level of antibody production by eliminating the immunogenic stimulus at an early stage.

CHRONIC DISEASE

If a parasite produces only minor damage in the host, a persistent, chronic relationship may develop and remain unnoticeable. Usually, however, **chronic infectious disease** is characterized by a rise and fall of symptoms referable to the activities of the parasite or its products and the host's responses to them. These responses often involve hypersensitivity reactions.

The symptoms may be acute each time the parasite is active; they may be of diminished character (subacute) if the host's defenses are more effective; or they may change in quality altogether and increase in intensity if new tissues become involved. Syphilis is an excellent example of the last situation. It is a long and slowly advancing disease that, if untreated, can progress through three stages, each with different symptoms. Tuberculosis also may be slowly or rapidly progressive. In both instances hypersensitive changes contribute to injury that is reflected by changing symptoms as time passes. Thus the pattern of chronic infections reflects the seesaw equilibrium between the two antagonists, just as it does in acute infections, although the level of reaction is different.

LATENT DISEASE

The most successful parasites do not produce lethal injury in their host or even evoke antimicrobial responses that result in their destruction. A number of microorganisms can live intracellularly in mononuclear cells for long time periods causing a **latent infectious disease** but emerge when the natural or acquired immune responses of the host become compromised. In addition to their role in acute and chronic disease, tuberculosis organisms can produce latent infections. Most tuberculosis seen in elderly individuals is the result of such infections, and the patients unknowingly may have carried the organism for years, usually in small lung foci.

The virus responsible for chickenpox may lie dormant in nerve cells after a childhood attack of the disease and reemerge years later, under conditions of stress, to produce shingles or herpes zoster. Similar latent viral infections may be responsible for chronic neurologic diseases known as "slow virus infections," but, as yet, little is known about their pathogenesis (see chap. 2, Prions).

LABORATORY DIAGNOSIS

SPECIMENS

Infectious diseases are diagnosed most accurately when the etiologic agents are isolated from patient material. The selection, timing, collection, and processing of patient specimens is of utmost importance. For this reason it is essential that there be an adequate liaison between the laboratory, physicians, and those health care personnel who collect patient specimens.

The laboratory depends on the physician for information about the possible nature of the infection so that microbiology technologists can use appropriate methods to isolate the pathogen. The physician in turn may depend on the laboratory for advice about selection and timing of the specimen. The nurse is frequently the essential person who collects the desired material at the right time, in a specified manner, and transports it promptly to the laboratory. Appropriate specimen collection requires the patient's cooperation and explicit directions must be provided, for example, when collecting a clean-voided midstream urine specimen or a sputum specimen (rather than saliva). In addition, the specimen must be collected in an appropriate transport container, labeled with the patient's name, and accompanied by a laboratory request slip identifying the patient, hospital or clinic location, type of material, time of collection, and nature of the studies desired (fig. 7.1). In many hospitals this patient and specimen information is transmitted from the outpatient and inpatient clinical units directly to the laboratory by computer entry (called computerized or automated order entry). The patient with an infectious disease depends on the physician, nurse, and laboratory to collaborate effectively so that prompt, accurate diagnosis and treatment may be assured. Neighboring hospitalized patients also depend on them to prevent cross-infection, particularly during the interval between correct diagnosis and appropriate treatment of the infectious case.

FIGURE 7.1 If the microbiologist is to be successful in the search for microorganisms of pathogenic significance in patients' specimens, the appropriate clinical material must be collected in suitable containers. Specimens and their accompanying request slips must be labeled correctly and completely.

Specimen Selection

Many kinds of specimens may be sent to the laboratory for the isolation and identification of significant pathogens, including blood, urine, feces, sputum, and pus from draining wounds. These types of specimen are the most common and most readily obtained. Infectious microorganisms may be found in a variety of tissues and body fluids, which may have to be taken by surgery, needle aspiration (bone marrow, cerebrospinal fluid, pleural or peritoneal exudates), or intubation (duodenal drainage). Clinical signs of infection should direct selection of the most appropriate specimen type. Thus a patient with symptoms referable to the urinary tract (urgency, frequency, pain on urination) should have urine cultures performed. Likewise, a patient with respiratory symptoms and an abnormal chest X ray requires cultures of sputum or material taken from lower areas of the respiratory tract. In a number of infections blood samples are frequently rewarding, particularly if the patient is febrile. Multiple blood specimens, usually two or three per febrile episode, are often necessary because organisms may be "showered" into the bloodstream at intervals from a local area of infection.

Specimen Timing

It is essential that specimens for laboratory culture be taken before antimicrobial therapy is initiated. Antimicrobial agents may not only affect the viability of microorganisms in the tissues, but they may also interfere with growth of organisms in culture, or change their metabolic characteristics.

Knowledge of the usual clinical course of a given infection, together with careful observation of the patient's signs and symptoms, guides the selection and timing of specimens sent to the laboratory. If the responsible pathogen is to be recovered in a viable, active state, not significantly affected by host defenses, specimens must be collected during the acute rather than convalescent stages of disease. Blood samples drawn during the prodromal period may yield microorganisms being disseminated silently from an original point of entry, but the stage of acute illness is most likely to reveal organisms not only in the blood but in specimens from other body areas where the organisms may be localized (e.g., a swab from the throat, an excised lesion, sputum, feces, aspirated spinal fluid, pus from an abscess). Similarly, throughout the course of a chronic infection, specimens collected from involved tissues at the height of recurring symptoms offer the laboratory the best opportunity for isolation and identification of the responsible agent.

Correct timing of the collection of blood samples for culture is critical because, as discussed, organisms may be released from the infection source at intervals. Since chills, followed by fever, indicate the presence of microorganisms in the bloodstream, cultures are most likely to be positive when they are collected during such episodes. On the other hand, in patients with heart valve (infective endocarditis) or blood vessel infections, organisms are present continuously in the bloodstream and multiple positive blood cultures may help to confirm the clinical diagnosis.

Timing of blood samples collected for serologic studies is also important because immunologic diagnosis often requires the demonstration of a rising level of antibody as infectious disease proceeds from the acute to the convalescent phase. The presence of antibody in a single serum sample may reflect either current infection or some previous exposure to the specific antigen; but an increasing quantity of the antibody in serum specimens drawn as the disease progresses provides strong evidence that the antigen is now involved in the infectious process. Since antibody formation can be slow, particularly on first exposure, it is important that blood samples be spaced to reflect the change in the patient's symptoms.

Specimen Collection

The proper collection of specimens involves not only their sterile or aseptic transfer into sterile containers but a number of equally important factors: a quantity adequate for the necessary studies; material representative of infection rather than of the commensal, contaminating flora of the region; and prompt examination by the laboratory. The volume of the specimen obtained for testing is often critical. For example, small numbers of significant organisms may not be detected unless a large specimen sample is available for concentration. Sputum specimens from patients suspected of having tuberculosis and spinal fluid from patients with signs and symptoms of meningitis are always concentrated by centrifugation before they are cultured. At certain stages of these diseases only a few organisms may be present in the material submitted for culture. A generous volume is also necessary when several culture methods or serologic tests must be performed on the same specimen.

If the specimen is taken from a source that is colonized with commensalistic organisms, or must pass through such a site (e.g., sputum coughed up through the oropharynx), it is important to collect material that is as free from contaminating organisms as possible and is representative of the infectious process. Thus, a draining abscess should not be sampled at its surface where normal flora of the skin may predominate. Instead the sample should be taken from its deep interior where the pathogenic agent can be found. In addition, the skin surface at the sample site should be cleansed thoroughly beforehand and treated with an effective antiseptic such as an iodophor (see chap. 9). The specimen should be delivered as quickly as possible to the laboratory so that microbiologic examination can begin promptly. If a delay occurs, fastidious pathogens may not survive in the presence of hardier organisms, or other components of the specimen may be deleterious—phagocytic action may continue in a sample of pus, urine may be too acid or alkaline, antimicrobial agents may continue to act.

Specimen containers should always be sterile so that environmental organisms are not introduced into cultures (fecal specimen containers are an exception since feces are plated on highly selective media). For specimens collected on swabs, special transport systems containing non-nutritive medium are available to protect any pathogens present from drying and overgrowth with commensals. When it is inoculated into blood cultures, an anticoagulant known as *sodium polyanethol sulfonate* also prevents continued activity of phagocytes and inactivates some antimicrobial agents. Since blood is commonly inoculated directly into blood culture broth at the patient's bedside, the effects of sodium polyanethol sulfonate help to ensure survival of organisms during their journey to the laboratory and throughout the incubation process. A knowledge of appropriate methods for collection and transport of specimens helps to ensure isolation of infectious agents and has direct patient benefit. All health care workers including nurses, physicians, and microbiology laboratory technologists must be aware of these methods and keep each other informed of new developments. Table 7.1 illustrates some examples of specimen collection for laboratory diagnosis.

Specimen Handling

Clinical specimens are selected and sent to the laboratory for culture because they may contain viable pathogenic agents of infectious disease. For this reason, every such specimen must be considered as a potential source of the disease for all who handle it. Careless manipulation of the specimen when it is placed in its container may contaminate the hands of the physician or nurse, the outer surfaces of the vessel, or the request slip accompanying it. The messenger who transports the container is then also subject to the risk of contaminated hands or clothing, as are laboratory personnel who subsequently work with the specimen. Each of these people in turn may have many continuing contacts with others—patients, other personnel, friends and family at home—all of whom may form a communicating chain for transfer of infection. Therefore every specimen must be handled with thoughtful care, placed in a container of adequate size to retain it without spilling, and fitted with a tight closure. If adjacent surfaces are contaminated by the specimen during collection, they should be promptly and carefully disinfected with an effective germicidal agent. In the microbiology laboratory all techniques and equipment used in handling specimens and cultures are designed to reduce risk of cross-infection to a minimum, both within the laboratory and beyond it. The same principles of aseptic care must be applied at the patient's bedside and throughout the hospital.

Functions of the Laboratory

The laboratory has three essential functions with regard to clinically significant infections: it can (1) establish or confirm the etiology by isolating and identifying the agent, (2) perform antimicrobial susceptibility testing on the implicated bacterial organisms, and (3) conduct serologic tests to determine the patient's specific antibody response. Thus, the laboratory can provide either a microbiologic or a serologic diagnosis, or both, together with valuable information as to the appropriate course of treatment for the patient. These functions are described in detail in chapters 4 and 6.

Within recent years, new technologies have been developed for the noncultural diagnosis of several infectious diseases. These systems permit the detection of microbial antigens (see chap. 6, Antigen Detection Methods) or specific nucleic acid sequences (using DNA probes, see chap. 4) directly in clinical specimens. The use of these systems precludes the need to isolate and subsequently identify a

TABLE 7.1 SOME EXAMPLES OF SPECIMEN COLLECTION FOR MICROBIOLOGIC DIAGNOSIS

Suspected Clinical Diagnosis	Specimen of Choice	Collection of Specimen	Comments or Precautions
Septicemia	Blood	Directly into blood culture broth or sterile tube with sodium polyanethol sulfonate	Disinfect venipuncture site thoroughly before collecting blood
Pharyngitis	Throat swab	Place swab in transport medium for delivery to laboratory	Swab throat areas showing redness or whitish patches
Pneumonia	Sputum; transtracheal aspirate; bronchoalveolar lavage	Sterile specimen container with leakproof lid or cap	Collect sputum, *not* saliva; aspirate from transtracheal catheter with syringe
Tuberculosis			
Pulmonary	Sputum	Same as above	First morning specimen; send at least 10 ml to laboratory
Disseminated	Urine	Same as above	First morning specimen; send 5–10 ml to laboratory
	Spinal fluid; tissue	Same as above	Specimen must be large enough to permit recovery of organisms if they are few
Gastrointestinal infection:			
Salmonellosis Shigellosis Campylobacteriosis	Feces	Clean specimen container with leakproof lid or cap	Specimen must be freshly collected and delivered promptly to laboratory
Meningitis	Spinal fluid	Sterile screw-capped tube or vial	Immediate delivery to laboratory is essential; send first portion collected for culture
Wound, abscess	Pus	Swab or aspirate in anaerobic transport system	Disinfect skin site thoroughly; collect pus from deep interior of wound; avoid surface contaminants; deliver to laboratory promptly
Urinary tract infection	Urine	"Clean-catch" (midstream) or catheterized specimen in sterile container with leakproof cap	Speed of delivery to laboratory essential to prevent growth of contaminants; refrigerate if delay cannot be avoided
Viral hepatitis	Blood for serologic tests	Clean tube with leakproof stopper	Microbiologic tests not readily available

microbial pathogen and allows test results to be obtained in much shorter times than by culture methods. These tests are most useful for microorganisms that are slow or difficult to grow in culture (for example, *Legionella,* mycobacteria, chlamydiae) but are available for detecting some common pathogens as well.

Currently, these techniques do not determine the antimicrobial susceptibility of the detected microorganism; therefore, when appropriate drugs for treating the particular bacterial species cannot be predicted, a culture must be done also to retrieve the organism for susceptibility testing. In addition, because these rapid methods may not detect the presence of small numbers of a pathogen in the specimen, negative test results may require a culture to be performed when the suspicion of the pathogen's presence is high.

Finally, the laboratory must report to the physician the results of all findings, both positive and negative, in a timely fashion. The introduction of computerized ordering and reporting has helped to speed the flow of information, and computers are an essential part of the hospital and laboratory environment (fig. 7.2).

QUESTIONS

1. List the clinical stages of acute infectious disease.
2. Describe the difference between latent and chronic diseases?
3. Define and give examples of the following terms: pyrogen, bacteremia, viremia, and septicemia.
4. What causes fever during the course of an infectious disease? How does fever affect normal physiology?
5. Name three essential functions a laboratory has in the diagnosis and treatment of clinically significant infections?
6. What are important considerations in the collection of specimens for the microbiology laboratory?

BIBLIOGRAPHY

August, M. J. 1985. Practical aspects of viral laboratory diagnosis. *J. Med. Technol.* 2 (8): 500–506, 514–515.

Baer, D. M. 1993. Urine screening. *MLO.* 25 (1): 12–14.

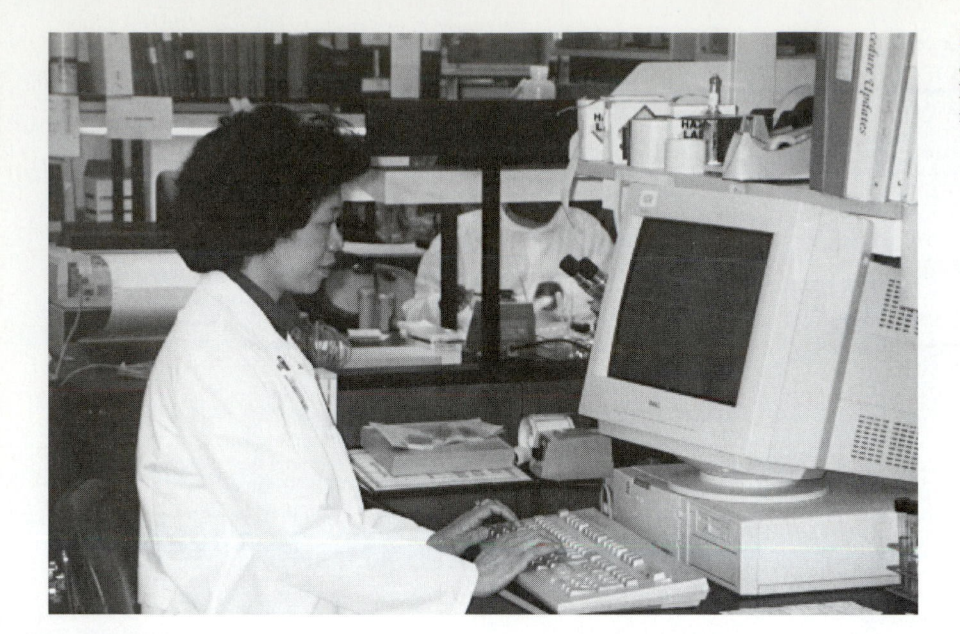

FIGURE 7.2 A technologist records in a computer the completed findings of a laboratory culture to be entered in the patient's record.

Baron, E. J., L. R. Peterson, and S. M. Finegold. 1991. *Bailey and Scott's diagnostic microbiology,* 9th ed. St. Louis: Mosby.

Bartels, J. H., and M. P. Weinstein. 1987. Appropriate use of the microbiology laboratory in the diagnosis of infectious disease. *Am. J. Infect. Control* 15 (5): 187–195.

Bottone, E. J., R. Girolami, and J. M. Stamm, eds. 1982. *Schneierson's atlas of diagnostic microbiology,* 8th ed. No. Chicago: Abbott Laboratories.

Flourney, D. J., and L. Adkins. 1986. Understanding the blood culture report. *Am. J. Infect. Control* 14 (1): 41–46.

Fuchs, P. C. 1993. Urine culture. *MLO.* 25 (10): 11–12.

Gurevich, I. 1988. How to make every culture count. *RN* 51 (8): 19–55.

Kohles, M. K., and P. L. Barry. 1989. Clinical laboratory and nursing personnel: Collaboration in improving patient care. *J. Nurs. Qual. Assur.* 3 (2): 1–10.

Kunz, L. J. 1990. Rapid diagnostic techniques: How to get the most from your microbiology laboratory. *Asepsis.* 12 (1): 14–17.

McFarlane, A. 1989. Using the laboratory in infection control. *Prof. Nurse* 4 (8): 393–394, 396–397.

Morello, J., H. Mizer, and M. Wilson. 1997. *Laboratory manual and workbook in microbiology: Applications to patient care.* 6th ed. Dubuque, Iowa: McGraw-Hill.

O'Bannon, R. H. 1988. The effects of improper specimen handling on lab tests. *MLO* 20 (11): 42–47.

Wasserman, M., and E. L. Keller. 1989. Fever, white blood cell count, and culture and sensitivity: Their value in the evaluation of the emergency patient. *Top. Emerg. Med.* 10 (4): 81–88.

SECTION III

A laboratory assistant surveys the latest insect catch to assess its capacity for spreading disease.
World Health Organization. Photo by R. Da Silva.

Epidemiology and Infection Control

CHAPTER 8

Sources, Routes, and Patterns of Infectious Diseases

OUTLINE

OBJECTIVES

On completing this chapter the student will be able to:

1. discuss how infectious microorganisms are transmitted from their sources in humans, animals, or the environment.
2. cite factors that influence the communicability, or transfer, of infectious diseases in human and animal populations.
3. define the meaning of the term *reservoir* with reference to infection.
4. describe differences in the means by which microorganisms are transmitted from living and inanimate reservoirs.
5. discuss the role of insect vectors in the transmission of infectious disease.
6. show how the life cycles of animal parasites affect how they are transmitted from one host to another.
7. cite the factors that influence the communicability of infectious diseases in human or animal populations.
8. describe the major patterns of infectious diseases and the factors in host populations that affect the frequency and spread of such diseases.
9. define epidemic, endemic, and pandemic diseases.
10. describe conditions in the human environment that influence the patterns and incidence of infectious disease.

KEY WORDS

Communicable
Endemic
Endogenous (end-**AH**-gen-us) sources of infection
Epidemic
Epidemiology
Pandemic
Reservoir of infection
Transmission routes of infection
Vectors of infection

EPIDEMIOLOGY

Epidemiology is the science that deals with the determinants, incidence, distribution, and control of health and disease in a population. It constitutes the sum of all known factors that determine the presence or absence of a disease in a given group. For infectious diseases, these contributing factors include the pathogen, host, and environmental conditions that together determine the occurrence of disease in a specified population over a stated course of time. In its original sense the word *epidemiology* (from the Greek "upon the people") applied to the study of **epidemics,** that is, of infectious communicable diseases that fell upon large numbers of people simultaneously or in rapid succession. After the microbial agents of many infectious diseases were recognized, it became apparent that they do not always produce disease in epidemic patterns but may be implicated also in occasional, sporadic cases. Furthermore, an epidemic outbreak may be limited to very small numbers of people or may be **pandemic,** spreading across continents to affect a larger number of people. The size of the outbreak depends in part on the level of resistance to the responsible microorganism in the exposed population.

It has also become apparent that a specific parasite may persist in a host population causing numerous subclinical infections. Here numbers of patients with clinical illness are few but relatively uniform from one time to another. When this occurs the disease is said to be **endemic,** meaning "in the people," as is the case with the common cold, for example.

The modern concept of epidemiology includes the study of interlocking factors that determine the occurrence, distribution, and effect of any disease on human welfare and health. Thus, we may speak not only of the epidemiology of a particular infection (such as tuberculosis or measles) but also of a noninfectious disease (diabetes or arthritis), or even of defined social problems (alcoholism, drug addiction, or poverty).

In the preceding chapters emphasis was placed on the major contributions of parasite and host to the causation and course of individual infectious diseases. In this section we shall discuss influences exerted on the host and microorganism by their natural, shared environment, examining the intersecting role of the latter in the epidemiology of infection. With knowledge of the epidemiologic basis of infectious disease, we can then consider how it can best be prevented or controlled by measures designed to destroy the parasite, limit its environmental distribution, or improve host resistance, or by any feasible combination of these approaches. Epidemiologic principles form the basis for all methods of controlling infection. They are as applicable to the single case of communicable disease as to large-scale problems of public health. Epidemiologic factors affecting patterns of infectious diseases are described at the end of this chapter.

COMMUNICABILITY OF INFECTION

Many microorganisms can survive under a wide range of environmental and host conditions because they have versatile metabolic mechanisms. Though widespread, they cause only a low incidence of infectious diseases in humans and animals because (1) relatively few microorganisms possess the factors needed to injure the host (pathogenicity), and (2) the healthy host has a remarkable ability to resist microbial attack (immunity). Even when a pathogenic microorganism is present in susceptible human or animal hosts, whether or not it will be **communicable** (transmitted directly or indirectly from one host to another) depends in part on the sources of infection from which disease may arise and in part on how easily the pathogen can be transmitted from source to host.

SOURCES OF INFECTION

A parasite that kills its host also dies unless it can be transferred to another environment where it can continue to multiply. The site of colonization within the host determines whether or not transmission is possible, and if so, how easily it can be accomplished. An organism localized in deep tissues without access to the body's excretions may not be able to exit under normal circumstances. As a result, it will not be an epidemiologic threat unless some intermediary agent transfers it to a new host. A blood-sucking insect, for example, may carry certain microorganisms from the bloodstream of one host to the tissues of another. For many human and animal pathogens, secretions from the mouth and upper respiratory tract or discharges from the intestinal or genitourinary tracts offer the simplest means of escape. In these instances, however, the organisms must be transferred directly to a new host or must be able to survive in the environment until one is found.

A local environment or host that supports the survival and multiplication of pathogenic microorganisms is spoken of as a **reservoir of infection.** An infected, though not necessarily diseased, human, animal, or insect host may constitute a *living* reservoir. Many environmental media (soil, air, food, milk, and water) may serve as *inanimate* reservoirs of infection.

The reservoir may or may not be the source from which the pathogen is transmitted to a new, susceptible host. For example, measles virus is transmitted directly from an infected to a susceptible (nonimmune) person. The infected individual serves both as the animate reservoir and source of the virus for the susceptible individual, who may in turn become a new reservoir and source of infection for others. On the other hand, legionella organisms growing in a hospital water supply may be spread to susceptible (immunocompromised) patients by shower heads in their rooms. The water supply is the inanimate reservoir but the source of the patients' infections is the shower head. Recognition of both the reservoir and source is important for controlling infection.

LIVING RESERVOIRS

Humans

The chief reservoirs of many important human infections are persons who, whether or not they display symptoms of clinical disease, harbor pathogenic organisms and transmit them to others. The most obvious human sources of infection are those people who are recognizably ill with a communicable disease. When it is promptly and accurately diagnosed, this source is also the most readily controlled, for sick individuals can be confined at home or in the hospital. The potential reservoir they represent is then restricted so that transmission of the pathogen is limited or impossible. At the same time, the patient is treated and successful therapy destroys the parasite.

The largest continuing human reservoir, and the most difficult to recognize or control, is comprised of infected individuals with only mild symptoms of illness, or none at all, who circulate freely in the community. Infected persons who display few or no symptoms are a potential source of infection to others and are referred to as *carriers*. In addition to those with inapparent disease, persons may be carriers before or after they suffer a known infectious process; that is, they are incubatory or convalescent carriers, respectively. When the carrier state persists for a long time, the person is referred to as a *chronic* carrier. The cook, "Typhoid Mary," who infected many individuals with typhoid bacilli during the early 1900s, is a notorious example of a chronic carrier.

The epidemiologic control of carriers of serious communicable diseases is of great importance. Efforts are directed primarily at preventing contamination of food and water supplies to avoid spreading disease from these environmental sources to large numbers of people.

Animals

Animal reservoirs of infection are a frequent threat to human health in all parts of the world. The infectious diseases of animals that can be transmitted to humans are called *zoonoses* (from the Greek words *zoon* for animal and *nosos* for disease). Zoonotic pathogens are found in every major group of microorganisms: fungi, bacteria, rickettsiae, viruses, protozoa, and helminths. Humans may acquire these infections through direct handling of animals and their products, by eating meat or other products from diseased animals, or through the bite of insects that prey on both animals and humans. There are a few zoonoses for which infected human beings also serve as reservoirs, capable of transferring their infection directly to other people (e.g., many *Salmonella* and streptococcal infections) or indirectly through another host or vector (e.g., the systemic protozoan diseases, leishmaniasis and malaria, transmitted by biting insects). For the most part, however, human beings are accidental victims of most zoonoses. If infected, they do not constitute a reservoir for such diseases as rabies, anthrax, brucellosis, or trichinosis because humans have no significant role in their transfer to other hosts.

Insects

The insects commonly involved in the transfer of infectious agents from one reservoir to another belong in the invertebrate phylum Arthropoda (literally, "jointed feet," from the Greek words *arthron* for joint and *pod* for foot). There are several classes of arthropods, but only two, the *Insecta* and the *Arachnida,* contain members that are important as **vectors,** or transmitters, **of infection.** The class *Insecta* includes flies, mosquitoes, fleas, lice, and the true bugs; the class *Arachnida* includes ticks and mites. Examples of important insect vectors are shown in figure 8.1a–f. Their association with disease is given in table 8.1.

Most of these arthropod vectors are parasitic themselves, preying on vertebrate animals to obtain blood, a rich protein nutrient for the insect. An enormous variety of animals—wild birds and domestic poultry, rodents and small game, wild and domestic animals—provide parasitic insects a source of support. When a biting or blood-sucking insect preys on an infected host, it may pick up infectious microorganisms in its blood meal and later deposit them on or inoculate them into the skin or mucous membranes of

TABLE 8.1 MEDICALLY IMPORTANT ARTHROPOD VECTORS

Class	Vector	Diseases Caused
Arachnida	Ticks	Rocky Mountain spotted fever, Q fever, relapsing fever, tularemia, Lyme disease
	Mites	Tsutsugamushi fever (scrub typhus), rickettsialpox, scabies
Insecta	Body louse	Epidemic typhus, relapsing fever
	Head and pubic ("crab") louse	Infestations known as pediculosis
	Reduviid ("assassin") bugs	Chagas' disease (South American trypanosomiasis)
	Flies (other than housefly)	Tularemia, bartonellosis, onchocerciasis, loiasis, leishmaniasis, African trypanosomiasis
	Mosquitoes	Yellow fever, viral encephalitis, malaria, filariasis
	Fleas	Bubonic plague, endemic murine typhus

FIGURE 8.1 (a) Flea. This arthropod transmits the plague bacillus from rats to humans. (b) Mite. The adult mite is involved in the transmission of scrub typhus, a rickettsial disease of the Far East. (c) Tick. A species of this insect transmits Lyme disease. (d) Louse. The human body louse is the vector of epidemic typhus fever, a rickettsial disease. (e) Mosquito. This mosquito of the genus *Aedes* transmits yellow fever. (f) Horsefly. A mechanical vector, this arthropod contaminates the external parts of its body or its digestive tract with excreta or secretions that contain the infectious agent. In the fly digestive tract the infectious organism may multiply. Hours or days later, the inoculum is deposited on food intended for human consumption, thus transferring the disease.
Centers for Disease Control, Atlanta, GA.

Flea
(a)

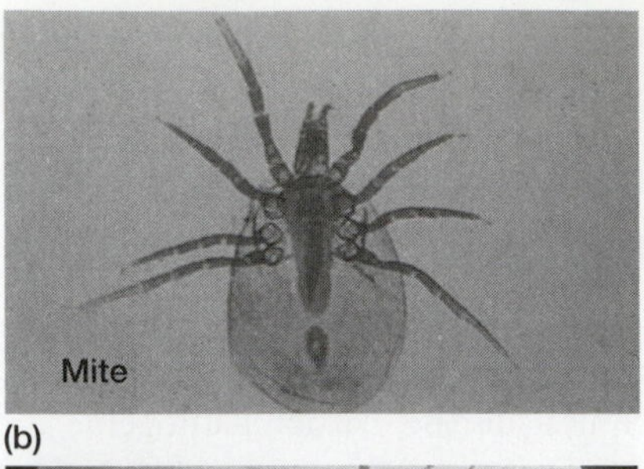

(b)

Tick
(c)

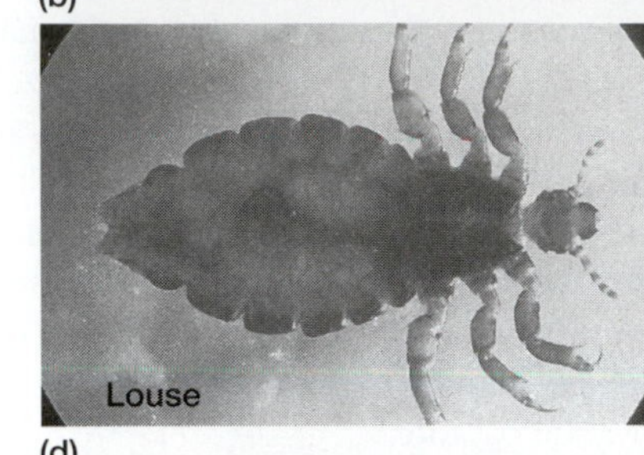

(d)

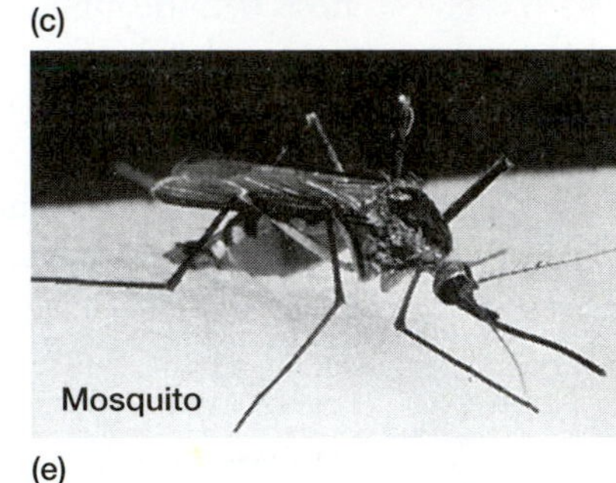

(e)

(f)

another individual of the same or different species. Thus a number of unrelated animal species may be linked together in the epidemiology of an infectious disease through the activities of an insect vector.

Often the parasitic arthropod becomes a true host to the pathogenic microorganism; that is, the insect is itself infected, affording the microorganism another site of multiplication or of development. There are some instances where a microbial parasite actually requires an arthropod host in which to undergo developmental changes that are essential to maintaining its life cycle. This is the case with the malarial parasite, which has a sexual stage of development in its mosquito host. Some pathogenic species of bacteria, viruses, and rickettsiae live within an insect host throughout its lifetime. The insect itself is not necessarily diseased by this parasitism and in some instances may even transmit infecting microbes through its eggs to its own succeeding generations. Such arthropod vectors can be considered reservoirs and sources of infection when they maintain and transfer the microbial agents of infectious disease to humans, animals, or both. Insects that serve as host and reservoir for infectious agents are often referred to as *biologic vectors* to distinguish them from those that function merely as mechanical carriers of microorganisms from one reservoir to another.

A number of nonparasitic arthropods are mechanical, or passive, vectors of infection. Among the more important are houseflies and cockroaches, which obtain their nutrients by scavenging. Flies, in particular, feed on and breed in all types of organic wastes such as human and animal excrement and garbage, rather than feeding directly on humans or animals. They do, however, visit intimately with them and can passively transport their microorganisms back and forth. Flies may ingest microbial parasites and later deposit them by defecation or regurgitation on any surface they alight upon. They also carry microorganisms on their hairy legs, depositing them in an invisible trail.

INANIMATE RESERVOIRS

The principal inanimate, or environmental, reservoirs for pathogenic organisms are soil, air, food, and water. By allowing the survival, if not the growth, of pathogens, such reservoirs may play a large part in the epidemiology of many infectious diseases.

Soil

The soil teems with microorganisms, most of which are not pathogenic for human beings. In general, the only natural inhabitants of the soil that may be human pathogens are fungi and the anaerobic bacterial endospore-former *Clostridium botulinum*. The conidia (spores) of fungi can survive for long periods on dry surfaces, but the growth and reproduction of these simple organisms require moist nutrient soil. Most fungi are saprophytic, but a few species of yeasts and molds are low-grade parasites. These may cause infections in humans through direct contacts with the soil, which constitutes the primary reservoir of fungi, or with dried earth blown about as dust. The virulent and invasive properties of the pathogenic fungi are not great and resistance of the human host to them varies, so contact with the reservoir and the ease of transmission do not alone assure disease. The fungous diseases are discussed in chapter 14.

Clostridium botulinum, an endospore-forming anaerobic bacterial species responsible for botulism (see chap. 16) also has a primary reservoir in the soil. The organism can be isolated from the intestinal tracts of fish and from the feces of domestic animals that graze on soil containing it in abundance, but it is rarely found as a commensal in the human intestine.

The soil may also serve as a secondary reservoir for some types of organisms whose primary source is the intestinal tract of human beings or animals, or both. However, this is true only for those pathogenic agents that can survive—if not multiply—in the soil until they enter a new host. The intestinal viruses (e.g., poliomyelitis virus) cannot remain alive for long outside of viable host cells; consequently, their transfer from contaminated soil must be very rapid. On the other hand, endospore-forming bacteria of the intestinal tract, protozoan cysts, and the eggs or larvae of intestinal helminths do survive in soil, retaining their infectivity long enough to be communicable from a secondary soil reservoir.

In the sense that soil generally does not support multiplication of many pathogenic organisms, but merely offers conditions that permit their survival (or, as in the case of some helminths, developmental changes), it is a limited, rather than a true, reservoir. The route of transmission of pathogenic bacteria, protozoa, and helminths from soil to a human host may be direct, as for the fungi, or quite indirect by way of other animal hosts.

Air

In its pure state, air does not supply nutrients for any living organisms, but in practical terms air is never pure. It is constantly being contaminated by dust from the soil, by various volatile products of human activities (e.g., those carried by smoke), and by respiratory secretions expelled from the mouth, nose, and throat of human beings while talking, laughing, or clearing their throats.

When their respiratory tracts are irritated by infection or by other agents, human beings may cough and sneeze, spraying the air over great distances with large quantities of finely scattered mucus and its content of commensalistic or potentially pathogenic microorganisms (see the illustration on p. 235). This mucous spray takes the form of droplets of various sizes containing protein and heavy with moisture. Heavier droplets settle out quickly, contaminating objects and surfaces on which they happen to fall, including nearby individuals. Lighter ones hang in the air for a longer time, gradually losing moisture by evaporation until they are mere droplet nuclei of flaked, dry protein, usually with microorganisms still clinging to or held within them. Many pathogenic microorganisms within such droplet nuclei survive long enough in air, or on the surfaces they settle upon, to remain infectious for other persons.

Microorganisms discharged from the human or animal body by other routes may also reach the air. Discharges from the intestinal and genitourinary tracts or from open wounds contain protein and other microbial nutrient, moisture, and often many potentially infectious organisms. Such material may contaminate the soil or any of the myriad objects with which humans surround themselves, including their clothing. As it dries, its protein particles become less sticky and readily flake off from the surfaces to which they have adhered. They can be floated into the air when disturbed, or picked up on moist hands, carrying any microorganisms that have survived the drying process.

Pathogenic microorganisms cannot multiply in air, but many remain viable within droplet nuclei or soil dust. The air thus becomes an intermediate source of many infectious diseases although, like contaminated soil, it is a limited secondary reservoir. The limiting factors in airborne infection include the capacity of microorganisms to survive drying; the speed and direction with which they are carried by air to new hosts; the availability of vectors such as insects or contaminated hands; the numbers and virulence of specific organisms; and the susceptibility of exposed hosts.

Food

Food for human and animal consumption is also a source of nutrient for many microbes. Raw food is probably always contaminated by microorganisms from the soil or from hands but is not necessarily a threat to health for two reasons: (1) the average contaminant from such sources lacks pathogenicity for humans, and (2) the microorganisms are normally not given an opportunity to multiply actively in food that is eaten fresh, or is adequately preserved before and after cooking. When food is protected from the effects of microbial growth and spoilage, it may be eaten raw or cooked without fear of infection. Such protection is afforded by refrigeration or freezing; adequate canning (fig. 8.2); smoking, salting, or drying methods; or pasteurization, as appropriate.

Unfortunately, contaminated food is all too frequently a primary or secondary reservoir for a variety of serious infections. Foods such as meat, fish, shellfish, and milk that are obtained from infected animal sources are a primary reservoir for helminthic and bacterial diseases acquired by persons who eat these foods raw or imperfectly cooked or treated. When originally clean food is contaminated by pathogenic microorganisms, dangerous secondary reservoirs of infectious disease are created, particularly if such contaminated food is improperly preserved, handled, or prepared so that microbial multiplication occurs before it is eaten. Some bacteria that contaminate food produce highly poisonous toxins as they multiply.

FIGURE 8.2 Home canning must be done carefully, using proper sterilization techniques to prevent botulism or other contamination.

Control of Microbes
Physical Control: Moist Heat

Food As a Primary Reservoir

Foods derived from infected animals or their products afford a direct route of transmission from the animal to the human host's intestinal tract. The infection that results is not necessarily limited to or even primarily localized in the intestinal tract, but the individual may then also become a reservoir capable of transmitting the agent to other persons or to animals.

Diseases acquired by ingesting infected meat, fish, poultry, eggs, or milk and maintained by an infected human reservoir fall chiefly into two groups: those of helminthic or those of bacterial origin (table 8.2, group 4).

Helminthic Diseases The helminthic diseases include those caused by pork, cattle, and fish tapeworms, ingested as larval forms encysted in the muscles of these animals, and the trematode (or fluke) infestations acquired by eating fish, crustaceans, or water plants infected with larvae. Although the pathogenic agents and the clinical symptoms of these diseases differ in many respects, their epidemiology is similar with regard to sources and routes of infection. Humans and animals both play a part in maintaining these helminths in nature by disseminating eggs into the environment. Their geographic distribution is limited by environmental influences that affect the life cycle stages of the parasite and its hosts. The social and eating habits of the human host also have their effect. In acquiring these infections, humans are often the victims of their taste for raw food or of carelessness in cooking it. Professional and home cooks may find these diseases an occupational hazard if they tend to check the taste of food before it is completely cooked.

Bacterial Diseases An increasing number of bacterial diseases has been found to be acquired from infected food. These include salmonellosis, campylobacteriosis, and yersiniosis. Before adequate government inspection and control of meats, herds, and milk in the United States and other countries, diseases such as bovine tuberculosis, brucellosis, and streptococcal infections were common. Except in some developing countries or areas where unpasteurized milk is consumed, these diseases are now rare.

Salmonella infections are widespread among animals including rodents, poultry, and other birds. They are also entrenched in many human populations. The transfer of infection among these reservoirs often occurs through the contamination of food supplies, but animal meat products, notably eggs, can be directly infectious. Fresh or dried eggs are a frequent source of *Salmonella* outbreaks, particularly when they are only lightly cooked or eaten raw, as in eggnogs. Eggs served as a valuable nutrient for patients and convalescing invalids may instead expose these particularly susceptible individuals to infection. For this reason, hospitals and nursing homes must choose suppliers of eggs from infection-free poultry farms and, when in doubt, serve only pasteurized or thoroughly cooked eggs.

Along with *Salmonella* infections, *Campylobacter* infections are commonly caused by consumption of undercooked poultry. The organism is found in the gastrointestinal tract of many animals and fowl; therefore, meat from these infected animals becomes contaminated during the slaughtering process. Infected animals also contaminate soil or water with their excreta, providing additional sources for human infection.

Yersinia infections occur more commonly in northern Europe than in the United States. The organism is found in a variety of animals including pigs, sheep, and cattle. Epidemics of foodborne yersiniosis have been linked to consumption of contaminated chocolate milk, bean sprouts, and chitterlings.

Food-Acquired Diseases Not Transmitted by Humans There are a few diseases that humans acquire from infected animal food sources but do not transmit to others. For these the infected individual is not a reservoir, but an accidental victim of infection (table 8.2, group 5).

Trichinosis, for example, is a roundworm disease of many wild animals, rodents, and swine. It is a systemic disease, induced by the encystment of larvae in muscles and organs. A person acquires it by eating the flesh of infected animals (domestic pork, wild game animals), ingesting larvae that have survived inadequate cooking. The human disease is also systemic and is the end of the line for the parasite, which has no exit from the host.

TABLE 8.2 THE RESERVOIRS OF SOME IMPORTANT INFECTIOUS DISEASES

Group 1. Humans are the only reservoir.

Diseases	
Viral	Measles; rubella; mumps; influenza; herpes; poliomyelitis; chickenpox; mononucleosis; hepatitis B; hepatitis C; AIDS
Bacterial	Shigellosis; whooping cough; diphtheria; gonorrhea; syphilis; lymphogranuloma venereum; trachoma; pneumonic plague; actinomycosis; chlamydial urethritis; pneumonia
Fungal	Candidiasis (the agent is an endogenous commensal of human alimentary mucosa)
Transmission	Direct, person to person; no durable environmental reservoirs

Group 2. Humans are the only living reservoir. Secondary reservoirs exist in the inanimate environment.

Diseases	
Viral	Hepatitis A
Bacterial	Typhoid fever; cholera; tuberculosis (human); staphylococcal and streptococcal infections (of human origin)
Protozoan	Amebiasis
Helminthic	Hookworm disease; ascariasis; enterobiasis
Transmission	Direct from person to person, or indirect from environmental reservoirs; food harvested from contaminated water (oysters, clams) or soil (garden vegetables) and eaten raw or insufficiently cooked; water for drinking, washing, cleaning, or swimming; soil contacts; surfaces

Group 3. Humans and one or more species of arthropods are the reservoirs.

Diseases	
Viral	Yellow fever (mosquito)
Rickettsial	Epidemic typhus fever (louse)
Protozoan	Malaria (mosquito)
Transmission	Arthropod to human; human to arthropod; no direct person-to-person transmission

Group 4. Humans and one or more species of animals are the reservoirs. Arthropods are involved as biologic or mechanical vectors for some. Secondary environmental reservoirs may exist for some.

Diseases		
Fungal	The superficial mycoses: ringworm of hair, skin, nails	
Bacterial	Salmonellosis; staphylococcal and streptococcal infections (of animal origin); tuberculosis (bovine); clostridial diseases (tetanus, gas gangrene); listeriosis; campylobacteriosis; yersiniosis	
Protozoan	Leishmaniasis; trypanosomiasis; giardiasis	
Helminthic	Filariasis (tissue nematodes); taeniasis (intestinal tapeworms); schistosomiasis and other trematode diseases	
Transmission	Direct from animal to human or animal Direct from person to person Indirect from environment, especially food, milk, water	
Except		
Clostridial diseases	Endogenous or indirect from fecally contaminated soil	
Leishmaniasis Trypanosomiasis Filariasis	Arthropod to humans or animals	No direct transmission
Taeniasis	Ingestion of infected beef or pork	No direct transmission
Trematode diseases	Ingestion or penetration of larvae from aquatic sources	No direct transmission

Group 5. Various animal or arthropod species, or both, are the reservoirs.

(Humans may be accidental victims of these infections but not the reservoir for them; they are rarely, if ever, transmitted from infected human beings to others of any species.)

Diseases	
Viral	Equine encephalitis; rabies; hantavirus infection
Rickettsial Bacterial	Scrub typhus; murine typhus; Rocky Mountain spotted fever; Q fever; psittacosis; anthrax; brucellosis; tularemia; bubonic plague; leptospirosis; relapsing fever; Lyme disease
Helminthic	Echinococcosis; trichinosis
Transmission	
Equine encephalitis	Horses to mosquito to humans No direct person-to-person transmission
Hantavirus infection	Deer mice to humans No direct person-to-person transmission
Rabies	Bite of a rabid dog or other animal Direct person-to-person transmission possible but not documented
Rickettsial diseases (except Q fever) and Lyme disease	Animal to arthropod (ticks, mites, fleas) to humans No direct person-to-person transmission
Q fever	Indirect from animals to humans via airborne dust or ingestion of raw milk Direct transmission from animals to humans possible No direct transmission from person to person

Continued

TABLE 8.2 *Continued*

Psittacosis	Direct contact with infected birds, or indirect through airborne dust Person-to-person transmission direct but rare
Anthrax Brucellosis Tularemia Leptospirosis	Direct or indirect from animals to humans Usual transmittal agents: dust from infective hides or tissues (anthrax); raw milk (brucellosis); ingestion of infective meat (tularemia); contaminated swimming water (leptospirosis)
Bubonic plague	Rat to rat flea to humans No direct person-to-person transmission except in cases of bubonic plague terminating with plague pneumonia
Relapsing fever	Lice and ticks transmit to humans; no direct transmission from person to person
Echinococcosis	Ingestion of infective ovum from intestinal tract of an animal, usually dogs
Trichinosis	Ingestion of pork containing encysted larvae
Group 6. Soil is the primary reservoir. Animals (including birds and fish) may be secondary reservoirs.	
Diseases	
Bacterial	Botulism; nocardiosis
Fungal	The systemic mycoses: blastomycosis; coccidioidomycosis; cryptococcosis; histoplasmosis; sporotrichosis
Transmission	Usually by contact with contaminated soil; through skin contact, inhalation of airborne dust; or, in the case of botulism, through foods contaminated with endospores

A few bacterial diseases, such as brucellosis and tularemia, also belong in this group of zoonoses that may be transmitted to, but not by, human beings. The reservoirs for brucellosis are domestic animals such as cattle, swine, sheep, and goats. Human infections may be airborne, from the dust of stables and pastures; they may be acquired through direct contacts with infected animals; or they may be transmitted through milk, cheese, and other dairy products derived from infected animals and distributed to people who would otherwise have no contact with the disease. Tularemia is a natural infection of many wild, and some domestic, animals. It is particularly frequent in rabbits. Several arthropod reservoirs also exist for this organism, commonly the wood tick and in some areas a biting fly or a mosquito. The disease is transmitted by these insects but may also be acquired through handling infected animals or by ingesting game meat (especially rabbit) that has not been fully cooked.

Food As a Secondary Reservoir

Food may become contaminated in a number of simple ways: directly from the infected intestine (via feces), skin, throat, or other superficial area to human hands, food, or milk; indirectly through fecal pollution of water supplies from which fish and shellfish are harvested or in which foods are washed; or through mechanical vectors such as houseflies trafficking from feces to food. Food is a most dangerous reservoir when it permits active multiplication of a contaminating pathogen. The best protection against growth of contaminants that have found their way into foods lies in immediate thorough cooking or prompt refrigeration, since this will kill or inhibit the growth of microbial pathogens.

Multiplication of pathogens in foods before they are eaten sets the stage for bacterial "food poisoning." When such contaminated food is ingested, one of two possible effects may ensue: (1) an active infection may result if organisms such as *Salmonella, Shigella,* and *Campylobacter* were present, or (2) symptoms of toxicity may occur if bacterial toxins were produced in the food during the period of active microbial growth prior to the meal. Examples of this type include toxins produced by *Staphylococcus aureus* and *Clostridium botulinum.* Food whose origin is unknown and whose handling and storage conditions are likely to be inadequate should never be consumed (fig. 8.3). Foodborne disease is discussed in more detail in chapter 16.

Milk

Milk is a special food, rich in protein, carbohydrate, fat, and other nutrients particularly favorable for growth of contaminating, pathogenic microorganisms. Although milk is sterile when produced by the healthy, lactating animal, it flows through mammary ducts that have a normal commensalistic flora. From that point it passes over surfaces and into containers that may be contaminated with a wide variety of bacteria and other microbes. For these reasons milk can be expected to contain a mixed microbial flora, most of which is of environmental origin and of no pathogenic importance. The most common bacterial species that contaminate and grow well in milk belong to the genera of *Streptococcus, Lactobacillus,* and *Micrococcus.* Species of these groups abound in milk but cause only souring when storage temperature and time permit their overgrowth. The potential threat offered by milk is that it provides similar opportunity for pathogenic contaminants to multiply.

Milkborne diseases originate either from infected animals or from the hands and methods of milk handlers. The important diseases of animals transmitted to human beings through their milk are tuberculosis and brucellosis. Proper pasteurization of commercially distributed milk has reduced the incidence of these diseases to low levels in developed countries. Other infections include yersiniosis, Q fever (a respiratory disease of rickettsial origin), and foot-and-mouth disease (a viral disease that can devastate cattle herds but to which humans are, fortunately, largely resistant).

FIGURE 8.3 Food prepared and served by street vendors, as in this Southeast Asian capital, is especially prone to contamination by pathogenic microorganisms.
World Health Organization. Photo by Pepas.

Human streptococcal infections may originate from animal sources or from the cross-infection of animal milks by infected handlers, as previously described.

Water

Like air, water from natural sources is seldom pure. Whether it comes to us from ground sources (springs, wells), from surface accumulation (rivers, lakes, reservoirs), or directly from the rain and snow that maintain these supplies, water always contains organic and inorganic material, suspended or dissolved. Rain washes down particulate matter suspended in the air as well as volatile chemicals from the atmosphere, many of which go into solution in rainwater. Surface and ground pools of water are not only fed by rain but are also exposed to soil. Consequently, water may normally contain many varieties of microorganisms derived from air or soil as well as the chemical substrates they need to maintain themselves and to multiply. The microbial or chemical contents of water may be removed by several methods (see chaps. 9 and 10). Such treatment is indicated whenever there is a chance that water has been contaminated by pathogenic organisms or harmful chemicals, or when the water has been polluted with materials that give it an obviously unpleasant odor, taste, or appearance.

The normal microbial flora of water is derived from soil or air. It does not include human pathogens but is composed of many heterotrophic and autotrophic microorganisms. Humans are responsible for the most frequent and hazardous contamination of water supplies by permitting it to be polluted with their sewage. The water drained from kitchen sinks, laundry tubs, bathtubs, and toilets not only contains high concentrations of soaps and detergents but teems with pathogenic and nonpathogenic microorganisms that can multiply in impure water. Disease can be transmitted to humans when they consume untreated water or food contaminated with such sewage. Diseases of enteric origin are most commonly transferred in this way. Microbial pathogens excreted in feces from an intestinal infection or from a lesion draining to the intestinal tract from the liver or gallbladder include the bacterial agents of typhoid and paratyphoid fever, shigellosis, yersiniosis, campylobacteriosis, and cholera; the viruses responsible for poliomyelitis and hepatitis A; the protozoan agents of amebic dysentery, giardiasis, and cryptosporidiosis; and a number of parasitic helminth ova.

Pathogenic organisms excreted in urine may also survive in and be transferred by water. This is true for the systemic *Salmonella* infections such as typhoid and paratyphoid fever, for hemorrhagic jaundice caused by the spirochete *Leptospira,* and for *Schistosoma* infections of the bladder.

For many pathogens, water is a limited reservoir because they cannot multiply in it. Although fish and shellfish that become infected with waterborne parasitic organisms may continue the chain of disease, water itself does not remain infectious indefinitely unless it is continuously supplied with the organic nutrient required by pathogenic organisms or is constantly recontaminated.

ENDOGENOUS SOURCES

The term *endogenous* means that something is originating, growing, or proceeding from within. In biology, "within" may refer to an entire living organism, such as the human body, or to a tissue segment or an individual cell. An infectious microorganism growing within certain cells or tissues of the body may cause no harm unless it gains access to other types of cells or tissues. For example, a *Salmonella* infection of the intestinal tract only may be asymptomatic, or it may produce symptoms resulting from the toxic effects of the organism within the intestine. If *Salmonella* organisms spread to body tissues beyond the intestinal tract, infection at different sites produces new and different symptoms. Thus, a systemic disease distinct in many ways from the original has been acquired from an **endogenous source of infection** rather than from an outside, or exogenous, reservoir, even though the first site of infection may have resulted from ingesting organisms from the outside reservoir.

It is equally important to remember that microorganisms living a commensalistic or saprophytic kind of life on the skin or mucosal linings may damage internal tissues to which they are foreign. Endogenous bacteria such as enteric bacilli and staphylococci normally living within the intestinal tract or on the skin, where they may reside unnoticed, can injure other tissues. Accidental trauma, skin or bowel sites eroded by ulcerative conditions or tumors, and surgical procedures are the most common means by which microorganisms are transferred from endogenous sources to parenteral tissues and establish themselves at these sites. Since the organisms involved seldom have well-marked properties of virulence, these infections may be subacute and slow to progress, depending also on the nature, function, and reactivity of the infected tissues (see chap. 5).

SUMMARY: DISEASES AND RESERVOIRS

Table 8.2 groups many of the common communicable diseases of humans or animals according to their major reservoirs.

ROUTES OF TRANSFER OF INFECTION

The communicability of infection depends not only on the nature and accessibility of reservoirs but also on the routes and mechanisms available for transmission of the infectious agent to new hosts. We have seen that there are four major **transmission routes of infection** through which microorganisms may enter the human body: respiratory tract, gastrointestinal tract, skin and mucosal surfaces, and other parenteral routes (see chap. 5).

These same avenues may also serve as exit routes, depending on the location of the organisms in the infected body and the availability of a transporting mechanism. Respiratory and gastrointestinal discharges are expelled from the body with ease and frequency. They may spread infection far and wide through droplet contamination of air; through contamination of surfaces, hands, and handled objects; and through establishment of secondary reservoirs of microbial growth in water, milk, or food supplies. Infectious discharges from exposed lesions of the skin and mucosa may also be widely disseminated by hands and other mechanical vectors or by person-to-person contacts. Deep tissue infections that are not exposed to the surface cannot be transmitted naturally unless the microbial agents find their way into the body's discharges or into an insect host as it takes a meal.

The transfer of an infectious microorganism from its exit point from one host to its portal of entry into another may be direct and immediate, or it may be delayed and sometimes extremely devious (see table 8.3). Intermediate, inanimate reservoirs may link the travel routes of the organism between living hosts, sometimes at close range, but in some instances, over great distances. Some of the parasitic protozoa and helminths display complex developmental cycles that require their transfer in sequence from one type of host or reservoir to another.

TRANSMISSION ROUTES FROM LIVING RESERVOIRS

There are two basic modes of transmission of infectious agents from infected human, animal, or insect hosts to susceptible human beings: (1) by direct contact with an infected, living host, or (2) after indirect contact with infectious material derived from a diseased host.

Direct Contact

This type of transfer involves close or frequent physical encounters between susceptible and infected hosts through which the infectious agent is transmitted directly to an appropriate site of entry to a new host.

Infected human hosts may transfer their microbes to other persons through physical contacts such as handshaking, kissing, and sexual intercourse. Infectious respiratory droplets may also be transmitted directly among people in close proximity. The most usual exit and entry sites for infections that are spread among and by humans are the respiratory, oral, and genital mucosa. Certain diseases are spread almost exclusively by sexual contact. However, the bacterial agents of gonorrhea and chlamydial disease may infect the conjunctiva of a newborn infant when it is delivered through the birth canal of an infected mother whose cervical discharge contains one or both of these organisms. Syphilitic mothers or those infected by the human immunodeficiency virus or hepatitis B virus may also transmit their disease to the fetus in utero. Children sometimes acquire sexually transmitted diseases from infected adults through fondling and kissing. Other infections may also be spread by contaminated hands, either those of the infected individuals or those of persons in close contact with them. Surface infections from suppurating wounds, pimples, or boils are readily transferable by hands as well.

Nurses, physicians, and others who care for infectious patients, particularly those whose diseases are directly transmissible, must take every precaution to protect others as well

TABLE 8.3 TRANSMISSION ROUTES

From Living Reservoirs		From Inanimate Reservoirs	
Direct	**Indirect**	**Direct**	**Indirect**
Kissing Handshaking Droplet transmission during talking, sneezing Sexual intercourse Insect vectors Animal contacts	Contact with infectious material derived from diseased host: sputum, feces, exudate from wound, animal tissues Usual route is oral following hand-to-mouth transfer of infectious material from contaminated surfaces.	Ingestion of contaminated food, milk, water Skin penetration by parasite from intermediate reservoir Injection of contaminated biologic products Soil contacts Dust inhalation	Via contaminated intermediary objects: utensils, clothing, bedding, skin Usual route is oral following hand-to-mouth transfer of infectious material from contaminated surfaces.

as themselves from contact spread of infection. Medical personnel may carry infectious organisms from one patient to another on their hands, on their clothing, or even on their faces. Acceptable patient care by all health care workers must include frequent, thorough hand washing and other aseptic techniques designed to prevent the spread of infection. These techniques are of particular importance to patients whose susceptibility to infection is high: surgical and obstetric patients; the very old or the very young; and those with underlying, debilitating, immunosuppressive disorders that undermine the normal resistance mechanisms (see chap. 9).

Infected animals may also transfer their infections to susceptible human hosts who come into direct contact with them. Animal breeders, farmers, dairyworkers, butchers, hide and leather workers, poultry raisers, pet lovers, and even meat eaters—all are subject to exposure to a number of animal infections that are transmissible by a direct route. The most frequent transfers occur by the hand-to-mouth route; by ingestion of infected meat or animal products such as milk or cheese; by direct inhalation of contaminated airborne excretions; or through abraded skin or mucosal surfaces.

Infected insect hosts or vectors may directly introduce infection into human tissues or blood. However, the insect-borne diseases are not transmissible between human or animal hosts. Their agents depend on the contacts, transport, and support of their arthropod hosts for their continuation in nature and their access to other host species.

Indirect Contact

Human infections may be acquired from other infected hosts even without immediate intimate contact. Microorganisms that can survive on surfaces contaminated by sick persons, animals, or infected insects may later find an entry route into susceptible human hosts who touch or handle such items. Clothing, bedding, surgical dressings, instruments, bedpans, paper or linen handkerchiefs, toys, eating utensils, food, drinking glasses, or their contents—all may provide a bridge between infected and new susceptible hosts. Insect vectors can play an indirect role in the spread of disease by depositing infectious material on food, skin, clothing, or other contact surfaces.

The usual route of entry for infections acquired by indirect contact is oral, following a hand-to-mouth transfer of infective matter from a contaminated surface. Occasionally, entry may occur through the skin or mucosa, particularly if a local injury or lesion is present.

Proper care of hospitalized patients includes the routine sterilization of many items that could be an indirect source of infection on contact. Items such as water carafes, glasses, dishes, bedding, bedpans, and thermometers are either disinfected or sterilized after use to ensure that neither acute nor clinically inapparent infections are transferred by them. Whenever possible, disposable materials are used instead; however, these items must be disposed of in a manner that does not permit infection to continue. Incineration is generally the method of choice but many medical wastes are buried in landfills, perpetuating an environmental problem.

TRANSMISSION ROUTES FROM INANIMATE RESERVOIRS

We have seen that most inanimate media (soil, air, food, milk, water) serve as secondary reservoirs, allowing certain microorganisms to survive and be transported between living, primary reservoirs. The route of transmission depends on the way humans are exposed to or use these secondary reservoirs. Infectious water, food, or milk may reach the gastrointestinal tract directly, by ingestion, or by an indirect pathway. For example, water used for washing clothes or utensils, for bathing or swimming, or for irrigating crops may contaminate food, hands, or clothing. Some of the helminthic parasites that undergo a phase of development in water or wet soil directly penetrate normal human skin (the larvae of hookworms and *Schistosoma* species), while other pathogenic organisms enter the body only through injured skin or the alimentary mucosa. Air containing infectious droplets or dust particles may enter the respiratory tract directly by inhalation, but often the particles are deposited on surfaces from which hand-to-mouth transfer can occur later.

Biologic products, including serum, plasma, soluble nutrients (e.g., dextrose or vitamin solutions), drugs, or vaccines, may also become contaminated with microorganisms capable of surviving and multiplying in them. When such

TABLE 8.4 ROUTES OF TRANSFER OF SOME IMPORTANT HUMAN PARASITES

Type of Transfer	Representative Organisms	Disease Produced	Other Host Needed in Life Cycle	Infective Form for Human[b]
Human-to-human[a]	*Entamoeba histolytica*	Amebic dysentery	None	Cysts in food, water
	Cryptosporidium parvum	Cryptosporidiosis	None	Oocysts in water
	Giardia lamblia	Giardiasis	None	Cysts in water
	Enterobius vermicularis	Pinworm infection	None	Eggs on hands, perianal skin, clothing, or in dust
Human-to-soil-to-human[a]	*Trichuris trichiura*	Whipworm infection	None	Mature eggs in soil
	Ascaris lumbricoides	Ascariasis	None	Mature eggs in soil
	Necator americanus	Hookworm infection	None	Larvae hatched from eggs in soil penetrate skin
	Strongyloides stercoralis	Strongyloidiasis	None	Larvae mature in soil then penetrate skin
Intermediate-host-to-human	*Trichinella spiralis*	Trichinosis	Pig Wild game	Larvae in infected muscle, usually pork or bear
Definitive-host-to-human	*Echinococcus granulosus*	Hydatid cyst	Dog	Eggs from dog feces
Human-to-intermediate host-to-human[a]	*Diphyllobothrium latum*	Fish tapeworm	Fish	Larvae in fish
	Taenia solium	Pork tapeworm	Pig	Larvae in pork
	Taenia saginata	Beef tapeworm	Cattle	Larvae in beef
	Schistosoma spp.	Schistosomiasis	Snails	Larvae in water penetrate skin
Insect-to-human-to-insect	*Plasmodium* spp.	Malaria	Mosquito	Sporozoites inoculated by bite
	Trypanosoma spp.	Sleeping sickness	Tsetse fly	Trypomastigotes inoculated by bite
	Leishmania donovani	Leishmaniasis	Sand fly	Promastigotes inoculated by bite

[a]Animals other than humans may serve as the definitive host for some of these parasites (see chap. 18).
[b]Except where indicated, the infective form is ingested.

contaminated materials are injected into the human body, serious infections may arise even though the agents involved may display little or no ability to produce disease by entering through the usual routes.

Table 8.3 summarizes the routes of transfer of infection.

ROUTES OF TRANSFER OF ANIMAL PARASITES

Animal parasites pass through cyclic, developmental stages of growth, maturation, and reproduction (see chap. 2). The full circle of changes involved is referred to as a *life cycle.* The continuation of parasitic protozoa and helminths in nature depends on completion of their life cycles, and their transfer to new hosts depends in turn on the transmissibility at a given developmental stage.

Life Cycles of Protozoans and Helminths

Many animal parasites cannot complete their life cycles in one host or reservoir. A human or animal host may be parasitized either by their adult, reproductive forms or by their intermediate, immature forms, sometimes both. The host in which the adult parasite lives and reproduces is called the *definitive,* or *final,* host, and the host in which immature stages of development occur is defined as the *intermediate* host. More than one type of intermediate host may be required, in sequence, to complete the life cycles of some parasitic worms. In some instances the host can support both adult and intermediate forms, but an *alternate* host is also needed to provide continuity for the organism. Animal parasites vary widely in complexity of development, but their types of transfer can be summarized (table 8.4). The life cycles of major parasites are detailed in chapters 18, 20, and 22 along with descriptions of the diseases they cause. A few are described here as examples of the sometimes highly complex ways in which these pathogens have ensured their survival.

Human-to-human transfer

The simplest type of life cycle is displayed by parasites such as *Entamoeba histolytica* and *Giardia lamblia.* The hardy infective cyst form of the organisms is ingested and survives passage through the stomach. In the intestinal tract, the trophozoite form emerges from the cyst and its activities are responsible for symptoms of disease. During infection, some trophozoites transform into the cyst forms, which are passed in feces. If these find their way into food or water, they are infective for others who may ingest them. Trophozoites do not survive long outside the body or during passage through the stomach, therefore they are not infective after ingestion. *Cryptosporidium parvum* oocysts undergo more complex development than do *E. histolytica* and *G. lamblia,* but the oocytsts are also excreted in the feces of infected humans and animals.

Enterobius vermicularis, the pinworm, also has a simple life cycle. Eggs are ingested and larvae hatch in the intestinal tract where they mature to adult males and females. The worms mate and, once their eggs are mature, gravid females migrate to the perianal region where the eggs are laid. The eggs, which contaminate bedding, clothes, hands, and dust, are easily ingested to begin the cycle again.

Human-to-soil-to-human

The eggs of roundworms pass out in the feces of the infected host and require a period of maturation in soil before they become infective. After mature *Trichuris* and *Ascaris* eggs are ingested, they hatch in the intestine. *Trichuris* larvae develop into egg-laying adults in the intestinal tract. *Ascaris* larvae, however, undergo a journey through the bloodstream to the lungs from where they pass up the epiglottis to the mouth and are swallowed. In the intestine again, they develop into adult, egg-producing worms.

The larvae of *Necator americanus* (the hookworm) hatch from eggs deposited in the soil and penetrate the skin of barefooted victims. From here, they are carried through the bloodstream to the lungs where their cycle follows that described for *Ascaris.* In another variation, the larvae of the roundworm *Strongyloides stercoralis* hatch in the intestinal tract to a noninfective form that becomes infective when deposited in soil. These larvae penetrate the skin and follow a cycle similar to that of *Necator.* If the noninfective *Strongyloides* larvae develop into the infective form in the human intestinal tract, a recurrent cycle of autoinfection is set up that can last for 30 or 40 years.

Intermediate-host-to-human

The larval form of *Trichinella spiralis,* the agent of trichinosis, is eaten by humans, usually in undercooked, infected pork meat. The larvae develop into adults in the intestinal tract and the adult female deposits live larvae into the bloodstream where they are carried to and encyst in striated muscle. Thus, humans are both the definitive and intermediate host. Humans represent a dead end for the parasite, but their cycle continues in carnivorous pigs, rats, and wild game, which eat infected animal flesh.

Definitive-host-to-human

Echinococcus granulosus (dog tapeworm) is uncommon in the United States but imported cases of echinococcosis are seen. The dog is the definitive host that carries the adult parasite in its intestinal tract. When humans accidentally ingest the eggs, the larval form penetrates the intestinal tract and develops into a hydatid cyst in one or more organs, especially liver, lung, and brain. Dogs acquire the infection by eating the organs of animals (usually sheep) infected with hydatid cysts.

Human-to-intermediate host-to-human

The larvae of several tapeworms (*Diphyllobothrium, Taenia*) are found in the flesh of animal foods eaten by humans, including fish, pork, and beef. When ingested, the larvae mature to adults in the intestinal tract, growing up to several feet in length. A special danger associated with the pork tapeworm, *Taenia solium,* is that its eggs may hatch in the human intestinal tract whereupon the larvae travel to and encyst in various body organs, just as *Echinococcus* larvae do. Serious consequences result when larvae encyst in brain tissue.

Schistosoma eggs in the feces and urine of infected humans are deposited with these excreta in soil or water. The larvae hatch and find their way into snails where they develop into another larval stage. These larvae exit into water and eventually penetrate the skin of humans as they wade, bathe, or participate in other activities in contaminated water. The larvae mature to adults in major blood vessels but the eggs of different species migrate to the intestinal tract or urinary bladder where they are excreted in feces or urine.

Insect-to-human-to-insect

The final type of life cycle is one in which insects ingest the parasite by feeding on blood or tissues (usually skin lesions) of humans infected with the parasite. The parasite undergoes development in the insect and is transmitted back to humans when the insect takes a blood meal or a bite.

Control of Parasitic Diseases

The control of diseases caused by animal parasites uses knowledge of their life cycles to interrupt their routes of transfer. Parasite control includes (1) proper disposal of human feces to prevent water or soil contamination; (2) protection of farm animals from infected food, destruction of rodents, and public education on the need for proper preparation of meat, especially pork, for the table; and (3) the destruction of insects, control of their breeding, and personal protection from insect bites. The recognition and treatment of humans and animals with parasitic diseases may remove some of the sources of infection, but inapparent infections often remain as continuing reservoirs.

PATTERNS OF INFECTIOUS DISEASES

Epidemiology is the study of the natural occurrence of diseases within a community. In human populations the occurrence of infectious diseases is characterized by patterns that are pandemic, epidemic, endemic, or sporadic. These terms reflect the numbers of persons who are clinically ill during a given period of time as well as those who have inapparent infections, the frequency with which overt illness occurs, and the rate of its spread through a population. The factors that influence these three features of community disease (numbers, frequency, spread) can be considered in three major categories: (1) the degree of effective resistance in the host population to a particular microbial agent of disease; (2) the sources, transmissibility, and incubation period of the infectious microorganism; and (3) the physical and social conditions of the human environment.

Resistance in Host Populations

When the number of cases of communicable disease occurring within a stated time period (week, month, year) exceeds expectations based on previous experience in similar periods, the situation is termed an *epidemic*. The extent of the epidemic is determined by the ratio of susceptible versus resistant individuals in the population, and this in turn is influenced by previous contact of the community with the disease. If the disease is a new one, or if the agent is a new strain of a mutating pathogen, its introduction may result in a large epidemic because the population contains a high proportion of susceptible people. Epidemics may also result when a disease is reintroduced to a community that has not experienced it for one or more generations. As the number of susceptibles declines during the course of an epidemic, spread of the disease becomes difficult and finally ceases unless new susceptibles are continuously introduced. Conversely, as the proportion of immune individuals rises, transmission is steadily reduced, even among the nonresistant. For this reason artificial active immunization of large numbers of people is often an effective block to the epidemic spread of infection.

On the other hand, even small numbers of cases of infectious disease may constitute an epidemic when they are seen in communities that are normally free of the disease in question. Small outbreaks may reflect a limited source of infection or a restricted opportunity for exposure of susceptibles, or they may be the forerunners of larger events to come.

The frequency of epidemic disease often reflects the rise, duration, and waning of immunity in a population. Before the widespread use of effective vaccines, many viral diseases of childhood such as measles and mumps, which provide a lifelong immunity, were seen primarily in young children rather than in immune adults. They often occurred in epidemic patterns, following two- to four-year cycles, as succeeding groups of susceptibles were born into a community or entered it from outside. Cyclic epidemics of influenza, however, usually occur among people of all ages, as their immunity wanes and different strains of the virus to which they are not immune become prevalent. Diseases such as the common cold are seen in fairly high incidence because many antigenically different agents cause the cold syndrome, and persons are immune only to those previously encountered.

Pandemic is a term that refers to epidemics that occur over widespread geographic regions. Within each region, however, the same factors apply as to epidemics.

Sources, Transmissibility, and Incubation Period

The rate at which an infectious disease spreads is primarily a function of the sources, transmissibility, and incubation period of the agent in question. The respiratory diseases that spread from person to person travel with great speed, particularly when people are exposed to each other in closed areas such as theaters, schools, barracks, and other communal places. The shorter the incubation period required for manifest disease, the more explosive the epidemic will appear, as in the case of influenza, streptococcal sore throat, or scarlet fever. Similarly, food-, water-, and milk-borne outbreaks may appear explosively because large numbers of people may be exposed to the same source of infection simultaneously, or within a short period of time. Diseases such as *Salmonella* gastroenteritis, staphylococcal food poisoning, and amebic dysentery have short incubation periods. They can appear in a population within a matter of 6 to 48 hours following ingestion of contaminated material. Other diseases such as infectious hepatitis, the systemic fungal diseases, or parasitic infestations take more time to develop. When these occur in epidemic patterns among people exposed to a common source (those sharing a meal containing virus-infected shellfish or trichina-infected pork, for example, or those working in an area where dust inhalation involves the common risk of fungous infection), the slow development of overt disease may make it difficult to relate individual cases to their point of origin. Arthropodborne infections may also appear in epidemic distribution when seasonal breeding increases the numbers of insect vectors that have access to an animal reservoir of infection. Outbreaks of equine encephalitis in the United States, for example, are usually associated with a spring or summer increase in the density of mosquito populations.

Diseases become endemic in populations with which infectious agents have established some balance. Clinically inapparent infections become the rule in such communities. Sporadic cases of disease occur from time to time, but epidemics do not occur while the equilibrium lasts. The balance can be disturbed by sudden changes in the proportion of susceptible persons, as when armies enter an area of endemic disease, or by environmental stresses that involve large numbers of people, such as those reduced by war to poverty and hunger. The ability of an infectious agent to establish endemicity in a human or animal population depends on the availability of suitable reservoirs and transmission routes and on the responses of susceptible individuals to infection.

Physical and Social Conditions

The immediate conditions of the human environment may influence enormously the patterns and incidence of infectious disease. Low-income groups and the unemployed experience the constant stresses of poor nutrition and crowded and unsanitary living conditions, and they lack information about preventive social habits. Exposure of such populations to a new and highly virulent infectious agent may have disastrous consequences of epidemic proportions. More often, persistent, low-grade infections are the rule in poor communities, and these endemic diseases contribute to a decreased resistance to any new diseases that may be introduced. Social customs, particularly in the matter of food and drink, sanitary standards, and even the political stability of a society, can play a role in the nature and incidence of communicable diseases. We have also seen (chap. 5) how large an influence the physical environment exerts on human interrelationships with microorganisms. Climate, weather, and geography determine the ecology of human, animal, plant, and microbial populations. For microorganisms, such vitally important factors may ensure the availability of reservoirs and vectors for their continuation in nature. For human beings, the total environment presents a constant challenge to discover the means of prevention or control of infectious disease, at its source or along the route of transmission.

QUESTIONS

1. What must an epidemiologist know to determine whether a disease is prevalent in a group or community?
2. Why is it necessary for a health professional to understand the epidemiology of an infectious disease?
3. What determines whether a disease is communicable?
4. Define reservoir of infection. Give examples of inanimate and living reservoirs. List reservoirs that may not be sources of infection.
5. Who are human carriers of infection? Why are they of great concern to the epidemiologist?
6. How do humans become infected with zoonoses? Give several examples of zoonoses and their transmission routes.
7. Why are flies and cockroaches referred to as mechanical vectors?
8. How do food, air, water, and soil contribute to disease transmission?
9. What are endogenous sources of infection?
10. What are the four major routes of transfer of infection to the human body?
11. Give examples of disease transmission by direct contact with infected humans.
12. How is disease transmitted by indirect contact?
13. Explain the differences between the infective forms of the following parasites: *Giardia lamblia, Necator americanus,* and *Trichinella spiralis.* Why is it important to understand life cycles of animal parasites?
14. Give an example of a common disease that is endemic. When is an epidemic disease referred to as a pandemic?

BIBLIOGRAPHY

Berlinguer, G. 1992. The interchange of disease and health between the Old and New Worlds. *Am. J. Public Health.* 82 (10): 1407–1413.

Buehler, J. W., O. J. Devine, R. L. Berkelman, and F. M. Chevarley. 1990. Impact of the human immunodeficiency virus epidemic on mortality trends in young men, United States. *Am. J. Public Health.* 80 (9): 1080–1086.

Cartwright, F. F., in collab. with M. D. Biddiss. 1991. *Disease and history.* New York: Dorset Press.

Centers for Disease Control and Prevention. *Morbidity and mortality weekly reports, MMWR* (A weekly report that discusses infectious disease). Atlanta: Centers for Disease Control and Prevention.

Cliff, A., and Haggett, P. 1984. Island epidemics. *Sci. Am.* 250 (5): 138–147.

Cohen, I. B. 1984. Florence Nightingale. *Sci. Am.* 250 (3): 128–137.

Fenner, F., et al. 1988. *Smallpox and its eradication.* World Health Organization: Geneva.

Garcia, L. S., and D. A. Bruckner. 1997. *Diagnostic medical parasitology.* 3d ed. Washington, D.C.: ASM Press.

Heyward, W. L., and J. W. Curran. 1988. The epidemiology of AIDS in the United States. *Sci. Am.* 259 (4): 72–81.

Kohles, M. K. and P. L. Barry. 1989. Clinical laboratory and nursing personnel: Collaboration in improving nursing care. *J. Nurs. Qual. Assur.* 3 (1): 1–10.

Last, J. M. 1989. Ethics and epidemiology. *World Health,* 22–24 June.

Mann, J. M., J. Chin, P. Piot, and T. Quinn. 1988. The international epidemiology of AIDS. *Sci. Am.* 259 (4): 82–89.

Marshall, M. M., D. Naumovitz, Y. Ortega, and C. R. Sterling. 1997. Waterborne protozoan pathogens. *Clin. Microbiol. Rev.* 10 (1): 67–85.

Noble, E., G. A. Noble, G. A. Schad, and A. J. MacInnes. 1989. *Parasitology: The biology of animal parasites,* 6th ed. Philadelphia: Lea and Febiger.

Patterson, K. D. 1986. *Pandemic influenza, 1700–1900: A study in historical epidemiology.* Totowa, N.J.: Rowman and Littlefield.

Peterman, T. A., and L. R. Petersen. 1990. Stalking the HIV epidemic: Which tracks to follow and how far? *Am. J. Public Health.* 80 (4): 401–402.

Pomeroy, C., and J. A. Englund. 1987. Cytomegalovirus: Epidemiology and infection control. *Am. J. Infect. Control* 15 (3): 107–119.

Roueche, B. 1991. *The medical detectives.* New York: Plume.

Schaberg, D. R. 1989. How infections spread in the hospital. *Respir. Care* 34 (2): 81–84.

Snider, S. 1991. Childhood diseases: New epidemics? *Emerg. Med. Serv.* 20 (1): 39–40.

CHAPTER 9

Prevention and Control of Infectious Diseases

OUTLINE

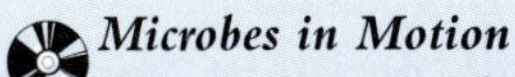

Microbes in Motion

This chapter covers the principles of prevention and control of infectious diseases in the hospital environment. Included are physical and chemical methods and antimicrobial agents used to treat infections in patients. The following books and chapters of the Microbes in Motion CD-ROM may be a useful preview or supplemental study aid to your reading: Control of Microbes: Terminology; Physical Control; Chemical Control. Antimicrobial Action: Cell Wall Inhibitors; Protein Synthesis Inhibitors; Agents Affecting Enzymes; Agents Affecting Nucleic Acids. Concepts of Epidemiology of Infectious Disease: Transmission.

OBJECTIVES

On completing this chapter the student will be able to:

1. cite several physical and chemical methods used to destroy microorganisms.
2. differentiate between disinfection and sterilization.
3. outline the modes of action of physical and chemical agents used to destroy microorganisms.
4. differentiate between medical and surgical asepsis.
5. discuss the nature and uses of antimicrobial agents in the treatment of patients with infectious disease.
6. describe the practical approach to treating infectious diseases in hospitals.
7. explain the need for infection control in hospitals and how these controls are applied.
8. define the role of the microbiology laboratory in providing effective information concerning the nature, treatment, and control of infectious patients.

KEY WORDS

Alcohols
Aldehydes
Alkylating agents
Antibacterial (anti-back-**TIER**-ial)
Antibiotic (anti-by-**OUGHT**-ic)
Antimicrobial (anti-my-**CROW**-bial) agent
Antisepsis (anti-**SEP**-sis)
Asepsis (a-**SEP**-sis)
Autoclave (auto-clave)
Beta lactamase (**BAY**-ta **LACK**-ta-mase) enzyme
Chemoprophylaxis (key-mo-pro-fill-**AX**-is)
Chemotherapy (key-mo-**THARE**-apy)
-cidal (**SIDE**-al)
Cleaning
Communicable disease
Communicable period
Contamination
Disinfection
Drug allergy
Drug resistance
Dry heat sterilization
Ethylene oxide (**ETH**-ill-ene **OX**-ide)
Filtration
Halogens
Immunoprophylaxis (imm-you-no-pro-fill-**AX**-is)
Immunotherapy (imm-you-no-**THARE**-apy)
Infection
Infectious disease
Isolation precautions
Nosocomial (no-zo-**COMB**-ial)
Pasteurization (past-er-iz-**A**-shun)
Penicillinase (pen-i-**SILL**-in-ase)
Phenols
Prophylaxis (pro-fill-**AX**-is)
Quarantine
Sanitization
-static (**STAT**-ic)
Steam sterilization
Sterilization
Surface-active agents
Ultraviolet radiation

GENERAL PRINCIPLES

Three basic approaches are used to prevent or control **infection:** (1) destruction or control of microbial pathogens; (2) elimination or control of the sources, routes, or agents of infection transmission; and (3) protection of the human host from the serious effects of disease using immunization, specific treatment of infections, or both.

The successful application of these approaches has led to some dramatic changes in the epidemic patterns of many infectious diseases. Smallpox, for example, has been eliminated within the past two decades through the success of a global Smallpox Eradication Program coordinated by the World Health Organization (WHO). Begun in 1967, this program of case detection, quarantine, and immunization has eradicated a once dreadful epidemic disease. Diphtheria has been controlled, although not totally eliminated, in countries where large-scale immunization programs have been maintained. Improved sanitation and the availability of effective vaccines have similarly reduced the incidence and spread of poliomyelitis to a level in which eradication can be anticipated. Where prevention has not been possible, the sources of transmitting agents of infection have been controlled, as in the cases of insectborne diseases caused by protozoa (malaria), viruses (yellow fever), and rickettsiae (typhus fever). Some infectious diseases can be controlled by preventing their severe effects. Viral diseases such as hepatitis A can be prevented or aborted if nonimmune persons receive hepatitis A vaccine or those who are exposed to the virus are given gamma globulin promptly to provide passive protective antibody. Specific chemotherapy administered quickly after the appearance of symptoms is now usually effective for controlling serious diseases caused by pneumococci and streptococci. Sexually transmitted diseases (chlamydial infection, gonorrhea, syphilis) remain largely uncontrolled at their human sources but can be successfully cured by early case detection and adequate chemotherapy. Thus, infections can be controlled significantly when diseases are aborted by either immunologic or chemotherapeutic means. In these instances not only is the individual case resolved, but further spread of virulent infection is prevented.

In this chapter, attention will be given to the principles involved in each of the three important approaches for prevention and control of infectious diseases. The emphasis here is on problems of individual infection. Application of the principles to protection of the public health is discussed in chapter 10.

DEFINITIONS

Discussion of the measures used to destroy or suppress microorganisms, or to break the chain of their transmission from one host to another, requires the use of technical terms whose meanings should be clear to all who use them. The student should review them before continuing to study the principles of control and should refer to them frequently in subsequent reading of the text to become thoroughly familiar with their proper use. Table 9.1 lists terms referring to the destruction or suppression of microorganisms; table 9.2 lists those used in the management of infectious diseases.

CONTROL OF INFECTIOUS MICROORGANISMS

The effectiveness of chemical and physical antimicrobial agents has been determined by extensive laboratory tests conducted in vitro and in vivo. On the basis of these examinations the following generalizations apply:

1. Microorganisms can be irreversibly "killed" or reversibly inhibited by a number of physical agents as well as by chemical methods.
2. Various microbial groups differ greatly in their response to sterilizing or inhibiting agents. The cells of a particular bacterial species may also vary in this respect; young, actively growing cells are more susceptible than older, resting ones, whereas endospores formed by some bacterial species are much more resistant to these agents than are their vegetative forms.
3. **Sterilization** is always a function of time, regardless of the agent employed or microorganism involved. Even in homogeneous cultures individual organisms vary in their susceptibility to sterilizing agents, and not all of them die simultaneously. In addition, the time for sterilization is directly related to the numbers of microorganisms present; longer time periods are required to destroy larger microbial populations.
4. The rate of sterilization also depends on the nature, concentration, or intensity of the sterilizing agent.
5. In choosing a method of sterilization, disinfection, or inhibition for a given situation, other factors that may limit the success or practicality of the procedure must be taken into account. Some of these factors are the nature and function of the material to be processed (fabrics, rubber, glassware, etc.) and the conditions that may affect the activity of the antimicrobial agent or the response of the microorganism (presence of proteins, lipids, salts, the pH, etc.).

The following section describes the more useful physical and chemical agents employed for sterilization, disinfection, or microbial inhibition, together with the reasons for and the limitations on their use.

ANTIMICROBIAL ACTIVITY OF PHYSICAL AGENTS

Heat

The application of heat is the simplest, most reliable, and least expensive way to sterilize materials that are not damaged by it. It is a rapid process to which all living protoplasm is susceptible.

TABLE 9.1 TERMS INDICATING DESTRUCTION OR SUPPRESSION OF MICROORGANISMS

Destruction	Suppression
Sterilization: Complete destruction of all forms of microbial life. An absolute term; there are no "degrees" of sterility.	**Asepsis:** Literally, without infection. Refers to techniques that prevent entry of living microorganisms.
Disinfection: Destruction of pathogenic microorganisms (except bacterial endospores) on inanimate objects by chemical or physical agents.	*Surgical asepsis:* Use of techniques designed to exclude **all** microorganisms.
Concurrent: Continuing disinfection of infectious discharges, or of objects soiled thereby.	*Medical asepsis:* Use of techniques designed to exclude agents of communicable disease but not necessarily **all** others.
Terminal: Final disinfection of surfaces and objects in an area previously occupied by an infectious patient.	**Sanitization:** Removal of pathogenic microorganisms from inanimate objects by mechanical and chemical cleaning. Does **not** imply sterilization or complete disinfection.
Antisepsis: Chemical disinfection of skin or other living tissue.	**Cleaning:** Removal of foreign matter such as soil or organic material. Never implies disinfection or sterilization.
-cide: Noun suffix meaning "killer" or "killing." Adjectival suffix = **-cidal.**	**Antibiotic:** Literally, against life. A microbial product that suppresses or destroys other microorganisms.
Germicide: Germ-killing agent.	**Antibacterial:** Against bacteria. Refers to agents that suppress or destroy bacteria.
Bactericide: Kills bacteria.	**Antimicrobial:** Against microbes. Refers to agents that suppress or destroy any microorganism.
Sporicide: Kills bacterial endospores.	**-stasis:** Noun suffix meaning "halted" or "arrested." Adjectival suffix = **-static.**
Fungicide: Kills fungi.	*Bacteriostasis:* Bacterial growth arrested; multiplication halted, but bacteria not killed and may resume growth when bacteriostatic agent is removed.
Virucide: Kills viruses.	*Fungistasis:* Fungal growth halted.
	Virustasis: Viral growth halted.

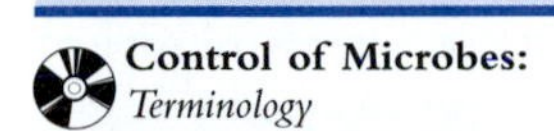
Control of Microbes:
Terminology

TABLE 9.2 TERMS USED IN THE MANAGEMENT OF INFECTIOUS DISEASES

Term	Denotes
Contamination	Presence of viable microorganisms on animate or inanimate surfaces; in water, food, or milk; in human or animal discharges.
Infection	Entry and multiplication of an infectious agent in the tissues of living hosts.
Infectious disease	Disease in a living host resulting from infection.
Communicable disease	Disease caused by an infectious agent or its toxic products, directly or indirectly transmissible from an animate or inanimate reservoir to a susceptible host. (Not every infectious disease is communicable between hosts of the same species.)
Nosocomial infection	Infection developing in a patient of a hospital or other health care facility. Includes infections not present at the time of admission, lingering infections acquired during a previous admission, and such infections among staff.
Isolation	Separation of infected persons or animals for the period of communicability in places and under conditions that prevent direct or indirect transmission of the infectious agent to susceptible persons who may spread the agent to others.
Quarantine	Formerly used to limit freedom of movement of healthy persons exposed to a communicable disease—for a time equal to the longest incubation period of the disease—to prevent effective contact with others not so exposed. Has been replaced by active surveillance of exposed persons through local health authorities although quarantine may still be applied to domestic animals.
Communicable period	Period during which an infectious agent may be transferred directly or indirectly from an infected person to another person, animal, or arthropod; or from an infected animal to humans. In diseases such as diphtheria and streptococcal infection, in which the first entry of the pathogen involves mucous membranes, the communicable period is from the date of first exposure until the infecting agent is no longer disseminated from those membranes (i.e., from before prodromal symptoms until termination of the carrier state, if such develops). In diseases such as tuberculosis, syphilis, and gonorrhea, communicability may occur at any time over a long, intermittent period when unhealed lesions permit discharge of infectious agents through body orifices or from skin surfaces. In diseases transmitted by arthropods (e.g., malaria and yellow fever), communicability exists when the infectious agent occurs in the blood or tissues of the infected person in infective form and in numbers sufficient for vector infection. The communicable period for the arthropod vector is that time during which the agent is present in the arthropod tissues in an infective form that can transmit infection.

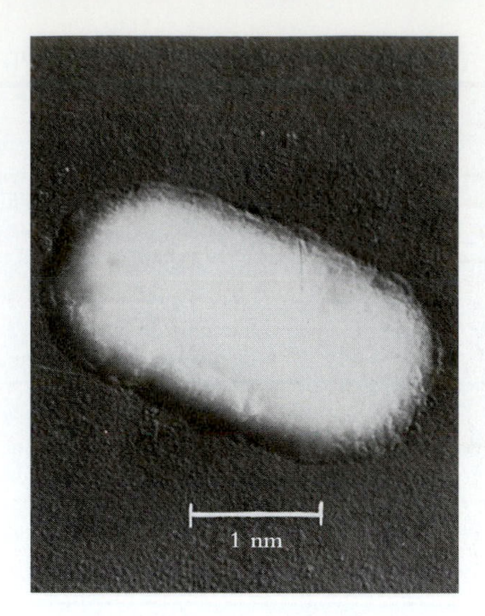

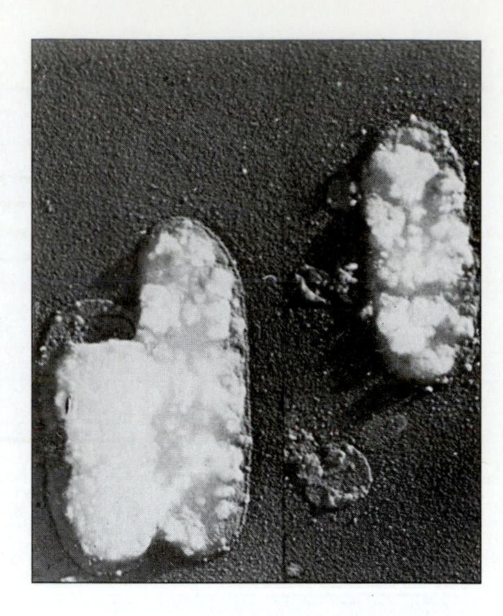

FIGURE 9.1 The effect of moist heat on *Escherichia coli*. (a) An electron micrograph of a typical cell of *E. coli* from a one-hour broth culture. Bar = 1 nm. (b) Three cells of *E. coli* that have been heated at 50° C in saline for 10 minutes. The coagulated cytoplasm becomes granular and shrinks away from the cell wall.

Reproduced from C. G. Heden and R. W. G. Wyckoff, "The electron micrography of heated bacteria," *Journal of Bacteriology* 58:153, 1949. American Society for Microbiology.

Control of Microbes
Physical Control: Moist Heat

Mode of Action

The efficiency of heat as a sterilizing agent is a function of both the intensity, or temperature, and the time for which a given temperature is maintained. As the temperature is increased, the exposure time needed for materials to be sterilized is decreased. Mathematical formulas are available to aid in determining the correct exposure times needed to kill specific microorganisms, but these calculations are used primarily in the food industry where killing foodborne disease agents is imperative. In medical use, exposure time is usually determined by the type of heat applied, the material to be sterilized, and whether bacterial endospores are likely to be present.

The antimicrobial effects of heat depend on whether the heat is moist or dry. Many microbial proteins are enzymes that function only within a narrow temperature range. When heat is applied in the presence of moisture, the structure of cellular proteins is altered, they are coagulated (as the white of an egg is coagulated when it is boiled), and their enzymatic function is destroyed (fig. 9.1). When the temperature of dry heat is raised, cell proteins do not coagulate, but they and other components of the microbial cell are oxidized. At low temperatures this process is much slower than protein coagulation in the presence of moisture. Moist-heat sterilization is more efficient, therefore, because it can be accomplished more rapidly and at lower temperatures than dry-heat methods. It is also far less injurious to those materials that would be oxidized by the high temperatures required for dry-heat sterilization.

Uses of Moist Heat

Most pathogenic microorganisms are killed when exposed for 10 minutes to temperatures between 50 and 70° C. However, the heat-resistant endospores of some species of pathogenic bacteria can survive at 100° C for one or two hours under particular conditions, while those of certain saprophytic species can resist up to 24 hours of boiling. The choice of a moist-heat method of sterilization depends therefore on the nature of the job to be done and the temperature that can be produced by the method.

Pasteurization **Pasteurization** is a process used primarily in the food and dairy industry. The minimum degree of moist heat that can be expected to kill nonsporing pathogenic bacteria in milk or other potable liquids is applied over a time that does not spoil the flavor or quality of the beverage. The method was first devised by Pasteur, who found that wine spoilage could be prevented by heating it to 60° C for a time sufficient to kill fermentative and oxidative organisms that cause souring. Pasteurization of milk has resulted in effective control of a number of important milkborne diseases, notably, tuberculosis, brucellosis, salmonellosis, and streptococcal diseases of bovine origin. In the United States these diseases are no longer common in cows; therefore, pasteurization serves primarily to retard spoilage by destroying organisms such as streptococci and lactobacilli that cause unpasteurized milk to sour. Other nonpathogenic heat-resistant bacteria that occur naturally in milk are not killed, however.

In the conventional pasteurization technique, the temperature of milk is raised to 62.9° C (145° F) for 30 minutes. Higher temperatures and shorter times, such as 71.6° C (161° F) for 15 seconds or 141° C (286° F) for 2 seconds, followed by rapid cooling are currently used to improve flavor and extend the shelf life of the product (fig. 9.2).

Pasteurization at 75° C (180° F) for 30 minutes is sometimes used in the hospital for rubber and polyethylene tubing and catheters that cannot withstand high temperature exposure. These items must be thoroughly cleaned with a disinfecting agent before pasteurization to remove residual organic matter from patient secretions that may protect microorganisms from the effects of heating.

Boiling Boiling water provides a simple means of disinfection under many circumstances. The temperature of boiling water and the steam it evolves is 100° C. This is well

FIGURE 9.2 A modern pasteurizer. The proper handling of milk before and after pasteurization reduces the possibility of disease transmission.
APV Crepaco, Inc., Tonawanda, NY.

above the point at which vegetative bacteria, fungi, and most viruses are killed, so boiling for 10 to 30 minutes kills all except heat-resistant endospores and viruses. Contaminated materials may be disinfected in this way if pathogenic endospores and resistant viruses are not involved in the infectious process. However, detergent agents must be added to the water when dirty items are boiled to prevent the coagulation of extraneous protein matter (blood, mucus, pus); bacteria trapped within coagulated particles are protected from the killing effects of temperature and moisture. Clean instruments, syringes, needles, and other metal or glass items may be disinfected by complete immersion in boiling water for not less than 30 minutes. Since the method cannot be guaranteed to produce sterility, safer methods must be used to prepare equipment when sterilization is essential.

Hot Water Water at temperatures from 60° C to boiling accomplishes sanitization, and at least partial disinfection, particularly if it is applied with force and agitation in washing and rinsing items to be cleaned. Modern laundry and dishwashing equipment provides far better conditions for disinfection than manual washing. Detergent soaps added to wash water also play an active role in inhibiting or destroying microorganisms at elevated temperatures.

Steam under Pressure Pressurized steam is the most reliable form of moist heat for sterilization. The temperature of freely flowing steam is 100° C. It is sometimes used in this form as a sanitizing or disinfecting agent, as in the steam flushing of bedpans in the hospital or of animal cages in the laboratory. Steam becomes a sterilizing medium when its temperature is increased by subjecting it to increasing positive pressure.

A steam pressure sterilizer is essentially a chamber that permits, in sequence, (1) flowing, saturated steam to enter and displaced air to exit, and then (2) the air outlet to close so that the continued admission of steam produces a rising pressure. As the pressure increases, the temperature of steam rises, and therefore shorter time periods can be used to accomplish sterilization. Temperature and time are the essential factors in the **steam sterilization** process; the increased pressure simply provides the means for raising the temperature of steam. Table 9.3 illustrates the influence of pressure on the temperature of steam and on the time needed to ensure destruction of heat-resistant organisms.

The **autoclave** is the most common equipment used for steam sterilization in hospital operating rooms and in microbiology laboratories. The most important factor for successful autoclaving is the complete elimination of air from the sterilizing chamber. Air is cooler, dryer, and heavier than steam. Although an air-steam mixture can be brought to any desired pressure, the mixture will have a lower temperature than that of pure steam at the same pressure. The temperatures stated in table 9.3 are accurate only for saturated steam unmixed with air.

When steam saturated with moisture enters the autoclave chamber, it quickly condenses on cold surfaces within. This condensation of steam releases a large amount of latent heat (about 540° per cubic foot) and at the same time wets materials exposed to it. The rapid release of heat and moisture within the sterilizer quickly raises the temperature of inner, as well as outer, surfaces of items to that of the surrounding steam. Vessels or containers having hollow air spaces must be covered loosely, however, or placed in horizontal positions (if they do not contain liquid material) to permit air displacement and steam condensation. At the end

TABLE 9.3 PRESSURE-TEMPERATURE-TIME RELATIONSHIPS IN STEAM PRESSURE STERILIZATION

Steam Pressure Pounds per Square Inch (Above Atmospheric Pressure)	Temperature Celsius	Fahrenheit	Time (Minutes Required to Kill *Exposed* Heat-Resistant Endospores)
0	100°	212°	—
10	115.5°	240°	15–60
15	121.5°	250°	12–15
20	126.5°	260°	5–12
30	134°	270°	3–5

FIGURE 9.3 A load of carefully packaged and arranged articles is being pushed into an autoclave (steam sterilizer). The articles are placed so that steam can circulate and permeate the packaged contents. The time necessary for sterilization depends on the size and nature of the load.

Putnam Hospital Center, Carmel, NY. Photo: B. Riley, Danae Associates.

Control of Microbes
Physical Control: Moist Heat

of the sterilizing period, steam is slowly released from the chamber. When the pressure within the autoclave returns to normal, a drying period is required for fabrics, clean glassware, and other items that must be free of moisture before use.

Since it takes a few minutes for steam pressure to rise within a closed chamber, the timing of an autoclave procedure cannot begin until sterilizing temperature has been reached. The time required to sterilize a particular item or group of materials placed together in a full loading of the autoclave chamber varies with the nature of the load (fig. 9.3). Steam penetration of thick, bulky, porous articles such as linen packs for the operating room takes longer than steam condensation on the exposed, impenetrable surfaces of metal or glass instruments, which are quickly raised to sterilizing temperatures. Microorganisms buried within the depths of porous materials are not reached as rapidly as those lying on exposed surfaces. For these reasons it is critical that the timing of autoclave procedures be judged carefully for each load.

Materials and items that must be sterilized vary in their susceptibility to heat. Nonporous materials such as glass or metal withstand it very well, but cloth, rubber, and plastic may be damaged. Time and temperature adjustments must be made with this in mind. Furthermore, the steam penetrability of wrapping materials must be borne in mind when the autoclave is used for sterilization. Rubber and some plastic surfaces can be successfully sterilized without deterioration under steam pressure, but they can never be used as covering for other items to be sterilized because they are impenetrable to steam.

The requirements and limitations of steam pressure sterilization may be summarized as follows:

1. The temperature of the steam chamber must reach a sterilizing level (121° C). This is possible only if all air originally present has drained out through the discharge outlet.
2. Items to be sterilized must be packaged and arranged within the autoclave to permit full steam penetration of each package. (Instruments or other objects to be used promptly after sterilization may be autoclaved without covering, provided they are not exposed to undue risk of air contamination upon removal from the sterilizer.)

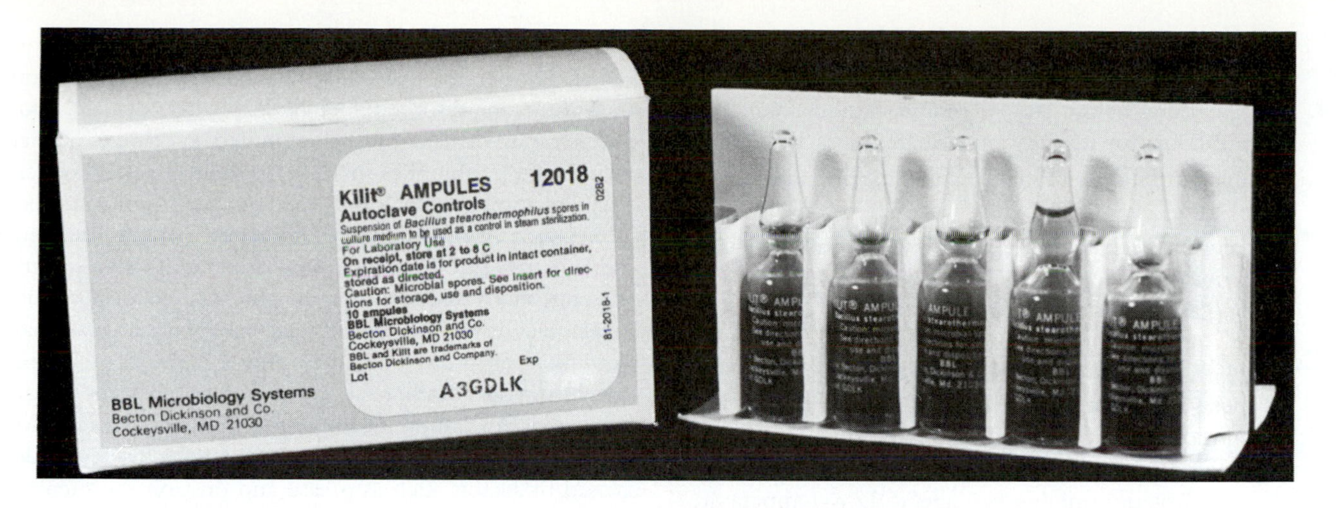

FIGURE 9.4 The final criterion for the safety of steam pressure sterilization is the bacteriological demonstration of its efficiency in killing heat-resistant organisms. A suspension of *Bacillus stearothermophilus* spores in culture medium can be used as a control on the procedure.
Courtesy of Becton Dickinson Microbiology Systems, Division of Becton Dickinson and Company, Cockeysville, MD.

3. The timing of the autoclave procedure must begin when the sterilizing temperature is reached, and sterilization must continue for a time period dictated by the nature and size of the load. It is dangerous to underestimate time, but it is inefficient and wasteful to prolong it unnecessarily and possibly injure valuable materials.

Modern autoclaves are equipped with regulating devices for the control of air discharge, pressure, temperature, and timing, which reduce the work of the operator to a minimum. Like all mechanical equipment, however, they require continuing intelligent monitoring. The safe operation of an autoclave is the responsibility of the person who uses it on each occasion.

The final criterion of the safety of steam pressure sterilization is the bacteriological demonstration of its efficiency in killing heat-resistant organisms (fig. 9.4). Suitable materials for conducting such tests (usually a preparation of living, heat-stable bacterial endospores) are available commercially, and the microbiology laboratory can perform the necessary culture work. The supervision and use of such controls, however, are the responsibility of those who consistently operate the autoclave. When properly carried out, bacteriologic tests can provide information about the mechanical efficiency of the autoclave; the adequacy of techniques for packaging individual, typical items; and the proficiency of the operator in arranging the total load and regulating conditions to permit steam penetration throughout the chamber.

Uses of Dry Heat

Dry heat sterilization has an oxidative effect not only on microorganisms but on the materials they contaminate. If the contaminated materials are expendable, they may be sterilized by incineration; if they are heat resistant, flaming or hot-air baking may be appropriate.

Incineration This is an effective technique for disposing of many contaminated objects that cannot be used again, provided the temperature within the incinerator is kept sufficiently high to ensure that materials are ashed promptly after being fed into it. Overloading a dry-heat incinerator may cause the partially burned material to "lump," condensing its moisture content, coagulating the protein material, and protecting the microorganisms trapped within. Incinerating chambers must be of adequate size for the maximum load introduced and capable of maintaining temperatures that will burn their largest loads promptly and completely, or sterilization cannot be guaranteed. The use of incinerators in the hospital for destroying bandages, syringes, and other refuse from infectious patients is open to serious question: the design of such equipment is frequently inadequate to the need; its use is turned over to untrained personnel; and the safety and efficiency of incineration are seldom subjected to critical review.

Hot-Air Baking Baking in an insulated, thermostatically controlled oven is an effective method for sterilizing materials that can withstand high temperatures. Since dry heat does not penetrate well, sterilization with dry heat requires higher temperatures for longer periods than it does with moist heat and may be destructive to some materials. Rubber and plastic goods are quickly destroyed by dry heat, but glass and metal withstand it very well. Some materials, such as powders, waxes, oils, glycerin, and petroleum jelly, are not penetrable by moist heat or would be damaged by moisture; these require hot-air sterilization.

The requirements and limitations of hot-air sterilization may be summarized as follows:

1. Most items that must be baked can withstand 160 to 165° C for 2 hours, or 170 to 175° C for 1 hour.
2. Clean, dry glassware, metal items, glass syringes, and needles can be baked at 170° C for 1 hour, without injury.
3. Oils, waxes, and jellies are not easily penetrated by moderate heat but can be damaged by intense heat. If

prepared in small, shallow units, they can be sterilized by baking for 2 hours at 160° C.

4. Temperatures below 160° C require periods longer than 2 hours for sterilization and are impractical.
5. Temperatures above 180° C cause charring in a very short time.
6. Whatever the temperature and time, each item in the load must be placed so that heated air flows freely among all objects to be sterilized.

Cold

As temperatures decrease, the metabolic activities of microorganisms are slowed until, under freezing conditions, their growth is completely inhibited. Cold therefore is a good bacteriostatic (inhibitory) agent. Low-temperature storage can be used to preserve microbial cultures for long periods. Freezing and drying under vacuum (lyophilization) preserves the viability of many bacteria for years.

Refrigeration or freezing effectively delays microbial spoilage of foods without greatly affecting the quality and flavor of stored items. The growth of spoilage organisms is very slow at normal refrigerator temperatures (0 to 8° C), or inhibited altogether at home freezer temperatures (−5 to −25° C). Frozen foods must be kept in the freezer until prepared for the table. It is unwise to refreeze food that has thawed because microbial growth may produce metabolic products or induce changes in food that could be unpleasant or harmful at the next thawing.

In the natural world, cold inhibits and sometimes even kills some water and soil microorganisms but does not sterilize these environmental reservoirs. As a matter of fact, ice from contaminated water has frequently been incriminated as a source of infection when used to chill drinking water, other beverages, or food.

Drying

Many microorganisms are inhibited but not killed by drying. In the absence of moisture, a microbial cell cannot grow and reproduce but may remain viable, resuming growth when moisture and nutrient are restored. This fact is applied in the laboratory to preserve cultures by drying them under vacuum from the frozen state. (Freeze drying is also used to preserve serum and other blood products.)

In the hospital situation it is most important to remember the survival capacity of dried microorganisms. Bacterial endospores are extraordinarily resistant to drying, but many vegetative bacteria, viruses, fungi, protozoa, and helminth ova may withstand it also, although to a lesser degree. Dust, clothing, bedding, and dressings may contain dried remnants of urine, feces, sputum, or pus contaminated with viable, infectious microorganisms. Careless shaking of soiled fabrics, sweeping, dry mopping, or dry dusting all contribute to wide dissemination of these potential pathogens.

Ultraviolet Radiation

The wavelengths of radiant energy that comprise the ultraviolet spectrum range from approximately 10 to 400 nm (1 nm = 10^{-9} m). Those with the strongest destructive effect on microorganisms are in the band from 250 to 280 nm since these are absorbed specifically by the nucleic acids. The resulting damage can be attributed to changes in their purines and pyrimidines that impair their normal function.

Ultraviolet radiation by mercury vapor lamps has been used extensively in a number of situations: to control airborne infection in critical hospital areas (operating rooms, nurseries, infectious diseases wards) or in confined areas of public use (elevators, schoolrooms, barracks); to decontaminate working areas in the laboratory; and to reduce contamination during the commercial preparation of sterile biologicals or of foods. An important limitation of ultraviolet radiation is that it penetrates very poorly and its effectiveness is decreased by factors such as grease and dust on the mercury lamps or extraneous material covering the microorganisms.

Although many studies have demonstrated that ultraviolet radiation does decrease the level of airborne microorganisms, it is not widely used for this purpose in hospitals. The airborne route of infection transfer is less important than direct contact with contaminated animate or inanimate objects.

Filtration

Coarse materials such as cotton, gauze, and paper can effectively filter bacteria from air when used correctly. These materials can be used to wrap items or packs of equipment to be heat-sterilized (fig. 9.5) because they will subsequently act as filters when packages are exposed to air, hands, or shelf storage. Fabrics and paper are not effective filters if they are moist or if their physical structure is damaged by overheating. Moisture-filled pores act as filters-in-reverse, for they promote passage of bacteria and other particles along their wet surfaces to the interior with a kind of capillary attraction. This is why a drying period is essential after steam pressure sterilization of wrapped equipment.

Masks used in surgical procedures (or in other situations requiring asepsis) are designed to prevent outward passage of microorganisms into a sterile area from the mouth, nose, and throat, or to protect the wearer from inhalation of infectious organisms. In either case they are effective only for the period during which they remain unsaturated with exhaled moisture. After that, they are not only useless as filters but dangerously capable of shedding contaminated droplets of moisture in either direction and should be discarded at once.

Modern forced-air ventilation systems also employ **filtration** devices to purify the air of dust. Hospital or laboratory areas where infection is of critical concern can be ventilated with equipment designed to filter out particles of submicroscopic size. However, the efficacy of such equipment is strictly limited by the extent to which it is monitored. Filters must be changed when they become clogged with dust and moisture, or their function is reversed and their action dangerous.

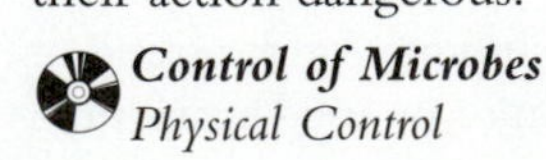
Control of Microbes
Physical Control

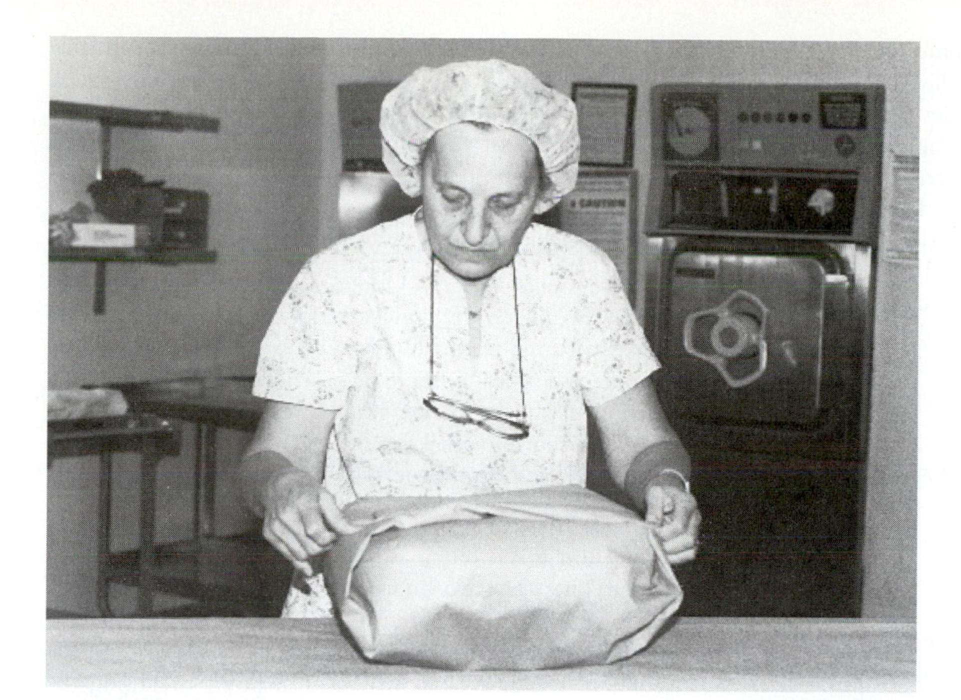

FIGURE 9.5 Instruments and articles are prewashed and carefully wrapped for sterilization in an autoclave.
Putnam Hospital Center, Carmel, NY. Photo: B. Riley, Danae Associates.

ANTIMICROBIAL ACTIVITY OF CHEMICAL AGENTS

Principles of Chemical Disinfection

Disinfection has been defined as a process that destroys pathogenic microorganisms. It is a chemical process involving interactions between disinfectant agents and various constituents of microbial cells. The result may be inhibitory to further growth of cells while they remain in contact with the chemical agent, but reversible upon its removal. In this case, the agent is said to be **static** in its effect (i.e., bacteriostatic, fungistatic). If the reaction produces irreversible changes in vital cellular components, the result is lethal, or **cidal,** and the cells are unable to grow again even when the agent is removed (table 9.1).

Like all chemical reactions, disinfection is influenced by the nature and concentration of the reactants and by time. Other factors may also influence the result, notably, the physical and chemical properties of contaminating substances that may be present (organic soil, inorganic salt) as well as pH and temperature. The presence of extraneous organic and inorganic materials may chemically inactivate the disinfectant or physically protect the microorganisms from destruction. Many chemical agents are lethal to pathogenic organisms under controlled conditions in a test tube, but under actual conditions of use in the hospital environment their efficacy may be limited by practical circumstances. It is essential, therefore, that those who apply chemical disinfection to the control of infectious disease understand both its principles and its limitations.

Microorganisms versus Disinfectants

The major groups of microorganisms are not uniformly susceptible to chemical disinfection. Among the animal parasites, protozoa and the ova and larvae of parasitic helminths are susceptible to germicide action. The vegetative forms of fungi are also readily killed by some disinfectants, but their spores are more resistant. The vegetative forms of most bacterial species succumb quickly upon exposure to active disinfectants. The tubercle bacillus is more resistant, but of all microbial forms, bacterial endospores display the greatest degree of resistance to chemical, as well as physical, agents of disinfection. Many viruses can be killed as easily as vegetative bacteria, and by the same germicides, but some, such as hepatitis viruses, are much more resistant.

The resistance of bacterial endospores to chemical action is related to their thick, relatively impervious walls and especially to the fact that their protoplasmic contents are extremely dehydrated. Thus, little of the water that is essential for reactivity with chemical agents is available in them. Tubercle bacilli are rich in lipids, and their waxy cell walls resist wetting (they are hydrophobic); hence, they are not readily penetrated by germicides in aqueous solutions but are more easily destroyed by those carried in lipid solvents or by hydrophilic "wetting" agents (soaps and detergents). Viruses that contain lipid, such as influenza and herpes viruses, are more susceptible to detergent germicides than are lipid-free viruses such as the poliomyelitis virus. The latter is destroyed by formalin and by alcohol. The resistance of hepatitis viruses has not been well defined. Decontamination with a 0.05 to 0.5% sodium hypochlorite solution appears to be effective in some instances (e.g., laboratory spills); however, whenever feasible, sterilizing methods are generally preferred for their destruction. The human immunodeficiency virus (HIV) is susceptible to several agents including 0.05% hypochlorite solution.

Since microorganisms differ in their response to disinfectants, the choice of an agent for a particular situation must take into account the type of microbial contamination that must be dealt with. To disinfect nondisposable mouth thermometers, for example, all pathogenic vegetative forms of bacteria that might be present in saliva or sputum, including the tubercle bacillus, must be destroyed. When destruction of bacterial endospores must be ensured, such as on instruments to be used for surgery and in dressings removed from a *Clostridium*-infected wound, heat sterilization must be used because disinfection procedures cannot guarantee killing of all endospores. Chemical sterilization with strong germicides such as formalin or glutaraldehyde is possible with a long exposure period (several hours). This may be feasible for heat-sensitive items (e.g., lensed instruments) provided they are not damaged by the germicide.

Destruction versus Inhibition The static or cidal effects of disinfectants are an important issue in their choice and use. In many instances the actual conditions for use of a chemical agent may lead only to inhibition of growth—a dangerous result if microbial destruction is planned or assumed. Microbiologic techniques can be used to demonstrate whether or not the use conditions for a disinfectant (its concentration, time of exposure, the presence of extraneous protein) are sufficient not only to prevent the growth of pathogenic microorganisms while they are in contact with the disinfecting agent, but to destroy them so that they cannot subsequently grow and multiply if and when the agent is removed.

It must be emphasized that the effectiveness claimed for a disinfectant in killing vegetative bacterial species (e.g., strains of *Staphylococcus* or *Salmonella*) may have no bearing on its ability to destroy fungal spores or bacterial endospores, tubercle bacilli, or some viruses. It should also be recognized that, while most classes of microorganisms can be killed within a 10 to 15-minute period by many of the available disinfectants, this may be true only for optimal, rather than average, conditions. Commercially available products bear labels that state their effective use conditions and the microbial agents that will be destroyed when such conditions are met.

Disinfectants versus Microorganisms

The chemical nature of a disinfectant determines its mode of action and its efficacy against different kinds of organisms. Some chemical agents coagulate the microbial cell protein, the effect being rapid and irreversible. The average disinfectant has a less powerful action, however, and denatures cellular proteins by more subtle, slower processes. Examples of disinfectant action are alteration of the chemical structure of enzymes by iodine compounds and lysis of cells or leakage of their contents induced by phenolics (or their detergent carriers).

Unfortunately, the activity of chemical disinfectants is not restricted to microbial proteins but is destructive for all living protoplasm to some degree. The stronger and most rapidly effective antimicrobial compounds are limited in their usefulness by their toxicity to human tissues. Strong germicides may be used under prescribed conditions for disinfection or chemical sterilization of some critical items, but less toxic compounds must be chosen for large-scale disinfection, as in hospital housekeeping or for surgical asepsis.

Other limitations on chemical disinfection include (1) the concentration of microbial cells to be inactivated, (2) the concentration of germicide that will do the job effectively without environmental damage, (3) inactivation of disinfectants by chemical contaminants from the environment, and (4) factors influencing the exposure of microorganisms to contact with disinfectants.

Microbial Concentration Under a given set of circumstances, the greater the concentration of microbial cells, the longer the time needed for disinfection. In practice, it is always necessary to observe this principle by thoroughly cleaning all items and surfaces before germicide treatment. With reduced microbial density, germicides are more efficient within a shorter time.

Germicide Concentration In general, the time required for disinfection is shortened as germicide concentration is increased. There are two important limitations to this principle, however. First, the activity of germicides in aqueous solutions is dependent on the water present. Decreasing the "water of reactivity" below a critical point by raising the germicide concentration may destroy some of the disinfectant activity of the solution. For example, concentrations of ethanol from 10 to 80% are increasingly bactericidal, the most efficient action being obtained in the 60 to 80% range. Ethanol solutions of 80 to 100% volume concentration are little more efficient than 40% solutions. Second, strong disinfectant solutions are corrosive to metals, fabrics, plastics, and skin; yet some may have little antimicrobial activity in weak solution.

The concentration of choice for a given disinfectant is that which displays the most rapid antimicrobial activity without simultaneous damage to materials exposed to it.

Inactivation of Disinfectants Germicides may be inactivated by organic or inorganic contaminants on items to be disinfected. The active portions of some disinfectant molecules may combine with human proteins of blood, mucus, or pus, and thus they are no longer free to react with microbial constituents. Some agents are more readily bound than others, by virtue of their chemical structure. The use of strong concentrations of some disinfectants, if compatible with materials to be treated, may ensure an excess of free, active molecules despite extensive protein contamination.

Inorganic compounds such as salts, metals, acids, or alkalis can also interfere with disinfectant activity by affecting the pH of the reaction or by combining with the germicide as proteins do. Volatilization may reduce the available concentration of some disinfectants, particularly those with an alcoholic base or those that depend on the activity of a halogen such as chlorine or iodine. Temperature is not an important factor in hospital disinfection, for most germicides are effective at temperatures ranging from just above to slightly below room temperature.

Other Factors Influencing Disinfection The presence of protein **contamination** on items to be disinfected may not only inactivate the germicide but also protect microorgan-

Table 9.4 LEVELS OF DISINFECTANT ACTION

	Bacteria				Viruses	
	Vegetative[a]	Tubercle Bacillus	Endospores	Fungi[b]	Lipid and Medium Size	Nonlipid and Small
High	I	+	+	+	+	+
Intermediate	+	+	−	+	+	+
Low	+	−	−	+	+	−

Source: From E. H. Spaulding, *Proceedings of the International Conference on Nosocomial Infections* (Washington, D.C.: USDHEW, Centers for Disease Control, 1970), p. 248.
Note: + indicates that cidal effect can be expected.
[a]Common forms of bacterial cells, e.g., staphylococci.
[b]Includes asexual "spores," but not necessarily dried chlamydospores and sexual spores.

isms from full contact with it. This protective effect offers another major reason for careful scrubbing and **cleaning** of objects to be chemically disinfected. Situations sometimes arise, however, when it may be unsafe to wash items grossly contaminated with infectious organisms. Surgical instruments covered with pus from infected wounds and heat-sensitive equipment similarly contaminated should not be handled. Instruments can be boiled with detergent disinfectants or placed in washer-sterilizer equipment, then carefully cleaned and lubricated before being sterilized again for another use.

The physical properties of objects and surfaces to be chemically disinfected must also be considered, for they influence the speed and ease with which effective results can be achieved. Hard, smooth surfaces are relatively easy to disinfect, but porous materials are not readily penetrated by many germicides. Grooved or hinged instruments and needles, catheters, and irrigation tubing of small bore offer particular difficulties, for their inner as well as outer surfaces must be in contact with the disinfectant. The choice of germicide and its concentration must be balanced against the type of material it is expected to disinfect, and the technique of application must ensure its efficacy under each set of circumstances.

Antisepsis

The chemical disinfection of skin is usually referred to as **antisepsis.** Many useful chemical disinfectants are too irritating to skin and mucous membranes to be used as antiseptics. When applied to skin, antiseptics reduce the numbers of surface organisms markedly and rapidly but do not penetrate readily to deeper layers, hair follicles, or glandular crypts. The mechanical action of scrubbing, followed by the flushing effect of running water, is an important factor in cleansing the skin surface and in removing some resident flora from deeper layers.

Soaps, alcohol, alcohol and iodine combinations, iodophor preparations, and chlorhexidine gluconate are among the most widely used antiseptics in surgical practice. All except soap have rapid activity.

Hexachlorophene is a skin disinfectant but is not recommended for repeated, routine use because it accumulates in an active residue on skin and has toxic effects, especially in infants. In addition, it is effective only against gram-positive bacteria and has little activity against other microorganisms.

None of these antiseptics can prevent restoration of the resident flora on gloved hands during lengthy surgical procedures. For this reason gloves that are nicked or punctured in the course of a surgical operation should be changed, particularly if enough time has elapsed to permit some bacterial recolonization on the hands.

A word of caution must be added regarding the storage and handling of antiseptics. Antiseptic solutions ready for use in the operating room for surgical scrub or for the preoperative preparation of patients' skin may become contaminated with microorganisms resistant to their chemical action. Sometimes this may happen also with strong disinfectants. If they are actively supporting microbial growth, such solutions constitute a dangerous reservoir of infection, particularly if they are used in surgical or other hospital situations where asepsis is essential. The assumption that the antimicrobial nature of disinfectants and antiseptics guarantees their sterility is not warranted. Working solutions, as well as stocks from which they are prepared for final use, should be stored and handled under aseptic conditions.

Major Classes of Disinfectants and Their Uses

Disinfectants can be classified on the basis of their general chemical nature and activity. One approach is to group them by level—high, intermediate, or low—according to their effects on different types of microorganisms (table 9.4). Agents that have a high level of activity essentially achieve chemical sterilization by destroying all microbial forms including highly resistant bacterial endospores. It must be realized that, in contrast to vegetative cells, which often are killed by disinfectants within only 10 minutes, considerably longer contact time (10 hours or longer) is required to destroy endospores. Intermediate level agents do not kill bacterial endospores, and therefore cannot be used when the disinfection process requires their destruction. Most antiseptics are low-level germicides and have activity only against vegetative bacteria, fungi, and lipid-containing viruses.

The following pages briefly describe the major classes of disinfectants and their uses. Table 9.5 summarizes their applications, properties, and important characteristics.

TABLE 9.5 SOME COMMON DISINFECTANTS WITH THEIR USE-DILUTIONS AND PROPERTIES

Germicide	Use-Dilution	Level of Disinfection	Inactivates[a]: Bacteria	Inactivates[a]: Lipophilic Viruses	Inactivates[a]: Hydrophilic Viruses	Inactivates[a]: *M. tuberculosis*	Inactivates[a]: Fungi	Inactivates[a]: Bacterial Endospores	Important Characteristics: Shelf Life >1 week	Important Characteristics: Corrosive/Deleterious Effect	Important Characteristics: Residue	Important Characteristics: Inactivated by Organic Matter	Important Characteristics: Skin Irritant	Important Characteristics: Eye Irritant	Important Characteristics: Respiratory Irritant	Important Characteristics: Toxic	Important Characteristics: Easily Obtainable
Isopropyl alcohol	60–95%	Int	+	+	−	+	+	−	+	±	−	+	±	+	−	+	+
Hydrogen peroxide	3–25%	CS/High	+	+	+	+	+	±	+	−	−	±	+	+	−	+	+
Formaldehyde	3–8%	High/Int	+	+	+	+	+	±	+	−	+	−	+	+	+	+	+
Quaternary ammonium compounds	0.4–1.6% aqueous	Low	+	+	−	−	±	−	+	−	−	+	+	+	−	+	+
Phenolic	0.4–5% aqueous	Int/Low	+	+	±	+	±	−	+	−	+	±	+	+	−	+	+
Chlorine	100–1000 ppm free chlorine	High/Low	+	+	+	+	+	±	+	+	+	+	+	+	+	+	+
Iodophors	30–50 ppm free iodine	Int	+	+	+	±	±	−	+	±	+	+	±	+	−	+	+
Glutaraldehyde	2%	CS/High	+	+	+	+	+	+	+	−	+	−	+	+	+	+	+

Modified from Laboratory Biosafety Manual, Geneva: World Health Organization, 1983.
Abbreviations: Int, intermediate; CS, chemical sterilant; + , yes; − , no; ± , variable results.
[a]Inactivates all indicated microorganisms with a contact time of 30 min or less, except bacterial endospores, which require 6–10 hours contact time.
Source: Modified from Rutala, W. A. 1996. Selection and Use of Disinfectants in Health Care, pp. 913-936. In C. Glen Mayhall, ed. *Hospital Epidemiology and Infection Control.* Williams & Wilkins, Baltimore.

Phenols

The **phenols** have been popular disinfectants ever since Lister first used carbolic acid in his operating room. Because it is irritating and corrosive for skin, carbolic acid is seldom used today as an antiseptic or disinfectant, but some of its derivatives have a safer application. Most phenolics currently employed contain a phenol molecule, chemically altered to reduce its toxicity or increase its antibacterial activity, in combination with a soap or detergent. The latter serves to disperse protein, lipid, and other organic matter, reducing the possibility that such material may bind and inactivate the disinfectant.

The best-known phenolic derivatives combined with soaps or detergents are the cresols and the bis-phenols. Among the cresol group are Lysol, Staphene, and Amphyl, these being proprietary names for different phenols used as environmental disinfectants. The bis-phenols, such as hexachlorophene, are so called because the chemical alteration involves the linkage of two phenol molecules.

The hospital applications of hexachlorophene have been strictly limited in recent years because of the potentially toxic hazards that accompany its continuous use, particularly for cleansing infant skin through which it is readily absorbed. It may be used in some instances to control nursery outbreaks of staphylococcal skin infections but never in concentrations exceeding 3%.

In high concentrations the phenols coagulate protein and effectively kill the vegetative cells of bacteria. When combined with detergents they are also good tuberculocides, virucides, and fungicides. They are not sporicidal under ordinary conditions of use as surface disinfectants, for they can kill endospores only in heated solutions applied for prolonged periods of time. When combined with soaps or detergents, the phenolics are not very soluble in water but remain active in aqueous concentrations as low as 1 to 5%. This combination of qualities makes them inexpensive to use but, more important, confers residual bactericidal or bacteriostatic effects when their solutions are left to dry on treated surfaces (walls, floors, furniture, skin). As water evaporates from phenolic films, the disinfectant molecule remains active for a time, deterring bacterial growth. These disinfectants can also be used to treat discarded cultures, sputum, feces, and heavily contaminated objects because of their low affinity for organic matter and their antibacterial activity, even in high dilution.

Alcohols

Low-molecular-weight **alcohols** that are water soluble may be bactericidal under proper conditions of use. Methyl alcohol, the lightweight of this family of compounds, is not a good disinfectant, but ethyl and propyl alcohols have antimicrobial activity in the presence of water. In absolute concentrations (100%) and those below 50%, alcohols have little or no effect on microorganisms. However, in 50 to 80% aqueous solution, ethyl, propyl, and isopropyl alcohols are lethal for the tubercle bacillus and for vegetative forms of other bacteria. At these concentrations alcohols can penetrate cell membranes and precipitate proteins, but they cannot penetrate bacterial endospore walls, and they are ineffective against certain viruses.

Alcohols are good lipid solvents and are frequently used to cleanse and disinfect skin surfaces, clearing it of oil, debris, and microorganisms. Other disinfectant compounds are sometimes dissolved in alcohol, rather than in water, the resulting tincture having added disinfectant properties as well as better grease solvency. (Tincture of iodine is a well-known example.) Alcoholic solutions dry the skin since they remove oil and thus damage one of the natural skin barriers. When used on traumatized tissues, they may add further injury by coagulating cellular proteins. The use of alcohols and tinctures as antiseptics is limited to normal skin surfaces (in preparation for venipuncture or minor surgery) and small, hard-surfaced objects such as thermometers.

Surface-Active Agents (Detergents and Soaps)

Surface-active agents decrease the tension that exists between molecules that lie on the surfaces of liquid suspensions or solutions. Liquids that have a high surface tension (e.g., oils) are not easily penetrated by other materials, nor do they spread into them. They tend, rather, to separate from surrounding substances because strong attracting forces hold their surface molecules together.

When the surface force of liquids is low, they spread, or "run," more easily. They are said to be "wet" because they flow across and penetrate into adjacent surfaces as pure water does. Liquids with low surface tension are also more "miscible"; that is, they mix and can be mixed with other fluids. Surface-active agents accumulate at interfaces of liquids (i.e., at their adjacent surfaces), lowering the surface tension of molecules on both sides of the interface so that they can run together. The liquids become more miscible, and the agent is said to have a "wetting" action.

The antibacterial action of soaps and detergents is partly due to their ability to lower the surface tension of liquid interfaces. Bacterial membranes themselves constitute interfaces between the intracellular cytoplasm and the moist, aqueous surroundings in which the bacteria usually lie. When surface-active agents come in contact with microbial interfaces, they remove the molecular barriers that hold cells and surrounding liquid media apart; they form a bridge between the interior contents of cells and the external milieu. Sometimes the bridging compound itself denatures microbial cell contents and induces cellular injury or death. If not, disinfectant chemicals can be mixed with surface-active agents to aid their entry into cells.

Soaps Soaps are surface-active agents that improve the miscibility of oils with water. For this reason they are very useful cleansing agents for skin and clothing. Soaps also have a lethal action on a few fastidious pathogenic bacteria, notably streptococci, pneumococci, and *Haemophilus* species, but have no chemical effects on most skin microorganisms.

When disinfectant chemicals are added to soaps (as in such proprietaries as Lifebuoy, Dial, Safeguard, and certain liquid soaps, which contain phenolic disinfectants or other compounds), their antibacterial efficacy is increased. The combination of soaps with disinfectants must be carefully gauged, however, to avoid neutralizing the antibacterial properties. For this reason the sequential use of soaps and disinfectants should be carefully considered, for even traces of soap left on skin or other objects may bind disinfectants subsequently applied and thus neutralize, or inactivate, their chemical effects on microorganisms.

Detergents Natural or synthetic soaps with a high degree of surface activity are called *detergents.* Depending on their chemical structure, they may or may not ionize in water solution. Those that do not ionize (neutral detergents) have no usefulness as disinfectants, although they may be good cleansers. Those that do form ions in aqueous solution have bactericidal activity, probably because their charged ions concentrate on bacterial cell membranes and disrupt their normal functions. Anionic detergents are those that form negatively charged ions (anions). These include the natural soaps (the sodium or potassium salts of fatty acids) and bile salts.

Cationic detergents form positively charged ions (cations). This group includes the quaternary ammonium compounds, or "quats." The most familiar proprietary name in this group is Zephiran. Although these compounds have many useful qualities (they are bacteriostatic in low concentrations, nontoxic, soluble in water, and inexpensive), they are severely limited by their affinity for many proteins and for soaps. They are also readily bound and inactivated by plant fibers, such as those in gauze and cotton pads. Aqueous solutions of quats bound to such materials can thus become reservoirs for bacterial growth and have been incriminated in numerous serious nosocomial (hospital-acquired) infections. For this reason, and because they lack the broad spectrum of activity (they are neither tuberculocidal nor sporicidal) of other useful disinfectants, their use in hospitals has been generally disapproved. Newer modifications of the quats, however, have resulted in germicidal detergents (T.B.Q., Cavicide) that appear to be significantly more effective, with microbicidal activity against bacteria, fungi, viruses, and the tubercle bacillus.

Halogens

The **halogens**—iodine, chlorine, and their derivatives—have bactericidal properties, probably because they damage protein enzymes and nucleic acids, inhibiting their activity.

Iodine This halogen is an effective antiseptic for normal skin and for superficial wounds. It is also very useful in environmental disinfection, for it is active against many vegetative bacterial species, including the tubercle bacillus, fungi, and some viruses. It is not an effective sporicide because concentrations that are employed must avoid toxicity. Pure iodine is not readily soluble in water, but dissolves relatively easily in solutions of sodium or potassium iodide. The release of free iodine from the iodine-iodide complex accomplishes disinfection. A 3.5% tincture of iodine in alcohol containing potassium iodide is a very effective skin antiseptic. Tinctures may also be used for disinfecting thermometers and small instruments, provided volatilization of these solutions is prevented.

Tinctures may damage skin if the iodine concentration has increased because of alcohol evaporation. This factor, as well as the staining effects of iodine, has tended to discourage the use of tinctures on skin. Both these limitations are counterbalanced by the rapidly effective action of iodine tinctures as compared with other antiseptics. For minor procedures such as venipuncture, the careful preparation of small areas of skin with iodine-alcohol solutions can produce safe results quickly. When large areas of skin must be treated, however, the irritating effects of iodine and alcohol outweigh the advantages. In this case, less toxic iodine preparations or other types of antiseptic agents are used, but more time must be allowed for their action and mechanical scrubbing becomes more important.

Iodine forms loosely bound complexes with organic molecules. In aqueous solution or suspension, the bound iodine carried by such a complex slowly dissociates and is freed. The combination of iodine and the carrier molecule is called an *iodophor.* Such compounds retain the useful properties of inorganic iodine solutions, but they are less toxic to the skin. The iodophors commonly used for environmental disinfection contain iodine combined with a nonionic detergent (some of the proprietary names include Wescodyne and Klenzade). In a concentration of 75 ppm of available iodine, the iodophors may be used for the general disinfection of floors, walls, and furniture. A concentration of 450 ppm has tuberculocidal activity. Formulations of iodine with polymeric molecules, such as polyvinyl pyrrolidone (povidone iodophors), are used for skin antisepsis. Betadine, Prepodyne, and Isodine are examples of this type of iodophor.

The iodophors have a wide range of antimicrobial activity. They should be used in fresh solution, for iodine slowly dissipates from dilutions that are allowed to stand. Fortunately, however, this change is easy to recognize as the original deep-brown iodine color fades to yellow with the loss of iodine. Fabrics and skin are not permanently stained by iodophors, but plastic materials treated with them remain discolored. The chief limitation of iodophor detergents is their reactivity with protein and other organic matter. In combining with such material, they lose disinfectant activity. It is important therefore that surfaces to be treated with iodophors be free of extraneous contamination.

Chlorine In gaseous or liquid form, chlorine is a useful and inexpensive bactericidal agent for water and sewage. A number of chlorine compounds release it readily as free or available chlorine in aqueous solution. Chlorine itself then combines with water to form hypochlorous acid, a potent oxidizing and bleaching agent with a rapid bactericidal action. A few parts per million of available chlorine in drinking water or swimming pools are disinfectant. Larger concentrations are required in sewage because some of the chlorine also combines with organic substances and is then no longer free to react with bactericidal cell proteins.

The most commonly used inorganic compounds of chlorine are hypochlorites of potassium, sodium (e.g., Clorox), or calcium (chlorinated lime). Sodium or potassium hypochlorite in aqueous solution (where it forms hypochlorous acid) can be used to disinfect small objects and surfaces and is used to decontaminate laboratory spills suspected of containing hepatitis and human immunodeficiency viruses. It is a convenient disinfectant for the sickroom at home but is seldom used for this purpose in hospitals because, like most other chlorine compounds, it is bleaching, corrosive, and malodorous. Chlorinated lime in solutions of 0.5 to 5% is used to disinfect dairies, barns, abattoirs, and similar installations. Inorganic chlorine compounds are also used to sanitize dishes and laundry.

Ions of Heavy Metals

Metallic ions such as mercury and silver readily combine with the proteins of living cells and are highly inactivated by organic matter. They inhibit microbial enzyme activity, but their effects are reversible. Therefore, these agents are bacteriostatic rather than bactericidal and have only limited applications. A 1% eyedrop solution of silver nitrate was once used in the eyes of newborn babies to prevent gonorrheal infection, but an ointment containing the antimicrobial agent erythromycin is now used instead. Silver nitrate is used to treat extensive burns as is silver combined with the antimicrobial agent sulfadiazine. Some microorganisms that cause burn wound infections are now resistant to killing by this latter preparation.

Chlorhexidine

Chlorhexidine, which belongs to a group of compounds known as biguanides, has been used for many years in England, but only recently in the United States. It has rapid bactericidal and some fungicidal activity. At low concentrations, chlorhexidine damages microbial membranes, resulting in leakage of cell contents; at high concentrations, it probably precipitates cellular proteins and nucleic acids. Chlorhexidine gluconate in a 4% emulsion (Hibiclens) is a highly effective antiseptic for use as a surgical scrub, hand wash, and cleanser for minor wounds. It rapidly reduces the number of organisms present on skin and is not affected by the presence of organic matter such as pus or blood.

Alkylating Agents

The antimicrobial activity of **alkylating agents** is due to the substitution of their alkyl groups (alkylation) for the reactive hydrogen on the chemical radicals of microbial proteins, nucleic acids, or enzymes. Alkyl groups are univalent hydrocarbon radicals such as occur in aldehydes (—CHO) and similar compounds, including ethylene oxide and beta-propriolactone. Alkylation disrupts microbial metabolic pathways resulting in destruction of microorganisms. Some of these alkylating agents are active as liquid disinfectants, some are used as gases, and some may be used in either the gaseous or liquid phase.

Aldehydes Formaldehyde and glutaraldehyde are the best-known examples of the group of disinfectants called **aldehydes.** Formaldehyde has a long history of use both as a gaseous fumigant and as a liquid germicide, but its use has been largely abandoned because of its irritating and toxic properties. It is damaging to skin and mucous membranes, and its odor, both as a gas and as a liquid, is unpleasant and penetrating—qualities that limit its usefulness as an environmental disinfectant. Formalin, a 37% aqueous solution of formaldehyde gas, is commonly used in high dilution to prepare microbial antigens for vaccines. It is valuable for this purpose because it can kill microorganisms or inactivate their toxins without destroying their antigenic properties.

Glutaraldehyde, a liquid compound related to formaldehyde, has come into prominence more recently. This disinfectant, as it is commonly used in a 2% buffered aqueous solution, is as effective as formalin without being noxious. It is cidal for most bacteria within five minutes, kills tubercle bacilli and viruses in ten minutes, and is sporicidal in ten hours. It can be considered a chemical sterilizer when used for a minimum of ten hours, or a disinfectant when applied for ten minutes. It is recommended as a soak for lensed instruments, rubber or plastic items, anesthesia or inhalation therapy equipment, sharp instruments and thermometers, and even for soiled instruments from septic cases.

Cidex is the proprietary name for the commercially available glutaraldehyde prepared with a buffer-activator that brings the pH to 8.0. This activated solution remains stable for 14 days if stored at cool room temperature. A newer formulation, acid glutaraldehyde (trade name, Sonacide) has properties similar to those of alkaline glutaraldehyde, but it is said to destroy endospores in one hour if the temperature is raised to 60° C. Although nonirritating when used correctly, glutaraldehyde should not be allowed to come in contact with the skin or eyes, but if this happens, it should be flushed away promptly with a copious flow of water.

Ethylene Oxide (ETO) The gaseous compound **ethylene oxide** is widely used for gas sterilization within a closed chamber similar to an autoclave. ETO is particularly suitable for this purpose because it is lethal for all varieties of microorganisms, is penetrative when under increased pressure, and does not permanently bind to items or materials exposed to it but dissipates slowly when these are removed from the sterilizer. It is a flammable, explosive gas, however, and for this reason it is used in a mixture with inert gases such as carbon dioxide or freon. In this condition it requires three to four hours under pressure to penetrate shallow layers of materials and to sterilize all surfaces within a chamber load. A minimum period of 24 hours at room temperature must be allowed following sterilization for ETO to dissipate from treated materials.

The time required, toxicity, flammability, and expense of the equipment needed are limitations to be weighed against its efficacy as a sterilizing agent for sensitive materials. Its use is generally reserved for objects that would be damaged by more efficient methods of heat sterilization or chemical disinfection. Lensed instruments, plastic items, catheters made of synthetic materials, Lucite and plastic components of heart-lung pumps, artificial heart valves and vessel segments, mattresses and other bedding, and many similar articles can be safely sterilized with ethylene oxide, this being the method of choice despite its expense.

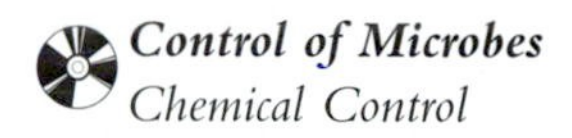
Control of Microbes
Chemical Control

Selection of the Appropriate Sterilization or Disinfection Method

When determining which method of sterilization or disinfection to use in health care settings, the potential for the item to transmit infection to the patient must be considered. Patient-care items have been divided in three categories according to the patient risk they present: critical, semicritical, and noncritical. Critical items are those that enter sterile tissues or the bloodstream, therefore requiring that they be *sterilized* before use, preferably in an autoclave or using ethylene oxide. Such items include surgical instruments, cardiac and urinary catheters, and intravenous fluids. Semicritical items are those that come in contact with mucous membranes or broken skin. These materials require treatment with disinfectants that have either high-level or intermediate-level action (see table 9.4). For example, respiratory therapy and anesthesia equipment and endoscopes require a high-level disinfectant, whereas intermediate-level disinfection only may be sufficient for other semicritical items such as reusable thermometers. Finally, noncritical items such as bedpans, patient furniture, and electrocardiogram leads may be cleaned and treated with low-level disinfectants. Consideration of how an instrument or device will be used in patient care followed by the use of an appropriate sterilization or disinfection method will ensure its safe use and prevent infection.

Control of Environmental Sources of Infection

"It may seem a strange principle to enunciate as the very first requirement in a Hospital that it should do the sick no harm." These words of Florence Nightingale should remain uppermost in the minds of all who are involved in patient care. Safety from infection is of critical importance in hospitals where, unfortunately, it is often most difficult to guarantee. It requires that all of the principles of asepsis, disinfection, and sterilization be continuously integrated to provide infection control and to prevent the occurrence of **nosocomial infection.**

The minimum goal for hospitals is cleanliness. This should be the coordinated responsibility of the heads of all major hospital departments including administration, housekeeping, maintenance, and nursing. Each area of the hospital, from its garbage disposal facility to the operating room, should be subjected to rigorous, daily cleaning to preclude survival and transmission of pathogenic organisms. Cleaning equipment should be chosen and applied with a knowledge of the microbiologic needs of hospitals. Dusting cloths, mops, brooms, and dry vacuum cleaners can contribute to the spread of microorganisms when they are used without regard for the principles of disinfection.

With a basic state of cleanliness existing throughout the hospital, aseptic conditions can then be established and maintained with relative ease wherever they are indicated. Medical **asepsis** has been defined as a condition excluding the infectious agents of **communicable disease,** while surgical asepsis implies the exclusion of all microorganisms (table 9.1). More often than not, asepsis is a goal rather than a realized condition, because it depends on the coordination of many techniques applied by many people (housekeepers, maintenance staff, nurses, physicians—even patients' visitors on occasion). Not all of these techniques can be fully effective under all circumstances, but when they are conscientiously integrated, they can provide the safe environment patients have a right to expect in modern hospitals.

In learning aseptic techniques it would be helpful to recall Joseph Lister's advice to his students:

> *. . . you must be able to see with your mental eye the septic ferments as distinctly as we see flies or other insects with the corporeal eye. If you can really see them in this distinct way with your intellectual eye, you can be properly on your guard against them; if you do not see them, you will be constantly liable to relax in your precautions.*

Medical Asepsis

Medical asepsis includes the fundamental concepts of good housekeeping (dust removal, scrubbing, laundering, disinfecting) with additional techniques applied to special conditions created by infectious illness. Some of these techniques are part of the general public education concerning communicability of infection: the personal hygiene of careful hand washing and bathing; sanitary handling of food and dishes; proper methods for cooking or refrigerating foods; hygienic attitudes toward body functions and excretions (toilet habits, covering the mouth and nose when coughing or sneezing, adequate disposal of facial tissues and handkerchiefs). More specialized methods are added to these by nurses, physicians, and other health care workers directly involved in the care of sick patients. These include all the precautionary measures taken to prevent direct transmission of infection from person-to-person or indirect transfer of pathogens by way of instruments, equipment, or any inanimate objects present in the sickroom.

Surgical Asepsis

Surgical asepsis adds the last links to this chain of techniques designed to eliminate the danger of transferring infection. The aseptic techniques that surround surgical procedures must provide the best guarantees that all microorganisms have been eliminated from any scene of action intimately involving the body's deep tissues (fig. 9.6). When injections are given, superficial wounds are sutured, or deep surgery is performed, the body's primary defense mechanisms are bypassed, and susceptibility to infection is increased at the site of the injury induced by the procedure.

Surgical asepsis is intended to prevent the introduction of any contaminating microorganisms. The area of the patient's skin to be entered is thoroughly cleansed and scrubbed with antiseptic, and if the procedure is to be extensive, surrounding areas are covered with sterile drapes so that a sterile field is established at the site of surgery. The surgeon and all personnel involved similarly prepare their own skin, scrubbing hands and forearms, and then with sterile gloves, gowns, caps, and masks they cover all areas from which microorganisms might be shed onto the operative field. Instruments, dressings, and sutures are all sterilized for use in surgery, and any equipment that cannot be presterilized is treated with an appropriate disinfection technique. The operating room walls, floors, and equipment are cleaned and disinfected before use, and every pattern of activity within the room is designed to maintain the aseptic atmosphere. The ventilation of operating rooms should be such that with doors closed, fresh, filtered air is delivered, circulated, and exhausted continuously.

Once asepsis has been adequately established in a closed, properly ventilated operating room, the number of microorganisms present in the air or on surfaces is reduced to the barest possible minimum. From that point on, as surgery proceeds, microbial contamination of the atmosphere progressively increases, its principal source being the people present in the room. The degree to which such contamination rises to pose potential danger to the patient depends largely on the technique and skill of attending personnel in controlling their own activities so that the major sources of their microorganisms (nose and throat, hands, uncovered skin and hair) remain under continuous, effective restraint. Close guards are kept against breaks in technique: damaged gloves are discarded and replaced; saturated masks are exchanged for fresh ones; backup supplies of sterile instruments, solutions, and other essential items are kept at hand.

It should be emphasized that surgical asepsis is a concept to be applied not only in operating rooms but in any situation requiring close protection of the patient when the risk of infection is high. It should be maintained for surgical wounds during the first few postoper-

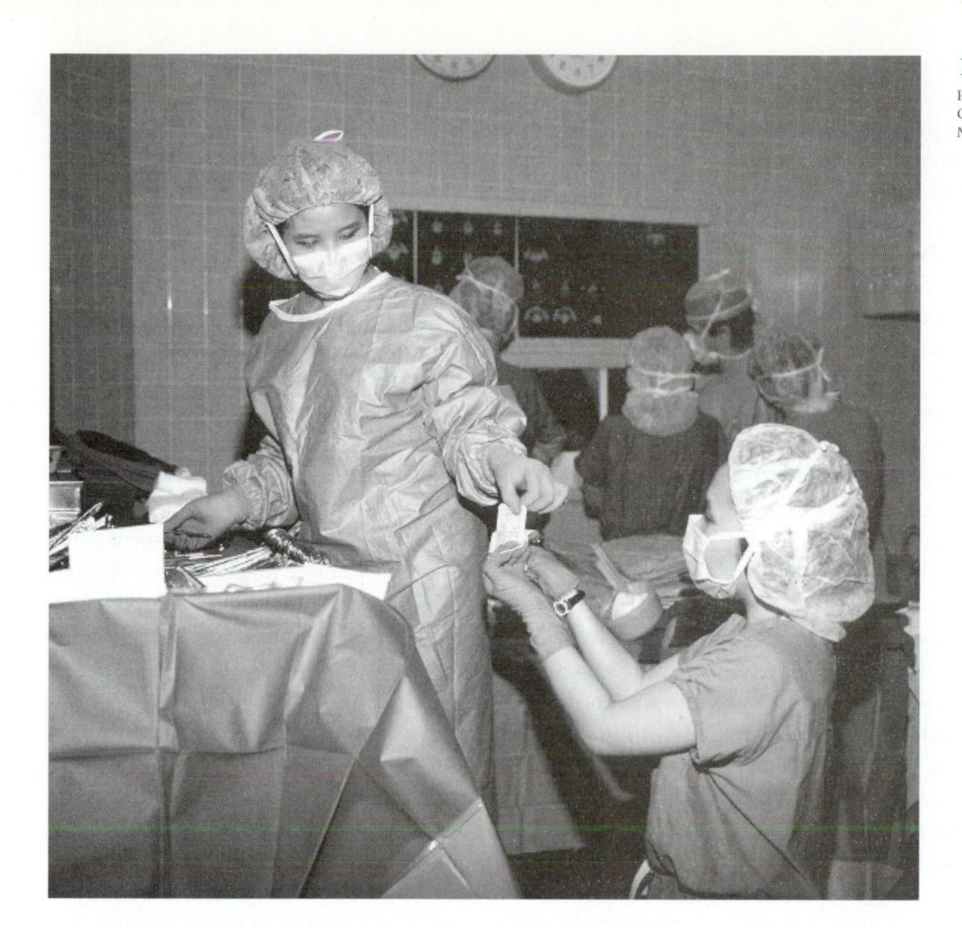

FIGURE 9.6 Modern operating room.
Reproduced from *Nursing Spectrum* 2: 8, April 9, 1990, Cover. O. R. Long Island College Hospital, Brooklyn, NY. Photo by Meryl Alexander-Tihany.

ative days, or until tissues are sufficiently healed to manage their own defenses. It is applied in delivery rooms to prevent sepsis of injured mucous membranes and in nurseries to protect susceptible newborns from any risk of infection. Wherever it is established, surgical asepsis represents a crowning touch to a basic structure of techniques that begin with sanitary cleanliness and end by disinfecting and sterilizing the environment to the fullest extent possible.

APPLICATION OF ASEPTIC TECHNIQUES IN HOSPITALS

In many hospitals, general policy with respect to infection control is established by a committee composed of representatives from the medical and nursing staffs, housekeeping and maintenance departments, the laboratory, and administration. The infection control committee can set up uniform guidelines for procedures throughout the hospital, basing these on sound principles of asepsis specifically applied to each local situation. Key members of this committee are the hospital epidemiologist and the infection control practitioner. The former is often a physician with special training in epidemiology or a clinician specialized in infectious diseases. This individual generally serves as the committee's chairperson, forming a liaison with the medical staff and authorizing final policy. The infection control practitioner (often a nurse) monitors the hospital for infection among patients or staff, supervises isolation and precaution techniques, and provides consultation and training for hospital personnel involved in problems associated with infection.

The function of the microbiology laboratory in infection control is to consult and provide surveillance to determine the efficacy of aseptic procedures. Appropriate laboratory tests can demonstrate whether or not hospital practices accomplish the desired result. When necessary for educational or epidemiological purposes, in-use testing can be applied to equipment, solutions, floors, furniture, or evaluation of techniques. In many situations it can provide final information concerning the value of preventive measures. The laboratory should also be prepared to combine its diagnostic function with surveillance methods so that outbreaks of particular infections, if they occur, can be quickly recognized and traced to their source. In this connection, the assistance of available epidemiologic experts from city or state health departments should be used. The microbiologist also plays an important role in providing information on the microorganisms that are most prevalent in the hospital or in a given service, and the patterns of their responses to antimicrobial agents. This can form the basis of medical policy for the clinically judicious use of antimicrobial agents and also provide an index of the efficiency of infection control policies.

Infection Control Principles

Decisions regarding the types of precautions to be taken in caring for infected patients are based on differences in communicability of their infections and possible routes of spread. Effective procedures must be established for isolating the disease (not necessarily the patient) in order to prevent transmission of infection between patients. These procedures are described more fully in chapter 11.

Since isolation of patients presents problems both for the patient and for the hospital, it should be used only when necessary to control transmission of infection. **Isolation precautions** should be discontinued promptly when the **communicable period** has passed or when specific treatment has removed the threat of further transmission. The procedures involved in carrying out isolation precautions are time consuming and expensive, and may even interfere with some aspects of good patient care. For example, the need to put on and take off gowns, masks, and gloves may discourage hospital staff and family from making frequent visits, adding to the patient's feeling of solitude.

Concurrent Disinfection

The techniques of concurrent disinfection are practiced throughout the course of the hospital stay, or until the physician considers the patient no longer infectious. To ensure destruction of pathogenic microorganisms, these techniques are applied as rapidly as possible to infectious patient discharges (excretions, secretions, blood, body fluids) or objects soiled by these discharges (bed linens, clothing, instruments, dressings). Such contaminated objects are frequently associated with transmission of infection to other patients. Nondisposable items are first disinfected, with a minimum of handling, then cleaned thoroughly or resterilized. If the patient's infection involves endospore-forming bacterial pathogens, any contaminated articles to be reused must be sterilized by steam under pressure. Disposable items may be incinerated, or autoclaved if this is more convenient.

Terminal Disinfection

When the patient has been discharged, terminal disinfection of the room is required. As in concurrent disinfection, special attention must be given to items that have been in direct contact with the patient or the patient's infectious material. All nondisposable equipment is either sterilized or carefully disinfected. All items that can be laundered are cleaned using good laundry techniques. Horizontal surfaces of furniture and plastic covers on pillows and mattresses can be scrubbed with liquid disinfectant. Floors are wet-vacuumed or mopped with a disinfectant-detergent solution. Washing of walls, blinds, and curtains is not indicated unless they are visibly soiled since these objects, as well as other environmental surfaces, rarely serve as vehicles for transmitting infections.

The Surgical Patient

Preparation of the patient scheduled for surgery requires adequate disinfection of the skin area to be incised. Unless the hair near the operative site is so thick that it will interfere with the surgical procedure, it should not be removed. If removal is necessary, it should be done as near the time of operation as possible, preferably immediately beforehand. Clippers or a depilatory rather than a razor should be used. This reduces the opportunity for skin injury and subsequent increased bacterial colonization. The area around and including the operative site should be scrubbed with a detergent solution; then an antiseptic solution such as chlorhexidine gluconate, iodophors, or alcohol should be applied. This area should be large enough to include the entire incision and an adjacent area large enough for the surgeon to work during the operation without contacting unprepared skin.

Patients being prepared for abdominal surgery are sometimes treated for one day beforehand with antimicrobial agents intended to "sterilize" the bowel contents or to reduce normal intestinal flora to a very low level temporarily. This procedure decreases the risk of infection to surrounding tissues of the peritoneum when the bowel must be opened, particularly if prolonged surgery is anticipated or if the patient's defenses are unusually low. Prophylactic antimicrobial agents may be administered to other patients undergoing surgical procedures associated with a high risk of infection.

The surgical patient who presents no evidence of overt infection before or after surgery does not require isolation, although the wound itself is isolated by a good dressing technique until normal healing occurs. On the other hand, when an infective lesion is incised in the operating room, particular precautions must be taken for concurrent and terminal disinfection of the surgical room and all its equipment. Infective postoperative patients must be managed in such a way that they offer no threat of infection transfer to other patients with unhealed surgical wounds. If the infection is severe and is not covered by a dressing, or the dressing does not adequately contain the pus, the patient should be placed in a private room and treated with full precautions.

Under less dangerous circumstances, techniques may be adapted to the needs of the situation. Careful dressing techniques are indicated, with immediate destruction of contaminated dressings. Good hand-washing techniques and sterile gloves are used when infected wounds are handled (fig. 9.7), or dressings may be changed with sterile transfer forceps, one for dirty use and one for clean. "Occlusive" dressings are sometimes applied to infected wounds for topical isolation of infectious drainage. This technique employs a waterproof lamina over the dressing, which adheres to the patient's skin around the full circumference of the area.

Personnel Techniques

Nurses and physicians who care for patients, and others (including their visitors) who come in contact with them, are sometimes responsible for continuing the chain of transmission of infection. The most obvious and frequent transportation offered to pathogenic microorganisms is by way of unwashed hands, but clothing and hair may also serve. It should also be remembered that medical personnel with colds, sore throats, or infectious skin lesions can transfer their own infections to patients, whose resistance may be lowered by their illness. The basic techniques that can prevent this transmis-

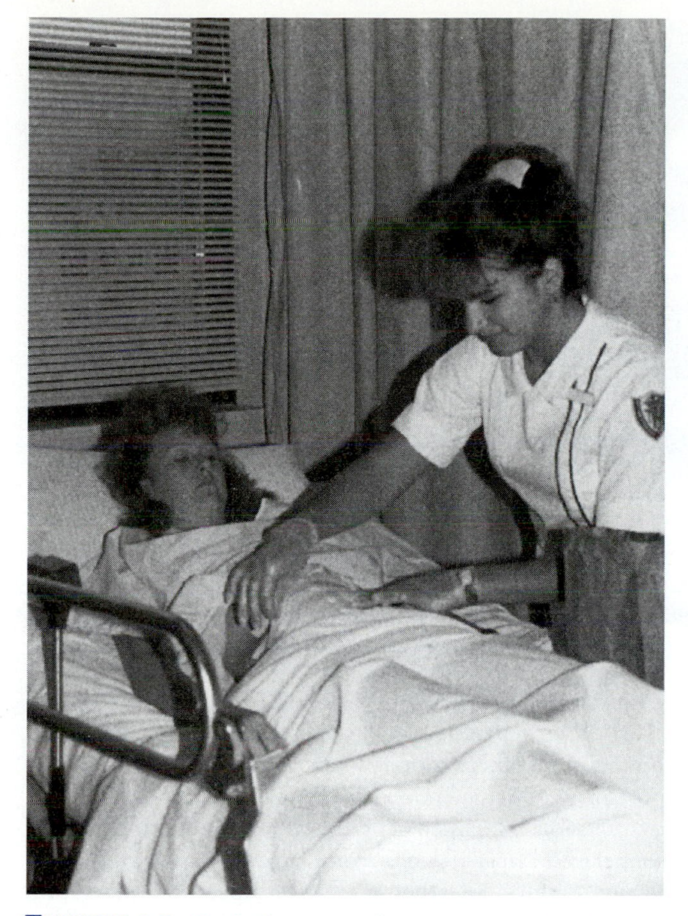

FIGURE 9.7 Sterile gloves are used to remove contaminated dressings and a new pair is then used to apply a fresh sterile dressing.

Concepts on Epidemiology of Infectious Disease
Transmission: Horizontal Transmission-Contact

sion in large measure are relatively simple, but they should be conscientiously and consistently practiced.

Hand Washing

Hand washing is the single most important means of preventing infection spread. Adequate hand-washing facilities should be available in all areas of the hospital for personnel and patients alike. Good antiseptic agents should be kept stocked in clean containers and changed frequently. Brush scrubbing is irritating to the hands and should be avoided, except in surgery, by those who must wash frequently.

The principles of antisepsis must be built in to hand-washing procedures if they are to be effective (fig. 9.8). Unless they have just been washed, the hands always carry their own normal flora together with contamination from other sources. Even when gloves are used, hands should be washed after taking care of an infected patient, after touching secretions and excretions, before performing invasive procedures, before touching wounds or patients who are especially susceptible to infection, and between patient contacts in intensive care units and newborn nurseries.

Hand washing should be thorough each time, with attention to the back as well as the palms, wrists, and each side of the fingers. Soap or detergent should be rubbed carefully for at least 30 seconds, holding the hands down from the elbow so that contaminated water will not run up the arm. Rinsing should be complete, preferably under running water with fingertips held up. Paper towels for drying are preferable to cloth, particularly in the sickroom, for a wet used towel may retain viable microorganisms. Unless an automatic device is available, the water should be turned off by using the paper towel rather than freshly washed hands. Hand lotion applied after each washing prevents chapped hands.

The fingernails should be kept short and scrupulously clean. Rings and wristwatches should not be worn by those caring for infectious patients.

Gown Technique

The use of gowns is indicated under many conditions, especially when soiling is likely. A clean gown that is ample in size and can be fastened securely protects the clothing and arms from contamination with infectious microorganisms. A fresh gown should be available for each person who enters the isolation room on each occasion, and it should be discarded after each use. Gowns are meant to be protective and, therefore, should be properly fastened and not be worn if wet or soiled. Moist areas on the gown may form a capillary bridge for the transfer of microorganisms to the clothing beneath.

Masks, Eye Protection, Face Shields

Masks, goggles, and face shields are worn during procedures and patient-care activities when splashes or sprays of blood, body fluids, secretions, or excretions might be generated. These items protect the mucous membranes of the eyes, nose, and mouth from transmission of pathogens. Surgical masks are generally worn when necessary to protect hospital personnel against spread of large, infectious droplets from infected patients who are coughing or sneezing (see chapter 11, droplet transmission). With the recent concern over spread of tuberculosis organisms that are resistant to multiple antimicrobial drugs, guidelines have been issued about specific types of masks that must be worn when caring for patients with tuberculosis. The high-efficiency particulate (HEPA) filter respirator is one of the best types for this purpose, but hospital personnel often have difficulty adjusting it properly over the nose and mouth. In addition, persons with compromising respiratory conditions such as asthma may have trouble breathing through it. New products for this purpose are continuing to be made available. The limitations of the mask should be understood and masks should be worn correctly and discarded promptly after use. When a used mask has been touched or handled, the hands should be washed thoroughly.

Personal Hygiene

Health care workers, because of their close association with sick people, must be particularly aware of personal hygiene to protect their own health as well as to avoid transmitting infection to others. Their clothing, hair, and skin should be

FIGURE 9.8 Proper and frequent hand washing, a medical aseptic technique, is an essential factor for preventing disease transmission. Soap, water, and friction are necessary to cleanse hands that have been contaminated in the care of patients and other activities.
Photography by Rachel Bloom.

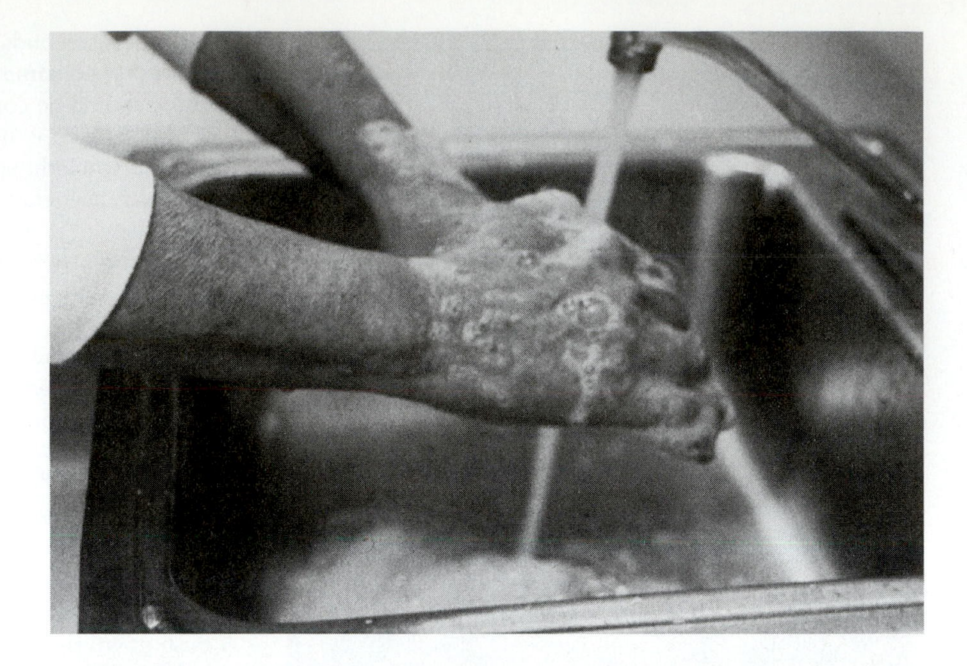

kept clean and in good condition. For nurses, one of the major reasons for wearing a uniform and traditional white duty shoes is that this kind of clothing is easy to clean. Soiling is immediately recognized and remedied. The practice of wearing this clothing outside of the hospital is a poor one because the nurse should come to the patient's bedside in clothes as freshly clean as possible.

In addition, soiled clothing may transmit infection outside of patient care areas. For the same reasons of hygiene, other health care professionals should avoid wearing laboratory coats, surgical scrub suits, or other hospital uniforms outside of their immediate work area.

Personal care of skin and hair should include use of effective soaps or shampoos together with nongreasy lubricants that help to avoid excessive drying. Hair should be shampooed frequently, particularly when caring for an infectious patient. The hair style should be simple and not require frequent arrangement with hands that may be contaminated. If hair is long or unruly, it should be controlled by a hairnet.

Above all, the habit of thorough, frequent hand washing should be developed early and continued throughout patient-care service. It should be a part of personal as well as professional life to guard the hands against contamination with potentially pathogenic microorganisms derived from one's own body or from others. These principles apply to all hospital workers, especially those who have direct patient contact.

Infections among Personnel

All medical personnel should have a preemployment medical evaluation and reevaluation at regular intervals. This is of particular importance for physicians, nurses, and dietary personnel who should be examined for evidence of active infections of the respiratory and intestinal tracts or the skin. When indicated, they should be isolated and treated as any other patient would be. When personnel infections are not severe and can be treated on an ambulatory basis, the employees should either be removed from duty altogether or be placed on duty in hospital areas where patient contacts are impossible or unimportant, until the infection has been cured or all danger of its transmission has passed. Appropriate immunization of personnel (see chap. 10) and ongoing practical instruction in infection control are other important aspects for protecting their health as well as that of their patients.

Control of the Hospital Environment

The entire physical environment of the hospital must be considered in planning and maintaining effective procedures for infection control. Increasing emphasis is now placed on the architectural design of hospitals to prevent access and transfer routes of infection from physical or human sources. In the meantime, many hospitals have fundamental faults of design that make it difficult to control traffic patterns, route essential hospital services (garbage and waste removal, the flow of clean and dirty laundry, food service), provide adequate isolation facilities for infectious patients, or protect areas such as operating rooms and intensive-care units, where infection control is of critical importance. In each hospital organization those responsible for providing an aseptic atmosphere must consider such indigenous problems and devise effective methods to cope with them at every level.

The principles of asepsis should form the basis of procedures applied by every hospital department. These departments include engineering and maintenance, housekeeping, and dietary (fig. 9.9). The continuing health of food handlers, as well as all other personnel, and their freedom from infection are the responsibility of the hospital's health service physician, with assistance from the microbiology laboratory.

In the medical and surgical care of patients, hospital personnel working together must consistently apply a pro-

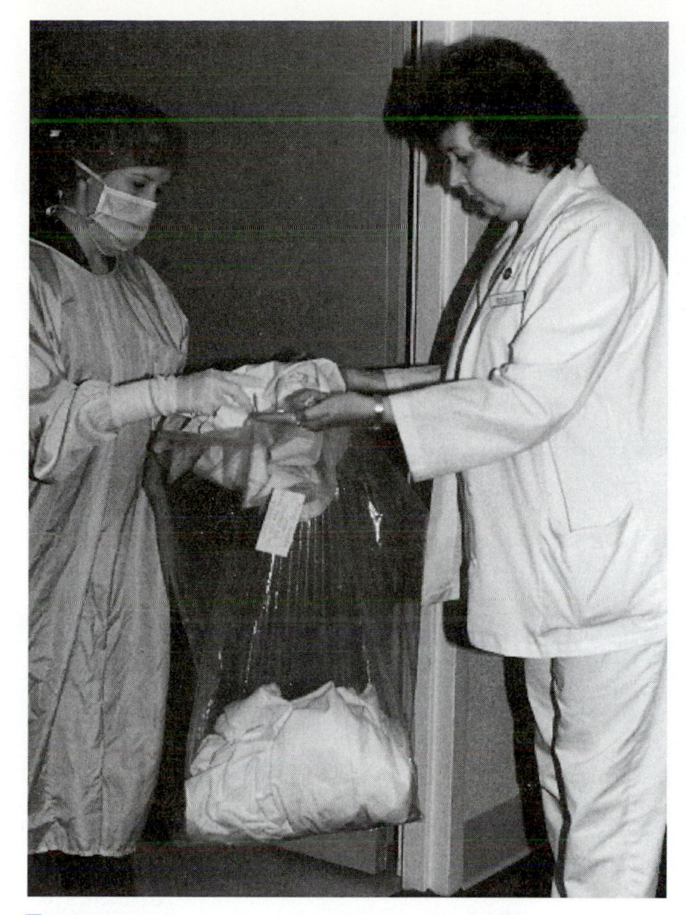

FIGURE 9.9 Double bagging by these nurses covers the contaminated linen inside the inner bag with another bag to protect laundry workers and others from contamination in the course of their work.
Putnam Hospital Center, Carmel, NY. Photo: B. Riley, Danae Associates.

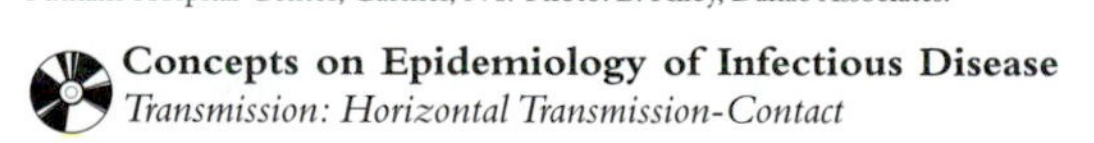

Concepts on Epidemiology of Infectious Disease
Transmission: Horizontal Transmission-Contact

cedural knowledge of asepsis. Patients being treated for noncommunicable diseases must not be exposed to risks of complicating infections, while the particular emotional as well as physical needs of those treated under isolation precautions must not be neglected. Maintaining aseptic conditions is the responsibility of each hospital employee.

CONTROL OF ACTIVE INFECTIOUS DISEASE

Confirming the diagnosis of an active infectious disease process on the basis of developing clinical symptoms and laboratory findings often requires some time—a few hours, a day or two, or even longer. Meantime, the physician must decide, based on probabilities, how to care for and treat the patient. If the disease could be communicable, the patient may be isolated either at home or in the hospital with appropriate precautions that depend on the clinical and epidemiologic possibilities (see chap. 11).

In the interim, the question of specific patient treatment, as well as any prophylactic protection for susceptible contacts, may be urgent. Therapy may have to be decided initially on the basis of clinical suspicion, then adjusted when the diagnosis is clear. Two kinds of agents are available for the control of many active infectious diseases: chemical agents (i.e., drugs or antimicrobial agents) for chemotherapy or chemoprophylaxis; and immune serums for immunotherapy or immunoprophylaxis.

For the sick patient, promptly instituted chemotherapy is generally most effective. However, in those diseases characterized by the action of potent microbial exotoxins (e.g., tetanus, botulism, or diphtheria), immunotherapy with specific antitoxin may be imperative early in the course of illness.

On the few occasions when it is necessary to protect the contacts of a patient ill with a communicable disease, chemoprophylaxis may be useful in some instances; in others, it may be necessary to consider passive immunization with immune serums (immunoprophylaxis).

TREATMENT OF THE PATIENT

Chemotherapy

The practice of treating disease with chemical agents may have begun with the earliest human beings. The scientific approach to **chemotherapy** came only after the etiology of infectious diseases was discovered in the late nineteenth century. Paul Ehrlich's work in the early years of the twentieth century laid the foundation for the systematic use of chemotherapeutics to cure microbial diseases. His experiments with arsenical drugs to treat syphilis (number 606 in a series of compounds tested was the most successful and famous) led him to recognize the selective activity of drugs on microorganisms as opposed to host cells with which they are associated in disease. He also encountered the development of resistance to drugs by microbial cells and explored means for countering this with combinations of compounds acting in different ways on microbial components.

The next important impetus to the chemotherapy of infection came in the 1930s with the discovery that a family of simple organic compounds known as sulfonamides possesses antimicrobial usefulness. In 1940 the demonstration that penicillin, an antibacterial substance produced by the mold *Penicillium,* is a therapeutic agent of great practical value opened a new age in the management of infectious diseases.

The dramatic success of penicillin in the treatment of infection led to the development of many useful drugs derived from fungi and other soil microorganisms. Although enormous numbers of antimicrobial agents have now been discovered, only a few have proved to be safe and effective in chemotherapy. Many are too damaging to host tissues in concentrations that are effective against microorganisms, whereas others are too limited in their antimicrobial range of action. As with other classes of drugs, the criteria for selecting antimicrobial agents to be used in chemotherapy are (1) low toxicity for host tissues, (2) lack of allergenic properties giving rise to host hypersensitivity, (3) a broad spectrum of activity against many different types of microorganisms,

(4) cidal rather than static action on microorganisms, and (5) antimicrobial effects to which microorganisms do not easily develop resistance. Not all of these properties are displayed by every useful antimicrobial agent, but the best possible combination is sought in the purification and preparation of such agents, which must meet minimum regulatory standards.

Most available chemotherapeutic agents are effective in bacterial infections only. Progress in developing useful antifungal, antiprotozoan, and antiviral compounds has been slow because most of the drugs investigated do not meet the five criteria listed and are especially toxic for human cells.

Mode and Range of Action of Antibacterial Drugs

Antimicrobial agents act by chemically interfering with the function or synthesis of vital microbial cell components. These agents may be classified by their modes of action, as shown in table 9.6. **Antibacterial** agents may act by inhibiting one or more of the steps involved in the synthesis of bacterial cell walls (fig. 9.10), or they may interfere with the permeability of the cell membrane. Recall from chapter 2 that bacterial ribosomes mediate the synthesis of cellular proteins. Some antimicrobial agents exert their effects because they attach to portions of the ribosome molecule causing the messenger RNA code for protein synthesis to be misread. Certain antimicrobial drugs act by interfering with nucleic acid metabolism, and others are inhibitory because they become substitutes, or analogues, for compounds normally used by bacterial cells in synthesis or enzyme function. Antimicrobial agents that act as analogues are commonly called antimetabolites. The sulfonamides, for example, act as analogues for a vitamin (para-aminobenzoic acid, or PABA) that many microorganisms require for synthesis of folic acid, an enzyme involved in the formation of purines and the pyrimidine, thymine. Substitution of a sulfonamide for PABA interferes with the cell's construction of properly functional folic acid and therefore prevents further growth (fig. 9.11). Bacterial cells that do not require folic acid for their growth or that can use preformed folic acid are not inhibited by the sulfonamides.

To be useful in chemotherapy, an antimicrobial agent must exert its effects on the pathogenic microorganism without seriously damaging host tissues. The toxicity of a particular drug for host cells as well as for microbial cells depends in part on whether or not they share the same kinds of components affected by the drug, or whether a given component has the same function in both. For example, some drugs that interfere with nucleic acid metabolism, such as actinomycins, are not selective for bacterial nucleic acids and are too toxic for host cells to be useful in chemotherapy. On the other hand, penicillin, which acts on bacterial cell-wall peptidoglycan, has little or no toxicity for human or animal cells. Since mammalian cells do not synthesize peptidoglycan, penicillin is not toxic for them. The difference in composition of the walls of gram-positive and gram-negative bacteria may also account for the greater activity of penicillin against the former.

TABLE 9.6 MECHANISM OF ACTION OF ANTIBACTERIAL DRUGS

Mechanism	Drug or Drug Class
Inhibit bacterial cell-wall synthesis	Aztreonam
	Bacitracin[a]
	Cephalosporins
	Cycloserine[b]
	Isoniazid[b]
	Penicillins
	Vancomycin
Affect permeability of microbial cell membranes	Polymyxins
Affect ribosomal protein synthesis	Aminoglycosides
	Azithromycin
	Chloramphenicol
	Clarithromycin
	Clindamycin
	Erythromycin
	Neomycin[a]
	Spectinomycin
	Tetracyclines
Interfere with nucleic acid replication or metabolism	Metronidazole
	Quinolones
	Rifampin
Act as antimetabolites	Aminosalicylic acid[b]
	Sulfonamides
	Sulfones[c]
	Trimethoprim

[a]Topical use only
[b]Used primarily for treating tuberculosis
[c]Used primarily for treating Hansen disease (leprosy)

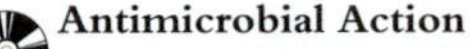
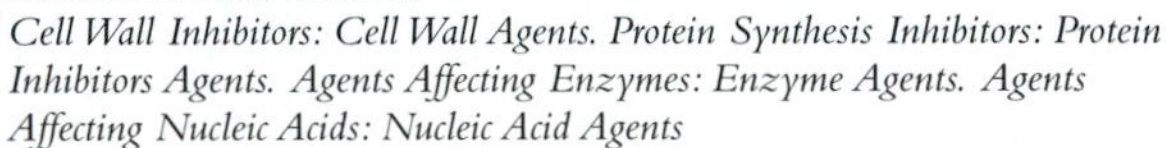

Antimicrobial Action
Cell Wall Inhibitors: Cell Wall Agents. Protein Synthesis Inhibitors: Protein Inhibitors Agents. Agents Affecting Enzymes: Enzyme Agents. Agents Affecting Nucleic Acids: Nucleic Acid Agents

Chemotherapeutic drugs may also be classified according to their range or spectrum of action against various microorganisms. Thus, some agents such as penicillin may have a narrow spectrum, affecting gram-positive bacteria primarily, and are the drugs of choice for many gram-positive infections. The tetracyclines and cephalosporins, however, are drugs with a broad spectrum of activity against both gram-positive and gram-negative organisms and the former are effective against rickettsiae. This clinically useful classification of antimicrobial drugs is shown in table 9.7.

Today many thousands of antimicrobial agents are known, some from natural sources, some synthesized in the laboratory, and many derived semisynthetically from a natural compound. Often when the active portion of an antimicrobial molecule is isolated and purified chemically, it is possible to add side chains to the basic structure and change its spectrum or mode of action. This process has produced semisynthetic compounds with greater, or different, antimicrobial activity than the parent substance. Only those agents derived from natural sources are properly called **antibiotics** so the term **antimicrobial agent** is often used to refer to both natural and synthetic drugs.

The "synthetic" penicillins have been very important in this regard because they enlarge the clinical usefulness of this antibiotic. The initial penicillin derivatives (e.g., penicillins

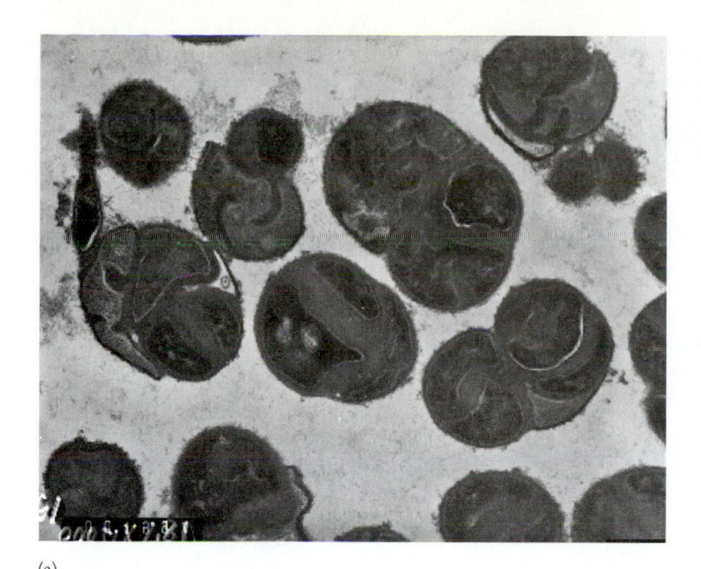

(a)

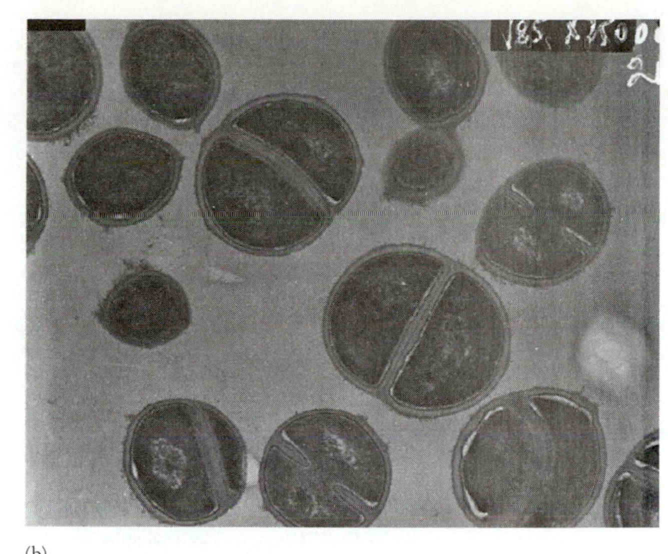

(b)

FIGURE 9.10 The effect of penicillin on susceptible staphylococci. (a) Staphylococci grown on agar containing one-third of the minimum inhibitory concentration (MIC) of penicillin show irregular shapes and numerous thick crosswalls. (b) The controls are staphylococci grown on nutrient agar without penicillin (×32,000). Compare the normal corsswalls in the controls with those in experiment (a).

Reproduced from V. Lorian, and B. Atkinson, "Some Effects of Subinhibitory Concentrations of Penicillin on the Structure and Division of Staphylococci." *Antimicrobial Agents and Chemotherapy,* 7: 864, 1975. American Society for Microbiology.

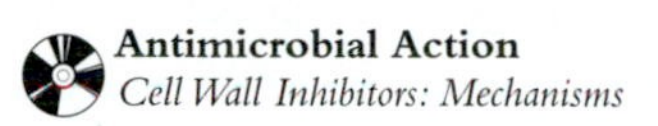

Antimicrobial Action
Cell Wall Inhibitors: Mechanisms

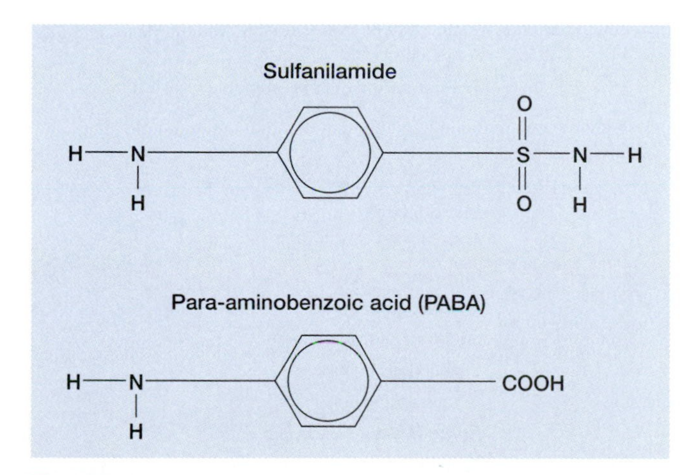

FIGURE 9.11 Chemical structures of sulfanilamide and para-aminobenzoic acid (PABA). Note the close similarity that makes it possible for the drug to substitute for the vitamin in bacterial syntheses, thus interfering with cellular metabolism.

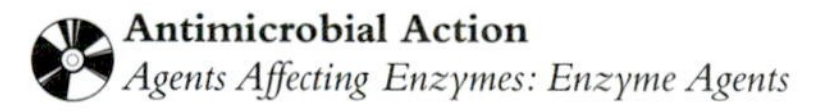

Antimicrobial Action
Agents Affecting Enzymes: Enzyme Agents

G and V) are destroyed by the action of a **beta-lactamase enzyme** that splits the beta-lactam ring portion of the penicillin molecule, thus rendering it inactive. This enzyme, often called **penicillinase,** is produced by many staphylococci and some gram-negative bacilli and is responsible for much of their resistance to this drug. Some synthetic penicillin molecules, such as oxacillin, methicillin, and nafcillin, are not destroyed by penicillinase because they have a slightly different chemical structure that protects them from the action of the beta-lactamase enzyme. Therefore, they can be used effectively to treat infections caused by penicillinase-producing staphylococci. Amoxicillin is a synthetic penicillin that is destroyed by penicillinase activity, but it has a broad spectrum of activity against gram-negative bacteria unlike the original antimicrobial agent (fig. 9.12).

The beta-lactamase ring formation also occurs in the cephalosporins, a group of drugs chemically similar to the penicillins with a similar mode of action. Chemical modifications of the parent molecule have produced a great many modified cephalosporins, some of which are resistant to beta-lactamase enzymes (in this case, cephalosporinase) and therefore more useful clinically against microorganisms possessing this enzyme (fig. 9.13). Parent drugs and their modifications are referred to as a *drug class;* for example, penicillin, amoxicillin, and methicillin are all antimicrobial agents of the penicillin class, whereas cephalothin, cefoxitin, and cefotaxime are drugs belonging to the cephalosporin class (see table 9.6).

Although the antimicrobial activities of a large number of natural and synthetic agents have been investigated in individual research laboratories and in pharmaceutical companies, only a limited number are ever marketed. Most of these agents are too toxic for human tissues, too expensive to produce in large quantities, or easily inactivated by existing microbial resistance mechanisms.

Problems in Antibacterial Chemotherapy

The principal result of antimicrobial drug action is to slow the rate of growth of young, actively multiplying microbial cells. The use of such agents alone does not necessarily cure the infection. Generally, the desired effect is to suppress the invading microorganisms entirely, or long enough for the body's nonspecific and specific resistance

TABLE 9.7 CHOICE OF ANTIBACTERIAL DRUGS FOR THERAPY

Causative Organism	Disease	Drug(s) of First Choice
Gram-negative cocci		
Neisseria gonorrhoeae	Gonorrhea	Ceftriaxone
Neisseria meningitidis	Meningitis	Penicillin
Gram-positive cocci		
Enterococcus spp.	Endocarditis Urinary tract infection	Ampicillin or vancomycin ± gentamicin Ampicillin, amoxicillin
Staphylococcus aureus	Furuncles, abscesses, pneumonia, meningitis, osteomyelitis, bacteremia, endocarditis	
Nonpenicillinase-producing		Penicillin
Penicillinase-producing		Methicillin, oxacillin
Methicillin-resistant		Vancomycin
Streptococcus pneumoniae	Pneumonia	Penicillin[a]
Streptococcus		
group A	Strep throat, scarlet fever, postpartum fever, erysipelas	Penicillin
group B	Neonatal meningitis, bacteremia	Penicillin, ampicillin
viridans group	Endocarditis	Penicillin ± gentamicin
other (including anaerobic) streptococci		Penicillin
Gram-positive bacilli		
Actinomyces spp.	Actinomycosis	Penicillin
Bacillus anthracis	Anthrax	Ciprofloxacin, doxycycline
Clostridium perfringens	Gas gangrene	Penicillin + clindamycin
Clostridium tetani	Tetanus	Metronidazole
Corynebacterium diphtheriae	Diphtheria	Erythromycin
Listeria monocytogenes	Meningitis, bacteremia	Ampicillin
Nocardia spp.	Nocardiosis	Trimethoprim-sulfamethoxazole
Gram-negative bacilli		
Bacteroides	Anaerobic infections	
mouth strains		Penicillin
gastrointestinal strains		Metronidazole
Brucella spp.	Brucellosis	Doxycycline + rifampin or gentamicin
Enterobacter spp.	Urinary tract infection	Imipenem
Escherichia coli	Urinary tract infection	A cephalosporin
Francisella tularensis	Tularemia	Streptomycin, gentamicin
Haemophilus influenzae	Meningitis, epiglottitis	Cefotaxime, ceftriaxone
Klebsiella spp.	Pneumonia, urinary tract infection	A cephalosporin
Legionella spp.	Pneumonia	Erythromycin + rifampin
Proteus mirabilis	Urinary tract infection	Ampicillin
Proteus, Providencia spp.	Urinary tract infection	Cefotaxime, ceftizoxime, ceftriaxone
Pseudomonas aeruginosa	Urinary tract infection Other infections	A fluoroquinolone Ticarcillin, mezlocillin + an aminoglycoside
Salmonella spp.	Typhoid, paratyphoid, enteritis	Ceftriaxone, a fluoroquinolone
Shigella	Dysentery	A fluoroquinolone
Vibrio cholerae	Cholera	Tetracycline
Yersinia pestis	Plague	Streptomycin, gentamicin
Acid-fast bacilli		
Mycobacterium tuberculosis	Tuberculosis	Isoniazid + rifampin ± pyrazinamide
Mycobacterium leprae	Hansen disease (leprosy)	Dapsone + rifampin ± clofazimine
Spirochetes		
Borrelia burgdorferi	Lyme disease	Ceftriaxone, doxycycline, amoxicillin
Borrelia recurrentis	Relapsing fever	Tetracycline
Leptospira interrogans	Leptospirosis	Penicillin
Treponema pallidum	Syphilis	Penicillin

[a]Penicillin-allergic patients are often treated with erythromycin or clarithromycin. Resistant strains are treated with vancomycin ± rifampin.
Source: Adapted from *The Medical Letter,* with permission.

TABLE 9.7 *Continued*

Causative Organism	Disease	Drug(s) of First Choice
Other bacteria		
Mycoplasma pneumoniae	Pneumonia	Erythromycin, azithromycin, clarithromycin
Rickettsia spp.	Rocky Mountain spotted fever, typhus and other fevers	Tetracycline
Chlamydia psittaci	Psittacosis	Tetracycline
Chlamydia trachomatis	Inclusion conjunctivitis, pneumonia	Erythromycin
	Urethritis, cervicitis	Doxycycline or azithromycin
	Lymphogranuloma venereum	Tetracycline
Chlamydia pneumoniae	Pneumonia	Doxycycline

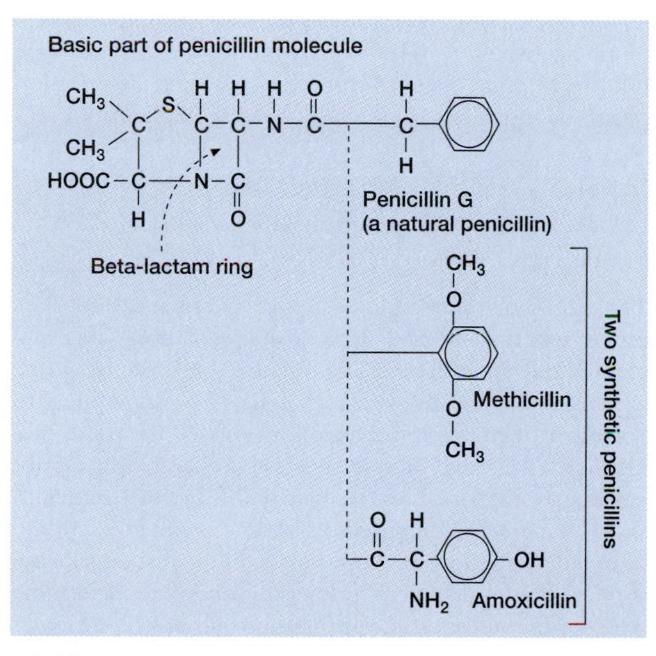

FIGURE 9.12 Two derivatives of penicillin. Substitution of different side chains on the basic part of the molecule at the point shown produces different penicillins with altered activities.

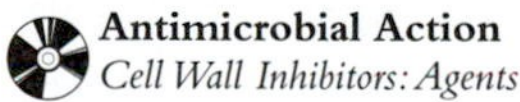
Antimicrobial Action
Cell Wall Inhibitors: Agents

FIGURE 9.13 Chemical structures of some cephalosporins. Cephalothin is referred to as a first generation cephalosporin, and cefoxitin and cefotaxime as second and third generation cephalosporins, respectively.

mechanisms to intervene and ensure complete recovery. Although the dramatic success of chemotherapy has reduced the incidence, morbidity, and mortality of many infectious diseases, certain problems must be emphasized.

Drug toxicity for the host may seriously interfere with the drug's value in therapy. Some drugs, such as penicillin, have few toxic effects and may be given in greater concentrations than the effective antimicrobial level required without inducing tissue damage. The dose of others must be carefully adjusted because their effective concentrations are not much different from their toxic concentrations, allowing little leeway. Different antimicrobial drugs may have characteristic toxicities for certain human tissues: gentamicin, an aminoglycoside, can induce renal damage and can also injure inner ear cells, resulting in partial or complete deafness; chloramphenicol in prolonged, high dosages may cause depression of bone marrow with aplastic anemia or other blood dyscrasias; certain sulfonamides may produce renal injury because they are insoluble and tend to precipitate out when excreted through the kidneys. These are some of the reasons why chemotherapeutic drugs should be issued only by a physician's order, prescribed with discriminating care, and taken only under a physician's supervision.

Drug allergies, or hypersensitivities, are frequent complications of the use of antimicrobial compounds. For example, gentamicin, vancomycin, and penicillin often produce allergic skin rashes. The effects of **drug allergy** are the result of previous sensitization of the patient with the drug; the timing and severity of the reaction depend on the route of administration of a repeated dose, the extent of the hypersensitive mechanism, and the tissues involved by it (see chap. 6).

Penicillin, although it is nontoxic, has a serious allergenic potential for hypersensitized individuals. Reactions range from skin rashes (manifested as a phase of the serum-sickness type of immediate response) to anaphylaxis, which may be fatal. These hypersensitivities are another important reason for using antimicrobial therapy with caution. Treatment of insignificant infections with antimicrobial agents should be discouraged, particularly if the drug's effect on the specific infectious agent being treated has not been established by laboratory tests.

Development of microbial resistance to drugs presents another increasingly important problem in chemotherapy. The genetic basis for the appearance of antimicrobial-agent-resistant mutants in susceptible bacterial populations was discussed in chapter 3. **Drug resistance** develops as a result of continuous use of drugs, particularly in closed environments such as hospitals. Continuous drug use tends to suppress susceptible microorganisms and to foster the survival of resistant strains. This situation is further encouraged by the constant personal contact between hospital patients and personnel during which resistant strains are directly or indirectly exchanged. This is the reason for the increase and persistence of drug-resistant strains of staphylococci and many species of gram-negative enteric bacilli indigenous to the human body. Finding an adequate drug with which to treat infections caused by resistant bacterial strains often poses a severe problem.

In the race between bacteria, which are developing resistance to even the newest antimicrobial agent formulations, and pharmaceutical companies, which are attempting to develop additional effective drugs, the microbes are clearly ahead. As strains of bacteria resistant to many, or even all, agents arise, the end of the twentieth century brings concern about the continued effectiveness of antimicrobial therapy to combat **infectious diseases.**

Combined therapy is sometimes useful in preventing the emergence of resistance in a bacterial strain. When administered together, two antimicrobial agents acting by different mechanisms on the causative organisms of the infection but not interfering with each other may act synergistically to prevent the appearance of resistant mutants.

Changes in the normal flora of the body frequently occur as a result of antimicrobial therapy. The balance maintained among the microorganisms of the skin and mucous membranes may be seriously disturbed if one or more drug-susceptible species is suppressed during the course of treatment. Under these circumstances, drug resistant organisms whose growth may have been inhibited by competitive microbial activities multiply, and a new and different infection may begin. These secondary, complicating infections, or *superinfections* as they are also called, occur most frequently on the mucosal surfaces of the mouth and throat, the intestinal tract, and the genitourinary system, but they may extend systemically as well. Strains of yeasts (notably, *Candida albicans*), staphylococci, and enteric bacteria are most often involved and are particularly difficult to treat because of their drug resistance. Antimycotic drugs such as nystatin or an imidazole agent may be administered to prevent outgrowth of yeasts if the normal balance is upset during antimicrobial therapy.

Surgical Treatment of Infection

When an infectious process is well localized by the inflammatory response and encapsulated by fibrosis as in abscess formation, chemotherapy may not be successful without surgery. The cellular barriers thrown up against the causative microorganism often preclude penetration of the area by bactericidal, or even bacteriostatic, drug concentrations. Under these circumstances surgical excision or drainage of the lesion is indicated if it lies in an accessible region of the body and if surgery does not compromise surrounding tissues. Antimicrobial agents given at the same time serve as a "cover" to prevent new sites of infection caused by organisms scattered by surgical manipulation.

Role of the Microbiology Laboratory in Chemotherapy for Bacterial Infections

The diagnostic microbiology laboratory can assist the clinician in selecting effective chemotherapeutic drugs for treatment of infectious disease. The causative organism must first be isolated in pure culture and then tested for its response to a number of antimicrobial agents, usually by the paper disk method. If the results of treatment with a chosen drug are not satisfactory, the laboratory may test the original organism again by more exacting methods to find a more useful drug or to indicate a more effective dosage of the first one. It may also test fresh specimens from the patient to determine whether the infectious agent has become drug resistant or whether a superinfection has arisen. Assays of the patient's serum, urine, or other body fluids for antimicrobial content may also be useful in guiding the physician in the choice of dosage or route of administration of the drug.

Laboratory methods for susceptibility testing of bacteria are described in chapter 4.

Antimicrobial Agents for Microorganisms Other than Bacteria

Drug manufacturers have not had the same success in developing large numbers and different classes of antimicrobial agents for microorganisms other than bacteria (e.g., fungi, parasites, and viruses), but some progress has been made. Unlike procaryotic bacteria, which have a unique cell wall and ribosomal structure and lack membrane-bound vesicles, eucaryotic fungal and parasite cell structures and metabolism more closely resemble those of higher animals. As a result, drugs that destroy these microbes are more likely to be toxic for humans. Viruses present an entirely different problem because, as described in chapter 2, they use the host's metabolic machinery to replicate. Thus, the action of antiviral agents must depend on specific inhibition of those few enzymes that viruses induce in order to reproduce but that are neither natural to nor inhibited in the host. Table 9.8 summarizes the most common antimicrobial agents available for nonbacterial infectious agents, their mode of action, and their preferred use. These agents must be used cautiously and the patient carefully monitored for adverse reactions.

Immunotherapy of Infectious Disease

Prior to the advent of antimicrobial agents, serums (usually from immunized animals) containing antibodies were frequently used to treat certain acute infections, notably, pneumococcal pneumonia and bacterial meningitis. Today, however, effective chemotherapy and large-scale active immunization programs have markedly reduced the incidence and serious consequences of such diseases. **Immunotherapy** is now largely limited to those few situations where bacterial infection is accompanied by the production of highly injurious or lethal toxins. These include diphtheria and the clostridial diseases, tetanus and botulism. Ingested botulism toxin, or diphtheria or tetanus toxin produced by bacteria growing in the tissues, is not affected by antimicrobial agents. In these situations, specific antitoxin given promptly on diagnosis prevents irreparable toxin damage to tissues. The concurrent administration of antimicrobial agents inhibits the further growth of the invading organisms and the toxins they produce.

CHEMOPROPHYLAXIS

Treatment of the Patient's Contacts

In general, antimicrobial agents should not be used to prevent spread of bacterial infection among the contacts of an active case. This procedure, called **chemoprophylaxis,** tends to encourage the emergence of resistant microorganisms that are not susceptible to the drug. One of the exceptions to this general rule, however, is the case of the sexually transmitted diseases caused by *Neisseria gonorrhoeae* and *Chlamydia trachomatis.* Patients with these infections and their contacts are examined and treated with ceftriaxone for gonorrhea and doxycycline or azithromycin for the chlamydial disease, and they are usually treated for both infections at once because they commonly coexist. Newborn babies are susceptible contacts of mothers who may have gonorrhea. The prophylactic measure in this case is treatment of the babies' eyes at birth with erythromycin or tetracycline ointment to prevent gonococcal conjunctivitis. Prophylactic measures are not effective in preventing chlamydial conjunctivitis, however, and infants who develop chlamydial disease after birth require therapy.

Persons in close contact with patients who have meningitis or bacteremia caused by another neisserial organism, *Neisseria meningitidis,* are treated prophylactically with sulfonamides (if laboratory tests demonstrate that the organism is susceptible to this drug) or with rifampin. Many strains are resistant to sulfonamides because of their previous widespread prophylactic use. Meningococci are transmitted under crowded conditions by droplets spread from the upper respiratory tract, and infection can reach epidemic proportions when persons live in close contact, as among military recruits, for example. Individuals who have only casual contact with the patient or the microorganism (e.g., laboratory technologists) do not require prophylaxis. As in meningococcal disease, close contacts of patients with *Haemophilus influenzae* meningitis may receive prophylactic treatment with rifampin, especially if they have not been immunized against this microorganism.

Among nonbacterial infectious diseases, chemoprophylaxis is used only for preventing the parasitic infection, malaria, and the influenza A virus infection. The source of malaria is not human but is a mosquito vector. In areas of the world where the disease and therefore infective mosquitoes are common, human exposure is frequent and often unavoidable. The drug chloroquine is effective in preventing most episodes of malaria, but in areas of the world where chloroquine-resistant malaria organisms have evolved, both prophylaxis and treatment may be difficult. Amantadine and rimantadine have been used successfully in prophylaxis of influenza A (see table 9.8). Their use, however, is recommended only for high risk individuals such as immunocompromised children and adults, the elderly, and persons with vital occupations whose activities would be curtailed by the disease or who might come in frequent contact with infected patients (e.g., health care workers). Use of these drugs is complicated by the need to confirm that an outbreak of viral disease is caused by the influenza A virus because amantadine and rimantadine are ineffective against other viruses. For all persons who would be candidates for receiving antiviral prophylaxis, immunization with the most current influenza vaccine is the preferred method of preventing disease.

Prophylaxis in Surgery and Medicine

Antimicrobial agents are used extensively in hospitals in the prophylactic treatment of surgical patients. **Prophylaxis** is indicated when the surgery involves nonsterile areas (e.g., colon, gallbladder, lower genital tract, burn wounds) or when devices are inserted that may become infected with saprophytic skin bacteria (e.g., heart valves, prosthetic joints). Surgical wounds most often are seeded with bacteria at the time of surgery, therefore prophylactic treatment of the patient with a broad spectrum agent for a short time period (24 to 48 hours) before and after the procedure is highly effective in preventing development of infection.

Short-term antimicrobial prophylaxis is often used at the time of dental work for persons with prosthetic heart valves or valves damaged by rheumatic fever (see chap. 15). During dental manipulation, mouth bacteria can escape into the patient's bloodstream, lodge on the heart valve surface, and multiply extensively to produce infective endocarditis, a potentially fatal illness. As oral bacteria are usually highly susceptible to antimicrobial agents, they are quickly

TABLE 9.8 ANTIMICROBIAL AGENTS FOR FUNGI, PARASITES, AND VIRUSES

Agent	Mechanism of Action	Susceptible Microorganisms	Comments
Antifungal agents			
Amphotericin B	Binds to sterols in fungal cell membranes increasing their permeability and causing leakage of cellular constituents	Systemic fungi including *Paracoccidioides brasiliensis, Coccidioides immitis, Cryptococcus neoformans, Histoplasma capsulatum, Sporothrix schenckii, Blastomyces dermatitidis, Candida* spp., *Aspergillus fumigatus*	Toxic for mammalian cells that contain sterols, such as certain kidney cells and erythrocytes. Used only for severe systemic fungal infections. May have activity against some protozoa (*Leishmania, Naegleria*).
Flucytosine	Degrades in cell to an antimetabolite; also inhibits DNA synthesis	Some *Cryptococcus, Candida* spp.	Used only for serious infections caused by susceptible strains of *Cryptococcus* and *Candida,* or topically in ophthalmic preparations.
Griseofulvin	Interrupts fungal cell mitosis; causes defective DNA production	Ringworm fungi including species of *Trichophyton, Microsporum,* and *Epidermophyton*	Used to treat ringworm infections of the hair, skin, and nails only. May need treatment for a year or longer for fungal nail infections.
Ketoconazole[a]	Alters cellular membranes and increases cell permeability	Active against a wide range of systemic and ringworm fungi (see above) and *Candida* spp.	Given orally for systemic and superficial fungal infections; not recommended for fungal meningitis. Topical preparation available. May cause liver toxicity.
Miconazole	Alters cellular membranes and interferes with intracellular enzymes	Same as ketoconazole	Given orally or intravenously for severe systemic fungal infections and intravenously for fungal meningitis. May be effective for eye infections caused by the parasite *Acanthamoeba.*
Nystatin	Binds to sterols in fungal cell membranes and alters their permeability	*Candida* spp.	Used orally or topically for local, but not systemic, *Candida* infections.
Terbinafine	Kills fungal cells by inhibiting an enzyme important for synthesis of the fungal cell wall and membrane	Ringworm fungi, yeasts	Used to treat ringworm infections, especially of the nails. Requires 6 weeks of therapy for fingernail infections and 12 weeks for toenail infections. May cause liver dysfunction and skin reactions.
Antiparasitic agents			
Albendazole	Kills worms and their larvae by interfering with ATP production and thus causing depletion of energy	Larval forms of *Taenia solium* and *Echinococcus granulosus*	The first drug approved in the United States for treating neuro-cysticercosis and hydatid cyst disease, the infections caused by these organisms. Effective in 70–80% of patients.
Chloroquine	Binds to nucleoproteins and interferes with protein synthesis	*Plasmodium* spp.	Effective against circulating forms of malarial parasites, but not *P. falciparum* gametocytes; many *P. falciparum* strains now resistant.
Emetine	Causes nuclear degeneration; interferes with trophozoite multiplication	*Entamoeba histolytica* trophozoites	Used in severe amebic dysentery only; used in combination with iodoquinol to completely eradicate disease.
Iodoquinol	Unknown	*Entamoeba histolytica*	Used alone for treating mild or asymptomatic intestinal disease or in combination with emetine for severe disease.
Mebendazole	Inhibits uptake of glucose by helminths depleting glycogen stores and interfering with absorptive and secretory functions	Pathogenic roundworms including hookworm, pinworm, whipworm, threadworm; some tapeworms (beef, pork, dwarf) *Echinococcus* larvae (hydatid cyst)	Highly effective in curing infection caused by susceptible worms.

Continued

[a]A number of antifungal imidazole drugs similar to ketoconazole in mechanism of action and spectrum of antifungal activity are available as topical preparations. These agents, which include clotrimazole, econazole, and terconazole, are not used to treat serious systemic disease.
[b]Rimantadine, a similar agent for prophylaxis of influenza A, may produce fewer side effects than does amantidine.

TABLE 9.8 *Continued*

Agent	Mechanism of Action	Susceptible Microorganisms	Comments
Niclosamide	Inhibits mitochondrial oxidative phosphorylation in worms	Most tapeworms, pinworm	Does not act on larval or tissue stage of tapeworm infection (cysticercosis).
Piperazine	Causes paralysis of worm by inhibiting nerve transmissions at neuromuscular junctions	*Ascaris lumbricoides* (roundworm), *Enterobius vermicularis* (pinworm)	Effective only when given in multiple doses.
Praziquantel	Unknown, but result is paralysis of worm suckers causing worm to dislodge from intestine	*Schistosoma* spp., flukes, some tapeworms, larval stage of *T. solium*	Highly effective against susceptible worms.
Primaquine	Interferes with DNA function of malarial parasites	*Plasmodium* spp.	Active against tissue forms of malarial parasites and *P. falciparum* gametocytes; not active against other erythrocytic forms, therefore must be used in conjunction with chloroquine to eradicate all parasite forms.
Pyrimethamine	Folic acid inhibitor similar to the antibacterial trimethoprim	*Toxoplasma gondii, Plasmodium* spp.	Used with a sulfonamide to treat toxoplasmosis; less effective against malarial parasites than other drugs.
Quinacrine	Unknown	*Giardia lamblia,* tapeworms, blood forms of malarial parasites (*Plasmodium* spp.)	Drug of choice for giardiasis; used for chemoprophylaxis against malaria; not effective for circulating gametocytes of *P. falciparum*.
Thiabendazole	Unknown	Most intestinal roundworms	Drug of choice for *Strongyloides* infections; variable results against other roundworms.
Antiviral agents			
Acyclovir	Interferes with DNA synthesis and inhibits viral replication of herpes simplex and varicella-zoster viruses; others unknown	Herpes virus family (including herpes simplex, varicella-zoster, Epstein-Barr, cytomegalovirus)	Used for treating initial and recurring herpes simplex infections in immunocompromised children and adults. In the immunocompetent host, used only against severe genital herpes. Some herpes viruses now resistant.
Amantadine[b]	Prevents virus from penetrating host cell or inhibits uncoating so that virus cannot replicate	Influenza A virus only	Virustatic drug; effective primarily for prophylaxis or symptomatic treatment of early influenza A. Most effective when given on exposure before symptoms develop.
Didanosine	Competitively inhibits reverse transcriptase enzyme; causes premature termination of viral replication	Retroviruses, including human immunodeficiency virus (HIV)	Used for treatment of patients with advanced HIV infection (AIDS) who cannot tolerate zidovudine therapy or who deteriorate clinically and immunologically during zidovudine therapy.
Ganciclovir	Competes with guanosine compound for uptake into viral DNA, interferes with DNA synthesis	Herpes virus family	Used almost exclusively to treat human cytomegalovirus infections, especially in immunosuppressed organ or bone marrow transplant patients. Some viruses already resistant.
Lamivudine	Interferes with viral DNA synthesis by acting as a competitive inhibitor of the viral reverse transcriptase enzyme	HIV	Approved for use in combination with zidovudine; may delay viral development of resistance to zidovudine or improve its effectiveness.
Ribavirin	Interferes with RNA and DNA synthesis, inhibiting protein synthesis and replication	In vitro, many RNA and DNA viruses	Available in United States only for aerosol therapy; used in infants to treat severe lower respiratory infections caused by respiratory syncytial virus.

[a]A number of antifungal imidazole drugs similar to ketoconazole in mechanism of action and spectrum of antifungal activity are available as topical preparations. These agents, which include clotrimazole, econazole, and terconazole, are not used to treat serious systemic disease.
[b]Rimantadine, a similar agent for prophylaxis of influenza A, may produce fewer side effects than does amantidine.

TABLE 9.8 *Continued*

Agent	Mechanism of Action	Susceptible Microorganisms	Comments
Saquinavir	Blocks action of a protease vital to the final stages of viral replication; the first of a type of drug known as a protease inhibitor to gain FDA approval	HIV	Approved for use in combination with zidovudine; does not work well alone. Ritonavir and indinavir are newer such drugs now in use in combination with zidovudine and other anti-HIV drugs.
Zalcitabine	Same as didanosine	HIV	Approved for use only in combination with zidovudine for adults with advanced AIDS.
Zidovudine	Inhibits reverse transcriptase enzyme in retroviruses	Retroviruses, including HIV	An "orphan" drug used only to treat symptomatic human immunodeficiency virus infections (AIDS).

destroyed in the bloodstream by appropriate antimicrobial agents taken before the dental procedure.

Immunoprophylaxis

For many infectious diseases the most effective means of control is mass active immunization of the population (**immunoprophylaxis**). Sporadic cases of diseases such as poliomyelitis, diphtheria, and pertussis may occur from time to time, however, even where immunoprophylaxis has been accomplished. Therefore, the immunity of exposed contacts must be considered. Renewed active immunization with vaccines may be needed to boost lagging immunity or to protect new susceptible members of the community. When exposure to tetanus infection threatens a nonimmune individual, passive immunization with antitoxin may be necessary to prevent disease, but active immunization with toxoid should then be given.

Human gamma globulin is used as a source of antibodies after susceptible individuals, for whom the risk of certain diseases is great, are exposed. For example, gamma globulin is administered if the disease is frequently accompanied by serious consequences (poliomyelitis, hepatitis A, mumps in adults) and active immunity has waned, or if no active artificial immunization is available. The prophylactic use of gamma globulin may also be advisable for individuals to whom any infectious disease may represent unusual risk, such as the very young, the very old, or others whose resistance is seriously compromised by medical or surgical problems.

QUESTIONS

1. Identify the similarities and differences between the terms in each of the following pairs: disinfection and sterilization; medical asepsis and surgical asepsis; contamination and infection; sanitation and antisepsis.
2. Define the following terms: infectious disease, communicable disease, nosocomial infection, communicable period.
3. Describe the action of heat on microorganisms.
4. What factors must be considered when sterilizing materials for medical use?
5. When is the sterilizing temperature in the autoclave reached? How is the temperature maintained?
6. How does ultraviolet radiation affect microorganisms? When is ultraviolet radiation used?
7. What is the action of chemical disinfection on microorganisms? Why are microorganisms not uniformly susceptible to chemical disinfection?
8. In sterilizing and disinfection procedures, what determines whether a patient care item is critical, semicritical, or noncritical?
9. When is concurrent disinfection warranted? When is terminal disinfection applied?
10. Why is hand washing considered the single most important means of preventing the spread of infection?
11. Describe the mode of action of the drugs penicillin, chloramphenicol, rifampin, and sulfonamides.
12. What are some of the problems encountered in the use of antimicrobial therapy?
13. Why are the 1990s referred to as being the postantibiotic era?

BIBLIOGRAPHY

Amantea, M. A., D. J. Drutz, and J. R. Rosenthal. 1990. Antifungals: A primary care primer. *Patient Care* 24 (15): 58–62, 64–65, 68.

American Hospital Formulary Service. 1994. *Drug information 94.* Bethesda, Md.: American Society of Hospital Pharmacists, Inc.

Ansari, S. A., V. S. Springthorpe, S. A. Sattar, et al. 1991. Comparison of cloth, paper, and warm drying in eliminating viruses and bacteria from washed hands. *Am. J. Infect. Control* 19 (5): 243–249.

Baumgardner, C. A., C. S. Maragos, M. J. Walz, and E. Larson. 1993. The effects of nail polish on microbial growth of fingernails. *J. Am. Oper. Room Nurs.* 58 (1): 84–88.

Benenson, A. S., ed. 1995. *Control of communicable diseases in man,* 16th ed. Washington, D.C.: American Public Health Association.

Block, S. S., ed. 1991. *Disinfection, sterilization and preservation* 4th ed. Philadelphia: Lea and Febiger.

Brady, M. T., J. Evans, and J. Cuartas. 1990. Survival and disinfection of parainfluenza viruses on environmental surfaces. *Am. J. Infect. Control* 18 (1): 18–23.

Brajtburg, J., and J. Bolard. 1996. Carrier effects on biological activity of amphotericin B. *Clin. Microbiol. Rev.* 9 (4): 512–531.

Burgert, S. J., and J. P. Burke. 1994. Antibiotic resistance: Will infection control meet the challenge? *Am. J. Infect. Control* 22 (4): 193–194.

Centers for Disease Control and Prevention. Guidelines for prevention of transmission of human immunodeficiency virus and hepatitis B virus to health-care and public-safety workers. 1989. *Morbid. Mortal. Weekly Report* vol. 38, 23 June, no. S–6.

Cohen, M. L. 1992. Epidemiology of drug resistance: Implications for a post-antimicrobial era. *Science* 257 (5073): 1050–1055.

Crow, S. 1989. Asepsis—an indispensible part of the patient's care plan. *Crit. Care Nurs. Q.* 11 (4): 11–16.

Crow, S. 1993. Sterilization processes: Meeting the demands of today's health care technology. *Nurs. Clin. N. Am.* 28 (3): 687–695.

D'Arco, S. H., and M. Hargreaves. 1995. Needlesticks: A multidisciplinary concern. *Nurs. Clin. No. Am.* 30 (1): 61–76.

Data show lab technicians, nurses most at risk. 1995. *Hosp. Employee Health* 11 (3): 35–36.

Douglas, G. R. Jr., and P. L. Chiodini. 1993. Antimicrobial agents: Viral/parasitic. *Curr. Opin. Infect. Dis.* 6 (6): 765–823, 853–861.

Favero, M. S., and E. A. Bolyard. 1995. Disinfection and sterilization strategies and the potential for airborne transmission of bloodborne pathogens, pp. 1071–1089. In *Prevention of transmission of bloodborne pathogens* edited by R. S. Rhodes and D. M. Bell. *Surg. Clin. N. Am.* Philadelphia: Saunders.

Garner, J. S. 1996. Guidelines for isolation precautions in hospitals. *Infect. Control Hosp. Epidemiol.* 17 (1): 53–80.

Goodman, H. S., R. D. Carpenter, and M. R. Cox. 1994. Sterilization of dental instruments and devices: An update. *Am. J. Infect. Control* 22 (2): 90–93

Gruendemann, B. J. 1990. Surgical asepsis revisited. *J. Ophthal. Nursing. Technol.* 9 (6): 250–253.

Gurevich, I., B. Yanelli, and B. A. Cunha. 1990. The disinfectant dilemma revisited. *Infect. Control. Hosp. Epidemiol.* 11 (2): 96–100.

Jackson, M., and P. Lynch. 1990. Infection control: In search of a rationale approach. *Am. J. Nurs.* 90 (10): 65–74.

Jackson, M., P. Lynch, D. C. McPherson, M. J. Cummings, and N. C. Greenawalt. 1987. Why not treat all body substances as infectious? *Am. J. Nurs.* 87 (9): 1137–1139.

Kunin, C. M. 1993. Resistance to antimicrobial drugs: A worldwide calamity. *Ann. Intern. Med.* 118 (7): 557–561.

Larson, E. L. 1995 APIC guidelines for handwashing and hand antisepsis in health care settings. *Am. J. Infect. Control* 23 (4): 251–253.

Larson, E. 1989. Handwashing: It's essential even when you wear gloves. *Am. J. Nurs.* 89 (7): 934–939.

Mathewson, H. S. 1990. Systemic antifungal agents. *Respir. Care* 35 (10): 987–989.

Mayhall, C. G., ed. 1996. *Infection control and hospital epidemiology.* Baltimore: The Williams and Wilkins Co.

McFarlane, A. 1989. Using the laboratory in infection control. *Prof. Nurse.* 4 (8): 393–394, 396–397.

Neu, H. C. 1992. The crisis in antibiotic resistance. *Science* 257 (5073): 1064–1073.

Nightingale, F. 1976. *Notes on Hospitals.* New York: Gordon Press.

Perkins, J. J. 1983. *Principles and methods of sterilization in health sciences,* 2d. ed. Springfield, Ill.: Charles C Thomas.

Pottinger, J., S. Burns, and C. Manske. 1989. Bacterial carriage by artificial versus natural nails. *Am. J. Infect. Control* 17 (6): 340–344.

Rutala, W. A. 1990. APIC guidelines for selection and use of disinfectants. *Am. J. Infect. Control.* 58: 99–117.

Rutala, W. A. 1996. Selection and use of disinfectants in health care. In *Infection control and hospital epidemiology,* edited by C. G. Mayhall, 913–936. Baltimore: The Williams and Wilkins Co.

Schaberg, D. R. 1989. How infections spread in the hospital. *Respir. Care.* 34 (2): 81–84.

Tuomanen, E. 1993. Breaching the blood-brain barrier. *Sci. Am.* 268 (2): 80–84.

Weinstein, S. A., N. M. Gantz, and D. Hibert. 1989. Bacterial surface contamination of patients' linens: Isolation versus standard care. *Am. J. Infect. Control* 17 (5): 264–267.

Wenzel, R. P. and M. A. Pfaller. 1991. Handwashing-efficacy versus acceptance: A brief essay. *J. Hosp. Infect.* 18 (suppl. B): 65–68.

Woodham-Smith, C. 1951. *Florence Nightingale.* New York: McGraw-Hill.

Young, H. K. 1994. Do nonclinical uses of antibiotics make a difference? *Infect. Control Hosp. Epidemiol.* 15 (7): 484–487.

CHAPTER 10

Communicable Disease Control in the Community

OUTLINE

Microbes in Motion

This chapter covers the concepts of disease control and prevention with emphasis on public health measures such as control of reservoirs of infection and immunization. The following books and chapters of Microbes in Motion CD-ROM may be a useful preview or supplemental study aid to your reading: Concepts of Epidemiology of Infectious Disease: Terminology, Disease Acquisition. Vaccines: Targets.

OBJECTIVES

On completing this chapter the student will be able to:

1. identify the functions of public health agencies at city, state, national, and worldwide levels in controlling communicable disease.
2. cite methods used by governmental agencies to monitor and control environmental sources of infectious disease.
3. discuss the importance of immunization programs in the control and prevention of human or animal diseases, and give examples of such programs.
4. outline general public health methods for controlling reservoirs of communicable disease in humans, animals, and insects.

KEY WORDS

Arthropod (**ARE**-throw-pod) vector
Bacteriologic standards for milk
Bacteriologic standards for water
Carriers of disease
Case finding and reporting
Food sanitation
Immunization schedules
Insect control
Isolation
Meat inspection
Milkborne diseases
Pasteurization (past-er-iz-**A**-shun)
Public health services
Quarantine
Reportable diseases
Rodent control
Sewage disposal
Shellfish control
Vaccine
Water purification
World Health Organization (WHO)
Zoonoses (zo-o-**NO**-sees)

PUBLIC HEALTH AGENCIES AND SERVICES

Prior to the twentieth century, federal, state, and city health departments in the United States experienced a slow and difficult history of organizational development. Not only was there uncertainty about the direction **public health services** should take and whether they had any value, but political and financial support were lacking as well. In the 1890s and 1900s, young American medical graduates who traveled to Europe for their postgraduate training came home with news of some revolutionary answers to old problems of epidemics and their possible control. These answers resulted from findings in the laboratories of Pasteur in France and Koch in Germany regarding the nature and transmission of infectious diseases. Since the turn of the twentieth century, governments around the world have taken increasing responsibility for controlling disease and offering other supportive health services.

In the United States today, health facilities are provided by government at every level. Most federal agencies involved in public health affairs are coordinated within the Department of Health and Human Services (HHS). Under HHS administration, the U.S. Public Health Service (whose origins date back to 1798) conducts research in the field of preventive medicine, provides hospital facilities for military personnel, and gives assistance to state and local health departments. Most research and epidemiologic investigation is conducted through divisions of the Centers for Disease Control and Prevention (CDC) in Atlanta, Georgia. The CDC Center for Infectious Diseases is made up of divisions for bacterial, viral and rickettsial, and parasitic infections. The Division of Surveillance and Epidemiology functions under the Epidemiology Program Office.

State Health Departments

In the United States, state health departments have authority over health regulations and services within their geographic areas. Many cities also assume local responsibility for health problems. For example, the New York City Department of Health is one of the oldest of such organizations. It was established in 1866 and was active in some of the most vigorous and successful battles fought in the United States against such public health problems as diphtheria, smallpox, cholera, and sexually transmitted diseases. The work of local community health departments led to the development of public health services and administration in this country. The Public Health Service now functions as an advisory leader in this field, offering assistance to the states when emergencies or special problems arise or new programs are to be developed.

International Health Organizations

Several organizations operate on an international level to protect public health, the most important of which is the **World Health Organization (WHO)** headquartered in Switzerland. The services of WHO, which are supported by many nations of the world, include distribution of technical information, standardization of drugs, and development of international regulations for controlling epidemic diseases. In many countries, WHO experts have helped governments develop programs for control or eradication of diseases such as smallpox, malaria, yaws, and trachoma. With increased control over the epidemic spread and endemic incidence of infections, improved standards of public health have contributed greatly to advancing the development of the world's resources.

Public Health Nursing

The first formal nursing training in the United States was given at New York Hospital in 1798. The first visiting nurse service was provided some 15 years later, in 1813, in Charleston, South Carolina. An epidemic of yellow fever in that city spurred the Ladies' Benevolent Society to organize a nursing service for the home care of the sick. By 1862, the year Florence Nightingale established a district nursing association in England (see Perspective 10.1), the Civil War in the United States had created an urgent need for medical and nursing care of sick and wounded soldiers. The Women's Central Relief Association, later incorporated into the Sanitary Commission, was organized to provide nursing care, and Miss Nightingale herself was an advisor to this group.

The eventual establishment of public health nursing as a profession dates from the late 1800s in New York. During that time, the New York City Mission provided trained nurses to care for the sick in their own homes; the Henry Street District Nursing Service was established by Lillian Wald and Mary Brewster (Miss Wald is considered the founder of public health nursing in the United States); and community nursing projects were sponsored by the New York City Department of Health. Public health nursing as we know it today is an essential function of local, state, and federal health services, and plays an international role as well.

PUBLIC HEALTH CONTROLS OF ENVIRONMENTAL RESERVOIRS OF INFECTION

The major environmental reservoirs of infectious disease in the public domain are water supplies, sewage, food, and milk. The sanitary standards and regulations developed during this century to monitor and control the purity of water and food have markedly reduced disease arising from these sources. Food- and waterborne epidemics, once so prevalent and devastating, now seldom occur in developed countries.

PERSPECTIVE 10.1 *Florence Nightingale: The Lady with the Lamp*

For all of her adult life, Florence Nightingale (1820–1910) believed that nursing should and could be a noble profession and worked diligently to make it so. Until she exerted her insight and determination, nursing was a job undertaken only by the lowliest of women who mopped up after the sick in prisons and hospitals, where filth and neglect hastened the demise of the sickest.

Born into a wealthy English family, Florence constantly had to defy the constraints of a rigid social order. Nonetheless, she pursued her idea of nursing, visiting hospitals at home and abroad, studying and learning about medical practice, and focusing on the needs of patients.

In 1854, when she was 34, England entered the Crimean War, joining France in its support of Turkey against Russian invaders. The war was bloody and the casualties were enormous. Ten times as many British soldiers died from cholera and typhus fever (transmitted by lice) than from battle wounds. By February of 1855 the mortality rate at the British military hospital at Scutari in the Crimea had reached 42.7 percent of the cases treated.

Florence Nightingale (1820-1910) documented the change in the mortality rate among British soldiers in the Crimea by her careful and assiduous use of statistics.

Courtesy The New York Public Library

When word of these casualties reached England, Miss Nightingale, by then well known for her efforts in nursing, was commissioned to take a contingent of 38 other young nurses to the Crimea. By the end of the war her name was legend. She had overcome the hostility of the military establishment to see that essential facilities and equipment for the care of the sick were provided. A laundry was installed with boilers for heating water; towels, soap, operating tables, scissors, bedpans, and stump pillows were supplied; and strenuous efforts to destroy lice were instituted. Sick soldiers came to know her as "the lady with the lamp" from her nightly visits to the hospital, lantern in hand. These late night rounds not only gave rise to poetic references to her as "the ministering angel of the Crimea" but also resulted in the dismissal of some of her nurses for delinquent behavior!

The statistics that Miss Nightingale kept meticulously showed that, within six months of her arrival in the Crimea, the mortality rate at Scutari had dropped from more than 42 percent to 2.2 percent. Thus her contention that sanitation and cleanliness are vital in the care of the sick was supported. Her renowned dictum helped to establish a new attitude and remains a compelling directive: "It may seem a strange principle to enunciate as the very first requirement in a hospital that it should do the sick no harm."

She returned to England in 1856, a world-famous and revered figure (see Perspective fig. 10.1). Yet she shunned publicity, saying that some 9,000 soldiers lay "in their forgotten graves," dead "from causes which might have been prevented." For the rest of her life she dedicated her efforts to establishing a commission to investigate military medical care and preventing the tragedy of needless deaths among fighting soldiers. Although she knew nothing of the causes of infectious diseases, she established standards of cleanliness and asepsis that saved lives then and today.

Ref: Cohen, I. B. 1984. Florence Nightingale. *Sci. Am.* 250:128–134.

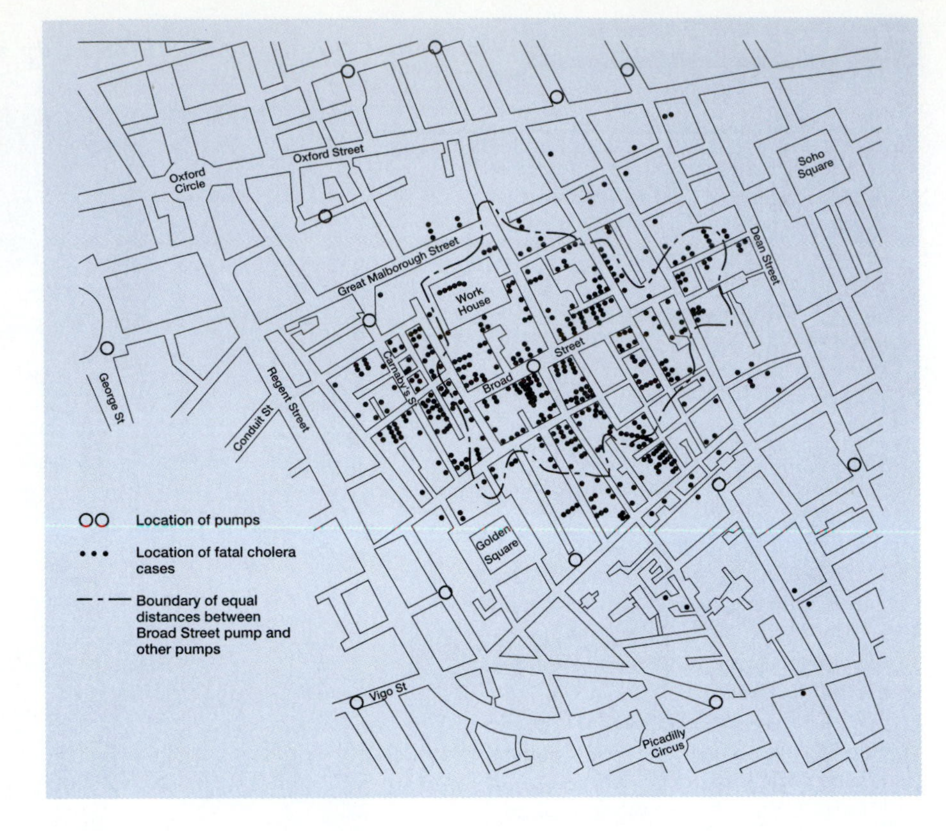

FIGURE 10.1 Snow's Broad Street pump map, one of the earliest examples of disease mapping. By marking each new case of cholera on the map, Dr. John Snow was able to prove that the Broad Street pump was a central point in the London cholera epidemic of 1854. This mapping also helped to establish the fact that cholera is a waterborne disease.

Courtesy of Pfizer, Inc.

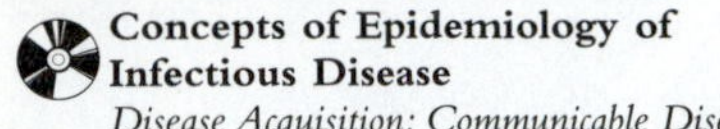
Concepts of Epidemiology of Infectious Disease
Disease Acquisition: Communicable Diseases

WATER AND SEWAGE SANITATION

The principal waterborne diseases are salmonelloses, shigellosis, amebic dysentery, giardiasis, and cholera. Sporadic outbreaks caused by other infectious agents such as the parasite, *Cryptosporidium,* also occur. Viruses excreted in feces may also be spread through contaminated water, as may some helminth ova. Fecal contamination of water supplies by humans and animals constitutes the major infectious hazard and the main need for water purification (see chap. 8).

Before the development of methods to safeguard public water supplies, epidemics of waterborne intestinal disease occurred frequently. Usually they were of short duration, beginning explosively among people sharing a contaminated water supply.

Most pathogenic organisms do not multiply or live long in water. Therefore, epidemics of waterborne infection are usually characterized by the rapid appearance of large numbers of cases (the number depending on the extent to which a water supply is shared) and an almost equally precipitous fall in incidence.

In some instances continuous undetected contamination of water from infectious sewage may cause prolonged epidemics. An epidemic of cholera that occurred in a London parish in the year 1854 provides a dramatic example. During a 10-day period in the autumn of that year, more than 500 people fell ill and died of cholera. There was a great deal of cholera elsewhere in the city at that time, but these cases were concentrated in one neighborhood. As the epidemic continued, two men, John Snow and John York, studied the area and proved, by epidemiologic methods only (the bacteriologic nature of the disease was not understood at that time), that the outbreak stemmed from a particular water source known as the Broad Street well. It was also discovered that sewage from the cesspool of a house on Broad Street was polluting the well and that there had been a case of undiagnosed intestinal illness in that house shortly before the cholera outbreak began. Snow traced many cases to water from the Broad Street well and also explained why people who lived in the same neighborhood, but did not drink from the polluted well, did not become ill (fig. 10.1).

The incidence of cholera, typhoid fever, and other intestinal infections rarely reaches epidemic proportions today in those countries or communities that have established standards for protection of water and disposal of sewage. These diseases persist in endemic form, however, in many parts of the world (fig. 10.2). Asymptomatic carriers of intestinal pathogens persist in all populations and are a possible source of new epidemics whenever there is a break in sanitation methods or systems. Such breaks may be minor, involving one well or one city apartment house. However, when major disasters such as war, floods, or earthquakes disrupt the functions of an entire city or a large area, the danger of large epidemics becomes real.

Sources of Public Water Supplies

Community water supplies are usually obtained from surface water collected from brooks, streams, and rivers and

FIGURE 10.2 Drinking from the same pool used for sewage disposal and clothes washing is a dangerous practice, which may result in the spread of many diseases.
World Health Organization. Photo by P. A. Pitter.

Concepts of Epidemiology of Infectious Disease
Terminology: Reservoirs. Transmission: Horizontal-Common Vehicle

stored in reservoirs. Surface water sources are fed by rainfall or by runoff of excess rain, and water often runs over large areas before accumulating in reservoirs. Consequently, the water is subject to pollution along much of its course and it must be purified before delivery to the community.

Local governments find it increasingly difficult to restrict industrial use of or home building on land involving their watersheds. Neighboring communities frequently create problems for each other by polluting shared water supplies. An "upstream" town, for example, may obtain clean water from one point in its river and use a lower point of the same river for sewage disposal. Other towns, located downstream from this point, are then subject to possible water contamination. Chemical pollution of rivers by industrial wastes similarly affects many communities sharing a common water supply. The need for firmer state and local government controls on the sources of water pollution is now widely recognized.

Groundwater accumulations from rain and snow can be tapped by digging wells for individual home use. The safety of well water depends on its protection from surface contamination or from the sewage disposal system employed. Wells must be deep enough to ensure good filtration of water seeping from the surface. Outdoor latrines, septic tanks, or ducts that run sewage off to a distant collection point must be located at sites that ensure protection of wells and their water sources.

Water Purification

To prepare it for consumption water is pumped from reservoirs or rivers into purification plants (fig. 10.3). It is first screened for large debris, such as leaves, twigs, or dead fish, then pumped into storage reservoirs where it is subjected to the following **water purification** processes:

1. *Aeration* increases the oxygen content of the water and speeds oxidative metabolism of organic material by bacteria. It also displaces other dissolved gases that contribute unpleasant odors and tastes to polluted water.
2. *Coagulation* of many soluble chemical contaminants is accomplished by adding compounds such as aluminum sulfate to create aggregates called *floc.* These colloidal complexes eventually settle out of solution.
3. *Sedimentation* of coagulated or insoluble materials removes remaining coarse contamination.
4. *Filtration* removes fine particles, including bacteria. Water seeping down through filter beds first passes slowly through closely packed sand filters, then more rapidly through looser sand filters into coarse gravel beds. Below the gravel, successive layers of pebbles, larger rocks, and tile continue the filtration, delivering relatively pure water into collecting tanks. The upper sand layers of a filtration system accomplish most of the purification. About 98% of bacteria and most suspended solids are trapped there. Oxidative processes remove organic matter, and many bacteria are destroyed by the activities of protozoa. Approximately six million gallons of water are filtered per day through an acre of sand bed, a rather slow process. The filters must be cleaned from time to time by replacing the top inch of sand and backflushing to clear accumulated materials.
5. *Chlorination* is the final step. The bacterial action of chlorine (see chap. 9) in a concentration of 0.2 mg/L ensures disinfection of filtered water and its safety along the piped route to the consumer. Chlorinated lime or sodium hypochlorite may be used as sources of chlorine, but the gaseous form is least expensive and the usual choice.

Close adherence to this water purification process, especially for surface waters, is needed to maintain the safety of potable water (i.e., water suitable for drinking). In spite of regulations and standard methods for evaluating water supplies, waterborne outbreaks continue to occur, presumably because of breaks in the treatment process. From 1993 through 1994, CDC received reports of 30 outbreaks involving 405,366 persons consuming water intended for drinking; 403,000 persons were afflicted during one outbreak (the largest ever documented in the United States) and 2,366 during the other 29. Most of these outbreaks resulted from deficiencies in the water treatment process. *Cryptosporidium parvum,* a protozoan parasite (see chap. 18) was involved in the large outbreak and in four others. *Giardia lamblia,* another protozoan parasite, was also involved in 5 outbreaks, but in an additional 5 outbreaks a causative agent was not found. Bacteria involved in a smaller number of waterborne disease outbreaks included *Campylobacter jejuni, Vibrio cholerae,* and species of *Shigella* and *Salmonella.* These organisms are discussed further in chap. 16.

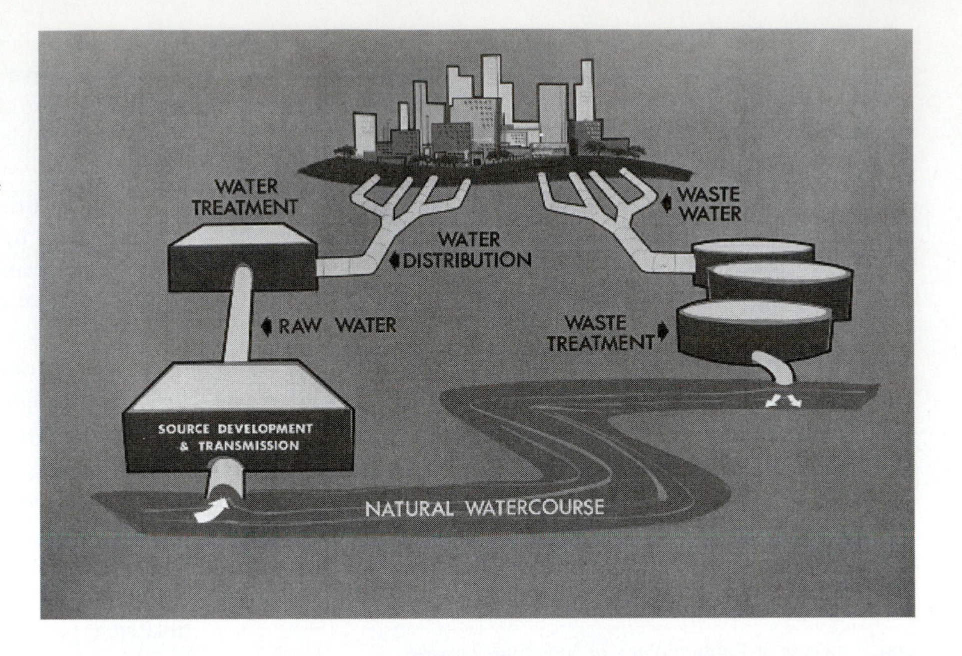

FIGURE 10.3 Water drawn from a natural watercourse must be treated in a purification plant to prepare it for consumption since upstream cities and towns may have polluted it with their wastes. Sewage and waste water should be treated before returning to the watercourse to avoid damaging indigenous aquatic life and to protect downstream cities.

Courtesy Centers for Disease Control, Atlanta, GA.

Sewage Disposal

Raw sewage, teeming with potentially pathogenic intestinal microorganisms, is the most dangerous threat to water supplies. It may have disastrous consequences for fish and wildlife as well as for human beings.

Sewage disposal on a community scale involves its collection in a treatment center. Bulky waste is screened out first, then the sewage is stirred and agitated to permit aeration. Large particles are broken up to promote bacterial decomposition of organic material. Solid matter settles out (fig. 10.4); the sludge that collects contains products of decomposition, which may be reduced in volume, dried, and sold for fertilizer (but is not recommended as fertilizer for plant foods that may be eaten raw).

The liquid component drawn off from sludge is treated with chlorine and can be pumped off into a river or the ocean as harmless effluent (see fig. 10.3).

In a septic tank the solid sludge of sewage settles to the bottom and must be pumped out periodically. The fluid effluent passes out of the tank into a soil drainage (leaching) bed where organic matter is decomposed by soil bacteria. In airplanes and trains, chemical methods are used to treat sewage in small collecting tanks.

Bacteriologic Standards for Water

Early drinking water standards were defined by the U.S. Public Health Service, but the quality of public water supplies is now subject to the terms of the U.S. Environmental Protection Agency 1989 National Primary Drinking Water Regulations. These **bacteriologic standards for water** are legal requirements which specify that testing for compliance must be done in a certified laboratory. Many health department laboratories maintain surveillance of community water supplies, testing them for chemical as well as microbiologic purity. Potable water should be free from odor, taste, color, harmful chemicals, nitrate, and other decomposition products of organic matter.

The bacteriologic analysis of water includes examination for the presence of contaminating coliform bacteria, an index of fecal pollution. The current Environmental Protection Agency standards call for a maximum contaminant level of zero coliforms in each 100 ml of water sampled. If coliforms are detected, repeat samples must be collected and analyzed for a subset of coliforms referred to as *fecal coliforms* or for the common intestinal inhabitant *Escherichia coli*. The regulations also stipulate that if fecal coliforms or *E. coli* are found, the public must be notified. Bacteriologic methods can also be used to detect the presence of enterococci (formerly known as fecal streptococci), another indicator of excretory water pollution.

When bacteriologic tests reveal unacceptable levels of fecal contamination, appropriate checks are made for possible sources of pollution or for failures in the purification system, and corrective action is recommended.

Chemical analyses are also essential in monitoring the quality of drinking water. Water that is microbiologically safe may not be potable because of chemical pollution with such substances as pesticides, detergents, metallic ions, alkalis, or similar contaminants.

FOOD SANITATION

Basic Control Measures

Strict cleanliness is important in all stages of food preparation but must be coupled with adequate refrigeration to prevent spoilage bacteria or pathogens from growing. Proper **food sanitation** methods are particularly important for foods that have been partially or fully cooked and for frozen foods. Cooked foods provide better culture media for bacteria than many uncooked foods. Frozen foods are not sterilized by the freezing process, and when they thaw, the

FIGURE 10.4 A primary settling tank for sewage. Before entering this tank, sewage is screened to remove large objects, then passed through a chamber where smaller particles are removed. In the tank, organic matter slowly settles out. This sediment is removed as sludge, treated in holding tanks, and finally dried for use as a fertilizer except on plants that will later be eaten without cooking.

Courtesy New York State Department of Health, Albany, NY.

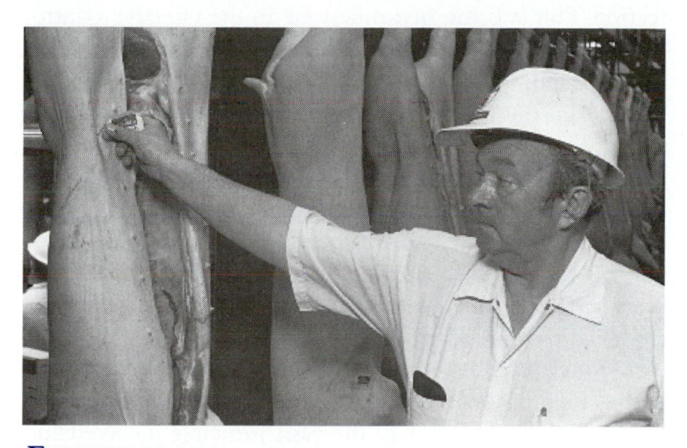

FIGURE 10.5 A government meat inspector examines meat in a commercial meat-packing plant.

Courtesy U.S. Department of Agriculture, Washington, D.C.

rupture of cells and fibers may make them more subject to bacterial growth. Proper cooking of food is necessary to ensure its safety from infectious organisms (or their toxins) that it may contain.

Public health regulations require inspection of food-processing plants, abattoirs, restaurants, dairies, and all other establishments involved in food preparation for adequacy of basic sanitation, including refrigeration methods. In addition, health service agencies at all levels conduct a continuous campaign to educate the general public in the sanitary principles involved.

Meat Inspection

Meat inspection for bacterial spoilage or other deterioration is part of the government controls on the safety of meat or poultry before it is distributed for retail sale (fig. 10.5). More important, certain primary infections of meat can be detected by visual inspection for tissue lesions. This is particularly true for encysted tapeworm larvae in beef or pork muscle, which can be seen and sometimes felt as small, white, hard nodules in the tissue. Unfortunately, fish tapeworms are not so easily detected, nor are larval encystments of the roundworm *Trichinella,* which may infest pigs.

A serious food sanitation problem in the United States is the high incidence with which chickens and their eggs are contaminated with *Salmonella* organisms. Species of this bacterium cause serious gastroenteritis and sometimes systemic disease, both of which can be life-threatening for very young, elderly, and immunocompromised persons. From 1988 through 1995, approximately 365,000 cases of salmonellosis were reported to the Centers for Disease Control and Prevention. Reported cases are thought to represent only a fraction of persons who develop the disease. In some states, 60% or more of chickens sold in supermarkets are contaminated with salmonellae. Once introduced into a flock, the organism becomes endemic among the birds and is difficult to eradicate. The speed and manner in which chickens are handled in processing plants causes contamination to spread, even to birds coming from uninfected flocks. Coupled with a shortage of government inspectors, the problem has become serious enough to warrant new regulations and recommendations from the U.S. Department of Agriculture and the Food and Drug Administration. Their focus is on the destruction of infected flocks and they recommend methods for refrigerating and pasteurizing eggs before cooking, especially in food service establishments. Fortunately, thorough cooking of all poultry products, including eggs, destroys the organisms and makes the foods safe for consumption. A decrease in the reported cases of salmonellosis in the United States from 21 out of 100,000 people in 1986 to 17 out of 100,000 in 1995 may indicate that these measures are effective in lowering the incidence of disease.

Control of Milkborne Diseases

The important diseases that are transmissible by milk are tuberculosis, brucellosis, streptococcal infections, and campylobacteriosis. The first three of these **milkborne diseases** are now uncommon in the United States because of better controls at the source of disease (i.e., inspecting, vaccinating, and eliminating infected cattle) and better controls on food handling. Pasteurization and sanitary bottling methods are also essential in controlling milkborne disease. Consumption of unpasteurized milk is the usual route by which campylobacteriosis is acquired. Campylobacteria are commonly found as normal flora in the gastrointestinal tract of cattle and can contaminate milk as it is collected. During the past few years, this organism has been responsible for almost all milk-associated outbreaks of gastroenteritis in the United States.

Pasteurization is an effective precautionary measure for assuring safe public milk supplies. It is a heat-disinfection procedure (see chap. 9) that destroys vegetative forms of all pathogenic microorganisms, with the possible exception of hepatitis virus. It does not destroy bacterial endospores, but these have no apparent importance in milk. Dairy sanitation must be strictly maintained if pasteurization is to be fully effective. This includes grooming of cows, sterilizing or disinfecting all equipment, and delivering milk into a closed presterilized system.

Bottling milk in presterilized glass, plastic, or waxed cardboard containers keeps its content of nonpathogenic organisms at a low level, but milk must be kept under constant refrigeration thereafter. Certain heat-resistant species of *Lactobacillus* and *Streptococcus* that are normally found in milk are not completely eliminated by pasteurization. Their subsequent multiplication, with degradation of nutrient in milk, is responsible for milk souring. Refrigeration, however, delays this process.

Bacteriologic standards for milk are determined by laboratory methods that (1) count total numbers of organisms surviving per milliliter of milk, (2) test for the presence of coliform bacteria, (3) determine adequacy of pasteurization by testing for the heat-sensitive enzyme phosphatase, normally present in raw milk (its absence from pasteurized milk indicates that heating conditions were sufficient for enzyme inactivation), and (4) test for the presence of antimicrobial agents. Milk is rated as grade A if the total bacterial count does not exceed 20,000 per milliliter after pasteurization and if the coliform count is less than 10 organisms per milliliter. In addition, the test for antimicrobial agents must be negative. No other grades of milk may be marketed for public consumption.

PUBLIC HEALTH CONTROLS OF LIVING RESERVOIRS OF INFECTION

The important principles involved in the prevention and control of infectious disease have been discussed previously in several contexts. They are summarized briefly here, together with a description of their broad applications to the protection of the public health.

CONTROL OF HUMAN DISEASE

Artificial Active Immunization

Probably the most important factor in preventing infectious disease in humans and animals is the development of active immunity. This may occur naturally following infection with a pathogenic microorganism, but the protective effect may vary in extent and duration. Artificial active immunization through the administration of available **vaccines** is therefore a valuable tool for preventing or modifying many human and animal infections.

The use of vaccines for the specific protection of human beings dates back to 1796 in England where Dr. Edward Jenner first experimented with the idea that people could be protected from serious disease by injection with material derived from others suffering from less severe infection. This was the first milestone in the effective control of smallpox (see Perspective 10.2).

Smallpox had been a terrible human scourge for centuries: transmitted readily by person-to-person contact, it had decimated military and civilian populations alike, lost and won wars, and sometimes threatened the survival of afflicted communities. (In Jenner's day, about 30% of children born in England died of smallpox before reaching maturity, and on the American continent many Indian populations were virtually wiped out by the inroads of this disease introduced by European settlers.)

Jenner's success with fluid collected from the less dangerous lesions of cowpox as a preventive agent for smallpox eventually led to a study of the basic principle involved and to its controlled application in preventing many infectious diseases (see chap. 6). Since then, smallpox has been eliminated from its last endemic strongholds in Bangladesh and Ethiopia and eradicated completely, thanks to concerted efforts of the WHO.

Mass immunization programs have also led to control over such diseases as tetanus, diphtheria, yellow fever, pertussis, and poliomyelitis. Unlike smallpox, these diseases have not been completely eliminated, but in countries where the vaccines are used routinely, disease incidence is low.

Successful control of infectious diseases with vaccines depends on a number of factors such as the global distribution of the pathogenic organism, its entrenchment in animal or arthropod reservoirs, the characteristics of its antigens, and the ease with which potent immunizing preparations can be made. Diseases such as measles, mumps, rubella, influenza, and the sexually transmitted infections—distributed worldwide and readily transmissible through direct contacts—are particularly difficult to control without effective vaccines. The problem has been solved for measles, mumps, and rubella with the availability of vaccines that appear to induce a solid immunity if administered during childhood with suitable dosage schedules. By contrast, influenza vaccines

PERSPECTIVE 10.2 *Smallpox: How It Was Controlled by Vaccination*

Resistance to the concept of vaccination diminished when smallpox epidemics began to vanish. This etching, "The Cow Pock, or the Wonderful Effect of the New Inoculation," by James Gilray (1802), is a caricature of Edward Jenner administering his vaccine to some reluctant people.

Courtesy National Library of Medicine.

Throughout recorded history, until recent times, smallpox was a disease that killed, blinded, or disfigured enormous numbers of people around the world. A continuing endemic disease in Europe, the incidence of smallpox would reach epidemic proportions on that continent about every five or ten years. In the seventeenth century, European peasants began the practice of "inoculating" their children. Fearful parents would seek out a mild case of smallpox and deliberately have their child touch one of the skin lesions so characteristic of the disease. Using this primitive method, some parents saved their children from all but a mild case of the disease and endowed them with a natural immunity that protected them from future epidemics. When this "hit-or-miss" procedure failed to work, however, the child would either become seriously infected and die or endanger others by contributing to the pool of infection. Nonetheless, the procedure was publicized by the Royal Society of London as a hopeful step forward.

In April of 1721, an epidemic of smallpox broke out in Boston. Theologian Cotton Mather, aware of the Royal Society's publication about inoculation, tried to organize the physicians of Boston to use this method to combat the disease. He was rebuffed by all except Dr. Zabdiel Boylston. Dr. Boylston inoculated three people: two of his own sons and a slave, none of whom suffered ill effects. Throughout Boston, however, the epidemic affected nearly one-half the population, and 844 people died.

Later, the practice of inoculation was succeeded by that of vaccination, instituted by Dr. Edward Jenner, in England, in the late 1700s. Jenner had observed that milkmaids in his farming area rarely, if ever, contracted smallpox, although the cows they milked often displayed a kind of skin lesion known as cowpox. He noted that cowpox among milkmaids was a mild form of smallpox that seemed to protect them from serious disease, and on that basis, he developed an inoculum from cows with cowpox to protect humans from smallpox. The Latin word for cow is *vacca,* so Jenner's technique became known as *vaccination.* In 1798, Jenner published a paper in which he noted that vaccinating people with cowpox material was far less dangerous than inoculating them with the virus of smallpox, but provided as much, or more, protection against the disease (see Perspective fig. 10.2).

Jenner's success with vaccination became widely known. Back in Boston in about 1800, Dr. Benjamin Waterhouse asked Jenner for cowpox material to vaccinate his children and also told his friend, then Vice President Thomas Jefferson, of Jenner's successes with vaccination. On this advice, Jefferson imported enough cowpox material to vaccinate his entire family and 60 members of his household, including the slaves.

When he became president, Jefferson influenced people throughout the new republic to use vaccination as a protection against smallpox. By the time he retired to Monticello in 1809, millions of people in America and Europe were vaccinated, and the threat of epidemics in those areas was vastly diminished, although it remained real in other parts of the world.

The smallpox virus now has been thoroughly studied, its relationship with the cowpox virus has been clarified, and the epidemiology of both diseases is understood. Vaccine has been administered so universally and successfully in all parts of the world that smallpox today is considered a disease that has been totally eradicated. As a result, smallpox vaccination is no longer required in any country.

Ref: Langer, W. I. 1976. Immunization against smallpox before Jenner. *Sci. Am.* 234: 112–117.

provide only a short-lived protection because of the changing antigenic nature of influenza viruses. The agents of gonorrhea and syphilis have defied efforts to prepare their antigens in a form suitable for mass immunization.

Diseases such as cholera, plague, typhus, and typhoid fever have been controlled partially through curbs on their sources (improved sanitation, insect control) and partially through selective immunization. Vaccines against these diseases do not induce solid, long-lasting immunity but have been useful for protecting persons at high risk of exposure (military people or travelers entering endemic areas).

Many research efforts continue in the field of vaccine development to improve antigenicity in available vaccines and to produce new ones. A number of microbial agents continue to be a challenge, either because they are difficult to cultivate (e.g., *Treponema pallidum,* the agent of syphilis) or because they do not possess a useful degree of antigenicity. For example, *Haemophilus influenzae* type b, an important agent of meningitis in young children (see chap. 15), has a polysaccharide capsule that is not antigenic in children younger than two years old. Because antibodies against this capsule protect against infection, researchers have sought ways to induce active immunity to the polysaccharide. Efforts have led to development of a so-called *conjugate vaccine* in which the *H. influenzae* type b polysaccharide capsule material is attached (conjugated) to an altered nontoxic protein toxin from a mutant diphtheria strain. In this form, the polysaccharide becomes immunogenic. This conjugate vaccine is now in routine use and has resulted in a remarkable decrease of over 90% in serious *H. influenzae* infections (see table 10.1).

Another exciting development resulting from recombinant DNA technology is the ability to insert selected genes from a variety of foreign microorganisms, including bacteria, viruses, and protozoa, into the nucleic acid of vaccinia virus (the cowpox virus once used to protect against smallpox). When the live recombinant vaccinia virus is inoculated into animals and multiplies in cells, antigens from the foreign microorganism are expressed and stimulate protective immunity. The vaccines are called *vaccinia virus vectors* because they are the means for transmitting the antigen into the animal cell. Animal studies have shown that these vaccines are highly successful in inducing both humoral and cellular immunity to microorganisms that cause diseases such as hepatitis, malaria, and streptococcal pharyngitis. However, they have not yet been approved for use in humans.

As more vaccines become available, the question remains: Who should be immunized and when? In general, everyone who is susceptible and likely to be exposed should be protected against diseases that pose a threat to individual or public health. In the United States, specific recommendations are made by the Public Health Service Immunization Practices Advisory Committee (APIC) and are published periodically.

Children

Immunization in childhood is indicated for those diseases that have a wide distribution, are readily transmissible, and for which effective vaccines are available. Current immunization programs for children include protection against hepatitis B, diphtheria, pertussis, tetanus, poliomyelitis, measles, mumps, rubella, and *H. influenzae* type b. In 1996, APIC recommended for the first time the use of a live attenuated varicella (chickenpox) vaccine for the routine immunization of healthy persons older than 12 months. The vaccine is not given to persons with a history of chickenpox because they are considered immune. Nor is the vaccine given to immunocompromised people because they risk the development of serious widespread varicella infection. **Immunization schedules** are based on the capacity of individual vaccines to induce solid immunity after one or more injections. Combined vaccines containing several different organisms or antigens simplify the procedure, but the length of the series of injections required then depends on the antigenicity of each of the individual components. It is important to remember that artificial immunization usually does not confer lifelong protection and that booster doses are necessary to maintain the effect, the interval depending on the characteristics of each vaccine preparation currently available (table 10.1).

TABLE 10.1 RECOMMENDED IMMUNIZATION SCHEDULE FOR CHILDREN

Recommended Age	Vaccine (Antigen) Administered
2 months	HB, DTP, IPV, Hib
4 months	HB, DTP, IPV, Hib
6 months	HB, DTP, OPV, Hib
12 to 15 months	DTP, Hib, MMR, Varicella
4 to 6 years	DTP, OPV, MMR
14 to 16 years	TD (repeat every 10 years)

Note: HB = hepatitis B; DTP = diphtheria-tetanus-pertussis; IPV = trivalent inactivated polio; OPV = trivalent oral polio; Hib = *Haemophilus influenzae* type b conjugate; MMR = measles-mumps-rubella; Varicella = live, attenuated varicella virus; TD = tetanus-diphtheria

Vaccines
Vaccine Targets: Bacterial Targets; Viral Targets

Travelers and Military Personnel

People who travel may be exposed to diseases unfamiliar to their own regions or to which they are not immune for other reasons (e.g., neglect of vaccination, no vaccine available). In this age of rapid and frequent world travel, controls on the international transmission of infectious disease must be maintained. Public health services of each country work to protect their people, and governments that support the WHO cooperate in extending this protection across national borders. In the United States, the Public Health Service publishes a booklet annually titled *Health Information for International Travel,* which outlines the vaccinations that may be required by international health regulations. The information is also available on the Internet (http://www.cdc.gov). The immunization schedule shown in table 10.1 would protect an individual from almost all diseases for which effective vaccines are available. In the 1990s, proof of yellow fever vaccination was required for travel to only a few countries, primarily in Africa. No other vaccines, including those for cholera and typhoid fever, were required, and no vaccinations were required to return to the United States.

To protect against some diseases that are prevalent in certain countries outside of the United States (malaria and hepatitis, for example), travelers need to adhere to certain prophylactic measures. In the case of malaria, chemotherapeutic agents should be taken before, during, and after likely exposure. For hepatitis B, hepatitis B vaccination is recommended; persons traveling to areas where the risk of hepatitis A is high should receive immune globulin or the hepatitis A vaccine. Routine, practical measures such as taking precautions against insect bites and avoiding uncooked foods and untreated local water are important for preventing infection and maintaining good health during travel. Because regulations change with disease prevalence, it is wise to refer to the latest edition of the *Health Information for International Travel* or consult a physician specializing in travel medicine before taking a trip, particularly to an underdeveloped country.

Medical Personnel

In hospitals and clinics, medical personnel are important candidates for controlled immunization programs. Physicians, nurses, microbiologists and other laboratory personnel, aides, orderlies, and all others who come into close contact with patients should be fully immunized against those diseases for which effective vaccines are available and which they may encounter in their work. This immunity is needed to protect themselves as well as to prevent transmission of avoidable diseases to susceptible patients. The basic program includes the vaccines listed in table 10.1, and immunity should be carefully maintained by a regular schedule of booster doses. All health-care workers should ensure that they are immune to varicella, either through natural infection or vaccination. One dose of vaccine appears adequate when given between 12 months and 13 years of age but two doses given 4 to 8 weeks apart is recommended for persons older than 13 years. In addition, influenza vaccine should be taken in the fall of each year; the antigenic nature of the influenza virus changes constantly (see chap. 13), so protection does not carry over from year to year.

Control of Human Carriers of Infection

It must be emphasized that human beings can be dangerous **carriers of disease**—never more so than when they go unrecognized. The normal appearance of a healthy carrier of pathogenic organisms, or of a person in the asymptomatic incubation period of an infectious disease, challenges epidemiologic detection.

Detection of the Incubating Disease

Until patients become symptomatic, little can be done to recognize them as reservoirs of infection. Under limited circumstances, as in closed populations (barracks, schools, prisons, even immigration offices), the suspicion of infection may be confirmed by microbiologic methods. Streptococci or meningococci, for example, may be recognized in throat cultures, or enteric pathogens in stool specimens; but in the absence of symptoms of active infection, such findings do not distinguish immune or resistant carriers from susceptible individuals who are incubating a disease. When there is serious risk of individual or epidemic disease, prophylactic measures are taken.

Detection of the Healthy Carrier

When carriers of specific infections have been recognized, they may be placed under suitable control. If their carrier state represents a threat to others, they may be treated with antimicrobial drugs in an effort to eliminate the infection, or they may be removed from areas or activities where transmission of the organisms is most likely.

Carriers of *Salmonella,* for example, should not be permitted to prepare food for others or to come in close contact with medical or surgical patients. It is sometimes difficult to eliminate such infections by chemotherapy, and the carrier state remains a difficult problem because it may persist in some individuals for long periods despite medical and surgical attempts at eradication.

Staphylococcus aureus carriers represent perhaps the most difficult problem of all, especially when they are found among hospital personnel. Staphylococcal cross-infections acquired by hospital patients can be serious, particularly in nurseries and other critical patient areas. Not all staphylococcal carriers are active shedders of the organism, and they do not always represent a dangerous threat to the patient. When the source of a patient's infection has been traced to an employee, however, attempts are made to eradicate the organism. In many instances the strains are difficult, if not impossible, to eliminate from some carriers, although a patient, systematic attack from several directions may succeed. Bactericidal antimicrobial therapy, maintained at an active level for a sufficient period of time, combined with consistent use of disinfectant soaps for skin care may prove successful. Beyond this, strategic job placement of hospital personnel carriers and their careful instruction in personal hygiene techniques can help to provide a reasonable solution to the double problem of patient protection and personnel employment.

Food Handlers

Health department sanitary codes generally have very specific requirements regarding preemployment examination of food handlers and their frequent surveillance. Routine preemployment tests include complete physical examination for evidence of communicable disease; a skin test to rule out tuberculosis; and stool examination and culture to screen for enteric pathogens. These examinations must be repeated at regular intervals after employment. Any symptoms of illness should be reported and treated promptly and the individual temporarily removed from the job if indicated. Sanitary toilet and washbasin facilities must be provided for kitchen personnel and continuing education provided emphasizing the need for careful hand washing. Other matters of personal hygiene should also be stressed, such as the cleanliness of hair, skin, and clothing; short, clean fingernails; and handkerchief protection of coughs and sneezes. Cuts and breaks in the skin of the hands should be covered with waterproof dressings when food is being handled, and the hands, arms, and face should be examined frequently for the appearance of infected pimples, boils, rashes, or other infectious lesions.

Communicable Disease Control

Definitions of the terms *communicable disease, isolation, quarantine,* and *communicable period* are given in table 9.2. A brief study of these definitions provides a review of the current principles associated with infection transfer.

Communicable diseases are not always directly transmissible between hosts of the same species, as we know. Their control depends on knowledge of their sources and routes of transfer. Because we now possess much more specific information than was available 100 years ago when the infectious nature of many diseases was first being discovered, the rules of control read very differently. **Isolation** and **quarantine** once involved entire communities in dreaded procedures of social imprisonment. Alarming notices were tacked on the doors of the quarantined; distress of physical segregation was added to the suffering of illness and death; the stench of fumigation and strong chemicals blended with the odors of sick and unsanitary cities—but the results were often effective. Today, isolation precautions are quietly applied in selective situations, primarily to prevent transfer of infection to susceptible people. The Centers for Disease Control and Prevention has issued a *Guideline for Isolation Precautions in Hospitals* that allows physicians or hospital infection control committees to use judgement in applying isolation precautions to prevent infection transmission (see chap. 11). The advice of local health officials may be sought as well.

The need for quarantine has diminished today because of the effective mass use of vaccines and a corresponding reduction in the number of persons susceptible to diseases controlled by this method. More adequate controls on animal diseases also have changed the situation as have systematic attacks on the arthropod vectors of some infectious diseases. Nonetheless, quarantine may be applied selectively: to travelers who fail to obtain appropriate artificial immunization; and to school-children and others in closed groups with many susceptibles (such as military personnel).

In the United States, the Public Health Service is responsible for controlling communicable diseases that might be disseminated through international trade and travel. This agency maintains quarantine stations at all ports, sea and air, where foreign commerce enters the country. All ships and planes arriving from ports in other countries are required to obtain health clearance from these quarantine stations. Radio contacts with ships or planes entering debarkation ports usually provide advance warning of overt illness aboard. These precautions are designed to prevent importation of such diseases as cholera, yellow fever, and plague. Currently, these are the only diseases subject to quarantine internationally. Passengers who arrive obviously ill with symptoms suggesting these diseases are quarantined in hospitals where they can be promptly treated.

As part of their request for permission to enter a port, vessel masters of cruise ships must report all persons who visited the ship's physician with a diarrheal illness. If 3% or more of passengers had a gastrointestinal illness during a one-week cruise, a quarantine officer boards the ship, and an epidemiologic investigation may be conducted.

Quarantine restrictions are strictly enforced, under penalty of fine, and can be lifted only by the responsible health agency.

Case finding and reporting within the community are important aspects of the preventive epidemiologic work of health departments. All local and state health boards require that physicians, hospitals, and similar institutions report certain communicable diseases. Some endemic diseases, such as chickenpox, are characterized by seasonal increases and cyclic epidemics every three to four years. Health agencies want to be forewarned of these rising incidence rates. Other diseases are reported so that the source of infection can be traced, with a view to preventing further transmission. Most such reports can be submitted by mail, but in the case of certain acute diseases, the local health department requests notification by telephone. Immediate reports should be made when cases of anthrax, cholera, diphtheria, foodborne illness, measles, meningococcal disease, plague, poliomyelitis, rabies, and typhoid fever are diagnosed. The **reportable diseases** at the national level beginning in 1996 are listed in table 10.2. The list reflects the changing pattern of important infectious diseases in the United States. It has been updated in two ways: (1) to exclude diseases that are now rarely seen and that do not pose epidemic threat (leptospirosis, rheumatic fever, tularemia) and (2) to add newly recognized diseases, such as cryptosporidiosis and hantavirus infection, or new types of infections produced by familiar pathogens such as invasive infections caused by group A streptococci ("flesh-eating bacteria").

Health departments also require that new cases of tuberculosis or sexually transmitted diseases be reported. Reporting sexually transmitted diseases such as gonorrhea and chlamydia urethritis is especially important because of the ease of transmission of such infections. All contacts of new cases should be investigated to ensure early diagnosis and treatment of developing disease.

Methods of prophylaxis for susceptible persons exposed to communicable diseases are discussed in chapter 9. Both chemoprophylactic and immunologic measures may be useful on a large scale to prevent epidemics. Mass im-

TABLE 10.2 DISEASES REPORTABLE AT THE NATIONAL LEVEL AS OF 1996

AIDS	Lyme disease
Anthrax	Malaria
Botulism	Measles
Brucellosis	Meningococcal disease
Chancroid	Mumps
Chlamydia trachomatis (genital infections)	Pertussis
Cholera	Plague
Coccidioidomycosis	Poliomyelitis, paralytic
Congenital rubella syndrome	Psittacosis
Congenital syphilis	Rabies, animal
Cryptosporidiosis	Rabies, human
Diphtheria	Rocky Mountain spotted fever
Encephalitis, California	Rubella
Encephalitis, eastern equine	Salmonellosis
Encephalitis, St. Louis	Shigellosis
Encephalitis, western equine	Streptococcal disease, invasive, group A
Escherichia coli O157:H7	*Streptococcus pneumoniae* (resistant)
Gonorrhea	Streptococcal toxic-shock syndrome
Haemophilus influenzae (invasive)	Syphilis
Hansen disease (leprosy)	Tetanus
Hantavirus infection	Toxic shock syndrome
Hemolytic uremic syndrome (post-diarrheal)	Trichinosis
Hepatitis A	Tuberculosis
Hepatitis B	Typhoid fever
Hepatitis C/non-A, non-B	Yellow fever
HIV infection, pediatric (less than 13 years old)	
Legionellosis	

munization is sometimes appropriate when diseases such as influenza threaten to involve large segments of the population. The expense and the risk of potentially dangerous side effects of the vaccine must be weighed against the public health benefit. Passive immunization on a large scale, as once applied in epidemics of poliomyelitis, is inappropriate today. Much of the population has either natural or artificial active immunity to diseases of epidemic significance. With the wane of certain pathogenic microorganisms from an immunized population, however, this natural stimulus to antibody production declines so that it becomes correspondingly more important to maintain active immunity by artificial means. Community immunization campaigns must emphasize the need for vaccine booster doses to help maintain a basic level of protective immunity.

CONTROL OF ANIMAL DISEASE

The principles of infectious disease control for animals are the same as for human beings. The basic premise in either case is that the best way to control disease is to prevent it using three major approaches: immunization whenever possible, sanitary protection of the environment, and protection of the healthy from exposure to the sick.

Control of Diseases of Domestic Animals

In some instances successful control of domestic animal infection is achieved by methods that cannot be used for humans. "Isolation of infection," for example, sometimes means destruction of infected, sick, or dying animals and vigorous methods for disinfection of carcasses. Quarantine may be applied to an entire herd of animals, without selecting those actually infected. Prophylactic antimicrobial agents may be used to prevent the chance of infection in healthy animals or fowl, as well as to protect exposed individuals. Use of drugs for such purposes, however, is thought to be an important factor in relation to the increasing number of antimicrobial-resistant bacteria that have emerged in animal and human infections. Controlling sources of animal diseases often involves control of insect vectors that perpetuate many of these infections as well as the protection of animals from infectious human beings. The chain of transmission of many infections links humans and animals by direct contact, and later through human consumption of animal products, notably milk, eggs, and meat.

Important **zoonoses** of domestic animals (diseases that are readily transmissible to humans) are tuberculosis, brucellosis, anthrax, taeniasis, trichinosis, and echinococcosis. Many of these diseases can be successfully controlled by adequate milk pasteurization and thorough cooking of animal products before ingestion. Vaccines are available to induce animal immunity for tuberculosis, brucellosis, and anthrax. Skin testing may also be of help in detecting infected animals. Sanitation and disinfection of animal quarters and of equipment used for preparing animal food products are critically important in preventing infection and controlling its spread. Other zoonoses such as cat-scratch disease, psittacosis, and toxoplasmosis may be acquired from household pets. Knowledge of their transmission routes is important for avoiding infection. The control and treatment of zoonoses will be discussed in detail in later chapters.

Rabies Control

Immunization of dogs is an essential control for a number of infectious diseases such as distemper, canine hepatitis, leptospirosis, and rabies. Of these, the last is also extremely dangerous to humans. Dogs, as well as a number of wild animals including foxes, wolves, raccoons, skunks, and bats are the natural reservoir of rabies virus (fig. 10.6). Transmission among animals or to humans is usually accomplished by the bite of a rabid animal. Rabies control depends on elimination of rabid animals and on protection of healthy dogs and cats by immunization with rabies vaccine. Immunization of humans is not recommended unless there is evidence of exposure to rabies.

When suspicion of rabies exists, biting dogs and cats must be quarantined for 10 days and observed for symptoms of disease. They must be destroyed if rabies is proven, and efforts must be made to find other animals possibly exposed. These animals also must be quarantined (for 3 months) and vaccinated or, if necessary, destroyed. Stray dogs and cats should be eliminated, and others immunized, with booster doses at regular intervals (every 3 years). Domestic dogs and cats should not be permitted to run freely, particularly in areas with endemic foci of rabies among wild animals.

FIGURE 10.6 This captured bat was submitted to a public health laboratory for examination for rabies (at least three bat varieties are known carriers).

Courtesy Centers for Disease Control, Atlanta, GA.

Concepts of Epidemiology of Infectious Disease
Terminology: Reservoirs; Zoonoses

Rabies control in wild animals is under supervision of federal and state fish and wildlife agencies. Hunting, trapping, and baiting methods are used to find and dispose of sick animals and to bring epidemics under control.

Rodent Control

Rodent control is directed at rats and mice, which are reservoirs for a number of infectious diseases that affect humans. In the past, diseases such as bubonic plague and murine typhus, both of which are transmitted by fleas from rodents to humans, were important problems. An understanding of the transmission route of these diseases has allowed preventive measures to be taken which, along with effective antimicrobial agents for therapy, have greatly decreased their occurrence. Currently, the deer mouse is known to be a reservoir for the agents of hantavirus pulmonary disease (chap. 13) and Lyme disease (chap. 22). Hantavirus infection is contracted by inhaling infective saliva or excreta produced directly from the animal, but Lyme disease is transmitted to humans by the bite of an infected tick that also feeds on deer mice.

For diseases in which rodents are directly responsible for transmission, prevention efforts must be aimed at eliminating rodents inside the home and reducing their access. Measures include covering all openings to the outside that might provide access, setting rodent traps, and eliminating potential food sources for rodents near the home. Persons engaged in outdoor activities such as camping or hiking should take precautions to avoid exposure to rodent infested areas and potentially infectious material.

Shellfish Control

Shellfish harvested from sewage-polluted waters may be infectious reservoirs of enteric diseases, notably cholera, vibrioses, and infectious hepatitis. Sanitary regulations for **shellfish control** in the shellfish industry are provided and enforced by local and state health departments, with the cooperation of the Public Health Service.

CONTROL OF ARTHROPOD VECTORS

Diseases transmitted solely by **arthropod vectors** may be controlled by eliminating the insect reservoir or preventing its human contacts. Principal members of this group of diseases are yellow fever, epidemic typhus, and malaria (see group 3 in table 8.2). Since human beings and the insect host are the only reservoirs for these diseases, significant control has been achieved by insect eradication programs coupled with immunization or chemoprophylaxis.

Diseases transmitted by arthropods from several animal reservoirs are much more difficult to control. Complete eradication of the vector would be required to eliminate them. Widely based arthropodborne infections include viral encephalitis, many rickettsial diseases, leishmaniasis and trypanosomiasis (both protozoan), filariasis (the tissue roundworm), and Lyme disease and bubonic plague (both bacterial). Arthropod control and protection of human beings from insect contacts are important, but other means of prevention must also be found.

Those diseases in which insects serve only as mechanical vectors do not depend on this means of transmission alone. Flies and other nonparasitic insects are often incriminated in transmission of poliomyelitis, salmonellosis, and other bacterial infections, but these diseases may be transmitted by other routes as well. Insect control therefore would not lead to elimination of these diseases, although it may help to reduce their incidence.

Control of Insects

The most effective methods of eradicating insects are those that prevent breeding. Mosquitoes are among the most important vectors of infectious diseases and the most usual targets of **insect control** programs. They breed in stagnant water: in swamps, still pools, unused rain barrels, or any forgotten container that collects and holds water for long periods (fig. 10.7). Mosquitoes can be eradicated or well controlled only if all these areas are found and drained or treated. Oily films spread on the surface of water prevent larvae from getting oxygen, or insecticides may be used, but draining is the most effective method of control.

Sanitary disposal of garbage and sewage and protection of manure or compost remove some of the breeding places of flies. These and other arthropods, such as lice, bedbugs, and mites, that breed in and around human dwellings are best controlled by strict sanitation and effective personal hygiene, reinforced by use of chemical insecticides.

FIGURE 10.7 Health workers engaged in mosquito control. In large parts of Africa, present methods of vector control are inadequate to check the spread of malaria.
World Health Organization. Photo by J. Mohr.

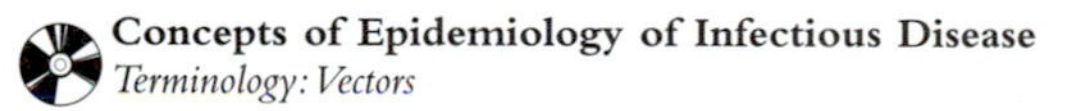

Concepts of Epidemiology of Infectious Disease
Terminology: Vectors

Chemical compounds may be used in sprays that kill insects on contact, but these ordinarily are applied only to limited areas. Residual insecticides are more effective in eradication because they can be applied to many surfaces where insects light or crawl. An unfortunate ecological side effect of the use of residual insecticides is their increasing concentration (biomagnification) to potentially toxic levels in the tissues of plants, animals, and, eventually, humans. Hydrocarbon compounds (chlordane, dieldrin) and organophosphates (malathion) have been used, but the industry continues to search for new compounds that not only are nontoxic, but to which insects have not developed resistance.

Protection of Human Beings from Insects

Control of insectborne diseases for which humans are a reservoir includes careful screening of known infectious cases from the vector. All possible precautions must be taken during the infective period of malaria for example, to screen patients from mosquitoes. In the case of epidemic typhus, patients must be deloused and prevented from acquiring further infestation from others. Personal hygiene and cleanliness of clothing or bedding are effective in controlling lice and other ectoparasites. Protective clothing is indicated when exposure to insects such as ticks or mites is anticipated, and frequent careful examination of the body should be made following such exposure. Above all, knowledge of the important vectors of infectious diseases is critical for avoiding risk of infection.

Questions

1. List the four major environmental reservoirs that public health agencies monitor and control. How have these agencies reduced diseases arising from these reservoirs?
2. What organisms are considered an index of fecal pollution of drinking water supplies?
3. What can persons do to protect the health of their family when preparing food?
4. What is the value to the community of immunization for individuals?
5. What immunizations should be completed and up to date by the time a child enters school at age 6?
6. What precautions should an international traveler take to maintain good health?
7. What immunizations are indicated for health care professionals? Why?
8. What diseases are reported to health departments? Why?

Bibliography

Abramov, V. 1989. Food irradiation—scientist's toy or everyman's joy? *World Health* 28–29 April.

Benenson, A. S., ed. 1995. *Control of communicable diseases in man,* 16th ed. Washington, D.C.: American Public Health Association.

The birth of WHO. 1989. *World Health.* 28–29 May.

Brock, T. D. 1989. Microorganisms from smallpox to Lyme disease. *Readings from Sci. Am.* New York: W. H. Freeman.

Burgess, W. 1993. The great white plague and other epidemics: Lessons from early visiting nursing. *Home Health Care Pract.* 6 (1): 12–17.

Centers for Disease Control and Prevention. 1994. General recommendation on immunization. Recommendations of the Immunization Practices Advisory Committe (ACIP). *Morbid. Mortal. Weekly Rep.* 43 (no. RR-1).

_______ 1995. *Health information for international travel.* HHS Publication No. (CDC) 95-8280. Washington, D.C.: U.S. Department of Health and Human Services.

_______ 1995. Licensure of inactivated Hepatitis A vaccine and recommendations for use among international travelers. *Morbid. Mortal. Weekly Rep.* 44 (29): 559.

_______ 1996. Prevention of varicella. *Morbid. Mortal. Weekly Rep.* 45 (no. RR-11).

_______ 1995. Progress toward global poliomyelitis eradication 1985–1994. *Morbid. Mortal. Weekly Rep.* 44 (4): 73

_______ 1996. Summary of notifiable disease, United States 1995. *Morbid. Mortal. Weekly Rep.* 44 (53).

_______ 1996. Surveillance for waterborne disease outbreaks 1993–1994. In CDC surveillance summaries. *Morbid. Mortal. Weekly Rep.* 42 (no. SS-1).

Cohen, I. B. 1984. Florence Nightingale. *Sci. Am.* 250 (3): 128–137.

Epstein, P. R. 1995. Emerging diseases and ecosystem instability: New threats to public health. *Am. J. Public Health* 85 (2): 168–172.

Frenkel, L. D. 1990. Routine immunizations for American children in the 1990s. *Pediatr. Clin. N. Am.* 37 (3): 531–548.

Glittenberg, J. E. 1990. Problems of global control of tuberculosis. *J. Prof. Nurs.* 6 (2): 73.

Greenberg, A. E., ed. 1995. *Standard methods for examination of water and wastewater.* 19th ed. Washington, D.C.: Am. Public Health Assoc.

Hager, W. D., M. Bird, J. C. Callahan, A. L. Frank, M. David, and J. Engelberg. 1990. Promiscuity and public health. *Hosp. Prac.* 25 (9): 63–78.

Hruby, D. E. 1990. Vaccinia virus vectors: New strategies for producing recombinant vaccines. *Clin. Microbiol. Rev.* 3 (2): 153–170.

Jacobson, C., and L. J. Strausbaugh. 1990. Incidence and impact of infection in a nursing home care unit. *Am. J. Infect. Control.* 18 (3): 151–159.

Jurgrau, A. 1990. Why aren't we protecting our children? Childhood immunization. *RN* 53 (11): 30–35.

Mann, J. M., J. Chin, P. Piot, and T. Quinn. 1988. The international epidemiology of AIDS. *Sci. Am.* 259 (4): 82–89.

Mayhall, G. C. 1995. *Hospital epidemiology and infection control.* Philadelphia: Williams and Wilkins.

McFeters, G. A., ed. 1990. *Drinking water microbiology: Progress and recent developments.* New York: Springer Verlag.

Nightingale, F. 1976. *Notes on Hospitals.* New York: Gordon Press.

Otterstetter, H., and A. Flores. 1987. Water: Source of life. *World Health* 11–13 October.

Poland, G. A. 1994. Adult immunization practice: The new paradigm. *Am. J. Infect. Control* 22 (1): 39–41.

Rodts, B., and S. Meister. 1990. Home care: Infection control takes top priority. *RN* 53 (12): 59–62.

Roueche, B. 1991. *Medical detectives.* New York: Plume.

Water, water everywhere—and not a drop to drink. 1993. *Hosp. Infect. Control* 20 (8): 112.

Woodham-Smith, C. 1951. *Florence Nightingale.* New York: McGraw-Hill.

World Health Assembly. 1995. Global eradication of poliomyelitis by the year 2000. Geneva: World Health Organization, January 1995.

CHAPTER 11

Communicable Disease Control in the Hospital

OUTLINE

Microbes in Motion

This chapter covers the concepts of disease control and prevention with emphasis on precautions to be taken in the hospital setting. The following books and chapters of the Microbes in Motion CD-ROM may be a useful preview or supplemental study aid to your reading: Concepts on Epidemiology of Infectious Disease: Disease Acquisition; Transmission.

OBJECTIVES

On completing this chapter the student will be able to:

1. state the major portals of entry and exit for important infectious diseases and indicate the type of hospital care such diseases require.
2. define the following terms and give an example of a disease to which each applies: airborne precautions, contact precautions, droplet precautions, and standard precautions.

KEY WORDS

Airborne precautions
Airborne transmission
Contact precautions
Contact transmission
Droplet precautions
Droplet transmission
Standard precautions
Transmission-based precautions

General principles

The chapter 9 section titled "Application of Aseptic Techniques in Hospitals" described important isolation precautions and personnel techniques. In the hospital environment, these precautions and techniques must be used to prevent transmission of infection among patients and between health care personnel and patients. Part 2 of the text will address specific infections transmitted by the respiratory, alimentary, and parenteral routes and through the skin and mucosae. First, however, it will be helpful to review the portals of entry and exit of the microorganisms, which affect the methods used for infection control. Refer to chapters 5 and 8 for greater detail regarding microbial entry routes and sources, and routes and patterns of infectious diseases.

The skyrocketing costs of medical care have led to a change in the way hospital health care costs are reimbursed. Regardless of whether patients have private insurance, are enrolled in a managed care program, or are covered by federal- or state-funded programs, only those who need the specialized treatment or procedures unique to the inpatient setting are admitted to the hospital. Once admitted, patients are discharged at the earliest possible time. While this trend may seem unjustified or unreasonable to some, it has benefits for many patients. Early return to the ambulatory state avoids complications encountered by the bedridden (such as pneumonia and bedsores). More importantly, prompt discharge removes the patient from the infection-laden hospital environment! Nonetheless, more patients with infectious diseases are now treated as outpatients and others may not show signs of an infection they acquired in the hospital until after discharge. As a result, it must be remembered that the basic procedures for preventing infection apply to any patient with an infectious disease, whether hospitalized or not.

Epidemiology and control of infectious diseases

Spread of infection within a hospital requires a source of the infection, a susceptible host, and a route of transmission. The source may be a patient, hospital personnel, or even a visitor. The source may not be obvious because he or she may not appear visibly ill but rather be in the incubation stage of disease or be an asymptomatic carrier of an organism that, once transmitted, can produce disease in a susceptible host. Hospitalized patients are especially vulnerable to contracting infections because of their underlying disease, drugs that they may be receiving such as antimicrobial and immunosuppressive agents, and procedures that allow the important skin and mucous membrane barriers to be breached (surgery or indwelling urinary or intravenous catheters, see chap. 21). Transmission of pathogens can occur by five main routes but only three—*contact, droplet,* and *airborne*,—are important in nosocomial (hospital-acquired) infections. The other two transmission routes, known as *common vehicle* (for example, an outbreak of food poisoning caused by contaminated food and drink) and *vectorborne* (primarily by arthropods and rodents), are more important in community-acquired infections but may also be important in hospitals in underdeveloped countries.

Transmission routes for nosocomial infections

Contact transmission is the most important and frequent route of infection transfer in hospitals. This route may be *direct,* in which microorganisms are transferred by bodily contact between an infected and uninfected person, such as during bathing or turning a patient. *Indirect* contact occurs when infectious agents are transmitted by inanimate, intermediate objects such as contaminated instruments or needles, or most importantly, by contaminated hands. Proper handwashing and changing of gloves between patient contacts are key in preventing infection transfer by contact transmission.

Droplet transmission occurs when the activities of an infected person (the source) cause contaminated respiratory secretions to be expelled into the air. When the person coughs, sneezes, talks, or is subjected to procedures such as suctioning and bronchoscopy, infected respiratory droplets are propelled short distances through the air, usually three feet or less. When they land on upper respiratory mucous membranes or conjunctiva of a susceptible host, infection may result.

Airborne transmission involves dissemination of small dried particles (usually 5 μm or less) of dust or droplet nuclei that contain infectious microorganisms. In contrast to droplet transmission in which particles travel only short distances, in airborne transmission the evaporated particles may remain suspended in the air for long periods and can be widely dispersed on air currents. When they are inhaled by susceptible persons, their small size allows them to reach the deepest recesses of the lung from where they can initiate infection. Because these droplets are disseminated so widely, special air handling and ventilation are required to contain infection transmission. The organisms that cause tuberculosis, measles, and chicken pox are examples of those transmitted in this manner.

The infection control measures in hospitals discussed in chapter 9 include proper handwashing technique and use of gowns, gloves, and masks. In addition, the Centers for Disease Control and Prevention (CDC) has developed two types of guidelines for isolation precautions to be used within hospitals. These include **standard precautions,** (formerly known as universal precautions), which are intended for use with every hospitalized patient, and **transmission-based precautions,** to be used for patients with known or suspected infections caused by pathogens transmitted by the three routes described above.

Standard precautions

In 1985, the CDC developed a strategy known as universal precautions, primarily to address concerns about transmission of the human immunodeficiency virus (HIV) but also applicable to the hepatitis B virus (HBV). The concept of universal precautions stressed that *"all patients should be assumed to be infectious for HIV and other blood-borne pathogens."* Thus, in the hospital or other health care setting, these precautions were followed when workers were exposed to blood and certain other body fluids especially if visibly contaminated with blood. This strategy has now been replaced by one known as "standard precautions," which also applies to **all** patients receiving care in hospitals, regardless of their diagnosis or whether they are known or presumed to have an infection.

Standard precautions apply not only to blood but to all body fluids, secretions, and excretions *except sweat,* regardless of whether they contain visible blood; nonintact skin; and mucous membranes. These precautions are designed to reduce infectious transmission from both recognized and unrecognized sources of infection in the hospital. They include the following: (1) washing hands after touching blood, body fluids, and contaminated items whether or not gloves are worn and washing hands immediately after removing gloves, between patient contacts, and whenever necessary to avoid transfer of microorganisms from one site to another on a single patient and to other patients and environments; (2) wearing gloves when touching blood, body fluids, secretions, excretions, and contaminated items; removing gloves promptly after use and washing hands immediately to avoid transfer of microorganisms to other patients or environments; (3) wearing a mask and eye protection or a face shield to protect mucous membranes of the eyes, nose, and mouth during patient-care activities that are likely to generate splashes or sprays of blood, body fluids, secretions, and excretions; (4) wearing a clean gown to protect skin and to prevent soiling of clothing during activities that are likely to generate splashes or sprays of blood, body fluids, secretions, or excretions; (5) handling used patient-care equipment soiled with blood, body fluids, secretions, and excretions in a manner that prevents skin and mucous membrane exposures, contamination of clothing, and transfer of microorganisms to other patients and the environment; and proper cleaning or disposal of such equipment; (6) ensuring that the hospital has adequate procedures for routine care, cleaning, and disinfection of environmental surfaces, furniture, and equipment; (7) handling, transporting, and processing used linen soiled with blood, body fluids, secretions, and excretions in a manner that prevents skin and mucous membrane exposures and contamination of clothing and that avoids transfer of microorganisms to other patients and environment. In addition, care must be taken to prevent injuries when needles, scalpels, and other sharp instruments are used, cleaned, or disposed of. Recapping needles is the most common cause of needle-stick injuries and should be avoided. Used disposable syringes, needles, scalpel blades, and other sharp items should be placed in puncture-resistant containers for disposal or reprocessing. Finally, patients who contaminate the environment or cannot maintain appropriate hygiene should be placed in a private room whenever possible. If a private room is not available, it might be permissible to place them in a room with a person who has the same type of infection, a procedure know as cohorting.

Transmission-Based Precautions

Transmission-based precautions are used *in addition to* standard precautions whenever patients are known or suspected to be infected by highly transmissible or epidemiologically important pathogens. The diseases to which each applies are listed in table 11.1.

TABLE 11.1 EPIDEMIOLOGY AND CONTROL OF INFECTIOUS DISEASES

Portal of Entry	Portal of Exit	Type of Precaution[a]	Diseases (chapter reference)
Acquired through Respiratory Tract			
Inhalation through mouth, nose; via contaminated fingers, apparatus for assisted ventilation	Respiratory droplets from saliva, sputum, throat secretions	Airborne	Measles (13) Tuberculosis (pulmonary or laryngeal) (12) Zoster (disseminated) (13)
		Contact	Diphtheria (cutaneous) (12) Pneumonia, viral, in infants and young children (13) Rubella (congenital) (13) Varicella (chickenpox) (13)[b]
		Droplet	Acute viral infections in infants and young children (13) Diphtheria, pharyngeal (12) Epiglottitis, *Haemophilus influenzae* (12) Erythema infectiosum (13)

TABLE 11.1 *Continued*

Portal of Entry	Portal of Exit	Type of Precaution[a]	Diseases (chapter reference)
			Hantavirus pulmonary syndrome (13) Influenza (13) Meningitis, bacterial (15) Meningococcemia (15) Mumps (13) Pertussis (whooping cough) (12) Plague, pneumonic (12) Pneumonia (*Haemophilus influenzae* in children) (12) Rubella (13) Streptococcal pharyngitis (12)
Acquired through Alimentary Tract			
In contaminated food, water, hands, soil; other contaminated objects placed in mouth	Infective human feces	Contact	Diarrhea with suspected infectious etiology (16–18) Enterocolitis, *Clostridium difficile* (16) Enteroviral infections (children) (17) Herpangina (17) Pleurodynia (13, 17)
		Standard	Amebiasis (amebic dysentery) (18) Cholera (16) Enterocolitis, *Staphylococcus aureus* (16) Enteroviral infections (adults) (17) Gastroenteritis caused by[c] *Campylobacter* spp. (16) *Cryptosporidium* spp. (18) *Escherichia coli* (16) *Giardia lamblia* (18) *Salmonella* spp. (16) *Shigella* spp. (16) *Vibrio parahaemolyticus* (16) Viruses, including Norwalk agent, rotavirus (17) *Yersinia enterocolitica* (16) Hepatitis, viral, type A (17)[c] Necrotizing enterocolitis (16) Poliomyelitis (17) Typhoid fever (*Salmonella typhi*) (16)
Acquired through Skin and Mucosa			
Through minor breaks in skin by direct contact with human or animal source of infection; through deep and penetrating injury such as arthropodborne disease, or accidental or surgical injury	Infectious lesions of skin and mucous membranes	Contact	Furunculosis, staphylococcal in newborns (19) Herpes simplex, disseminated (19) Impetigo (19) Skin, burn, and wound infection, major (19, 23)
		Standard	Abscesses, skin, burn, and wound infection, minor (19, 23) Acquired immune deficiency syndrome (AIDS) (19) Gonococcal conjunctivitis in newborns (19) Leptospirosis (20) Scalded skin syndrome (19) Syphilis, all stages (19)
Acquired Parenterally			
Introduced directly into tissues as a result of injury to normal skin and mucosal barriers; usually a result of medical or surgical techniques, arthropod bites, accidental injuries	Not directly transmissible among humans under usual conditions	Standard	Arthropodborne viral fevers (22) Babesiosis (22) Endometritis, group A *Streptococcus* (21) Hepatitis, viral, types B and C (21) Malaria (22) Rat-bite fever (22) Relapsing fever (22)

[a]Standard precautions are always used *in addition to* airborne, droplet, and contact precautions. Multidrug resistant organisms from any source require contact precautions.
[b]Requires airborne as well as contact precautions
[c]Contact precautions required for diapered or incontinent children under 6 years old

Airborne precautions, designed to reduce the risk of infectious transmission on airborne droplet nuclei or dust particles, require a private room with special air handling and ventilation; and the door must be kept closed. Persons entering the room are required to wear masks if the patient has tuberculosis, but not if the patient has measles or chickenpox and the person entering is immune. Patient transport should be limited but when it is required, the patient should wear a mask. Cohorting is allowed if a private room is not available.

Droplet precautions, designed to reduce the risk of droplet transmission, require a private room or cohorting, wearing a mask when working within three feet of the patient, and limiting movement and transport of the patient or masking patients who require transport.

Contact precautions are used for patients with infections that can be transmitted by either hand- or skin-to-skin contact with the patient or indirect contact with environmental surfaces or patient-care items. Contact precautions require a private room or cohorting, wearing gloves and washing hands after glove removal, ensuring ungloved, washed hands do not touch contaminated items before caring for another patient, wearing a gown if soiling is likely, limiting movement and transport of the patient or taking precautions to minimize the risk of transmission of infectious material if the patient must be moved, and avoiding sharing of equipment among patients.

Table 11.1 lists the common portals of entry and exit for infectious disease acquired by different routes, and the category of precautions recommended by the CDC.

QUESTIONS

1. Why are hospitalized patients vulnerable to infections?
2. What routes of transmission are most common in nosocomial infections?
3. How do direct and indirect transmission differ? What procedures are necessary to prevent infection transfer by contact transmission?
4. How does droplet transmission result in infection?
5. What procedures are required to prevent airborne transmission of droplet nuclei?
6. List two types of precautions recommended by the CDC for hospitals? When is each indicated?
7. List the type of precaution indicated for the following diseases: tuberculosis; viral pneumonia; diarrhea with infectious etiology; hepatitis A.

BIBLIOGRAPHY

Benenson, A. S., ed. 1995. *Control of communicable disease in man,* 16th ed. Washington, D.C.: American Public Health Association.

Bennett, J. and V. Brachman, eds. 1992. *Hospital infections* 3d. ed. Boston: Little, Brown.

Crow, S. 1989. *Asepsis: The right touch.* Something old is now new. Bossier City, La.: Everett Pub.

Garner, J. 1996. Guidelines for isolation precautions in hospitals. *Infect. Control Hosp. Epidemiol.* 17 (1): 53–80.

Gerberding, J. L., F. R. Lewis, and W. P. Schecter. 1995. Are universal precautions realistic? pp. 1091–1104. In Prevention of transmission of bloodborne pathogens edited by R. S. Rhodes and D. M. Bell. *Surg. Clin. N. Am.* Philadelphia: Saunders.

Goldmann, D., and E. Larson. 1992. Handwashing and nosocomial infections. *N. Engl. J. Med.* 327 (2): 120–122.

Larson, E. 1985. Handwashing and skin: Physiologic and bacteriologic aspects. *Infect. Control* 6 (1): 11–23.

Lowbury, E. J. 1992. *Control of hospital infection.* 3d ed. New York: Chapman & Hall.

Mandell, G. L., G. R. Douglas, and R. Dolin. 1995. *Principles and practice of infectious diseases.* 4th ed. New York: Churchill Livingstone.

Olsen, R. J., P. Lynch, M. B. Coyle, J. Cummings, T. Bokete, and W. E. Stamm. 1993. Examination gloves as barriers to hand contamination in clinical practice. *JAMA* 270 (3): 350–353.

Perceval, A. 1993. Wash hands, disinfect hands, or don't touch? Which, when, and why? *Infect. Control Hosp. Epidemiol.* 11 (5): 273–275.

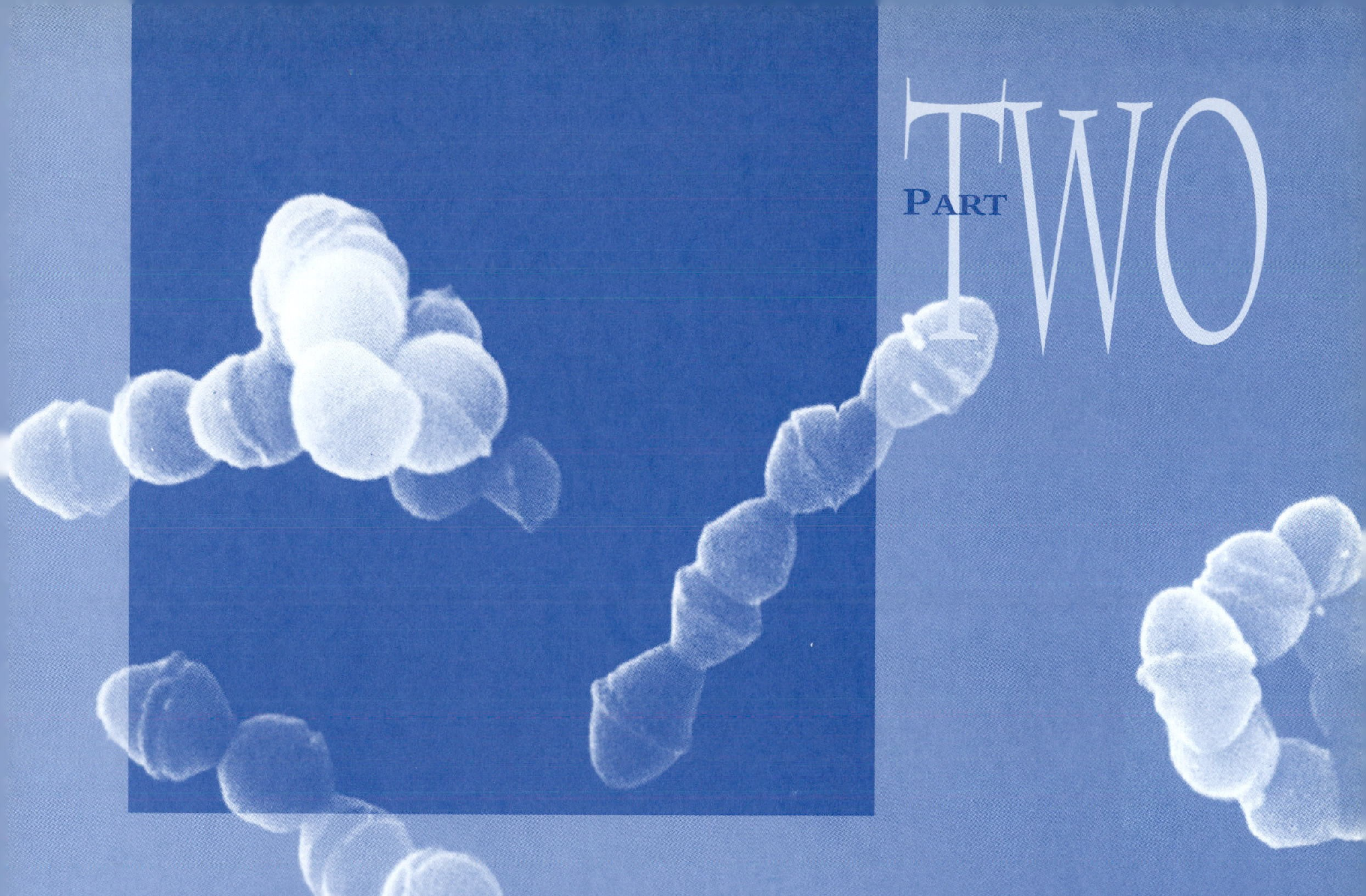

PART TWO

MICROBIAL DISEASES AND THEIR EPIDEMIOLOGY

INTRODUCTION TO PART 2

The preceding sections of this text have dealt with basic concepts concerning the nature of microbial diseases. The contributions of pathogenic microorganisms and the human host to the pathogenesis of infectious disease have been discussed, and epidemiologic principles involved in treatment, control, and prevention of infection have been reviewed.

In part 2 specific infectious diseases of human importance are presented in a manner designed to emphasize application of epidemiologic principles to their management. The sequence of approach is based on the practical considerations that confront those involved in patient care with each case of infectious disease: recognition of its important clinical features, establishment of the laboratory diagnosis, and institution of appropriate control measures.

Each disease is discussed with respect to its clinical appearance and pathogenesis, the availability of laboratory methods for confirming the diagnosis (what specimens are of value; how and when they should be collected), and the most important aspects of its epidemiology. Necessary control measures are indicated in each case.

Some infectious diseases are easily spread by several routes of transmission and entry; others are self-limited but nonetheless can spread widely among the patient's contacts; some are restricted by developmental and physiologic requirements of the infectious microorganism as well as by opportunities it may have for entry to or exit from the human body; and still others may be acquired by some fortuitous means but are not transmissible under ordinary circumstances. Measures for controlling further spread of infection must apply to the demands of a specific situation and be based not only on some knowledge of the infectious microorganism and how it is transferred but also on the probability of exposure of susceptible human hosts. Diseases encountered in the private home, or in rural situations, sometimes have a very different epidemiologic potential from those occurring in crowded urban districts or in hospitals.

With these considerations in mind, discussions of specific infectious diseases have been grouped primarily on the basis of the most common route of entry to the human host available to the microbial agent. They are arranged in three sections (IV through VI) dealing, respectively, with entry through the respiratory tract, the gastrointestinal route, and skin and mucosal surfaces or parenteral routes. Repeated emphasis is placed on the fact that the site of entry of an infecting organism is not always the focus of developing infection, nor is it necessarily associated with the microorganism's route of exit and transmission. When appropriate, within each section diseases are grouped in chapters according to the nature of their causative agents (e.g., bacteria, viruses, fungi, animal parasites).

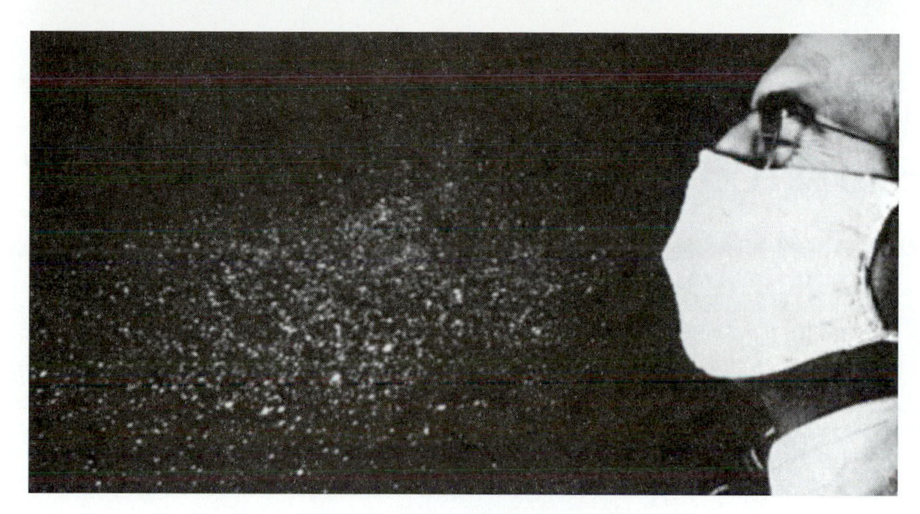

Effect of a mask on the aerosol from a sneeze.

Jenninson, Marshall W. "The Dynamics of Sneezing—Studies in High Speed Photography" *Science Monthly,* 52: 24–33, 1941.

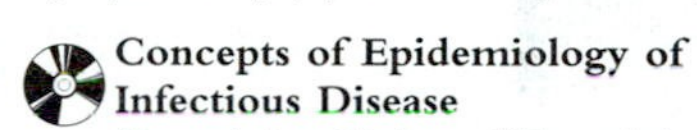

Concepts of Epidemiology of Infectious Disease
Transmission: Horizontal Transmission—Airborne

Infectious Diseases Acquired through Inhalation

12. Bacterial Diseases Acquired through the Respiratory Tract

13 Viral Diseases Acquired through the Respiratory Tract

14. Systemic and Subcutaneous Mycoses: Fungal Diseases Acquired through Respiratory, Parenteral, and Endogenous Routes

15. Nonrespiratory Diseases Acquired through the Respiratory Tract

CHAPTER 12

Bacterial Diseases Acquired through the Respiratory Tract

OUTLINE

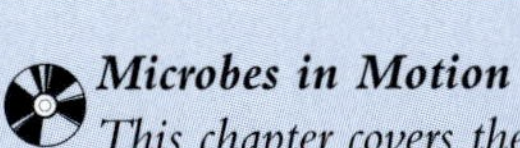

Microbes in Motion

This chapter covers the important bacterial diseases acquired through the respiratory tract including the clinical characteristics, causative organisms, laboratory diagnosis, epidemiology, and control. The following books and chapters of the Microbes in Motion CD-ROM may be a useful preview or supplemental study aid to your reading: Gram-positive Organisms in Human Hosts: Gram-positive Cocci-Streptococci; Gram-positive Cocci-Enterococci. Gram-negative Organisms in Human Hosts: Gram-negative Bacilli-Aerobes. Miscellaneous Bacteria in Human Hosts: Mycoplasma; Mycobacteria. Microbial Pathogenesis in Infectious Diseases: Toxins. Vaccines: Vaccine Targets.

OBJECTIVES

On completing this chapter the student will be able to:

1. describe the most important bacterial diseases acquired through the respiratory tract.
2. cite the clinical characteristics of each disease.
3. name the causative organism(s) of each disease.
4. indicate the specimens required for laboratory diagnosis.
5. describe important features of the epidemiology of each disease.
6. list methods for controlling active infection.
7. describe methods for preventing the disease.

KEY WORDS

Arcanobacterium haemolyticum (ar-**CANE**-oh-back-tear-i-umhemo-**LIT**-i-cum)
Antistreptolysin (anti-strep-toe-**LICE**-in)
Bacterial pneumonia
BCG vaccination
Beta hemolysis (**BAY**-ta he-**MOLL**-i-sis)
Bronchopneumonia
Chlamydial (clam-**ID**-ial) pneumonia
Corynebacterium (co-**RHINE**-ee-back-**TIER**-i-um)
Coxiella (cox-i-**ELLA**)
Deoxyribonuclease (dee-oxy-ribe-o-**NUKE**-lee-ase)
Diphtheria (diff-**THEER**-ia)
Enteric bacilli (ent-**TEAR**-ic ba-**SILL**-eye)
Enterococcus (ent-er-o-**COCK**-us)
Erythrogenic (e-ryth-ro-**GEN**-ic) toxin
Fusospirochetosis (fuse-oh-spire-oh-key-**TOE**-sis)
Ghon (**GO**-n) complex
Glomerulonephritis (gla-**MER**-u-lo-neph-**RITE**-is)
Gonococcal pharyngitis
Haemophilus (he-**MOFF**-ill-us) *influenzae*
Hemolytic (hemo-**LIT**-ic) toxin
Hyaluronic (high-al-ure-**ON**-ic) acid
Hyaluronidase (high-al-ure-**ON**-i-dase)
Infant pneumonitis
Klebsiella (kleb-si-**ELLA**)
Lancefield serologic grouping
Laryngitis (lahr-in **JITE-is)**
Legionella (legion-**ELLA**)
Legionnaires disease
Leukocidin (loo-co-**SIDE**-in)
Lipoteichoic (lip-o-tie-**CO**-ic) acid
Lobar (**LOW**-bar) pneumonia
M protein
Mantoux (man-**TOO**) test
MOTT (mycobacteria other than tuberculosis)
Mycobacterium (my-ko-back-**TIER**-i-um)
Mycoplasma (my-ko-**PLAZ**-ma)
Nocardia (no-**CARD**-ia)
Nocardiosis (no-card-i-**OH**-sis)
Otitis (oh-**TITE**-is) media
Peptostreptococcus (pep-toe-strep-toe-**COCK**-us)
Pertussis (per-T**USS**-is)
Pharyngitis (fahr-in-**JITE**-is)
Pneumococcal (new-mo-**COCK**-al) pneumonia
Pneumococci (new-mo-**COCK**-si)
PPD (purified protein derivative)
Primary atypical pneumonia
Primary pneumonic plague
Pseudomonas (soo-do-**MO**-nas)
Psittacosis (sit-a-**CO**-sis)
Pyrogenic (pie-row-**GEN**-ic) exotoxin
Q fever
Quellung (**KWELL**-ung) reaction
Rheumatic (room-**AT**-ic) fever
Scarlet fever
Sinusitis (sign-us-**ITE**-us)
Staphylococcus (staff-i-lo-**COCK**-us) *aureus*
Streptococcal (strep-toe-**COCK**-al) sore throat
Streptococcal virulence factors
Streptococci (strep-toe-**COX**-eye)
Streptococcus (strep-toe-**COCK**-us)
Streptokinase (strep-toe-**KINE**-ase)
Streptolysin (strep-toe-**LICE**-in)
Tonsillitis
Toxic shock syndrome (TSS)
Tubercle bacilli
Tuberculosis
TWAR agent
URI (upper respiratory infection)
Vincent's angina
Viridans (**VEER**-i-dans) streptococci
Whooping cough

INTRODUCTION

Bacterial diseases acquired through the respiratory tract include some of the most important and serious human infections. The respiratory route of access makes their control difficult and assures their widespread distribution. Thus, low-grade but debilitating infections with some of these agents are very common, and the threat of epidemics of acute infections must be continuously controlled.

The diseases described in this chapter are diverse in their nature and symptomatology, but bacterial infections share a few common features. Some pathogenic bacteria are called *pyogenic* because the body responds to them with the defensive production of pus. Purulent exudates in the tissues are one of the most frequent signs of bacterial infection, whether it be pharyngitis, pneumonia, meningitis, or a focal abscess. When infection is severe, a general leukocytic response occurs, with elevated white blood cell counts (usually at least 15,000 cells per cubic millimeter). High fever (102° F [38.9° C] and above) and pain at the localizing site of infection (acute sore throat, chest pain, headache) are also common signs of developing bacterial disease. In some instances, disseminating bacterial toxins cause distinct signs and symptoms as in diphtheria and some streptococcal infections.

Many acute bacterial diseases that were once accompanied by crippling or fatal effects (meningococcal meningitis, scarlet fever) can now be controlled by prompt, effective antimicrobial therapy. Early clinical recognition and rapid laboratory confirmation of the diagnosis are essential for choosing appropriate treatment and for controlling active infection.

Effective vaccines are not available for many diseases, thus for these, prompt antimicrobial therapy of the patient remains the best control measure. For diphtheria and whooping cough, the threat of epidemic disease in the United States has been eliminated by mass immunization, begun in childhood. Such diseases can be prevented only through continued immunization programs for susceptible populations.

UPPER RESPIRATORY INFECTIONS (URI)

The Clinical Disease

The symptoms, pathology, and etiology of many common bacterial **upper respiratory infections (URI)** are superficial and nonspecific. These infections may be acute, chronic, or recurrent. Febrile reactions are mild or do not occur. The basic picture is one of inflammation and hyperemia (congestion). Terms such as **pharyngitis, laryngitis, tonsillitis,** and **sinusitis** indicate the regional mucous membranes involved. **Otitis media,** or inflammation of the middle ear, results from extension of pharyngeal infection through the eustachian canal. Mucosal injury following viral infections, inhalation of toxic vapors or smoke, excessive dryness, or pollen or dust allergies may permit certain members of the indigenous bacterial flora to grow in the compromised tissue. The result of these secondary infections may be the formation of purulent or serous exudates, localizing abscesses, and hyperplasia of tissues chronically involved (adenitis, adenoiditis, follicular tonsillitis). Persistent infections are frequently marked by hypersensitivity immune responses to bacterial cells or their products. More specific symptoms and pathology may be produced by truly pathogenic invaders.

The Organisms

Infections of the upper respiratory mucosa may be caused by a variety of microorganisms. The most important agent of bacterial pharyngitis is the group A beta-hemolytic ***Streptococcus.*** Other less common causes of bacterial pharyngitis include ***Arcanobacterium haemolyticum*** and ***Neisseria gonorrhoeae.*** Although other potential pathogens such as ***Haemophilus influenzae,*** **pneumococci,** ***Staphylococcus aureus,*** gram-negative **enteric bacilli,** and ***Pseudomonas*** species are often isolated from pharyngeal cultures, these bacteria are thought not to play a role in infection at that site. Rather they are involved in other upper respiratory infections such as otitis and sinusitis.

LABORATORY DIAGNOSIS

Smears and cultures may be unrewarding in distinguishing normal microbial flora of upper respiratory mucosa from those responsible for nonspecific infections. However, cultures taken with thin swabs aimed at the principal lesions may reveal the causative microorganisms with some precision.

Care must be taken to avoid contaminating the swab with saliva or the mucous secretions of uninvolved tissues. The collected material is usually placed immediately in supportive "transport" medium, or is streaked out on a blood agar plate, pending its delivery to the laboratory. Swabs should never be replaced in empty tubes where they may dry out before laboratory culture can be initiated because fastidious species may die, and even the hardier organisms (staphylococci or enteric bacilli) may be greatly reduced in numbers so that they cannot be detected or their significance in culture is missed. In the laboratory the swabs are streaked on appropriate agar plates and incubated. Isolated colonies appearing on incubated plates are identified and, if indicated, tested in pure culture for antimicrobial susceptibility. In general, pneumococci and other streptococci appearing on such plates are not routinely tested with antimicrobial agents. Pneumococci are usually susceptible to penicillin. However, the incidence of penicillin-resistant pneumococci is increasing. Most other streptococci are susceptible to penicillin but in any case their association with clinical disease should first be confirmed, particularly if they are not beta-hemolytic strains (see later section, Streptococcal Diseases).

Serologic tests are not done.

Epidemiology

I. Communicability of Infection

A. *Reservoir, Sources, and Transmission Routes* Human beings are the reservoir of these infections, and the source may be endogenous. When person-to-person transmission is involved, the route is by airborne droplets or direct or indirect contact with secretions from the nose and throat of infected persons.

B. *Incubation Period* Unknown for endogenous infection; probably very short (1 to 2 days) when transmissible.

C. *Communicable Period* During time of active inflammation, excessive mucus production, or drainage of purulent exudates from nose and throat.

D. *Immunity* Specific immune responses may not occur naturally or be protective, but they may be produced against the specific antigens of some invading strains. The nonspecific resistance of normal, healthy tissues appears to constitute a major defense mechanism. Local injury to mucosal surfaces predisposes to infection by endogenous microorganisms and increases susceptibility to exogenous infections, should exposure occur.

II. Control of Active Infection

A. *Transmission-based Precautions* Not warranted, but patient should avoid directly exposing those who may be unusually susceptible to infection (infants, aged people, surgical or obstetric patients).

B. *Other Precautions* Infected persons should practice careful personal hygiene: cover coughs and sneezes, collect nasal discharges in disposable tissues and discard these in closed paper or plastic bags for incineration, wash hands frequently and carefully.

C. *Treatment* Nonspecific supportive measures designed to remove surface accumulations of exudates, promote blood supply, and reduce allergic reactivity of tissues, if any. Specific antimicrobial agents are sometimes effective but should not be used until adequate clinical observation and laboratory studies have demonstrated their need.

D. *Control of Contacts* Case finding: Indicated only when risk of infection to exposed persons is great, as in hospital nurseries, operating rooms, or intensive-care units. Infectious personnel working in these areas should be removed from this duty until their infections are no longer communicable.

III. Prevention

Immunization Limited value only.

STREPTOCOCCAL DISEASES

GENERAL COMMENTS

The **streptococci** are a large and diverse family of microorganisms, among which several groups are of great clinical importance. Collectively, they are associated with a wide variety of infectious diseases ranging from superficial infections of the skin or mucous membranes to serious systemic diseases often characterized by the toxic effects of the cellular or extracellular products of the infecting strain. The classification of streptococci on the basis of their properties and the diseases they produce is discussed here.

The important routes of entry for streptococci are the respiratory tract, the skin and superficial membranes, and traumatized tissues. Streptococcal diseases associated with respiratory entry include streptococcal sore throat (pharyngitis-tonsillitis) and scarlet fever. These and the important sequelae of infections caused by group A streptococci, notably rheumatic fever and glomerulonephritis, are described in this section. Certain group A streptococci have earned the name "flesh-eating bacteria" because they cause a severe invasive skin infection that is discussed in chapter 19. Endocarditis, frequently associated with streptococci found among the indigenous oral flora, is discussed in chapter 15 as a nonrespiratory infection.

Two other specific streptococcal diseases with a different entry route and epidemiology are erysipelas and postpartum endometritis. These are discussed in chapters 19 and 21, respectively, the former being a skin contact disease, the latter of iatrogenic (caused by medical treatment) origin. In addition, under appropriate circumstances, streptococci may be responsible for diseases of the bone (osteomyelitis), ear (mastoiditis), lymph nodes (lymphadenitis), and peritoneal cavity (peritonitis). In skin diseases such as impetigo, or in wound infections, hemolytic streptococci are often found together with other bacterial pathogens such as staphylococci.

Streptococcal Virulence

Little is known of the pathogenicity and virulence of streptococci except for group A. For group A streptococci, such properties are related to cellular and extracellular substances that have a toxic effect on human cells (see also chap. 5).

Capsular and Cell Wall Substances

The capsule of group A streptococci is an envelope of **hyaluronic acid.** This mucoid substance is chemically identical to the hyaluronic acid of human connective tissue. Therefore, it is treated as "self" by the immune system during infection and does not elicit an antibody response. Since the capsule is not coated with specific antibodies, phagocytosis is not enhanced by opsonization as it is with the pneumococcus, for example (see later section, Pneumococcal Pneumonia). The chemical nature of the capsule thus helps to protect the organism against the accelerated phagocytosis that occurs when opsonins coat bacterial surfaces.

Beneath the capsule and extending into it as minute projections or fimbriae is a component called **M protein.** This antiphagocytic substance prevents ingestion of the organism by phagocytes. This property makes the M protein an important **streptococcal virulence factor.** It is also antigenic, conferring type specificity on group A strains.

Another important virulence factor found in the group A streptococcal wall is a substance known as **lipoteichoic**

acid. This material helps the organisms initiate infection by binding them to the surface of human epithelial cells so that they are not washed away by body secretions such as saliva.

The cell wall constituent of greatest serologic significance, especially among hemolytic strains, is a carbohydrate. The serologic classification of streptococci is based on the antigenic specificity of this cell wall carbohydrate, as first described by Dr. Rebecca Lancefield in the early 1930s. The **Lancefield serologic grouping** method recognizes antigenic differences in the carbohydrates of streptococcal strains separating them into groups designated by letters A through V. Further classification of group A strains into numbered types is then based on differences in their M protein antigens. Since the M protein is the important antiphagocytic factor, immunity to group A streptococcal infections depends on anti-M antibodies, and thus is type specific. For example, an individual host may be resistant to a strain of type 12 but not to type 4 streptococci.

Extracellular Products

The **hemolytic toxins** (**streptolysins**) of group A streptococci distinguish them from other species found more usually among commensal flora of the upper respiratory tract. These streptolysins completely destroy red blood cells in culture media, a property referred to as **beta hemolysis.** (Strains of **viridans streptococci,** which occur most frequently as normal flora in the throat, effect an incomplete hemolysis of blood cells termed *alpha* hemolysis. It is often accompanied by a green discoloration of the zone of incomplete lysis surrounding viridans colonies. Nonhemolytic streptococci are referred to as *gamma* strains.)

Streptolysin O, one type of group A streptococcal lysin, is antigenic. Specific antibodies produced in response to its presence in the body during active streptococcal infection are called **antistreptolysins.** Laboratory detection of antistreptolysin O in a patient's serum indicates present or past infection with group A beta-hemolytic streptococci.

In addition to hemolytic toxins, virulent streptococci often produce the enzymes hyaluronidase, deoxyribonuclease, and streptokinase, as well as a leukocidin. **Hyaluronidase** production accounts for some of the invasive properties of these organisms, assisting in their spread through local tissues by breaking down the connective-tissue matrix; thus, it is called a *spreading factor.* **Deoxyribonuclease** helps break down host cell DNA. **Streptokinase** activity assists invasive spread because its defibrinating effect dissolves clots that may entrap organisms. **Leukocidin** destroys phagocytes.

A streptococcal **pyrogenic exotoxin** (formerly known as **erythrogenic toxin**) produced by some strains of beta-hemolytic streptococci is responsible for the skin rash of scarlet fever. The soluble toxin is distributed through the body by the bloodstream from a site of streptococcal infection in the throat or elsewhere. Its toxicity for blood vessels and cells in the skin results in the characteristic diffuse reddening of scarlatinal rash. This substance is antigenic and stimulates formation of an antibody capable of neutralizing its effects on dermal tissues. Only streptococcal strains that are infected by specific bacterial viruses produce streptococcal pyrogenic exotoxin (see chap. 3, Bacteriophages). Streptococcal pyrogenic exotoxin is also thought to be responsible for initiating other severe complications of group A streptococcus infection as discussed in chapter 19.

Distribution of Serologic Groups of Streptococci

Most strains of pathogenic streptococci belong to group A, whereas both human and animal pathogenicity is associated with group B. In recent years group B strains have been increasingly detected as colonizers of the female genital tract. Their presence at that site is a potential hazard to infants born of colonized mothers because they may cause serious, often fatal, neonatal group B streptococcal meningitis or bloodstream infections. In some cases, the infected mothers have also developed serious postpartum group B disease (see chap. 21).

Group C strains are largely restricted to animal hosts but have been isolated from the pharynx of humans with pharyngitis as well as from occasional patients with abscesses and pneumonia. Organisms with the group D antigen belong to two distinct subsets, one of which has now been reclassified into its own genus, ***Enterococcus.*** Previously known as *fecal streptococci* because they are part of the normal flora of the intestinal tract, enterococci are often associated with urinary tract or wound infections as well as endocarditis (see chap. 15). Nonenterococcal group D streptococci may also cause endocarditis, bacteremia, or urinary tract infection. Differentiation of the group D antigen-bearing organisms is important clinically because *Enterococcus* species are more difficult to treat than are nonenterococal strains. One *Enterococcus* species in particular (*Enterococcus faecium*) has become resistant not only to the usual therapy of ampicillin and an aminoglycoside (see table 9.7) but also to the alternative therapy of vancomycin and an aminoglycoside. Vancomycin-resistant enterococci emerged in the late 1980s with the widespread use of the drug. The appearance of such highly resistant microorganisms is of major significance to health-care professionals because, currently, no therapeutic agents in use are available for effective treatment of patients infected with vancomycin-resistant enterococci.

Other serologic (Lancefield) groups are sometimes involved in human disease. Some important properties of streptococci, and the human diseases with which they are associated, are shown in table 12.1. Not all streptococci can be grouped by serologic methods, for some lack a group-specific carbohydrate in their cell walls. This is true of some of the alpha, or viridans, streptococci found normally on the respiratory or oral membranes. Anaerobic streptococci (***Peptostreptococcus*** species) of the normal mouth, intestinal tract, and vagina cannot be grouped serologically for the same reason. These organisms are not primarily pathogenic but may cause opportunistic infections.

TABLE 12.1 CLASSIFICATION OF STREPTOCOCCI AND ENTEROCOCCI AND THEIR DISEASES

Lancefield Group	Organism	Hemolysis	Cellular Products	Extracellular Products	Clinical Diseases
A	*S. pyogenes*	Beta	Hyaluronic acid capsule Group A carbohydrate M protein Lipoteichoic acid	Streptolysins Hyaluronidase Deoxyribonuclease Streptokinase Leukocidin Streptococcal pyrogenic (erythrogenic) exotoxin	Pharyngitis-tonsillitis Skin infections (impetigo) Erysipelas Scarlet fever Postpartum endometritis Rheumatic fever Glomerulonephritis Endocarditis
B	*S. agalactiae*	Beta	Group B carbohydrate	Streptolysins	Neonatal sepsis and meningitis Postpartum endometritis
C	*S. equisimilis*	Beta	Group C carbohydrate	Streptolysins	Impetigo Pharyngitis
D	*Enterococcus* (*E. faecalis*)[a] Nonenterococcus (*S. bovis*)[a]	Gamma	Group D carbohydrate—amino acid complex		Endocarditis Urinary tract infections
Not grouped	Viridans streptococci[b]	Alpha		Glucan or glucan-like substance	Dental caries Endocarditis
Not grouped (anaerobic strains)	*Peptostreptococcus* species	Gamma			Brain abscess Lung abscess Postpartum endometritis

[a]Other species of group D streptococci or enterococci may be alpha, beta, or gamma hemolytic.
[b]Includes species such as S. salivarius, S. mitis, and S. mutans.

Gram-Positive Organisms in Human Hosts
Gram-Positive Cocci-Streptococci: Taxonomy; Antigenic Structures; Virulence Factors; Diseases. Gram-Positive Cocci-Enterococci: Taxonomy; Antigenic Structures; Virulene Factors; Diseases

Latex agglutination tests for grouping streptococci are readily available commercially and the tests can be performed rapidly in the clinical microbiology laboratory. Therefore, beta-hemolytic streptococci are routinely identified to species in many laboratories, especially when they have been isolated from serious patient infections. If these methods are not available or not used, most clinically important strains can be identified presumptively by certain biochemical properties. These include susceptibility to low concentrations of bacitracin (characteristic of group A), esculin hydrolysis (group D), or production of the CAMP factor (group B). The CAMP test takes advantage of the enhanced (synergistic) hemolysis produced when the CAMP factor from group B streptococci interacts with a hemolysin from *Staphylococcus aureus* on a blood-containing agar medium (colorplate 3e).

Serologic identification can also be useful in epidemiologic tracking of sporadic or epidemic streptococcal disease. Since infection with group A strains may be followed by serious sequelae (e.g., rheumatic fever, acute glomerulonephritis), early recognition of these organisms and effective chemotherapy are very important.

CLINICAL STREPTOCOCCAL DISEASES

Streptococcal Sore Throat

Streptococcal sore throat is an infection of pharyngeal tissues characterized by edema and reddening of posterior and anterior surfaces of the throat and soft palate. (They have a swollen, "beefy" look.) Petechial lesions are sometimes seen. When tonsils and other lymphoid tissues are involved, an acute exudative response occurs. Cervical lymph nodes at the angle of the jaw may become tender and enlarged; fever and malaise are frequent.

Scarlet Fever

Scarlet fever results when (1) the strain causing streptococcal sore throat produces streptococcal pyrogenic exotoxin, and (2) the patient is *not* immune to the toxin. The lesions in the throat or at other sites appear together with the characteristic skin rash. (The name *erythrogenic toxin* was changed to *streptococcal pyrogenic,* or "feverproducing" *exotoxin* because fever and other toxic effects accompany the rash.) As the febrile reaction intensifies, nausea and vomiting may occur, and the white blood cell count rises. The rash is distributed to the neck and chest, the axillary and inguinal folds, and the soft skin of the inner sides of the arms and thighs. (Desquamation of these skin areas occurs during convalescence.) The tongue becomes swollen and reddened and the papilla protrude, giving an effect described as "strawberry tongue" (colorplate 5a).

Streptococcal pyrogenic exotoxin is antigenic, stimulating production of specific antitoxin. There are three antigenically distinct types of this exotoxin. As antitoxin levels rise in circulating blood, the skin reaction begins to subside. Antitoxin levels produced during the course of disease protect the individual from the effects of future streptococcal throat infections caused by strains producing the same antigenic type of exotoxin. Prompt therapy of initial throat infection effectively prevents development of scarlet fever by also aborting production of this pyrogenic exotoxin by the infecting strain.

An increase in serious group A streptococcal diseases has taken place since the late 1980s, including some cases of a toxic-shock-like syndrome that have been associated with the streptococcal pyrogenic exotoxin. Previously, *S. aureus* alone was thought to be responsible for this syndrome. The chemical structure of the streptococcal exotoxin is closely related to that of the *S. aureus* toxin that has been implicated in some cases of **toxic shock syndrome (TSS).**

In 1990, the group A *Streptococcus* gained widespread public attention when Jim Henson, creator of the popular Muppet characters, died of disseminated streptococcal infection. He had been ill with a respiratory viral infection, which presumably predisposed him to infection by the bacterium, although such fatal streptococcal infections have been rare since the advent of effective antimicrobial agents. Several states in the United States have also reported an increased incidence in rheumatic fever, one of the sequelae of group A streptococcal infection and a disease that had almost disappeared in this country. Currently, microbiologists and epidemiologists are charting the incidence of disease and examining properties of the associated microorganisms to better understand the factors responsible for this resurgence of streptococcal disease.

LABORATORY DIAGNOSIS

I. Identification of the Organism

(Streptococcus pyogenes, group A*)*. The microscopic, cultural, and serologic characteristics of virulent streptococci may be summarized as follows:

A. They are gram-positive cocci that grow in chains (see colorplate 1b).

B. They are facultative aerobic organisms that grow best on blood-enriched media. Isolated colonies on such media are characteristically surrounded by zones of beta hemolysis (colorplate 3a). Microaerophilic incubation (decreased atmospheric O_2 and increased CO_2 provided in a "candle jar" or CO_2 incubator) enhances growth but may affect the characteristic hemolysis. Cultures for isolating these organisms therefore are often incubated under anaerobic conditions, which enhances hemolytic activity of streptolysin O.

C. They may be serologically identified by grouping and by typing methods, usually latex agglutination, when epidemiologic considerations warrant.

II. Identification of Patient's Serum Antibodies

A. Antibacterial (somatic) antibodies are type specific. They have no practical value in diagnosis of current infection because they arise in late stages of the disease.

B. Antistreptolysin O (ASO) titers are indicative of current or past pharyngeal infection with beta-hemolytic strains of streptococci. The titer (level of antibody) is a clue to the recency of infection.

C. Tests to detect antibodies against other streptococcal enzymes (e.g., hyaluronidase, deoxyribonuclease) may be run in parallel with the ASO test. These are sometimes positive when the ASO test is negative.

III. Specimens for Diagnosis

A. Throat swabs in transport medium for culture.

B. Blood for culture.

C. Whole blood (or serum) for antistreptolysin or other antistreptococcal enzyme titers. At least two samples should be submitted: one collected during the acute stage of illness (when the titer may be low) and another taken in the late or convalescent stage to determine whether the antibody level has increased.

IV. Special Laboratory Test

Nonculture, immunologic methods are available for direct detection of group A streptococci in throat swab specimens. Group A specific antigen can be extracted into a solution from bacteria on the swab. The extract solution is then tested for the presence of group A antigen using a latex agglutination test, or more usually, an EIA (see fig. 6.17). These tests are usually performed in clinics and physicians' offices because they are rapid (taking from 10 to 30 minutes). Results must be interpreted carefully, however, because falsely negative tests can occur if low numbers of group A streptococci are present in the specimen. If the results for patients with clinical evidence of pharyngitis are negative, a throat swab for "strep" culture should always be sent to the laboratory.

Epidemiology

I. Communicability of Infection

A. *Reservoir, Sources, and Transmission Routes* Humans are the reservoir for group A streptococci. Asymptomatic carriers as well as patients in acute and convalescent stages of disease shed these organisms in their respiratory secretions. Transmission is by droplet spread, direct contact, and indirect contact with temporarily contaminated environmental sources, including milk or food. Contaminated hands carry these organisms from handkerchiefs to clothing, skin, and food, as well as objects handled. Droplet spray contaminates many nearby surfaces also, including the faces of close contacts.

B. *Incubation Period* Rarely more than 3 days.

C. *Communicable Period* Persists during the incubation period and throughout clinical illness, or longer, unless adequately treated. The carrier state may last for many months, and untreated persons with exudative lesions are a constant source of virulent organisms. Adequate antimicrobial therapy eliminates the communicability of disease, if not all group A streptococci, within 24 to 48 hours.

D. *Immunity* Susceptibility to streptococcal sore throat and scarlet fever is general, but immunity against specific M-protein type strains often develops as a result of inapparent infections. Subsequent infections are caused by different M-protein types to which the person is not immune. Antitoxin immunity to streptococcal pyrogenic exotoxin prevents scarlet fever but not streptococcal pharyngitis. A satisfactory vaccine for immunization is not yet available.

II. Control of Active Infection

A. *Transmission-based Precautions* Droplet precautions for streptococcal pharyngitis. For streptococcal skin infections (chap. 19), contact precautions are recommended. If patient hygiene is poor, the patient may be placed in a private room for 24 hours from the start of antimicrobial therapy. Treatment should be continued for at least ten days. Infection control may be verified by repeated throat cultures.

B. *Other Precautions* Concurrent disinfection of respiratory discharges and all contaminated items. Terminal disinfectant-cleaning.

C. *Treatment* Penicillin is the usual drug of choice in patients who are not allergic to it. Streptococci are also susceptible to erythromycin, cephalosporins, and clindamycin.

D. *Control of Contacts*

1. *Isolation* Not indicated.

2. *Prophylaxis* Immunization not available. Antimicrobial prophylaxis may be indicated for patients with recurrent infections.

3. *Case finding* Not indicated except in outbreaks.

E. *Epidemic Controls* Reports of streptococcal epidemics may be required so that efforts can be made to find the source and mode of transmission. Identification of the group and type of streptococcal strain involved in epidemics is important for detecting sources. Suspected persons, milk, or food can then be checked by bacteriologic and immunologic techniques. Contact precautions are used for patients involved in epidemic outbreaks and they are treated with antimicrobial agents. Contaminated commercial food or milk supplies must be removed from the market. Sources may be pinpointed by examining the distribution of infected patients and their possible exposure to common transmission routes.

III. Prevention

A. Public education on the ease and hazards of transmission of streptococcal diseases; the need for personal hygienic controls and for prompt medical control of developing illness.

B. Control of food and milk handlers. Inspection of dairy cows for mastitis or other infection, sanitation of dairy establishments, pasteurization of milk (boil or discard if evidence of streptococcal contamination exists).

C. Penicillin prophylaxis for children and others for whom risk of repeated infection is great. The group includes those who have had rheumatic fever or other diseases as an aftermath to streptococcal infection.

Sequelae of Group A Streptococcal Infections

Late manifestations or consequences of streptococcal throat infections include development of rheumatic fever or acute glomerulonephritis in a small proportion of patients. These diseases are neither infectious nor communicable. They are believed to develop as a consequence of both humoral and cellular responses to streptococcal antigens.

Rheumatic Fever

Rheumatic fever is the most important and frequent of the poststreptococcal diseases. Following upper respiratory tract infection with group A streptococci, about 3% of untreated patients (the incidence is highest in childhood) develop rheumatic fever after an average interval of three weeks. Major clinical symptoms include fever, carditis, and polyarthritis (inflammatory involvement of joints, usually migratory in pattern). The severity of symptoms varies from mild and scarcely noticeable to acute illness with irreversible cardiac damage. The lesions of the disease may include skin involvement (subcutaneous nodules) and central nervous system injury (chorea). Rheumatic fever may persist as a chronic disease for years, or it may be a recurrent acute or subacute problem following repeated streptococcal infections. Many serologic types of the group A strains have been demonstrated in the predisposing throat infections. When numerous cases of rheumatic fever are seen following a streptococcal epidemic, epidemiologic studies can relate them to the specific infecting strain.

During the 1960s and 1970s, rheumatic fever virtually disappeared from developed countries. Beginning in 1985, however, an increased incidence has been seen throughout the United States. Many of these rheumatic fever patients had previously been infected with highly mucoid strains of group A streptococci belonging to only a few immunologic M types. Studies are underway to examine both the unusual streptococcal strains and factors affecting patient predisposition to this disease.

Acute Glomerulonephritis

Hemorrhagic lesions in the glomeruli may develop following streptococcal infection with certain strains. Unlike rheumatic fever, which occurs only after pharyngeal streptococcal infection, **glomerulonephritis** can appear after either pharyngeal or skin infections. The clinical symptoms include blood or albumin in the urine (hematuria, albuminuria), generalized edema, and hypertension. The average interval between infection and this type of sequel is shorter than for rheumatic fever. Glomerulonephritis is usually an acute disease from which most patients recover without residual damage, and recurrent episodes of this type are rare. Occasionally, symptoms of rheumatic fever may occur with those of glomerulonephritis. The disease is thought to be initiated by deposition of circulating streptococcal antigen-antibody (immune) complexes on the basement membranes of glomerular capillaries.

The laboratory diagnosis of these diseases is by previous bacteriologic evidence of streptococcal infection. Current demonstration of significant serum antibody titers for streptococcal enzymes, such as streptolysin O or deoxyribonuclease, supports the clinical diagnosis.

The epidemiology of streptococcal sequelae does not include considerations of their control since they are not communicable. The primary concern is prevention, particularly of rheumatic fever with its potential for permanent cardiac damage. Prompt, effective antimicrobial therapy of active streptococcal disease eliminates the organism and may prevent the distribution of sensitizing streptococcal antigens through the body, thus reducing the incidence of rheumatic fever. Penicillin prophylaxis for those who have had rheumatic fever can successfully prevent its recurrence.

OTHER BACTERIAL CAUSES OF PHARYNGITIS

Arcanobacterial Pharyngitis

The Clinical Disease

Arcanobacterium haemolyticum is being increasingly recognized as a cause of exudative pharyngitis, particularly in children and young adults. The clinical presentation of disease is similar to the pharyngitis caused by group A streptococci and is often associated with the production of a skin rash on the trunk and extremities indistinguishable from the rash of scarlet fever. Arcanobacterial pharyngitis can be treated succesfully with penicillin or erythromycin.

LABORATORY DIAGNOSIS

I. Identification of the Organism
(Arcanobacterium haemolyticum).

Microscopic and Cultural Characteristics The organisms are gram-positive bacilli that are nonencapsulated, nonmotile, and nonspore-forming. They are facultatively anaerobic organisms that grow best on blood-enriched media. Colonies on these media are characteristically surrounded by small zones of beta hemolysis, usually after 48 hours of incubation. The presence of 5% CO_2 in the atmosphere stimulates bacterial growth.

II. Specimens for Diagnosis

The specimen of choice for culture is a pharyngeal swab in transport medium.

Gonococcal Pharyngitis

The sexually transmitted disease, gonorrhea, is the most common manifestation of infection with *Neisseria gonorrhoeae* (the gonococcccus, see chap. 19). In addition to this urogenital infection, the organism can also cause **gonococcal pharyngitis,** an infection that is increasing in incidence. Orogenital contact is the major risk factor for acquiring gonococcal pharyngeal infection. Most infections are asymptomatic but mild cases of pharyngitis have been reported. Because *N. gonorrhoeae* can be carried in the oropharynx, infection can be transmitted to others by close contact such as kissing. A pharyngeal swab sample in transport medium is the specimen of choice to identify individuals who may be infected. Laboratory methods for the cultural diagnosis and treatment of gonococcal infections are discussed in detail in chapter 19.

Vincent's Angina (Fusospirochetosis)

Vincent's angina or **fusospirochetosis** is also known as trench mouth because it was common among soldiers who served "in the trenches," especially during the two world wars. It is a mixed bacterial infection caused by at least two anaerobes, *Fusobacterium necrophorum* and a spirochete. The disease is characterized by the production of an exudate or pseudomembrane on the affected tissue. Patients usually have a very fetid breath, which suggests the presence of the disease. Vincent's angina occurs most frequently in persons who have poor oral hygiene (as was the case with soldiers in the field during wartime). The diagnosis is established by examining a Gram-stained smear of pharyngeal or tonsillar exudate and seeing both spindle-shaped (fusiform), gram-negative bacilli and spirochetes. Culture of the sample is not indicated and patients respond promptly when treated with penicillin or erythromycin.

LOWER RESPIRATORY INFECTIONS: BACTERIAL PNEUMONIAS

PNEUMOCOCCAL PNEUMONIA

The Clinical Disease

Community-acquired **bacterial pneumonias** are most frequently of pneumococcal origin. **Pneumococcal pneumonia** is an acute infectious disease involving alveoli or bronchi of the lungs, or both. The onset is sudden and marked by chills, fever, chest pain, difficult breathing (dyspnea), and cough. The sputum is characteristically bright red or rusty in color, reflecting the presence of blood.

In acute **lobar pneumonia,** the alveolar spaces of an involved lobe fill with a fibrinous exudate containing red blood cells and polymorphonuclear leukocytes (PMN), and the lung becomes consolidated. **Pneumococci** are present in large numbers in the exudate; they may also be in the bloodstream during early stages of disease, or persistently if infection is not arrested. These are extracellular, capsulated organisms. The capsules prevent ingestion of pneumococci by phagocytic cells and hence contribute to their virulence. The capsular material is antigenic, however, stimulating the production of antibodies that combine with it, change its surface properties, and thus promote phagocytic uptake of the bacterial cells. Once taken up by the PMN and macrophages, pneumococci are readily killed.

Antibody production reaches a peak between the fifth and tenth days and may bring about a dramatic and sudden recovery of untreated cases referred to as "crisis." Before penicillin was available, patients either recovered at the time of crisis or were treated by immunotherapy, that is, by administration of a serum specific for the capsular antigen type of the causative strain. Today, pneumococcal pneumonia is treated promptly, usually with penicillin, which prevents or arrests lobar consolidation and terminates the disease rapidly. Pulmonary damage is not permanent.

Bronchopneumonia is more diffuse and less localized than the lobar variety, involving the bronchial tree rather than the alveoli.

Pneumococci are members of the upper respiratory flora in 5 to 40% of healthy persons, and in some instances they may be transmitted to others. Pneumococcal pneumonia is seldom a primary infection but occurs secondary to injury induced by other causes. The organisms enter the alveoli by aspiration of oral secretions as a result of impairment of normal defenses by virus infection, chemical irritation, allergic damage, alcoholism, or impaired pulmonary circulation (as in chest surgery or cardiac disease), for example.

The factors that predispose to pneumonia may also lead to extension of infection along mucosal surfaces of the upper respiratory tract. Pneumococcal sinusitis and otitis media may occur, and involvement of mastoid surfaces may lead to pneumococcal extension to the meninges with resulting meningitis. Pneumococci circulating in the bloodstream may reach the brain by this route as well, then localizing on the meninges or within the brain.

LABORATORY DIAGNOSIS

I. Identification of the Organism
(Streptococcus pneumoniae).

A. *Microscopic Characteristics* The pneumococcus is characteristically a gram-positive diplococcus; it occurs in pairs or in short chains of pairs. The cocci are lancet-shaped or elliptical in morphology; the broad ends of each pair are adjacent, the thinner ends point away from each other (colorplate 1a).

B. *Culture Characteristics* Pneumococci are somewhat fastidious and do not survive well in competition with hardier organisms of the mixed throat flora present in expectorated sputum specimens. For this reason sputum specimens should be submitted promptly to the laboratory so that appropriate culture techniques can be initiated.

Pneumococci grow well on enriched culture media containing blood (see fig. 4.3) incubated in an atmosphere of 5 to 10% CO_2. The colonies growing on blood agar plates are surrounded by a zone of incomplete (alpha) hemolysis, usually accompanied by greening. In this respect they resemble alpha-hemolytic streptococci but can be differentiated from the latter by differences in colony morphology, some biochemical reactions to carbohydrates, and their greater susceptibility to bile salts and other surface-active agents. Optochin is a chemical substance that inhibits pneumococcal growth. It can be used in a culture method as a disk susceptibility test to identify the organism presumptively.

C. *Immunologic Methods of Identification* Pneumococcal capsules are composed of an antigenic polysaccharide that stimulates antibody production. When anticapsular antibody is formed, it combines with the poly-saccharide so that phagocytosis of the opsonized bacteria proceeds normally. When the reaction of antibody with capsulated pneumococci is viewed under the microscope, the capsule appears to swell and its outer edge becomes sharply demarcated (fig. 12.1). This **quellung reaction** (*quellung* is the German word for "swelling") can be used in the laboratory to identify pneumococci and to distinguish among their antigenic types.

At least 84 immunologic types of pneumococci have been distinguished on the basis of differences in the chemical structure of their antigenic capsules as detected by specific antiserums. Only a small number of types is consistently associated with pneumococcal disease: types 1 to 8 account for most cases of adult pneumococcal pneumonia; types 1 to 8 plus types 14 and 19 are responsible for most cases in children.

The immunologic identification of pneumococcal strains was a matter of clinical urgency before antimicrobial drugs were developed for therapeutic use because specific antiserums were used to treat pneumococcal infections. Chemotherapy has obviated the clinical need for pneumococcus typing, but the procedure remains valuable in tracing the epidemiology of pneumococcal disease.

Immunologic reagents for latex agglutination (see chap. 6) are available for rapidly identifying pneumococci grown in culture or for detecting their soluble capsular antigens in body fluids (urine, serum, and cerebrospinal fluid) of infected persons. These reagents consist of latex particles coated with a polyvalent pneumococcal antiserum. This antiserum is made up of pooled antisera containing antibodies to the most common pneumococcal capsular types.

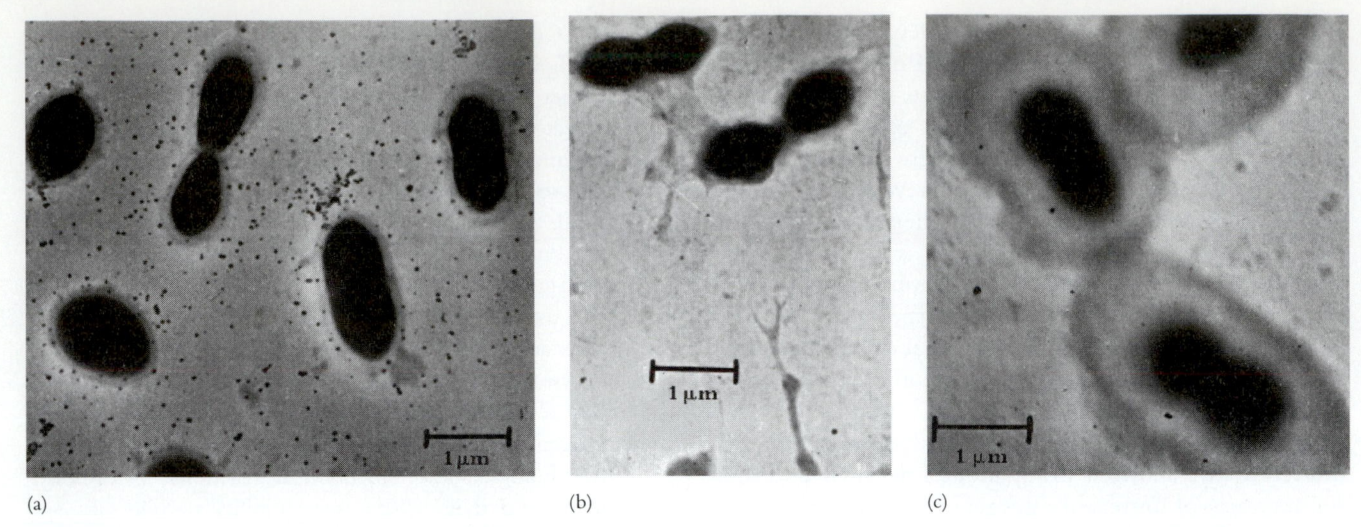

FIGURE 12.1 Pneumococcus typing by capsular swelling reaction. (a) *Streptococcus pneumoniae* type 3 not exposed to pneumococcus antiserum. Their capsules are visible only as faint halos surrounding the bacteria. (b) *S. pneumoniae* type 3 in type 1 serum. These organisms were exposed to a nonspecific antipneumococcal serum prepared with type 1 organisms. The appearance of the capsules does not differ from those seen in (a). (c) *S. pneumoniae* type 1 were exposed to a specific type 1 antipneumococcal serum. A strong quellung reaction occurred as a result of specific antibody combination with capsular antigen. Note lancet shape of the organism.

II. Identification of Patient's Serum Antibodies

Not done.

III. Specimens for Diagnosis

(Submitted, whenever possible, before antimicrobial therapy is started.)

A. Blood samples drawn in the early febrile stage of disease may yield pneumococci promptly upon appropriate culture.

B. Sputum specimens for smear and culture: Gram-stained smears can be made and examined immediately for the presence of typical gram-positive diplococci suggestive of pneumococci. Culture confirmation is usually possible within 24 to 48 hours.

C. Spinal fluid (if pneumococcal meningitis or other CNS involvement is suspected) for smear, culture, and latex agglutination tests.

D. Urine specimens may sometimes contain excreted antigen detected by latex agglutination tests even though no viable organisms are present in the specimen.

Epidemiology

I. Communicability of Infection

A. *Reservoir, Sources, and Transmission Routes* Human beings are the reservoir; the organisms are present in respiratory secretions of asymptomatic carriers and patients with active infection. Transmission is by direct contact with infected persons, by droplet spread, and by indirect contact with infectious secretions on recently contaminated surfaces.

B. *Incubation Period* Unknown for infection of endogenous origin; probably 1 to 2 days for transmitted infection.

C. *Communicable Period* Probably during time when pneumococci are present in large numbers in nasal and oral secretions. Prompt chemotherapy reduces these numbers sharply and shortens the communicable period.

D. *Immunity* The nonspecific resistance of normal healthy tissues constitutes the major defense mechanism, holding most infections to an inapparent or subclinical level. The immunity following an attack is type specific; its duration depends on the level of antibody reached before antimicrobial therapy eliminated the antigenic stimulus.

II. Control of Active Infection

A. *Transmission-based Precautions* Not indicated.

B. *Other Precautions* During the communicable period, concurrent disinfection of respiratory secretions and of contaminated items. Terminal cleaning includes disinfection of furniture and floor.

C. *Treatment* Penicillin is the usual drug of choice but penicillin-resistant strains are now being reported worldwide and are increasing in incidence. The resistant strains are treated with vancomycin, with or without rifampin.

Susceptible pneumococci can also be treated with a wide range of antimicrobial drugs (erythromycin, cephalosporins, clindamycin, and clarithromycin). The route of administration (oral, intramuscular, or intravenous) depends on the severity of symptoms, localization of infection, characteristics of the drug, and response of the patient.

D. *Control of Contacts* None indicated.

E. *Epidemic Controls* Generally not necessary unless outbreaks threaten in closed populations with low resistance (pediatric wards, geriatric institutions, military hospitals, prisons). Vigorous sanitary and hygienic measures are indicated with chemoprophylaxis for the unusually susceptible. Epidemic outbreaks must be reported to the local health department; individual case reports are not required.

III. Prevention

Immunization A vaccine that protects against the 23 most common antigenic types of pneumococci is available. It is recommended primarily for persons at high risk, including the elderly, alcoholics, or individuals with chronic diseases.

COMMON BACTERIAL PNEUMONIAS OTHER THAN PNEUMOCOCCAL

The Clinical Disease

In nonhospitalized patients about 20% of bacterial pneumonias are caused by organisms other than the pneumococcus. They are acute febrile infections that may localize in any segment of the bronchopulmonary system. They are often sequels to virus infections (influenza, viral pneumonia) or other injury and are caused most often by opportunistic bacteria that are, like the pneumococcus, normal inhabitants of the upper respiratory tract (e.g., *Staphylococcus aureus, Streptococcus pyogenes, Klebsiella pneumoniae,* and *Haemophilus influenzae*).

The pathogenesis of these pneumonias is similar to that caused by *S. pneumoniae.* They occur most frequently in persons whose normal resistance has been lowered in some way. The degree of permanent damage done to lung tissues varies, however, with the capacities of the causative organism to induce tissue injury. Unlike the pneumococcus, some of these bacterial invaders may produce necrosis of local tissues or blood vessel walls through the action of exo- or endotoxins, or as a result of yet unidentified factors. *S. aureus, S. pyogenes,* and *K. pneumoniae* (a gram-negative bacillus) induce hemorrhagic consolidation and eventual necrosis of lung parenchyma if untreated. This damage resolves slowly and affected tissue is often replaced by permanent scar tissue (fibrosis).

K. pneumoniae and *H. influenzae* share with the pneumococcus the pathogenic feature of a polysaccharide capsule that protects them from phagocytosis. The capsular polysaccharides of these organisms are also antigenic and can be identified by serologic methods. There are 77 known antigenic types of *K. pneumoniae* but only six of *H. influenzae,* the most pathogenic of the latter being type b. The incidence of these pneumonias is not high, but their morbidity and fatality depend on the patient's resistance and on the adequacy of chemotherapy. In ***Klebsiella*** pneumonia, pulmonary abscess formation, often associated with bronchial obstruction, is a common complication.

H. influenzae infection, usually caused by type b strains, is seen in nonimmunized children. It often begins in the upper respiratory tract as a pharyngitis, sinusitis, or epiglottitis. In the latter case, severe infection can lead to laryngeal obstruction, sometimes requiring tracheotomy to prevent asphyxia. Invasion of the lower respiratory tract, with development of *H. influenzae* pneumonia, occurs in adults as well as children, often as a sequel to virus infection. (The organism acquired its name because it was isolated frequently from cases of pneumonia that occurred during the 1918 viral influenza epidemic and was wrongly concluded to be the cause of the primary disease.) *H. influenzae* frequently invades the bloodstream from respiratory sites of infection, and in nonimmunized children meningitis is a common result (see Bacterial Meningitis, chap. 15). The infection can also extend directly to the central nervous system from a local site of infection such as in the inner ear (otitis media). Before the recent successful use of *H. influenzae* type b vaccine (see table 10.1), this organism was the usual cause of meningitis in infants and very young children, probably because they had not yet developed antibodies as a result of colonization and subclinical infection. The widespread use of this vaccine has reduced the incidence of *Haemophilus influenzae* meningitis in the pediatric age group by more than 95%. *Haemophilus* infections are now occurring more commonly in nonimmunized adults, perhaps because early treatment of their minor childhood respiratory diseases prevented them from producing protective antibody or because their antibody levels have declined with age.

Secondary bacterial pneumonias can also be caused by a variety of gram-negative commensalistic bacteria, particularly those commonly associated with the intestinal tract or environmental sources (e.g., *E. coli, Proteus* and *Pseudomonas* species, *Serratia marcescens*). These are almost always nosocomial infections seen in hospitalized patients, especially those who have undergone chest or abdominal surgery or who have underlying pulmonary disease and are receiving inhalation therapy. Prior chemotherapy of these patients may have contributed to the pneumonia by modifying the normal flora of the superficial mucosae. Treatment with inappropriate antimicrobial agents such as penicillin, to which gram-negative bacilli are resistant, results in colonization of these organisms in the upper respiratory tract and increases the risk of further infection. In addition, the water in ventilator reservoir nebulizers and room humidifiers can easily become contaminated with *Pseudomonas,* and a number of pneumonia outbreaks linked to such sources have been reported.

LABORATORY DIAGNOSIS

I. The Organisms

A. *Streptococci* Hemolytic strains of serologic group A are the primary human pathogens. Microscopically, they are gram-positive cocci occurring in short to long chains. In culture they are somewhat fastidious but are supported by blood-enriched media incubated aerobically or anaerobically. Colonies on blood agar are surrounded by a zone of complete (beta) hemolysis. Immunologic identification is based on differentiation of their polysaccharide antigens in the cell wall. Numerous groups, designated by letters, are recognized but most of the strains associated with human disease fall into group A. (See the discussion of streptococcal diseases earlier in this chapter.)

B. *Staphylococci* Microscopically, gram-positive cocci in clustered arrangements, resembling grapes. These organisms grow readily in most simple nutrient media. Strains most frequently associated with human disease produce a plasma-coagulating enzyme, coagulase. When this characteristic is seen, strains are reported as *Staphylococcus aureus.* (See the discussion of staphylococcal diseases in chapter 19.)

C. *Klebsiella pneumoniae* Microscopically, a gram-negative plump, short bacillus with a large capsule. They are nonsporulating and nonmotile. Klebsiellae grow with ease on common laboratory media, their colonies being quite mucoid because of capsular polysaccharide production. The capsules are antigenic and can be identified by immunologic methods such as the quellung reaction. There are many antigenic types but most strains that cause human respiratory infection fall into capsular types 1 and 2. (Other types are associated with urinary infections.)

D. *Haemophilus influenzae* Microscopically, a gram-negative bacillus, slender but pleomorphic. Most are short and coccoid, but long filamentous rods may also be seen. This organism is nonmotile and nonsporulating. It requires enriched media and microaerophilic incubation for growth. *H. influenzae* produces antigenic capsular polysaccharides identifiable by the quellung and latex agglutination techniques. There are six antigenic types, lettered a through f. Type b causes most human systemic infections.

E. *Gram-Negative Bacilli* Most of the gram-negative bacilli associated with the intestinal and upper respiratory mucous membranes are aerobic or facultative, nonsporulating rods, indistinguishable from each other microscopically. The major genera include *Escherichia, Enterobacter, Proteus, Serratia,* and *Pseudomonas.* They are identified by genus and species on the basis of biochemical properties, particularly characteristic enzymic pathways in the use of carbohydrates. Except for *Escherichia,* members of these genera are also common in the environment (soil, water).

II. Identification of Patient's Serum Antibodies

Except for streptococcal infection, immunologic diagnosis is not routinely practiced.

III. Specimens for Diagnosis

A. Sputum specimens for smear and culture.

B. Blood specimens for culture.

C. Throat or nasopharyngeal swabs in transport medium may be useful for culture, particularly if concurrent URI exists.

D. Specimens should be collected for laboratory diagnosis before antimicrobial therapy is started, whenever possible. Prompt transport to the laboratory is essential for successful isolation of fastidious pathogens such as *Haemophilus* and streptococci.

Epidemiology

The epidemiology of these infections is similar to that of pneumococcal pneumonia. Specific immunity plays a minimal role, the best defense and preventive control being the maintenance of healthy, nonspecific resistance.

Secondary bacterial pneumonias occur sporadically, never in epidemic patterns. Except for *Haemophilus influenzae* infections, which require droplet precautions, nursing care does not require transmission-based precautions. Appropriate care should be taken with the patient's secretions, however. Concurrent disinfection of inhalation therapy equipment with high-level agents is of particular importance.

Specific antimicrobial therapy is based on the identification of the infectious agent and antimicrobial susceptibility testing. Penicillin is the drug of choice for streptococcal infections and for those rare staphylococci that do not produce penicillinase. Staphylococci that are penicillinase producers are generally susceptible to synthetic penicillins that are not inactivated by penicillinase (e.g., methicillin and nafcillin). Importantly, some strains of staphylococci, called methicillin-resistant *Staphylococcus aureus* or MRSA, have developed resistance to these synthetic penicillins in which case, the therapeutic drug of choice is vancomycin. For *H. influenzae* pneumonia or meningitis, ampicillin has been the drug of choice. However, 30 to 40% of *H. influenzae* strains also produce a penicillinase that inactivates this agent, and for these the drug of choice is a third-generation cephalosporin such as ceftriaxone. *K. pneumoniae* is characteristically resistant to penicillin and is treated with either a cephalosporin or gentamicin, or a combination of these two drugs. These variations in microbial responses to antimicrobial agents underline the need for adequate susceptibility testing in the laboratory.

Legionnaires Disease (Legionellosis)

During the summer of 1976 an epidemic of respiratory illness swept through a group of people who had attended a state convention of American Legionnaires in Philadelphia. Before it ended, this epidemic took 34 lives and hospitalized some 170 persons of the 5,000 or so who had gathered for the convention. A large-scale search using every available method for microbial isolation eventually revealed that a "new" bacterium, subsequently named *Legionella pneumophila,* was the etiologic agent of this disease. This organism is now known to have been responsible for a number of other outbreaks, as well as sporadic cases, of pneumonia dating back at least as far as 1947. In most instances the retrospective diagnosis could be made by detecting antibodies against the organism in patients' sera that had been stored frozen. With the isolation and further study of this organism, diagnostic laboratory tests have been refined so that it can be grown in culture and detected by fluorescent antibody (see chap. 6), enzyme immunoassay, and nucleic acid probe tests (see chap. 7). More than 30 other species of ***Legionella*** have been identified, but *L. pneumophila* is responsible for 90% of infections caused by this organism group.

L. pneumophila is a fastidious gram-negative bacillus with an aquatic natural habitat including lakes, rivers, streams, and, in particular, thermally polluted water (water containing discharges that raise the water temperature). The organism is found in many water distribution systems including in homes and hospitals. **Legionnaires disease** (or legionellosis) is presumed to result from inhaling contaminated aerosols or perhaps aspirating contaminated water. Infection is reported most commonly in immunocompromised hospitalized patients, especially those who have undergone organ transplant.

The Clinical Disease

Legionnaires disease may take several forms ranging from a mild, flulike illness to a severe infection that affects many body systems. Initially, patients have mild upper respiratory symptoms, headache, muscle pain, fever, and commonly, diarrhea. This may progress to pneumonia with chest pains, lung consolidation, cough, and sputum production. Many patients have some type of underlying factor that predisposes to legionellosis such as a chronic disease, cigarette smoking, alcoholism, and immunosuppressive therapy. Disease is more common and more severe in the elderly and in males. Patients with mild infections usually recover spontaneously, but others, especially those with predisposing conditions, may develop kidney failure, intravascular coagulation, and, eventually, death due to pulmonary insufficiency or shock. A high fever continues until appropriate therapy is instituted or spontaneous resolution of the disease occurs.

LABORATORY DIAGNOSIS

The diagnosis of legionellosis can be made in a number of ways. These include by isolating the organism in culture, by demonstrating the agent in patient tissues with direct fluorescent antibody or DNA probe tests, by detecting antigen in the patient's urine with latex agglutination or enzyme immunoassay tests, or by serologic methods.

I. Identification of the Organism

Legionella pneumophila may be isolated on a variety of specially prepared bacteriological culture media: one of the best being a charcoal-yeast extract medium. It does not grow on routine laboratory media; therefore, it is extremely important to notify the laboratory that a diagnosis of legionellosis is suspected so that appropriate culture methods can be used. The organism grows more slowly than many respiratory pathogens, requiring three to seven days for colonies to appear. *L. pneumophila* produces no reaction in common laboratory biochemical tests; thus, identification is usually made by the direct fluorescent antibody technique using specific, fluorescent dye-labeled antiserum. Since there are many antigenically different strains of *L. pneumophila,* each specimen may have to be examined with several antisera or pools of antisera. Lung tissue obtained at biopsy or autopsy, and other specimens, may also be treated with the fluorescent dye-labeled antiserum and examined microscopically to search for brightly fluorescing bacilli in the material (see colorplate 2b). Although the organism is gram-negative, it stains faintly and cannot be distinguished morphologically from a variety of other gram-negative bacilli. DNA probes have also been used for identification.

II. Identification of Patient's Serum Antibodies

As with other infectious diseases, serologic diagnosis is best made by testing both an acute and convalescent phase serum for a fourfold or greater rise in antibody titer. Indirect fluorescent antibody or EIA tests are used for this purpose. It is essential that paired specimens be examined as many persons have antibodies to *Legionella* in the absence of overt disease (for example, persons who worked in the lobby of the Philadelphia hotel where the outbreak of Legionnaires disease occurred).

III. Specimens for Diagnosis

A. Lung tissue obtained at biopsy and pleural fluid are preferred for *Legionella* isolation. The organism has been isolated also from tracheal aspirates, bronchoalveolar lavage, and occasionally sputum when cultured on medium with antimicrobial agents added that are inhibitory to normal respiratory flora, but not to *Legionella.*

B. Acute and convalescent serum samples drawn at appropriate intervals and examined by the indirect fluorescent antibody or EIA method. A single titer of 256 may indicate active disease if an acute serum is not available.

C. Urine for antigen detection by EIA or latex agglutination (see chap. 6).

Epidemiology

I. Communicability of Infection

A. *Reservoir, Sources, and Transmission Routes* The reservoir for the organism is water and appears to provide a risk when it is subjected to aerosolization and subsequently inhaled. Legionellae can survive routine chlorine treatment of water supplies and even the elevated temperatures of hot water (when below 140° F). It is readily found at the bottom of hot water storage tanks where environmental bacteria apparently provide it with the nutrients essential for its survival. The exact mode of transmission remains uncertain but is assumed to be by inhaling aerosols or by aspirating water or oropharyngeal contents contaminated by the organism. In hospitals, the tap water used to rinse ventilatory assistance equipment is a potential source of legionellae that would be instilled directly into the lungs. In community-acquired infections, the source is thought to be organisms aerosolized in ventilation-dehumidification systems. Dissemination by air conditioners, evaporative condensers, and whirlpools has also been reported.

Transmission between human beings has not been definitively documented, and animal infections are not known. Important factors appear to be an underlying predisposition and sufficient environmental exposure to the organism.

B. *Incubation Period* The incubation period is 2 to 10 days and is probably related to the dose of organisms inhaled as well as the physical condition and immune status of the host.

C. *Communicable Period* Although the organism can be isolated from respiratory secretions, the disease is not known to be communicable among human beings.

D. *Immunity* Antibody titers to *L. pneumophila* have been demonstrated to last for 10 years in some instances, but they usually decrease by 6 to 18 months. Antibody is generally specific for the serotype causing infection, but some cross-protection among strains may occur.

II. Control of Active Infection

Although legionellosis was originally considered a highly contagious disease and strictest precautions were recommended, the absence of human-to-human transmission has led to relaxation of the standards. Currently only standard precautions are required. When a cluster of cases occurs, epidemiologic tracing is needed to pinpoint and eliminate the source, but the affected persons cannot be predicted.

Treatment with erythromycin plus rifampin is highly effective in most instances. Azithromycin, clarithromycin, and trimethoprim-sulfamethoxazole are useful as alternative drugs. Laboratory studies do not correlate with the organism's susceptibility in vivo; therefore, they are not recommended at present.

III. Prevention

No immunization is available. In institutions with high rates of nosocomial legionellosis, high-risk patients may be treated with prophylactic antimicrobial agents. When spread of infection can be traced to aerosolization from air conditioning systems, eliminating the organism from the system by chemical means may prevent further spread of infection. Respiratory therapy equipment that acts to humidify or distribute air should be thoroughly disinfected and serviced with care. Further study is needed to define the environmental reservoirs and natural distribution of this organism.

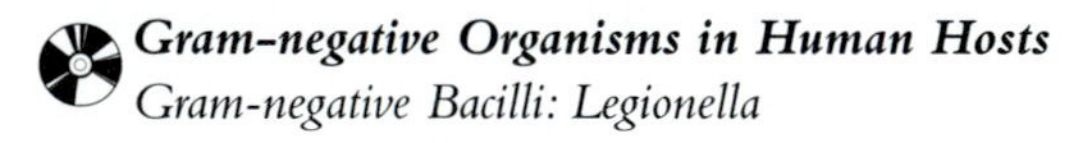
Gram-negative Organisms in Human Hosts
Gram-negative Bacilli: Legionella

PRIMARY ATYPICAL PNEUMONIA

The Clinical Disease

The term **primary atypical pneumonia** refers to a clinical syndrome rather than to any one of the several infectious agents that may induce it. Clinically, the phrase implies an acute lower respiratory infection of primary origin. The disease is characterized by fever and other constitutional symptoms, and pulmonary infiltration that is patchy, irregular, or diffuse, but seldom if ever consolidated. There is usually not a marked leukocytic response, the white blood cell count remaining within normal limits. This picture is "atypical" with respect to most bacterial infections, which characteristically induce exudative reactions in the lungs or bronchi.

Nonbacterial infectious pneumonias of the infiltrative type are caused by a number of different microbial agents including a large, heterogeneous group of viruses (see table 13.1). In addition, an unusual type of bacterium called *Mycoplasma pneumoniae* is characteristically associated with primary atypical pneumonia.

Mycoplasmas are unlike other bacteria in that they do not possess a rigid cell wall. Consequently, their protoplasmic structures are very plastic and pleomorphic. Because of their plasticity and extremely narrow width, even the longest filaments of ***Mycoplasma*** are filterable. The organisms are prevalent in human beings in both health and disease, as well as in many species of animals with which they have contact. *Mycoplasma* strains are found among the normal flora of the respiratory and urogenital membranes. They are also associated with inflammatory conditions of these tissues (cervicitis, urethritis, prostatitis). They have been identified in pus from deep tissue abscesses and associated with a variety of systemic illnesses.

Mycoplasma pneumoniae produces a spectrum of respiratory disease in humans that varies from mild upper respiratory tract infections to severe bilateral pneumonia. Primary atypical pneumonia due to *M. pneumoniae,* however, is usually a mild, self-limited infection that occurs most often in children and young adults. The clinical picture is variable, with an insidious onset. Within about 2 weeks of exposure the symptoms begin with fever and chills, headache, malaise, and cough. The sputum is mucoid or purulent, but the cough is not always productive. Usually one lower lobe of the lung is involved in a patchy consolidation, but bilateral lower lobe involvement may also occur. Upper lobe pneumonia, pleural effusion, and pulmonary abscess are sometimes seen, but these are rare.

Untreated primary atypical pneumonia usually resolves within 2 or 3 weeks, but antimicrobial therapy with erythromycin or tetracycline can bring about prompt resolution of symptoms.

LABORATORY DIAGNOSIS

I. Identification of the Organism

(Mycoplasma pneumoniae).

Microscopic and Cultural Characteristics Mycoplasmal pneumonia may be difficult to distinguish clinically from pneumonia due to other infectious agents, particularly viruses, and the laboratory diagnosis may require some time. Often a presumptive diagnosis is made by clinical and X-ray findings and the patient's response to trial therapy with tetracycline or erythromycin. However, a laboratory diagnosis can be made with cultures and serologic methods. A DNA probe for detecting the organism is also available.

Mycoplasmas require special enriched media for growth, and culture methods are not routinely available in most clinical laboratories. On agar media, the colonies are difficult to see without a lens because they are extremely small and because the delicate, protoplastic cells extend

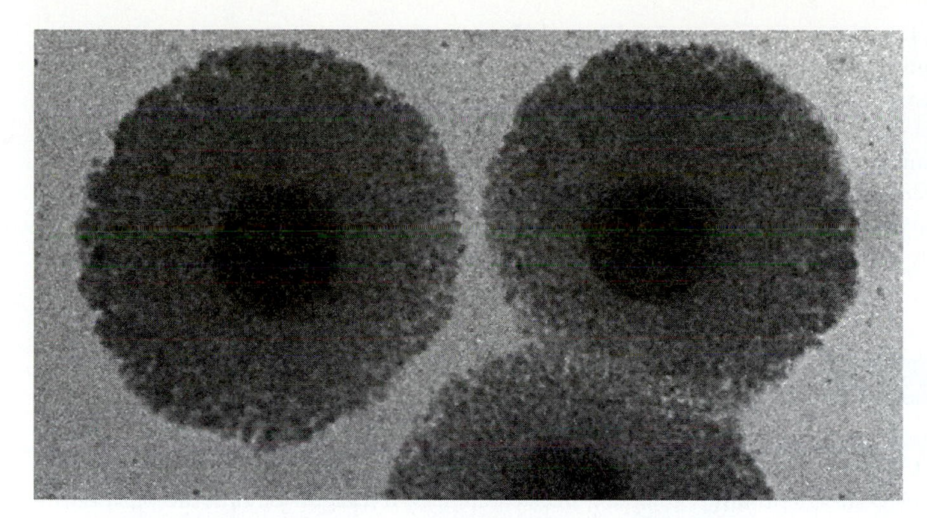

FIGURE 12.2 Mycoplasma colonies growing on agar are typically dense in the center with a delicate perimeter, giving them a "fried egg" appearance.

Courtesy Dr. Yvonne Faur, Public Health Laboratories, Department of Health of the City of New York.

Miscellaneous Bacteria in Human Hosts
Mycoplasma: Growth and Metabolism

through and within the medium rather than being heaped on the surface as are other bacterial colonies. When the agar surface is viewed through the low-power lens of a microscope, *Mycoplasma* colonies have a typical "fried egg" appearance, being dense in the center where filaments of growth extend downward into the agar and then spill over to the surface to form a delicate perimeter (fig. 12.2). Colonies of *M. pneumoniae* display only the dense "yolk." Microscopically, these organisms display many different forms: long filaments, bacillary and coccal shapes, disks, and granular bodies.

II. Identification of Patient's Serum Antibodies

Mycoplasma pneumoniae was identified in the early 1960s as the cause of those cases of atypical pneumonia that are associated with development of "cold agglutinins" in a patient's serum. This term refers to the fact that the serum agglutinates human erythrocytes in the cold, at 4° C but not at 37° C (group O cells must be used). Such activity is not displayed by normal serum, but increases markedly in about 50% of cases of mycoplasmal pneumonia.

In addition to a rising titer of cold agglutinins, patients develop specific antibodies to *Mycoplasma pneumoniae.* These can be demonstrated by the immunofluorescent staining technique.

III. Specimens for Diagnosis

A. Sputum, pharyngeal swabs, and pleural fluid for isolation of *M. pneumoniae.*

B. Acute and convalescent serum specimens.

Epidemiology

I. Communicability of Infection

A. *Reservoir, Sources, and Transmission Routes* Humans are the reservoir of mycoplasmal infection, its source being nasal and oral secretions of infected persons. Infection is transferred by droplet inhalation or by direct and indirect contacts.

B. *Incubation Period* Usually about 2 weeks, but may be from 7 to 21 days.

C. *Communicability* Probably highest during the first week of illness.

D. *Immunity* Immunologic evidence, as well as the incidence of *M. pneumoniae* in healthy persons and in patients with primary atypical pneumonia or febrile upper respiratory illness, indicates that this organism is widely disseminated. The incidence of disease in the general population is low, but the attack rate is higher among children and young adults. Epidemics occur most frequently among military and institutional populations. The incidence is highest in the fall and winter, but outbreaks within families or closed groups may occur throughout the year.

II. Control of Active Infection

Control of mycoplasmal pneumonia does not appear to hinge on specific preventive measures. The hospitalized patient requires only standard precautions, although suitable care should be taken with respiratory secretions.

Since *M. pneumoniae* has no cell wall, it is resistant to penicillin but susceptible to erythromycin, azithromycin, and clarithromycin. These drugs are useful in severe cases of mycoplasmal infection. In mild cases of mycoplasmal infection, especially in the absence of a specific laboratory diagnosis, antimicrobial therapy may be avoided because the disease is self-limited and also because it is difficult to distinguish clinically from viral infections for which antimicrobial agents are useless.

III. Prevention

No immunizing or other specific preventive measures are available.

Miscellaneous Bacteria in Human Hosts
Mycoplasma: Virulence Factors; Diseases

Q FEVER

The Clinical Disease

Q fever was first recognized as a clinical entity in the mid-1930s in Australia. It was named by the letter Q to indicate "query," pending the demonstration of its agent, soon considered as a rickettsia. Q fever is an acute systemic disease with a sudden onset of fever and chills, myalgia, and severe headache. The upper respiratory tract is not commonly involved, but an interstitial pneumonia is frequent, with chest pain and cough. Nausea and vomiting often occur, and in some cases symptoms may be referable to hepatic or meningeal involvement or endocarditis. Chronic infection may persist for a few months but usually recovery is complete in about 3 weeks.

LABORATORY DIAGNOSIS

Clinical and X-ray diagnosis reveals signs of viral (atypical) pneumonia, but these are not specific for Q fever. Laboratory studies such as blood counts and tests for cold agglutinins (negative in Q fever but positive in mycoplasmal pneumonia) are also nonspecific. Laboratory diagnosis is best provided by identification of the organism or the demonstration of specific antibodies in the patient's serum, or both.

I. Identification of the Organism

The causative agent, *Coxiella burnetii,* is an obligate intracellular parasite long thought to be a rickettsia; however, because of important differences, it is undergoing reclassification. Its morphology is irregular, with forms as large as usual bacteria, or so minute as to be barely visible in stained preparations examined by the light microscope. It is filterable and resistant to physical and chemical agents. Unlike the rickettsia, it does not produce a cutaneous rash and does not require an arthropod vector to infect humans. The organism infects ticks, lice, mites, and other arthropods, and these probably have a role in transmission among animals, but human infection with ***Coxiella*** is acquired directly by inhalation, by ingestion in unpasteurized milk, through the conjunctiva, or through minor lesions in skin. This distinction from the rickettsiae, all of which are transmitted to humans by insect vectors (see chap. 22), together with the other differences noted, are the basis for the reclassification.

II. Identification of Patient's Serum Antibodies

Tests for specific antibody in the patient's serum, early and late in the disease, may be performed by a variety of methods including agglutination, immunofluorescence, and EIA. Titers are significantly high by the end of the second or third week.

III. Specimens for Diagnosis

A. Specimens of blood taken during the febrile period, as well as sputum, urine, tissue, or spinal fluid (as indicated by symptomatology), are injected into small laboratory animals or chick embryos for isolation of the organism. Laboratory facilities for this work should include special safeguards against the high risk of infection of laboratory personnel and others who work in the same building.

B. Serologic diagnosis is simpler, safer, and preferred by the diagnostic laboratory. Acute and convalescent samples of serum are submitted for testing.

Epidemiology

I. Communicability of Infection

A. *Reservoirs, Sources, and Transmission Routes* Domestic animals (cattle, sheep, and goats), ticks, and other arthropod vectors are the natural reservoirs of this infection, humans being an incidental victim and unimportant to its natural maintenance. As a zoonosis, Q fever is typically an inapparent infection with a worldwide distribution in domestic livestock. Animals with inapparent infections may shed large numbers of the organisms in milk, urine, and feces, and in placental tissues and blood. Raw milk from infected animals is probably a source of infection for humans, but the richest common source of the organisms is the dust of pasture soil and of barns housing infected animals. Inhalation of airborne organisms is thought to be the most important route of transmission. Q fever may be an occupational hazard for workers in abattoirs, in plants where wool or hides are processed, and also in laboratories where the organism is handled.

Direct person-to-person transmission does not ordinarily occur, but Q fever has been reported as a hospital cross-infection and as a contact infection for nurses and physicians, particularly those conducting postmortem examinations.

B. *Incubation Period* From 2 to 4 weeks, with an average time of 18 to 21 days.

C. *Communicable Period* Not ordinarily transmissible from person to person (see I, A above).

D. *Immunity* Susceptibility to Q fever appears to be general in areas where the organism is endemic in animals. Recovery from illness is thought to be accompanied by lifelong immunity.

II. Control of Active Infection

A. *Transmission-based Precautions* Person-to-person transmission is unusual therefore only standard precautions are needed.

B. *Other Precautions* The organism can survive for months on clothing, on farm implements, in dust and dirt, and presumably, on hospital equipment. Concurrent disinfection techniques should be practiced in the hospital unit where the Q fever patient is cared for, and terminal disinfection-cleaning is advisable as well.

C. *Treatment* Tetracyclines are the drugs of choice. Chloramphenicol is also useful.

D. *Control of Contacts* Human contacts are usually not infected, but a search should be made for the source of infection in animals, unpasteurized milk, or the laboratory.

E. *Epidemic Controls* Case reports are indicated in areas where Q fever is endemic. Persons exposed in an outbreak are observed and promptly treated with antimicrobial agents if symptoms occur.

III. Prevention

A. Immunization can be effected with an inactivated vaccine. This procedure is recommended for laboratory workers and others whose occupational risk of infection is high.

B. Pasteurization of milk prevents dissemination of this infection.

C. Control depends on limiting spread of infection from animals, regulations on shipment of livestock, and vaccination of animals where economically feasible.

PRIMARY PLAGUE PNEUMONIA

The Clinical Disease

Plague is an endemic infection of rats and other wild rodents, maintained by a flea vector. When humans are bitten by this flea and infected, they develop the bubonic form of plague (see chap. 22). In late or terminal stages of bubonic plague, the invading organism (*Yersinia pestis*) may disseminate to the lungs, where it localizes and multiplies in large numbers. The sputum and respiratory secretions of such patients are extremely infectious and can transmit the organism directly to others. The respiratory entry of the infectious agent then results in **primary pneumonic plague.** This disease may occur in epidemic patterns among persons closely exposed to a plague source and to each other. It is a highly fatal infection, characterized by rigor, severe headache, generalized pain, difficult breathing, productive cough, and high fever. Sputum is watery, frothy, and easily projected in a droplet spray for several feet by the coughing patient. Droplet precautions must be instituted during patient care. Prompt diagnosis and antimicrobial therapy may be lifesaving.

The laboratory diagnosis and the epidemiology of plague are discussed in chapter 22.

CHLAMYDIAL PNEUMONIAS

As discussed in chapter 2, chlamydiae are bacteria and can be included as causes of bacterial pneumonia. They are given separate treatment here because each of three species, *Chlamydia psittaci, Chlamydia trachomatis,* and *Chlamydia pneumoniae,* produces a pneumonia that is transmitted in a unique way or affects a distinct patient population.

Psittacosis (Ornithosis, Parrot Fever)

The Clinical Disease

Psittacosis is a respiratory and systemic infectious human disease transmitted by birds such as parrots, parakeets, and lovebirds (known as psittacine birds), pigeons, and farmyard fowl. When the human clinical disease was first recognized and associated with psittacine birds, it was called psittacosis to emphasize this relationship. Later it became obvious that a large variety of birds and domestic poultry may transmit the disease, and the more general term *ornithosis* was introduced to describe these infections when they occur in, or are acquired from, birds other than the psittacine variety.

Human infection may occur without clinical symptoms or as a mild disease, but it may also be seen as a severe pneumonia. The infectious agent, *Chlamydia psittaci,* gains entry to the body through the respiratory tract. It localizes primarily in cells of the mononuclear phagocyte system. After a period of multiplication the organism spreads through the body in the bloodstream, signaled by a sudden onset of fever, chills, headache, and malaise. The respiratory tract involvement may resemble influenza or primary atypical pneumonia. In mild cases this may be the extent of symptoms, with recovery in about a week. Pulmonary infection may be severe, however, with a cough productive of mucopurulent sputum. Fatalities seldom occur if the patient is treated, but when they do, they are attributable to vascular damage. A skin rash may be seen as a result of capillary flushing, nosebleeds are common, and thrombophlebitis may occur, with threat of embolism. Early recognition and adequate antimicrobial therapy usually ensure recovery from psittacosis without circulatory or other complications.

LABORATORY DIAGNOSIS

The clinical diagnosis of psittacosis can be confirmed in the laboratory by isolation and identification of the agent from human cases or suspected bird sources, and by serologic methods.

I. Identification of the Organism

Chlamydia psittaci is an obligate intracellular parasite that can be readily cultivated in chick embryos, small laboratory animals, and cell culture. It is identified by its growth in infected cells where it forms "reticulate bodies" (see chap. 2) and by its reactivity with specific antiserum. Culture is a hazardous procedure that should be carried out only in specialized laboratories.

II. Identification of Patient's Serum Antibodies

Serologic diagnosis is made by testing acute and convalescent samples of the patient's serum. Antibodies can be demonstrated by immunofluorescence or EIA techniques, or by neutralizing infectivity of the agent for animal cells. However, antibody production may be delayed in patients treated with antimicrobial agents, and a 2-week serum sample may not show a significant rise in titer over the acute serum in such cases. Repeated serologic testing done at 3 and 4 weeks after onset may confirm the diagnosis in treated cases.

A single titer of 32 is presumptive evidence of infection in a patient with an illness suggestive of ornithosis.

III. Specimens for Diagnosis

A. Specimens of blood, sputum, or spleen and liver biopsy are submitted for isolation of psittacosis agent. Suspected bird tissues also yield psittacosis agent.

B. Acute and convalescent serum samples drawn at appropriate intervals.

C. A brief clinical history and tentative diagnosis should accompany specimens for laboratory study.

Epidemiology

I. Communicability of Infection

A. *Reservoir, Sources, and Transmission Routes* *C. psittaci* is transmitted to humans primarily from a reservoir in infected birds. Person-to-person transmission occurs rarely. Birds bred for sale as pets, such as parakeets, canaries, and thrushes, are frequently involved. Other important reservoirs for ornithosis are domestic fowl, particularly turkeys, ducks, and geese, but also pigeons and some water birds. Infection rates are high among workers in pet shops and aviaries, poultry farmers, pluckers, and processors. Infected birds may show symptoms such as diarrhea, droopiness, and ruffled feathers; but often they do not appear to be ill. The infectious agent is present in the respiratory secretions of infected birds and also in droppings. It spreads rapidly among birds by contact and through the air. The dust from dried, infected droppings is widely disseminated on air currents created by the motion of wings and feathers. Young birds may acquire the infection from their parents while still in the nest, and although the infection is often inapparent, they continue to excrete the agent for long periods.

Human infection is contracted through inhalation of infectious droplets or dust particles. When direct person-to-person transmission occurs, the source of infection is the sputum of infected persons, the agent being spread in airborne droplets or possibly through indirect contacts with sputum-contaminated articles. Hospital personnel who care for psittacosis patients are sometimes involved in outbreaks of the disease.

B. *Incubation Period* About 10 days are required for symptoms to appear following exposure, the range being 4 to 15 days.

C. *Communicable Period* Human patients are infectious during the acute stage of illness. Birds may be infectious for weeks or months.

D. *Immunity* An attack of the disease confers incomplete protective immunity, and second attacks occur. Psittacosis has a worldwide, nonseasonal distribution. The number of cases reported annually is directly related to the volume of traffic in birds. In the general population it is a sporadic disease, with occasional family outbreaks stemming from an infected pet bird.

II. Control of Active Infection

Standard precautions only are required although careful disposal of discharges is prudent.

Treatment with tetracycline antibiotics is often effective, particularly if it is begun early and continued for several days after fever and other symptoms have subsided. Relapses occur when specific therapy is inadequate. Tetracyclines are also used to treat sick birds.

Contacts should be investigated, but neither quarantine nor immunization is recommended. Identification of the source of infection in diseased birds is essential to control, not only to prevent further spread of human cases from a current source, but to prevent a widening flock epidemic. Infected pet birds are traced to their pet shop or aviary origin; domestic flocks are examined closely if they are involved. Microbiologic diagnosis is made at autopsy of infected birds, and appropriate flock controls are instituted. These may include mass dosing with tetracycline, elimination of sick birds and incineration of their bodies, and thorough cleaning and disinfection of premises, nests, and roosts.

To ensure effective application of control measures at the source of infection, every human case must be reported to the local health authority.

III. Prevention

Effective immunizing vaccines are not available for prevention of psittacosis. Prevention is best achieved by regulations on the import of and traffic in psittacine birds, with quarantine of imports as indicated or of pet shops associated with cases. Improved methods for raising birds for pet sales and prophylactic feeding of tetracyclines have helped to prevent spread of infection in aviaries. Implementation of these measures has resulted in a marked decrease of disease incidence in the United States. In 1995, 64 cases of psittacosis were reported to the CDC.

Chlamydia pneumoniae (TWAR Agent)

The Clinical Disease

Like the Legionnaires disease bacillus, *Chlamydia pneumoniae* has long been responsible for respiratory disease but was definitively identified as a causative agent only during the 1980s. As a result of better procedures for growing chlamydiae, the organism was isolated in cell culture from eight patients with acute respiratory infections. These patients also had antibodies against the agent detected by microimmunofluorescence tests, thus confirming the relationship of the organism with **chlamydial pneumonia.** Known to be a chlamydia, the agent was first thought to be a strain of *C. psittaci*. It was called the **TWAR agent:** TW from "**T**ai**w**an," where an unidentified organism (later shown to be *C. pneumoniae*) had been isolated in chick embryo yolk sacs in 1965, and AR from "**a**cute **r**espiratory," the designation given to a strain isolated from a student with pharyngitis. Because it has distinctive pear-shaped elementary bodies in cell culture and is not related to other *Chlamydia* species by chromosomal DNA analyses, it is considered a distinct species rather than a strain of *C. psittaci*.

Clinically, *C. pneumoniae* produces a variety of respiratory illnesses including pneumonia, bronchitis, pharyngitis, sinusitis, and a febrile illness resembling influenza. Symptoms include fever, cough (often productive), sore throat, hoarseness, and pain on swallowing. On chest X ray, pulmonary involvement is usually confined to part of one lobe in the middle or lower lung. The clinical and radiologic appearance of the disease most closely resembles that produced by *Mycoplasma pneumoniae*.

Infection with *C. pneumoniae* occurs worldwide and serologic tests show that 30 to 50% of the population has antibodies to the organism. Disease occurs both endemically and epidemically; epidemics have been reported, par-

Bacterial Diseases of Humans

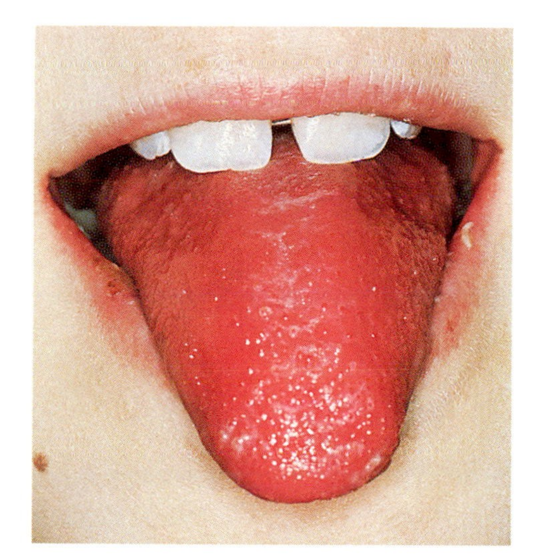

a) A "strawberry tongue" is typically seen in scarlet fever.
Armed Forces Institute of Pathology, Washington, D.C.

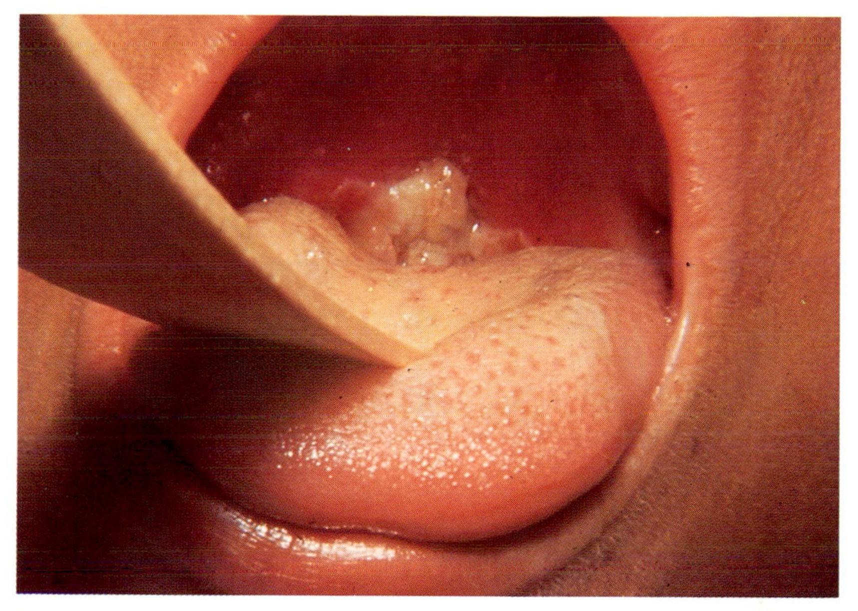

b) Diphtheria is characterized by the growth of a thick, grayish pseudomembrane in the oropharynx.
Centers for Disease Control, Atlanta, GA.

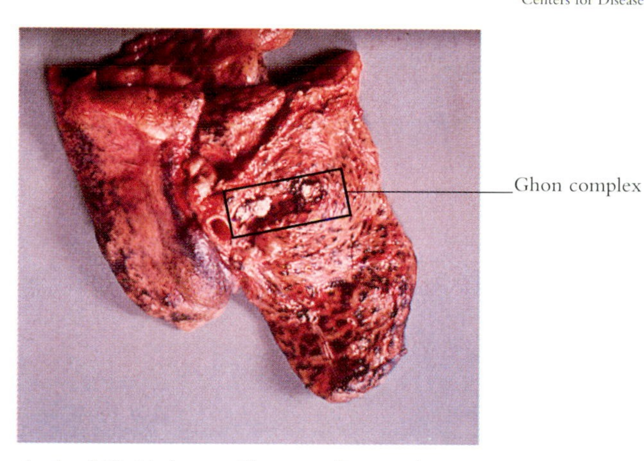

c) A calcified lesion, or Ghon complex, seen in a tuberculous lung.
Dept. of Veterans Affairs, Washington, D.C.

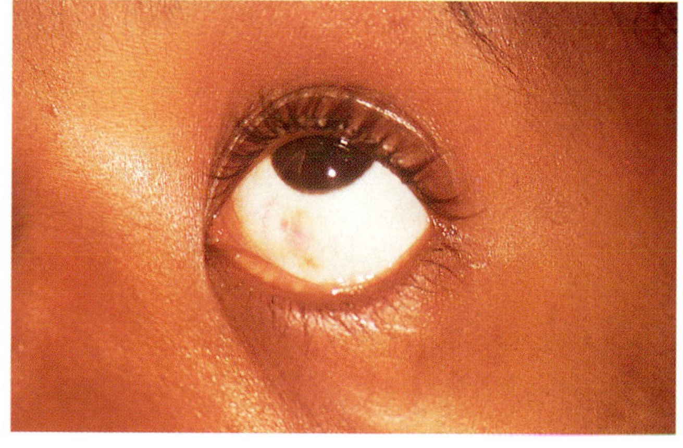

d) Conjunctival lesions in a patient with disseminated gonococcal infection.
Centers for Disease Control, Atlanta, GA. Photo by Dr. M. F. Rein.

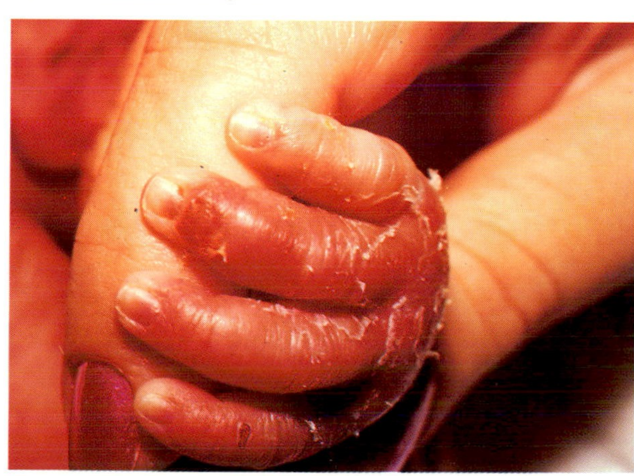

e) Scalded skin syndrome in a newborn infant.
© Charles Stoer/Camera M.D. Studios.

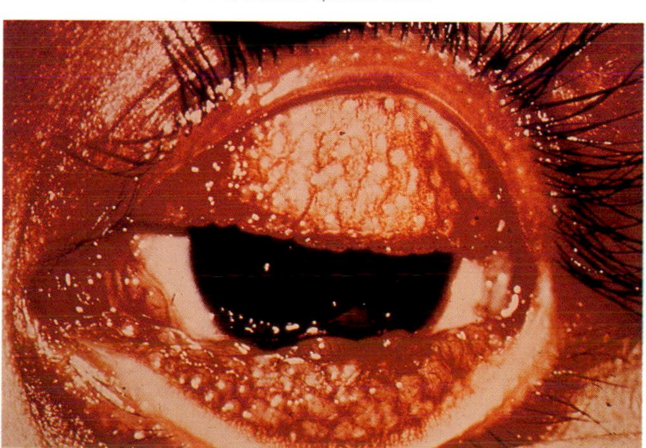

f) Trachoma of the eye. Both eyelids are marked by follicular hypertrophy and the inflammatory nodules cover the conjunctiva.
Armed Forces Institute of Pathology, Washington D.C.

Viral Diseases of Humans

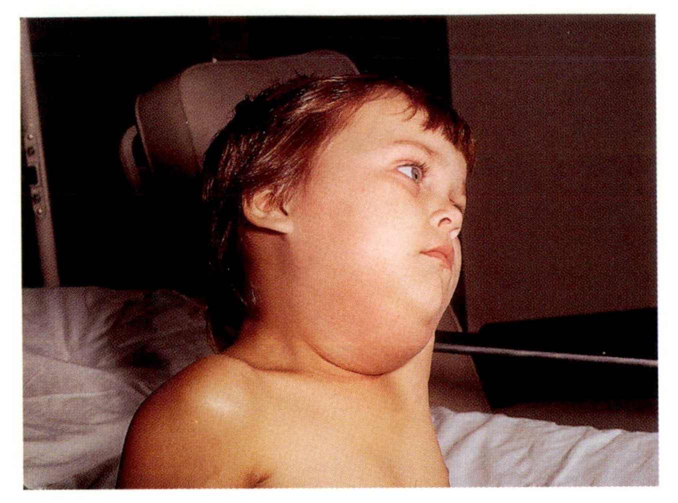

a) Mumps is a generalized viral infection that usually affects the parotid glands. This child has a diffuse lymphedema of the anterior neck due to a mumps virus infection of the thyroid gland.

Centers for Disease Control, Atlanta, GA.

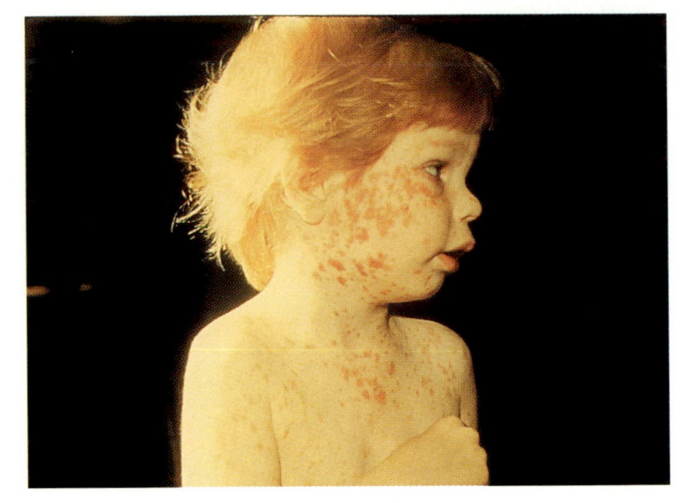

b) A small boy with a severe case of measles.

Centers for Disease Control, Atlanta, GA.

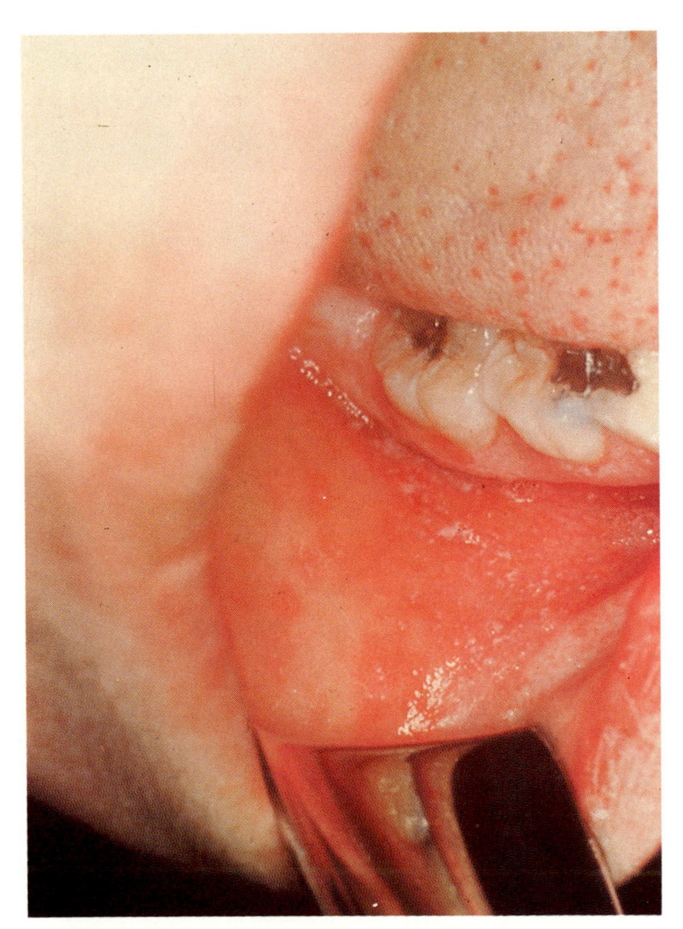

c) Koplik spots on the inside cheek of a child with 5-day-old measles.

Centers for Disease Control, Atlanta, GA.

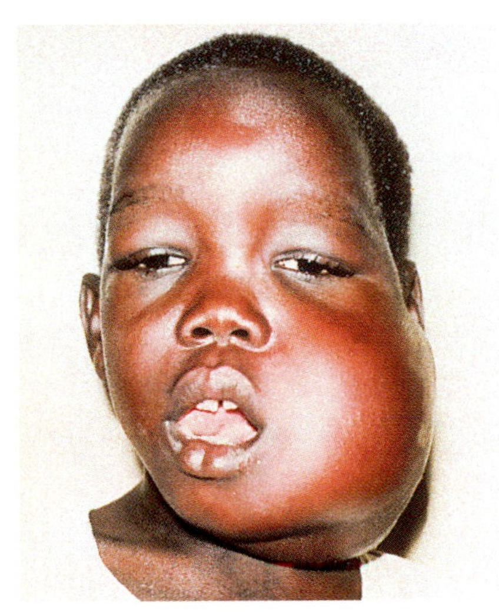

d) The Epstein-Barr virus causes Burkitt's lymphoma.

Centers for Disease Control, Atlanta, GA.

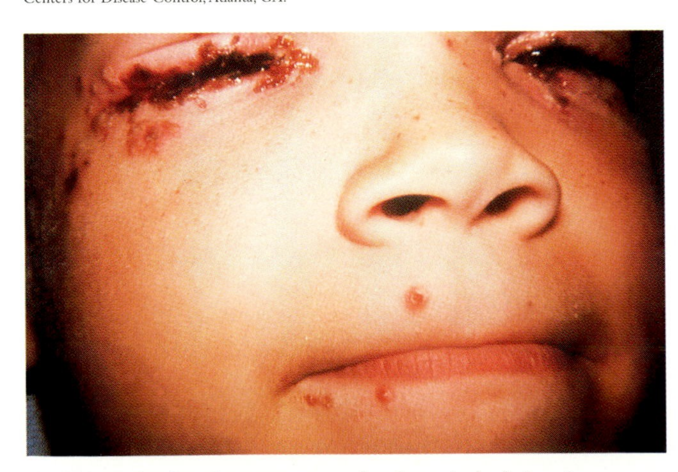

e) Herpes simplex virus recurrent periocular vesicular lesions in a 7-year-old with a history of labial herpes.

Centers for Disease Control, Atlanta, GA. Photo by E. L. Herrman.

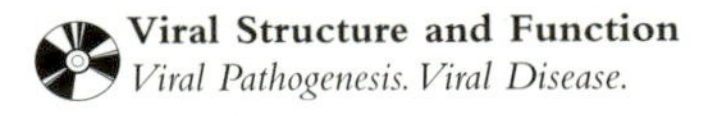

Viral Structure and Function
Viral Pathogenesis. Viral Disease.

ticularly in Scandinavian countries where many early studies of this organism were performed. No bird or animal reservoirs have been found. Spread appears to be from person to person.

LABORATORY DIAGNOSIS

I. Identification of the Organism
Chlamydia pneumoniae has been isolated from patient specimens in chick embryo yolk sac and cell cultures, but the latter method is easier to perform in the laboratory. Diagnosis is made by testing the cell culture with specific monoclonal antibody to detect *C. pneumoniae* antigen. Only specialized laboratories routinely perform these tests.

II. Identification of Patient's Serum Antibodies
A microimmunofluorescence test is performed but only in specialized laboratories. Rises in both IgM and IgG antibody can be detected using appropriate fluorescein reagents. In primary disease, IgM antibody appears after approximately 3 weeks and IgG after about 8 weeks. If only IgG antibody is detected early, it may indicate a reinfection.

III. Specimens for Diagnosis
Sputum specimens or pharyngeal swab specimens in special chlamydia transport medium may be sent to a reference laboratory for cell culture isolation of the organism. Pharyngeal specimens are preferable because sputum is often toxic to the cells used for culture. Acute and convalescent serum specimens may also be referred for microimmunofluorescence and complement-fixation tests.

Epidemiology

I. Communicability of Infection

A. *Reservoir, Sources, and Transmission Routes* Infection occurs worldwide and is limited to humans. No animal or bird host has been identified and therefore spread is presumed to be by human-to-human transmission. The mechanism of spread is not yet known.

B. *Incubation Period* Relatively long; approximately 10 to 30 days following exposure.

C. *Communicable Period* Unknown, may be as long as 8 months.

D. *Immunity* Antibodies detectable, but second attacks have been demonstrated in military recruits.

II. Control of Active Infection
Personal hygiene methods such as covering mouth and nose when coughing and sneezing; sanitary discharge of objects contaminated with infectious secretions. Standard precautions only are required. Tetracycline and erythromycin are used for treatment.

III. Prevention
As with other respiratory diseases, avoiding crowded living conditions or close contact with ill persons. Close contacts may be examined for signs of the disease and treated. No immunization is available.

Infant Pneumonitis

The Clinical Disease

Infant pneumonitis is caused by a third chlamydial species, *Chlamydia trachomatis,* and occurs more commonly than ornithosis. Infants acquire the organism at birth from the infected maternal genital tract (see chap. 19). They become ill at 4 to 16 weeks of age with respiratory symptoms of wheezing and coughing; fever or other toxic signs are usually absent. Diagnosis is made by culturing respiratory secretions in cell culture or by microimmunofluorescence tests. Treatment is with erythromycin or a sulfonamide..

OTHER LOWER RESPIRATORY BACTERIAL DISEASES

WHOOPING COUGH (PERTUSSIS)

The Clinical Disease

Whooping cough (or **pertussis**) is a highly communicable, acute respiratory disease, particularly prevalent in nonimmunized children. It begins with a catarrhal stage that resembles a cold, except that it persists for 10 to 12 days, the initial mild cough becoming progressively irritating, and finally paroxysmal. At this time the organism is extending downward from the upper respiratory tract along the mucosal surfaces of the larynx, trachea, and the bronchial tree. The epithelial tissue along the way is injured and a leukocytic exudate is formed. Pneumonia may develop if the organisms multiply in the interstitial lung tissues. Secondary opportunists may proliferate in alveolar exudates, but the pneumonia of pertussis is interstitial. Excessive mucus production in the bronchial tract may plug the lower airways. This is the cause of the paroxysm, which is characterized by a series of struggling coughs on a single inhalation ending with an involuntary "whoop" of inspiration through narrowed, mucus-filled passages. The mucus is clear but tenacious and difficult to expel. Its obstruction of airways, together with thickening of interstitial tissue in the lung, reduces the amount of oxygen available for the blood. Convulsions may ensue if hypoxia continues. Antimicrobial drugs may reduce the severity of the disease, but they do not change its course to any marked degree. The paroxysmal stage may remain acute for 2 to 3 weeks and persist for 1 or 2 months.

The mortality rate is high for infants: about 80% of the deaths occur in children under one year of age. Active immunization programs have greatly reduced the incidence of this disease in the past 30 years (fig. 12.3), but epidemic cycles continue every 2 to 4 years. Pertussis may occur in all seasons, but the epidemic incidence is highest in the winter and early spring. Because immunity after vaccination is not lifelong, many adults are susceptible to pertussis and probably represent a high proportion of infected patients. Disease is often not diagnosed in adults because symptoms may be atypical.

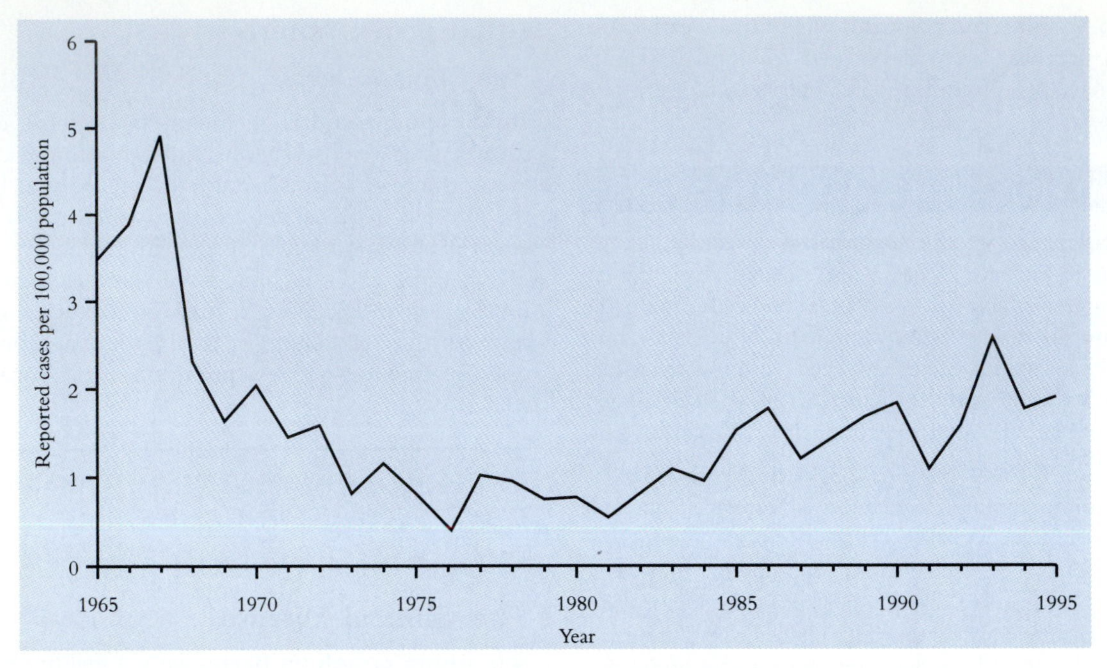

FIGURE 12.3 Pertussis (whooping cough) cases in the United States from 1965 to 1995. Immunization programs have been responsible for the great reduction in the incidence of pertussis. The DTP vaccine was licensed in 1949.

Centers for Disease Control and Prevention *MMWR Summary of Notifiable Diseases, United States, 1995.* (October 25, 1996, Vol. 44: 53).

Gram-Negative Organisms in Human Hosts
Gram-Negative Bacilli—Aerobes: Bordetella

Vaccines
Vaccine Targets: Bacterial Targets. Immunization Schedules

LABORATORY DIAGNOSIS

I. Identification of the Organism

(Bordetella pertussis).

A. *Microscopic Characteristics* The organism is a small gram-negative bacillus closely resembling *Haemophilus.*

B. *Culture Characteristics* The pertussis bacillus requires complex media for its isolation and growth. Regan-Lowe agar is commonly used. It is a mixture of horse blood and charcoal agar. Cephalexin is added to the medium to inhibit growth of other respiratory flora. The pertussis bacillus, which is resistant to these drugs, may then grow as small transparent colonies on the medium.

C. *Methods of Identification B. pertussis* possesses several components that are antigenic. Fluorescent antibody tests are used to detect the organisms directly in nasopharyngeal material or in culture. The agglutination technique can also be used to confirm the identity of isolated organisms.

II. Identification of Patient's Serum Antibodies

Serologic tests are of little diagnostic value.

III. Specimens for Diagnosis

A. Nasopharyngeal swabs in transport medium for culture.

B. Throat swabs may be collected in addition to nasopharyngeal swabs. "Cough" plates, which were once recommended for pertussis diagnosis, are not satisfactory and their use should be discouraged.

Epidemiology

I. Communicability of Infection

A. *Reservoir, Sources, and Transmission Routes* Humans are the only reservoir. Infectious respiratory secretions transmit the infection directly by droplet spread or indirectly by contamination of objects in the environment, including hands and handkerchiefs.

B. *Incubation Period* Not more than 21 days, usually 7 to 10 days.

C. *Communicable Period* Communicability is highest in the late days of the incubation period and during the catarrhal stage. Familial contacts continue to run a high risk of infection until the paroxysmal stage begins to subside. In the schoolroom or out of doors, communicability is reduced by distance and by the susceptibility of the organism to the effects of drying. For general control, the disease should be considered communicable for a period extending from the seventh day following exposure to the end of the third week of the paroxysmal stage.

D. *Immunity* Resistance to pertussis depends on specific immunity. The incidence of clinical disease and of inapparent or atypical infections is high in childhood, especially in preschool years. Active immunity acquired during an attack of the disease in childhood is durable, but occasional second attacks occur in exposed adults. Artificial active immunization is effective and should be started in infancy.

II. Control of Active Infection

A. *Transmission-based Precautions* Droplet precautions are recommended for 7 days after the start of effective therapy. Children and other patients hospitalized with pertussis must be placed in a private room. Precautions are most urgent in hospital nurseries and pediatric wards. At home the sick child should be separated from susceptible children (particularly young infants) insofar as practical. The school-age patient must be kept at home for the duration of the recognized communicable period (usually about 4 weeks).

B. *Other Precautions* Concurrent disinfection of respiratory secretions and all articles contaminated with these discharges. Masks are worn by those in close contact with the patient. Hospital units are cleaned and disinfected terminally. At home frequent cleaning and disinfection of the sickroom should be practiced. The protection of susceptible children includes use of disposable dishes or separating and boiling dishes used by the infectious patient.

C. *Treatment* Antimicrobial therapy is not markedly effective but may help to modify the most severe symptoms, prevent secondary pulmonary infection, and shorten the communicable period. Erythromycin is the drug of choice.

D. *Control of Contacts*

1. *Quarantine* Nonimmune children exposed to whooping cough should be excluded from school or other public contacts with susceptible children for a 2-week period. This time corresponds with the incubation and communicability of the disease. Exceptions may be granted for public schools that have adequate surveillance programs for the detection of early illness in children. Younger children and infants must be guarded from infection with particular care.

2. *Prophylaxis* Unvaccinated, exposed preschool children should be given pertussis vaccine (in combination with tetanus and diphtheria toxoids) as soon as possible, as should children who have not received vaccine within 3 years. Erythromycin prophylaxis for close contacts is also recommended.

3. *Case finding* The obvious case may be found and isolated easily. Attempts to find atypical cases or those that have been misdiagnosed must also be made.

E. *Epidemic Controls* Individual cases must be reported to local health departments. Beyond this, accurate case finding of unrecognized active infections is the key to controlling further spread.

III. Prevention

A. *Immunization* Artificial active immunization during infancy affords the best protection against whooping cough. Bacterial vaccines that effectively stimulate the production of protective antibody are available. Pertussis vaccine (usually combined with diphtheria and tetanus toxoids [DTP]) may be given to infants 6 to 8 weeks of age. Two doses are then given at 2-month intervals, followed by additional booster doses 6 to 9 months later and again at 4 to 6 years of age. The vaccine is estimated to be more than 80% effective in preventing pertussis.

In rare instances, administration of pertussis cellular vaccine (a preparation of whole, killed pertussis organisms) has been associated with convulsions and permanent neurologic damage. Although conclusive evidence that these effects are caused by the vaccine is lacking, in some countries pertussis vaccine is no longer administered routinely. The result has been a resurgence of disease in those countries, with an increased number of deaths due to pertussis. Consequently, work has been in progress on an acellular vaccine containing *B. pertussis* toxins and other components thought to be important in disease production. These acellular vaccines showed promise in early studies conducted primarily in Japan. They are now licensed in the United States for use in the fourth and/or fifth doses of the diphtheria, tetanus, pertussis (DTP) vaccine series (see table 10.1).

B. *Control of Reservoir, Sources, and Transmission Routes* Control of the human reservoir and sources of infection is accomplished most effectively by specific immunity actively acquired through infection or by immunization.

Diphtheria

The Clinical Disease

Large epidemics of **diphtheria** no longer occur throughout most of the world, although in 1994, an epidemic that affected approximately 48,000 people was reported from the former Soviet Union. The disease remains a potential threat, however, as evidenced by the outbreak in the Soviet Union and by sporadic cases and small outbreaks that occur among persons who lack naturally or artificially acquired active immunity against the toxin produced by the responsible organism. Such outbreaks occurred in the United States in San Antonio, Texas, in 1970 and Seattle, Washington, in 1975 (fig. 12.4). While the incidence of clinical diphtheria has sharply declined since the introduction of nearly universal artificial immunization (fig. 12.4), the case fatality rate of 5 to 10% has not changed significantly.

Diphtheria begins as an acute infection of the mucous membranes of the upper respiratory tract. The growing diphtheria bacilli produce an exotoxin that destroys the local epithelial cells and induces an inflammatory response at the site. The necrotic surface epithelium becomes enmeshed in a fibrinous and cellular exudate that lies like a membrane over the involved area. This pseudomembrane formation advances as the organisms continue to grow. It may cover tonsils and pharynx (colorplate 5b) or extend downward to the larynx and trachea. Laryngeal diphtheria may cause death by suffocation, particularly in infants or young children (fig. 12.5). The pseudomembrane is tough and cannot be lifted without tearing the structures beneath, leaving an exposed, bleeding undersurface.

Patients often appear extremely toxic, even though fever is moderate (100 to 102° F [37.8 to 38.9° C]). Cervical lymph nodes become enlarged and tender, and there

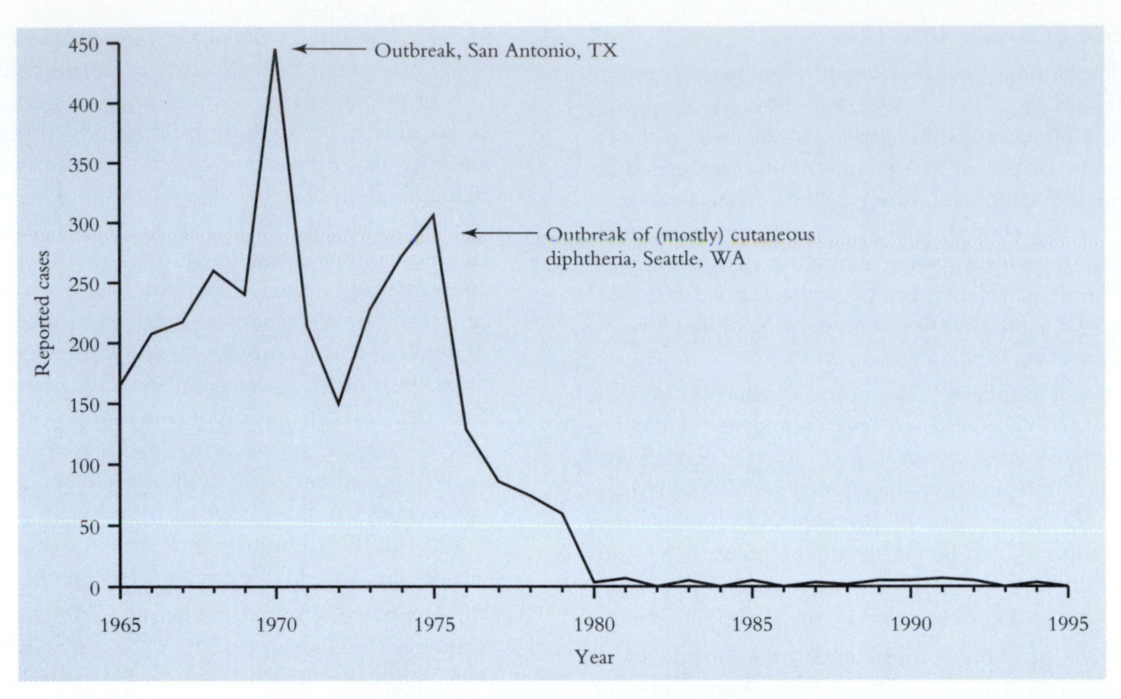

FIGURE 12.4 Diphtheria cases in the United States from 1965 to 1995. There has been a steady decline of diphtheria since 1975 and the disease has stabilized at low levels since 1980. DPT, the diphtheria, pertussis, tetanus vaccine, was licensed in 1949.

Centers for Disease Control and Prevention. *MMWR Summary of Notifiable Diseases, United States, 1995.* (October 25, 1996, Vol. 44: 53).

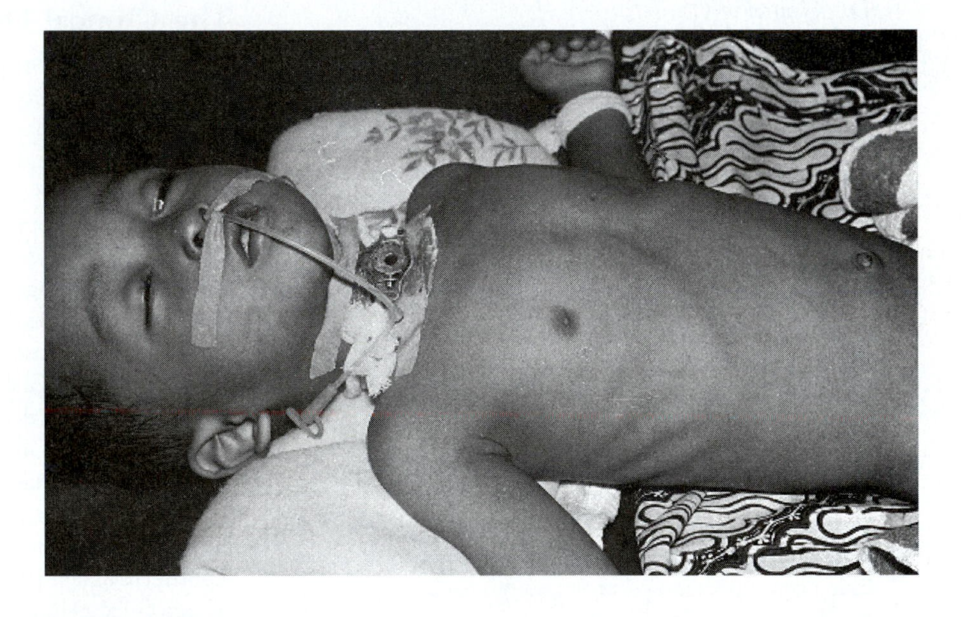

FIGURE 12.5 Diphtheria is still one of the most lethal communicable diseases. A tracheotomy keeps this Indonesian child alive.

World Health Organization. Photo by L. Matlovski.

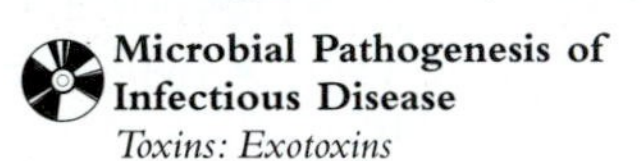

Microbial Pathogenesis of Infectious Disease
Toxins: Exotoxins

may be extensive edema of the neck. The organisms continue to proliferate in the pseudomembrane and to produce toxin, which is absorbed and distributed through the body. The exotoxin is responsible for the systemic manifestations of the disease. It is composed of two parts, A and B. B binds the toxin to the surface of the host cell after which A is cleaved off and enters the cell where it acts by inactivating protein synthesis. It does so by inhibiting an enzyme known as elongation factor 2. This enzyme prevents messenger RNA from interacting with transfer RNA so that amino acids can no longer be added to the developing polypeptide chains (see fig. 2.10). The toxin affects all eukaryotic cells, but cardiac, neural, and renal cells are particularly susceptible. Diphtheria toxin is extremely potent. A single molecule can completely inactivate protein synthesis in a cell within a few hours. For this reason, infection with a toxigenic strain of *C. diphtheriae* is often referred to as a *toxemia* or *intoxification*.

Because of its effects, the exotoxin may cause severe injury to parenchymal cells of the liver, adrenals, kidneys, and (a more immediate threat) the myocardium. Cardiac deaths are frequent among patients succumbing in the stage of systemic toxin involvement. Cranial and peripheral nerves, both motor and sensory, may also be damaged, with resulting paralysis of ocular muscles, the soft palate, or the extremities, depending on the nerve affected.

Aside from this picture of classical diphtheria, the organism is sometimes associated with wound or skin infections. These superficial infections may also be marked by local membrane formation and absorption of exotoxin with resulting systemic toxicity. Diphtheritic skin or wound infections are common in tropical countries, but recently these infections have been recognized in the United States, especially in southern areas. The lesions may persist without healing for weeks or months and represent an important source of organism dissemination.

Prompt clinical diagnosis of diphtheria is the most important factor in preventing crippling or fatal toxicity. Immediate passive immunization with antitoxin provides circulating antibody to neutralize toxin before it can attach to cells in the myocardium or elsewhere and cause irreparable damage. Simultaneous therapy with an appropriate antimicrobial agent then eliminates the infectious agent from the throat. The diagnosis is confirmed in the laboratory by demonstrating the organism in smears and cultures of the throat exudate.

LABORATORY DIAGNOSIS

I. Identification of the Organism

(*Corynebacterium diphtheriae*). The diphtheria bacillus is one of several species belonging to the genus ***Corynebacterium,*** but it is the major pathogen. The others are common commensals of the skin and mucous membranes that occasionally may cause infective endocarditis in patients with damaged heart valves, or after cardiac surgery (see chap. 15). The microbiologist must distinguish these "diphtheroid" species from *C. diphtheriae* in cultures of material from the nose and throat.

A. *Microscopic Characteristics* The corynebacteria are gram-positive bacilli, nonspore forming, noncapsulated, and nonmotile. *C. diphtheriae* is slender with granular contents that stain irregularly and give it a beaded or barred appearance. Individual rods are often club-shaped. They lie in characteristic "V" and "Y" formations or in groups lined up like fence posts (palisaded). Diphtheroids appear in this arrangement also, but they are usually shorter, thicker rods. The distinction may be difficult to make in direct smears of throat exudate, but the smear report is evaluated by the physician in the light of clinical findings. If these suggest diphtheria, the decision to give antitoxin is made at once, with or without culture confirmation.

B. *Culture Characteristics* All corynebacteria grow aerobically on blood agar, but certain special media are useful for selecting and differentiating the diphtheria bacillus from other organisms present in throat cultures. Loeffler's medium is used because it does not support the growth of most commensal contaminants. The addition of potassium tellurite to blood agar inhibits development of many throat organisms and also distinguishes colonies of *C. diphtheriae,* which become black, from some other bacteria that are capable of growing in the presence of this chemical.

The morphology of *C. diphtheriae* is confirmed by Gram-stained smears, and the toxigenicity of the isolate is then tested by the Elek plate method (see fig. 6.15). If toxin precipitation by antitoxin is demonstrated in this test, the need for prompt immunotherapy is underscored.

Toxin-producing strains of *C. diphtheriae* (called toxigenic strains) are infected with a temperate bacteriophage (see. chap. 3). This virus carries a gene that directs the production of exotoxin by the bacterial cell. Uninfected bacterial cells (nontoxigenic strains) do not produce toxin although they retain their ability to invade and multiply in human epithelial tissue.

II. Identification of Patient's Serum Antibodies

The demonstration of circulating antitoxin is not appropriate in the diagnosis of diphtheria because active production of antibody in the sick individual occurs much too late to have diagnostic value.

III. Specimens for Diagnosis

In pharyngeal diphtheria, material from the suspected lesion in the throat and a nasopharyngeal swab are submitted for smear and culture. These are the only specimens of diagnostic value, since diphtheria bacilli do not spread from those sites. Throat material is best collected by running a thin swab under the lifted edge of the pseudomembrane. The organisms may be present in nearly pure culture in the area beneath, whereas many saprophytic organisms may be living on the dead tissue at the surface of the membrane. In wound diphtheria, specimens are taken from the base of the lesion after thorough cleansing of the area and removal of crusted material.

Epidemiology

I. Communicability of Infection

A. *Reservoir, Sources, and Transmission Routes* Clinically or inapparently infected humans are the reservoir. Discharges from the normal respiratory membranes of carriers are a source of infection as well as those from active throat or skin lesions. The organism is transmitted in direct contacts, by droplet spread, and through indirect contacts with objects contaminated by respiratory secretions.

B. *Incubation Period* Usually only 2 to 5 days.

C. *Communicable Period* This varies with the persistence of bacilli in the lesions or on recovering mucous membranes. The carrier state may persist for many weeks following convalescence. Antimicrobial therapy reduces the number of carriers and shortens the carrier state, but these measures have not eliminated the problem completely because the carrier state sometimes recurs after therapy.

D. *Immunity* Resistance to clinical disease depends on the presence of circulating antitoxin, either passively or actively acquired. An attack of diphtheria does not always confer immunity; therefore, patients should receive immunizing toxoid during the convalescent stage of this disease. The passive antibody acquired through antitoxin administration in the acute stage of infection persists for only 2 to 3 weeks. Infants born of immune mothers acquire passive immunity congenitally, but this lasts only 5 or 6 months. A complete course of active immunization with diphtheria toxoid provides the most effective immunity.

II. Control of Active Infection

A. *Transmission-based Precautions* Patients ill with pharyngeal diphtheria should be treated with droplet precautions and those with wound diphtheria should be kept on contact precautions until two cultures taken on consecutive days following the completion of antimicrobial therapy are bacteriologically negative. When laboratory evidence is difficult to obtain, isolation may be terminated after the patient has received 14 days of appropriate antimicrobial therapy.

B. *Other Precautions* Concurrent disinfection of all articles involved in droplet or contact transfer of infection. Terminal cleaning-disinfection of the room should be thorough.

C. *Treatment* Antitoxin is given in a single dose of 20,000 to 100,000 units as indicated by the duration and severity of symptoms. Because the antitoxin is prepared in horses, the patient should first be carefully evaluated for hypersensitivity to horse serum proteins. Erythromycin and penicillin are the drugs of choice for therapy, but they have no effect on diphtheria toxin or the toxic symptoms.

D. *Control of Contacts*

1. *Isolation* Adult contacts who care for children or handle food should be removed from these occupations until bacteriologic evidence shows them not to be carriers. Contacts with inapparent infections should be given antimicrobial therapy.

2. *Prophylaxis* Nonimmune close contacts of an active case should be given the first dose of toxoid and an appropriate antimicrobial agent and should then receive complete immunization. A physician can determine the need for active immunization of individuals who have an accurate history of previous immunization. Immunization may be particularly important for older people whose immunity has waned.

3. *Case finding* In communities where the population has been universally immunized, carriers and atypical cases are rare. When an unusual incidence of clinical diphtheria begins to appear in any community, however, a careful search for possible sources should be made.

E. *Epidemic Controls* Immediate surveys are made to determine the general level of immunity in the population involved. Active immunization is conducted on the scale indicated, concentrating on the youngest children but extending to all age groups. The search for contacts and sources is intensified. Individual case reporting generally is required in this country and others.

III. Prevention

Immunization Artificial active immunization with diphtheria toxoid on a mass scale has provided effective control of the human reservoir of this disease. Immunity must be maintained with booster doses of toxoid at appropriate intervals (see table 10.1).

Diphtheria exposure is particularly likely for hospital personnel in contact with patients, for military personnel traveling to or stationed in endemic areas, and for professional travelers (such as airline flight personnel), teachers, and others in contact with many people from different parts of the world.

Future control of diphtheria depends on the continuing education of people everywhere as to the necessity for adequate artificial active immunization.

TUBERCULOSIS

Tuberculosis is a chronic infectious disease caused by one member of a large group of bacteria associated with numerous human and animal infections. These organisms are members of the genus ***Mycobacterium,*** several of which may infect humans, producing pulmonary disease similar to tuberculosis or more superficial infections of the skin or lymph nodes. Some of these are primarily pathogens of cattle, fowl, reptiles, or fish, whereas others are nonpathogenic saprophytes of soil and water that may also be found among the commensal flora of the human skin or superficial membranes.

The pathogenic mycobacteria are intracellular parasites that primarily infect macrophages. Most can be cultivated on artificial media and are therefore not obligate parasites. Some members of the group, notably the causative agent of Hansen disease (leprosy), *Mycobacterium leprae,* cannot be cultivated but grow only in certain animals or in cultures of macrophages. Some of these obligate intracellular parasites are associated with disease in rodents, but *M. leprae* is the only one of clinical importance to humans.

Mycobacterium tuberculosis, the most usual cause of human tuberculosis, has a respiratory route of entry, as do other strains associated with pulmonary disease. Hansen disease (leprosy) is also spread by this route (see chap. 19). Cutaneous mycobacterial infections are acquired by contact with infected animal or marine sources. The most important mycobacterial species are shown in table 12.2, together with the diseases they cause.

Although the pulmonary disease caused by several mycobacterial species is clinically indistinguishable from that due to *M. tuberculosis,* the discussion of tuberculosis that follows refers to this organism, the host responses it induces, and its epidemiology.

The Clinical Disease

Tuberculosis is a chronic disease characterized by a continuing interplay between the invading organism and host mechanisms for resistance, especially hypersensitivity responses. Specific immunity is primarily of the cell-mediated type, but nonspecific resistance is also important.

TABLE 12.2 MYCOBACTERIA IN INFECTIOUS DISEASE

Disease	Species	Host(s)	Route of Entry
Tuberculosis	*Mycobacterium tuberculosis*	Human	Respiratory
	Mycobacterium africanum	Human	Respiratory
	Mycobacterium bovis	Cattle and human	Alimentary (milk)
Pulmonary disease resembling tuberculosis (mycobacterioses)	*Mycobacterium avium*[a]	Fowl and human	Respiratory
	Mycobacterium intracellularea[a]	Human	Environmental contacts (water and soil)?
	Mycobacterium kansasii	Human	
	Mycobacterium szulgai	Human	
	Mycobacterium xenopi	Human	
Lymphadenitis (usually cervical)	*Mycobacterium tuberculosis*	Human	Respiratory
	Mycobacterium scrofulaceum	Human	Environmental contacts (water and soil)?
	Mycobacterium avium complex	Fowl and human	
Skin ulcerations	*Mycobacterium ulcerans*	Human	Environmental contacts?
	Mycobacterium marinum	Fish and human	Aquatic contacts
Soft tissue	*Mycobacterium fortuitum*	Human	Environmental contacts?
	Mycobacterium chelonae	Human	
Hansen disease (leprosy)	*Mycobacterium leprae*	Human	Respiratory
Saprophytes: water, soil; human skin and mucosae	*Mycobacterium smegmatis*		
	Mycobacterium phlei		
	Mycobacterium gordonae		

[a]*Mycobacterium avium* and *M. intracellulare* are so closely related that they are often identified as the *Mycobacterium avium* complex.

Miscellaneous Bacteria in Human Hosts
Mycobacteria: Taxonomy; Habitat; Virulence Factors; Diseases

Pathogenesis

The virulence of **tubercle bacilli** is associated with their level of capacity for establishing and multiplying in the human host and with their degree of resistance to human defenses. After the bacilli first enter into a susceptible host, production, progression, or healing of tuberculosis lesions depends on the numbers of invading organisms and their success in multiplying locally.

When lesions are produced, they may be one of two types, depending on local host reaction. *Exudative lesions* are formed by an acute inflammatory reaction, with engulfment of tubercle bacilli in a fluid exudate containing polymorphonuclear and monocytic cells. This type of lesion may subside and heal; progress and lead to necrosis of local tissue; or develop into the second type of lesion—a *tubercle.* This second type of lesion is called *productive* (i.e., producing new tissue) because it consists of cells (without exudate) organized in a fairly definite way around the offending organisms. It is called a *granuloma* because it resembles a tumor. Its central part contains a compact mass of large multinucleated cells (giant cells) formed by coalescence of mononuclear cells. Surrounding it is a zone of "epithelioid" cells (which are modified macrophages), and an outer zone composed of lymphocytes, monocytes, and fibroblasts. The tubercle bacilli reside in the center of the tubercle, principally intracellularly, within giant cells or mononuclear cells. Eventually the tubercle may become fibrous at its periphery and calcified within. The organisms may remain viable for years even under these conditions, or they may slowly dwindle and die. Often the central area of the tubercle becomes soft and cheesy in a process called *caseation* or *caseation necrosis.* Caseous tubercles sometimes break open, spilling their contents into surrounding tissue and forming a cavity that then may heal by fibrosis and calcification.

Spread of the organisms through the body may occur in a number of ways. From exudative lesions or cavitating tubercles they may extend directly into adjacent tissue, enter the bloodstream or the lymphatic flow, or find their way along mucosal surfaces. The cavitation of a caseous tubercle in the lung, for example, may spill organisms into a bronchus. From here they may pass downward into the lung or be coughed up into the throat, swallowed, and passed into the gastrointestinal tract. When tubercle bacilli initially enter the body through respiratory or alimentary routes, they are spread from their original site through lymphatic ducts into regional lymph nodes.

Patterns of Disease

Two types of tuberculosis are recognized: *primary,* or first-infection type, and *reinfection* tuberculosis.

In primary tuberculosis the initial lesion is of the acute exudative type and spreads rapidly by way of lymphatics to regional nodes. Within a few days the organisms reach the thoracic duct and the general circulation, disseminating to all parts of the body. In most individuals, however, as the hypersensitivity response develops, the exudative lesions quickly resorb and heal, although with some residual calcification, fibrosis, and scarring. The healed lesion is referred to as a **Ghon complex** (colorplate 5c) and can be detected on X ray.

Primary tuberculosis occurs most frequently in childhood but may also be seen in adults who have not acquired infection previously. In first infections the primary lesion usually occurs in the lower two-thirds of the lungs.

Reinfection tuberculosis is a more chronic disease characterized by lesions of the productive type. Reinfection may be acquired from new exogenous sources but, more commonly, is caused by organisms that survive in and extend from primary lesions (endogenous reinfection). Regional lymph nodes are seldom involved in reinfection. The prominent lesions of this type of disease almost always begin in the apex of a lung and progress downward. Tubercles form, become caseous, and cavitate; the organisms pass down the bronchial tree to new sites where tubercles form again to resist their spread.

Endogenous reinfection may occur many months or years after first infections appear to be healed. Fibrosis and calcification are not always sufficient to suppress and finally destroy the organism. Lowered host resistance may permit organisms to break through from a primary focus or a regional lymph node, with resulting spread. Occasionally, organisms from a cavitating focus enter the blood or lymph stream in sufficient numbers to assure widespread distribution throughout the body. General infection then ensues, with tubercle formation in many organs. This distribution is called *miliary tuberculosis* because the tubercles are small and hard and resemble millet seed.

Resistance, Hypersensitivity, and Immunity

The speed and extent with which reinfection tuberculosis progresses are partly determined by the numbers of organisms released from infected foci and their route of spread. Progress of the disease is also determined by the increasing capacity of the host to resist, that is, to localize bacilli and retard their growth or destroy them and thus limit their dissemination. The nonspecific resistance of the inflammatory process and of mononuclear phagocytic cells plays a primary role in first infection. Cell-mediated immunity (CMI) also develops, increasing the reactivity of mononuclear cells. They become more adept at phagocytosis, and tubercle formation becomes more rapid. Although the CMI response inhibits the multiplication of tubercle bacilli within mononuclear cells, it is unable to bring about destruction of all of the organisms. Tubercle bacilli within macrophages located in a tubercle somewhere in the lung or lymphatic system may remain in check for long periods but can multiply again if local conditions and lowered resistance permit. At the same time delayed hypersensitivity to the proteins of the tubercle bacillus may lead to a destructive inflammatory reaction as new multiplication occurs. Resultant necrosis of adjacent tissues can then contribute to the spread of the organisms and progressive disease. Hypersensitivity can thus be involved in a new cycle of caseation, cavitation, and spread.

The tuberculin skin test for hypersensitivity to the proteins of tubercle bacilli becomes positive during the course of first infection and remains so for many years. Methods for testing and the interpretation of reactions are described later.

Clinical Symptoms

The wide variety of symptoms in tuberculosis reflects the fact that the organism may involve any organ to which it spreads. The intensity of symptoms depends on the numbers and location of foci and the host response to them. There may be no symptoms at all in a first infection, or the primary course may be rapidly progressive and sometimes fatal. The signs of reinfection tuberculosis may also be minimal or prominent. Common symptoms include fever and malaise, easy fatigue, and loss of weight. In pulmonary disease there may be a productive cough yielding blood as well as sputum. Tuberculosis of kidney, liver, bone, genital organs, or other tissues may occur simultaneously by hematogenous spread, usually from a pulmonary focus. Or, another of these sites may be involved separately, with symptoms characteristic of the location.

Tuberculous meningitis may occur as a result of bacteremia. It is more commonly seen in children following first infection than in adults. The symptoms are more chronic and slowly progressive than those of the acute bacterial meningitides (see Meningitis, chap. 15), with lower cerebrospinal fluid cell counts that are predominantly lymphocytic. The clinical form of tuberculous meningitis is similar in some respects to that seen in fungal meningitis (see chap. 14).

LABORATORY DIAGNOSIS

I. Identification of the Organism

(Mycobacterium tuberculosis). The tubercle bacillus belongs to the genus *Mycobacterium,* whose most important members are shown in table 12.2.

A. *Microscopic Characteristics* The mycobacteria stain with difficulty. Once they take an appropriate stain, however, they resist decolorization, even by acids, and are referred to as "acid-fast" bacilli because of this property. The Ziehl-Neelsen (or Kinyoun) stain (see chap. 4) is commonly used to identify them, the organisms appearing red against a blue background. Another stain, using the fluorescent dyes auramine and rhodamine, permits rapid detection of the organisms as bright objects against a dark background. This technique is used in many laboratories today.

Tubercle bacilli are slender straight or slightly curved rods, often granular or beaded in appearance. Like the corynebacteria (see Diphtheria, this chapter), they are often seen in palisade formations, or in "V" and "L" positions that suggest a snapped stick. (The colloquial phrase "red snapper" is often quite appropriate.) (colorplate 2d)

B. *Culture Characteristics* The growth rate of tubercle bacilli and most mycobacteria is much slower than that of other bacteria. In general, these mycobacteria have a generation time of 18 to 24 hours. As such, several days or weeks are required for primary isolation, with an average incubation time of 3 to 6 weeks on solid culture media. They are aerobic, but their growth is enhanced with increased atmospheric CO_2. Incubation is at 35° C; however, duplicate cultures of material from skin lesions are also incubated at 32° C, the temperature preferred by certain environmental species that may produce cutaneous infections.

A number of media are available for culture of tubercle bacilli. Some of these contain complex nutrients such as egg, potato, and serum. The most commonly used egg medium is Lowenstein-Jensen (colorplate 3f). Other media, such as Middlebrook 7H-11 or 7H-12, are formulated to provide the factors best known to promote mycobacterial growth. Many laboratories use two or more different media to ensure successful cultures and some use a radiometric method to detect the presence of mycobacteria in patient specimens. This method relies on a special instrument to monitor the release of radioactive $^{14}CO_2$ from a selective Middlebrook 7H-12 broth medium. The broth contains nutrients labeled with the radioisotope ^{14}C. If mycobacteria are present in the specimen, they metabolize the radioactive substrates, thus releasing increased amounts of $^{14}CO_2$. This method is more sensitive than other culture methods; therefore, specimens positive for mycobacteria can often be detected within several days to 2 weeks rather than the usual 3 to 6 weeks. Similar methods that avoid the use of radioactive nutrients are now being introduced.

Since tubercle bacilli may produce lesions in various parts of the body, many types of specimens are appropriate for culture. Some of these, however, contain commensalistic organisms indigenous to the region (sputum, voided urine, etc.). Many bacterial species also grow on the media used for isolation of tubercle bacilli, and they grow much faster. Their multiplication rapidly depletes nutrient and also makes the search for tubercle bacilli virtually impossible in the mixed culture. For these reasons it is necessary, prior to culture, to treat such specimens with chemicals that destroy organisms other than mycobacteria.

The chemical treatment, often with sodium hydroxide, is carefully controlled to avoid undue suppression of tubercle bacilli. Fluid specimens are then concentrated by centrifugation, and the sediment is inoculated onto appropriate culture media. Cultures are routinely examined every week for developing growth of *M. tuberculosis* (colorplate 3f) or more often in cases of clinical urgency. Negative cultures are not discarded until they have been incubated for at least 6 to 8 weeks.

The lipid, hydrophobic nature of the surface of tubercle bacilli accounts for their heaped, clumped colonial growth on the surface of solid media. Certain compounds with a detergent or wetting action (e.g., Tween 80) are used in liquid media to permit more dispersed and rapid growth. The waxy character of the surface of this organism contributes to its resistance to chemical and physical agents. Tubercle bacilli are not as readily killed by antiseptics or disinfectants as are other vegetative microorganisms. They are also more resistant to drying and can survive for long periods in dried sputum or other infectious discharges. For this reason special care must be taken in selecting chemical disinfectants, adjusting their concentrations, and timing the disinfection period when tubercle bacilli are to be killed by chemical methods.

C. *Species Identification* Because they grow so slowly, mycobacteria are more difficult to identify to the species level than are most other human pathogens. Initial observations include rate of colony growth (visible in greater or less than 7 days), colony pigmentation (buff to orange in color), growth at low (32° C) and high (42° C) temperatures, and reaction in a few biochemical tests. Among the most important tests for distinguishing *M. tuberculosis* from other potential human mycobacterial pathogens is the one that detects accumulation of niacin in culture medium on which colonies of the organism are growing. Because it displays this characteristic, *M. tuberculosis* is said to be niacin-positive. (With rare exceptions, other species afflicting humans are niacin-negative.) *M. tuberculosis* also grows slowly in culture, produces buff-colored colonies, and grows at 35 to 37° C but not at 32 or 42° C.

A rapid DNA probe test can be used to identify *M. tuberculosis* and some other mycobacterial species growing in culture. The test is highly accurate but expensive to perform. It has found greatest use in reference and other laboratories with high rates of positive mycobacteria cultures.

D. *Immunologic Methods of Identification* Such tests are performed primarily on a research basis or in reference laboratories.

II. Identification of Patient's Antibodies

A. *Circulating (Humoral) Antibodies* These are of no diagnostic value.

B. *Skin Tests* Tuberculin hypersensitivity is associated with immune T lymphocytes. It can be detected by skin tests using protein antigens from filtrates of tubercle bacillus cultures. When the antigen is injected into the skin of a hypersensitive individual, an inflammatory reaction occurs. The cellular response consists primarily of activated macrophages (responding to lymphokine stimulation) with a few lymphocytes. This cellular infiltrate, along with an increased blood vessel permeability, results in a reddened, slightly raised, hard area of induration at the skin site where the antigen was injected. The degree of reaction depends on the level of individual immunity; severe reactions may be accompanied by blood vessel damage, thrombosis, and tissue necrosis, with sloughing at the skin test site. This delayed reaction is usually fully developed by 48 to 72 hours, in contrast to immediate reactions such as anaphylaxis that occur within minutes after antigen exposure (see chap. 6).

Since the hypersensitivity of tuberculosis is long lasting, a positive reaction indicates past infection that may or may not be currently active. (The possibility of current infection must then be determined by clinical findings, X ray, and bacteriologic methods.) A negative tuberculin reaction is displayed by individuals who have never been infected, by those who are in the preallergic early stage of first infection, and often by those with advanced miliary tuberculosis.

1. *Skin-testing antigens* Old tuberculin (OT) is a concentrated filtrate of a broth culture of tubercle bacilli. It contains, in addition to the tuberculoproteins derived from the organism, a number of irrelevant constituents of the bacilli and of the medium that may induce nonspecific reactions in the skin. A **purified protein derivative (PPD)** is obtained by chemical fractionation and purification of OT. This purified antigen is preferred for skin testing.

2. *Skin-test doses* The usual dose administered has been carefully standardized to distinguish infected from noninfected persons. A dose of 5 tuberculin units (TU), equivalent to 0.0001 mg PPD, is given first; if the test is negative and suspicion of disease is high, a subsequent injection with 250 TU is administered.

3. *Skin-testing methods* A number of techniques have been used to introduce tuberculins into the skin, but the **Mantoux test** is most accurate and therefore preferred. This test requires the intracutaneous injection of 0.1 ml of fluid containing 5 TU. The material is injected into the ventral surface of the forearm, just within the epidermal layer. The need for individual syringes and needles is a disadvantage when performing mass tuberculin test surveys. In these instances, the less accurate tine test, which employs dried OT on multiple metal tines in a round plastic head, is used for intracutaneous inoculation.

4. *Reading and interpreting tuberculin skin tests* Skin tests should be read in 48 hours, or no later than 72 hours (colorplate 4g). They are interpreted as positive if an area of induration with at least a 10-mm diameter can be felt (the test is "read" with fingers lightly pressed over the injection site). Edema and erythema may extend beyond the indurated area, but are disregarded. Strongly positive reactions may persist for several days, but weak ones disappear quickly after 72 hours. Nonspecific reactions may appear during the first 24 hours but do not persist to the second day.

Tuberculin reactivity begins within 6 to 14 weeks after infection and persists for long periods. It may be lost in the presence of overwhelming tuberculosis infection or during the course of such diseases as measles, sarcoid, and Hodgkin's (anergy). As pointed out previously, a positive test does not necessarily denote present infection, nor does it imply immunity. Reinfection from exogenous sources, however, is much less common than endogenous reactivation of latent disease.

III. Specimens for Bacteriologic Diagnosis

A. In pulmonary tuberculosis, sputum is the most appropriate specimen for smear and culture. Three specimens collected on consecutive or alternate days increase the cultural diagnostic rate. If the patient is unable to produce sputum, gastric washings or sputum induced by saline nebulization may be useful. In disseminated, or miliary, forms of the disease any specimen from a suspected focus of infection may be relevant: pleural fluid, synovial fluid, peritoneal fluid, or spinal fluid; urine, aspirated bone marrow, or exudate from lesions in bone, biopsied tissue, and other suspected material. Blood specimens cultured with special methods are sometimes rewarding if blood is drawn during a period of active bacteremia. Stool specimens have the least value and offer the greatest technical difficulties. They are warranted only in rare instances of suspected involvement of intestinal lymphatic tissue, when this cannot be confirmed by other means.

B. Sputum should be collected in the morning when the patient first awakens. Sputum pools in the bronchi during sleep and thus contains many more bacilli than casual specimens coughed up during the day. However, any specimen that appears to contain caseous particles should be submitted for culture. If the patient is nebulized in the early morning, a more representative specimen can often be obtained. Propylene glycol should not be used in the nebulizer, however, since this substance can inhibit or kill tubercle bacilli.

Twenty-four-hour sputum collections are to be discouraged because they are contaminated with oral bacteria, which may multiply still further in specimens held at room temperature throughout the collection period.

C. A single collection of the first morning urine specimen is more satisfactory than a 24-hour pooled specimen to which the same objections apply as with sputum collections.

D. The precautions described for the collection of sputum should be intelligently applied to the collection and transport of any specimen. Collections are made in sterile containers without added preservative. They must be kept tightly capped. The patient should be instructed to avoid the collection of saliva and to handle the container hygienically. Containers must not be overfilled, lest their contents leak through the cap. When the containers are ready for transport to the laboratory, the outer surfaces should be wiped clean with disinfectant, the cap again tested for fit, and the transporter instructed to carry the container in an upright position within a paper or plastic bag.

E. Smears can be made directly from clinical specimens or from their concentrates. A smear report indicating the presence of acid-fast bacilli must be considered presumptive evidence of tuberculosis, but it is important to remember that acid-fast saprophytic organisms may be found in sputum, urine, gastric washings, and similar specimens. In addition, immunocompromised patients, especially those with AIDS, are commonly infected with mycobacteria other than *M. tuberculosis.* Growth characteristics of the organisms must be identified in culture to confirm laboratory diagnosis.

F. Cultures are made directly or on centrifuged concentrates of fluid or other specimens that do not have a mixed, commensalistic flora (see previous section Culture Characteristics).

IV. Special Laboratory Tests

A. *Antimicrobial Susceptibility Testing* is of particular value in determining resistance of strains from AIDS patients and treated cases of tuberculosis (see later section Treatment).

Epidemiology

I. Communicability of Infection

A. *Reservoir, Sources, and Transmission Routes* Humans are the most common reservoir of human tuberculosis,

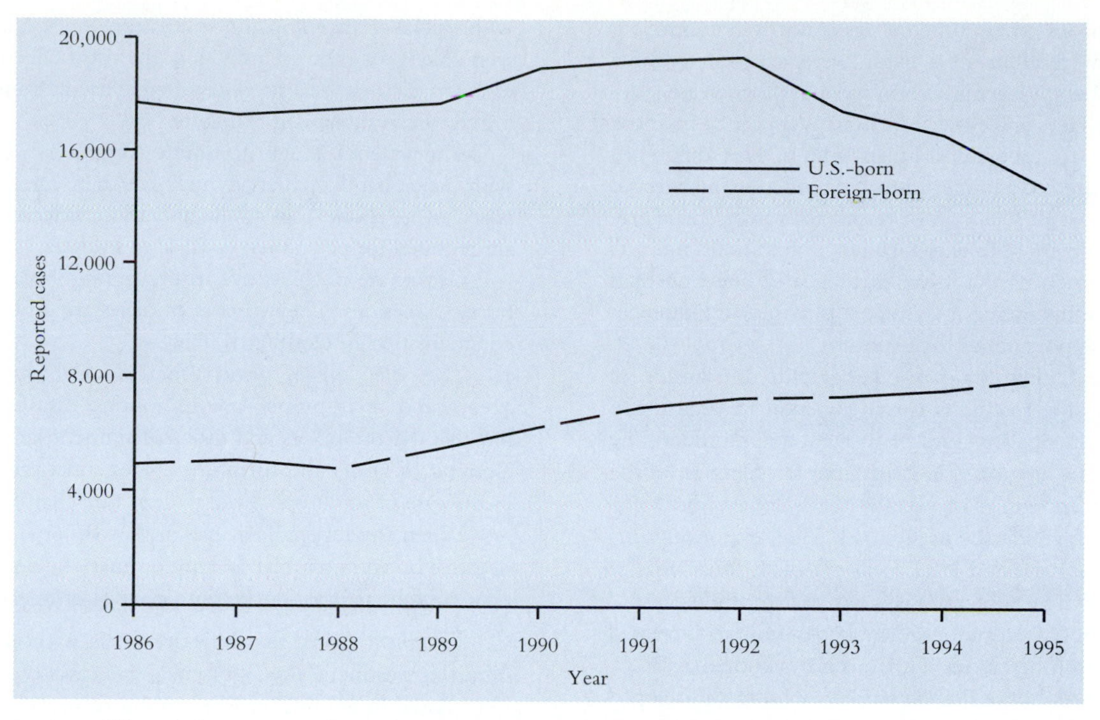

FIGURE 12.6 Tuberculosis cases among foreign-born persons and persons born in the United States from 1986 to 1995. There has been a slight decrease in the number of cases in persons born in the United States, but tuberculosis cases among the foreign-born has increased from 4,925 (21.6%) in 1986 to 7,930 (34.7%) in 1995.

Centers for Disease Control and Prevention *MMWR Summary of Notifiable Diseases, United States, 1995.*(October 25, 1996, Vol. 44: 53).

the source of infection being the respiratory secretions of persons with active pulmonary lesions. Transmission is almost always by inhalation of airborne droplet nuclei. The closed living conditions in families, the military, and institutions may permit prolonged exposure to an active case, especially an unrecognized one. This type of situation most frequently contributes to spread of infection and the incidence of active cases.

Diseased cattle are a reservoir for the bovine strain *Mycobacterium bovis.* Humans are infected by consumption of milk (or other dairy products) from tuberculous cattle. The lesions acquired through gastrointestinal entry are usually extrapulmonary, and this type of infection is not directly transmitted by the infected individual. Bovine tuberculosis may also be acquired through inhalation of airborne organisms in and around barns housing diseased animals or by handling infectious animal products. Human pulmonary infection with bovine strains is communicable. In developed countries, the cattle reservoir has been controlled through continuous tuberculin testing of herds and pasteurization of milk.

B. *Incubation Period* The period from known effective exposure to the appearance of a primary lesion of first infection is usually about 6 to 12 weeks. The time required for development of new lesions following reinfection varies with host resistance and with the source and numbers of organisms. The interval between first infection and reinfection tuberculosis may be of many years' duration.

C. *Communicability* The communicable period persists for as long as the patient discharges tubercle bacilli. The degree of communicability depends on the intensity of contamination of air with infectious droplets. This is influenced by the coughing habits of the patient, as well as by the dynamics of air circulation and fallout. Actual transmission to another person during the communicable period depends on inhalation of at least one infectious droplet nucleus. Inhaled tubercle bacilli must be carried far down the airways to a point in the lung where fixed mononuclear phagocytes can pick them up and their intracellular multiplication can begin. The sputum of an infected patient must carry fair numbers of bacilli to ensure this process. It becomes a most dangerous source of communicable infection when it is sprayed into the air by forceful coughing or sneezing. In a closed environment, shared by a number of people, communicability of infection from an active case may be high.

D. *Immunity* Susceptibility to infection is general but is influenced by age, sex, race, ethnicity, nutrition, and general health. Children under 3 years of age are highly susceptible, but the incidence of tuberculosis in older children declined sharply after controls on bovine tuberculosis were improved. In 1991, the Department of Health and Human Services reported a disturbing increase of tuberculosis (9.4% between 1989 and 1990) in all age groups (fig. 12.6). The rise was highest in some racial and ethnic minorities, children, and foreign-born persons. AIDS patients, in whom latent disease frequently reactivates, and immigrants from countries where tuberculosis is not well controlled, along with children who likely were exposed to these groups of people, accounted for most of the increase.

Black races and American Indians appear to be more susceptible to tuberculosis than whites. This difference may be related to a longer history of endemic infection in Caucasians, with selection of resistant survivors, or it may be explained by differences in living conditions and nutrition. Women have a lower incidence of infection than men. These differences may reflect physiologic influences or varied opportunities for exposure.

In the United States, the geographic distribution of new cases of tuberculosis is much greater in large urban areas than in smaller cities or in rural communities. The prevalence of new cases in large cities is a function of the dense concentrations of people of the lowest socioeconomic status and at the highest risk. High-risk groups include black males and legal or illegal aliens from areas of the world where tuberculosis is less well controlled.

A bacterial vaccine is available that confers increased resistance to tuberculosis. This vaccine is known as BCG, in recognition of the French workers who first developed it ("bacille de Calmette et Guérin"). BCG is a living, attenuated strain of bovine tubercle bacilli that protects against naturally acquired human tuberculosis in 60 to 80% of vaccinated individuals. Because the risk of tuberculosis infection is so low in the United States, **BCG vaccination** is recommended only under very special circumstances: for tuberculin-negative infants in close contact with active, poorly controlled cases of tuberculosis; for hospital workers in hospitals with a high rate of tuberculin conversion among staff members; and in other circumstances in which the infection rate is high and the usual tuberculin skin test monitoring is not feasible or successful. One reason the vaccine is not popular in the United States is that BCG-vaccinated persons become tuberculin-positive and so cannot be tested in this way for signs of active infection.

II. Control of Active Infection

A. *Transmission-based Precautions* The communicability of tuberculosis can be brought under control by specific antimicrobial therapy coupled with effective hygienic practices. Patients with active infection should be hospitalized for treatment, at least until smears of sputum (or other discharge) stained by an acid-fast method are negative. During their stay in a hospital, patients are placed on airborne precautions. When patients are treated at home, these precautions should be observed while they remain in a communicable stage of disease. This is particularly important if there are young children or other highly susceptible individuals in the family. Public health nursing supervision should be provided with instruction for all family members, as well as the patient, in the hygiene of tuberculosis prevention.

B. *Other Precautions* During the communicable period, hospitalized patients should be cared for in a private room with special air handling and ventilation. Since tuberculosis is chiefly an airborne infection, the most effective precautions are those that prevent tubercle bacilli from reaching the air in significant numbers.

The patient is taught to muffle coughing or sneezing with paper handkerchiefs, which are then carefully disposed of in plastic bags for incineration. Effective masks are required for personnel, visitors, and others.

Gowns are not effective in preventing spread of airborne infections. Their use is unnecessary unless gross contamination of clothing is likely.

The selection of good tuberculocidal disinfectant agents, and their proper use in routine disinfection of reusable thermometers and other equipment, or for daily cleaning of floors and furniture, are essential precautions in the care of the hospitalized patient (see chap. 9).

When the tuberculosis patient has become sputum-negative on antimicrobial therapy, ordinary hygienic practices are sufficient for continuing control of infection.

C. *Treatment* Most first infections heal without recognition or treatment. Recognized, active cases of tuberculosis should be hospitalized to prevent secondary cases of infection and to give the patient immediate treatment.

Currently, initial therapy of tuberculosis is with isoniazid (INH) and rifampin. These drugs are administered daily for 9 months, or for only 6 months if pyrazinamide is added during the first 2 months. The drugs are used in combination because drug resistance to any one of them may develop quickly in infecting strains of tubercle bacilli. Laboratory testing of drug susceptibility may be done on initial cultures of tubercle bacilli but is required if the patient's sputum remains bacteriologically positive or reverts after being negative. Changes in the therapeutic regimen may be necessary if laboratory studies demonstrate drug resistance. Since 1991, strains resistant to multiple drugs have become common among AIDS patients and their contacts.

D. *Control of Contacts*

1. *Quarantine* Not indicated.

2. *Prophylaxis* Immunoprophylactic use of BCG may be warranted for those close contacts of active cases who are tuberculin-negative. Chemoprophylactic use of isoniazid may be indicated for contacts of active cases for whom risk of infection is particularly high. Alternative prophylaxis should be considered if the active case is likely to be infected with an isoniazid-resistant strain of *M. tuberculosis.*

3. *Case finding* Tuberculin test surveys of contacts are of particular value. In view of the recognized hazards of repeated X irradiation, tuberculin skin testing is preferred for continuing surveillance of exposed negative reactors. This should be done on a semiannual basis for children and at least annually for adults (e.g., nursing and other health-care personnel). If negative skin reactors convert to a positive tuberculin response, chest X ray and other intensive studies are useful in detecting early infection. Prompt chemotherapy may then abort development of serious disease.

E. *Epidemic Controls* Case reports to local health departments are obligatory in most communities. This procedure permits epidemiologic analysis of the distribution of cases so that unusual clusters and incidence of cases may be promptly recognized and an immediate search made for the origins of infection. International control calls for X-ray screening of immigrants.

III. Prevention

A. *Immunization* BCG vaccination provides protection in specific instances. When properly performed, BCG vaccination converts 60 to 80% of tuberculin-negative persons to a state of skin sensitivity. The protection afforded is partial, and its duration remains undetermined, although in one study protection continued for at least 10 years. Mass vaccination with BCG is applicable only in situations where risk of infection and number of tuberculin-negative contacts of active disease are both high.

B. *Control of Reservoirs, Sources, and Transmission Routes* Methods for prevention of further infection are summarized below:

1. Find and treat active tuberculosis as rapidly as possible. New cases are most likely to be found among contacts of recently diagnosed patients, members of racial and ethnic minorities, and the foreign-born.
2. Find the source (or sources) of each active case, isolate the individual, and apply chemotherapeutic methods as indicated.
3. Continue the search for bovine tuberculosis. Eliminate tuberculin-reactive cattle. Pasteurize milk.
4. Provide BCG vaccination for tuberculin-negative persons who are closely or continuously exposed to risk of infection.
5. Provide chemotherapy for tuberculin-positive exposed individuals for whom the risk of infection is great.
6. Conduct frequent community surveys by skin testing.
7. Provide public health nursing supervision of active cases under treatment at home.
8. Provide continuing public education concerning the importance, origin, and control of tuberculosis.

The availability of these methods plus the anticipated development of new technologies for treatment, diagnosis, and prevention, prompted the United States Public Health Service to set a goal of eradicating tuberculosis by the year 2010. Although the incidence of disease has begun to decline again after the upsurge in 1989 to 1992 (fig. 12.6), it is unlikely that the target date will be hit. The problem of control is further intensified by the organism's increasing drug resistance.

Other Mycobacterioses

Mycobacteria other than *M. tuberculosis*—*M. bovis, M. avium,* and *M. leprae*—have a wide distribution in nature. Many of them are found in soil and water and could easily infect humans or animals through inhalation or ingestion. The epidemiology of the diseases caused by these organisms has not been well defined, however. Except for *M. leprae,* (see chap. 19), there is no evidence that infection can be spread from person to person. There is a good deal of evidence, however, that those species associated with human pulmonary disease are widely distributed in the population. This is indicated by the fact that their protein extracts (equivalent to PPD from tubercle bacilli) produce skin reactions in many persons tested. Since even tuberculin-negative persons may react to extracts from species other than *M. tuberculosis,* this skin reactivity appears related to specific subclinical infection with these other species. In addition, *M. avium* complex has become a significant cause of opportunistic disease in AIDS patients in the United States. *Mycobacterium kansasii* afflicts other immunocompromised patients as well as do newly recognized strains that have been cultivated only recently. These include *Mycobacterium genavense* and *Mycobacterium haemophilum,* which have special growth requirements for isolation.

Microbiologists used to refer to these strains as "atypical," "unclassified," or "anonymous" mycobacteria. When first identified as agents of pulmonary disease similar to tuberculosis, it was recognized that these mycobacteria are culturally and biochemically different from *M. tuberculosis.* Since then they have been classified into species based on the cultural and biochemical characteristics described for *M. tuberculosis.* A name more recently used to refer to this group of microorganisms is **MOTT** (***m**ycobacteria **o**ther **t**han **t**uberculosis*).

Certain MOTT bacteria are relatively resistant to the antituberculosis drugs in common use. A high proportion are isoniazid resistant, but many can be successfully treated by using combinations of three or more drugs.

Nocardiosis

Nocardiosis is a disease produced by several filamentous bacterial species that once were thought to be fungi or an intermediate form between bacteria and fungi. Their principal characteristics indicate that they are true bacteria, however, because of their narrow diameter (approximately 1 μm), absence of a nuclear membrane, and susceptibility to antibacterial antimicrobial agents and bacteriophages. These organisms are classified in the genus ***Nocardia*** of the order *Actinomycetales,* which also includes the mycobacteria and an anaerobic genus *Actinomyces.* Another major group in this order are the streptomycetes, soil organisms from which many useful antimicrobial agents have been derived. Like the streptomycetes, nocardiae are responsible for a subcutaneous, nodular disease known as mycetoma (see chap. 14) but they cause respiratory disease primarily.

Causative Organism

Nocardia asteroides, Nocardia brasiliensis, Nocardia otitidiscavarium (also known as *N. caviae*).

The Clinical Disease

Nocardiosis is a chronic pulmonary disease characterized by suppuration and abscess formation. The organisms are

SUMMARY TABLE 12.3 BACTERIAL DISEASES ACQUIRED THROUGH THE RESPIRATORY TRACT

Clinical Disease	Causative Organism	Other Possible Entry Routes	Incubation Period	Communicable Period
I. Upper Respiratory Infections				
Pharyngitis	Usually *Streptococcus pyogenes* (group A)		1–3 days	During time of active inflammation
Epiglottitis (children)	*Haemophilus influenzae*			
Sinusitis *Otitis*	*Streptococcus pneumoniae* *Streptococcus pyogenes* *Haemophilus influenzae* *Staphylococcus aureus* Gram-negative enteric bacilli			
II. Streptococcal Diseases				
A. *Strep sore throat*	*Streptococcus pyogenes*		1–3 days	While throat secretions are bacteriologically positive
B. *Scarlet fever*	*Streptococcus pyogenes*		1–3 days	As **II** *A*
C. *Sequelae of group A strep infections:*				
Rheumatic fever			3 weeks postinfection	Not communicable
Acute hemorrhagic glomerulonephritis			1–3 weeks postinfection	Not communicable
III. Lower Respiratory Infections: Bacterial Pneumonias				
A. *Pneumococcal pneumonia*	*Streptococcus pneumoniae*		1–2 days	Probably during time sputum is bacteriologically positive
B. *Other bacterial pneumonias*	*Streptococcus pyogenes* *Staphylococcus aureus* *Klebsiella pneumoniae* *Haemophilus influenzae*		1–2 days	As above
C. *Legionnaires disease*	*Legionella pneumophila*	Unknown	Unknown	Not communicable
D. *Primary atypical pneumonia*	*Mycoplasma pneumoniae*		7–21 days	During first week of illness
E. *Q fever*	*Coxiella burnetii*	Ingestion (milk), through conjunctiva and minor skin lesions	2–4 weeks	Ordinarily not communicable from person to person
F. *Primary plague pneumonia*	*Yersinia pestis*	Bubonic plague is arthropodborne (see chap. 22)	3–4 days	Probably during time sputum is bacteriologically positive

Continued

Specimens Required	Laboratory Diagnosis	Immunization	Antimicrobial Agents for Treatment	Precautions[a]
Throat swab	Culture	Limited value	Choice depends on testing of isolated organisms; penicillin for *S. pyogenes*	Principles of personal hygiene, support of healthy resistance
Discharge from site	Smear and culture			
Throat or nasopharyngeal swabs	Culture, rapid antigen-detection test	None	Penicillin Erythromycin Clarithromycin Azithromycin Clindamycin Cephalosporins	Droplet precautions; concurrent disinfection of respiratory secretions
As **II** *A*	As **II** *A*	None	As **II** *A*	As **II** *A*
Serum	Antistreptolysin titer	None	Penicillin to prevent further strep infection	None
Serum	Antistreptolysin titer, antideoxyribonuclease titer	None	Penicillin to prevent further strep infection	None
Urine	Microscopic analysis (no culture)			
Sputum Blood	Smear and culture Culture	Pneumovax, vaccine against 23 common types	Penicillin vancomycin ± rifampin[b]	Concurrent disinfection of respiratory secretions
Sputum Blood	Smear and culture Culture	None except *H. influenzae*	Choice depends on testing of isolated organisms	Concurrent disinfection of respiratory secretions
Lung tissue, tracheal aspirate, pleural fluid, bronchoalveolar lavage Serum	Culture, DNA probe Serology	None	Erythromycin + rifampin Trimethoprim-sulfamethoxazole Azithromycin Clarithromycin	Concurrent disinfection of respiratory secretions
Pharyngeal swabs, sputum and pleural fluid Acute and convalescent serums	Culture, DNA probe Rising antibody titer	None	Erythromycin Azithromycin Clarithromycin	Concurrent disinfection of respiratory secretions
Blood, spinal fluid, sputum, urine Acute and convalescent serums	Isolation of organism in animal or chick embryo Rising antibody titer	Inactivated vaccine	Tetracyclines Chloramphenicol	Concurrent disinfection of respiratory secretions
Sputum Blood Serum	Smear and culture Culture Serology	Killed vaccine	Streptomycin Gentamicin	Droplet precautions

Continued

SUMMARY TABLE 12.3 *Continued*

Clinical Disease	Causative Organism	Other Possible Entry Routes	Incubation Period	Communicable Period
IV. Chlamydial Pneumonias				
A. Psittacosis	*Chlamydia psittaci*		4–15 days	Not highly communicable
B. TWAR pneumonia	*Chlamydia pneumoniae*		ca. 10–30 days	Unknown, may be months
C. Infant pneumonitis	*Chlamydia trachomatis*		2–4 weeks	Probably during time secretions contain organisms
V. Whooping Cough				
	Bordetella pertussis		7–10 days; not more than 21	Highest during late incubation and catarrhal stages
VI. Diphtheria				
	Corynebacterium diphtheriae		2–5 days	Carrier state may be persistent
VII. Tuberculosis				
	Mycobacterium tuberculosis	Gastrointestinal (bovine strains)	6–12 weeks for primary infection	While sputum is bacteriologically positive
	Mycobacterium species	Human, animal, and environmental contact		Not communicable person-to-person?
VIII. Nocardiosis				
	Nocardia asteroides *Nocardia brasiliensis* *Nocardia otitidiscavarium (caviae)*	Parenteral	Unknown, probably few days to weeks	Not usually communicable

[a]Standard precautions apply to all patients with infectious diseases. Other transmission-based precautions indicated when applicable.
[b]For strains highly resistant to penicillin.

weakly acid-fast filamentous bacilli. Unlike *M. tuberculosis,* they are matted, branching, and display many coccoid forms as well. Dissemination of the organism from pulmonary lesions results in scattered abscesses in subcutaneous tissues, the peritoneum, and the brain. The disease has a high mortality rate unless diagnosed before widespread metastases have occurred.

Incubation Period

Not known; probably a few days to a few weeks.

LABORATORY DIAGNOSIS

Specimens for smear and culture include sputum, pus, exudates, biopsies. Examination of clinical specimens may reveal yellow-pigmented flecks or sulfur granules.

Microscopically, granules appear as tangled masses of filamentous, branching bacillary forms. The organisms are gram-positive and acid-fast (weak). A modified acid-fast stain must be used when their presence is suspected.

In culture the organism grows slowly on ordinary media at room temperature or 37° C. It is aerobic. The colony

Specimens Required	Laboratory Diagnosis	Immunization	Antimicrobial Agents for Treatment	Precautions
Blood, sputum, spleen and liver biopsy	Isolation of organism	None	Tetracyclines Chloramphenicol	Concurrent disinfection of discharges; terminal disinfection
Acute and convalescent serums	Rising antibody titer			
Sputum	Isolation of organism	None	Tetracyclines Erythromycin Clarithromycin	Concurrent disinfection of respiratory secretions
Acute and convalescent serums	Rising antibody titer			
Nasopharyngeal swab Acute serum	Cell culture IgM antibody titer	None	Erythromycin Sulfonamides	Concurrent disinfection of respiratory secretions
Nasopharyngeal swabs	Smear and culture	Killed whole-cell or acellular vaccine	Erythromycin	Droplet precautions Concurrent disinfection of respiratory secretions and contaminated articles
Blood	White blood cell count			
Swab of suspected lesion in throat Nasopharyngeal swab Wound swab (if appropriate)	Smear and culture Elek test	Toxoid	(Diphtheria antitoxin) Erythromycin Penicillin	Droplet precautions until bacteriologically negative; concurrent disinfection of nasal and oral secretions and contaminated articles; contact precautions for wound diphtheria
Sputum Gastric washings Body fluids Exudates from localized lesions	Smear and culture Smear and culture Smear and culture Smear and culture	BCG vaccine for persons at high risk	Isoniazid + rifampin + pyrazinamide ± ethambutol or streptomycin Ciprofloxacin	Airborne precautions while sputum or other discharges are bacteriologically positive Concurrent disinfection of respiratory secretions and contaminated articles
As above	As above	None	Varies depending on species	Concurrent disinfection as above
Sputum, pus, exudates, biopsies	Smear and culture	None	Sulfonamides Trimethoprim-sulfamethoxazole	Concurrent disinfection

is often waxy and folded. Orange, yellow, and red pigments are produced by different strains.

Epidemiology

Contaminated soil or dust is the source of infection. Airborne organisms are presumably the source of pulmonary disease; local tissue entry through minor wounds of skin is the origin of mycetoma. This disease is sporadic in all parts of the world.

Nocardiosis may be a primary disease or an opportunistic one in the host compromised by malignancy or immunodeficiency. The prognosis depends on early diagnosis before metastasis to the brain can occur. Surgical incision of abscesses with drainage of pus is important whenever possible. Treatment with sulfonamides for prolonged periods is helpful if instituted early. Trimethoprim-sulfamethoxazole is an alternative drug.

Questions

1. Which factors contribute to mucosal injury and predispose the tissues to secondary bacterial infections?

2. How are streptococci classified? Why is this classification useful?
3. How do extracellular products of streptococci contribute to their virulence?
4. Why is immunological identification useful in the epidemiologic tracking of streptococcal disease?
5. What do toxic shock syndrome and scarlet fever have in common?
6. Why is rheumatic fever a poststreptococcal disease? How is this disease treated?
7. Which specimens are required to find the causative microorganism of pneumonia in the laboratory?
8. Which part of the population should receive the pneumococcal vaccine?
9. Which organisms are the most common cause of pneumonia?
10. What is the reservoir for the organism that causes Legionnaires disease? How is the disease usually transmitted?
11. Why is *Mycoplasma pneumoniae* resistant to penicillin? What other drugs can be used to treat the pneumonia caused by this organism?
12. What is the reservoir for psittacosis? Why is ornithosis a more accurate name for this disease?
13. Why has the incidence of psittacosis decreased?
14. What causes the cough characteristic of pertussis?
15. Why is it important that children be immunized against pertussis early in life?
16. Which laboratory specimen is required for the laboratory diagnosis of pharyngeal diphtheria? Why?
17. Which precautions are necessary when caring for a patient ill with pharyngeal diphtheria?
18. How is diphtheria controlled?
19. How does primary tuberculosis differ from reinfection tuberculosis?
20. Why is the tuberculosis organism referred to as "acid-fast"?
21. How does hypersensitivity contribute to tubercular disease?
22. What does a positive tuberculin test signify? Which of the available tests is preferred?
23. List the most useful antituberculosis drugs. Why are they often given in combination? How long is the course of treatment?
24. Why is there a new concern about the epidemiology and treatment of tuberculosis?
25. What precautions are necessary when caring for a patient with tuberculosis?

BIBLIOGRAPHY

Bass, J. 1989. The face of TB changes again. *Hosp. Prac.* 24 (4a): 81–100.

Bradley, O. M., and J. L. Norris. 1989. Bacterial nosocomial pneumonia. *Physician Assist.* 13 (6): 35–38, 40.

Caldwell, M. 1988. *The last crusade: The war on consumption. 1862-1954.* New York: Macmillan.

Calubrian, O. V., G. S. Rosenbaum, and N. C. Klein. 1989. Atypical pneumonia. *Emerg. Med.* 21 (4): 166–168, 171–172.

Caruthers, D. D. 1990. Infectious pneumonia in the elderly. *Am. J. Nurs.* 90 (2): 56–60.

Centers for Disease Control and Prevention. 1995. Diphtheria acquired by U.S. citizens in the Russian federation and the Ukraine. *Morbid. Mortal. Weekly Rep.* 44 (12): 237.

Centers for Disease Control. 1994. Guidelines for preventing the transmission of *Mycobacterium tuberculosis* in health care facilities. *Morbid. Mortal. Weekly Rep.* 43 (RR-13): 132.

Centers for Disease Control and Prevention. 1995. Pertussis—United States, January 1992–1995. *Morbid. Mortal. Weekly Rep.* 44 (28): 525–529.

Centers for Disease Control and Prevention. 1995. Progress toward elimination of *Haemophilus influenzae.* type b disease among infants and children—United States, 1993-1994. *Morbid. Mortal. Weekly Rep.* 44 (29): 545.

Centers for Disease Control and Prevention. 1995. Tuberculosis morbidity—United States, 1994. *Morbid. Mortal. Weekly Rep.* 44 (20): 387.

Cornell, C. 1988. Tuberculosis in hospital employees. *Am. J. Nurs.* 88 (4): 484–486.

Daly, J. A. 1989. Resurgence of group A streptococcal disease: Rheumatic fever. *Clin. Microbiol. Newsl.* 11 (21): 161–163.

Daugherty, J. S., M. D. Hutton, and P. M. Simone. 1993. Prevention and control of tuberculosis in the 1990s. *Nurs. Clinics N. Am.* 28 (3): 599–611.

Dubos, R., and J. Dubos, eds. 1987. *The white plague: Tuberculosis, man and society.* New Brunswick, N.J.: Rutgers University Press.

Hoge, C. W., and R. F. Breiman. 1991. Advances in the epidemiology and control of *Legionella* infections. *Epidemiol. Rev.* 13: 329–340.

Invasive strep increasing in children. 1991. *Emerg. Med.* 23 (11): 161–162.

Janai, H., H. R. Stutman, and M. I. Marks. 1990. Invasive *Haemophilus influenzae* type b infections: A continuing challenge. *Am. J. Infect. Control* 18 (3): 180–186.

Klugman, K. P. 1990. Pneumococcal resistance to antibiotics. *Clin. Microbiol. Rev.* 3 (2): 171–196.

Lowry, P. W., and L. S. Tompkins. 1993. Nosocomial *legionellosis:* A review of pulmonary and extrapulmonary syndromes. *Am. J. Infect. Control* 21 (1): 21–27.

Madsen, L. 1990. Tuberculosis today. *RN* 53 (3): 44–51.

Mandell, G. L., J. E. Bennett, and R. Dolin. 1995. *Principles and practice of infectious diseases,* 4th ed. New York: Churchill Livingstone.

The Medical Letter. 1994. *Handbook of antimicrobial therapy.* New Rochelle, N.Y.: The Medical Letter, Inc.

Moore, P. S., and C.V. Broome. 1994. Cerebrospinal meningitis epidemics. *Sci Am.* 271 (5): 38–45.

Neville, K., A. Bromberg, R. Bromberg, S. Bonk, B. A. Hanna, and W. N. Rom. 1991. The third epidemic-Multidrug resistant tuberculosis. *Chest* 105 (1): 15–18.

Rothman, S. M. 1994. *Living in the shadow of death: Tuberculosis and the social experience of illness in American history.* New York: Basic Books.

Stratton, C. W. 1993. Tuberculosis, infection control, and the microbiology laboratory. *Infect. Control Hosp. Epidemiol.* 11 (8): 181–187.

Tablan, O. C., L. J. Anderson, W. H. Arden, et al. 1994. CDC Hospital Infection Control Practices Advisory Committee. Guidelines for prevention of nosocomial pneumonia. *Am. J. Infect. Control* 22 (4): 247–292.

TB. A global emergency. 1993. *World Health* 46 (4): entire issue.

Williams, J., N. Schneider, and M. E. Gilligan. 1995. Implementing a tuberculosis control program. *Am. J. Infect. Control* 23 (1): 152–155.

Winn, W. C. 1988. Legionnaires disease: Historical perspective. *Clin. Microbiol. Rev.* 1 (1): 60–81.

CHAPTER 13

Viral Diseases Acquired through the Respiratory Tract

OUTLINE

Microbes in Motion

This chapter covers the important viral diseases acquired through the respiratory tract including the clinical characteristics, causative organisms, laboratory diagnosis, epidemiology, and control. The following books and chapters of the Microbes in Motion CD-ROM may be a useful preview or supplemental study aid to your reading: Viral Structure and Function: Viral Disease. Vaccines: Vaccine Targets.

OBJECTIVES

On completing this chapter the student will be able to:

1. cite the clinical characteristics of important viral diseases acquired through the respiratory tract.
2. name the organism(s) responsible for each disease.
3. outline methods most commonly used for their laboratory diagnosis.
4. discuss the epidemiology of these diseases.
5. describe usual methods for controlling active infection.
6. discuss methods for preventing such diseases.

KEY WORDS

Adenovirus (**AD**-en-o-virus)
Antigenic (anti-**GEN**-ic) drift
Antigenic shift
Chickenpox
Common cold
Coronavirus (cor-**OH**-na-virus)
Coxsackievirus (cox-**SACK**-ie-virus)
Cytomegalovirus (site-o-**MEG**-a-lo-virus)
Echovirus
Epstein-Barr (EBV) virus
Erythema infectiosum (ere-i-**THEME**-a infect-she-**OH**-sum)
Fifth disease
German measles
Hantavirus (**HAN**-ta-virus)
Hemagglutinin (he-ma-**GLOOT**-in-in)
Herpes zoster (**HER**-peas **ZOSS**-ter)
Heterophile (**HET**-er-o-file) titer
Infectious mononucleosis
Influenza
Measles
Mumps
Myxovirus (**MIX**-o-virus)
Neuraminidase (new-ram-**IN**-i-dase)
Parainfluenza virus
Respiratory syncytial (sin-**SISH**-ial) virus (RSV)
Reye (**RAY**) syndrome (**SIN**-drome)
Rhinovirus (**RHINE**-o-virus)
Rubella (rube-**ELLA**)
Rubeola (rube-e-**OH**-la)
Smallpox
Varicella-zoster (vary-**CELL**-a-**ZOSS**-ter) virus
Variola (vary-**OH**-la) major
Zoster (**ZOSS**-ter)

INTRODUCTION

Viral diseases acquired by respiratory transmission and entry include some of the most contagious infections known. **Influenza,** for example, often occurs in rapidly spreading epidemic form, and the common viral diseases of childhood, such as **mumps** and **measles,** are so highly communicable that few who have not been immunized escape infection.

The diseases of this chapter are presented in two sections: (1) those whose agents infect the upper or lower respiratory tract following respiratory entry and (2) those that are systemic infections induced by agents entering by the respiratory route (including mumps, measles, and the pox diseases).

A few systemic viral diseases, for example, those caused by poliomyelitis or **coxsackieviruses,** have either alimentary or respiratory tract entry. Because the gastrointestinal route appears to be most common for these diseases, they will be discussed in chapter 17 with other viral infections of intestinal tract entry. It should be noted, however, that respiratory transmission is also possible, and appropriate precautions should be taken.

Unlike many of the bacterial infections considered in chapter 12, particularly those of the upper respiratory tract where bacteria reside chronically, the viral diseases described in this chapter are always acute and self-limited, although they may leave residual damage. A few antiviral agents have become available, but in general, viral diseases do not respond to antimicrobial therapy and are resolved instead by the body's defense mechanisms. Specific immunity, either natural or by vaccination, plays a large role, notably in **chickenpox,** measles, **rubella,** and mumps. Viral diseases of the respiratory tract itself, however, induce only transient immunity either because the virus constantly changes its surface antigens (as with influenza) or because so many different antigenic types exist (as with the **common cold**). Vaccination is not useful in controlling the common cold, but it has been of major importance in controlling systemic viral diseases.

The easy and frequent communicability of these infections by the respiratory route accounts for their epidemic character and presents a difficult challenge in patient care, especially in hospitals. The general principles of control of respiratory infections should be reviewed (see chap. 11).

VIRAL DISEASES OF THE RESPIRATORY TRACT

INFLUENZA

Epidemic influenza has been called the "last great uncontrolled plague of mankind." Control is difficult because the virus has a remarkable tendency to change the antigenic nature of its surface proteins, thereby producing new strains to which the population is not immune. Three major types of influenza virus, A, B, and C, are recognized based on their nucleocapsid-antigen composition. Type A strains are found among animals such as swine, horses, and domestic fowl, as well as in humans. Types B and C appear to be restricted to humans. The A and B strains are of major importance in human epidemics, whereas C strains occur only sporadically, usually in children. Type C influenza is sufficiently different from types A and B that it may eventually be grouped separately.

The two surface proteins primarily responsible for immunity to and virulence of the influenza virus are a **hemagglutinin** and a **neuraminidase.** The hemagglutinin confers on the virus the ability to hemagglutinate (clump) red blood cells, but its most important function is to attach the virus to host cells to initiate infection. The neuraminidase is an enzyme that breaks down viscous protective respiratory mucus so that the virus hemagglutinins can attach more readily. Neuraminidase may also help bring about the release of the virus from the host cell's surface.

Antigenic changes in hemagglutinins and neuraminidases can occur in two ways: (1) through a process of gradual mutation of the RNA that codes for production of these proteins (see chap. 2), referred to as **antigenic drift,** and (2) through a process known as **antigenic shift,** in which completely new influenza virus subtypes are produced to which the human population has little or no immunity. Antigenic shift is thought to occur when two strains of the virus simultaneously infect a host and the genes coding for surface protein antigens become rearranged. The result is a "new" virus strain that contains genes from both of the "old" strains. In this case, the virus RNA codes for hemagglutinins and neuraminidases whose protein structures are antigenically different from those of previous strains. Antigenic shift is most common with type A influenza virus and is thought to take place mostly in animals and birds. Antigenic shift, though a rare occurrence, is responsible for the many deaths resulting from intermittent influenza pandemics that have swept across the world.

The Clinical Disease

Influenza is an acute respiratory infection accompanied by fever, chills, and general malaise. Headache, muscular aches and pains, and a feeling of exhaustion that borders on prostration are common. The tissues of the upper respiratory tract are inflamed and have a bright red, mucoid appearance. Lymphoid tissue is often enlarged, but there is no purulent exudate. Coryza is not so prominent in influenza as it is in "head colds" or in bacterial infections of the upper respiratory tract. A dry, hacking cough is usual, and there may be laryngitis with hoarseness. The disease is usually self-limited, with fever for about 3 days and recovery in a week or so.

Complications

The chief complication of influenza is pneumonia, resulting from initial viral injury extending to the epithelium of the bronchial tract and alveoli and frequently followed secondarily by bacterial invasion of these tissues. The most common secondary invaders are *Staphylococcus aureus, Haemophilus influenzae,* and *Streptococcus pneumoniae* (see chap. 12).

When influenza occurs in epidemic form, it spreads rapidly throughout large segments of the communities involved. Attack rates are usually highest in school-aged children because of increased opportunities for transmission. Elderly individuals and those with chronic debilitating diseases or other reasons for lowered resistance are also especially susceptible, and for these groups, the risk of complication and death is greatest. During epidemic years, the mortality rate is much higher than usual in the community.

In rare instances, central nervous system involvement may also occur as a postinfluenza complication. **Reye syndrome,** for example, is an acute encephalitis accompanied by fatty degeneration of the liver and other viscera. This syndrome may follow other virus infections as well, especially chickenpox. It is seen in children and young adolescents and has a mortality rate of 10 to 40%. Permanent brain damage in survivors has also been reported. Epidemiologic evidence suggests a relationship between Reye syndrome and the ingestion of salicylates (e.g., aspirin), therefore, the use of salicylates in children with influenza and chickenpox is contraindicated.

Guillain-Barré syndrome (GBS), usually a self-limited but generalized paralysis, is another rare central nervous system disease linked with influenza and possibly other virus infections. The mechanism by which viruses induce this syndrome is not clear.

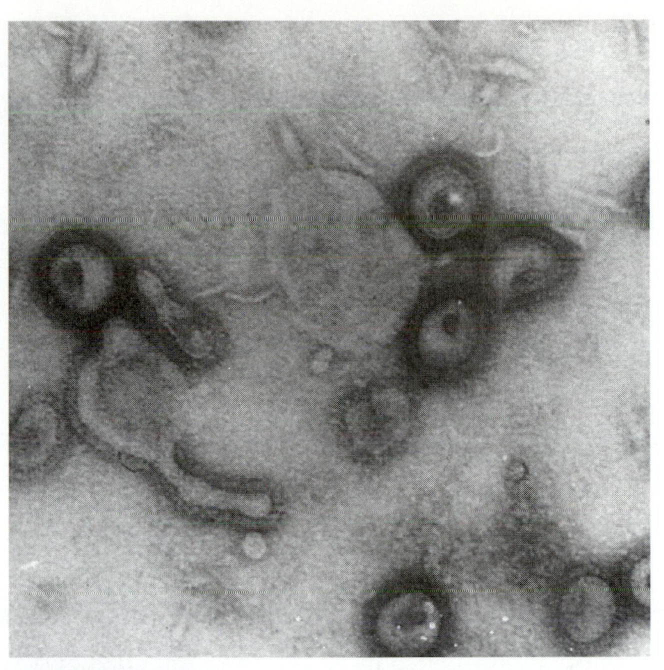

FIGURE 13.1 Electron micrograph of Asian influenza virus shows the fine structure of the virus particles (×80,000).
Courtesy Lederle Laboratories, Division of American Cyanamid, Wayne, NJ.

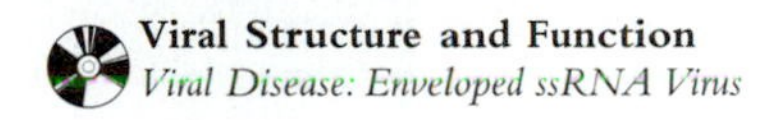

Viral Structure and Function
Viral Disease: Enveloped ssRNA Virus

LABORATORY DIAGNOSIS

I. Identification of the Organism

During epidemic periods influenza is usually diagnosed clinically but sporadic cases can be identified only by laboratory findings. Mild or asymptomatic infections often go unrecognized.

A. Influenza virus (fig. 13.1) can be recovered from throat and nasal secretions by inoculating these into embryonated chicken eggs. The virus multiplies in the amniotic and allantoic fluids, which can be harvested from the eggs in 3 to 4 days and tested for virus activity (fig. 13.2). Cultures of monkey kidney cells are also used for isolation. One of the properties of this virus that is most easily recognized is its ability to agglutinate the red blood cells of several animal species. Influenza, like mumps and measles viruses, belongs to a group called **myxoviruses.** The prefix *myxo* means "slime" and refers to an affinity for mucin. Red blood cells have a surface mucoprotein to which the virus hemagglutinin attaches, causing hemagglutination. Because viruses other than influenza also have hemagglutinating properties, harvested egg fluids and cell cultures must be tested with specific influenza antiserums. If influenza virus is present, specific antibody will combine with it and prevent hemagglutination from occurring. This is called a *hemagglutination-inhibition (HI) test.*

B. Immunologic identification of influenza virus requires that the antigenic type of the nucleocapsid (A, B, or C) be determined. Each strain is then given a designation that includes the specific antigenic type, the geographic origin of an isolate (e.g., Texas, Beijing); the strain number; and the year of its isolation. Thus, A/Texas/36/91 (H1, N1) indicates an A variant isolated during an epidemic in Texas in 1991, whereas B/Beijing/32/92 (H3, N2) identifies the Beijing strain of 1992. Influenza subtypes are based on hemagglutinin (e.g., H1, H2, H3) and neuraminidase (N1, N2) antigens. The identification of type specific variants in epidemics is important for preparing immunologically effective vaccines so that future epidemics may be prevented (see Prevention, III, A).

II. Identification of Patient's Serum Antibodies

Antibodies are produced during a clinical attack of influenza. Antibodies appear in the circulating blood in increasing concentration following onset, reach a peak level in about 2 weeks and persist at high levels for about 4 weeks. Thereafter the level gradually falls until a new infection is acquired.

The patient's antibodies are usually identified by the hemagglutination-inhibition method previously described, using laboratory-propagated virus. For diagnostic purposes, it is important that two specimens of serum be tested: one obtained during the acute stage of illness and the other taken about 2 weeks later in the convalescent stage. The serologic diagnosis is made on the basis of a fourfold or greater rise in antibody titer during the course of the disease. This pairing of serums is necessary because low levels of influenza antibody persist in healthy persons between infections and do not signify current disease.

FIGURE 13.2 Fertile eggs being inoculated with seed virus, the first step in the production of many virus vaccines. After injection, the egg is sealed with collodion and incubated to permit maximum virus growth.
Courtesy Wyeth-Ayerst Laboratories, Philadelphia, PA.

III. Specimens for Diagnosis

A. Swabs of the pharynx and nasal passages and saline nasal washings are the recommended specimens for virus isolation. Material on swabs is placed into transport medium (buffered saline or broth) and, along with the washings, is sent immediately to the virus laboratory or refrigerated until their transport can be arranged.

B. An acute serum taken within a day or two of the onset of illness.

C. A convalescent serum taken 2 weeks after onset.

Note: The first serum specimen can be retained in the refrigerator until the second blood is drawn, and the two can be sent to the laboratory as a pair. To prevent hemolysis in storage or transit, serum should be separated from whole blood and stored in a sterile tube using aseptic technique to prevent bacterial contamination and deterioration of the serum.

D. A brief clinical history and tentative diagnosis specifying the dates of onset and of specimen collection should accompany every patient's specimens.

Epidemiology

I. Communicability of Infection

A. *Reservoirs, Sources, and Transmission Routes* The reservoir for human influenza is humans, although animal reservoirs (swine, horses, fowl) may be possible sources of new human variant strains. It is thought that genetic recombinations may occur among human and animal virus strains if they occasionally find their way from one species to another. The frequency of such human-to-animal-to-human transmissions and their epidemiologic significance have not been firmly established, however.

Among infected human beings, transmission is by airborne droplets, direct contact, and indirect contamination of fomites with respiratory secretions. The virus persists for several hours in dried respiratory secretions.

B. *Incubation Period* Very short, usually only 1 or 2 days.

C. *Communicable Period* Probably does not extend past the acute febrile stage (3 to 4 days).

D. *Immunity* Specific active immunity confers resistance, but an effective concentration of IgA in respiratory secretions (the site of virus entry) as well as circulating antibodies are needed. Antibody can protect against reinfection with the same or a closely related strain for several years. Repeated infections with different influenza virus variants broaden the immunity. Protective immunity can also be obtained with influenza vaccines, although the extent of protection depends on the antigenic types incorporated in the immunizing material.

In the absence of type specific immunity, human beings are universally susceptible. In general, the incidence of infection is highest in schoolchildren, lower in children under 6, and decreases progressively among adults of advancing age. When populations encounter new strains or when old ones recur after some time, age incidence becomes more uniform. The disease is generally most severe, and mortality from complications highest, in elderly adults.

II. Control of Active Infection

A. *Transmission-based Precautions* Droplet precautions are used during the acute stage of infection.

During epidemics, patients admitted to the hospital with suspected influenza should be placed in private rooms with negative air flow or grouped together whenever possible.

B. *Other Precautions* For hospitalized patients, precautions could include concurrent disinfection of respiratory discharges and items or equipment contaminated by these secretions during the acute stage. The patient should be taught handkerchief control of coughing and sneezing, whether at home or in the hospital.

C. *Treatment* The usual antimicrobial drugs are ineffective in primary influenza and should be used only to treat secondary bacterial complications. Amantadine or rimantidine may be used early in influenza A infections (see Chemoprophylaxis, III, B).

D. *Epidemic Controls* Individual contact controls are not feasible in influenza, and sporadic cases are not reported unless the specific strain type has been determined. Health departments maintain surveillance programs to monitor the prevailing types and patterns of epidemics. This information is exchanged at state, national, and international levels so that the antigenic types expected to produce the next epidemic are incorporated into current vaccines. Vaccines must be available and administered before epidemics get underway because the disease develops and spreads rapidly, outstripping the immune response to active immunization.

During epidemics, unnecessary public gatherings should be avoided, but work, school, or other institutional functions need not be disrupted. Hospitalization of influenza cases is indicated only for patients at high risk or those with complications. Those who have not been immunized should be considered for chemoprophylaxis with amantadine or rimantidine (see III, B). Elective admissions should be curtailed and visiting discouraged. Hospital rules should include routine immunization of personnel before the epidemic season (see Perspective 13.1).

III. Prevention

A. *Immunization* Mass immunization against influenza in periods between epidemics has not generally been considered feasible because of the limited duration of vaccine protection and also because for healthy persons it is usually a mild disease. Efforts to prevent or control influenza have been aimed at protecting those at the greatest risk of developing serious complications and dying. Such persons, including chronically ill adults and children as well as people over age 65, should be vaccinated annually, regardless of the amount of influenza in their communities. Vaccination is also advisable for all hospital personnel and for others in public services (fig. 13.3).

In the United States, federal agencies regularly review the antigens present in influenza vaccines and recommend changes or additions according to the strains likely to be prevalent during the year. For example, the vaccine recommended for 1996–1997 contained antigens similar to those of strains A/Wuhan/359/95 (H3, N2), A/Texas/36/91 (H1, N1), and B/Beijing/184/93.

B. *Chemoprophylaxis* Amantadine and rimantidine are drugs that have been studied extensively in clinical trials for their prophylactic value in preventing influenza A (but not influenza B) infections. When a documented influenza A epidemic is present in the community, the use of either drug is recommended for patients at high risk who have not been vaccinated. Because amantadine and rimantidine do not interfere with antibody production, influenza vaccine should also be administered. The drugs act by inhibiting entry of virus into host cells or inhibiting replication of influenza A strains. Clinical studies also indicate that they have therapeutic value in uncomplicated influenza. If given early in the course of illness (within 48 hours of onset), they can shorten the duration and relieve the severity of symptoms.

THE COMMON COLD AND OTHER ACUTE VIRAL RESPIRATORY DISEASES

Clinical Diseases

The **common cold** and a number of other acute infections of the upper, and sometimes the lower, respiratory tract are caused by viruses. These infections seldom have

Perspective 13.1 *Influenza: A Modern-Day Plague*

Influenza ranks as one of the world's most devastating diseases, gaining that notoriety primarily because of its performance in the years 1918 and 1919 (Perspective fig. 13.1). In just two years a great pandemic of influenza traveled around the world in three waves, killing more than 20 million people and leaving economic and social upheaval in its wake. For a single infectious disease to cause such devastation in such a short time is unprecedented in history, and to this day the episode remains largely unexplained.

Influenza is by no means a disease new to the twentieth century. One of its earliest epidemics was recorded in 412 B.C. by the Greek physician Hippocrates. Influenza takes its name from the fifteenth century Italians who attributed its epidemic recurrence to the *influence* of the stars.

At the height of the influenza epidemic of World War I policemen were often required to wear masks to protect themselves.

Culver Pictures, Inc.

Today we think of influenza as a winter nuisance. Its viral nature is understood and vaccines are available, but it remains an endemic threat, especially to the elderly and other vulnerable groups. The history of its inroads in 1918 reminds us of its lethal potential. The statistics were enormous: 500,000 people died in the United States; in India, 20 million. Forty-three thousand people died of influenza on the Western Front during World War I, nearly 80 percent of the total number of American deaths during that war.

In early studies performed in search of the cause of this disease, the bacillus *Haemophilus influenzae* was often isolated from the sputum of influenza patients, and so this organism was at first believed to be, and was named as though it were, the primary agent of the disease. Later, it became apparent that secondary bacterial pneumonias often take over in the wake of viral infection of the lungs, and it was recognized that influenza is viral, not bacterial, in nature.

The influenza virus has been thoroughly studied, and its nature and properties are now understood; however, no definitive explanation has ever been given for the pandemics of 1918 and 1919. Vigorous efforts to provide effective vaccines have so far prevented further episodes of such widespread nature, holding the disease in check from year to year. The drugs amantadine and rimantidine have been found to prevent serious influenza A infection if administered soon after exposure, and antimicrobial agents are used to treat the secondary bacterial pneumonias that contribute to its morbidity and mortality.

The question remains. What happened in 1918?

Ref: Kaplan, M. M., and Webster, R. G. 1977. The epidemiology of influenza. Sci. Am. 237:88–92.

distinct clinical features, and their viral agents fall into diverse classifications. Clinically, two types of infection are recognized: (1) nonfebrile illnesses with symptoms localized in the upper respiratory tract and (2) more severe involvement of lower respiratory structures with constitutional symptoms including fever, general malaise, and sometimes gastrointestinal disturbances. Virologically, differences are noted in the location and type of cellular injury produced by different viruses as well as their antigenic and biologic properties. Epidemiologically, these diseases are characterized by an easy distribution from the human reservoir by respiratory transmission routes, general population susceptibility, and high incidence rates with varying degrees of immune response.

1. *The common cold* is a catarrhal infection of the nose and throat, characterized by local congestion and excessive mucus secretion as well as general malaise and fatigue. Fever is rare, and the duration is usually limited to a few days or a week. Secondary bacterial infections of involved mucous membranes are frequent, inducing sinusitis, laryngitis, tracheitis, bronchitis, otitis media, or combinations of these syndromes.

   One hundred antigenic types of **rhinovirus** produce the common cold, and thus the frequent recurrences that characterize this illness are due to infections by antigenically different strains to which individuals have not become immune. Rhinovirus infections are probably the most common acute

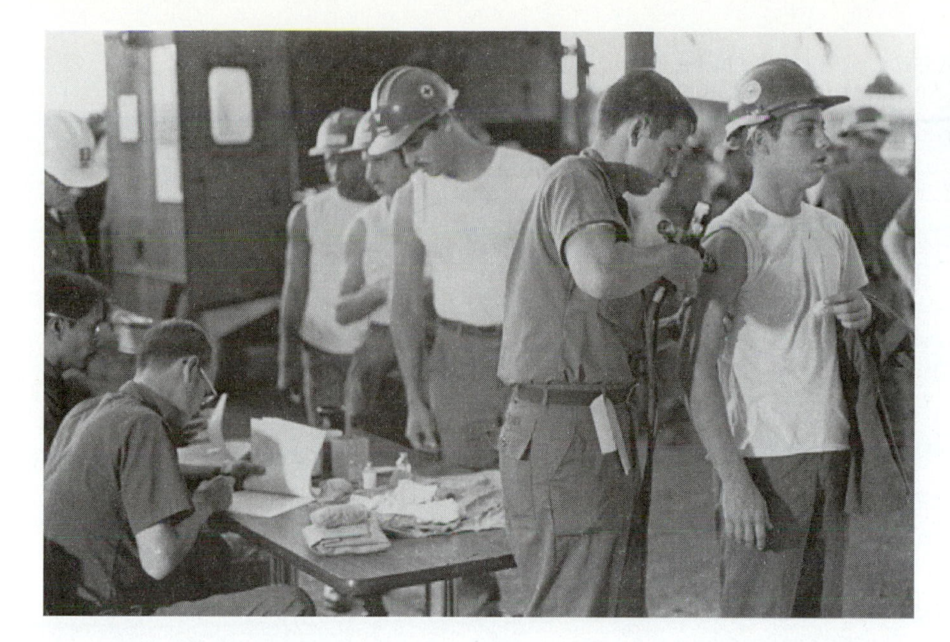

FIGURE 13.3 The battalion medical department of the United States Navy administers influenza vaccinations to Navy Seabees on active duty

Reproduced from D. A. Ice, "Improving medical care in naval mobile construction battalions," *Navy Medicine* 79 (6): Nov.-Dec., 1988.

Vaccines
Vaccine Targets: Viral Targets

human infections, resulting in many days lost from school and work. Other respiratory viruses may produce common cold symptoms that are indistinguishable from those caused by rhinoviruses (see tables 13.1 and 13.3).

2. *Febrile viral diseases* of the respiratory tract generally involve either the underlying lymphoid tissue and the superficial membranes of the upper respiratory tract or the tissues extending downward from the larynx and trachea into the bronchial tree and lungs. A number of localizing signs are related to the nature and site of viral injury: lymphoid swelling; hoarseness and croup; the dry cough of laryngeal or tracheal irritation; the productive cough of bronchitis or pneumonia; or deep pulmonary infection, which may lead to interstitial reactions or production of consolidating alveolar exudates.

VIROLOGIC DIAGNOSIS

The common cold and other acute respiratory diseases are usually mild, self-limited infections and, therefore, diagnosis is primarily clinical. When more severe syndromes are produced or when epidemics are suspected, documentation of the responsible agent is useful. During the past few years, cell culture lines and immunologic reagents have become commercially available so that clinical microbiology laboratories can accomplish viral diagnosis more readily. As with all infections, collecting the appropriate specimen at the time when the viral agent or serum antibodies are most likely to be present increases the likelihood of making a specific diagnosis. Swab or fluid specimens for virus culture are usually placed in a buffered viral transport medium containing a protein such as gelatin. The specimens are taken immediately to the laboratory or, if necessary, shipped on ice. Different viruses prefer to grow in different cell culture types, and some require incubation at special temperatures (e.g., rhinoviruses at 33 to 34° C). Thus, the laboratory technologist must inoculate the specimen into several cell types and perhaps incubate at more than one temperature.

Immunologic diagnosis for some viruses can be made by direct specimen examination using fluorescent antibody and enzyme immunoassay methods. In the latter instance especially, specimen transport is less critical because the viral antigen rather than a viable virus is detected. Available serologic tests include neutralization, hemagglutination-inhibition (for hemagglutinating viruses), and enzyme immunoassays. Acute and convalescent serums are preferred to document a fourfold rise in antibody titer.

Table 13.1 summarizes available methods for diagnosing common cold and acute respiratory disease viruses.

Epidemiology

Communicability of Infection

So far as is known, human beings are the only reservoir for these viruses, which are transmitted by droplets spread on airborne routes, by direct contacts, or by indirect contamination of the immediate environment. Some may also be spread by the fecal-oral route. The common cold is most likely transmitted by contaminated hands touching the nasal mucosa or conjunctiva, thereby inoculating these sites with virus.

The incubation period for the common cold is generally 1 to 2 days. The infectious agent is probably communicable for a day or so before symptoms begin and throughout the course of active infection. In the more severe acute diseases involving lower respiratory tissues, the incubation period is longer, from a few days to a week or more. The infections are communicable while active symptoms persist.

Susceptibility to these infections is general, although young children and infants are more frequently and seriously

TABLE 13.1 VIRUSES ASSOCIATED WITH HUMAN RESPIRATORY DISEASES AND THEIR LABORATORY DIAGNOSIS

Virus (no. of members)	Associated Disease(s)	Appropriate Specimen(s)	Direct Detection[a]	Culture	Serologic Tests[a]
Rhinovirus (>100)	Common colds (adults and children)	Nasopharyngeal washings	EIA and nucleic acid probe under development	Virus grows in human cell cultures at 33° C	Neutralization test available but not commonly used
Adenoviruses (47)	Undifferentiated acute respiratory diseases (ARD) Acute pharyngitis Epidemic follicular conjunctivitis Pharyngoconjunctival fever Epidemic keratoconjunctivitis Bronchitis and bronchial pneumonia Acute hemorrhagic cystitis Diarrhea	Pharyngeal swabs Conjunctival scrapings Sputum Stool	Electron microscopy FA	Human cell cultures	EIA HI Neutralization
Parainfluenza viruses (4)	Mild or acute upper respiratory infection Laryngotracheobronchitis (croup) Bronchitis Pneumonia	Nasal washings Nasopharyngeal aspirates Lower respiratory secretions Lung tissue	IF EIA	Human and monkey kidney cell culture	EIA HI Neutralization
Coronaviruses[b] (2)	Upper respiratory illness, possibly infant gastroenteritis	Nasal swabs and washings Stool	FA EIA	Human embryonic lung cell culture	EIA HI Neutralization
Coxsackieviruses[c] (30)	Febrile upper respiratory illness Aseptic meningitis Encephalitis Paralysis Herpangina (vesicular pharyngitis) Pleurodynia (epidemic, febrile thoracic myalgia) Myocarditis Exanthems (rashes) Generalized newborn disease	Stool Throat swab Others according to symptoms (e.g., CSF, vesicle fluid, pleural fluid)	Not yet available	Monkey kidney cell culture	Neutralization
Echoviruses[d] (33)	Febrile upper respiratory illness Aseptic meningitis Encephalitis Paralysis Exanthem Neonatal diarrhea Generalized newborn disease	Stool Throat swabs Others according to symptoms	Not yet available	Human embryonic lung cell culture	Neutralization

[a]Abbreviations: FA = fluorescent antibody test; EIA = enzyme immunoassay; HI = hemagglutination-inhibition
[b]Many others are found in animals; only two have been well studied in human beings.
[c]Named for a town in New York state where the virus was first isolated
[d]Echo = **E**nteric**c**ytopathogenic **h**uman **o**rphan virus

involved than adults. Specific immunity is induced by these diseases, but it is often short-lived. Reinfections are common but may be milder than preceding attacks because of lingering immunity. As mentioned previously, the antigenic diversity of rhinovirus is also responsible for repeated episodes of respiratory disease.

Control and prevention of the common viral respiratory infections are a difficult problem whose importance can be judged by the frequency of attack and the extent to which disability interferes with school or work. These viral infections often predispose to secondary bacterial complications that may become chronic. Personal and public attention to hygienic and sanitary measures that limit respiratory and oral spread of infection is the chief control in families and groups. Antimicrobial agents have no effect on viral infections and should be used only when bacterial complications arise. In those situations, drugs should be selected on the basis of laboratory isolation of clinically significant bacterial strains individually tested for susceptibility.

Human interferon, produced by recombinant DNA technology, has been used successfully as a prophylactic agent against rhinovirus infection. This substance, applied nasally for short periods, prevents the development of the common cold in exposed persons. Unfortunately, the effect is limited to colds produced by rhinoviruses and use for more than 1 week often results in nasal discomfort and small areas of nasal ulceration.

RESPIRATORY SYNCYTIAL VIRUS

Although it causes acute upper respiratory disease in older children and adults, **respiratory syncytial virus** (**RSV**) is treated separately here because the virus is also the major cause of lower respiratory tract illness in infants and young children. The disease is cyclical, recurring each year during the winter and spring months, and attack rates are high. Immunity to RSV infection is only partial; therefore, subsequent attacks are common, although milder, in older persons.

Clinically, RSV is associated with pneumonia, bronchiolitis, tracheobronchitis, or upper respiratory tract illness. Fever and otitis media are common especially in the lower respiratory form of the illness. Because it is highly contagious, it spreads rapidly among family members. Of more serious consequence, RSV is an important cause of nosocomial (hospital-acquired) infection (see chap. 21) and spreads to both children and their adult caretakers who are exposed on a hospital ward. It is presumed that infected adults spread the disease from infant to infant in hospitals either directly from their respiratory secretions or on items contaminated with infected secretions. The virus can survive on contaminated tissues, hospital gowns, or other materials for up to 1 hour and on counter tops for up to 24 hours. Disease is more severe in immunocompromised patients and death may result.

LABORATORY DIAGNOSIS

Once the laboratory documents the arrival of "RSV season," diagnosis is usually clinical. Nasal washings are the preferred specimen although a nasopharyngeal swab may also be used. RSV grows in several human cell culture lines producing the characteristic syncytial cytopathic effect from which it derives its name. Virus can be detected directly in respiratory washings by examining the material using fluorescent antibody and EIA methods. For epidemiologic purposes, antibodies in paired sera are detected by EIA and neutralization assays. Serologic tests are of limited value in diagnosis.

Control and Prevention

Awareness of the easy transmissibility of RSV infection is especially important for controlling the disease in the hospital setting. Patients are placed on contact precautions for the duration of their illness. Gowns may be worn if soiling is likely, but masks are not of value. Careful hand washing, especially after contact with infected respiratory secretions is mandatory.

The antiviral agent ribavirin (see table 9.8) is approved for treating infants with RSV lower respiratory tract infection. This virustatic agent decreases the clinical severity of the infection as well as the amount of virus shed in respiratory secretions. It is administered as an aerosol for 2 to 5 days. In some patients, viral shedding recurs when ribavirin administration is stopped. No RSV vaccine is yet available.

HANTAVIRUS PULMONARY SYNDROME

In 1993, an outbreak of a serious pulmonary illness occurred on a Navajo Reservation in the "four corners" area of the United States bordering Arizona, Colorado, New Mexico, and Utah. Approximately 50 people became ill with this mysterious pulmonary syndrome and about 50% of them died. After intensive epidemiological and laboratory investigations, a newly discovered infectious agent, at first called Muerto Canyon virus but now known as Sin Nombre (without name) virus, was identified as the agent of this disease. The name change reflects the fact that no geographic area wished to claim responsibility for the virus. This agent belongs to the Hantavirus group of viruses and, therefore, the infectious pulmonary process it causes is called **hantavirus** pulmonary syndrome.

The Clinical Disease

Patients with hantavirus pulmonary syndrome initially present with a mild, flulike illness that can develop rapidly into severe pulmonary disease resulting in respiratory arrest. The first symptoms of disease are the sudden onset of fever, headache, and generalized myalgia or muscle aches. These symptoms may be accompanied by abdominal pain and, within four to five days, a dry, nonproductive cough and shortness of breath (dyspnea) may develop. Within several hours, this mild respiratory illness can become an acute syndrome of respiratory distress. Fluid accumulates in the lungs as a result of the increased permeability of blood vessels in the lungs, which leads to respiratory failure and death.

LABORATORY DIAGNOSIS

The diagnosis of hantavirus pulmonary syndrome is established most reliably by serologic tests using an enzyme immunoassay. The presence of specific IgM antibodies in a patient's serum sample or the demonstration of a significant rise in IgG antibodies in acute and convalescent serum specimens is diagnostic of hantavirus infection. Hantavirus can be recovered only with difficulty from clinical specimens such as throat washings, pulmonary secretions, or lung tissue. These culture attempts are made in specially equipped virus reference laboratories.

Epidemiology

The *Hantavirus* genus belongs in the family Bunyaviridae and includes several species. All hantaviruses are parasites of wild rodents. However, each species of *Hantavirus* has a single major rodent host. For instance, Hantaan virus, the cause of severe hemorrhagic fever with a renal syndrome, is carried by the striped field mouse, *Apodemus agrarius.* Hantaan virus produces disease only in certain areas of the world where this species of mouse is indigenous, such as Korea, China, eastern Russia, and the Balkans. Two species of *Hantavirus,* Prospect Hill and the newly discovered Sin Nombre virus, are indigenous to rodents in the United States. Prospect Hill virus has been recovered from the meadow vole, *Microtus pennsylvanicus,* but is not known to cause human disease. Sin Nombre virus, the agent of hantavirus pulmonary syndrome, is carried by the deer mouse, *Peromyscus maniculatus.* These rodents become chronically infected and excrete the virus in urine, feces, and saliva for indefinite periods.

Humans become infected with Sin Nombre virus by the respiratory route from aerosols contaminated with infectious rodent urine or feces. Less commonly, people may acquire disease from the bite of an infected rodent. The incubation period following exposure is typically one to two weeks but may range from a few days to six weeks. There is no evidence that the disease can be transmitted by person-to-person contact.

In the 1993 outbreak of hantavirus pulmonary syndrome, large mouse populations were present in the southwestern United States. These infected deer mice entered human dwellings and outbuildings where subsequent aerosolization of their infected urine and feces resulted in the outbreak of disease. Since the discovery of Sin Nombre virus as the cause of hantavirus pulmonary syndrome, more than twenty states, including several eastern states, have reported human infections.

There is no specific therapy for hantavirus pulmonary syndrome. Infected persons have been treated with the antiviral agent, ribavirin, but its efficacy has not been clearly established. No vaccine is available to prevent the disease.

Control and Prevention

Control of mouse populations is an important measure in reducing the incidence of this disease. Dwellings with large amounts of potentially infectious rodent droppings should be aired out while unoccupied. Before the buildings are cleaned, the rodent droppings should be wet down thoroughly with a household disinfectant solution consisting of detergent and 1.5 cups of bleach per gallon of water. An old spray bottle is ideal for applying this disinfectant. Rodent droppings should be wiped up and placed in plastic bags for disposal together with any material used for the clean up such as paper towels. Vacuum cleaners or brooms that might create unnecessary aerosols should not be used. Wearing gloves, face masks, long-sleeve clothing, and protective eyewear during the clean-up process is recommended to minimize the risk of personal exposure. Patients are cared for using droplet precautions.

SYSTEMIC VIRAL DISEASES ACQUIRED THROUGH THE RESPIRATORY ROUTE

MUMPS

The Clinical Disease

Mumps is a familiar disease involving the parotid glands. The infection is usually generalized because viremia frequently occurs after the virus has multiplied in the parotid gland or in the superficial epithelium of the upper respiratory tract. The virus may localize subsequently in other salivary glands; the testes or ovaries, especially in adolescents or adults; the thyroid gland; and occasionally the central nervous system.

Mumps begin typically as a parotitis, with sudden onset of fever and swelling and tenderness of one or both parotids, followed sometimes by enlargement of sublingual or submaxillary glands (colorplate 6a). Tissue damage is not great, but the swelling causes pain, particularly with mouth and throat movements. The swelling reaches a peak in about 2 days and persists for a week to 10 days, outlasting the fever. When infection at other locations occurs, it generally follows the salivary gland inflammation in about a week, but it can occur simultaneously or even without parotitis. Tissue injury is not severe or permanent except in the testis, which may be compressed by the swelling within its limiting membrane and later atrophied. If bilateral orchitis occurs (which is not usual), the damage may lead to sterility. Ovarian infection does not result in sterility because the ovaries can swell without compression by a limiting anatomic sheath. Meningoencephalitis occurs in fewer than 10% of cases; fatality is rare.

LABORATORY DIAGNOSIS

Mumps parotitis is diagnosed clinically, without need for laboratory studies, but when other tissues are involved without the characteristic initial syndrome, the virus laboratory may be needed to establish the diagnosis.

I. Identification of the Organism

Mumps virus can be propagated in chick embryos or in monkey kidney cell culture. It is identified as a paramyxovirus (see table 2.6) by its hemagglutinating property. It is also characterized by its ability to lyse red blood cells as well as to agglutinate them. Serologic identification of virus is made by demonstrating specific neutralization of these properties by mumps antiserum.

II. Identification of Patient's Serum Antibodies

A serologic diagnosis of mumps can be made by demonstrating a rising titer of specific antibody following an attack of mumps. The titer of the convalescent serum must be at least four times higher than that of the acute serum in order to confirm a mumps diagnosis.

Hemagglutination-inhibition, EIA, and neutralization techniques are employed. The last method involves neutralization by antibody of mumps virus infectivity for eggs or for cell cultures.

III. Specimens for Diagnosis

A. Saliva, spinal fluid, or urine may be collected for virus isolation, as indicated by localization of symptoms, a few days after onset.

B. Acute and convalescent serums.

C. A single serum taken by the fifth day of illness (or up to 3 months after) for demonstrating mumps-specific IgM antibody.

D. A brief clinical history and tentative diagnosis with recorded dates of onset and specimen collection must accompany the specimens.

Epidemiology

I. Communicability of Infection

A. *Reservoirs, Sources, and Transmission Routes* Mumps is a human disease, the virus source being the saliva of infected persons. It is transmitted primarily by airborne droplets or by direct contacts and articles freshly contaminated by infectious saliva. The transfer must be rapid to be effective, apparently through close contacts.

B. *Incubation Period* May be as short as 12 days or as long as a month, but the average time is 18 days to 3 weeks.

C. *Communicable Period* Mumps is most communicable at the time symptoms begin, but virus may be present in saliva from about 6 days before onset until 9 days after the first swelling.

D. *Immunity* Mumps is generally a mild disease of childhood and confers a lasting immunity. Inapparent infections are frequent and also lead to antibody production as well as skin hypersensitivity. Mumps skin testing however is not considered reliable for detecting immune persons.

II. Control of Active Infection

A. *Transmission-based Precautions* The patient is cared for with droplet precautions and isolated from new nonimmunized contacts for as many days as indicated by the persistence of active symptoms of parotitis, usually 7 to 9 days. Contacts of the previous week presumably were exposed and should be watched for signs of developing infection.

B. *Other Precautions* Respiratory and oral secretions are infectious; therefore, concurrent disinfection of contaminated articles should be carried out. The hospitalized patient should have individualized care.

C. *Treatment* There is no specific treatment for mumps; the disease is usually benign and self-limited.

D. *Control of Contacts* The routine investigation or isolation of contacts is not warranted, especially if they are children.

E. *Epidemic Controls* When epidemics are under way, there are no recognized control measures of practical value.

III. Prevention

Active Immunization Live attenuated vaccine is available either alone or in combination with rubella and measles live virus vaccines. It provides solid immunity and is safe for children from 1 year of age and older. The recommended time for mumps immunization, together with measles and rubella vaccines (MMR injection), is at 12 to 15 months of age with a booster dose given at 4 to 6 years of age. Nonimmune older children and adolescents should receive the vaccine before they reach maturity, but this and other live virus preparations are contraindicated for immunocompromised or immunodeficient individuals. Widespread mumps vaccination has resulted in a dramatic decrease in the incidence of this disease (fig. 13.4). For school admission, many states now require mumps vaccination, along with others such as diphtheria, polio, rubella, and measles.

MEASLES (RUBEOLA)

The Clinical Disease

Measles, or **rubeola,** is a common but sometimes severe infectious disease of childhood. It begins with fever and the sneezing, coughing, and conjunctival irritation typical of colds or other upper respiratory infections, except that the patient often appears sicker than a mere cold would warrant. In 3 or 4 days, the characteristic blotchy, dark-red rash appears on the face to confirm that something more serious than a cold is developing. The skin rash, or exanthem (colorplate 6b), progresses dramatically within the next 24 to 48 hours, extending to the neck, the chest, and finally the entire trunk and extremities, including the palms and soles. In the meantime, systemic features of disease intensify, with fever reaching 104 to 105° F (40 to 40.5° C), the cough deepening to an obvious symptom of bronchitis, the patient becoming progressively irritable and dyspneic. Physical examination of the mouth and throat reveals bright red lesions with whitish centers (enanthem) on the buccal mucosa and palate. These mucosal lesions, called Koplik spots, often appear before the skin rash is visible and may provide the first clue to clinical diagnosis (colorplate 6c).

In about a week or 10 days the rash begins to lose its angry red look and slowly becomes brownish. Fever and respiratory symptoms subside rapidly, barring complications, and an uneventful recovery ensues. The symptoms of measles are caused by the rubeola virus, which enters the body by the respiratory route, localizes, and multiplies in lymphoid tissue of the pharynx. After about 10 days of

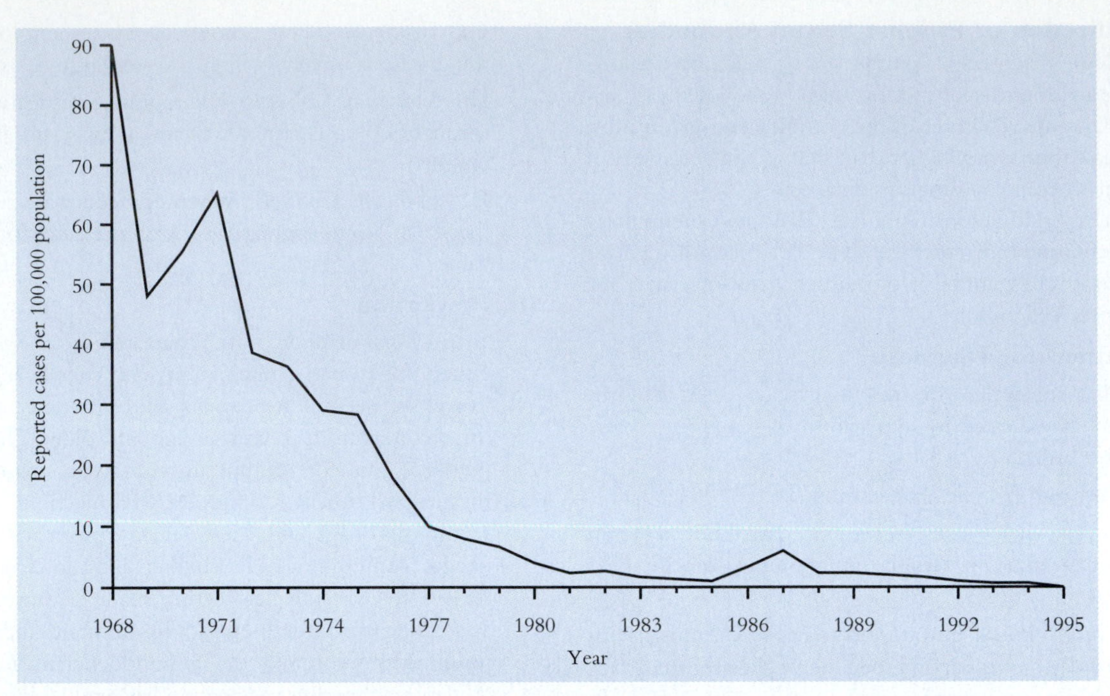

FIGURE 13.4 Reported mumps cases in the United States from 1968 to 1995. Mumps became nationally notifiable in 1968. In 1995, 906 mumps cases were reported—the lowest number ever reported in the United States during one year. The mumps vaccine was licensed in 1967.

Centers for Disease Control and Prevention, *MMWR Summary of Notifiable Diseases, United States,* 1995. (October 25, 1996, Vol. 44: 53.)

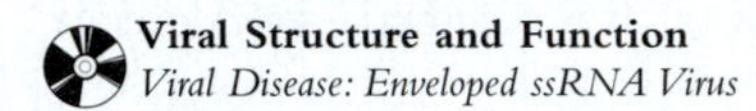

Viral Structure and Function
Viral Disease: Enveloped ssRNA Virus

Vaccines
Vaccine Targets: Viral Targets. Immunization Schedules

incubation, as the catarrhal period begins, viremia occurs. Virus can be found not only in respiratory discharges but also in conjunctival secretions and the blood for at least 2 days after the rash appears. Superficial capillaries are probably injured by hypersensitivity reactions to measles antigen in the lining cells of the blood vessels. This pathology leads to the skin rash as well as Koplik spots on oral mucous membranes. A generalized lymphoid tissue reaction and a decreased number of circulating lymphocytes also occur in response to virus activity.

When the virus infection involves the lower respiratory tract, a severe complication called *measles pneumonia* results. In addition, both upper and lower respiratory tracts become more susceptible to secondary bacterial infection, another frequent complication of measles. These infections can usually be controlled with antimicrobial therapy. A rare (1 in 2,000 cases) but more serious complication arises if the virus enters the central nervous system. The encephalitis that results is fatal in up to 30% of cases displaying this syndrome, and about 40% of survivors are left with some degree of brain damage.

Subacute sclerosing panencephalitis (SSPE) is another unusual disease also linked to measles on the basis of a measleslike virus isolated from brain tissue of affected individuals who also have high titers of measles antibody. This fatal, progressive nerve disease usually begins about 5 years after the measles infection and is referred to as a "slow" virus disease. Since the widespread use of measles vaccine, SSPE has virtually disappeared in the United States.

LABORATORY DIAGNOSIS

Measles is usually diagnosed on the basis of its characteristic clinical appearance. With the widespread use of measles vaccine in children, however, more cases are being seen in unimmunized adults, as well as atypical cases in immunized children. When the rash is atypical or Koplik spots fail to appear, differentiation from other virus exanthems may be difficult, and laboratory confirmation of the diagnosis may be valuable.

I. Identification of the Organism

Measles virus can be detected directly in patient specimens by immunofluorescence tests with specific measles antibody. The organism can be propagated in cell cultures such as human and monkey kidney. It produces characteristic changes in these cultures, and its typical inclusion bodies can be seen in the nuclei of invaded cells. When harvested from cell cultures, rubeola, a paramyxovirus, can be demonstrated to have hemagglutinating properties. Serologic identification of virus is obtained with measles antiserum, which specifically inhibits the hemagglutinating activity of rubeola or neutralizes its infectivity for human cell cultures. Fluorescent antibody tests are also used.

II. Identification of Patient's Serum Antibodies

Specific measles antibody appears in the patient's serum during the course of disease, rising in titer from the acute to the convalescent period. These antibodies may persist for many years following an attack of measles and are sufficiently protective to make second attacks infrequent.

The patient's serum is tested for its ability to react with known strains of measles virus by neutralization of infectivity, hemagglutination inhibition, or enzyme immunoassay. The usual fourfold rise in antibody titer is required for diagnosis, or detection of IgM antibody.

III. Specimens for Diagnosis

A. Blood samples, nasopharyngeal and conjunctival secretions, and urine can be submitted for virus isolation in the early acute phase, within 24 hours of the first appearance of rash. Blood should be defibrinated or anticoagulated; nasopharyngeal and conjunctival secretions should be placed in holding medium provided by the laboratory. These materials should be sent immediately for culture or frozen pending their transport.

B. Acute and convalescent serum samples.

C. A single serum specimen taken no later than 30 days after onset for demonstrating measles-specific IgM antibody.

D. A brief history and tentative clinical diagnosis with recorded dates of onset and specimen collection must accompany the specimens.

Epidemiology

I. Communicability of Infection

A. *Reservoirs, Sources, and Transmission Routes* Measles is one of the most readily transmissible of infectious diseases with a human reservoir. It is spread by airborne droplets derived from respiratory discharges of infected persons, who may remain infectious from the time of onset of symptoms throughout the catarrhal period and for 4 or 5 days of the eruptive stage. Infectious droplets are transferred directly or indirectly by contamination of clothing or other articles with secretions from the nose and throat. Subsequent drying releases into the air droplet nuclei containing viable rubeola virus.

B. *Incubation Period* From exposure to the time of first symptoms of fever and "cold," usually about 10 days. Development of the rash requires another 3 to 4 days.

C. *Communicable Period* See Epidemiology, I, A above.

D. *Immunity* Although natural measles infection is now uncommon in the United States because of widespread immunization, a single attack confers lifelong protective immunity. Following a mild or aborted case, immunity may be less durable, and second attacks of measles may occur. Maternal antibody is transferred in utero, but congenitally acquired immunity does not persist past the first year of life.

Measles is now endemic only in those parts of the world where vaccination is not routine and is most prevalent in winter and spring. Natural disease is most common and widespread among children so that about 90% of unvaccinated people reaching age 20 have acquired immunity through infection. The immune state of the population is the principal factor determining the frequency of epidemics, which may occur every 2 years.

When an effective, attenuated, live measles virus vaccine was developed in the 1960s, measles was targeted for eradication in the United States by 1982. Instead, a series of localized measles outbreaks occurred, primarily on college campuses and in urban areas, during the middle and late 1980s (see fig. 13.5). This resurgence of a disease thought to have been conquered led to a reevaluation not only of levels of compliance with the vaccination requirements but of the schedule for administering the vaccine (see later section, Prevention, III).

II. Control of Active Infection

A. *Transmission-based Precautions* The patient should be placed in a private room on airborne precautions for the duration of illness in order to minimize the risk of further exposure of susceptible contacts, particularly if these include infants or small children for whom the disease may be serious. Patients should be protected against secondary bacterial complications, particularly if they are hospitalized.

B. *Other Precautions* Patient care techniques are directed at control of airborne spread. Concurrent disinfection of all articles contaminated by nose-and-throat secretions is indicated. Hospitals should be particularly careful to protect very young and immunocompromised patients from infection acquired through possible measles admissions.

C. *Treatment* There is no specific treatment for measles. The use of antimicrobial drugs is aimed at preventing secondary bacterial infections (especially in patients with underlying debilities) or treating complications as they arise.

D. *Control of Contacts* Live attenuated vaccine will protect against measles if given to nonimmune contacts within 72 hours of exposure. If given later in the incubation period, it may prevent severe illness without producing adverse effects. If vaccine is contraindicated (immunodeficiency, immunosuppression, leukemia, tuberculosis, pregnancy, infant less than 1 year old), measles immunoglobulin (human) may be given within the first 6 days after exposure. The earlier immunoglobulin is administered, the better the chance of preventing the disease altogether.

E. *Epidemic Controls* Case reporting is required by health departments because it provides the best opportunity for the prompt protection of vulnerable contacts. When cases are concentrated in institutions or isolated villages, public health measures should include, as indicated by the local situation, isolation of the sick, prompt immunization programs aimed at all potentially susceptible individuals, and immunoprophylaxis for those who should not receive live virus.

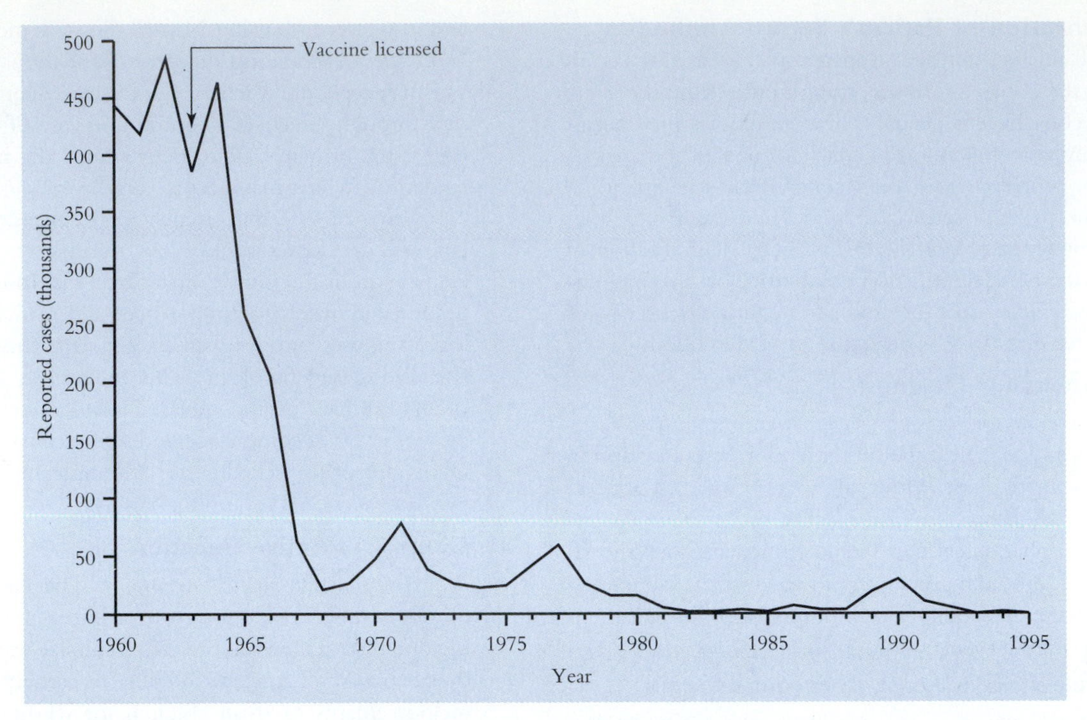

FIGURE 13.5 Measles (rubeola) cases in the United States, from 1960 to 1995. The dramatic fall of measles cases since the introduction of the vaccine in 1963 is illustrated. In 1995, 309 cases of measles were reported—the lowest annual total ever. The largest outbreaks were reported among communities that do not accept vaccination.

Centers for Disease Control and Prevention, *MMWR Summary of Notifiable Diseases, United States,* 1995 (October 25, 1996, Vol. 44: 53.)

III. Prevention

Immunization Live attenuated measles vaccine provides the most effective preventive control. Since the vaccine was licensed in 1963, the incidence of measles has decreased by 99% (fig. 13.5). The outbreaks that occurred during the late 1980s and into 1990 (see previous section, Immunity, I, D) were found to affect primarily unvaccinated preschool-aged children, including those younger than the recommended age for vaccination of 15 months and students and personnel on college campuses, many of whom had received the then standard single dose of vaccine before they were 15 months old. Because maternal antibody persists in some children beyond 1 year of age, the vaccine is considered less effective when administered before 15 months of age.

These findings have produced new recommendations to help achieve the goal of eliminating measles in the United States, including the administration of two doses of vaccine (rather than only one). The first dose is given at 12 to 15 months of age and the second between 4 and 6 years upon entry to kindergarten or first grade. The first dose should be given at 12 months of age in areas where cases of measles continue to be reported and in counties that have a large inner-city urban population. When immunoglobulin has been given to an unvaccinated person after measles exposure, vaccine should be given 3 months later if the person is at least 15 months old. After 3 months, the passively acquired measles antibodies should have disappeared.

All persons vaccinated before their first birthday with live virus vaccine as well as those who were vaccinated with the previously available killed measles vaccine need to be reimmunized with two doses of vaccine given at least 1 month apart. Documented evidence of measles immunity is now required for employment in most hospitals. People are considered immune if (1) they were born before 1957, when few escaped natural infection, (2) they have a physician's confirmation of a clinical case of measles, or (3) they demonstrate the presence of measles antibody by serologic testing.

The immunity gained by both natural infection and vaccination with live measles vaccine is considered long-lived, perhaps lasting for the life of the individual. Among the rare side effects of vaccine administration are fever and a transient rash. The vaccine may be administered alone or in combination with mumps and rubella vaccines, the so-called *MMR vaccine.*

RUBELLA (GERMAN MEASLES)

The Clinical Disease

Before the availability of an effective vaccine, **rubella,** or **German measles,** was one of the frequent but mild viral exanthems of childhood. Its chief importance lies in the effect it may have on infants born of mothers who contract the infection during early months of pregnancy.

Rubella virus probably enters the body through the respiratory tract but is disseminated by the bloodstream.

Symptoms begin as a mild upper respiratory infection followed by enlargement of cervical lymph nodes, characteristically those of the back of the neck, and finally include the eruption of a generalized macular rash. The rash is distributed first on the face and head, spreads rapidly to the neck and trunk, but subsides in 2 or 3 days. Rapid recovery is the rule, but in adults occasional complications result from virus localization in joints or the central nervous system.

When rubella occurs in pregnant women, transfer of the virus across the placenta to the fetus may cause serious and permanent damage to developing embryonic tissues, particularly if they are infected during the earliest, most rapid stage of formation. The danger of such effects diminishes after the fourth or fifth month of pregnancy. Stillbirths and spontaneous abortions may occur, and babies brought to term may display one or more serious anomalies such as heart defects, liver and spleen enlargements with dysfunction, deafness, cataracts, mental retardation, or other brain damage.

LABORATORY DIAGNOSIS

Clinically, rubella must be distinguished from measles, scarlet fever, and a number of rash-producing virus infections.

For virus isolation in the early, acute stage of disease, nasal and pharyngeal swabs, blood, urine, or cataract material and lens fluid should be submitted to the laboratory.

For serologic diagnosis, the hemagglutination inhibition test has been commonly used; however, neutralization, indirect fluorescent antibody, latex agglutination, and EIA methods are now available. The laboratory should be consulted by the physician requesting rubella serology. The choice of test and the timing of blood collection depend on the patient's age, exposure risk, clinical condition, and history of rubella vaccination.

Decisions concerning pregnancy termination often rest on diagnostic laboratory tests, and it is essential that they be adequately planned and properly interpreted.

Epidemiology

I. Communicability of Infection

A. *Reservoirs, Sources, and Transmission Routes* So far as is known, humans are the only reservoir for rubella. The nasopharyngeal secretions of infected persons contain the virus and transmit it when sprayed into the air. Airborne droplet nuclei and direct or indirect contacts are the usual transferral routes.

B. *Incubation Period* In adults, symptoms become apparent about 14 to 21 days after exposure. The average incubation time is 18 days.

C. *Communicable Period* This disease is very communicable during the week before and the week after the appearance of the rash.

D. *Immunity* A single attack of rubella appears to provide a permanent immunity, but in the absence of natural infection or active immunization, susceptibility is general. Infants may be protected for the first few months of life by maternal antibody acquired in utero.

In natural infection, rubella antibodies begin to appear in the serum soon after the rash subsides and rise to a peak within 2 or 3 weeks. Thereafter the persistence of antibody provides lifelong immunity. Demonstration of antibody by serum assay is therefore a more reliable index of immunity than is a history of rubella.

When symptoms suggest current, active rubella infection, the diagnosis can be confirmed by demonstrating a fourfold rise in antibody titer in a "convalescent" serum, compared with an "acute" serum, or the presence of IgM antibody (primary response to active infection) as distinct from IgG antibody (secondary and persistent after previous infection).

It is particularly important that the laboratory diagnosis be established in the first trimester of pregnancy for those women who have symptoms suggestive of rubella or who think they have been exposed to the disease. In many American states and elsewhere in the world, documentation of rubella in early pregnancy provides legally acceptable grounds for abortion to prevent the birth of babies with congenital deformities. Conversely, demonstration of an unchanging, positive rubella titer in an asymptomatic pregnant woman exposed to rubella signifies that she is immune and not infected and that abortion on this basis is not indicated. Routine premarital screening for rubella antibodies has become common and detects those women who are candidates for rubella vaccine.

II. Control of Active Infection

A. *Transmission-based Precautions* Droplet precautions are recommended for 7 days after onset of the rash. Susceptible persons should stay out of the room if possible.

The epidemiology of congenital rubella is related to the fact that rubella outbreaks can occur after childhood. In the spring of 1964 the United States experienced a large epidemic among young adults which resulted in congenital disaster for thousands of babies in the form of cataracts, heart defects, nerve damage resulting in deafness, and mental retardation. Improved laboratory techniques made it possible to study many of those infants extensively. It was discovered that they remain infectious for many months after birth, the virus being recoverable from throat secretions or spinal fluid up to 18 months postnatally. It is thought that this continuing infection with rubella virus may account for the progression of symptoms observed in affected babies, especially those with central nervous system defects.

It is also important to recognize that congenitally infected infants can transmit virus to susceptible adults who care for them. The hazard is great only for women in the first trimester of pregnancy. If they contract the disease, a 20 to 25% risk of congenital infection can be expected. Furthermore, apparently normal babies born of infected mothers may shed virus, and the fetus of a mother with an asymptomatic infection may be affected to the same degree as one whose mother had a rash.

For these reasons, it is advisable for susceptible pregnant women to avoid unnecessary contacts with newborn babies, especially during the first trimester.

B. *Other Precautions* Concurrent disinfection of respiratory secretions is advisable but not required.

C. *Treatment* No specific treatment for rubella is available. If the patient is pregnant and the diagnosis of rubella is confirmed by the laboratory, therapeutic abortion is strongly advised. The use of gamma globulin as an alternative to abortion is not indicated because it does not prevent viremia in the mother or the transfer of virus in the blood to the fetus.

D. *Control of Contacts* Susceptible women of childbearing age who are in contact with babies or others suspected of having rubella infection should be identified by serologic tests. If antibody cannot be demonstrated in their serum, they should be immunized with rubella vaccine, provided the restrictions described in the following section on prevention are noted.

E. *Epidemic Controls* None that is effective.

III. Prevention

Immunization In 1969 a live attenuated rubella vaccine became available in the United States. Two doses of this vaccine are recommended for all children. It is given in combination with mumps and measles vaccines at 12 to 15 months of age and again at 4 to 6 years of age. It may also be given to adolescent or young adult females, but because it is a live vaccine the following restrictions must be carefully observed: (1) it should not be given to a pregnant female; (2) it should not be given to persons who already possess rubella antibody, as demonstrated serologically; and (3) it should not be given to any female who is likely to become pregnant within 3 months of vaccination. With the last restriction in mind, rubella vaccine may be given to a susceptible woman during the postpartum period, when the possibility of an unrecognized pregnancy is remote.

The vaccine produces transient arthritis and arthralgia in about 30% of teenage or older females. Its use is contraindicated for persons whose immune mechanisms are compromised, although it appears that it can be safely given to persons with acquired immune deficiency syndrome (AIDS). Although the vaccine should be avoided during pregnancy, some pregnant women have been vaccinated by accident. No adverse effects on their fetuses were demonstrated even though evidence of infection by the rubella vaccine strain was documented. Therefore, inadvertent vaccine administration during pregnancy is not a cause for alarm that congenital rubella syndrome will result.

Currently, vaccination is recommended for all children 12 months of age or older, schoolchildren not vaccinated in infancy, and susceptible adults—particularly females, persons in the military, and hospital personnel. The vaccine may be given alone or in combination with mumps and measles vaccines (MMR vaccine) at 15 months of age. If these recommendations are followed, rubella may eventually be eliminated in the United States. From 1969 when the vaccine was introduced until 1995, the number of reported cases of rubella infection decreased from 60,000 to 200 (see fig. 13.6). With the decline in infection has come a decrease in the number of infants with congenital rubella syndrome; 7 cases were reported in 1995. Approximately 10% of women who receive premarital rubella serologic screening have no protective antibodies; therefore, rubella eradication remains a number of years away.

Erythema Infectiosum (Fifth Disease)

The Clinical Disease

Erythema infectiosum is a usually mild viral illness that may be accompanied by upper respiratory tract symptoms. The disease for many years was known as one of the childhood illnesses accompanied by a rash, but it was not until the 1970s that it became associated with a small DNA virus called *parvovirus B-19.* The infection has been referred to as **fifth disease** because it was the fifth of a group of six exanthematous childhood diseases that also includes measles, rubella, and scarlet fever.

Children are most often affected by this virus. After a short febrile period, they develop a brilliant-red facial rash characterized by its "slapped cheek" appearance. The rash is enhanced by exposure to sunlight, bathing, and a variety of stress factors. Eventually the rash spreads to the extremities but usually not the trunk. As the rash resolves, it develops a lacy pattern of redness that is another characteristic feature of the disease. In addition to the fever and rash, symptoms may include abdominal pain, nausea, and pharyngitis. The length of illness averages 11 days.

Although the infection is mild in healthy children, adults may develop transient arthritis and arthralgias (joint pains). The most serious effects are seen in persons who have red blood cell defects that make their cells more susceptible to hemolysis (e.g., sickle cell anemia, hemolytic anemia) and in the fetuses of women who contract the disease during pregnancy. The virus apparently attacks the red blood cell precursor cells in both types of persons, leading to a condition known as *aplastic crisis* (failure to produce red blood cells) in the former and fetal hydrops (an edematous accumulation of fluid in the tissues) in the latter. Spontaneous abortion and stillbirth of the fetus are other possible consequences.

Precautions

Patients with erythema infectiosum should be placed on droplet precautions for 7 days after the onset of illness. Susceptible pregnant women and those with hemolytic diseases should be especially careful to avoid contact with infected persons. Hospital workers should take special care in handling infected respiratory secretions because nosocomial outbreaks have been reported. Diagnosis is best made by de-

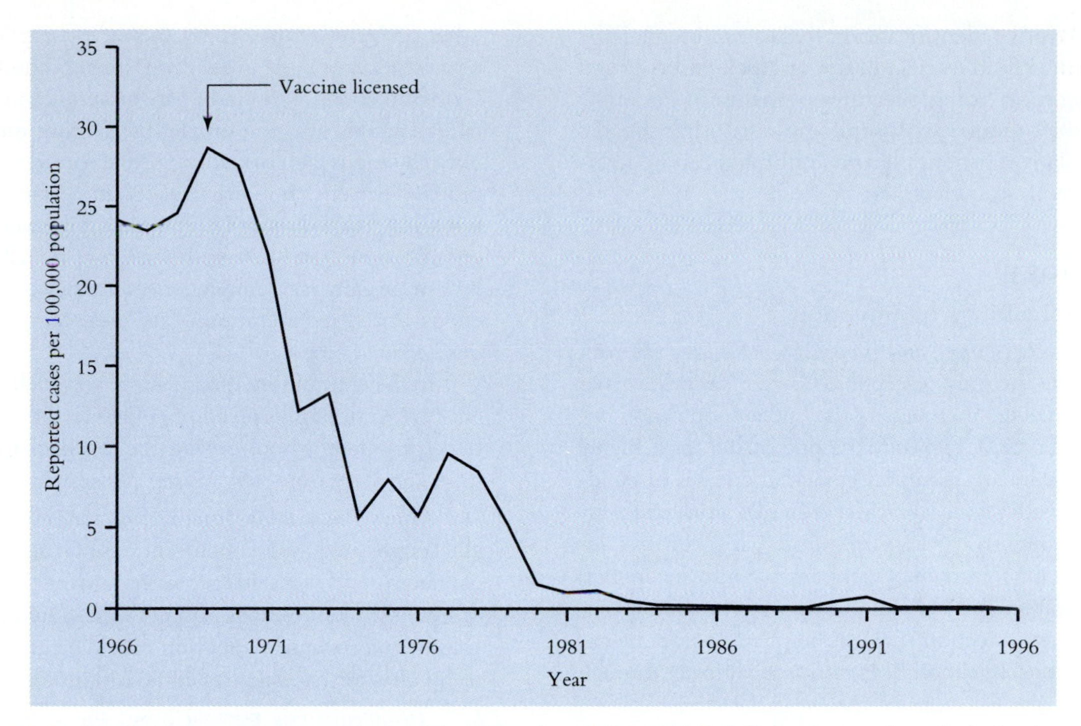

FIGURE 13.6 Rubella (German measles) cases in the United States from 1966 to 1995. Rubella has shown a dramatic decline since the vaccine was licensed in 1969; only 128 cases of rubella were reported in the United States in 1995, the lowest number ever reported.
Centers for Disease Control and Prevention *MMWR Summary of Notifiable Diseases, United States,* 1995 (October 25, 1996, Vol. 44: 53.)

tecting specific IgM antibodies against the virus, but the test is not widely available. Treatment is primarily supportive. No vaccine is available.

CHICKENPOX AND HERPES ZOSTER

The Clinical Disease

Chickenpox (varicella) and shingles (also known as **herpes zoster**) have existed for many centuries, but only recently has their relationship become apparent. Chickenpox is a mild but highly communicable disease of childhood, characterized by fever, malaise, and an itching vesicular skin eruption. Zoster, or shingles, bears little clinical resemblance to chickenpox but is a recurrent infection with the same virus, which has persisted in latent form in nerve ganglia. It occurs sporadically in adults, who develop an incapacitating, painful inflammation of posterior nerve roots and ganglia. The path of the affected sensory nerves is marked by crops of vesicular eruptions of the skin of the area. These vesicles are morphologically identical with those of chickenpox, but the clinical appearance of zoster is distinct from the childhood disease.

Varicella-zoster virus, a herpes virus, is believed to enter the body through the respiratory mucosa and, after a period of viremia, to localize in the cells of the skin. The first symptoms of fever and malaise appear about 12 to 14 days after exposure. The rash appears first on the trunk and then progresses to the face, limbs, and sometimes the mucosa of the mouth and throat. The lesions are often more abundant on covered than on exposed areas of the body, and they appear in successive crops for 3 to 4 days, so that different stages of pock development may be seen at the same time. The pocks begin as papules and then develop into weepy vesicles, which become crusted as they begin to heal. The lesions resolve without scarring as a rule, and recovery is usually rapid and uneventful. The disease has an extremely low mortality rate and only occasionally may lead to dangerous complications such as viral pneumonia in adults or encephalitis in children.

Zoster also begins with fever and malaise as virus multiplication proceeds, and extreme tenderness develops along the dorsal nerve roots where virus localization has occurred. In a few days crops of vesicles appear on the skin area supplied by the sensory nerves of the affected root. The eruption is generally unilateral, appearing most commonly on the trunk along the lower rib line, sometimes on the shoulders, neck, or head.

LABORATORY DIAGNOSIS

Clinically, the appearance and sequence of distribution of skin lesions usually distinguish chickenpox from other "pox" diseases. When required for epidemiologic reasons, the laboratory can identify these viruses or perform tests to measure the patient's specific immune response. Before the eradication of smallpox (see later section on Smallpox), these were the only methods to distinguish definitively between the two diseases.

Morphologically, varicella and smallpox viruses have much in common, but distinctions can be made in the location of inclusion bodies within infected host cells or in cell cultures. Positive identification of virus can be made by cultivating vesicle fluid in cell culture or chick embryos and by confirming morphologic features of inclusion bodies or the damage they induce with serologic tests using specific antibody. Similarly, a patient's serum antibodies can be identified in tests with known strains of viruses.

Epidemiology

I. Communicability of Infection

A. *Reservoirs, Sources, and Transmission Routes* Humans appear to be the only susceptible host for varicella-zoster virus, the infected members of the human population being the reservoir. Chickenpox is one of the most highly communicable and common epidemic diseases of childhood. It spreads rapidly, the chief source of virus being respiratory secretions or discharges from early vesicular lesions of the mucous membranes or skin. Airborne droplets as well as direct and indirect contacts with infectious secretions carry infection into the respiratory tract of exposed contacts. Infection does not occur directly through the skin.

B. *Incubation Period* Two to three weeks are commonly required following exposure. The average case becomes symptomatic in about 14 to 16 days.

C. *Communicable Period* Infectivity is greatest from a day or so before the rash appears to the fifth or sixth day of the eruption. Scabs are not infective, but the disease remains communicable while fresh crops of vesicles are appearing. Zoster is not readily communicable because the virus is not present in the respiratory tract, but this infection can be the source of varicella in children and has been known to initiate epidemics.

D. *Immunity* Infection with varicella-zoster virus confers a long-lasting immunity to chickenpox. Persons who have recovered from varicella have serum antibodies against the varicella-zoster virus. Zoster usually occurs in adults who have had a natural varicella infection at an early age.

Susceptibility is universal among those who have not had chickenpox. The disease in adults is often more severe than it is in children, and in past times this has contributed to the difficulty of distinguishing it from smallpox. In temperate zones, chickenpox is more prevalent in the winter and spring, but zoster occurs sporadically without seasonal incidence.

The viruses isolated from chickenpox and from zoster are identical. During a chickenpox infection, the virus sometimes infects nerve cells and remains latent and inactive within them for long periods. A stress or physiologic insult later seems to provide an opportunity for activation of the latent virus, with the characteristic result known as zoster. The exact mechanism for virus reactivation is unknown.

II. Control of Active Infection

A. *Transmission-based Precautions* Children should not be hospitalized with chickenpox unless complications or other problems require it. By the time complications appear, chickenpox is usually no longer communicable. Cases sometimes develop in patients already in the hospital, and in this situation droplet precautions must be used until all lesions are crusted to prevent spread to other susceptible persons. It is particularly important to protect other patients who may be immunosuppressed for a variety of reasons. Children with leukemia are subject to severe or possibly fatal consequences if they contract chickenpox. Airborne precautions are used for patients with disseminated zoster.

In the community, the isolation of sick children is not effective in stopping spread of epidemics, but they should be kept out of school during the week of the eruptive stage and away from very young susceptible contacts.

B. *Other Precautions* In the hospital, nursing care of chickenpox requires concurrent disinfection of objects contaminated by respiratory secretions or the discharges of fresh skin lesions. The patient's care should be individualized, and particular attention should be given to proper techniques for handling soiled bedclothing and linens.

C. *Treatment* For normal hosts, supportive treatment only is recommended for chickenpox or shingles. Aspirin should not be given to children because, as with influenza, a link between chickenpox, aspirin, and Reye syndrome has been established. Scratching of the itchy eruptions may lead to secondary bacterial infection of open vesicles; therefore, the patient's skin should be kept scrupulously clean and fingernails short. Hand and nail scrubbing is also essential for health care personnel who have close contact with the patient.

Immunocompromised hosts may develop severe complications of chickenpox such as pneumonia and widespread, generalized eruptions. For these patients, treatment with the antiviral agent acyclovir (see table 9.8) has been successful in decreasing the severity and length of illness.

D. *Control of Contacts* Isolation of contacts is ordinarily not effective and prophylactic measures are not indicated, except possibly for exposed susceptibles for whom chickenpox might be a serious disease (children or adults on immunosuppressive or cytotoxic drugs or patients with AIDS). Administration of serum (varicella-zoster immune globulin, or VZIG) obtained from convalescent patients with high varicella-zoster antibody titers has been effective in preventing or modifying disease in susceptible children. The serum is in short supply, and therefore its use is restricted primarily to high-risk children such as those with immunodeficiency diseases.

III. Prevention

A vaccine has been tested successfully in Japan on both normal and immunocompromised children. A high rate of antibody production to varicella-zoster virus was observed in both groups, although a common side effect of the vaccine in immunocompromised children was the development of a rash. In 1996, the vaccine was recommended for routine administration to children 12 to 18 months of age.

SMALLPOX (VARIOLA MAJOR)

Smallpox, a severe, acute infectious disease also known as **variola major,** was recognized and feared for many centuries in all parts of the world. Recorded descriptions of its epidemic effects extend from the writings of Galen in the second century. Wherever it was introduced into noninfected susceptible populations, it left a trail of suffering, disfigurement, and death. In 1966, the World Health Organization (WHO) embarked on a program to eradicate smallpox from the world. Several features of the disease made this goal feasible: smallpox afflicts human beings only; there are no animal reservoirs and no carrier state; case finding is aided by the appearance of skin lesions in all who contract the disease; and most important, a highly effective vaccine is available (see Perspective 10.2).

For a number of years, surveillance and control programs as well as the strict monitoring of travelers arriving from previously infected areas kept most countries free of smallpox. By intensive case finding and vaccination of direct and indirect contacts, the disease was eradicated from pockets of infection in India, Bangladesh, and Africa. In October 1977, the last naturally acquired case was found in Somalia, and on December 9, 1979, the Global Commission for the Certification of Smallpox Eradication certified that smallpox had been eradicated from the world.

The virus now exists only in the frozen state in two WHO-approved laboratories (Atlanta and Moscow), and the clinical disease is discussed here primarily for historical reasons. The dreaded possibility remains, however, that undetected foci of disease are present somewhere in the world or that some nonapproved laboratories have not destroyed their stock of smallpox virus cultures. In the latter case, this viral agent could someday be used inappropriately, for example, in germ warfare. Because smallpox vaccination is no longer recommended throughout the world, the results could be devastating to future susceptible populations.

FIGURE 13.7 A smallpox victim in Africa, a sight the world may never see again.
World Health Organization. Photo by J. Breman.

The Clinical Disease

Variola virus enters the susceptible host through the upper respiratory mucosa. During the incubation period, the virus multiplies in lymphoid tissue and is distributed by the bloodstream to many parts of the body. The virus localizes first in mononuclear phagocytic cells, is released again into the circulating blood, and then localizes in skin cells. Symptoms begin when the second stage of viremia becomes intense. The skin rash begins to appear after one or several days of fever, erupting first on the face and then symmetrically on the arms, trunk, and legs (fig. 13.7). The exanthem progresses through macular, papular, and vesicular stages to the formation of pustules on about the eighth or ninth day. Crusts harden on the pustules and fall off after another 10 days or 2 weeks.

The nature and full development of the rash provide an index of the severity of the disease and the extent of the patient's immunity. In severe cases, the rash may become confluent and even hemorrhagic. There is a direct correlation between death rates and the extent and character of the rash.

Recovery is signaled by drying of the pustular lesions. Successive crops of eruptions do not occur in smallpox, so that individual pocks all begin to heal at about the same time. As the scabs fall away, healing is usually accompanied by scarring. Complications of smallpox infection include secondary bacterial invasion of skin lesions, sometimes with resultant septicemia.

LABORATORY DIAGNOSIS

Most cases of smallpox were diagnosed clinically on the basis of the appearance, development, and distribution of skin eruptions, together with a history of sudden febrile illness preceding exanthem. The clinical suspicion of smallpox called for immediate preventive measures to be instituted for all contacts, even before laboratory confirmation was obtained. Smallpox had to be distinguished from chickenpox, drug eruptions, meningococcemia, and other viral diseases accompanied by skin eruption.

Should new cases ever be suspected, only a few diagnostic virus laboratories can now confirm clinical smallpox or assist in doubtful cases.

INFECTIOUS MONONUCLEOSIS

The Clinical Disease

Infectious mononucleosis is an acute infectious disease characterized by fever, pharyngitis, tonsillitis, cervical lymphadenopathy, and, in some instances, involvement of the spleen and liver (with jaundice). It is rarely fatal, but death has been associated with rupture of the spleen.

In recent years, attempts have been made to link another syndrome referred to as *chronic fatigue syndrome* with the virus causing infectious mononucleosis. The patients are usually young adult women who have symptoms similar to infectious mononucleosis and a chronic, incapacitating fatigue. No evidence exists that chronic fatigue syndrome and infectious mononucleosis are related; therefore, this theory has been discounted. The etiology of chronic fatigue syndrome remains unknown but some evidence indicates the disease is caused by an increased, damaging immunological response to a virus mediated by cytokines (see chap. 6).

LABORATORY DIAGNOSIS

I. Identification of the Organism

The agent is a herpes virus called the **Epstein-Barr virus (EBV)**. It is associated not only with infectious mononucleosis but also with nasopharyngeal carcinomas and with a lymphoma known as *Burkitt's tumor* (indigenous in children of Central Africa) (colorplate 6d).

II. Identification of Patient's Serum Antibodies

Antibodies against EBV arise early in the acute stage of disease, reach high levels during convalescence, and persist for years. The indirect immunofluorescence technique is used to demonstrate them.

Heterophile antibodies, or sheep cell hemagglutinins, also develop during the course of infectious mononucleosis. The serologic test for identifying them has been in use for many years and is still the simplest diagnostic technique routinely available for the disease. Serum from some normal persons also agglutinates sheep red cells, but agglutinin absorption distinguishes heterophile antibody of mononucleosis. The normal antibody is absorbed when the serum is mixed with guinea pig kidney tissue but not when mixed with beef red cells. Heterophile antibody of mononucleosis is absorbed by beef erythrocytes but not appreciably by guinea pig kidney tissues. Thus, two serum samples giving the same sheep cell hemagglutination titer can be differentiated as shown in table 13.2.

Horse cells used in place of sheep cells have been found to provide more accurate results. Most laboratories now use a rapid "spot" test in which drops of patient serum are absorbed directly with the beef cells and guinea pig kidney tissue on a slide. The test can be completed in only a few minutes and is as accurate as tests performed in tubes.

Small numbers of infected persons do not develop heterophile antibodies. If an accurate diagnosis is needed, specific serological tests that detect Epstein-Barr virus IgM and IgG antibodies are available, but these are usually performed only in specialized laboratories.

III. Specimens for Diagnosis

Serum samples for heterophile agglutination or for EBV antibodies.

Blood counts show marked lymphocytosis with many abnormal lymphocytes.

Chemical analysis of blood reveals dysfunction of liver parenchyma.

Epidemiology

I. Communicability of Infection

Humans appear to be the reservoir of this disease, although studies of EBV have shown that it also occurs among other primates. It is distributed among humans in many parts of the world. In some areas studied, half of the 1-year-old children and 90% of adults and children over 4 years of age displayed EBV antibody.

Clinically apparent infectious mononucleosis occurs most commonly among children and young adults. Among these, the major route of transmission is probably through intimate oral contact with exchange of saliva. (Mononucleosis has been called the "kissing disease" because of this communicability.)

The incubation period among contacts is from 4 to 6 weeks with a communicable period beginning before symptoms appear and extending during convalescence for many months. The virus may be shed in respiratory secretions for 1 year or more after infection.

II. Control of Active Infection

Standard precautions only apply to cases of infectious mononucleosis. Concurrent disinfection of oral and respiratory discharges or articles soiled by them is advisable.

There is no specific therapy. Clinical trials with acyclovir (see table 9.8) have shown some success but it is not sufficiently effective to recommend its routine use.

CYTOMEGALOVIRUS INFECTION

The Clinical Disease

Serologic surveys have shown that **cytomegalovirus (CMV) infection** is extremely common and usually asymptomatic. Depending on the population studied, antibodies have been found in from 20 to 100% of young adults, with the higher prevalence in lower socioeconomic groups. The infected individual sheds virus in a number of body secretions and excretions. When pregnant women are infected during their last trimester, virus present in their bloodstream (viremia) may cross the placenta, producing

TABLE 13.2 SEROLOGIC DIFFERENTIATION OF INFECTIOUS MONONUCLEOSIS

Heterophile Titer before Absorption	Titer after Absorption with: Beef RBC	Guinea Pig Kidney Tissue	Diagnosis
Yes	Yes	No	Normal serum
Yes	No	Yes	Infectious mononucleosis

congenital fetal disease. In a small number of cases, this congenital infection results either in intrauterine fetal death or in birth of an infant with one or all of the following signs: hepatomegaly, splenomegaly, jaundice, purpura, a petechial rash, and various neurologic deficits. Even when the disease is not apparent at birth, mental retardation may become evident at a later time. Neonatal infections may be acquired at birth either from virus present in maternal cervical secretions or by subsequent contact with other infectious maternal secretions. Approximately 1% of newborns are infected with CMV, but most are asymptomatic, even though they shed large amounts of virus, especially in urine. In children and adults, clinically apparent infection may take several forms, including pneumonia, hepatitis, and mononucleosis. In the last instance, the clinical signs are similar to those of EBV mononucleosis, but the heterophile antibody test is negative. CMV infections also occur in post-transfusion and post-transplant patients. Presumably, the virus is present in the donor blood or organ, although reactivation of latent disease (which is characteristic of all herpes viruses, including CMV) in the recipient cannot always be ruled out. CMV disease is especially severe in immunocompromised patients, including those with acquired immune deficiency syndrome (see chap. 19).

LABORATORY DIAGNOSIS

I. Identification of the Organism

CMV is a herpes virus belonging to the same group as the Epstein-Barr virus, the herpes simplex viruses (see chap. 19), and the varicella-zoster virus. Patient specimens are inoculated into human fibroblast cell cultures and observed for the appearance of characteristic swollen cells with intranuclear inclusions. Specific fluorescent antibody reagents are used to confirm the identity of the virus. Direct examination of stained smears of tissue cells present in a variety of body fluids and secretions may reveal CMV inclusions, but the cell culture technique is more sensitive and therefore preferable.

II. Identification of Patient's Serum Antibodies

Detectable antibodies to CMV arise during both clinical and subclinical infections. The indirect fluorescent antibody test is preferred, but neutralization and indirect hemagglutination methods are also available. Detecting IgM antibody in the serum of neonates may help distinguish antibody produced as a result of active infection from passively transferred maternal IgG.

III. Specimens for Diagnosis

A. Virus can be recovered from urine, saliva, tears, milk, semen, stools, vaginal or cervical secretions, and the buffy coat of blood (the white blood cell layer that accumulates between the red blood cells and plasma when anticoagulated blood is centrifuged). For maximum viral recovery, specimens must be inoculated into cell cultures as quickly as possible, but refrigeration for short time periods is acceptable if processing is delayed.

B. Acute and convalescent phase serums.

C. Body fluids and secretions for cytologic demonstration of viral inclusions (not as sensitive as culture).

Epidemiology

I. Communicability of Infection

CMV is highly communicable, as evidenced by the large number of individuals with CMV antibody titers. Most postnatal primary CMV infections are acquired by droplet or possibly fecal-oral spread. Infection may also occur by means of blood transfusion, organ transplantation, or possibly sexual contact. Neonates most likely acquire the virus during birth from infected cervical secretions or later from infected breast milk. The incubation period ranges from 2½ to 7 weeks. Virus is shed for extended periods of time, even though the infected individual is asymptomatic and may have a high CMV antibody titer.

II. Control of Active Infection

Control is difficult because of the high incidence of inapparent infection. With known cases, control measures in the hospital include standard precautions for respiratory secretions, urine, blood, and feces. Pregnant personnel should be informed of the infection risk and excluded from contact with known cases of the disease. The antiviral agents ganciclovir and foscarnet (see table 9.8) have proven effective in cases of severe CMV infection in immunocompromised patients.

QUESTIONS

1. What is influenza? What organisms may complicate the disease? Why?
2. When is influenza immunization indicated?
3. What are postinfluenza complications? Are they preventable?
4. Why is it important for health departments to monitor the types of influenza infections and patterns of influenza epidemics?

SUMMARY: TABLE 13.3 VIRAL DISEASES ACQUIRED BY THE RESPIRATORY ROUTE

Clinical Disease	Causative Organism	Other Possible Entry Routes	Incubation Period	Communicable Period	Specimens Required
Influenza	Influenza virus		1–2 days	Acute febrile stage (3–4 days)	Pharyngeal and nasal swabs Nasal washings Acute and convalescent serums
Common cold or coldlike	Rhinoviruses Adenoviruses Parainfluenza viruses Coronaviruses Coxsackieviruses Echoviruses	By fecal-oral route	1–2 days	During time of active symptoms	Same as above
Respiratory syncytial disease	Respiratory syncytial virus		2–8 days	During time of active symptoms	Nasal washings Nasopharyngeal swab Acute and convalescent serums
Hantavirus pulmonary syndrome	Sin Nombre Virus	Rodent bite	1–10 days	Not communicable person-to-person	Acute and convalescent serums
Mumps	Mumps virus		18 days to 3 weeks	6 days before onset of symptoms until 9 days after first swelling	Acute and convalescent serums
Measles (rubeola)	Measles virus		About 10 days	From onset of symptoms through catarrhal period and for 4–5 days of eruptive stage	Blood serums Blood Urine Conjunctival and nasopharyngeal secretions
Rubella (German measles)	Rubella virus	Transfer of virus across placenta	14–21 days	Week before and after appearance of rash	Blood serums Throat and nasal swabs, blood, urine
Erythema infectiosum	Parvovirus B-19	Transfer of virus across placenta	4–20 days	Before onset of rash to about 1 week after symptoms	Blood serums
Chickenpox and herpes zoster	Varicella-zoster virus		2–3 weeks	Greatest from a day before rash appears to 5th or 6th day of eruption	Vesicle fluid Blood serums
Infectious mononucleosis	Epstein-Barr (EBV) virus	Congenital?	4–6 weeks	Before onset, during clinical illness, and throughout convalescence	Blood serums
Cytomegalovirus infection	Cytomegalovirus	Transfer of virus across placenta	2.5–7 weeks	Virus shed for long time period	Urine, saliva, tears, milk, semen, stools, vaginal or cervical secretions Acute and convalescent serums

*Standard precautions apply to all patients with infectious diseases. Other transmission-based precautions indicated when applicable.

5. What is the particular concern associated with respiratory syncytial virus (RSV)? How is this disease controlled in the hospital?
6. What are the clinical complications of measles?
7. What is Reye syndrome?
8. What type of vaccine is used to control measles? When is it administered?
9. What is the clinical significance of rubella? Why should health care professionals, including men who care for pregnant women, be immunized for rubella?
10. What is hantavirus pulmonary syndrome? What causes this disease?
11. What is the reservoir for hantaviruses? How do humans become infected with Sin Nombre virus?
12. How is hantavirus disease controlled?
13. Why is smallpox immunization no longer required in the United States?

Laboratory Diagnosis	Immunization	Treatment	Precautions*
Virus isolation Antibody titer	Killed vaccine	None specific Rimantidine or amantadine may shorten course if given early	Droplet precautions are used during the acute stage of infection
Virologic diagnosis usually not indicated Virus isolation Direct antigen detection Antibody titer	Killed vaccine for adenoviruses, use limited	None specific	None required but principles of personal hygiene, support of healthy resistance protective
Virus isolation Direct antigen detection Antibody titer	None	Ribavirin for children with serious disease	Contact precautions Careful hand washing mandatory
Antibody titer (4-fold rise in IgG or single IgM)	None	None specific	Standard precautions only
Antibody titer (4-fold rise in IgG or single IgM)	Live vaccine (2 doses)	None specific	Droplet precautions; concurrent disinfection of contaminated articles; hospitalized patient in private room
Antibody titer (IgG or IgM) Virus isolation Direct antigen detection	Live vaccine (2 doses) Gamma globulin for high-risk individuals	None specific	Airborne precautions; concurrent disinfection of articles contaminated by nose-and-throat secretions; protect young and immunocompromised persons from measles cases
Antibody titer Virus isolation	Live vaccine (2 doses)	None specific	Droplet precautions (susceptible persons should stay out of room if possible)
IgM antibody titer	None	None specific	Droplet precautions; concurrent disinfection of articles contaminated with respiratory secretions
Virus isolation to distinguish from smallpox virus Antibody titer	Live vaccine	Acyclovir for immunocompromised patients	Droplet precautions in hospital to protect susceptible individuals
Heterophile agglutination test EBV antibodies	None	None specific	Standard precautions; concurrent disinfection of oral and respiratory discharges
Virus isolation Antibody titer	None	Ganciclovir for severe infections in immunocompromised patients	Standard precautions; concurrent disinfection of articles contaminated with infectious secretions and excretions; terminal disinfection

14. What are the serious effects of erythema infectiosum, or fifth disease?
15. What are the similarities and differences between chickenpox and shingles?
16. What drug is contraindicated for children with chickenpox? Why?
17. What is the etiology of infectious mononucleosis? Why is it called the "kissing disease"?
18. What laboratory tests are used to diagnose infectious mononucleosis?
19. Why is cytomegalovirus of concern to pregnant women?

BIBLIOGRAPHY

Benenson, A. S., ed. 1995. *Control of communicable disease in man,* 16th ed. Washington, D.C.: American Public Health Association.

Beveridge, W. I. B. 1977. *Influenza: The last great plague.* New York: Neale Watson Academic Publications.

Brady, M. 1994. Epstein-Barr virus infection in children: Implications for the treatment of infectious mononucleosis. *J. Pediatr. Health Care* 8 (5): 233–235.

Caruthers, D. D. 1990. Infectious pneumonia in the elderly. *Am. J. Nurs.* 90 (2): 56–60.

Centers for Disease Control and Prevention. 1989. December 29. Rubella prevention: Recommendations of the immunization practices advisory committee. *Morbid. Mortal. Weekly Rep.* 38 (no. S-9).

Centers for Disease Control and Prevention. 1990. Measles prevention: Recommendations of the immunization practices advisory committee. *Morbid. Mortal. Weekly Rep.* 39 (no. RR-15).

Centers for Disease Control and Prevention. 1995. Recommended childhood immunization schedule—United States, 1995. *Morbid. Mortal. Weekly Rep.* 44 (no. RR-5).

Centers for Disease Control and Prevention. 1995. Update: Influenza activity—Worldwide 1995. *Morbid. Mortal. Weekly Rep.* 44 (35): 644.

Cormier, D. P., and D. R. Mayo. 1988. Parvovirus B-19 infections. *Clin. Microbiol. Newsl.* 10 (7): 49–52.

Davis, D. M. 1988. Reye's syndrome, 1988. *J. Emerg. Nurs.* 14 (2): 110–111.

Evans, A. S., ed. 1991. *Viral infections of humans: Epidemiology and control* 3d ed. New York: Plenum.

Faoagali, J. L., and D. Darcy. 1995. Chickenpox outbreak among staff of a large, urban adult hospital: Costs of monitoring and control. *Am. J. Infect. Control.* 23 (4): 247–250.

Filippell, M. B., and T. Rearick. 1993. Respiratory syncytial virus. *Nurs. Clin. N. Am.* 28 (3): 651–671.

Henderson, D. A. 1976. The eradication of smallpox. *Sci. Am.* 235 (4): 25–33.

Kahn, A. S. et al. 1996. Hantavirus pulmonary syndrome: The first 100 U.S. cases. *J. Infect. Dis.* 173 (6): 1297–1303.

Kaplan, M. M., and R. G. Webster. 1977. Epidemiology of influenza. *Sci. Am.* 237 (6): 88–92.

Langer, W. L. 1976. Immunization against smallpox before Jenner. *Sci. Am.* 234 (1): 112–117.

Lennette, E. H., P. Halonen, and F. A. Murphy, eds. 1988. Laboratory diagnosis of infectious diseases: Principles and practice, vol. II. *Viral, rickettsial and chlamydial diseases.* New York: Springer-Verlag.

Mandell, G. L., J. E. Bennett, and R. Dolin. 1995. *Principles and practice of infectious diseases,* 4th ed. New York: Churchill Livingstone.

Patterson, K. D. 1986. *Pandemic influenza 1700–1900.* Totowa, N. J.: Rowland & Littlefield.

Rashotte, J. 1989. The seasonal invader—Respiratory syncytial virus. *Can. Nurse.* 85 (10): 29–32.

Sanderson, D. 1993. The worst disaster in recorded history. *Old Farmers Almanac* 201: 114–117.

Thomas, R. 1991. Hantavirus: Quick and deadly. *RN* 57 (8): 11–18.

Todd, B. 1989. Getting ready for flu season: Vaccines for elders? *Geriatr. Nurs.* 10 (5): 227–228.

Walker, J. H. 1993. Commentary on respiratory syncytial virus and the use of ribavirin. *Enas. Nurs. Scan. Emerg. Care* 3 (2): 9.

Wood, P. J. 1993. Measles, mumps, and mud: Childhood *epidemics* at the turn of the century. *Nurs. Prax. NZ.* 8 (3): 21–29.

CHAPTER 14

Systemic and Subcutaneous Mycoses: Fungal Diseases Acquired through Respiratory, Parenteral, and Endogenous Routes

OUTLINE

Microbes in Motion

This chapter covers the important fungal diseases acquired through the respiratory, parenteral, and endogenous routes including the clinical characteristics, causative organisms, laboratory diagnosis, epidemiology, and treatment. The following book and chapter of the Microbes in Motion CD-ROM may be a useful preview or supplemental study aid to your reading: Fungal Structure and Function: Diseases.

OBJECTIVES

On completing this chapter the student will be able to:

1. differentiate between and give an example of systemic, subcutaneous, and opportunistic fungal infections.
2. explain the pathogenic qualities of fungal agents of human disease.
3. discuss the nature of important human mycotic diseases in terms of their (a) causative organism(s), (b) clinical characteristics, (c) methods for laboratory diagnosis, (d) epidemiology, and (e) treatment.

KEY WORDS

Arthrospore (**ARTH**-ro-spore)
Aspergillosis (ass-per-jill-**OH**-sis)
Aspergillus (ass-per-**JILL**-us)
Blastomyces (blast-o-**MY**-sees)
Blastomycosis (blast-o-my-**CO**-sis)
Blastospore (**BLAST**-o-spore)
Candida (**CAN**-did-ah)
Candidiasis (can-did-**EYE**-a-sis)
Chlamydospore (clam-**ID**-o-spore)
Chromomycosis (chromo-my-**CO**-sis)
Coccidioides (cock-sid-i-**OID**-ease)
Coccidioidomycosis (cock-sid-i-**OID**-o-my-**CO**-sis)
Cryptococcosis (crypt-o-cock-**OH**-sis)
Cryptococcus (crypt-oh-**COCK**-us)
Dimorphic (di-**MORPH**-ic) fungus
Endogenous (end-**AH**-gen-us)
Germ tube
Histoplasma (histo-**PLAZ**-ma)
Histoplasmosis (histo-plaz-**MO**-sis)
Interstitial plasma cell pneumonia
Macroconidia (mac-ro-co-**NID**-ia)
Maduromycosis (ma-**DURE**-o-my-**CO**-sis)
Microconidia (my-crow-co-**NID**-ia)
Mucor (**MEW**-core)
Mycetoma (my-seat-**OH**-ma)
Mycoses (my-**CO**-sees)
Mycotic (my-**COT**-ic) diseases
Nocardia (no-**CARD**-ia)
North American blastomycosis
Opportunistic mycosis
Paracoccidioides (para-cock-sid-i-**OID**-ease)
Paracoccidioidomycosis (para-cock-sid-i-**OID**-o-my-**CO**-sis)
Pneumocystis (new-mo-**CYST**-is)
Pneumocystosis (new-mo-cyst-**OH**-sis)
Rhizopus (**RISE**-o-pus)
South American blastomycosis
Sporothrix (**SPORE**-o-thrix)
Sporotrichosis (spore-o-trick-**OH**-sis)
Subcutaneous mycoses
Superficial mycoses
Systemic mycoses
Thrush
Verrucous (ve-**RUKE**-ous) dermatitis
Yeast
Zygomycosis (zi-go-my-**CO**-sis)

Mycotic infections: clinical and epidemiologic distinctions

Clinically, human **mycotic diseases** fall into three distinct patterns: **superficial mycoses** involving only the outermost epithelial structures of the body (skin, hair, and nails), **systemic mycoses** arising from fungal infections of deep tissues, and **subcutaneous mycoses** deriving from organisms that penetrate through skin and generally remain localized in subcutaneous tissues. Superficial fungal infections are usually chronic and difficult to treat but are not a threat to general health because the causative agents do not invade deeper tissues and are not disseminated. Fungi responsible for systemic infections, on the other hand, can invade internal body tissues and be disseminated widely. They may remain inactive or be held quiescent by the body's defense mechanisms, producing few if any symptoms of importance, or they may cause progressively advancing disease. Subcutaneous mycoses are generally acquired through trauma to the skin, their agents being introduced into deep dermal tissues. Although such infections may involve extensive subcutaneous areas and induce marked regional lymphadenopathy, generally they are not disseminated to internal organs.

Important epidemiologic distinctions can also be made for these three types of **mycosis.** Fungi of superficial infections have a natural reservoir in soil, human beings, and some animals and are directly transmissible by close or indirect contacts among infected individuals. Most agents of both the deep and subcutaneous mycoses, however, appear to have a reservoir only in soil, where they live as saprophytes. They are capable of parasitism but are not directly transmissible from one person to another. The systemic mycoses are generally acquired through inhalation of fungal spores or conidia (see chap. 2) from a soil source, whereas subcutaneous fungal infections are acquired through injury to the skin, with introduction of spores or conidia into underlying tissues. The term *spores* will be used throughout this chapter in general reference to both spores and conidia.

In keeping with these distinctions, the superficial fungal infections are described in Section VI, together with other diseases acquired and transmitted through skin contacts (see chap. 19). Systemic and subcutaneous mycoses are grouped together here because, irrespective of their routes of entry, the basic patterns of their pathogenesis and epidemiology are similar.

Systemic and subcutaneous mycoses: sources, entry routes, pathogenesis, epidemiology

Sources

Many species of fungi are associated with systemic and subcutaneous disease in humans. Most of these are derived from an exogenous soil reservoir and can cause a primary mycosis following effective entry into the healthy body. Others are more often associated with opportunistic infection in persons with underlying disease or an immunodeficiency. One important yeast genus, ***Candida,*** has several species that live commensally on human mucosal surfaces. From this **endogenous** source, *Candida* species (particularly *C. albicans*) can induce serious infection, often in infants, aged people, or those debilitated by other diseases. The exact source in nature is not known for some fungi, the most prominent example being ***Blastomyces.***

Free-living fungi are not ubiquitously distributed in nature but often are concentrated in particular areas where soil conditions permit their growth. Some fungi like ***Aspergillus*** occur in widely scattered parts of the world, others such as ***Coccidioides*** and ***Paracoccidioides*** in limited geographic regions. Little is known about the natural life of these fungi or about the factors that determine their geographic distribution. At least two pathogenic genera, ***Cryptococcus*** and ***Histoplasma,*** are often associated with birds such as pigeons, chickens, and starlings. They may be spread in bird droppings (which act as an enrichment medium) or be carried on their contaminated feathers and distributed widely on air currents. *Histoplasma* has also been isolated from bat caves.

Entry routes

In soil, these microorganisms produce many spores that can survive for long periods, with or without moisture, in earth or on wood or plant surfaces. When spore-laden dust is stirred up by feet, shovels, bulldozers, or wind currents, the airborne spores may be inhaled in sufficient numbers to reach the lungs and induce primary pulmonary mycoses. This is a common entry route for diseases such as **histoplasmosis** and **coccidioidomycosis.**

In other instances, fungal spores such as those of ***Sporothrix*** may be introduced subcutaneously through minor skin injuries incurred by people working with contaminated soil, plants, or wood. Splinters and thorns are common agents of such injuries. The primary infection in these cases is extrapulmonary, as in **sporotrichosis** or **chromomycosis.** Although most subcutaneous mycoses do not disseminate to deep, internal tissues, sporotrichosis sometimes does. Rarely, sporotrichosis occurs as a primary pulmonary disease that can be misdiagnosed as tuberculosis.

In general, human beings have a high natural resistance to fungal infections. Opportunistic fungal infections, however, have increased with the widespread use of broad-spectrum antimicrobial agents that destroy competing bacteria, corticosteroid and immunosuppressive therapy and diseases such as AIDS that decreases normal cellular defense mechanisms, and the self-administration of contaminated injectable drugs such as heroin (or injection with contaminated paraphernalia). Cardiac surgery, indwelling intravenous catheters, hyperalimentation, organ transplantation, and underlying malignancies (leukemia, lymphoma) are also associated with opportunistic mycotic infection.

Fungal Infections of Humans

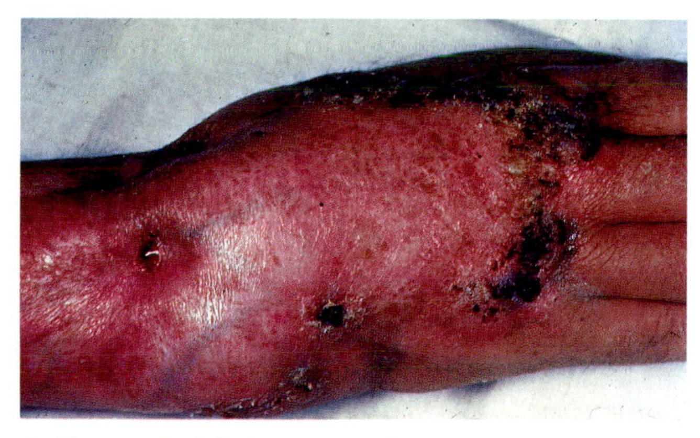

a) Blastomycosis of the forearm caused by *Blastomyces dermatitidis.*

From E. S. Beneke et al., *Human Mycoses,* 1984. Reproduced by permission of the Upjohn Company and E. S. Beneke.

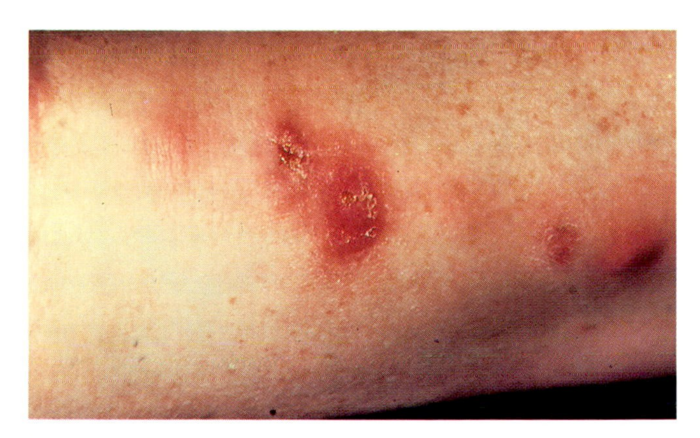

b) Sporotrichosis of the arm caused by *Sporothrix schenckii.*

© Everett S. Beneke/Visuals Unlimited.

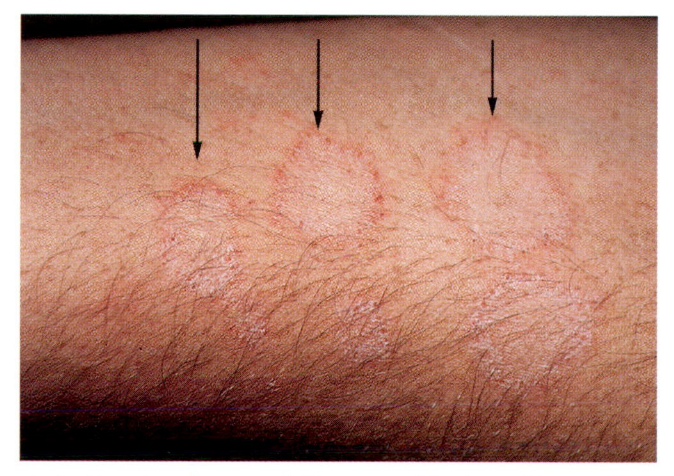

c) In this patient tinea corporis (ringworm of the forearm) is caused by the fungus *Trichophyton mentagrophytes.* The circular rash is typical of this infection.

© Carroll H. Weiss/Camera M. D. Studios.

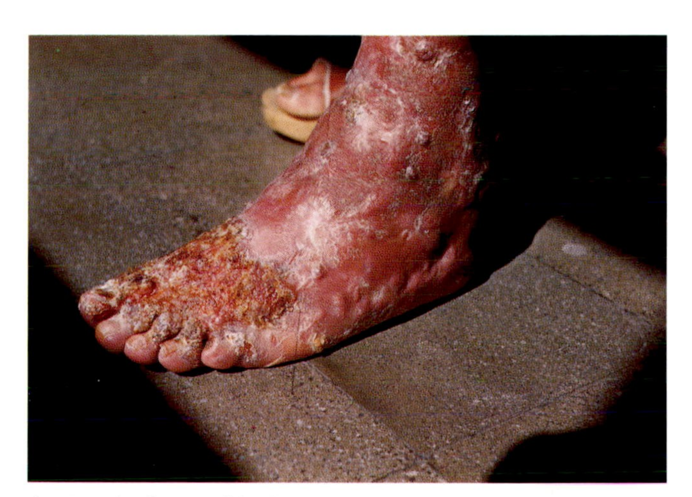

d) Fungal infection of the foot (mycetoma) caused by *Madurella mycetomatis.*

From E. S. Beneke et al., *Human Mycoses,* 1984. Reproduced by permission of the Upjohn Company and E. S. Beneke.

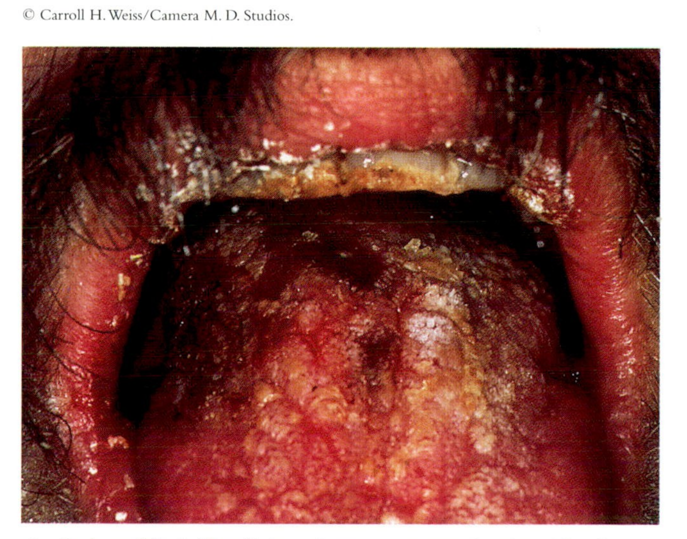

e) Oral candidiasis (thrush) in an immunosuppressed patient. The disease is characterized by white patches on the tongue.

Centers for Disease Control, Atlanta, GA and John Molinaire Ph.D. University of Detroit, Detroit, MI.

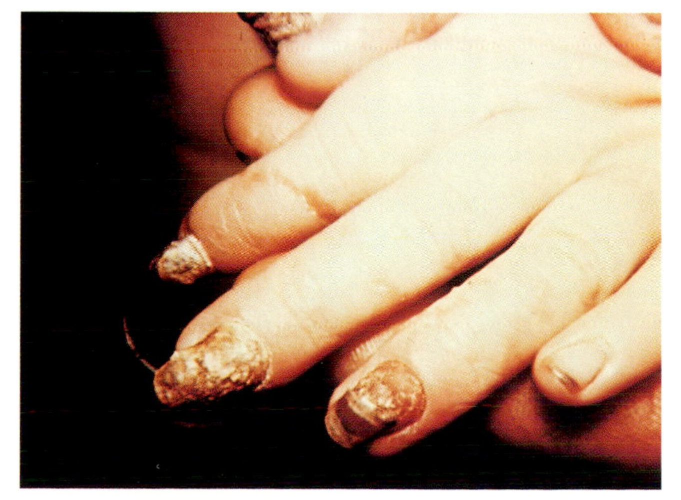

f) Paronychia and onychomycosis of the fingernails caused by *Candida.*

© Everett S. Beneke/Visuals Unlimited.

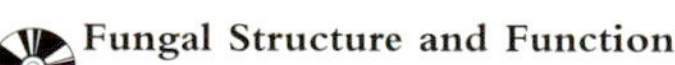

Fungal Structure and Function
Diseases: Environmental Diseases Occupation; Inhalation; Penetration

The Appearance of Two Parasitic Diseases in Humans

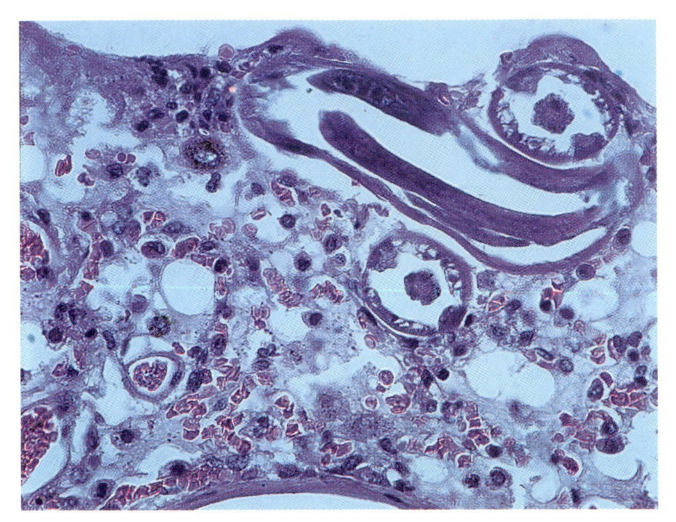

a) *Ascaris lumbricoides* in lung tissue.
Courtesy Dr. Paul Beaver, Tulane University, New Orleans, LA.

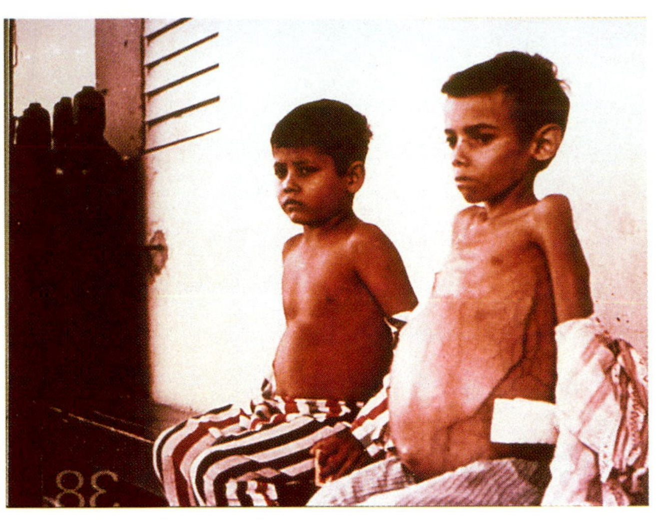

b) Schistosomiasis in two young boys. The eggs lodged in the liver interfere with the portal circulation and result in liver enlargement.
World Health Organization.

Examples of Parasitic Protozoa

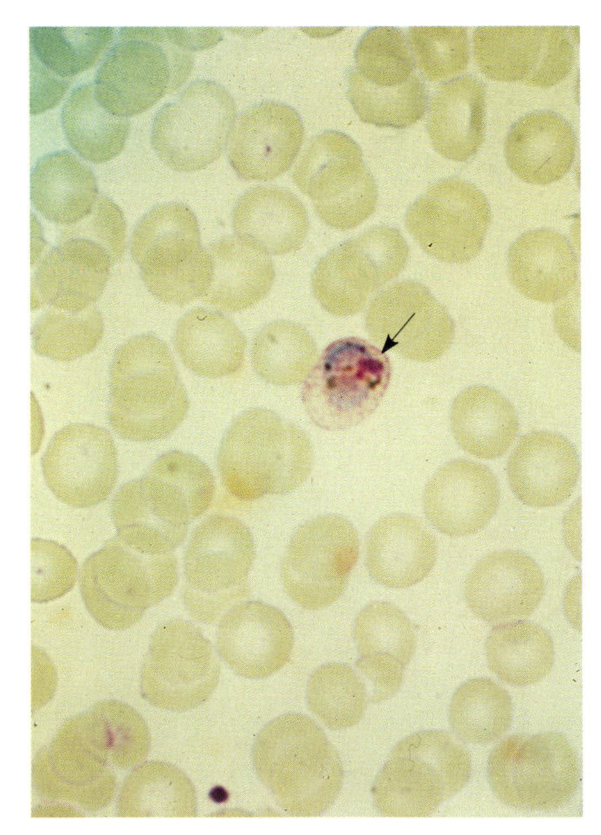

c) *Plasmodium vivax* trophozoite in a red blood cell (arrow). The red and blue malarial parasite causes the red blood cell to enlarge and to show characteristic stippling (Schüffner's dots).
Courtesy Dr. E. J. Bottone, Mt. Sinai Hospital, New York, NY.

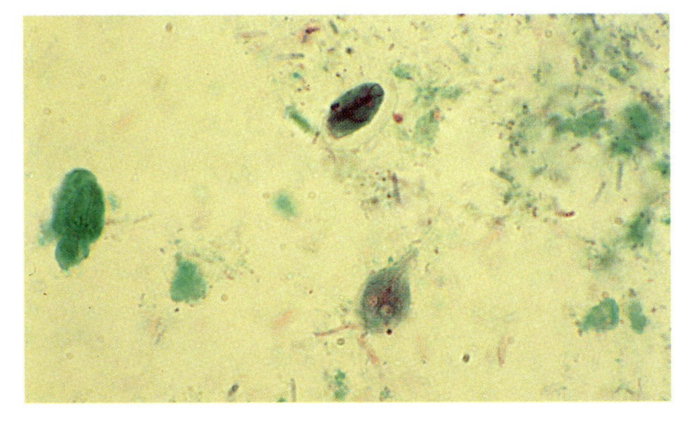

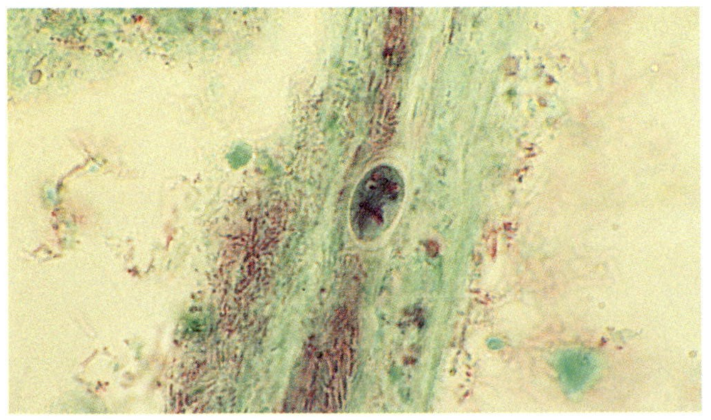

d) *Giardia lamblia* (top) trophozoite and cyst seen in stool specimen of patient with giardiasis. (bottom) The characteristic features of the cyst are revealed more clearly in this view (×1,000).

TABLE 14.1 CLASSIFICATION OF SYSTEMIC AND SUBCUTANEOUS MYCOSES

Type	Sources	Entry Routes	Primary Infection	Disease	Causative Organism(s)
Primary Systemic Mycoses	Exogenous	Respiratory or parenteral	Pulmonary or extrapulmonary	Histoplasmosis	*Histoplasma capsulatum*
				Coccidioidomycosis	*Coccidioides immitis*
				Blastomycosis (North American)	*Blastomyces dermatitidis*
				Paracoccidioidomycosis (South American blastomycosis)	*Paracoccidioides brasiliensis*
				Cryptococcosis	*Cryptococcus neoformans*
Subcutaneous Mycoses	Exogenous	Parenteral	Extrapulmonary	Sporotrichosis	*Sporothrix schenckii*
				Chromomycosis	*Phialophora, Fonsecaea, Cladosporium, Rhinocladiella* species
				Mycetoma (Madura foot)	*Madurella, Pseudallescheria* species, and others
Opportunistic Mycoses	Endogenous	Skin, mucosae, or gastrointestinal tract	Superficial or disseminated	Candidiasis	*Candida albicans* and other species
	Exogenous	Respiratory	Pulmonary	Aspergillosis	*Aspergillus fumigatus* and other species
	Exogenous	Respiratory or parenteral	Pulmonary or extrapulmonary	Zygomycosis	*Mucor, Rhizopus, Absidia,* and others

The distinction between primary and opportunistic mycoses is sometimes difficult. Diseases such as histoplasmosis and **cryptococcosis** have occurred with increasing incidence among patients on prolonged corticosteroid therapy. Conversely, although nearly all patients with **aspergillosis** or **zygomycosis** have some underlying predisposing disease, rare cases may occur as primary infections. In healthy individuals, however, opportunistic fungi of low pathogenicity must gain entry in some extraordinary way (be inhaled in large numbers from a rich source in soil or be introduced traumatically) to establish primary disease.

Similarly, endogenous fungi such as *Candida* species require a predisposing opportunistic condition to induce damaging infection in superficial or deep tissues. *Candida* strains live commensally on normal skin and on oral, vaginal, and intestinal mucosae. Alteration of the normal bacterial flora by antimicrobial therapy may permit their overgrowth with resultant damage to local tissues. Lowered host resistance, trauma, or other circumstances may contribute to further extension of *Candida* infection. Superficial **candidiasis** is much more frequent than systemic disease, but *Candida* may localize in lungs, meninges, the endocardium, or other tissues.

Table 14.1 shows a classification of the primary systemic, subcutaneous, and opportunistic mycoses, indicating source and entry routes.

PATHOGENESIS

The typical tissue reaction to deep fungal infections is a chronic granuloma with necrosis or abscess formation. This is the same general type of lesion that develops in tuberculosis, with certain differences characteristic of the proliferating organism or of its location in the body. Mononuclear, epithelioid, and giant cells surround and attempt to contain the organism, and this group of cells is surrounded peripherally by dense accumulations of lymphocytes and fibroblasts. The different fungal agents germinate and reproduce in different ways. Under the pressure of human tissue opposition, proliferation may be halted or held to a minimum for long periods of time, or it may advance progressively as conditions permit. A necrotizing or suppurating granuloma may release fungi into adjacent tissues or into lymphatic channels or blood vessels that distribute them widely in the body. Infection may then progress at new sites.

In primary pulmonary mycoses resulting from inhalation of fungal spores, granulomatous lesions may be distributed through any portion of the lungs. The extent of disease varies from a single, small lesion to multiple foci involving large areas of both lungs. Areas of consolidation can resemble bacterial pneumonia or extensive tumor formation. Cavities may be formed, or there may be miliary, or fine or coarse nodular patterns. Infection may progress extensively within the lung or may spread hematogenously to other tissues (liver, spleen, kidneys, brain and meninges, bone, and skin) from a primary pulmonary site.

Skin and subcutaneous lesions sometimes develop by dissemination from systemic (usually pulmonary) foci, or they may represent the primary form of disease at the site of entry of the agent. **Paracoccidioidomycosis** (South American blastomycosis), for example, is a primary pulmonary mycosis with early lesions in the lung and later involvement of subcutaneous tissues, especially at the mucocutaneous borders of the mouth and nose, with extensive lymphadenopathy. It is possible, however, that such facial lesions may sometimes represent the primary site of entry of the fungus. In the subcutaneous mycoses, granulomas are formed in the deep tissues of the skin. Often these nodular lesions tend to break down and ulcerate, discharging through the surface. In sporotrichosis, minor injuries to the skin of the hands or arms admit the spores from thorns, splintered wood, or soil. The organism then slowly advances from the site of entry, where a subcutaneous nodule usually forms along the course of the regional draining lymphatics. Disseminated forms of sporotrichosis occur but are not common. As pointed out previously, this disease may develop as a primary pulmonary infection, though this is rare. The lesions of chromoblastomycosis and **mycetoma** are characteristically limited to tissues of the extremity originally exposed to infection (usually the feet and legs), but they can be extensive and disabling.

Hypersensitivity, a manifestation of cell-mediated immunity, is the characteristic immunologic response to mycoses. This specifically enhanced reactivity to components of the invading fungus probably contributes in many instances to necrosis of granulomatous lesions and to progressive advance of infection, as it does in tuberculosis of the reinfection type (see chap. 6 and chap. 12). Protective humoral immunity does not develop in mycotic diseases, although circulating antibodies measurable by serologic methods do appear in many instances. These have some diagnostic and prognostic value but do not appear to have a role in resistance.

The clinical symptoms depend on the location of the organism and the degree of nonspecific as well as hypersensitive reactivity evoked by the different fungi. Pulmonary mycosis often begins as a subacute respiratory infection, with low-grade fever, dyspnea, and nonproductive cough. As the disease progresses, these symptoms intensify, but the cough becomes productive of purulent sputum. Pain in the chest, loss of weight, fatigue, and night sweats may ensue. Disseminated disease produces symptoms referable to hypersensitivity and to the involvement of other organs.

The diagnosis is sometimes difficult because the clinical characteristics of systemic mycotic lesions are similar to those of tuberculosis and other granulomatous diseases. Laboratory identification of the causative fungus in smears and cultures of sputum, body fluids, or pus from localized lesions usually provides a specific diagnosis.

The microscopic characteristics of certain pathogenic fungi as they are seen in tissues or in direct smears and wet mounts made from clinical specimens differ from those that appear when the organisms grow in the environment or in laboratory culture. This difference in the structure of tissue and nontissue phases of the same fungus is of some epidemiologic interest because infected human beings do not transmit the parasitic fungi directly to others. It is only when the parasitic forms have converted to their environmental type of growth outside of the body that spores or other structures with an infectious potential are produced. A fungus that exists in two forms, one in tissues and the other in the environment, is referred to as a **dimorphic fungus.**

Skin tests and serologic methods are sometimes helpful in establishing the presumptive diagnosis of mycotic disease or in indicating the prognosis. Tuberculin-type skin tests, using sterile, somewhat purified culture filtrates of the organisms, produce delayed hypersensitivity reactions if they are positive. This may indicate either currently active or past (possibly healed) infection. However, skin testing usually has more epidemiologic and prognostic value than diagnostic usefulness. For example, in areas where histoplasmosis is endemic, more than half the population may have positive skin tests, yet be clinically asymptomatic. Conversely, negative skin tests are not uncommon in the presence of clinically active histoplasmosis or coccidioidomycosis, and such findings cannot rule out the diagnosis. Prognostically, a reversion from a positive to a negative skin test often indicates anergy, the failure of the immune mechanisms, and the opposite of allergy or hypersensitivity. This may happen in the terminal stages of overwhelming, progressive infection, signaling a grave prognosis. Presently, skin testing is seldom done because skin test antigens are not widely available. Serologic tests are preferred.

Serologic methods used in the diagnosis of mycotic disease include latex agglutination, fluorescent antibody techniques, enzyme immunoassays, and precipitin tests such as tube precipitation and immunodiffusion (ID). A titer of 32 or greater on a single serum specimen is highly suggestive of infection. Demonstrating a fourfold or more increase or decrease in titer is diagnostic provided that care is taken in test performance and interpretation to rule out cross-reactions. In general, procedures using fungal antigens tend to lack either specificity or sensitivity, or both. Fungal serologic tests are usually performed only in specialized reference or large hospital laboratories.

Epidemiology

Systemic mycoses are frequently limited to particular geographic areas of the world. Within these areas a majority of people may acquire the fungal infection, but a very small minority of these develop serious progressive disease. Some mycoses, such as histoplasmosis and coccidioidomycosis, appear to be increasing in importance. This may reflect improved diagnosis, an increased rate of exposure among the ever-larger traveling public, or a greater number of people with heightened susceptibility. Some of the factors that lower resistance to mycotic infection are (1) other persistent, debilitating disease (cancer, diabetes, tuberculosis, malnutrition), (2) widespread use of chemotherapeutic drugs and hormones that alter human metabolism and disturb balances among commensalistic flora of the body, (3) immunodeficiencies such as AIDS or the use of immunosuppressive drugs, (4) drug abuse, and (5) local acute or chronic injury that permits fungi to enter deep tissues (accidental or surgical trauma, ulcerative lesions of skin or of oral or intestinal membranes, lesions of avitaminosis, X irradiation).

The incubation period of systemic mycoses varies with the numbers of organisms introduced and the ease of their access to sites where they can proliferate. Fungi do not multiply rapidly, and as a general rule the first symptoms of infection do not appear for at least 10 days to 2 weeks. In many instances, the incubation period may extend from 3 to 5 weeks, or it may be impossible to define.

Control of active mycotic infection does not require isolation or any special measures to protect the patient's contacts. Care should include application of concurrent disinfection techniques to sputum or other discharges (as clinically appropriate) and to contaminated articles. Terminal disinfection-cleaning should be provided. These precautions are necessary to prevent possible conversion of the fungus from the parasitic to infectious spore forms.

When epidemics of airborne mycoses occur (coccidioidomycosis, histoplasmosis), recognition of their sources makes it possible to apply dust control measures or to eliminate heavy concentrations of fungus in limited areas (chicken houses, barns, and starling roosts may be sources of *Histoplasma* infection). These procedures may be of preventive value when applied on a routine basis in endemic areas.

Treatment of deep fungal infections varies with the disease. Amphotericin B is an effective broad-spectrum antifungal agent used to treat most systemic mycoses. It is not equally effective in all pulmonary infections, but it appears to be the most potent drug available for each of them. Unfortunately, its use is complicated by toxic and undesirable side effects which include fever, nausea and vomiting, anemia, and renal failure. Alternative drugs may be used to avoid these toxic effects; however, amphotericin is the most effective therapeutic agent available for blastomycosis, coccidioidomycosis, and histoplasmosis.

Other drugs such as 5-fluorocytosine (flucytosine) and imidazoles (ketoconazole, miconazole; see table 9.8) have been used, with varying degrees of success, to treat fungal infections. Sporotrichosis is most commonly treated with an oral solution of potassium iodide. Cryptococcosis is treated with flucytosine and amphotericin, a combination that makes it possible to reduce the amount and, therefore, the toxicity of the amphotericin. Systemic candidiasis is treated with amphotericin or ketoconazole. The antimicrobial agents used for bacterial infections have no value in treating fungal diseases because the drugs do not affect the eucaryotic fungal cell structures.

PRIMARY SYSTEMIC MYCOSES

Individual mycotic diseases are briefly described in outline form on the following pages. The preceding section should be consulted for general information when a particular disease is being studied. Specific details are inserted only when they represent an exception to the general features of mycosis.

HISTOPLASMOSIS

Causative Organism

Histoplasma capsulatum, a dimorphic fungus.

The Clinical Disease

Primary pulmonary infection is frequent in endemic areas but usually asymptomatic. Small granulomas in the lung heal with calcification, as evidenced by chest X ray and histoplasmin-positive and tuberculin-negative skin tests. (Calcified lung lesions cannot be distinguished from healed tuberculosis or coccidioidomycosis by X ray alone.) With heavy exposure to airborne infection, clinical pneumonia develops. Illness may be prolonged but can subside with spontaneous healing or disease may continue to disseminate in a small minority of cases. When dissemination occurs, the mononuclear phagocyte system is particularly involved, with localizations in lymph nodes, spleen, and liver. Granulomatous lesions may be found in many organs. Progressive histoplasmosis is usually fatal, and its course is more rapid in children than in adults.

Incubation Period

When infection can be traced, about 5 to 18 days.

LABORATORY DIAGNOSIS

I. Skin Tests

A positive intradermal test with histoplasmin (a culture filtrate of the mycelial phase of the fungus) denotes either past or recent exposure to *H. capsulatum* or an antigenically related organism. The skin test may be negative in the late, disseminated stage of disease, but this test is not in common use at present.

II. Laboratory

Specimens of sputum, urine, blood; pus aspirated from lesions; biopsies from bone marrow, liver, lymph nodes, skin lesions.

In tissue sections, *Histoplasma* appears as an intracellular, single-celled yeast, 2 to 4 μm in diameter, often packed within giant cells or macrophages (fig. 14.1).

In culture the organism is dimorphic: at 37°C on enriched media the organism is yeastlike, as it is in tissues. At atmospheric temperature on ordinary media the mold form appears, with mycelium, **microconidia,** and characteristic spiny ("tuberculate") **macroconidia** (fig. 14.2).

Blood for serologic tests is useful. A rising titer is strong evidence of active disease. Histoplasma skin test antigen provokes a rise in antibody titer; therefore, blood for serology should be drawn before skin tests are done.

Epidemiology

The organism is found in soil enriched with bird droppings, as in chicken coops, pigeon cotes, barns, and also in caves inhabited by bats.

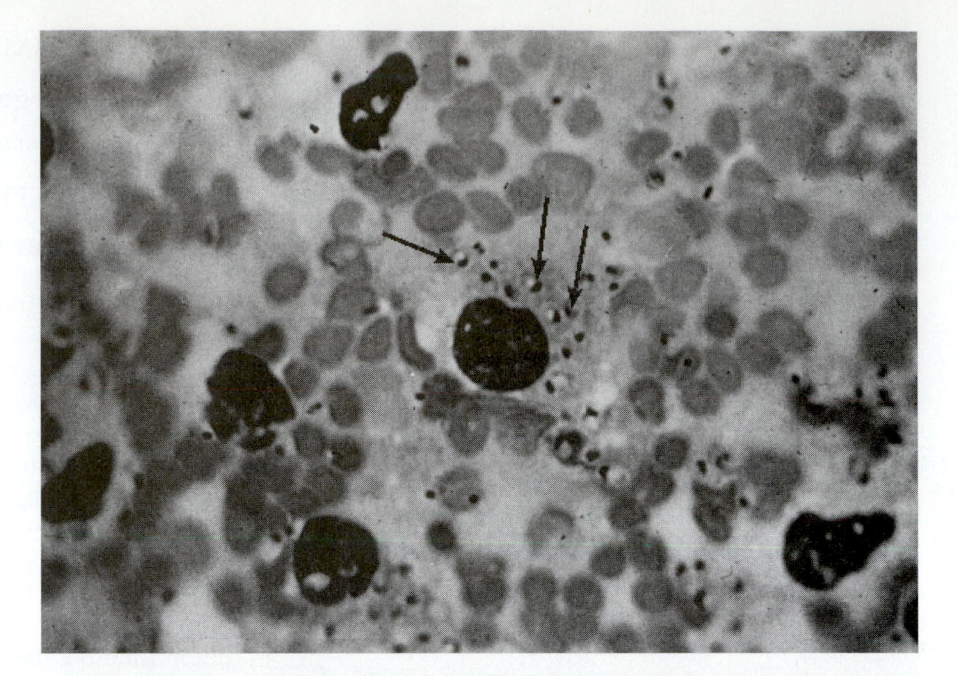

FIGURE 14.1 Photomicrograph of *Histoplasma capsulatum* in tissue section. Small yeastlike, single-celled organisms (arrows) are seen within macrophages (×1,500).

Centers for Disease Control, Atlanta, GA.

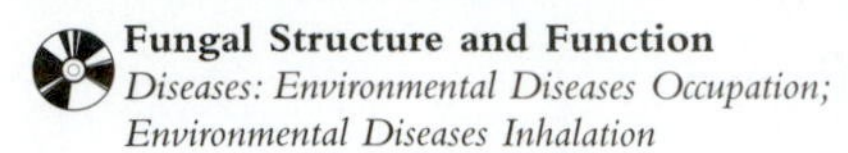

Fungal Structure and Function
Diseases: Environmental Diseases Occupation; Environmental Diseases Inhalation

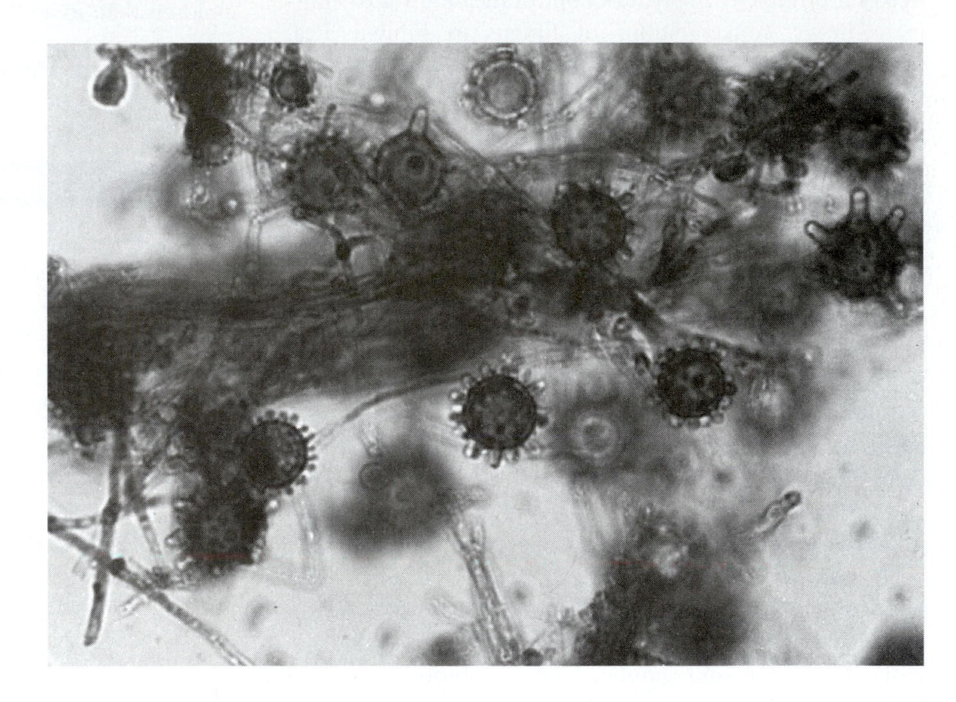

FIGURE 14.2 Photomicrograph of the mycelial phase of *H. capsulatum* (stained with lactophenol cotton blue), showing large, spiny (tuberculate) macroconidia (×800).

Reproduced from E. L. Hazen, M. A. Gordon, and F. C. Reed. *Laboratory Identification of Pathogenic Fungi Simplified*. 3rd ed., 1970. Courtesy Charles C. Thomas, Publisher, Springfield, IL.

Endemic areas are found throughout the world, in temperate and tropical zones. In the United States, localized distribution in enriched soils occurs in many central and eastern states, including Tennessee, Kentucky, Illinois, Michigan, Georgia, Louisiana, Ohio, Indiana, Missouri, and Arkansas.

People of all ages are susceptible to histoplasmosis and become infected when they inhale *Histoplasma* microconidia. The highest incidence of disseminated, fatal disease occurs in infants and in the aged, but in endemic areas up to 90% of the population may display evidence of exposure (positive skin tests) without signs of active infection. There are no significant racial or sex differences, except that males predominate among adult infected patients, probably because of greater exposure. Disseminated histoplasmosis is associated more than casually with leukemia, lymphoma, AIDS, and Hodgkin's disease.

Histoplasmosis is often, but not strictly, a rural disease. People from urban areas who visit the country may be exposed to windblown spores, contaminated garden soil fertilized with chicken manure, or caves where bat guano (excrement) is abundant (spelunkers run the risk of such exposure). Urban sources of exposure also exist, as in areas where starlings roost or in soil around old houses and other buildings inhabited by bats.

Amphotericin B is the drug of choice. Ketoconazole may be used as an alternative drug.

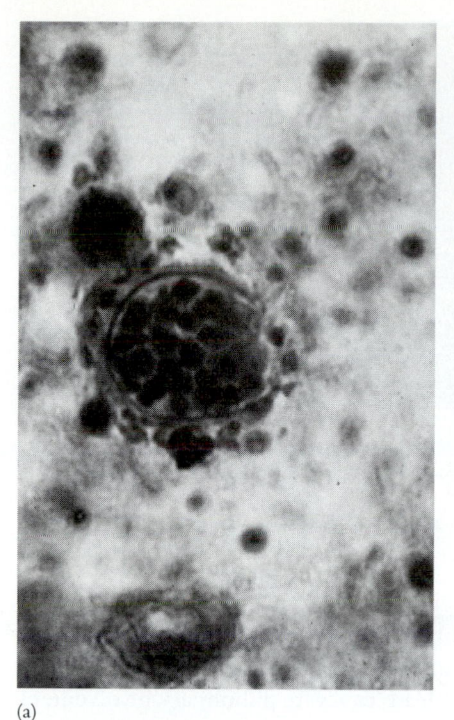
(a)

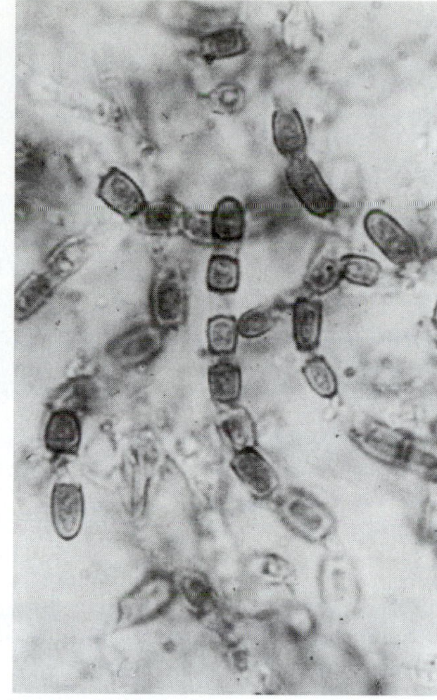
(b)

FIGURE 14.3 *Coccidioides immitis.* (a) Mature spherule seen in infected tissues (×800). (b) Fragmenting arthrospores develop in cultures at room temperature (×1,000).

Courtesy Dr. John W. Rippon.

Fungal Structure and Function
Diseases: Environmental Diseases Inhalation

COCCIDIOIDOMYCOSIS

Causative Organism

Coccidioides immitis, a dimorphic fungus.

The Clinical Disease

Coccidioidomycosis typically begins as a respiratory infection. The primary infection may be asymptomatic or resemble acute influenzal illness, with fever, chills, cough, and chest pain. These symptoms may subside without residual traces, or healing may be accompanied by fibrosis and calcification of granulomatous lesions in the chest. Rarely, the disease progresses to a disseminated form resembling tuberculosis, with lesions in lungs, bones, joints, subcutaneous tissues, skin, internal organs, brain, and meninges. With widespread dissemination, symptoms are acute, and prostration and death may occur within a few weeks.

Incubation Period

For primary infection, between 10 days and 3 weeks. Development of granulomas is often gradual and inapparent.

LABORATORY DIAGNOSIS

Specimens of sputum, pus from aspirated lesions, biopsies, spinal fluid if indicated by symptoms.

Microscopic examination reveals thick-walled, spheric structures (spherules) containing many endospores (fig. 14.3a). In tissue sections or biopsies of granulomas, the spherules may be seen within giant cells or in the acellular areas of central necrosis.

In culture it is difficult to reproduce the in vivo conditions that encourage spherule production (the endospores resemble single yeast cells but do not reproduce by budding). At atmospheric temperature on ordinary media the mold growth consists of an abundant aerial mycelium. The branching hyphae develop into chains of fragmenting **arthrospores,** which are easily disseminated and highly infectious (fig. 14.3b). In the laboratory, cultures are always examined under a biological safety cabinet.

Blood for serologic tests is useful. Precipitins (precipitating antibodies) appear in the course of symptomatic disease. The latter may be delayed for some weeks, then persist for a few months. A rising titer of antibody indicates the probability of disseminated disease and a poor prognosis. Other useful serologic tests include latex agglutination, immunodiffusion, and fluorescent antibody.

Epidemiology

Soil and spore-laden dust are sources of infection for humans and a number of domestic animals. In the United States the disease is endemic in the Southwest, particularly the San Joaquin Valley of southern California. Other endemic areas are found in parts of Central and South America.

C. immitis grows in soil, producing large numbers of arthrospores like those seen in cultures. These are easily disseminated on air currents. It has been demonstrated that in the hot summer the fungus disappears from the desert floor but persists in rodent burrows where it is protected from solar heat and radiation. After the winter and spring rains, it spreads again over the desert floor. During natural events such as dust storms and earthquakes, when clouds of arthrospores are aerosolized, the incidence of disease increases dramatically. In the two months following an

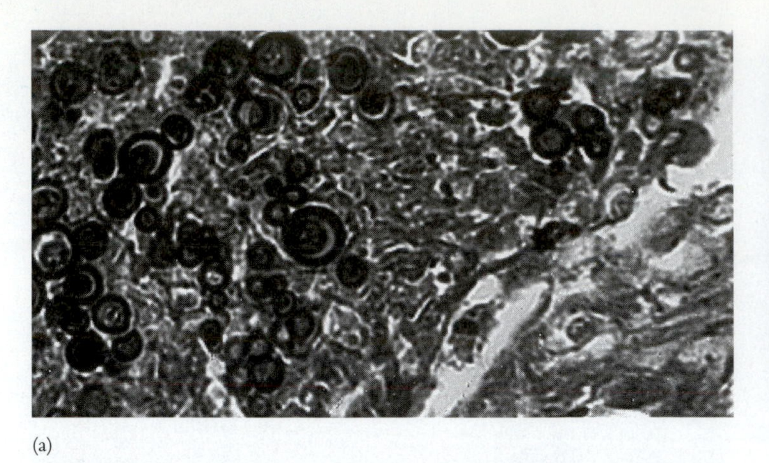
(a)

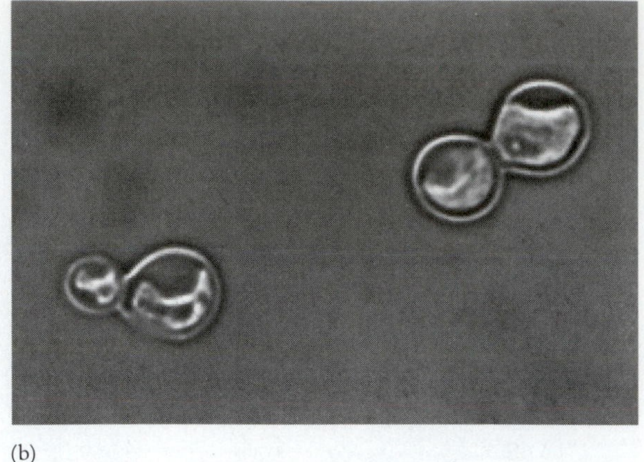
(b)

FIGURE 14.4 *Blastomyces dermatitidis.* (a) Large, spheric, thick-walled budding cells in human lung tissue (×800). (b) Budding cells from growth on blood agar at 37° C (×1600).

earthquake in southern California in the 1990s, 170 persons were diagnosed with coccidioidomycosis as compared with only 52 cases during the entire previous year.

The arthrospores of the culture and soil phase are extremely infectious. In an unusual example of person-to-person transmission from a hospitalized patient with osteolytic coccidioidomycosis, six members of a medical staff were infected by inhaling arthrospores that had been produced on the patient's plaster cast. The tissue spherule form of the fungus had converted to the arthrospore-forming mold phase in the plaster.

People of all ages are susceptible, but in an endemic area the disease is usually a mild one of early childhood or frequent in adults who have migrated to the area. Disseminated fatal disease occurs most often among infected pregnant women, dark-skinned males, and immunocompromised persons such as those with AIDS or who are on immunosuppressive therapy. Adult white females more frequently display allergic lesions (erythema nodosum, erythema multiforme), presumably because of an inherent resistance and an ability to develop a rapid, cell-mediated, immunologic defense. Such allergic lesions are uncommon in dark-skinned men and immunocompromised persons.

Amphotericin B, ketoconazole, and miconazole are effective in treating coccidioidomycosis but relapses have often been seen with the last two drugs. Therapy is often prolonged for disseminated infection. Mild pulmonary disease usually resolves without therapy.

BLASTOMYCOSIS (NORTH AMERICAN)

Causative Organism

Blastomyces dermatitidis, a dimorphic fungus.

The Clinical Disease

Systemic **blastomycosis** (or **North American blastomycosis**) usually begins as a primary pulmonary infection acquired through inhalation of **blastospores.** Initial symptoms are those of an upper respiratory viral infection, but progressive infection leads to severe pulmonary involvement and dissemination to subcutaneous tissues, bones, genital and visceral organs, and the central nervous system. Cutaneous blastomycosis is often the first sign of dissemination in systemic disease. Rarely, skin lesions may be a primary event resulting from local entry of the organisms into the skin (colorplate 7a).

Incubation Period

From 4 to 6 weeks, an estimate based on several outbreaks of disease originating from a common source.

LABORATORY DIAGNOSIS

Specimens of sputum, pus, and exudates from lesions, biopsy of skin granulomas.

Microscopic examination reveals the tissue form to be a large, thick-walled yeast whose budding forms characteristically have broad bases at the point of attachment of two cells (fig. 14.4a).

In culture the organism grows in the yeast form on blood agar incubated at 37° C (fig. 14.4b). At room temperature on ordinary fungal growth media, the mold form is white to tan in color and grows into a colony with single-celled smooth conidia.

Serologic tests for blastomycosis are not as useful as for other fungal diseases. Immunodiffusion and enzyme immunoassay tests are of some value for aiding diagnosis, but these tests are performed in only a few laboratories. Skin tests are not available.

Epidemiology

The incidence of blastomycosis is unknown because many cases are asymptomatic or mild. The reservoir is thought to be in moist soil enriched with a high organic content from

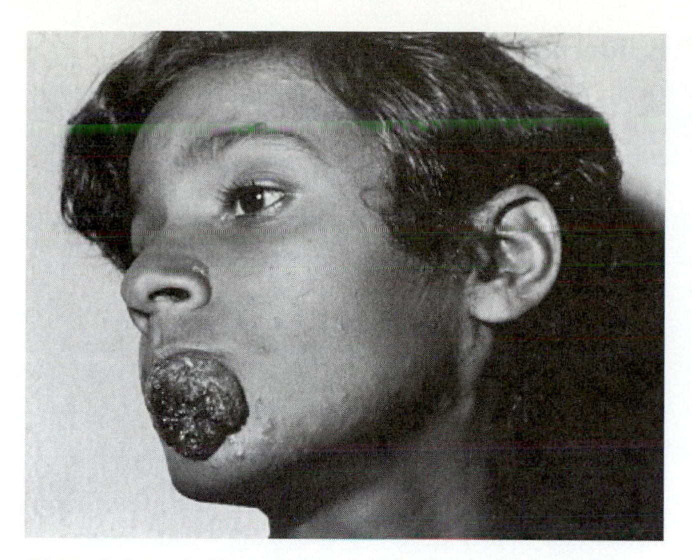

FIGURE 14.5 Mucocutaneous facial lesion caused by the fungus *Paracoccidioides brasiliensis.* Note also the enlargement of cervical lymph nodes.

Courtesy Dr. Jose Lisboa Miranda, Universidade Gama Filho, Rio De Janeiro, Brazil.

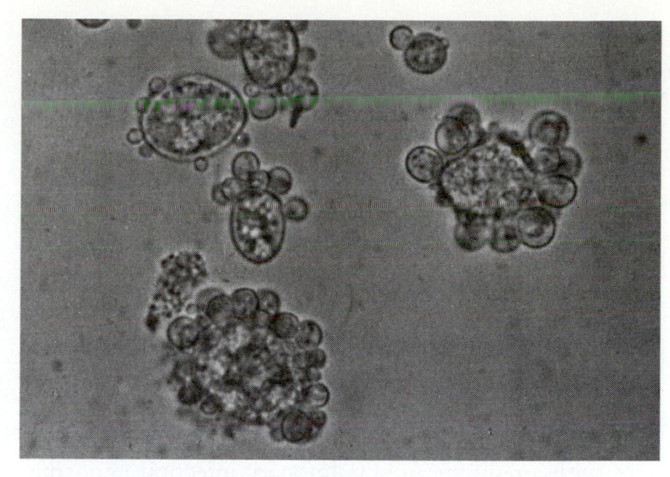

FIGURE 14.6 *Paracoccidioides brasiliensis.* Photomicrograph of yeast cells showing multiple budding (wet mount made from a culture grown on blood agar at 37° C) (×2,500).

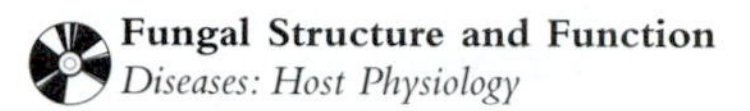

Fungal Structure and Function
Diseases: Host Physiology

decaying vegetation and decomposed wood. The mode of transmission is by inhalation of microconidia from the environment. The disease occurs primarily in Canada and the United States, most commonly in eastern and midwest portions along waterways such as the Mississippi and Ohio rivers and the Great Lakes. A number of reports indicate that it may also be found in Africa, India, and the Middle East.

This sporadic disease occurs among people of all ages with a somewhat higher incidence in middle age. It is more frequent in males than females, but no occupational associations have been demonstrated. Blastomycosis also occurs in dogs, cats, and horses.

Amphotericin B is most effective in severe, life-threatening cases. Ketoconazole is used in milder cases of the disease and is given to patients who have relapsed after amphotericin B therapy or who cannot tolerate amphotericin B toxicity.

PARACOCCIDIOIDOMYCOSIS (SOUTH AMERICAN BLASTOMYCOSIS)

Causative Organism

Paracoccidioides brasiliensis, a dimorphic fungus.

The Clinical Disease

Paracoccidioidomycosis (or **South American blastomycosis**) is endemic in Latin America, particularly Brazil. It was first called *pseudococcidioidal granuloma* because the tissue form of the organism was thought to resemble *Coccidioides.* It produces a chronic granulomatous disease of the lung, the mucocutaneous membranes of the face (fig. 14.5), oral membranes, lymph nodes, and visceral organs. Cervical lymphadenopathy may resemble that of Hodgkin's disease. Disseminated infection is highly fatal; localized disease may persist for years if untreated.

Incubation Period

A few weeks.

LABORATORY DIAGNOSIS

Skin Tests Limited diagnostic value but useful epidemiologically in detecting early cases and in delimiting endemic areas.

Laboratory Specimens of pus from lesions, biopsies. Sputum if pulmonary disease occurs.

Microscopically, the tissue form of the organism is a large, thick-walled budding yeast, 5 to 25 μm in diameter with a thick, double refractile wall. Many cells show multiple buds so that the appearance resembles a pilot wheel (fig. 14.6).

In culture, the tissue form grows at 37° C on blood agar. At room temperature on ordinary media the organism is a mold.

Serologic diagnosis is usually made by immunodiffusion tests. Counterimmunoelectrophoresis and enzyme immunoassays are available in specialized laboratories.

Epidemiology

The organism is thought to be associated with vegetation, wood, or soil, especially that along riverbanks in subtropical regions of Latin America. The infection is primarily acquired through inhalation, but it may possibly be introduced through the skin or oral membranes, presumably by direct contact with contaminated vegetative material. The incidence of this mycosis is highest among young male adults whose manual labor brings them in contact with the fungus, but the disease is sporadic rather than epidemic.

This mycosis is uniformly fatal if not treated. Sulfonamides can be effective in the mildest forms of the disease, but amphotericin B is recommended for patients who do not respond to sulfonamides, for those who relapse after treatment with sulfonamides, and for those with advanced disease. Ketoconazole appears to be an excellent drug for treating this infection.

CRYPTOCOCCOSIS

Causative Organism

Cryptococcus neoformans, a yeast.

The Clinical Disease

Infection in humans usually occurs through the respiratory tract, with a resulting mild pulmonary infection frequently undiagnosed. The most common clinical extension of **cryptococcosis** in humans is the slow development of a chronic meningitis. The course of cryptococcal meningitis may extend over several years, resembling degenerative central nervous system disease, syphilitic or tuberculous meningitis, or brain tumor. The organism may occasionally gain entry to the body through the skin or the mucosa of the upper respiratory or intestinal tracts, with isolated lesions in subcutaneous tissues, lymph nodes, tongue, muscles of the back, and other areas. (Oral entry is especially frequent in animals.) Lesions of this type may remain localized or extend to the brain and meninges.

Incubation Period

Not defined; meningitis usually follows months or years after inapparent pulmonary infection.

LABORATORY DIAGNOSIS

Specimens for wet mounts and culture include spinal fluid, sputum, and exudates from cutaneous lesions if they occur.

Microscopic examination of wet mounts of centrifuged spinal fluid or other specimens reveals a single or budding yeast cell surrounded by a large capsule. This polysaccharide capsule can also be seen against a dark background obtained by suspending the material to be examined in India ink. Special stains such as mucicarmine are used to highlight the capsule in tissue preparations (fig. 14.7).

In culture the organism grows as a yeast on all media at temperatures ranging from 20 to 37° C. Capsule production can be demonstrated on special media or by mouse inoculation.

Serologic diagnosis can be made with good precision using the slide latex agglutination method (the laboratory must take steps to ensure elimination from the patient's serum of a factor present in rheumatoid arthritis that also agglutinates the latex antigen). In central nervous system infection, cryptococcal antigen can be demonstrated by latex agglutination in spinal fluid and serum.

Skin tests are not available.

Epidemiology

The organism grows saprophytically in the external environment. It can be isolated consistently from pigeon nests and droppings and from soil in many parts of the world. Transmission is thought to occur through inhalation of contaminated dust, but the development of disease depends on the intensity of exposure and the numbers of organisms reaching the lower respiratory tract.

Sporadic cases occur on a worldwide basis, in domestic and wild animals as well as in humans. The disease is more frequent in human males than in females. Although there are no significant differences related to age, race, or occupation, the disease is associated more than randomly with leukemia, lymphosarcoma, Hodgkin's disease, diabetes, and prolonged steroid therapy. It has been estimated that approximately 8% of AIDS patients in the United States and Africa have developed cryptococcosis. For these reasons, some consider it to be an opportunistic infection, even though not all affected persons are immunocompromised.

The prognosis is good for pulmonary and cutaneous forms of the disease, but disseminated or meningeal cryptococcus may be fatal even when treated. The combined use of amphotericin and flucytosine is the treatment of choice for cryptococcal meningitis as well as for other severe forms of the infection. Mild pulmonary disease may resolve spontaneously without therapy.

SUBCUTANEOUS MYCOSES

Like the systemic mycotic diseases, those with a parenteral entry route are not transmissible from person to person. Subcutaneous mycoses tend to be chronic and slowly progressive, but they usually do not disseminate to systemic tissues. Three important diseases of this type are described briefly next.

SPOROTRICHOSIS

Causative Organism

Sporothrix schenckii, a dimorphic fungus.

The Clinical Disease

Sporotrichosis is a chronic infection usually initiated by entry of the organism through the skin of an extremity (frequently a hand or arm) and characterized by development of nodular lesions in subcutaneous tissues along the path of the regional lymphatic drainage. The nodules frequently soften, ulcerate at the surface, and discharge this pus (colorplate 7b). The clinical picture is highly characteristic. Dissemination of the organism to other parts of the body occurs rarely.

Incubation Period

The initial lesion may be seen within 3 weeks, or it may require up to 3 months to appear.

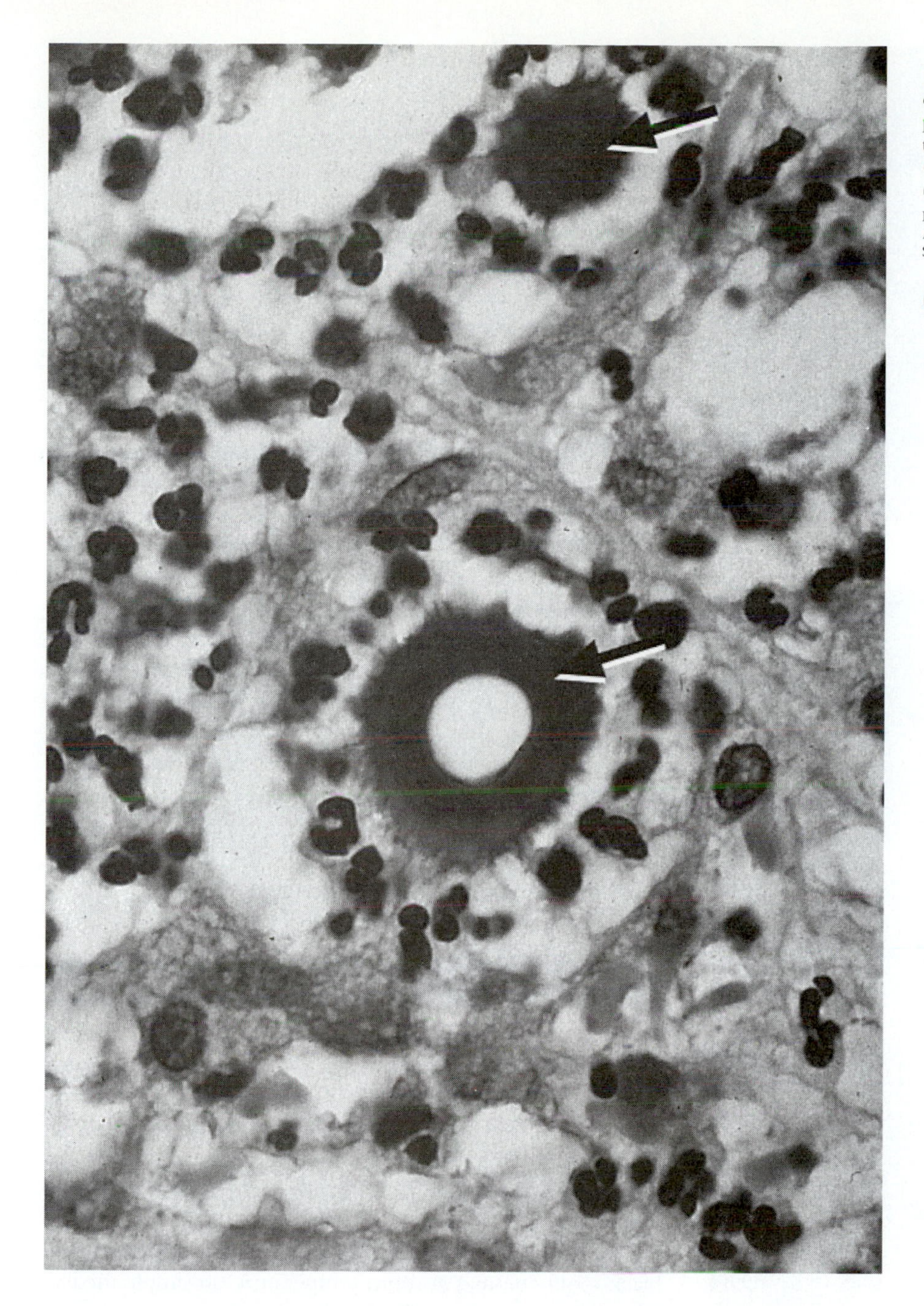

FIGURE 14.7 *Cryptococcus neoformans* yeast cells in a preparation made from the skin of a patient with disseminated cryptococcal infection. The preparation has been stained so that *Cryptococcus* capsules (arrows) are readily visible (×1,000).

Fungal Structure and Function
Diseases: Host Physiology

LABORATORY DIAGNOSIS

Specimens Pus, aspirates, or biopsy material from subcutaneous lesions for smear and culture.

Microscopic examination of pus does not always reveal the tissue phase of the organism, which is a small, gram-positive, cigar-shaped, budding cell. Specific fluorescent antibody reagent can detect the organisms in smears more readily than other stains.

In culture the yeastlike tissue form grows on enriched media at 37° C (fig. 14.8a). At room temperature on simple media the organism is a mold, often darkly pigmented. Delicate hyphae on the aerial mycelium carry small, ovoid spores in petallike clusters (fig. 14.8b).

Epidemiology

The organism grows in soil, on wood, and on vegetation. Infection is a sporadic occupational hazard of gardeners and farmers. An epidemic was reported from South Africa involving about 3,000 gold-mine workers infected in common by contaminated mine timbers. The disease has been reported from all parts of the world and has been seen in animals as well as humans.

The cutaneous lymphatic form of the disease responds well to treatment with an oral solution of potassium iodide. This therapy is less satisfactory in disseminated infection, therefore amphotericin B is used instead.

In the care of patients with sporotrichosis, special attention should be given to the disposal of contaminated

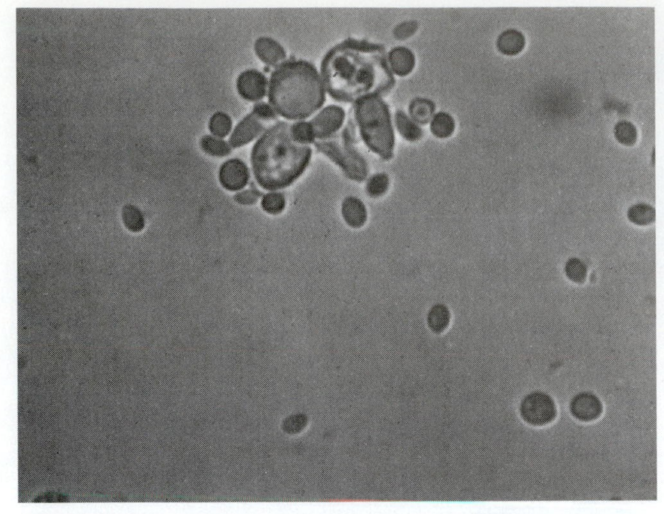

(a)

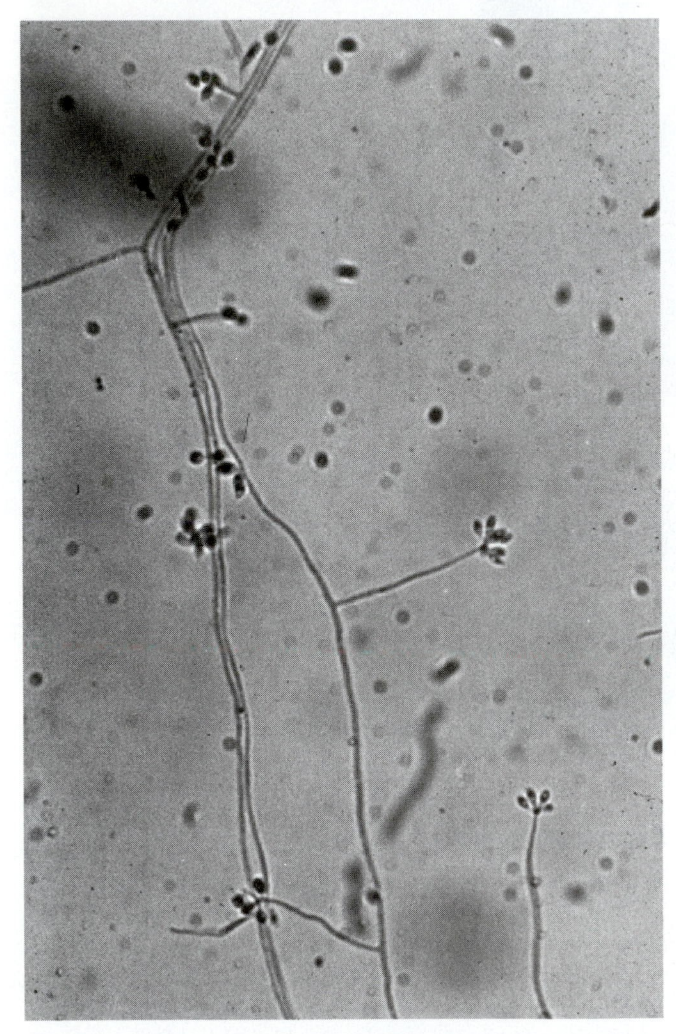

(b)

FIGURE 14.8 *Sporothrix schenckii.* (a) Yeastlike cells taken from a culture grown on blood agar at 37° C (×800). (b) A microscopic view of the mycelial growth obtained at room temperature (note the delicate petallike spores) (×400).

Reproduced from J. W. Rippon, *Medical Mycology: The Pathogenic Fungi and the Pathogenic Actinomycetes,* 3rd, 1988. W. B. Saunders, Philadelphia, PA.

Fungal Structure and Function
Diseases: Environmental Diseases Penetration

dressings, as they may be infectious for persons with open skin lesions.

CHROMOMYCOSIS (VERRUCOUS DERMATITIS)

Causative Organism

Several species of darkly pigmented mold fungi, notably of the *Cladosporium, Fonsecaea, Phialophora, Exophiala,* and *Rhinocladiella* genera.

The Clinical Disease

Chromomycosis, or **verrucous dermatitis,** is a chronic, slowly progressive, granulomatous infection of the skin and lymphatics. Warty cutaneous nodules develop very slowly and become prominent vegetations that do not always ulcerate. The lesions are located as a rule on an exposed extremity and are almost always unilateral. The feet and legs are generally involved, possibly because they are frequently exposed to the soil source of the organism. Surgical excision of early lesions and treatment with antifungal agents such as amphotericin B or itraconazole help to limit the slow progression of the disease. No specific drug(s) administered alone or in combination has been completely successful in its cure. The infection does not become generalized.

Incubation Period

Unknown but probably requires weeks or months.

LABORATORY DIAGNOSIS

Specimens The crusts of lesions or biopsy material.

Microscopic examination of the tissue reveals brown-pigmented, branching hyphal strands and dark-brown, thick-walled septate bodies about the size of a leukocyte. These structures reproduce by splitting rather than budding. The several fungi that cause this disease produce the same form in tissues.

In culture these fungus species produce dark-brown or black mold colonies at room temperature on simple media. They do not grow at 37° C. The specific, responsible organism is identified by its characteristic spore structures.

Epidemiology

The organisms probably exist in wood, soil, or vegetation and are introduced through traumatized tissue. Sporadic cases occur in many parts of the world, but this is primarily a disease of rural tropical regions (the West Indies, Central and South America, Asia). It is seen frequently in male laborers.

MYCETOMA (MADUROMYCOSIS)

Causative Organism

Numerous soil fungi including *Madurella, Pseudallescheria, Curvularia, Exophiala,* and *Acremonium* species (also the bacteria *Nocardia, Streptomyces,* and *Actinomadura*).

The Clinical Disease

Mycetoma, also called **maduromycosis** or Madura foot, is a chronic infection of subcutaneous tissues characterized by swelling and nodule and abscess formation, with suppurative extension through sinus tracts or fistulae. The foot is commonly involved, especially in tropical areas where people go without shoes (colorplate 7d). Deep extension of the infection involves muscles and bones. The affected area swells to a globular shape, losing its structural form.

Therapy with streptomycin, dapsone, or trimethoprim-sulfamethoxazole is useful if the infecting organism is *Streptomyces, Actinomadura,* or ***Nocardia.*** The true fungi that cause this disease are not affected by these drugs but may respond to amphotericin therapy. Intractable disease may require amputation to save the patient from death by secondary bacterial infection.

Incubation Period

The first lesions may require months to develop; progression takes place over a period of years.

LABORATORY DIAGNOSIS

Specimens Draining pus from sinus tracts or biopsy material.

Microscopic examination and culture may reveal bacillary forms or fungi. Culture techniques should include methods appropriate for the growth of either.

Epidemiology

Mycetomas are seen most frequently in tropical and semitropical areas where people acquire the infection through the injured skin of bare feet. The disease is seen occasionally in scattered areas of the United States and in Mexico, South America, Africa, and Asia.

OPPORTUNISTIC MYCOSES

Many different fungi can produce an **opportunistic mycosis** in a host with inadequate cellular or humoral resistance or other underlying disease. Unfortunately, the number of such hosts has increased considerably during the last two decades as a result of the AIDS epidemic and because many more therapies prolong life in persons with diseases that were fatal not many years ago. Health care personnel in most hospital and clinic settings today routinely see patients with opportunistic infections.

The organisms responsible for infection may be acquired from environmental sources to which many persons are exposed. Under ordinary circumstances, these fungi lack the qualities needed to produce disease; that is, they are not innately pathogenic and cause no harm to healthy hosts. A few yeasts are endogenous in the human body, living as commensals among the normal skin flora and on mucous membranes. They too may invade tissues and cause disease under circumstances in which host defenses are lowered.

Among the **yeasts,** species of the genus *Candida* are the most common opportunists. Some of the outstanding features of candidiasis are described next. Two environmental fungal diseases that are often seen in compromised hosts are also discussed briefly. These are aspergillosis and zygomycosis, the latter referring to infection caused by members of the phylum Zygomycota, notably ***Mucor*** and ***Rhizopus.*** Many other fungi, including the yeast *Malasezzia furfur* (which ordinarily causes a mild skin condition known as *pityriasis versicolor*) and environmental species in the mold genera *Fusarium* and *Pseudallescheria,* can invade and cause fatal disseminated infections in immunocompromised patients. Microbiology laboratory technologists and other health care personnel must be aware of the clinical significance of such fungi when they are isolated from compromised patients and not simply dismiss them as laboratory contaminants.

CANDIDIASIS

Causative Organism

Candida albicans (also *C. tropicalis, C. stellatoidea, C. parapsilosis, C. guillermondii,* and others), a yeast.

The Clinical Disease

Candidiasis is usually a superficial mycosis of skin or mucous membranes in which the organism is ordinarily a member of the normal flora. A number of factors causing injury to or resulting in unusual susceptibility of tissues may permit local entry of the organism or development of systemic progressive disease (see Epidemiology).

Several types of clinical candidiasis may be described briefly as follows:

Oral candidiasis (**thrush**) is particularly frequent in infants. Loosely adherent, white, pseudomembranous patches of fungal growth occur on the tongue and buccal mucosa (colorplate 7e).

Vaginal infection is common in pregnancy, diabetic women, and those on antimicrobial therapy or oral contraceptives. It involves vulvar surfaces as well as the vagina, producing irritation, intense pruritus, and discharge.

Cutaneous candidiasis often involves smooth skin of the intertriginous folds, particularly in axillary, inguinal, and inframammary regions. The interdigital folds of the feet and hands may be involved, and in infants, the diaper area. The cutaneous form may also be quite generalized, and these areas become reddened and weepy with exudate from small vesicles or pustules.

Nail infections, onychia and paronychia, are common. Swellings develop around or in the nailbed, and the nails become thick, hard, and deeply grooved (see colorplate 7f).

Esophageal candidiasis is common in patients receiving therapy for blood and lymphatic malignancies and in AIDS patients. Swallowing is painful, chest pains may occur, and white patches resembling thrush are seen on endoscopic examination of the esophagus.

In some immunocompromised patients the organisms disseminate through the bloodstream from a site of local infection or colonization, usually in the gastrointestinal tract. The resulting systemic candidiasis can involve almost any organ including the brain, kidney, and endocardium.

Incubation Period

Infant thrush develops within 2 to 5 days of exposure. In other forms of infection the incubation period is variable and difficult to ascertain.

LABORATORY DIAGNOSIS

Specimens Scrapings or pus from surface lesions, sputum, blood, and spinal fluid, as indicated by clinical symptoms.

Microscopic examination of clinical specimens in wet mounts or stained smears reveals an oval, budding yeast. Very often the pseudomycelium produced by these organisms in culture is also seen in the specimen, indicating an established colonization or infection (see colorplate 2e).

In culture, the organism grows well on common culture media both at room temperature and at 37° C. Under both conditions it is a yeast, sometimes with a pseudomycelium produced by elongation and pinching off of germinating cells. When grown on media deficient in nutrient, *Candida albicans* produces clusters of thick-walled **chlamydospores** (resting spores). This feature, as well as a "**germ tube**" sprout produced after two hours of incubation in serum, is often used to distinguish this species from other members of the *Candida* genus that are not frequently associated with mycosis.

Serologic methods are not useful in the diagnosis of candidiasis.

Epidemiology

Factors that commonly predispose to *Candida* infections include the following:

1. Antimicrobial therapy, which disturbs the normal bacterial flora that holds *Candida* in check under ordinary circumstances.
2. Debilitating organic disease, particularly diabetes.
3. Pregnancy, which appears to affect the carbohydrate content of the vagina, leading to an increased population of *Candida*.
4. Malignant tumors and leukemia.
5. Nutritional deficiencies.
6. Continuing maceration of tissues caused by frequent or prolonged immersion of the hands in water; or excessive perspiration in the frictional, intertriginous folds (axillary, inguinal, inframammary, interdigital).
7. Drug abuse. Dirty needles may introduce *Candida* and other organisms that settle into deep tissues.
8. Cardiac and gastrointestinal surgery, procedures requiring indwelling intravenous catheterization, hyperalimentation, and peritoneal dialysis may all open a path for the exogenous or endogenous spread of *Candida*.
9. AIDS, which increases susceptibility to oral, esophageal, and dermal *Candida* infections.
10. Administration of immunosuppressive drugs for a variety of conditions including organ transplantation.

Skin and oral infection is also seen in newborn infants, particularly premature babies who acquire the yeast from the mother's colonized vagina during passage through the birth canal. Thrush is frequently communicable in the nursery situation (colorplate 7e). It may be transmitted to babies by nursing mothers with breast infection and may also appear as a nursery outbreak arising from a common environmental source such as inadequately sterilized bottle nipples.

Control

Standard precautions are required. When nursery epidemics occur, every effort should be made to find the source of infection. (Bottle nipples should be considered as the first possibility.) Concurrent disinfection and terminal cleaning techniques must be carefully maintained.

Treatment

Some forms of candidiasis, especially superficial infections, may clear without therapy. Vulvovaginitis in pregnant women, for example, usually improves following delivery. In superficial candidiasis, control of underlying factors or disease is important and may obviate the need for further treatment, for example, good diabetic therapy or judicious changes in or discontinuance of antimicrobial therapy.

Superficial lesions are usually treated topically. The antifungal agent nystatin may be used as a cream or an ointment for skin infections or as a solution for oral and vaginal douches. Topical clotrimazole and miconazole are also of therapeutic value.

Systemic candidiasis is a serious disease with a poor prognosis if untreated. Amphotericin B must be used in these cases, alone or in combination with flucytosine.

ASPERGILLOSIS

Causative Organism

Aspergillus fumigatus (also *A. flavus;* rarely *A. niger* and *A. terreus*), a mold.

The Clinical Disease

Aspergillosis is a granulomatous disease of the lung or other tissues. Pulmonary disease may be necrotizing, with cavitation and hematogenous spread to other organs. Other forms of aspergillosis include nasal-orbital and cutaneous involvement. Pulmonary disease with this fungus is often secondary to tuberculosis, silicosis, tumor, immunodeficiency, or steroid therapy.

A dilated bronchus or old tuberculosis cavity may become colonized, with the formation of a fungus ball or tightly matted mass of mycelial fragments. This form of infection, called *aspergilloma,* rarely produces symptoms because it is merely a saprophytic growth in the cavity, with no tissue invasion. However, hemoptysis (spitting of blood) may occur and be serious enough to require surgical intervention.

Healthy persons who are frequently exposed to the spores in an environmental source (soil, compost, wood chips) may develop allergic aspergillosis. Inhalation of spores and colonization of the respiratory tract without invasive disease may so sensitize such individuals that each subsequent exposure may be followed by asthmatic symptoms.

Incubation Period

A few days to weeks following exposure.

LABORATORY DIAGNOSIS

Specimens Sputum, bronchial aspirates or lavage, draining pus, or other material, depending on suspected localization.

Microscopic examination and culture reveal the typical structures of this fungus, which grows as a mold at room and higher temperatures. Species identification is made by the morphology of the spore-bearing structure and the color of the spores.

Serologic methods and skin tests may be of little or no diagnostic value in invasive aspergillosis because the immune mechanisms of such patients have been undermined by predisposing disease. In aspergilloma and allergic aspergillosis, skin tests and serologic techniques (immunodiffusion) may be diagnostically useful.

Epidemiology

Healthy persons have a high degree of resistance to this fungus found ubiquitously in the environment.

Invasive pulmonary aspergillosis has an extremely poor prognosis. Amphotericin B alone or in combination with flucytosine is required, but the underlying disease itself often has a poor prognosis as well. Any immunosuppressive therapy in use should be reduced or discontinued in the face of superimposed mycosis.

Bronchial colonization without invasion may be treated supportively, particularly with measures designed to improve pulmonary function and drainage. Surgical resection is recommended whenever feasible.

ZYGOMYCOSIS (MUCORMYCOSIS)

Mycoses of this type include infections caused by members of the Zygomycota phylum of fungi. These infections are variously called **zygomycosis** or mucormycosis. The former term refers to the characteristic formation of a zygospore in the sexual reproduction of fungi of this group (see chap. 2).

Causative Organisms

Species of *Mucor, Rhizopus, Absidia,* and other molds.

The Clinical Disease

Zygomycosis is characterized by massive invasion of blood vessels (with formation of emboli), pulmonary lesions, infection of orbital and nasal sinus tissues, gastrointestinal ulcerations, and terminal central nervous system involvement. Disseminated infection is rapidly fatal, often occurring even before the laboratory diagnosis can be made.

LABORATORY DIAGNOSIS

The fungi can be recovered readily from appropriate clinical specimens, growing rapidly as coarse, wooly molds at 37° C. However, the mere isolation of *Mucor, Rhizopus,* or *Absidia* species from sputum does not constitute a diagnosis of zygomycosis, as patients with bronchiectasis may cough up fungal spores, especially after exposure to a dusty environment. If pulmonary invasion has occurred, the sputum will be repeatedly positive for the fungus species involved, but cultures will be negative after a few days if the colonization was transitory or noninvasive. Curetted tissue for direct microscopic examination is most important for confirming the diagnosis, for hyphal segments of the fungus actively growing among the tissue cells can then be visualized.

In making the diagnosis it is important for the laboratory to report these organisms, even though they may often be mere contaminants. The physician may then weigh the significance of the laboratory report with the clinical findings in recognizing an invasive zygomycosis.

Epidemiology

These fungi are common environmental saprophytes to which healthy individuals have a high degree of resistance. A few members of the group are associated with primary mycoses seen largely in tropical areas, but most cases are opportunistic in patients with predisposing illnesses such as diabetes, malignancy, or immunodeficiency. Therapeutic control of the underlying disease coupled with administration of amphotericin B constitutes the best control.

PNEUMOCYSTOSIS

Causative Organism

Pneumocystis carinii, an organism of uncertain classification, most likely a fungus.

The Clinical Disease

Formerly included in the protozoan parasite phylum Apicomplexa, *P. carinii* is now thought to be a fungus. It is the causative agent of diffuse, interstitial pneumonia in infants, persons with AIDS, and those undergoing immunosuppressive therapy.

SUMMARY TABLE 14.2 MYCOTIC INFECTIONS

Clinical Disease	Causative Organisms	Other Possible Entry Routes	Incubation Period	Communicable Period
Primary (Pulmonary) Systemic Mycoses				
Histoplasmosis	*Histoplasma capsulatum*	Parenteral	5–18 days	Not usually communicable from person to person
Coccidioidomycosis	*Coccidioides immitis*	Parenteral	10 days to 3 weeks	Same as above
Blastomycosis (North American)	*Blastomyces dermatitidis*	Parenteral	4–6 weeks	Same as above
Paracoccidioidomycosis (South American blastomycosis)	*Paracoccidioides brasiliensis*	Intestinal Parenteral?	A few weeks or undefined	Same as above
Cryptococcosis	*Cryptococcus neoformans*	Parenteral	Not defined	Same as above
Subcutaneous Mycoses				
Sporotrichosis	*Sporothrix schenckii*	Parenteral	3 weeks to 3 months	Not usually communicable from person to person
Chromomycosis	Dark-pigmented fungi: *Cladosporium, Exophiala Fonsecaea, Phialophora*	Parenteral	Unknown	Same as above
Mycetoma	*Madurella* *Pseudallescheria* *Curvularia* *Exophiala* *Acremonium* Bacteria: *Nocardia* *Streptomyces* *Actinomadura*	Parenteral	First lesions may take months to develop	Same as above
Opportunistic Mycoses				
Candidiasis	*Candida albicans* and other species	Parenteral Endogenous Skin or mucosa	Infants (thrush), 3–5 days	During time of lesions
Aspergillosis	*Aspergillus fumigatus, A. flavus,* and other species		Few days to weeks	Not communicable
Mucormycosis	*Mucor* *Rhizopus* *Absidia* and other species		1–2 days	Not communicable
Pneumocystosis	*Pneumocystis carinii*		1–2 months	Unknown

Specimens Required	Laboratory Diagnosis	Immunization	Treatment	Precautions
Sputum, pus, bone marrow, spinal fluid Biopsies of liver, spleen, lung, lymph nodes Skin test Blood serum	Wet mounts Cultures Fungal stains for tissues Cultures Positive for histoplasmin Serologic tests	None	Amphotericin B Ketoconazole	Standard precautions; concurrent disinfection of contaminated articles prevents conversion of fungus to environmental form
Sputum, pus from lesions, spinal fluid Skin test Blood serum	Wet mounts Cultures Positive for coccidioidin Serologic tests	None	Amphotericin B Miconazole Ketoconazole	Same as above
Sputum, pus and exudates Biopsies of skin granulomas	Wet mounts Cultures Fungal or FA stains for tissues Cultures	None	Amphotericin B Ketoconazole	Same as above
Pus from lesions, biopsies, sputum Blood serum	Wet mounts Cultures Serologic tests	None	Sulfonamides Amphotericin B Ketoconazole	Same as above
Sputum, spinal fluid, exudates from lesions Blood serum Spinal fluid	India ink mounts Cultures Serologic tests for antigen and/or antibody	None	Flucytosine + Amphotericin B	Same as above
Pus, aspirates, or biopsy from subcutaneous lesions	Wet mounts Fungal or FA stains for tissues Cultures	None	Potassium iodide Amphotericin B	Standard precautions; concurrent disinfection of discharges and contaminated articles prevents conversion of fungus to environmental form
Crusts of lesions Biopsy material	Same as above	None	Surgical excision early Amphotericin B Itraconazole	Same as above
Draining pus or biopsy material	Same as above	None	Amphotericin B for fungi Streptomycin, dapsone, trimethoprim-sulfamethoxazole for bacteria	Same as above
Scrapings or pus from surface lesions, sputum, blood, spinal fluid, as indicated	Wet mounts Smears Cultures	None	Nystatin Imidazoles Amphotericin B Flucytosine	Standard precautions; concurrent disinfection and terminal cleaning techniques must be carefully maintained; find source of infection in nursery
Sputum, bronchial aspirates and lavage, pus, blood Allergy: skin tests Blood serum	Wet mounts Cultures Serologic tests	None	Amphotericin B ± Flucytosine	Standard precautions; concurrent disinfection
Curetted tissue, sputum, pus from areas involved	Wet mounts Cultures	None	Amphotericin B	Same as above
Lung biopsy Aspirated mucus Sputum	Microscopic examination for cysts stained with special stains or fluorescent antibody reagents	None	Trimethoprim-sulfamethoxazole Pentamidine	Standard precautions

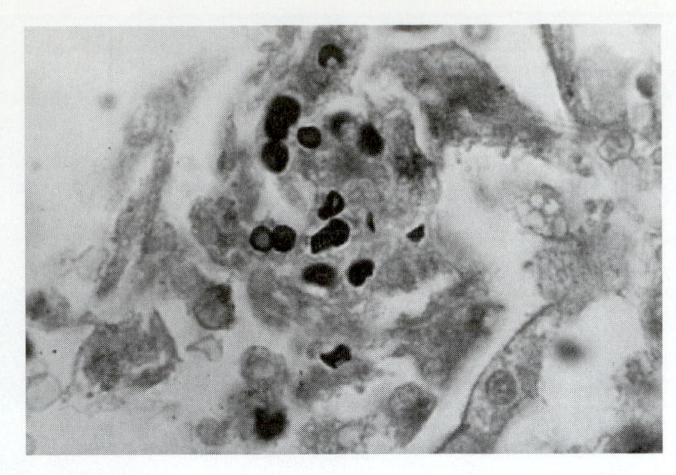

FIGURE 14.9 Pulmonary pneumocystosis: *Pneumocystis carinii* in lungs of a hemophilia patient with AIDS.
Centers for Disease Control, Atlanta, GA.

Pneumocystosis is an acute and often fatal disease. The organism is widespread among humans and is acquired by inhalation of the cyst form. Primary infection is usually subclinical, but the cysts remain dormant in lung tissue. In most persons, disease results only after the latent infection is reactivated by an immunosuppressive condition. In premature and other compromised infants, an unusual infiltration of plasma cells, thickening of the alveolar walls, and a frothy, alveolar exudate (referred to as **interstitial plasma cell pneumonia**) are seen. The exudate is filled with ***Pneumocystis*** organisms and host cells (fig. 14.9). At autopsy, the lungs are characteristically heavy and airless.

The clinical symptoms of *Pneumocystis* pneumonia are not specific; however, when an immunocompromised patient is febrile and has a dry, nonproductive cough and shortness of breath, the X-ray finding of interstitial pneumonia is highly suggestive. In severe cases, organisms can disseminate from the lung to involve almost any other organ including liver, spleen, bone marrow, and skin.

Incubation Period

Unknown. Usually 1 to 2 months after the immunosuppressive event begins.

LABORATORY DIAGNOSIS

Specimens Biopsied lung tissue, aspirated bronchial and alveolar washings (bronchoalveolar lavage), or occasionally, sputum.

Microscopic examination of specimens stained with silver impregnation fungal stains or fluorescent antibody reagents reveals the characteristic extracellular cysts. They are rounded or oval, have a diameter of 6 to 9 μm, contain up to eight internal bodies, and are enclosed by a viscous wall (fig. 14.9). A PCR assay for detecting this organism in clinical specimens is under development.

Culture has been successful in some research laboratories, but culture methods for diagnostic laboratories are not yet routinely available.

Serologic methods are not useful because most healthy adults have antibody as a result of subclinical infection. Immunocompromised patients may not demonstrate circulating antibody even when overt disease is present.

Epidemiology

Pneumocystis species infect a wide variety of animals, but animal strains appear to be different from those that infect humans. The relationship of animal-to-human disease is unknown. Person-to-person spread has not been documented, but outbreaks among children in nurseries and orphanages and hospitalized patients on organ transplant wards suggest direct respiratory spread from infected or asymptomatic persons. Most children in the United States develop antibody as a result of asymptomatic infection by the time they are 4 years old. An environmental source of *P. carinii* has been suggested, but the existence of such a reservoir is unproven.

In the early 1980s, the unusually high number of young homosexual men developing *Pneumocystis* pneumonia was one of the first clues to the impending AIDS epidemic (see chap. 19). In North America, up to 80% of patients with AIDS develop pneumocystosis at least once, and repeated infections occur in those persons who are not maintained on prophylactic therapy.

Trimethoprim-sulfamethoxazole (primarily an antibacterial agent) is the drug of choice for treating pneumocystosis. It is also widely used as a prophylactic agent for preventing infections in organ transplant patients and persons undergoing cancer chemotherapy. Pentamidine is used for therapy and prophylaxis of *Pneumocystis* pneumonia in AIDS patients because they have a high incidence of adverse reactions to trimethoprim-sulfamethoxazole (e.g., skin rashes, decreased white blood cells, and changes in liver function). When used for prophylaxis, pentamidine is administered in aerosolized form.

QUESTIONS

1. How are the human mycotic diseases classified clinically? What are the epidemiologic distinctions of these three types of mycoses?
2. Why have the numbers of opportunistic fungal infections increased?
3. What are the typical tissue reactions to deep fungal infections?
4. Why is it difficult to diagnose systemic mycoses?
5. What factors commonly predispose to candidiasis?
6. Describe the five types of clinical candidiasis.
7. What is the drug of choice for *Candida* infections?
8. Describe the management of the patient with a mycotic infection.

BIBLIOGRAPHY

Amantea, M. A., D. J. Drutz, and J. R. Rosenthal. 1990. Antifungals: A primary care. *Patient Care.* 24 (18): 58–68.

Balows, A., W. J. Hausler, and E. H. Lennette, eds. 1988. *Laboratory diagnosis of infectious diseases: Principles and practice. Vol. 1. bacterial, mycotic and parasitic diseases.* New York: Springer-Verlag.

Bartlett, M. S., and J. Smith. 1990. *Pneumocystis carinii: An opportunist in immunocompromised patients. Clin. Microbiol. Rev.* 4 (2): 137–149.

Beneke, E. S. 1984. *Human mycoses.* Kalamazoo, Mich.: Upjohn.

Brummer, E., E. Castaneda, and A. Restrepo. 1993. Paracoccidioidomycosis: An update. *Clin. Microbiol. Rev.* 6 (1): 89–117.

Centers for Disease Control and Prevention. 1995. Histoplasmosis—Kentucky 1995. *Morbid. Mortal. Weekly Rep.* 44: (38): 701.

Chessin, L. N., T. Hoffman, and L. D. Sabath. 1988. Targeting systemic mycoses. *Patient Care.* 22 (13): 121–132.

Fungal infections: Deadly in the immunocompromised. 1989. *Emerg. Med.* 21 (15): 18–26.

Fungal infections late in life. 1988. *Emerg. Med.* 20 (19): 28–31, 41–42.

Fuerst, M. L. 1993. Current trends in the treatment and prevention of opportunistic infections. *AIDS-Patient Care.* 7 (1): 38–39.

Hammarstein, J. E., and J. F. Hammarstein. 1990. Histoplasmosis—Recognition and treatment. *Hosp. Prac.* 25 (6): 95–126.

Hazen, K. C. 1995. New and emerging yeast pathogens. *Clin. Microbiol. Rev.* 8: 462–478.

Hess, P. 1991. Catching the fever in L.A—*coccidioidomycosis* . . . an outbreak of valley fever. *Emerg. Med. Serv.* 23 (5): 39.

Histoplasmosis in AIDS. 1988. *Emerg. Med.* 20 (17): 65–66.

Hogan, L. H., B. S. Klein, and S. M. Levitz. 1996. Virulence factors of medically important fungi. *Clin. Microbiol. Rev.* 9 (4): 469–488.

Hughes, W. T. 1990. The prevention of *P. carinii* pneumonia. *Hosp. Prac.* 25 (4A): 33–43.

Kauffman, C. A. 1993. Fungal infections in the elderly. *Emerg. Med.* 25 (6): 21–27; 30–32.

Larone, D. 1995. *Medically important fungi.* 3d ed. Herndon, Va.: ASM Press.

Mandell, G. L., J. E. Bennett, and R. Dolin. 1995. *Principles and practice of infectious diseases,* 4th ed. New York: Churchill Livingstone.

Mathewson, H. S. 1991. Drugs for pulmonary mycoses. *Respir. Care.* 39 (6): 652–656.

Rippon, John. 1988. *Medical mycology: The pathogenic fungi and the pathogenic actinomycetes,* 3d ed. Philadelphia: W. B. Saunders.

Roccograndi, J. F., and K. S. Clements. 1993. Managing AIDS-related meningitis. *RN* 56 (11): 36–39.

Roueche, B. 1991. The Liberace room. In *The medical detectives.* New York: Plume.

Stringer, J. R. 1996. *Pneumocystis carinii:* What is it, exactly? *Clin. Microbiol. Rev.* 9 (4): 489–498.

What you need to know about crypto—An emerging infection. 1991. *Hosp. Infect. Control* 21 (11): 157–158.

CHAPTER 15

Nonrespiratory Diseases Acquired through the Respiratory Tract

OUTLINE

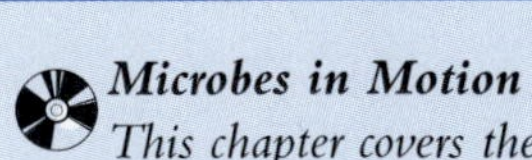

Microbes in Motion

This chapter covers the agents of microbial meningitis and infective endocarditis and their pathogenesis, epidemiology, control, and prevention. The following books and chapters of the Microbes in Motion CD-ROM may be a useful preview or supplemental study aid to your reading: Gram-positive Cocci: Streptococci; Microbial Pathogenesis of Infectious Diseases: Toxins.

OBJECTIVES

On completing this chapter the student will be able to:

1. name the agents of microbial meningitis and infective endocarditis.
2. discuss the mechanisms by which such organisms may reach the central nervous system or heart valves.
3. describe the methods used for the laboratory diagnosis of meningitis and endocarditis.
4. explain the epidemiology of these diseases.
5. describe methods for controlling active infection.
6. discuss current methods for preventing these diseases.

KEY WORDS

Amebic meningitis
Bacterial meningitis
Chemical meningitis
Fungal meningitis
Infective endocarditis (**END**-o-card-**EYE**-tis) (IE)
Meningitis (men-in-**JITE**-is)
Meningococcal meningitis
Meningococcus (men-in-jo-**COCK**-us)
Shunt-associated meningitis
Viral encephalitis (en-seff-a-**LITE**-us)
Viral meningitis

INTRODUCTION

Meningitis and infective endocarditis are diseases affecting the central nervous system and the endocardium of the heart valves, respectively. They are included in this section because most often the microbial agents involved either enter the body through the respiratory tract and initiate infection there, or they are members of the normal flora of the upper respiratory tract that cause disease when they accidentally find their way into the bloodstream and become lodged on a compromised heart valve.

MENINGITIS

GENERAL COMMENTS

The term **meningitis** signifies an acute inflammation of the membranes covering the brain and spinal cord (fig. 15.1). It may be associated with a variety of nonspecific injuries but is also commonly caused by bacterial and viral infection of the meninges (table 15.1). Fungi and protozoa may also be involved. The clinical as well as the epidemiologic pattern of infectious meningitis varies greatly with the causative organism.

Bacterial meningitis is clinically acute, characterized by an outpouring of purulent exudate over the membrane surfaces. The term *septic meningitis* is sometimes used to refer to this purulence. The exudate is largely composed of polymorphonuclear cells which appear in great numbers in the spinal fluid. Epidemiologically, most types of bacterial meningitis are associated with respiratory routes of entry. The bacterial agents are often commensals of the respiratory mucosa. From this local site they may reach the bloodstream and infect other organs, with meningitis a secondary result of bacteremia. In some instances, there may be direct penetration from the upper respiratory tract and invasion of the central nervous system. Meningitis may also be secondary to head injury or surgical or diagnostic procedures involving the central nervous system. Spinal anesthesia, lumbar puncture, and pneumoencephalography may allow opportunistic pathogens to invade directly.

The age and immune status of the host are correlated with the specific types of bacteria involved in meningeal infections. The major defense mechanism against bacteremia and the spread of infection to the central nervous system is thought to be the activity of opsonizing serum antibodies.

In newborn infants less than 2 months of age whose immune mechanisms are undeveloped, meningitis may be the result of colonization of the intestinal tract or skin by organisms encountered in the birth canal, with ensuing spread through the infant's bloodstream. Direct invasion by commensalistic bacteria of the birth canal may occur, but this is rare. The organisms most frequently involved in neonatal meningitis are group B streptococci, *Listeria monocytogenes,* and *E. coli,* all found as members of the vaginal flora.

In older children from 2 months to 5 years of age, meningitis results from bacteremia or direct extension with organisms colonizing the upper respiratory tract. Until recently, when a specific vaccine became widely available, the most frequent agent was *Haemophilus influenzae* type b. *Streptococcus pneumoniae* and *Neisseria meningitidis* are the other common causes of bacterial meningitis in this age group.

In most adults, these are also the common modes of meningeal infection, with *N. meningitidis* (meningococcus) the usual agent in young adults and *S. pneumoniae* (pneumococcus) in the older age group. *H. influenzae* meningitis is now also seen in older patients who have lost their serum antibody for type b strains. Infections following head injuries, or those of iatrogenic origin, most often involve staphylococci, group A streptococci, pneumococci, and *Pseudomonas aeruginosa.*

The clinical picture of tuberculous meningitis is an exception to that of other bacterial infections of the meninges. This is a subacute or chronic progressive disease, infection being spread through the blood from a primary lesion in the lung. There is no purulent polymorphonuclear exudate, and the local lesion, wherever it occurs, is a granuloma composed of lymphocytes and other mononuclear cells.

Viral meningitis (sometimes called *aseptic meningitis* because it is nonbacterial and not purulent) is clinically a more benign disease, usually of short duration. Cellular reaction is less marked than in bacterial infections and is characterized by a predominance of lymphocytes. The viral agents of meningitis may come and go on either respiratory or oral routes of entry and transfer, reaching the meninges by way of the bloodstream, as do bacteria.

Viral encephalitis, or infection of brain tissue itself, is a much more severe disease involving cerebral dysfunction as well as all the manifestations of meningitis.

Fungal meningitis is very different clinically and epidemiologically from bacterial or viral meningeal disease. The fungal organisms are airborne and usually enter the body by the respiratory route, but they come from soil reservoirs. Humans are not a reservoir and do not transmit them directly. Clinical infections induced by fungal agents are chronic, persistent, and slowly progressive. From primary lesions in the lungs or mononuclear phagocyte system, fungi may disseminate to other organs or to the meninges. When meningitis occurs it is also chronic, progressive, and often fatal. The cellular reaction is predominantly lymphocytic.

Amebic meningitis is more correctly a meningoencephalitis because brain tissue as well as the meninges are involved. Patients are usually children or young adults who almost always have a recent history of swimming in warm, freshwater lakes or ponds. The motile amebas, whose natural habitat is in the environment, gain access to the central nervous system through the nasal mucosa and the cribriform plate. The disease has an acute onset and is marked by a purulent meningeal reaction as well as hemorrhages and edema in brain tissue. Death usually ensues within 1 week.

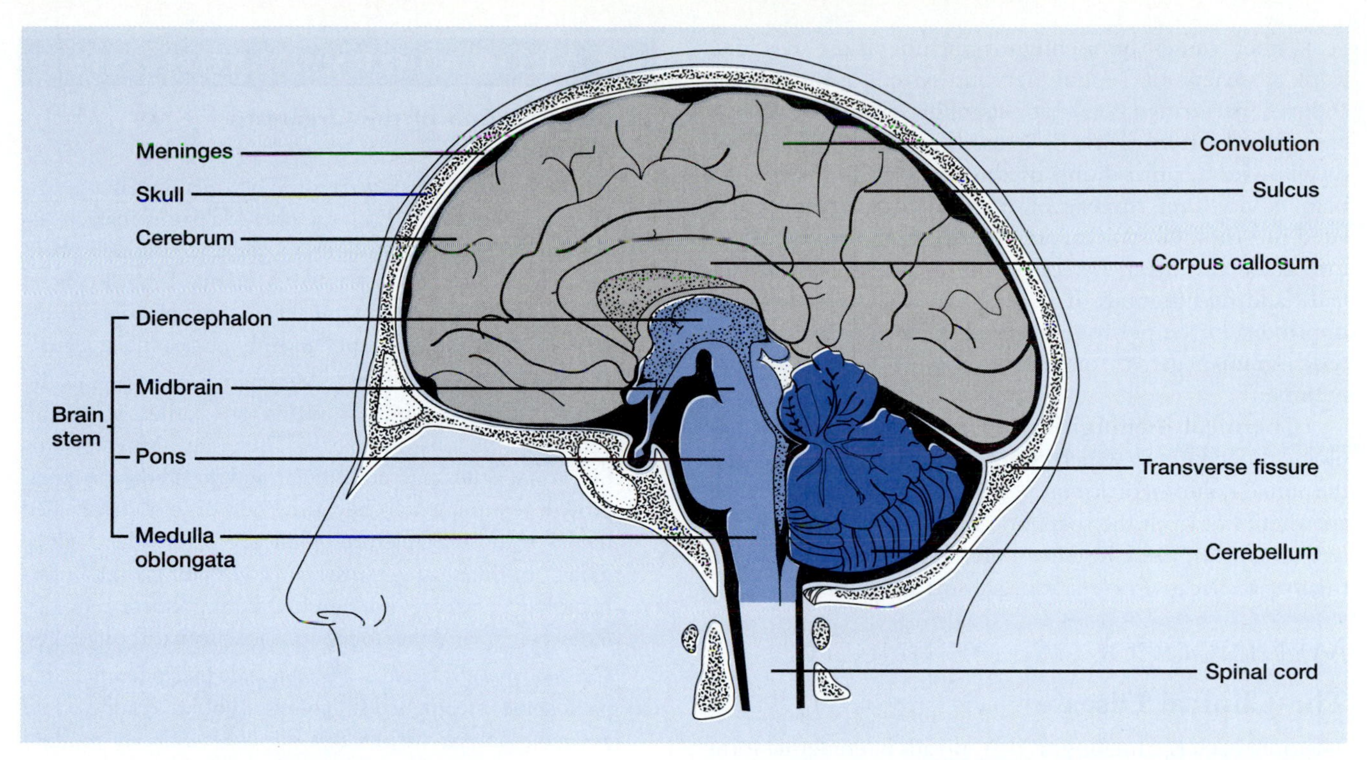

FIGURE 15.1 The major parts of the brain: the cerebellum, cerebrum, and brain stem. The meninges are the membranes under the skull bones that cover and help protect delicate brain and spinal cord tissues.

TABLE 15.1 IMPORTANT MICROBIAL AGENTS OF MENINGITIS

Type of Infection	Mode of Spread to Meninges	Organisms Most Commonly Involved
Bacterial Meningitis		
Infants under 2 months	From intestinal tract or skin by way of bloodstream Direct from birth canal	*Streptococcus,* group B *Escherichia coli* *Listeria monocytogenes*
Children, 2 months to 5 years	From upper respiratory tract by way of bloodstream Direct invasion	*Haemophilus influenzae* type b* *Streptococcus pneumoniae* *Neisseria meningitidis*
Adults	From upper respiratory tract by way of bloodstream Direct invasion	*Neisseria meningitidis* *Streptococcus pneumoniae* *Haemophilus influenzae* type b (rare)
Any age	Direct invasion after head injury, neurosurgical or diagnostic procedure	Staphylococci *Streptococcus,* group A *Streptococcus pneumoniae* *Pseudomonas aeruginosa*
Tuberculous meningitis	From lesion in lung by way of bloodstream	*Mycobacterium tuberculosis*
Viral (Aseptic) Meningitis	From upper respiratory or intestinal tract by way of blood	Mumps virus Enteroviruses
Viral Encephalitis	From upper respiratory or intestinal tract by way of blood Arthropod vector	Mumps virus Enteroviruses Herpes viruses Arboviruses
Fungal Meningitis	From lesion in lung or other organ by way of blood	*Cryptococcus neoformans*
Amebic Meningitis	Direct invasion through nasal mucosa and cribriform plate after entry from fresh water	*Naegleria fowleri*

*In nonimmunized children.

Other causes of meningitis include those resulting from a variety of central nervous system invasive procedures performed on patients. For example, **shunt-associated meningitis** is caused by infections of indwelling intracranial shunts used to draw off fluid from the brain cavity after surgery or trauma. The organisms isolated in these instances are usually skin organisms of low-grade pathogenicity, for example, *Staphylococcus epidermidis* and diphtheroids. If the second end of the shunt is implanted in the peritoneum (a common procedure), enteric gram-negative organisms may sometimes cause meningitis.

Chemical meningitis may be caused by special radiologic procedures in which contrast material is injected into the spinal column, or it may result from injection of spinal anesthetics or from the soaps and disinfectants used to treat instruments. In these instances, the illness resembles aseptic meningitis because no microorganisms are isolated.

MENINGOCOCCAL MENINGITIS

The Clinical Disease

The only bacterial meningitis that spreads in epidemic form is that caused by *Neisseria meningitidis,* the **meningococcus. Meningococcal meningitis** is an acute disease characterized by sudden onset of fever, severe headache, painful rigidity of the neck, nausea, and vomiting. Convulsions are often seen in children. Delirium or coma is frequent. The meninges are involved in an acute inflammatory reaction characterized by a purulent exudate. If the disease is not arrested before this exudate becomes abundant, the meninges later become thickened as inflammation progresses. Hemorrhage and thrombosis of small blood vessels contribute to the damage, which may be residual if treatment is delayed or ineffective.

Meningococci enter the body through the nasopharynx and localize there. They may remain there without any injury to tissues, or they may induce a local inflammation. In a small proportion of cases, meningococci enter the bloodstream. They may be filtered out and killed by the body's defense mechanisms, or meningococcemia may ensue, with distribution of the organisms to many foci throughout the body, notably the skin, joints, lungs, adrenal glands, and central nervous system.

Damage of capillary walls by meningococcal endotoxin leads to small petechial hemorrhages in the skin and other affected tissues. Fulminating infections may lead to adrenal hemorrhage, circulatory collapse, and shock (Waterhouse-Friderichsen syndrome). The clinical appearance of meningococcemia is marked by fever, chills, acute malaise, petechial rash, and prostration. The disease may be rapidly fatal, with death occurring in a few hours before meningitis can develop.

Although the usual route of spread to the meninges involves bacteremia, meningococci may penetrate the thin bony lamina posterior and superior to the nasopharynx and invade the central nervous system directly.

LABORATORY DIAGNOSIS

I. Identification of the Organism

Neisseria meningitidis.

A. *Microscopic Characteristics* The organism is a gram-negative diplococcus. Each member of a pair is characteristically flattened on its adjoining side and rounded on its outer edge so that it resembles a kidney bean. In direct stained smears of purulent spinal fluid the diplococci are often seen intracellularly in polymorphonuclear phagocytes.

B. *Culture Characteristics* Meningococci belong to the genus *Neisseria,* which contains one other important pathogen (*Neisseria gonorrhoeae*). These organisms are aerobic, nonsporulating, nonmotile, and fastidious in their growth requirements. They grow best on blood-enriched media incubated microaerophilically with added CO_2. Their colonies are transparent, glistening, and nonhemolytic. They are distinguished from gonococci by differences in their ability to produce acid from carbohydrates.

C. *Immunologic Methods of Identification* Like pneumococci, pathogenic meningococci possess capsules composed of antigenic polysaccharides that stimulate antibody production. Their recognition by capsular swelling or agglutination techniques may be of value in diagnosis and in epidemiologic tracking of strains involved in outbreaks of meningitis.

Latex agglutination is used to detect capsular antigens in cerebrospinal fluid or urine (see chap. 6). The capsular antigens fall into several serologic types. Groups A, B, and C were once the most prominent. Group A was responsible for worldwide epidemics, whereas groups B and C were associated with sporadic or endemic disease. In recent years, new serotypes including those labeled X, Y, Z, 29-E, and W-135 have been identified. Of these, group Y has assumed clinical importance as the cause of small outbreaks of meningococcal disease. Group A still causes epidemics in some parts of the world.

II. Specimens for Diagnosis

A. Blood cultures drawn as soon as possible after onset of symptoms.

B. Spinal fluid samples for immediate examination of Gram-stained smears and initiation of culture. Demonstration of capsular polysaccharide antigens in spinal fluid or urine, if successful, provides a quick, confirmed diagnosis. White blood cell counts and chemical analysis for glucose and protein levels are also essential.

C. Nasopharyngeal swabs in transport medium may detect carriers.

D. Petechial lesions on the skin may be scraped or aspirated and submitted for smear and culture.

E. Identification of patient's serum antibodies is seldom necessary.

III. Special Considerations

A. Meningococcal meningitis is a medical emergency requiring prompt, accurate diagnosis and therapy. Permanent tissue damage or death may result from delay. Pretreatment specimens, aseptically collected in adequate quantity and promptly examined, are essential to laboratory diagnosis.

B. Spinal fluid must be analyzed by chemical and cytologic methods as well as by culture techniques. Collection in three separate sterile tubes is recommended for speedy and efficient distribution to appropriate sections of the laboratory. The quantity per tube should be determined by consultation with the laboratory so that the volumes are adequate for the tests to be performed. If only one tube is used for collection, it should be submitted first to the microbiology laboratory for aseptic withdrawal of an aliquot for smear and culture.

C. Any material submitted for culture of meningococci should be clearly identified for this purpose so that appropriate examinations may be made promptly by the laboratory.

D. Meningococci lyse rapidly in extravasated body fluids without nutrient and are very sensitive to temperature changes. For these reasons specimens for meningococcus culture should be delivered without delay to the laboratory and placed in the hands of the microbiologist, not in incubators, in refrigerators, or on unattended laboratory benches.

Epidemiology

I. Communicability of Infection

A. *Reservoirs, Sources, and Transmission Routes* Humans are the only known reservoir. The nasopharyngeal secretions of infected persons are the principal source. Epidemics are spread primarily by asymptomatic carriers rather than by infectious cases and have occurred most frequently in closed populations such as military groups. Meningococci do not survive long in the environment; transmission is by direct contact or droplet spread. Occasional meningococcal isolates from the genitourinary tract have been reported, but the implications for transmission are not known. At least one case of fatal neonatal meningococcal meningitis has been reported, the mother's cervicovaginal cultures being positive for *N. meningitidis.*

B. *Incubation Period* The extremes are 2 and 10 days, the average being 3 to 4 days.

C. *Communicable Period* The disease is potentially transmissible while meningococci are present in respiratory discharges. Effective chemotherapy generally clears the nasopharyngeal focus within 24 to 48 hours.

D. *Immunity* Asymptomatic carriers greatly outnumber cases of infectious disease, indicating generally low susceptibility to clinical illness. Specific immunity following recovery from the disease is of uncertain importance in providing future protection. Persons who lack any one of certain components of complement, primarily C5-C8 (see fig. 6.13), are at increased risk for acquiring infection.

Capsular polysaccharide vaccines for artificial active immunization have been developed and are being used for high-risk groups, for example, military recruits. The vaccine contains groups A, C, Y, and W-135 polysaccharides. Group B is not included because its polysaccharide capsule is not sufficiently immunogenic to produce an antibody response. This problem is serious because *N. meningitidis* group B is responsible for more than 50% of cases of meningococcal meningitis in the United States.

II. Control of Active Infection

A. *Transmission-based precautions* Droplet precautions for at least 24 hours following initiation of specific therapy or until the nasopharynx is culturally negative.

B. *Other Precautions* During the communicable period, concurrent disinfection of respiratory secretions and of contaminated items. Terminal cleaning should include disinfection of furniture and floor.

C. *Treatment* Intravenous penicillin in high dosage is the regimen of choice for either meningitis or meningococcemia. Chloramphenicol is useful for those patients who are hypersensitive to penicillin or related drugs.

D. *Control of Contacts*

1. *Isolation* None.

2. *Prophylaxis* Immunization with the A, C, Y, W-135 polysaccharide vaccine appears to be a safe, effective method for providing artificial immunity and preventing disease caused by organisms of these groups. The durability of artificially acquired immunity has not yet been determined.

The choice of drugs for chemoprophylaxis depends on the susceptibility of the strain of *N. meningitidis* to which contacts have been exposed. Although sulfadiazine was formerly the drug of choice for this purpose, certain strains (especially groups B and C) are now resistant to it. Rifampin is used in the United States for close family contacts of cases of meningococcal meningitis. Penicillin is ineffective in eliminating the carrier state unless administered intramuscularly in high doses, and it is not generally recommended for the treatment of contacts. Health care workers receive prophylaxis only if they have had direct, intimate contact with a patient's respiratory secretions (e.g., during mouth-to-mouth resuscitation).

E. *Epidemic Controls* Individual cases must be reported by telephone to local health departments. In 1996, approximately 3,200 cases of meningococcal disease were reported. Chemoprophylaxis of close contacts and surveillance of their health and living conditions limit the spread of an outbreak by reducing the number of carriers and permitting early segregation of the clinically ill. Outbreaks occur most frequently in closed populations (military barracks, schools, camps, institutions). Physical separation, increased ventilation of quarters, and insistence on good personal hygiene help to limit outbreaks. The last major epidemic of meningococcal disease in the United States occurred in 1943 when about 13 cases per 100,000 population were reported but the rates have remained

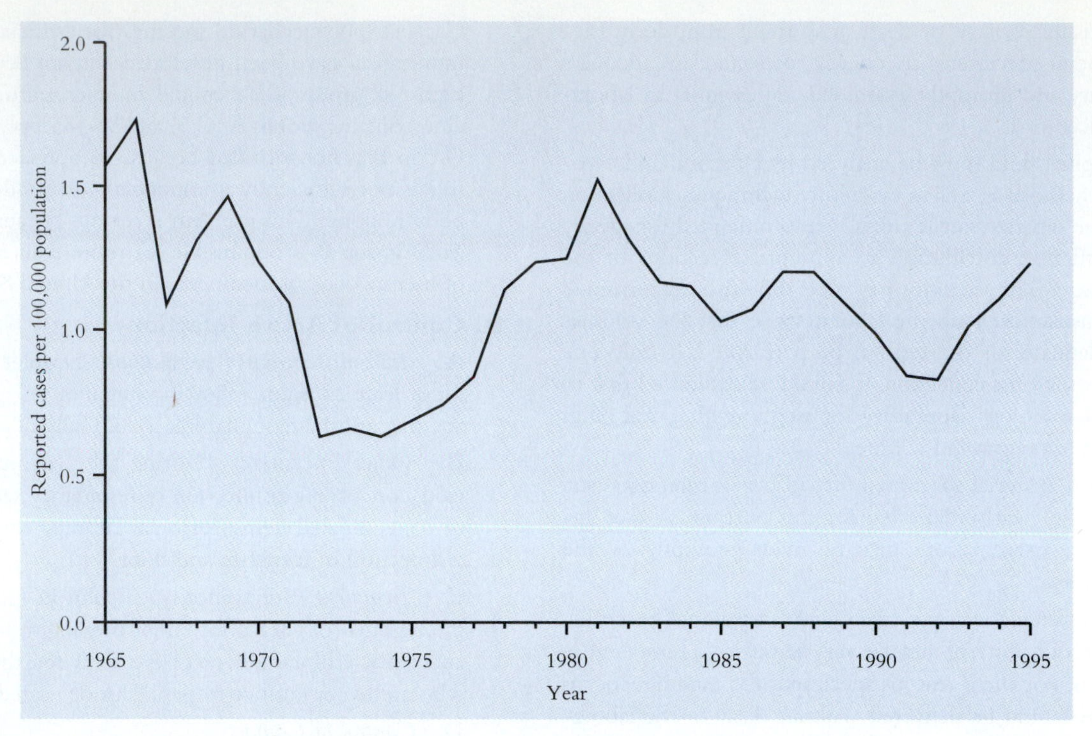

FIGURE 15.2 Meningococcal disease in the United States from 1965 to 1995. Although overall rates of meningococcal disease have remained relatively constant, serogroup C outbreaks are occurring more frequently in the United States, and vaccine use for outbreak control has increased. The meningococcal vaccine licensed in the United States provides protection against serogroups A, C, Y, and W-135, but not against serogroup B.
Centers for Disease Control and Prevention *MMWR Summary of Notifiable Diseases, United States, 1995.* (October 25, 1996, Vol 44: 53).

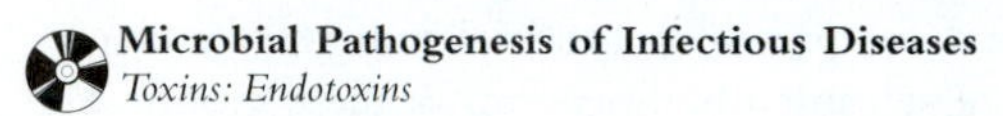

relatively constant since then at 1 to 2 cases per 100,000 poulation (fig. 15.2).

III. Prevention

A. *Immunization* Not routine. In the future, immunization may prove to be the best means of control for selected groups at risk.

B. *Control of Reservoirs, Sources, and Transmission Routes* Public education concerning the source of infection, respiratory hygiene, avoidance of direct contacts and crowded living, working, or traveling conditions.

OTHER BACTERIAL AGENTS OF MENINGITIS

The Clinical Disease

Acute bacterial meningitis may be secondary to respiratory tract infections caused by organisms of the indigenous flora. Entry of these organisms into the bloodstream from the site of active lesions in the upper or lower respiratory tract probably is their chief route of access to the meninges. The latter event does not occur frequently, presumably because the blood's defensive barriers are effective in eliminating most bacteremic infections. Bacteria of enteric origin are also sometimes responsible for acute meningitis if they find a disseminating route from colonizing sites in the bowel. The clinical disease produced in these situations resembles meningococcal meningitis in every way, except that petechial lesions characteristic of meningococcal injury are usually not seen in other bacterial infections.

Fractures of the skull or other head injuries may create possibilities for direct invasion of the subarachnoid space by bacteria. Surgical and other procedures involving even minor trauma to the central nervous system may also open a pathway for invading microorganisms and may result in meningitis.

LABORATORY DIAGNOSIS

I. Identification of the Organisms

Differentiation of these meningitides requires bacteriologic recognition of their agents. The organisms most frequently involved are listed in table 15.1. Some of these, notably *Haemophilus influenzae* type b and *Streptococcus pneumoniae* are also associated with bacterial pneumonias. A variety of other bacteria may also be responsible for meningitis and are usually seen in patients compromised by a physical or immunological predisposing condition. These organisms include a number of commensal bacilli of the intestinal tract such as *E. coli, Klebsiella,* and *Proteus,* or environmental bacteria such as *Pseudomonas.*

The bacteriology of enteric bacilli is described in chapter 16.

Tuberculous meningitis is an important disease that may occur as a sequel to pulmonary tuberculosis. This disease is discussed in chapter 12.

II. Identification of Patient's Serum Antibodies
Serologic diagnosis is seldom helpful.

III. Specimens for Diagnosis

A. Blood for culture, particularly in the early stage of disease.

B. Spinal fluid for smear and culture.

C. Nasopharyngeal swabs or sputum specimens sometimes help in recognizing endogenous sources of meningeal infection.

D. Spinal fluid, serum, or urine for direct antigen detection by latex agglutination (*H. influenzae, S. pneumoniae,* group B streptococci). The test is helpful primarily for patients who have received antimicrobial agents before the spinal tap is performed, in which case Gram stain and culture may be negative.

IV. Special Considerations
Acute infectious meningitis is a medical emergency (see previous section, Meningococcal Meningitis, III. Special Considerations, items A, B, C).

Epidemiology

I. Communicability of Infection

A. *Reservoirs, Sources, and Transmission Routes* Humans are the reservoir. Endogenous infectious lesions in the respiratory tract or other organs are often an immediate source of meningeal infection. Nasal, oral, and intestinal discharges of infected persons may be a source for susceptible individuals. Transmission is by direct contact, by droplet spread, or by the fecal-oral route. In the case of group B streptococcal meningitis, the organism is acquired by the neonate during passage through the colonized maternal vagina.

B. *Incubation Period* Unknown.

C. *Communicable Period* Organisms are readily transmissible, but communicability depends on susceptibility of contacts.

D. *Immunity* General resistance to the pathogenic potential of these organisms is high, but children are often more susceptible than adults, particularly in the case of *H. influenzae.* This organism is one of the most common causes of meningitis in infants and children who are not vaccinated. Antibody response may lower susceptibility to severe *H. influenzae* infection, but in general the role of specific immunity to other organisms of this group is not significant.

II. Control of Active Infection

A. *Transmission-based precautions* Droplet precautions for *H. influenzae* and *N. meningitidis;* otherwise standard precautions are sufficient for other agents.

B. *Other Precautions* Concurrent disinfection of respiratory secretions or other discharges as indicated by possible systemic foci of infection.

C. *Treatment* Antimicrobial drugs are effective when given promptly with onset of symptoms. Speed is essential in identifying the causative organism and determining its susceptibility to antimicrobial agents. Pending preliminary and confirmed information from the laboratory, antimicrobial agents are given empirically to control the disease.

D. *Control of Contacts* Generally not indicated.

E. *Epidemic Controls* Generally not applicable. Case reports should be sent to local health departments.

III. Prevention

A. *Immunization* Pneumococcal vaccine (Pneumovax) is given to persons at special risk because of underlying disease (diabetes, asplenia) and is recommended for persons over 50 years old. *H. influenzae* conjugate vaccine (see table 10.1) is now administered as part of routine childhood immunizations. The widespread use of this vaccine has resulted in a 95% reduction in the incidence of *H. influenzae* meningitis in young children. No vaccines are available for other agents.

B. *Prophylaxis* for *H. influenzae* meningitis is with rifampin. The drug is administered to the same groups of people as those who receive it for meningococcal meningitis prophylaxis. Prevention of meningitis caused by other agents rests with avoidance of infectious contacts by susceptible persons, particularly children. Support and maintenance of healthy resistance are also important.

For prevention of group B streptococcal meningitis, recent recommendations call for collecting rectal and vaginal swabs from all pregnant women at 35 to 37 weeks' gestation and culturing them for group B streptococci. Women who have positive cultures are given penicillin during delivery. For penicillin-allergic patients, clindamycin or erythromycin are given instead. Antibiotic is also given to women whose culture results are not known, who deliver prematurely before cultures are taken, who have had ruptured membranes for longer than 18 hours but have not yet delivered, or have a temperature higher than 38°C. These recommendations are expected to reduce significantly the occurrence of group B streptococcal meningitis.

Infective Endocarditis

Infective endocarditis (IE) was formerly referred to as either *subacute* or *acute* endocarditis, based on the severity of disease and the speed with which it progressed. Because there is significant overlap between the two forms of disease, the term infective endocarditis is commonly used instead. Certain microorganisms, such as streptococci indigenous to the respiratory and oral flora, and other commensalistic organisms of the normal mucosal surfaces, however, tend to be associated with a more chronic, slowly progressing disease.

Summary Table 15.2 NONRESPIRATORY DISEASES ACQUIRED THROUGH THE RESPIRATORY TRACT

Clinical Disease	Causative Organism	Other Possible Entry Routes	Incubation Period	Communicable Period
I. Bacterial Meningitis				
A. Meningococcal meningitis	*Neisseria meningitidis*		3–4 days	While nasopharynx is bacteriologically positive
B. Other bacterial agents of meningitis	*Haemophilus influenzae* *Streptococcus pneumoniae* *Streptococcus* group B *Staphylococcus aureus* *Staphylococcus* species Gram-negative bacilli *Listeria monocytogenes*	Group B streptococci from maternal vagina Enteric bacteria possibly enter by fecal-oral route *Listeria* by congenital transfer	Unknown	Respiratory transmission does not necessarily lead to meningeal infection
II. Infective Endocarditis				
	Streptococcus species *Staphylococcus* species Gram-negative bacilli and others	Endogenous sources	Unknown	Not communicable

Infective endocarditis is not communicable. Infection is of endogenous origin, involving localization of normal flora organisms at some site of previous injury on the endocardium, usually a valve surface. At least two factors appear to be essential to development of the more chronic, subacute form of disease: (1) a preexisting defect of the valvular endocardium (though this may not be clinically detectable before the onset of disease) and (2) the bacteremic distribution of microorganisms from a resident focus on a skin or mucosal surface disrupted by trauma, surgery, or some manipulation.

Most patients with the disease (about 75%) have a history of rheumatic fever as the most probable cause of preexisting cardiac damage. Congenital cardiac defects and valvular injuries induced by arteriosclerosis or syphilis may also predispose to infection.

Minor dental surgery is often considered to be the initiating cause of transient bacteremias that lead to IE. Many other possibilities exist as well: nonsurgical manipulation of genitourinary tissues (catheterization, cystoscopy, prostatic massage), ulcerative erosions or perforating lesions of the bowel, or surgical transection of surfaces with a normal flora—all represent possible mechanisms by which commensalistic microorganisms may be introduced into the bloodstream.

Ordinarily these bacteremias are terminated rapidly, the organisms being incapable of localizing in normal systemic human tissues. IE develops if the microorganisms localize on damaged valvular surfaces and multiply there. The site may characteristically be involved in a fibrinous response to injury, and small aggregates of platelets in fibrin may be formed. Microorganisms can grow in this milieu protected from the action of phagocytes or other bactericidal components of blood. Their presence accelerates fibrinous tissue reaction, and "vegetations" are built up around their proliferating growth. After such infection has been established, bacteremia is usually steady and continuous.

Increasing valvular dysfunction leads to symptoms of cardiac distress. If the situation is not arrested by antimicrobial treatment, the valve may be occluded by vegetative growth or fail to function because of loss of elasticity, with fatal results in either case. These friable vegetations break off in small pieces called emboli in the surging flow of blood propelled through the valve by ventricular contraction. Such infectious emboli then travel down the arterial vessels, finally lodging in and occluding vessels that are too small to pass them further. The seriousness of this situation depends on the nature of the tissue thus deprived of its blood supply and also on the capacity of the organisms to continue multiplying at the new site.

LABORATORY DIAGNOSIS

The laboratory diagnosis of IE depends on successful isolation of the causative organism from blood. Since there is a fairly continuous release of organisms from the valvular lesions, blood specimens taken at any time are likely to be positive in culture. However, only small numbers of organisms are usually present, and a few days may be required for their growth and identification. Whenever possible, four to six samples of blood for culture should be drawn over the course of two or three days before antimicrobial therapy is instituted. Isolation of the same organism from a majority of them is strong evidence for a diagnosis of IE.

Strains of alpha-hemolytic *Streptococcus* are most frequently isolated from blood cultures in the subacute form

Specimens Required	Laboratory Diagnosis	Immunization	Antimicrobial Agents for Treatment	Precautions
Spinal fluid Spinal fluid Blood	Smear and culture Latex agglutination Culture	Quadrivalent vaccine (groups A, C, Y, W-135)	Penicillin Rifampin (contacts or carriers)	Droplet precautions for first day after therapy initiated
Spinal fluid Spinal fluid, urine Blood	Smear and culture Latex agglutination for *H. influenzae,* pneumococcus, and group B streptococci Culture	Pneumovax *H. influenzae* conjugate vaccine	Choice depends on testing of isolated organism	Droplet precautions for *H. influenzae* for first day after therapy initiated
Blood	Culture	None	Same as above	Standard precautions

of endocarditis. Enterococci are also common, whereas enteric bacilli and anaerobic organisms are less common.

It should be noted that modern methods of heart surgery, including the use of arterial catheters, open easy avenues for infection of damaged or prosthetic valves. Rigorous aseptic precautions are required to protect patients undergoing cardiac surgery or catheterization from directly implanted infection. Successful surgical techniques may be compromised or negated by microbial contamination that develops into infection. The organisms most frequently involved in this situation are staphylococci of the skin and mucosa. Some are benign coagulase-negative strains of the resident skin flora (*Staphylococcus epidermidis*), which generally induce subacute but still serious problems; others are of the virulent coagulase-positive variety (*Staphylococcus aureus*) capable of causing acute or overwhelming infection, particularly of traumatized or malfunctioning tissues (see chap. 21).

The treatment of streptococcal endocarditis generally is penicillin and gentamicin in combination, or vancomycin, when organisms of the alpha-hemolytic (viridans) group are involved. Enterococci may be treated with penicillin or ampicillin in combination with gentamicin. Vancomycin may also be used, especially if the patient is allergic to penicillin. Antimicrobial susceptibility testing of the isolated organisms is essential in indicating the choice of drugs. The physician may find it of even greater value to know what dosage is required to kill, rather than merely suppress, the organism. Quantitative tube dilution or microdilution assays of the antimicrobial agent are performed with the organism isolated from the patient so that the dose, frequency, and route of administration of the drug may be adjusted to maintain a bactericidal level of antimicrobial agent in the bloodstream and tissues. The patient's blood serum can also be assayed for its bactericidal activity against the causative agent as a further measure of the adequacy of therapy and of response to it (see chap. 4).

The epidemiology of endocarditis points to the need for preventive measures (not for communicable infection controls): prevention of the causes of endocardial injury (rheumatic fever is the primary target) and control of transmission of infecting organisms from their sources. Early administration of penicillin for group A streptococcal infections has helped to reduce the incidence of rheumatic fever and chemoprophylaxis is useful in decreasing the occurrence of bacteremia after surgery or other provocative procedures. In these cases, antimicrobial agents are given before, during, and for a short time after surgery. Antimicrobial-induced changes in the normal flora and the emergence of resistant bacterial strains counterbalance the prophylactic use of these agents, therefore, their use must be reserved for carefully selected situations.

In recent years, the acute form of IE has been seen primarily in drug abusers who presumably inoculate contaminated material directly into their skin or bloodstream or who use contaminated needles and syringes for injection. *Staphylococcus aureus* and *Pseudomonas aeruginosa* are isolated most commonly from the blood cultures of these patients. These organisms frequently produce disease on previously normal heart valves, and the infection progresses more rapidly than that caused by streptococci. It is also more likely to be fatal.

Table 15.2 summarizes the characteristics of meningitis and infective endocarditis.

Gram-Positive Organisms in Human Host
Gram-Positive Cocci-Streptococci: Taxonomy; Habitat; Diseases

Questions

1. What is meningitis? What is the major defense mechanism against bacteremia and the spread of infection to the central nervous system?
2. What microorganisms cause meningitis?
3. How does bacterial meningitis differ from viral meningitis?
4. Which type of meningitis is spread in epidemic form?
5. What specimens are required for the diagnosis of meningitis?
6. What forms of immunization are available for meningitis?
7. How does infective endocarditis develop?
8. What factors appear to be essential in the development of the more chronic subacute form of endocarditis?
9. How can endocarditis be prevented?

Bibliography

Acute meningitis: Toward a more accurate diagnosis. 1990. *Emerg. Med.* 22 (8): 99–108.

Bradford, A. 1994. Meningococcal meningitis. *Intensive. Crit. Care. Nurs.* 10 (3): 199–208

Klein, N. C., and B. A. Cunha. 1994. Bacterial meningitis. *Emerg. Med.* 26 (3): 14–18.

Mandell, G. L., J. E. Bennett, and R. Dolin. 1995. *Principles and practice of infectious diseases.* 4th ed. New York: Churchill Livingstone.

Marks, M. I. 1990. Meningitis: A management challenge. *J. Am. Acad. Physician Assist.* 3 (6): 442–452.

Mickles, L. I., and D. P. Mickles. 1994. Listeria meningitis: A case study. *Crit Care Nurse* 14 (4): 22, 25–32.

Moore, P. and C. V. Broome. 1994. Cerebrospinal meningitis epidemics. *Sci. Am.* 271 (5): 38–45.

Purdie, K., and A. Peterson. 1994. Managing meningitis: What causes it, how can it be treated and how can nurses protect themselves? *Nurs. New Zealand* 2 (10): 18–19.

Tunkel, A. R., and W. M. Scheld. 1993. Pathogenesis and pathology of bacterial meningitis. *Clin. Microbiol. Rev.* 6 (2): 118–136.

Section V

A United States Department of Agriculture meat inspector is checking the temperature in smoked salami.

U.S. Department of Agriculture, Washington D.C.

Infectious Diseases Acquired through Ingestion

CHAPTER 16

Bacterial Diseases Acquired through the Alimentary Route

OUTLINE

Microbes in Motion
This chapter covers the important bacterial agents acquired through the alimentary route including those that cause disease by invasive and toxigenic mechanisms. The following books and chapters of the Microbes in Motion CD-ROM may be a useful preview or supplemental study aid to your reading: Anaerobic Bacteria: Gram-positive Bacilli; Gram-positive Organisms in Human Hosts: Gram-positive Bacilli; Gram-negative Organisms in Human Hosts: Bacilli–Facultative Anaerobes; Miscellaneous Bacteria in Human Hosts: Zoonoses; Microbial Pathogenesis of Infectious Diseases: Toxins.

OBJECTIVES

On completing this chapter the student will be able to:

1. name several acute bacterial infections acquired through the gastrointestinal tract.
2. identify bacterial species that produce toxins responsible for systemic disease or local disturbance in the gastrointestinal tract.
3. differentiate between bacterial infections and bacterial toxic diseases acquired through the gastrointestinal tract.
4. describe the pathogenic, invasive, toxigenic, and hemorrhagic types of *Escherichia coli.*
5. explain the action of ingested preformed bacterial exotoxins in causing disease.
6. cite the types of systemic diseases induced by those pathogenic bacteria or their toxins that enter through the gastrointestinal tract.

KEY WORDS

Bacillary (**BA**-sill-airy) dysentery
Bacillus cereus (**SER**-ious)
Bacterial food poisoning
Botulism
Brucellosis (**BRUCE**-ell-**OH**-sis)
Campylobacter (**CAMP**-y-lo-bact-er)
Cholera
Clostridium (closs-**TRID**-i-um) *botulinum* (bot-u-**LINE**-um)
Clostridium difficile (diff-ee-**SEAL**)
Clostridium perfringens (per-**FRINGE**-ens)
Escherichia (esh-er-**ICH**-ee-ah) *coli* (**CO**-lie)
Gastrointestinal infections
Helicobacter (**HELL**-ee-co-back-ter) *pylori* (pile-**OR**-ee)
Listeriosis (list-**EAR**-ee-**OH**-sis)
Paratyphoid fever
Salmonellosis (**SAL**-men-el-**OH**-sis)
Shigellosis (**SHE**-gill-**OH**-sis)
Staphylococcal enterotoxin
Typhoid fever
Vibrio (**VIB**-ri-oh)
Yersinia (yer-**SIN**-ee-a)

Introduction

Bacterial **gastrointestinal infections** acquired through the alimentary route fall into three categories: (1) infections of the gastrointestinal tract caused by ingestion of viable bacteria, usually in contaminated food or water, with resulting acute gastroenteritis; (2) intoxications caused by ingestion of preformed toxins (also referred to as enterotoxins) produced by bacteria growing in contaminated foods; and (3) systemic infections caused by dissemination of ingested microorganisms from intestinal sites to internal tissues and organs.

The first category may be further subdivided into two disease types: (1) those caused by organisms that multiply in and destroy intestinal mucosa and submucosa and (2) those that multiply and produce harmful toxins in the gastrointestinal tract.

Infectious gastroenteritis may be caused by a number of bacterial agents. The basic clinical syndrome common to most of these infections is abdominal distress and diarrhea, often with nausea and vomiting. Clinical distinctions, except in the case of **shigellosis** and **cholera,** are usually difficult or impossible, and specific diagnosis depends on the laboratory identification of the organism in the patient's feces or vomitus or in suspect food or water.

In many instances a specific diagnosis may not be made at all; these infections are usually self-limited and recovery is rapid and complete. Many cases of gastroenteritis are probably caused by viruses whose isolation and identification are time-consuming and expensive. Also, a variety of noxious chemical agents other than bacterial toxins may induce similar symptoms (such agents include medications, allergens, and poisonous plant and animal substances). However, when the cause is infectious and the disease becomes epidemic because it is transmitted directly from infected individuals or indirectly from contaminated food or water sources, every effort should be made to establish the definitive diagnosis so that the source of the agent may be found and controlled if possible. Bacterial diseases that develop abruptly after the ingestion of contaminated food are referred to as *foodborne diseases* (sometimes called *food poisoning*). They often occur as outbreaks among many persons who shared the food.

Public control of all the diseases discussed in this chapter depends on the protection of food and water supplies and the sanitary disposal of sewage. The same principles apply in hospitals. Food must be protected from environmental contamination and from handlers who may be carriers of infection. Standard precautions sometimes coupled with contact precautions for young children must be followed carefully in the management of patients.

Acute Gastrointestinal Disease

The most prominent symptom of most gastrointestinal infections is diarrhea, often accompanied by crampy abdominal pain, nausea, and vomiting. Mild fever, chills, and headache are usually, but not always, present. Many bacterial species have been incriminated in acute foodborne gastroenteritis. The most important of these are discussed here.

Infections Caused by Microbial Multiplication

Salmonellosis

The *Salmonella* genus is remarkable because of its size, wide distribution in nature, and nearly uniform pathogenicity of its many species for humans, animals, and birds. Some strains are more frequently associated with one type of host than with another, but almost all are pathogenic for humans, causing a disease referred to as **salmonellosis.** Only *Salmonella typhi,* the agent of typhoid fever, appears to be restricted to a human reservoir.

Morphologically the salmonellae are indistinguishable from those members of the normal intestinal flora collectively called gram-negative enteric bacilli. The latter include, among others, *Escherichia coli* and species of *Proteus, Providencia, Klebsiella, Enterobacter,* and *Citrobacter.* All of these and the salmonellae are gram-negative, facultatively anaerobic, non-spore-forming rods, many of them motile. They are distinguished from one another by differences in their biochemical properties, possessing enzymes for metabolism of carbohydrates such as lactose, dextrose, and sucrose, and by differences between their specific antigenic components. The biochemical characteristics of the salmonellae are fairly uniform but they are antigenically heterogeneous (more than 1,800 different serotypes are recognized). Taxonomic tools that use genetic comparisons have indicated that, despite their antigenic diversity, all salmonellae are of a single species; however, common practice is to refer to each serotype as a distinct species.

Salmonella infections are most often acquired by ingesting food or water fecally contaminated with these bacilli. Soiled hands often play a role in implanting these organisms in food or directly into the mouth. Meat and other products of infected animals and shellfish are also involved in the wide distribution of salmonellosis.

A dramatic increase in salmonellosis during the past few years has been traced to eggs with intact shells. Presumably, salmonellae spread to the ovaries of infected hens whose eggs become a source of infection when they are laid. Inapparent human infection also contributes to sporadic or epidemic disease.

Salmonella infections may be one of three types and sometimes a combination of these: enteric fever, bacteremia, or gastroenteritis. The first two are systemic infections resulting from spread of organisms through the blood following intestinal tract multiplication, whereas the third (and most common) is intestinally localized infection.

Most *Salmonella* serotypes are agents of gastroenteritis. Although they are enormously diverse antigenically, their clinical and epidemiologic potential is very similar.

Gastroenteritis (Salmonella Food Poisoning)

The symptoms of this infection usually begin from 8 to 48 hours following ingestion of the organisms, with sudden onset of nausea, vomiting, abdominal pain, and diarrhea, often accompanied by fever and chills. In severe cases, the patient may be prostrate. The organisms invade the intestinal mucous membranes, and an acute inflammatory reaction originating in the subepithelial tissues results. The organisms usually do not enter the bloodstream but can be recovered from stool specimens.

Occasionally more than one strain of *Salmonella* may be found in a patient's stool, and a case that begins as a gastroenteritis may sometimes develop into enteric fever.

LABORATORY DIAGNOSIS

I. Identification of the Organism

The organisms are gram-negative bacilli. They grow well in laboratory culture, but their isolation from the mixture of bacteria present in stools may be difficult. This problem is overcome by using selective agar media that suppress the growth of other enteric organisms but permit the growth of *Salmonella* species. The incorporation of lactose into these media, together with an indicator that responds with a color change when acid is produced, serves to differentiate the colonies that grow. With rare exceptions, *Salmonella* strains do not break down lactose into acid end products and are considered lactose nonfermenters. Their colonies, therefore, appear colorless on differential media. Normal enteric bacilli, on the other hand, characteristically ferment lactose. Their colonies take on the indicator color as the pH decreases with the breakdown of the sugar to acid products. This color differentiation makes it possible to select suspicious, colorless colonies and inoculate them as pure cultures into other media necessary for their identification. Certain other organisms normally found in stool specimens, such as *Proteus* and *Pseudomonas* species, are also lactose negative and must be differentiated from *Salmonella* by recognizing a combination of biochemical characteristics (for example, fermentative or oxidative enzymes for other carbohydrates and use of certain amino acids).

Serologic identifications of isolated organisms biochemically typical of *Salmonella* are made with specific antiserums by simple slide agglutination techniques (see chap. 6). Antiserums that detect specific antigens present in the bacterial outer membrane (O antigens) and on flagella (H antigens) are used to determine the organism's exact serotype and species name.

II. Identification of Patient's Serum Antibodies

In gastroenteritis without bloodstream invasion there is usually no antibody response and serologic tests are not helpful.

III. Specimens for Diagnosis

In gastroenteritis, multiple (usually three) stool specimens should be submitted for culture during the acute stage. When positive stool cultures are obtained, specimens should be submitted to the laboratory a week or so after the patient has returned to normal health as a check for establishment of the intestinal carrier state. Other cultures and serologic testing are not indicated unless signs of enteric fever or bacteremia develop.

Epidemiology

I. Communicability of Infection

A. *Reservoirs, Sources, and Transmission Routes* *Salmonella typhi* is found only in humans but other strains have a worldwide distribution in domestic and wild animals, some reptiles such as turtles, and poultry. The source of infection is the feces of infected animals or persons, and the environmental reservoir is usually (1) food derived from an infected animal (meats, eggs, milk and other dairy products); (2) food contaminated during storage by the feces of infected animals, especially rodents; (3) food contaminated by an infected person during its processing or preparation; (4) water contaminated by infectious sewage; (5) food obtained from contaminated water (shellfish); or (6) direct or indirect contact with persons having clinical or inapparent infections. When infection is derived from any of the first five of these sources, the result is likely to be an outbreak among persons sharing the contaminated food or water supply. The numbers of cases resulting from contact with an infected individual depend on the nature and closeness of contacts and on the food handling activities of the infected person, especially of food for family or public consumption.

B. *Incubation Period* Gastroenteritis usually begins in about 8 to 12 hours but may be delayed for 48 hours or more, depending on the numbers of organisms ingested and the degree of multiplication in the intestinal tract.

C. *Communicable Period* Communicability continues throughout active infection and for as long as salmonellae are shed in excreta. Adults generally do not become permanent intestinal carriers, but babies and young children may shed the organisms for months.

D. *Immunity* Gastrointestinal infection produces little or no immunity, and new attacks may occur after each exposure.

II. Control of Active Infection

A. *Transmission-based Precautions* Children may be isolated in private rooms and cared for in hospitals with contact precautions, but this is not necessary for cooperative adults who are cared for with standard precautions. Patients sharing toilet facilities must be instructed in hand washing techniques.

B. *Other Precautions* These include concurrent disinfection of all infectious excreta and of objects contaminated by them (bedpans, rectal thermometers, enema equipment, clothing). Toilet seats and other bathroom facilities should be disinfected frequently, especially when they are shared. Individual bedpans should be provided for bedridden patients. Gowns and gloves should be worn by all persons in direct contact with the patient and by those who must handle contaminated articles. Hand washing must be practiced scrupulously by the patient and by personnel in attendance.

These precautions should be continued until the patient's stool is free of salmonellae. This assurance is usually based on three negative stool cultures obtained on successive days during convalescence. Patients must be excluded from food handling while culture-positive or until three negative stool cultures are obtained.

C. *Treatment* Supportive and symptomatic therapy is of greatest value in *Salmonella* gastroenteritis. Antimicrobial agents are not useful and may even prolong the period of fecal excretion of the organisms.

D. *Control of Contacts* Food handlers are the particular targets of efforts to prevent spread of *Salmonella* infection. The close contacts of infected patients may be examined in a search for unreported cases or inapparent carriers.

E. *Epidemic Controls* Individual case reports in epidemics of *Salmonella* gastroenteritis guide the efforts of epidemiologists to find active cases and carriers, as well as food or water sources. Special situations may occur in hospitals, as in the case of a *Salmonella*-contaminated dye that was inadvertently administered through a gastric tube to patients undergoing studies of abdominal tumors. Another hospital outbreak was traced to the practice of serving raw eggs in milk shakes, or other forms, to nutritionally debilitated patients. In this instance, eggs marketed by local poultry farmers were the source of infection.

These incidents point to the need for continuing awareness of the many living and environmental reservoirs of *Salmonella* infection and for controlling obvious sources. When outbreaks occur, controls are tightened with exclusion of suspected foods from the market; recommendations for pasteurization, boiling, or exclusion of milk; thorough cooking of eggs, poultry, and meats; and chlorination or boiling of water, depending on the epidemiologic evidence.

III. Prevention

A. *Immunization* Bacterial vaccines do not confer protection against *Salmonella* gastroenteritis but may help prevent typhoid fever.

B. *Control of Reservoirs, Sources, and Transmission Routes* The many factors involved in prevention of salmonellosis may be summarized as follows:

1. Public water supplies must be protected, purified, and chlorinated (see chap. 10). Private water supplies must also be protected. When doubt exists concerning the safety of water for drinking or washing, it should be either boiled or disinfected with chlorine or iodine preparations. Disinfectant tablets are available for this purpose.

2. Adequate disposal of sewage is essential.

3. Meats, eggs, and milk must be processed by sanitary procedures. Infection should be eliminated at the animal source insofar as possible. Stored foods should be safeguarded from possible contamination by rodent feces.

4. Shellfish should be protected by control of sewage disposal in waters from which they are harvested.

5. Milk should be pasteurized, and meats and shellfish should be cooked when questions arise concerning their purity.

6. Eggs should be kept refrigerated until used and cooked until both yolk and white are firm before they are eaten. Pasteurized eggs are recommended for use in restaurants and in institutional settings such as nursing homes and hospitals.

7. Prepared foods should be refrigerated and cross-contamination between raw and cooked foods prevented.

8. Food handlers and methods for preparing foods in public places should be under the continuing supervision of local health departments.

9. Flies should be controlled by adequate screening and judicious use of insecticides at their breeding places or entry routes.

10. Carriers must be supervised and restricted from food handling.

11. Food handlers and the public must be continuously educated as to the sources and transmission of *Salmonella* infections.

SHIGELLOSIS (BACILLARY DYSENTERY)

The Clinical Disease

Shigellosis, or **bacillary dysentery,** is an acute intestinal disease characterized by diarrhea, abdominal cramps, and fever. The liquid stools often contain blood and mucus and are passed with difficulty because of rectal spasm (tenesmus). The large bowel is inflamed, and mucous membrane damage leads to ulceration and bleeding. The infection is usually self-limited, but when it subsides, the intestinal ulcers heal with scar tissue formation. Uncomplicated recovery is usual, although the clinical severity varies with the causative strain and the patient's age and general condition. Ordinarily the case fatality rate is less than 1%, but in epidemics in tropical countries in which living conditions and sanitation are poor, the disease may be rapidly fatal to as many as 20% of the patients.

The injury from *Shigella* infection results from microbial multiplication in the intestinal mucosa. At least one species, *Shigella dysenteriae,* also secretes a toxin that most likely plays a role in tissue destruction and may be responsible for the watery diarrhea.

Unlike salmonellae, some strains of which may invade the bloodstream and cause enteric fever, shigellae rarely enter parenteral tissues but, rather, remain in the bowel.

LABORATORY DIAGNOSIS

I. Identification of the Organism

A. *Microscopic Characteristics* Shigellae closely resemble other enteric bacteria in appearance. They are gram-negative, non-spore forming, nonmotile, facultatively anaerobic bacilli.

B. *Cultural Characteristics* Methods for isolating *Shigella* species from stool specimens are similar to those described for salmonellae. *Shigella* species are lactose nonfermenters (one species ferments it very slowly); therefore, their colonies are colorless on selective and differential agar plates. They are distinguished from *Salmonella* and other species by lack of motility, by fermentation of specific carbohydrates (most often without gas production), and by other biochemical parameters.

C. *Serologic Identification* Final identification of isolated *Shigella* species is accomplished by slide agglutination tests with specific *Shigella* antisera. The genus contains four antigenic groups: group A (*S. dysenteriae*), group B (*S. flexneri*), group C (*S. boydii*), and group D (*S. sonnei*). Some of these groups contain more than one serotype.

II. Identification of Patient's Serum Antibodies

Serologic tests are not helpful in diagnosis.

III. Specimens for Diagnosis

Feces or rectal swabs only are suitable for culture of *Shigella* species. Specimens should be received by the laboratory in the freshest possible condition because these organisms do not survive well outside the human body.

Organisms are often concentrated in flecks of mucus and blood present in diarrheic stools, and these flecks should be selected for culture. If the patient is examined by sigmoidoscopy, cultures may be taken from the ulcerative lesions visualized through the instrument.

Epidemiology

I. Communicability of Infection

A. *Reservoirs, Sources, and Transmission Routes* The reservoir for shigellae is the human intestinal tract, and infected feces are always the source. Transmission may occur by the direct fecal-oral route or indirectly through contaminated objects, hands, or mechanical vectors, but widest distribution is through contamination of water or food.

B. *Incubation Period* Usually short, within 1 to 4 days, but not more than 7 days.

C. *Communicable Period* For as long as the organisms are present in feces. The carrier state does not develop so frequently in shigellosis as it does in salmonellosis. Most convalescents do not shed the bacilli for more than 2 or 3 weeks, but occasionally, persons remain chronically infected and may suffer relapses.

D. *Immunity* Specific antibodies develop during the course of disease but are not protective against new infections. People of all ages are susceptible, but shigellosis is often more serious in children and debilitated adults. Although it has a worldwide distribution, epidemics are more common and severe in tropical and other areas in which nutrition and living conditions are poor, and sanitation is inadequate. Outbreaks often occur in institutions, particularly in orphanages, convalescent homes for children, mental hospitals, and jails.

II. Control of Active Infection

A. *Transmission-based Precautions* Standard precautions are sufficient unless the patient is diapered or incontinent. In these instances, contact precautions should be applied.

B. *Other Precautions* Concurrent disinfection of feces and contaminated objects (bedpans, toilet seats, bed linens, rectal thermometers, sigmoidoscopes, enema equipment, clothing) must be thorough. If the sewage system is adequate, feces may be disposed of directly into it; if not, fecal matter should first be disinfected chemically. Gowns should be worn by all who come in contact with the patient if soiling is likely, and scrupulous attention should be given to hand washing, both by the patient and by personnel.

C. *Treatment* Many *Shigella* strains are resistant to a number of antimicrobial drugs that were once used for treatment. For this reason, antimicrobial susceptibility testing should be done to determine the most effective agent for individual strains. Trimethoprim-sulfamethoxazole, ampicillin, or ciprofloxacin may be useful. Effective antimicrobial drugs quickly eliminate shigellae from the feces, reducing fever and diarrhea within 12 to 18 hours after treatment is begun. Antidiarrheal drugs, on the other hand, may prolong fever, diarrhea, and excretion of shigellae, so they are not used in treating shigellosis.

The most important supportive therapy is control of dehydration, with fluid and electrolyte replacement.

D. *Control of Contacts* The control of food handlers is important in shigellosis as it is in salmonellosis. If the contacts of active cases are food handlers, they should not be allowed to work until their fecal cultures are found to be negative for shigellae. A search is made among contacts for other cases of active shigellosis, atypical cases, and carriers.

E. *Epidemic Controls* All cases must be reported so that the source of infection may be found in the population or the environment. Investigations should be centered on common water, food, and milk supplies.

III. Prevention

A. *Immunization* Effective vaccines are not available.

B. The prevention of shigellosis is based on control of its human reservoir and sanitary control of its environmental sources, through chlorination of water, adequate sewage disposal, fly control, and protection of food, water, and milk from human or mechanical vectors.

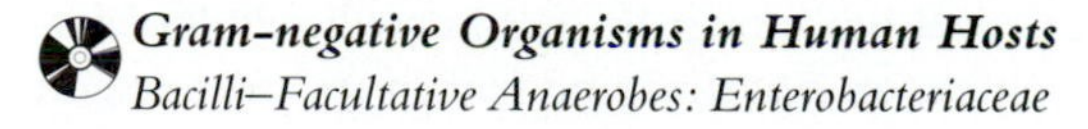

Gram-negative Organisms in Human Hosts
Bacilli–Facultative Anaerobes: Enterobacteriaceae

CAMPYLOBACTER JEJUNI

The Clinical Disease

Campylobacteria are primarily agents of zoonotic diseases, but several species also infect humans. Among these is *Campylobacter jejuni,* which may be the most common bacterial agent of diarrhea in children. In some countries, including the United States, it is isolated more frequently from patients with gastroenteritis than are *Salmonella* and *Shigella,* especially during the summer months. It has relatively strict growth requirements, and therefore, its true significance was not established until the 1970s when effective methods for its cultivation were widely used. The incubation period of ***Campylobacter*** enteritis is relatively short, approximately 2 to 4 days. The onset of illness is sudden, with abdominal pain, nausea, fever, headache, and muscle pain. Watery, bloody diarrhea may be present at onset of the disease or may not occur until several days after other symptoms appear. Intestinal biopsies show a loss of surface epithelium and infiltration of the mucosa with neutrophils, plasma cells, and lymphocytes. In some instances the tissue appearance is identical with that seen in patients with ulcerative colitis. The illness rarely lasts longer than one week, although 25% of patients may have a relapse. The organism invades the intestinal mucosa and elaborates several toxins, but the relationship of these toxins to disease production is not yet known.

LABORATORY DIAGNOSIS

I. Identification of the Organism

A. *Microscopic Characteristics* The organism is a slender, curved, gram-negative bacillus that may have one or more spirals and sometimes appears in arrangements resembling gull wings, commas, or the letter *S.* Because the organisms are present in large numbers in the stool, Gram stain or direct examination of the specimen under phase-contrast microscopy may reveal the morphologically characteristic organisms.

B. *Cultural Characteristics* Conventional stool culture methods used for isolating *Salmonella* and *Shigella* do not detect *Campylobacter.* Therefore, specimens are plated on special selective culture media containing antimicrobial agents and incubated at 42° C in an atmosphere of 5% O_2, 10% CO_2, and 85% N_2. After 24 to 72 hours, gray, flat, often watery colonies grow, containing characteristic curved, motile bacilli. When necessary, a few biochemical tests are available to differentiate *C. jejuni* from other *Campylobacter* species.

C. *Immunologic Identification* Immunologic diagnosis is possible, but the antiserums are not widely available.

II. Identification of Patient's Serum Antibodies

Patients develop specific antibodies against *C. jejuni,* but no serologic tests are available for routine use.

III. Specimens for Diagnosis

Feces or rectal swabs must be sent immediately for laboratory culture, or swabs may be sent in special transport media. The organisms will survive for several days when refrigerated in the transport medium.

Epidemiology

I. Communicability of Infection

A. *Reservoirs, Sources, and Transmission Routes* Campylobacteriosis is primarily a zoonotic infection. The organisms are commensals in the gastrointestinal tract of many animals, especially cattle, sheep, and poultry. Most humans acquire the infection by ingesting contaminated food or water. Outbreaks have been traced to unpasteurized milk, untreated surface water, and undercooked meats, especially poultry which is thought to be the source in more than 50% of cases. Cats and dogs also carry the organism and may transmit infection directly to humans. Similarly, persons exposed to farm animals are at risk for infection. Person-to-person transmission by the fecal-oral route may occur.

B. *Incubation Period* From 1 to 7 days but usually 2 to 4 days after exposure.

C. *Communicable Period* While patients are shedding organisms in their feces. This period may vary from a few days to up to a few weeks if the person is untreated.

D. *Immunity* Not well understood but specific antibodies appear to protect against disease caused by immunologically related strains.

II. Control of Active Infection

A. *Transmission-based Precautions* Patients are treated under standard precautions. They may be placed in private rooms if their hygiene is poor.

B. *Other Precautions* Gowns are worn if soiling is likely and gloves are used when touching infective material. Proper hand washing is imperative. Feces and articles soiled with feces should be treated with concurrent disinfection methods. If sewage facilities are adequate, feces may be discharged directly into sewers without disinfection. Symptomatic patients should be excluded from food handling and from close contact with potentially susceptible persons such as children in day care and hospitalized patients.

C. *Treatment Campylobacter* enteritis is usually self-limited and treated only with fluid replacement. In severe cases, erythromycin and ciprofloxacin are used. Tetracycline and gentamicin are alternative drugs.

III. Prevention

Campylobacteriosis is prevented in a manner similar to salmonellosis. All food, especially poultry, should be thoroughly cooked and efforts should be made to prevent recontamination of cooked foods. Milk should be pasteurized and water supplies chlorinated. The threat posed by farm animals and domestic pets should be realized. Puppies and kittens with diarrhea can be treated with antimicrobial agents to reduce their risk to children. No immunization is available.

Campylobacter fetus

Another *Campylobacter* species, *C. fetus,* is associated with a variety of systemic illnesses including bacteremia, meningitis, arthritis, pneumonia, and peritonitis. These infections occur most commonly in patients with underlying illness such as diabetes, malignancies, and cardiovascular disease. In these instances, the organism is considered to be an opportunistic pathogen. *C. fetus,* which grows at 35° C and requires an increased CO_2 atmosphere, can be isolated from the bloodstream, sterile tissues, or body fluids of infected patients. Gentamicin, ampicillin, or chloramphenicol (for central nervous system disease) are used for therapy.

HELICOBACTER PYLORI

Helicobacter pylori organisms were first isolated from humans in 1982. Because the isolates resembled strains known as *Campylobacter*-Like-Organisms, they were originally called *Campylobacter pylori.* However, subsequent taxonomic studies showed major genotypic and phenotypic differences between this organism and other *Campylobacter* species, which led to the establishment of a new genus, *Helicobacter,* in 1989. Currently the genus *Helicobacter* contains at least nine different species of which three are significant human pathogens. ***Helicobacter pylori*** is a major cause of human gastritis and peptic and duodenal ulcer disease, whereas *H. fennelliae* and *H. cinaedi* are causes of proctitis, enteritis, and systemic infections in homosexual men.

The Clinical Disease

Recent evidence has clearly established that *H. pylori* is an important, though not exclusive, cause of acute and chronic gastritis, peptic ulcer disease, and duodenal ulceration. People with these syndromes usually have acute upper gastrointestinal illness with nausea and upper abdominal pain. Patients may also experience fever, vomiting, heartburn, belching, epigastric pain, and diarrhea. If the illness is not properly diagnosed and treated, most people with acute gastritis go on to develop chronic infection. The symptoms of acute gastritis do not usually return but these individuals develop a chronic, diffuse superficial gastritis that usually remains asymptomatic. Importantly, chronic gastritis is a well-known risk factor for the development of gastric carcinoma and gastric lymphoma, which emphasizes the importance of properly diagnosing and treating this disease.

LABORATORY DIAGNOSIS

I. Identification of the Organism

H. pylori is a microaerophilic, helical, or curved gram-negative bacillus that measures 0.3 to 0.5 μm wide and 1.5. to 5.0 μm long. The organism produces an extremely potent enzyme, urease, which hydrolyzes urea to ammonia and carbon dioxide. The presence of this enzyme is a reliable marker for detecting the organism in tissue samples.

A. *Characteristics in Tissue* Diagnosis of *H. pylori* infection can be made by histologic examination of biopsy tissue from the antral portion of the stomach. The biopsy sample is collected through an endoscope. Sections treated with silver stains reveal the presence of *H. pylori* as black, curved, or spiral-shaped organisms adjacent to the gastric mucosa.

B. *Cultural Characteristics* Diagnosis can be made by detecting the organism's potent urease activity in the CLO (*Campylobacter*-Like-Organisms) culture test, which is performed by placing a small piece of the biopsy specimen in a plug of urea agar covered with plastic on a special slide. The organism's potent urease activity can be detected by a change in the color of the indicator in the agar from yellow (acid) to pink (alkaline) within 24 hours. Usually, however, this reaction is positive within 15 to 30 minutes and provides a rapid and reliable test to detect the organism's presence.

C. *Other Identification Method* Another test for detecting the potent urease activity of *H. pylori* is the carbon-labeled breath test. After the patients drink a solution of ^{13}C- or ^{14}C-labeled urea, their breath is sampled to detect the presence of ^{13}C- or ^{14}C-labeled carbon dioxide, a product of the breakdown of urea by urease.

II. Identification of Patient's Serum Antibodies

Enzyme immunoassay tests for establishing the serologic diagnosis of infection are also available. These tests are used to detect the presence of IgA antibody in serum or to demonstrate a significant change in IgG antibody titer in paired serum specimens.

III. Specimens for Diagnosis

Biopsy specimens from the stomach obtained by endoscopy for histologic examination and culture tests for urease production. Single or paired serum specimens for enzyme immunoassay to detect *H. pylori* antibodies.

Epidemiology

Humans are the natural reservoir for *H. pylori* and the organism can be detected in persons from all parts of the world. It can be recovered from dental plaque, saliva, and feces, which serve as convenient reservoirs of transmission to others, particularly family members. However, the major habitat of *H. pylori* is the human gastric mucosa. The organism can survive at the low pH of the stomach by producing urease, which converts urea to the alkaline ammonia and carbon dioxide, thereby neutralizing the stomach acids.

In developing countries, infection is acquired in childhood. Up to 90% of children are infected by age 5. In the United States, children do not usually become infected. However, by age 60, 50% of the population of the United States has evidence of infection. In both developing and developed countries, the incidence of disease is highest in low socioeconomic groups for whom crowding and poor sanitation are important risk factors. Currently, no vaccines are available to prevent infection.

Because the organism grows slowly and to minimize the development of antimicrobial resistance, a 14-day course of therapy with three drugs, bismuth subsalicylate, metronidazole, and tetracycline, is recommended. Cures are seen in up to 90% of infected patients, but increasing resistance of the organism has led to a search for alternative agents.

YERSINIA ENTEROCOLITICA

Yersinia enterocolitica is a gram-negative bacillus that causes severe enterocolitis in humans. The first human cases were diagnosed in New York in 1939, but before the 1960s ***Yersinia*** had been known primarily as an agent of epidemic illness in domestic and wild animals (pigs, sheep, rabbits). Improved diagnostic methods have indicated that yersiniosis is common in some areas of the world, primarily Northern Europe.

In humans, the most common form of the illness is acute enterocolitis, occurring primarily in children under 7 years of age. The symptoms of diarrhea, abdominal pain, and sometimes fever are like those caused by salmonellae and shigellae. Occasionally, the organisms infect the mesenteric lymph nodes producing symptoms suggestive of acute appendicitis. In immunocompromised patients, yersiniae may invade past the intestinal tract to produce disseminated infection, with focal abscesses in the liver, spleen, kidneys, or other organs. Invasive disease may develop in previously healthy patients as well. About 20% of patients develop erythema nodosum or arthritis following an initial bout of diarrhea and fever.

LABORATORY DIAGNOSIS

The diagnosis of *Y. enterocolitica* infections depends on isolating and identifying the organism from feces in uncomplicated enterocolitis. Surgery for suspected appendicitis may reveal mesenteric lymphadenitis, and culture of excised lymph nodes then establishes the diagnosis. In disseminated infection, cultures of blood or other body fluids and tissues may be diagnostic. *Y. enterocolitica* grows well on enriched media, but on the selective and differential media normally used for stool cultures, it grows best at room temperature rather than 35° C. Such media are generally incubated first at 35° C to rule out salmonellae or shigellae and then held at room temperature for another 24 hours. Alternatively, the media may be inoculated in duplicate with one set incubated at 35° C for 24 hours and the other set incubated at room temperature for 48 hours. Selective media specific for growth of *Y. enterocolitica* are also available. Suspected colonies of *Y. enterocolitica* are then tested for characteristic biochemical features.

The serologic diagnosis of yersiniosis involves testing the patient's serum with a battery of strain-specific antiserums. The procedure is done only in specialized laboratories.

Most *Yersinia* infections are self-limited, and therefore therapy is used only for severe cases. The antimicrobial drugs that may be used include trimethoprim-sulfamethoxazole, ciprofloxacin, gentamicin, and cefotaxime.

Y. enterocolitica appears to be widespread in nature and has been isolated from water, fruits, and vegetables as well as several animal species. In humans, primary entry from environmental or animal sources most likely is through the digestive tract. In a large, interstate outbreak of yersiniosis in the south central region of the United States, foodborne transmission was documented. The source was pasteurized milk supplied by a dairy in Memphis. Most of the 172 patients required hospitalization, and 17 underwent appendectomies before the nature of the illness was clarified. About 14% of patients had extraintestinal infections of the throat, blood, urinary tract, central nervous system, or of a wound. *Y. enterocolitica* does not usually survive pasteurization, but when large numbers are present in milk, viable organisms may persist after pasteurization and multiply at refrigerator temperature. Other well-documented foodborne outbreaks have occurred: one caused by contaminated chocolate milk, another by contaminated tofu, and a third by raw chitterlings (pork intestine). In the last case, the organism was transferred primarily to children by contact with the contaminated hands of the food handlers.

Control and prevention are similar to those for *Salmonella* infections.

INFECTIONS CAUSED BY MICROBIAL MULTIPLICATION WITH ENTEROTOXIN PRODUCTION

CHOLERA

The Clinical Disease

Cholera is an acute diarrheal disease caused by an exotoxin (enterotoxin) produced by *Vibrio cholerae.* Following their ingestion in fecally contaminated water, the ***Vibrio*** organisms colonize the small bowel, secreting enterotoxin that binds rapidly to the intestinal epithelial cells. Its effect is to activate an enzyme in mucosal cells, which in turn increases the production of a specific compound (cyclic adenosine monophosphate or cAMP) by the epithelium. The rising concentration of this compound markedly alters the permeability of the mucosa, resulting in an outpouring of fluid and salts. There is a profuse diarrhea, with "rice water" stools containing mucus and mucosal epithelial cells. Onset is sudden and is accompanied by nausea, vomiting, abdominal pain, and severe dehydration. Collapse, shock, and death may follow if the patient is not promptly and continuously rehydrated until the infection subsides. This disease may occur in explosive epidemic patterns in susceptible populations, with death rates as high as 75%. In endemic areas the fatality rate varies from 5 to 15%.

Cholera has been endemic in India and other parts of Asia for the past century and a half and has caused several major pandemics. The latest one, which began in 1961, spread throughout Asia, Africa, and Eastern Europe and entered South America in 1991 where it caused widespread illness and death. Most cases in this pandemic have been caused by a *V. cholerae* strain referred to as *El Tor.* The El Tor variant more commonly produces asymptomatic than symptomatic infection, results in a prolonged carrier state, and is better able to survive in the environment than the classical cholera organism. These characteristics promote its survival and ready passage among humans. In 1992, a large outbreak of cholera occurred in southern India followed by an epidemic in Bangladesh. This epidemic has continued into the late 1990s and involves great numbers of people in neighboring countries. A different strain of *V. cholerae,* called *V. cholerae* O139 "Bengal," has caused these recent outbreaks. This organism is genetically related to the El Tor pandemic strain but has a different antigenic composition. As a result, people living in areas of endemicity who previously have been infected with the El Tor biotype are not immune to the new strain.

In endemic areas, 4 to 30 times as many persons have asymptomatic infections as clinical disease, depending on the infecting strain. In these areas cholera is predominantly a disease of children, but when it spreads to previously uninvolved sections, the attack rates are at least as high in adults as in children.

Cholera has not been a major concern in the United States because of the high level of sanitation and hygiene. *V. cholerae* strains isolated in the United States between 1911 and 1973 were invariably acquired in other parts of the world.

Since 1973, however, more than 50 cases of nonimported infections with toxigenic *V. cholerae* have occurred in this country, primarily along the Gulf Coast in Texas and Louisiana. In one outbreak, 11 persons acquired the infection by eating inadequately cooked crabs from a Louisiana marsh. The exact source of the organism in most of the other infections is not clear. These cholera cases are not associated with the current pandemic because the isolated strains are similar to one another but differ from the pandemic strain. Cases of cholera caused by the pandemic strain began to appear in the United States in 1991. Most were in people entering the country from areas of South America where the pandemic was raging. Other cases resulted from eating contaminated shellfish (especially crabs) illegally brought into the United States from South America. Additional cholera cases are likely to occur, especially along the Gulf Coast and in travelers from South America who may have contracted the disease. Medical personnel must remain alert to this possibility and notify the laboratory of suspected cases so that appropriate procedures can be used to isolate the organism from patient specimens. In the eight-year period, 1988 to 1995, 223 cases of cholera were reported in the United States.

Cholera is strictly an intestinal disease; the organisms do not reach the bloodstream or establish systemic foci. Consequently, the disease is ordinarily self-limited. Recovery, when it occurs, is usually quite rapid, with no signs of residual damage. Convalescents may shed vibrios in feces for 2 or 3 weeks, but the carrier state does not appear to be permanent.

LABORATORY DIAGNOSIS

I. Identification of the Organism

(Vibrio cholerae). Microscopically, vibrios are small, gram-negative, aerobic, nonsporulating bacilli, curved in the shape of a comma. They have a single polar flagellum which gives them very rapid motility.

Culturally, vibrios grow on nonselective laboratory media, but in addition, a selective medium, thiosulfate-citrate-bile salt-sucrose agar (TCBS) is recommended. The organisms are recognized initially by their characteristic morphology on TCBS and their ability to grow in a medium with a strongly alkaline pH (8.5 to 9.5).

Serologically, vibrios can be identified by slide agglutination tests using specific antiserum.

II. Identification of Patient's Serum Antibodies

Specific antibodies against the vibrios and their toxins appear in the patient's serum during cholera infection. Serodiagnosis is not applied routinely for acute cholera but is used for epidemiologic purposes or retrospective diagnosis.

III. Specimens for Diagnosis

Stool specimens or rectal swabs (during the acute stages of illness) should be sent for bacteriologic culture. If the specimen will not be processed immediately, an appropriate transport medium should be used.

Epidemiology

I. Communicability of Infection

A. *Reservoirs, Sources, and Transmission Routes* Cholera is a human disease transmitted by organisms in the feces or vomitus of infected persons. Contaminated water supplies are the usual vehicle for epidemic spread. Food is not ordinarily involved in epidemics, but shellfish especially may be the source of sporadic cases in endemic areas where mild cases of cholera continue. In the latter situation, water is less important as a transmitting agent than are direct contacts of the household variety. Hands, utensils, clothing, and flies may contaminate food or carry infection directly to the mouth.

B. *Incubation Period* Usually 2 to 3 days, but in rapidly spreading cholera, many cases begin within a few hours of exposure.

C. *Communicable Period* Cholera is transmissible as long as vibrios are present in vomitus or feces, usually several days after recovery. They may, however, persist in the intestinal tract for several weeks.

D. *Immunity* Active immunity results from an attack of disease or from artificial immunization with bacterial vaccine, but the latter is less effective than previous illness in preventing subsequent disease.

Both circulating immunoglobulins and local secretory IgA are produced by the body. These antibodies act on the bacteria themselves, but antibodies to the enterotoxin are also produced. Secretory IgA developed in the small bowel is thought to function by preventing attachment of the vibrios to intestinal epithelial cells. Antitoxic antibodies most likely intercept enterotoxin before it binds to mucosal cells.

Human susceptibility to cholera is not consistent.

II. Control of Active Infection

A. *Transmission-based Precautions* Cholera patients may be cared for with standard precautions but contact precautions are required for diapered or incontinent children less than six years of age. Effective hand washing and scrupulous cleanliness are basic to the management of cholera wards in hospitals.

B. *Other Precautions* These include concurrent disinfection of feces, vomitus, and contaminated equipment and bedding. Feces must be chemically disinfected when adequate sewage systems are not available. Personnel in contact with the patients must wear clean gowns if soiling is likely and practice careful hand washing.

C. *Treatment* Adequate rehydration of the patient with restoration of electrolyte balance is critical. Tetracycline antibiotics may dramatically reduce the duration and volume of diarrhea as well as speed the elimination of vibrios from the feces. Trimethoprim-sulfamethoxazole or ciprofloxacin is useful but somewhat less effective than tetracycline for therapy.

D. *Control of Contacts* Exposed contacts should be kept under surveillance for several days. If fecal cultures are positive for vibrios, long periods of observation may be required. A search should be made for other cases of cholera and for possible sources in common water supplies or among household contacts.

A prolonged gallbladder carrier state may develop in up to 3% of convalescents, particularly adults, but the role of such carriers in transmission is not clear. Surveillance of these individuals is maintained whenever feasible.

Vaccination of contacts is not warranted, as it cannot prevent current infection within the short incubation period. Passive immunization is not available.

E. *Epidemic Controls* Case reporting is required by the International Health Regulations of WHO. Control measures just described are expanded to the entire community when epidemics threaten. When the general water supply cannot be immediately purified, drinking water must be boiled. Patients with poor hygiene should be isolated, and supervisory controls tightened on members of households in which cholera has occurred and on public eating places and food supplies.

III. Prevention

Cholera prevention depends on the continuing application of the controls outlined. Immunization has a short-term value only. In endemic areas cholera vaccine must be given repeatedly to persons who run the risk of frequent exposure. Most authorities, including WHO and the U.S. Public Health Service, do not recommend the routine use of vaccine to prevent cholera. Proof of vaccination is no longer required for entry into any country. Personal and public sanitary measures remain the best protection.

Gram-negative Organisms in Human Hosts
Bacilli–Faculative Anaerobes: Vibrionaceae

VIBRIO PARAHAEMOLYTICUS

Vibrio parahaemolyticus is a microorganism related to *V. cholerae* that has received increasing recognition as an agent of bacterial diarrhea. This vibrio was first isolated in an outbreak of food poisoning in Japan in 1951 and is the most common cause of bacterial gastroenteritis in that country. In recent years, *V. parahaemolyticus* has been documented as the cause of several outbreaks in the United States, particularly along the East Coast. Fish or shellfish are most commonly involved, including raw fish or sashimi in Japan and improperly cooked seafood (steamed crabs) in the United States. The organism's survival in frozen shrimp has also been described. Vibrio gastroenteritis is primarily a disease of the summer months when there are large numbers of vibrios in coastal seawaters and when bacterial multiplication in unrefrigerated foods is encouraged.

V. parahaemolyticus food poisoning appears to be both an infectious and a toxic disease. Symptoms usually begin about 12 hours after ingestion of the contaminated food, but the incubation period ranges from 2 to 48 hours. The disease may be clinically indistinguishable from *Salmonella* gastroenteritis, with abdominal pain, diarrhea, nausea, and vomiting. Mild fever, chills, and headache are also characteristic. The diagnosis must be established bacteriologically by isolating *V. parahaemolyticus* from the patient's feces or vomitus. These vibrios are not found in the feces of asymptomatic persons and appear to be pathogenic only for humans. They can be cultured on ordinary media to which 1 to 3% sodium chloride has been added (they are halophilic) and on TCBS agar. Enteropathogenic strains are hemolytic on blood agar.

Vibrio gastroenteritis is not a serious disease as a rule. Recovery is usually complete within 2 to 5 days and antimicrobial therapy is rarely indicated. The very low fatality rate is associated with infection in aged, debilitated patients.

CLOSTRIDIUM PERFRINGENS

Clostridium perfringens is a gram-positive, anaerobic endospore-forming bacillus responsible for about 3% of reported outbreaks of food poisoning and for about 11% of individual reported cases. There are numerous species of *Clostridium,* but only a few are involved in disease production. These pathogenic clostridia are noted for producing exotoxins that can seriously injure human tissues. *C. perfringens,* which typically is associated with gas gangrene (see chap. 23), elaborates four major toxins. It is classified into five types, A through E, on the basis of its production of these potent substances. Type A is primarily responsible for diseases in humans, and types B, C, D, and E cause a variety of diseases in animals. All of these types occur naturally in the intestinal tracts of animals, and type A can be found normally among human intestinal flora.

Clostridial food poisoning is associated with the ingestion of foods such as meat, poultry, or gravy that are contaminated with *C. perfringens.* The organism produces heat-resistant endospores that may survive the temperatures at which contaminated meat is cooked. Cooking causes anaerobic conditions by driving off dissolved oxygen. Then if the cooked food is allowed to stand at room temperature, the spores germinate and the bacilli grow rapidly. Once contaminated food is ingested, the clostridia begin to form endospores in the intestinal tract, and at this time, the enterotoxin is formed by the sporulating organisms.

Symptoms usually appear within 8 or 10 to 24 hours after eating contaminated food. The disease is mild as a rule, with crampy abdominal pain and diarrhea the only symptoms. Nausea, vomiting, and fever are seldom prominent. The illness is brief, with symptoms usually subsiding within 24 hours. Chemotherapy is not warranted, and treatment is supportive only.

Individual cases probably go undiagnosed in many instances because of the mild nature and short duration of this type of food poisoning. When outbreaks occur, the diagnosis is established by identifying the organism in appropriate anaerobic cultures of the incriminated food. Fecal cultures may be useful if *C. perfringens* is isolated from affected persons in higher-than-normal numbers and is of the same serotype as isolates from the implicated food.

Clostridial food poisoning can be prevented by avoiding long periods in which cooked foods are held at moderate temperatures. If foods must stand for several hours before they are served, they should be kept at very warm or very cool temperatures (above 60° C or below 5° C).

BACILLUS CEREUS

Bacillus cereus is another gram-positive, endospore-forming bacillus, but unlike the clostridia, it is an aerobic organism. It has been incriminated in several outbreaks of food poisoning. The incubation period is about 10 hours, after which symptoms of abdominal pain, profuse watery diarrhea with rectal spasm (tenesmus), and nausea occur. Fever is rare, and the illness persists for only about 12 hours. Experimental evidence indicates that large numbers of the organism must be ingested to produce disease. *B. cereus* produces at least two enterotoxins, one of which behaves in a manner similar to cholera toxin.

Foods most likely to be contaminated with *B. cereus* include grain and vegetable dishes. These should not be allowed to remain at room temperature for long periods after cooking, as endospores may germinate and the bacilli multiply rapidly.

CLOSTRIDIUM DIFFICILE

Patients receiving antimicrobial agents often develop mild to severe diarrhea with or without intestinal pathologic changes. During the 1950s, *Staphylococcus aureus* was thought to be responsible for this phenomenon, even though staphylococci were not always found in the patients' stool specimens. In the late 1970s, it was discovered that the primary pathogen in this syndrome is the gram-positive, anaerobic endospore-forming bacillus, ***Clostridium difficile,*** which produces toxins responsible for the disease (colorplate 12b).

Administration of almost any antimicrobial agent may result in this syndrome. Presumably, the mechanism by which these agents incite disease is to suppress the growth of normal intestinal flora thus allowing *C. difficile,* which tends to be more resistant to antimicrobial agents, to flourish. Alternatively, the normal intestinal flora may somehow prevent the production of, or inactivate, *C. difficile* toxins. Typically, patients have diarrhea, fever, and abdominal pain. In severe cases, sigmoidoscopy reveals yellow-white, raised, exudative plaques or pseudomembranes; hence, the disease is sometimes referred to as *pseudomembranous colitis* (colorplate 12a). Not all patients develop pseudomembranes, however, in which case the disease may be referred to as *antibiotic-associated colitis.* Two toxins, A and B, are produced by *C. difficile.* Both are found in disease, but toxin A is thought to be responsible for the most severe symptoms. Toxin A is an enterotoxin and toxin B is a cytotoxin.

LABORATORY DIAGNOSIS

C. difficile has been isolated from the stools of approximately 4% of healthy individuals, but from 90% of those of individuals with pseudomembranous colitis. To isolate the organism in culture, selective media containing antimicrobial agents to inhibit normal stool flora are needed. Isolation of the organism alone does not confirm the diagnosis; instead, assays for *C. difficile* toxin(s) must be performed. In one assay, an extract of the stool specimen is inoculated into a cell culture that is then observed for up to 48 hours for characteristic cell damage. The assay, which detects toxin B, is confirmed as positive if the same stool extract mixed with a specific antitoxin (for neutralizing the toxin) displays no cytotoxicity. An enzyme immunoassay is available to detect toxin A in stool culture extracts within 4 hours of receiving the specimen, and a latex agglutination assay detects a *C. difficile* protein antigen in stool within 3 to 10 minutes. Positive tests for either toxin A or toxin B are thought to be the best indicators of disease.

In patients suspected of having *C. difficile* colitis, antimicrobial therapy is immediately stopped. Many persons require no further treatment. For severe cases, however, metronidazole or vancomycin are administered to eradicate *C. difficile* from the intestinal tract.

The appearance of clusters of hospitalized patients with *C. difficile* colitis led to the discovery that large numbers of *C. difficile* endospores may contaminate the environment of individuals ill with this disease. The organism was isolated from toilets, bedpan hoppers, dustpans, used bedding, washbasins, bathroom floors, and other areas in proximity to the patient. Most important, it was isolated from the hands and stools of asymptomatic hospital personnel, suggesting that they may have been involved in disease transmission. Strict attention to contact precautions, scrupulous cleaning of potentially contaminated objects, and careful hand washing may prevent spread of the organism among hospitalized persons who are more susceptible to infection.

ENTERIC INFECTIONS CAUSED BY *ESCHERICHIA COLI*

Certain strains of ***Escherichia coli*** have long been known to cause severe or lethal diarrhea in newborn domestic animals and in human infants. More recently, this intriguing microorganism, once thought to be simply a harmless commensal in the intestinal tract, has been shown to cause serious enteric disease in adults as well. Enteropathogenic strains are distinct from *E. coli* strains causing infections of parenteral tissues, particularly of the urinary tract (see chap. 21).

Different strains of *E. coli* produce at least five types of diarrheal disease: (1) enterotoxigenic *E. coli* (ETEC), largely responsible for traveler's diarrhea (turista), elaborate several toxins, at least one of which acts in the same manner as cholera toxin; (2) enteroinvasive *E. coli* (EIEC) invade the surface layers of the intestinal tract and cause disease similar to shigellosis; (3) enteropathogenic *E. coli* (EPEC) cause epidemic diarrhea in hospital nurseries; and (4) enterohemorrhagic *E. coli* (EHEC) elaborate a toxin similar to that produced by one species of *Shigella* (Shiga toxin), which stimulates production of a watery, bloody diarrhea, and (5) enteroaggregative *E. coli* (EAggEC) cause a watery diarrhea in children and adults characterized by vomiting, dehydration, and occasionally, fever, abdominal pain, and bloody stools. Factors responsible for the different ways by which *E. coli* strains produce these diseases are not completely known. In most instances, the bacteria harbor plasmids (extrachromosomal DNA fragments, see chap. 3) that contain the genetic instructions for microbial attachment to intestinal epithelial cells and for subsequent invasion and/or elaboration of toxins. Some strains also have characteristic O (outer membrane somatic antigens) and H (flagella) antigens that can be detected by serologic studies. For example, EHEC almost always have the O157 and H7 antigens.

ENTEROTOXIGENIC *E. COLI*

The Clinical Disease

Approximately 50% of cases of travelers' diarrhea have been attributed to ETEC. The disease most often develops in persons who visit underdeveloped or developing countries and who do not take proper precautions with regard to food and drink. The onset of disease is usually within 1 or 2 weeks after arrival in the country. The patient develops abdominal pain, malaise, loss of appetite, and sudden onset of watery diarrhea. Although ETEC toxin resembles cholera toxin, the illness is much less severe. Nausea, vomiting, and low-grade fever occur in one-half to one-third of patients. Symptoms last from 3 to 5 days.

LABORATORY DIAGNOSIS

Diagnosis is based primarily on the history of travel and on clinical findings. Immunoassay and DNA probe tests that detect the specific *E. coli* toxins are available in research laboratories, but these require the tedious isolation of virulent strains from among normal flora *E. coli*.

Control and Prevention

Contaminated food and water are the sources of infection, therefore travelers should eat only well-cooked foods and water treated by chemicals or boiling. Even bottled water may not be safe to drink. Prophylactic administration of antimicrobial agents is discouraged because of the risk of developing disease with an antimicrobial-resistant organism. Treatment is primarily with fluid and electrolyte replacement, although drugs such as trimethoprim-sulfamethoxazole and ciprofloxacin may shorten the period of illness. Known cases are cared for with standard precautions. Persons living in areas where ETEC are prevalent apparently develop immunity to these strains. Vaccines are under development.

ENTEROINVASIVE *E. COLI*

The Clinical Disease

EIEC invade tissue and therefore provoke a marked inflammatory response. Patients have severe abdominal cramps, malaise, fever, and bloody stools, a clinical picture similar to shigellosis. The strains are uncommon in the United States but are endemic in developing countries where they are responsible for approximately 1 to 5% of diarrheal illness.

LABORATORY DIAGNOSIS

Bloody diarrhea in the absence of other enteric pathogens (*Shigella, Entamoeba histolytica*) may provide a clue to the presence of EIEC. DNA probes that detect the plasmid responsible for tissue invasiveness and an animal model for invasiveness are available in specialized laboratories.

Control and Prevention

Control and prevention are similar to those described for ETEC. Food, dairy products, and water have been reported as sources of infection by EIEC. In severe cases of dysentery, antimicrobial agents such as ampicillin may be given to lessen the duration of illness.

ENTEROPATHOGENIC *E. COLI*

The Clinical Disease

EPEC are classically associated with epidemic infant diarrhea seen in hospital nurseries and in community epidemics, especially during the summer. The epidemic can spread swiftly from one baby to another. EPEC strains attach to intestinal epithelial cells and cause disruption of the cells' surface microvilli. Clinically, the result is severe, watery diarrhea with dehydration and shock. This disease was prevalent throughout the world during the 1940s and 1950s but since the 1960s has occurred primarily in developing countries, including South America and parts of Africa and Asia. It is uncommon in North America and Europe.

LABORATORY DIAGNOSIS

Stool specimens from infants with diarrhea are cultured on selective media that allow detection of typical *E. coli* colonies. These colonies are then tested in agglutination tests with a battery of antiserums against the O antigens common to EPEC strains. When a recognized group is found, it is reported by the numbers assigned to the particular antigen (e.g., O55, O111, O119). An EIA is also available for detecting EPEC. Serologic tests are not widely available.

Prevention and Control

Infants acquire the infection from contaminated infant formula and weaning food or by transmission on contaminated hands of hospital personnel. In hospitals, contact precautions must be adhered to, with special attention to scrupulous hand washing. Ill infants may need to be cared for together, but separated from healthy infants in private rooms (cohorted). Individual equipment and bathing facilities should be provided for each infant. Infant formula must be prepared and handled as aseptically as possible. Breast-fed babies are less likely to develop infection because maternal milk provides protective maternal antibodies. Successful therapy depends on the prompt replacement of lost fluids and electrolytes. When necessary, antimicrobial therapy is guided by the susceptibility pattern of isolated organisms.

ENTEROHEMORRHAGIC *E. COLI*

The Clinical Disease

The association of EHEC with hemorrhagic colitis resulted from the investigation of a multistate outbreak of disease in the United States in 1983. Illness was associated with ingestion of hamburgers at several outlets of a well-known fast-food chain. A specific *E. coli* serotype, O157:H7, was isolated from about half of the patients, but was found in no healthy controls. Subsequently, other investigations of patients with bloody diarrhea revealed that the organism commonly causes this syndrome. Patients with the disease initially experience abdominal cramps and watery diarrhea, followed by a bloody intestinal discharge. The diarrheal fluid contains no inflammatory cells nor do the patients have fever. In some patients, especially children, diarrhea is followed by a syndrome referred to as *hemolytic uremic syndrome* (*HUS*). Patients with HUS have acute renal failure, decreased platelets, and a hemolytic anemia. Modern therapies such as renal dialysis have reduced fatalities caused by this disease from 50% to only 10%. In a 1993 EHEC outbreak, again associated with undercooked hamburger served by restaurants of a different fast-food chain, at least two children died of HUS. EHEC produce colitis by elaborating two Shiga-like toxins, so called because they are almost identical to the toxins produced by one species of *Shigella*. These EHEC toxins are called Shiga-like toxin 1 (also known as verotoxin 1) and Shiga-like toxin 2 (verotoxin 2).

LABORATORY DIAGNOSIS

Unlike most other *E. coli* strains, O157:H7 EHEC do not ferment sorbitol, therefore a selective medium with sorbitol (sorbitol-MacConkey agar) is used to distinguish these strains from others. Because other *E. coli* serotypes may produce hemorrhagic colitis and HUS, although less commonly, tests for detecting the Shiga-like toxins in stool filtrates have been developed. These are not widely used except in specialized laboratories. The tests are similar to the immunologic and cytotoxic assays used to detect *C. difficile* toxins A and B, respectively (see previous section, *C. difficile,* Laboratory Diagnosis). DNA probes to detect genes for toxin production are in development in research laboratories. Latex agglutination tests are also available to screen sorbitol-negative isolates of *E. coli* for the presence of the O157 and H7 antigens.

Control and Prevention

The reservoir of infection is in animals, particularly cattle, deer, and sheep, although evidence from some outbreaks indicates that disease may be transmitted by direct contact among humans. Poorly cooked beef, raw milk, and unpasteurized apple juice have been implicated as the source of disease, therefore these should be avoided. This disease can be easily prevented by cooking hamburgers until the interior is no longer pink and the juices run clear. Patients are treated under standard or contact (if incontinent) precautions with rehydration and administration of electrolytes.

ENTEROAGGREGATIVE *E. COLI*

The Clinical Disease

Enteroaggregative *E. coli* strains, also known as enteroadherent *E. coli,* are an important cause of travelers' diarrhea in developing countries, particularly Mexico and North Africa. Persons infected with EAggEC develop an intestinal illness characterized by watery diarrhea, vomiting, and dehydration. In addition, some patients may have fever, abdominal pain, and bloody stools. EAggEC can also cause a chronic type of diarrhea that has been reported in many parts of the world.

LABORATORY DIAGNOSIS

The diagnosis of EAggEC is achieved by testing *E. coli* isolates recovered from patients' stool specimens for their ability to adhere in an aggregative way to specialized cells, Hep-2 cells, in culture. A DNA probe test is also available for detecting specific genes unique to this organism. Both tests are available only at special reference or research laboratories.

Prevention and Control

The risk factors for acquiring EAggEC infection are unknown. However, by following the recommended control measures for ETEC, the risk of acquiring diarrhea caused by EAggEC will likely be reduced.

DISEASES CAUSED BY INGESTION OF PREFORMED BACTERIAL TOXINS

Foodborne disease caused by preformed bacterial toxins falls into two clinical patterns: acute gastroenteritis, as induced by staphylococcal enterotoxin, and the systemic intoxication typical of botulism.

STAPHYLOCOCCAL ENTEROTOXIN POISONING

Acute food poisoning due to **staphylococcal enterotoxin** is one of the principal forms of bacterial foodborne disease in the United States. This illness is an intoxication induced by preformed toxin produced by staphylococci multiplying actively in food before it is consumed. Symptoms begin within 1 to 6 hours, usually about 2 to 4 hours, after the contaminated food is eaten. The onset is abrupt and often violent, with nausea and vomiting, cramps, and diarrhea. The patient may be prostrate, with a depressed temperature and blood pressure. Severe as these symptoms are, they pass quickly, usually within a day or so, as the toxin is eliminated and the injured intestinal mucosa recovers.

The **diagnosis** of staphylococcal intoxication can be made tentatively from clinical symptoms. *Staphylococcus aureus* may be isolated in significantly large numbers from the suspected food if it is still available for testing. The presence of many staphylococci in vomitus or feces is also significant. When staphylococci are isolated from food or gastrointestinal contents, they may be tested for enterotoxin production or typed with bacteriophages (bacterial viruses, see chap. 2).

The source of staphylococcal contamination of food is usually a person with an infected lesion on hands, arms, or face, although these organisms may also reside on healthy skin and respiratory mucosa. A great variety of foods may support growth of staphylococci, but those most often involved are custards and pastries made with cream, salads and salad dressings, and meats that have been sliced or ground. When staphylococci are implanted in foods they require a few hours to grow and elaborate toxin. Growth and multiplication occur most rapidly in warm rooms. Cooking kills the organisms, but if toxin has already been formed, it may survive, for it is stable to heat. Refrigeration does not kill the organisms or destroy the toxin but inhibits microbial growth and thus prevents enterotoxin formation. For this reason, meats and foods prepared in advance of meals must be kept refrigerated until they are eaten or cooked.

Control of staphylococcal food poisoning depends on the recognition of cases, the food source, and the infected human reservoir. Bacteriologic cultures are made from suspected food, from lesions or from the nose of handlers, and from patients' vomitus. Strains of staphylococci isolated from these sources are then compared as to phage type, enterotoxin production, and other properties to establish their relationship to the case and its source. Control of food handlers includes education in personal hygiene and supervision in maintaining sanitary techniques for handling, preparing, and refrigerating foods. Persons with obvious skin infections should not be permitted to handle foods until the condition has been treated and cured. In many areas the public health laws require refrigeration of foods prepared for public consumption.

BOTULISM

Botulism is an acute and often fatal disease that results from ingesting a toxin preformed in foods by the anaerobe ***Clostridium botulinum.*** Unlike other forms of **bacterial food poisoning,** botulism is a systemic intoxication. The powerful toxin is absorbed from the intestinal tract and causes motor paralysis of cranial and peripheral nerves, effects that are sometimes irreversible and fatal. Nausea, vomiting, and diarrhea may occur early in the disease, although the first signs are usually related to neurotoxin activity: dizziness, difficulty in swallowing, and ocular disturbances (double vision). Neuromuscular involvement of the intestinal wall may lead to severe constipation; death usually results from respiratory paralysis. An incubation period of 12 to 36 hours, rarely longer, reflects the time required for the toxin to be absorbed from the intestinal tract, disseminated through the body, and attached to peripheral nerves. The amount of toxin ingested may also affect the length of the incubation period, although this is a very powerful substance, capable of exerting effects in very small doses. Patients who recover do so slowly, as the toxin is inactivated and eliminated and the inhibitory effects at the myoneural junctions slowly regress. They may require artificial respiration for some time. Immunotherapy with botulinal antitoxin is sometimes lifesaving.

The **diagnosis** of botulism can often be made on clinical grounds but requires bacteriologic confirmation and identification of its source. The organism may be isolated from the suspected food if it is available for testing, but the best proof lies in demonstrating (1) that the food is toxic for mice and (2) that this toxicity can be specifically neutralized by botulinum antitoxin. Patients' sera may also be tested for toxin in this way. When canned foods are involved, related lots or batches should be tested.

PERSPECTIVE 16.1 *The Case of the Deadly Soup*[a]

Mr. Roger C., a 60-year-old executive, was admitted to a New York City hospital in acute distress with neurological symptoms including dizziness, difficulty swallowing, and double vision. He had been well and apparently had no underlying organic disease. In spite of all immediate medical efforts, his condition worsened rapidly, and he was placed on a respirator as he became unable to breathe. He died within 10 hours of admission.

At the same time, his wife was suffering nausea and vomiting, and she too had some positive neurological findings, but her symptoms were milder. The family physician, suspicious of the symptoms, immediately questioned the wife about any food they might both have eaten within the last few days.

Mrs. C. reported that they had both recently enjoyed a new brand of canned gourmet soup. This soup was vichysoisse, a French potato soup usually served cold. Mrs. C. had merely chilled the can before opening it, and her husband had eaten a much larger portion than she because the soup was one of his favorites.

Suspecting botulism, whose symptoms reflect the action of a bacterial toxin on the central nervous system, the physician notified public health officials of the cases. Inspectors promptly went to Mrs. C's house and recovered samples of the vichysoisse from the can, not yet discarded. These samples were delivered to the public health laboratory, where routine procedures to prove botulism were followed. Laboratory mice were injected. Those that received a sample of the soup, but no treatment, died. Mice given soup and injected with botulinum antitoxin survived. On the basis of these conclusive findings, Mrs. C. was given a botulinum antitoxin, and she recovered.

Other inspectors from the local health department then visited the canning factory where the vichysoisse was produced. There they found that the processing methods could allow the production of botulinum toxin in the cans. The factory was closed down.

Had Mrs. C. heated the soup for 10 minutes, the heat-sensitive toxin would have been destroyed and Mr. C. would be alive. But vichysoisse is intended to be served cold and heating destroys the flavor.

Without question, the fault belongs to the canning factory that failed to take the precautions necessary to ensure that the endospores of *Clostridium botulinum* (an anaerobe) could not survive and germinate under the anaerobic conditions of the sealed can. The result was production of the deadly toxin (Perspective fig. 16.1).

Poor commercial and home-canning techniques have long been incriminated as sources of botulism, but better public understanding has decreased the number of cases such as these. Nonetheless, sporadic cases continue to appear, so physicians, nurses, and public health workers must remain alert to its symptoms and possible sources.

[a]Based on an actual case history.

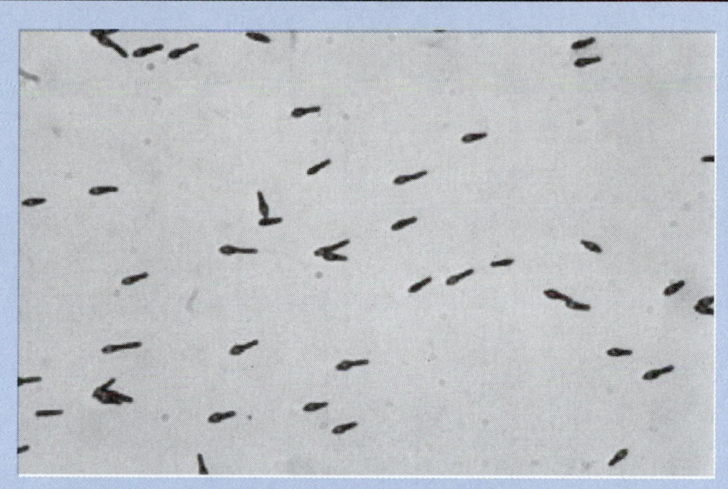

The organism Clostridium botulinum *produces a deadly toxin under the anaerobic conditions that occur in cans and jars that are inadequately processed to kill the endospores.*
Centers for Disease Control, Atlanta, GA.

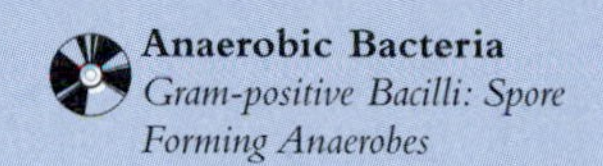

Anaerobic Bacteria
Gram-positive Bacilli: Spore Forming Anaerobes

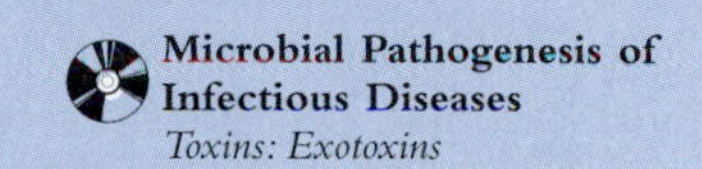

Microbial Pathogenesis of Infectious Diseases
Toxins: Exotoxins

C. botulinum is an anaerobic, endospore-forming, motile, gram-positive bacillus, normally found in soil and the intestinal tracts of animals and fish. Its endospores may contaminate vegetables and other foods. When these are preserved by canning techniques, endospores may survive inadequate processing and germinate later under the anaerobic conditions provided in a can or jar. The vegetative bacilli then proliferate and elaborate toxin. Canned foods are frequently eaten with little or no subsequent cooking, and under these circumstances active toxin is ingested (see Perspective 16.1). Boiling foods for at least 10 minutes will inactivate the toxin. The chief danger lies in the fact that these foods may display little or no change in appearance or odor related to the presence of toxin so that the need for boiling or discarding them may not be apparent. The foods most commonly involved are home-canned vegetables, olives, mushrooms, corn, spinach, and string beans. Commercial canning processes have largely overcome the problem, but outbreaks have been related to commercially packed tuna fish, smoked fish, and potato soup.

Control of botulism hinges on safe measures for processing foods (especially those likely to harbor *C. botulinum*), prompt recognition of the disease and its source, and recall from the market of suspected foods. Cases must be reported so that a search for the source can be made without delay. Persons known to have eaten the incriminated food may be treated prophylactically with botulinal antitoxin. This procedure is most effective if antitoxin reaches the bloodstream before the toxin does so. There are seven types of *C. botulinum,* each with its own distinct antigenic type of toxin (A through G). Types A, B, and E are most prevalent in human disease. Prophylactic antiserum must contain antibodies specific to the type of toxin involved in the disease, if this can be determined; otherwise polyvalent antiserum containing several type-specific antibodies must be used.

Although government controls provide for inspection of commercially canned foods, unforeseen problems arise occasionally. Cooks should use safe methods to can foods, as endospores of *C. botulinum* may survive long periods of boiling, especially in neutral or alkaline foods. Unless the food is acidic, it cannot be safely processed except in a pressure cooker. If there is the slightest reason for doubt, home-canned nonacid foods should be boiled for at least 10 minutes before they are served or even tasted. The contents of swollen or damaged cans (fig. 16.1) should never be eaten.

Figure 16.2 illustrates the reported cases of foodborne botulism between 1975 and 1995 in the United States. Except in outbreak situations almost all of these were related to home-canned or home-processed food.

Occasionally, wounds may become contaminated with soil that contains endospores of *C. botulinum.* If the injury provides suitable anaerobic conditions, the endospores germinate, and the vegetative bacilli multiply and produce neurotoxin. The resulting disease is indistinguishable from foodborne botulism but is often unrecognized because there is no link to contaminated food.

Table 16.1 shows some of the important distinctions between bacterial food poisoning caused by the ingestion of infectious microorganisms and that resulting from the consumption of preformed bacterial toxins.

FIGURE 16.1 This can's swelling is due to the activities of microbial contaminants within. Its contents should not be eaten.

U.S. Food & Drug Administration, Washington, D.C.

Infant Botulism

Some evidence has indicated that botulism is responsible for certain cases of sudden infant death syndrome (SIDS), sometimes referred to as "crib death." SIDS has been a baffling event, most likely to occur in infants up to 3 months of age. Because the typical victims have been apparently healthy babies and have often been found dead in their cribs, many of these deaths were formerly attributed to accidental suffocation in bedclothes. This has been shown to be seldom, if ever, the true cause of death, but the basic etiology has remained undocumented. A number of cases of infant botulism have been recognized, however, and a small percentage of these have died. The age pattern matches that of SIDS, as do the symptoms of respiratory paralysis. It has been postulated that endospores of *C. botulinum* reaching the digestive tracts of infants may germinate there. This does not happen in the adult bowel, with its teeming microbial flora, but conditions in the infant may permit it. Infants may then be highly susceptible to minute quantities of botulinal toxin produced by the multiplying bacilli. *C. botulinum* organisms and toxin have been found in the stools of some infants with SIDS. Presently, honey is the only food that has been implicated in this syndrome. *C. botulinum* type B was isolated both from infants with the disease and from honey fed to them, but not in all cases. Botulinal endospores also have been isolated from vacuum cleaner dust and from frontyard and backyard soil samples of several patients' homes, thus implicating environmental sources as well. Other etiologies, as yet not well-defined, are thought to account for most cases of SIDS, however.

INFECTIOUS SYSTEMIC DISEASES

TYPHOID AND PARATYPHOID FEVERS AND SALMONELLA BACTEREMIA

Typhoid Fever

Typhoid fever is a systemic bacterial disease of humans caused by *Salmonella typhi.* Like other salmonelloses, this infection is transmitted by the fecal-oral route via contaminated food and water. Unlike most others, however, it does not lead to an acute gastroenteritis but rather to the involvement of parenteral sites.

When *S. typhi* is ingested in contaminated food or water, it first localizes in the deep lymphoid tissue of the intestinal wall (Peyer's patches). The lymph flow carries the organisms into the thoracic duct, and from there they reach

TABLE 16.1 SUMMARY OF BACTERIAL FOODBORNE DISEASES

Disease	Agent	Incubation Period	Communicable Person to Person	Symptoms	Control
Infectious gastroenteritis	*Salmonella* spp.	8–48 hours	Yes	Fever, N, V, D,[a] for several days	Prevent contamination of food, water; control carriers, food handlers; cook and store foods at proper temperatures
	Shigella spp.		Yes		
	C. jejuni		Yes		
	Y. enterocolitica		Yes		
	Vibrio spp.		No		
	C. perfringens		No		
	B. cereus[b]		No	N, V, D, for several hours	
	E. coli		Depends on infecting strain	Depends on infecting stain	
Staphylococcal intoxication	Preformed enterotoxin of *S. aureus*	1–6 hours	No	Acute abdominal cramps, N, V, D 1 or 2 days; subnormal temperature	Prevent contamination and multiplication of staph in food; control food handlers
Botulism	Preformed neurotoxin of *C. botulinum*	12–36 hours	No	Central nervous system intoxication; double vision; flaccid paralysis	Prevent contamination and germination of endospores in canned foods; reheat canned foods before serving (boil 10 min); antitoxin for all who have eaten suspected food

[a]N, V, D = nausea, vomiting, diarrhea. All or some of these symptoms may be present in each infection.
[b]Preformed toxin involved as well as active infection.

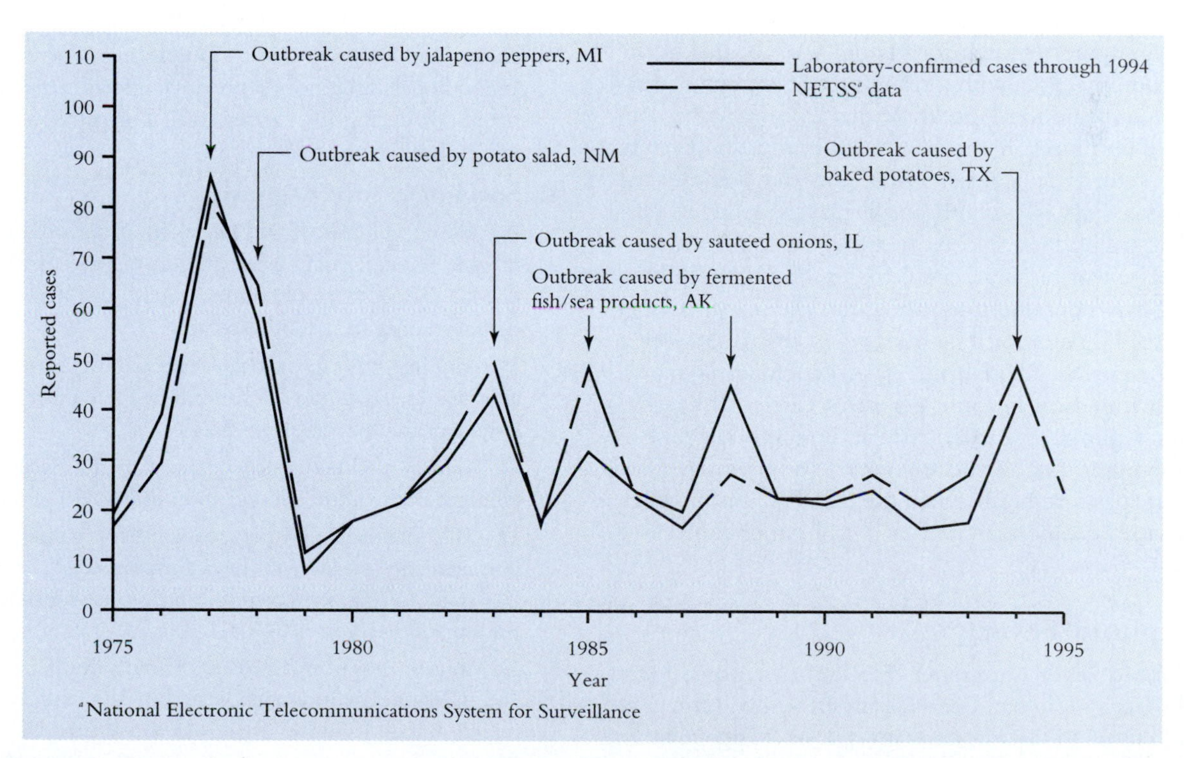

FIGURE 16.2 Botulism (foodborne) cases in the United States from 1975 to 1995. Although cases of foodborne botulism remain uncommon, outbreaks (including restaurant-associated ones) continue to occur and because of the severity of the illness, constitute public health emergencies

Centers for Disease Control and Prevention *MMWR Summary of Notifiable Diseases, United States, 1995* (October 25, 1996, Vol. 2. 44: 53).

the bloodstream. They may be widely disseminated but usually localize in the liver and gallbladder. The kidneys may also be infected, as well as the spleen, bone marrow, or lungs. In the liver and kidneys there is focal necrosis of parenchymal cells at the site of colonization, and lymphoid tissue becomes enlarged and more cellular. The gallbladder is always infected and is a major source of organisms that appear in the stool during the course of the disease and often continuing into convalescence. Organisms in the lungs produce a bronchitis or pneumonitis. The cellular response is mononuclear rather than polymorphonuclear, and the total number of white blood cells in the circulating blood is often depressed (leukopenia).

The symptoms of typhoid fever include a gradual, insidious onset, with fever rising in steps to an average height of 104° F (40° C), with a relatively slow pulse. There is often abdominal tenderness, but constipation is more frequent than diarrhea. Rose spots often appear on the trunk during the first or second week, together with symptoms of pulmonary involvement. In severe cases, the patient may become delirious or stuporous (the Greek word for stupor is *typhos*).

The organisms may be recovered from the blood during the first week to 10 days when bacteremia and dissemination occur. Stool cultures may be positive from the onset, although the greatest numbers of organisms are found after 2 or more weeks. They may remain positive long after the convalescence. Urine cultures may yield the organisms in the second and third weeks and continue to do so for long periods thereafter. The organisms may be found in bone marrow, bile drainage, or sputum if localized in the bones, gallbladder, or lungs.

The immune response to typhoid fever begins in the early days of infection, with a rising titer of antibodies during the second and third weeks. At this time, organisms can no longer be found in the bloodstream; the temperature gradually returns to normal levels; and the convalescent stage begins. Foci of typhoid bacilli may nonetheless persist in the gallbladder, spleen, or kidney within mononuclear cells, out of reach of circulating antibody. From such sites they may continue to be shed into the intestinal tract (by way of the common bile duct when the gallbladder is involved) or excreted in urine. Patients remain asymptomatic, but they may become persistent carriers. In recovery from typhoid fever, the cellular immune response is more important than circulating antibody. Intracellular organisms are destroyed only when the macrophages in which they survive become activated by T cell lymphokines (see chap. 6).

Paratyphoid Fever

Paratyphoid fever is a milder disease than typhoid fever. It usually has a more rapid onset, with fever, involvement of lymphoid tissue in the intestine, and spleen enlargement. In more severe cases there may be rose spots and other symptoms similar to those of typhoid. The duration of symptoms may be from 1 to 3 weeks. *S. paratyphi* A and B, or closely related serotypes, are the causative organisms.

Salmonella Bacteremia

Salmonella gastroenteritis is sometimes followed by bloodstream invasion (see earlier section on salmonellosis). Although many *Salmonella* strains can invade in this way, some species (e.g., *S. choleraesuis* and *S. enteritidis*) are more likely to be associated with severe illness. The organisms spread to various body sites and can produce perineal and pelvic abscesses and, rarely, meningitis, endocarditis, pneumonia, pyelonephritis, arthritis, or osteomyelitis. Such infections with salmonellae are often associated with other chronic disease or defective host defenses. Patients with sickle cell anemia, for example, appear to be more susceptible, as do those with metastatic cancer. Peripheral vascular grafts and foreign bodies are also subject to colonization by salmonellae, presumably from a low-level source of infection in the bowel, with periodic subclinical bacteremia.

LABORATORY DIAGNOSIS

I. Identification of the Organism

See previous section, Salmonellosis.

II. Identification of Patient's Serum Antibodies

The Widal test may be useful in demonstrating antibody titers during systemic *Salmonella* infection but is useful in only about 50% of patients. This is an agglutination method using known *Salmonella* O (somatic) and H (flagellar) antigens. A rising titer of agglutinins occurring from the onset through the course of the disease signifies active infection. A finding of typhoid or paratyphoid agglutinins in only one serum sample may not distinguish antibody persisting from a previous infection, from vaccination, or during the carrier state. A negative result does not exclude the disease.

III. Specimens for Diagnosis

A. Blood cultures should be taken at periodic intervals (two to three per day) especially during the first week of disease. Bone marrow cultures may also be useful in demonstrating bacteremic spread.

B. Multiple stool specimens (collected on consecutive or alternate days) should be submitted for culture, particularly after the first week of disease.

C. Urine specimens may be positive for *Salmonella* by culture, usually in the second and third weeks of infection.

D. Bile drainage may be collected for culture, during convalescence or later, to check for persistence of the carrier state and to demonstrate biliary tract localization of the organisms.

E. Sputum, pus from abscesses, synovial fluid, spinal fluid, and similar collections as indicated by the clinical signs of infection may yield the organisms in culture.

F. An acute serum taken during the first week of disease, as close to onset as possible, may be submitted for the

PERSPECTIVE 16.2 *Typhoid Fever: An Early Explanation of Its Transmission*

In 1853, an outbreak of typhoid fever occurred in a small town in Wales (Perspective fig. 16.2). A few months later, an investigation by English physician William Budd (1811–1880) proved beyond doubt the source and route of transmission of this small but deadly epidemic. His account of his findings and deductions is a masterpiece of scientific thinking:

In the month of November, 1853, it being Cowbridge Race week, two balls were held on two alternate nights at the principal hotel of that little Welch town.

These balls were attended by about 140 persons, the greater number from different parts of Wales, but some from Gloucester, Somerset and other distant counties.

Almost immediately afterwards, a number of these persons were seized with typhoid fever, and as many as eight died of it. Among the sufferers there was a considerable proportion who had never been in one another's company, except in the Cowbridge ballroom. It is clear, therefore, that they had in some way contracted the fever there, and that the typhoid poison was present at this hotel in no common degree of force and intensity. It is not recorded that fever was specially prevalent at the time in the neighbourhood; and, with the exception of one or two persons who lived in the house, those who attended the balls appear to have been alone attacked.

The small town of Cowbridge, Wales, was the site of a typhoid fever outbreak in 1853.

An occurrence so painfully striking, and in all ways so remarkable as this, naturally attracted great attention at the time, and an inquiry was held, with a view to discover the cause of the calamity. The only sanitary defect elicited by this inquiry, in explanation of so terrible an outbreak, consisted in the fact 'that the supper-room was merely a temporary transformation of a loft over a seven-stalled stable,' and that the passage between it and the ball-room was partly built over a large tank which collected the water from the roof of the house.

About fifteen months after this outbreak occurred I was called to Cowbridge to a case of typhoid fever which had come down by direct lineal succession from one of the original sufferers; and I took the opportunity of ascertaining as well as I could, both from the people of the hotel and from the medical man consulted on the occasion, the leading circumstances of it.

First in order came the all-important fact, which had not been disclosed to the gentleman who originally reported on the outbreak, that there had been a case of typhoid fever in the hotel immediately before the balls were held.

The disease occurred in the person of a gentleman visiting the hotel, and who was laid up there for some time with it. A day or two before the balls, although not yet fully convalescent, he left the house on account of the approaching festivities.

As none of the ball-goers had been in the presence of the sick man, it was obvious that they did not contract the fever from direct personal infection. There was no reason to believe that the infection was communicated through the air, as no offensive smell in the ball-room, or, indeed, anywhere in the house, was noticed by the guests. From this and other considerations I was led to infer that drinking water was the most probable vehicle of it.

Continued

PERSPECTIVE 16.2 ***Continued***

A visit to the courtyard of the hotel left in my mind no doubt that this was the true view of the disease. The cesspool and drain, which I was informed received the bulk of the diarrhoeal discharges from the fever patient, was at the time of the outbreak so near to the well, that, under the conditions of soil and locality, percolation from one to the other was almost inevitable. I further learnt, from persons who were present at the balls, that, as usual on such occasions, many drinks—lemonade among others—were largely supplied there and freely drunk.

This much, then, was sure—that a considerable number of the persons who attended the balls drank freely water from a well in close proximity to a receptacle which, for a considerable time, had received the specific excreta from the diseased intestine of a fever patient.

To complete the demonstration, nothing more seems needed than the fact within a few days a considerable number of these persons were found to be infected with the same fever. It is seldom that, in the difficult work of tracing the causes of disease, we can succeed in bringing fundamental relations so close together as these.

Source: William Budd, *Typhoid Fever: Its Nature, Mode of Spreading, and Prevention*. London, 1873.

Widal (*Salmonella* agglutinin) test followed by a convalescent serum collected after 10 days to 2 weeks to demonstrate any rising antibody response to active infection.

Epidemiology

I. Communicability of Infection

A. *Reservoirs, Sources, and Transmission Routes* Human beings are the only reservoir for *S. typhi* and a primary one for the *S. paratyphi* serotypes. Family contacts may be transient carriers. The carrier state is most common among persons who acquire their infection during middle age and is especially frequent among women. One of the most famous typhoid carriers, nicknamed "Typhoid Mary" (her name was Mary Mallon), is known to have been responsible for 10 outbreaks of typhoid fever between 1896 and 1906, involving 51 cases and 3 deaths through her activities as a food handler.

B. *Incubation Period* For typhoid fever, usually 1 to 2 weeks, sometimes 3 weeks; for paratyphoid fever, about 1 to 10 days; for bacteremia, difficult to define.

C. *Communicable Period* While the organisms persist in excreta, usually from the first week through convalescence. About 10% of typhoid fever convalescents shed bacilli for varying periods up to 3 months, and about 2 to 5% become permanent carriers. The carrier rate is variable in paratyphoid fever and other systemic infections.

D. *Immunity* Human populations show a general susceptibility to *Salmonella* infections, with more severe reactions among the very young and the very old. Exposure to systemic infection generally results in immunity, but the degree and duration are variable. Second attacks occur but are not usual after age 30 or 40.

II. Control of Active Infection

A. *Transmission-based Precautions* A private room is not required, but standard precautions must be followed. Contact precautions are used for diapered or incontinent children less than 6 years old. Typhoid fever patients should not be cared for at home unless adequate sanitary conditions exist, and good nursing care is available. Their rooms should be screened from flies and kept scrupulously clean. The patients may not be released to resume general activities, especially if these include food handling, until their stools and urine appear to be free of salmonellae. This assurance is generally based on three negative cultures of stool (and urine, if indicated) obtained on successive days during convalescence.

B. *Other Precautions* These include concurrent disinfection of all infectious excreta and objects contaminated by them (bedpans, bed linens, rectal thermometers, enema equipment, clothing). When community sewage disposal systems are adequate, toilet facilities may be used directly or for bedpan contents. Toilet seats and other bathroom facilities should be disinfected frequently, especially when shared. In hospitals, individual bedpans or toilets should be provided for patients with enteric infections, or the bedpans should be sterilized if they are reused. When good sewage disposal is not available, preliminary treatment of bedpan contents with chemical disinfectants is necessary. Gowns should be worn if soiling is likely, and hand washing should be practiced with scrupulous care by the patient as well as by personnel.

C. *Treatment* Ciprofloxacin, and ceftriaxone are the drugs of choice for typhoid fever and other systemic infections. Antimicrobial susceptibility testing with the patient's strain should be performed. If it is resistant to the above drugs, chloramphenicol, ampicillin, or trimethoprim-sulfamethoxazole may be of value.

D. *Control of Contacts* Close contacts should be excluded from food handling until cultures prove them to be free of inapparent infection.

The source of every case of typhoid or other enteric fever is investigated. The possibilities of unreported cases or asymptomatic carriers are explored, together with the question of contaminated food or water as a source of common exposure in outbreaks (see Perspective 16.2).

E. *Epidemic Controls* Individual case reports are required by local health authorities. Suspected food is discarded, and milk must be boiled, pasteurized, or discarded until proven safe. Suspected water supplies (especially after floods) must be treated with chlorine or iodine or boiled before use. Vaccine is not recommended during outbreaks because it complicates the serologic diagnosis of suspected illness.

III. Prevention

A. *Immunization* Vaccine is available for typhoid fever only and is only 50 to 70% effective. Current recommendations indicate vaccination of persons subject to unusual exposure, through either occupation or travel. Vaccine is also used in areas of high endemicity or in institutions in which sanitation cannot be maintained.

Vaccine elicits the production of agglutinating antibodies, which appear in the circulating blood and persist for some time. These antibodies may act protectively during the early bacteremic stages of typhoid fever, before the organisms invade mononuclear phagocytes. When serum from patients with febrile disease of suspected *Salmonella* origin is tested for typhoid agglutinins, a positive titer must be interpreted in the light of a history of vaccination or repeated natural exposure to, or active infection with, these organisms.

B. *Control of Reservoirs, Sources, and Transmission Routes* Prevention of typhoid fever depends on adequate protection of water supplies (travelers in endemic areas should take care to use water that is boiled, chlorinated, or disinfected with iodine tablets); sanitary disposal of human excreta; pasteurization of milk and dairy products; fly control; sanitary supervision of food processing in public eating places, with special attention to foods eaten raw and to the provision and use of hand washing facilities; and limitations on the collection and marketing of shellfish from clean waters.

The identification and continued supervision of typhoid carriers is an essential preventive measure. Carriers sometimes become impatient and uncooperative under surveillance, and this necessitates authenticating specimens yielding negative cultures. Antimicrobial therapy with amoxicillin, ciprofloxacin, or trimethoprim-sulfamethoxazole, if prolonged, may end the carrier state. If this fails, removal of the gallbladder may be considered.

Patients convalescing from enteric fevers must be warned of the significance of the carrier state, if they remain culture-positive. They should be given special instruction in personal hygiene, hand washing after using the toilet and before meals, and the sanitation of excreta when adequate public sewers are not available. They must be excluded from food handling until satisfactorily demonstrated to be free of continuing infection.

Brucellosis

The Clinical Disease

Brucellosis is a systemic infection characterized by variability—of entry route, tissue localizations, and onset and symptoms. It is really a disease associated with animals. Called Bang's disease when it affects cattle, it also occurs in hogs, sheep, goats, horses, and other domestic or wild animals. It is sometimes transmitted to humans through infectious animal tissues or milk. Ingestion of contaminated dairy products is one of the major entry routes, but the organisms also may enter the body through minor skin breaks or inhalation.

Whatever their portal of entry, the organisms usually find their way through lymphatic channels to regional lymph nodes, the thoracic duct, and the bloodstream, with eventual dissemination to many organs. They are localized most frequently in the mononuclear phagocyte system, being taken up by mononuclear phagocytes of organs and tissues such as spleen, liver, bone marrow, and lymph nodes. The cellular reaction around them is granulomatous, as noted so often in chronic infectious diseases. The nodules frequently undergo central necrosis, and if they break down, the infection extends to other tissues. Cholecystitis, bone marrow invasion, and meningitis sometimes occur. Persons repeatedly exposed to brucellae or their antigens develop a delayed-type hypersensitivity response.

The onset of brucellosis is usually insidious, with malaise and weakness, assorted muscle aches and pains, and an intermittent fever that rises late in the day but falls during the night; the patient is drenched with sweat at that time. When onset is acute, fever and chills are more pronounced, and the patient may have severe headache, backache and exhaustion. There may be gastrointestinal symptoms, enlarged lymph nodes and spleen, hepatitis with jaundice, and mental depression. These generalized symptoms may subside for several weeks, but if untreated, the disease often becomes chronic, with persisting or extending lesions. Endocarditis or pyelonephritis may develop, endometritis may occur in women, and various forms of nervous system disease may be seen.

Recovery may be gradual in untreated patients, often with residual allergy or damage to bones and other tissues, or the infection may become latent and asymptomatic, with chills and fever recurring over the years as old lesions are reactivated.

Laboratory Diagnosis

I. Identification of the Organisms

There are four major species, each of which has a distinct primary reservoir: *Brucella abortus* (cattle), *Brucella melitensis* (goats), *Brucella suis* (swine), and *Brucella canis* (dogs). Any of these strains may infect humans or animals other than the principal host.

Microscopically, brucellae are very small gram-negative coccobacilli, nonmotile and non-spore-forming. They are much smaller and more delicate in appearance than the enteric bacilli of the human intestinal tract.

Culturally, these organisms are quite fastidious. They require enriched culture media, and some will not grow unless additional atmospheric CO_2 is provided during 35° C incubation. They are differentiated from other gram-negative bacilli and from one another by a few biochemical reactions and by serologic agglutination with strain-specific *Brucella* antisera.

II. Identification of Patient's Serum Antibodies

Agglutinating antibodies appear in the circulating blood during early stages of infection and persist for some time, especially in chronic disease. Diagnosis by serologic testing is complicated by difficulties distinguishing between those antibodies that persist even after successful therapy and those that result from ongoing chronic infection. In addition, some patients develop so-called *blocking antibodies* that block the agglutination reaction unless high dilutions of serum are tested. The results of the tests must be interpreted in light of the patient's history with regard to time of possible exposure, onset, and current stage of symptoms.

III. Specimens for Diagnosis

A. Repeated blood cultures are useful during early stages of acute infection or during febrile periods of chronic disease. Cultures of bone marrow may yield the organism also.

B. Biopsies of lymph nodes, spleen, liver, or other tissues apparently involved in infection; body fluids, such as spinal fluid; or sections of nodules in tissues removed by surgery may be bacteriologically diagnostic.

C. Serum samples for serologic tests may be submitted.

Epidemiology

I. Communicability of Infection

A. *Reservoirs, Sources, and Transmission Routes* Brucellosis is transmitted to humans from domestic animal reservoirs through direct contacts with animal tissues (placentas, aborted fetuses, slaughterhouse blood and meat), inhalation of contaminated dust of barns or abattoirs, or ingestion of unpasteurized dairy products from infected animals. It is rarely transmitted from person to person, even when the organisms are present in discharges from the infected human case.

B. *Incubation Period* When the onset is slow and insidious, the incubation period is difficult to determine but is estimated at 2 to 3 weeks.

C. *Communicable Period* Brucellosis is not ordinarily communicable among humans.

D. *Immunity* Patients who recover from brucellosis become immune to it. Reinfections may occur if exposure is massive, but a milder illness results. Judging by the numbers of people who can be infected through ingestion of contaminated milk or cheese, it would appear that most human beings are susceptible to *Brucella* infection. Pasteurizing milk prevents this form of exposure on any large scale, and brucellosis remains primarily an occupational hazard for farmers, veterinarians, and slaughterhouse workers exposed to infected animals. Artificial active immunization is not available for humans, but a bacterial vaccine is used for calves in endemic areas.

II. Control of Active Infection

A. *Transmission-based Precautions* Patients with brucellosis are cared for with standard precautions ensuring safe disposal of excreta and secretions.

B. *Treatment* of diagnosed *Brucella* infection with antimicrobial drugs generally must be prolonged to eradicate organisms from protected intracellular sites in granulomatous nodules, and relapses may be common. The drug of choice is doxycycline, administered with streptomycin rifampin or gentamicin.

C. *Contact Controls* are applicable only to infected animal reservoirs or to detection of sources of infection in unpasteurized dairy products from infected herds. Suspected animals are tested for delayed-type skin hypersensitivity and for serum antibodies. Reactive animals are slaughtered.

D. *Epidemic Controls* depend on good case reporting of human infections, which may be mild or unrecognized because of difficulties in bacteriologic or serologic diagnosis. When brucellosis is suspected or confirmed, local health agencies can investigate sources and set up appropriate controls on infected animals or their products.

III. Prevention

Prevention of brucellosis depends on recognizing its sources in animals; limiting its spread among them; and controlling occupational hazards, infected meats, milk, and other animal foods.

Vaccines for human use are not available. They are sometimes used for the immunization of calves but only in areas in which the disease is prevalent.

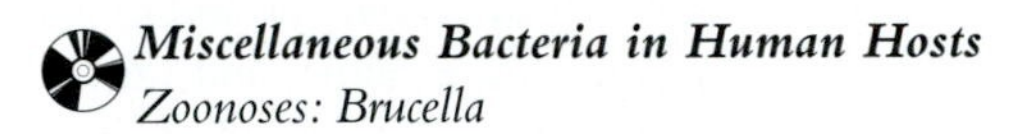

LISTERIOSIS

The Clinical Disease

Listeriosis has attracted new interest in recent years because of its association with several outbreaks of foodborne disease. The organism has a widespread distribution in soil and among many wild and domestic animals in which it also causes disease. It is acquired by humans primarily through ingestion of contaminated foods or congenital transfer. Most likely, healthy human adults are resistant to infection, but the organism lives commensalistically in their intestinal tracts until they become more susceptible to the disease. Listeriosis is most common in pregnant women, newborns, adults with underlying debilitating or immunosuppressive conditions such as diabetes and malignancies, and those receiving immunosuppressive drugs. It is seen more frequently in patients with acquired immunodeficiency syndrome or AIDS (see chap. 19) than in the general population.

Infection in pregnant women is usually mild and flu-like. However, during the bacteremic stage of maternal infection, the fetus may become infected through the placenta. Without prompt treatment, spontaneous abortion or stillbirth may result. Some infants may have an overwhelming listerial sepsis at birth or shortly thereafter, whereas others develop meningitis several days to weeks after birth. The mode of organism transmission in these latter cases is not understood. In some nursery epidemics of disease, transmission has been caused by inadequate hand washing or by breaks in appropriate patient care techniques (e.g., improperly cleaned rectal thermometers, contaminated mineral oil in a container used to bathe babies).

Foodborne listeriosis occurs in epidemic and sporadic patterns. The largest epidemic in North America was traced to ingestion of a specific brand of Mexican-style soft cheese. Most cases of disease in that outbreak occurred in pregnant Hispanic women or their offspring. Other foods implicated in transmission include unpasteurized milk, ice cream, raw vegetables, salami, undercooked chicken, and uncooked hot dogs.

LABORATORY DIAGNOSIS

I. Identification of the Organism

Listeria monocytogenes.

A. *Microscopic Characteristics* The organism is a gram-positive, nonsporulating, motile, aerobic bacillus.

B. *Cultural Characteristics* *Listeria* is sometimes difficult to isolate from foods and body sites that are not normally sterile. Isolation is usually from blood, spinal fluid, placenta, and amniotic fluid. The colonies are beta-hemolytic and resemble those of group B streptococci, organisms that cause very similar perinatal disease. A Gram stain of colonies reveals the rodlike rather than coccal nature of the listerial cells. Microscopically, *Listeria* resembles nonpathogenic members of the coryneform group (diphtheroids), which are common commensals on skin and mucous membranes. In contrast to the diphtheroids, which are nonmotile, *Listeria* organisms have a characteristic end-over-end, "falling leaf" type of motility when viewed in a wet preparation.

Listeria survives at cold temperatures and can sometimes be isolated from foods or specimens containing mixed flora by storing them at refrigerator temperatures for days or weeks. Most other pathogens do not withstand this so-called "cold enrichment" technique.

C. *Immunologic Identification* Not useful.

II. Identification of Patient's Serum Antibodies

Not routine.

III. Specimens for Diagnosis

A. Blood for culture in early stages of maternal disease.

B. From infected infants, blood for culture and spinal fluid for smear and culture.

C. From mothers of infants with perinatal listeriosis, amniotic fluid and placenta.

D. Suspect foods in cases of foodborne epidemics.

Epidemiology

I. Communicability of Infection

A. *Reservoirs, Sources, and Transmission Routes* The major reservoir is in soil. Listeriosis in food and dairy animals may lead to infection of humans who ingest improperly cooked foods or contaminated unpasteurized milk and dairy products. Approximately 5% of people are asymptomatic intestinal carriers of *L. monocytogenes.*

B. *Incubation Period* Variable, from 3 to 70 days have been reported.

C. *Communicable Period* Most communicable for fetuses during maternal bacteremic stage of infection.

D. *Immunity* Fetuses and newborns are highly susceptible; most adults are resistant unless afflicted with a debilitating or immunosuppressive condition. Antibodies do not appear to be protective. Protective immunity may be related to cellular rather than humoral defenses.

II. Control of Active Infection

A. *Transmission-based Precautions* Standard precautions are required although most humans are resistant to infection and the disease is not readily communicable.

B. *Other Precautions* No other special precautions are recommended, but concurrent disinfection of items used during care of infected infants is advisable.

C. *Treatment* Ampicillin is the drug of choice; trimethoprim-sulfamethoxazole is an alternative for ampicillin-allergic patients.

D. *Control of Contacts* Not indicated.

E. *Epidemic Controls* Outbreaks of infection should be investigated to identify the source and prevent further spread of infection.

III. Prevention

No immunizing agents are available. Persons at risk should avoid eating unpasteurized milk products and raw or partially cooked meats. Foods should be prepared in a manner to prevent cross-contamination between raw and cooked foods.

Gram-positive Organisms in Human Hosts
Gram-positive Bacilli: Listeria

QUESTIONS

1. How are bacterial diseases acquired through the alimentary route classified?
2. Why is it difficult to distinguish clinically between infectious gastroenteritis diseases? How is the diagnosis made?
3. What type of precautions are required by the CDC for hospitalized patients?
4. Why is it necessary to send stool specimens to the laboratory without delay?
5. How does the laboratory contend with the normal flora of the stool while attempting to isolate a pathogen from feces?
6. Why is campylobacteriosis considered a zoonotic disease? How is the disease transmitted to humans? How can the disease be prevented?

SUMMARY TABLE 16.2 BACTERIAL DISEASES ACQUIRED THROUGH THE ALIMENTARY ROUTE

Clinical Disease	Causative Organism	Other Possible Entry Routes	Incubation Period	Communicable Period
Acute Gastrointestinal Disease				
Salmonellosis	*Salmonella* species (many serotypes)		8–10 hours, may be delayed for 48 hours or more	For duration of active infection and as long as salmonellae are shed
Shigellosis (bacillary dysentery)	*Shigella dysenteriae* *Shigella flexneri* *Shigella boydii* *Shigella sonnei*		1–4 days, not more than 7 days	As long as organisms are in feces
Gastroenteritis	*Campylobacter jejuni*		2–4 days	Not known; probably while organisms are present in feces
Gastritis, peptic ulcer	*Helicobacter pylori*		Unknown	Not known if communicable
Gastroenteritis	*Yersinia enterocolitica*		2–7 days	As long as organisms are in feces
Infections Caused by Microbial Multiplication with Enterotoxin Production				
Cholera	*Vibrio cholerae*		2–3 days, few hours during outbreak of cholera	While organisms are present in vomitus and feces; may persist in intestinal tract for several weeks after recovery
Gastroenteritis	*Vibrio parahaemolyticus*		8–48 hours	Not communicable from person to person
	Clostridium perfringens		2–4 days	Not communicable
	Bacillus cereus		As above	Not communicable
Pseudomembranous colitis	*Clostridium difficile*		Undefined	As long as patients shed organisms and endospores contaminate environment

[a]Contact precautions are used in addition to standard precautions if the patient is diapered or incontinent.

7. Why is it important that an infection with *Helicobacter pylori* be properly diagnosed and treated promptly?
8. What biochemical activity of *H.pylori* assists in the diagnosis of infection?
9. What is pseudomembranous colitis? What organism is implicated in this disease?
10. What is the course of therapy if pseudomembranous colitis is diagnosed? How can the disease be controlled in the hospital?
11. What is the effect of the enterotoxin of *V. cholerae* on the small intestines? How is the disease spread in epidemic form?
12. What characteristic of the El Tor variant of *V. cholerae* made it possible for the disease to spread in recent epidemics? What is the significance of the antigenic difference between the El Tor variant and the 0139 Bengal strain of *V. cholerae*?
13. What five strains of *E. coli* cause diarrheal diseases?

Specimens Required	Laboratory Diagnosis	Immunization	Treatment	Transmission-based Precautions[a]
Stools	Culture	None	None (unless dissemination occurs). Use of antimicrobial agents may prolong fecal excretion of organisms	Standard precautions; concurrent disinfection of all excreta and contaminated objects; individual toilets or bedpans; chemical disinfection of excreta if sewage disposal is inadequate; gowns; hand washing by patient and personnel
Bowel contents Rectal swabs Material taken from ulcerative lesions during sigmoidoscopy	Culture Culture Culture	None	Ciprofloxacin or other fluoroquinolone	Same as above
Stools	Culture	None	Fluid replacement Erythromycin Tetracycline Ciprofloxacin	Standard precautions
Tissue from stomach biopsy Serum	Culture IgG and IgA assays	None	Bismuth sulfate + metronidazole, amoxicillin	Standard precautions
Stools Lymph nodes Blood Body fluids Tissues	Culture Culture Culture Culture Culture	None	Trimethoprim-sulfamethoxazole Gentamicin Cefotaxime Ciprofloxacin	Standard precautions
Stools Vomitus Rectal swabs in acute stage	Culture	Killed vaccine	Fluid and electrolyte replacement Tetracyclines	Same as salmonellosis
Stools	Culture	None	None	Standard precautions
Suspect food Stools	Culture Culture	None	Supportive	Standard precautions
Suspect food	Culture	None	Supportive	Standard precautions
Stools Stool extract	Culture Tissue culture with antitoxin control Enzyme immunoassay Latex agglutination assay	None	Stop antimicrobial agents Metronidazole or vancomycin if severe	Standard precautions

Continued

14. How can enterohemorrhagic *E. coli* infection be prevented?

15. What is the difference between staphylococcal food poisoning and botulism? How can staphylococcal food poisoning be prevented?

16. Under what conditions does *C. botulinum* produce exotoxin?

17. How can botulism be prevented?

18. How does *S. typhi* infection differ from that of other salmonelloses?

19. Where do *S. typhi* localize in the body? What implications does this have for public health?

20. How is brucellosis transmitted to humans? What control measures are necessary for cases of active brucellosis?

21. Why is listeriosis of concern for pregnant women?

SUMMARY TABLE 16.2 *Continued*

Clinical Disease	Causative Organism	Other Possible Entry Routes	Incubation Period	Communicable Period
Enteric Infections Caused by *Escherichia coli*				
Travelers' diarrhea	Enterotoxigenic *Escherichia coli*		12–72 hours	Generally not communicable from person to person
	Enteroaggregative[b] *Escherichia coli*		1–3 days	Same
Gastroenteritis, dysentery	Enteroinvasive *Escherichia coli*		2–3 days	As long as organisms are in feces
Epidemic diarrhea in nurseries	Enteropathogenic *Escherichia coli*		2–4 days; maximum of 3 weeks	As long as babies remain symptomatic and adult carrier or environmental source remains undetected
Hemorrhagic colitis	Enterohemorrhagic *Escherichia coli*		3–5 days	As long as organisms are in feces
Diseases Caused by Ingestion of Preformed Toxin				
Staphylococcal intoxication	*Staphylococcus aureus* enterotoxin		1–6 hours	Not communicable from person to person
Botulism	*Clostridium botulinum* toxin		12–36 hours	Not communicable from person to person
Infectious Systemic Diseases				
Typhoid fever	*Salmonella typhi*		1–2 weeks, sometimes 3 weeks	Through active infection, and as long as salmonellae are shed in excreta; 10 percent of convalescents shed bacilli up to 3 months, 2–5 percent become permanent carriers
Paratyphoid fever	*Salmonella enteritidis var paratyphi* A or B		1–10 days	Through active infection and for as long as salmonellae are shed in excreta; carrier rate is variable
Salmonella septicemia	*Salmonella choleraesuis*		Same as above	Same as above
Brucellosis	*Brucella abortus* *Brucella melitensis* *Brucella suis* *Brucella canis*	Inhalation Skin contacts	1–3 weeks	Not ordinarily communicable from person to person
Listeriosis	*Listeria monocytogenes*	Congenital	3–70 days	To fetus, during maternal bacteremic phase

[b]May also cause a chronic diarrhea.

Specimens Required	Laboratory Diagnosis	Immunization	Treatment	Transmission-based Precautions[a]
Stools	Assays for toxin	None	Primarily supportive Trimethoprim-sulfamethoxazole	Standard precautions
Stools	Culture + assays for cell adherence	None	Ciprofloxacin Same	Standard precautions
Stools	DNA probe Animal invasiveness assays	None	Ampicillin	Standard precautions
Stools from sick and exposed babies Stools from mothers and personnel	Culture Virus isolation (to exclude viral etiology) Culture	None	Fluid and electrolyte replacement Antimicrobial therapy depends on testing of isolated organism	Symptomatic babies placed on contact precautions; set up clean and contaminated nurseries; consult infection control committee; review techniques in obstetric service and nursery; survey mothers and personnel for symptoms of illness; survey recently discharged babies
Stools Stool extract	Culture Toxin assay	None	Supportive	Standard precautions
Vomitus Suspected food (if available)	Smear and culture Culture for staph Phage typing Toxicity test	None	Supportive	Standard precautions
Suspected food Serum	Smear and culture Toxicity for mice Toxicity for mice	None	Antitoxin	Standard precautions
Blood Stool specimens Urine specimens Bile drainage Sputum Pus from abscesses Synovial fluid Spinal fluid (as appropriate clinically) Acute and convalescent blood serums	Culture Culture Culture Smear and culture Agglutinin titers	Heat-killed vaccine	Ciprofloxacin Ceftriaxone Chloramphenicol Ampicillin Trimethoprim-sulfamethoxazole	Standard precautions; concurrent disinfection of all contaminated objects; scrupulous attention to hand washing
Same as above	Same as above	None	Same as above	Standard precautions
Same as above	Same as above	None	Same as above	Standard precautions
Blood Bone marrow Biopsies of lymph nodes, spleen, liver, spinal fluid Serum samples	Culture Smear and culture Agglutinin titers	Not available for humans	Doxycyclines + rifampin or gentamicin	Safe disposal of excreta and secretions
Blood CSF Amniotic fluid Placenta Suspect food	Culture Smear and Culture Culture	None	Ampicillin Trimethoprim-sulfamethoxazole	Standard precautions; concurrent disinfection during care of infected infants

22. What precautions are required when caring for patients with acute gastrointestinal diseases?

Bibliography

Benenson, A. S. ed. 1995. *Control of communicable diseases in man.* 16th ed. Washington, D.C.: American Public Health Association.

Blumenthal, D. 1990. *Salmonella enteritidis:* From the chicken to the egg. *FDA Cons.* 24: 6–10.

Centers for Disease Control and Prevention. 1990. Recommendations for collection of laboratory specimens associated with outbreaks of gastroenteritis. *Morbid. Mortal. Weekly Rep.* 39 (RR-14).

Centers for Disease Control and Prevention. 1995. Cholera associated with food transported from El Salvador-Indiana 1994. *Morbid. Mortal. Weekly Rep.* 44 (20): 385.

Centers for Disease Control and Prevention. 1995. *Escherichia coli* 0157:H7 outbreak linked to commercially distributed dry-cured salami-Washington and California 1994. *Morbid. Mortal. Weekly Rep.* 44 (9): 157.

Centers for Disease Control and Prevention. 1995. Outbreak of acute gastroenteritis attributable to *Escherichia coli* serotype 0104:H21-Helena, Montana, 1994. *Mortal. Morbid. Weekly Rep.* 44 (27): 501.

Centers for Disease Control and Prevention. 1996. Surveillance for foodborne-disease outbreaks—United States, 1988-1992. *Morbid. Mortal. Weekly Rep.* 45 (SS-5).

Centers for Disease Control and Prevention. 1996. Surveillance for waterborne disease outbreaks—United States 1993–1994. *Morbid. Mortal. Weekly Rep.* 45 (SS-1).

Ching-Lee, M. R., A. R. Katz, D. M. Sasaki, and H. P. Minette. 1991. Salmonella egg survey in Hawaii: Evidence for routine bacterial surveillance. *Am. J. Public Health* 81 (5): 761–766.

Ewald, P. W. 1991. Waterborne transmission and the evolution of virulence among gastrointestinal bacteria. *Epidemiol. Infect.* 106 (1): 83.

Holmes, S. 1989. Careful food handling reduces the risk of listeria. *Prof. Nurse* 4 (7): 322–324.

Kaper, J. B., J. G. Morris, Jr., and M. M. Levine. 1995. Cholera. *Clin. Microbiol. Rev.* 8 (1): 48–86.

Kilgore, P. E., R. C. Holman, M. J. Clarke, and R. I. Glass. 1995. Trends of diarrheal diseases—Associated mortality in U. S. children, 1968 through 1991. *JAMA* 274 (14): 1143–1148.

Lovett, J., D. W. Francis, and J. M. Hunt. 1987. *Listeria monocytogenes* in raw milk: Detection, incidence and pathogenicity. *J. Food Prot.* 50: 188–192.

Mandell, G. L., J. E. Bennett, and R. Dolin. 1995. *Principles and practice of infectious diseases.* 4th ed. New York: Churchill Livingstone.

Marshall, B. J. 1995. *Helicobacter pylori:* The etiologic agent for peptic ulcer. *JAMA* 274 (13): 1064–1066.

Midura, T. F. 1996. Update: Infant botulism. *Clin. Microbiol. Rev.* 9 (2): 119–125.

Roueche, B. 1991. A game of wild Indians. In *The medical detectives.* New York: Plume.

Schuchat, A., B. Swaminathan, and C. V. Broome. 1991. Epidemiology of human listeriosis. *Clin. Microbiol. Rev.* 4 (2): 169–183.

Skoutelis, A. T., et al. 1994. Hospital carpeting and epidemiology of *Clostridium difficile. Am. J. Infect. Control* Aug. 22 (4): 212–217.

Suppaiah, L. 1988. Pseudomembranous colitis induced by *Clostridium difficile. Crit. Care Nurse.* 8: 65–72.

Swan, P. 1983. Cholera: A visitation on Victorian society; The first outbreak of cholera in Britain 150 years ago. *Nurs. Mirror* 157 (15): 30–34.

Thorne, G. M. 1991. Salmonella: The chickens and the eggs. *Clin. Microbiol. Newsl.* 13 (9): 65–68.

Yablon, S. A., R. Kretenberg, K. Fruhmann. 1993. *Clostridium difficile-* related disease: Evaluation and prevalence among inpatients with diarrhea in two freestanding rehabilitation hospitals. *Arch. Phys. Med. Rehabil.* 71 (1): 9–13.

CHAPTER 17

Viral Diseases Acquired through the Alimentary Route

OUTLINE

Microbes in Motion

This chapter covers the important viral agents acquired through the alimentary route including diseases they cause, laboratory diagnosis, epidemiology, control, and prevention. The following books and chapters of the Microbes in Motion CD-ROM may be a useful preview or supplemental study aid to your reading: Vaccines: Vaccine Targets.

OBJECTIVES

On completing this chapter the student will be able to:

1. distinguish among the important viral agents acquired through the alimentary route.
2. describe the clinical characteristics of these viral diseases.
3. identify the causative agents of the diseases.
4. cite the methods used for their laboratory diagnosis.
5. discuss the epidemiology, control, and prevention of these diseases.

KEY WORDS

Adeno (**AD**-en-oh) viruses
Aseptic meningitis
Astroviruses
Calici (cal-**ISS**-ee) viruses
Coxsackie (cox-**SACK**-ie) viruses
Echoviruses
Encephalitis
Entero (**ENTER**-oh) viruses
Hepatitis A virus
Hepatitis E virus
Herpangina (her-pan-**GINE**-ah)
Inactivated polio vaccine (IPV)
Neonatal myocarditis
Norwalk viruses
Oral poliovirus vaccine (OPV)
Picorna (pe-**CORN**-ah) viruses
Pleurodynia (plu-roe-**DIN**-i-ah)
Poliomyelitis
Polioviruses
Postpoliomyelitis syndrome
Rota (**ROW**-ta) viruses
Sabin vaccine
Salk vaccine

INTRODUCTION

Several important human viral diseases are acquired through the fecal-oral route. Some of the viral agents involved enter the body and localize in the intestinal tract but then disseminate and exert their pathogenic effects in other tissues. Such viruses include the enteroviruses and the agent of hepatitis A (infectious hepatitis).

Other viruses remain localized in the intestinal tract and are responsible for many cases of enteric disease, primarily in children but also in adults. These include rotaviruses, certain adenoviruses, Norwalk virus, caliciviruses, and astroviruses, some of which cause disease in other animals as well. Because they do not grow (or grow only poorly) in cell cultures, the importance of these viruses in gastroenteritis has been known only since the 1970s, when each was visualized by electron microscopy in fecal filtrates. Serologic studies have confirmed that the infections they cause are widespread throughout the world.

This chapter describes the principal viral diseases transmitted through the alimentary route and their clinical and epidemiologic patterns.

ENTEROVIRUS DISEASES

The **enteroviruses** (polioviruses, coxsackieviruses, and echoviruses) are members of a group called **picornaviruses** because they are the smallest of the RNA viruses (*pico* = very small; *rna* indicates the type of nucleic acid they possess). These viruses are spheric, have icosahedral symmetry, lack an envelope, and measure only 20 to 30 nm in diameter (see table 2.6). Members of a second subgroup of the picornaviruses, the rhinoviruses, are associated with common colds. Acute respiratory disease is also caused by some enteroviruses such as certain types of coxsackievirus and echovirus (see table 13.1).

The enteroviruses are widely distributed in human populations and are far more commonly associated with asymptomatic infections than with clinically apparent disease. When they induce injury, their clinical and epidemiologic patterns vary. The upper respiratory tract may be their primary site of entry, multiplication, and subsequent transmission, but they usually implant more permanently in the bowel. From there, they may disseminate to parenteral tissues of the infected host, causing cell damage and host reaction, and they may be transmitted by the fecal-oral route to the oropharynx of humans in close contact.

Enterovirus infections differ in severity, tissues involved in disease, and type of lesions produced. They range from simple upper respiratory syndromes, such as the common cold, to serious illnesses, including paralytic poliomyelitis, myocarditis, and meningitis. Some coxsackievirus and echovirus types have been associated only with diseases of the respiratory tract, but others are known to produce several kinds of systemic infections. Conversely, a particular clinical entity (meningitis, bronchitis, pneumonitis) may be caused by any one of several different enteroviruses. The clinical classification of these viruses is, therefore, not very satisfactory, except for their common relationship to the alimentary route of entry, implantation, and transmission. Virologically, they have some physical and chemical properties in common: they are small viruses, in the range of 20 to 30 nm in size; their architecture is similar, as revealed by electron micrography; and their nucleic acid structures are closely related. Antigenically, the three families of poliovirus, coxsackievirus, and echovirus represent distinct groups. The poliovirus group contains three major types, whereas the other two contain many more.

POLIOMYELITIS

The Clinical Disease

Poliovirus infection is usually inapparent. As a disease, it may take one of three forms or a combination of these. (1) Most commonly, **poliomyelitis** is a mild, nonparalytic illness characterized as a gastrointestinal disturbance, with fever, nausea and vomiting, headache, sore throat, and drowsiness. The course is short, with recovery in a few days. This syndrome can be diagnosed as an abortive poliomyelitis only by laboratory identification of the virus or of specific antibody response. (2) Patients with nonparalytic poliomyelitis may develop symptoms of meningitis, with stiffness and pain in the neck and back. This is an aseptic meningitis, typical of the kind produced by many viruses (see chap. 15). These symptoms may continue up to about ten days and be followed by rapid, complete recovery. The diagnosis again depends on laboratory findings, although clinical diagnosis may be based on the prevalence of typical poliomyelitis among the patient's contacts. (3) Paralytic poliomyelitis may develop without preliminary warning but more usually follows the initial, mild form of febrile illness. Viral damage to lower motor neurons results in a flaccid paralysis, usually asymmetric and commonly involving lower extremities, although this varies with the site of virus localization. In bulbar polio, there may be brainstem invasion with paralysis of respiratory muscles (fig. 17.1) or spasm and incoordination of nonparalyzed muscles.

Poliovirus enters the body through the mouth, transmitted by fecally contaminated objects or hands. It implants and multiplies in the lymphoid tissue of the intestinal wall. The multiplying virus then escapes into the bloodstream and travels to the oropharynx and other body tissues. It can be isolated both from the throat and from feces during this time. Virus continues to be excreted in feces for a number of weeks and is often identified in stools of persons with inapparent infection. Virus can be isolated from the blood of patients with nonparalytic polio and of others in the preparalytic stage of central nervous system disease.

Circulating viruses probably cross the blood-brain barrier through transcapillary diffusion and multiply in the neurons, causing variable degrees of injury. The anterior horn cells of the spinal cord are frequently but not exclusively the sites of intracellular localization and multiplication. Posterior horn cells and sensory ganglia may be

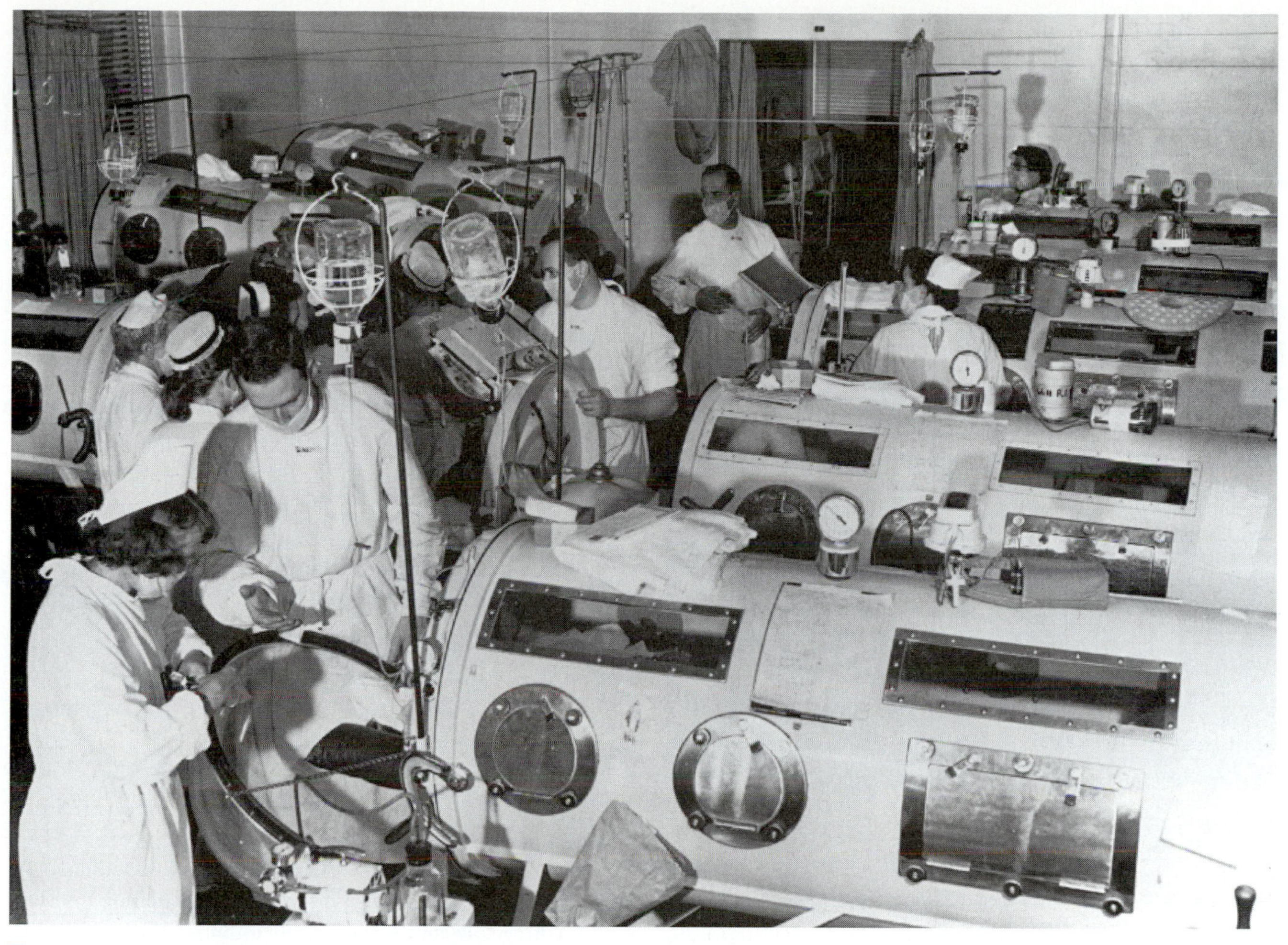

FIGURE 17.1 A hospital ward filled with iron lungs was a familiar sight during the years before 1956 when mass inoculation against poliomyelitis began.

March of Dimes, Birth Defects Foundation, White Plains, NY.

Vaccines
Vaccine Targets: Viral Targets

involved, as well as intermediate gray matter. (The term *poliomyelitis* means that the gray matter [*polio-*] of the spinal cord [*-myel-*] is inflamed [*-itis*].) Damage to these cells may be severe and may proceed rapidly to complete destruction. Cells that are not destroyed may later recover their function. Inflammation appears at the site where neurons are attacked, with focal collections of mononuclear and some polymorphonuclear cells. The virus does not affect peripheral nerves and muscles directly, but these undergo changes reflecting motor neuron damage. Muscle involvement usually reaches its peak quickly, within a few days. Recovery from paralytic polio is slower than in most acute infections, requiring several months, and residual damage often is crippling for a lifetime.

LABORATORY DIAGNOSIS

I. Identification of the Organism

Poliovirus (fig. 17.2) can be propagated in human or monkey cell cultures and identified by antibody neutralization of its infectivity. There are three antigenic types of poliovirus: 1, 2, and 3.

II. Identification of Patient's Serum Antibodies

Antibodies to poliovirus appear early in both abortive and paralytic disease. They may also be detected in the blood of persons with inapparent infection, vaccinated persons, and contacts of the latter; therefore, serologic diagnosis of active poliomyelitis requires paired acute and convalescent serum samples. Serums are tested with the three antigenic types of poliovirus by neutralization techniques. The titer of the second sample must be at least four times greater than that of the first to be considered diagnostically significant of active disease.

III. Specimens for Diagnosis

A. Throat washings and swabs collected within a few days of onset of symptoms may be submitted for virus isolation. Washings should be refrigerated at 4° C until they can be transported to the laboratory. Swabs are kept in viral transport medium obtained from the laboratory.

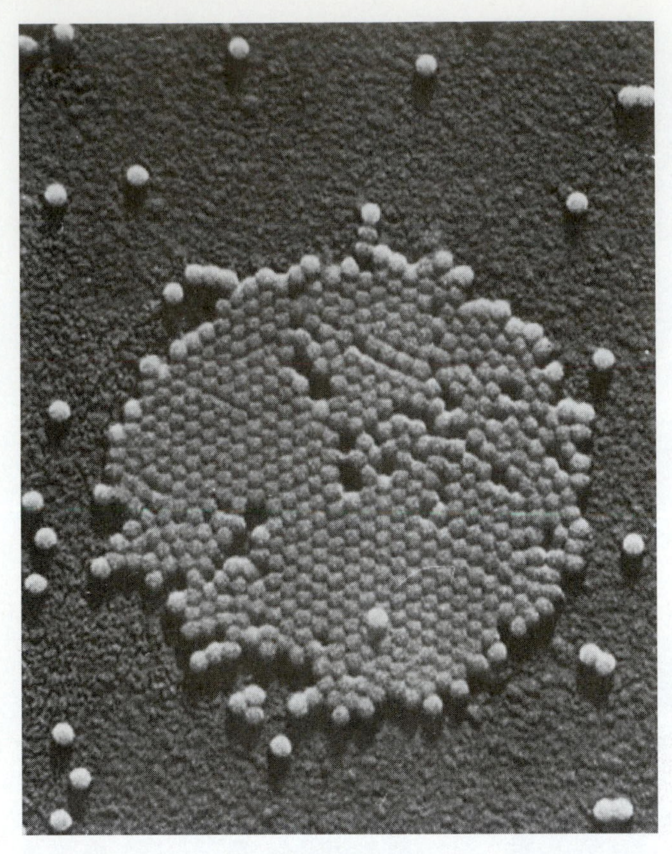

FIGURE 17.2 Electron micrograph of the poliomyelitis virus (×148,000).
Courtesy Lederle Laboratories, Division of American Cyanamid, Wayne, NJ.

B. Fecal specimens in sterile containers or rectal swabs in viral transport medium may be collected up to 4 weeks after infection.

C. Acute and convalescent serums (see Laboratory Diagnosis, II).

Epidemiology

I. Communicability of Infection

A. *Reservoirs, Sources, and Transmission Routes* Humans are the only known reservoir of poliovirus infection. The disease is transmitted by the fecal-oral route, the virus being present in the feces of infected persons. Contacts must be fairly close and direct since infection rates are highest among persons living together in households. Spread from fecal sources is enhanced by a warm climate and is greatest during the summer seasons in temperate climate zones. During epidemic periods, poliovirus has been recovered from flies, roaches, and sewage, but the role played by contaminated vectors, food, or water does not appear to be prominent. Direct human contacts are the most usual factor in transmission.

B. *Incubation Period* Usually between 1 and 2 weeks but may range from 3 days to 4 or 5 weeks.

C. *Communicable Period* Virus is present in feces before the onset of illness and usually persists for varying periods, up to about 6 weeks.

D. *Immunity* Type specific antibodies induced by clinically inapparent and recognizable infections confer resistance to infection. Second attacks are related to infection with a different poliovirus serotype. Immunization with poliovirus vaccine provides immunity to all three types of virus. Immunity is durable provided that the initial and booster doses of vaccine are adequate. Many unvaccinated, close contacts of live, oral vaccine recipients develop protective immunity because the vaccine strains are shed in feces and transmitted in a manner similar to natural disease.

In the United States, many children from disadvantaged backgrounds remain unvaccinated and approximately 21% of young adults are susceptible to at least one poliovirus type. Nevertheless, no cases of disease caused by natural type viruses (nonvaccine strains) have been reported since 1979. Thus, polio vaccination appears to have eradicated the natural disease in this country. The goal of the World Health Organization is to eradicate poliomyelitis globally by the year 2000, but many doubt whether this goal will be achieved.

II. Control of Infection

A. *Transmission-based Precautions* When patients suspected of poliomyelitis are hospitalized, they should be handled with standard precautions.

B. *Other Precautions* During the first week of illness, concurrent disinfection techniques should be followed for fecal excreta or articles exposed to fecal contamination. When adequate sewage disposal systems are not available, bedpan collections must be chemically disinfected before discarding. Terminal disinfection is necessary.

C. *Treatment* There is no specific antimicrobial therapy for poliomyelitis.

D. *Control of Contacts*

1. *Isolation* Not justified because the virus is usually widespread before active cases appear. In individual situations, the activities of family members may be limited to prevent contacts with persons not previously exposed.

2. *Case finding* Important to prevent further spread.

E. *Epidemic Controls* Case reporting is required so that community protection from epidemic polio can be provided without delay. In countries that advocate mass immunization, the incidence of paralytic polio has been strikingly reduced. Paralytic cases now occur chiefly in preschool-aged children who have not been vaccinated or have not received adequate doses of vaccine, usually children whose families live under the poorest socioeconomic conditions.

When the incidence of reported cases indicates the possibility of an outbreak, the involved community should be immunized as rapidly as possible. Oral vaccines can be administered to large numbers of people in a very short time.

III. Prevention

A. *Immunization* Two kinds of poliovaccine are available, and both provide durable active immunity.

The **inactivated polio vaccine (IPV),** or **Salk vaccine,** contains the three types of poliovirus, which have been formalin-inactivated. In the United States as of 1997, IPV is given in a series of two injections at 2 and 4 months of age followed by administration of oral poliovirus vaccine (see below). In other countries, IPV is sometimes used for the full vaccination series, given also at 18 months and 4 to 6 years of age. In this latter case, immunity lasts for at least 5 years. The killed virus stimulates production of circulating antibodies that can neutralize any infective viruses that may reach the bloodstream from intestinal sites. Shedding of natural virus from the intestinal tract is not prevented, nor can the killed vaccine be transmitted through the fecal-oral route to promote more widespread immunity among unvaccinated contacts.

Oral poliovirus vaccine (OPV), or **Sabin vaccine,** contains live, attenuated strains of all three antigenic types. The live viruses in OPV multiply in intestinal sites and induce production of secretory antibodies (sIgA) that provide resistance to later infection. They also stimulate the production of humoral antibodies that prevent infective virus from localizing in the CNS. The dose schedule for vaccine administration is at 6 months, and 4 to 6 years of age, following two previous doses of IPV (see table 10.1). Reinforcement immunization should be given to all persons who are planning to travel to countries where polio is endemic.

Mass immunization with oral poliovirus vaccines has reduced the incidence of paralytic poliomyelitis to extremely low levels. However, about 1 in approximately 3 million persons vaccinated develops paralytic or fatal disease within a month of taking OPV. Most of these are children less than four months old or persons who are immunocompromised. For this reason, some countries prefer to administer IPV instead, and the recommendation in the United States has been changed from four doses of OPV to two initial doses of IPV followed by OPV. During the past few years, fewer than ten cases of paralytic polio per year have been reported in the United States; all cases documented by culture were caused by vaccine strains.

As the incidence of poliomyelitis decreases, the public incentive to maintain effective levels of immunity also wanes. If immunization programs decline, the numbers of susceptible persons will increase again. For the individual victim of paralytic polio, the consequences of failure to provide vaccine can last a lifetime (fig. 17.3).

About 20 to 30% of people who have recovered from paralytic poliomyelitis experience pain and muscle weakness beginning many years after the acute illness. The cause of this condition, known as **postpoliomyelitis syndrome,** is unknown. It is thought to be related to further loss of already compromised nerve control of the affected muscles as the persons age.

Coxsackievirus Diseases

Clinical Syndromes

The **coxsackieviruses** were first recognized in 1948 in a study of an outbreak of a disease that simulated paralytic poliomyelitis. The disease, however, was milder and did not leave residual paralysis. Clinically, this syndrome was one of aseptic meningitis with some muscle weakness, or paresis. In the laboratory, virus isolated from these cases was shown to be distinct from poliovirus in its antigenic and infective properties, and so it was named for the town (Coxsackie, New York) in which it had caused an epidemic. Since that time, coxsackieviruses have been found to have a worldwide distribution and to be associated with a number of acute illnesses, some relatively mild and of short duration, others of more serious consequence.

Two antigenic groups of coxsackieviruses, A and B, are recognized, each containing several serologic types. Coxsackieviruses are similar to echoviruses in their physical properties, clinical illnesses produced in humans, and epidemiology. Laboratory identification of causative agents isolated from patient specimens has shown that they are responsible for the syndromes discussed below.

Aseptic Meningitis and Encephalitis

These diseases are among the most common manifestations of coxsackievirus infection, caused by serotypes of both groups A and B. **Aseptic meningitis** caused by coxsackieviruses cannot be distinguished clinically from that induced by other viral agents. **Encephalitis** may produce paralysis like that seen in poliomyelitis but is usually less extensive and more transient when caused by coxsackieviruses.

Herpangina

Herpangina is an acute, brief illness characterized by fever, sore throat, difficulty in swallowing, vomiting, and abdominal pain. The pharynx typically shows many vesicular lesions. Recovery occurs within 2 to 4 days. Herpangina is often a summer illness of children, associated with group A coxsackieviruses, but older persons may also contract it. Aseptic meningitis has been reported as an uncommon complication of infection.

Pleurodynia

Fever and acute muscle pain in the chest wall characterize **pleurodynia.** Malaise, headache, and other ill-defined symptoms occur but disappear in a week or two, with complete recovery. Group B coxsackieviruses are the chief cause of this syndrome, which sometimes occurs in epidemic form, most often in the summer and fall.

Respiratory Infections

Some group A viruses may produce the rhinopharyngitis typical of the common cold (see chap. 13). Some strains may also induce nodular lesions of the lymphatic tissue of the pharynx, the syndrome being similar to herpangina, but milder. Group B coxsackievirus has been associated with primary atypical pneumonia in infants and young children.

Neonatal Myocarditis

Neonatal myocarditis or encephalomyocarditis is associated with coxsackie group B viruses, but it is not known

FIGURE 17.3 Proper immunization to prevent polio would have avoided permanent physical handicap.
World Health Organization. Photo by E. Schwab

how babies acquire this infection. During the first week of life, lethargy, diarrhea, vomiting, and feeding problems, sometimes with fever, may occur. These symptoms may disappear, or there may be rapid development of cardiac and respiratory distress, the baby being cyanotic and dyspneic, with evidence of myocarditis. The disease may be fatal, but infants who recover do not display residual damage.

Other clinical syndromes sometimes associated with coxsackieviruses include hepatitis, acute conjunctivitis, gastroenteritis, and febrile illnesses that are accompanied by rashes of the face, neck, and trunk.

LABORATORY DIAGNOSIS

I. Identification of the Organism

These viruses can be propagated in cell cultures or in suckling mice and recognized by distinct pathogenic effects. A number of antigenic types in both groups A and B have been identified by specific antisera that neutralize infectivity of the viruses.

II. Identification of Patient's Serum Antibodies

Rising titers of neutralizing antibodies appear in patients' serums during the course of these viral infections. Their identification with known strains of viruses provides serologic confirmation of coxsackievirus disease. In practice, these tests are usually limited to use in outbreak situations.

III. Specimens for Diagnosis

Throat or nasopharyngeal washings and stool specimens are submitted for viral isolation. Spinal fluid may yield the virus in aseptic meningitis and fecal specimens are positive in this instance also.

Acute and convalescent serum samples are useful for serologic diagnosis.

Epidemiology

I. Communicability of Infection

Coxsackieviruses appear to have a human reservoir only. Their frequent presence in throat washings or feces and the easy spread of infections such as herpangina among familial contacts suggest transmission by fecal-oral or oral-oral direct or indirect contacts. The incubation period is short (a few days), and the communicable period is probably limited to the acute stage of illness. Encounters with these viruses appear to be frequent, resulting in the formation of antibodies that persist for years. Many adults possess antibodies to a wide variety of antigenic types.

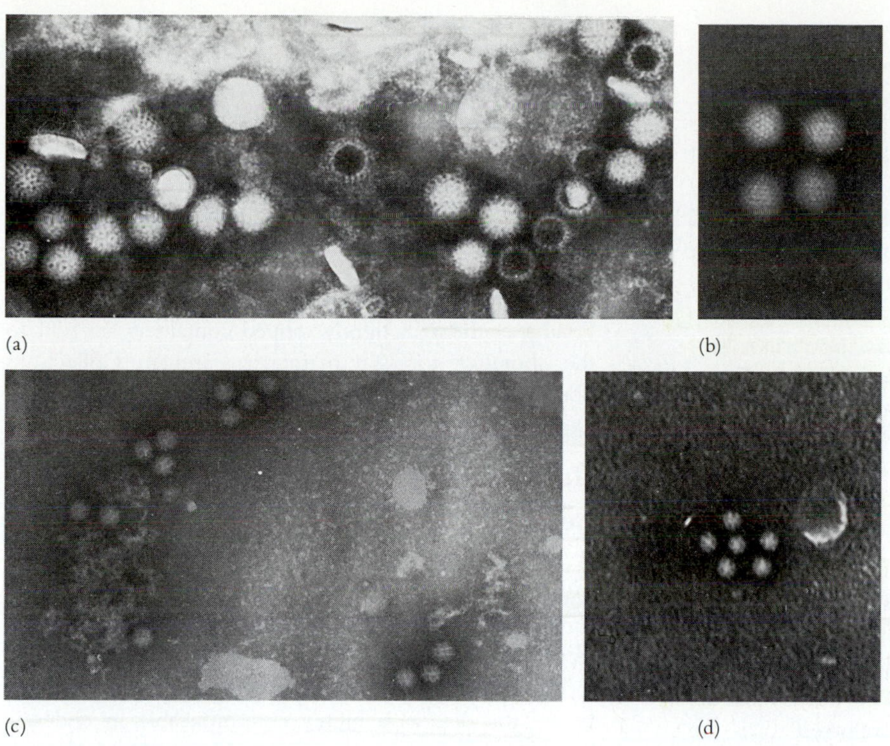

FIGURE 17.4 Common gastroenteritis viruses from stool specimens of pediatric patients with acute gastroenteritis (×140,000). (a) Rotaviruses; (b) adenoviruses in stool; (c) caliciviruses; (d) astroviruses.

Reproduced from M. L. Christensen, "Human Viral Gastroenteritis," *Clinical Microbiology Reviews* 2:51–89, 1989. Photo by Cynthia Howard. American Society for Microbiology.

II. Control of Active Infection

The question of isolation and precautionary care of these infections is often academic because virologic diagnosis is seldom complete until long after the patient recovers, and clinical diagnosis is difficult or inconclusive. When the disease is suspected, patients are treated with standard precautions except that contact precautions are used for infants and young children for the duration of the illness.

ECHOVIRUS DISEASES

Clinical Syndromes

Echoviruses are frequently isolated from the human intestinal tract but are less often documented as causative agents of disease. Diagnosis depends entirely on laboratory tests in which virus is isolated and identified or serum antibodies are detected in significant titer or both.

Diseases that echoviruses produce are similar to those caused by coxsackieviruses. They range from common colds to febrile respiratory illnesses, gastroenteritis, and aseptic meningitis. The latter syndrome may include muscle weakness and spasm and must be differentiated from poliomyelitis. This type of disease in young children is often accompanied by a rash.

LABORATORY DIAGNOSIS

Echoviruses grow well in monkey kidney cells. Their isolation from throat washings or feces (spinal fluid, in aseptic meningitis) affords the best diagnosis. More than 30 antigenic types of echoviruses have been recognized, and it is not uncommon to find more than one type in a given specimen. The size of this group makes serologic diagnosis difficult under ordinary circumstances, for the patient's serum must be tested against the entire battery, often by more than one serologic method.

Epidemiology

The echovirus group appears to have a worldwide distribution. Infections occur frequently during infancy and childhood, and the viruses have a rapid spread among intimate contacts. Echovirus infections probably occur far more widely in the community than is recognized on the basis of incidence seen in hospitalized patients, but there are no effective means of controlling community spread. The protection of young infants from older children or adults with acute febrile illnesses is always advisable.

VIRAL GASTROENTERITIS

Beginning in the early 1970s, several groups of viruses were seen by electron microscopy in the stools of patients with diarrhea (fig. 17.4). However, these viruses, which include rotaviruses, enteric adenoviruses, Norwalk and Norwalk-like viruses, caliciviruses, and astroviruses, could not be isolated in routine cell cultures. Most infect infants and young children, sometimes in epidemic form, but some, especially the Norwalk agents, also infect adults. Discovery of these viruses has expanded knowledge of the etiology and epidemiology of nonbacterial diarrheal illness in all age groups. The current difficulties in isolating these agents in culture has hindered attempts at their control, especially by vaccination.

ROTAVIRUSES

The Clinical Disease

Rotaviruses were first described in 1973 when they were seen in electron micrographs of duodenal biopsies and fecal extracts from children with nonbacterial gastroenteritis. Worldwide, rotaviruses are responsible for 50% or more of the acute diarrheal disease of infants and young children (less than two years of age) who require hospitalization. Nosocomially acquired rotavirus infection is common in children hospitalized for other reasons, and the virus is frequently responsible for epidemics in day-care centers. Every year, approximately 3.5 million cases occur in the United States, 140 million cases are reported worldwide, and almost 1 million deaths result from these infections.

The disease is characterized by acute onset of vomiting and watery diarrhea, symptoms that last for approximately 6 days. These signs are accompanied by fever and abdominal pain. The virus replicates in epithelial cells on the tips of villi of the small intestine. Loss of fluid and electrolytes can lead to severe dehydration and death if replacement therapy is not initiated. In temperate climates, most disease occurs during the cooler months (it was once referred to as the "winter vomiting" disease).

Subclinical human rotavirus infections apparently are common. By the end of the third year of life, almost all infants have developed antibody to the virus, and high antibody levels continue into adulthood, suggesting frequent subclinical reinfection. Rotavirus is transmitted by the fecal-oral route and is highly contagious. Many parents may be subclinically infected at the time their children are ill. In addition, ingestion of large numbers of rotavirus particles can overcome immunity in adults. Foodborne and waterborne outbreaks of disease in adults have been reported.

Rotaviruses have also been detected in the stools of many young animals with diarrhea, including calves, piglets, rabbits, poultry, and kittens, but animal-to-human transmission of the virus has not been documented. Currently, five groups of rotaviruses are known: A, B, C, D, and E. Group A causes most human infections in North America and Europe, whereas group B has been responsible for large outbreaks in adults of all ages in China. Groups B through E are found primarily in animals.

LABORATORY DIAGNOSIS

I. Identification of the Organism

The agent is a 70-nm RNA virus with a double-layered capsid that gives it the appearance of a wheel (*rota* is the Latin word for "wheel"). The organism can be seen in fecal extracts examined by electron microscopy (see fig. 17.4a), but most laboratories diagnose the disease by using an enzyme immunoassay to detect rotavirus antigen in stools. A latex agglutination assay is also available but is not as sensitive as immunoassay. Several PCR techniques for detecting rotavirus have been developed in research laboratories but these tests are not currently available for routine use in the clinical laboratory. Special tests and procedures have been used to grow the organism in culture, but these are not routinely available either.

II. Identification of Patient's Serum Antibodies

Neutralization and enzyme immunoassay tests can detect IgA, IgM, and IgG antibodies, but these are not routinely performed.

III. Specimens for Diagnosis

Fecal specimens in tightly capped containers are sent to the laboratory for assay of rotavirus antigen. Collections obtained during the first 2 to 3 days of illness are most likely to yield positive results.

Epidemiology

I. Communicability of Infection

A. *Reservoirs, Sources, and Transmission Routes* Humans are the reservoir of group A rotavirus. Disease is presumably spread by the fecal-oral route. Up to 1 trillion infectious particles per milliliter of stool are shed, and the infectious dose in children may be as small as 10 particles; hence, the disease is highly communicable in endemic form.

B. *Incubation Period* The incubation period is approximately 24 to 72 hours but usually is less than 48 hours.

C. *Communicable Period* During the acute stage of disease while the virus is shed in feces, usually 6 to 8 days.

D. *Immunity* After 3 years of age, most persons are immune to the disease. Infants 1 to 2 years old and immunocompromised children of all ages are most susceptible to infection. Malnourished children in underdeveloped countries are also highly susceptible.

II. Control of Active Infection

A. *Transmission-based Precautions* Patients are cared for with standard precautions except that contact precautions are used for diapered or incontinent children. A private room is needed if patient hygiene is poor.

B. *Other Precautions* Because of the high infectivity of the virus, gowns and gloves should be worn whenever caring for patients and handling their feces and vomitus. Scrupulous hand washing is mandatory.

C. *Treatment* Oral water and electrolyte replacement are the primary means of therapy to avoid potentially fatal dehydration. No specific antiviral therapy is available.

D. *Control of Contacts* Protect infants and young children from persons with acute gastroenteritis, especially in hospital and day-care settings.

E. *Epidemic Controls* Search for and eliminate possible food and water sources. These are unusual in comparison with person-to-person transmission.

III. Prevention

A. *Immunization* An oral vaccine containing attenuated rotaviruses is currently being tested in the United States, but results have been inconsistent.

ADENOVIRUSES

The Clinical Disease

Adenoviruses of the serologic types that cause primarily respiratory disease (see table 13.1) were isolated in cell culture from patients with viral gastroenteritis in the 1960s. In the late 1970s, however, adenovirus particles that could not be cultured were seen by electron microscopy in the stools of children with diarrhea (see fig. 17.4b). These viruses, at first called *fastidious enteric adenoviruses,* belong to specific serotypes numbered 40 and 41, and they are the primary agents of adenovirus gastroenteritis. They are responsible for 5 to 20% of hospitalizations for childhood diarrhea in developed countries. The viruses cause disease primarily in children less than two years of age, but older children and adults may also be infected, often asymptomatically. The most prominent symptom is diarrhea, with or without vomiting and fever, often accompanied by respiratory symptoms. The incubation period is between 3 and 10 days, and illness may last more than 1 week.

Infections occur throughout the year without seasonal incidence, tend to be endemic rather than epidemic, and spread person to person. Outbreaks occur primarily in hospitals or day-care settings; no food- or waterborne transmission has been reported.

LABORATORY DIAGNOSIS

I. Identification of the Organism

Adenoviruses are 70 to 75-nm DNA viruses. Forty-one serotypes cause human disease, but types 40 and 41 are most commonly associated with gastroenteritis. Specialized culture methods for viral isolation have been developed, but routine diagnosis depends on electron microscopy or, more usually, enzyme immunoassays performed on fecal specimens. Serologic tests are not routinely available.

II. Specimens for Diagnosis

Stool specimens collected early in the course of illness for electron microscopic examination or antigen detection by enzyme immunoassays.

Epidemiology

I. Communicability of Infection

A. *Reservoirs, Sources, and Transmission Routes* Humans are the reservoir of infection, and transmission is by the fecal-oral route.

B. *Incubation Period* Generally between 3 and 10 days.

C. *Communicable Period* As long as viruses are shed in feces. Illness is longer with adenoviruses than with other enteric viral pathogens, sometimes up to 2 weeks.

D. *Immunity* Long-term immunity most likely is acquired during childhood infection.

E. *Treatment* Fluid replacement provides supportive therapy.

II. Control of Active Infection

Control is the same as for rotavirus gastroenteritis except that food- and waterborne disease have not been described.

III. Prevention

No vaccine is available for adenovirus gastroenteritis.

NORWALK AND NORWALK-LIKE VIRUSES

In 1972 a virus that is now called the **Norwalk virus** was seen by electron microscopy in the stool of a patient involved in a gastroenteritis outbreak in Norwalk, Ohio. This organism is now known to be involved in epidemics of viral gastroenteritis affecting primarily older children and adults. In addition, several 25- to 30-nm viruses detectable in the stools of patients with gastroenteritis have properties in common with the Norwalk virus and thus are called Norwalk-like viruses.

The Norwalk viruses are associated with outbreaks of viral gastroenteritis occurring in families, schools, communities, and institutions and have been responsible for outbreaks on cruise ships, in resort camps, in colleges, and in nursing homes. On the basis of antibody surveys, the Norwalk virus group appears to have a worldwide distribution. It probably causes disease during all seasons of the year.

Infection with the Norwalk viruses is most likely spread from person to person by the fecal-oral route. Outbreaks have occurred after ingesting contaminated ice, oysters, and drinking water and after swimming in a contaminated lake. The incubation period is approximately 24 to 48 hours. The most common symptoms are nausea, vomiting, abdominal cramps, lethargy, and diarrhea lasting for approximately 12 to 24 hours. The illness is typically mild and self-limited, and patients rarely require hospitalization. Virus is shed maximally at onset of disease and for 2 to 3 days following onset.

The clinical diagnosis of Norwalk virus disease is difficult because the symptoms are nonspecific. Laboratory diagnosis is hampered because the virus does not grow in cell culture or in experimental animals. It was found initially by immune electron microscopy (IEM), a procedure in which virus can be identified by adding convalescent patient's serum to stool extracts and observing the preparation under the electron microscope for the presence of antibody-coated virus. This method is primarily a research tool and not adaptable to the routine clinical laboratory. Radioimmunoassays and enzyme immunoassays for antigen detection have been developed but are not widely available.

Serologic tests have permitted the study of a large number of sera and revealed that approximately 50% of adults have antibodies against the Norwalk agent.

Treatment consists primarily of fluid replacement. Effective hand washing, proper disposal or disinfection of contaminated materials, and alertness to the purity of drinking water or water in swimming pools can help to limit the number of outbreaks caused by these agents.

No vaccine is available.

Caliciviruses

In 1976, the first two reports of **calicivirus** infections in babies with diarrhea appeared. Like other enteric viruses, the viruses were first seen by electron microscopic examination of diarrheic stools and could not be grown in cell culture. The virus derives its name from the 32 cup-shaped depressions on the virus surface (*calyx* = cup or chalice). At certain rotations of the virus, these depressions resemble a six-pointed star like the Star of David (fig. 17.4c). Some virologists consider the Norwalk virus to be a calicivirus.

Studies in England and the United States have revealed that approximately 3% of children hospitalized for diarrhea excrete caliciviruses. The incubation period is from 1 to 3 days, with illness lasting approximately 4 days. The predominant symptoms are vomiting and diarrhea and resemble those of rotavirus infection. Antibody studies show widespread immunity in persons older than 12 years.

Person-to-person transmission is presumed to be responsible for most endemic disease, but some cases have been traced to contaminated shellfish and drinking water. Several outbreaks in England and Japan have occurred in institutional settings (e.g., an orphanage, a school, and nursing homes), but no outbreaks have been reported in the United States.

Treatment and control are the same as for other viral enteric diseases. No vaccine is available because the virus cannot be grown in vitro.

Astroviruses

Astroviruses are 28- to 30-nm viruses with a star-shaped configuration on their surface (fig. 17.4d). They are responsible for approximately 5% of cases of gastroenteritis in children less than seven years of age. Adults can also be infected but experience a milder disease. Most adults have antibodies against the virus.

The incubation period is 24 to 36 hours, with symptoms lasting 1 to 4 days. Watery diarrhea, vomiting, fever, and abdominal pain are common. Transmission is primarily person to person in children and in institutional settings, but contaminated water and shellfish have been responsible for a few disease outbreaks. No specific therapy other than fluid and electrolyte replacement is available.

Hepatitis Viruses

Viral hepatitis is a major cause of acute liver disease in the United States and other parts of the world. It is caused by several viruses including five referred to alphabetically as hepatitis viruses A, B, C, D, and E. Of these five, only hepatitis viruses A and E are transmitted by the alimentary route. The others are transmitted parenterally, primarily by blood transfusion, and are discussed in chapter 21.

Hepatitis A Virus (HAV)

Hepatitis A virus is a small, 27-nm picornavirus morphologically similar to a number of other small, spheric viruses. It can be demonstrated in the feces of approximately 50% of patients with acute hepatitis A using immune electron microscopy (IEM) along with a specific HAV antiserum to produce antigen-antibody aggregates that can be seen with the electron microscope. HAV is shed in feces late in the incubation period and in the early days of illness, corresponding to the time of known fecal infectivity.

Enzyme immunoassays have been developed for detecting HAV antigen in stool specimens and HAV antibodies in patients' serums. IEM has been used to demonstrate clumping of the fecal virus particles by acute and convalescent serums from patients with hepatitis A.

Hepatitis A is known also as infectious hepatitis or epidemic hepatitis. Its primary transmission is by the fecal-oral route, and therefore its spread may occur in an epidemic as well as sporadic pattern. It may also be transmitted by parenteral injection, as shown in volunteers, but serologic studies indicate that it is rarely spread by blood transfusion. Contaminated needles and syringes may play a role among drug addicts, but for practical purposes the natural spread of hepatitis A appears to be from person to person. The route may be direct or by way of contaminated food or water. The incubation period ranges from 15 to 45 days followed by abrupt onset of symptoms.

The Clinical Syndrome

Clinically, infection with HAV may result in a spectrum of syndromes. Often it begins as many other infections do, with fever, malaise, vomiting, anorexia, and discomfort. These symptoms subside after a few days and are succeeded by jaundice. The liver may be enlarged and tender, and bile may appear in the urine. Injury appears to be restricted to the liver, the degree of destruction to parenchymal cells determining the severity and mortality of the disease. In the most fulminating cases, which are unusual, death may occur within about 10 days; in more slowly advancing cases death occurs in 1 or 2 months. The fatality rate is low, however, and the majority of cases recover, though residual liver damage may take some time to repair. The disease in children is mild and probably occurs more frequently than is recognized. The infection is more severe in adults with underlying liver disorders and in pregnant women.

Clinically inapparent infection is more common than symptomatic disease, the proportion being estimated at 50 to 90%.

LABORATORY DIAGNOSIS

I. Identification of the Organism

In general, the agents of viral hepatitis have not been cultivated in cell culture or in experimental animals other than primates. Although now available, cell culture and IEM are not practical or available on a routine basis for identifying HAV. The diagnosis of hepatitis A rests with laboratory tests for liver function and HAV serum antibodies. Clinical findings and epidemiologic information are also important.

II. Identification of Patient's Serum Antibodies

Serologic diagnosis of hepatitis A is made primarily by IEM and enzyme immunoassay. At the time of onset of symptoms, IgM antibodies are already present in the serum. The IgM titer drops off after 3 to 6 months, and IgG antibody is then detectable.

III. Specimens for Diagnosis

Until convenient methods for virus isolation are developed, the diagnosis is made on clinical and epidemiologic grounds, supported by tests for abnormal liver function and serologic evidence of the disease. Blood for detecting antibodies to HAV and to other viral hepatitis agents should be collected.

Epidemiology of Hepatitis A

I. Communicability of Infection

A. *Reservoirs, Sources, and Transmission Routes* Humans and some nonhuman primates are the reservoir. There are three general sources of orally transmitted infection: (1) water supplies contaminated by human sewage; (2) food such as shellfish from naturally contaminated sources, especially if not cooked at temperatures that inactivate the virus; and (3) food prepared or handled by infectious persons who practice poor hygiene. Cooking in the last instance is irrelevant if the food is handled just before serving.

Intimate contact with infected persons may result in transmission, and in families or institutions it may lead to epidemics. The most explosive epidemics are likely to originate from drinking water, particularly in camps during summer months.

Parenteral transmission is possible (but uncommon) through injection of contaminated blood, either by transfusion or through the use of contaminated medical, dental, or tattooing instruments.

B. *Incubation Period* Varies from 15 to 45 days, with an average of 1 month.

C. *Communicable Period* Studies in human volunteers and epidemiologic evidence indicate that the disease is communicable from the latter half of the incubation period and for a few days after onset of jaundice. Studies on fecal excretion in humans by using immune electron microscopy, however, suggest that the person who already has symptoms of hepatitis A is unlikely to transmit the infection to others.

D. *Immunity* Immunity is widespread in the adult population as a result of childhood infection. Second attacks of clinically apparent HAV disease are rare. Early infections are probably largely inapparent or so mild as to escape recognition. The persistence of antibodies in adult serum and their concentration in preparations of gamma globulin make it possible to provide passive immunoprophylaxis for contacts.

II. Control of Active Infection

A. *Transmission-based Precautions* Hospitalization of uncomplicated cases is discouraged because it exposes other patients and medical personnel to the risks of infection. Isolation at home is also unwarranted, as family contacts have been fully exposed before the patient's symptoms begin. In the hospital, a private room is not required except for children or uncooperative adults.

B. *Other Precautions* Standard or contact (for children) precautions are emphasized in the care of viral hepatitis patients. Concurrent disinfection techniques should be applied to the disposal of feces, urine, and blood. Only sterile, disposable needles and syringes should be used.

C. *Treatment* There is no specific therapy for infectious hepatitis.

D. *Control of Contacts* Isolation of contacts is not indicated, but case finding is important so that close contacts may receive passive immunization with immune globulin. IgG antibodies present in the immune globulin provide protection for about 2 months if given within the first week after exposure. Later immunization may also be effective, and if exposure is prolonged a second dose may be indicated.

E. *Epidemic Controls* Hepatitis A is a reportable disease to protect contacts and to prevent possible epidemic outbreaks. A search is made for the source of infection when the incidence of cases is unusually great. Mass administration of immune globulin is sometimes indicated when institutional outbreaks occur. Sanitary and hygienic standards are reviewed and strengthened as indicated, with particular regard to sewage disposal and the protection of food and water supplies.

III. Prevention

In 1995, an inactivated hepatitis A vaccine was licensed for use in persons two years of age or older. Two doses are given to children between 2 and 18 years of age: a first dose and a booster dose 6 months later. For persons immunized at 18 years of age or older, a higher concentration of vaccine is given, also in two doses spaced 6 months apart. Studies have indicated that almost all children, adolescents, and adults develop protective levels of antibody after completing the immunization series. Protection is established by 4 weeks after the initial dose and is estimated to last for at least 20 years. The vaccine is recommended for all susceptible persons traveling to or working in countries with a high or intermediate incidence of hepatitis A virus disease. People who might be at risk of contracting or transmitting the disease should also receive vaccine. These include persons working with HAV-infected animals, persons with chronic liver disease, and food handlers.

In addition, passive immunization is useful for travelers, medical personnel, and others under immediate risk of exposure and children less than two years of age. When immune globulin is given to exposed contacts early in the incubation period, only mild or asymptomatic infections result. Although passive immunity prevents clinical disease from developing in these exposed persons, they develop long-term, protective, active immunity in response to the viral infection.

SUMMARY TABLE 17.1 VIRAL DISEASES ACQUIRED THROUGH THE ALIMENTARY ROUTE

Clinical Disease	Causative Organism	Other Possible Entry Routes	Incubation Period	Communicable Period
Poliomyelitis	Poliovirus		Usually 1–2 weeks, may range from 3 days to 4–5 weeks	Before onset of symptoms and during infection
Coxsackievirus Diseases	Coxsackieviruses	Respiratory tract	3–5 days	During acute, febrile stage of illness
Echovirus Diseases	Echoviruses	Respiratory tract	Unknown	During acute, febrile stage of illness
Viral Gastroenteritis	Rotaviruses		24–48 hours	Through first week of illness
	Adenoviruses	Respiratory tract	3–10 days	May be up to 2 weeks
	Norwalk virus group		24–48 hours	2–3 days after onset of illness
	Caliciviruses		1–3 days	
	Astroviruses		24–36 hours	
Viral Hepatitis	Hepatitis A virus	Parenteral route	15–45 days	During incubation period and possibly through first week of symptoms
	Hepatitis E virus		4–8 weeks	Same as above

HEPATITIS E VIRUS (HEV)

Hepatitis E virus is one of a group of viruses formerly referred to as *non-A, non-B hepatitis viruses* because their exact identity was unknown. Within the past few years, two members of this group, hepatitis C virus (see chap. 21) and hepatitis E virus, have been identified. Much progress has been made in understanding their molecular structures, modes of transmission, and methods for control, but much remains to be learned.

Hepatitis E virus was not characterized until 1990. It is most closely related to the calicivirus family and is unrelated to hepatitis A virus, a picornavirus. Like HAV, however, it causes viral hepatitis transmitted by the fecal-oral route. Several large waterborne outbreaks of HEV have been reported, primarily in tropical and subtropical countries with inadequate public sanitation and with poorly nourished populations. Pregnant women are at increased risk for severe infection; a mortality rate of 20% has been seen in some outbreaks. In otherwise healthy persons, disease is mild and self-limited. In developed countries, infection occurs in persons who have traveled to areas where the disease is endemic. Good sanitation and better hygienic practices presumably limit the spread of HEV in developed countries. Diagnosis of infection is usually established by detecting a significant rise in IgG titer between acute and convalescent serum specimens.

QUESTIONS

1. What is the reservoir for poliovirus? How is the disease transmitted?
2. How does the oral polio vaccine (OPV) protect an individual from acquiring poliomyelitis?
3. What are the most common manifestations of coxsackievirus disease?
4. How is rotavirus transmitted? How is the disease diagnosed? Who is most susceptible?
5. How are adenoviruses transmitted? How is the disease diagnosed?
6. What hepatitis viruses are transmitted by the alimentary route?
7. How is infection with hepatitis A virus (HAV) diagnosed?
8. What type of immunization is available for hepatitis A? When is it indicated?
9. What prevents the spread of hepatitis E virus (HEV) in developed countries?

Specimens Required	Laboratory Diagnosis	Immunization	Treatment	Transmission-based Precautions
Throat washing or swab	Virus isolation	IPV (Salk vaccine, killed) OPV (Sabin vaccine, live)	None specific	Standard precautions
Rectal swab in transport medium				
Acute and convalescent serum	Rising antibody titer			
Throat and nasopharyngeal washings	Virus isolation	None	None specific	Standard precautions except contact precautions for infants and young children
Stool specimens				
Spinal fluid				
Acute and convalescent serum	Rising antibody titer			
Throat washings, feces, spinal fluid	Virus isolation	None	None specific	See above
Stool specimen	EIA for viral antigen	None	None specific	Standard precautions
Acute and convalescent serums	EIA, neutralization			
Stool specimen	EIA for viral antigen	None	None specific	Standard precautions
Stool specimen	IEM (if available)	None	None specific	Standard precautions
Acute and convalescent serums	Rising antibody titer			
Stool specimen	EM (if available)	None	None specific	Standard precautions
Stool specimen	EM (if available)	None	None specific	Standard precautions
No microbiologic tests available	None	Inactivated vaccine	None specific	Standard precautions
Blood serum	EIA for IgM and IgG against hepatitis A	Passive		
Blood serums		None	None specific	Standard precautions

BIBLIOGRAPHY

Benenson, A. S., ed. 1995. *Control of communicable diseases in man.* 16th ed. Washington, D.C.: American Public Health Association.

Centers for Disease Control and Prevention. 1990. Recommendations for collection of laboratory specimens associated with outbreaks of gastroenteritis. *Morbid. Mortal. Weekly Rep.* 39: RR-14.

Centers for Disease Control and Prevention. 1990. Viral agents of gastroenteritis: Public health importance and outbreak management. *Morbid. Mortal. Weekly Rep.* 39: RR-5.

Centers for Disease Control and Prevention. 1995. Progress toward global poliomyelitis eradication 1985-1994. *Morbid. Mortal. Weekly Report* 44 (4): 273.

Centers for Disease Control and Prevention. 1996. Prevention of hepatitis A through active or passive immunization. *Morbid. Mortal. Weekly Rep.* 45 (RR-15).

Centers for Disease Control and Prevention. 1996. Surveillance for waterborne disease outbreaks–United States, 1993-1994. In CDC surveillance summaries. *Morbid. Mortal. Weekly Rep.* 45: SS-1.

Christensen,. M. L. 1989. Human viral gastroenteritis. *Clin. Microbiol. Rev.* 2 (1): 51–89.

Dock, N. L. 1991. The ABC's of viral hepatitis. *Clin. Microbiol. Newsl.* 13 (3): 17–20.

Greenberg, A. E., ed. 1995. *Standard methods for the examination of water and wastewater.* 19th ed. Washington, D.C.: American Public Health Association.

Lennette, E. H., P. Halonen, and F. A. Murphy, eds. 1988. *Laboratory diagnosis of infectious diseases: Principles and practice, Vol. II., Viral, rickettsial, and chlamydial diseases.* New York: Springer-Verlag.

Mandell, G. L., J. E. Bennett, and R. Dolin. 1995. *Principles and practice of infectious disease,* 4th ed. New York: Churchill Livingstone.

Melnick, J. L. 1996. Current status of poliovirus infections. *Clin. Microbiol. Rev.* 9 (3): 293–300.

Paul, J. R. 1972. *A history of poliomyelitis.* New Haven, Ct.: Yale University Press.

Polio eradication: Part and parcel of EPI (expanded programme on immunization). 1989. *World Health* December: 6–9.

Sepples, S. B. 1992. Polio nursing: The fight against paralysis—In the years before an effective vaccine was available. *Nursing Connections* 5 (3): 31–38.

Smith, J. S. 1990. *Patenting the sun.* New York: William Morrow.

CHAPTER 18

Parasitic Diseases Acquired through the Alimentary Route

OUTLINE

Microbes in Motion

This chapter covers the important parasites acquired through the alimentary route including their life cycles, diseases they cause, laboratory diagnosis, epidemiology, treatment, and control. The following books and chapters of the Microbes in Motion CD-ROM may be a useful preview or supplemental study aid to your reading: Parasitic Structure and Function: Protozoa Classification; Intestinal Protozoa; Tissue and Blood Protozoa; Cestodes; Nematodes; Trematodes.

OBJECTIVES

On completing this chapter the student will be able to:

1. describe the life cycles of several animal parasites.
2. discuss the major clinical features of parasitic diseases.
3. describe the epidemiology of parasitic diseases.
4. differentiate between intestinal and invasive parasitic infections.
5. relate the clinical treatment and management of intestinal and invasive parasitic diseases to the life cycles of the parasites involved.

KEY WORDS

Amebiasis (am-ee-**BI**-ah-sis)
Ascariasis (ask-are-**EYE**-ah-sis)
Balantidiasis (bal-an-ti-**DIE**-ah-sis)
Blastocystosis (blasto-oh-sis-**TOE**-sis)
Cestodes (**SESS**-toads)
Clonorchiasis (clahn-ork-**EYE**-ah-sis)
Cryptosporidiosis (cryp-toe-spore-id-i-**OH**-sis)
Cyclosporiasis (si-klo-spore-**EYE**-ah-sis)
Cysticercosis (sis-ti-circ-**OH**-sis)
Diphyllobothriasis (di-phil-oh-bahth-**RYE**-ah-sis)
Echinococcosis (ee-kine-oh-cock-**OH**-sis)
Enterobiasis (entero-**BI**-ah-sis)
Fascioliasis (fash-ee-oh-**LIE**-ah-sis)
Fasciolopsiasis (fash-ee-oh-lop-**SIGH**-ah-sis)
Flukes
Giardiasis (gee-are-**DIE**-ah-sis)
Hymenolepiasis (hi-men-oh-lep-**EYE**-ah-sis)
Intestinal parasites
Invasive parasites
Microsporidiosis (my-cro-spore-id-i-**OH**-sis)
Nemathelminths (knee-mat-**HELL**-minths)
Nematodes (**NEEM**-ah-toads)
Paragonimiasis (para-gone-ee-**MY**-ah-sis)
Platyhelminths (plat-ee-**HELL**-minths)
Protozoans (pro-toe-**ZO**-ans)
Roundworms
Taeniasis (teen-**EYE**-ah-sis)
Tapeworms
Toxocariasis (tox-oh-car-**EYE**-ah-sis)
Toxoplasmosis (tox-oh-plaz-**MO**-sis)
Trematodes (**TREE**-ma-toads)
Trichinosis (trick-in-**OH**-sis)
Trichuriasis (trick-ure-**EYE**-ah-sis)

THE GENERAL NATURE OF PARASITIC DISEASES

In general, animal parasites that live in or on human tissues induce injury at the site where they are located in the body. A few protozoans are commensal inhabitants of the human bowel, and some helminths (roundworms and tapeworms) infect the intestinal tract as saprophytes. For the most part, however, these organisms are truly parasitic. They depend on one or more hosts to maintain them in nature and damage the host as a result of their parasitic activities. Some of the injury is simply due to their size. Even though small protozoans and immature forms of helminths are microscopic, they can induce structural damage within tissues. In addition, because they are living, metabolizing organisms, they compete for nutrient with the host's own cells. Many use host materials and excrete metabolic end products that are toxic to cells around them or, if disseminated, to distant tissues.

EPIDEMIOLOGY

Animal parasites can enter the human body through one of three routes: the alimentary tract, the skin or mucosa, or the direct parenteral route. Most are taken in through the mouth, either from the feces of other infected persons or eaten in infected animal or aquatic plant foods. These and the diseases they cause are described in this chapter.

The control and prevention of parasitic diseases are based on knowledge of their communicability and an understanding of the developmental requirements of parasites and their life cycles. All major taxonomic groups of parasites (**protozoans, nematodes, cestodes,** and **trematodes**) can produce disease after ingestion. Their subsequent routes of transfer are identified in chapter 8 (see table 8.4) and the life cycles of each are described in this chapter.

Most parasites that enter by the alimentary route localize in the intestinal tract, but a few produce disease by invading adjacent tissues or disseminating larvae to systemic locations.

Patients with active parasitic infections are treated with standard precautions and special care should be taken with feces or other excretions in which infective forms may leave the body. People with active amebiasis should not be involved in food handling until the disease has been treated because amebae may be transmitted by this route.

The source of any parasitic infection should be sought, whether it is an infected human contact or an animal food, to prevent further spread of the disease. Preventive measures are based on interrupting the chain of transfer of parasites to other hosts in which they can develop. Sanitary disposal of feces and urine, sewage control, and protection of drinking-water supplies are always important (fig. 18.1). Animal farming should include measures that prevent infection of animals raised for meat, and meat inspection provides further preventive control. Education of the public regarding the nature of these diseases, the need for careful selection of government-inspected meats, and proper methods of cooking these meats are the final links in the chain of prevention in some cases. In others it rests on public understanding of the role played by certain fish and aquatic plants that are eaten raw or without thorough cooking.

INTESTINAL INFECTION

Protozoans and helminths are **intestinal parasites** that parasitize the bowel producing varying degrees of injury to the intestinal lining. The most invasive protozoan, *Entamoeba histolytica,* actually erodes, penetrating into intestinal submucosa and producing undermined, ulcerative lesions. One ciliated protozoan, *Balantidium coli,* (fig. 18.2), also induces ulcerative lesions, but these are usually not deeply erosive. Some species of amebae and flagellates live as saprophytes in the bowel, inducing no injury.

Helminth infection of the bowel is usually not serious unless the worm actually parasitizes the intestinal tissues or its larval forms migrate beyond the bowel. The hookworm (acquired through larval penetration of the skin; see chap. 20) is a notable example of a helminth adult that parasitizes the body from an intestinal site, attaching itself to the mucosa with a cutting mouth through which it ingests the host's blood. This situation is debilitating and can lead eventually to severe anemia if many worms are involved. In trichinosis and sometimes ascariasis, larval migration into parenteral tissues, rather than intestinal infection, leads to the serious aspects of these diseases.

Many intestinal **roundworms (nemathelminths)** and **tapeworms (platyhelminths)** live saprophytically in the bowel. Some attach with hooklets and suckers but derive nutrient from bowel contents, not from living tissue. Some have no means of attachment, but their activities or products may cause local injury. If they are numerous and large enough (some species of tapeworm may reach up to 20 feet, or 6 meters, in length), they can create intestinal obstruction. The small intestinal roundworms that are not equipped for holding on, for example, the pinworm (*Enterobius*) or whipworm (*Trichuris*), often find their way or are pushed into the cul-de-sac of the appendix. This seldom causes difficulty but may be an incidental finding in appendixes that have been surgically removed for some other reason.

INVASIVE INFECTION

Animal parasites enter tissues, either from intestinal sites or directly through skin (by way of insect vectors or their own penetrative capacity). The body then attempts to deal with these **invasive parasites** as it does any foreign object, by marshaling phagocytic cells and fibrinous exudates to contain them. If this succeeds, scar tissue formation and calcium deposit will complete the job. The resulting degree of impairment of local tissue function depends on the extent of infection: it may be minor and unnoticeable or permanently handicapping. The total effect on health depends on the nature of tissue involved and the extent to which it contributes to the body's vital functions.

FIGURE 18.1 Children in impoverished areas are more prone to parasitic infestations due to poor sanitary conditions
World Health Organization. Photo by R. Blessing.

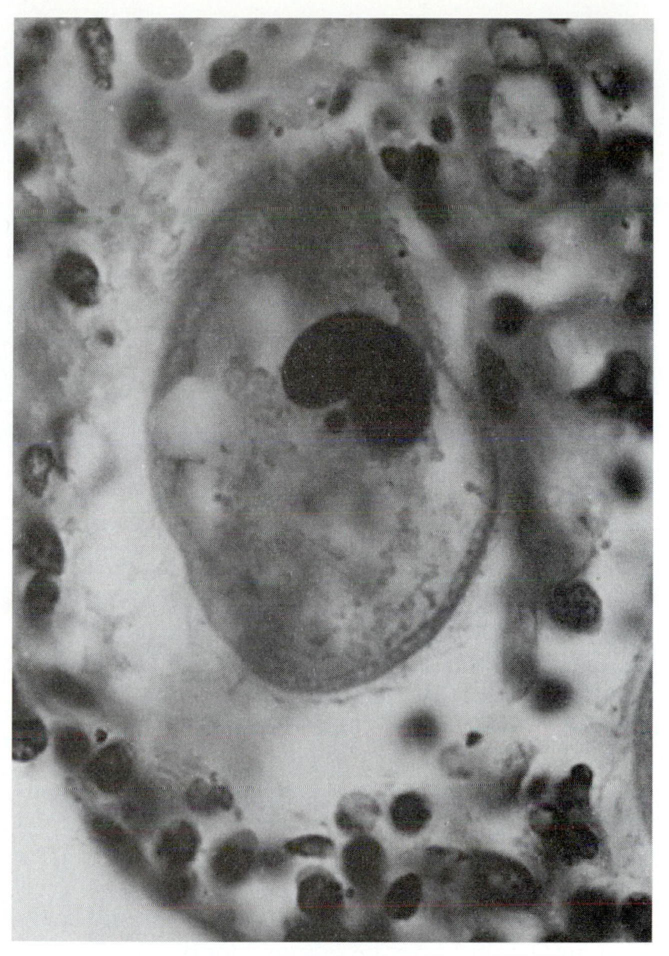

FIGURE 18.2 Trophozoite of *Balantidium coli* with both large macronucleus and small micronucleus visible.
Courtesy Dr. Paul Beaver, Tulane University, New Orleans, LA.

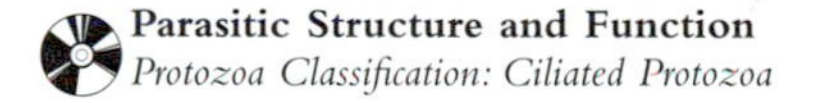
Parasitic Structure and Function
Protozoa Classification: Ciliated Protozoa

The immune response to antigenic components of animal parasites or their products usually is one of cell-mediated immunity and hypersensitivity. This response may occur when antigenic products of intestinal parasites are absorbed from the bowel, but the allergic reaction is most marked in parasitic diseases that are characterized by parenteral localizations of the parasite in either adult or immature stages. Eosinophilic polymorphonuclear cells, which frequently play a role in allergic responses, are characteristically seen in increased numbers in circulating blood in several parasitic diseases. Delayed-type hypersensitivity can sometimes be demonstrated by using skin-testing antigens extracted from parasites. Circulating antibodies are generally not evoked by intestinal parasites but may be produced in response to those that invade parenteral tissues. Detection of humoral antibodies by serologic techniques can be of diagnostic value, therefore, in some invasive infections. Artificial immunization procedures have not been developed, however, because of the generally poor humoral immunity induced by animal parasites and because of the antigenic complexity of these organisms.

LABORATORY DIAGNOSIS

Identification of Parasites

Parasitic infections can seldom be diagnosed on clinical grounds alone. They almost always require laboratory confirmation by microscopic examination of appropriate specimens for ova, larvae, or adult forms of the suspected parasite. Stool specimens are required when intestinal infections by protozoa or helminths are suspected. Small, selected portions of feces may be suspended in saline or water for direct examination as wet-mount cover slip preparations. The numbers of ova or of the parasite in other developmental stages may be few in proportion to the quantity of feces passed at any one time. For this reason, it is often necessary to use concentration techniques to increase the numbers of parasitic forms in the fecal residue.

For diagnosing parenteral infections, specimens of urine, sputum, or blood may be relevant, depending on the clinical indications and the nature of the suspected parasite. In some instances, the location of parasitic forms does not permit them to migrate into excreta, secretions from superficial tissue, or the bloodstream, but biopsies of involved tissue may reveal the organisms.

The various forms of animal parasites that may be present in clinical specimens are usually easily identified by microscopic examination. The ova, larvae, and adult forms of helminthic parasites are large, readily visualized (even without staining, as a rule), and identified by characteristic morphologic features.

Protozoan parasites are also recognizable by microscopic techniques, but they are smaller and sometimes more difficult to find in clinical specimens. A few protozoans can be cultivated on special laboratory media (e.g., *Entamoeba histolytica, Trichomonas, Leishmania, Trypanosoma* spp.) or propagated in laboratory animals (*Toxoplasma gondii*), but these techniques are not always available.

Serology

When laboratory techniques for demonstrating the parasite itself are unsuccessful, serologic methods may be a useful diagnostic adjunct, especially for invasive parasitic diseases. The specificity and sensitivity of current serologic methods are quite reliable for parenteral diseases such as invasive amebiasis, toxoplasmosis, trichinosis, and others. Some of the most useful techniques include indirect hemagglutination (IHA), indirect fluorescent antibody (IFA), bentonite flocculation (BF), and enzyme immunoassays (EIA).

TREATMENT

The treatment of parasitic diseases is often difficult because many antiparasitic drugs are also toxic for the host. Because the goal must be either complete eradication of the parasite or reduction in its numbers, the physician must first know the identity of the parasite, its location, and the relative intensity of the infection. These factors must be weighed against the difficulties inherent in the planned treatment. The physical condition of the patient, particularly one who has been compromised by other diseases, may contraindicate the use of some drugs. In some instances, if more than one developmental stage is involved in the infection, it may be necessary to use a combination of agents. The patient's care may also require supportive measures as well as effective procedures to prevent reinfection.

Treatment of protozoan diseases usually is aimed at complete elimination of the organisms, because protozoa multiply in the host. In the case of helminths, however, unless additional infective stages are ingested, the number of infecting organisms generally does not increase. For this reason, antihelminth therapy may be designed to reduce the density of worm infection rather than eliminate the last worm.

In recent years, a number of broad-spectrum drugs (metronidazole, mebendazole, praziquantel, and others) have been developed for treating both protozoan and helminthic intestinal parasitic diseases. These are less toxic than some of the older drugs and are well tolerated in most instances. The most serious problems in chemotherapy are encountered in invasive infections, particularly those that produce chronic, inflammatory lesions (filariasis, schistosomiasis, the trematode diseases, and others). Extensive fibrosis surrounding the parasite in tissue may render chemotherapy virtually useless, and surgical intervention or supportive treatment may be the only recourse. The problem of drug resistance may also arise, as with some strains of the malaria parasite that now fail to respond to the drugs of choice and must be treated with alternative medications (see chap. 22).

INTESTINAL PROTOZOAN DISEASES

GIARDIASIS

Organism

(*Giardia lamblia.*) A flagellated protozoan (phylum Zoomastigina) with two developmental stages, trophozoite and cyst, causes **giardiasis.** The trophozoite has a tumbling, vibrant motility imparted by four pairs of flagella located on one side of the pear-shaped body, a sucking disc that attaches to the intestinal microvilli to help it resist peristalsis, and a pair of symmetrically located nuclei that look like eyes staring back from microscopic preparations (colorplate 8d). It is about 10 to 20 μm in length and 5 to 15 μm in diameter. The cyst is nonmotile, is oval in shape, has four nuclei when mature (colorplate 8d), and is 11 to 14 μm long and 7 to 10 μm wide. The organism is often shrunken inside the cyst wall so that it appears to be surrounded by a halo.

Life Cycle (fig. 18.3)

I. Definitive Host

Humans are the primary reservoir, but beavers and other wild and domestic animals may be infected.

II. Intermediate Host

None.

III. Human Infection

A. *Source of Infection* Infected human or animal feces contaminating water and food; occasionally person-to-person contact especially among children.

B. *Infective Form* Cysts in contaminated food, water, and feces.

C. *Localization and Development* These organisms reproduce in the upper small intestine. They attach to epithelial cells but do not invade intestinal tissue.

D. *Exit Route* Feces. Trophozoites and cysts may be found, but cysts are typically seen.

IV. Cyclic Development Outside of Human Host

None.

Characteristics of Disease

This organism ordinarily lives saprophytically in the duodenum and jejunum, producing no symptoms until it reaches sufficient numbers to involve large surface areas of the upper intestine. Chronic diarrhea, dehydration, excessive mucus secretion, and flatulence result. The stools are pale in color, most likely because of interference with fat absorption.

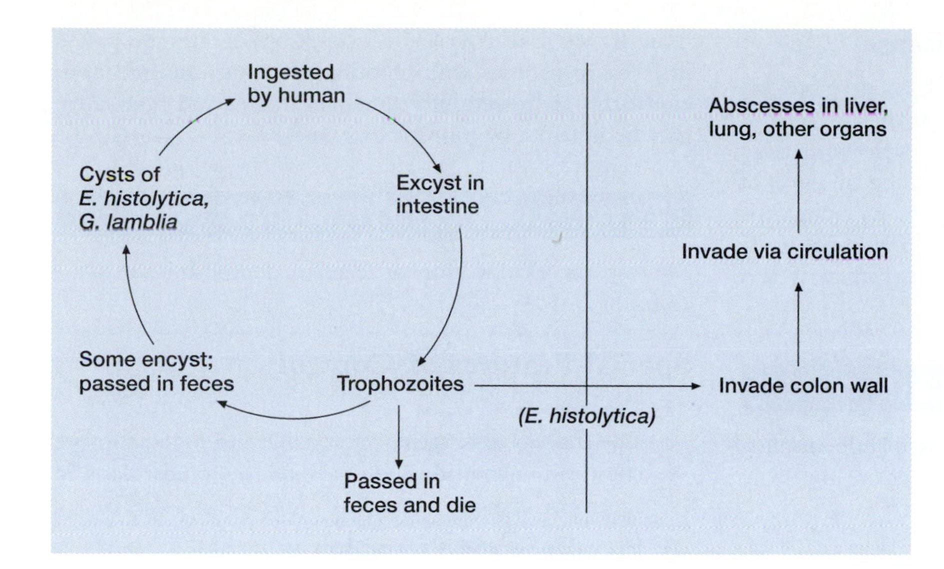

FIGURE 18.3 Human-to-human route of transfer characteristic of *Entamoeba histolytica* and *Giardia lamblia*. Only *E. histolytica* trophozoites may invade the intestinal mucosa and be carried to various body organs, where they cause abscess formation.

Parasitic Structure and Function
Intestinal Protozoa: Intestinal Protozoa Life Cycles

DIAGNOSTIC METHODS

Identification of large numbers of trophozoites or cysts, or both, in feces. Antigen assays to detect *G. lamblia* directly in stool specimens using immunofluorescence and EIA are also available. Both assays, but particularly the EIA, are highly sensitive and are useful in detecting low numbers of organisms.

Special Features of Control

I. Treatment

The drug of choice is metronidazole (Flagyl), but quinacrine (Atabrine) is also widely used. Quinacrine frequently produces headache, dizziness, and vomiting, whereas these effects are uncommon with metronidazole. The latter drug is teratogenic for experimental animals and therefore contraindicated for some patients, especially pregnant women.

II. Prevalence and Prevention

Giardiasis has a worldwide distribution but is more prevalent in areas of poor sanitation. A high incidence is sometimes associated with contaminated tap water. Tour groups traveling to the former Soviet Union often experience a high incidence of giardiasis associated epidemiologically with drinking tap water in St. Petersburg (Leningrad). Children are more frequently infected than adults. In the United States, where outbreaks also occur, the carrier rate varies from 2 to 20%, depending on the age group and community surveyed.

Standard precautions should be followed for infected individuals.

Prevention depends on the same hygienic measures applicable to other diseases transmitted by the fecal-oral route. Individual sources of infection and vehicles of transmission should be sought, especially when outbreaks occur among families, in institutions, or among travelers.

BLASTOCYSTOSIS

Organism

(Blastocystis hominis). A protozoan parasite formerly thought to be a commensal intestinal yeast causes **blastocystosis.** Its taxonomy is uncertain but it may belong in the phylum Rhizopoda along with the amebae. Several morphologic forms have been seen, but in most clinical samples the organism appears spheric with a large central body resembling a vacuole that fills 50 to 95% of the cell. The cytoplasm is seen as a narrow band at the periphery of the cell, clumped into two or four opposing pods that bulge the central body membrane inward. The cell diameter is widely variable, ranging from 4 to 15 μm. A cyst form has been described recently.

Life Cycle

I. Definitive Host

Humans are the primary reservoir but *B. hominis* has been isolated from monkeys, apes, pigs, and guinea pigs.

II. Intermediate Host

None.

III. Human Infection

A. *Source of Infection* Most likely, transmission is by the fecal-oral route either directly from person to person or by contaminated food and water.

B. *Infective Form* Not clearly established; either trophozoites or cysts in contaminated food, water, and feces.

C. *Localization and Development* The organisms develop in the large bowel, producing inflammation of the intestinal lining in some, but not all, patients.

D. *Exit Route* Feces. Trophozoites are most commonly found.

IV. Cyclic Development Outside of Human Host

Probably none.

Characteristics of Human Disease

Not all persons with *B. hominis* in their feces develop gastrointestinal symptoms, therefore some physicians do not consider it a pathogen. However, most people with large numbers of organisms in their stool and the absence of other known pathogens develop symptoms including abdominal discomfort, diarrhea or constipation, anorexia, nausea, and weight loss. Blood, mucus, and leukocytes may also be present in the feces.

DIAGNOSTIC METHODS

Identification of large numbers of characteristic trophozoites in feces. No culture or serologic tests are routinely available.

Special Features of Control

I. Treatment

Metronidazole and iodoquinol (Yodoxin) are used for therapy, but they must often be administered for long periods before the trophozoites are eradicated. Some persons develop chronic disease requiring repeated treatment. Trimethoprim-sulfamethoxazole, diiodohydroxyquin, and emetine may also be effective.

II. Prevalence and Prevention

Reports of the disease have come from many countries worldwide, especially those in which blastocystosis is endemic and where living conditions and hygiene are poor. The incidence in the United States is low compared with that in tropical countries.

Preventive measures for giardiasis apply also to blastocystosis.

BALANTIDIASIS

Organism

(Balantidium coli.) A ciliated protozoan in the phylum Ciliophora, with both a trophozoite and cyst stage causes **balantidiasis.** The surface of the trophozoite is covered with many cilia, the beating motion of which enables the trophozoites to move about rapidly. The trophozoite and cyst contain two nuclei, a large macronucleus and a small micronucleus (fig. 18.2). This organism, which is the only pathogenic ciliate, is also the largest parasitic protozoan, measuring 50 to 200 μm in length and 40 to 70 μm in breadth. Although *B. coli* infects primarily swine, humans are occasional victims when they accidently ingest cysts in infected human or swine feces.

Characteristics of Human Disease

Balantidia produce ulcers or subsurface abscesses in the mucosa and submucosa of the large intestine. Infection may be acute, chronic, or asymptomatic. In acute disease, severe dysentery with bloody, mucoid stools may develop. Extraintestinal infection has been reported with involvement of the peritoneum, kidney, and vagina, but these are uncommon. In active diarrheal infections, trophozoites are passed in feces. In patients with chronic infections, the infectious cyst forms are passed into the environment and from there may be ingested by humans or swine.

DIAGNOSTIC METHODS

Microscopic examination of feces to demonstrate trophozoites or cysts.

Special Features of Control

I. Treatment

The antimicrobial agents tetracycline and iodoquinol are effective in eliminating this parasite. Metronidazole is an alternative choice.

II. Prevalence and Prevention

Although it has a worldwide distribution, the incidence of balantidiasis in humans is low and primarily related to occupational exposure. Outbreaks have been traced to fecally contaminated water, but sporadic transmission is by soiled hands, food, or flies. Prevention depends on good sanitation and hygiene as well as the identification of any source of common infection.

INVASIVE PROTOZOAN AND COCCIDIAN DISEASES

AMEBIASIS (AMEBIC DYSENTERY)

Organism

(Entamoeba histolytica.) An amebic protozoan (phylum Rhizopoda) with two developmental stages; the trophozoite, or vegetative form, and a cyst stage, causes **amebiasis.**

The trophozoite (18 to 20 μm in diameter) is two and one-half to three times the size of an erythrocyte. It has an active ameboid motion and can ingest red blood cells when it parasitizes the human body (see fig. 2.21a,b). The trophozoite has one nucleus and multiplies by either fission or encystation. It can be cultivated through several generations on special artificial laboratory media.

In human infection, encystation of the trophozoite occurs in the cecum and colon. The trophozoite rounds up, extrudes undigested food particles, shrinks in size, and secretes a cyst wall. Young cysts at first have one nucleus, but as they mature two consecutive mitotic divisions take place, producing four nuclei in each fully developed cyst.

When viable mature cysts are swallowed, they pass unchanged through the stomach. When the acid environment around them becomes neutral or slightly alkaline in the small intestine, excystation takes place. The cyst wall weakens and the multinucleate ameba within squeezes out. Another mitotic division of the four nuclei ultimately leads to the formation of eight small trophozoites, or amebulae, each with one nucleus in its cytoplasm.

The amebulae do not colonize in the small intestine but are carried along into the cecum, where they may make contact with the mucosa and become lodged in the glandular crypts. Once they begin to feed and grow, the amebulae develop into mature trophozoites, and the developmental cycle is complete.

When examined microscopically, *E. histolytica* trophozoites and cysts are distinguished from those of other intestinal amebae by the details of their nuclear structure, their cytoplasmic inclusions, and the motility of trophozoites.

Life Cycle (fig. 18.3)

I. Definitive Host

Humans are the only reservoir.

II. Intermediate Host

None.

III. Human Infection

A. *Source of Infection* Infected human feces. Can be spread by direct fecal-oral contact or through contaminated food or water. Water may be a source of epidemics.

B. *Infective Form* Cysts, which can survive in environmental sources.

C. *Localization and Development* Trophozoites parasitize the large bowel. They enter the cyst stage when fecal material becomes more solid and unfavorable for the trophozoites. Cyst forms may persist and be responsible for an asymptomatic carrier state. Erosion of capillaries in intestinal lesions may lead to dissemination of the organisms to the liver, lung, brain, or other organs.

D. *Exit Route* Feces. Trophozoites are found in active intestinal infections; cysts are present in chronic intestinal disease or the carrier state. When infection at other sites occurs, either form may be found in pus from the localized lesion.

IV. Cyclic Development Outside of Human Host

A. *In Environment* Trophozoites may pass into the cyst stage but are easily destroyed by chemical or physical pressures before this occurs. Cysts may survive for some time but do not evolve into trophozoites until ingested.

B. *In Hosts Other Than Humans* Does not establish disease in other hosts.

Characteristics of Human Disease

Amebiasis may be acute, chronic, or asymptomatic. Amebic trophozoites invade the wall of the large bowel, feeding on erythrocytes. They may erode the local area, creating ulcerative lesions that can extend deeply through the submucosa under the entry point and undermining the mucosal surface to produce characteristic flask-shaped ulcers. In acute dysentery, there may be profuse bloody diarrhea, with mucus and pus in the stool. Chronic amebic dysentery is characterized by intermittent mucoid diarrhea, sometimes with blood and pus cells, or there may be only occasional discomfort with mild diarrhea or constipation. When infection extends to extraintestinal sites, focal abscesses form; these are most often found in the liver but may also form in the lung, brain, or other body areas. The organism may also extend to new sites if these abscesses rupture.

DIAGNOSTIC METHODS

Microscopic examination of feces is indicated when intestinal infection is suspected. Trophozoites remain active for a short time only, and for this reason stool specimens should be kept warm during their transport to the laboratory or be placed in a polyvinyl alcohol (PVA) preservative solution to maintain the trophozoite morphology. Saline-purged stool specimens may show the organisms when they are not detected in routinely passed stools.

Aspirated pus from parenteral lesions or sputum may reveal the organisms when systemic disease is suspected.

Culture techniques are sometimes useful when microscopic examination fails to reveal the amebae or cysts.

Serologic tests are useful in diagnosing all forms of amebiasis except the cyst carrier state. They are particularly useful in extraintestinal infections. The most commonly used, highly sensitive method is the indirect hemagglutination test, which is positive in approximately 95% of patients with extraintestinal amebiasis.

Special Features of Control

I. Treatment

Metronidazole is the drug of choice for both intestinal and extraintestinal forms of amebiasis but is contraindicated for pregnant women. This drug is somewhat more effective against trophozoites than cysts. Iodoquinol is used for patients with asymptomatic infections.

Surgical aspiration may be necessary for the treatment of abscesses, but chemotherapy should be given first to inhibit dissemination of the infection during the surgical procedure.

II. Prevalence and Prevention

E. histolytica infection is prevalent in many parts of the world but is found particularly in areas of poor sanitation, such as the tropics. However, even usually well-controlled water supplies can become contaminated, as in a large Chicago outbreak that claimed about 100 lives during the World's Fair of the 1930s. Even in temperate zones, the incidence of amebiasis is estimated at about 5%, whereas that of asymptomatic infection is thought to be much higher.

Patients with active amebiasis are not isolated but are treated with standard precautions. They, as well as asymptomatic cyst passers, should be removed from food-handling occupations until freed of the organism. Sources of infection among the patients' contacts or in their environment should be found if possible.

Preventive measures on a public scale involve sanitary controls of feces and sewage disposal, the protection of water supplies, and health agency supervision of restaurants.

FIGURE 18.4 *Toxoplasma gondii* may be acquired by humans in two ways: ingesting oocysts shed from the definitive host, the cat, or ingesting tissue cysts in uncooked or poorly cooked tissues of intermediate hosts. The organisms excyst in the human and develop into tissue cysts in numerous organs. If pregnant women become infected, they may transmit the parasite to the fetus.

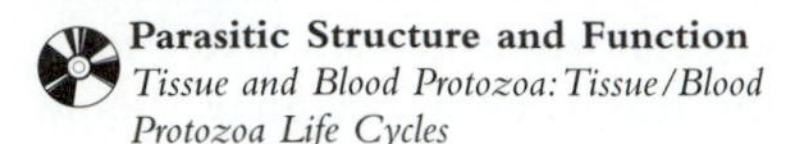
Parasitic Structure and Function
Tissue and Blood Protozoa: Tissue/Blood Protozoa Life Cycles

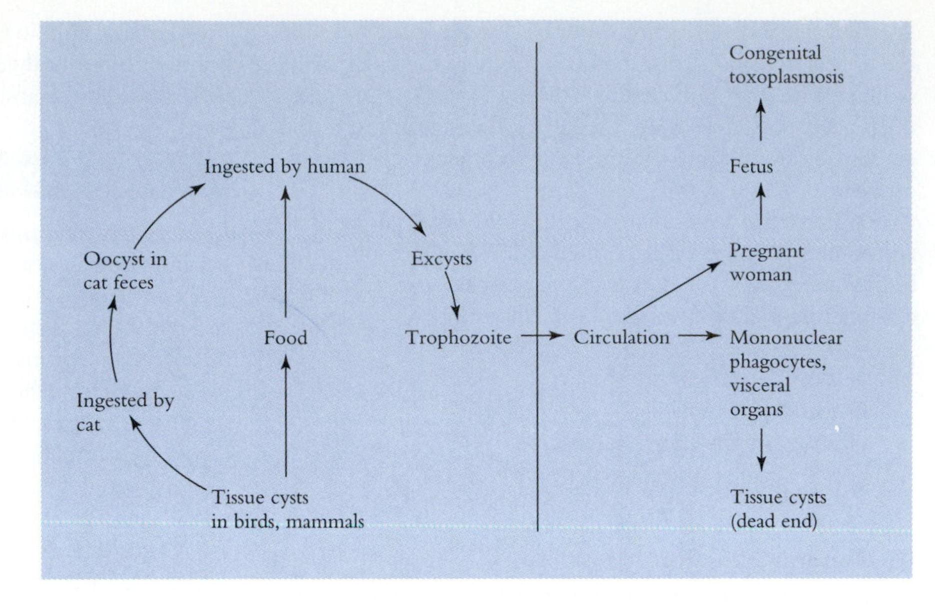

TOXOPLASMOSIS

Organism

Toxoplasma gondii.

Life Cycle (fig. 18.4)

An intracellular protozoan belonging to the phylum Apicomplexa, classified in the order *Coccidia*. *T. gondii* has both sexual and asexual phases in its life cycle, the former taking place in the intestinal epithelium of cats. Forms called *oocysts* are shed in cat feces and develop further in soil, becoming infective for humans and other animals in about 1 to 3 days. When this form is ingested, the trophozoite develops. The latter is a crescent-shaped organism, 4 to 8 μm in length and 2 to 3 μm in width. It can invade and multiply in cells of the mononuclear phagocyte system, blood vessels, and visceral organs, eventually forming tissue cysts with many viable *Toxoplasma* organisms. Cats, which are the principal reservoir and definitive host of the oocystic form, acquire the infection by eating small, infected animals (mice) and birds. Humans and other warm-blooded animals (dogs, cattle, sheep, rodents, chickens) are intermediate hosts, harboring the invasive stage in parenteral tissues.

Characteristics of Human Disease

Human **toxoplasmosis** is acquired through either oral or congenital routes of transmission. Orally transmitted disease has a source either in cat feces or in the inadequately prepared meat of infected animals. Congenital toxoplasmosis is transmitted through the placenta from an infected mother who is usually asymptomatic.

The primary infection, orally acquired, seldom induces serious illness and may often go unrecognized, being marked only by fatigue and muscle pains. More acute disease is rare, with a range of symptoms from fever and lymphadenopathy to generalized muscle involvement or cerebral infection. The symptoms are usually nonspecific, resembling those of aseptic meningitis, hepatitis, pneumonia, or myocarditis, depending on the localization of the parasite.

The organisms may remain dormant in cysts for many years and then reactivate when the host becomes immunocompromised. The AIDS epidemic has brought about a resurgence of cerebral toxoplasmosis, which is a leading cause of death in persons with AIDS.

Congenital toxoplasmosis is also severe, sometimes resulting in death of the fetus. Surviving infants may display a maculopapular rash, liver disease, splenomegaly, or brain damage. Abnormalities include hydrocephaly, microcephaly, jaundice, or chorioretinitis. Cerebral calcifications contribute to mental retardation.

DIAGNOSTIC METHODS

The organisms can sometimes be demonstrated in body fluids or biopsied tissues, but diagnosis depends more frequently on serologic tests. Indirect hemagglutination, indirect fluorescent antibody, and enzyme immunoassay techniques are most useful. Interpretation of results may be difficult because antibody levels, even in persons with subclinical infections, remain high for long periods.

Special Features of Control

I. Treatment

Sulfonamides combined with pyrimethamine are used for treatment but are not completely satisfactory. They are more effective in acute than in chronic forms of toxoplasmosis. Spiramycin is an alternative drug.

II. Prevalence and Prevention

Infection is common in humans and animals, but clinical disease is uncommon. Preventive measures are important primarily for pregnant women who may transmit crippling or fatal disease to their fetuses. They should be alert to the possibilities of acquiring infection through handling or eating raw meat or through contacts with cats and their litterboxes.

CRYPTOSPORIDIOSIS

Organism

Cryptosporidium spp. (usually *C. parvum* in humans).

Life Cycle (fig. 18.5)

Like *Toxoplasma* species, *Cryptosporidium* species are classified as coccidian parasites and have similar life cycles. When ingested by a susceptible host, the *Cryptosporidium* oocysts are immediately infective. Sporozoite forms are released from oocysts in the intestinal tract where they infect the brush border of microvilli. The sporozoites become surrounded by the host cell membrane to form a vacuole at the cell surface. After several asexual divisions, they develop into sexual forms that reproduce into thin- and thick-walled oocysts, 4 to 6 μm in size, containing four sporozoites. The thick-walled oocyst is passed in the feces and is infective for the next host. Thin-walled oocysts may rupture in the intestinal tract to release sporozoites which invade new epithelial cells. This last process leads to persistent **cryptosporidiosis,** especially in immunocompromised patients.

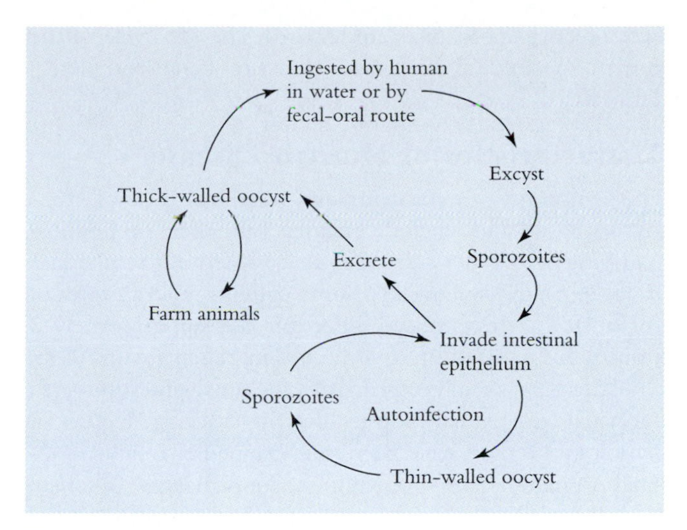

FIGURE 18.5 *Cryptosporidium parvum* can reach high numbers in the intestinal tract because during one stage of its life cycle it can reinfect the human intestine without passing to the outside. Humans ingest it in water contaminated with oocysts or by the direct fecal-oral route.

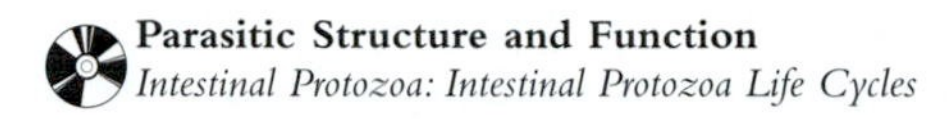
Parasitic Structure and Function
Intestinal Protozoa: Intestinal Protozoa Life Cycles

Characteristics of Human Disease

Cryptosporidium species are a widespread cause of diarrhea among many animals (especially calves and other young farm animals). The disease was once thought to be uncommon in humans, but infection in healthy persons is self-limited, lasting 1 to 2 weeks. Greater awareness of the human disease was brought about when many AIDS patients developed severe diarrhea that was impossible to control. Cryptosporidiosis is but one of many opportunistic infections that afflicts these immunocompromised patients; others, for example, are pneumocystosis, toxoplasmosis, and cytomegalovirus and *Mycobacterium avium* complex disease.

Cryptosporidiosis is transmitted by the fecal-oral route from person to person, animal to person, or in contaminated water. The largest waterborne-disease outbreak ever recorded occurred in Milwaukee, Wisconsin, in 1993 as a result of inadequate removal of *C. parvum* oocysts by the filtration and chlorination treatment of the drinking water. More than 400,000 persons became ill, and 4,400 of these were hospitalized. Clinical symptoms include nausea, low-grade fever, abdominal cramps, anorexia, and several watery, frothy bowel movements a day. In immunocompromised patients who are unable to eradicate sporozoites from the intestinal tract, the disease becomes chronic and may result in death.

DIAGNOSTIC METHODS

Diagnosis is made by examining bowel biopsy material or detecting the parasite in stool specimens. A special type of concentration technique for examining stool specimens must be used; therefore the laboratory should be notified of the suspected diagnosis. The concentrate is examined by acid-fast or direct fluorescent antibody stains for the presence of characteristic thick-walled oocysts. A highly sensitive enzyme immunoassay is also available for detecting *Cryptosporidium* antigen in fecal specimens.

Special Features of Control

Effective antiparasitic therapy for cryptosporidiosis is not yet available. Treatment is primarily supportive with replacement of fluid loss. If possible, any immunosuppressive therapy the patient is receiving should be discontinued. Hospitalized patients are treated with standard precautions. Persons working with animals (e.g., farmers, veterinarians) must be aware of the opportunity for transmission and take precautions when discarding fecal material.

CYCLOSPORIASIS

Organism

Cyclospora cayetanensis.

Life Cycle

Like *Toxoplasma* and *Cryptosporidium* spp., members of the genus *Cyclospora* are classified with the coccidian parasites. They infect several animal species and their life cycles vary with the host. Some *Cyclospora* species complete both their sexual and asexual development in cells of the host's intestinal tract, whereas for others, sexual and asexual development take place in different organs. Oocysts excreted from the host require a period of maturation in the external environment ranging from 4 to 14 days. As they mature, the oocysts develop two sporocysts from each of which two infective sporozoites are released. The life cycle of the human species, *C. cayetanensis,* has not been completely defined.

Characteristics of Human Disease

The symptoms of **cyclosporiasis** resemble those of cryptosporidiosis and include nausea, anorexia, abdominal cramping, and watery diarrhea that sometimes results in a 15 to 20 pound weight loss. Some patients report a fever of up to 102° F. If untreated, infection lasts approximately 2 months but may last up to 4 months in patients with AIDS. Disease seems to affect primarily the small intestine with duodenal and jejunal biopsies showing varying degrees of pathology. Oocysts have been seen trapped in mucus in jejunal aspirates. In biopsy samples, abnormalities observed have included mild to moderate acute inflammation of the intestinal epithelial surface.

Epidemiology

Disease has been reported in the Americas (including the United States), the Caribbean, Africa, Bangladesh, Southeast Asia, England, and Eastern Europe. Although most cases in the United States had been reported by tourists who visited countries with high rates of diarrhea, in the late 1990s, a series of sporadic cases and outbreaks occurred in at least 11 states and in Canada. Some of the outbreaks were related to ingestion of strawberries from California and raspberries, but not all persons who became ill had ingested these fruits.

Epidemiological evidence suggests that the source of the organism is in contaminated water. Organisms resembling *Cyclospora* oocysts have been found in Lake Michigan, in water from a rooftop reservoir of a hospital where 20 employees became ill, and in chlorinated drinking water from a river and municipal supply in Nepal where 14 people developed illness. Further studies of this organism are needed to more completely define its epidemiology.

DIAGNOSTIC METHODS

Cyclospora oocysts can be seen in wet mounts of freshly collected fecal specimens or fecal specimens stained with acid-fast stains (see chap. 4). When viewed under ultraviolet fluorescence using appropriate filters, unstained organisms autofluoresce a green color. The oocysts are 8 to 10 μm in diameter, which differentiates them from those of *Cryptosporidium parvum* (4 to 6 μm). Monoclonal antibodies and nucleic acid probes for use in diagnosis are under development. A low increase in the IgM but not IgG titers between acute and convalescent sera was found in children in Peru. Serologic tests are not readily available and therefore have little use in diagnosis.

Special Features of Control

Unlike other coccidian infections, cyclosporiasis is readily treated with the antibacterial combination of trimethoprim and sulfamethoxazole. Patients generally respond to this therapy within 1 to 3 days. Because widespread occurrence of the disease is a newly recognized phenomenon, transmission-based precautions have not yet been recommended. The disease and its mode of transmission suggest that the control measures used for cryptosporidiosis should serve for control of cyclosporiasis as well. Thorough washing of raw fruits and vegetables before ingestion is a wise precaution.

MICROSPORIDIOSIS

Organism

The term microsporidia encompasses five genera known to cause human **microsporidiosis:** *Enterocytozoon, Septata, Nosema, Encephalitozoon,* and *Pleistophora. Enterocytozoon* and *Septata* are principally pathogens of the gastrointestinal tract whereas the other genera usually cause extraintestinal infections of the lung, kidney, eye, or skeletal muscle. Because these organisms can be identified to the genus level only in specially equipped reference or research laboratories, they are often referred to collectively as microsporidia.

Life Cycle

The microsporidia are classified in the order *Microsporidia* of the phylum *Microspora,* which contains more than 100 genera and at least 1,000 species. The organisms are ubiquitous in nature and infect nearly all invertebrates including mosquitoes, spiders, fleas, grasshoppers, and bees as well as vertebrate animals. Microsporidia are obligate intracellular parasites that have a unique life cycle. Unlike some other intestinal protozoa, their infective stage is a spore rather than a cyst. Their exact mode of transmission is unknown but is presumed to be by contact with infected persons or animals or their excretions. A respiratory route of infection has also been considered.

The spores are extremely small (1 to 2 μm), ovoid in shape, and highly specialized. They contain a coiled tubular filament called a polar tubule that, when extended, attaches to epithelial cells in the small intestine and allows the spore contents, or sporoplasm, to pass through it into the host target cell (fig. 18.6). The sporoplasm undergoes development into multiplying stages called meronts and sporonts. Sporonts develop into spores that are released when the cell eventually ruptures. The spores may infect other intestinal cells to continue the disease process, or they may be passed to the outside in feces, urine, and respiratory secretions.

Characteristics of Human Disease

Human microsporidiosis, which was first described in 1985, is most commonly a gastrointestinal disease of severely immunocompromised persons, such as persons with AIDS. Moreover, up to 30% of AIDS patients with cryptosporidiosis may be infected with microsporidia as well. The typical clinical presentation of intestinal microsporidiosis is in a patient who is severely immunocompromised and has a 6- to 10-month history of chronic, intermittent diarrhea. The diarrhea worsens when the patient

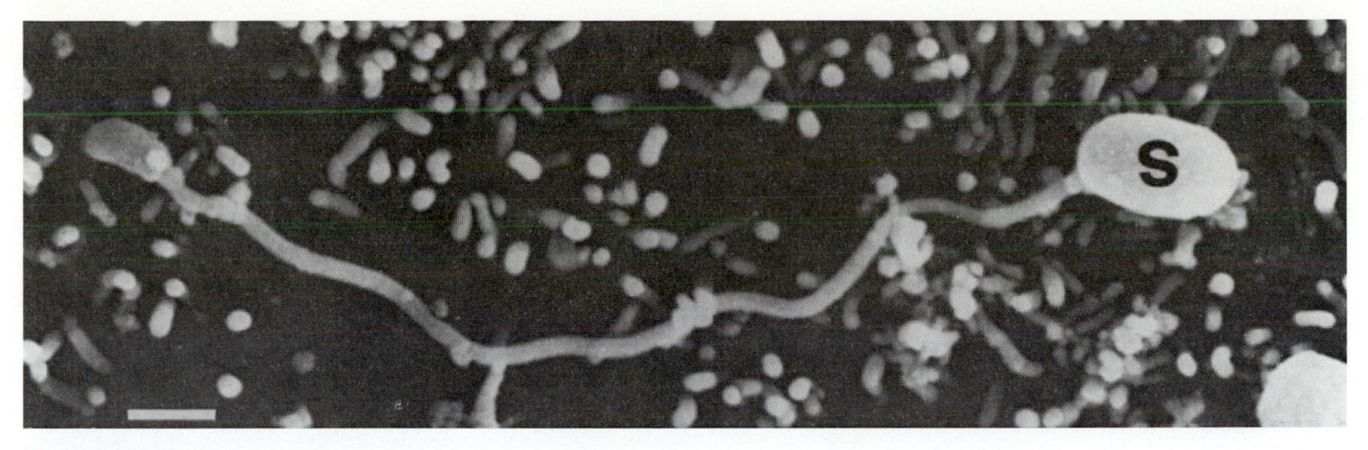

FIGURE 18.6 Scanning electron micrograph of a microsporidial spore (S) with an extended polar tubule and a sporoplasm at the end of the tubule. When the tubule attaches to an epithelial cell, the sporoplasm enters the cell, multiplies, and eventually forms new spores. This parasite was obtained from the urine of a patient with AIDS who had disseminated microsporidial infection.

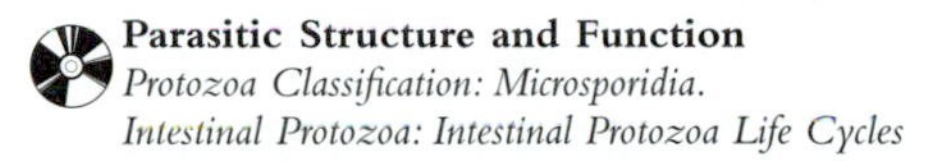

Parasitic Structure and Function
Protozoa Classification: Microsporidia.
Intestinal Protozoa: Intestinal Protozoa Life Cycles

eats meals that are high in fat. Fever is usually not present and because the diarrhea is not continuous, patients may not seek immediate medical attention. Over the course of many months, they may experience a wasting syndrome characterized by significant weight loss and emaciation.

DIAGNOSTIC METHODS

Diagnosis of microsporidiosis is achieved by visualizing the organisms in bowel biopsy specimens by electron microscopy or detecting the organisms in stool specimens by conventional light microscopy. In the latter case, because of the organisms' small size, the specimens must be stained with special stains in order to be seen.

Special Features of Control

Currently, there is no effective therapy for the treatment of microsporidiosis. Albendazole may be effective in some cases of infection caused by *Enterocytozoon* and *Septata.* Standard precautions should be used when caring for patients with microsporidiosis, and their discharges should be handled with special care.

INTESTINAL NEMATODE (ROUNDWORM) DISEASES

ENTEROBIASIS

Organism

(*Enterobius vermicularis* [pinworm].) The smallest of the parasitic intestinal **roundworms** and the cause of **enterobiasis,** the females measure about 13 mm in length ($^{1}/_{2}$ to $^{3}/_{4}$ in.) and the male about 5 mm. The gravid female produces eggs containing well-developed larvae, many of them motile within their shells. The ova are broadly oval, are flattened on one side, and measure about $60 \times 30\ \mu m$ (fig. 18.7).

Life Cycle (fig. 18.8)

I. Definitive Host
Humans.

II. Intermediate Host
None.

III. Human Infection

A. *Sources of Infection* Ova on the perianal skin of infected persons, rarely in feces. Direct or indirect transfer via clothing, bedding, toys, or other objects, and sometimes food.

B. *Infective Form* Ingested ova.

C. *Localization and Development* Following ingestion of ova, the larvae hatch and develop into adults in the intestinal tract. After copulation, the female usually migrates to the anus where she produces thousands of eggs. The adults die, but reinfection and transfer of infection from the ova are common.

D. *Exit Route* Anus and perianal region.

IV. Cyclic Development Outside of Human Host

A. Environmental development of the embryonated ova occurs within a few hours, usually while the eggs are still clinging to the skin. These ova are relatively resistant to drying and to disinfectants and may remain infective in dust or on surfaces for many days, although most die within 2 to 3 days.

B. Hosts other than humans are not involved in the cycle.

Characteristics of Human Disease

Intestinal infection with adult worms produces few, if any, symptoms. Perianal migrations of the female and oviposition in that region cause an intense local pruritus. Itching is often particularly irritating at night, inducing restlessness, insomnia, and incessant scratching. Children are frequently

FIGURE 18.7 *Enterobius vermicularis* (pinworm). (a) A female pinworm extruding a cloud of eggs (×20). (b) Pinworm eggs (×500) are characteristically deposited on the perianal skin. Each female discharges an average of about 11,000 eggs

Hugo Terner, Montefiore Hospital and Medical Center, Bronx, NY.

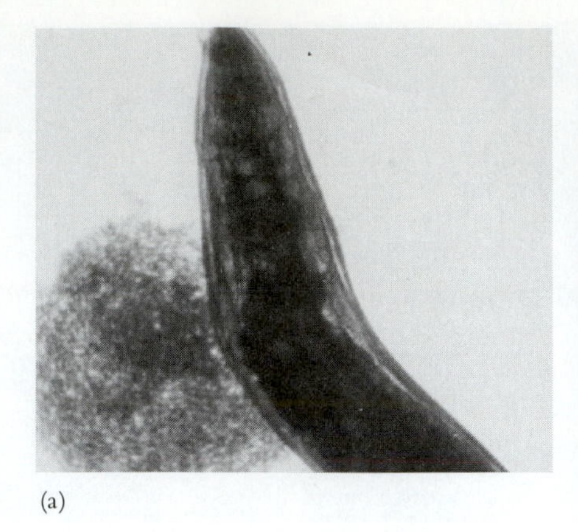

(a)

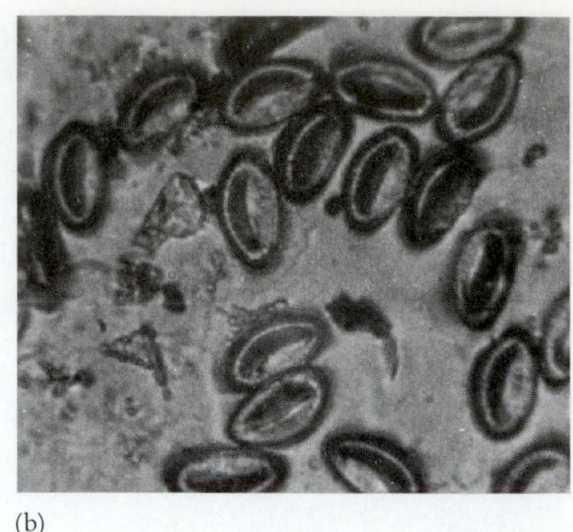

(b)

infected by other youngsters and can transfer the infection to an entire household.

DIAGNOSTIC METHODS

Pinworm ova may be seen by microscopic examination of Scotch tape applied to the perianal skin. The eggs may also be present in feces or in material scraped from beneath the fingernails. Adult worms, females and sometimes males, can also be found in the skin of the perianal region.

Special Features of Control

I. Treatment

The major problem in the treatment of pinworm infection is the prevention of reinfection, difficult to accomplish even with major hygienic efforts. Treatment with anthelminthic drugs, such as pyrantel pamoate (Antiminth) and mebendazole (Vermox), has a high degree of success when the drugs are administered as a single oral dose followed by a second dose after two weeks. They are relatively nontoxic, palatable, and usually well tolerated.

When pinworms become entrenched in families or institutions, it may be necessary to treat entire groups two or three times a year.

II. Prevalence and Prevention

Enterobiasis has a worldwide distribution with high infection rates in some areas. It is the most common helminth infection in the United States, with 5 to 10% of the population involved. Prevalence is highest among children and their mothers. Because of its easy transmissibility, pinworm infection is usually a familial or institutional infection.

Chemotherapy should be combined with vigorous methods of household disinfection, including bathrooms, bed linens, and bedroom dust. Careful personal hygiene is also important, particularly with regard to cleanliness of hands and fingernails.

TRICHURIASIS

Organism

(*Trichuris trichiura* [whipworm].) A slender, little roundworm shaped like a whip causes **trichuriasis.** The posterior end is thickened, the anterior slim and drawn out. It is about twice the size of the pinworm, measuring up to 5 cm, the male a little shorter.

The characteristic ovum is barrel-shaped and thick-shelled, with a bulging protuberance at each pole (fig. 18.9).

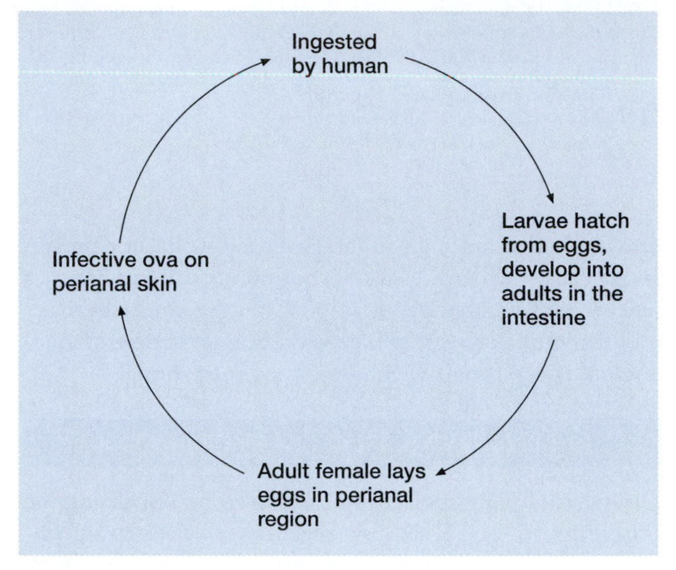

FIGURE 18.8 Human-to-human route of transfer characteristic of *Enterobius vermicularis.*

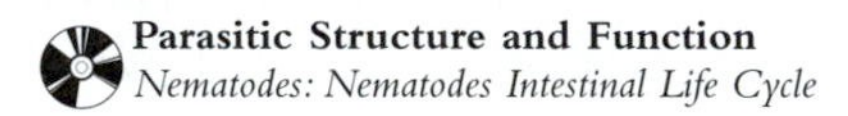
Parasitic Structure and Function
Nematodes: Nematodes Intestinal Life Cycle

Life Cycle (fig. 18.10)

I. Definitive Host

Humans.

II. Intermediate Host

None.

III. Human Infection

A. *Source of Infection* Soil in which eggs from human feces have matured.

B. *Infective Form* Embryonated egg.

C. *Localization and Development* After ingestion, the shell of the ovum is digested away, and the larva emerges and develops into an adult in the small intestine. Adults attach to the wall of the cecum by "threading" their ante-

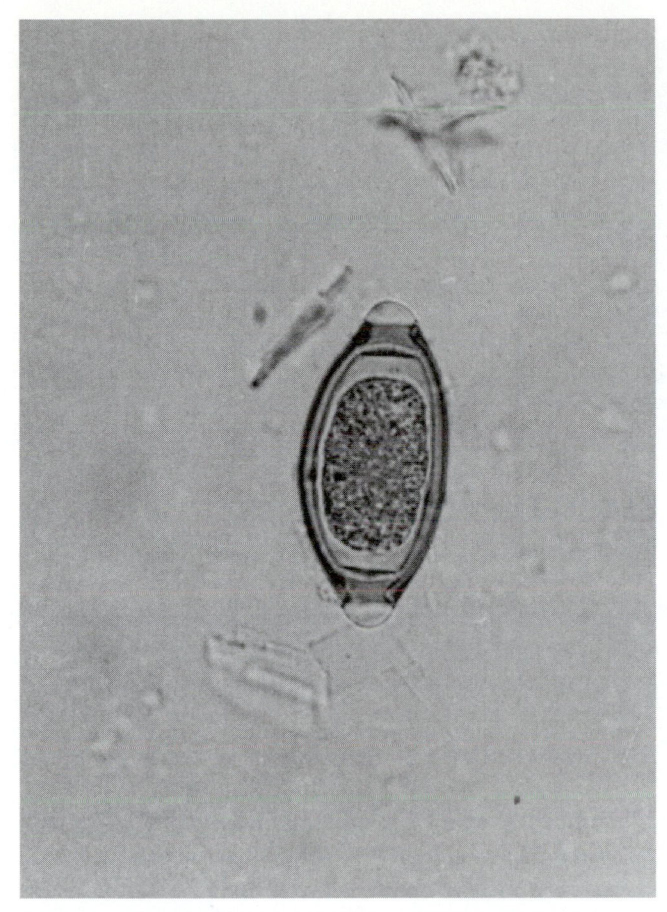

FIGURE 18.9 A nonembryonated egg of the whipworm *Trichuris trichiura.* The barrel-shaped ovum with a transparent plug at each end is characteristic of this species.

Courtesy Dr. Paul Beaver, Tulane University, New Orleans, LA.

rior end into the mucosa. Each female can produce five or six thousand eggs per day.

D. *Exit from Body* Eggs discharged in feces.

IV. Cyclic Development Outside of Human Host

In Environment Ova require 10 to 21 days in soil for further development of the embryo before they are infective. Other living hosts are not involved.

Characteristics of Disease

Light infections produce few, if any, symptoms. Massive infections may occur in children, with worm attachments extending far down the colon. Inflammation of the mucosa, mucous diarrhea, systemic toxicity, and anemia may result in severe, untreated cases. These symptoms resemble those of hookworm disease. Adult infections are usually asymptomatic.

DIAGNOSTIC METHODS

Demonstration of *Trichuris* ova in feces, by microscopic examination.

Special Features of Control

I. Treatment

The firm attachment of adult worms to the intestinal wall makes trichuriasis difficult to treat. Mebendazole, a broad-spectrum anthelminthic, is the drug of choice. It is given in two daily doses for a period of three days and appears to be both safe and effective.

II. Prevalence and Prevention

The whipworm has a wide distribution but is especially prevalent in warm, moist climates where sanitation is poor and crowding contributes to the ease of transmission.

Prevention requires good personal hygiene, sanitary disposal of feces, and careful washing of vegetables or other foods that could be contaminated with infective eggs from the soil.

ASCARIASIS

Organism

(*Ascaris lumbricoides.*) The largest of the roundworms, cylindric, tapered, measuring 20 to 30 cm in length, causes **ascariasis.**

The ovum is large (50 × 75 μm), spheric to ovoid, with a very thick, rough outer shell (fig. 18.11). Some infertile eggs, with a thinner shell, are produced also.

Life Cycle (fig. 18.10)

I. Definitive Host

Humans.

II. Intermediate Host

None.

III. Human Infection

A *Source of Infection* Soil in which eggs from human feces have matured.

B *Infective Form* Embryonated egg.

C *Localization and Development* After ingestion, larvae hatch from the ova in the duodenum. They penetrate the wall of the small intestine, reach the portal circulation, and are carried through the right side of the heart into the lungs. They may remain in the lungs for several days (colorplate 8a), then penetrate the pulmonary capillaries, reach the alveoli, and work their way up the bronchial tree. They are coughed up past the epiglottis, swallowed, and on rearrival in the small intestine develop into adult males and females.

D *Exit from Body* Ova are discharged in feces. (Each female can produce approximately 200,000 eggs per day.)

IV. Cyclic Development Outside of Human Host

In Environment Ova require 2 to 3 weeks of incubation in the soil for maturation of the embryo before they are infective. Other living hosts are not involved.

Characteristics of Disease

Intestinal infection may be mild and asymptomatic or produce abdominal pain, vomiting, diarrhea, and indigestion.

FIGURE 18.10 Human-to-soil-to-human route of transfer characteristic of *Trichuris trichiura* and *Ascaris lumbricoides*. *Ascaris* larvae travel through the circulatory system to the lungs and migrate to the pharynx. They are then swallowed and develop into adults in the intestinal tract.

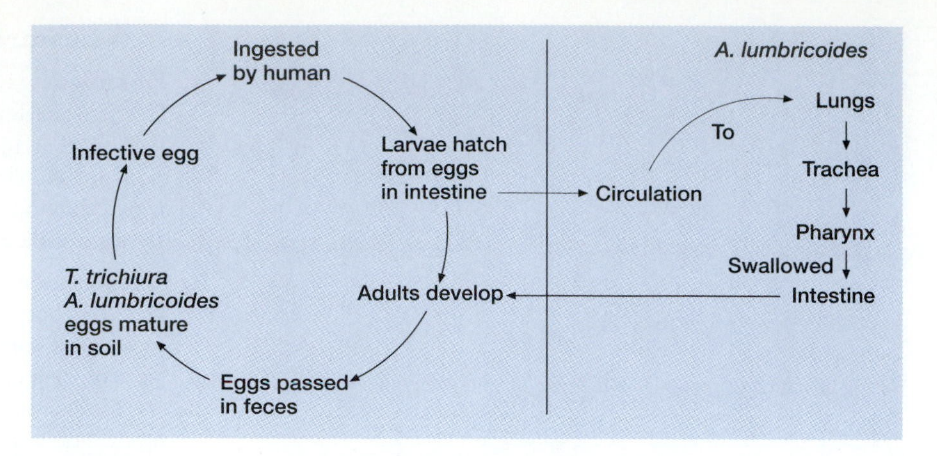

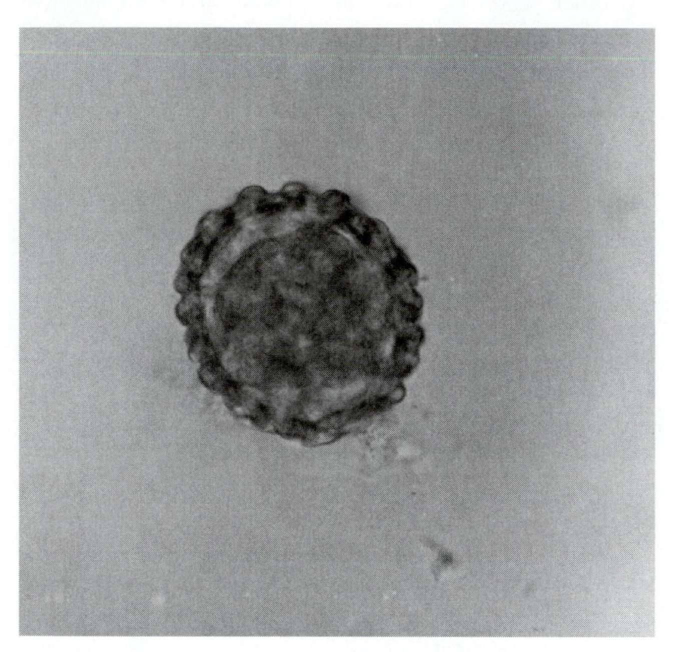

FIGURE 18.11 The ovum of *Ascaris lumbricoides* (×400).
Courtesy Dr. Paul Beaver, Tulane University, New Orleans, LA.

Masses of worms can cause intestinal obstruction, and systemic toxicity may occur.

Pulmonary infection with migrating larvae may induce a severe pneumonitis, sometimes with secondary bacterial infection. Fever, spasms of coughing, and asthmatic breathing are characteristic, and allergic reactions indicate hypersensitivity to the ascarids. These may include hive-type allergic rashes and eosinophilia.

Larvae sometimes reach the general circulation and are lodged in other tissues, producing severe symptoms if they reach organs such as the kidney or brain.

DIAGNOSTIC METHODS

Ascaris pneumonitis is diagnosed clinically and by radiologic techniques. Occasionally, larvae may be seen in sputum. Later, the recognition of ova in feces during intestinal infection confirms the diagnosis. Live adult worms may also be passed in feces.

Special Features of Control

I. Treatment

Mebendazole or pyrantel pamoate are used to treat intestinal ascariasis. Piperazine is an alternative drug. Intestinal obstruction caused by adult ascarids may necessitate intubation or surgery. There is no specific therapy for migrating larvae.

II. Prevalence and Prevention

Ascariasis is a common, worldwide infection. In moist tropical areas, the prevalence may exceed 50% of a population. Infection is heaviest and most frequent in young children of early or preschool age (fig. 18.12). Control depends on preventing contamination of soil with infective feces and educating children and adults in techniques of sanitation and hygiene.

INVASIVE NEMATODE (ROUNDWORM) DISEASES

TRICHINOSIS

Organism

(*Trichinella spiralis*.) A minute worm that is smaller than the pinworm and just visible to the unaided eye causes **trichinosis.** The female is viviparous; that is, she deposits viable larvae rather than eggs.

Life Cycle (fig. 18.13)

I. Definitive Hosts

Swine, rats, and wild animals are definitive hosts. Humans become a definitive host by eating infected meat but do not maintain the parasite in nature.

II. Intermediate Host

Each host serves as both the definitive and intermediate host, harboring both adult and larval forms of the parasite.

III. Human Infection

A *Source of Infection* Meat of infected animal (usually pork or bear meat).

B *Infective Form* Viable larvae encysted in animal's muscle tissue.

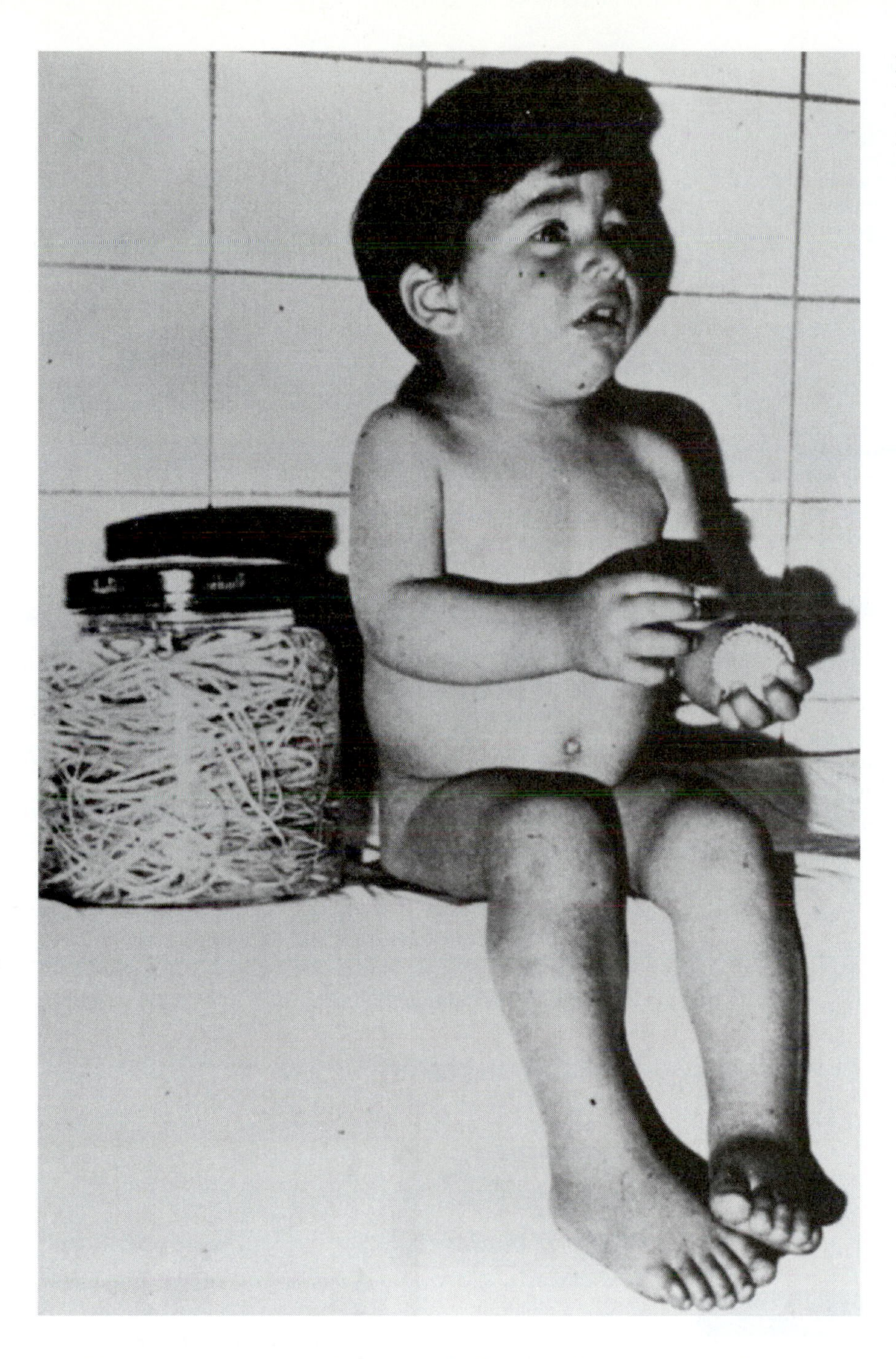

FIGURE 18.12 Massive ascariasis in a two-year-old child. The bottle contains 278 worms expelled in one day with the help of the drug piperazine.

Reproduced with the permission of the author and editor from P. Garaguso, "Intestinal Parasitic Disease," *Annales Nestlé* 46/47: 2–47, 1981.

Parasitic Structure and Function
Nematodes: Nematodes Disease

C *Localization and Development* After being swallowed, the larvae excyst during digestion of the meat in the stomach. They pass into the duodenum and within 4 or 5 days develop into male and female adults. The females burrow into the mucosa of the villi, or even into submucosal layers or mesenteric lymph nodes, and begin to deposit larvae, which reach the lymphatic and blood circulation and are distributed through the body. Larvae may lodge temporarily in a variety of tissues from which they reenter the circulation. When they reach striated muscle tissue, they become encysted. The muscles most frequently involved are the diaphragm, tongue, biceps, deltoid, and those of the thoracic and abdominal walls. Larvae cannot develop into adults unless they subsequently are ingested by another host.

D *Exit from Human Body* None. Not communicable by humans.

IV. Cyclic Development Outside of Human Host

A *In Environment* None.

B *In Hosts Other Than Humans* Carnivorous animals maintain this parasite (see Perspective 18.1). Swine are infected from garbage scraps containing infected meat. Adult and larval trichinae both develop in one host, but each worm requires two hosts to maintain itself.

Characteristics of Human Disease

Trichinosis often appears as a small, sporadic outbreak among family or other groups who have eaten the same

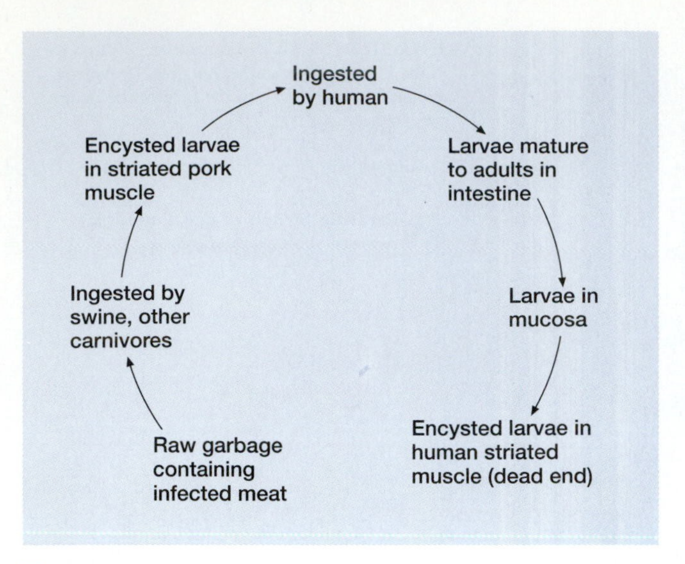

FIGURE 18.13 Intermediate host-to-human route of transfer characteristic of *Trichinella spiralis.* This cycle in swine and other animals is the same as that in humans.

Parasitic Structure and Function
Nematodes: Nematodes Tissue Life Cycle

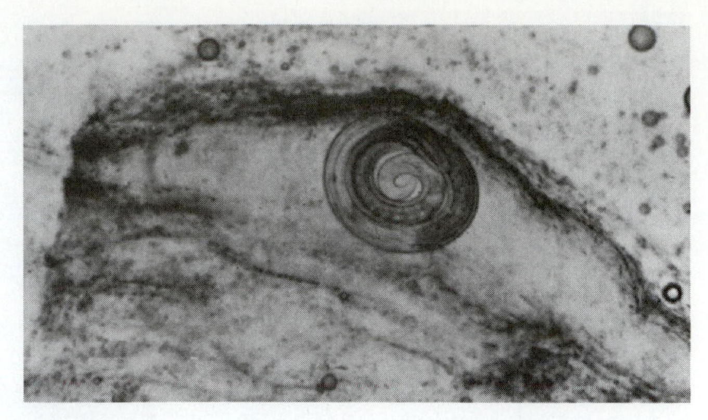

FIGURE 18.14 *Trichinella spiralis* larva in a preparation of digested muscle tissue (×128). The patient from whom the muscle biopsy was obtained had eaten infected bear meat while on a hunting trip. The meat was cooked over a campfire and not heated sufficiently to kill the viable larvae inside the meat. (See Perspective 18.1.)

infected, undercooked roast pig or bear meat. (Sausages and other products may also contain infective larvae.) The severity of resulting disease depends on the number of viable larvae eaten. Trichinosis may be a mild infection or a serious and threatening disease. Initially there may be some gastrointestinal discomfort as the larvae invade the intestinal mucosa and develop into adults. During larval migration there is fever, often a characteristic periorbital edema, and, in a few days, muscle pain, chills, weakness, and sometimes prostration. Eosinophilia is usually marked. There may be widespread muscle injury due to the presence of larvae. Respiratory distress and myocardial involvement are the most severe events and may cause death. Recovery begins when larval migration ceases, varying between 3 and 8 weeks. Residual handicap to muscle function depends on the numbers and location of encysted larvae, which are slowly calcified but may remain viable for many months.

DIAGNOSTIC METHODS

Visualization of the parasite in the laboratory requires examination of muscle tissue taken by biopsy (fig. 18.14). Bits of muscle may be sectioned and stained by histologic techniques or can be digested by proteolytic enzymes and examined microscopically for freed, motile larvae.

Circulating antibodies appear in the patient's serum after 2 to 3 weeks and provide a useful basis for serologic diagnosis. The bentonite flocculation test is a sensitive, specific method used for diagnosis. For this test, trichinella antigen is bound to bentonite particles that flocculate (precipitate) when they are mixed with patient serum containing trichinella antibodies. Enzyme immunoassay or immunofluorescence methods may also be used for serologic diagnosis.

Special Features of Control

I. Treatment

Mebendazole is given to eliminate adult *T. spiralis.* Corticosteroids reduce the inflammatory response and thus help relieve the symptoms caused by larval encystment.

II. Prevalence and Prevention

Trichinosis has a wide distribution, but its prevalence depends on local practices of preparing and eating pork. Control and prevention depend on eliminating sources of infection for hogs (feeding cooked garbage, eliminating infected animals detected by skin testing, destroying rats), and processing meat by adequate cold storage or freezing and final thorough cooking.

TOXOCARIASIS (VISCERAL LARVA MIGRANS)

Organisms

Toxocara canis, Toxocara cati.

Life Cycle

Toxocariasis, also known as *visceral larva migrans,* is a disease of young children caused by the larvae of animal ascarids. *Toxocara canis,* a canine ascarid, and *Toxocara cati,* its feline counterpart, normally infect dogs and cats, which maintain their host-to-soil-to-host life cycles in nature. In animals, this nematode infection is like that of ascariasis in humans. Humans are an accidental host to whom the parasite is not fully adapted, and when human infection occurs, the life cycle is aborted in the stage of larval migration through viscera (see fig. 18.15).

Humans are infected by ingestion of *Toxocara* eggs produced by adult worms residing in the intestinal tract of a dog or cat, passed in animal feces, and matured in the soil. After ingestion, the embryonated eggs hatch in the intestine. Larvae penetrate the wall of the bowel and migrate to liver and lungs in the blood and lymph systems. The immature larvae may

PERSPECTIVE 18.1 *Trichinosis and the Bearburger*[a]

Jack M., a 55-year-old postal worker from New York City, took a week's vacation to hunt bear in Maine. Several weeks after his return, he complained of muscle pain, especially in his chest. Fearing a heart attack, Jack called his physician, who promptly admitted him to a local hospital. Mr. M's symptoms seemed to support a provisional diagnosis of myocardial infarction, but his electrocardiogram and other admission tests were all within normal limits—except for an elevated eosinophil count.

Aware that eosinophils are elevated in parasitic infestations, Jack's physician questioned him about his activities in the past several months. Jack reported that he and his friends had spent an entire week tracking bear in the woods of Maine without success. Unwilling to be denied the meal they had anticipated for so long, however, the group stopped at a local butcher shop that was advertising fresh-ground bear meat, bought a few pounds, and took the meat back to the campsite for supper. Jack chose to have his bearburger rare, while the others preferred theirs well-done.

Bears and other mammals may be reservoirs of trichinosis infection.

After hearing this history and suspecting a parasitic infestation, Jack's physician ordered a muscle biopsy which, sure enough, revealed the larvae of *Trichinella spiralis*. Jack's symptoms were related to the trichinosis he had acquired from infected bear meat because he had chosen to eat it undercooked. He was treated with an antiparasitic agent to eliminate any adult worms and with corticosteroids to reduce the inflammatory response induced by the larvae encysted in his muscles. Involvement of the diaphragm, in particular, had caused the chest pain and Jack's feeling that he was having a heart attack. Although Jack's condition improved, the invasion of *Trichinella* larvae into his muscles continued to handicap him because they can be killed but not eliminated. He never regained full use of the invaded muscles.

How do wild bears acquire trichinosis? Even in the wilderness, they are frequent visitors to human garbage dumps and are scavengers of refuse. If bears eat infected meat along with other scraps they find, they may become an alternate host to the parasite (Perspective fig. 18.1), which encysts in their muscles just as it did in those of Jack M.

Fortunately, such occurrences are rare today. Government inspection of domestic meats ensures the safety of most butchers' customers, but wild animals are another story!

[a]Based on an actual case history.

wander from the lungs to other tissues, continuing their migrations for years. They do not mature or make their way back to the intestinal tract for adult development, as do the larvae of *Ascaris lumbricoides*.

Visceral larva migrans (VLM) is most common in very young children who have close contact with dogs and cats and also in those with the habit known as pica (eating unnatural foods such as soil, clay, starch, chalk, ashes). Animals are also infected by ingesting eggs from the soil. Even more important, however, a pregnant female dog can transmit infection to her fetuses if the immature larvae migrate into their tissues.

Characteristics of Human Disease

VLM is usually a self-limited, benign disease. Most patients show complete recovery over a period of years. The severity of disease depends on the intensity of the infection. Clinically, it is characterized by fever, hepatomegaly, and pulmonary symptoms such as coughing and wheezing. The white blood cell count is often elevated, and there is a marked eosinophilia (80 to 90%). Endophthalmitis, caused by larval migration to the eye, is sometimes seen, and cerebral or myocardial symptoms occur occasionally.

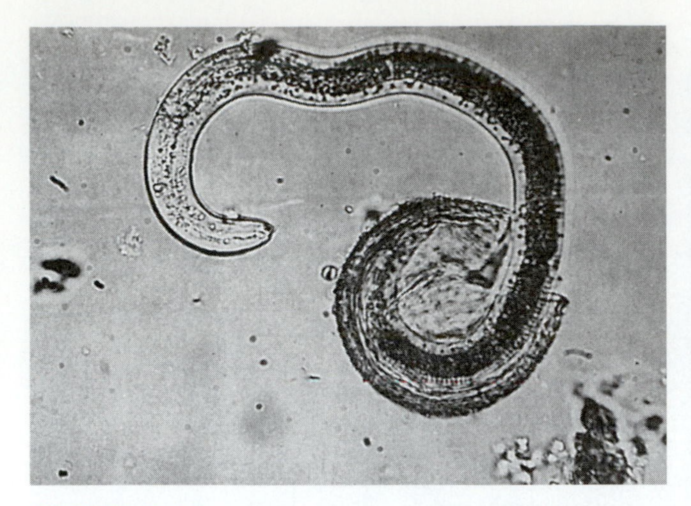

FIGURE 18.15 A larva of *Toxocara canis* (×500) involved in human visceral larva migrans (toxocariasis).
Courtesy Dr. Paul Beaver, Tulane University, New Orleans, LA.

DIAGNOSTIC METHODS

Positive serologic tests using *Toxocara* and *Ascaris* antigens (using indirect hemagglutination, fluorescent antibody, and EIA tests) suggest the diagnosis but are not conclusive. A definitive diagnosis can be made only by liver biopsy and demonstration of the larvae in the tissue sections, but this procedure is rarely performed.

Special Features of Control

Treatment of VLM is difficult and should be instituted only in severe cases. Diethylcarbamazine (Hetrazan) or thiabendazole is used but must be given in multiple daily doses for up to ten days.

Prevention is best achieved by controlling contact of children with dogs and cats and eliminating animal feces from play areas. Children's sandboxes, for example, are attractive to cats and should be covered when not in use. Pets should be adequately dewormed, and children should be discouraged from pica. All family members should be taught the importance of hand washing after handling soil and before eating.

INTESTINAL CESTODE (TAPEWORM) DISEASES

TAENIASIS

Organisms

(*Taenia saginata* [beef tapeworm], *Taenia solium* [pork tapeworm].) The adult tapeworms that cause ***taeniasis*** are similar morphologically, differing in structure of the scolex (*T. solium* possesses a crown of hooklets [fig. 18.16], *T. saginata* is unarmed) and internal features of the proglottids (fig. 18.17). Adults are seldom less than 2 meters in length and may reach 6 meters (fig. 2.27). The ova of the two species are morphologically identical. They are spheric (35 to 40 μm in diameter), have a striated outer shell, and the embryo within possesses hooklets.

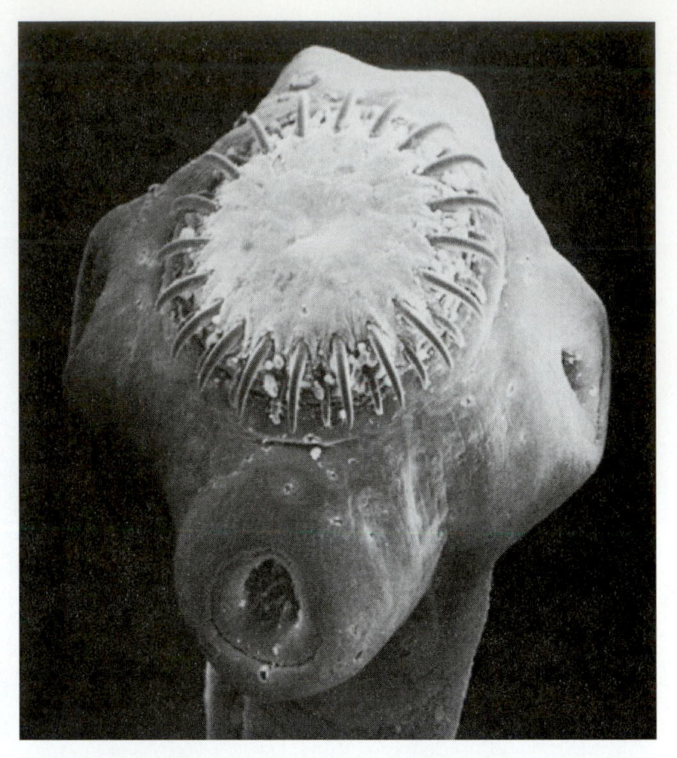

FIGURE 18.16 Scanning electron micrograph of the scolex of the tapeworm *Taenia solium* (×500).
Kessel & Shih/Peter Arnold, Inc.

Life Cycle (fig. 18.18)

I. Definitive Host

Humans.

II. Intermediate Host

Cattle (*T. saginata*) and swine (*T. solium*). Humans can also serve as an intermediate host for *T. solium*.

III. Human Infection

A *Source of Infection* Infected, inadequately cooked flesh of beef or pork.

B *Infective Form* Viable larvae encysted in muscle tissue of animals.

C *Localization and Development* Following ingestion, larvae excyst in the stomach and pass into the small intestine where the scolex attaches to the intestinal wall, developing into a mature worm within 5 to 12 weeks. The hermaphroditic proglottids produce many eggs.

D *Exit from Human Body* Eggs are discharged in the feces. Gravid proglottids break off easily and may also be passed in the stool.

IV. Cyclic Development Outside of Human Host

Animal hosts ingest eggs derived from infective human feces deposited on soil where animals graze. When the eggs reach the animal's intestinal tract, the larvae hatch, penetrate the intestinal wall, and are distributed through the

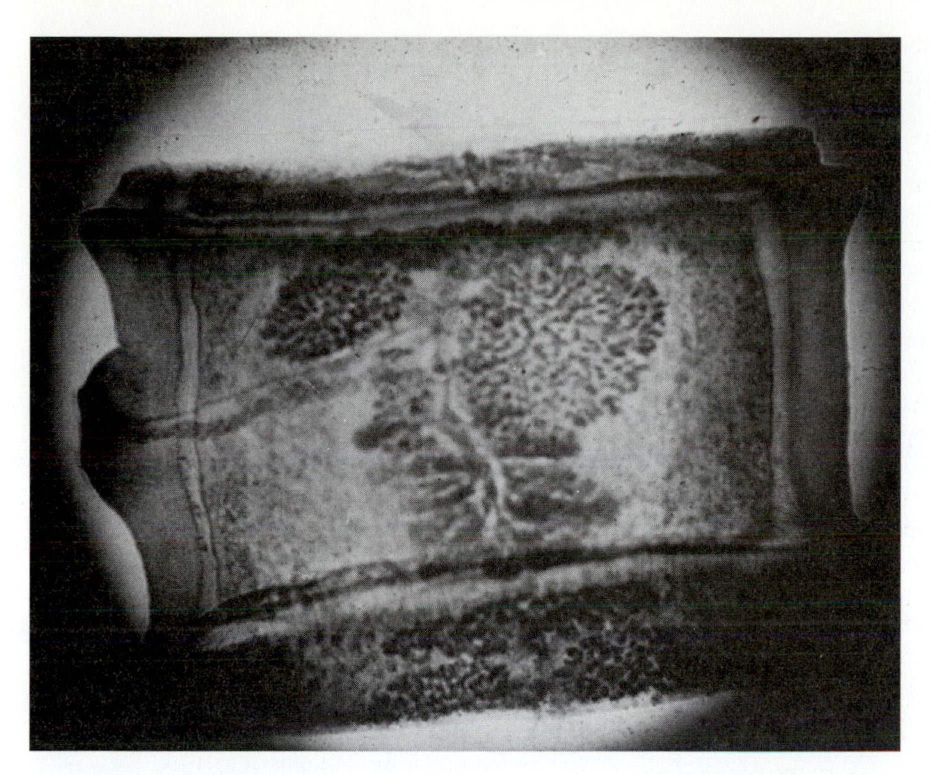

FIGURE 18.17 The proglottid of *Taenia solium.* Note the central uterine canal and its few bilateral branches. Mature ova pass from the uterus along the genital canal (seen extending to the left in the photograph) and escape through the prominent lateral genital pore (×8).

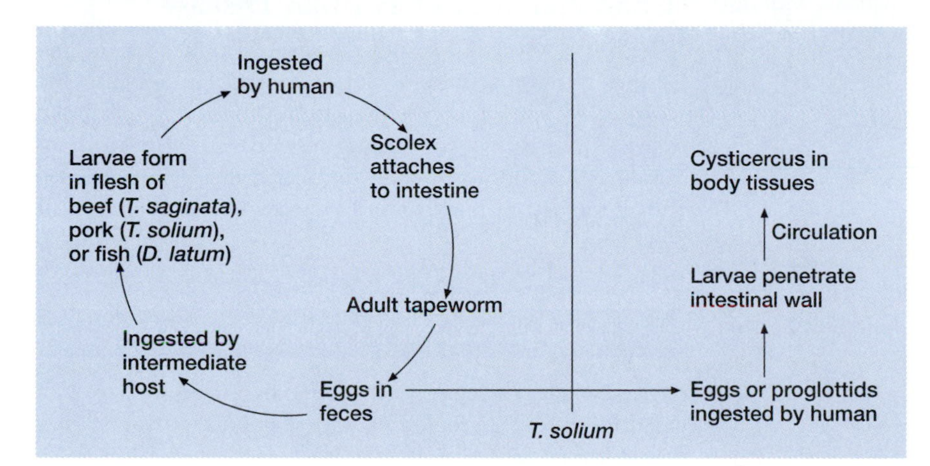

FIGURE 18.18 Human-to-intermediate host-to-human route of transfer characteristic of *Diphyllobothrium latum, Taenia solium,* and *Taenia saginata.* The eggs of *T. solium* only are infective for humans, and accidental ingestion of these eggs or proglottids containing them results in cysticercosis.

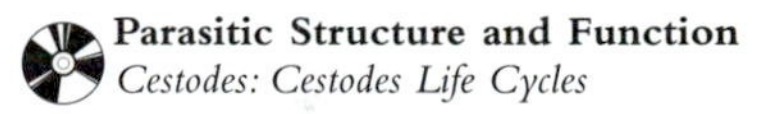
Parasitic Structure and Function
Cestodes: Cestodes Life Cycles

body by the bloodstream. They localize and encyst in skeletal muscle, in forms referred to as *bladder worms,* or *cysticerci.* In the case of *Taenia solium,* this invasive form of infection may also occur in humans (see Cysticercosis below).

Characteristics of Human Disease

Intestinal infection with either of these species is relatively mild. In persistent infections there may be diarrhea, epigastric distress, vomiting, and weight loss, and sometimes anemia and eosinophilia. Although these worms are quite long, they rarely cause intestinal obstruction. Most patients with *Taenia* infections display no symptoms. *T. solium* is potentially dangerous, however, because it can cause the invasive, systemic form of disease called *cysticercosis* described later.

DIAGNOSTIC METHODS

Laboratory diagnosis of intestinal taeniasis is made by recognition of proglottids passed in feces or the microscopic identification of *Taenia* ova. Species identification is made on the basis of characteristic features of the proglottids.

Special Features of Control

I. Treatment

To eradicate the infection, the scolex must be eliminated or a new adult worm will be generated. Niclosamide, given in a single dose, is the drug of choice but one dose of praziquantel is also effective. In treating *T. solium* infection, it is important to prevent the patient from vomiting, because gravid proglottids and eggs may be brought into the stomach by reverse peristalsis. If this happens, the eggs may hatch in the upper small intestine, penetrate, and cause cysticercosis.

II. Prevalence and Prevention

These tapeworm infections are prevalent wherever beef or pork is eaten raw or lightly cooked. The incidence is highest in Latin America, eastern Europe, Africa, and Asia, as well as some tropical countries of the western hemisphere. *T. saginata* is more common than *T. solium,* and the latter is rare in the United States and Canada.

The control and prevention of *Taenia* infections depend on the sanitary disposal of human feces and protection of animal feeding grounds. Government meat inspectors reject grossly infected meats in abattoirs (bladder worms can be seen and felt in the tissues), but thorough cooking provides the final safeguard. Patients with *T. solium* infections are cared for with standard precautions to prevent accidental ingestion of infective eggs by the patients or their contacts.

DIPHYLLOBOTHRIASIS

Organism

(*Diphyllobothrium latum* [fish tapeworm].) This tapeworm, like the *Taenia* species, can reach great lengths in patients with **diphyllobothriasis.** The structure of the scolex differs from that of the taeniae (fig. 18.19), the proglottids are much broader than long, and the ovum is larger (about 50×75 μm). The ovum is broadly oval and has a thin shell with a lidded opening (operculum) at one end. The embryo within has hooklets.

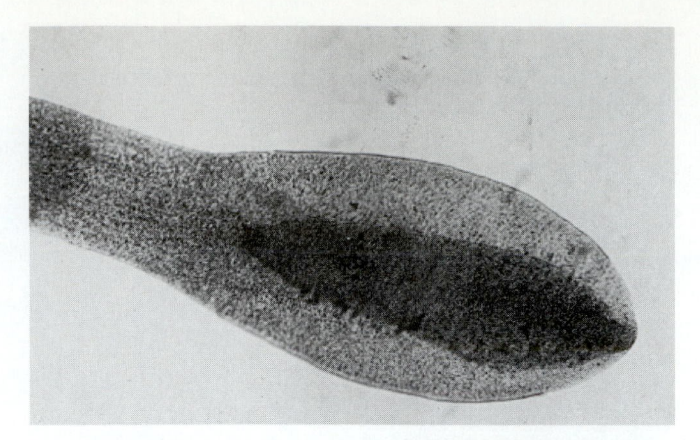

FIGURE 18.19 Scolex of *Diphyllobothrium latum* (×25). Note the long, deep muscular cleft by which this tapeworm attaches itself to the intestinal wall.

Courtesy Dr. Kenneth Phifer, Rockville, MD.

Life Cycle (fig. 18.18)

I. Definitive Host

Humans and many animals (e.g., cats, dogs, bears, foxes, walruses).

II. Intermediate Hosts

Copepods (aquatic arthropods) and fish.

III. Human Infection

A *Sources of Infection* The flesh of infected freshwater fish.

B *Infective Forms* Viable larvae in fish tissues.

C *Localization and Development* After they are swallowed, larvae pass through the stomach into the small intestine. The scolex attaches to the wall and develops into a mature, egg-producing worm.

D *Exit from Human Body* Eggs are discharged in feces. They are not infective for humans.

V. Cyclic Development Outside of Human Host

When eggs passed in human feces are deposited in water or contaminate water from draining sewage, ciliated embryos are released and swim about until ingested by the first intermediate host, the copepod. The parasite develops within the arthropod for 2 or 3 weeks. If the copepod is then eaten by a fish, the larval form develops further and migrates into the fish's muscles (second intermediate host). Fish-eating marine or land animals, including humans, then acquire the intestinal infection by eating the fish raw. People can avoid infection by cooking fish thoroughly or freezing it for 24 hours at $-10°$ C.

Characteristics of Human Disease

The fish tapeworm does not ordinarily produce serious intestinal symptoms unless it causes obstruction. Systemic intoxication sometimes results from absorbed metabolic wastes from the worm. In rare instances a severe primary anemia is produced because the worm in the intestine deprives the host of the vitamin B_{12} needed for red blood cell formation.

DIAGNOSTIC METHODS

The characteristic eggs are identified by microscopic examination of feces. Proglottids and segments of worms are sometimes passed as well.

Special Features of Control

I. Treatment

As in taeniasis, if the scolex is not eliminated, a new worm will be generated. Treatment is the same as for the taeniae, with niclosamide or praziquantel given in a single dose.

II. Prevalence and Prevention

Diphyllobothriasis is endemic in northern and central Europe, parts of the Near and Far East, and, on this continent, in Canada, the Great Lakes region, and Florida—all areas where freshwater fish are an important food and where rural systems for sewage disposal may fail to protect fresh water from contamination. Infected human beings are the primary source of fish infection in endemic areas. Control and prevention require protection of fresh water from human fecal pollution, as well as the education of cooks in the preparation of fish and the risks of sampling partially cooked fish foods.

Hymenolepiasis

Organisms

(*Hymenolepis nana* [dwarf tapeworm], *Hymenolepis diminuta* [rat tapeworm].) These are the smallest of the intestinal tapeworms, measuring only about 2.5 to 5 cm in length.

Hymenolepiasis caused by *Hymenolepis nana* is common, particularly in the southern United States. Its life cycle is unusual, for it is the only tapeworm that does not require an intermediate host. The adult worm lives in the human intestinal tract, producing ova that are discharged in the feces and are directly infective for others if swallowed. Ova remaining in the bowel may also hatch into larval forms in intestinal villi and continue development into adult forms in the bowel lumen. This means that a single host may harbor the parasite indefinitely as well as pass it along to others. Dwarf tapeworm infection is generally mild until large numbers of adult worms are formed, causing intestinal irritation and systemic intoxication from absorption of worm products. Treatment is with praziquantel (the drug of choice) or niclosamide. Control requires breaking the chain of transfer among human contacts by means of improved sanitation and hygiene.

Hymenolepis diminuta infects humans infrequently but is primarily an intestinal infection of rats, kept going by fleas that ingest ova passed in rat feces. (Fleas are scavengers as well as bloodsucking insects.) The larval stage of development occurs in the arthropod host, which is then ingested by the rat host it parasitizes, and the intestinal cycle begins again. Human intestinal infection results from accidental swallowing of an infected flea or other arthropod. This situation can arise under squalid, rat-infested living conditions. Rat extermination, relief of poverty, and improvement of living conditions reduce the possibility of human exposure to this parasite.

Invasive Cestode (Tapeworm) Diseases

Cysticercosis

Human **cysticercosis** may result from ingesting the eggs of *Taenia solium* derived from the feces of a person with an intestinal infection with this worm. The eggs hatch in the small intestine, and the larvae penetrate the intestinal wall and are carried by the bloodstream into striated muscle tissue, subcutaneous foci, and sometimes to vital organs, where development of a cysticercus may have severe consequences. Autoinfection is also possible if eggs from an intestinal infection reach the stomach or upper small intestine by reverse peristalsis. When human infection of this type occurs, no further development of the parasite is possible, for it can find no exit route by which it can reach another host.

The symptoms of cysticercosis are referable to the location of the bladder worms. They have been reported most frequently in subcutaneous tissues, but also in the eye, brain, muscles, and visceral organs. Diagnosis requires excision of a cyst and its microscopic examination. The appearance of the cysts in computerized tomographic (CT) scans is often highly suggestive of the diagnosis. Effective therapy requires praziquantel or albendazole, or surgery.

Echinococcosis (Hydatid Cyst)

Organism

(*Echinococcus granulosus*.) A small tapeworm having only four segments: a scolex, an immature proglottid, one maturing proglottid, and one gravid proglottid is the cause of **echinococcosis** (fig. 18.20c). The ovum is indistinguishable from that of the *Taenia* species.

Life Cycle (fig. 18.21)

I. Definitive Host

Domestic dogs and wild canines.

II. Intermediate Hosts

Sheep, cattle, swine, and sometimes humans.

III. Human Infection

A *Source of Infection* Feces from infected animals, usually domestic dogs.

B *Infective Form* Egg derived from adult worm living in canine intestine.

C *Localization and Development* After being swallowed, the eggs hatch in the duodenum; the embryos migrate through the intestinal wall, enter the portal bloodstream, and eventually lodge in capillary filter beds in the liver and in various other tissues. A fluid-filled sac, or hydatid cyst, forms at the sites of tissue localization. Within the cysts many immature scolices develop but cannot mature (fig. 18.20b). The cysts enlarge, creating pressure and causing damage to tissue architecture. They may rupture and spill their contents into adjacent areas where new cysts then form (fig. 18.20a).

D *Exit Route from the Human Body* None. Not communicable by humans.

IV. Cyclic Development in Nature

Dogs and sheep (or other domestic animals raised for meat) ordinarily maintain this parasite. Sheep ingest the ova when they graze in pastures contaminated by infective dog feces, and they develop the larval form of tissue infection, as described for humans. In rural areas, dogs usually have access to discarded portions of slaughtered animals. If so, they ingest the larvae in these tissues and develop an intestinal infection, which completes the cycle.

Characteristics of Human Disease

Echinococcosis is usually quite serious, although symptoms depend on the size and location of the cysts. They

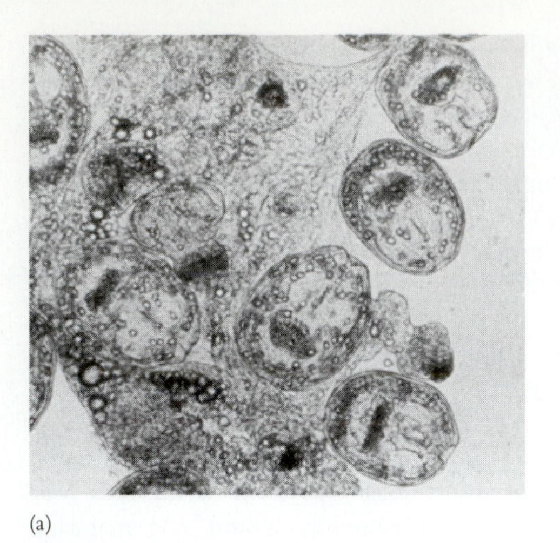

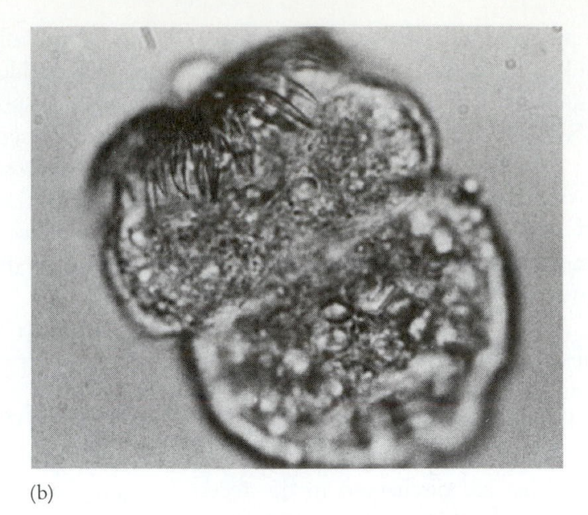

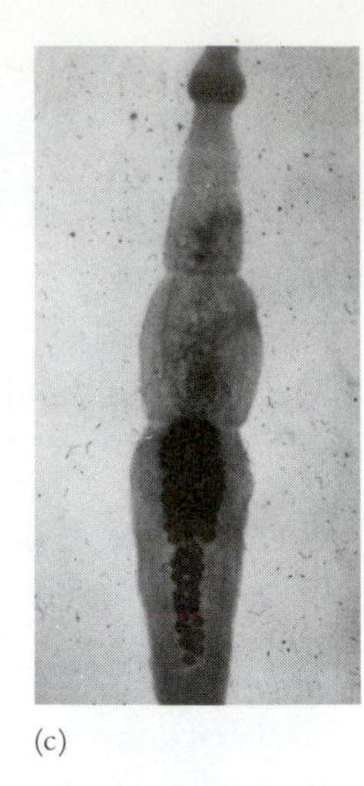

(a) (b) (c)

FIGURE 18.20 *Echinococcus granulosus.* (a) Microscopic appearance of unstained hydatid sand aspirated from a cyst (×100). (b) Evaginated scolex of *Echinococcus* larva found in the same microscopic preparation (×400). (c) Adult tapeworm (×10).

(a, b) Reprinted by permission of the publisher from R. Gaynes and P. Gardner, "*Echinococcus* Liver Abscess," *Clinical Microbiology Newsletter* 7:5, 36–37, 1985. Copyright 1985 by Elsevier Science Publishing Co., Inc. (c) Hugo Terner, Montefiore Hospital and Medical Center, Bronx, NY.

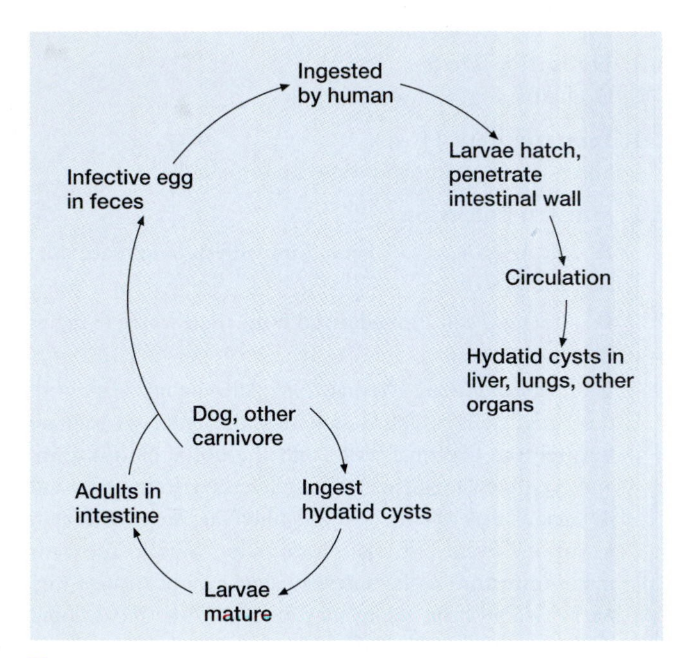

FIGURE 18.21 Definitive host-to-human route of transfer characteristic of *Echinococcus granulosus.* The dog and other carnivores acquire the parasite by eating infected viscera of sheep, swine, cattle, etc. The larval forms in the hydatid cysts mature into adult egg-laying tapeworms in the intestine of the carnivore.

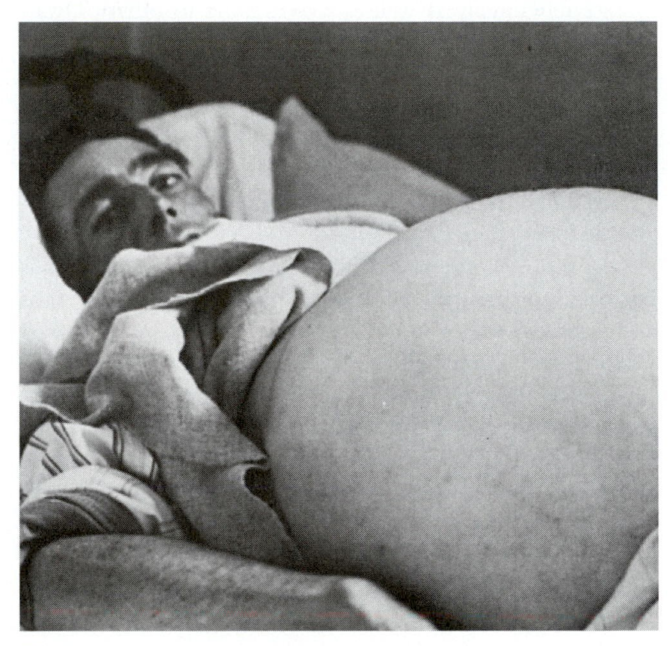

FIGURE 18.22 This man has large hydatid cysts in his abdomen. Treatment of echinococcosis used to be surgical and very complicated; now WHO is coordinating clinical trials for a new drug that may bring relief to patients without the need for surgery.

Courtesy World Health Organization. Photo by Medical Journal of Australia.

sometimes cause death, but they may also persist throughout life, producing little difficulty. Multiple cysts in the liver or other abdominal sites may result in abdominal distension (fig. 18.22).

DIAGNOSTIC METHODS

Diagnosis is usually made on the basis of clinical and radiologic findings. The cysts and the scolices within them can be identified in the laboratory only after their surgical removal. If cysts rupture and discharge their contents to an orifice (bowel, urinary tract, bronchial tree), examination of feces, urine, or sputum may reveal immature scolices.

Serologic methods such as indirect fluorescent antibody, indirect hemagglutination, enzyme immunoassay, and bentonite or latex flocculation can help establish the diagnosis.

Special Features of Control

I. Treatment

Surgical techniques are useful if the cysts lie in operable sites and if they have not extended too widely. Cautious surgical dissection is important because cysts are easily nicked or broken and may spill their infectious contents into the surrounding tissues. Preoperative injection of the cyst with alcohol, formalin, or aqueous iodine is recommended to kill scolices before surgery. Albendazole in large doses is effective if surgery is not possible.

II. Prevalence and Prevention

Echinococcosis is most common in countries in which grazing animals are herded by dogs that also have close contact with humans: South America, the Mediterranean countries of Europe and Africa, and Australia. The disease is rarely acquired in the United States.

Control and prevention depend on breaking the chain of transfer, particularly from sheep or cattle to dogs. The disease declines when strict controls keep dogs away from slaughtered animal tissues. Human disease can be prevented by good household and personal hygiene, particularly when dogs are free-roaming members of the household in an endemic area.

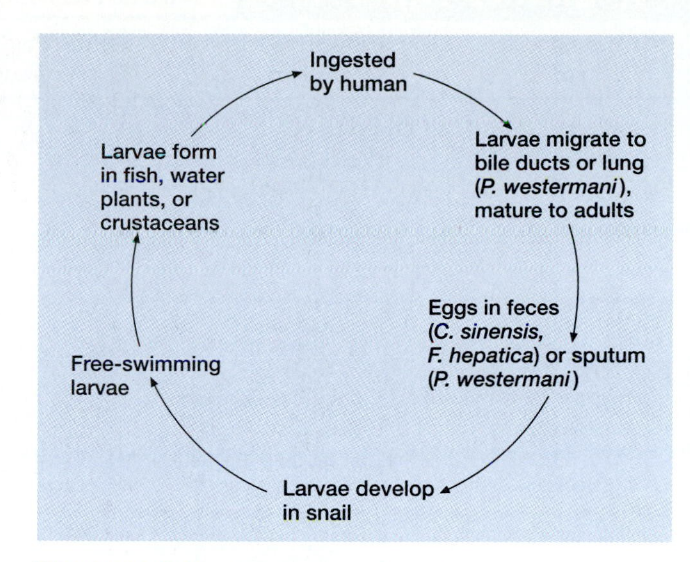

FIGURE 18.23 Human-to-intermediate host-to-human route of transfer characteristic of *Clonorchis sinensis, Fasciola hepatica,* and *Paragonimus westermani*. The snail is the intermediate host, but for humans, the infective form is found in fish, water plants, or crustaceans, respectively.

INVASIVE TREMATODE DISEASES (LIVER, INTESTINAL, AND LUNG FLUKES)

The *trematode* diseases described here are caused by a small group of animal parasites with many biologic features in common. They share the same group of definitive hosts (humans and some domesticated animals such as dogs, cats, sheep, and swine) that acquire them by ingestion, and their cyclic development in intermediate hosts is similar but not identical. They differ from blood **flukes** (*Schistosoma* species) in that their adult forms are hermaphroditic, whereas the schistosomal flukes are bisexual, but live in such close association that this distinction is not very important. The more useful epidemiologic distinction is that the larval forms of *Schistosoma* find their way into the body from their environmental sources by penetrating exposed skin or mucosa (see chap. 20), whereas the flukes discussed here have an alimentary route of entry.

Life Cycle (fig. 18.23)

I. Definitive Hosts

Humans, dogs, cats, sheep, and swine.

II. Intermediate Hosts

Two are usual: snails are the first intermediate host, and aquatic animals (crayfish, crabs, or fish) or plants (water chestnuts and others) act as the second intermediate hosts, which are ingested by humans and other land animals.

III. Human Infection

A. *Source of Infection* Infective aquatic life eaten raw or partially cooked.

B. *Infective Form* Larvae (metacercariae) encysted in aquatic animals or plants.

C. *Localization and Development* After they are eaten, the digestive process frees the encysted larvae. *Fasciolopsis buski,* the intestinal fluke, fastens to the intestinal wall and develops into an adult at that site. The larvae of other species of this group penetrate the intestinal wall and travel through body cavities or ducts to reach their respective sites of choice. The liver flukes (*Clonorchis* and *Fasciola*) usually settle in the bile ducts in the liver, whereas the lung fluke (*Paragonimus*) finds its way from the abdominal cavity through the diaphragm into the pleural cavity, the lungs, or, finally, the bronchioles. The development of larvae into adult flukes is completed in these locations, usually after several weeks, and the hermaphroditic adult worms begin to produce eggs. Depending on the location of the adult worm, ova may find their way directly into the feces (*Fasciolopsis*), down the bile ducts into the bowel and the feces (*Clonorchis* and *Fasciola*), or up the bronchial tree with sputum (*Paragonimus*). Some eggs may be deflected from these paths and be caught in surrounding parenchymal tissues, where defensive host reactions trap them in cellular exudates that hold the ova but disrupt local tissue functions (as the adult worm may also do).

D. *Exit Routes from Human Body* Ova of liver and intestinal flukes exit through fecal discharges, and lung fluke ova exit in sputum. Other sites are possible if adult worms locate in aberrant positions. These ova are not infective for human beings.

IV. Cyclic Development Outside of Human Host

A. *In Environment* The ova of these parasites must reach water to continue their developmental cycles. Ciliated embryos (miracidia) escape from their shells and swim about until they are taken up by snails.

B. *In Hosts Other Than Humans* Snails are the first intermediate host for each of these species (also for the schistosomal flukes). The parasites go through maturation stages in snails and emerge again as free-swimming forms called *cercariae*. These larval forms then encyst again in a

SUMMARY TABLE 18.1 PARASITIC DISEASES ACQUIRED THROUGH THE ALIMENTARY ROUTE

Clinical Disease	Causative Organism	Other Possible Entry Routes	Incubation Period	Communicable Period
Intestinal Protozoan Diseases				
Giardiasis	*Giardia lamblia*		1–4 weeks, usually 7–10 days	Throughout infection
Blastocystosis	*Blastocystis hominis*		Unknown, probably same as above	Same as above
Balantidiasis	*Balantidium coli*		Unknown	Same as above
Invasive Protozoan and Coccidian Diseases				
Amebiasis	*Entamoeba histolytica*		Usually 2–4 weeks	Same as above
Toxoplasmosis	*Toxoplasma gondii*	Via placenta	Usually 1–3 weeks	Probably no human-to-human transmission except congenital
Cryptosporidiosis	*Cryptosporidium*		Probably 1–12 days	Oocysts shed for several weeks
Cyclosporiasis	*Cyclospora cayetanensis*		Unknown	Not communicable person to person; oocysts must mature in environment
Microsporidiosis	Microsporidia (5 genera)	Respiratory	Unknown	Probably while shedding spores
Intestinal Nematode Diseases				
Enterobiasis	*Enterobius vermicularis*		3–6 week life cycle	Throughout infection
Trichuriasis	*Trichuris trichiura*		Long, variable	Not communicable from person to person; ova require period of maturation in soil
Ascariasis	*Ascaris lumbricoides*		Several months	Same as above
Invasive Nematode Diseases				
Trichinosis	*Trichinella spiralis*		Usually 8–15 days	Not communicable from person to person
Toxocariasis (VLM)	*Toxocara canis* *Toxocara cati*		Probably weeks or months	Same as above
Intestinal Cestode Diseases				
Taeniasis	*Taenia saginata* *Taenia solium* (see Cysticercosis below)		8–10 weeks	Same as above
Diphyllobothriasis	*Diphyllobothrium latum*		3–6 weeks	Same as above
Hymenolepiasis	*Hymenolepis nana* *Hymenolepis diminuta*		Unknown	*H. nana* communicable while parasite persists in intestines *H. diminuta* not communicable from person to person
Invasive Cestode Diseases				
Cysticercosis	*Taenia solium*			As long as *T. solium* persists in intestines
Echinococcosis	*Echinococcus granulosus*		Long, months to years	Not communicable from person to person
Invasive Trematode Diseases				
Clonorchiasis	*Clonorchis* spp.		Unknown	Same as above
Fascioliasis	*Fasciola hepatica*		Variable	Same as above
Fasciolopsiasis	*Fasciolopsis buski*		About 2 months	Same as above
Paragonimiasis	*Paragonimus westermani*		Unknown	Same as above

*Standard precautions apply to all patients with parasitic diseases. Other precautions indicated when applicable.

Specimens Required	Laboratory Diagnosis	Immunization	Treatment	Precautions[a]
Stools	Microscopic examination for trophozoites or cysts	None	Metronidazole Quinacrine	
Stools	Microscopic examination for trophozoites or cysts	None	Metronidazole Iodoquinol	Sanitary disposal of feces
Stools	Microscopic examination for trophozoites or cysts	None	Tetracyclines Iodoquinol	Personal hygiene
Warm or PVA-fixed stools Blood serum	Same as above Serology	None	Metronidazole Iodoquinol	Control food handlers; sanitary disposal of feces
Body fluids Biopsy Blood serum	Microscopic examination for trophozoites Serology	None	Sulfonamides with pyrimethamine Spiramycin	Instruct pregnant women about source in cats, raw meat
Stools Bowel biopsy	Concentration Microscopic examination	None	Supportive Replace fluid loss	Protect immunodeficient patients
Stools	Wet mount Acid-fast stains Examination under UV microscopy	None	Trimethoprim-sulfamethoxazole	Probably same as for *Cryptosporidium*
Stools Bowel biopsy	Microscopic examination Electron microscopic examination	None	Albendazole for some genera	Same as above
Scotch tape of perianal area	Microscopic examination for ova of adults	None	Mebendazole Pyrantel pamoate	In home: vigorous disinfection involving bathroom, bed linens, bedroom dust, hands, fingernails; in hospital: concurrent disinfection of fecally contaminated objects, careful hand washing
Stools	Ova	None	Mebendazole	Sanitary disposal of feces
Sputum Stools	Larvae Ova or adults	None	Mebendazole Pyrantel pamoate	Same as above
Muscle biopsy Blood serum	Larvae Serology	None	Mebendazole Corticosteroids	No special precautions necessary
Liver biopsy Blood serum	Larvae Serology	None	Diethylcarbamazine Thiabendazole	Same as above
Stools	Ova or proglottids	None	Niclosamide Praziquantel	Sanitary disposal of feces (see Cysticercosis)
Stools	Ova or proglottids	None	Same as above	Sanitary disposal of feces
Stools	Ova	None	Same as above	*H. nana:* sanitary disposal of feces, careful hand washing before eating and after defecation; *H. diminuta:* no special precautions necessary
Cyst excision	Larvae	None	Surgery Praziquantel Albendazole	No special precautions
Cyst excision Blood serum	Scolices Serology	None	Surgery Albendazole	Same as above
Stools	Ova	None	Praziquantel	Sanitary disposal of feces
Stools	Ova	None	Bithionol Surgery	Same as above
Stools	Ova	None	Praziquantel Niclosamide	Same as above
Sputum Stools	Ova Ova	None	Bithionol Praziquantel	Same as above

second aquatic host—fish, crustaceans, or water plants—until they are ingested by a mammalian host that offers suitable conditions for their maturation as adults. The second intermediate host differs for each of these species.

Clonorchiasis

The *Clonorchis* adult lodges in the bile ducts. Symptoms of **clonorchiasis** may result from obstruction of the ducts and include jaundice or liver disease. The liver may become enlarged, tender, and cirrhotic. Symptoms may persist for many years without therapy. Praziquantel is the drug of choice. The disease is acquired by ingesting larvae encysted in fish. Control requires sanitary disposal of human feces so that the ova of human infections do not reach water where snail and fish hosts may perpetuate the cycle. Thorough cooking of fish is another measure necessary in endemic areas (the Far East and Southeast Asia, including Hawaii, Korea, and Vietnam).

Fascioliasis

Fascioliasis occurs in sheep- and cattle-raising countries around the world, including the United States. The adult worm *Fasciola hepatica* develops from larvae ingested with infected water plants (such as watercress). It inhabits the bile ducts and produces symptoms referable to obstructive jaundice. The drug bithionol or surgical removal of the adult worm may be effective. Sheep and cattle are the principal animal reservoirs. Control depends on eliminating infected animals, eradicating snails, and protecting water greens cultivated for domestic animal or human consumption from contamination.

Fasciolopsiasis

The adult trematode *Fasciolopsis buski* causes **fasciolopsiasis** by infecting the intestinal tract, producing local irritation with resultant diarrhea, vomiting, anorexia, and other gastrointestinal discomfort. Absorption of toxic metabolites produced by the worm induces systemic reactions and allergy. The face becomes edematous, and the abdomen may be distended and painful with ascites and allergic edema of the wall. Treatment with praziquantel or niclosamide is effective in clearing the worms from their intestinal attachments, and recovery is good if the accompanying intoxication has not been too damaging. This is primarily a disease of the Orient. Human beings and some of their domestic animals maintain the parasite. Fecal contamination of water leads to the parasite's further development in snails and eventual encystment on water plants, particularly the popular water chestnut, which is often eaten raw. Control can be achieved by sanitary disposal of feces or their disinfection before use as a fertilizer, destruction of snail hosts in areas where water plants are harvested, or preliminary treatment of such vegetables by a few seconds' dip in boiling water.

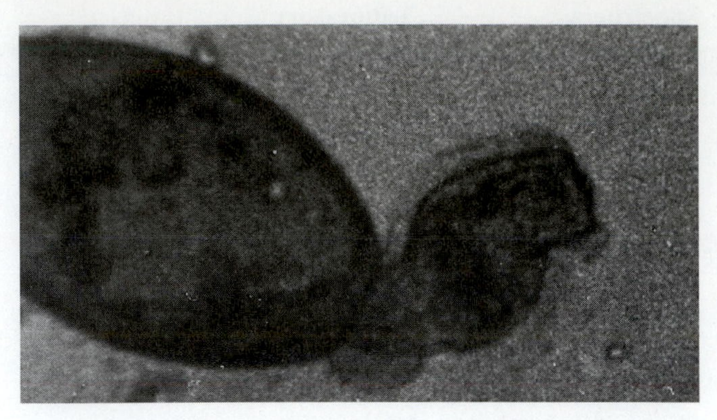

FIGURE 18.24 An ovum of *Paragonimus westermani* with an open operculum through which the embryo is emerging (×1,000).
Hugo Terner, Montefiore Hospital and Medical Center, Bronx, NY.

Paragonimiasis

The adult lung fluke *Paragonimus westermani* infects pulmonary tissues to cause **paragonimiasis.** The ova may be coughed out in sputum or swallowed and discharged in feces. When they reach water and a snail host, larval development proceeds (fig. 18.24). Cercariae emerging from snails find their way into crayfish or crabs and encyst in these second intermediate hosts. Humans and other animals (dogs, cats, wild carnivores) that eat these infected crustaceans perpetuate the parasite. The human infection can be treated with praziquantel or bithionol, the former being the drug of choice. It can be prevented by adequate cooking of crustacean foods. In areas of the world where this disease is endemic (the Far East and parts of India, Africa, and South America), control of the snail host is important.

Questions

1. What part of the human body is usually injured by animal parasites?
2. Describe the immune response to the antigenic components of animal parasites.
3. What are the inherent problems of drug therapy in parasitic diseases?
4. What is the infective form of *Giardia lamblia?* How is it transmitted?
5. What are the two developmental forms of *Entamoeba histolytica?* Where does the excystation take place?
6. How is amebiasis controlled?
7. What are the transmission routes of toxoplasmosis? How can the disease be prevented?
8. Why has cryptosporidiosis become more prevalent in recent years?
9. Why is cyclosporiasis thought to be a water-borne disease?
10. How is pinworm infection diagnosed? Why is this parasitic infection difficult to treat?

11. How is trichuriasis diagnosed? Why is this disease difficult to treat?
12. Briefly describe the development of the adult form of *Ascaris lumbricoides* following the ingestion of ova.
13. What are the pathological effects of ascariasis?
14. Can trichinosis be transmitted by humans?
15. In the treatment of taeniasis, why is it imperative to retrieve the scolex?
16. How is taeniasis prevented?
17. How does *Hymenolepis nana* differ from other tapeworms?
18. In what parts of the world is echinococcosis prevalent? Why?

BIBLIOGRAPHY

Amin, N. M. 1984. Giardiasis. *Hosp. Med.* 20 (5): 39–48.

Beaver, P. C., and R. C. Jung. 1985. *Animal agents and vectors of human disease.* 5th ed. Philadelphia: Lea & Febiger.

Benenson, A. S., ed.1995. *Control of communicable diseases in man.* 16th ed. Washington, D.C.: American Public Health Association.

Beutel, W. D., and J. G. Allison. 1983. Ascariasis. *Consultant* 23 (7): 213, 216–217, 220–222.

Bonner, A., and R. Dale. 1986. Giardia lamblia: Day care diarrhea. *Am. J. Nurs.* 86 (7): 818–820.

Current, W. L., and L. S. Garcia. 1991. Cryptosporidiosis. *Clin. Microbiol. Rev.* 4 (3): 325–328.

Desowitz, R. 1983. *New Guinea tapeworms and Jewish grandmothers: Tales of parasites and people.* New York: Avon.

Garcia, L. S., and D. A. Bruckner. 1997. *Diagnostic medical parasitology.* 3rd ed. Washington, D.C.: ASM Press.

Katzman, E. M. 1989. What's the most common helminth infection in the U.S? *MCN* 14 (3): 193–195.

Kaye, D. 1988. The spectrum of strongyloidiasis (case study). *Hosp. Prac.* 23 (10): 11–15, 118–119, 125–126.

Marshall, M. M., D. Naumovitz, Y. Ortega, and C. R. Sterling. 1997. Waterborne protozoan pathogens. *Clin. Microbiol. Rev.* 10 (1): 67–85.

The medical letter handbook of antimicrobial therapy. 1994. New Rochelle, N.Y.: The Medical Letter, Inc.

Ortega, Y., and C. R. Sterling. 1996. *Cyclospora cayetanensis:* Epidemiology and diagnosis. Clin. Microbiol. Newslett. 18 (22): 169–172.

Roueche, B. 1991. A pig from Jersey. In *Medical detectives.* New York: Plume.

Sand, P. 1984. Congenital toxoplasmosis. *Perinat. Press* 8 (8): 119–122.

Sheahan, S. L., and J. P. Scabolt. 1987. Management of common parasitic infections encountered in primary care. *Nurse Pract.* 12 (8): 19–20, 22, 25.

Spencer, I. M., and L. S. Monroe. 1982. *The color atlas of intestinal parasites.* 2d ed. Springfield, Ill.: Charles C. Thomas.

Stenzel, D. J., and P. F. L. Boreham. 1996. *Blastocystis hominis* revisited. *Clin. Microbiol. Rev.* 9 (4): 563–584.

Weller, P. F. 1983. Intestinal parasites: Amebiasis and giardiasis are indigenous—and spreading. *Consultant* 23 (4): 115–116, 126, 130–131.

Weber, R., R. T. Bryan, D. A. Schwartz, and R. L. Owen. 1994. Human microsporidial infections. *Clin. Microbiol. Rev.* 7 (4): 426–461.

Wolfe, M. S. 1992. Giardiasis. *Clin. Microbiol. Rev.* 5(4): 93–100.

Zierdt, C. H. 1991. *Blastocystis hominis*—Past and future. *Clin. Microbiol. Rev.* 4 (1): 61–79.

Section VI

Sexually active young adults are among the prime targets for sexually transmitted diseases.
World Health Organization. Photo by P. Almasy.

Infectious Diseases Acquired through the Skin and Mucosa or Parenterally

CHAPTER 19

Infectious Diseases Acquired through the Skin and Mucosa and Transmitted by Human Contact

OUTLINE

Microbes in Motion

This chapter covers the important bacterial, viral, and parasitic agents of sexually transmitted diseases along with other bacterial, viral, and fungal agents that are important in diseases of the skin and mucosa transmitted by human contact. The laboratory diagnosis, epidemiology, control, and prevention of these diseases is discussed. The following books and chapters of the Microbes in Motion CD-ROM may be a useful preview or supplemental study aid to your reading: Miscellaneous Bacteria in Human Hosts: Treponemes; Chlamydia; Mycobacteria. Gram-negative Organisms in Human Hosts: Gram-negative Cocci. Parasitic Structure and Function: Protozoa Classification. Viral Structure and Function: Viral Pathogenesis. Gram-positive Organisms in Human Hosts: Staphylococci; Streptococci.

OBJECTIVES

Upon completing this chapter the student will be able to:

1. describe the serious infections of the eye acquired through human contact.
2. cite other types of microbial diseases acquired through human contact and how they involve superficial or systemic tissues.
3. name the causative organisms of the important sexually transmitted diseases.
4. describe the methods used for the laboratory diagnosis of sexually transmitted diseases.
5. explain the epidemiology of these diseases.
6. discuss methods for controlling sexually transmitted diseases and for preventing those diseases.

KEY WORDS

Acquired immunodeficiency syndrome (AIDS)
Bacterial conjunctivitis
Bacterial vaginosis (vage-in-**OH**-sis) (BV)
Chancroid (**SHAN**-croid)
Chlamydia (clam-**ID**-ee-ah)
Chlamydial urethritis
Ebola (e-**BOWL**-uh) virus
Epidermophyton (epi-derm-**OFF**-ee-ton)
Erysipelas (**ERE**-ee-**SIP**-eh-las)
Genital herpes (**HER**-peas)
Genital warts
Gonorrhea
Granuloma inguinale (gran-u-**LOAM**-ah in-gwin-**ALLEY**)
Hansen disease (HD)
Hemorrhagic conjunctivitis
Herpes simplex virus (HSV)
Herpetic (her-**PET**-ic) disease
Human immunodeficiency virus (HIV)
Impetigo (imp-eh-**TIE**-go)
Inclusion conjunctivitis
Kerato (**CARE**-ah-toe) conjunctivitis
Leprosy
Lymphogranuloma venereum
Microsporum (micro-**SPORE**-um)
Necrotizing fasciitis (neck-crow-**TIE**-zing fas-she **EYE**-tis)
Neisseria (nice-**EAR**-ee-ah)
Reagin (ree-**AGE**-in)
Retrovirus
Ringworm
Scalded skin syndrome
Sexually transmitted diseases (STD)
Staphylococcal (**STAFF**-ill-oh-**COCK**-al) infections
Streptococcal (**STREP**-toe-**COCK**-al) infections
Superficial mycoses
Syphilis
Tinea (**TEEN**-ee-ah)
Toxic shock syndrome (TSS)
Trachoma (track-**OH**-ma)
Treponema (trep-oh-**NEEM**-ah)
Trichomoniasis (trick-oh-moan-**EYE**-ah-sis)
Trichophyton (tri-**COFF**-ee-ton)

Introduction

The major source of microorganisms that infect the skin and superficial mucosae appears to be the infected skin or mucosal lesions of other persons. A few organisms, such as staphylococci and streptococci, may have a reservoir in the normal respiratory tract, and the AIDS virus is carried in the blood and some body secretions; but the epidemiology of these infections characteristically indicates transmission from one person to another through direct contact. Fomites may play an indirect role in the spread of some diseases, but they are generally not a major factor. Prevention and control therefore depend primarily on principles of personal hygiene and adequate protection during sexual activity, as well as early diagnosis and adequate treatment. Effective vaccines and immunoprophylactic measures have proved difficult to develop and are not yet available for control of the contact diseases.

Sexually Transmitted Diseases

The term **sexually transmitted diseases (STD)** refers to those infectious diseases that are transmitted only, or most commonly, through intimate sexual contacts. The microbial agents of STD are limited to and transmitted by human hosts. Under normal circumstances they do not survive in the environment or in other hosts. Most can attach to, colonize, and even penetrate mucosal surfaces, but probably not normal intact skin. Their entry routes are generally the genital, anal, or oropharyngeal surfaces.

Ordinarily, STDs occur among adults and adolescents, but children can acquire infections through sexual or other close contacts with adults, and infants may be infected with the agents of sexually transmitted diseases at birth or in utero. Infection from environmental sources is not common, but accidental transfer from freshly contaminated objects, infectious discharges, or transfused blood is occasionally reported, particularly in medical or laboratory personnel or in infants.

A variety of potentially pathogenic organisms can be transferred by sexual contacts. Among these, several are consistently associated with disease in adults: the spirochete of syphilis; the bacterial agents of gonorrhea, chancroid, and granuloma inguinale; and the chlamydial agents of nongonococcal urethritis and lymphogranuloma venereum. Most of these agents cause acute or chronic reactions in tissues adjacent to the site of entry. In some instances, microorganisms transmitted through the genital route produce infections of little or no consequence in adults, but intrauterine infection of infants born of infected mothers may be severe or fatal.

Acquired immunodeficiency syndrome (AIDS) is transmitted by blood containing the **human immunodeficiency virus (HIV)** as well as by sexual contact. Therefore, it is also considered a parenterally acquired infection. Because sexual transmission is an important route for acquisition of infection, it is discussed in this section with other STDs.

Effective vaccines for sexually transmitted diseases have not been developed, but efforts to achieve this kind of control continue. In the meantime, prevention remains an unsolved problem. Public education along with prompt diagnosis and adequate treatment of each discovered case afford the only external controls possible. These in turn depend on the efficacy of community-wide efforts to inform the public of the nature and dangers of untreated STD. Physicians and laboratories are requested to make prompt reports of diagnosed cases, whereupon local health agencies send trained advisers to interview patients, find the sources of their infections if possible, and treat all contacts. Such treatment is usually offered at government expense in health department clinics.

Syphilis

The Clinical Disease

Untreated, noncongenital **syphilis** progresses through three stages extending over many years, with long intervening periods of latency. As a result of intimate, usually sexual, contact with an infected individual, the causative spirochetes (***Treponema** pallidum*) enter the body through small breaks in the mucous membranes or sometimes in the skin. Approximately 30% of persons exposed in this manner develops syphilis.

During the primary stage of disease, the spirochetes multiply locally, producing a superficial lesion called a *chancre* (colorplate 10a). Some may also reach the draining lymph nodes of the area and spread to the bloodstream. The chancre usually requires about 2 to 6 weeks to develop. It is a painless ulcer with a hard, firm base, located usually on the penis of the male and on the labia, vaginal wall, or cervix of the female. Depending on the mode of infection, the chancre may appear at other sites such as the rectum, lip, or palm of the hand, but only one lesion is formed. The regional lymph nodes are often enlarged but not tender. The chancre heals spontaneously, even if not treated, but without therapy the systemic infection continues.

After the chancre disappears, the lesions of the secondary stage develop from 4 to 6 weeks later (usually within 3 months of the time of infection) (colorplate 10b). Secondary syphilis commonly involves the skin, with lesions on the trunk, extremities, genitalia, palms, and soles of the feet. The rash may be macular, papular, or pustular, the lesions representing foci of disseminated spirochetes (colorplate 10c). Patchy lesions may also appear on the genital or oropharyngeal mucosa or in the anal canal. These mucous patches (like the primary chancre) contain many treponemes and are highly infectious. Generalized lymphadenopathy is common in this stage of disease, and there may be other manifestations such as arthritis or involvement of the liver, spleen, eye, or central nervous system.

As with primary syphilis, the lesions of secondary disease disappear with or without treatment but may recur in about 25% of patients. If the syphilitic patient is untreated or if therapy is inadequate, a latent, asymptomatic period ensues. In the early part of the latent stage, up to about 2 years from the time of infection, syphilis remains potentially infectious if secondary lesions recur. After 2 years, the late latent stage of the disease is less communicable by sexual contact but may be transmitted to a fetus by an infected mother.

Latency may continue for life, but about one-third of untreated cases advance to the tertiary stage of disease, or late syphilis. The onset of this stage may be as early as 3 years or as late as 30 years following the primary infection, but in most cases at least 10 years elapse before tertiary symptoms develop. Late syphilis is characterized by the formation of soft, granulomatous lesions called *gummas* (colorplate 10d), which may occur in a variety of tissues: skin, bones, blood vessels, liver, and central nervous system. Spirochetes may be difficult to demonstrate in these lesions, which probably represent a delayed-type hypersensitivity response to the organism or its products. The location, number, and clinical effects of gummatous lesions are quite varied, producing symptoms like those of many other chronic diseases (syphilis has been called the "great imitator"). Manifestations may be benign, or there may be cardiovascular disease (aortic aneurysm is not uncommon), CNS involvement, or a combination of these. CNS manifestations are common in patients with AIDS. Some of the symptoms of neurosyphilis include paresis (incomplete paralysis), tabes dorsalis (also called *locomotor ataxia,* a shuffling walk induced by sensory tract degenerations in the cord), deafness, blindness, and progressive mental deterioration. With vigorous campaigns for the discovery and treatment of early syphilis, the tertiary disease has become uncommon in the United States and elsewhere.

Congenital Syphilis

Untreated syphilis in pregnant women may produce congenital disease of the fetus, because the spirochete can cross the placental barrier and infect fetal tissues. Miscarriages and stillbirths occur, but the greatest tragedy involves live infants born with generalized (including CNS) lesions. Sometimes there are no obvious signs of syphilis at birth, but months or years later the congenital case develops blindness, deafness, or mental retardation. The superficial lesions of congenital syphilis are as infectious as those of the disease acquired in any other way. Syphilitic babies may transfer the disease to adults who fondle and kiss them, and conversely, normal babies may be infected in this way by persons in primary or secondary stages of infection. Spirochetes from infected mothers cannot cross the placenta during the first trimester of pregnancy. Thus, early diagnosis and treatment of the mother during this stage of pregnancy prevents infection of the fetus and the subsequent development of congenital defects in the newborn.

LABORATORY DIAGNOSIS

Syphilis may be diagnosed clinically from the appearance of characteristic lesions and a history of sexual contact with an infected person, but this disease is so variable in its symptomatology and so serious in its consequences that laboratory confirmation is always sought.

I. Identification of the Organism

(Treponema pallidum.)

A. *Microscopic Characteristics* Treponemal spirochetes are slender spirals with about 8 to 15 coils evenly spaced at 1-μm distances, the entire length being from 5 to 15 μm. They are very thin (about 0.2 μm in width), have a multilayered cell membrane beneath the cell wall, and six flagellalike fibrils between the cell wall and membrane. They have a distinctive drifting motility, their bodies curving in graceful, slowly undulating motions.

Treponemes do not take ordinary stains but can be seen in tissue sections treated with silver nitrate. Metallic silver precipitates on the surface of spirochetes, making them appear black against a light background of tissue (fig. 19.1).

They can also be visualized in wet preparations of material taken from chancres or mucous patches if the microscope is adjusted for dark-field illumination. This method is commonly used to make a rapid, presumptive diagnosis of syphilis. The microscope technique is described in chapter 4. Figure 19.2 illustrates the appearance of treponemes when examined by the dark-field technique: note that they are brightly illuminated against a very dark background, so that their coiled morphology can be clearly seen. The spacing of the coils and differences in motility help distinguish treponemes from other spirochetes.

T. pallidum-specific antibody labeled with a fluorescent dye is available for performing direct fluorescent antibody tests (DFA) on the specimen (see chapter 6). In this test, patient specimen dried on a microscope slide is exposed to the specific antitreponemal antibody. The slide is then observed microscopically with an ultraviolet light source for the presence of the characteristic spirochetes, which should fluoresce brightly.

Treponema pallidum has not been successfully cultivated on artificial laboratory media or in embryonated hen's eggs. The organism has been grown in cell culture, but this method is not widely used. The organism must be diagnosed on the basis of microscopic characteristics, in material obtained from active lesions, or by serologic testing of the patient's serum.

II. Identification of Patient's Serum Antibodies

Two types of antibodies develop during the course of syphilitic infection, and both can be detected in the circulating blood. One type is called *reagin:* the other is specific treponemal antibody.

Reagin is a nontreponemal antibody that, in the test tube, reacts with lipid antigens extracted from beef heart (cardiolipin) and other mammalian tissue, combining with them to produce flocculation. Although the antigen is not derived from the spirochete, it is thought to be similar to lipid material released in the body as a result of

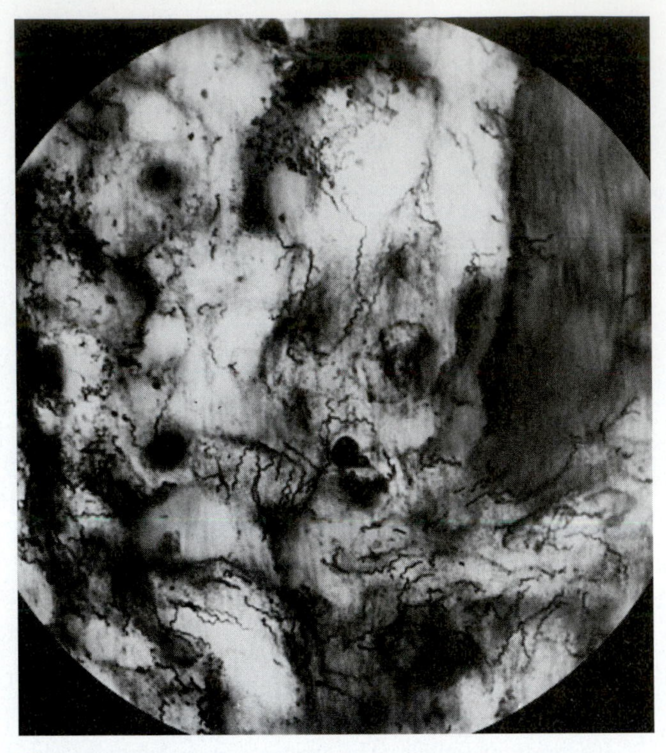

FIGURE 19.1 *Treponema pallidum* in a tissue section treated with silver nitrate (×1,000).
Courtesy New York State Health Department, Albany NY.

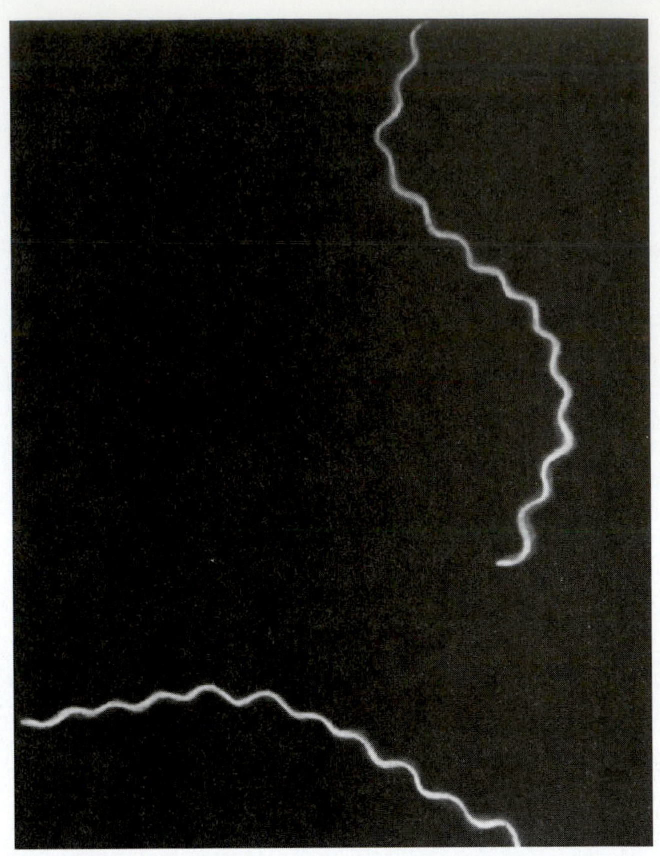

FIGURE 19.2 Dark-field photomicrograph of treponemes (×2,000).
Centers for Disease Control, Atlanta, GA.

tissue destruction during active infection or to resemble some antigenic component of the organism. In any case, the antibody that appears during the disease process gives positive reactions in several flocculation tests. The VDRL slide flocculation test (developed by the Venereal Disease Research Laboratory of the U.S. Public Health Service, hence its name) has been used most commonly. Current modifications of flocculation or agglutination techniques include such rapid reagin tests as the Rapid Plasma Reagin (RPR), used for screening large numbers of patients' serums, or the Automated Reagin Test (ART), an automated version of the RPR. The methods for detecting reagin may give biologic false-positive tests for syphilis, particularly in other infectious diseases such as malaria, tuberculosis, leprosy, infectious mononucleosis, viral hepatitis, and other spirochetal infections. Reagins also appear in spinal fluid during syphilitic infections of the central nervous system. In this case there is little likelihood of obtaining a false-positive test, but a negative result merely excludes CNS syphilis. It does not rule out syphilitic infections elsewhere in the body. Reagin tests are negative in approximately 25 to 30% of patients with untreated primary and late syphilis.

Specific Treponemal Antibody Specific humoral antibodies against *T. pallidum* also arise during the course of the disease. A number of serologic tests for these have been developed and are particularly useful when reagin tests are suspected of being falsely positive.

One of the most reliable antibody tests is the FTA-ABS (*fluorescent treponemal antibody-absorbed*) technique. This is an indirect fluorescent test in which the patient's serum is brought in contact with lyophilized (freeze-dried) *T. pallidum* that has been fixed to a glass slide. This mixture is then overlaid with a fluorescein-tagged conjugate of antihuman globulin. If the patient's serum contains specific treponemal antibodies, the organisms are coated with fluorescein and can be seen under an ultraviolet microscope (see chap. 6). If the patient's serum has no antibody, there is no coating, and the spirochetes are not visible. Preliminary treatment of the test serum with a related treponemal strain absorbs out group antibodies for commensal treponemes and makes the FTA test still more specific for *T. pallidum*.

Microhemagglutination tests for *T. pallidum* (MHA-TP) are the most convenient specific confirmatory tests now available, but they may be negative in about 35% of patients with primary syphilis. These tests employ sheep red blood cells sensitized with an antigen obtained from *T. pallidum* by ultrasonification. When the red cell-antigen complex is mixed with antitreponemal antibody, visible hemagglutination occurs (the term *micro-* refers to the small volumes of blood and reagents used for this test). Commercial availability of the antigen and the simplicity of the test procedure are among the advantages of the MHA-TP.

The confirmatory procedures all have some disadvantages in the serologic diagnosis of syphilis. For example, all give positive results with the nonsexually transmitted, cutaneous treponemal diseases, yaws, pinta, and bejel, which occur primarily in tropical and semitropical countries. This makes it impossible to diagnose latent syphilis in patients exposed to other treponematoses who present no clinical evidence of syphilis. Also, the FTA-ABS and MHA-TP tests infrequently give false-positive results in certain patients. The fluorescent method may be positive in some autoimmune diseases such as rheumatoid arthritis or systemic lupus erythematosus; the hemagglutination test may be positive in similar connective tissue disease syndromes and in infectious mononucleosis.

The specific treponemal tests remain reactive for many years following syphilitic infection, whether or not the disease is treated. Therefore, these tests do not provide information regarding the clinical activity of syphilis, nor are they useful in following response to treatment. Reagin tests, on the other hand, may give rising or falling titers with changes in the status of disease. A rising titer may indicate recent infection, reinfection, or relapse. A fall in titer signifies adequate treatment, and an unchanging low titer indicates inactive infection.

III. Specimens for Diagnosis

A. Tissue fluid expressed from primary or secondary superficial lesions is submitted for dark-field or DFA examination. Arrangements must be made with the laboratory for immediate dark-field examination of this material because spirochetes die rapidly when they leave the body and cannot be identified with certainty when they are dead. Timing is not as critical with the DFA test because test results depend on visualizing brightly fluorescing organisms rather than recognizing characteristic motility. Spirochetes at the lesion surface may be inactive in patients who have applied ointments, salves, and other home remedies, thus making dark-field examination unsuccessful. With careful cleaning of the lesion using bland soap and water, the organisms may again become microscopically recognizable within 24 hours. If this delay represents a risk that the patient will not return for a follow-up visit, the physician may perform the DFA test, if available, or choose to treat on clinical evidence alone. Spirochetes disappear from the superficial lesions within a few hours of antimicrobial therapy, and there is usually no later opportunity to demonstrate them. Presumptive serologic evidence can be obtained rapidly with the VDRL, RPR, or other screening test. If this test is positive, the diagnostic problem is solved. However, the primary lesion often appears a week or more before detectable levels of reagin, so that a negative test is not conclusive at this stage.

B. Late syphilis can be diagnosed by histologic examination of gummatous lesions removed by surgery or biopsied. Silver impregnation techniques are used to demonstrate the organisms in tissues.

C. Serologic tests for reagin in the serum generally become positive between the third and sixth weeks of infection and between the fourth and eighth weeks in spinal fluid from patients with CNS involvement. These antibodies may persist in low titer for months or years, even in treated cases, but in latent syphilis patients may become reagin-negative.

D. Serologic tests for specific treponemal antibodies in the serum remain positive for many years, even in treated cases.

Epidemiology

I. Communicability of Infection

A. *Reservoirs, Sources, and Transmission Routes* See page 410.

B. *Incubation Period* Average time from exposure to development of primary lesion is about 3 weeks, but this period may be as short as 10 days.

C. *Communicable Period* Untreated syphilis may be communicable throughout the primary and secondary stages and during the first years of latency. With adequate treatment, infectivity ceases after about 1 day.

D. *Immunity* There is no natural resistance to syphilis among humans, but infection slowly leads to some immunity against reinfection, probably because of the development of specific treponemal antibodies. (Reagins do not play a protective role.) When syphilis is treated adequately during its early stages, the individual does not have an opportunity to develop specific antibodies and again becomes fully susceptible.

II. Control of Active Infection

A. *Isolation* A private room is not necessary. Antimicrobial therapy eliminates the threat of communicability within 24 hours. To avoid reinfection, patients should not engage in sexual intercourse with previous partners who have not been treated.

B. *Transmission-based Precautions* Care of patients with active superficial lesions includes application of standard precautions (see chap. 11). Blood, discharges from open lesions, and objects freshly contaminated through direct contacts may be highly infectious.

C. *Treatment* *Treponema pallidum* is highly susceptible to penicillin. Long-acting formulations of this drug, such as benzathine penicillin G, are customarily used. Patients with hypersensitivity to penicillin may be treated with doxycycline or erythromycin, but these drugs are not as treponemicidal as penicillin, and a watch must be kept for remittent symptoms of active infection.

If syphilitic pregnant women are adequately treated during the first trimester of pregnancy, the risk of congenital disease of the fetus will be minimal. The infants should be followed closely after birth to make certain their serologic tests for syphilis are negative, or they should be treated at birth.

The Jarisch-Herxheimer reaction is a febrile response to specific therapy that may occur in patients in the

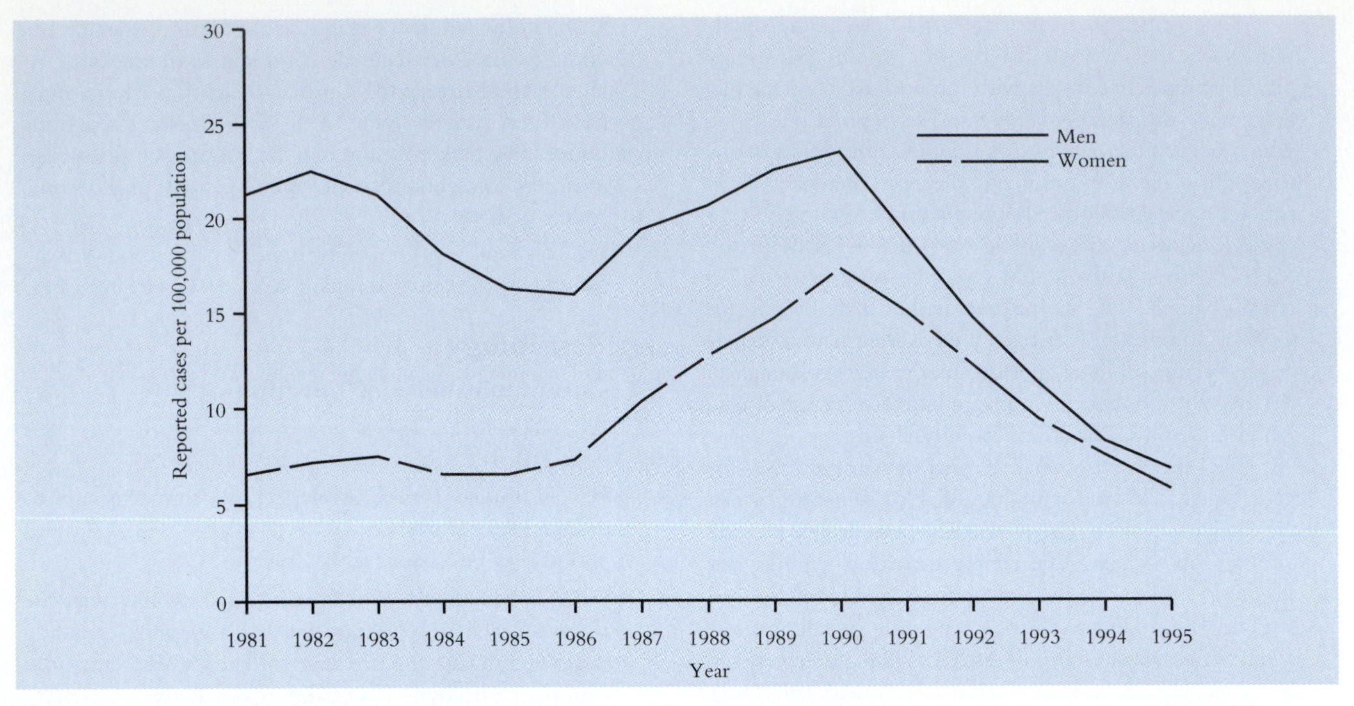

FIGURE 19.3 Syphilis (primary and secondary) in the United States, from 1981 to 1995. In 1995, the rate of primary and secondary syphilis among men decreased from 8.4 per 100,000 in 1994 to 6.8, and from 7.5 to 5.8 among women.

Centers for Disease Control and Prevention *MMWR Summary of Notifiable Diseases, United States, 1995* (October 25, 1996, Vol. 44:53).

Miscellaneous Bacteria in Human Hosts
Treponemes: Treponemes Virulence Factors, Treponemes Diseases

secondary or later stages of syphilis. It is thought to be related to the release of an endotoxinlike substance from the spirochetes killed by antimicrobial agents.

D. *Control of Contacts* Prophylactic methods are unreliable for controlling syphilis, although the use of condoms reduces the risk of infection to the sexual partner. Some spermicidal gels may have antitreponemal activity. Prostitution and other forms of sexual promiscuity ensure the continuation of this disease. Although the incidence of disease rose throughout the late 1980s, an impressive decline has been seen since 1990 (fig. 19.3). Case finding by investigating contacts of patients known to have syphilis remains the most effective form of control, as it offers an opportunity for treating otherwise undetected cases. Contact tracing is done by specially trained public health nurses or epidemiologists.

E. *Epidemic Controls* All state health departments require individual case reports so that efforts can be made to find and treat contacts who represent sources of infection. Efforts to detect and control syphilis include serologic tests for pregnant women, military personnel, and others. Personnel health services in hospitals and business corporations may also require that new employees receive a serologic test for syphilis.

In spite of countrywide and international measures to control it, the disease continues to spread, albeit at lower levels, particularly among adolescents and male homosexuals. Syphilis can be widely disseminated from a single active case. Each new transfer has the same potential for spreading widely, particularly among the sexually promiscuous (fig. 19.4). The optimism associated with the knowledge that syphilis can now be cured promptly and relatively easily with penicillin is perhaps misplaced. The early cure of infection is essential in controlling its spread, but it leaves little or no residual resistance to reinfection. Furthermore, primary infection may go unnoticed (particularly in women) and therefore untreated, and so may spread to large numbers of contacts, including the children congenitally infected. The renewed and intensified efforts being made by health agencies to maintain control through case finding and treatment appear to be having some effect in the United States (fig. 19.3).

III. Prevention

The prevention of syphilis and other sexually transmitted diseases depends on the efforts of an informed public, armed with a basic knowledge of sex and health, to promote the social and physical welfare of its young people, in particular. Working together, the general public, its public health services, and the medical community could reduce the problem. Better control and prevention await the development of an effective vaccine, and this in turn depends on new methods for the cultivation of the spirochete in vitro or in a practical animal model.

Gonorrhea

Gonorrhea has been known from the early ages, one of its first descriptions appearing in an Egyptian papyrus dating from 3500 B.C. The written records of every developing civilization, from China to Mesopotamia to early Rome, contain descriptions and prescriptions relating to a disease known for

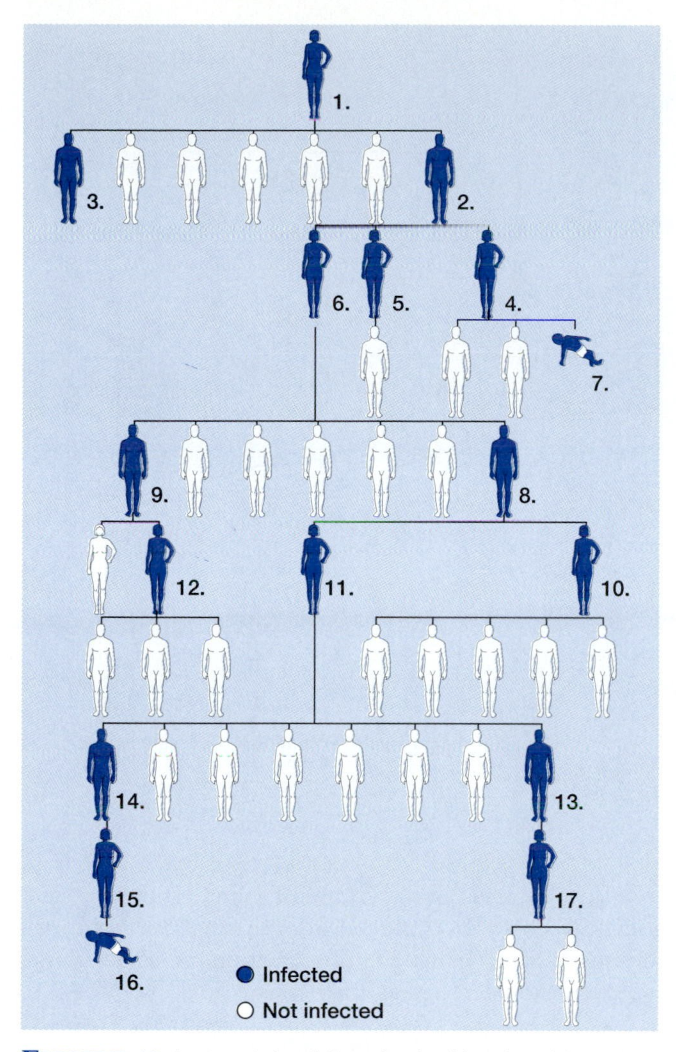

FIGURE 19.4 A typical syphilis outbreak. Although syphilis is a sexually transmitted disease, note that it can be transmitted to the fetus in utero resulting in a congenitally acquired infection.

Modified from *Today's VD Control Problem 1975.* Copyright © 1975 American Social Health Association. Reprinted by permission.

its characteristic urethral discharge, or "issue," and the strictures, or "strangury," that so often followed. In the third century B.C., Hippocrates defined its mode of transmission, listing it among the consequences of the "excesses of the pleasure of Venus" (STDs were formerly referred to as "venereal" diseases, a term derived from the name of this Roman goddess of love). The misnomer "gonorrhea" came from the Roman physician Galen in the second century A.D. Galen mistakenly assumed that the male urethral discharge represented an involuntary loss of semen and coined the term (*gonos* = seed; *rhoia* = flow), which has remained part of medical language ever since. Among the many vernacular and medical terms that have been applied to this disease, the phrase the "great sterilizer" perhaps holds the strongest implications of its consequences if untreated, especially in women.

The Clinical Disease

Gonorrhea is a sexually transmitted disease of worldwide distribution that reached epidemic proportions during the 1970s. Beginning in the early 1980s, however, its incidence began to level off and decrease (fig. 19.5). It is primarily a genitourinary tract infection of adults that may proceed to disseminated systemic disease. In young female children it may appear as a vulvovaginitis, and in newborn infants infected at birth it may be manifest as a severe ophthalmic infection.

Adult Gonorrhea

In the adult male, gonorrhea begins as an acute urethritis, manifested by a thick, purulent discharge. If untreated, it may extend to an epididymitis or prostatitis. Urethral scarring may produce strictures requiring surgical relief. Not uncommonly, systemic involvements such as arthritis, endocarditis, or meningitis may develop (see Disseminated Gonococcal Infection) in untreated cases.

In the adult female initial localization of infection is most often in the cervix but may occur in the urethra or anal canal. It is often mild or asymptomatic in the early stage, producing no discharge or one that goes unnoticed. If not treated, infection may extend upward into the Fallopian tubes and on into the pelvic peritoneum. The resultant salpingitis and pelvic inflammatory disease (PID) may heal by scarring when the suppurative process subsides, and this may lead to obstruction of the tubes and sterility. Silent, residual infection in untreated or unsuspected cases, particularly in women, is most responsible for continuing spread of the disease to new contacts.

Pharyngeal and anorectal infections have become increasingly common among heterosexual as well as homosexual adults. Pharyngeal infection is seldom symptomatic but may present as an exudative pharyngitis. Anorectal gonorrhea can produce a marked proctitis with rectal discharge, tenesmus, and bloody diarrhea. More commonly it is asymptomatic or causes only mild rectal burning or pruritus.

Gonorrhea is primarily an infection of the columnar and transitional epithelium. On entry into the body, the causative organism (***Neisseria** gonorrhoeae,* commonly called the *gonococcus*) attaches to mucosal cells by means of pili and surface antigens present on the cell wall (see fig. 5.6). Following attachment, gonococci reach the subepithelial connective tissue by penetrating the intercellular epithelial spaces. Sometimes they are engulfed by epithelial cells, but more usually they are taken up in large numbers by polymorphonuclear leukocytes as part of an inflammatory response to their presence. Spread of gonococci to adjacent tissues is probably facilitated by their multiplication in the film of secretions that covers the genitourinary surfaces. Studies in women indicate that spread into the endometrial cavity and Fallopian tubes is increased among those using an IUD (intrauterine device) for contraception and decreased in women using oral contraceptives. Menstruation also appears to facilitate spread to the upper genital tract, perhaps as a result of cervical dilatation. Gonococci attach to spermatozoa and thus can be carried on sperm into the upper tract.

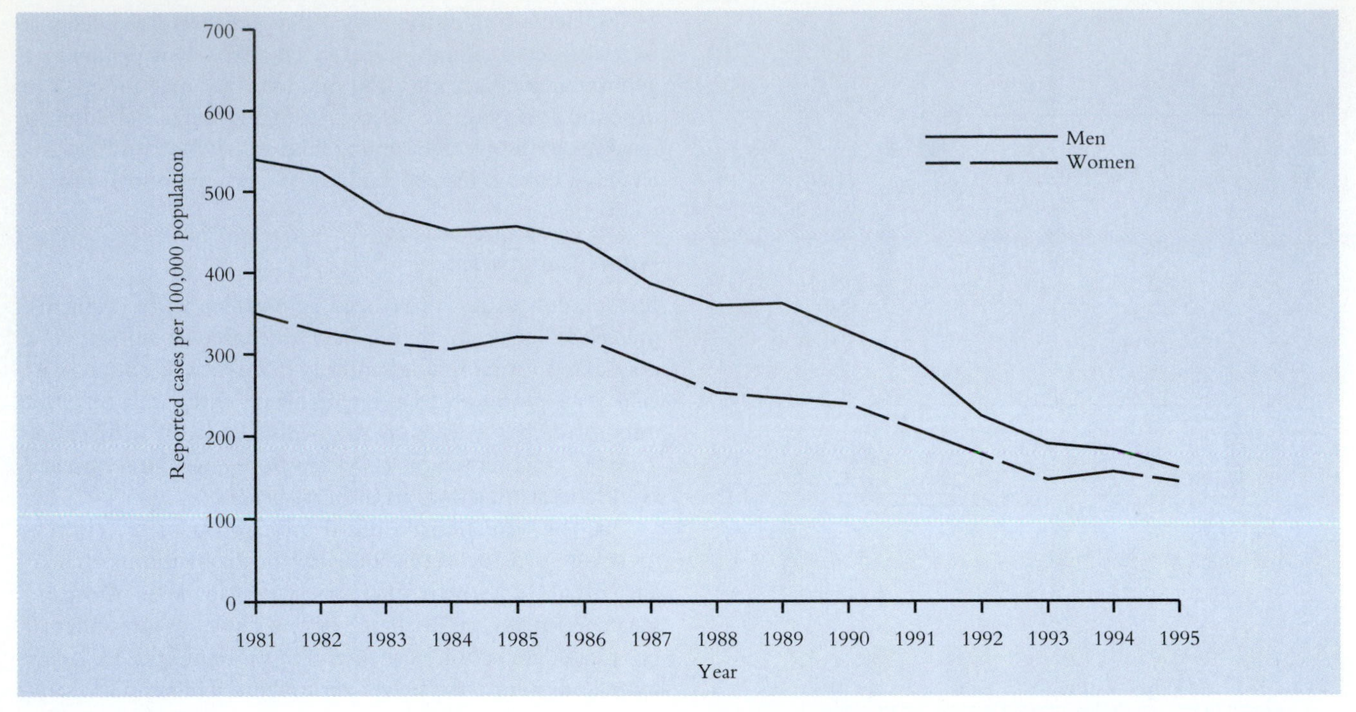

FIGURE 19.5 Gonorrhea cases in the United States from 1981 to 1995. In 1995, the rate of gonorrhea among men continued to decline. Although the rate in women had risen slightly from 1993 to 1994, it declined again in 1995.

Centers for Disease Control and Prevention *MMWR Summary of Notifiable Diseases, United States, 1995* (October 25, 1996, Vol. 44:53).

Secretory IgA antibody to gonococci appears in vaginal and urethral secretions in gonorrhea, but its role is not yet clear. It may inhibit gonococcal attachment to mucosal cells or promote opsonization by PMN. Conversely, gonococci produce an enzyme that can inactivate IgA and thus may protect themselves. Low levels of humoral antibodies may appear in genitourinary infections, but these tend to increase with repeated infections and in patients with disseminated gonorrhea.

For infection with *N. gonorrhoeae* to occur, the organisms must come into contact with susceptible epithelium. The only surfaces available for initial infection in adults are the urethra, endocervix, rectum, pharynx, and conjunctiva, as the organism cannot attach to and survive on cornified squamous epithelium. Coitus, fellatio, cunnilingus, and anal intercourse provide routes of transmission to all but the last of these sites. Transfer of gonococci to the eyes may be accomplished by unwashed hands contaminated either with an infectious genital discharge or accidentally in the laboratory.

Gonorrhea in Female Children

The vaginal epithelium of prepubescent girls is not cornified, so the entire vagina is susceptible to gonococcal infection. The genital labia may also be involved, and in severe cases infection can extend into the urethra and bladder. These mucous membranes become inflamed, and there is a profuse leukorrheal discharge. Sexual or other intimate contacts with infected adults appear to be responsible for most cases. There are some reports of transmission from mothers to infant daughters in the course of routine baby care, and a few cases may have resulted from sharing a bed with an infected parent. However, the question of child abuse should be considered in all such cases.

Gonococcal Infection of the Newborn

Ophthalmia neonatorum may be a serious consequence of untreated or undetected maternal infection. The eyes of a newborn baby may become infected at the time of delivery during passage through the birth canal if the mother has active gonorrhea. If the baby's eyes are not promptly treated, infection begins as an acute conjunctivitis, which rapidly involves the cornea and commonly ends in blindness (colorplate 10e). This problem can be readily avoided by treating the eyes of all newborn babies prophylactically with erythromycin ophthalmic ointment (0.5%), a procedure required by law in most states. Alternatively, either a 1% tetracycline ointment or a silver nitrate solution applied to the eyes immediately after birth prevents both gonococcal ophthalmia and chlamydial conjunctivitis. In another form of neonatal gonorrhea, infection of the chorioamnion may lead to prematurity, prolonged rupture of fetal membranes, and sepsis and orogastric colonization in the infant.

Gonococcal Conjunctivitis

This infection may be acquired by people of all ages and usually occurs through the transfer of genital infection by contaminated hands. After an incubation period of 1 to 3 days, the conjunctiva becomes intensely red and swollen, and there is a profuse purulent discharge. Prompt treatment brings about full recovery, but neglect can lead to corneal ulceration, panophthalmitis, and eventual blindness.

Disseminated Gonococcal Infection (DGI)

In about 1 to 3% of patients with gonorrhea, disseminated infection occurs through the entry of gonococci into the bloodstream from a site of localized infection. This is more common in women than men, occurring most frequently during pregnancy or menstruation. DGI may follow asymptomatic genital, pharyngeal, or rectal infection as well as more severe local disease.

There are two stages of DGI. In the first there is a bacteremia, followed by the appearance of skin and conjunctival lesions (colorplate 5d) and multiple joint involvement (polyarthralgia or polyarthritis). If DGI is untreated during this stage, the second phase may begin. Bacteremia ceases and polyarthritis disappears, but the infection localizes in a single joint. This is called the *septic joint stage,* because the affected joint develops a purulent effusion in which gonococci are usually present. There is a cellular infiltration of the synovium, which may lead to its destruction as granulation tissue is formed. Without treatment, this stage progresses to immobilization of the joint, with ankylosis and fibrosis.

Meningitis, endocarditis, myocarditis, and pericarditis all are possible complications of DGI, particularly in cases in which bacteremia persists.

LABORATORY DIAGNOSIS

The control of gonorrhea depends on its recognition and treatment, not only in obvious cases but also in patients with asymptomatic infections and their contacts. Therefore laboratory confirmation must be obtained whenever possible. In recent years the practice of taking cultures on a mass scale, particularly from women, who are often asymptomatic carriers of infection, has become a major part of a vigorous program to achieve better control.

I. Identification of the Organism

(Neisseria gonorrhoeae).

A. *Microscopic Characteristics* The gonococcus is a gram-negative, kidney-bean-shaped diplococcus, morphologically indistinguishable from the meningococcus (see chap. 15). In stained smears of exudates, gonococci are characteristically found in intracellular positions within polymorphonuclear leukocytes (colorplate 4e). They cannot be identified on the basis of morphology alone, however. Other gram-negative organisms that normally inhabit mucosal surfaces or that may be associated with infections may look very much like gonococci.

With this in mind, the laboratory sends only a presumptive report of the findings in a smear of exudate, with no commitment as to identification. When suspicious organisms are found, the report may read, "gram-negative intracellular diplococci seen."

B. *Culture Characteristics* The presumptive diagnosis of gonorrhea is confirmed by culture techniques. *N. gonorrhoeae* is a fastidious organism requiring enriched media containing hemoglobin, plasma, and yeast components. In addition, the media contain a battery of antimicrobial agents that do not affect *N. gonorrhoeae* but inhibit the growth of other organisms commonly found as commensals on the mucosal surfaces to be cultured. This permits gonococci, if they are present, to grow in relatively pure culture, free of competition or interference from other less fastidious organisms, and simplifies their isolation.

The requirement for CO_2 is particularly important for the primary isolation of the gonococcus. In addition, this organism is extremely sensitive to the effects of excessive temperatures and drying. For these reasons, the success of cultures depends on the care with which specimens are collected and the speed of their transport to the laboratory. Preferably, a suitable growth agar medium is inoculated directly and taken to the laboratory immediately so that it can be placed under CO_2 incubation without delay. If transport to the laboratory is to be delayed for some hours, a growth-transport agar medium system should be used. Such systems provide both the growth agar and the necessary CO_2 atmosphere. Whenever possible, cultures initiated in such transport systems should be incubated at 35°C during the time they are being held for transport to the laboratory.

Specimens taken directly to the laboratory may be streaked out there on a suitable growth medium and incubated in a candle jar or CO_2 incubator. After incubation, colonies appearing on the culture plates are tested for the presence of oxidase, an enzyme possessed by all *Neisseria* and some other microorganisms. Oxidase-positive colonies are then smeared, Gram stained, and examined microscopically. Those with typical morphology may then be distinguished from other *Neisseria,* including the meningococcus, and confirmed as *N. gonorrhoeae* based on their reactions in carbohydrate media. Identification systems that determine unique enzyme profiles of the organism are also available. Cultures from urogenital sources (i.e., cervix, vagina, or urethra, the last from either sex), may be reported as presumptively positive, without confirmatory techniques, if they meet the criteria of a positive oxidase reaction and typical morphology. Colonies grown from cultures of the anus or oropharyngeal sites, or those grown from any site in cases of suspected sexual abuse, must be fully differentiated from those of *N. meningitidis* or similar organisms that may also be isolated.

C. *Immunologic Methods for Identification* Direct fluorescent antibody reagents are available for confirming the identification of organisms grown in culture (this method is described in chapter 6). This test correlates well with biochemical identification methods and provides results more rapidly (in one to two hours as opposed to days).

D. *Nonculture Methods for Detection* Enzyme immunoassay and nucleic acid probe tests for detecting gonococci provide an alternative to culture for detecting *N. gonorrhoeae* directly in urogenital specimens. These tests are especially useful for screening specimens when culture facilities for *N. gonorrhoeae* are not readily available or when there may be significant delays in transporting specimens to the laboratory. These delays can adversely affect the recovery of the organism in culture. Under such circumstances, the use of these technologies provides a significant advantage over culture because they do not depend on the presence of viable gonococci to establish the diagnosis of gonorrhea. For the nonculture methods, special transport systems that differ from those used for culture are required.

Currently, the most widely used probe assay is a 2-hour test that detects gonococcal ribosomal RNA directly in urethral or cervical samples. The assay is highly sensitive and specific for gonococci. Because the nucleic acid is more stable than are whole organisms, delays in transport of the specimen to the laboratory do not affect the result. A major disadvantage of the nonculture tests is that they are not approved for testing specimens from extragenital sources (e.g., pharyngeal, rectal specimens). In addition, if antimicrobial susceptibility tests are needed, a culture isolate is not available for performing them. Nonculture tests should not be used when there is a question of sexual abuse, especially in children.

II. Identification of Patient's Serum Antibodies

Much effort and many trials with a variety of antigen preparations and serologic techniques have as yet failed to yield a clinically useful serologic test for gonorrhea.

III. Specimens for Diagnosis

Exudates from urogenital or other mucosal membranes for smear and/or culture. Urogenital specimens for EIA or nucleic acid probe tests.

In the male, Gram-stained smears of urogenital exudates are examined microscopically. If smears are positive for typical gram-negative diplococci, cultures are often not performed, especially in public health clinics or other facilities where large numbers of patients are screened for infection. In these settings, male patients are treated at once on the basis of positive smears. If smears are negative but there is clinical evidence or suspicion of gonorrhea, cultures are initiated. Exudates from the anus or pharynx should be cultured in any case, as smears may fail to reveal the organisms at these sites or distinguish morphologically similar organisms that may be present there normally. Commensal *Neisseria* species, and sometimes *N. meningitidis,* inhabit the throat; thus the identification of all suspicious oropharyngeal isolates must be confirmed as *N. gonorrhoeae* by biochemical or immunologic tests. In a few instances, *N. meningitidis* has been isolated from urethral exudate of males with signs and symptoms of gonorrhea.

In the female, exudates from all sources are routinely cultured. Smears may fail to reveal gonococci, especially in asymptomatic infections, and are not recommended except to confirm clinically obvious symptoms. Smears from females can be especially difficult to interpret because organisms morphologically similar to gonococci are often found normally on genital membranes. Cervical discharges are most likely to yield *N. gonorrhoeae,* but vaginal, urethral, anal, or pharyngeal cultures may also be positive. Isolates from rectal and throat cultures must be distinguished from *N. meningitidis* or other *Neisseria* species and culturally confirmed as gonococci. *N. meningitidis,* however, has been involved in a number of severe female genital tract infections.

If joint involvement is suspected in chronic gonorrhea, aspirated synovial fluid should be submitted for culture.

Cultures of blood or spinal fluid may yield the organism if systemic dissemination has occurred.

In gonococcal ophthalmia, pus from the conjunctiva is collected.

For proper collection and transport of specimens to the laboratory see Laboratory Diagnosis, I, B and D.

Whenever possible, antimicrobial susceptibility tests should be performed on isolates to detect emerging patterns of resistance to drugs.

Epidemiology

I. Communicability of Infection

A. *Reservoirs, Sources, and Transmission Routes* Gonorrhea is a human disease. Exudates from the involved mucous membranes of infected persons are the source of infection, transmitted by direct contact, usually sexual.

B. *Incubation Period* Usually 3 to 9 days, except in the case of ophthalmia neonatorum, which develops within 1 to 5 days.

C. *Communicable Period* If specific antimicrobial therapy is given, symptoms and communicability subside within a day or so. Without therapy, asymptomatic adult gonorrhea may remain communicable for months or years. Conjunctival membranes should be considered infectious until all discharge has ceased or for 24 hours after therapy is begun.

D. *Immunity* Susceptibility to gonorrhea is universal. Reinfection may readily occur because gonococci are antigenically diverse; therefore, cross-protection against different gonococcal strains does not develop.

II. Control of Active Infection

A. *Isolation* Hospitalized infants and children may be isolated for 24 hours following the administration of antimicrobial agents. No isolation is necessary for adults, but they should be instructed to avoid sexual contact with previous partners who have not been treated.

B. *Transmission-based Precautions* Standard precautions with concurrent disinfection of exudates from lesions or articles soiled by them. Health care workers should be aware of the susceptibility of the conjunctiva to gonococcal infection. Routine terminal cleaning of the unit is sufficient.

C. *Treatment* The gonococcus has a tendency to develop resistance to the antimicrobial agents used against it. Penicillin, once the drug of choice, can no longer be administered routinely because of the prevalence worldwide of beta-lactamase (penicillinase)-producing strains.

Instead, since 1976 when these resistant strains appeared, other drugs have been recommended for therapy. The cephalosporin ceftriaxone is now one of the most common drugs used. Coinfection with the sexually transmitted organism *C. trachomatis* has also become commonplace; therefore, doxycycline is now administered to eradicate that organism as well. This therapeutic regimen is also effective for treating most patients with unsuspected syphilis, but serologic follow-up for syphilis is recommended. Patients allergic to ceftriaxone are given a quinolone drug, such as ciprofloxacin, or spectinomycin instead, provided the organism is susceptible to the drugs.

Resistance to penicillin is conferred in some strains when the gonococci acquire a plasmid (see chap. 3) that enables them to produce penicillinase. Other strains are resistant because of chromosomal mutations, which most likely make the gonococcal outer membrane less permeable to penicillin. Plasmid-mediated tetracycline resistance has also been found in some strains. Strains resistant to other antimicrobial agents, including ceftriaxone, may soon emerge.

Disseminated gonococcal infections are treated initially with intravenous or intramuscular ceftriaxone for a few days followed by oral ciprofloxacin for a total of 7 to 10 days. Gonococcal meningitis and endocarditis require intravenous ceftriaxone administered for 10 days and 3 to 4 weeks, respectively.

D. *Control of Contacts* Rapid case finding to locate and treat the source of gonococcal infection is essential to control this disease. Case finding requires trained interviewers to identify those sexual partners with whom the patient has had contact within 10 days of the onset of infection. Full, curative doses of antimicrobial agents for gonococcal and chlamydial infection are given to all contacts, regardless of their symptoms, and serologic tests for syphilis are performed. Routine culture screening of asymptomatic, sexually active women in high risk populations is also important for control.

All infants born to infected mothers should be monitored for symptoms of developing infection. Orogastric and rectal cultures should be taken. Conjunctival cultures are not needed if adequate prophylaxis was given at birth to prevent ophthalmia neonatorum.

III. Prevention

Condoms provide some protection against transmission and acquisition of gonorrhea. The diaphragm, cervical cap, and topical spermicides (those containing nonoxynol-9) offer somewhat less protection; washing, urinating, or douching after exposure are ineffective. The use of low doses of antimicrobial agents immediately after exposure may reduce the risk of infection but also increases the possibility that the strain will be resistant to antimicrobial agents if infection is acquired.

To date, the search for gonococcal antigens of immunogenic value in a vaccine has met with little success, but the efforts continue.

CHLAMYDIAL (NONGONOCOCCAL) URETHRITIS

When urethritis in males cannot be diagnosed as gonococcal infection, it is referred to as *nongonococcal urethritis* (NGU) or **chlamydial urethritis.** Improved methods for cultivating and identifying chlamydiae in genital secretions have demonstrated that most patients with NGU are infected by *Chlamydia trachomatis.* This organism is the most common cause of STD, affecting millions of people in the United States each year. Some evidence links the mycoplasma organism *Ureaplasma urealyticum* with some cases of NGU, but these infections are believed to be much less common than chlamydial infection.

Chlamydiae are obligate intracellular parasites that must obtain energy-yielding compounds from their host cells (see chap. 2). They parasitize primarily columnar epithelial cells, especially those of the genital and respiratory tracts. Two of the three ***Chlamydia*** species, *Chlamydia psittaci* and *Chlamydia pneumoniae,* are usually involved in respiratory disease (see chap. 12). In addition to nongonococcal urethritis, strains of *C. trachomatis* are responsible for eye infections (trachoma and inclusion conjunctivitis); for lymphogranuloma venereum, another type of STD; and for infant pneumonia (see chap. 12).

The Clinical Disease

The symptoms of NGU in males are similar to those found in male patients with gonorrhea but are usually less severe. The urethral exudate in NGU is less purulent and more mucoid than in gonorrhea, but clinical distinction between the diseases is difficult. In Gram-stained smears of urethral exudate, the absence of intracellular gram-negative diplococci characteristic of gonococcal infection suggests chlamydial infection, but the two infections often coexist in patients. *C. trachomatis* is also responsible for epididymitis in sexually active young men and proctitis in homosexual men. Chlamydial infection is often asymptomatic in both men and women, as is gonorrhea.

In women, the primary sign of clinical infection is a mucopurulent cervical discharge containing increased numbers of polymorphonuclear leukocytes. Disease may extend upward in the reproductive tract to cause infection of the Fallopian tubes (salpingitis) and the endometrial lining of the uterus (endometritis). Salpingitis often leads to tubal scarring and blockage, resulting in infertility. The risk of infertility is 10% after a single chlamydial infection and more than 50% after three or more infections. Acute urethritis is another chlamydial syndrome in women. The symptoms resemble those of an acute urinary tract infection, i.e., painful urination (dysuria) and frequent urge to urinate. Microscopic examination of urine reveals many polymorphonuclear leukocytes, but no bacteria are seen or grown in culture. If infected women become pregnant and deliver vaginally, the risk to infants of developing inclusion conjunctivitis and/or interstitial pneumonitis is significant.

LABORATORY DIAGNOSIS

Formerly, a presumptive diagnosis of NGU was made if (or when) gonorrhea was excluded by Gram stain, culture, or both. The high incidence of the disease has resulted in the development of culture and nonculture methods for the more precise detection of chlamydial organisms.

I. Identification of the Organism

(Chlamydia trachomatis).

A. *Microscopic Characteristics* The organisms are too small to be seen with the ordinary light microscope unless special stains are used. Giemsa-stained smears of cell scrapings from the infected site may reveal chlamydial inclusion bodies, but this method requires intensive search and is often unrewarding.

B. *Culture Characteristics* Because chlamydiae are obligate intracellular pathogens, they cannot be grown on ordinary artificial culture media; instead, they must be grown in cell cultures. A cell line known as *McCoy cells* is used for this purpose in many laboratories. McCoy cells growing in a small screw-capped vial containing a cover slip (referred to as a *shell vial*) are inoculated with specimen material and are centrifuged at high speed for 1 hour to ensure contact between the chlamydiae and the McCoy cells and therefore to enhance infection in the cell culture. After 48 hours of incubation, the cover slip is removed and treated either by staining with an iodine solution or staining with a chlamydial fluorescent antibody reagent. Iodine stains the glycogenlike vacuoles produced only in *Chlamydia*-infected cells (colorplate 2f). Likewise, the fluorescent antibody reagent stains intracellular chlamydial elementary bodies (see chap. 2). The glycogen-stained preparations can be viewed under the ordinary light microscope, but a fluorescence microscope, that is, one with an ultraviolet light source, is needed to view the brightly fluorescing elementary bodies.

C. *Nonculture Methods for Diagnosis* Specimen material dried on microscope slides may be treated with a specific fluorescent antibody reagent and viewed with a fluorescence microscope. If the specimen contains chlamydial elementary bodies, they appear as brightly fluorescing particles against a dark background resembling a starry sky. Often, though, only a few infected cells are present in the preparation and the search is laborious. As with gonorrhea diagnosis, enzyme immunoassay and nucleic acid probe methods have been developed for detecting *C. trachomatis* directly in patient specimens. Because most clinical laboratories do not have cell culture facilities available and because the disease has become widespread, these assays have gained increasing acceptance. Most recently, a polymerase chain reaction (PCR) amplification assay has been approved for use for detecting *C. trachomatis* in clinical samples. In general, PCR is more sensitive than are enzyme immunoassays or nucleic acid probes for detecting the organism. The same precautions apply to use of these tests as to those for detecting *N. gonorrhoeae,* and culture is required to confirm the diagnosis in cases of suspected sexual abuse.

II. Identification of Patient's Serum Antibodies

Serologic assays cannot distinguish past from current infection nor which of the three *Chlamydia* species is responsible for the infection, and therefore are not used.

III. Specimens for Diagnosis

Cervical and urethral specimens obtained by vigorous scraping with a nontoxic swab are preferred, but rectal specimens may also be obtained. Rayon, cotton, or Dacron swabs on plastic shafts are better than swabs with wooden shafts. Specimens from women may be collected using a small, cytologic brush to increase the number of cells obtained. Because the organisms are located intracellularly, the specimen must contain cells from the site of infection; a specimen taken from mucopurulent discharge alone, for example, is unsatisfactory.

Epidemiology

I. Communicability of Infection

A. *Reservoirs, Sources, and Transmission Routes* Like gonorrhea, *C. trachomatis* is found only in humans, who usually transmit the infection by sexual contact. In the case of inclusion conjunctivitis and infant pneumonia, the organism is most likely acquired at birth during passage through the infected mother's birth canal.

B. *Incubation Period* Thought to be 1 to 2 weeks.

C. *Communicable Period* Throughout the duration of illness; probably not infectious within a few days after therapy.

D. *Immunity* Antibodies can be detected in serum, but because multiple infections are common, they may not be protective. Alternatively, repeated infections may be caused by different serotypes of the organism.

II. Control of Active Infection

A. *Transmission-based Precautions* Standard precautions are applied for the duration of illness. Gloves are worn when handling infectious discharges such as purulent genital exudates. Patients should refrain from sexual activity until treatment is completed.

B. *Treatment* The recommended treatment is doxycycline or azithromycin administered for 7 days. Erythromycin is preferred for pregnant women. Patients should be treated concurrently for gonorrhea because of the high incidence of coinfection.

C. *Control of Contacts* Case reporting is required in all states. Whenever possible, sexual contacts should be treated even in the absence of symptoms. Pregnant women should be tested early in pregnancy for syphilis, gonorrhea, and chlamydial infection and, if considered at high risk for acquiring infection, tested again during the third trimester.

III. Prevention

No vaccine is available. Young, sexually active patients with multiple partners are at highest risk for STD and should be tested for chlamydial infection by culture, antigen detection, or nucleic acid probe methods even if asymptomatic. Treatment of all confirmed cases and their partners is mandatory. Practice of good sex hygiene, as discussed for gonorrhea, is also recommended.

LYMPHOGRANULOMA VENEREUM

The Clinical Disease

Lymphogranuloma venereum has a worldwide distribution but is most common in tropical or semitropical areas. In the United States it is endemic in southern regions bordering the Gulf of Mexico. It is a sexually transmitted infection to which all ages and races are susceptible but is most common in sexually active young adults, including homosexuals. Lymphogranuloma venereum (LGV) begins with a small, painless primary lesion on the genital mucosa marking the site of entry. This may be a papule or an ulcerated lesion. Clinical evidence of systemic infection begins in about 2 weeks with enlargement of the regional lymph nodes, which may proceed to suppurate and drain, and with fever, general malaise, and sometimes involvement of membranous surfaces of the joints, eye, or brain. Chronic infection within pelvic lymphatics may be of long standing, eventually inducing blockade with resultant elephantiasis of genital tissues and strictures.

LABORATORY DIAGNOSIS

The agents of lymphogranuloma venereum are strains of *C. trachomatis* that are serologically distinct from those that cause nongonococcal urethritis and trachoma. With special staining techniques, the organism can be visualized within the mononuclear cells of exudates obtained from draining lymph nodes or in tissue biopsies. It can also be cultivated from these specimens in yolk sacs of embryonated eggs and in cell cultures. In preparations harvested from these sources, it is identified by microscopic morphology and by serologic identification of its antigenic properties.

Antibodies produced during the course of LGV infection can be identified by microimmunofluorescence tests, but positive tests must be considered along with clinical findings.

Special Features of Control

Specific treatment with doxycycline for 3 weeks results in prompt relief of fever and pain, but enlarged lymph nodes require several weeks or even months to heal. Aspiration of lymph nodes and surgical relief of elephantiasis or strictures may be required. Erythromycin is the alternative drug of choice.

The incubation period may require from 1 to 3 or 4 weeks. The primary lesion usually appears within 7 to 12 days, but if it fails to develop, the first sign of infection may be an inguinal bubo appearing about a month after exposure. Untreated LGV may be communicable for months or years, but even with treatment the lesions heal slowly.

Resistance to reinfection develops during the course of this disease, but the infecting organism may remain latent and viable within tissues, unaffected by antibody production.

LGV is primarily a sexually transmitted infection, but it is also sometimes transmitted to children through contacts with freshly contaminated environmental sources. Control and prevention depend on the detection and treatment of sexual contacts, as outlined for other STDs.

Miscellaneous Bacteria in Human Hosts
Chlamydia: Chlamydia Virulence Factors; Chlamydia Disease

CHANCROID

The Clinical Disease

Chancroid, a sexually transmitted disease, is relatively widespread but its incidence in the United States since the 1980s has increased markedly, especially in a few coastal states.

The chancroid is the initial lesion that occurs on the genitalia: a soft chancre with ragged edges. Unlike the hard chancre of syphilis, the ulcer is swollen and painful, as are the regional lymph nodes. The chancroid and the nodes become suppurative and necrotic, and pus-filled inguinal nodes may break down and suppurate to the surface. These "buboes" exude the causative bacteria, which may then be inoculated into new sites. Extragenital lesions may be seen on other mucosal and skin surfaces (tongue, lip, breast, umbilicus).

LABORATORY DIAGNOSIS

Haemophilus ducreyi is the responsible agent, a small, gram-negative bacillus, nonmotile and nonendospore-forming. It is classified in the same genus as *Haemophilus influenzae* (see chap. 12), but it is culturally more fastidious, requiring media with a high hemoglobin content as well as incubation in an atmosphere with increased CO_2 and high humidity. It can be visualized in Gram-stained smears of exudates from the chancroid or buboes as individual or sometimes chaining rows of bacilli.

The differential diagnosis of chancroid includes dark-field examinations of exudate and serologic testing to rule out syphilis.

Special Features of Control

Erythromycin and ceftriaxone are the drugs of choice. Sexual partners should be treated also.

Chancroid has a short incubation period of 3 to 5 days and is communicable through infectious discharges of the genital lesions and buboes.

Measures for the control and prevention of this disease are similar to those described for other STDs.

Granuloma Inguinale

The Clinical Disease

Granuloma inguinale is a chronic, sexually transmitted disease chiefly of tropical areas, but it is also found in temperate zones. A small nodule appears initially at the site of entry of the organism on the genital mucosa. When the nodule ulcerates, it may be mistaken for the hard chancre of syphilis or the soft one of chancroid. As disease progresses, the ulcer spreads peripherally and may superficially involve a large area of the mucous membrane.

LABORATORY DIAGNOSIS

The causative organism is *Calymmatobacterium granulomatis.* Formerly known as *Donovania granulomatis* or simply as "Donovan bodies," these organisms appear in biopsies of the lesions as short, gram-negative bacilli crowded into mononuclear cells. They have been cultivated in embryonated eggs and also on complex artificial media but these are not routine procedures.

Special Features of Control

Tetracyclines are the drugs of choice for this disease. Trimethoprim-sulfamethoxazole and chloramphenicol are useful alternatives.

This disease occurs more frequently among males than females. Many male patients are homosexuals, whereas the infection is uncommon in prostitutes or the heterosexual partners of infected persons.

Acquired Immunodeficiency Syndrome (AIDS)

During the summer of 1981, reports from the Centers for Disease Control described occurrences of *Pneumocystis* pneumonia (see chap. 14), an unusual form of cancer known as Kaposi sarcoma, and a number of other severe opportunistic infections, all in previously healthy, young homosexual men living in large metropolitan areas such as New York City and San Francisco. Shortly thereafter, the syndrome was recognized in other population groups throughout the United States. Members of these groups included individuals with hemophilia who had received treatment with clotting factors prepared from pooled human blood, intravenous drug users, Haitians residing in the United States, and a small number of people who had received blood transfusions. The disease came to be called **acquired immunodeficiency syndrome (AIDS)** because of the marked immune deficits that made afflicted persons susceptible to opportunistic infections. It became evident that AIDS was spread by sexual contact and by intravenous administration of contaminated blood.

During the next 15 years, the number of reported AIDS cases increased each year, the syndrome spreading to involve heterosexual men and women and newborn children. By 1996, more than 500,000 reports of persons with AIDS had been made to public health departments in the United States. Of these, more than 62% died, making AIDS a leading cause of death among young men and women as well as children less than 5 years of age. In addition, the World Health Organization estimated that by 1996 nearly 4.5 million cases of AIDS had occurred worldwide and that 18 million adults and 1.5 million children have been infected with HIV. Reports have come from virtually all countries among all races, ages, and social classes; but almost half of the reported cases have originated in sub-Saharan Africa.

Experts estimate that, by the year 2000, 30 to 40 million people will be infected with the human immunodeficiency virus; more than 90% of them in developing countries. In addition, the infection will have progressed to AIDS in almost 10 million people.

The early recognition and spread of the disease set off an unprecedented search for the causative agent. This search was characterized by much political and moral debate over government funding for research and was marred by an international scandal in which a French and an American researcher both claimed to be the first to isolate the AIDS virus. In 1983, Dr. Luc Montagnier and his colleagues at the Pasteur Institute in France described in the literature a virus isolated from an AIDS patient, a virus which they referred to as *LAV* (*l*ymphadenopathy *a*ssociated *v*irus; lymphadenopathy is an initial sign of AIDS). In 1984, Dr. Robert Gallo of the National Institutes of Health (NIH) in Washington, D.C., reported isolating HTLV-III from AIDS patients, a virus he considered to be a new member of a group of *h*uman *T* cell *l*ymphotrophic *v*iruses (the AIDS virus destroys specific human T cells). HTLV-III was declared to be the AIDS virus. Later, however, LAV and HTLV-III were found to be identical, and it was realized that HTLV-III was not an independently isolated strain but was actually an isolate of LAV that had been sent to the NIH for study.

The LAV/HTLV-III virus, renamed **human immunodeficiency virus** (**HIV**), was patented jointly by the French and Americans and used to prepare diagnostic reagents for a serologic test for HIV. This serologic test has helped to safeguard the purity of transfused blood supplies and to detect infected persons at risk for transmitting the virus, especially through sexual contact and shared use of intravenous drug paraphernalia. However, the unfortunate social stigma and the economic consequences experienced by persons with AIDS have created a new medical-legal climate in which right-to-privacy issues prevent the best use of information obtainable by serologic testing.

In spite of the intensive efforts that led to the early discovery of the etiologic agent and to the development of a diagnostic test for the disease, no antiviral agent cure for AIDS has yet been found, nor has an effective vaccine for prevention been developed. New antiviral drugs, however, show promise in greatly slowing progression of the disease.

TABLE 19.1 COMMON OPPORTUNISTIC INFECTIONS ASSOCIATED WITH THE AIDS DIAGNOSIS

Opportunistic Infection	Organism	Chapter Reference
Candidiasis (esophagus, trachea, bronchi, lungs, vagina)	*Candida albicans*	14
Cervical cancer	Human papillomavirus	19
Cryptococcosis (disseminated)	*Cryptococcus neoformans*	14
Cryptosporidiosis	*Cryptosporidium* spp.	18
Cytomegalovirus disease (disseminated)	Cytomegalovirus	13
Herpes simplex virus infection (disseminated)	Herpes simplex virus	19
Microsporidiosis	Microsporidia	18
Mycobacterium avium complex or *Mycobacterium kansasii* disease (disseminated)	*Mycobacterium avium* complex	12
	Mycobacterium kansasii	12
Pneumocystis carinii pneumonia	*Pneumocystis carinii*	14
Toxoplasmosis of the brain	*Toxoplasma gondii*	18
Tuberculosis	*Mycobacterium tuberculosis*	12

The Clinical Disease

AIDS is an illness that progresses slowly; it usually develops over the course of many years and, according to current knowledge, is invariably fatal. Primary HIV infection may be asymptomatic or may result in a mononucleosislike illness with fever, sweats, muscle pains, nausea, loss of appetite, diarrhea, and a nonexudative pharyngitis. Many patients also develop a rash, lymphadenopathy, and neurologic symptoms. Serologic tests for HIV antibody are usually negative at this stage (they become positive after 1 to 3 months), but antigen detection tests reveal a characteristic viral core antigen in the blood. The disease progresses at a slow rate until after 2 to 5 years, when the patient develops persistent infections such as herpes zoster (see chap. 13) and oral candidiasis, or thrush (see chap. 14), and experiences weight loss, fatigue, and persistent diarrhea. These signs and symptoms are sometimes referred to as the *AIDS-related complex* (*ARC*), and they indicate progression to full-blown AIDS. The most common clinical manifestations establishing a diagnosis of AIDS are infections with organisms that seldom cause disease in persons with normal cell-mediated immunologic defenses. Table 19.1 lists the most frequent of these opportunistic infections whose occurrence is usually the first signal of AIDS (colorplates 9a–f and fig. 19.6). In addition, several malignancies such as Kaposi sarcoma, cervical cancer, and cancers of the lymphatic tissue are associated with HIV infection.

An important defect in cell-mediated immunity in AIDS is a decrease of CD4 + helper T-lymphocytes and an increase in CD8 + suppressor T-lymphocytes (see chap. 6). CD4 + T-lymphocytes bear on their surface a glycoprotein substance known as CD4 which is a receptor for HIV. HIV attaches to the CD4 molecule and is taken into the cell where it multiplies and eventually kills the cell (fig. 19.7). In general, when fewer than 200 CD4 + T-lymphocytes per cubic millimeter are present in the circulation, the patient begins to experience the opportunistic infections characteristic of AIDS. Other body cells, including nerve and intestinal cells, bear the CD4 receptor molecule. HIV invasion of these cells may be responsible for the central nervous system and intestinal tract symptoms that patients with AIDS experience. HIV also has an effect on mononuclear blood cells and B lymphocytes, involving them in the complex immunologic suppression characteristic in persons with the disease. Without specific therapy, 80 to 90% of patients die 3 to 5 years after they are diagnosed with AIDS. Therapy slows progression of the disease but does not cure it.

HIV is a **retrovirus** belonging to the same family as HTLV-I and HTLV-II, but is distinct from them in being classified within the lentivirus subfamily. These RNA viruses possess an enzyme called *reverse transcriptase,* which allows the RNA genome to be transcribed into a DNA copy that can be integrated into the chromosome and persist in latent form within the host. (see chap. 2).

LABORATORY DIAGNOSIS

Indirect evidence of infection is gathered by monitoring the level of CD4 + T-cells and the ratio of CD4 + to CD8 + T-cells. Serologic and antigen-detection tests provide conclusive diagnosis. The progress of infection and response to antiviral therapy can be determined by assessing viral load; that is, the amount of HIV RNA circulating in the patient's bloodstream.

I. Identification of the Organism

Human immunodeficiency virus (fig. 19.8). The strain found in most cases of disease is known as HIV-1. A second strain, HIV-2, has been isolated from some patients in West Africa and from West Africans living in other countries.

A. *Microscopic Characteristics* HIV can be seen only by electron microscopy. It is a spherical virus measuring 80 to 130 nm in diameter with three layers: (1) an inner nucleocapsid core enclosed in (2) an icosahedral capsid surrounded by (3) a lipid envelope from which viral glycoprotein spikes project.

B. *Cultural Characteristics* The organism can be isolated from patients' blood by culturing their blood mononuclear cells in cell culture medium. Addition of lymphocytes from non-HIV-infected persons is needed for viral

FIGURE 19.6 The clinical spectrum of HIV infection.
Centers for Disease Control, Atlanta, GA.

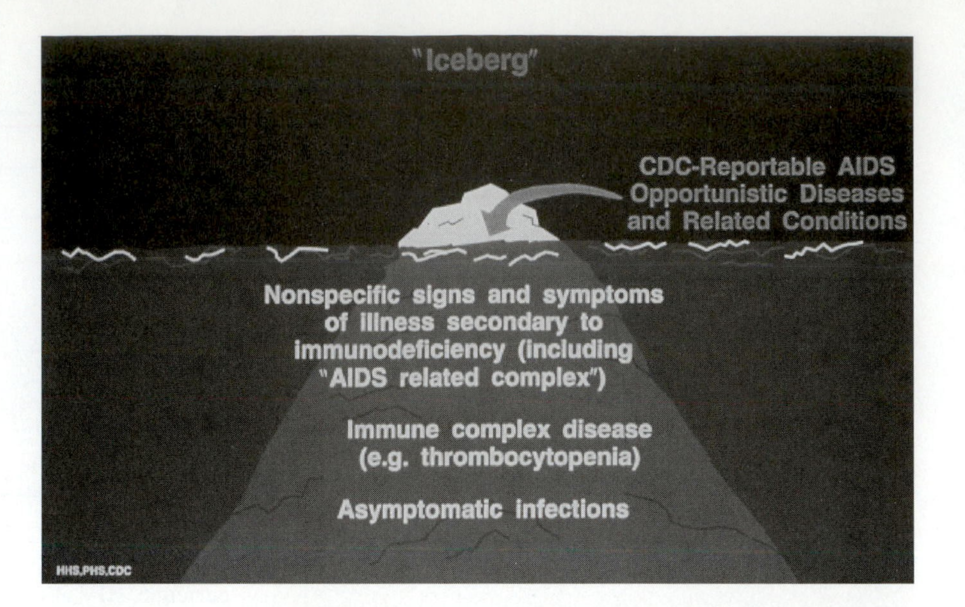

replication to occur in culture. To detect the presence of the virus growing in cell cultures, the fluid medium is assayed for the presence of the characteristic reverse transcriptase enzyme or for HIV antigen. Alternatively, cells can be examined for intracellular HIV antigen with a specific fluorescein-conjugated HIV antiserum in an immunofluorescence assay. Cell culture can be hazardous and should be carried out only by trained persons using strict isolation precautions.

C. *Immunologic Identification* HIV-1 can be serologically distinguished from HIV-2 by antibodies prepared against outer viral components. HIV-1 antigen is detected in serum or other patient body fluids by an enzyme immunoassay and in peripheral blood mononuclear cells by an immunofluorescence assay. Antigen can often be detected early in disease, before HIV antibody is present. A highly sensitive method has been developed in which a polymerase chain reaction (PCR) is used to amplify viral DNA. This method, which can detect one HIV DNA molecule in 100,000 lymphocytes, is becoming widely used.

D. *Viral load measurements* Several nucleic acid amplification tests are available to measure the quantity of HIV RNA in patient plasma. Viral RNA is extracted and isolated from the specimen and amplified using molecular technologies. The test is not used to diagnose infection but rather to indicate disease prognosis (higher levels are associated with shorter times to death and lower levels with a much longer time). With the availability of new, more effective drugs (see Treatment section), viral load measurements are used to assess patient response to therapy. RNA levels that decrease from an initial, baseline number are consistant with a,good response but increasing viral loads indicate active viral replication and treatment failure.

II. Identification of Patient's Serum Antibodies

Several methods are available for detecting antibodies in patients' serum. The enzyme immunoassay is the most common screening assay. The antigen used consists of purified, ruptured viral particles. Viral antigens can also be prepared synthetically or by recombinant DNA techniques. This sensitive enzyme immunoassay method is known to give false-positive results, therefore all positive tests must be confirmed with another method, for example, the indirect immunofluorescence assay using patients' sera and HIV-infected cell cultures. The most common test of confirmation is known as the immunoblot assay (see chap. 6). In this test, viral protein antigens are separated in an electric field and transferred to paper strips. Patient sera are then reacted with the strips. In a positive test, patient antibody binds to one or more HIV protein antigens on the strip.

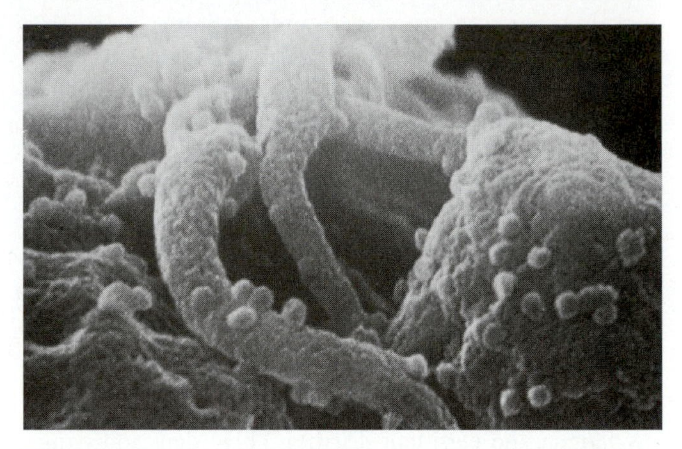

FIGURE 19.7 Scanning electron micrograph of HIV-infected CD4+T lymphocytes showing virus budding from their plasma membrane (low magnification).
Centers for Disease Control, Atlanta, GA. Photo by E. Palmer.

III. Specimens for Diagnosis

For cell culture, peripheral blood is the most common specimen, but other body fluids and tissues may be cultured (e.g., saliva, tears, urine, genital secretions, intestinal biopsies). Plasma is used for viral load measurements, and

Diseases Related to AIDS

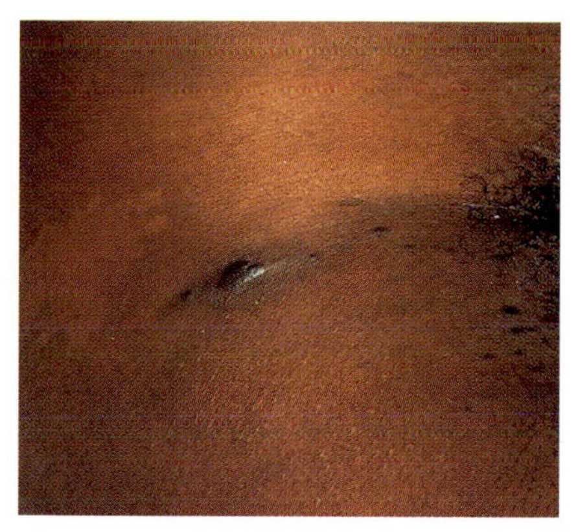

a) Kaposi's sarcoma on the arm.
Centers for Disease Control, Atlanta, GA.

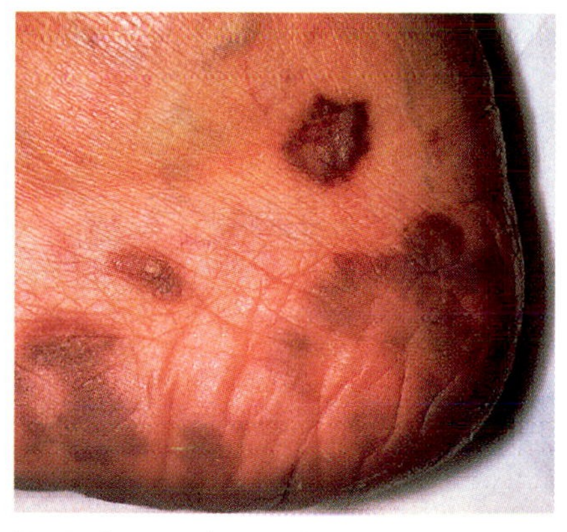

b) Violaceous plaques of Kaposi's sarcoma on the heel and foot.
Centers for Disease Control, Atlanta, GA.

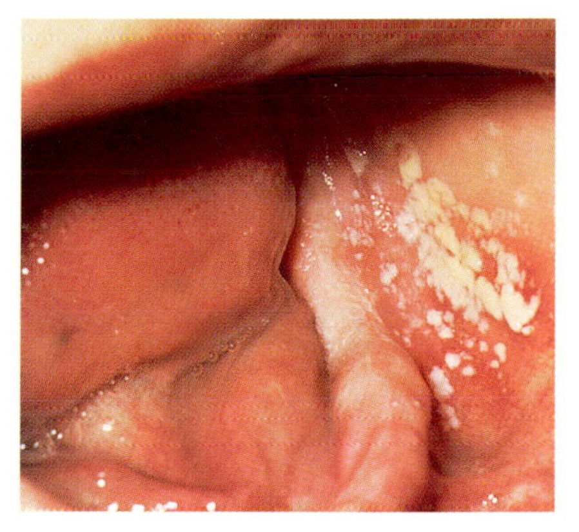

c) Acute pseudomembranous candida infection of the mouth.
Centers for Disease Control, Atlanta, GA. Photo by Dr. Sol Silverman, Jr. D.D.S. University of California, San Francisco.

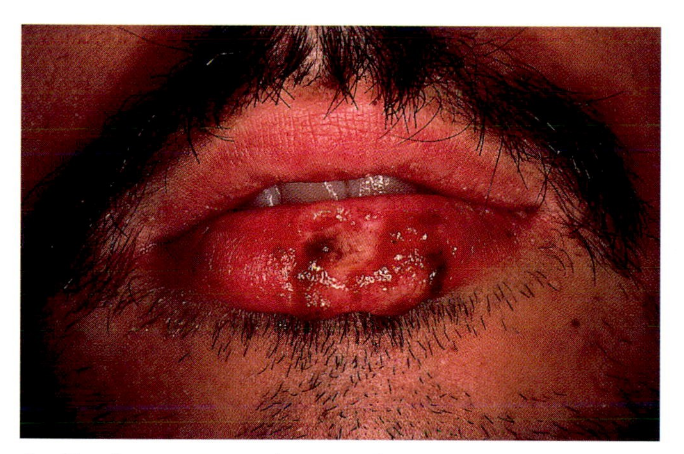

d) Chronic mucocutaneous herpes simplex, one month duration.
Centers for Disease Control, Atlanta, GA.

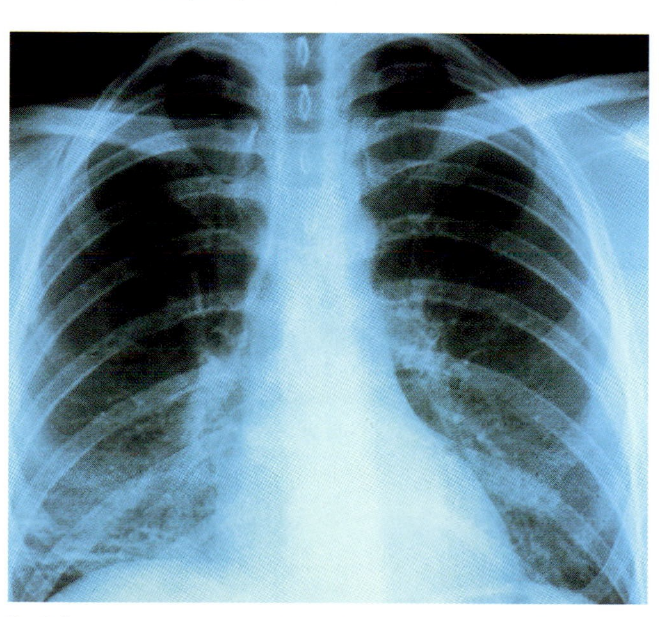

f) Pulmonary pneumocystosis (chest X ray showing interstitial infiltrate).
Centers for Disease Control, Atlanta, GA.

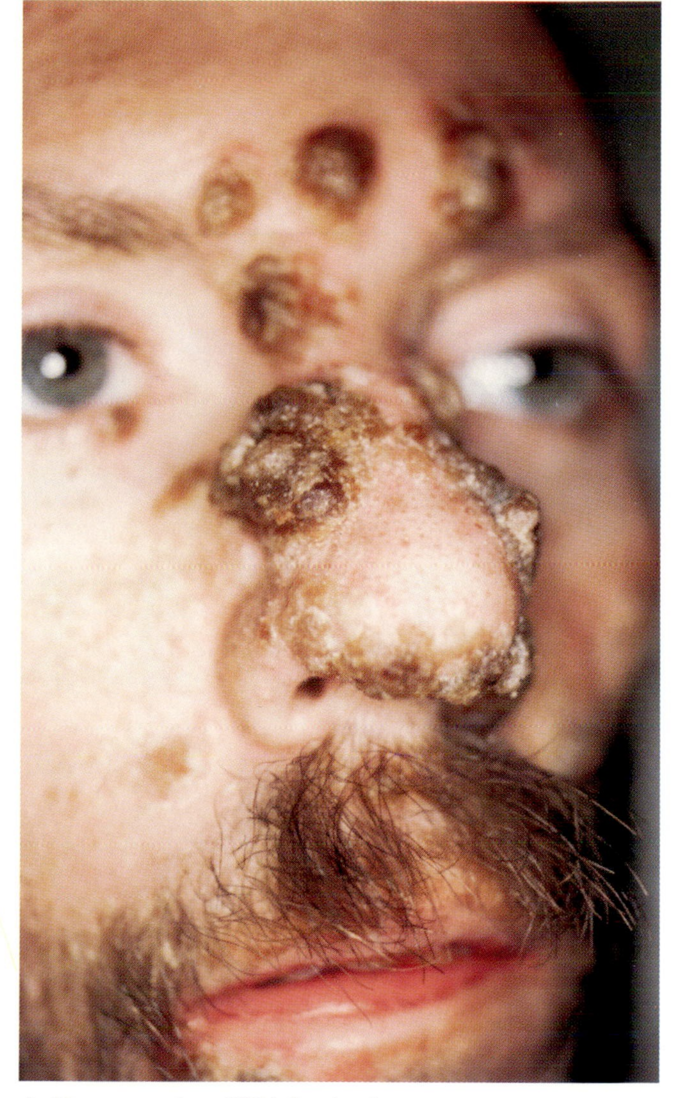

e) Herpes zoster in an HIV-infected patient.
From W. B. Hoppenjans, "Prolonged Cutaneous Herpes Zoster in AIDS," *Archives of Dermatology* 126: 1048–1049, 1990. Copyright 1990 American Medical Association. Photo courtesy William B. Hoppenjans, M.D., University of Cincinnati.

COLORPLATE TEN

Sexually Transmitted Diseases

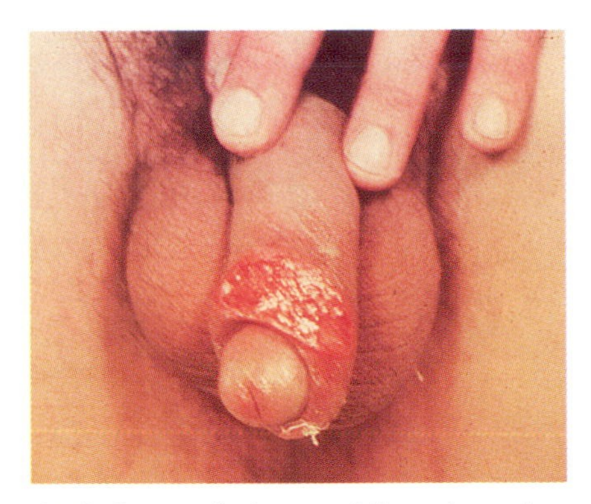

a) A chancre of primary syphilis on the penis.
Centers for Disease Control, Atlanta, GA.

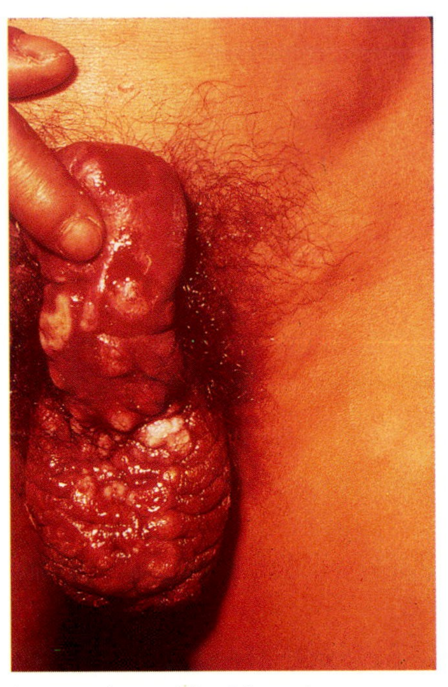

b) Secondary syphilis of the penis.
Centers for Disease Control, Atlanta, GA. Photo by Brian Hill, New Zealand

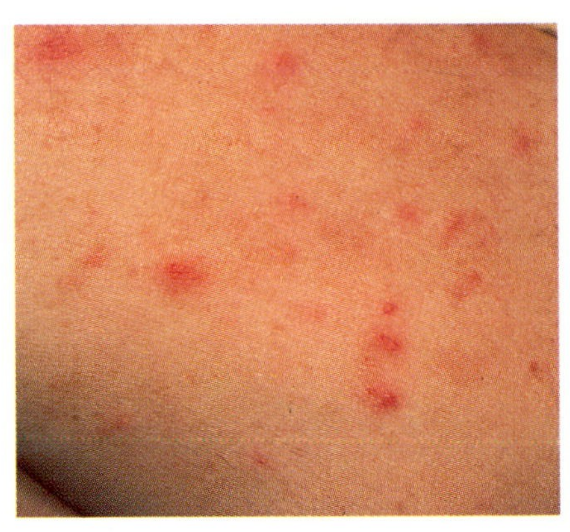

c) The skin rash of secondary syphilis.
Centers for Disease Control, Atlanta, GA.

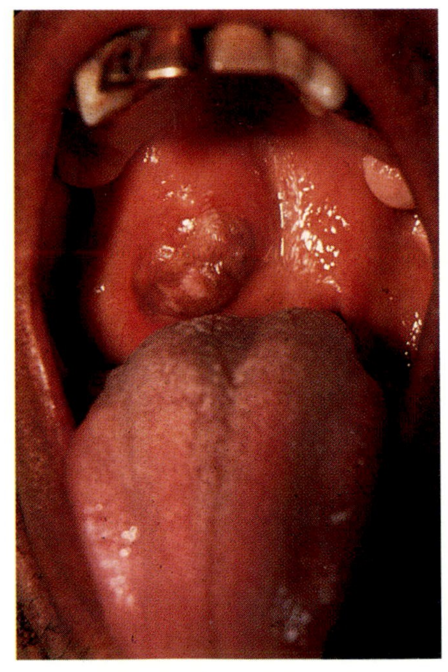

d) This gumma of the soft palate is typical of tertiary syphilis.
Centers for Disease Control, Atlanta, GA.

e) Ophthalmia neonatorum in a newborn. Prophylaxis with silver nitrate or antimicrobial agents has controlled a disease that was once the leading cause of blindness in children.
Centers for Disease Control, Atlanta, GA. Photo by Dr. M. R. Rein.

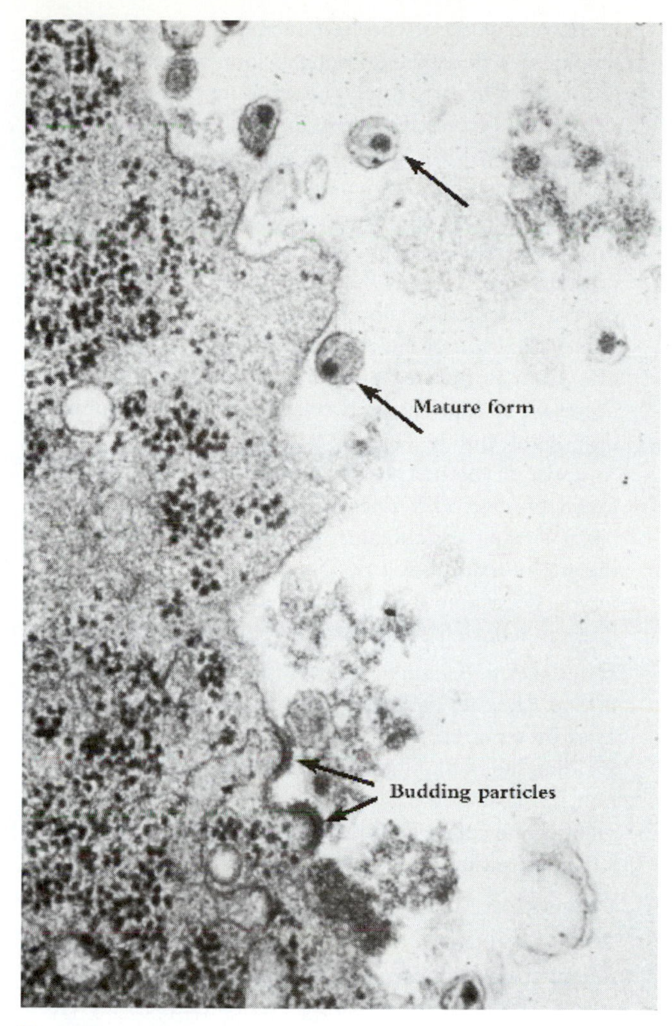

FIGURE 19.8 HIV virus found in hemophilia patient who developed AIDS (magnification of virus particles ranges in size from 90-120 nm).
Centers for Disease Control, Atlanta, GA.

plasma, serum, or other body fluids may be tested serologically for antigen or antibody. If necessary, specimens may be stored at –70° C before they are tested.

Epidemiology

I. Communicability of Infection

A. *Reservoirs, Sources, and Transmission Routes* Humans are the only reservoir although a virus similar to HIV, the simian immunodeficiency virus (SIV), has been found in African green monkeys. SIV and HIV may have been derived from a common precursor virus. HIV is transmitted by intimate sexual contact, contaminated blood and blood products, and passage from mother to child prenatally, during the birth process, or shortly after birth. Casual contact and insect vectors do not transmit infection. Rectal intercourse is a major route of sexual transmission, with greatest risk to the receptive partner. The presence of other sexually transmitted diseases increases transmission, especially ulcerative infections such as genital herpes and chancroid. Health care workers appear to be at low risk for acquiring infection even after sustaining injuries with needles or other sharp objects contaminated with HIV-infected blood. Transmission from a dentist to at least five patients has been reported, but evidence indicates that the dentist did not follow prescribed precautions to prevent possible exchange of blood or body fluids. Health care workers with AIDS are cautioned against performing invasive techniques on patients and are advised to tell their patients about their illness.

B. *Incubation Period* Variable. Although antibodies appear within 1 to 3 months, 10 years or more may elapse after infection takes place before a clinical diagnosis of AIDS is made. The incubation period in infants is shorter than in adults.

C. *Communicable Period* Most likely throughout disease beginning shortly after infection. Communicability may increase as the severity of immunosuppression increases and in the presence of other ulcerative sexually transmitted diseases.

D. *Immunity* Susceptibility appears to be general. Although HIV antibodies develop, they are not protective and no definitively diagnosed patient has been known to recover.

II. Control of Active Infection

A. *Isolation* Patients may be placed in a private room if their hygiene is poor, otherwise isolation is not required. HIV is of low transmissibility when appropriate precautions are taken.

B. *Transmission-based Precautions* In the hospital, patients are treated with standard precautions (see chap. 11). Gowns and gloves are worn if soiling is likely and when touching infective material. Only blood that has been tested for HIV antigen and antibody and found to be negative is used for transfusion. The HIV antibody status of organ and sperm donors should be determined, and persons suspected of engaging in high-risk practices should be excluded from donation. Blood products administered to patients with hemophilia are heated to inactivate the virus.

In the clinical laboratory, patient specimens are treated with standard precautions (see chap. 11). All specimens are considered to be HIV-infected and are handled with gloves. Depending on the degree of exposure, masks, protective eyewear, and protective clothing may also be worn. Procedures that may generate aerosols are carried out in a biological safety cabinet. Spills onto skin or inanimate objects are immediately cleaned and disinfected. HIV is susceptible to most common disinfectants including 0.05% Clorox. Precautions should always be taken to avoid needle-stick injuries.

In the community, education about activities that carry a high risk of acquiring AIDS (multiple sexual partners, intravenous drug use with needle sharing) must be stressed. Use of latex condoms during intercourse with an HIV-positive person appears to decrease the risk of infection but may not completely eliminate it.

C. *Treatment* While no effective drug is available to cure HIV infection or AIDS, two groups of antiviral agents, reverse transcriptase inhibitors and protease inhibitors, are used to prolong survival of patients by slowing the progress of HIV infection. Both groups of agents work by inhibiting key enzymes necessary for viral replication.

The reverse transcriptase inhibitors prevent the conversion of viral RNA to DNA. Two major classes of reverse transcriptase inhibitors, nucleoside analogues (nucleosides are building blocks of RNA and DNA) and nonnucleoside inhibitors are available. The nucleoside analogues, which include zidovudine (AZT), didanosine (DDI), and zalcitabine (DDC) (see table 9.8), were the first group of antiviral agents developed for treating AIDS. AZT slows the progression of disease and is most effective when administered early in infection whereas DDI is effective when given either in combination with AZT or to patients who no longer respond to AZT. DDC is used in combination with AZT for adults with advanced AIDS. In general, these nucleoside analogues are not very potent inhibitors of HIV replication and have been of only moderate clinical benefit to HIV-infected patients.

Nonnucleoside reverse transcriptase inhibitors, such as nevirapine and delaviridine, are more potent inhibitors of HIV replication than are AZT, DDI, or DDC. These agents have significant antiretroviral activity, but resistant viruses emerge rapidly if either drug is used alone. To reduce the incidence of drug resistance, the agents are used in combination with other antiviral drugs; for example, nevirapine given with both AZT and DDI shows prolonged and substantial antiviral activity.

The protease inhibitors are a relatively new and highly potent group of antiviral agents that show considerable promise in slowing or completely inhibiting HIV replication. These agents act by inhibiting aspartyl protease, an enzyme that is vital to the final stages of viral replication. In the absence of this essential enzyme activity, noninfectious virus particles are synthesized and the viral replication cycle is disrupted. Currently available protease inhibitors include saquinavir, indinavir, and ritonavir (see table 9.8). These agents are highly potent inhibitors of HIV replication and are well-tolerated by patients. In clinical trials, the use of the protease inhibitors either alone or in combination with a reverse transcriptase inhibitor, has significantly slowed the progression of disease in HIV-infected patients. In many instances, this therapy has been so successful that even patients with advanced disease have become well enough to return to school or work and resume normal activity.

D. *Control of Contacts* Contacts of persons with AIDS should take precautions against parenteral contamination with blood or secretions. No transmission through casual contact, tears, or saliva has been documented. The average risk for HIV infection from percutaneous (through the skin) exposure to HIV-infected blood is 0.3%, but the risk is higher in certain instances, such as after a deep injury or if blood is visible on the device causing the injury. For health care workers who have such a high-risk exposure, prophylactic therapy beginning as soon as possible is recommended. In early trials, postexposure prophylaxis with zidovudine decreased the risk for HIV seroconversion by approximately 79%. Currently, the three-drug combination of zidovudine, lamivudine, and indinavir is recommended. Prophylactic treatment is offered also to health care workers whose mucous membranes or skin is exposed to blood or body fluids of HIV-infected persons but the risk of seroconversion is low.

A clinical trial indicated that the risk of transmitting HIV from an infected pregnant woman to her newborn decreased by approximately two-thirds by the administration of zidovudine. Thus all pregnant women are advised to undergo voluntary testing for HIV and be treated if found positive. HIV-infected women should also avoid breast feeding their infants, a practice that may increase the rate of transmission by 10 to 20%. Studies are underway to determine the effectiveness of the newer types of drugs in reducing congenital HIV infection.

E. *Epidemic Controls* Currently, AIDS is a pandemic disease. All states and most countries require that known cases be reported. Prevention depends on knowing and avoiding high-risk practices. The World Health Organization has instituted a global prevention and control program. Some countries forbid entry of persons with AIDS, a highly controversial issue.

III. Prevention

A. *Immunization* No vaccine is currently available but intensive developmental efforts are ongoing. Different strains of the virus are antigenically distinct, therefore studies have been aimed at finding common antigenic components for use in vaccines. In one trial, AIDS patients were vaccinated with HIV antigens attached to the vaccinia virus (the virus once used to provide immunity against smallpox); the attempt resulted in the death of several patients, apparently from disseminated vaccinia virus infection.

OTHER SEXUALLY TRANSMITTED DISEASES

Genital transfer of infection may involve a number of organisms that ordinarily produce localized genitourinary tract symptoms. In males, urethritis not associated with *Chlamydia* or the gonococcus may be caused by several agents that can be sexually transmitted. In females, these organisms may cause a low-grade vaginitis or cervicitis. Most of these infections do not constitute a serious threat to the health of adults, although they may be responsible for reproductive failures. The most serious implications, however, may involve infants infected congenitally or at birth.

Trichomoniasis

The flagellated protozoan *Trichomonas vaginalis* (fig. 19.9) may cause a particularly irritating local infection in women. The organism can be found as a commensal of the genital

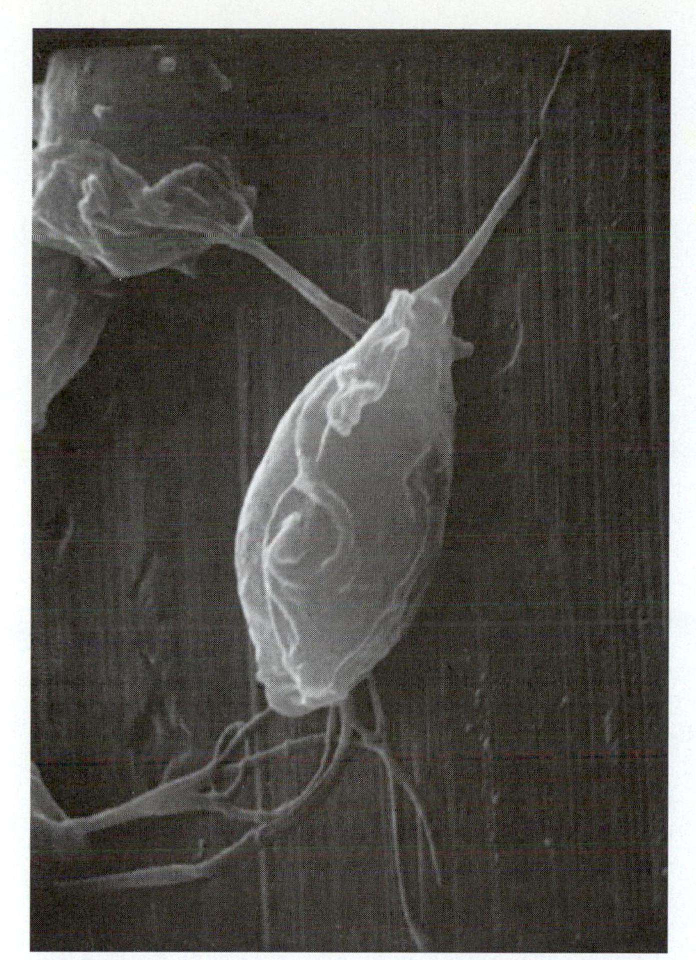

FIGURE 19.9 A scanning electron micrograph of the flagellated protozoan, *Trichomonas vaginalis* (×1,500).

William G. Barnes, Ph.D., John D. Arnold, M.D., and Arthur Berger, Department of Pathology, General Hospital and Medical Center and Harry S. Truman Research Laboratory, Kansas City, MO.

Parasitic Structure and Function
Protozoa Classification: Flagellated Protozoa

mucosa. In men, it lives on the urethral surfaces or in the prostate and seminal vesicles, where it seldom produces any symptoms. When transferred to women it may induce a chronic vaginitis characterized by a profuse malodorous discharge. Other bacteria (lactobacilli) that normally live in the vagina and maintain the acidity of the mucous membrane are displaced by trichomonads, which feed on bacteria and also on the leukocytes marshaled in response to their invasion. The membranes become superficially inflamed, and there is usually an intense pruritus of the vagina and vulva. **Trichomoniasis** is easily diagnosed by microscopic examination of the vaginal secretions or of seminal discharge from the male. The organisms are often present in urine also. The actively motile parasite can be seen in unstained wet mounts and can be cultured on a medium similar to that used for amebic protozoans. Trichomoniasis can be treated effectively with metronidazole given orally, but concurrent treatment of sexual contacts is necessary to prevent reinfection.

Genital Herpes

Herpes simplex viruses (HSV) are responsible for a spectrum of human contact diseases; these are discussed more fully later in this chapter. **Genital herpes** is a distinct clinical entity usually associated with HSV type 2 and occasionally with type 1. It is sexually transmitted among adults, in whom it may appear as a primary or recurrent disease. The characteristic lesion is a vesicle filled with clear fluid and surrounded by an area of inflammation, edema, and congestion. These vesicular lesions may break down, becoming ulcerative and extremely painful. In males, herpetic vesicles usually occur on the glans penis, the prepuce, or the penile shaft. In women, the principal site of genital herpes is the cervix, but lesions may also occur in the vagina or on the vulva. The distribution of lesions varies in both sexes, however, depending on sex practices. In recurrent infections, lesions are fewer in number and resolve in a shorter time than in primary disease. Severe to fatal infection may occur in patients with immunologic deficiencies or in newborn infants infected at the time of birth from the mother's genitalia.

The clinical diagnosis of genital herpes is confirmed by demonstrating HSV type 2 (or sometimes, type 1) in vesicular fluid using cell culture techniques. Microscopic examination of stained epithelial cells obtained from the base of a vesicle may reveal characteristic intranuclear inclusion bodies. Immunofluorescence methods can also be used to detect viral antigens in cells. This latter method is almost as sensitive as culture methods for diagnosing genital herpes. Humoral antibodies to HSV can be detected by a variety of serologic techniques and may provide a serologic diagnosis.

There is no known cure for HSV infection. Acyclovir taken orally for 7 to 10 days reduces viral shedding and the duration of disease in patients with primary infection. The drug is less effective in recurrent genital herpes infections. Continuous suppressive acyclovir therapy in patients with frequent episodes (more than six per year) may decrease the rate of recurrences.

Patients should be told about the natural history of genital herpes infection and be advised to abstain from sexual contact while lesions are present. The risk of transmitting the virus during asymptomatic periods is unknown but does occur. Early in pregnancy, a woman should inform her physician of her history of genital herpes.

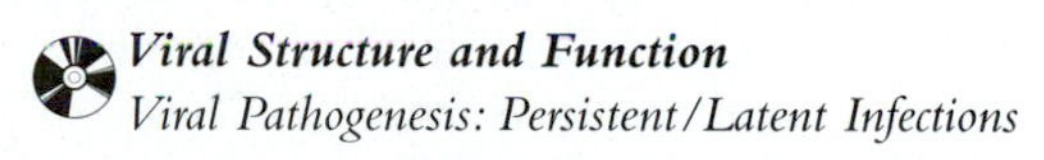

Viral Structure and Function
Viral Pathogenesis: Persistent/Latent Infections

Genital Warts

Genital warts, or condyloma acuminata, are the most common viral sexually transmitted disease. These growths, flesh to gray in color, are hyperkeratotic lesions that are either flat or attached by a broad, stalklike peduncle. In males, they are located most commonly on the penile shaft; in women, they are typically found over the perineal area, the vagina, and cervix. The perianal area may also be involved, especially in homosexual men. Most patients are asymptomatic, but itching, burning, and pain may be present.

Like warts in other body areas, genital warts are caused by a specific subset of the almost 60 types of human papillomaviruses (HPV). Recently, papillomavirus types 16, 18, and 31 have been strongly associated with the presence of cervical cancer in women. Epidemiologically, women with a history of genital warts are four times more likely to develop cervical carcinoma than women without warts. In studies of cervical tissue taken from women with carcinoma, HPV DNA was found to be present in almost 90% of the cases. Even when overt, wartlike growths are absent, application of dilute acetic acid to the cervix may reveal whitish patches that contain HPV DNA, but the rate at which these lesions develop into cervical carcinoma is not known. The increased incidence of cervical carcinoma seen in women with AIDS has caused HPV to be added to the list of agents causing opportunistic infections in AIDS patients (see table 19.1).

Current methods for identifying HPV depend on molecular hybridization techniques that are too cumbersome to use on a routine basis; therefore, widespread testing for HPV is not recommended at present. A variety of methods is available for removing the warts: application of a resinous plant extract called podophyllin, freezing with liquid nitrogen, and laser therapy. Therapy is recommended because of the association of warts with genital malignancies, but treatment is not always successful.

Herpes simplex virus type 2 was once thought to be associated with genital cancers, but careful epidemiologic studies have discounted its role in this type of malignancy. Whether papillomaviruses are indeed the inciting agent remains to be determined, although current evidence is highly suggestive.

Bacterial Vaginosis (BV)

Once referred to as nonspecific vaginitis, **bacterial vaginosis** (or **BV**), is a common vaginal infection that is presumed to be sexually transmitted. The disease is characterized by the presence of a moderate amount of grayish-white, adherent vaginal discharge that patients describe as having a "fishy" odor, especially after sexual intercourse. The vaginal pH is elevated above the usual pH of 4.5 and Gram-stain examination of vaginal fluid reveals characteristic "clue" cells. Clue cells are vaginal epithelial cells coated on their surfaces with gram-negative coccobacillary organisms. Straight and curved gram-negative and gram-variable (staining irregularly gram-positive) bacilli are often seen also. The large, gram-positive lactobacilli present in the healthy vagina are absent. Inflammation and perivaginal irritation are milder in BV than in trichomoniasis and candidiasis, and only an occasional pus cell is seen on the smear. In some pregnant patients, BV has been associated with premature labor.

The etiology of BV is unknown. *Gardnerella vaginalis,* a small gram-variable bacillus was once assumed to be the cause, but this organism is found in many asymptomatic women as well as those with BV. The straight and curved gram-negative bacilli seen on smears of a patient's discharge have been recently isolated and identified as anaerobic organisms in a new genus, *Mobiluncus* (motile hooks). Most likely, the infection is caused not by a single bacterial species but by the synergistic interaction of several species including *G. vaginalis* and the *Mobiluncus* species.

BV is diagnosed by finding at least three of the following four criteria: a characteristic discharge, a pH greater than 4.5, clue cells, and a positive "whiff" test, which is performed by adding a drop of 10% potassium hydroxide to vaginal secretions on a glass microscope slide. In a positive test, a fishy odor emanates from the discharge. The most effective treatment is oral metronidazole given for 7 days. Male sexual partners are not treated unless recurrent infections persist in the woman.

Sexually Transmitted Enteric Disease

A wide range of enteric organisms can be sexually transmitted. The mechanism in general is the fecal-oral passage of infectious agents, predominantly during oral-genital or oral-anal contact. Such transmission occurs predominantly in males between ages 20 and 40 because they appear to be more promiscuous and sexually active during this age period. Most such diseases have been recognized in homosexual men, but this may be related to the more intense surveillance by public health workers who treat large numbers of homosexual patients with sexually related diseases. Heterosexual individuals also acquire infection in this manner.

Any organism that has an infective form present in the feces can be sexually transmitted, including bacteria, protozoa, viruses, and helminths. Among the diseases seen most commonly are shigellosis, salmonellosis, amebiasis, giardiasis, hepatitis A and B, and pinworm infection. Signs, symptoms, and treatment of these diseases are the same as for infections acquired by the conventional fecal-oral route, as discussed in chapters 16 through 18. Control depends on patient counseling by the physician or health worker regarding sexual transmission of these infections.

Congenital and Neonatal Diseases Resulting from Sexually Transmitted Infections

Some of the organisms transferred among adults during sexual intercourse may have little consequence except for a developing fetus or an infant born of an infected mother, but for these the results may be disastrous.

Intrauterine infections fall into two epidemiologic categories. Those acquired by extension of nonsexually transmitted maternal infection include congenital rubella syndrome, cytomegalovirus infection, listeriosis, and toxoplasmosis. Respiratory or alimentary routes of entry are involved in the initial infection of the mother in these in-

TABLE 19.2 SUMMARY OF INFECTIOUS DISEASES OF INFANTS

	Diseases	Organisms	Chapter Reference
Congenital Infections	Syphilis	*Treponema pallidum*	19
	Gonococcal amniotic infection syndrome	*Neisseria gonorrhoeae*	19
	Listeriosis	*Listeria monocytogenes*	16
	Rubella syndrome	Rubella virus	13
	Toxoplasmosis	*Toxoplasma gondii*	18
	Cytomegalovirus infection	Cytomegalovirus	13
	AIDS	Human immunodeficiency virus	19
Infections Acquired at Birth	Gonococcal ophthalmia neonatorum	*Neisseria gonorrhoeae*	19
	Herpesvirus infections	Herpes simplex virus	19
	Inclusion conjunctivitis Infant pneumonitis	*Chlamydia trachomatis*	12, 19
	Pneumonia Septicemia Meningitis	Group B streptococci	12
	AIDS	Human immunodeficiency virus	19
Neonatal Infections	Epidemic diarrhea	Enteropathogenic *E. coli*	16
	Myocarditis Encephalomyocarditis	Coxsackievirus	13, 17
	Skin infections	*Staphylococcus aureus*	19
	Pneumocystosis	*Pneumocystis carinii*	14
	AIDS (breastfed infants)	Human immunodeficiency virus	19

stances. The fetus is usually infected through the placenta during a stage of disease when the organisms are in the maternal bloodstream. The second category includes those acquired by extension of the maternal sexually transmitted disease, the most notable example being congenital syphilis. Gonococcal infection of the chorioamnion also falls within this group and may be more common than previously recognized.

Neonatal disease acquired at birth from infected mothers with sexually transmitted disease include gonococcal ophthalmia neonatorum, inclusion conjunctivitis and infant pneumonitis of chlamydial origin, and herpes simplex virus infection. In each of these instances, the infant is infected during passage through the birth canal by exposure to infectious organisms on the mother's genital mucosa. The maternal infection may have been asymptomatic but in any case has been unrecognized and untreated. Neonatal infections may also occur with other organisms belonging to the mother's indigenous mucosal flora, such as *E. coli* and group B streptococcal meningitis.

The postnatal infectious diseases of infants generally are acquired through the usual routes of transfer from infected adult carriers or environmental sources. In some instances, however, infections acquired congenitally or at birth may not be immediately apparent. This is often the case in congenital syphilis, for example, in which infants appear healthy at birth and do not develop signs of infection until about the third week of life. Recognition of the nature and origin of infant infections is thus of great epidemiologic importance, requiring careful assessment of the need for treatment of the mother and any other adults involved. HIV may be transmitted at any time during the perinatal period by women not receiving AZT therapy.

Brief descriptions of the infectious diseases of infants are included in the discussion of each disease and are summarized in table 19.2.

INFECTIOUS DISEASES OF THE EYE

Infectious eye diseases acquired from infected persons fall into two groups: those of bacterial etiology (including chlamydial infections) and those caused by viral agents.

Numerous strains of pathogenic bacteria are associated with eye infections, but the clinical result is similar in each case: there is an acute conjunctivitis with production of a purulent exudate. Bacteria generally do not invade the epithelial cells of the conjunctiva or surrounding tissues but multiply extracellularly where they may be taken up by phagocytes in the exudate. The clinical and pathologic features of chlamydial or viral infection, on the other hand, are related to invasion of the conjunctival epithelium, or cells of other ocular tissues, and to the intracellular activities of the infectious agent. In most of these invasive infections, initial acute conjunctivitis may be succeeded by a chronic stage of slow destruction with scarring that can lead to blindness. Figure 19.10 illustrates the structure of the eye.

ACUTE BACTERIAL CONJUNCTIVITIS

The Clinical Picture

Bacterial conjunctivitis begins with symptoms of local irritation, tearing, and congestion of the small blood vessels of the conjunctiva over the cornea and under the eyelids. As infection proceeds, the eyelids become swollen and

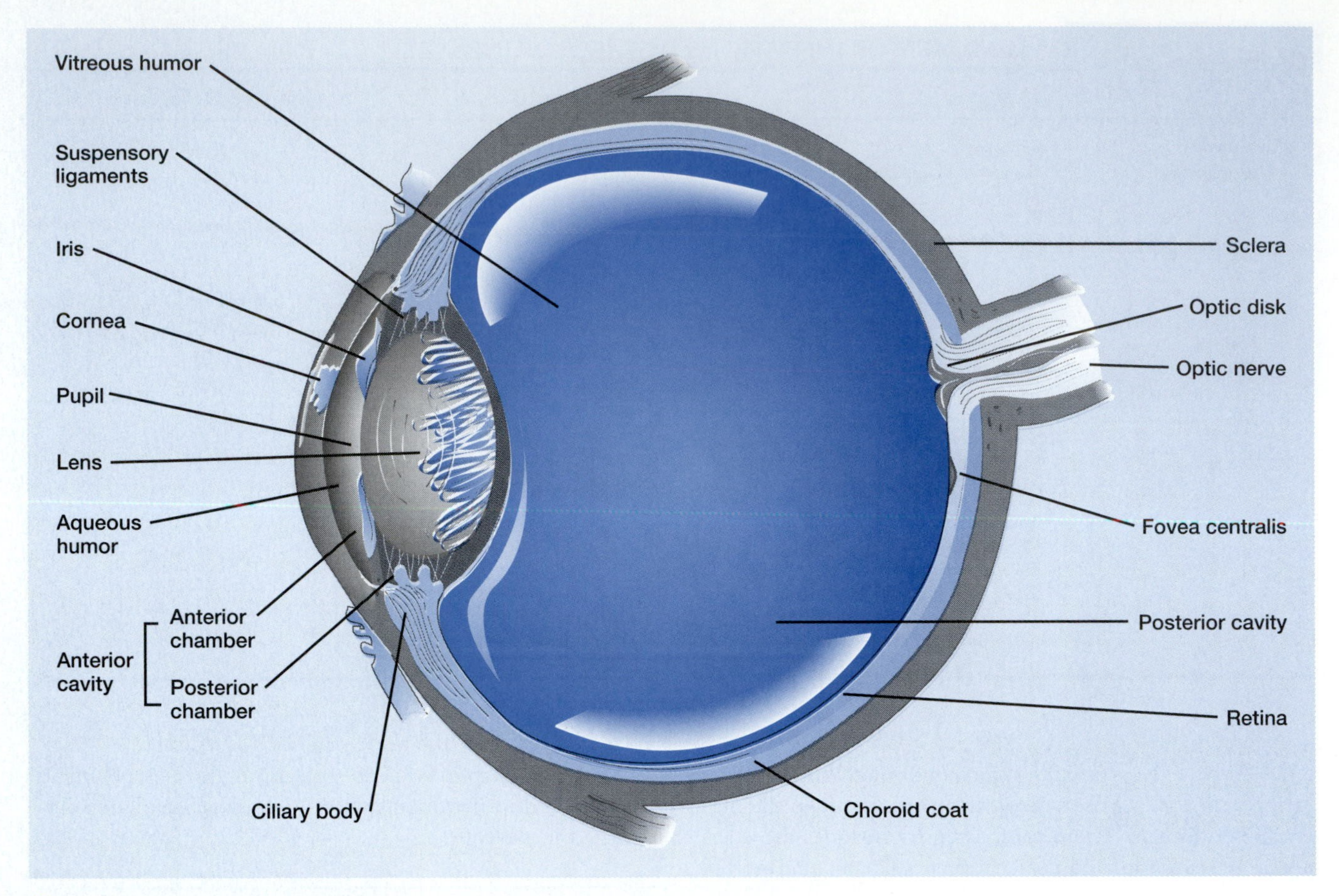

FIGURE 19.10 The structures of the eye.

reddened, and a reactive, cellular exudate is formed that may cause the eyelids to stick together. The affected eye (both or only one may be involved) becomes sensitive to light and feels hot and painful to the patient. These symptoms vary in degree, but neglect may lead to chronic infections with more serious involvement of deep tissues. As a rule, these infections respond well to treatment and heal without scarring or other residual damage. Because vascular injection is often a marked feature of bacterial conjunctivitis, the popular word for this type of infection is "pink-eye."

LABORATORY DIAGNOSIS

Microscopic examination of the exudate from eye infections caused by bacteria usually reveals the organism, and culture techniques provide final identification. The agents identified most commonly in children are strains of pneumococci or species of *Haemophilus* and *Moraxella*. Other organisms associated with the respiratory membranes, such as streptococci or staphylococci, may also be responsible for conjunctivitis in adults and children. It should be remembered also that gonococci from infected genital membranes can establish themselves on the conjunctival mucosa and produce severe infection in adults as well as in babies born of mothers with gonorrhea.

Epidemiology

These infections are most common among young children or others with lowered resistance. They are spread from infectious throat or conjunctival discharges and may pass rapidly from one small child to another in day-care centers, kindergartens, nursery schools, or playgrounds. They often yield to local antimicrobial therapy, but systemic dosage is sometimes required. Healing is usually rapid and without residual damage if treated early. The best control lies in prompt therapy and careful attention to hygienic care of healthy eyes. Children should be taught as early as possible to respect their eyes and to avoid rubbing them, especially with dirty hands. Parents, child-care workers, and teachers should be alert to children who rub their eyes to an unusual extent, for this may be the first symptom of a developing infection.

CHLAMYDIAL DISEASES OF THE EYE

Trachoma

Trachoma is a serious communicable disease of the eye prevalent in hot, dry areas of the world, particularly among people who live in overcrowded and unsanitary conditions. The infection is widespread in Mediterranean countries, parts of Africa, the Middle East, Asia, and South America. In the United States it is a public health problem on the Indian reservations of the Southwest.

Trachoma is caused by *Chlamydia trachomatis,* an intracellular bacterial agent. Different serologic types of this organism are associated with inclusion conjunctivitis and pneumonia of the newborn, chlamydial urethritis in adults, and lymphogranuloma venereum. These microorganisms are obligate intracellular parasites that can be seen as inclusion bodies in the cytoplasm of invaded cells. In trachoma and inclusion conjunctivitis they are found in epithelial cells of the conjunctiva.

Trachoma begins as an acute inflammatory conjunctivitis but is later characterized by excessive accumulations of tissue cells in tumorlike formations (colorplate 5f). Lymphocytes accumulating in and beneath the conjunctiva form lymphoid "follicles." This hyperplasia usually involves the conjunctiva of the upper lid and may extend under the mucosa covering the upper half of the cornea. Conjunctival scars may shrink, causing the eyelid to turn inward so that the lashes constantly abrade the cornea. This trauma, together with scarring and possibly, a toxin produced by the organism, deforms the eye and threatens blindness because of corneal opacity. Secondary bacterial infection is common and further damages the cornea, usually resulting in total loss of vision. Mild cases of trachoma are common, however, and rarely result in the loss of vision.

LABORATORY DIAGNOSIS

The laboratory diagnosis is made by microscopic demonstration of the typical intracytoplasmic inclusion bodies in epithelial cells of conjunctival scrapings. Special tissue stains or immunofluorescent staining is used. The agent may be cultivated in embryonated eggs or cell cultures and identified serologically. Identification of specific antibodies in the patient's serum can be made with immunofluorescent methods, and secretory IgA antibodies can be detected in tears. It is sometimes difficult to distinguish the agent from that of inclusion conjunctivitis by serotyping, but the clinical picture and epidemiology of the two diseases are quite different.

Epidemiology

Trachoma is spread by direct or indirect contacts with conjunctival discharges from infected persons, most commonly within families. It can be treated successfully if discovered early, so prompt diagnosis constitutes the best means of control. Tetracyclines are the drugs of choice, but sulfonamides are also effective. Children are more frequently affected than adults and if their recovery is to be complete, must be diagnosed and treated before the disease has advanced to the point of scar formation. In areas where trachoma is prevalent, mass treatment programs with topical antimicrobial agents are coupled with efforts to improve sanitary standards.

Inclusion Conjunctivitis

Inclusion conjunctivitis is an acute conjunctivitis of newborn babies resulting from maternal, sexually transmitted infection, as described previously. The organism, *C. trachomatis,* is rarely transmitted to adults from infected babies' eyes, but infectious genital secretions may be transferred on hands.

The disease is less severe than trachoma, but if untreated it may persist in a chronic stage for several months. The incubation period is from 5 to 14 days (in the newborn this distinguishes it from gonococcal ophthalmia, which usually begins within 36 hours of birth). Infant conjunctivitis is characterized by acute inflammation, particularly of the lower lids. Purulent exudate is produced, but there is no lymphoid follicle formation or scarring. In older children and adults, there is lymphoid hyperplasia, as in trachoma, but it is most marked on the lower lids and resolves without scarring. If properly treated, there is no corneal involvement in inclusion conjunctivitis.

Confirmation of the diagnosis can be obtained by laboratory demonstration of inclusion bodies in epithelial cells scraped gently from the conjunctival surface of the lower lid and also by culture methods and immunologic techniques. Appropriate tests to rule out gonococcal infection should also be performed. The infection responds rapidly to erythromycin therapy. The newborn disease is difficult to prevent, as the mother's infection is usually inapparent. The current practice of instilling tetracycline or erythromycin ointment in neonates' eyes should reduce the incidence of infection.

VIRAL DISEASES OF THE EYE

At least three viruses are associated with specific syndromes of acute conjunctivitis. Adenovirus (usually types 8 and 19) causes epidemic **keratoconjunctivitis,** an infection in which the conjunctivae are inflamed and the cornea may become temporarily keratinized and opaque. Systemic signs such as low-grade fever and headache may also be present. The virus is very hardy and resists treatment commonly used for disinfecting ophthalmologists' instruments. Therefore it may be spread in clinics or offices where strict asepsis is not maintained. There is no specific therapy for adenovirus conjunctivitis.

Herpes simplex virus type 1 and occasionally type 2 may infect the conjunctival membrane or cornea. Although the infections are usually superficial and resolve quickly, the cornea may be ulcerated and scarred, or keratinized. Antiviral drugs (such as trifluridine and vidarabine) may be useful for suppressing the organism in superficial cells, but deep infections may require ophthalmic surgery.

Enterovirus type 70 is associated with epidemics of a **hemorrhagic conjunctivitis** seen primarily in Africa, Asia, and European countries. The infection is characterized by pain, swollen eyelids, injection of the conjunctivae, and subconjunctival hemorrhage. Epidemics are most frequent in families or in situations of overcrowding and poor hygiene where it is transmitted by direct or indirect contact with discharge from infected eyes.

When necessary, diagnosis is made by isolating the responsible virus in cell cultures. Conjunctivitis may be a manifestation of other viral infections as well.

INFECTIOUS DISEASES INITIATED ON SKIN OR MUCOUS MEMBRANES

STAPHYLOCOCCAL INFECTIONS

Staphylococci are among the most common members of the normal flora of humans. Several species live as commensals on different surface areas of the body, primarily skin and mucous membranes, but three species are frequently associated with **staphylococcal infections.** *Staphylococcus aureus* has a number of virulence properties and is responsible for a wide variety of superficial and systemic infections as well as diseases caused by its exotoxins. *Staphylococcus epidermidis,* the most common aerobic bacterium on the skin surface, is of relatively low virulence. In patients with lowered defense mechanisms, and especially those with indwelling intravenous catheters, it can invade the bloodstream and cause serious disease. The third species, *Staphylococcus saprophyticus,* has an unusual capacity for causing urinary tract infections in young women, primarily between the ages of 16 and 25 years.

When *S. aureus* produces disease, the source is sometimes endogenous; resident staphylococci enter deep tissues through broken skin or mucosal barriers or multiply at a local site where normal physiologic barriers have broken down. The source of infection may also be the purulent discharge from active lesions of another person or organisms shed by an asymptomatic nasal carrier. The skin is the most frequent route of entry and the most common site of localized *S. aureus* infection. In the following discussion, the staphylococcal diseases are grouped as superficial, deep tissue, or toxic infections.

1. Superficial Staphylococcal Infections

The typical skin lesion produced by *S. aureus* is an abscess that may range in size and severity from the small but annoying pimple to larger and more painful furuncles (boils) and carbuncles. Furuncles arise from abscesses localized in hair follicles. They are uncommon on the scalp but may occur on many other hairy surfaces of the body, common sites being the axillae, the back of the neck, and the buttocks. (Abscesses in the follicles of the eyelashes are called *sties.*) Carbuncles are deep-seated abscesses in subcutaneous tissues. They are large, bumpy, and painful because there may be several closely associated, or intercommunicating, pockets of pus. These infections and others, such as staphylococcal paronychia (inflammation of the soft tissues around the nails) and decubitus ulcers (bedsores), usually arise at sites of minor but continuously irritated skin injuries. Such injuries may be caused by the constant pressure of clothing or the macerating effect of heat and moisture, especially in patients with lowered physiologic resistance. Bedridden patients who do not receive proper care probably suffer most from these infections, which add to the underlying debility that confines them to bed.

Acne is a condition associated primarily with certain anaerobic bacteria that live commensally on the skin, but staphylococci may be involved in a secondary role, producing pustular acne vulgaris. Acne is a problem of adolescence related to physiologic and hormonal changes as well as possible dietary factors. The irritating effect of fatty acids produced from sebum by the anaerobes is also thought to be important. Acne is characterized by the recurrent appearance of crops of comedones ("blackheads" and "whiteheads") and pimples on the face, often in such high density that the epithelium cannot recover and is replaced by scar tissue. It can be controlled by skin soaps that reduce normal bacterial flora, judicious hormone therapy, and careful use of antimicrobial agents. Tetracyclines applied topically may be helpful. In more severe cases, oral drugs such as erythromycin or clindamycin may be given.

2. Deep-Tissue Infections Caused by Staphylococci

Within the tissues staphylococci may localize anywhere, depending on the route of entry and their accessibility to the bloodstream. From the upper respiratory tract they may extend to sinuses or the middle ear and mastoid or downward to the bronchial tract and pulmonary beds. Staphylococcal pneumonia is one of the most serious bacterial pneumonias because the organism and its toxins produce hemorrhagic, necrotizing lesions in the lung parenchyma.

Staphylococci sometimes cause postpartum endometritis, gaining access to the pelvic tract from the superficial genital mucosa during childbirth. Streptococci are associated with this type of disease more frequently (see chap. 21).

S. aureus may gain access to the bloodstream through skin wounds or abscesses and be distributed to various sites. They can grow in any tissue. If local defenses are not adequate, abscesses can form in bone (osteomyelitis), in the covering tissue that sheathes bone (periostitis), in liver or spleen, the kidney pelvis (pyelonephritis), or brain (meningitis or encephalitis). Depending on the site of entry, focal lesions develop in one or more of these sites, but if septicemia becomes fulminating and widespread abscesses develop, the result may be fatal.

Generalized infections are more frequent and dangerous in the very young or old and those with problems that open the door to superimposed infection. The latter include surgical and obstetric patients, people with respiratory diseases, patients with renal or bladder dysfunctions, diabetics, and others with endocrine disorders that affect resistance. In cardiac patients requiring arterial catheterization or open-heart surgery, or patients with indwelling intravenous catheters, *S. epidermidis* may be introduced into the bloodstream or directly to the endocardium resulting in septicemia or endocarditis. In drug abusers, *S. aureus* injected directly from the skin or from contaminated paraphernalia is the more usual cause of staphylococcal endocarditis.

Even a minor abscess in the skin may be the source of disseminated infection if the fibrous capsule encasing it is broken by the pressure of squeezing so that microorganisms within the lesion are released to surrounding tissues and adjacent capillaries. It is particularly dangerous to squeeze pimples and boils that appear on the face because the draining lymphatics and blood capillaries may carry released staphylococci directly into areas contiguous with the brain.

Any of the three common *Staphylococcus* species, *S. aureus, S. epidermidis,* and *S. saprophyticus,* may be isolated from the urine of patients with significant urinary tract infection. Recovery of *S. aureus* usually indicates that the organism has spread through the bloodstream to the kidneys, with abscess production. The significance of *S. epidermidis* urinary tract infection is not clear, although such infections are usually confined to the bladder, producing only a cystitis. Because *S. epidermidis* can be a member of the normal perineal flora, this organism is considered to be a urinary pathogen only when approximately 10^4 to 10^5 or more organisms per milliliter of voided urine are recovered in culture. Currently, the only infection in which *S. saprophyticus* has been implicated is cystitis in young women. The source of the organism is unclear because it normally is not found on the skin or periurethral area. Infection is accompanied by typical symptoms of urgency, hesitancy in voiding, and dysuria.

3. Staphylococcal Toxic Disease

Certain *S. aureus* strains in localized foci of infection produce an epidermolytic toxin that causes separation of the upper layers of skin to form blisterlike lesions, a condition referred to as *bullous impetigo.* In adults, only a few of these lesions usually form and are limited to a specific body area. Widely disseminated disease, termed **scalded skin syndrome,** may occur in children (colorplate 5e). It is characterized by the formation of large superficial blisters that may peel off in "sheets." In neonates, the disease has been seen in association with nursery epidemics of staphylococcal disease. Staphylococci can be isolated from material obtained from the base of the lesions but are usually not found in the vesicle fluid.

Toxic Shock Syndrome (TSS)

Toxic shock syndrome (TSS), associated with *S. aureus,* was first described in 1978, although cases of patients with similar symptoms have been reported since 1927, when the disease was usually referred to as *staphylococcal scarlet fever.* The classic symptoms include sudden onset of high fever, hypotension, headache, severe myalgia, profuse vomiting and diarrhea, mucous membrane inflammation, diffuse neurologic abnormalities, and a desquamating skin rash. Serious complications include shock, cardiac arrhythmias, renal failure, and pulmonary edema with respiratory failure.

Of the initial cases of TSS reported to the Centers for Disease Control, 96% involved women and at least 92% of these experienced onset during a menstrual period. Because epidemiological data implicated use of high-absorbency tampons as a risk factor for TSS, a particular brand of high absorbency tampons was removed from the market. In addition, manufacturers lowered the absorbency and changed the fiber composition of other brands. Women are advised to wear the lowest absorbency tampons compatible with their needs. Even though the precise relationship between tampon use and TSS has not been determined, changes in the product as well as increased public awareness of early signs of illness have markedly reduced the incidence of menstrual TSS. Reports of nonmenstrual TSS, which occurs in men, women, and children of all ages and races, have not decreased. A wide array of predisposing infections have been responsible for nonmenstrual TSS, including empyema, septic abortions, and surgical wound infections.

Most *S. aureus* strains isolated from TSS patients produce a toxin referred to as toxic shock syndrome toxin-1 or TSST-1. In several animals, this toxin produces a TSS-like illness. Many persons who are colonized with TSST-1-producing strains of *S. aureus* have high levels of antibody against the toxin and never develop TSS. In contrast, patients with TSS lack preexisting antibody. Disease results when conditions favor the growth and multiplication of *S. aureus,* with subsequent toxin production. Therefore, it appears that the pathogenesis of TSS requires exposure of a nonimmune individual to a toxin-producing *S. aureus* strain. Many additional aspects of TSS and its pathogenesis remain unexplained.

Foodborne disease caused by staphylococcal enterotoxin is described in chapter 16.

LABORATORY DIAGNOSIS

I. Identification of the Organism

(Staphylococcus aureus) (Staphylococcus epidermidis) (Staphylococcus saprophyticus).

A. *Microscopic Characteristics* Species of *Staphylococcus* cannot be distinguished by microscopic examination. They are gram-positive cocci, typically arranged in grapelike clusters, nonmotile, and nonspore-forming.

B. *Culture Characteristics* Strains of *S. aureus* grow well under aerobic or anaerobic conditions on ordinary laboratory media. On blood agar they often possess the golden-yellow pigment from which their name was derived. They are often hemolytic and almost invariably produce a substance called coagulase, which causes clotting of plasma.

Strains of *S. epidermidis* grow under the same conditions, but they possess a white pigment. They are generally nonhemolytic and coagulase negative.

These findings represent the general rule, but there may be variations. Occasional strains of *S. epidermidis* may be hemolytic, and strains of *S. aureus* may be nonhemolytic. Variants of either type may be associated with human lesions, sometimes with minor superficial infections of surgical wounds (e.g., stitch abscesses) or with chronic, low-grade, but threatening systemic infections occurring most frequently as an aftermath of cardiac surgery or catheterization.

S. saprophyticus produces yellow- or white-pigmented colonies and is coagulase-negative. It is distinguished from

S. epidermidis by its resistance to low concentrations of novobiocin in laboratory tests.

Biochemical test kits are available to identify the three common species and other staphylococci. Unless they are isolated from serious infections, other species are referred to only as "coagulase-negative staphylococci." In addition to *S. aureus,* a few species isolated from animals produce the coagulase enzyme.

C. *Immunologic Methods of Identification* Immunologic methods are generally not applicable because the staphylococci are a large and heterogeneous group antigenically. The tools most useful in recognizing strains of common epidemiologic importance in hospital or community cross-infections are phage typing and antimicrobial susceptibility testing (see Special Laboratory Tests, IV).

II. Identification of Patient's Serum Antibodies

The serum of most healthy adults contains antibodies against numerous staphylococcal products and components, and therefore, serologic tests are not useful.

III. Specimens for Diagnosis

Pus or fluid from any lesion in any part of the body may be submitted for smear and culture. Blood, spinal fluid, pleural, abdominal, synovial fluid, or urine may be appropriate for culture depending on clinical circumstances.

Stained smears of the exudate are examined for the presence of clusters of organisms resembling staphylococci.

Cultures are examined for pigment, hemolysis, coagulase activity, and, for coagulase-negative urine isolates, novobiocin susceptibility.

Further biochemical tests may be done as appropriate. *S. aureus* isolates from patients with suspected toxic shock syndrome may be sent to special reference laboratories that have the ability to test for TSST-1 production.

IV. Special Laboratory Tests

When it is suspected that cross-infections have been incurred from a single source of infection, two types of laboratory information may support or confirm this:

A. Antibiograms of staphylococci isolated from a series of infected patients may be compared. An antibiogram is the pattern of susceptibility and resistance shown by a given strain to a battery of antimicrobial agents. When all isolated strains have similar or identical patterns, this constitutes presumptive evidence of their epidemiologic relationship.

B. Phage types of strains isolated from a series of infections may be compared. *S. aureus* strains are generally susceptible to bacteriophages, each being lysed by one or more particular phages. If the pattern of lysis among all strains tested is the same (or very similar), the epidemic probably is caused by a single strain. The common source of that strain should be identified and eliminated if possible.

Epidemiology

I. Communicability of Infection

A. *Reservoirs, Sources, and Transmission Routes* Staphylococcal infections have their primary reservoir in humans and are spread from colonizing sources on normal human membranes (usually in the anterior part of the nose) or in open furuncles, draining sinuses, tonsillar abscesses, and similar lesions. They may be transmitted by direct personal contacts or indirectly through contaminated hands and objects. Airborne transfers may be important for viable organisms in dust or in droplet nuclei from respiratory sources. In hospitals, dirty dry mops and dustcloths may contribute to their spread, as well as soiled bed linens, dressings, and unwashed hands of personnel.

B. *Incubation Period* This varies and depends on the route of entry, the resistance of the infected individual, and the numbers of virulent organisms introduced. In surgical wounds, infection may appear within a day or two. When organisms are introduced directly into the bloodstream, symptoms may ensue within hours. Infections of normal "intact" skin, such as boils or other abscesses, usually require from 4 to 10 days to appear.

C. *Communicable Period* Staphylococcal infection is transmissible throughout the time purulent discharges drain from open lesions or the nasal carrier state persists.

D. *Immunity* Resistance to staphylococcal infections is based primarily on nonspecific mechanisms. Antibody titers are not consistent and seem to have little relationship to immunity against reinfection. Susceptibility is fairly general, but those at greatest risk are the newborn, those debilitated by chronic diseases or age, surgical patients, and those who must undergo protracted medical or surgical therapy for underlying organic diseases.

Persons with immunity against TSST-1 do not develop toxic shock syndrome.

II. Control of Active Infection

A. *Transmission-based Precautions* It is seldom necessary or feasible to isolate patients being treated at home, but they should be kept away from infants or family members who are ill and debilitated. In the hospital, contact precautions are used for all patients with major skin, wound, or burn infections, but standard precautions are adequate for other syndromes. In the newborn nursery, isolation may be accomplished by cohorting with thorough attention to concurrent disinfection.

B. *Other Precautions* Dressings from discharging lesions should be placed in impervious plastic bags, sealed, and disposed of in a safe manner. Bed linens and clothing should be carefully placed in tightly closed bags marked for care in transport and handling prior to laundering. If routine laundering cannot assure disinfection, sheets and bedclothes should be autoclaved or washed in boiling water with detergent. Concurrent disinfection must be carried out for all items and equipment in the patient's room that might be contaminated by infectious discharges or by drying droplet nuclei disseminated from the patient's secretions or exudates. If soiling is likely, a clean gown should be worn by each person who enters the room, including visitors. Upon leaving, persons should discard their gowns in a hamper provided near the hand-washing facility, carefully wash their hands with disinfectant liquid soap, and dry them with paper towels. Terminal cleaning-disinfection must be thorough.

C. *Treatment* Staphylococci are notorious for their ability to become resistant to new antimicrobial agents in a short time. Because of the marked variability in their responses to useful antimicrobial drugs, isolated strains must be tested in the laboratory for their antimicrobial susceptibility. Therapy is guided by the results of these susceptibility tests. Topical therapy may be used for localized skin infections along with medical or surgical methods that promote drainage. Systemic agents are needed for deep-tissue infections or under conditions that may promote dissemination of the organism from a local source.

Although penicillin was once the drug of choice for *S. aureus* infections, almost all strains now produce the beta-lactamase enzyme, penicillinase, and are resistant to this agent. Beta-lactamase-resistant penicillins such as methicillin are used instead. Unfortunately, many strains, especially those isolated from hospitalized patients, are also methicillin-resistant, thus making it necessary to use other, more toxic and expensive, drugs such as vancomycin for treatment.

Genetic control of penicillinase production in staphylococci resides on a transferable extrachromosomal DNA element (plasmid). Their resistance to methicillin depends on a chromosomally induced change in proteins in the staphylococcal outer membrane (see chap. 3). Thus, *S. aureus* takes advantage of more than one resistance mechanism in its efforts to evade the bactericidal consequences of antimicrobial agents.

The antimicrobial susceptibility of *S. epidermidis* and *S. saprophyticus* also varies, so the physician depends upon the results of laboratory testing to aid in selecting the appropriate drug when therapy is indicated. *S. epidermidis* strains are often even more resistant to antimicrobial agents than is *S. aureus; S. saprophyticus* is usually less resistant.

D. *Control of Staphylococcal Infections in Hospitals* *S. aureus* was once a common cause of hospital-acquired infections, especially in the 1950s and 1960s, but the use of effective antimicrobial agents and appropriate isolation precautions has since decreased the threat of staphylococcal outbreaks. Emerging resistance to multiple antimicrobial agents, however, may bring recurrence of the problem, therefore, strict principles of personal hygiene should be applied to patients and practiced by personnel in all hospital situations. Personnel found to be active shedders of antimicrobial resistant strains should be removed from patient contact until therapy clears the carrier state; however, therapy is not always effective or may not result in permanent clearance. In addition, high-risk patients should be protected from infection by rigid adherence to contact or standard isolation precautions.

III. Prevention

No effective vaccines are available. Prevention of staphylococcal infections in hospitals requires a coordinated policy and the cooperation of all responsible for the safety of patients. When clusters of staphylococcal infection are identified, direct patient care personnel are screened by culture. Shedders are treated as described in the previous section. Intranasal administration of the agent mupirocin has been effective in eradicating nasal carriage of *S. aureus* in some people but many strains rapidly become resistant to this drug.

Gram-positive Organisms in Human Hosts
Staphylococci: Virulence Factors; Diseases

STREPTOCOCCAL INFECTIONS

The Clinical Disease

Erysipelas, impetigo, and **necrotizing fasciitis** are **streptococcal infections,** skin infections caused by strains of group A beta-hemolytic streptococci. In erysipelas, systemic infection also occurs, with bacteremia, fever, malaise, and marked leukocytosis. The skin lesions occur most frequently on the face and legs, beginning at an inconspicuous portal of entry of the organism in the skin (perhaps rubbed into an abrasion or a hair follicle). The lesion is tender and red and spreads rapidly, having a raised advancing margin. On the face, erysipelas often spreads bilaterally from the nose across the cheeks in a "butterfly" pattern. This highly contagious infection is more severe in patients with debilitating conditions.

Impetigo, or streptococcal pyoderma, is especially common in children and occurs when the streptococcus is inoculated directly into a superficial skin abrasion such as an insect bite, dermatitis, or eczema. The lesion evolves into a thick, raised, crusted area that is only slightly painful. Staphylococci as well as streptococci are sometimes present in the lesions. Acute glomerulonephritis is an important complication of streptococcal impetigo (see chap. 12). Necrotizing fasciitis is an old disease that has had a dramatic recurrence during the past few years. The term "flesh-eating bacteria" was coined to describe the invasive, tissue destructive lesions produced by the group A streptococci. Although the number of cases of this skin infection appears not to have increased, the severity of the current disease suggests that the streptococci have become more virulent than in the past. Recent cases have occurred primarily in young healthy persons who experienced a minor trauma to an extremity. The infection involves deep subcutaneous tissue with destruction of connective tissue and fat but usually not muscle or skin. In addition to the local wound infection, the patients experience a severe systemic illness that may result in shock, disseminated intravascular coagulopathy, or respiratory, liver, or kidney failure. Death results in from 20 to 60% of patients. The systemic signs are sometimes referred to as streptococcal toxic-shock syndrome because of their similarities to staphylococcal toxic-shock syndrome and are thought to be related to the action of streptococcal pyrogenic exotoxin.

The streptococcal diseases and the role of streptococci in skin infections are discussed in chapter 12, but a few

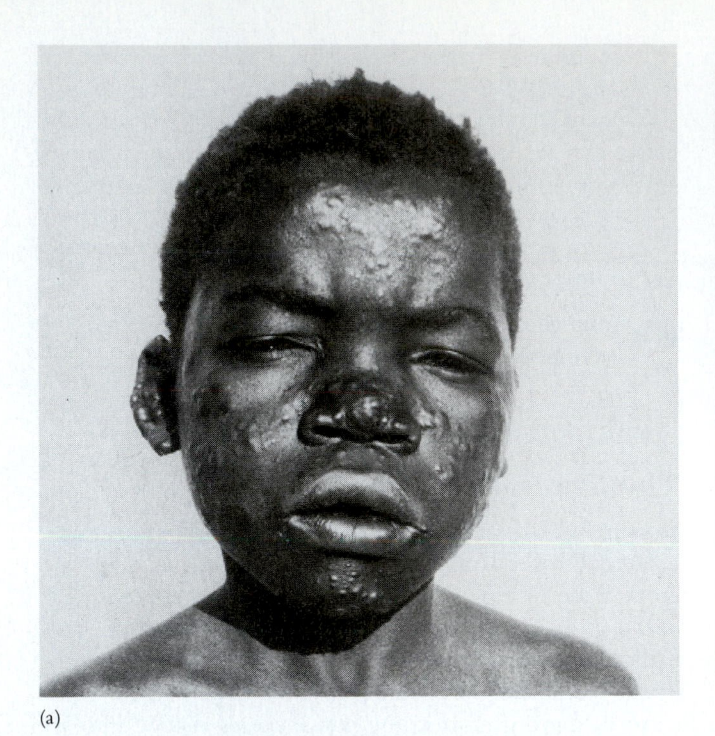

(a)

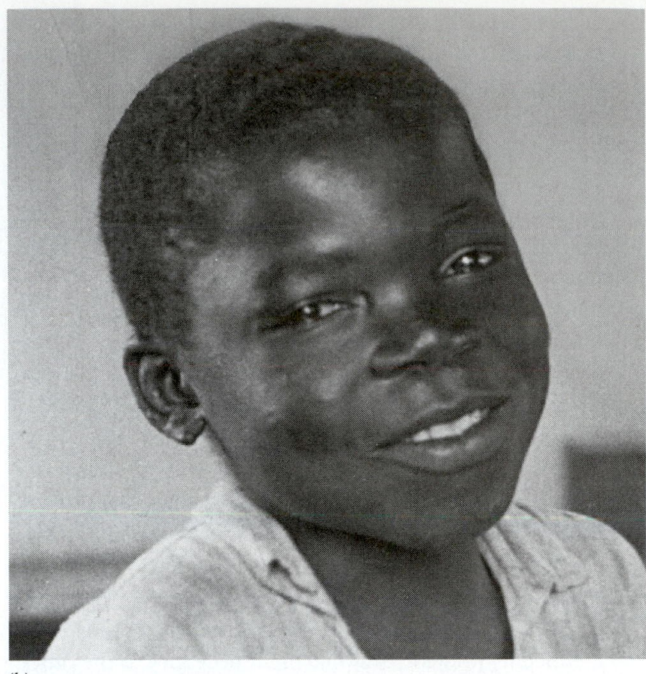

(b)

FIGURE 19.11 (a) An early case of cutaneous (lepromatous) Hansen disease. (b) The same child following treatment with Hansen disease drugs.
Dr. Roy Pfatzgraff, ALM International, Greenville, SC.

Miscellaneous Bacteria in Human Host
Mycobacteria: Mycobacteria Virulence Factors; Mycobacteria Diseases

details pertinent to their laboratory diagnosis and epidemiology are mentioned here.

LABORATORY DIAGNOSIS

In erysipelas, specimens for smear and culture should be taken from the raised peripheral border of lesions. In necrotizing fasciitis, specimens are taken from surgically obtained material. Blood cultures may yield the organisms if taken during febrile periods. The lesions of impetigo are so characteristic that diagnosis is most often made clinically.

Epidemiology

Erysipelas occurs most frequently in aged people and in infants. It is characterized by recurrences in the same individual, possibly because of hypersensitivity or reinfection from endogenous sources. Its epidemiology is similar to that of other streptococcal diseases described in chapter 12. Hospitalized patients should be isolated and contact precautions should be followed, especially if the newborn nursery is involved.

Impetigo occurs primarily in children during warm seasons in temperate climates or year-round in the tropics. It may spread epidemically among close contacts such as in families and schools. Prevention depends on good general hygiene and proper attention to insect bites and abrasions.

Gram-positive Organisms in Human Hosts
Streptococci: Virulence Factors; Diseases

HANSEN DISEASE (LEPROSY)

The Clinical Disease

Hansen disease (HD), or **leprosy,** is a chronic, very slowly progressive disease that requires many years to develop. Although HD is highly contagious, more than 90% of individuals are resistant to infection. Persons who develop the disease appear to have a specific defect of cell-mediated immunity that is not yet well understood. When HD is untreated, its progressive development leads to disfigurements that have been greeted for centuries with horror and fear. No social ostracism could be more complete and no physical exclusion more dreadful than that to which persons with HD have been subjected throughout human history. In recent years, medical knowledge has shed light and objectivity on the problem, relieving some of the pressure on the afflicted patient.

The clinical manifestations of leprosy are broad. They range from a single, localized, frequently self-healing patch on the skin to progressive, diffuse involvement of the whole skin, numerous peripheral nerves, and many organs (fig. 19.11a). In general, three forms of HD are recognized, each of which progresses slowly. In the lepromatous form, there is diffuse infiltration of bacilli in skin, mucous membranes, peripheral nerves, and some organs. Infiltration is generally most intense in peripheral areas where skin temperature is low and least intense in warmer, protected areas such as the axilla, groin, and popliteal fossae. Histologic examination reveals characteristic foam cells and granulomas. Acid-fast bacilli, seen in large numbers under acid-fast stains, com-

FIGURE 19.12 Armadillos are useful in drug research for the treatment of Hansen disease. They are one of the few animals in which the organism *Mycobacterium leprae* can be grown and that are naturally infected with a similar bacterium.

Courtesy The Star, Carville, LA.

pletely fill the cytoplasm of cells. No lymphocytes are seen and the immune response is largely ineffective in limiting progress of the disease.

The tuberculoid form of leprosy is characterized by decreased sensation in well-defined skin nodules and localized involvement of one or two nerve trunks. Bacilli are difficult to find, and those that are found are usually fragmented. In the affected areas, the whole dermis is infiltrated by the granuloma which is made up of nests of lymphocytes and epithelioid cells (specialized macrophages). Many of these epithelioid cells fuse into giant cells called *Langhan cells.*

The third form, borderline leprosy, has clinical and pathological characteristics of both of the other forms. Patients have large numbers of skin lesions of varying types including macules, nodules, and plaques, all of which may be found in the same patient. Sensory loss in the lesions is often easily demonstrated. Many patients suffer extensive peripheral nerve damage.

Any of the three forms of HD can result in sensory and motor loss, particularly in the limbs and on the face. Consequently, the afflicted area is exposed to injury and infection which can result in the characteristic deformities. The deformities may be limited to a single digit, or, over the years, may result in almost complete destruction of the hands and feet as well as severe damage to the eyes.

All patients with HD experience periodic acute inflammatory responses (referred to as *reactive phenomena*) in tissues where *M. leprae* is present. In lepromatous disease, the phenomena are characterized by the appearance of tender, erythematous nodules in the skin, a condition referred to as *erythema nodosum leprosum,* or ENL. Nerves, eyes, testes, joints, and lymph nodes may also be affected and the patient may be febrile. In tuberculoid and borderline HD, reactive phenomena appear as erythema in skin lesions, often with edema and sometimes necrosis. Pain, swelling, tenderness, and loss of function in nerve trunks are common manifestations of this reaction. Some borderline patients also suffer from ENL.

LABORATORY DIAGNOSIS

I. Identification of the Organism

(Mycobacterium leprae).

The agent of HD is an acid-fast bacillus in the same genus as the tubercle bacillus (see chap. 12). Unlike the latter, *M. leprae* (originally called *Hansen's bacillus*) is an obligate intracellular organism. It has never been satisfactorily cultivated in cell-free media or cell culture. In recent years, efforts to establish HD in experimental animals have succeeded, and the disease can now be studied in normal and immunologically deficient mice, some primates, and the nine-banded armadillo (fig. 19.12).

In the tissue of humans or animals with the lepromatous form of disease, aggregates of *M. leprae* are found almost completely filling characteristic foam cells. In contrast, bacilli are rarely found in tuberculoid leprosy.

II. Identification of Patient's Serum Antibodies

Reliable serologic tests for HD have yet to be developed. A skin-testing antigen, lepromin, has some value in determining the prognosis, as it can distinguish the forms of HD. The antigen is an extract of infected tissue that produces delayed-type hypersensitivity reactions in the skin of patients with the tuberculoid form of disease, but the test is negative in lepromatous cases.

It should be remembered that sera from patients with HD often give false-positive results in reagin tests for syphilis.

III. Specimens for Diagnosis

Scrapings or biopsies of skin lesions are submitted for microscopic examination for intracellular acid-fast bacilli. The nasal mucosa is a preferred area for sampling. Another is the active periphery of a nodule cut through its epidermal layer and scraped along the inner edge of the cut for cells representing the central portion of the granuloma. Acid-fast bacilli are seen in mononuclear cells.

Epidemiology

I. Communicability of Infection

A. *Reservoirs, Sources, and Transmission Routes* This disease has a human reservoir. Recently, it has been determined that organisms are more likely to be transmitted from the respiratory tract than from skin lesions. Patients with lepromatous disease shed large numbers of bacilli in their respiratory secretions. Presumably these are inhaled and spread through the body during a bacteremic episode. Infants and young children are the most susceptible, but their contacts with infectious parents or other adults must be long and intimate if they are to contract the disease.

B. *Incubation Period* Very long periods of exposure are required, the average time for appearance of symptoms being 3 to 5 years. Infants heavily exposed from birth have been known to develop the disease in as short a time as 7 months, or usually, 1 to 2 years.

C. *Communicable Period* HD is considered infectious if bacilli can be demonstrated in skin or mucosal lesions, especially of the upper respiratory tract. Organisms are always demonstrable in active lepromatous cases and often in untreated borderline cases. Organisms are rarely seen in tuberculoid cases. Infectiousness is lost within a few days of effective therapy.

D. *Immunity* Immunity is of the cell-mediated, hypersensitive type, as indicated by reactivity to lepromin. Patients with the more progressive lepromatous type of infection have no resistance and are lepromin-negative, whereas tuberculoid patients as well as infected but asymptomatic persons are lepromin-positive. The long incubation period and the closeness of contact required for transmission of the disease indicates a high natural resistance.

HD is found primarily in tropical and semitropical areas of the world. There are now an estimated 5 to 10 million cases worldwide. Most are found in Central Africa, Southeast Asia, and Central and South America. Approximately 6,000 cases are known in the United States, primarily in the south.

II. Control of Active Infection

A. *Transmission-based Precautions* People with HD need not be rigidly isolated. Caring for them with standard precautions is sufficient. Most can be successfully treated in clinics or out-patient departments. Patients may be admitted for a few days at the start of treatment to provide them with intensive health education, but admission is not required on clinical grounds. Many patients experience reactive phenomena during the course of disease and may require hospitalization to manage acute neuritis or fever.

B. *Other Precautions* Modern treatment is so effective that patients become noninfectious within a day or two after treatment is begun. No special disinfection procedures are recommended for their discharges.

C. *Treatment* Two or more drugs must be used concurrently. For tuberculoid disease, a combination of dapsone and rifampin is used. Only 6 months of therapy is required. However, many clinicians prefer to treat longer than 6 months because lesions often appear clinically active past this period and because reactive phenomena may occur for more than a year after treatment is initiated. For lepromatous and borderline disease, patients are normally treated for 2 years with drugs, preferably rifampin, clofazimine, and dapsone. Treatment does not need to be continued until all acid-fast material has disappeared from the skin. Experience has shown that the granulomas continue to clear for years after treatment has been stopped (fig. 19.11b).

Reactive phenomena in tuberculoid and borderline cases are treated with steroids. In lepromatous disease, ENL is suppressed with clofazimine or thalidomide. The latter drug is never used in women of childbearing age because it is known to cause birth defects. Steroids may also be administered, but only in conjunction with clofazimine.

D. *Control of Contacts* The chief method of control involves case finding, especially among family contacts of newly diagnosed individual cases. Close contacts should be examined yearly for at least 5 years after contact with an infectious patient and should be treated if necessary. Case reporting is mandatory in the United States.

III. Prevention

A vaccine that combines the antituberculosis vaccine (BCG) with inactivated *M. leprae* has been found not to be more effective than BCG alone and results with the latter have been variable.

HERPES SIMPLEX VIRUS

Herpesvirus Infection

Herpes simplex virus (HSV) belongs to the group of viruses that includes cytomegalovirus, varicella-zoster virus (chickenpox and herpes zoster), and Epstein-Barr virus (infectious mononucleosis). It produces vesicular eruptions of the skin or mucous membranes similar to those of herpes zoster (see chap. 13), but it differs antigenically and in its epidemiologic patterns of spread. HSV often establishes a lifelong association with infected human beings. It may involve various mucous membranes (oral, pharyngeal, conjunctival, genital) or mucocutaneous areas in recurrent infections interspersed with long periods of asymptomatic latency.

In the wide spectrum of herpetic infections, most clinical syndromes are mild. One of the most common is the recurrent labial herpes of adults, familiar to everyone as the "cold sore" or "fever blister" that crops up again and again in some individuals at a mucocutaneous margin of the lip (colorplate 6e). Primary acute stomatitis, an inflammation of the oral membranes, occurs in children, and genital herpes is a sexually transmitted disease of adults described earlier in this chapter.

More severe forms of herpetic disease occur less frequently in the healthy host. These include keratoconjunctivitis, chronic eczema, and meningoencephalitis. Generalized neonatal infection, acquired at birth from exposure to maternal genital lesions, is perhaps the most disastrous form of **herpetic disease** and is often fatal. Ordinarily the newborn child is protected from critical infection during the first months of life by maternal antibody produced in the course of low-grade infections experienced by the mother. If the mother acquires her first genital infection during pregnancy, however, and the child is born before sufficient protective antibody can be transferred placentally, neonatal infection may ensue. The virus is widely disseminated in infant tissues, producing necrotizing lesions in the skin and mucosa, liver, and brain.

Immunocompromised patients are at high risk for developing severe HSV infections, especially patients with AIDS (colorplate 9d), those with extensive burns, and those receiving immunosuppressive therapy.

Two serologic types of HSV are recognized, displaying biologic as well as antigenic differences. HSV type 1 is characteristically associated with most nongenital infections. HSV type 2 is the cause of genital lesions and neonatal disease and has also been recovered from cases of aseptic meningitis. These viruses can be distinguished by immunologic methods applied either directly to material from patient lesions or to cells from cultures inoculated with patient specimen.

LABORATORY DIAGNOSIS

HSV inclusions can be visualized by either light or fluorescence microscopy in appropriately stained material obtained from the base of lesions. The virus can be isolated from skin and mucosal lesions or from spinal fluid in central nervous system infections and identified immunologically. Serologic diagnosis can be made by demonstrating a rising titer of antibodies, usually in enzyme immunoassay tests.

Epidemiology

Humans are the reservoir. HSV may be present in saliva, respiratory secretions, conjunctival discharges, and on the genital mucosa. In males, the principal reservoir may be semen. The virus is transmitted by direct personal contacts of all kinds.

Labial herpes is the most common result of the reactivation of latent virus infection. Fever blisters and cold sores reappear on the face or lips, the vesicular lesions usually crusting in a few days and then healing spontaneously. Reactivated lesions may also involve other areas of skin or mucous membranes elsewhere on the body. This reactivated infection is usually the result of trauma (sunburn is a common cause), hormonal or other physiologic changes (e.g., menstruation), or concurrent infectious disease (respiratory infection, bacterial meningitis, malaria). Occasionally, reactivation may lead to serious CNS disease, such as meningoencephalitis, but the latter occurs most often as a primary herpetic infection acquired through direct contact with the virus in the saliva of carriers.

Health care workers and others who handle infants with skin lesions or patients with chronic eczematoid disease may be infected through direct contact. Extreme care should be used in such nursing care, with strict observance of contact precautions (see table 11.1).

Older patients with herpetic lesions should avoid contact with newborn infants, burn patients, or others with chronic skin diseases. Patients on immunosuppressive therapy or those with immunodeficiencies must also be protected from the risk of generalized herpetic disease.

The antiviral agent acyclovir, when administered intravenously, is effective in lowering mortality in patients with neonatal HSV and herpes encephalitis. Topical preparations of the drug trifluridine are helpful for HSV keratitis but not for cutaneous lesions. Acyclovir in oral or topical form is effective for treating primary, but not recurrent, genital herpes and also disseminated and central nervous system HSV infections. In addition, when administered prophylactically to immunocompromised patients who are at risk for recurrent infections, acyclovir appears to prevent these recurrences. There is concern, however, that with continuous prophylactic use, strains resistant to this agent may emerge. Preliminary HSV vaccine trials in animals have shown promise, but their use in humans is several years away.

SUPERFICIAL MYCOSES

Ringworm Infections

Superficial mycoses of the hair, skin, or nails are referred to collectively as **ringworm** infections. The clinical term for ringworm is **tinea** (derived from the Latin word for "worm"). In medical usage, this term is modified by another Latin word indicating the part of the body affected (e.g., *tinea capitis,* ringworm of the scalp; *tinea corporis,* ringworm of the body; *tinea pedis,* ringworm of the feet; *tinea unguium,* ringworm of the nails). Many other more specific anatomic terms may be used for fungous infections of particular areas of the body, such as *tinea barbae* (the bearded area of face and neck are involved) or *tinea cruris* (inguinal folds are infected).

These "dermatophytoses" are caused by members of a group of fungi called *dermatophytes* because they can invade skin or its keratinized appendages but do not infect systemic tissues. Three important genera of dermatophytes (***Epidermophyton, Microsporum,*** and ***Trichophyton***) are described briefly in the section on Laboratory Diagnosis. A particular species of dermatophyte may cause a variety of clinical lesions in different areas of the body, and conversely, different fungi may cause similar clinical manifestations.

Epidemiologically these infections have much in common. Some of the causative fungi have a reservoir in both humans and animals, some in humans only, but all are directly transmissible from person to person, unlike the systemic fungi (see chap. 14). The route of transmission may be through direct contact with infected persons (or animals) or from indirect sources of environmental contamination derived from active cases. Methods for control and treatment of these infections are similar and are described in outline form in table 19.3. The following is a brief description of the four major clinical forms of dermatophytosis and the causative fungi.

1. Tinea Capitis (Ringworm of the Scalp)

This infection usually begins with the appearance of a small, scaling papule which is red and itchy. Numerous papules may appear and spread peripherally. The hair of the affected area becomes brittle and is easily broken off leaving patches of baldness (alopecia) (fig. 19.13). Sometimes the underlying tissue becomes inflamed and ulcerated; this type of lesion is called a *kerion.* The hair shafts themselves may be infected with fungus, or fungal growth may form a sheath around

TABLE 19.3 EPIDEMIOLOGY OF SUPERFICIAL MYCOSES

Disease	Causative Fungus	Reservoir	Sources	Transmission	Incubation Period
Tinea capitis	*Microsporum* sp. *Trichophyton* sp.	Humans and animals	Lesions Combs Toilet articles Barber's tools Headrests of upholstered seats in theaters, trains, etc.	Direct or indirect contacts	10 to 14 days
Tinea favosa	*T. schoenleinii*	Humans	Crusts from scutula	Direct or indirect contacts (favored by crowding and filth)	10 to 14 days
Tinea corporis	*Epidermophyton* *Microsporum* sp. *Trichophyton* sp.	Humans and animals	Lesions Clothing Floors Shower stalls	Direct or indirect contacts	10 to 14 days
Tinea pedis	*Epidermophyton* *Trichophyton* sp.	Humans	Lesions Shoes and socks Floors Shower stalls	Direct or indirect contacts	10 to 14 days
Tinea unguium	*Trichophyton* sp.	Humans	Lesions	Person-to-person transmission unusual but may extend to other nails of infected individual	Source and time of exposure usually unknown

*See text discussion.

the shaft. Some of the *Microsporum* species fluoresce in ultraviolet light, which can be an aid to diagnosis. The suspected area of the scalp is examined under an ultraviolet ("Wood's") lamp, and fluorescing hairs are removed with forceps for laboratory examination.

Some *Trichophyton* species create a lesion known as "black dot" ringworm, the effect being created by dark broken stumps of hair shafts sticking up in an area of alopecia and scaling. These fungi do not fluoresce under the Wood's lamp.

Tinea favosa is a clinical lesion of the scalp caused by a particular species of *Trichophyton* (*T. schoenleinii*). Cup-shaped yellowish crusts, called *scutula,* are formed. The lesion is deep and may heal by scarring, with permanent destruction of hair follicles.

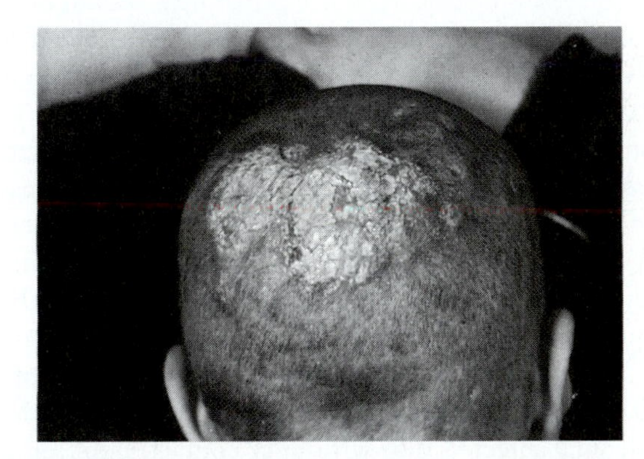

FIGURE 19.13 Tinea capitis, or ringworm of the scalp, often occurs in childhood. This infection is caused by *Microsporum audouinii.*

2. Tinea Corporis (Ringworm of the Body)

This term refers to fungal infections of the glabrous (smooth, nonhairy) skin anywhere on the body. The lesions are characteristically flat, spreading, and ring-shaped (colorplate 7c). The periphery of the lesion is always the most active, being raised, erythematous, sometimes vesicular and weeping, sometimes dry and scaling. Moist lesions may form crusts as they heal. The infection advances peripherally, leaving a central area of healing, normal skin in many instances.

3. Tinea Pedis (Ringworm of the Feet)

"Athlete's foot" is one of the most common and widespread superficial fungous infections. It may often appear as nothing more than a scaling or cracking of the skin, particularly between the toes, but it may become chronic and severe,

Communicable Period	Special Treatment[a]	Precautions	Other Controls
While lesions remain active	Daily shampoo Fungicidal ointment Protect with skullcap At night, pull out infected hairs and reapply ointment Griseofulvin by mouth, especially for *M. audouinii* infections	Keep head covered with cotton cap that can be sterilized frequently by boiling or autoclaving	Investigate household for other infected cases Look for infection among pet animals or farm animal contacts
While lesions remain active	Epilation usually essential Other therapy as above	As above	More common in Africa and the Middle East than in U.S.A.; should be treated before entry
While lesions remain active	Keep skin very clean by soap-and-water bathing Clean away crusts and scabs, if any, and apply fungicidal ointments Griseofulvin or terbinafine by mouth	Infected persons should not frequent public pools and gymnasiums Clothing, especially underclothes, should be disinfected if in contact with lesions	Look for source of infection among human or animal contacts
While lesions remain active	As above Keep feet clean and dry; expose to air as much as possible	Cotton socks or hose should be worn and disinfected between uses, preferably by boiling	Infected persons should not frequent public pools and gymnasiums unless foot covering is worn; the care of such places should include daily disinfectant-cleaning
While lesions remain active but little communicability for others	Nails are kept closely pared and filed Fungicidal soaks and ointments topically Griseofulvin or terbinafine by mouth	As above if toenails are involved When fingernails are infected, gloves should be worn only to maintain contact with fungicides; otherwise nails should be kept clean and dry	As above if toenails are involved

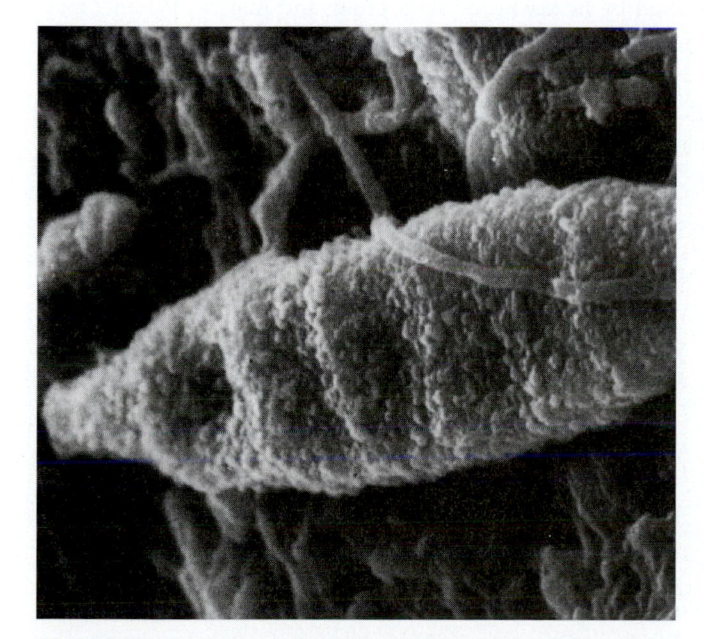

FIGURE 19.14 Scanning electron micrograph of *Microsporum gypseum* macroconidium (×800). The macroconidium is lying among hyphae of the mycelial mat. Macroconidia of this genus are spindle-shaped, septate, and have tough, thick walls.

William G. Barnes, Ph.D., John D. Arnold, M. D., and Arthur Berger, Department of Pathology, General Hospital and Medical Center and Harry S. Truman Research Laboratory, Kansas City, MO.

with vesicle formation and maceration of tissues. Scaling may extend over the heels and soles of the feet, sometimes with vesicular eruptions or acute eczematoid reactions. The tissues of the sole may be undermined by ulcerations, with resulting cellulitis, lymphangitis, and systemic reaction. This form of tinea pedis can be chronically disabling. The hands may also become infected, and allergic skin reactions to fungal products may appear on many parts of the body, but particularly the arms and legs.

4. *Tinea Unguium (Ringworm of the Nails)*

Nails of the hands or feet may become involved in chronic infections caused by dermatophytic fungi. (Infections with *Candida albicans* are also common but have a somewhat different epidemiology, being of endogenous origin.) The affected nails become discolored, thickened, brittle, and often deeply grooved or pitted. Caseous epidermal debris accumulates under the nail, and the top may separate. Eventually the whole nail may be destroyed if the fungus invades the entire plate. Unlike *Candida* or staphylococcal infections, there is usually no paronychial involvement (see colorplate 7f).

LABORATORY DIAGNOSIS

The characteristic features of the three genera of dermatophytes may be summarized as follows:

Epidermophyton
The colony is velvety and greenish yellow, with radiating furrows. Microscopically, macroconidia identify the fungus. These are large, club-shaped, septate, smooth bodies usually borne in clusters on the hyphae. The only important species is *E. floccosum*.

Microsporum
Colonies are velvety, woolly, or powdery, depending on the species. They also vary in color, one of the important species being light gray to brown and the other two, bright orange or yellow. Pigmentation is seen best on the underside of the colonies. The macroconidia are spindle-shaped, septate, thick-walled, and rough-surfaced (fig. 19.14). The three important species are *M. audouinii, M. canis,* and *M. gypseum*.

Trichophyton
There are many species displaying a variety of colony forms and pigments. On microscopic examination, these fungi are seen to produce many microconidia but often few macroconidia. The latter are thin, pencil-shaped bodies, multiseptate, thin-walled, and smooth. Some important species are *T. mentagrophytes, T. rubrum, T. tonsurans, T. violaceum,* and *T. schoenleinii*.

Most of these fungi grow slowly in the laboratory, requiring 2 to 4 weeks for maturation and identification of characteristic reproductive structures. They are strict aerobes that grow best at room temperature (22 to 28° C) rather than at body temperature.

Specimens for Diagnosis
Scrapings of involved areas of skin, vesicular fluid, exudates, or hairs may be submitted for microscopic examination and culture. In general, it is best to take material from the active margins of advancing lesions, scraping deeply under the lip of crateriform lesions or into the inflamed base of vesicles or papules. Hairs and skin scrapings may be examined under the microscope directly, using a 10% potassium hydroxide solution to clear cellular debris. When infected hair is examined, fungal conidia (spores) may be seen invading the shaft itself (endothrix) or clinging to its external surfaces (ectothrix). Skin scrapings usually reveal branching hyphal fragments without conidia.

Calcofluor white, a fluorescent stain that stains fungal cell walls, may also be used to visualize organisms directly in specimen material. When the smear is observed microscopically under an ultraviolet light source, bright, fluorescing hyphae are easily seen against a black background. Hair and skin scrapings materials may be satisfactorily cultured on a variety of simple media, the most common of which is Sabouraud's dextrose agar, usually containing antimicrobial agents to inhibit growth of contaminating bacteria.

Treatment

Superficial mycotic infections are treated locally with a variety of fungistatic or fungicidal solutions, ointments, lotions, or dusting powders. Some of the imidazole compounds have been found clinically useful in topical applications. These include miconazole, ketoconazole, econazole, and clotrimazole. In clinical use, the imidazoles are active against superficial *Candida albicans* infections as well as the dermatophytes. A newer agent, terbinafine, may be more effective than the imidazoles, especially in tinea pedis. Systemic therapy of dermatophyte infections is presently limited to oral terbinafine and griseofulvin.

VIRAL HEMORRHAGIC FEVER AND EBOLA VIRUS

Viral hemorrhagic fever (VHF) is an extremely uncommon yet serious human infectious disease associated with high mortality rates. Several viruses cause VHF including **Ebola virus,** Marburg virus, Lassa virus, and Crimean-Congo virus. Little is known about the natural reservoirs for these viruses but epidemics of VHF have been documented in various parts of the world, with most outbreaks occurring in Africa. In April of 1995, an outbreak of VHF caused by Ebola virus was documented in Zaire, Africa. A total of 296 persons developed symptomatic infection of whom 79% died. Some of the important aspects of this Ebola VHF outbreak follow.

Clinical Disease

The usual incubation period of Ebola VHF is 5 to 10 days but may range from 2 to 21 days. Symptomatic disease is characterized by an abrupt onset of fever often accompanied by headache, myalgia, chills, and malaise. Patients may also have severe abdominal pain, diarrhea, nausea, and vomiting. Within 5 to 7 days of onset of symptoms, most patients develop a maculopapular rash that is most prominent on the trunk. In the second week of illness, the patient either becomes afebrile and improves remarkably or dies from shock with multiorgan system dysfunction and disseminated intravascular coagulation. The mortality rates associated with documented outbreaks of Ebola VHF have ranged from 50 to 90%.

LABORATORY DIAGNOSIS

The laboratory diagnosis of Ebola virus infection is established by recovery of the organism in culture of appropriate specimens (e.g, blood, tissue). The specimens should be collected during the acute stage of disease whenever possible. Alternatively, an enzyme immunoassay to detect Ebola virus antigen directly in clinical specimens has been used successfully. An immunofluorescence assay is available to screen for Ebola virus antibodies in acute and convalescent serum specimens. However, false-positive and nonreproducible results are often obtained with this assay and serologic testing is considered unreliable for diagnosis.

Epidemiology

The natural reservior for Ebola virus is unknown. In cases of epidemics, the outbreak can usually be traced back to the index case (i.e., first known patient) but no farther. Monkeys, bats, ticks, and spiders have been incriminated as possible reservoirs of infection but the virus has never been recovered from any of these sources. Person-to-person transmission occurs by contact with blood or other infected body fluids that contain virus, including vomitus, urine, and stool. As such, health-care providers are at increased risk of acquiring disease by coming in contact with infectious secretions of hospitalized patients. In Africa, disease transmission has also been associated with the reuse of nonsterile needles and syringes. Airborne transmission of disease has never been documented in humans and is considered a rare possibility if patients are in the advanced stages of disease.

Control and Prevention

No antiviral therapy or vaccine is available to treat or prevent Ebola virus infection. Infected patients are treated with supportive care and careful management of fluid replacement. Disease transmission can be controlled effectively or prevented by avoiding exposure to infected body fluid secretions. Caregivers should follow the guidelines of contact precautions and wear protective eyewear, masks, gowns, and gloves when attending to infected patients (see chap. 11).

QUESTIONS

1. How are sexually transmitted diseases controlled?
2. What is congenital syphilis? How is the infant affected?
3. Briefly describe the three stages through which untreated, sexually transmitted syphilis progresses. Identify the communicable period in each stage.
4. Why is syphilis called the "great imitator"?
5. What tests are used for screening a patient's serum for syphilis?
6. What is the Jarisch-Herxheimer reaction?
7. What is the communicable period for gonorrhea?
8. What are the clinical consequences of gonorrhea in the male? in the female? Why is gonorrhea called the "great sterilizer"?
9. How are infants infected with *Neisseria gonorrhoeae?* How are young children infected?
10. What is the treatment for gonorrhea? Why has the treatment changed?
11. How can gonorrhea be prevented and controlled?
12. What is the most common cause of nongonococcal urethritis (NGU)?
13. What are the major complications of NGU in women and infants?
14. What is lymphogranuloma venereum? chancroid? granuloma inguinale?
15. List the congenital and neonatal diseases resulting from sexually transmitted diseases.
16. List three eye infections and the etiology of each.
17. What precautions are required when caring for patients with eye infections?
18. What conditions predispose to generalized staphylococcal infections?
19. What bacterial product produces symptoms of staphylococcal toxic shock syndrome (TSS)?
20. How can women reduce their risk of TSS?
21. What is the significance of bacteriophage typing in staphylococcal infections?
22. Why is it difficult to treat staphylococcal infections?
23. How does disease produced by "flesh-eating bacteria" resemble TSS?
24. What are the three clinical forms of Hansen disease? How do they differ?
25. How is Hansen disease treated?
26. What is AIDS? How is the disease transmitted?
27. What laboratory test is used for AIDS diagnosis?
28. What is the incubation period for AIDS?
29. Distinguish between the terms HIV positive, AIDS-related complex (ARC), and AIDS.
30. Why are the morbidity and mortality rates from HIV infections so high?
31. Why is AIDS a pandemic disease?
32. What methods are used to control AIDS?
33. What precautions are required when caring for a patient with AIDS?
34. What do the dermatophytoses have in common?
35. How are the superficial mycotic infections treated?
36. How is Ebola virus disease transmitted?

BIBLIOGRAPHY

Abel, L. J. 1994. *Chlamydia trachomatis* Prevalence and risk factors in a family planning setting. *Nurs. Pract.* 19 (8): 30–32.

AIDS: A special threat to women. 1990. *World Health* (whole issue), November-December.

Alter, M. J., P. J. Coleman, W. J. Alexander, et al. 1989. The importance of heterosexual activity in the transmission of hepatitis B and non-A, non-B hepatitis. *JAMA* 262 (9): 1201–1205.

Aral, S. O., and K. H. King. 1991. Sexually transmitted diseases in the AIDS era. *Sci. Am.* 264 (2): 62–91.

Balows, A., W. J. Hausler, and E. Lennette, eds. 1988. Laboratory diagnosis of infectious diseases: Principles and practice, vol. 1. *Bacterial, mycotic and parasitic diseases.* New York: Springer-Verlag.

SUMMARY TABLE 19.4 INFECTIOUS DISEASES ACQUIRED THROUGH SKIN AND MUCOSA AND TRANSMITTED BY HUMAN CONTACT

Clinical Disease	Causative Organism	Other Possible Entry Routes	Incubation Period	Communicable Period
I. Sexually Transmitted Diseases				
Syphilis	*Treponema pallidum*	Parenteral Congenital	10–21 days	Probably communicable during 2-4 years of primary and secondary stages and during first years of latency if untreated
Gonorrhea	*Neisseria gonorrhoeae*	Conjunctiva (newborn or adults) Chorioamnion	3–9 days; in ophthalmia neonatorum 1–5 days	In adults, asymptomatic disease communicable for years if not treated; conjunctival membranes are infectious until discharges cease or therapy initiated
Chlamydial (nongonococcal) urethritis	*Chlamydia trachomatis*		1–2 weeks after sexual contact	While infection persists untreated
Lymphogranuloma venereum	*Chlamydia trachomatis*		1–4 weeks	Untreated LGV may be communicable for months or years
Chancroid	*Haemophilus ducreyi*		3–5 days	Communicable through infectious discharges of genital lesions and buboes
Granuloma inguinale	*Calymmatobacterium granulomatis*		Unknown	While infective agent is found in lesions
Acquired immunodeficiency syndrome (AIDS)	Human immunodeficiency virus	Blood Congenital	Variable	Throughout infection

Beares, N. D., J. C. Robinson, L. Timpson, and J. M. Keller. 1990. *Chlamydia trachomatis* in a high-risk population. *J. Am. Acad. Physician Assist.* 3 (8): 614–617.

Benenson, A. S., ed. 1995. *Control of communicable diseases in man.* 16th ed. Washington, D.C.: American Public Health Association.

Black, D. 1988. *The plague years: A chronicle of AIDS, epidemic of our times.* New York: Simon and Schuster.

Brandt, A. M. 1987. *No magic bullet: A social history of venereal disease in the United States since 1880.* Oxford: Oxford University Press.

Bryceson, A., and R. E. Pfaltzgraff. 1990. *Leprosy.* New York: Churchill Livingston.

Caldwell, J. C., and P. Caldwell. 1996. The African AIDS epidemic. *Sci. Am.* 274 (3): 62–68.

Cates, W. Jr., and D. E. Toomey. 1990. Sexually transmitted diseases: Overview of the situation. *Prim. Care* 17 (1): 7–27

Centers for Disease Control and Prevention. 1989. Guidelines for prevention of transmission of human immunodeficiency virus and hepatitis B virus to health care and public safety workers. *Morbid. Mortal. Weekly Report* vol. 38, no. S-6.

Centers for Disease Control and Prevention. 1990. Reduced incidence of menstrual toxic shock syndrome—United States 1980-1990. *Morbid. Mortal. Weekly Report* 39 (25): 421–423.

Specimens Required	Laboratory Diagnosis	Immunization	Antimicrobial Treatment	Transmission-Based Precautions
Tissue fluid aspirations Gummatous lesion biopsy Blood serum	Dark-field or direct fluorescent antibody examination for spirochetes Silver impregnation demonstrates spirochete Positive for reagin (VDRL, RPR) Positive for treponemal antibodies (FTA-ABS, MHA-TP)	None	Penicillin Doxycycline Erythromycin	Private room not necessary (antimicrobial therapy eliminates threat of communicability in 24 hours); standard precautions
Exudate from lesions Blood, joint fluid in DGI	Smear, males Culture, females and smear-negative males Enzyme immunoassay DNA probe Culture	None	Ceftriaxone + doxycycline Ciprofloxacin or spectinomycin for allergic patients	Standard precautions
Urethral, cervical scrapings	Cell culture Direct fluorescent antibody DNA probe Enzyme immunoassay PCR	None	Doxycycline Azithromycin Erythromycin	If hospitalized, standard precautions
Exudate from lesions Tissue biopsy Blood serum	Smear and isolation in embryonated eggs or cell culture Rising titer of complement-fixing antibodies	None	Doxycycline Erythromycin	Standard precautions
Exudate from lesions	Smear and culture	None	Erythromycin Ceftriaxone	Same as above
Biopsy of lesions	Isolation of organism in chick embryos	None	Tetracyclines Trimethoprim-sulfamethoxazole Chloramphenicol	Same as above
Blood Blood serum	Cell culture Enzyme immunoassay Immunoblots		Reverse transcriptase inhibitors Protease inhibitors	Same as above

Centers for Disease Control and Prevention. 1991. The HIV/AIDS epidemic: The first 10 years. *Morbid. Mortal. Weekly Report* 40 (22): 357–369.

Centers for Disease Control and Prevention. 1995. Syringe exchange programs—United States, 1994-1995. *Morbid. Mortal. Weekly Report* 44 (37): 700.

Centers for Disease Control and Prevention. 1995. Update: Acquired immunodeficiency syndrome—United States 1994. *Morbid. Mortal. Weekly Report* 44 (4): 64.

Centers for Disease Control and Prevention. 1995. U. S. Public Health Service recommendations for human immunodeficiency virus counseling and voluntary testing of pregnant women. 44: RR-7.

Centers for Disease Control and Prevention. 1996. Update: Provisional public health service recommendations for chemoprophylaxis after occupational exposure to HIV. 45 (22): 468–472.

Cuzzell, J. Z. 1990. Clues: Pain burning and itching. *Am. J. Nurs.* 90 (7): 15–16.

Cuzzell, J. Z. 1990. Clues: Recurrent punched-out lesions. *Am. J. Nurs.* 90 (5): 21–22.

Dawkins, B. J. 1990. Genital herpes. *Prim. Care.* 17 (1): 95–113.

DesJarlais, D. C., and S. R. Friedman. 1994. AIDS and the use of injected drugs. *Sci. Am.* 270 (2): 82–88.

Dirubbo, N. E. 1987. The condom barrier. *Am. J. Nurs.* 87 (10): 1306–1307.

SUMMARY TABLE 19.4 *Continued*

Clinical Disease	Causative Organism	Other Possible Entry Routes	Incubation Period	Communicable Period
II. Eye Diseases				
Acute bacterial conjunctivitis	Streptococci Staphylococci Pneumococci *Haemophilus* species Gonococci	Respiratory	Short, 24–72 hours	During active infection
Trachoma	*Chlamydia trachomatis*		About 5–12 days	During active infection
Inclusion conjunctivitis (newborn)	*Chlamydia trachomatis*		Unknown	During active infection
Keratoconjunctivitis	Adenovirus (types 8 and 19) Herpes simplex virus (type 1)	Respiratory	Unknown	During active infection
Hemorrhagic conjunctivitis	Enterovirus (type 70)			
III. Skin Contact Diseases				
Staphylococcal infections (includes scalded skin syndrome, toxic shock syndrome)	*Staphylococcus aureus*	Respiratory Parenteral	Variable; in surgical wounds 1–2 days; impetigo, boils, 4–10 days	During time of purulent discharge of open lesion and as long as nasal carrier state persists
Erysipelas *Impetigo*	Beta-hemolytic streptococci, group A	Respiratory	Unknown, probably about 2 days	During active infection, no longer communicable 24 hours after institution of antimicrobial therapy
Hansen disease (leprosy)	*Mycobacterium leprae*	Respiratory	3–5 years	During time bacilli can be demonstrated in skin or mucosal lesions or in respiratory secretions, or for 3 months after initiation of therapy
Herpes simplex infection	Herpes simplex virus, type 1 and type 2	Respiratory	Up to 2 weeks	During active infection
Viral hemorrhagic fever	Ebola, Marburg, Lassa, Crimean-Congo viruses	?Respiratory	5-10 days	During active infection

Editorial. 1995. The evolving epidemiology of syphilis. *Am. J. Public Health* 85 (8): 1053.

Facts about Hansen's Disease. 1990. *The Star* 50: 18.

Fee, E., and D. Fox, eds. 1988. *AIDS: The burdens of history.* Berkeley, Calif: University of California Press.3

Fineberg, H.V. 1988. The social dimension of AIDS. *Sci. Am.* 259 (4): 128–134.

Grady, C. 1988. HIV: Epidemiology, immunopathogenesis, and clinical consequences. *Nurs. Clin. N. Am.* 23 (4): 683–696.

Specimens Required	Laboratory Diagnosis	Immunization	Agents for Treatment	Transmission-Based Precautions
Exudate or discharge from eye	Smear and culture	None	Antimicrobial	Standard precautions ophthalmic ointment
Epithelial cells scraped gently from eyelids Blood serum	Smear for cytoplasmic inclusion bodies Cell culture, eggs Rising titer of antibodies	None	Tetracycline Sulfonamides	Standard precautions
Epithelial cells scraped gently from eyelids	Smear for cytoplasmic inclusion bodies Cell culture, eggs	None	Erythromycin	Standard precautions; isolate infected babies in nurseries
Eye swab or secretions Blood serum	Cell culture Rising titer of antibodies	None	Nonspecific except trifluridine for herpes	Contact precautions; careful aseptic techniques in dispensaries and industrial clinics
Exudate from lesions Blood Pleural fluid Abdominal fluid Synovial fluid	Smear and culture Antibiograms Phage typing		Antimicrobial agents chosen on basis of susceptibility	Contact precautions; stringent aseptic techniques; concurrent disinfection of all articles possibly contaminated by infectious discharges; gown and *hand washing* with disinfectant soap; terminal disinfection; standard precautions for toxic shock syndrome
Scrapings from raised peripheral border of lesions Exudate Blood	Smear and culture Smear and culture Culture	None	Penicillin Erythromycin	Contact precautions; concurrent disinfection of discharges and dressings from lesions; thorough terminal disinfection
Scrapings or biopsies of skin lesions Skin tests	Acid-fast smears Lepromin test prognostic; + in tuberculoid form only	Vaccine in development	Sulfones + rifampin, clofazimine, or ethionamide	Standard precautions; concurrent disinfection of respiratory discharges; terminal cleaning
Exudates from lesions: HSV 1 (nongenital lesions) HSV 2 (genital and neonatal lesions) Spinal fluid Blood serum	Virus isolation Characteristic inclusions in stained cells from lesions Virus isolation Rising titer of antibodies	None	Trifluridine for conjunctivitis Acyclovir for genital herpes	Contact precautions
Exudates from lesions Blood Tissue	Virus isolation Antigen detection	None	None	Contact precautions

Gurevich, I. 1990. Counseling the patient with herpes. *RN* 53 (2): 22–28.

Hastings, R. C., D.V. Opremolla, eds. 1994. Leprosy. In *Medicine in the tropics.* New York: Churchill Livingstone.

Henig, R. M. 1995. The lessons of syphilis in the age of AIDS. *Civilization* 2 (6): 36–43.

Jensen, S. C., and S. Eaton. 1985. The herpes simplex viruses: The facts health professionals need to know. *Radiol. Technol.* 56 (5): 302–305.

Larson, E., and L. Bobo. 1992. Effective hand degerming in the presence of blood. *J. Emerg. Med.* 10 (1): 7–11.

Lennette, E. H., P. Halonen, and F. A. Murphy, eds. 1988. Laboratory diagnosis of infectious diseases: Principles and practice. *Viral, Rickettsial and Chlamydial Diseases.* Vol. II. New York: Springer-Verlag.

Louks, A. 1987. Chlamydia: The unheralded epidemic. *Am. J. Nurs.* 87 (7): 920–922.

Mandell, E. G., J. E. Bennett, and R. Dolin. 1995. *Principles and practice of infectious diseases,* 4th ed. Philadelphia: W. B. Saunders.

Morita, M. M. 1993. Methicillin-resistant *Staphylococcus aureus.* Past, present and future. *Nurs. Clin. N. Am.* 28 (3): 625–637.

Nettina, S. L. 1990. Syphilis, a new look at an old killer. *Am. J. Nurs.* 90 (4): 68–70.

Nowak, M. A., and A. McMichael. 1995. How HIV defeats the immune system. *Sci. Am.* 273 (2): 58–65.

Preston, R. 1994. *The Hot Zone.* N.Y.: Random House.

Reddy, R. E. 1988. Genital herpes. *Physician Assist.* 12 (7): 21–23, 27–29, 31–34.

Rolfs, R. T., M. Goldberg, and R. G. Sharrar. 1990. Risk factors for syphilis: Cocaine use and prostitution. *Am. J. Public Health.* 80 (7): 853–857.

Roman, A., and K. H. Fife. 1989. Human papillomaviruses: Are we ready to type. *Clin. Microbiol. Rev.* 2 (2): 166–190.

Roper, W. L. 1991. Current approaches to prevention of HIV infection. *Public Health Reports* 106 (2): 111–115.

Rosebury, T. 1973. *Microbes and morals: The strange story of venereal disease.* New York. Ballantine.

Sheahan, S. L., and J. P. Seabolt. 1989. Chlamydia trachomatis infections: A health problem of infants. *J. Pediatr. Health Care.* 3 (3): 144–149.

Solomon, M. Z., and W. DeJong. 1989. Preventing AIDS and other STDs through condom promotion: A patient education intervention. *Am. J. Public Health.* 79 (4): 453–458.

Spencer, L. 1990. What's wrong with this patient? *RN* 53 (4): 41–43.

Spiegel, C. 1991. Bacterial vaginosis. *Clin. Microbiol. Rev.* 4 (4): 485–502.

Stevens, D. 1995. Streptococcal toxic-shock syndrome. *Emerging Infect. Dis.* 1: 69–78.

Swanson, J. M., and W. C. Chenitz. 1990. Psychosocial aspects of genital herpes: A review of the literature. *Public Health Nurs.* 7 (2): 96–104.

Temkin, E. 1994. Turn-of-the-century nursing perspectives on venereal disease. *Image* 26 (3): 207–211.

Todd, J. K. 1988. Toxic shock syndrome. *Clin. Microbiol. Rev.* 1 (4): 432–446.

Touchstone, D. M., and D. D. Davis. 1992. Consider chlamydia. *RN* 55: 32–36.

Trees, D. L. and S. A. Morse. 1995. Chancroid and *Haemophilus ducreyi:* An update. *Clin. Microbiol. Rev.* 8 (3): 357–375.

U.S. paying high price for HIV and TB epidemics. 1995. *AIDS Alert* 10 (3): 13–14.

Valenti, W. M. 1994. AIDS: Problem solving in infection control—compliance with universal precautions: Getting to yes. *Am. J. Infect. Control* 22: 177–178.

Weiss, R. A. 1993. How does HIV cause AIDS? *Science* 260 (5): 1273–1279.

Weitzman, I., and R. C. Summerbell. 1995. The dermatophytes. *Clin. Microbiol. Rev.* 8 (2): 240–259.

CHAPTER 20

Infectious Diseases Acquired through Skin or Mucosa and Transmitted by Animals or from the Environment

OUTLINE

Microbes in Motion

This chapter covers the important bacterial and parasitic agents that cause infectious diseases transmitted from animal or environmental reservoirs. A discussion of their laboratory diagnosis, epidemiology, control, and prevention is included. The following books and chapters of the Microbes in Motion CD-ROM may be a useful preview or supplemental study aid to your reading: Gram-positive Organisms in Human Hosts: Gram-positive Bacilli. Miscellaneous Bacteria in Human Hosts: Treponemes. Parasitic Structure and Function: Nematodes; Trematodes.

OBJECTIVES

On completing this chapter the student will be able to:

1. cite several bacterial and parasitic diseases with animal or environmental reservoirs.
2. identify the sources of these diseases.
3. describe the causative organisms, laboratory diagnosis, epidemiology, and methods of prevention and control for each disease.

KEY WORDS

Anthrax
Bacillus anthracis (an-**THRACE**-sis)
Francisella tularensis (france-is-**ELL**-ah tool-are-**EN**-sis)
Hookworm
Leptospira interrogans (lep-toe-**SPIRE**-ah in-**TERR**-oh-gans)
Leptospirosis (lep-toe-spire-**OH**-sis)
Necator americanus (knee-**KAY**-tor am-er-i-**KAHN**-us)
Schistosoma (shiss-toe-**SO**-ma)
Schistosomiasis (shiss-toe-so-**MY**-ah-sis)
Strongyloides (strong-ill-**OID**-ease)
Strongyloidiasis (strong-ill-oid-**EYE**-ah-sis)
Tularemia (tool-are-**EEM**-ee-ah)

Introduction

The infective agents described in this chapter have little in common except the epidemiologic importance of their transmission to humans from animals or intermediate sources in water or soil. All have a common route of entry through human skin or mucosal surfaces.

Anthrax and tularemia are zoonoses that may also be transmitted through indirect contact with animal products or through entry routes other than skin and mucosa. For example, these diseases may have respiratory or alimentary routes of entry, and tularemia may also be transmitted by arthropods. The cutaneous route is emphasized here because it is the most direct and most common entry route.

Leptospirosis is acquired most often from environmental sources, although its reservoir is in infected animals. Other diseases such as brucellosis and ringworm (superficial mycoses) may also be acquired through the skin by contacts with infected animals, but alternative entry routes are more important for these infections. Brucellosis is discussed in chapter 16 as an alimentary route infection; ringworm is usually transmissible between infected humans and is discussed in chapter 19.

Hookworm and *Strongyloides* species are intestinal roundworms maintained in humans with intermediate periods of maturation in soil. The schistosomes are trematodes (blood flukes) with a cycle of development in humans or animals (definitive hosts) and in aquatic intermediate hosts. For each of these organisms, the infective form is found in water or soil and can establish human disease by penetrating the skin.

The infections produced by these organisms are usually systemic, having major effects in deep tissues. Although the skin is seldom involved in the disease process (except in anthrax), infections are acquired through small cuts and breaks in skin that go unnoticed, or by larval penetration. The control and prevention of these diseases depend, in part, on avoiding contact with infected animals and protecting humans and the environment (primarily water and soil) from sources of infection.

Diseases with an Animal Reservoir

The infections described in this section pose a special threat to persons with direct occupational or recreational exposure either to certain animals or to environmental sources contaminated with their products. Awareness of these animal sources and knowledge of microbial transmission routes are essential for preventing their related diseases.

Anthrax

The Clinical Disease

The **anthrax** bacillus is an aerobic endospore former that is primarily associated with severe disease of domestic animals, notably sheep, cattle, goats, and horses. The resistant endospore can survive for years in soil and in the hides, hair, and wool of infected animals. Human anthrax is an occupational disease in industrial workers who process animal products that may be contaminated with anthrax endospores. It also is a hazard for farmers and veterinarians who work directly with infected animals. This disease is now infrequent in the United States. Domestic animal herds have been largely freed of the infection, and imported animal products are sterilized either before processing or at final distribution for sale. Some effective controls have been instituted for handling raw products that would be damaged by sterilization, and immunizing agents are now available for protecting exposed workers. Anthrax persists, however, in the arid and semiarid regions of the Middle East, Asia, Africa, and South America.

Human anthrax usually begins as an infection of the skin, with formation of a lesion known as "malignant pustule." The endospores enter the skin, probably through minor abrasions at the point of pressure contacts with the hide or other product being handled. Within a day or so a papule appears at the site of entry, rapidly develops into a vesicle, and progresses to necrosis. This necrotic ulcer, or "eschar," is sometimes underlaid by hard, swollen tissues (colorplate 12c). The encapsulated vegetative bacilli that germinate from the endospores multiply in the wound and produce a toxin whose cell-damaging effect results in the edematous swelling of the surrounding area. The bacilli spread through adjacent lymph channels into the bloodstream, multiplying freely. The overwhelming septicemia that results, if untreated, leads to shock and death in a few days. A similar disaster may follow the inhalation or ingestion of endospores.

Laboratory Diagnosis

I. Identification of the Organism
(Bacillus anthracis).

A. *Microscopic Characteristics (colorplate 12d)* The genus *Bacillus* includes large gram-positive, endospore-forming, aerobic rods. Many species of this genus are saprophytes, such as *B. subtilis* or *B. cereus* which live in soil, dust, vegetation, or water and seldom cause disease in healthy people. ***Bacillus anthracis*** is the only member of the group that possesses properties that are almost always pathogenic for humans and animals. They can often be distinguished from other species of the genus because they are encapsulated and are nonmotile. The vegetative bacilli line up in chains, with individual ends squared off at regular intervals.

B. *Culture Characteristics B. anthracis* grows readily on blood agar incubated at 35° C under aerobic conditions. The colonies are nonhemolytic and dull gray. They sometimes have irregular margins, but some variants are smooth and discrete with a "ground glass" appearance. Short outgrowths sometimes extend from the colony edge, giving a "Medusa head" appearance. Colonies are mucoid and tenacious when touched with an inoculating needle.

To avoid accidental inhalation and contact with highly infectious endospores, all work with suspected cultures of *B. anthracis* should be performed in a biological safety cabinet and gloves should be worn (see fig. 4.6).

Immunofluorescent staining is available in specialized laboratories to identify the bacteria in smears made from cultures or in direct smears of exudate.

II. Identification of Patient's Serum Antibodies

Serologic diagnosis can be made by demonstrating a rising titer of antibodies in the patient's serum using an indirect hemagglutination technique, but this method is not widely available.

III. Specimens for Diagnosis

A. Exudate from the cutaneous lesion is submitted for smear, culture, and immunofluorescent staining.

B. Sputum specimens may reveal the organisms by smear or culture when inhalation anthrax is suspected.

C. Blood cultures should be collected as soon as the nature of the cutaneous lesion or systemic disease is suspected.

D. All specimens must be transported under stringent safety conditions to prevent skin contact or inhalation of endospores.

E. Acute and convalescent serums for serologic diagnosis.

Epidemiology

I. Communicability of Infection

A. *Reservoirs, Sources, and Transmission Routes* The reservoir of anthrax infection is in domestic animals such as cattle, sheep, horses, pigs, or goats. Infected animals die of anthrax, but their tissues may harbor endospores for very long periods. Pasture soil also becomes contaminated and may remain so for years, with infection being spread among animals, from carcasses or soil, by biting flies or other insects, and by vultures. No effective method is known for disinfecting the soil in areas where animal anthrax has been prevalent, and this environmental contamination remains a possible source of both human and animal infection. Anthrax may be transmitted to human beings from a great variety of animal products, including bone meal, shaving-brush bristles, hair or wool used in textile industries, and hides processed for leather goods production. Endospores from such sources may be inhaled or ingested (in contaminated meat) or enter through the skin.

The popularity of handicrafts and the importation of such items as goatskin drums, alpaca wool, and goat's hair have led to a few recent cases in the United States, each beginning as a cutaneous infection. One death resulted from dissemination of the disease before it was recognized clinically.

B. *Incubation Period* Usually 2 days but up to a week.

C. *Communicable Period* Anthrax has not been reported as transmissible between human beings, but the possibility of transfer of pulmonary infection exists. The infectivity of animal or soil sources may persist for years.

D. *Immunity* Natural resistance to anthrax varies among animal species. Humans are thought to be moderately resistant to infection. The annual world total of anthrax is thought to be between 20,000 and 100,000 cases, but in the United States only sporadic cases, associated with occupational hazards, are seen.

II. Control of Active Infection

A. *Transmission-based Precautions* Hospitalized patients with cutaneous lesions are placed under standard precautions (see chap. 11). Cases of inhalation anthrax or patients with disseminated pulmonary infection should be cared for in the same manner. These precautions are advisable to prevent contamination of the environment with resistant endospores. They should be continued until skin lesions or sputum are bacteriologically negative.

B. *Other Precautions* The organisms multiplying in human tissues are in the vegetative state. When discharged to the environment in exudates from cutaneous lesions or in sputum, they convert rapidly to endospores, which can be destroyed only by incineration or sterilization by steam or gas (ETO). Contaminated dressings should be placed in impervious bags and burned. Visibly soiled bed linens should be autoclaved before laundering. When soiling is likely, gowns should be worn to prevent contamination of the clothing of those who come in close contact with the patient, and hand-washing facilities must be provided in the unit. Terminal disinfection-cleaning of the patient area is also important.

C. *Treatment* Because anthrax can develop quickly into a fulminating septicemia, early diagnosis and treatment are essential. The drug of choice is penicillin, but erythromycin and tetracycline may also be used. The mortality rate in treated anthrax is essentially zero, as compared with a 5 to 20% fatality among untreated cutaneous cases or an even higher percentage among systemic cases.

D. *Control of Contacts* A case report of anthrax leads to an immediate search for the source of infection and the prompt diagnosis and treatment of any other human cases exposed to and infected by the same source. Outbreaks are associated either with industrial workers or with farmers directly involved in the raising, handling, or slaughter of animals. Handicrafters and hobbyists working with animal leathers, hairs, and wools may be at risk if these items are acquired from areas where infection is endemic in soil or animals.

III. Prevention

A. *Immunization* A vaccine is available for the active immunization of animals and persons exposed to high occupational risks. The material is a cell-free antigen obtained from an anthrax culture filtrate.

B. *Control* The control of natural reservoirs of anthrax requires special precautions with dead or dying animals to prevent contamination of others or the environment. Carcasses must be cremated or buried in deep lime pits. In areas where anthrax is prevalent, healthy animals and their human handlers should be vaccinated annually.

Animal products for commercial processing are sterilized insofar as is practical. Bone meal is autoclaved before incorporation into animal feeds, and fibers used in brushes are sterilized. Hair and wool can be washed with soap and exposed to formaldehyde. Safety measures for industrial workers include the provision of protective clothing and gloves, as well as immunization programs. Dust control in high-risk industrial areas may be necessary.

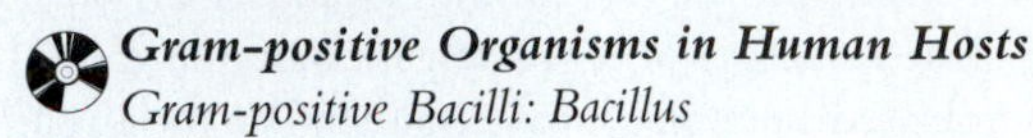

Gram-positive Organisms in Human Hosts
Gram-positive Bacilli: Bacillus

TULAREMIA

The Clinical Disease

Tularemia is an infectious disease of wild animals; hares and rabbits are among the chief sources of human disease. The organism may enter through the skin or mucous membranes, including the conjunctivae. Entry may also occur by respiratory, gastrointestinal, or parenteral routes if the organism is injected by the bite of infective insect vectors, primarily ticks.

In some instances infection is limited to tissues at the site of entry, with formation of a local ulcer and involvement of regional lymph glands. Cutaneous infection is frequent among those who handle or skin infected animals. The lesion takes an "ulceroglandular" form, characterized by regional lymphadenopathy and an ulcer at the entry site, usually on the hands, arms, or face. Primary ocular infection may also occur, accompanied by formation of papules on the eyelid, conjunctivitis, and swelling of adjacent lymphoid tissue; this is the so-called "oculoglandular" form of the disease.

Primary tularemic pneumonia may result from inhalation of the bacilli. When they are ingested, necrotizing lesions may be formed in the mouth, pharynx, or gastrointestinal mucosa. Submaxillary, cervical, or mesenteric lymph nodes may be involved, depending on the site of colonization. Occasionally, tularemia may begin without localizing signs, developing as a febrile systemic infection.

Whatever the initial site of infection, enlargement of regional lymph nodes may lead to their suppuration, dissemination of the organisms to various tissues of the body, and the formation of granulomatous nodules in systemic foci. These nodules may also break down, and the organisms may be distributed still further in a progressive disease pattern. Fatal septicemia and death may ensue without early diagnosis and treatment.

LABORATORY DIAGNOSIS

I. Identification of the Organism

(Francisella tularensis).

A. *Microscopic Characteristics* The organism is a short, gram-negative bacillus, nonmotile, nonendospore-forming, aerobic or microaerophilic. Morphologically it is similar to *Yersinia pestis,* the agent of bubonic plague (see chap. 22), with which it was formerly classified in the genus *Pasteurella.*

B. *Culture Characteristics* ***Francisella tularensis*** is difficult to grow in laboratory media. It requires blood or tissue enrichments and added cystine. Incubation at 35 to 37°C for 2 to 3 days yields minute colonies, appearing as transparent drops on the surface of blood agar plates.

The organism is usually identified and distinguished from *Pasteurella* and *Yersinia* species by agglutination reactions of the isolate with specific antiserum.

II. Identification of Patient's Serum Antibodies

Demonstration of a rising titer of agglutinating antibody is essential to serologic diagnosis of this disease, especially in persons who have been previously exposed and have a persisting titer of residual antibody.

III. Specimens for Diagnosis

Exudates from cutaneous or mucosal ulcers, material aspirated from suppurating lymph nodes, blood, or sputum may be appropriate for culture, as indicated by clinical signs and symptoms. These materials should be collected, transported, and handled with strict aseptic technique and full awareness of the infectivity of these organisms. Laboratory infections have been common, particularly in research settings where experimental animals are used in virulence studies. Therefore, all specimens and cultures in the laboratory should be handled only in a biological safety cabinet.

Preliminary reports on stained smears made from these specimens (except blood samples) may sometimes tentatively confirm the clinical diagnosis. For specific identification of *F. tularensis* directly from patient specimens, examination is best made by fluorescent antibody techniques. In subacute or chronic infections the organisms may be extremely difficult to see in smears or to propagate in culture. Serologic diagnosis with acute and convalescent serum specimens may be most useful in such cases.

Epidemiology

I. Communicability of Infection

In nature, tularemia is transmitted among animal hosts by several biting arthropods including deerflies, wood ticks, and rabbit ticks, and these insects may also transmit the disease to human beings. Humans are an accidental host for the infectious organism and do not transmit it to others. The incubation period following exposure to or ingestion of infected animal meat (or contaminated drinking water) is usually about 3 days, with a possible range of from 2 to 10 days. Animal meats may remain infective, even though frozen, for periods up to 3 years.

II. Control and Prevention

There is no need to isolate patients with active infection, as tularemia is not a communicable disease. However, stan-

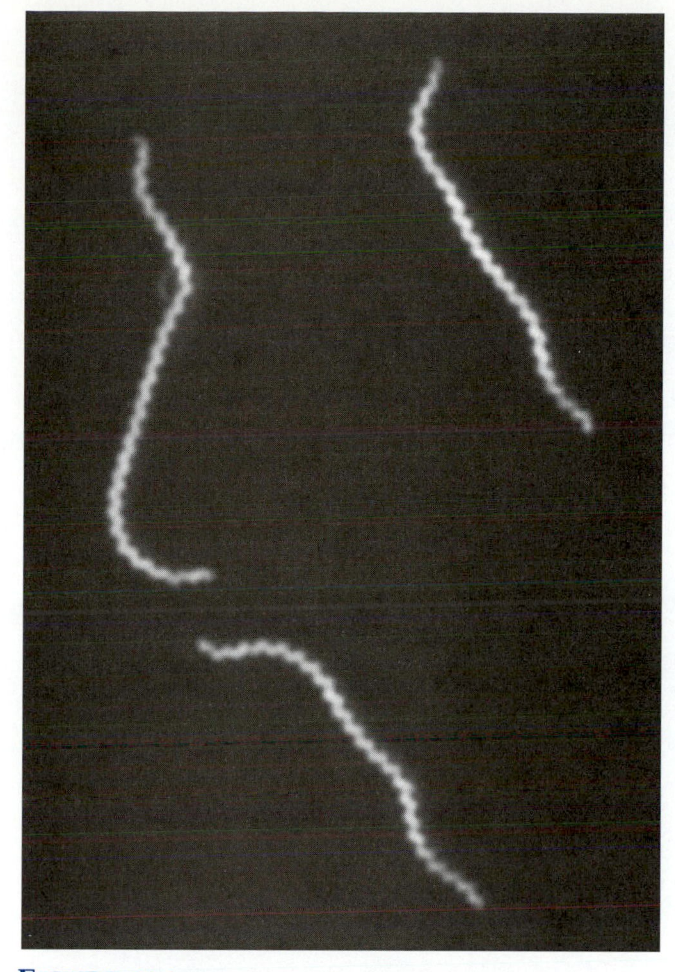

FIGURE 20.1 Dark-field photomicrograph of leptospires (×1,600)

Drs. J. D. Fulton and D. F. Spooner, London School of Hygiene and Tropical Medicine, London, England.

Miscellaneous Bacteria in Human Hosts
Treponemes: Treponemes Taxonomy; Treponemes Virulence Factors; Treponemes Diseases. Zoonoses: Leptospira

dard precautions should be applied. Mucosal or cutaneous discharges, which can sometimes be teeming with the infecting organism, should be disinfected in order to avoid the accidental transfer of infection. Dressings should be burned, and syringes and needles used to collect blood samples should be handled with particular care and sterilized promptly or incinerated.

Tularemia can be effectively treated with streptomycin, gentamicin, chloramphenicol, or tetracyclines. Cure is more rapid with early treatment, but specific therapy is continued for several days after fever and other symptoms have subsided in order to avoid the establishment of chronic, low-grade, persistent infection.

The source of infection should be sought in each case to prevent further transmission from a lingering reservoir, such as remaining portions of game meat stored in the refrigerator. The sale or shipment of infected animals or meats is a matter of concern to public health authorities who enforce controls on this possible mode of spread. Campaigns are conducted to educate the public, especially hunters, concerning the sources of infection and the techniques of prevention: wearing clothing protective against insect vectors, wearing rubber gloves when dressing wild game, thoroughly cooking such meat, and boiling water obtained in areas where infected animals may have contaminated it.

Immunizing vaccines are available, but their use in the United States is restricted to those who run a high risk of infection, notably laboratory workers who perform research with *F. tularensis.*

LEPTOSPIROSIS

The Clinical Disease

Leptospirosis is an acute infection characterized by fever, chills, headache, malaise, and often by jaundice. In severe disease, there may be hemorrhage, hemolytic anemia, and azotemia. The organisms localize chiefly in the kidneys, producing tissue necrosis and organ dysfunction. The leptospires disseminate through the bloodstream to other body areas such as the liver and central nervous system. Leptospiral meningitis is clinically of the benign aseptic type and may represent an immune complex reaction (see chap. 6) rather than infection. It usually occurs when antibodies, but only a few leptospires, are present in the CSF. The duration of the acute illness is from 1 to 3 weeks, with possible recurrences. Mortality is low, the outcome usually reflecting severity of damage to renal or hepatic tissues.

LABORATORY DIAGNOSIS

I. Identification of the Organism

(Leptospira interrogans). ***Leptospira interrogans*** is pathogenic for humans and includes many different antigenic types (serovars or serotypes), distributed among animal hosts in different parts of the world. They are morphologically indistinguishable but produce clinical disease that differs in degree and severity.

Leptospira are very tightly coiled spirochetes, about the same length as the treponemal agent of syphilis, but their spirals are much finer, being only about 0.1 to 0.2 μm wide (fig. 20.1). Unstained, the organism is too narrow to be seen with the ordinary light microscope but can be visualized by dark-field or phase-contrast microscopy. One or both ends of the organism are often bent into a hook shape. The spirochetes have a very active corkscrew rotation, and when viewed by dark-field microscopy, they sometimes appear to be chains of tiny cocci in rapid undulant motion. Like the treponemes, they do not stain well except with silver stains. Fluorescent antibody methods are also used to visualize the organisms microscopically.

Leptospira can be propagated in a serum-enriched semisolid medium incubated in the dark at 30°C or at room temperature.

II. Identification of Patient's Serum Antibodies

Agglutinating antibodies develop after the first week of leptospiral infection and reach a peak level within 1 month. Both microagglutination and macroagglutination tests are available, but the former method is preferred because it is more sensitive and specific. The antigens used in these tests are made from *Leptospira* organisms grown in culture and then treated with formalin.

III. Specimens for Diagnosis

Fresh blood or urine may be submitted for dark-field examination or for culture, the choice of specimen depending on the phase of disease. Organisms are present in the blood during the first week of infection (bacteremic phase) and in the urine after the first week (leptospiuric phase). Specimens are inoculated into culture media, and growth is usually detectable by dark-field or phase-contrast microscopic examination within 14 days.

Epidemiology

I. Communicability of Infection

A. *Reservoirs, Sources, and Transmission Routes* The reservoir of leptospirosis is in domestic and wild animals, including rats and other rodents. Infected animals excrete large numbers of spirochetes in their urine and may contaminate water used for drinking, swimming, or cultivating crops. People whose occupations may take them into contaminated water include workers in sugar cane and rice fields, sewers and mines, processors in fish plants, and soldiers dug into trenches or foxholes. The disease may also be acquired by people in direct contact with infected animals, such as farmers, veterinarians, and abattoir workers. The route of entry is probably through the skin or conjunctival, oral, or alimentary tract mucosa.

B. *Incubation Period* The average time required for development of symptoms following exposure is 10 days, with a range from 3 to 30 days.

C. *Communicable Period* Leptospirosis is not transmitted directly from person to person. Human beings are an accidental host for this organism.

D. *Immunity* Infection confers a long-lasting protective immunity, but otherwise human susceptibility appears to be universal.

II. Control and Prevention

Isolation is not necessary in the care of these patients. Standard precautions (see chap. 11) should be applied, especially to blood and urine. Treatment is largely supportive, although penicillin and tetracycline may have some therapeutic effect in the early stages of infection. Antimicrobial drugs may not eliminate the organisms.

Control measures consist of a search for the source of infected water, programs for extermination of rats and other rodents, and the use of protective clothing by workers exposed to infection. Contaminated waters should be closed to swimmers, and domestic animals should not be permitted access to such watering places. Vaccines for administration to livestock and dogs are available and contain leptospiral strains prevalent in the local area.

PARASITIC DISEASES WITH AN ENVIRONMENTAL RESERVOIR

The nature and epidemiology of the parasitic diseases are discussed at some length in chapter 18, together with a review of the major types of life cycles displayed by the animal parasites (see also chap. 8). Refer to these chapters for general discussions of each of the parasitic diseases outlined below.

HOOKWORM DISEASE

Organisms

Necator americanus (American **hookworm**), *Ancylostoma duodenale* ("Old World" hookworm), *Ancylostoma braziliense* (agent of "creeping eruption"). The small adult roundworms have mouths equipped with chitinous plates, or "teeth," by which they attach themselves to intestinal mucosa (fig. 5.4).

Hookworm ova measure about 40 to 60 μm in length, are oval in shape with rounded ends, and have a thin outer shell. Normally, when they are passed promptly in the feces, the developing embryo has not yet reached a differentiated stage but appears cellular. More developed eggs show distinct segmentation of the embryo (fig. 20.2).

Life Cycle (fig. 20.3)

I. Definitive Host

Humans.

II. Intermediate Host

None.

III. Human Infection

A. *Source of Infection* Soil in which larvae have matured after hatching from eggs derived from human feces.

B. *Infective Form* Larvae that penetrate exposed skin (fig. 20.4).

C. *Localization and Development* Larvae enter through the skin, reach lymphatics and the bloodstream, and are carried to the lungs. They migrate into alveoli, up the bronchial tree to the trachea, epiglottis, and pharynx, are swallowed, and finally reach the small intestine where they attach to the intestinal mucosa, develop into male and female adults, copulate, and produce eggs.

D. *Exit from Body* Hookworm eggs are discharged in feces.

IV. Cyclic Development Outside of Human Host

In Environment In moist, shaded, warm soil, eggs hatch in 1 or 2 days (fig. 20.5), and the emerging larvae feed on soil bacteria and organic matter. Larvae go through developmental changes and in about a week reach a stage that

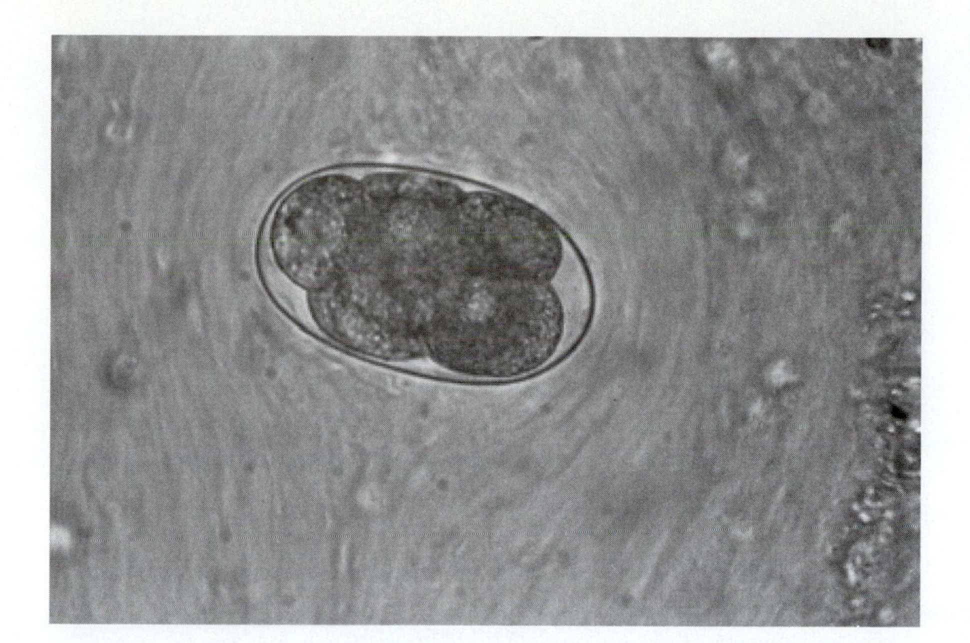

FIGURE 20.2 Photomicrograph of hookworm ovum (×1,000).
Courtesy Dr. Paul Beaver, Tulane University, New Orleans, LA.

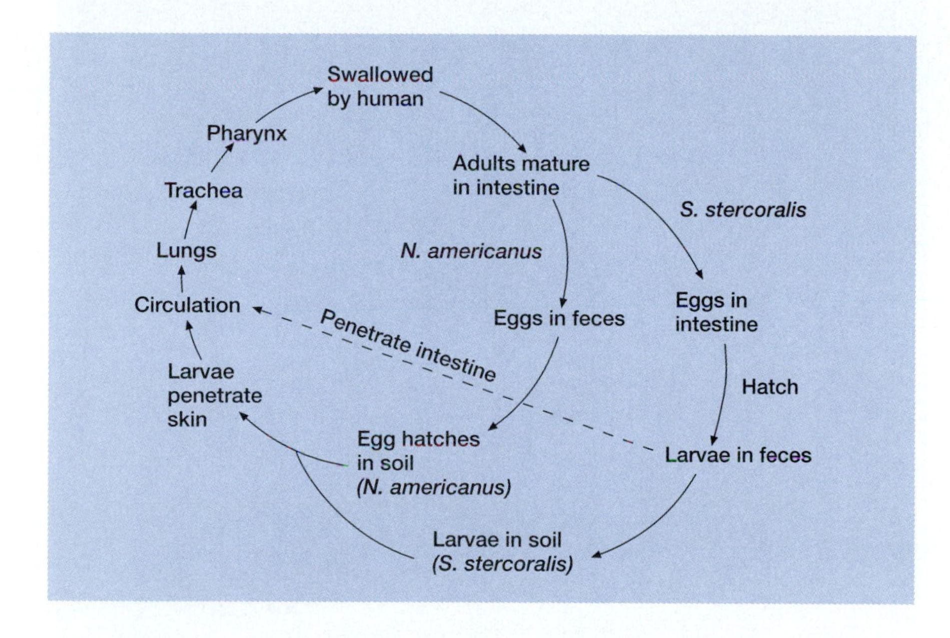

FIGURE 20.3 Human-to-soil-to-human route of transfer characteristic of *Necator americanus* and *Strongyloides stercoralis*. *S. stercoralis* larvae can penetrate directly into the intestinal mucosa or perianal skin to cause autoinfection.

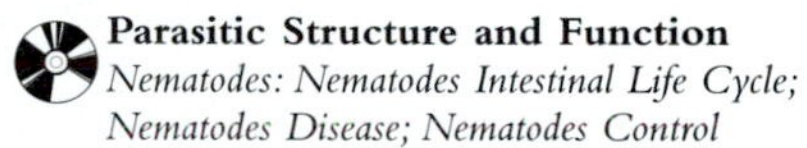

Parasitic Structure and Function
Nematodes: Nematodes Intestinal Life Cycle; Nematodes Disease; Nematodes Control

is infective for humans. This stage may remain viable in soil for about 2 weeks.

Characteristics of Human Disease Hookworm infestation produces a chronic disease with symptoms related to the blood-sucking activities of the intestinal worms. It is estimated that a single hookworm may remove nearly 0.1 ml of blood per day. When this loss is multiplied by thousands of worms and by all the days of a long infection, the debilitating effects of this disease may be well understood. Severe anemia is a characteristic result. Children afflicted with this disease are often malnourished to begin with, and their infection constitutes an added physical burden leading to symptoms of weakness, fatigue, pulmonary disability on exertion, and physical or even mental retardation.

Light infections may produce few, if any, symptoms, but the infected individual may nonetheless perpetuate the disease if eggs discharged in feces have an opportunity to mature in soil. Treatment may be indicated for this reason alone.

In the United States, hookworm disease is caused by species of the genus *Necator,* whereas *Ancylostoma* infestations are common in Europe, Southeast Asia, and the Far East.

Dogs and cats are infected with a species of hookworm, *Ancylostoma braziliense,* that has a similar cycle in the animal's intestine, with larval maturation in soil. These larvae may also penetrate human skin but cannot migrate further or develop in the human body. They may remain alive in epidermal tissues, creating serpentine tunnels and producing intense local tissue

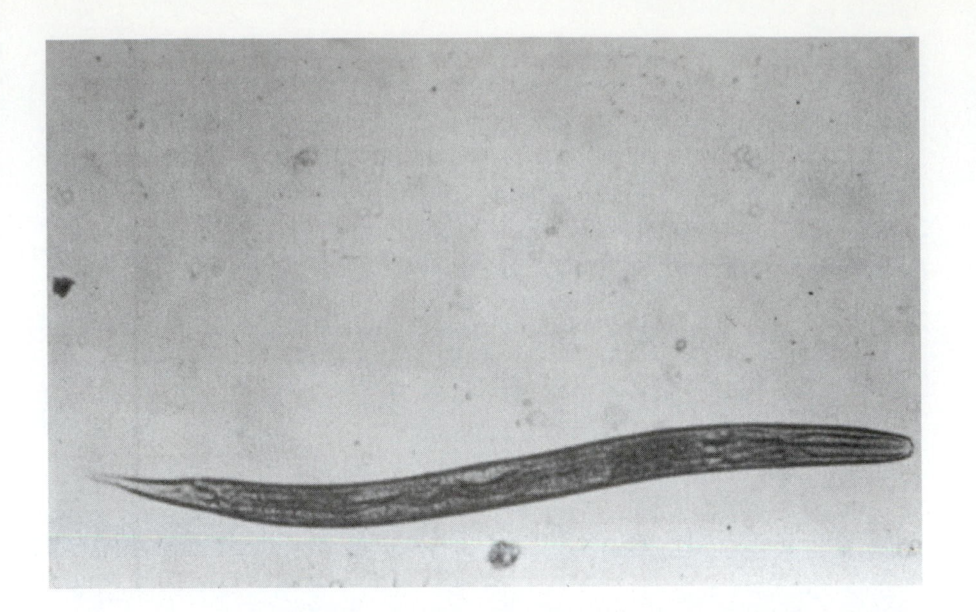

FIGURE 20.4 Larval stage of *Necator americanus* (×500).
Dr. Mark L. Dembert.

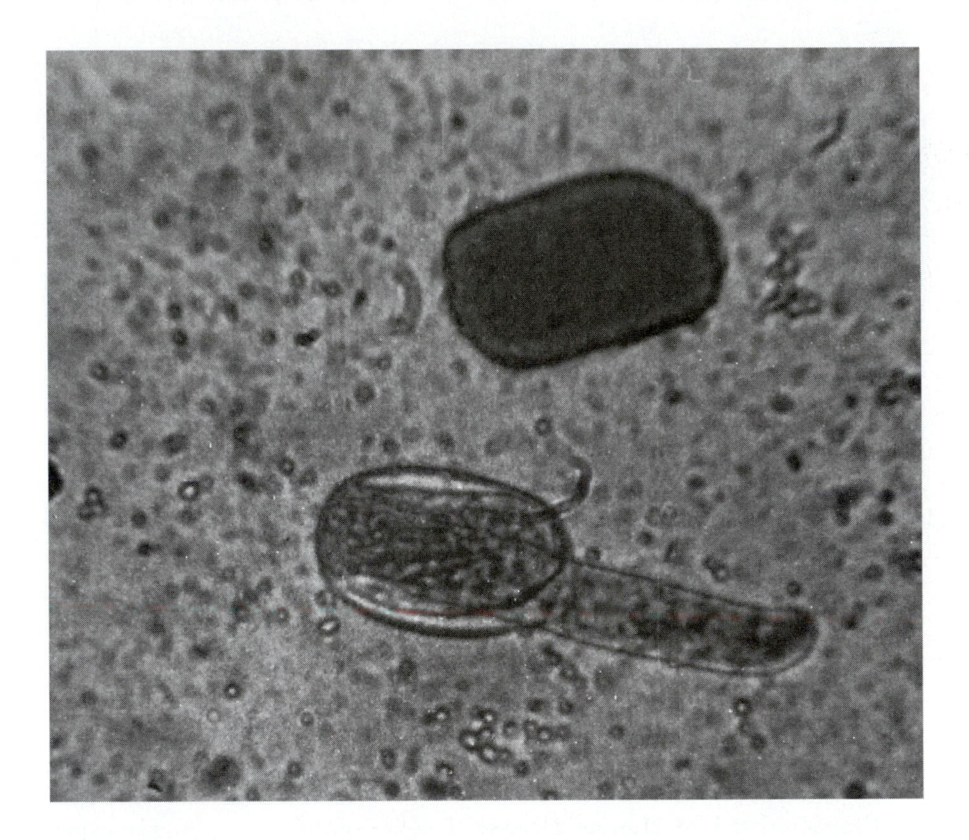

FIGURE 20.5 Photomicrograph of a hookworm larva emerging from an ovum (×500).
Hugo Terner, Montefiore Hospital and Medical Center, Bronx, NY.

reactions of erythema, cellular infiltration, induration, and sometimes vesicular eruptions. This form of epidermal hookworm infection is known as *creeping eruption, cutaneous larva migrans,* or *ground itch*.

DIAGNOSTIC METHODS

Diagnosis of intestinal hookworm disease is established by microscopic identification of hookworm ova in feces. An estimate of the number of eggs produced daily is sometimes useful to the physician in deciding whether or not to treat lightly infected patients. If the patient is unlikely, by virtue of living conditions and sanitary habits, to perpetuate the organism's life cycle, treatment may be unnecessary.

Creeping eruption of hookworm origin is diagnosed on the basis of its clinical appearance and a history of exposure to domestic animals who may harbor the intestinal worm.

Special Features of Control

I. Treatment

Hookworm disease is treated with mebendazole (Vermox) or pyrantel pamoate (Antiminth), the drugs of choice. Specific therapy should be supplemented by supportive treatment of the accompanying anemia. A balanced diet is also an essential part of the treatment.

Creeping eruption is treated locally with thiabendazole cream, or the drug may be given by mouth when many lesions are present.

II. Prevention

Prevention of hookworm disease is primarily a matter of establishing and maintaining sanitary systems for the disposal of feces so that the parasite cannot continue its cycle in soil. Campaigns to promote wearing shoes in rural areas where the organism is entrenched in the human population are a stopgap measure only. The disease can be eliminated by the installation and maintenance of adequate methods for feces disposal in privies or latrines that can be chemically disinfected or in sewerage systems.

STRONGYLOIDIASIS

Organism

(*Strongyloides stercoralis*). Adult ***Strongyloides*** are smaller than hookworms and do not possess cutting teeth or plates for attachment.

Life Cycle (fig. 20.3)

I. Definitive Host

Humans.

II. Intermediate Host

None.

III. Human Infection

A. *Source of Infection* Soil in which larvae have matured after discharge in human feces.

B. *Infective Form* Larvae that penetrate exposed skin.

C. *Localization and Development* Identical with those of hookworm, except that the ova usually hatch in the bowel and larvae are discharged in feces.

D. *Exit from Body* Larval forms are passed in the stool.

IV. Cyclic Development Outside of Human Host

In Environment The larvae mature in soil and may live freely for several generations. The infective form is similar to that of the hookworm and can penetrate human skin, migrate through the body, reach the lungs, and then move into the intestinal tract.

V. Autoinfection

An unusual feature of this parasite is that larval maturation may occur within the infected bowel. The infective larvae may then penetrate the wall of the colon or rectum, enter the tissues, migrate to the lungs, bronchial tree, and intestinal tract, and establish a new generation of adults. Infection may be continued in this way for many years, with frequent recurrences of symptoms. In patients whose resistance is lowered by immunosuppressive therapy, especially corticosteroids, such infections may become generalized, with larvae found in virtually all parts of the body.

Characteristics of Human Disease

The symptoms of **strongyloidiasis** vary according to the density of infection. Migration of larvae through the skin may induce local pruritus; their pulmonary passage sometimes induces pneumonitis or bronchial irritation; but the chief symptoms are related to the burrowing of developing larvae and adults into the villi of the duodenum and small intestine. Epigastric pain, nausea, vomiting, weight loss, weakness, and disturbances in bowel function are common symptoms, sometimes suggesting peptic ulcer. Allergic skin reactions and eosinophilia also are common. Light infections may produce mild variations of these symptoms or none at all.

DIAGNOSTIC METHODS

Microscopic examination of the stool usually reveals viable, active *Strongyloides* larvae. These must be distinguished morphologically from hookworm larvae, which occasionally hatch in the bowel before they are discharged in feces. When strongyloidiasis is suspected but cannot be confirmed by stool examination, aspirated duodenal drainage may reveal larvae.

Special Features of Control

I. Treatment

The drug of choice is thiabendazole; an alternative is ivermectin. These drugs may cause nausea, vomiting, and dizziness, and patients must be followed closely while under treatment.

II. Prevention

Measures for control and prevention of *Strongyloides* infections are the same as those required for hookworm disease. The most effective and permanent control is provided by installation of sanitary measures for disposal of human feces. Shoes are essential in endemic areas where the primary source of infection has not been controlled.

SCHISTOSOMIASIS

Organisms

Schistosoma mansoni (occurs in the Caribbean, South America, Africa), *Schistosoma haematobium* (Africa, Middle East, India), and *Schistosoma japonicum* (Far East, Philippines), all produce the disease **schistosomiasis.** The male and female adult ***Schistosoma*** worms live in close association (fig. 20.6a). The ova are quite large, measuring about 50 μm in diameter and from 100 to 175 μm in length, depending on the species. They have a transparent shell which in two species is equipped with a prominent spine (laterally placed on the *S. mansoni* egg [fig. 20.6b], terminal on the ovum of *S. haematobium*). The ovum of the third species (*S. japonicum*) has a very small, short, curved hook on one side. When the embryo in the shell is fully developed, it is ciliated and, upon being hatched, can swim

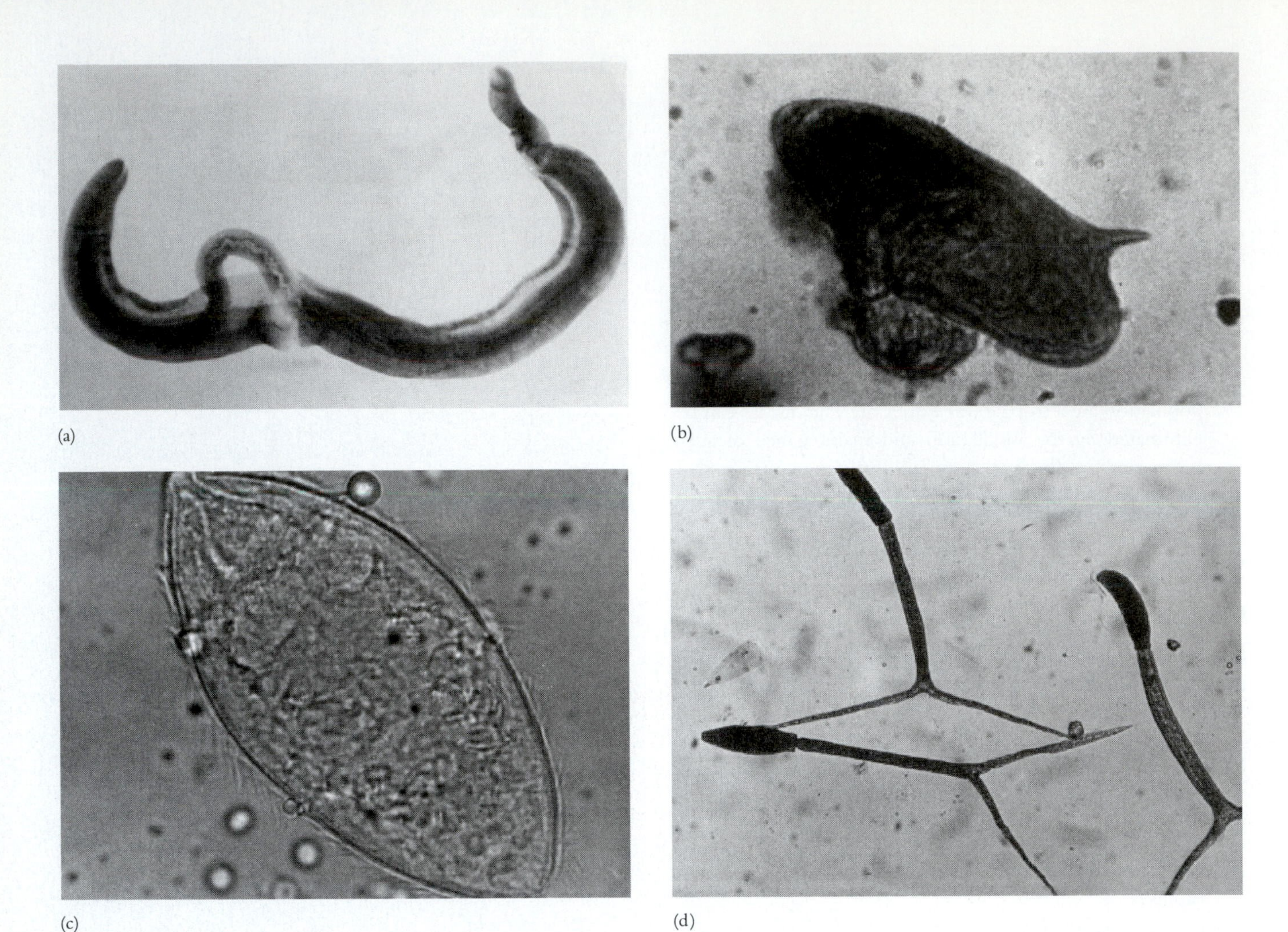

FIGURE 20.6 Four stages in the life cycle of *Schistosoma mansoni*. (a) Male and female adults. The slender female is interlocked with the larger male (×6). (b) *Schistosoma mansoni* egg from a human case of schistosomiasis (×150). (c) A free-swimming miracidium hatched from the egg (×240). Note the fine rim of cilia on the outer side of the cell membrane. (d) Fork-tailed cercariae (×150). This form emerges following cyclic development of the miracidium in a snail host, swims freely in water, and infects humans by penetrating their skin.

(b) Hugo Terner, Montefiore Hospital and Medical Center, Bronx, NY.

about freely. The hatched embryo is called a *miracidium* (fig. 20.6c).

Life Cycle (fig. 20.7)

I. Definitive Host

Humans and domestic or wild animals.

II. Intermediate Host

Snails.

III. Human Infection

A. *Source of Infection* Water infested with larval forms (cercariae) liberated from the intermediate snail host.

B. *Infective Form* Free-swimming larvae, also referred to as "fork-tailed cercariae" (fig. 20.6d).

C. *Localization and Development* Larvae penetrate the skin of persons swimming, wading, or working in infested water. They enter the bloodstream, are carried to the liver, mature to adult forms, and migrate through hepatic vessels or veins of the abdominal cavity. *S. mansoni* and *S. japonicum* remain in hepatic or mesenteric vessels, but *S. haematobium* usually finds its way into the venous complex of the pelvis. Deposited eggs find their way out of venules into the lumen of hepatic ducts or pass more directly into the bowel or urinary bladder, depending on the location of the adult worm. Many ova may lodge in tissues adjacent to the site where they are deposited and be immobilized there by local cellular reactions.

D. *Exit from the Body* The ova of *S. mansoni* and *S. japonicum* may be discharged in feces; those of *S. haematobium* usually emerge into the bladder and are excreted in urine but may also find their way into the colon and be passed in the stool.

IV. Cyclic Development Outside of Human Host

A. *In Environment* *Schistosoma* ova must reach water to begin their maturation. The ciliated miracidia break out of their enclosing shells and swim about in water until they

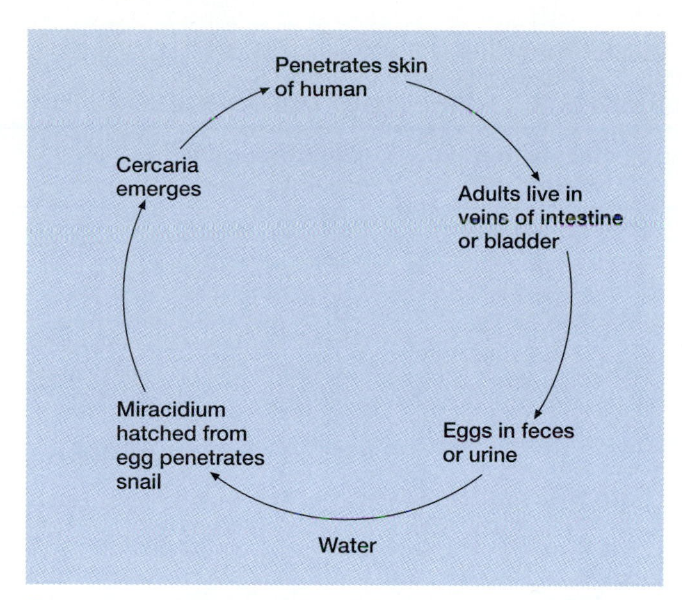

FIGURE 20.7 Human-to-intermediate host-to-human route of transfer characteristic of schistosomes. All species have the same life cycle, but the adults live in veins of intestine (*S. mansoni, S. japonicum*) or bladder (*S. haematobium*).

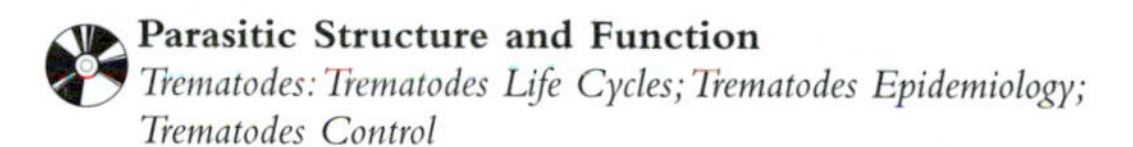

Parasitic Structure and Function
Trematodes: Trematodes Life Cycles; Trematodes Epidemiology; Trematodes Control

are ingested by snails, which serve as intermediate hosts for further development.

B. *In Snail Hosts* The parasites undergo several morphologic changes in snail tissues, emerging finally as microscopic larval forms that also can swim freely. The larvae are called *cercariae* and are characterized by their forked tails, the anterior ends of which are thickened and elongated. Glands in the anterior section of the larva produce enzymes that assist in penetration of human skin once the parasite has become attached.

Characteristics of Human Disease

The symptoms of schistosomiasis are related to the location of the adult worm and the damage inflicted on adjacent tissues by the migration of ova. The adults may cause rupture of small vessels as they migrate together through venules, but chronic injury is induced by ova wandering in parenchymal tissues. Polymorphonuclear exudates collect around the eggs, and abscesses form. When ova lodge in the lumen of hepatic ducts or in the bowel, small ulcers may form around them. The eggs sometimes dislodge and are carried by the bloodstream into the lungs where they may obstruct capillaries and give rise to abscesses in the pulmonary bed. In the liver they may obstruct the portal circulation leading to enlargement of both liver and spleen (colorplate 8b). From the venous complex in the pelvis, eggs may reach the bladder, genital organs, or colon. Further migration of adult worms through connective venules may lead to deposition of eggs in other parts of the body such as the skin, conjunctiva, or spinal cord and brain.

During the course of active infection and egg deposition, the disease is characterized by allergic reactions; fever, urticaria, and eosinophilia may be intense. When the eggs trapped in parenchymal tissues eventually die, granulomatous pseudotubercles are formed, with foreign-body giant cells surrounding and engulfing the dead ova. The eggs may be removed entirely or may become calcified and surrounded by scar tissue. Extensive scarring disrupts normal tissue architecture and interferes with its function. Obstructions of blood vessels and bile ducts result from the body's efforts to heal the lesions induced by *Schistosoma* ova.

DIAGNOSTIC METHODS

Diagnosis of schistosomiasis is confirmed by the demonstration of characteristic eggs in feces, urine, or biopsied tissue. Because both dead and live eggs may be excreted, the viability of the embryo should be confirmed by observing movement or encouraging hatching of the miracidium in a nonchlorinated, water-diluted specimen incubated at 35 to 37° C. The demonstration of viable eggs indicates the presence of active adults and the need for treatment.

Special Features of Control

I. Treatment
Schistosomiasis was once treated with toxic antimony compounds, but the current drug of choice for disease caused by all *Schistosoma* species is praziquantel (Biltricide). Administration of this drug has resulted in a high cure rate.

II. Prevention
Major control of schistosomiasis can be achieved only by the sanitary disposal of human excreta, which prevents infection of the snail host and perpetuation of the parasite's life cycle. Snail eradication and land reclamation programs afford some relief but cannot eliminate the disease while human infection and unsanitary practices persist.

QUESTIONS

1. Why is anthrax classified as an occupational disease?
2. Why has the incidence of anthrax decreased in the United States?
3. What are the characteristics of the genus *Bacillus?*
4. How do laboratory technologists protect themselves from highly infectious endospores of *Bacillus anthracis?*
5. What are the chief reservoirs of *Francisella tularensis?*
6. How do humans become infected with tularemia?
7. Why is it important to trace the source of infection for humans with tularemia?
8. How is an acute infection of leptospirosis characterized? What organs are ultimately affected by this disease?

Summary Table 20.1 INFECTIOUS DISEASES ACQUIRED THROUGH SKIN OR MUCOSA AND TRANSMITTED BY ANIMALS OR FROM INTERMEDIATE ENVIRONMENTAL SOURCES

Clinical Disease	Causative Organism	Other Possible Entry Routes	Incubation Period	Communicable Period
Antrax	*Bacillus anthracis*	Respiratory Gastrointestinal	Less than 1 week	Not communicable from person to person
Tularemia	*Francisella tularensis*	Respiratory Gastrointestinal Parenteral	1–10 days	Not transmissible from person to person
Leptospirosis	*Leptospira interrogans*		3–30 days	Not usually communicable from person to person
Hookworm disease	*Necator americanus* *Ancylostoma duodenale* *Ancylostoma braziliense*		Variable; ova are seen in stools 6 weeks after initial infection	Not communicable from person to person
Strongyloidiasis	*Strongyloides stercoralis*		About 17 days	Not communicable from person to person
Schistosomiasis	*Schistosoma mansoni* *Schistosoma haematobium* *Schistosoma japonicum*		4–6 weeks	Not communicable from person to person

9. What groups of people are particularly prone to leptospirosis?
10. What anatomical characteristics do hookworms possess?
11. How do hookworms affect the health of the infected person?
12. How is hookworm best controlled?
13. What aspect of the life cycle of *Strongyloides* makes it especially serious for immunocompromised patients?
14. What is the intermediate host for schistosomiasis? What is the infective form for this disease?
15. What form of *Schistosoma* species causes damage to human tissues?
16. How can schistosomiasis be prevented?

Bibliography

Balows, A., W. J. Hausler, and E. Lennette. eds. 1988. *Laboratory diagnosis of infectious diseases: Principles and practice.* Vol. 1, *Bacterial, mycotic and parasitic diseases.* New York: Springer-Verlag.

Benenson, A. S., ed. 1995. *Control of communicable diseases in man.* 16th ed. Washington, D.C.: American Public Health Association.

Desowitz, R. 1983. *New Guinea tapeworms and Jewish grandmothers: Tales of parasites and people.* New York: Avon.

Garcia, L. S., and D. A. Bruckner. 1997. *Diagnostic medical parasitology.* 3d ed. Washington, D.C.: ASM Press.

Hotez, P. J., and D. Pritchard. 1995. Hookworm infection. *Sci. Am.* 272 (6): 68–74.

Specimens Required	Laboratory Diagnosis	Immunization	Treatment	Transmission-Based Precautions
Exudates from lesions Sputum Blood Acute and convalescent serums	Smear and culture Immunofluorescence Culture Rising titer of antibodies	Cell-free vaccine	Penicillin Erythromycin Tetracycline	Standard precautions until lesions are bacteriologically negative; contaminated dressings and disposable items burned; visibly soiled bed linens autoclaved before laundering; steam sterilization of heat-stable items; gas sterilization for heat-sensitive items; gown technique and hand washing essential; terminal disinfection-cleaning important; cultures done in biological safety cabinet
Exudates from ulcers, aspirated material from suppurating lymph nodes Blood Sputum Acute and convalescent serums	Smear and culture Immunofluorescence Rising titer of antibodies	Attenuated live vaccine	Streptomycin Gentamicin Tetracyclines Chloramphenicol	Standard precautions; cultures done in biological safety cabinet
Blood Urine Spinal fluid	Culture	Killed vaccines for animals	Penicillin and other antimicrobial agents suppress but do not eliminate the organism	Standard precautions
Stools	Identification of ova		Mebendazole Pyrantel pamoate	Sanitary disposal of feces
Stools	Identification of viable, active strongyloid larvae		Thiabendazole Ivermectin	Sanitary disposal of feces
Stools Urine	Identification of ova		Praziquantel	Sanitary disposal of feces

Jahn, T. L., E. C. Bovee, and F. F. Jahn. 1978. *The protozoa.* 2d ed. Dubuque, Iowa: William C. Brown.

Katzman, E. M. 1989. What's the most common helminth infection in the U.S? *MCN* 14 (3): 193–195.

Kaye, D. 1988. The spectrum of strongyloidiasis (case study). *Hosp. Prac.* 23 (10): 11–15, 118–119, 125–126.

The medical letter handbook of antimicrobial therapy. 1994. New Rochelle, N.Y.: The Medical Letter, Inc.

Murray, P. R., ed.-in-chief. 1995. *Manual of clinical microbiology.* 6th ed. Washington, D.C.: ASM Press.

Nield, H. 1990. Leptospirosis: Weil's diseases. *Occup. Health* (London) 42: 140–142.

Roueche, B. 1991. A man named Hoffman. In *The medical detectives.* New York: Plume.

———. 1991. A rainy day on the Vineyard. In *The medical detectives.* New York: Plume.

———. 1991. A swim in the Nile. In *The medical detectives.* New York: Plume.

Sanford, J. P. 1990. Humans and animals: Increasing contacts, increasing infections. *Hosp. Prac.* 25 (2): 123–140.

Taylor, P. 1987. Popular technology beats schistosomiasis. *World Health* (July) 28–29.

Verghese, T. 1990. Guinea worm, the fiery serpent. *World Health* September-October, pp. 20–22.

Zygmunt, D. J. 1990. *Toxoplasma gondii. Infect. Control Hosp. Epidemiol.* 11 (4): 207–211.

Infections and Infectious Diseases Acquired through Medical and Surgical Procedures

OUTLINE

Microbes in Motion
This chapter covers the causative organisms, laboratory diagnosis, and epidemiology of infectious diseases acquired through medical and surgical procedures. The following book and chapter of the Microbes in Motion CD-ROM may be a useful preview or supplemental study aid to your reading: Bacterial Structure and Function: External Structures.

OBJECTIVES

On completing this chapter the student will be able to:

1. discuss the importance of the hospital environment in terms of its potential for transmitting nosocomial infections to patients.
2. describe the infectious risks of medical or surgical treatment in hospitalized patients.
3. identify the aseptic techniques essential in catheterization and surgical treatment of patients.
4. cite important infectious diseases associated with medical and surgical procedures.
5. describe the causative organism(s), laboratory diagnosis, and epidemiology of the diseases associated with medical and surgical procedures.
6. discuss the safety precautions required in blood banks.

KEY WORDS

Blood bank precautions
Catheterization techniques
Contaminated injectables
Dane particle
Glycocalyx (gly-co-**KALE**-icks)
Hepatitis B virus (HBV)
Hepatitis C virus (HCV)
Hepatitis D virus (HDV)
Hypodermic techniques
Lensed instruments
Nosocomial (nose-oh-**COMB**-ee-al) infection
Postpartum endometritis (end-oh-meh-**TRITE**-iss)
Puerperal (poo-**ERE**-per-al) fever
Slime layer
Surgical infection
Urinary tract infections

For many patients it could well be said that "hospitals are dangerous to your health." Reports from the Centers for Disease Control indicate that more than 2 million patients acquire infections in hospitals every year. Many of these **nosocomial infections** are more difficult to treat than those acquired elsewhere because the organisms producing them are resistant to a variety of antimicrobial agents. They are epidemiologically dangerous because of the ease with which they can be transmitted, directly or indirectly, and the speed with which they can spread through the wards, particularly those containing surgical patients, new mothers, and babies. Hospitalized patients are more susceptible to infection because their illness often makes them physiologically less capable of resisting it. The problem is compounded by institutional living conditions or by risks inherent in certain medical, surgical, and nursing procedures. Opportunities for exposure are multiple and may arise from a number of directions and sources.

INFECTIOUS RISKS OF MEDICAL AND SURGICAL PROCEDURES

This chapter is concerned with infections that arise as a result of medical or surgical interventions. Four major procedures are considered: (1) catheterization of the urinary tract or blood vessels, (2) surgery, (3) instrumentation through mucosal surfaces, and (4) insertion of hypodermic needles for injection or withdrawal of fluids. The nature and sources of infections that may be acquired through these routes are reviewed, and some diseases characteristically associated with hospital or other medical practices are discussed.

CATHETERIZATION OF THE URINARY TRACT OR BLOOD VESSELS

Catheters are frequently inserted into the long narrow passages of the urinary tract (ureters or the urethra) or into blood vessels for diagnostic or therapeutic purposes. **Catheterization techniques** may be a source of infection if equipment is not properly sterilized and if strict aseptic precautions are not used in the process of insertion.

The urinary tract is the most common site of infections acquired in the hospital. Urinary tract catheterizations may result in cystitis, pyelonephritis, or more disseminated systemic infections. Procedures most frequently associated with complicating urinary infections are urethral or ureteral catheterization, use of indwelling catheters for continuous bladder drainage, cystoscopy, and prostate or bladder surgery. These infections may have an endogenous source in the microbial flora of the patient's genital, perineal, or perianal mucosa, or they may be introduced from exogenous sources. The technique of catheterization should be conducted with strict asepsis and with great care to avoid injuring delicate mucosal linings with the catheter itself. Vascular catheters may introduce microorganisms directly into the blood. Organisms of low pathogenicity are generally removed by phagocytes and antibacterial components of blood, but some may localize and multiply, sometimes at sites where the vascular endothelium is injured. The result may be local inflammation at the site of catheter insertion (thrombophlebitis or cellulitis) or even bacteremia.

Catheters are placed through the skin for several purposes: to provide hyperalimentation feedings; to administer medications (particularly to outpatients with hematologic disorders); to monitor physiologic functions; or to perform peritoneal or vascular dialysis. Catheters may become infected with organisms of low virulence, such as *Staphylococcus epidermidis,* which are commonly found as part of the normal skin flora (see chap. 19). Even though *S. epidermidis* is regarded as being of low virulence in human disease, it has emerged as the leading cause of catheter-associated infections. Some strains of *S. epidermidis* produce an exopolysaccharide called a **glycocalyx,** or slime, that allows them to adhere to the smooth surfaces of catheter tips. Evidently, the staphylococci migrate from the skin along the exterior portion of the catheter. They may also be introduced into the catheter lumen by medical or nursing personnel after the catheter is disconnected from the hub. Regardless of how they enter, staphylococci, once attached to the catheter surface, can bind fibronectin, collagen, and other host proteins to form a glycoproteinaceous film that rapidly coats the biomedical device.

Firmly attached to the catheter implant, the organism becomes completely coated with slime (fig. 21.1). This biofilm provides the organism with a protective microenvironment in which growth conditions are optimal. The **slime layer** also serves as a mechanical barrier, protecting the organism from the action of antimicrobial agents and normal host defense mechanisms. Because the infection cannot be treated effectively by antimicrobial agents or cleared by body defenses, the catheter must be removed to eliminate the infectious source.

Staphylococcus aureus (coagulase-positive) is responsible for some infections that complicate the patient's postcatheterization course, but *S. epidermidis* (coagulase-negative) is more commonly involved. Gram-negative enteric bacilli such as *Klebsiella, E. coli,* and *Pseudomonas* are also known offenders. In addition to skin, the sources of these infections may be mucous membranes of the patient or even of a member of the medical team. Environmental contamination of catheters or instruments may occur because these procedures frequently are carried out in X-ray rooms or other areas where radiologic visualization of catheter placement is possible, but where strict asepsis is difficult to maintain. In dialysis procedures, especially, if aseptic precautions are not rigidly adhered to, equipment or supplies may become contaminated with microorganisms such as *Pseudomonas* and *Acinetobacter* species, which commonly inhabit water.

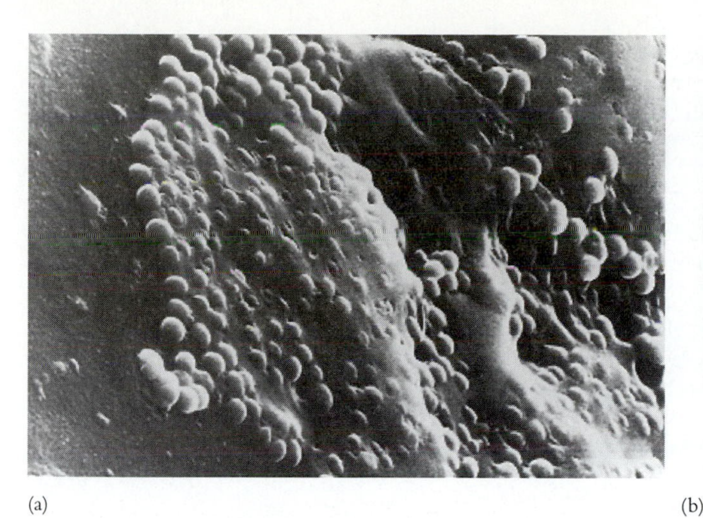

(a)

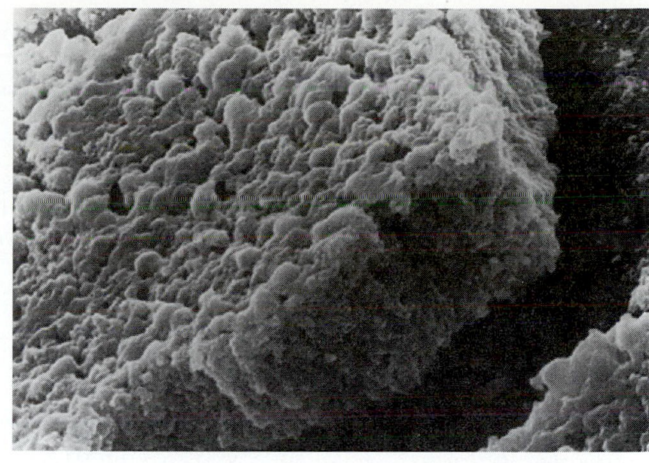

(b)

FIGURE 21.1 Coagulase-negative staphylococci (strain KH 11) produced this slimy material on the inner surface of a polyethylene catheter: (a) a thick layer of slime after 48 hours of incubation and (b) embedding of the bacteria after incubation.

From Peters, G., Locci, R., and Pulverer, G., "Adherence and Growth of Coagulase-Negative Staphylocci on Surfaces of Intravenous Catheters," Journal of Infectious Disease 146–4:479–482, 1982. University of Chicago Press.

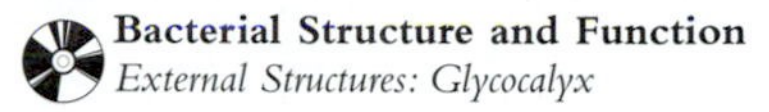

Bacterial Structure and Function
External Structures: Glycocalyx

SURGERY AND SURGICAL IMPLANTS

One of the most obvious risks of surgery lies in exposure of normally sterile tissues to exogenous or endogenous sources of infection. Major sources of exogenous **surgical infection** are the skin and respiratory secretions of the operating team and environmental contamination from objects in intimate contact with surgically exposed tissues (instruments, sponges, sutures, linen drapes, irrigating solutions). Contaminated air may also play a role in surgical infections. Endogenous infections arise when microorganisms disseminate into adjacent tissues or the blood from their usual habitat on the patient's skin or mucous membranes, such as the lumen of the intestinal tract, genitourinary membranes, and oral mucosa.

Surgical patients are often subjected to a variety of procedural risks in addition to those inherent in surgery. They may require intubation (insertion of an airway into the larynx) while under anesthesia; administration of intravenous fluids; urinary catheterization; or other management that can open the door to infection. Inhalation anesthesia also leaves such patients more vulnerable to respiratory infection.

The degree of trauma suffered by tissues exposed to surgery or to the ancillary procedures just mentioned is an important determining factor in the development of wound infections. Excessive or rough handling damages the architectural integrity of tissues. It may interfere with vascular structures leaving areas of devitalized, necrotic tissue. Bacteria can grow without opposition in these tissues, as well as in hematomas formed as a result of continued bleeding from severed vessels. Speed, skill, manual dexterity, and delicacy are essential factors in the control and prevention of surgical wound infections.

The nature and source of infecting microorganisms vary somewhat with the type of surgery performed and the region of the body involved. Table 21.1 shows some infectious complications that may result from surgery and indicates their possible origins. Sources of surgical infection are often difficult to prove conclusively, but note that the microorganisms involved are those commonly associated with human surface tissues. Typically, the best evidence incriminating surgical personnel, techniques, or supplies is demonstration of a high incidence of wound infections among surgical patients or those of a given service, especially if the same etiologic agents are involved.

Technological advances have also led to surgical procedures in which human joints are replaced by a variety of orthopedic devices and foreign bodies of other kinds are implanted. Whatever the device—mechanical or autologous heart valves, pacemaker wires and power packs, the lenses implanted following cataract extraction, or breast and penile implants—all predispose the patient to the risk of subsequent infection. If the device becomes infected, the infectious process typically is refractory to antimicrobial therapy, making it necessary to surgically remove the old device and implant a new one.

INSTRUMENTATION THROUGH MUCOSAL ORIFICES

Lensed instruments are frequently used for examining mucosal passageways and deep surfaces (endoscopy). These instruments consist, essentially, of a rigid or flexible tube equipped with magnifying lenses, a light source at the inserted end, and a series of prisms that direct light to the observer's eye, permitting visualization of the depths of the

TABLE 21.1 INFECTIOUS COMPLICATIONS THAT MAY RESULT FROM SURGERY

Surgical Area	Types of Infection and Their Possible Origins		Some Sources of Exogenous Infections
	Endogenous	Exogenous	
Mouth, nose, or throat	Infections of local soft tissues Regional lymphadenopathies Bacteremia and systemic infection		
	Organisms from upper respiratory tract or buccal mucosa: Streptococci (may induce infective endocarditis) Staphylococci *Actinomyces israelii* (cervicofacial or thoracic actinomycosis) Other anaerobes from upper respiratory tract	Streptococci (especially beta-hemolytic strains) Staphylococci (especially *S. aureus*)	Infected personnel Unsterile instruments, wound packing, sponges, irrigating solutions
Chest (including cardiac surgery)	Wound infections Pulmonary infection Systemic infection		
	Organisms from upper respiratory tract or skin: Streptococci Staphylococci (coagulase-negative as well as coagulase-positive strains sometimes involved in endocarditis following cardiac surgery) *Pseudomonas aeruginosa* Gram-negative enteric bacilli: *E. coli* *Klebsiella* species *Proteus* species Anaerobic organisms: *Actinomyces israelii* *Bacteroides* species *Fusobacterium* species Peptostreptococci	As in the endogenous column on left, except anaerobes are not derived from exogenous sources	Infected personnel Hands in punctured or torn gloves Contaminated gowns Contaminated instruments, sutures, solutions Contaminated fluids or blood for transfusion Contamination in heart-lung machine
Intestinal tract	Wound infections Infection of bowel wall Peritonitis Systemic infections (including endocarditis)		
	Organisms from intestinal tract or skin: Streptococci Staphylococci (usually *S. aureus*) Gram-negative enteric bacilli: *E. coli* *Klebsiella* species *Proteus* species *P. aeruginosa* *Salmonella* species (may be spread from inapparent bowel infection, inducing *Salmonella* septicemia) Anaerobic organisms: *Clostridium tetani* *Clostridium perfringens* and others of gas gangrene group *Bacteroides* species Peptostreptococci	As in the endogenous column on left, except salmonellosis is rarely derived from exogenous sources during surgery; anaerobes are not derived from exogenous sources except clostridial endospores	Infected personnel Hands in punctured or torn gloves Contaminated gowns Contaminated instruments, sutures, solutions Contaminated blood or fluids given intravenously

Continued

TABLE 21.1 *Continued*

Surgical Area	Types of Infection and Their Possible Origins: Endogenous	Exogenous	Some Sources of Exogenous Infections
Orthopedic surgery	Wound infections Osteomyelitis Systemic infections		
	Organisms from skin or intestinal tract: Streptococci (usually beta-hemolytic) Staphylococci (usually *S. aureus*) Anaerobic organisms: *Clostridium tetani* *Clostridium perfringens* and others of gas gangrene group Peptostreptococci *Bacteroides* species Gram-negative enteric bacilli *P. aeruginosa*	As in the endogenous column on the left except nonendospore-forming anaerobes are not derived from exogenous sources	Infected personnel Hands in punctured or torn gloves Contaminated gowns Contaminated instruments, sutures, solutions Contaminated blood or fluids given intravenously Contaminated prosthetic devices implanted in tissues Plaster dust
Ophthalmic surgery	Conjunctivitis Bulbar infections		
	Organisms from conjunctiva or skin: Pneumococci Streptococci Staphylococci *Haemophilus* species Herpes simplex virus	The bacterial species listed in the endogenous column on the left plus: *Pseudomonas* Fungi (various species, e.g., *Aspergillus, Mucor*) *Chlamydia trachomatis*	Infected personnel Contaminated hands or gowns Contaminated instruments Contaminated eye drops or anesthetizing solutions
Plastic surgery	Wound infections Infectious disruption of grafts		
	Organisms from skin or mucosa: Streptococci Staphylococci Gram-negative enteric bacilli *P. aeruginosa*	As in the endogenous column on the left	Infected personnel Hands in punctured or torn gloves Contaminated gowns Contaminated instruments, sutures, solutions
Gynecologic and obstetric surgery	Postpartum endometritis Urinary tract infections (cystitis, pyelonephritis)		
	Organisms from genital or intestinal tract mucosa: Streptococci Staphylococci Gram-negative enteric bacilli Anaerobic organisms: *Clostridium perfringens* and others of the gas gangrene group *Bacteroides* species *Fusobacterium* species Peptostreptococci	As in the endogenous column on the left Streptococci and staphylococci most commonly associated with postpartum endometritis; beta-hemolytic streptococci or *S. aureus* strains usually derived from obstetric personnel; nonendospore-forming anaerobes not exogenous	Infected personnel Hands in punctured or torn gloves Contaminated gowns Contaminated instruments, solutions Contaminated catheters

passageway. They are constructed for specific use on various anatomic structures and are named accordingly, each term having the suffix *-scope* to indicate a viewing instrument. The following are some examples:

Instrument	Examination of:
Nasopharyngoscope	Nasopharynx
Laryngoscope	Larynx
Bronchoscope	Trachea, bronchial tree
Gastroscope	Esophagus, stomach
Sigmoidoscope	Sigmoid, rectum
Cystoscope	Urethra, bladder
Endoscope	(A collective term)

These instruments present risk of infection for two reasons: (1) they are difficult to sterilize because of their construction; their lenses, lighting system, and prisms are mounted in cementing substances that may be dissolved by moist heat or chemical solvents; and (2) they must be large enough in diameter to permit an adequate view of the passageway, which means that they can disrupt delicate mucosal linings. If not properly disinfected, endoscopes may introduce extraneous microorganisms, including those from patients previously examined, and if not carefully inserted, they may irritate or break the mucosal barrier, permitting entry of local or exogenous organisms to deeper tissues.

HYPODERMIC NEEDLES, SYRINGES, AND INJECTABLE SOLUTIONS

Among the many tools of medical practice that may be responsible for the parenteral introduction of infection, the hypodermic needle is the most frequently, and sometimes the most carelessly, used. **Hypodermic techniques** can be directly or indirectly responsible for infection for any of the following reasons:

1. The needle itself is contaminated.
2. The syringe or other vessel containing fluid to be injected is contaminated.
3. The patient's skin is inadequately disinfected.
4. The patient's skin is contaminated with microorganisms growing in inactive disinfectants used to prepare the injection site.
5. The injected fluid contains contaminating organisms.

Sources of Infection by the Hypodermic Route

The sources of contamination for needles, syringes, injectable solutions, or intravenous infusion systems are quite varied (fig. 21.2). The most obvious of these and their serious implications are described next.

1. Reused Instruments

Pathogenic microorganisms may be derived from the blood, tissues, or skin of infectious patients for whom the needle or syringe was previously used, implying that the reused instruments were inadequately cleaned and sterilized. The infectious agents most frequently involved under these conditions are the viruses of hepatitis A and hepatitis B. These viruses are not only quite resistant to most useful agents of disinfection, but they are pathogenic if introduced directly into blood or tissues of susceptible, nonimmune persons, even if the numbers introduced are small. Proper heat sterilization of reusable needles or syringes, or the use (and discard) of disposable presterilized equipment, is the only guarantee for preventing the spread of viral hepatitis by this route.

2. Syringe and Needle Handling

Faulty techniques may result in contamination of needles or syringes after sterilization. If these items are not handled correctly, microorganisms from the operator's hands or respiratory secretions may reach the shaft of the needle, the inner surfaces of a disassembled syringe, or its tip. The shaft of a sterile needle must be continuously protected from contamination from the air, bench top, or other surface where it is laid pending final preparation of the patient's skin. Disposable needles packaged in plastic protectors should be left in place until the moment comes to fill the syringe with the material to be injected or to introduce the needle into the patient's skin. If a loaded syringe and needle must be momentarily set aside *before* an injection is given, the plastic needle protector should be replaced. The protector should *never* be replaced on a used needle because this recapping procedure is the most common cause of needle-stick injuries to health care personnel.

3. Skin Disinfection

Inadequate disinfection of the patient's skin before introducing hypodermic needles may also lead to infection. Organisms usually introduced from this source include streptococci, staphylococci, enteric bacteria, *Pseudomonas,* and yeasts. The degree of risk depends on the immunologic status of the host, the number of organisms injected, their virulence, or the toxicity of any preformed products of their growth in exogenous sources. Small numbers of avirulent organisms from the skin are easily removed by phagocytes, but the properties of virulent organisms such as hemolytic streptococci or coagulase-positive staphylococci provide them with greater resistance to the body's defenses. Such bacteria can be dangerous even in small numbers if introduced directly into blood tissues, or if the patient is immuncompromised.

The route of introduction of organisms from the skin or other sources also influences the outcome. Microorganisms injected into intradermal or subcutaneous tissues meet a great deal of opposition from local phagocytes and blood components supplied to well-vascularized tissues. In intramuscular sites, contaminating organisms may have more opportunity to establish themselves because muscle tissue contains fewer fixed phagocytes, but the blood supply soon brings polymorphonuclear cells and plasma components for opsonization. Intravenous injections or venipuncture for collection of

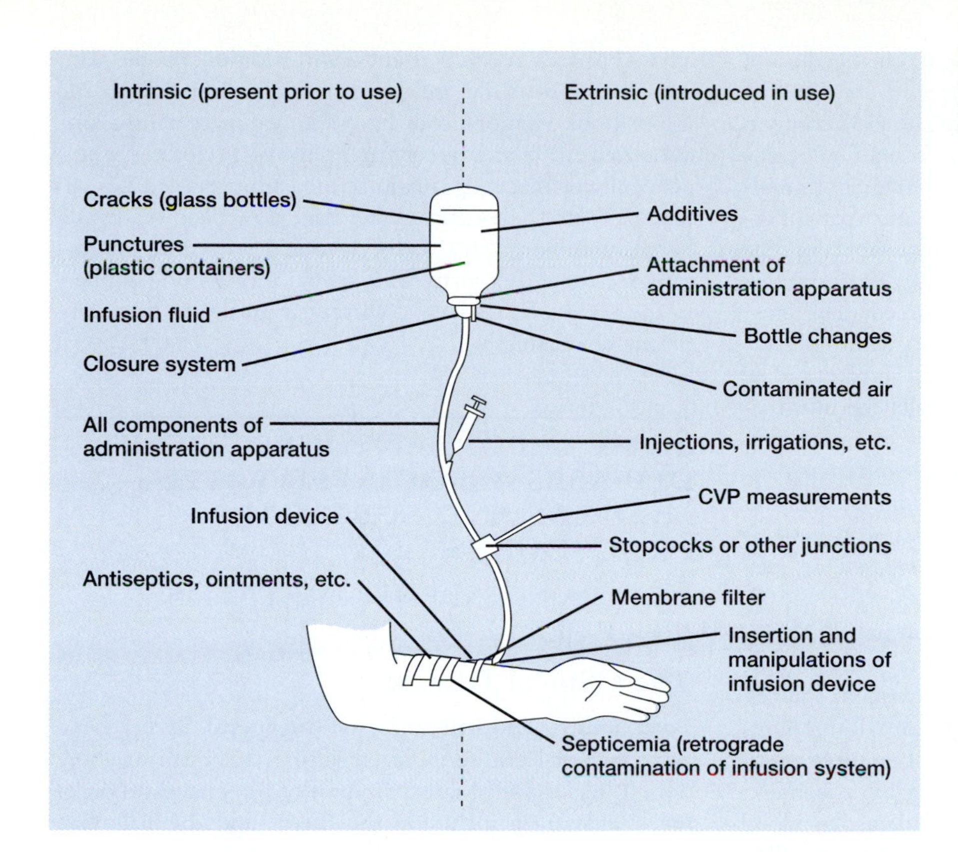

FIGURE 21.2 Potential mechanisms for contamination of intravenous infusion systems

Reproduced, with permission from Dennis G. Maki, D. A. Goldmann, and F. S. Rhame, in *Ann. Intern Med.* 1973; 79:867

blood samples affords the greatest risk of introduction and wide dissemination of contaminating microorganisms.

4. Inactivated Antiseptics

Another dangerous possibility of bacterial infection arises when the hypodermic technique introduces materials in which microbial multiplication has occurred before injection. This situation can happen when bacteria have been growing in skin antiseptics that have become inactive or to which the organism is resistant. When the patient's skin is prepared for injection by applying the "disinfectant," large numbers of contaminating organisms are essentially wiped onto the skin surface. The needle then passes through this teeming culture into the tissues below or into a vein.

Antiseptics and other chemical disinfectants may be inactivated in a number of ways. Some are readily bound by protein or other substances that combine with the active portions of their molecules (see chap. 9) and become useless after such exposure. Fresh working solutions should be prepared frequently and kept in clean, closed containers, protected from chemical or excessive microbial contamination. The practice of setting up boats with cotton or gauze pads soaked in antiseptic solution, ready to use, is not acceptable because cellulose fibers may bind with the antiseptic and inactivate it. This affinity for cellulose and protein is one of the deficiencies of quaternary ammonium compounds. Bacteria are continually introduced into such boats from each hand reaching for a new pad, and in a short time a rich culture of organisms is formed in a solution that can no longer suppress them.

5. Contaminated Injectable Fluids

Fluids injected by hypodermic methods may be the most dangerous source of infection, particularly if microorganisms have had an opportunity to grow in them and produce toxic substances. The patient who receives **contaminated injectables** is not only at risk of generalized systemic infection but also may suffer a severe febrile reaction to injected pyrogenic microbial by-products (gram-negative or endotoxin shock).

Any type of injectable solution may be involved in this type of accident, sometimes with disastrous results for the patient. Fluids given intravenously; vitamins; vaccines; antisera; anesthetizing solutions used to induce local skin or mucosal anesthesia in dental, ophthalmic, or other surgery or for such procedures as lumbar puncture, biopsies, or bone marrow aspiration—all these and many others may be the source of contaminating organisms capable of causing shock or establishing infection. Continual vigilance is essential to prevent these problems. Careful visual inspection of each ampule, vial, or bottle immediately before its use is helpful only if the material to be injected is in clear solution and is not hidden by a label or by markings on the container. Cracked or otherwise damaged vessels containing injectable material should never be used.

The use of vials containing multiple doses of fluids for injection should be discouraged because each entry into such a container affords a new opportunity for contamination. These vials are generally sealed with a cap containing a rubber diaphragm through which the hypodermic needle is inserted for withdrawal of a dose of material to be administered to a patient. The hole made by the needle as it passes through the rubber may be enlarged every time the vial is used, threatening contamination of the vial's contents by this route. The needle itself may introduce a few organisms. If contaminating bacteria have an opportunity to multiply in the vial before it is used again, the next patient receiving a dose of its contents may acquire a superimposed infection. Vials or bottles containing cloudy material that originally was clear should be discarded immediately. Those containing fluids that are not clear to begin with (e.g., microbial vaccines) present a more difficult problem. The manufacturer's label should be read carefully for information concerning the presence of an antibacterial preservative, date of preparation and sterilization, expiration date, and any instructions for the proper handling, administration, and storage temperature of the material. Whenever possible, injectable solutions should be prepared or purchased in single-dose vials.

Blood or blood products for transfusion (e.g., platelets, fibrinogen) require special emphasis in this regard. Not only are they subject to contamination from extraneous sources, but they may also transfer active infection from donor to recipient. Under well-controlled circumstances, most problems of this type can be avoided by the legal requirements for operation of blood banks, which stipulate that blood donors must be examined for evidence of current active infection and history of acute or chronic disease transmissible by blood. Unfortunately, however, a few diseases of this type exist in latent, asymptomatic form. Would-be donors may be unaware of their infection or may give an inaccurate history of past disease, unintentionally or willfully. Diseases most readily transmissible from "silent" donors are syphilis, malaria, viral hepatitis, and AIDS. In the case of syphilis, routine physical and serologic examination is generally successful in screening infectious donors. Malaria and hepatitis A are not readily diagnosed in their inapparent stages, but a well-taken history can provide evidence of clinically characteristic disease experienced in the past. Hepatitis B is a difficult problem because a number of people carry the viral agent of this disease without showing evidence of clinical illness. Serologic techniques for detecting such individuals have become more reliable in recent years and have greatly alleviated the risk of transfusion hepatitis (see discussion of hepatitis, below).

Acquired immunodeficiency syndrome is a more recently detected, serious disease whose transmission by blood is well documented. In an effort to eliminate the problem of transfusion-transmitted AIDS, blood collection services screen all prospective donors by taking careful and confidential medical histories and excluding all persons from the donor pool who have a history of high-risk behavior (e.g., intravenous drug abuse, homosexual activity, or recent exposure to a prostitute). In addition, all units of blood collected from the remaining pool of donors are screened for the presence of HIV antigen and antibodies. If the blood is found to contain antigen or antibodies to HIV, it is discarded and the donor is notified and advised to seek medical attention. In addition, people who test positive for HIV are permanently excluded from the donor pool. Hemophiliacs have also acquired transfusion-associated AIDS after receiving Factor VIII (clotting agent) concentrates that were manufactured from pooled human plasma. Factor VIII concentrates have now been made safe by heating them to 60° C for 1 hour; a procedure that inactivates the AIDS virus. The activities of blood suppliers in the United States have contributed greatly to the safety of our blood supply.

Some Important Infectious Diseases Associated with Medical and Surgical Procedures

Postpartum Endometritis (Puerperal Fever)

The Clinical Disease

Postpartum endometritis, or **puerperal fever,** is an acute bacterial disease incurred during childbirth or abortion. It begins with local symptoms of bacterial invasion of the genital tract, including the uterus and the Fallopian tubes. If not treated promptly or adequately, it may extend as a peritonitis, usually with septicemia.

Laboratory Diagnosis

I. Identification of the Organisms

Postpartum endometritis may be caused by a number of different bacterial agents, most of which are endogenous to the normal female gastrointestinal and lower genital tract. The organisms involved most frequently are aerobic streptococci (alpha-hemolytic, group B, and enterococci), anaerobic cocci, gram-negative enterics (*E. coli, Proteus, Klebsiella*) and the anaerobic, gram-negative bacillus, *Bacteroides fragilis.* Occasionally, group A beta-hemolytic streptococci acquired from an exogenous source cause serious infection. Mixtures of organisms are frequently found.

II. Identification of Patient's Serum Antibodies

In postpartum endometritis caused by group A streptococci, the serum antistreptolysin titer rises between the onset of acute symptoms and the convalescent stage. Serologic diagnosis has no value when other types of infection induce this syndrome.

III. Specimens for Diagnosis

Pus from the cervix or uterus is appropriate for smear and culture (aerobic and anaerobic).

Blood cultures may be useful if the disease has advanced to a septicemic stage.

When streptococcal infection is suspected, samples of blood should be submitted for acute and convalescent group A streptococcal serologic tests.

Diseases Transmitted by Arthropod Vectors

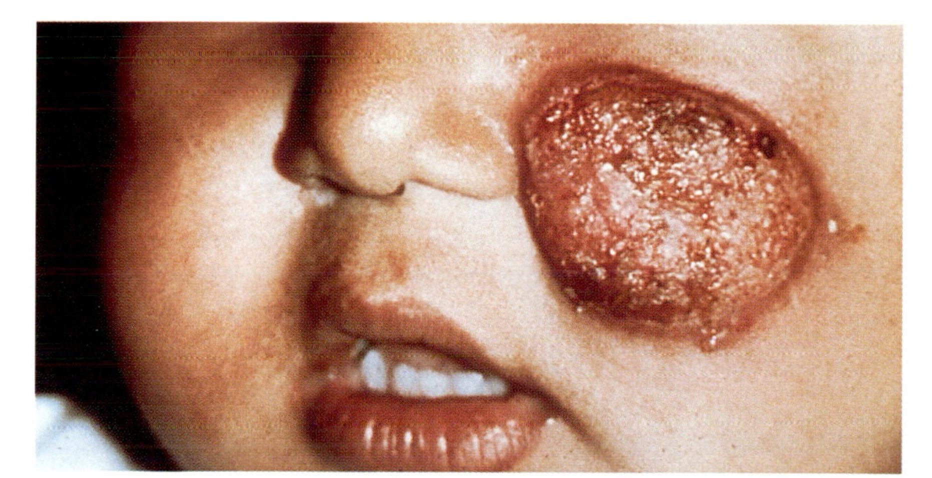

a) The lesion on the face of this child is a form of leishmaniasis.

World Health Organization.

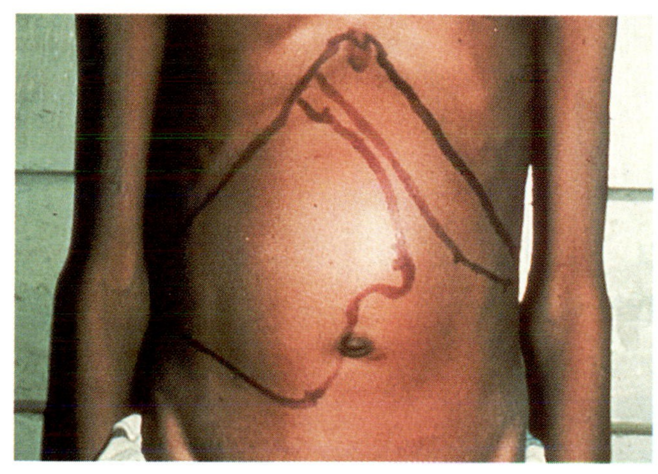

b) An enlarged liver is another sign of leishmaniasis.

World Health Organization.

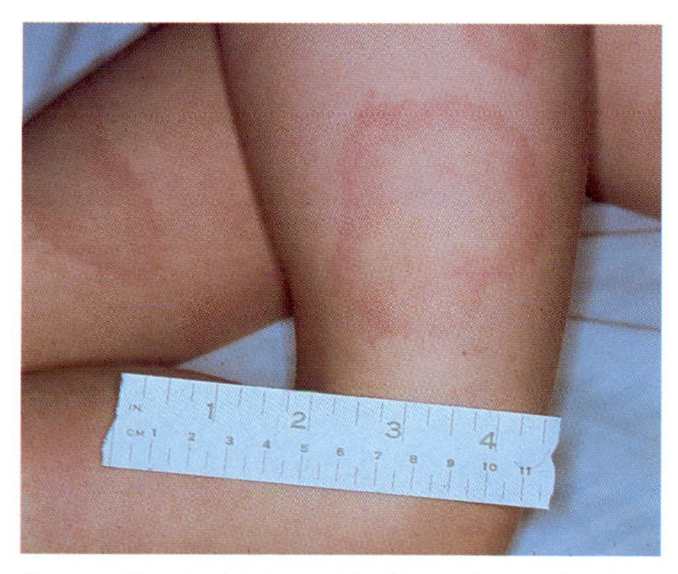

d) A spreading red rash is characteristic of many patients who have Lyme disease.

From A. C. Steere, *Annals of Internal Medicine* 86: 658–698, 1977, American College of Physicians. Courtesy Allen C. Steere, Tufts University.

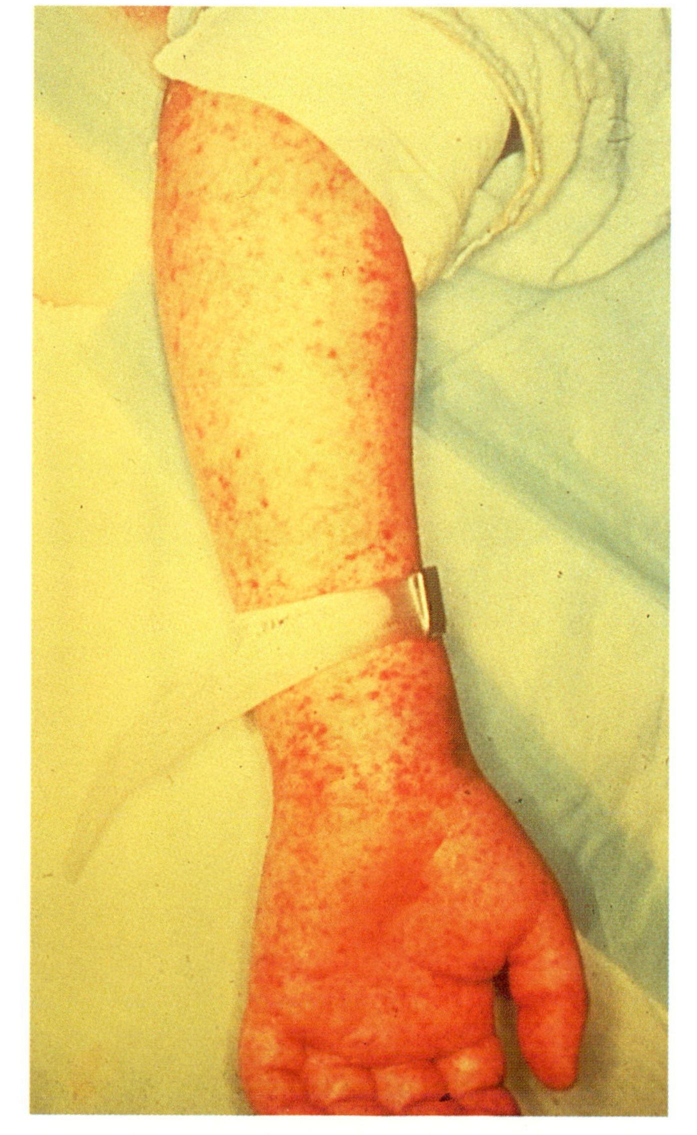

c) This rash is characteristic of Rocky Mountain spotted fever.

Centers for Disease Control, Atlanta, GA.

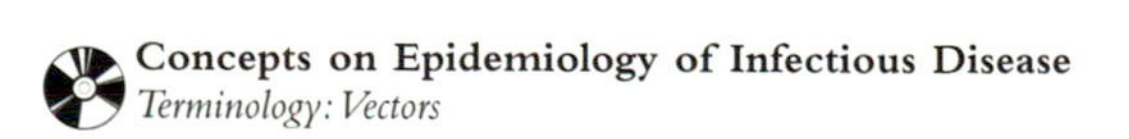

Concepts on Epidemiology of Infectious Disease
Terminology: Vectors

Diseases Caused by Endospore Forming Bacteria

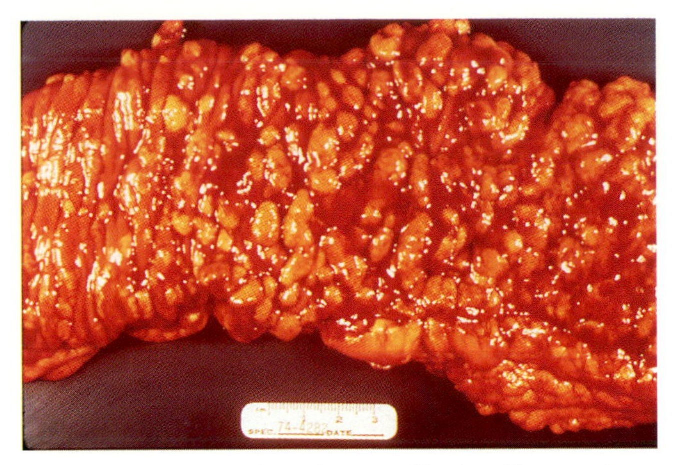

a) Pseudomembranous colitis is characterized by plaquelike mucosal lesions of the bowel.

Courtesy Dr. Stephen Allen, Indiana University Medical School, Indianapolis, IN.

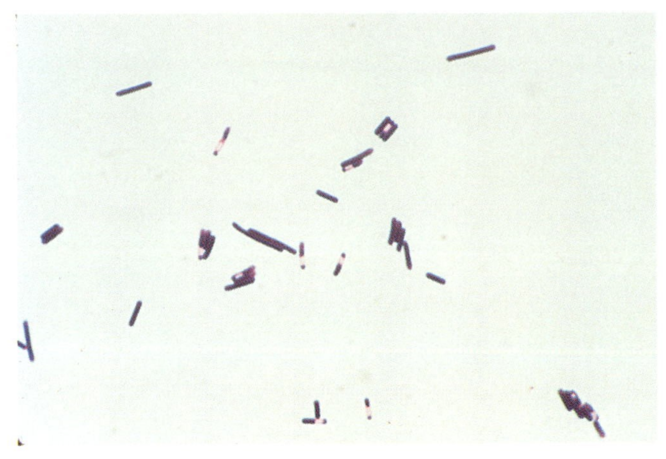

b) *Clostridium difficile,* the organism implicated in pseudomembranous colitis.

Centers for Disease Control, Atlanta, GA.

Anaerobic Bacteria
Gram-Positive Bacilli: Spore Forming Anaerobes

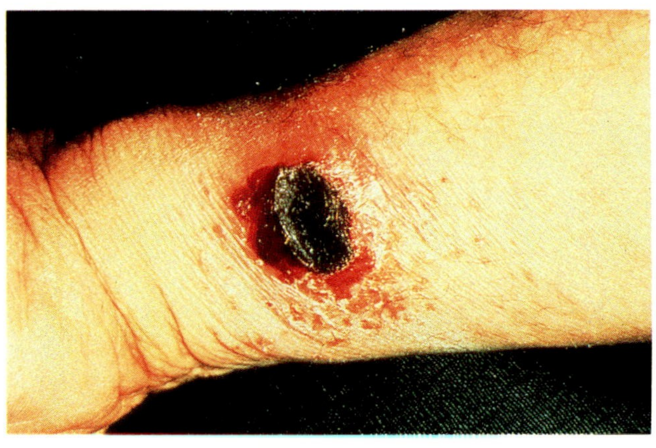

c) Anthrax of the forearm. Butchers who handle infected carcasses or gardeners using contaminated bonemeal can be infected with *Bacillus anthracis.*

From Emond and Rowland, A Colour Atlas of Infectious Disease, 2nd ed. 1987. Wolfe Publishing, London, England. Distributed by Mosby-Year Book.

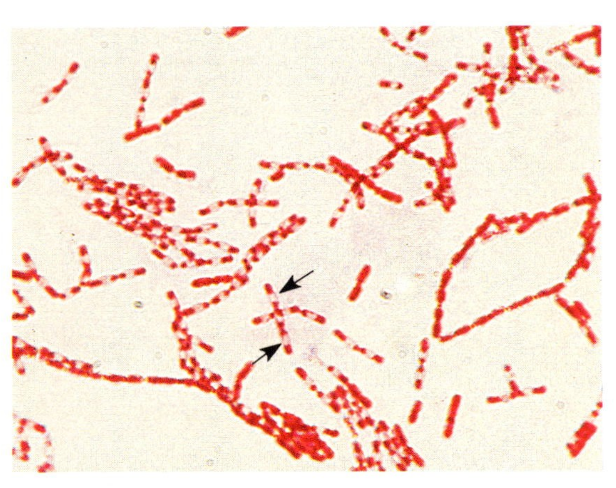

d) *Bacillus anthracis,* an aerobic, endospore-forming bacterium, causes anthrax. The endospores appear as clear areas (arrow) within gram-positive vegetative bacterial cells (×1,000).

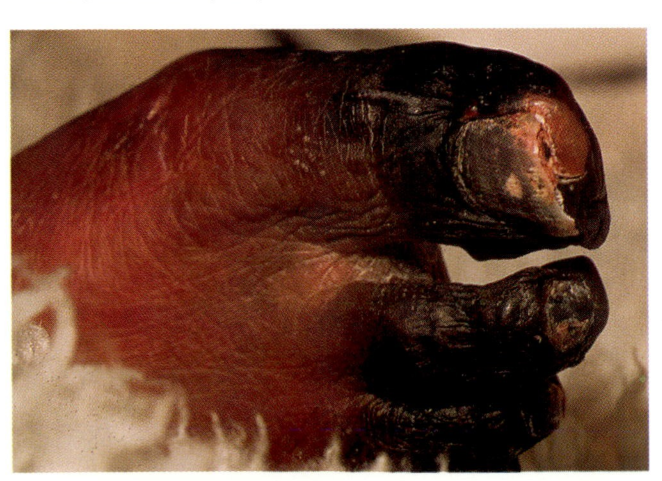

e) Gangrene of the toe. Gas is produced in the tissues by the organism *Clostidium perfringens.*

Armed Forces Institute of Pathology, Washington, D.C.

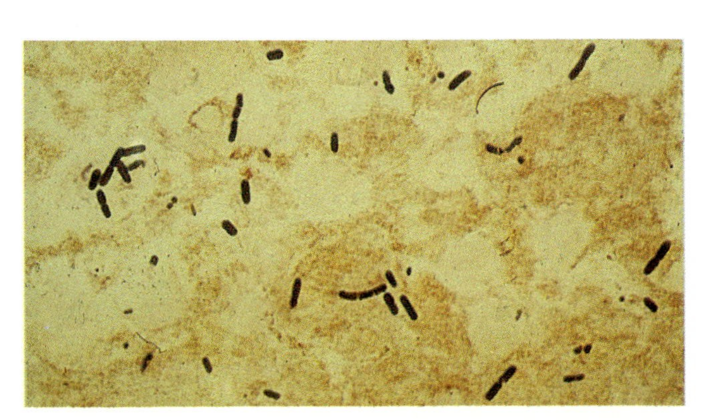

f) *Clostridium perfringens,* an endospore-forming anaerobic bacterium, produces cytotoxins that kill healthy tissue causing the spread of gangrene (×1,000).

Epidemiology

I. Communicability of Infection

A. *Reservoirs, Sources, and Transmission Routes* The human reservoir supplies the most common organisms encountered in postpartum endometritis, whereas the chief source of group A streptococcal infection is the nose, throat, or hands of obstetricians, midwives, nurses, or other attendants, or instruments contaminated by these sources.

Before the mid-1860s, when Joseph Lister first applied Pasteur's concepts of infectious disease to surgery, puerperal fever was a frequent complication of childbirth, and a dreaded one because of its high mortality rate. It was seen most often in hospital maternity wards, but mothers who delivered at home were not safe either, because their physicians or midwives brought the infection to them from other cases they had attended. Wiping their contaminated hands on lapels or aprons, these attendants would set to work, full of confidence, to deliver a new baby and to institute a new infection.

Fully two decades before Lister's aseptic approach to surgery began to impress his surgical colleagues with its success in preventing infection, the first important epidemiologic study of puerperal fever had been reported by an American physician, Oliver Wendell Holmes. In 1843 he presented a paper, "The Contagiousness of Puerperal Fever," to the Boston Society for Medical Improvement. This report created a furor among the leading obstetricians of the day because Holmes dared to present the thesis that puerperal fever was due largely to "the carelessness, ignorance, and negligence of the obstetrician and midwife."[1] A similar conclusion was being reached independently at about the same time by an Austrian physician, Ignaz Semmelweis (see Perspective 21.1). Holmes and Semmelweis did not know each other, but they had much in common. In their own obstetric practices they instituted rules of cleanliness, which included hand washing as well as other careful precautions regarding changes of clothing between infectious cases or after attending autopsies of patients who had died of puerperal sepsis. They managed to keep their patients relatively free of infection, but they were subjected to a great deal of ridicule and abuse from medical colleagues. Semmelweis suffered more from this than Holmes and in the end was so disturbed by his failure to persuade other physicians to accept methods he had used effectively to save women in childbirth that his sanity was affected, and he died in a mental hospital (of sepsis from a wound on his hand!). Holmes's reaction was different:

> *Whatever indulgence may be granted to those who have heretofore been the ignorant causes of so much misery, the time has come when the existence of a private pestilence in the sphere of a single physician should be looked upon not as a misfortune but a crime; and in the knowledge of such occurrences, the duties of the practitioner to his profession, should give way to his paramount obligations to society.*[2]

These words remain even more pointed today when the causes of infectious disease are so well understood, yet sometimes ignored or underestimated.

B. *Incubation Period* Usually between 1 and 3 days.

C. *Communicable Period* Infection caused by normal genital flora is not communicable. Group A streptococcal infection is communicable while purulent discharges continue. When treatment is prompt and adequate, communicability can be controlled in a day or two, but untreated patients may remain infectious for days or weeks.

D. *Immunity* Antibacterial immunity may develop against groups A and B *Streptococcus* and their extracellular products if one of these organisms is the cause of the infectious disease, but susceptibility to other kinds of infection remains general.

II. Control of Active Infection

A. *Transmission-based Precautions* Women who develop postpartum endometritis should be cared for using standard precautions. Appropriate and adequate antimicrobial therapy shortens the period of communicability to 24 hours for group A streptococcal infection. Precautions are maintained for the duration of illness when other organisms are implicated.

B. *Other Precautions* Careful concurrent disinfection is required for vaginal discharges and all equipment soiled by them (bedpans, items required for perineal care, specula). Soiled clothing and bed linens should be placed in individual bags and marked for special handling for transport to the laundry. When endospore-bearing bacteria (e.g., *Clostridium* species) have been demonstrated as the causative agents of postpartum endometritis, all equipment (including linens) must be bagged and autoclaved before cleaning. Disposable items are placed in closed containers and incinerated; heat-sensitive items may be soaked in sporicidal disinfectants or subjected to gas sterilization. When soiling is likely, a clean gown must be worn by each person who attends or examines the patient and careful hand-washing techniques must be observed.

C. *Treatment* Penicillin is the drug of choice in cases of hemolytic streptococcal infection. Erythromycin, a cephalosporin, or others may be effective. Therapy of other bacterial infections is adjusted on the basis of the antimicrobial susceptibility of isolated organisms suspected to be responsible for the patient's disease.

D. *Control of Contacts* Obstetricians, midwives, nurses, and other persons who attend abortions or deliveries should be examined if they have been associated with

[1] Wilson G. Smillie, *Public Health, Its Promise for the Future* (New York: Macmillan, 1955), p. 156.

[2] Oliver Wendell Holmes. The contagiousness of puerperal fever, *Medical Classics.* **1**:243, November 1936 (from *New England Quarterly Journal, Medicine & Surgery.* **1**, 1843, published in Boston by D. Clapp, Jr.).

Perspective 21.1 *Ignaz Semmelweis: A Pioneer in Medical Asepsis*

Ignaz Semmelweis (1818–1865) is remembered as one of the first practicing physicians to recognize the importance of hand washing, and to urge its use on scientific grounds, as a primary defense against the spread of disease, especially among hospitalized patients.

Semmelweis was a Hungarian obstetrician who practiced in a prominent Viennese hospital at the height of his medical career. Obstetrical practice at that time was plagued by puerperal (childbirth) fever which killed young mothers at a rate ranging from 3 to 25 percent, a horrifying statistic then and now. Even more disconcerting, the mortality rate was much higher among hospitalized women than it was among those who delivered at home.

Prevailing medical opinions as to the cause of this problem was uninformed. Medical pundits variously declared it to be the result of "lactation," poor ventilation, or more generally, "miasma." In short, they had no clue, much less a cure. While they debated, women continued to die in childbirth or soon thereafter—especially in hospital obstetrical wards.

Dr. Semmelweis began to take a closer look at the hospital where he worked. It had two obstetrical divisions: the First Division trained medical students; the Second Division trained midwives. He started to keep notes and observed almost immediately that the mortality rate from puerperal fever was four times greater in the First Division than in the Second Division. What could account for this enormous difference?

When a colleague died of a wound infection incurred while performing an autopsy on a woman who had died of puerperal fever, it occurred to Semmelweis that the autopsy had been attended by medical students on the First Division. He was struck with the possibility that the students might carry something from the autopsy room and infect their patients. But what could serve as the carrier? Semmelweis did not consider clothes; doctors in those days wore their frock coats from one patient to the other and even to the autopsy room. He did, however, consider hands—those common mediators. With this in mind, Semmelweis asked the medical students to wash their hands before examining patients. When this proved ineffective, he added chlorinated lime to the hand-washing basins (Perspective fig. 21.1); quickly and dramatically thereafter, the mortality rate on the First Division fell below that of the Second.

The washstand used by Ignaz Semmelweis in his efforts to lower the mortality rate among postpartum women in a maternity hospital.

World Health Organization. Photo by J. Faundez

Unfortunately, the Viennese medical profession ignored this innovative procedure and its implications. No one there, including Semmelweis, as yet understood the nature of infectious disease and the possible routes of its transmission. But Semmelweis had provided some clear evidence of one route that might have led to broader understanding, had anyone paid notice.

Disenchanted with Viennese medicine, Semmelweis finally returned to his native Hungary where he again worked hard to change attitudes. Midwives and a few Hungarian physicians began to practice hand washing. In 1861, some 13 years after leaving Vienna, he published his now-famous treatise: "The Etiology, Concept and Prophylaxis of Puerperal Fever." Still, no one paid attention to his theory.

In 1865, he was driven to insanity, haunted by thoughts of the many women who had died and would continue to die because his proven theory was being ignored. Within six weeks after its onset, he died of the type of septic infection he had fought so long to eradicate.

Had he been born a few years later and in a different place, the tragedy of his life might instead have been medical triumph. He would have heard of Pasteur's work in Paris and of Koch's in Germany, and he would have understood more about the nature and causes of transmissible diseases.

Semmelweis, working with incomplete knowledge, died in obscurity and defeat. Nevertheless, we salute him today as one who perceived and proved a fundamental and timeless medical truth: Hands can carry dangerous infection, but they can easily be prevented from doing so.

Source: "Four Against Infection." 1972. *Hospital Sepsis.* Medical Learning Systems.

patients developing group A streptococcal postpartum endometritis. Cultures of the nose, throat, skin, or active lesions should be taken to establish the possible source of the organism. Individuals harboring group A streptococci and those with other acute or chronic infections should be removed from obstetric duties and treated specifically before they are allowed to return to the floor. Delivery techniques should be carefully reviewed and environmental sources of infection sought, with particular attention to the sterility of instruments or other equipment. These investigations are of particularly urgent concern when there is an unusual incidence of cases within an individual hospital service or practice.

III. Prevention

The prevention of postpartum endometritis depends entirely on maintenance of strict asepsis in obstetric techniques and procedures. It is also essential that the patient be protected from possible sources of infection in the nose, throat, hands, or infectious lesions of those who attend or visit, or of other patients who have communicable diseases.

URINARY TRACT INFECTIONS

Clinical Diseases

Urinary tract infections frequently occur as a result of urologic procedures carried out in medical offices or hospitals. The procedures most commonly involved include operations on the bladder, transurethral resection of the prostate, cystoscopy, urethral or ureteral catheterizations, and the use of indwelling catheters for continuous bladder drainage. These infections may involve colonization and inflammation of the urethra (urethritis), bladder (cystitis), ureters (usually an ascending ureteritis), pelvis or the kidney, or renal parenchyma (pyelonephritis). Infections of the lower urinary tract may be mild, transient, and insignificant, but they may lead to acute or chronic kidney disease, sometimes with permanent damage to renal function. In some instances infectious disease may arise from endogenous sources, particularly when procedural damage inflicted on the epithelial lining of the tract (e.g., by catheterization) provides an opportunity for entry and establishment of commensal organisms from genitourinary mucosal surfaces. In many cases, infection is introduced by faulty technique, by inadequately sterilized or disinfected equipment, or by unwashed hands.

Urinary tract infections are not associated exclusively with such procedures. Kidney infection, for example, may be a complication of systemic dissemination of microorganisms, especially staphylococci, from another entry route. Other types of disease or anatomic problems may also predispose to infection of the urinary system. Stones or tumors may be important foci of bacterial localization.

The frequency with which urinary tract infection occurs in females may have several explanations. The shortness of the female urethra may be a factor, and with young girls the problem may be related to their tendency to retain urine until the bladder becomes quite distended. In adults, tumors sometimes cause urinary retention (fibroid tumors of the uterus, in women; prostatic tumors, in men; bladder tumors), and the resulting distention strongly predisposes to cystitis. Pregnant women may have a similar problem as the developing fetus places increasing pressure on surrounding pelvic structures. During labor, urinary retention is common, and when catheterization is necessary, another risk of infection is added to preexisting possibilities.

LABORATORY DIAGNOSIS

I. Identification of the Organisms

Common agents of urinary tract infections include many species of gram-negative enteric bacilli (*Escherichia coli, Klebsiella, Proteus*); *Pseudomonas aeruginosa; Staphylococcus aureus,* and sometimes coagulase-negative staphylococci; and enterococci. *Candida albicans* is sometimes involved, particularly in infections of older or debilitated persons.

A number of pathogenic organisms associated with systemic disease may cause primary or secondary lesions in the kidney, and these organisms may also be identified in the urine. They include the following:

Mycobacterium tuberculosis (see chap. 12).
Salmonella species (including *S. typhi*) (see chap. 16).
Leptospira interrogans (see chap. 20).

In addition, the blood fluke *Schistosoma haematobium* (see chap. 20) characteristically lodges in pelvic veins, producing ova that migrate through the bladder wall and are excreted in urine.

II. Identification of Patient's Serum Antibodies

Serologic diagnosis generally has no value in infections of the lower urinary tract.

III. Specimens for Diagnosis

Urine specimens for microscopic examination and quantitative culture (see Special Laboratory Tests, IV).

The practice of collecting catheterized urine specimens for bacteriologic studies has been largely abandoned because of the risk of superimposed infection inherent in the technique. This procedure should be reserved for those cases in which a conclusive demonstration of infection is difficult to obtain with voided specimens.

For patients with indwelling catheters, a recommended method for obtaining a urine specimen for laboratory examination is described in figure 21.3.

For routine purposes, clean-voided, midstream urine collections are satisfactory. When quantitative culture results are desired, urine should be delivered to the laboratory within 30 minutes of collection or refrigerated until transport can be arranged. If specimens are permitted to remain at room temperature for periods longer than 1/2 hour, contaminating organisms from external genitourinary mucosa may multiply freely, providing quantitative results that are erroneously high and open to misinterpretation.

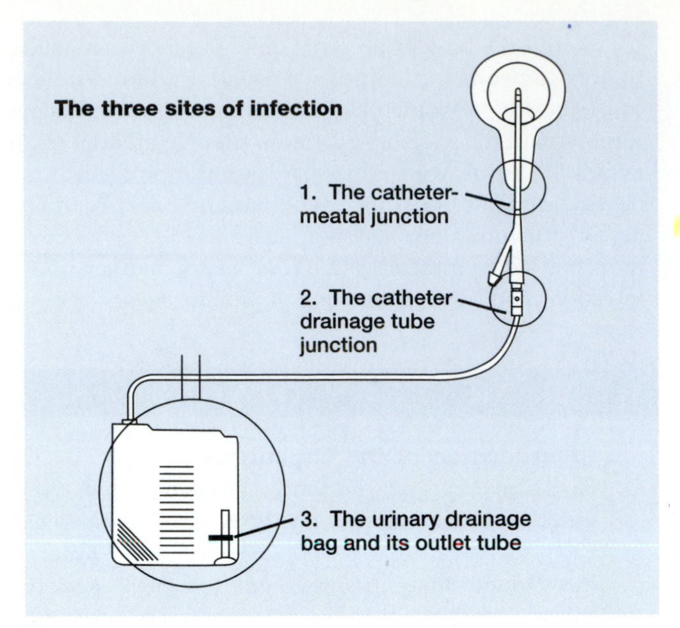

FIGURE 21.3 The three possible sites of infection when a urinary drainage bag is used. To obtain a urine specimen, a sterile syringe and needle are required, and the specimen is aspirated at the catheter drainage tube junction (2).

Courtesy Baxter Healthcare Corporation, Deerfield, Illinois.

Satisfactory specimens can be collected after careful cleaning of the meatus and external mucosa using gauze sponges moistened with tap water and liquid soap. The first and last portions of urine voided should be discarded and only the midstream portion collected in a sterile container with a cover. Male collections offer little difficulty, but when urine samples are collected from women, care must be taken not to permit contamination of the open lip of the vessel by surrounding unsterile surfaces. The labia should first be carefully cleansed and the first passage of urine discarded. The collecting vessel is held close to, but not touching, the meatus, and the patient positioned so that a vertical stream can be delivered directly into the container.

IV. Special Laboratory Tests

Qualitative techniques for examination or culture of urine are adequate when pathogenic species such as the tubercle bacillus or *Salmonella* are suspected. These organisms are of diagnostic significance regardless of their numbers in a given specimen.

Quantitative culture techniques are used to establish the significance of staphylococci, streptococci, or gram-negative enteric bacilli in urine (see chap. 4).

Antimicrobial susceptibility testing is of value only when performed on pure cultures of isolated organisms present in significant number in clean-voided urine specimens.

Epidemiology

I. Communicability of Infection

A. *Reservoirs, Sources, and Transmission Routes* See the initial discussion.

B. *Incubation Period* Generally short (1 to 5 days).

C. *Communicable Period* Continues for as long as urine contains infectious microorganisms such as tubercle bacilli or salmonellae; otherwise not communicable.

D. *Immunity* Susceptibility is general, and immunity does not usually develop.

II. Control of Active Infection

A. *Transmission-based Precautions* Precautions are usually not necessary for patients with urinary tract infections.

B. *Other Precautions* Concurrent disinfection or sterilization procedures should be carried out for all instruments or equipment used for urologic examination or treatment. This includes reusable bedpans, urometers, catheters, cystoscopes, and items used for perineal care. Gowns and gloves should be worn by personnel performing catheterizations or other urologic procedures, and hands should be washed before and after these techniques. Soiled bed linens and clothing should be placed in individual bags marked for special handling in their transport to the laundry.

C. *Treatment* Choice of antimicrobial therapy depends on the results of laboratory studies identifying significant organisms and testing their susceptibility to antimicrobial drugs. Supportive measures often include use of antiseptic solutions as bladder irrigants and drugs that maintain the pH of the urine at alkaline levels that are suppressive for most bacteria.

D. *Control of Contacts* The source of possible contamination of the urinary tract should be sought, particularly when the incidence of cases suggests extraneous reservoirs persisting within a urologic service or practice. Techniques should be carefully reviewed with all personnel and corrective procedures instituted as indicated. Sterilization and disinfection methods should be tested for efficacy using bacteriologic techniques.

III. Prevention

Prevention of urinary tract infections associated with urologic procedures depends on the adequate sterilization or disinfection of equipment, especially catheters and cystoscopes, and on careful maintenance of aseptic technique.

A. *Sterilization or Disinfection of Equipment*

1. *Cystoscopes* These instruments cannot be sterilized by the use of heat in autoclaves or ovens because such methods quickly damage their delicate lens and prism systems beyond repair. They can be sterilized by ethylene oxide gas, and this is the method of choice when gas-sterilizing equipment is available. The method is expensive and time-consuming, requiring an adequate supply of instruments to outlast the 4-hour period of sterilization and a matching period for dissipation of the toxic gas from instrument packs (see chap. 9). Gas sterilization is not destructive to cystoscopes, and this advantage must be weighed, together with safety factors, against the expense and inconvenience of the method.

Disinfection of cystoscopes is possible but requires meticulous attention to detail, including preliminary cleaning and proper application of germicides effective against common and uncommon microbial agents including the tubercle bacillus.

2. *Catheters* Most urethral and ureteral catheters can be successfully sterilized in the autoclave (20 minutes at 121° C, or 250° F). This is true of soft rubber, latex, and woven-base catheters. Hard rubber, "shellac," and "web" catheters may be chemically disinfected with agents used for disinfecting cystoscopes.

The problem of sterilization and disinfection of catheters can best be solved by using presterilized disposable polyethylene catheterizing equipment.

B. *Aseptic Technique in Urologic Nursing Practice* The technique of catheterization is frequently the responsibility of nurses or of orderlies working under nursing supervision. They must apply principles of asepsis throughout the chain of procedures involved in sterilization or disinfection of equipment and preparation of the patient. Precautions taken should be of the same order as those required for a surgical dressing. Before initiating the procedure, the nurse obtains sterile equipment and then positions the patient with proper illumination. After careful hand washing and donning sterile gloves, the nurse cleans the urinary meatus and surrounding mucosal surfaces with a suitable antiseptic, then inserts the catheter with maximum care to avoid contaminating it by contact with the patient's skin or nearby objects. A skillful touch must be developed for this procedure so that the risk of damage to delicate mucosal surfaces is minimized. When the procedure is completed, used materials are discarded into a container that can be closed for transport to an incinerator or an autoclave, and the nurse carefully washes his or her hands once more.

Indwelling catheters are often an infectious hazard because bacterial infection may spread in retrograde fashion from contaminated drainage bottles. This hazard can be reduced by using a closed system that does not permit air contamination of the drainage container or contact between the catheter and the contents of the vessel. Presterilized plastic bags and catheter tubing are available but require frequent attention to ensure that the tubing remains in the proper position. The catheter tubing should never be allowed to come loose from the collecting vessel and to hang free at its distal end where it can be readily contaminated. Also, the bag should not be permitted to fill to the point of contact with the delivery end of the tubing, or retrograde contamination of the bladder may occur. A safe method for collecting urine specimens from an indwelling catheter is illustrated in figure 21.3.

Viral Hepatitis

Viral hepatitis may be caused by hepatitis A virus (HAV); hepatitis B virus (HBV); hepatitis C virus (HCV), formerly known as non-A, non-B hepatitis; the delta agent, or hepatitis D virus (HDV); and hepatitis E virus (HEV). The five hepatitis viruses can be divided into two groups based on their mode of transmission. HAV and HEV are transmitted by the fecal-oral route (chap. 17) and the others are bloodborne.

Hepatitis B Virus

The **hepatitis B virus (HBV)** may well be the cause of the most serious and persistent viral infection on earth, infecting more than 200 million people worldwide. In the United States, HBV is of major public health importance as the number of cases reported continues to increase each year. The Centers for Disease Control estimates that 300,000 people in the United States contract HBV infection every year. Of these, 25% develop acute hepatitis and 6 to 10% become chronic carriers of HBV. Thus, an additional 18,000 to 30,000 people annually are added to the population of chronic carriers in the United States. Chronic carriers are the main reservoir for HBV, and they pose the greatest risk to health care workers because most are asymptomatic and may not be aware that they carry the virus. The epidemiology of HBV is discussed here because of the frequency with which it is indirectly transmitted through blood and needles. Importantly, health care workers can be protected from accidentally acquiring HBV from a contaminated needle-stick injury by receiving the safe recombinant HBV vaccine (see later section, III A).

Hepatitis B virus contains several antigens that are related to disease and infectivity. One is called *hepatitis B surface antigen,* or HBsAg, because it is derived from the outer surface layer of the hepatitis B virion. This antigen commonly occurs in chronic HBV infection and may be found in the blood for many months or years. It is not necessarily associated with active clinical disease, however.

The infectious viral particle (known as the **Dane particle** after its discoverer, D. S. Dane) carries HBsAg on its surface and the core antigen, HBcAg, in its core. During active infection, HBV core components are synthesized in the nuclei of liver cells, while the surface or coat antigens form in the cytoplasm. Completed viral cores pass from the nucleus into the cytoplasm where they are surrounded by surface antigen and are released, fully assembled, into the bloodstream. Some HBsAg is released "empty" into the bloodstream because it is produced in excess of the HBcAg; this is the antigen often found in the blood of patients with chronic HBV infection.

A third HBV antigen, HBeAg, is probably associated with the core of the Dane particle also. The presence of this antigen in circulating blood is highly significant because it tends to correlate well with infectivity. Thus, HBsAg-positive persons who are also positive for HBeAg are usually highly infectious and often transmit this infection to intimate contacts. Persons lacking HBeAg or who have antibody to HBeAg (anti-HBe) usually have low levels of circulating virus and are less likely to spread this disease to their contacts.

Epidemiology of Hepatitis B

I. Communicability of Infection

A. *Reservoirs, Sources, and Transmission Routes* Humans are the only known natural reservoir for the viruses of hepatitis. At one time it was thought that hepatitis B, or serum hepatitis, could be transmitted only by parenteral routes such as inoculation or the transfusion of blood or blood products. It is now known that HBV can be spread in other ways as well, for its surface antigen, HBsAg, can be found in saliva and other body fluids such as semen, urine, cerebrospinal fluid, tears, and breast milk. Transmission may therefore be either direct, between persons in close contact, by virus entering the body through mucous membranes or percutaneously; or indirect, through transfused blood, contaminated needles, or other parenteral means.

Serologic tests for the detection of the surface antigen of HBV in blood indicate that HBV has a significant distribution in the human population but that the risk of infection for most people is small. Among adult volunteer blood donors, for example, about 1 to 5 per 1,000 have HBsAg and are considered potentially infectious. Although blood donors do not represent the entire population, studies of this group and others indicate that the highest risk of infection is among people whose occupation or household exposure results in close, continued contact with HBsAg-positive persons. Nonimmunized hospital staff and other health care personnel are more likely to acquire HBV infection, and within hospitals, the patients or staff in hemodialysis units, surgery units, the laboratory, and blood banks are at greater risk. Sexually active male homosexuals and intravenous drug abusers are also at great risk for contracting hepatitis B.

The principal mechanism of indirect transmission of HBV is by blood transfusion or other parenteral means of transferring virus from the blood of an HBV-positive person. Self-injection of drugs with contaminated needles, use of multiple-dose syringes, tattooing, ear piercing, and handling bloody objects have all been implicated. Drug abusers provide a large reservoir for HBV and are a particular problem because of the frequency with which they offer themselves as paid blood donors. In the United States, prohibition on payment for blood donation as well as federal regulations requiring the testing of all donor blood for HBsAg and the sensitive test methods available for this purpose have markedly reduced the incidence of transfusion-associated hepatitis.

B. *Incubation Period* 40 to 180 days, usually 12 to 14 weeks.

C. *Communicable Period* In experiments with human volunteers, blood taken from infected persons 12 to 13 weeks before the onset of symptoms and up to 8 days following the appearance of jaundice has produced hepatitis. Infected persons who have served as blood donors have been demonstrated to be capable of transmitting infection for many years.

HBsAg or antibody to HBcAg (core antigen) can be demonstrated in the blood of many, but not all, carriers, and many of the blood-positive carriers have never had clinically recognizable hepatitis.

D. *Immunity* Susceptibility to HBV is general, but infection usually confers immunity. Reinfection occurs at a low rate; however, this may be caused by a related virus such as hepatitis C virus.

II. Control of Active Infection

A. *Transmission-based Precautions* Standard precautions are used for hospitalized patients.

B. *Other Precautions* Special care should be taken with syringes, needles, and any other equipment contaminated with the HBV patient's blood. These should be discarded or carefully sterilized before reuse. Suitable sterilization techniques are

1. Autoclaving for 15 minutes at 121° C (250° F).
2. Dry heat for 2 hours at 160° C (320° F).
3. Boiling in water for 10 minutes (100° C, 212° F).

The following chemical disinfectants are presumed to be effective alternatives to heat:

1. 0.05 to 1.0% sodium hypochlorite (500 to 10,000 ppm available chlorine) for 30 minutes.
2. 40% aqueous formalin (16% aqueous formaldehyde) for 12 hours.
3. 20% formalin in 70% alcohol for 18 hours.
4. 2% aqueous acid or alkaline glutaraldehyde for 10 hours.
5. Gas sterilization with ethylene oxide, following manufacturer's recommendations.

C. *Treatment* There is no specific therapy for hepatitis B. After nonimmunized people are exposed to HBsAg-positive inocula (e.g., needle sticks or contact with mucosal surfaces), human hepatitis B immune globulin (HBIG) is administered intramuscularly as soon as possible within a 7-day period, preferably within 24 hours. A second dose is sometimes given 4 weeks after the first although its usefulness has not been established. This immunoprophylaxis is usually successful in protecting against infection but has no effect once the disease is established.

D. *Control of Contacts* A search is made for the source of infection, particularly if such source is associated with blood banks, clinics, medical offices, or hospitals where large numbers of people are bled as donors, treated with parenterally administered medications, or given blood transfusions. Case reports are helpful in determining the incidence of the problem and the techniques required to control it.

III. Prevention

A. *Immunization* An effective recombinant DNA hepatitis B vaccine has been available since 1986 and universal vaccination of children was recommended in 1991. The vaccine consists of a highly purified HBsAg produced by a yeast into which a plasmid containing the gene for hepatitis B surface antigen has been inserted. The vaccine is administered in a series of three intramuscular doses, with the first given between birth and 2 months of age and the second and third doses given at 4 and 6 months, respectively. If they have not been previously immunized, vaccination is recommended for those persons at high risk of acquiring the disease (see previous section, I, A). The vaccine field trials suggested that the vaccine is partially effective even when given early in the course of established infection. Presumably this is related to the long incubation period of HBV disease. At present the only patients for whom postexposure vaccination is recommended are infants born to HBsAg-positive mothers who should receive both HBIG and vaccine. Passive immunization with HBIG does not interfere with the active immunity induced by vaccine.

B. *Control of Blood Donors* As part of **blood bank precautions,** persons who have themselves received a blood transfusion within the past 6 months are not accepted as donors. Persons with a history of clinical hepatitis at any time are rejected, as are those known or suspected to be narcotics addicts.

Prospective blood donors are screened serologically for HBsAg or anti-HBcAg in their blood and rejected if positive.

The names of persons developing posttransfusion hepatitis are reported to local health agencies, together with the identification numbers of donors who supplied blood in each case. The health agency then circulates to all blood banks a current list of patients and donors involved so that the banks can check all future donors against the names of known carriers.

C. *Control of Blood or Blood Products* The transfusion of unscreened whole blood is limited to those for whom it represents an immediate therapeutic necessity. The use of unscreened, pooled blood products is particularly discouraged because pooling increases the chance of including virus-positive blood.

D. *Control of Equipment* Disposable, presterilized needles, syringes, lancets, tubing, and blood bottles should be used whenever possible. When nondisposable items must be used, they should be sterilized or disinfected adequately before reuse. Fresh, sterile items must be used for every injection or finger puncture.

E. *Personal Hygiene* Good personal hygiene is fundamental in the prevention of hepatitis B infection. Careful hand washing is the single most important practice, but prevention rests on avoiding the most obvious sources of HBV.

Hepatitis C Virus

With the availability of serologic tests for diagnosing hepatitis A and B, patients with clinical, but not immunologic, evidence of hepatitis were identified. This condition was named non-A, non-B hepatitis because the diagnosis was made primarily by *excluding* the infections caused by the two viral agents then known. In 1988, one agent responsible for non-A, non-B hepatitis was discovered; it was called **hepatitis C virus (HCV).**

HCV generally produces a mild disease, and more than 50% of patients may be symptom-free. The incubation period ranges from 42 to 103 days. In acute disease, the manifestations are similar to those of acute hepatitis B. The patient may become a chronic carrier of the virus and, thus, an important vehicle for its spread. Transmission is by both systemic and nonsystemic routes, although exposure by transfusion of blood or blood products and needle sticks have been the most clearly documented. One report implicates this virus in a hepatitis A-like foodborne outbreak. HCV hepatitis is currently recognized as the most common form of posttransfusion hepatitis in the United States (accounting for 80 to 90% of the cases) and is seen frequently in dialysis patients and drug abusers. A serologic test is available to identify patients infected with HCV.

Hepatitis D Virus

Hepatitis D virus (HDV), also known as the *delta agent,* is found only in people who are also infected with HBV. HDV is transmitted parenterally and the disease of patients coinfected by HDV and HBV is more severe. Because HDV cannot cause infection in the absence of HBV, the HBV vaccine prevents both types of infection.

Questions

1. Why are hospitalized patients particularly susceptible to infection?
2. Why are nosocomial infections frequently difficult to treat?
3. What are the sources associated with the hypodermic route of infection?
4. What are the sites of infection with the use of a urinary drainage bag? What techniques are used to minimize the risk of infection?
5. What is postpartum endometritis? What organisms are responsible for this disease? How do women become infected?
6. What organisms cause viral hepatitis? Why is HBV infection a serious problem?
7. How is HBV infection prevented and controlled?

Bibliography

Benenson, A. S., ed. 1995. *Control of communicable diseases in man.* 16th ed. Washington, D.C.: American Public Health Association.

Brosnan, K., A. M. Parham, B. Rutledge, D. J. Baker, and J. S. Redding. 1988. Stopcock contamination. *Am. J. Nurs.* 88 (3): 320–323.

Centers for Disease Control and Prevention. 1988. Changing patterns of groups at high risk for hepatitis B in the United States. *Morbid. Mortal. Weekly Report* 37: 429–432, 437.

Centers for Disease Control and Prevention. 1990. Nosocomial transmission of hepatitis B virus associated with a spring-loaded fingerstick device—California. *Morbid. Mortal. Weekly Report* 39 (35): 610.

Conti, M. T., and L. Eutropius. 1987. Preventing UTIs: What works? *Am. J. Nurs.* 87 (3): 307–309.

Culpepper, M. M., and P. G. Adams. 1988. Nursing in the Civil War. *Am. J. Nurs.* 88 (7): 981–984.

Efforts continue to unravel mysteries of enigmatic HCV. 1993. *Hosp. Infect. Control* 20 (8): 105–109.

Gould, D. 1991. Keeping on tract: An overview of recent research on the management of indwelling catheters, indicating areas where further work is needed. *Nurs. Times* 90 (10): 58–62.

Grau, P. A. 1991. Are you at risk for Hepatitis B? *Nursing* 21 (3): 44–46.

Hamory, B. H. 1989. Hepatitis B vaccines: Who, when, why and how much? *Am. J. Infect. Control* 17 (3): 119–120.

Jacobson, E. 1990. Hospital hazards, Part 2: How to protect yourself. *Am. J. Nurs.* 90 (4): 48–53.

Kaspar, L. N., and P. M. Peter. 1991. Wound care and dressings. In *Basic nursing skills and concepts.* Mosby Year Book. St. Louis: Mosby.

Khatib, R., et al. 1994. A common source outbreak of gastroenteritis in a teaching hospital. *Infect. Control Hosp. Epidemiol.* 15 (8): 534–535.

Lott, J. W., et al. 1994. Keeping up with neonatal infections: Designer bugs, Part 1. *Matern. Child Nurs.* 19 (4): 207–213.

Nystrom, B. 1994. Impact of handwashing in intensive care: Examination of the evidence. *Infect. Control Hosp. Epidemiol.* 15 (7): 435–436.

Rhame, F. S. 1989. Nosocomial aspergillosis: How much protection for which patients? *Infect. Control Hosp. Epidemiol.* 10 (7): 296–299.

Roueche, B. 1991. Three sick babies. In *The medical detectives.* New York: Plume.

________. 1991. The west branch study. In *The medical detectives.* New York: Plume.

Smith, F. 1994. Pseudomonas infection. *Nurs Times* 90 (46): 55–56.

Technique with multidose vials leads to pseudomonad infections. 1990. *Hosp. Infect. Control.* 17 (6): 73–74.

Thomas, D. J. 1989. Neonatal nosocomial infections. *Crit. Care Nurs. Q.* 11 (4): 83–94.

Venezia, R. A., M. D. Agresta, E. M. Hanley, K. Urquhart, and D. Schoonmaker. 1994. Nosocomial legionellosis associated with aspiration of nasogastric feedings diluted in tap water. *Infect. Control Hosp. Epidemiol.* 15 (8): 529–533.

CHAPTER 22

Infectious Diseases Acquired from Arthropod Vectors

OUTLINE

Microbes in Motion

This chapter covers the important disease-causing bacterial, viral, and parasitic agents that are acquired from arthropod vectors. A discussion of their laboratory diagnosis, epidemiology, control, and prevention is included. The following books and chapters of the Microbes in Motion CD-ROM may be a useful preview or supplemental study aid to your reading: Concepts on Epidemiology of Infectious Disease: Terminology. Miscellaneous Bacteria in Human Hosts: Zoonoses. Parasitic Structure and Function: Tissue and Blood Protozoa; Nematodes.

OBJECTIVES

On completing this chapter the student will be able to:

1. cite the major types of infectious diseases transmitted by biting insects.
2. describe the role of infected humans in the epidemiology of these diseases.
3. outline the clinical nature of Lyme disease, typhus fever, yellow fever, and malaria.
4. cite the causative organisms of these diseases.
5. name the insect vectors associated with each disease.
6. describe the methods used for their laboratory diagnosis.
7. discuss the epidemiology and methods for controlling and preventing the diseases.

KEY WORDS

Arbovirus diseases
Arbovirus encephalitides (en-seff-ah-**LIT**-ah-dees)
Arthropod vectors
Babesiosis (bab-ease-ee-**OSE**-iss)
Borrelia burgdorferi (bore-**EEL**-ee-ah burg-**DORF**-er-eye)
Borrelia recurrentis (reck-er-**ENT**-iss)
Ehrlichiosis (air-lick-ee-**OSE**-iss)
Filariasis (fill-are-**EYE**-ah-sis)
Leishmania (leash-**MANE**-ee-ah) species
Leishmaniasis (leash-man-**EYE**-ah-sis)
Loiasis (lo-**EYE**-ah-sis)
Lyme disease
Malaria
Onchocerciasis (on-co-serk-**EYE**-ah-sis)
Plague
Plasmodium (plaz-**MO**-dee-um) species
Relapsing fever
Rickettsial (rick-**ETT**-si-al) diseases
Spotted fever group
Trypanosoma (trip-**PAN**-oh-so-mah)
Trypanosomiasis (trip-**PAN**-oh-sohm-**EYE**-ah-sis)
Typhus fever group
Yellow fever
Yersinia pestis (yer-**SIN**-ee-ah **PEST**-iss)

Introduction

A large group of human or animal diseases is perpetuated in nature by arthropod vectors that transmit infectious agents to mammalian hosts and play a vital role in the developmental cycle of the microorganisms. The microbial agents of this group fall in each of the major classes of pathogenic microorganisms except the fungi. The human diseases are presented here in groups according to the etiologic agent.

The epidemiology of these infections varies and holds different implications for human beings. They may be accidental hosts and victims of parasites maintained primarily by animals and arthropods, or they may be the primary or only mammalian reservoir, sustaining the parasite sequentially with insect hosts. Human beings usually acquire these diseases only from **arthropod vectors,** and they are not directly transmissible from one person to another. Plague is an exception to this rule. As described in chapter 12, the pneumonic phase of plague is directly transmissible from person to person.

Arthropods and their role as hosts of infectious disease are described in chapter 8 and methods for their control are discussed in chapter 10.

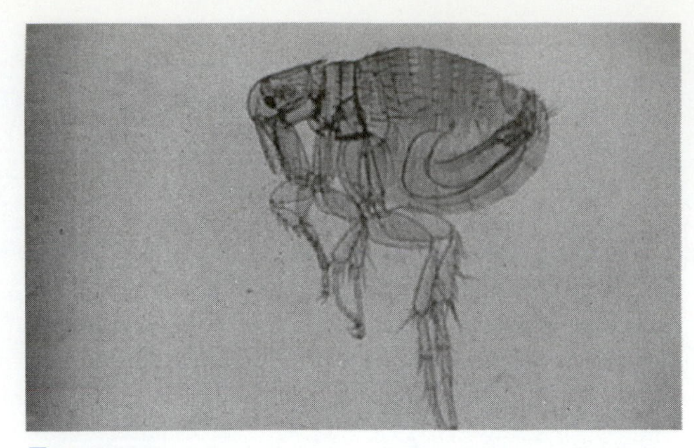

FIGURE 22.1 Bubonic plague results from the bite of a flea, *Xenopsylla cheopis.* The flea carries the microorganism *Yersinia pestis* that it acquires by biting an infected rodent.

Centers for Disease Control, Atlanta, GA.

Concepts on Epidemiology of Infectious Disease
Terminology: Vectors

Miscellaneous Bacteria in Human Hosts
Zoonoses: Yersinia

Bacterial diseases

Plague

The Clinical Disease

Plague is a virulent disease that has been a threat to human societies for centuries. During the course of human history, epidemics of the Black Death have devastated populations in various parts of the world. Some of the most ravaging epidemics occurred during the Middle Ages in the crowded cities of Europe, India, and Asia, sometimes killing as many as two-thirds of the population (see Perspective 22.1).

The last pandemic of plague began in Hong Kong in 1894. As it spread through large segments of the world, it left an estimated 43 million people dead. It was during that outbreak that the causative organism was discovered, and the epidemiology of the disease began to be understood. Since then plague has come under control in most parts of the world and is rarely seen in epidemic form today. It remains endemic in the western United States and in other areas of the world where the wild rodent population harbors the organism, but human infection is limited to sporadic cases involving unusual exposure to rodents or their fleas. Public health controls have prevented further spread from such cases.

The plague bacillus produces two principal forms of human illness: bubonic and pneumonic plague. Other clinical phases, such as septicemic or meningeal plague, sometimes occur.

The bubonic form results from the bite of a flea (fig. 22.1) that is carrying the organisms it acquired by previously biting an infected rodent. The injected bacilli are carried along the lymphatic channels into regional lymph nodes, the thoracic duct, and the bloodstream, which disseminates them throughout the body. The most usual sites of dissemination are to the spleen, liver, and lungs. Symptoms begin suddenly, with fever, chills, severe headache, and exhaustion. The patient may have vomiting and diarrhea, and often becomes delirious. Damage to lymph nodes results in hemorrhagic inflammation, swelling, and pain. The swollen lymph nodes are called "buboes" and occur most frequently in the axilla or groin. They may become necrotic and suppurate to the surface. Hemorrhagic lesions in other parts of the body may also become necrotic. In the skin these necrosing areas appear very dark, hence the term *Black Death.* If septicema ensues, there may be destructive lesions in every part of the body, with pleural or peritoneal effusions, pericarditis, or even meningitis. This form of plague is not directly transmissible to other persons but may be carried to others by fleas.

The pneumonic form of plague results when organisms localize in the lungs of the victim of bubonic plague. This form is highly contagious, being spread by the coughing of the infected patient. Cases of primary plague pneumonia may then occur among people in contact with the patient (see chap. 12).

Laboratory Diagnosis

The clinical diagnosis of plague is sometimes difficult before characteristic buboes appear. In the early stages the symptoms may resemble those of other diseases, so laboratory diagnosis may be of critical importance.

I. Identification of the Organism

(Yersinia pestis).

A. *Microscopic Characteristics* The plague bacillus is a short gram-negative rod, nonmotile and nonendospore-

PERSPECTIVE 22.1 *The Black Death*

Today, safe as we are from the great waves of infection that once carried large numbers of people to their deaths, it is difficult to imagine life during the huge epidemics that once took place. Among those epidemics, bubonic plague reigns as one of the greatest killers of all time (Perspective fig. 22.1).

Following are two descriptions that may give us a clearer picture. The first describes some of the horrors of the infection and the death that resulted. The second provides some sense of the times from the point of view of those whose job was to clean up after a sweeping epidemic.

1. In her book *A Distant Mirror,* Barbara Tuchman presents a chapter titled "This Is the End of the World: The Black Death" in which she describes the main features of bubonic plague. This was an epidemic that wiped out one-quarter to one-half of the population of Europe in the 1300s, and Tuchman depicts in detail the symptoms of those who became infected. Black swellings, about the size of an egg or apple, appeared in the armpit and groin. The swellings oozed blood and pus, caused severe pain, and were followed by spreading boils and black blotches on the skin from internal bleeding. These patients died quickly, often within five days of the first symptoms. She also characterized the pneumonic form of plague: the fever, the coughing and spitting of blood, the heavy sweating, and the even more rapid death. Of both the bubonic and pneumonic forms of plague, Tuchman had this to say: "In both types everything that issued from the body—breath, sweat, blood from the buboes and lungs, bloody urine, and blood-blackened excrement—smelled foul. Depression and despair accompanied the physical symptoms, and before the end 'death is seen seated on the face.'"

2. Nearly 400 years after the epidemic described by Tuchman, plague again threatened England and Europe. Daniel Defoe published his *Journal of the Plague Year* in 1722 in which he attempted to caution the people by describing some of the events of the earlier epidemic of 1665. The following excerpt details some of the harsh problems of dealing with massive numbers of the dead:

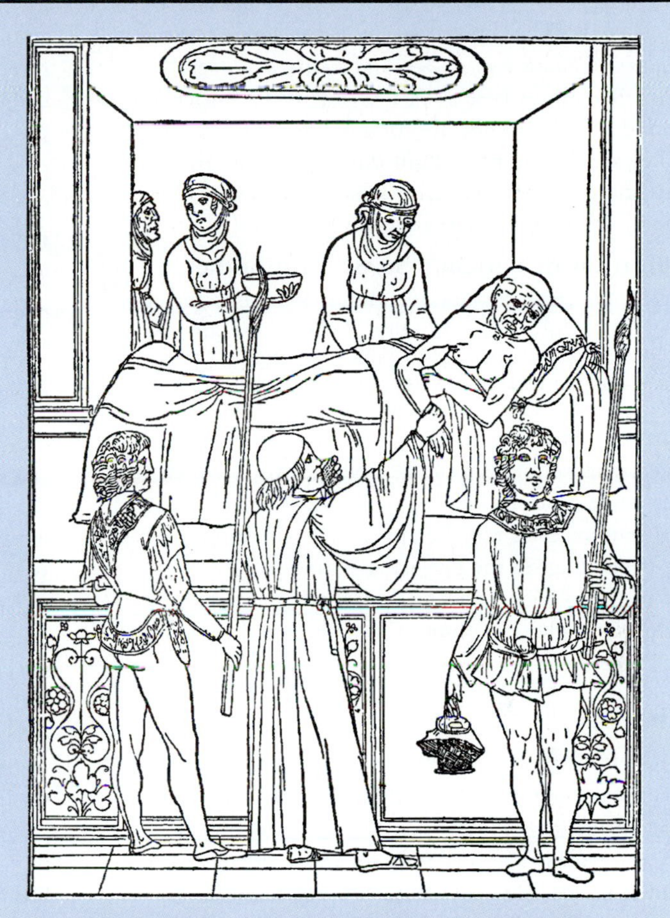

A physician tends a victim of bubonic plague, the Black Death.
Culver Pictures, Inc.

"This brings these two men to a further remembrance. The name of one was John Hayward who was at that time under-sexton of the parish of St. Stephen, Coleman Street. By under-sexton was understood at that time gravedigger and bearer of the dead. This man carried, or assisted to carry, all the dead to their graves which were buried in that large parish, and who were carried in form, and after that form of burying was stopped, went with the dead-cart and the bell to fetch the dead bodies from the houses where they lay, and fetched many of them out of the chambers and houses; for the parish was, and is still, remarkable particularly, above all the parishes in London, for a great number of alleys and thoroughfares, very long, into which no carts could come, and where they were obliged to go and fetch the bodies a very long way; which alleys now remain to witness it, such as White's Alley, Cross Key Court, Swan Alley, Bell Alley, White Horse Alley, and many more. Here they went with a kind of hand-barrow and laid the dead bodies on it and carried them out to the carts; which work he performed and never had the distemper at all, but lived about twenty year after it, and was sexton of the parish to the time of his death. His wife at the same time was a nurse to infected people, and tended many that died in the parish, being for her honesty recommended by the parish officers, yet she never was infected neither.

"He never used any preservative against the infection, other than holding garlick and rue in his mouth, and smoaking tobacco. This I also had from his own mouth. And his wife's remedy was washing her head in vinegar, and sprinkling her head-cloaths so with vinegar as to keep them always moist; and if the smell of any of those she waited on was more than ordinary offensive, she snuffed vinegar up her nose and sprinkled vinegar upon her head-cloaths, and held a handkerchief wetted with vinegar to her mouth."

Source: Barbara Tuchman, *A Distant Mirror,* 1978, New York: Alfred A. Knopf, Inc. and Daniel Defoe, *Journal of the Plague Year,* 1960, New York: Signet Classics.

forming. With special stains such as Giemsa's, the organisms characteristically show bipolar staining, which gives them the appearance of safety pins. They can be positively identified by immunofluorescent techniques applied to smears of exudates.

B. *Culture Characteristics* ***Yersinia pestis*** is an aerobic organism that grows best at 30° C on media enriched with blood or tissue fluids. Its colonies are gray and viscous. They are distinguished from other species of *Yersinia* or from *Pasteurella* (with which they were formerly classified), by their biochemical reactions.

II. Identification of Patient's Serum Antibodies

The passive hemagglutination technique demonstrating at least a fourfold rise in antibody titer during the course of disease is most commonly used to establish a serologic diagnosis.

III. Specimens for Diagnosis

Multiple blood samples should be collected for culture.

Pus aspirated from buboes, paracentesis fluids, spinal fluid, or sputum each may be appropriate for smear and culture, depending on the localizing symptoms.

Blood should be collected for serologic testing as soon as possible after the onset of symptoms and again after 2 or 3 weeks.

Epidemiology

I. Communicability of Infection

A. *Reservoirs, Sources, and Transmission Routes* Humans are an accidental host for the plague bacillus, which has a primary reservoir in rats and wild rodents (squirrels, field mice, and others). Rodent fleas transmit the infection among animals. When the flea ingests organisms in a blood meal taken from an infected rodent, the organisms multiply and obstruct its feeding passage. The next time the flea bites, the new meal cannot be swallowed, so that the blood eddies over the obstructing mass of bacilli and is regurgitated back into the bite wound of the current victim, injecting a mass of organisms. Fleas are most likely to bite human beings present at the time the rodent host dies, leaving the insect parasite in search of a new source of food.

Plague can also be acquired through direct contact with infected animal tissues, and accidental infections have been reported among laboratory workers. The most hazardous situations arise in cases of pharyngeal or pneumonic infection, transmissible by inhalation of droplet aerosols from infected patients. This can occur whether the pneumonic phase is primary or secondary to bubonic plague.

Plague in humans is relatively rare in the United States today. Where it exists, the disease is associated with wild animal reservoirs in the western mountain regions of the country, especially New Mexico and Colorado.

Urban plague has been controlled throughout the world, but sylvatic (wild rodent) plague remains endemic in many areas, e.g., South America, parts of Africa, the Near East, Indonesia, and Asia, as well as the western American states.

B. *Incubation Period* Two to 6 days. Average time for development of pneumonic plague is 1 to 6 days.

C. *Communicable Period* Rat fleas may remain infected for months. Bubonic plague is not directly transmissible among people, but pneumonic plague is highly communicable until infection is successfully cleared by antimicrobial therapy.

D. *Immunity* Recovery from plague is usually accompanied by a durable immunity. This acquired immunity offers the only natural resistance to the human disease.

II. Control of Active Infection

A. *Transmission-based Precautions* If hospitalized, the patient with bubonic plague should be treated immediately for flea infestation and cared for with standard precautions.

Patients with pneumonic plague must be cared for with droplet precautions against airborne spread of infection (see chap. 11). Such precautions must be maintained until at least the third day following the start of antimicrobial therapy or until the patient clinically responds to treatment.

B. *Other Precautions* Health care workers attending pneumonic plague patients must observe stringent precautions to protect themselves and to avoid spreading disease to others. Masks, gowns, and gloves should be worn and the patient should be cared for in a private room. Prophylactic doses of antimicrobial agents offer further protection for personnel. The patient's excretions, purulent discharges, sputum, contaminated equipment, and bed linens must be disinfected concurrently. Terminal disinfection-cleaning must be thorough. Dead bodies must be handled with strict asepsis.

C. *Treatment* Antimicrobial therapy is effective if instituted early. In pneumonic plague prognosis is best when treatment is begun within 24 hours of onset. Streptomycin and tetracyclines are the drugs of choice and are often given in combination. Chloramphenicol is an effective alternative.

D. *Control of Contacts* When cases of plague are reported, the immediate concern is to protect contacts and prevent spread by rodent fleas or by airborne routes if pneumonic plague develops. Close contacts of active cases are dusted with insecticide powders, and observed closely during a 6-day period for signs of infection. Antimicrobial therapy is instituted with the first evidence of febrile illness. Where these procedures are not feasible, chemoprophylactic drugs are administered to contacts daily throughout the 6-day incubation period.

III. Prevention

A. *Immunization* Killed vaccines are available that provide protective immunity for at least 1 year. Immunization is recommended for persons at high risk. Laboratory workers, medical personnel in areas where infection is endemic in the rodent population, travelers, and others whose occupation may expose them to the animal source of infection should be vaccinated.

B. *Control of Reservoirs* The most effective method for preventing urban plague is destruction of rats and their fleas. Rat eradication campaigns attempt to eliminate unsanitary conditions that encourage rat breeding. When rodent controls are applied on a mass scale, simultaneous measures to eliminate rat fleas are necessary to prevent spread of hungry insects into the nearby human population.

Public health measures also include the ratproofing of buildings in sea- and airports, quarantine of ships from areas where plague has been reported, fumigation of such vessels as indicated, and frequent surveys of the native rodent population for evidence of plague infection.

LYME DISEASE

The Clinical Disease

Lyme disease was first recognized in 1975 when a cluster of children living in Lyme, Connecticut, developed an arthritic illness. Intensive epidemiologic investigations revealed that the disease is a tickborne spirochetal infection, widespread in the United States and also occurring in Europe, the Far East, the former Soviet Union, and Australia.

The disease has several stages, some of which mimic other illnesses and make diagnosis difficult. The initial infection is a lesion that appears after 3 to 14 days at the site of a tick bite. This hallmark lesion, called *erythema chronicum migrans,* (or erythema migrans), is bright red and spreads out from a central papular area that later clears. In many patients, multiple lesions appear at other sites after several days (colorplate 11d), some of which may coalesce. During this time, the patient may have a low-grade fever, headache, arthralgias and myalgias, and a stiff neck. If the infection is not treated, the disease progresses to the second and third stages after a few weeks or months. These stages involve the musculoskeletal system, the nervous system, and the cardiovascular system. Arthritis occurs in one or more joints, may last for weeks, and sometimes recurs several times. The patient has symptoms of meningeal irritation accompanied by headache, sharp pains in the trunk or extremities, facial palsy, and meningitis with lymphocytic cells in the cerebrospinal fluid. The heart disorder is a diffuse inflammation of the heart muscle and lining (myopericarditis) and is almost always self-limited, lasting from a few days up to 6 weeks. In some patients, defects in electrical conduction (heart block) interfere with normal cardiac function, but heart failure is rare.

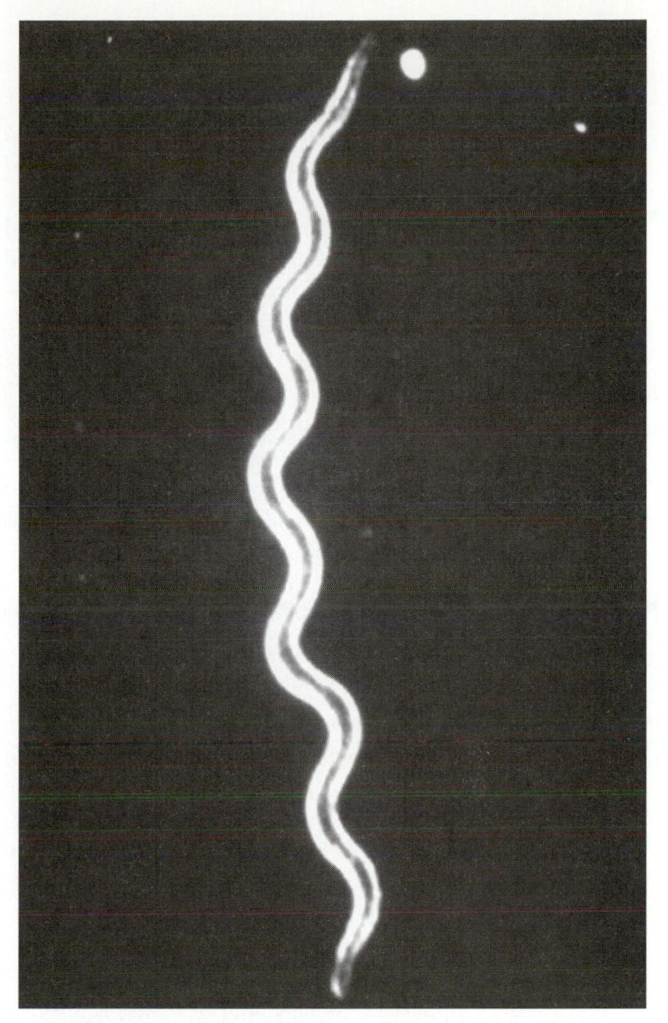

FIGURE 22.2 *Borrelia burgdorferi*, the agent of Lyme disease.

Miscellaneous Bacteria in Human Hosts
Zoonoses: Borrelia

LABORATORY DIAGNOSIS

I. Identification of the Organism

(Borrelia burgdorferi).

The spirochete is about 10 to 30 μm long with 7 to 11 flagella wound around the bacterial cell (fig. 22.2). By phase-contrast or dark-field microscopy, it can usually be distinguished from other borreliae by its looser, more irregular coiling. Although borreliae are gram-negative, they are difficult to see with the Gram stain, therefore Giemsa or silver stains are preferred. Direct fluorescent antibody tests are also used to visualize the organism, and nucleic acid probe detection tests are under development.

Borrelia burgdorferi can be grown in special culture media, but these are not widely available and the yield of positive cultures is low, even in confirmed cases of disease. Laboratory animals, including hamsters, rats, and mice, can be infected.

II. Identification of Patient's Serum Antibodies

Enzyme immunoassays and immunofluorescence assays to detect IgG and IgM anti-*B. burgdorferi* antibodies are available for laboratory diagnosis. Because many of these tests are not well standardized among different laboratories, interpretation of results is often difficult. False-positive serologic tests may be obtained in patients with infectious mononucleosis, rheumatoid arthritis, or other spirochetal diseases. Immunoblot tests may be used to confirm positive enzyme immunoassay tests in patients who do not have a suggestive clinical history.

III. Specimens for Diagnosis

Blood specimens are obtained for serologic diagnosis. IgM antibodies reach a peak after 3 to 6 weeks of infection, and IgG antibodies are highest months to years later. Because of the problems of nonstandardized serologic tests and false-positive reactions, these tests should be performed only when clinical evidence strongly suggests the diagnosis of Lyme disease.

Epidemiology

I. Communicability of Infection

Lyme disease is the most commonly reported vectorborne disease in the United States. The organism is transmitted only through the bite of several tick species in the genus *Ixodes* and is not communicable from person to person. The ticks feed primarily on wild rodents and deer. Cases have been reported from at least 47 states, but most occur in the northeastern, mid-Atlantic, and north central states, and the northern Pacific coastal areas.

II. Control and Prevention

Control depends on avoiding sites suspected to be infested with ticks, using tick repellents, and wearing protective clothing. When it is necessary to spend time in tick-infested areas, skin should be inspected frequently and any ticks found should be removed promptly with a tweezers, taking care to remove their mouth parts.

Ceftriaxone is the drug of choice, but amoxicillin and tetracyclines are also effective, especially in the later stages of disease. Prophylactic therapy after tick bites is not recommended.

RELAPSING FEVER

The Clinical Disease

Relapsing fever is a spirochetal infection characterized by repeated, alternating febrile and afebrile episodes that last 2 to 3 days. This syndrome of relapse and remission may continue for 2 or 3 weeks. Symptoms are related to the presence of spirochetes in the bloodstream. Fever, chills, headache, and malaise mark the release of organisms from the spleen, liver, kidneys, and other visceral tissues where they are localized. There may be three to ten such relapses of decreasing severity, before symptoms of active infection subside. In epidemic situations the mortality rate may be high (50%) among debilitated victims of this disease, but ordinarily it is in the range of 2 to 10%.

LABORATORY DIAGNOSIS

I. Identification of the Organism

(Borrelia recurrentis).

The spirochete of relapsing fever is ***Borrelia recurrentis,*** a long, irregularly coiled, spiraled organism (twice the dimension of species of *Treponema* or *Leptospira*). The coils are haphazardly spaced. The organism rotates in the characteristic corkscrew fashion of spirochetes but is more flexible than other species in its twisting and lashing motions.

Borrelia species can be stained with several dyes, such as methylene blue, Giemsa and Wright stains, and the safranin of the Gram stain.

Borreliae can be cultivated in liquid media enriched with blood or serum, or in chick embryos, mice, and rats.

II. Identification of Patient's Serum Antibodies

Agglutinating, lysing, and *Borrelia*-immobilizing antibodies arise during the course of infection, but their identification is difficult because cultivated spirochetes used to identify them possess unstable antigenic components. These organisms also appear to undergo antigenic variation during the course of human infection, possibly as a result of interactions with antibody produced in vivo. The multiplication of new variants may account for the relapsing character of the infection, remissions occurring as specific antibodies for each variant reach a neutralizing level.

III. Specimens for Diagnosis

Multiple blood samples should be collected during the febrile periods for examination by stained smear, darkfield microscopy, culture, and animal inoculation.

Serologic tests with spirochetal antigens may be inconclusive for reasons outlined in the preceding section (II).

Epidemiology

I. Communicability of Infection

Relapsing fever has two epidemiologic patterns of spread. (1) Human epidemics arise when humans and one of their common ectoparasites, the body louse, become reservoirs for the organism. The louse rapidly transmits disease from one person to another, especially when overcrowded conditions occur among people who are unable to maintain hygienic standards or adequate nutrition. Wars, famines, floods, and other disasters favor louseborne epidemics, now restricted to parts of Asia, Africa, and South America. (2) Under natural circumstances, this infection has a reservoir in wild rodents (squirrels, prairie dogs, and others) and is transmitted among them or to humans by ticks. The endemic, tickborne disease occurs sporadically in humans in many parts of the world, including the western part of the United States, Mediterranean Europe and Africa, the Near East, and South America.

The incubation period following the bite of a tick or louse varies from 3 days to about 2 weeks, with an average time of 1 week.

The disease is not communicable from person to person. Ticks may remain infective throughout their lifetime of many years. Lice become infective within a few days of ingestion of blood from infected persons and remain so within their short life span of 5 to 6 weeks.

TABLE 22.1 ARTHROPODBORNE RICKETTSIAL DISEASES

Diseases	Insect Vector	Animal Host	Causative Organism
Typhus Group			
Epidemic typhus	Louse	Humans	*Rickettsia prowazekii*
Murine typhus (endemic typhus)	Flea	Small rodents	*R. typhi*
Scrub typhus (tsutsugamushi fever)	Mite	Wild rodents	*R. tsutsugamushi*
Spotted Fever Group			
Rocky Mountain spotted fever	Tick	Wild rodents Dogs	*R. rickettsii*
Boutonneuse fever	Tick	Wild rodents Dogs	*R. conorii*
Rickettsialpox	Mite	Mice	*R. akari*

The duration of immunity following an attack of relapsing fever is unknown. Acquired immunity provides the only form of resistance.

II. Control and Prevention

Patients with active infection do not require isolation if free of insect parasites, but their blood should be handled in accordance with standard precautions. Control of this disease depends on eliminating lice as the source of epidemic infection or suppressing ticks in areas where the disease is endemic among native rodents. Human infection can be treated with tetracyclines or penicillin.

RICKETTSIAL DISEASES

GENERAL EPIDEMIOLOGY

The rickettsiae are described in chapter 2 as coccobacillary, nonmotile, extremely small bacteria. They are obligate intracellular organisms that parasitize arthropods, notably ticks, mites, lice, and fleas. These insects in turn parasitize animals, such as rodents, rabbits, and dogs. Together the insect and animal hosts maintain the reservoir and cycle of rickettsial infection. In most instances, human beings are occasional, accidental victims of **rickettsial diseases** transmitted to them by infective arthropods. The outstanding exception to this general epidemiologic pattern is the form of typhus fever known as *epidemic typhus.* In this case, human beings are the only known reservoir for the rickettsial agent. When infected human beings are parasitized by lice, these insects acquire infection and transmit it from person to person. Rapid epidemic spread may occur among masses of people in overcrowded, unhygienic conditions.

Q fever has been considered another exception to the usual route of transfer of rickettsial infection. The infective agent *Coxiella burnetii* has been considered a rickettsia but displays certain biologic differences from other organisms of the group and is undergoing reclassification. It is maintained in nature by ticks, which transmit it to domestic animals. The human disease is acquired as a respiratory infection transmitted on airborne routes from sources associated with infected animals (chap. 12). Unlike rickettsioses, Q fever is not transferred by an infective arthropod.

CLINICAL RICKETTSIOSIS

The arthropodborne rickettsial infections are clinically and epidemiologically similar. In the human body these organisms parasitize endothelial cells of small blood vessels. After introduction into the skin, they may disseminate through the bloodstream, localizing in many peripheral vessels, as well as those of the heart or brain. The incubation period ranges from a few days to about 3 weeks, with an average time of 2 weeks. Clinical symptoms include fever, headache, and a typical skin rash. Depending on the virulence of the organism or the severity of infection, there may be prostration, delirium or stupor, myocarditis, or peripheral vascular collapse. These diseases respond well to antimicrobial therapy, although antimicrobial drugs merely suppress the organisms and do not eliminate them from the body. Antibodies arise during the course of infection, and the immune response leads to final recovery, usually without residual tissue damage. The rickettsial agent of epidemic typhus sometimes persists in its human reservoir for many years following recovery from the initial disease. These infected persons display no symptoms until some change in individual resistance permits a recrudescence of infection. This recurring form of typhus is called *Brill-Zinsser disease.*

The rickettsial diseases are associated with particular types of arthropod vectors. They can be classified in three groups on the basis of insect transmission and certain clinical features as shown in table 22.1.

TYPHUS FEVER GROUP

Epidemic Typhus

Epidemic typhus is a severe disease belonging to the **typhus fever group** and has a high mortality rate (6 to 30%). During the Middle Ages it swept in epidemic form through armies and among the massed refugee victims of the wars. The louse is a human ectoparasite that thrives when human conditions are at their worst (fig. 22.3). Unsanitary, crowded, malnourished people living in refugee camps, prisons, or slums are the usual hosts for body and head lice. In central and eastern Europe, Africa, Asia, Mexico, and

FIGURE 22.3 The louse is the vector of epidemic typhus.
Centers for Disease Control, Atlanta, GA.

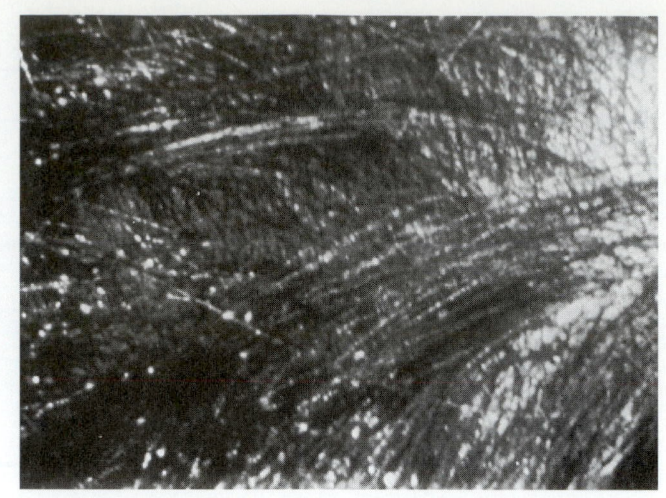

FIGURE 22.4 Lice breed on human hosts, leaving eggs (nits) on the skin and hair.
Centers for Disease Control, Atlanta, GA.

South America, these same people also become endemic reservoirs for the rickettsial parasite, transmitting it to one another by way of their lice. The infected louse excretes the organism in its feces, defecating on the skin at the time it bites. When the bite is scratched, organisms are rubbed into the minor wound, penetrating to the tissues below the skin. It is small comfort to the human victim that the louse also dies of this infection.

Special features of the control of active typhus infection require early treatment of patients: delousing them, their clothing, surroundings, and contacts and protecting the medical personnel who care for them. Isolation is not necessary, provided the patients are protected from further infestation by lice. Tetracycline or chloramphenicol is given as specific treatment. In epidemic areas, vaccines and sometimes antimicrobial drugs are used prophylactically for those who run a high risk of infection. Vigorous efforts are made to eliminate the louse from the resident population. An effective insecticide should be used to delouse the skin, clothing, and bedding.

Lice breed on their human hosts, leaving eggs (nits) on the skin or hair, and must be carefully removed (fig. 22.4). Bathing, shampooing, and frequent washing of clothes are highly effective in discouraging louse infestation.

Typhus vaccines have proved to be of great value in preventing this disease and are recommended for medical personnel in areas of risk, as well as for travelers or military units entering such regions.

Brill-Zinsser Disease

This modified form of typhus may occur in an individual who has recovered from a previous attack of the disease. Following recovery, the organisms remain latent for many years, the disease recrudescing only when some condition of stress shifts the patient's balance of resistance. If this happens in an area where human louse infestation is widespread, the recrudescing case may be the starting point of a new epidemic of unmodified typhus.

Murine Typhus

Murine typhus, also called *endemic typhus,* is a sporadic disease occurring in many parts of the world where people are exposed to infected rats and their fleas. It was once an urban disease associated with granaries and other areas in cities or seaports where rats congregate and breed. In the United States, continuing rat eradication campaigns have greatly decreased the incidence of this disease. Murine typhus is clinically milder than epidemic typhus but runs a similar course, conferring active immunity on those who recover. The fatality rate is less than 1%.

Scrub Typhus (Tsutsugamushi Fever)

Scrub typhus is a rickettsial infection limited to the eastern part of the world. Mites and wild rodents that maintain the natural infection are found in Southeast Asia, Australia, many Pacific islands, and some sections of China, Japan, and Korea. There are numerous antigenic types of the rickettsial species. The immunity resulting from disease is not cross-protective, so second and third attacks are common in resident populations. Military operations in these regions have been marked by epidemics occurring among susceptible persons newly exposed to mite infestation.

Scrub typhus and other miteborne rickettsial infections are characterized by a primary lesion, or eschar, that forms at the skin site where an infective mite attached and inoculated the organisms. Fever, headache, and rash follow, persisting for about 2 weeks unless treated with antimicrobial agents. Control of scrub typhus depends on protection against mites, together with chemoprophylaxis. Military measures include stripping vegetation from camp sites, spraying insecticide, applying rodent controls, impregnating

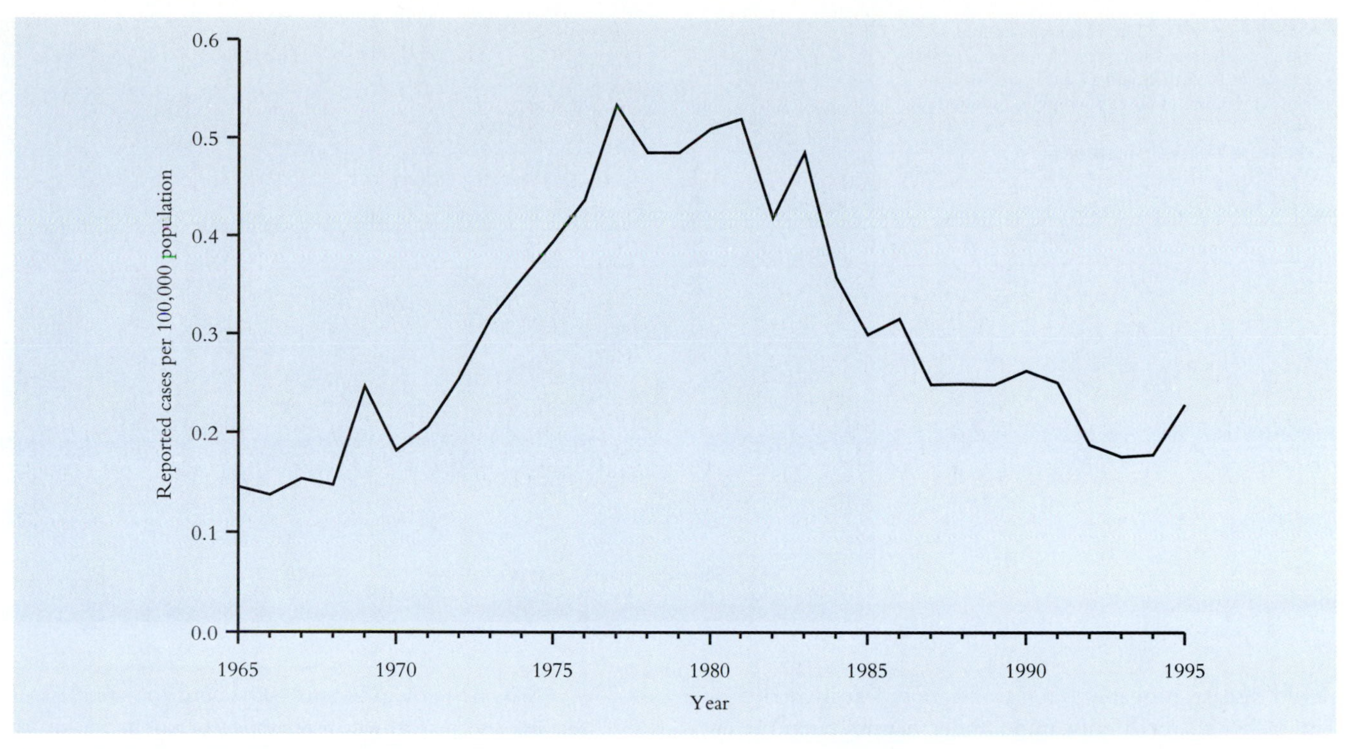

FIGURE 22.5 Rocky Mountain spotted fever (RMSF) cases in the United States from 1965 to 1995. In the United States, Rocky Mountain spotted fever is now more prevalent in the southeast than in the Rocky Mountain area. After a steady decline, an upsurge of the disease was seen through the 1970s. The upsurge may have been related to increased exposure to infected ticks as a result of farmlands converted into housing developments and increased recreational activities in wooded areas. Rocky Mountain spotted fever is the most common fatal tickborne disease in the United States, with a case fatality rate of 4%.

Centers for Disease Control and Prevention *MMWR Summary of Notifiable Diseases, United States, 1995* (October 25, 1996, Vol. 44: 53).

clothing and bedding with miticidal chemicals, and using insect repellents on the skin.

SPOTTED FEVER GROUP

Rocky Mountain Spotted Fever

A member of the **spotted fever group,** Rocky Mountain spotted fever was first recognized in patients who acquired their infection in the Rocky Mountain area (colorplate 11c). Since then it has become most prevalent in the southeastern part of the United States, but has been reported in almost every state. The disease also occurs in the western provinces of Canada and in Mexico and South America. Its prevalence in the United States is illustrated in figure 22.5.

The tick reservoir differs in various regions (fig. 22.6). The dog tick (*Dermacentor variabilis*) maintains the infection in the east; the wood tick (*Dermacentor andersoni*), in northwestern sections. Animals (rabbits, field mice, dogs) parasitized by ticks help maintain the infection in nature. Ticks themselves are an important reservoir because they pass the organisms along, transovarially, to the eggs of succeeding generations.

Special features in the control of rickettsial spotted fever include antimicrobial therapy and the prompt removal and destruction of ticks from the patients' bodies, if any remain at the time they become ill. Isolation and special precautions are not necessary. Prevention of this disease depends primarily on protecting individuals who may be exposed to ticks in endemic areas (hunters and campers, sheepherders, people engaged in land clearance or forest protection work). Protective clothing is important, but persons at risk learn to examine themselves closely for ticks and to remove them promptly. Dogs should also be examined frequently and carefully so that ticks can be removed before they have embedded their biting mouthparts in the skin. It is not practical to attempt tick eradication except in limited areas where the land has been cleared. Various insecticides, such as chlordane or lindane, can control the vectors in these areas.

Boutonneuse Fever

This rickettsial spotted fever is endemic in Mediterranean countries, though similar diseases also occur in other parts of the world, notably Australia, India, and Siberia. Each of them is associated with a particular tick reservoir and certain animal hosts parasitized by the arthropod.

Rickettsialpox

This disease appears to be limited to some cities of the former Soviet Union and of eastern and central United States (New York, Boston, Hartford, Philadelphia, Cleveland). It is a

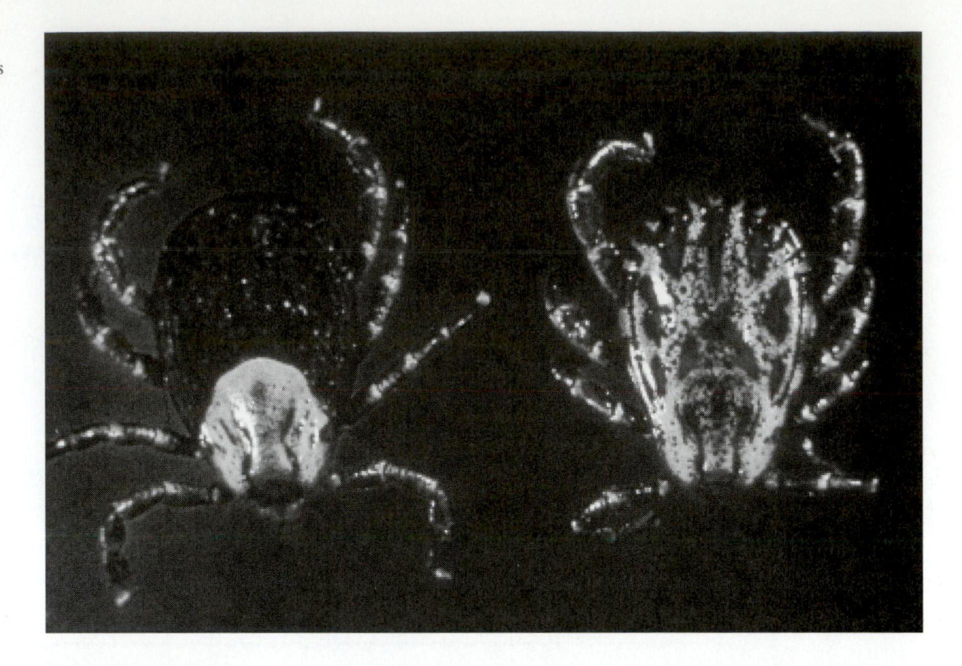

FIGURE 22.6 The *Dermacentor* genus of ticks is well-known since several species, such as *D. variabilis* (dog tick) and *D. andersoni* (wood tick), are vectors of Rocky Mountain spotted fever.
Centers for Disease Control, Atlanta, GA.

mild disease, probably often undiagnosed. It is carried by a mite that shares it with mice under natural circumstances. The human disease begins with an eschar at the site of the mite's attachment to the skin and proceeds as a febrile illness with a rash that lasts about a week. Rickettsialpox may appear in small outbreaks involving people in apartment houses or tenements infested with mice that carry the infection and maintain it with mites, the latter also parasitizing the human tenants. Control is readily accomplished by clearing basements and incinerator closets of trash that encourages mouse habitation and breeding. Insecticides are also helpful.

LABORATORY DIAGNOSIS

I. Identification of Organisms

Microscopic and cultural identification of rickettsiae is technically difficult and hazardous for the laboratory worker. Biopsies of skin lesions may be stained by special techniques, including direct fluorescent antibody, to reveal the organisms in endothelial cells. Specimens of whole blood (or macerated blood clots) can be inoculated into embryonated chick eggs, cell cultures, guinea pigs, or mice to propagate the infectious agents, but these techniques are usually available only in public health reference laboratories.

II. Identification of Patient's Serum Antibodies

Specific antibodies arise during the course of rickettsial infections which can be identified by indirect fluorescent antibody, indirect hemagglutination, enzyme immunoassay, or latex agglutination techniques using rickettsial antigens.

III. Specimens for Diagnosis

The organisms may be isolated from blood and tissue specimens in special reference laboratories. These laboratories should be contacted about appropriate shipping conditions.

Specific serological tests with acute and convalescent sera are also performed in reference or public health laboratories.

Control and Prevention

I. Isolation

These diseases are not directly communicable from person to person. The first step required in their control is the removal and destruction of insects on the patient's body. Continuing measures must be taken to protect patients and personnel from insects, particularly in louseborne typhus. When this is assured, isolation is not necessary.

II. Precautions

The blood of patients with rickettsial diseases may be infectious. Syringes and needles used for blood collection should be handled with caution and safely discarded or sterilized promptly.

III. Treatment

Tetracyclines and chloramphenicol are the drugs used in rickettsial infections. They have sharply reduced mortality rates formerly seen, especially those associated with Rocky Mountain spotted fever and epidemic typhus.

IV. Chemoprophylaxis

Chloramphenicol has been used prophylactically to control scrub typhus in endemic areas. It does not avert *infection* but prevents the development of *disease* if continued for a month.

V. Immunity

An attack of rickettsial disease usually confers permanent immunity against the specific infecting strain.

VI. Immunization

For epidemic typhus, formalinized vaccines are available containing rickettsiae cultivated in chick embryos. They induce an immunity that remains effective for a few months to a year. Persons exposed to risk of infection should receive booster doses at regular intervals as indicated by the local situation. No vaccines are available for other rickettsial diseases.

EHRLICHIOSIS

Species of the genus *Ehrlichia* are classified in the family *Rickettsiaceae* but have some characteristics resembling chlamydiae; that is, they have a complex developmental cycle and grow in membranous vacuoles in host cells. They are obligate intracellular parasites that live primarily in host leukocytes. Most are of veterinary importance, but at least 3 species, *Ehrlichia sennetsu, Ehrlichia chaffeensis,* and an agent closely resembling *Ehrlichia equi,* infect humans. *E. sennetsu* is responsible for a form of **ehrlichiosis** called *sennetsu fever,* an acute febrile illness characterized by lymphadenopathy, lethargy, and white blood cell abnormalities. This disease has been recognized in Japan since the late nineteenth century and most reported cases have come from that country. The mode of transmission is unknown. Diagnosis is made by inoculating the patient's blood into mice, which are highly susceptible to the disease. Infected mice die 3 to 4 weeks after inoculation showing enlarged lymph nodes, liver, and spleen; ehrlichial organisms can be demonstrated in peritoneal cells using indirect fluorescent antibody or other special staining methods.

E. chaffeensis was first isolated in 1986 from an infected soldier at Fort Chaffee, Arkansas. The disease it causes is known as human monocytic ehrlichiosis because the organism infects mononuclear phagocytic cells. Humans become infected by a tick bite, most frequently in late spring or early summer. The Lone Star tick (*Amblyomma americanum*) and the American dog tick (*Dermacentor variabilis*) are the principle vectors of human transmission. Ehrlichiae are not transferred transovarially in ticks; thus, another host reservoir is suspected, most likely the white-tailed deer.

Ehrlichiae enter the skin through the tick bite and disseminate hematogenously. Clinical signs and symptoms of the disease resemble those of Rocky Mountain spotted fever and include fever, headache, myalgia, arthralgia, anorexia, nausea, cough, pharyngitis, diarrhea, and abdominal pain. In about 36% of patients, a nonhemorrhagic rash may appear late in the illness. Most patients also have leukopenia, thrombocytopenia, and elevated concentrations of liver enzymes in their sera. Most human cases occur in western, south central, and south Atlantic states; the mortality rate is 3 to 5%. The diagnosis of infection is usually established by demonstrating a significant rise in antibody titer to antigens of *E. chaffeensis* in an indirect immunofluorescence assay.

Human granulocytic ehrlichiosis was first described in 1994 by Johan Bakken, an infectious disease physician in Duluth, Minnesota. He observed that several of his patients had inclusions that resembled intracellular ehrlichiae in their neutrophils, or granulocytes (in contrast to mononuclear phagocytes infected by *E. chaffeensis*). These inclusions are referred to as morulae. The illness was characterized by fever, chills, headache, and myalgia. Severely ill patients developed disease of multiple systems including interstitial pneumonia, hepatocellular necrosis, leukopenia, anemia, thrombocytopenia, and hemorrhage. The illness had a mortality rate of 5 to 10%. The as yet unidentified *Ehrlichia* species responsible for this disease closely resembles the veterinary pathogen of horses, *Ehrlichia equi.*

Human granulocytic ehrlichiosis is associated with a tick bite and is most likely transmitted by the deer tick, *Ixodes scapularis.* The human infections documented in the eastern United States and the known distribution of *E. equi* suggest that human granulocytic ehrlichiosis has a wide geographic distribution throughout the United States. Diagnosis of human infection can be established presumptively by detecting morulae in granulocytes in blood smears stained with Romanowsky stain. Definitive diagnosis requires the use of PCR, serologic methods, or special immunological tissue stains which, at present, are available only in research laboratories.

Both human and animal ehrlichial disease is treated with tetracycline. Much work is required to determine the pathogenesis and mode of transmission of these unusual microbes.

VIRAL DISEASES

Arthropodborne viruses are often called *arboviruses* to distinguish them from others that are transmissible from person to person. The arboviruses characteristically multiply in arthropod hosts without damaging them and are transmitted by their bites to humans or other animals, including birds and reptiles. In some instances, the arthropod is itself a reservoir, passing a virus along to its offspring, but some arboviruses must cycle between mammalian reservoirs and the arthropod host.

The arbovirus group includes a large number of viruses, usually classified in several groups on the basis of antigenic relationships, arthropod vectors, and the types of disease they cause in humans (table 22.2). Mosquitoes, biting flies, and ticks are the arthropods that most often play host to viruses. The biologic requirements for survival of an arthropod species may sharply limit geographic distribution of the virus or the human disease associated with it.

The **arbovirus diseases** are presented here in two groups based on clinical distinctions: (1) systemic infections caused by viscerotropic viruses (those with a predilection for cells of visceral organs) and (2) primary infections of the brain and spinal cord caused by neurotropic viruses (those with an affinity for neural cells). The characteristic clinical syndrome of the latter group of diseases is encephalitis (plural, encephalitides).

SYSTEMIC ARBOVIRUS DISEASES

Yellow Fever

The Clinical Disease

Yellow fever is an acute infectious disease with symptoms that may be mild or extremely severe. The fatality rate is

TABLE 22.2 CLASSIFICATION OF SOME ARTHROPODBORNE VIRUS DISEASES[a]

Virus Group	Name of Virus	Vector	Human Disease	Where Found
Togaviruses				
Alphaviruses (group A)	Eastern equine	Mosquito	Encephalitis	Americas
	Venezuelan equine	Mosquito	Encephalitis	South and Central America, Florida
	Western equine	Mosquito	Encephalitis	Americas
Flaviviruses (group B)	Dengue	Mosquito	Fever, rash, hemorrhagic fever	Africa, Asia, South America, Caribbean Islands, India
	Japanese B encephalitis	Mosquito	Encephalitis	Asia
	Louping ill	Tick	Encephalitis	Great Britain
	Murray Valley	Mosquito	Encephalitis	Australia, New Guinea
	Russian spring-summer	Tick	Encephalitis	Europe, Asia
	St. Louis	Mosquito	Encephalitis	Americas, Jamaica
	Yellow fever	Mosquito	Hemorrhagic fever	Africa, South and Cental America
Bunyaviruses				
(group C)				
California group	California encephalitis	Mosquito	Encephalitis	U.S., Canada
Sandfly fever group	Several	Sandfly (*Phlebotomus*)	Fever	South and Central America, Europe, Africa, Asia
Reoviruses				
Colorado tick fever group	Colorado tick fever	Tick	Fever	U.S., Canada
Rhabdoviruses				
Vesicular stomatitis group	Vesicular stomatitis	Sandfly	Fever	Americas

[a]Partial list, adapted from Abram S. Benenson, ed. *Control of Communicable Diseases in Man,* 15th ed., (Washington, D.C.: A.P.H.A., 1990).

low (less than 5%) among people who live in tropical areas where the disease is endemic but may be as high as 30 to 40% among susceptible people not previously exposed. At one time epidemic yellow fever was one of the great plagues of tropical and subtropical areas of the Americas. The epidemic disease came under control in 1900 when Dr. Walter Reed demonstrated the causative agent in the blood of patients and in a mosquito vector, *Aedes aegypti* (see Perspective 22.2). Mosquito eradication programs promptly eliminated the disease from large urban centers, but a jungle form persists in mosquito vectors of the tropical rain forests of South America and Africa. Continuous control depends on vigilant efforts to keep the *Aedes* vector suppressed, as well as on active immunization of persons visiting or residing in areas where yellow fever is endemic.

When yellow fever virus is injected into human skin through the bite of an infective mosquito, the organism spreads to the lymph nodes, enters the bloodstream, and localizes in the liver, spleen, kidney, and bone marrow. After a few days, symptoms begin with fever, chills, headache, and malaise. Foci of virus multiplication in the viscera become necrotic. The liver parenchyma is often hardest hit, symptoms of jaundice being prominent (hence *yellow* fever). In severe cases the gastric mucosa may be involved, hemorrhage from the stomach producing black vomitus. Death may result if hepatic and renal damage is excessive. Patients who recover do so completely, without residual damage.

LABORATORY DIAGNOSIS

Virus is present in the bloodstream during the first 3 to 5 days of illness. It can be recovered and identified by inoculating mice or cell cultures with blood or tissue samples collected from a suspected case.

As disease progresses, antibodies appear in the patient's serum in rising titer. Paired sera are tested by neutralization, enzyme immunoassay, and hemagglutination-inhibition methods, although cross-reactions occur between yellow fever and other flaviviruses.

Epidemiology

I. Communicability of Infection

A. *Reservoirs, Sources, and Transmission Routes* In urban areas, human beings and the *Aedes aegypti* mosquito are reservoirs.

In jungle areas, mosquitoes maintain the natural infection with monkeys but can transmit it to susceptible human beings.

B. *Incubation Period* Three to 6 days.

C. *Communicable Period* Virus is present in the bloodstream just before symptoms begin and for the next few days. During this time humans are infective for the mosquito. Following the mosquito's blood meal, the virus requires about 10 to 12 days to develop and multiply before the mosquito becomes infectious. Mosquitoes remain infected while they live but do not pass virus along to their progeny.

D. *Immunity* During infection antibodies are produced rapidly and following recovery provide permanent immunity.

II. Control of Active Infection

A. *Transmission-based Precautions* It is not necessary to isolate patients, but standard precautions should be maintained during their hospitalization. It is essential that they be screened from mosquitoes, especially during the first few days of illness and viremia.

B. *Other Precautions* The patients' living quarters and all those nearby are immediately sprayed with insecticides.

C. *Therapy* There is no specific antimicrobial therapy.

D. *Control of Contacts* Nonimmune persons in the immediate vicinity of a new case of yellow fever are vaccinated promptly, and a campaign is conducted against mosquito species that might carry and transfer the infection.

E. *Epidemic Controls* Yellow fever is a matter of international concern, and all governments work through the World Health Organization to report the disease and control it.

III. Prevention

A. *Immunization* A live attenuated strain (17D) of yellow fever virus is used for artificial active immunization. Routine use of vaccine is desirable for persons living in or traveling to areas where the jungle type of the disease is endemic and for laboratory workers who might be exposed to the virus.

B. *Continuing control and surveillance* programs are necessary to prevent recurrence of yellow fever as an epidemic disease.

Other Arbovirus Fevers

Many other systemic virus infections are carried by arthropods in different parts of the world. Some are mosquitoborne, and others are carried by ticks or biting flies. A group of these diseases is characterized, like yellow fever, by hemorrhagic lesions in muscosal areas (enanthems), but petechial or purpuric skin lesions also occur (exanthems). These hemorrhagic fevers, as they are called, occur in scattered areas (Crimea, Southeast Asia and the Pacific Islands, the former Soviet Union and Siberia and South America), usually limited by the range of the arthropod host harboring the virus.

Some arboviruses do not produce hemorrhagic injury but characteristically elicit a maculopapular rash like that of measles. The fever lasts for a few days and is accompanied by lymph node involvement, joint and muscle pains, pain in the eyes, and conjunctivitis. Viruses of this group do not localize in the liver or other vital organs, and these fevers are rarely fatal. They are distributed geographically according to the range of their vectors.

When viruses are successfully introduced into new arthropod hosts, they may cause outbreaks of disease and become endemic in the geographic area of the new vector. In the Americas, dengue is a well-known example of an arbovirus disease introduced into the Caribbean area by the *Aedes aegypti* mosquito vector. Dengue (also known as "breakbone fever" because of the severe muscle and joint pain experienced by many patients) was first recognized in the Caribbean in 1963, when several large outbreaks occurred. It is now endemic in that area and in South America and is seen with increasing frequency among residents as well as vacationing travelers. The latter are warned by the U.S. Public Health Service to take precautions to avoid mosquito bites and to report any acute febrile illness they may experience within two weeks of returning home. This usually mild virus disease is also endemic in parts of Asia, the southwest Pacific, and West Africa.

Other examples of arbovirus fevers include Colorado tick fever, seen in western areas of the United States, and sandfly fever. Several viruses of the latter group are vectored by biting flies of the genus *Phlebotomus.* (This name and the general term *phlebotomine* refer to the fact that sandflies are bloodsucking insects.) The diseases of this group have a scattered distribution in the Americas, as well as the hot, dry parts of Europe, Africa, and Asia, as shown in table 22.2.

ARBOVIRUS ENCEPHALITIDES

A rather large group of encephalitides is associated with arboviruses. Most **arbovirus encephalitides** are carried by mosquitoes, but a small group is also transmitted by ticks. Birds and other animals (rodents, bats, reptiles, and domestic animals such as horses and sheep) appear to play a role in maintaining these infections in nature, with the assistance of ectoparasites.

In the United States only mosquitoborne encephalitides are known, the tickborne variety being restricted to spotty areas of northern and central Europe and the former Soviet Union. The clinical diseases recognized in the United States and other parts of this hemisphere are Eastern equine encephalitis (EEE), Western equine encephalitis (WEE), Venezuelan equine encephalitis (VEE), and St. Louis encephalitis (SLE). (In the Far East, Japanese B and Murray Valley encephalitis are arbovirus infections carried by mosquitoes.)

Eastern Equine, Western Equine, Venezuelan Equine, and St. Louis Encephalitis

The Clinical Disease

These encephalitides are acute, self-limited diseases characterized by inflammation of the brain, spinal cord, and meninges. They may occur as mild cases of aseptic meningitis, or they may be quite severe, with a sudden onset of high fever, chills, nausea, and malaise and localizing symptoms of meningitis and encephalitis occurring within the next day or two. Pain and stiffness of the neck, drowsiness or stupor, and disorientation are frequent. In severe cases, there may be aphasia, convulsions, and coma. Spastic paralysis may occur. There may be sequelae, such as impaired

PERSPECTIVE 22.2 *The Yellow Plague*

Yellow fever, a mosquitoborne viral disease of the tropics, descended upon the colonies of the New World in the late 1700s. The disease remained a problem throughout the nineteenth century, particularly in the Atlantic and Gulf ports of America and in the Caribbean, and especially during the hot months of summer.

Yellow fever made its way to North America from Africa aboard the slave ships and from South America on trading vessels, but this was neither known nor understood at the time. The sailing ships, which had to carry large supplies of fresh water, offered excellent conditions for breeding mosquitoes and for their transport from port to port. When the disease ravaged the city of Philadelphia during the summer of 1793, little or nothing was known of the nature of infectious diseases, much less that some could be transmitted by insects.

The epidemic in Philadelphia was chronicled by one of the city's eminent physicians, Dr. Mathew Carey. In his work titled "A Short Account of the Malignant Fever Lately Prevalent in Philadelphia," Carey wrote that the general opinion was that "the disorder probably originated from some damaged coffee or other putrefied vegetable and animal matters." Accordingly, efforts to control the disease included "cleaning the streets properly and purified by the scavengers and all the filth immediately hauled away"; of course, these measures were fruitless.

Within two months of the appearance of yellow fever in Philadelphia, 10 percent of the population died (the disease may have a fatality rate as high as 85 percent among those stricken). Many fled the city as the death toll increased. The College of Physicians recommended that the tolling of the church bells be discontinued because it was having a demoralizing effect on the people of the city. Dr. Carey, Dr. Benjamin Rush, and other physicians worked day and night tending the sick, using the treatment of vigorous purging and bleeding that Dr. Rush believed to be most effective. This regimen, however, simply hastened the demise of most victims.

Before the epidemic ended, 11,000 people were stricken and Philadelphia was in complete chaos. The government no longer functioned, schools closed, and businesses failed. Finally, in the fall of the year following the first frost, this plague came to an abrupt end, but not before it had killed more than 4,000 people.

Similar outbreaks continued to harass the New World until the start of the twentieth century. In the early 1800s, most of Napoleon's army of 33,000 men in Haiti was wiped out within a month after yellow fever struck. In 1853, New Orleans suffered an outbreak that killed 7,849. The Panama Canal building project had to be abandoned when one-third of the work force placed there by the French government between 1881 and 1889 died of yellow fever.

The American army suffered high casualties from yellow fever during the Spanish-American War in 1898. After the war Major Walter Reed, based in Cuba, was commissioned to study the disease. A Havana physician, Dr. Carlos Juan Finlay, believed that an agent, specifically a mosquito, transmitted yellow fever. The malarial parasite was, by that time, known to be transmitted by a mosquito, so Dr. Reed's army commission proceeded to work with mosquito larvae provided by Dr. Finlay.

Because no animal was known to have contracted yellow fever, human volunteers were sought for the study. The volunteers were housed in two sets of small, tightly screened cottages (Perspective fig. 22.2). In one set of cottages, mosquitoes that had recently acquired a blood meal from a yellow fever patient were released. In the other, subjects slept on bloodstained sheets and wore the clothes of victims of the disease. Results were conclusive: only those who were in contact with the infected mosquitoes developed yellow fever. Clara Maass, a volunteer nurse, was among those bitten and who died of the disease in 1901. She was buried with full military honors.

As a result of these efforts the mosquito vector of yellow fever was identified conclusively as a member of the species *Aedes aegypti,* and extensive efforts to exterminate it were begun. In February 1901 Major William Gorgas, chief sanitary officer of the U.S. Army, sent inspectors to check cisterns (household receptacles for storing water) for breeding mosquitoes. Orders were given for the tops of barrels to be covered, and spigots were used to drain off water from the bottom. Old bottles and cans where mosquitoes could breed were removed from yards, and fines were imposed on citizens who did not comply with the efforts. Within a year, the city of Havana was free of yellow fever for the first time in 150 years. Through similar extermination campaigns the disease disappeared from the United States and Panama, where construction of the canal resumed, this time by the United States Government.

In 1901, Dr. Reed and Dr. Finlay published their proof that yellow fever is transmitted to humans from a mosquito

These shacks at Camp Lazear in Quemados, Cuba, were the site of Dr. Walter Reed's research to prove that the mosquito transmitted yellow fever.
Otis Historical Archives, National Museum of Health and Medicine, Armed Forces Institute of Pathology

vector that has taken its last meal from an individual sick with yellow fever. Reed, studying the filtered blood of victims of the disease and discerning nothing unusual in these specimens, concluded that the agent itself must be so small as to be "ultramicroscopic," what is now referred to as a filterable virus. Dr. Max Theiler, a South African-American research physician and virologist then working at the Rockefeller Institute in New York City, developed a preventive vaccine for which he received a Nobel prize in 1951.

Yellow fever is no longer a plague, but it remains endemic in certain parts of Africa and South America where preventive measures have kept it under control.

mental function, blindness, deafness, or epilepsy, and paralysis may also be residual. The severity and mortality rate of these "American" encephalitides are not so great as those seen in Japanese B encephalitis, but they take their worst toll among infants and young children.

LABORATORY DIAGNOSIS

The viruses of these diseases are seldom isolated from the blood of patients because viremia usually occurs before the onset of symptoms. By the time illness becomes apparent, virus is localized within the central nervous system and is identified only in brain tissue taken at autopsy. These viruses can be cultivated in mice and embryonated eggs, then identified by neutralizing antibodies.

Serologic diagnosis can be obtained by identifying the patient's serum antibodies using neutralization, enzyme immunoassay, and hemagglutination-inhibition techniques. A rising titer of antibody must be demonstrated.

When outbreaks of these encephalitides occur, epidemiologists search for mosquito and animal reservoirs. Virus may be isolated from these sources and tested with patient serum as well as with known antiserums.

Epidemiology

I. Communicability of Infection

A. *Reservoirs, Sources, and Transmission Routes* The identity of year-round reservoirs for these viruses is not known, but wild birds, rodents, and reptiles are thought to be important in the annual cycle. During the summer mosquito season, birds are the source of infection for mosquitoes, and the latter transfer it to human beings. The equine viruses are also transferred to horses, and large epizootics may occur among these animals. Neither horses nor human beings are a likely source of infection for mosquitoes because the viremic stage is brief, and virus is not heavily concentrated in the blood. The virus of St. Louis encephalitis is not known to infect horses. A number of mosquito vectors may be involved. The most important of them belong to the genus *Culex*.

B. *Incubation Period* Varies from 4 days to 3 weeks, with an average time of 10 to 15 days.

C. *Communicable Period* These diseases are not transmissible from person to person, and humans probably do not infect the mosquito. Viremia persists in birds for several days, and the mosquito remains infective throughout its life.

D. *Immunity* In areas where these diseases occur frequently, most adults are immune because of frequent inapparent infection. Active disease results in solid immunity against the infecting virus strain. Infants, young children, and aging people are among the most susceptible because their immunity is incomplete or waning. See table 22.2 for the geographic distribution of these encephalitides.

II. Control of Active Infection

Isolation and special patient precautions are not indicated. The disease must be clinically differentiated from others that may represent extension of systemic infection (poliomyelitis, mumps, rabies, postinfectious viral encephalitides, as well as those of bacterial or mycotic origin). There is no specific treatment.

Control and prevention of these infections depend on the recognition of human cases, as well as cases of related disease in horses or birds, and the elimination of vector mosquitoes.

Vaccines are generally not available, but investigational EEE and WEE vaccines can be obtained for persons at high risk of infection (e.g., laboratory workers and epidemiologists).

PARASITIC DISEASES

Most arthropodborne parasitic diseases are caused by protozoan species. There are three important genera of protozoa transmitted by arthropods to humans and animals, each genus containing several species: *Plasmodium,* the agent of malaria; *Leishmania,* which is associated with three types of disease; and *Trypanosoma,* the cause of African sleeping sickness and of a South American infection called *Chagas' disease.*

In addition, there is a family of nematodes called *filarial worms* (filaria meaning "thread"), whose adult forms live in human tissue or body cavities. Mosquitoes are intermediate hosts for these parasites.

All organisms of this group require an arthropod host in which to complete part of their developmental cycle. Definitive and intermediate hosts alternate in maintaining natural infection. The life cycles and the characteristic human diseases produced by these parasites are outlined here, providing highlights of the developmental patterns, clinical syndromes, and control methods for each.

MALARIA

Organisms

(*Plasmodium vivax, Plasmodium malariae, Plasmodium falciparum, Plasmodium ovale.*)

The ***Plasmodium*** *species* that cause **malaria** parasitize human red blood cells in which they progress through stages of asexual development. In the process, precursors of sexual cells (gametocytes) are also formed. When the gametocytes are taken up by a mosquito, they proceed through stages of sexual division in the arthropod.

Life Cycle (fig. 2.20f)

I. Definitive Host

Mosquitoes of the genus *Anopheles.*

II. Intermediate Host

Humans (a number of animals may be hosts for *Plasmodium* species that are not infective for humans).

III. Human Infection

A. *Source of Infection* The bite of an infective female anopheline mosquito.

B. *Infective Forms* Sporozoites congregated in the salivary gland of the mosquito and injected into the skin when she bites.

C. *Localization and Development* Sporozoites are first carried by the bloodstream to a localizing point in the liver where they undergo several cycles of asexual multiplication during the following week or two. Eventually basic units called *merozoites* emerge from the liver and invade red blood cells. Within invaded erythrocytes, each merozoite (there may be more than one in a cell in some types of infection) initiates a new asexual cycle of division. Successive forms that develop in the red blood cell are called (1) *rings* (these are young, circular trophozoites); (2) *trophozoites* (larger than rings and often irregular in shape as seen in colorplate 8c); (3) *schizonts* (display many segments formed by the division and redivision of nuclear material); (4) *mature schizonts,* which fill and eventually rupture the erythrocytes. The scattered segments, again called merozoites, enter new cells and the cycle repeats with a growing density of parasites. This process is called *schizogony.*

In time, some of the parasites develop into specialized precursors of gametes, or sex cells. The precursors are called *microgametocytes* and *macrogametocytes.* These do not combine sexually in the human body but must wait for the conditions provided by the mosquito host.

In *P. falciparum* infections, the cell membranes of parasitized erythrocytes become sticky and adhere to the walls of capillaries in internal organs. Consequently, only young ring forms and gametocytes are found in smears of the circulating blood; intermediate developmental forms are not seen.

In *P. vivax* and *P. ovale* malaria only, another parasite stage called the *hypnozoite* is found in liver cells. In this form, the organism remains quiescent for a variable time and then continues development to the merozoite stage, invading peripheral blood erythrocytes. As a result, the patient experiences a relapse of disease.

D. *Exit from Human Body* The female anopheline mosquito must take a blood meal from an infected person, and the blood removed must contain both micro- and macrogametocytes. Any asexual forms ingested do not develop further in the insect.

IV. Cyclic Development in Mosquito Host

The microgametocyte fertilizes the macrogametocyte, and a zygote is formed. The latter enters the wall of the mosquito's stomach, forms an oocyst, and sporozoites develop. This process is known as *sporogony* and culminates with the liberation of hundreds of sporozoites into the mosquito's body cavity. Most migrate to the salivary or other glands of the insect where they can develop no further until injected into a human host. The life cycle of the malaria parasite is shown diagrammatically in figure 2.20f.

Characteristics of Human Malaria

Malaria is an old disease that has killed millions of people since the ancient Chinese, Assyrians, and Egyptians first made note of it. It is a chronic disease directly responsible for deaths in about 1 to 2% of adult cases, but the rate is much higher among infants. Endemic malaria has been controlled and virtually eliminated from many countries, but it remains a serious public health problem in Central and South America, Africa, the Near and Far East, Southeast Asia, and the southwest Pacific.

Humans are the only known host for *P. vivax, P. malariae, P. falciparum,* and *P. ovale,* although other species have been recognized in animal infections. *P. vivax* and *P. falciparum* are the most common species afflicting humans and *P. falciparum* produces the most severe clinical disease. *P. malariae* has a wide distribution, but the incidence of infection is low. *P. ovale* causes a relatively mild human disease acquired only in West Africa. Throughout the "malaria belt" that stretches across tropical and subtropical areas of the globe, the female *Anopheles* mosquito is the responsible vector. The spread of human malaria is restricted chiefly by her requirement for warm, moist climates in altitudes of less than 6,000 feet. Human populations living within the range of this mosquito ensure perpetuation of the parasite until the vector is eradicated.

The asexual cycle in human peripheral blood cells is responsible for many of the clinical symptoms of malaria and for the cyclic recurrences. When erythrocytes containing large numbers of maturing schizonts simultaneously rupture, the infected person experiences paroxysmal fever and shaking chills.

These symptoms may last for a few minutes to an hour and are often accompanied by nausea, vomiting, and severe headache. Fever may persist for several hours because liberated merozoites circulate before invading new red blood cells. The episode concludes with a drenching sweat that leaves the patient afebrile and exhausted. For the next two or three days the patient feels reasonably well, until another cycle of parasite maturation and a new paroxysmal stage occur. These episodes of fever and chills are irregular at first but later develop a rhythm characteristic for the parasite species. *P. falciparum* and *P. vivax* malaria are usually tertian: that is, cyclic fever and chills recur every third day (at 48-hour intervals), but the cycle is quartan for *P. malariae,* with paroxysms recurring every fourth day (72 hours).

Malaria eventually burns itself out if the patient is not reinfected by another mosquito. Hepatomegaly, splenomegaly, anemia, and in severe *P. falciparum* malaria, hemorrhagic damage to the kidneys and liver ("blackwater fever") can occur. "Cerebral malaria" is another possible consequence of *P. falciparum* infections, resulting in excessive fever, delirium or coma, and death when the vascular supply to the brain is seriously compromised by parasitic occlusions of cerebral blood vessels.

DIAGNOSTIC METHODS

Blood smears are examined microscopically. Thin blood smears are made by spreading a drop of blood thinly across a slide as is done for a white blood cell differential count. Thick smears are also important because the organisms may be scarce and hard to find. A thick smear is made by placing a drop of the patient's fingertip blood in the center of a glass slide. A wooden applicator is used to spread the drop over an area about the size of a dime. Thin smears are stained to demonstrate parasites in red blood cells. Thick smears are first treated so that erythrocytes are lysed, freeing their parasites. When properly stained, the organisms can be seen lying among white blood cells and platelets left on the slide.

When blood smears are reported negative for malarial parasites, but the clinical picture and history of exposure are suggestive, it is important to collect blood for additional smears. Maturing organisms are present in the blood (except in the case of *P. falciparum*) during the few hours preceding the paroxysms of fever and chills. When the latter can be predicted, therefore, specimens should be taken while the fever is rising and during the first chills. Differentiation of *Plasmodium* species depends on finding organisms in varying stages of maturity and requires an expert eye.

Serologic diagnosis can be made with the indirect fluorescent antibody technique or indirect hemagglutination, but these tests are not routinely available.

Special Features of Control

Treatment

Malaria was treated with extracts of cinchona bark long before its etiologic agent was known. (The parasite was discovered in 1880 by Charles Louis Alphonse Laveran, a French army doctor.) The active ingredient of cinchona bark is quinine, which was the standard drug until more effective synthetic compounds were developed.

Currently, chloroquine is the drug of choice for treating malaria caused by all *Plasmodium* species. However, some *P. falciparum* strains are resistant to this drug, and therefore, in these patients, a quininelike drug, quinidine gluconate, is relied on for therapy. It is used in combination with another drug such as tetracycline. To prevent malarial relapses caused by *P. vivax* and *P. ovale,* primaquine is given in addition to chloroquine to kill the residual hypnozoite forms in the liver.

Chloroquine is also of value as a *prophylactic* agent to prevent the development of clinical malaria. The drug must be taken from two weeks before entry until six weeks after leaving a malarious area. People traveling to areas in which chloroquine-resistant malaria is prevalent may need to take a combination of drugs for prophylaxis.

Control

Control depends on coordinated efforts to treat and cure malaria in humans, thus eliminating the intermediate reservoir, and to eradicate the mosquito host. While being treated, patients should be cared for with standard precautions and screened from mosquitoes until the disease is under control.

Chemoprophylaxis is of particular importance for persons entering endemic areas. The increasing volume of tourism and foreign immigration has been correlated in recent years with an increase in the number of malaria cases imported into the United States (fig. 22.7).

Mosquito eradication programs include community application of residual insecticides around dwellings but the most effective control is achieved when mosquito breeding places are eliminated by draining and filling stagnant swamps or water holes or by chemical treatment that kills larvae (fig. 10.7).

Efforts by the World Health Organization to eradicate malaria from the world have not been successful, in part because of increased resistance of the mosquito vector to certain insecticides, parasite resistance to antimalarial drugs, and the high expense of the eradication campaign. The recently devised method for in vitro cultivation of malaria parasites may eventually lead to effective vaccine production.

BABESIOSIS

Babesia species are obligate intraerythrocytic protozoan parasites transmitted by ticks (the same *Ixodes* ticks that transmit Lyme disease). Most disease occurs in various wild mammals and domestic animals such as cattle, goats, pigs, horses, and dogs. The first human case of **babesiosis** was described in 1957, although forms resembling the organism had been described in blood films before then. Several cases have been reported from Europe and the former Soviet Union, primarily in splenectomized patients. In the United States, the disease is limited almost exclusively to islands off the New England coast, especially Nantucket and Martha's Vineyard, and parts of Long Island. These infections are caused primarily by *Babesia microti,* a species normally found in rodents. Only a few cases have been seen in splenectomized patients.

Symptoms of disease are generally similar to those of malaria, except that the fever is not cyclic. Fatigue, myalgia, headache, nausea, arthralgia, splenomegaly, and mental depression may also occur. Results of serologic studies indicate that the disease is often mild and self-limited, although it is generally more severe when it occurs in older patients and those who have been splenectomized. Diagnosis is made by finding the malarialike trophozoites in stained blood films. During schizogony, the organism divides into no more than four trophozoites, which appear in a characteristic tetrad or "Maltese cross" formation. Serologic diagnosis can be made by using an indirect immunofluorescent test on acute and convalescent sera. Combination therapy with clindamycin plus quinine has been successful for most, but not all, patients.

LEISHMANIASIS

Organisms

(*Leishmania donovani, Leishmania tropica, Leishmania mexicana, Leishmania braziliensis,* and other species.)

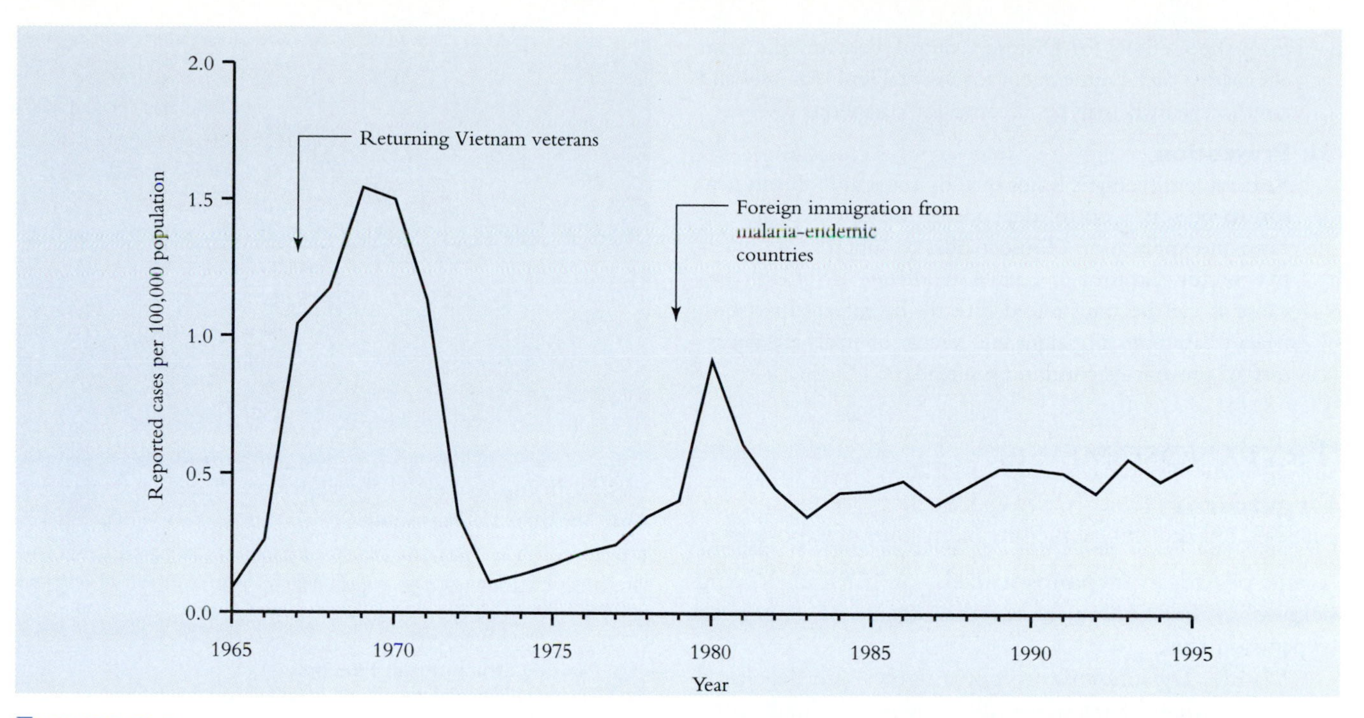

FIGURE 22.7 Malaria cases in the United States from 1965 to 1995. Most cases of malaria are found among foreign civilians who emigrate from endemic areas. In 1994, *Plasmodium vivax* was diagnosed in three persons in Houston, Texas. These persons had not traveled to areas where malaria is endemic nor did they have a recent bloodborne exposure. This strongly suggests mosquitoborne transmission of malaria in the United States.

Centers for Disease Control and Prevention *MMWR Summary of Notifiable Diseases, United States, 1995* (October 25, 1996, Vol. 44: 53).

Leishmanias are protozoan hemoflagellates. **Leishmaniasis** is caused by the presence of these intracellular parasites in their human host where they multiply as nonmotile, round forms (amastigotes) in mononuclear phagocytic cells. The individual species multiply in different body tissues and, therefore, produce different clinical syndromes. The human is the intermediate host; each species requires a definitive host, a sandfly, to complete its life cycle. Amastigotes ingested by the sandfly develop into flagellated, motile forms (promastigotes) that multiply in the insect's intestine. Later, the promastigotes migrate to the mouth parts of the insect and are inoculated into a new host by regurgitation when the sandfly bites another host.

Characteristics of Human Disease

Visceral leishmaniasis, (kala-azar), caused by *L. donovani,* is a disease of tropical and semitropical countries that occurs primarily in children and young adults. It is a chronic, systemic disease that is progressive and fatal when untreated. Following entry into the skin the organisms are engulfed by mononuclear phagocytes and carried through the blood to various organ sites. They are found in the liver (colorplate 11b), spleen, bone marrow, lymph nodes, and many other organs.

Cutaneous leishmaniasis is caused by different ***Leishmania*** *species* in different geographic locations. One form, called *oriental sore,* is caused by *L. tropica* and is found primarily in tropical and semitropical areas of Europe, Africa, and the Near and Far East. Similar disease is produced by *L. mexicana,* which occurs especially in Latin America. The organisms are found in mononuclear phagocytes and lymphoid tissue of the skin where they produce one or more ulcerating lesions (colorplate 11a).

Mucocutaneous leishmaniasis, also called *American leishmaniasis* because it occurs in Latin America, is caused by *L. braziliensis.* The organism parasitizes cells of the skin and mucosa of the mouth, nose, and pharynx. The patient may develop multiple skin lesions as a result of autoinoculation, or the disease may be transmitted from person to person by direct contact.

DIAGNOSTIC METHODS

Visceral leishmaniasis is diagnosed by demonstrating intracellular amastigotes in samples of spleen, liver, lymph nodes, bone marrow, or blood. Clinical material can be inoculated into special culture media to propagate the promastigote form. Nucleic acid probes are available in a few research laboratories. Serologic diagnosis can be made with an indirect immunofluorescence test.

Cutaneous leishmaniasis is diagnosed by demonstrating intracellular *Leishmania* amastigotes in cells of skin scrapings taken from the edges of exposed lesions or by culturing the promastigote form.

Special Features of Control

I. Treatment

Antimony compounds provide effective therapy for leishmaniasis, but these are available only for investigational use from the Centers for Disease Control. Pentamidine is an alternative therapeutic agent for visceral leishmaniasis, and amphotericin B may be effective for cutaneous disease.

II. Prevention

Visceral leishmaniasis is not directly transmitted from person to person. Control depends on treatment of human cases and application of insecticides to suppress the infective vector. Control of cutaneous disease is difficult because it can be transmitted directly by infected persons. Mass treatment campaigns and vector controls are necessary to prevent its continuing spread.

TRYPANOSOMIASIS

Organisms

(*Trypanosoma brucei gambiense, Trypanosoma brucei rhodesiense* [agents of African **trypanosomiasis,** or "African sleeping sickness"], *Trypanosoma cruzi,* the agent of American trypanosomiasis.)

Mature ***Trypanosoma*** (trypomastigotes) are flagellated protozoans that characteristically possess an undulating membrane. The African species appear in this form in the blood of human hosts (fig. 22.8). The American species may be found in the trypomastigote form in human blood but also in an intracellular amastigote stage in the mononuclear phagocytic cells of many organs. The insect stage of *T. cruzi* and the African species is called the *epimastigote.*

As with leishmania organisms, trypanosomes have humans and animals as intermediate hosts and insects as definitive hosts. The tsetse fly is the definitive host for the African species; for the American species, the definitive host is the "kissing" bug (reduviid bug), so called because of its tendency to bite at the mucocutaneous junction of the lips. When the insect bites or takes a blood meal from an infected human, it ingests the parasite which subsequently develops in the intestinal tract. In the tsetse fly, the organisms migrate to the salivary glands and are inoculated into a new host when the fly bites its next victim. In the reduviid bug, the parasite matures in the intestinal tract and is passed in the insect's feces. When the insect bites humans or animals, infected feces are deposited on skin and are rubbed into the bite wound or other skin abrasion by scratching.

Characteristics of Human Disease

African trypanosomiasis may begin with the formation of a lesion at the site of inoculation. As extracellular multiplication proceeds, the organism reaches the bloodstream and is deposited in many organs. When the central nervous system is invaded, the "sleeping-sickness" stage of infection develops, characterized by headaches, mental apathy, and neuromuscular disorders. As the disease progresses, sleepiness becomes pronounced and the patient passes into a deep coma. Death results from heart failure, meningitis, or secondary bacterial or parasitic infections, especially pneumonia.

American trypanosomiasis, or Chagas disease, occurs in Latin America. It begins with the development of a swollen lesion (the chagoma) at the site of inoculation, often around the lips or eyes. The latter site is common in children. In adults, the disease may be mild or asymptomatic, but it is often fatal for children who develop fever and malaise. Organisms can be found in the mononuclear phagocytes of many organs. Death may result with chronic progression to the central nervous system or with myocardial failure.

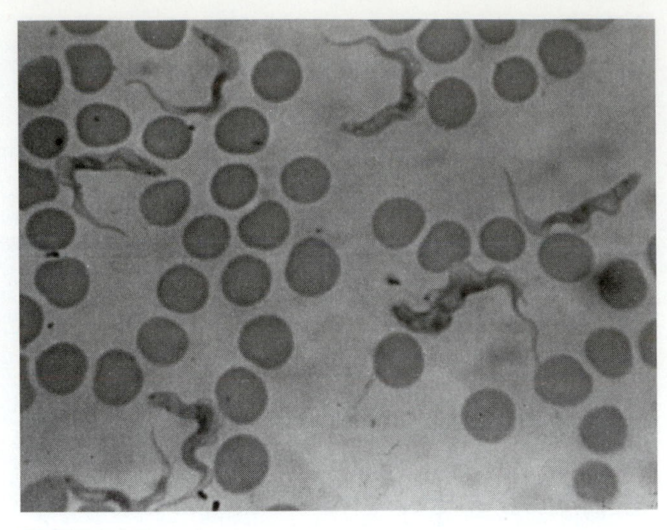

FIGURE 22.8 *Trypanosoma brucei gambiense* in a blood smear. Note the terminal flagellum and the undulating membrane that extend along the organism's body (×850).

Courtesy of Dr. Thomas W. M. Cameron

Parasitic Structure and Function
Tissue and Blood Protozoa: Tissue/Blood Protozoa Life Cycles

DIAGNOSTIC METHODS

African trypanosomiasis is usually diagnosed by demonstrating the organisms in smears of peripheral blood (fig. 22.8) or lymph node aspirates. During stages of CNS involvement they may be seen in spinal fluid. Special culture and animal inoculation methods are available. Serologic diagnosis can be made by fluorescent antibody and enzyme immunoassay tests, but these are not well standardized. Diagnostic methods for American trypanosomiasis are similar to those for the African disease. Microscopic examination of specimens, culture methods, and animal inoculations establish the diagnosis if positive. Sensitive serologic methods include fluorescent antibody tests and agglutination procedures.

Special Features of Control

I. Treatment

The early stages of African trypanosomiasis are treated with pentamidine or suramin and advanced disease with arsenical compounds such as melarsoprol or tryparsamide. Once the disease reaches the somnolent stage, the prognosis is generally poor. Treatment of American trypanosomiasis is particularly difficult when chronic, progressive stages have been reached. Two drugs, nifurtimox and benzimidazole, are used for therapy but must be administered for up to 4 months.

TABLE 22.3 HUMAN FILARIAL INFECTION

Organism	Sources of Infection	Infective Form	Adult Localization	Geographic Transmission	Distribution
W. bancrofti	*Culex* *Aedes* *Anopheles* mosquitoes	Larvae matured in mosquito	Large lymph vessels, especially in the pelvis	Larvae called "microfilariae" circulate in blood, especially at night; these are transferred to mosquito	Tropical and subtropical regions, especially: Africa Asia Central and South America Pacific Islands
B. malayi	*Mansonia* *Anopheles* *Aedes* mosquitoes	Larvae matured in mosquito	As above	As above	Asia Malaysia Indonesia Philippines
O. volvulus	Blackfly	Larvae matured in fly	Subcutaneous tissues, especially those of head and shoulders, lower trunk, and legs	Microfilariae remain in skin and lymphatics adjacent to adult and are picked up by biting fly	Guatemala Mexico Venezuela Brazil Ecuador Central Africa
L. loa	Deerfly Mango fly	Larvae matured in fly	Adult migrates through subcutaneous tissues	Microfilariae in circulating blood are picked up by biting fly	West and Central Africa The Sudan

II. Prevention

Control of trypanosomiasis is difficult. Measures include intensive efforts to clear the tsetse flies and reduviid bugs from villages and homes while conducting mass treatment campaigns in local populations.

FILARIASIS

Organisms

Species of roundworms. In the case of *Wuchereria bancrofti* and *Brugia malayi,* the adults live in lymphatic vessels. In infections caused by *Onchocerca volvulus* and *Loa loa,* the adults live in skin and subcutaneous tissues.

Life Cycles and Characteristics of Human Disease

Humans are the definitive host for the parasites causing **filariasis.** Various species of mosquitoes and biting flies serve as intermediate hosts (table 22.3). After the larval parasites are inoculated into a human by an insect bite, they mature into bisexual adults. The adults copulate and produce microfilariae which, depending on the species, either circulate in the blood (fig. 22.9) or remain trapped in tissues. Insects become infected with microfilariae while feeding on infected persons.

Allergic reactions including hives, fever, and local tissue inflammation are prominent in filarial infections.

Bancroftian and Malayan Filariasis

Filariasis caused by *Wuchereria bancrofti* or *Brugia malayi* is characterized by lymphadenopathy, which may lead to obstruction of lymph flow and elephantiasis of the region drained by affected vessels. Adult worms of *W. bancrofti* localize most commonly in lymphatic vessels of the pelvis, blocking flow from the genitalia and lower limbs. Genital involvement is rare with *B. malayi.* The microfilaria of *Wuchereria* and *Brugia* species display "nocturnal periodicity"; that is, they are found most commonly in peripheral blood during the night rather than during the day. Interestingly, circulation of the microfilariae coincides with the feeding habits of their insect host.

Onchocerciasis

Onchocerciasis, caused by *Onchocerca volvulus,* is characterized by the appearance of firm nodules in subcutaneous tissues of the head, shoulders, or lower part of the body. The microfilariae do not enter the blood but migrate through the tissues. If they migrate from lesions on the head to the eye, they may produce a local reaction, ocular disorders, and sometimes blindness.

Loiasis

Loiasis, caused by *Loa loa,* is also an infection of subcutaneous tissues, but the adult worm migrates, producing extensive, probably allergic, reactions at each new site. As the worm migrates, so called "fugitive" or "calabar" nodules form and then resolve, a characteristic of the disease. Sometimes the worm may wander through the conjunctival membrane across the ball of the eye causing pain and swelling for the afflicted person and great surprise for any observer. Microfilariae circulate in the peripheral bloodstream, where they may be found during the day as well as at night. Its vector is *Chrysops,* a tabanid fly and a daytime feeder.

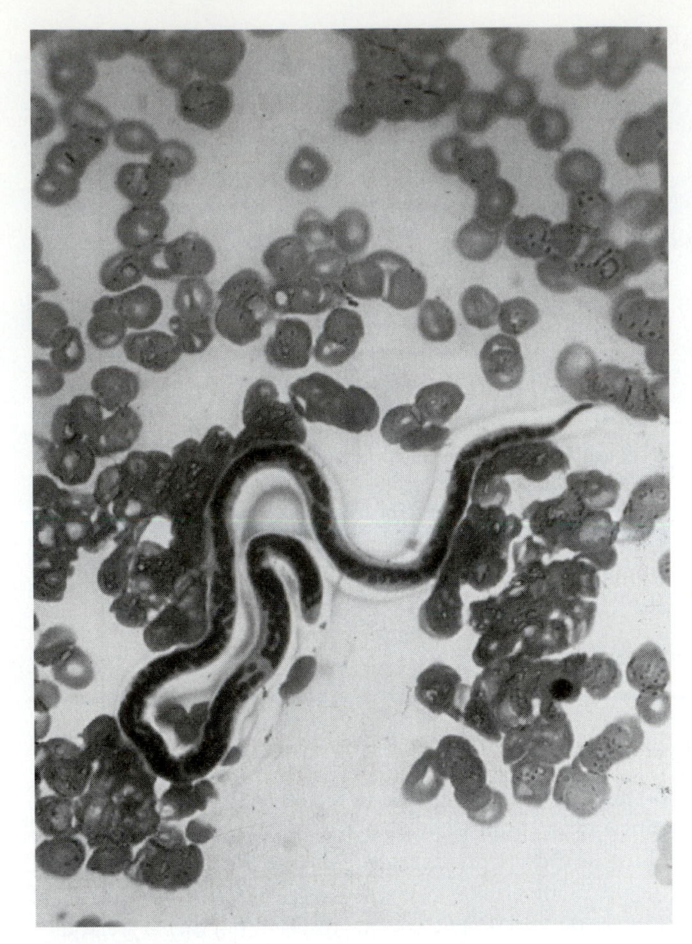

FIGURE 22.9 Microfilaria *(Loa loa)* in a thick smear of blood (×246).
Courtesy Dr. Stephen Lerner, Wayne State University, School of Medicine.

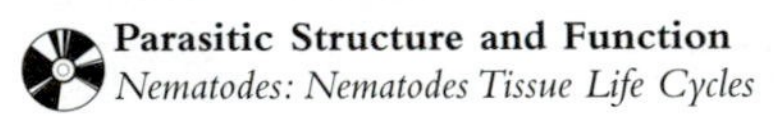

Parasitic Structure and Function
Nematodes: Nematodes Tissue Life Cycles

DIAGNOSTIC METHODS

Diagnosis is made in the laboratory by demonstrating viable microfilariae in wet-mount preparations of blood or by staining them in dried smears of blood (fig. 22.9). Filtration and concentration techniques are available to enhance detection.

Onchocerciasis is diagnosed by examining tissue obtained by biopsy or by surgical removal of a nodule. Either microfilariae or adult worms may be found.

Treatment and Prevention

Diethylcarbamate and ivermectin are used to eradicate microfilariae, but elimination of adult filarial worms is difficult. Surgical removal of adult worms may be the most feasible cure of onchocerciasis and loiasis, but the method is not practical in bancroftian or Malayan filariasis. Antihistamines or corticosteroids may help relieve allergic reactions.

Control and prevention require concomitant efforts to suppress insect vectors and to treat all infected human hosts who may perpetuate the life cycles of the parasites.

QUESTIONS

1. What vector transmits the plague bacillus?
2. What two kinds of illness can the plague bacillus produce? What precautions are required for each? Why do these precautions differ?
3. What organism causes Lyme disease? What is the vector? the intermediate host? the accidental host?
4. Describe the early symptoms of Lyme disease. What chemotherapy is used to treat the disease?
5. What organism causes relapsing fever? What arthropod transmits this disease?
6. What vector transmits epidemic typhus fever? What steps are necessary to control this disease?
7. How is Rocky Mountain spotted fever maintained in nature? What vector transmits this disease?
8. How are rickettsial diseases diagnosed in the laboratory?
9. Differentiate between the two types of ehrlichiosis in the United States.
10. What is the reservoir for the yellow fever virus? the vector? How is the disease controlled?
11. What is the definitive host for the *Plasmodium* species? the intermediate host? the vector?
12. What causes the symptoms of malaria?
13. What is the goal of malarial therapy?
14. What is a hemoflagellate? Name a disease caused by a hemoflagellate.
15. What is the vector for leishmaniasis?
16. Differentiate between the two types of leishmaniasis.

BIBLIOGRAPHY

Anonymous. 1950. Clara L. Maass. *Am. J. Nurs.* 50 (6): 343.

Benenson, A. S., ed. 1995. *Control of communicable diseases in man,* 16th ed. Washington, D.C.: American Public Health Association.

Carey, M. 1958. A short account of malignant fever lately prevalent in Philadelphia. In *Curiosities of medicine.* New York: Berkley Medallion.

Cartwright, F. F. in collaboration with M. D. Biddiss. 1991. The Black Death. In *Disease and history.* New York: Dorset Press.

———. 1991. General Napoleon and General Typhus. In *Disease and history.* New York: Dorset Press.

Centers for Disease Control and Prevention. 1995. Arboviral disease–United States 1994. *Morbid. Mortal. Weekly Report* 44 (35): 641.

Centers for Disease Control and Prevention. 1995. Imported dengue–United States 1993–1994. *Morbid. Mortal. Weekly Report* 44 (18): 292.

Centers for Disease Control and Prevention. 1995. Local transmission of *Plasmodium vivax* malaria–Houston, Texas, 1994. *Morbid. Mortal. Weekly Report* 44 (15): 292.

Centers for Disease Control and Prevention. 1995. Lyme disease–United States 1994. *Morbid. Mortal. Weekly Report* 44 (24): 459.

Culpepper, M. M., and P. G. Adams. 1988. Nursing in the civil war. *Am. J. Nurs.* 88 (7): 981–984.

Defoe, D. 1960. *Journal of the plague year.* New York: Signet Classics.

Desowitz, R. S. 1991. *The malarial capers: More tales of parasites and people, research and reality.* New York: W. W. Norton.

Dunelson, J. E., and M. J. Turner. 1985. How the trypanosome changes its coat. *Sci. Am.* 252 (2): 44–51.

Ehrich, R. 1994. Current therapies and prophylaxis of malaria. *Nurs. Pract.* 19 (9); 49–51,55–56.

Eron, C. 1981. The sting of death. In *The virus that ate cannibals.* New York: Macmillan.

Estes, M. E. Z. 1900. Chagas' disease. *Crit. Care Nurse.* 9 (7): 48–62.

Friedman, M. J., and W. Trager. 1981. The biochemistry of resistance to malaria. *Sci. Am.* 244 (3): 154–165.

Guinther, L. 1932. A nurse among the heroes of the yellow-fever conquest. *Am. J. Nurs.* 32 (2): 173–176.

Habicht, G. S., G. Beck, and J. L. Benach. 1987. Lyme disease. *Sci. Am.* 257 (1): 78–83.

Hawking, F. 1970. The clock of the malaria parasite. *Sci. Am.* 222 (6): 123.

Kantor, F. S. 1994. Disarming Lyme disease. *Sci. Am.* 271 (3): 34–39.

Magnarelli, L. A. and J. S. Dumler. 1996. Ehrlichioses: Emerging infectious diseases in tick-infested areas. *Clin. Microbiol. Newsl.* 18 (11): 81–83.

Malaria: The unwelcome immigrant. 1989. *Emerg. Med.* 21 (13): 128–144.

Malaria: 1991. *World Health* September/October 2: 29.

Marbit, M. D., and E. Willis. 1990. Lyme disease: Implications for health educators. *Health Educ.* 21 (2): 41–43.

Massey, K. 1982. Rocky Mountain spotted fever: A national disease. *MCN* 7 (2): 104–109.

McEvedy, C. 1988. The bubonic plague. *Sci. Am.* 258 (2): 118–123.

McNeill, W. H. 1976. *Plagues and peoples:* Garden City, N.Y.: Anchor Press.

Monath, T. P. 1988. Yellow fever gains ground. *World Health* July, 20–23.

Nanduri, J., and J. W. Kazura. 1989. Clinical and laboratory aspects of filariasis. *Clin. Microbiol. Rev.* 2 (1): 39–50.

Overturf, O. 1989. Bacterial and rickettsial zoonosis associated with tick and flea bites. *TOP Emerg. Med.* 10 (4): 67–79.

Powell, J. H. 1949. *Bring out your dead: The great plague of yellow fever in Philadelphia in 1793.* Philadelphia: University of Pennsylvania Press.

Rahn, D. W., and S. E. Malawista. 1991. Lyme disease: Recommendations for diagnosis and treatment. *Ann. Intern. Med.* 114 (6): 472–481.

Rikihisa, Y. 1991. The tribe *Ehrlichieae* and ehrlichial diseases. *Clin. Microbiol. Rev.* 4 (3): 286–308.

Roueche, B. 1991. The foulest and nastiest creatures that be. In *The medical detectives.* New York: Plume.

Sanford, J. P. 1990. Humans and animals: Increasing contacts, increasing infections. *Hosp. Pract.* 25 (2): 123–140.

Shryock, R. H. 1947. *American medical research: Past and present.* New York: Commonwealth Fund.

———. 1960. *Medicine and society in America: 1660–1860.* Ithaca, N.Y.: Cornell University Press.

Silverstein, D. B. 1989. Lyme disease: Easy to treat, easy to miss. *J. Am. Acad. Nurse Pract.* 1 (3): 73–76.

Strode, G. K. 1951. *Yellow fever.* New York: McGraw-Hill.

Teglia, M. Q., O. Teglia, S. Santiago, and B. A. Cunha. 1991. Case studies in infections: Falciparum malaria. *Emerg. Med.* 23: 165–173.

Verhoef, H. and R. Bos. 1992. Development and vector-borne diseases. *World Health* November/December.

Walker, D. H. 1989. Rocky Mountain spotted fever: A disease in need of microbiological concern. *Clin. Microbiol. Rev.* 2 (3): 227–240.

Ziegler, P. 1991. *The Black death.* Dover, N. H.: Alan Sutton, Inc.

Zinsser, H. 1975. *Rats, lice and history.* Boston: Little, Brown.

SUMMARY TABLE 22.4 INFECTIOUS DISEASES TRANSMITTED BY ARTHROPOD VECTORS

Clinical Disease	Causative Organism	Other Possible Entry Routes	Incubation Period	Communicable Period
I. Bacterial Diseases Transmitted by Arthropods				
Plague	*Yersinia pestis*	Respiratory	1–6 days	Not transmissible among human beings except when plague pneumonia develops; then communicable during time sputum is bacteriologically positive
Lyme disease	*Borrelia burgdorferi*		3–32 days	Not transmissible among human beings
Relapsing Fever	*Borrelia recurrentis*		5 days to 2 weeks	Not communicable among human beings
II. Rickettsial Diseases Transmitted by Arthropods				
Epidemic typhus fever, Brill-Zinsser disease	*Rickettsia prowazekii*		6–15 days, usually about 12 days	Not communicable among human beings
Murine (endemic) typhus	*Rickettsia typhi*		Same as above	Not communicable among human beings
Scrub typhus (tsutsugamushi fever)	*Rickettsia tsutsugamushi*		6–21 days	Not communicable among human beings
Rocky Mountain spotted fever	*Rickettsia rickettsii*		3–10 days	Not communicable among human beings
Boutonneuse fever	*Rickettsia conorii*		5–7 days	Not communicable among human beings
Rickettsialpox	*Rickettsia akari*		Probably 10–24 days	Not communicable among human beings
Ehrlichiosis	*Ehrlichia sennetsu* *Ehrlichia chaffeensis* Unidentified species		5–28 days	Not communicable among human beings
III. Viral Diseases Transmitted by Arthropods				
Yellow fever	Yellow fever virus		Less than a week, as a rule 3–6 days	Not transmissible from person to person
Eastern equine *Western equine* *Venezuelan equine* *St. Louis encephalitis*	Alphaviruses		4–5 days to 3 weeks, average 10–15 days	Not transmissible from person to person
IV. Parasitic Diseases Transmitted by Arthropods				
Malaria	*Plasmodium* species		12–30 days, depends on species	When gametocytes circulate in blood[a]
Babesiosis	*Babesia* species		About 2 weeks	
Leishmaniasis	*Leishmania* species		From a few days to 24 months, depends on species	While leishmaniae circulate in blood[a]
Trypanosomiasis	*Trypanosoma brucei* subspecies		1–3 weeks, depends on species	While trypanosomes circulate in blood[a]
Filariasis	*Wuchereria bancrofti* *Brugia malayi* *Onchocerca volvulus* *Loa loa*		About 9 months before microfilariae are found in blood	While microfilariae circulate in blood[a]

[a]Transmissible to insect vector only.
[b]For *P. vivax* and *P. ovale* hypnozoites only.

Specimens Required	Laboratory Diagnosis	Immunization	Treatment	Transmission-Based Precautions
Blood Pus aspirated from buboes, paracentesis fluids; spinal fluid Sputum Acute and convalescent serum	Culture Smear and culture Rising titer of antibodies	Killed vaccines for people at high risk	Streptomycin Tetracyclines Chloramphenicol	Standard precautions; droplet for pneumonic plague Concurrent disinfection of discharges; thorough terminal disinfection-cleaning
Acute and convalescent serum	As above	None	Tetracyclines Amoxicillin Ceftriaxone	None required
Blood	Smear and culture Dark-field exam Animal inoculation	None	Tetracyclines Penicillin	Standard precautions
Blood	Isolation of rickettsiae in specialized facilities	Killed vaccines		
Acute and convalescent serum	Antibody titers by: Indirect fluorescense Indirect hemagglutination Enzyme immunoassay Latex agglutination	None	Tetracyclines Chloramphenicol	Isolation not necessary if patient is thoroughly devectored
Blood Blood Acute and convalescent serum	Isolation in mice Special stains PCR Rising titer of antibodies	None	Tetracycline Chloramphenicol	None required
Blood Acute and convalescent serum	Virus isolation in mouse Rising titer of antibodies	Live attenuated vaccine	None specific	Standard precautions; patient must be screened from mosquitoes
Blood Acute and convalescent serum	Virus isolation Rising titer of antibodies	None except EEE and WEE vaccine (inactivated)	None specific	Isolation and special precautions not indicated
Blood smears, thick and thin	Recognition of parasitic forms in red blood cells		Chloroquine Quinine plus sulfonamide+ Pyrimethamine Primaquine[b]	Standard precautions; protect patient from vector Same as above
Same as above	Same as above		Clindamycin plus quinine	
Biopsies of spleen or liver, blood Aspirated bone marrow Acute and convalescent serum	Culture and microscopic recognition of intracellular forms Rising titer of antibodies		Stibogluconate Antimony compounds Amphotericin B	
Spinal fluid Blood Aspirated lymph	Microscopic recognition of intracellular or extracellular forms; culture		Nifurtimox Benzimidazole	Isolation and special precautions not indicated; protect patient from vector
Acute and convalescent serum	Rising titer of antibodies		Diethylcarbamazine	
Blood	Recognition of microfilariae		Ivermectin	

Infectious Diseases Acquired through Accidental Injuries

OUTLINE

Microbes in Motion

This chapter covers the important bacterial and parasitic agents that cause infectious diseases transmitted from animal or environmental reservoirs. A discussion of their laboratory diagnosis, epidemiology, control, and prevention is included. The following books and chapters of the Microbes in Motion CD-ROM may be a useful preview or supplemental study aid to your reading: Vaccines: Vaccine Targets. Gram-negative Organisms in Human Hosts: Bacilli-Facultative Anaerobes. Anaerobic Bacteria: Gram-positive Bacilli; Normal Flora; Gram-positive Cocci; Gram-negative Bacilli; Gram-negative Cocci. Microbial Pathogenesis of Infectious Diseases: Exotoxins.

OBJECTIVES

On completing this chapter the student will be able to:

1. cite several diseases acquired through animal bites or accidental injury.
2. discuss the clinical management of burn wounds.
3. define the term street wound.
4. describe the clinical nature, laboratory diagnosis, precautions, and epidemiology of rabies, tetanus, and gas gangrene.

KEY WORDS

Anaerobic infections
Bartonella henselae (bar-ton-**ELL**-a hen-**SEL**-ee)
Bite wound infections
Burn wound infections
Cat scratch disease
Clostridial (closs-**TRID**-ee-al) infections
Clostridium species
Gas gangrene
Pasteurellosis (past-ure-ell-**OH**-sis)
Rabies
Rabies immunoprophylaxis (imm-u-no-pro-fill-**AX**-iss)
Rat bite fever
Street wound
Tetanus (lockjaw)

INTRODUCTION

This chapter covers infectious diseases characteristically acquired through injury to the skin and mucous membranes such as animal bites, anaerobic infections, "street wounds," and burns. Diseased animals sometimes transmit their infections through injuries they inflict on human beings, rabies being a notable example. More usually, animals transmit viral or bacterial agents that are commensalistic for them but virulent for human beings when introduced into their subcutaneous tissues by a bite or scratch. The microbial species involved may induce severe local or even systemic infection. Similar problems may arise as a result of human bites.

The term **street wound** is used to distinguish accidental injuries exposed to environmental sources of infection from those inflicted by animals, burns, or surgery (discussed in chapter 21) or contaminated with endogenous microorganisms. Animals transmit parasitic or commensalistic microorganisms, whereas environmental reservoirs teem with saprophytic bacteria and fungi, many of which cannot establish themselves in human tissues. Among the most serious diseases acquired from environmental sources are those of clostridial origin (tetanus and gas gangrene) and those caused by fungi. Vegetative clostridial cells are found in the intestinal tract of human beings and animals, but their resistant endospores survive and are ubiquitous in the external environment. The fungi that cause subcutaneous or systemic mycoses normally live as saprophytes in the soil (see chap. 14).

Other organisms most likely to infect injured soft tissues include species of staphylococci, streptococci, and gram-negative bacteria such as *Pseudomonas aeruginosa,* a notorious troublemaker in wounds and burns. Anaerobes thrive in soft tissues when local conditions meet their requirement for reduced oxygen tension. The development and severity of wound infections depend primarily on conditions in traumatized tissues and their exposure to microorganisms that have a pathogenic potential.

INFECTIOUS DISEASES ACQUIRED THROUGH ANIMAL BITES

RABIES

The Clinical Disease

Rabies is a viral infection occurring primarily in wild carnivorous animals (skunks, foxes, raccoons) and bats. Transmission to human beings is rare but almost invariably fatal. Human and animal infection is usually acquired through the bite of a rabid animal whose saliva contains the virus. Transmission through the respiratory route, by airborne virus, has occurred under unusual circumstances, as in spelunkers exploring a bat-infested cave and aerosol transmission to a laboratory worker. Human-to-human transmission has been documented only in a few instances when patients received corneal transplants from individuals who died of an unknown neurologic disease that later proved to be rabies.

When introduced into tissues, the rabies virus travels along sensory nerves from the point of inoculation to the central nervous system, where it multiplies. In the brain, the virus localizes primarily in the hippocampus and cerebellum. From the CNS, the virus travels along nerves in a centrifugal pattern to invade a variety of organs and tissues. The chief clinical manifestations are those of an acute encephalomyelitis, with both physical and psychologic symptoms of deranged cerebral function. The onset of fever, malaise, and headache is followed by sensory disturbances, particularly at the site of the bite wound, spasms in the muscles required for swallowing, convulsive seizures, and respiratory paralysis. The mental suffering of the patient is associated with a fear of swallowing because of painful throat spasms that can bring on attacks of convulsive choking. (This disease has been called *hydrophobia* because of the patient's fear of swallowing water or liquids.) Apprehension, excitability, and delirium then give way to a progressive, general, flaccid paralysis, and the patient gradually lapses into coma. Death, usually within 6 days, may result from respiratory paralysis, peripheral vascular collapse, or both.

LABORATORY DIAGNOSIS

I. Identification of the Organism

The rabies virus forms inclusion bodies within the nucleus of the affected brain cells. These characteristic inclusions, called *Negri bodies,* are readily visualized under the light microscope in stained impression smears but are absent in 20 to 30% of patients. Fluorescein-tagged rabies antiserum is also used to identify viral antigen in infected tissue cells. The virus is isolated in white laboratory mice or cell cultures. It is identified serologically by known antiserum that specifically neutralizes its infectivity.

II. Identification of Patient's Serum Antibodies

Antibodies that develop during the course of infection or following vaccination can be detected by neutralization tests using infected mice or cell cultures. Enzyme immunoassay and immunofluorescent techniques may also be used. Serologic diagnosis is seldom helpful in rapidly fatal cases because significant levels of antibody do not have time to develop.

III. Specimens for Diagnosis

During life, rabies virus may be isolated from saliva, throat swabs, conjunctival and nasal secretions, spinal fluid, and urine. At autopsy, possible sources include brain, spinal cord, submaxillary salivary glands, lacrimal glands, muscle tissue, lung, kidney, pancreas, and adrenals. Such specimens may be examined by the fluorescent rabies antibody (FRA) technique. A positive FRA test is diagnostic, but if it is negative, specimens are inoculated into mice or cell cultures. Rabies virus, if present, causes death of mice within a few days. Mouse brains and cell cultures are examined for virus by the FRA technique. Direct examina-

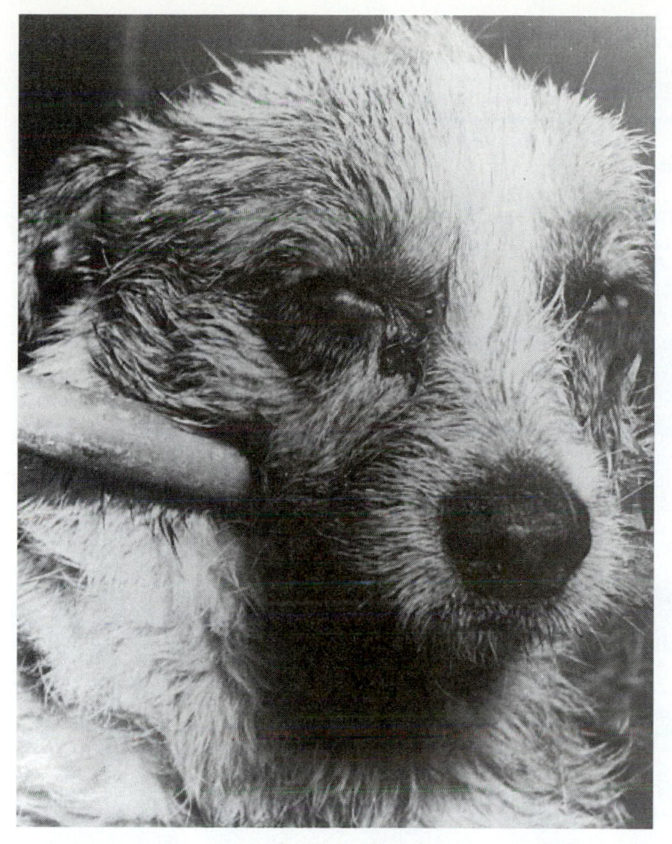

FIGURE 23.1 The paralytic form of rabies is seen in the closeup of this dog's face.
Centers for Disease Control, Atlanta, GA.

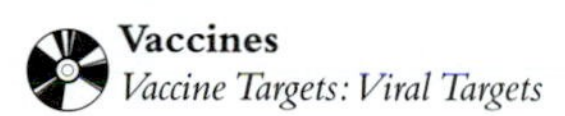

Vaccines
Vaccine Targets: Viral Targets

tion of neck skin biopsies by FRA methods is also a sensitive diagnostic test.

All specimens collected for virus isolation or FRA tests should be frozen as soon as possible, pending their transport to the laboratory. Most clinical laboratories are not equipped for handling rabies virus, and such specimens must be forwarded to the nearest reference or health department laboratory.

An acute serum for serologic studies should be collected as soon as the diagnosis of rabies is suspected. This should be followed by additional serum collections in 5 and 10 days, and later, if possible. Serologic diagnosis, when it can be made, is generally retrospective at best.

IV. Examination of Suspect Animal

When a bat, dog, or other possibly rabid animal has bitten a human being, it is of immediate importance to isolate and examine the animal if possible. Appropriate decisions must be made promptly as to the need for prophylactic vaccination of the victim (see Perspective 23.1).

Captured bats are killed and examined at once for rabies infection. Selective judgment should be applied to the disposition of other animals, however. A veterinarian or physician competent to diagnose rabies should be called to examine a captive dog or wild animal that has bitten someone.

A healthy dog or cat that bites a person should be quarantined for 10 days and observed for clinical signs of developing rabies, but it is best to sacrifice a captured wild animal at once and examine its brain for the disease. If it was infective at the time of the bite, a quarantined dog or cat will generally show signs of rabies (e.g., increasing excitability or paralysis) within a week (fig. 23.1). Animals that appear to be rabid at the time of the bite or later during quarantine should be killed and examined without delay. The animal's head must not be destroyed because only the brain or salivary secretions can provide conclusive evidence of rabies infection.

Rapid diagnosis of the animal infection is critical if the human bite victim is to have time to be actively immunized against the disease. **Rabies immunoprophylaxis** is effective only because the infection generally has a prolonged incubation period. However, immunization itself requires multiple doses of vaccine given over a 4-week period and must be started promptly if it is to be successful. When the possibility of exposure is reasonably certain or has been confirmed, vaccination is initiated without question. The decision is more difficult when the biting animal cannot be properly diagnosed because it escaped, was killed prematurely, or was destroyed in such a way that the head cannot be examined. Failure to properly preserve the intact head of a suspect animal (it should be packed in ice or kept frozen for transport to the laboratory) may contribute to the problem of animal diagnosis (see table 23.1 for recommendations concerning immunoprophylaxis).

Epidemiology

I. Communicability of Infection

A. *Reservoirs, Sources, and Transmission Routes* Animals, especially wild ones, constitute the largest reservoir for rabies. In the United States, seven types accounted for almost all of the 7,800 confirmed animal cases in 1995: skunks, raccoons, bats, cattle, cats, dogs, and foxes, and a new outbreak is emerging in coyotes in Texas. Domestic dogs acquire the disease from wild animals. In the United States, dogs and indigenous bat species are the usual vectors of human rabies. From 1980 through 1994, 13 of 16 human cases acquired in the United States resulted from exposure to bats. All of the infections were fatal.

It is possible, but not unusual, for human beings to acquire the infection from rabid animals without being bitten. Scratches or other minor skin wounds, or mucosal surfaces exposed to infectious saliva, may provide a route of entry. Even the airborne transmission of rabies has been documented. In one episode the virus was acquired through exposure to the dust of a cave harboring millions of bats (fig. 23.2). In another instance, the source of virus was an aerosol generated by a defective machine used for spraying virus in an animal inoculation experiment. The fact that the laboratory worker had been immunized but nonetheless acquired rabies from a modified virus suggests that airborne transmission increases the risk of infection

PERSPECTIVE 23.1 *Joseph Meister: A Lucky Boy*

In 1885, Joseph Meister, a 9-year-old boy from the Alsace region of France, was bitten severely by a rabid dog. Had this episode occurred any earlier time in history, he would most certainly have died. As it was, the news from Paris was that Dr. Louis Pasteur had developed an effective rabies vaccine for dogs, and Joseph was taken to Paris in the desperate hope that this vaccine might save him. The boy was the first human to receive the vaccine.

Pasteur had been cultivating the rabies virus in rabbits, attenuating it, and injecting it into dogs. After three years of such experiments, he had successfully immunized more than 50 dogs against rabies.

Pasteur presented a dramatic account of the case in a paper published in the *Comptes Rendus Academie des Sciences.* Following are excerpts from this original work:

> *The dog has now been rendered refractory to rabies. It may be inoculated with the virus of rabies under the skin, or even after trephining, on the surface of the brain, without any subsequent development of rabies.*
>
> *Never having once failed when using this method, I had in my possession fifty dogs, of all ages and every race, refractory to rabies, when three individuals from Alsace unexpectedly presented themselves at my laboratory, on Monday the 6th of last July.*
>
> *Theodore Vone, grocer, of Meissengott, near Schlestadt, bitten in the arm, July 4th, by his own dog, which had gone mad.*
>
> *Joseph Meister, aged 9 years, also bitten on July 4th, at eight o'clock in the morning by the same dog. This child had been knocked over by the dog and presented numerous bites, on the hands, legs, and thighs, some of them so deep as to render walking difficult. The principal bites had been cauterized at eight o'clock in the evening of July 4th, only twelve hours after the accident, with phenic acid by Dr. Weber, of Ville.*
>
> *The third person, who had not been bitten, was the mother of little Joseph Meister.*
>
> *At the examination of the dog, after its death by the hand of its master, the stomach was found full of hay, straw, and scraps of wood. The dog was certainly rabid. Joseph Meister had been pulled out from under him covered with foam and blood.*
>
> *M. Vone had some severe contusions on the arm, but he assured me that his shirt had not been pierced by the dog's fangs. As he had nothing to fear, I told him that he could return to Alsace the same day, which he did. But I kept young Meister and his mother with me.*

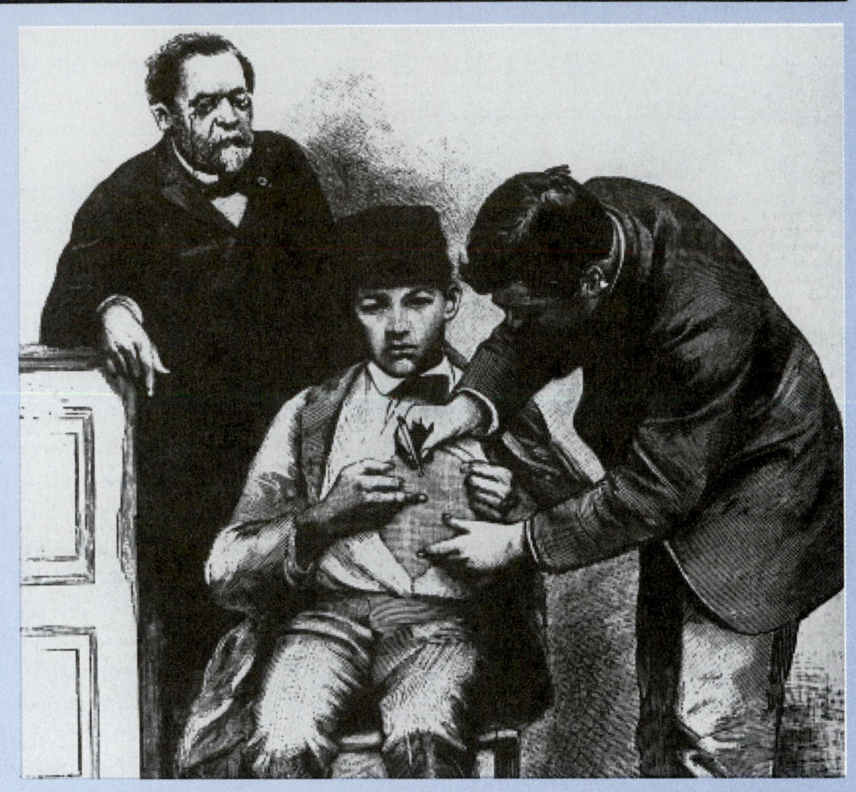

Louis Pasteur observes as his experimentally developed rabies vaccine is administered to a human for the first time. Joseph Meister's life was saved by the decision to try the vaccine. Wood engraving "An Inoculation for Hydrophobia" from Harpers Weekly, *1885.*

National Library of Medicine.

Meantime, Pasteur consulted with colleagues at a weekly meeting of the Academy of Sciences on July 6th. Telling them of this case, two of them

> *". . . M. Vulpian and Dr. Grancher . . . had the goodness to come and see little Joseph Meister at once, and to take note of the condition and the number of his wounds. There were no less than fourteen."*

Combined opinion was that Joseph was certain to develop rabies and die, but Pasteur told them of his latest experiments and successes with dogs. His account continues:

> *The death of this child appearing to be inevitable, I decided, not without lively and sore anxiety, as may well be believed, to try upon Joseph Meister the method which I had found constantly successful with dogs . . .*

The upshot was that young Meister received multiple inoculations of Pasteur's rabies vaccine over the next 10 days (Perspective fig. 23.1) and survived his ordeal.

Pasteur concluded his account:

> *"Since the middle of August I have looked forward with confidence to the future good health of Joseph Meister. At the present time, three months and three weeks have elapsed since the accident, his state of health leaves nothing to be desired . . ."*

So did the Meister story end happily, and so did the successful history of rabies vaccination begin.

From L. Clendening, *Source Book of Medical History.*

TABLE 23.1 RECOMMENDATIONS FOR RABIES IMMUNOPROPHYLAXIS[a]

Nature of Exposure	Status of Animal (Irrespective of Previous Vaccination): At Time of Exposure	During 10 days[b]	Recommended Treatment
I. Contact, but no lesions Indirect contact No contact	Rabid	—	None
II. Licks of the skin; scratches or abrasions; minor bites (covered areas, or arms, legs, trunk)	Suspected as rabid[c]	Healthy	Start vaccine. Stop if animal remains healthy for 5 days[b,c,d]
		Rabid	Start vaccine; upon positive diagnosis of animal give antiserum, complete vaccine course
	Rabid domestic animal; wild animal[e]; or animal unavailable for observation	—	Antiserum + vaccine
III. Licks of mucosa, major bites (multiple or on face, head, neck, finger)	Rabid or suspect[c] domestic animal; wild animal[e]; or animal unavailable for observation	—	Antiserum + vaccine. Stop treatment if animal remains healthy for 5 days[b,c,d]

[a]Adapted from guidelines suggested by the Expert Committee on Rabies of the World Health Organization.
[b]Holding period applies only to dogs and cats.
[c]All unprovoked bites in endemic area should be suspect unless animal proved negative by brain FRA.
[d]Or if its brain is proved negative by FRA tests.
[e]Exposure to rodents or rabbits almost never requires specific antirabies treatment.

FIGURE 23.2 A colony of bats in flight. These cave dwellers are an important reservoir of rabies, transmitting the virus through their bite. The dust in caves colonized by bats may also be infectious if inhaled.
Centers for Disease Control, Atlanta, GA.

and/or that the respiratory mucosae offer a particularly effective route of entry.

Person-to-person transmission of rabies by infectious secretions may be possible because human saliva contains virus during the course of disease. Such transfer, however, has not been documented. Persons dying of unknown neurologic disease are not used as corneal transplant donors because of the unfortunate transfer of fatal rabies infection by this route to a few individuals.

B. *Incubation Period* The average time is 3 to 12 weeks but may be shorter or greatly prolonged. The duration of the incubation is influenced by the length of the path that the virus must travel from the site of inoculation to the brain. Bites inflicted on the face and neck, particularly those involving mucous membranes, may result in infection, with onset of symptoms in 1 or 2 weeks. The amount of virus inoculated and the distribution of the local nerve supply also affect the length of the incubation period.

C. *Communicable Period* Virus is present in the salivary secretions of biting animals for several days before their symptoms appear and throughout the course of their illness. Bats may shed virus for weeks without displaying symptoms of illness.

D. *Immunity* Natural immunity to rabies virus does not exist among human beings or animals, but antibody production can be stimulated by virus vaccines.

II. Control of Active Infection

A. *Transmission-based Precautions* The patient should be hospitalized and treated with standard precautions (see chap. 11) throughout the course of the illness.

B. *Other Precautions* The patient's saliva should be considered infectious, and all items contaminated with it should be disinfected. The nursing care of these patients is difficult because of their apprehension, excitability, and disorientation. Particular caution must be exercised by those in close attendance to protect their skin and clothing from salivary contamination by wearing gowns, masks, and gloves.

C. *Treatment* There is no specific treatment for rabies after symptoms have started and the disease is established. However, immediate attention should be given to the bite wound. It should be thoroughly cleaned by irrigating and washing with soap and water. After washing, wounds are treated by installing half the dose of hyperimmune antirabies serum into the skin and muscle tissue beneath the wound and the other half intramuscularly in the gluteal area.

Intensive medical care is given to support the patient, with particular attention to pulmonary and cardiac monitoring. There are at least three documented cases of human survivors of clinical rabies, two in the United States and one in Argentina. Although one of these patients suffered severe residual damage, their survival suggests that continuous intensive care, started early in the disease and maintained at a high level, may be lifesaving.

D. *Control of Contacts* Human contacts of a rabies patient need not be vaccinated unless the patient's saliva has made direct contact with an open wound or mucous membrane. A search is made for the rabid animal source of the case, if it has not already been captured, and for other persons or animals that may have been bitten.

III. Prevention

Following an animal bite, the prevention of human rabies depends on appropriate treatment of the wound (see II, C) to kill the virus locally if possible and on prompt immunologic procedures. The latter include active immunization with vaccine and passive immunization with hyperimmune antirabies serum. Vaccination is also the chief preventive measure for people at high risk of exposure to rabies.

A. *Active Immunization* The original rabies vaccine prepared by Pasteur contained virus that he had passed through rabbits (fig. 23.3). The spinal cords of these animals were dried to reduce still further the virulence of the organisms for human beings. Since then, this vaccine has been modified a number of times, for example, the virus has been grown in duck embryos as well as in neural tissue, but these vaccines often produced severe allergic reactions. Currently, a human-diploid-cell-strain rabies vaccine (HDCV) is used in the United States. It consists of purified virus grown in cell culture and inactivated with beta-propriolactone. The vaccine produces an excellent antibody response, with only minimal adverse effects.

Humans are vaccinated with HDCV after injury has been inflicted by an animal known or suspected to be rabid (table 23.1). The procedure requires a series of five intramuscular injections on days 0, 3, 7, 14, and 28. Dogs are immunized by a single intramuscular injection of a live-virus vaccine, repeated once yearly or every 3 years, depending on the vaccine (fig. 23.4).

B. *Passive Immunization* Hyperimmune antirabies human or horse antiserum also provides effective protection when used in conjunction with active immunization. Antiserum is given intramuscularly, a portion being injected in and around the bite wound, the remainder in the buttocks. Passive immunization should be administered immediately after a bite exposure to rabies.

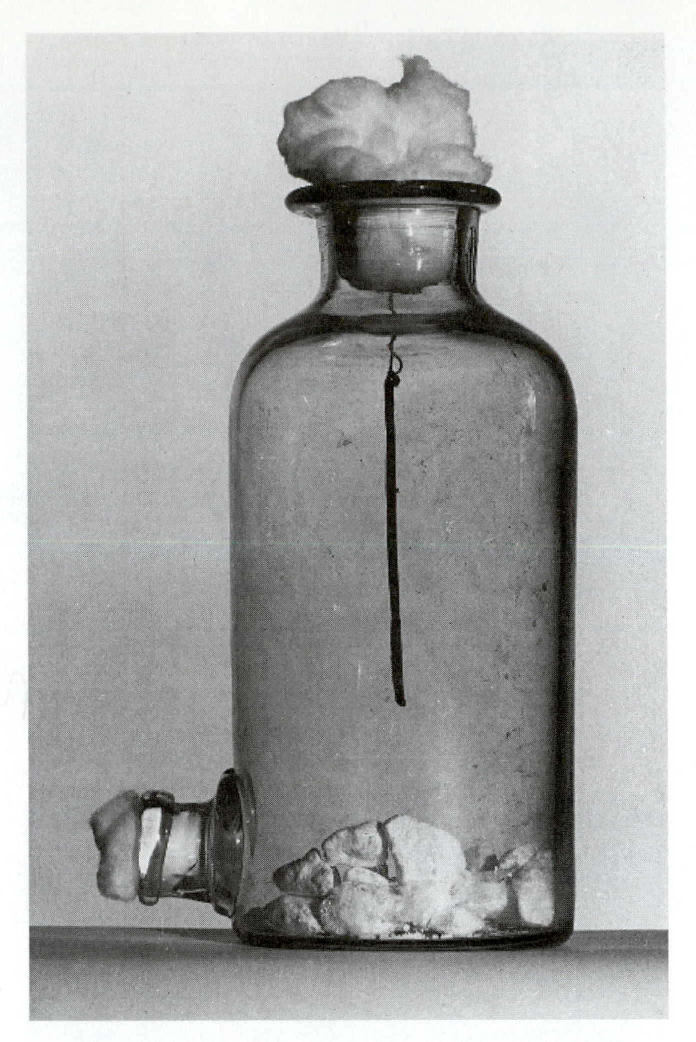

FIGURE 23.3 This flask was used by Louis Pasteur in his experimental work with rabies. It contains a rabbit spinal cord suspended for drying over a desiccant. Pasteur's rabies vaccine contained virus "fixed" by passage through rabbits. and dried by this method to attenuate its virulence for humans

Conservateur du Musée Pasteur, Paris, France.

C. *Preexposure Immunization* Rabies vaccine is usually given to persons at risk of exposure, such as laboratory workers, veterinarians, dog- and wildlife-control personnel, spelunkers, and other field workers serving in areas where rabies is endemic and prevalent. HDCV is given intradermally or intramuscularly on days 0, 7, and 21 or 28, followed by booster injections every 6 months or 2 years as needed to maintain a neutralizing titer of 5 in serum.

IV. Control

Rabies control requires suppression of the disease in the animal population. Immunization of dogs and cats and enforced restrictions on stray animals are essential features of control and are required by most local or state governments. Control of rabies in wild animals is the responsibility of state and federal wildlife conservation agencies, who work in cooperation with public health authorities when the threat of rabies exists (see chap. 10).

FIGURE 23.4 Rabies in dogs can be prevented by an annual rabies vaccination.
Courtesy Dr. Theodore Hoch, D.V. M., Yorktown, NY.

Apparently healthy dogs and cats that have bitten people need not be destroyed hastily. Instead, they should be quarantined for 10 days and observed for possible onset of disease.

The widespread incidence of rabies in the bat population of the United States appears to justify the vaccination of all persons bitten by bats. The necessity for treatment is often underscored by the fact that the offending bat may not be captured and hence is unavailable for examination. Furthermore, it is often difficult to demonstrate virus in bats by rapid smear examination, and the time required for virus cultivation may be more than the patient can safely afford. An oral vaccination program for rabies control in wildlife is being evaluated in the United States.

CAT SCRATCH DISEASE

Cat scratch disease is characterized by regional lymphadenopathy associated with a scratch or bite wound usually (but not necessarily) inflicted by a cat. There is an initial mild inflammatory reaction at the site of the wound, sometimes with induration or minor ulceration. Nodes that drain the area then become swollen with a thick, gray pus and eventually suppurate. This reaction requires at least 2 or 3 weeks to subside but may persist for months. There is often a recurring fever, sometimes with chills and general malaise. The eyes are sometimes involved, displaying the form of "oculoglandular" syndrome that may be seen in tularemia infections (see chap. 20). Recovery is always complete but requires variable periods of time.

The previously unknown causative agent of cat scratch disease has now been identified as ***Bartonella henselae,*** a rickettsialike organism. This same bacillus also causes lesions in the blood vessels of skin and other organs in patients with AIDS.

The diagnosis of cat scratch disease is made by isolating the organism from lymph node biopsy tissue on special culture media or by detecting significant levels of antibodies to *B. henselae* by EIA or immunofluorescence procedures. Bacillary organisms are also seen on lymph node biopsy specimens stained with silver stains. PCR techniques for diagnosis are available in specialized reference laboratories.

A skin-testing antigen prepared from heat inactivated suspensions of lymph nodes or pus has been used to aid the diagnosis of cat scratch fever. This antigen produces a delayed type hypersensitivity reaction in patients with the disease. It is not commercially available, however, and is not recommended for routine use because it does not differentiate past from present infection.

The lesions of cat scratch disease resemble those of other infections incurred through animal bites, notably pasteurellosis (discussed under Bacterial Infections of Bite Wounds, below). Good diagnostic bacteriologic techniques are required to differentiate such infections. Cat scratch disease occurs sporadically in all parts of the world among people of all ages but is seen most often among children and young adults. This age distinction may be due to frequency of exposure or to decreasing susceptibility in older adults following repeated encounters with the infectious agent. Cats do not become ill with this infection and are thought to be merely mechanical vectors of the agent. Infected kittens may have organisms in their bloodstreams for many months. Human cases have been acquired from sources other than cats, such as wood splinters or thorns, insect bites, and wounds incurred while slicing meats. This chronic, benign lymphadenopathy affects approximately 24,000 people annually.

Treatment is with ciprofloxacin or trimethoprim-sulfamethoxazole for cat scratch disease or erythromycin or doxycyline for skin lesions in patients with AIDS. Hospitalized patients are treated with standard precautions.

BACTERIAL INFECTIONS OF BITE WOUNDS

Resident bacteria of the mouths of animals or humans may produce serious local or even systemic infections when introduced deeply into tissues by biting teeth. The normal flora of human and animal mouths include both aerobic and anaerobic microorganisms (see table 5.1), although there are some differences in bacterial species. The anatomy and physiology of animal mucous membranes differ in many respects from those of human mucosae and therefore provide different conditions for commensalistic microorganisms. Animals often harbor transient microbial species from the soil, feces, their own hair or fur, or other sources, depending on their nature and habits.

Pathogenesis

Development of **bite wound infections** is greatly influenced by the nature of the injury. An open tear or laceration inflicted by teeth can be easily cleansed and kept aerated while healing proceeds so that microorganisms have little opportunity to establish themselves. Puncture wounds may implant bacteria deeply in positions that are inaccessible unless the wound is laid open by surgical incision.

Infected bites often contain mixed flora, but sometimes a particular species may predominate if it possesses more of the properties required for growth in human tissues. The process may be promoted still further if tissue-defense mechanisms are handicapped by injury to blood supply, with local hemorrhage, clotting, and devitalization of surrounding cells. This not only interferes with the marshaling of phagocytes and antibacterial components of plasma but also prevents adequate oxygenation of the local area, which becomes progressively anaerobic.

Many bacteria that live aerobically on mucous membranes or on culture media in the laboratory are, in reality, "facultative" anaerobes capable of growth and multiplication under conditions of reduced oxygen tension or anaerobiasis. Facultative anaerobes of the resident oral flora include species of staphylococci and streptococci, microaerophilic organisms, such as species of *Haemophilus,* and obligate anaerobes of the *Actinomyces* and *Bacteroides* genera.

Clinical Appearance

The character of the infectious process in a bite wound reflects, to some extent, the nature and properties of the multiplying organism as well as the tissue response to it. Staphylococci characteristically induce formation of localizing abscesses, or suppuration (chap. 19), whereas streptococci are often more invasive because their enzymes and toxins destroy the cells and components of connective tissue (chap. 12). Cellulitis, which is an advancing inflammatory involvement of subcutaneous and connective tissues, is frequently induced by streptococcal infection. Other invasive organisms may also cause this type of infection (see the discussion of clostridial gas gangrene in the following section). *Bacteroides, Fusobacterium,* and related species are gram-negative, nonendospore-forming anaerobic rods normally found in the mouth or alimentary tract but are capable of causing suppurative, inflammatory reactions in soft tissues. Mixed (synergistic) infections with anaerobic streptococci and staphylococci can produce severe gangrenous infections that may require amputation.

If local tissue reaction fails or if prompt medical treatment is not obtained, dissemination through draining lymph channels or the bloodstream may ensue. The spread of infection from a wound on an extremity is often marked by red streaks extending proximally. Such streaking is a sign of lymphangitis, inflammation of the lymph vessels draining the affected part to proximal nodes, which may also become involved. Regional lymphadenopathy with swelling and tenderness may then follow. Immediate treatment is indicated at the first signs of lymphangitis in order to prevent bacteremia.

Systemic Diseases Incurred through Bite Wounds

Aside from local trauma and infection, animal bites may lead to certain characteristic clinical syndromes typically associated with specific microorganisms. These include tetanus and gas gangrene, syndromes arising from particular clostridial infections; pasteurellosis; and rat bite fever, a term that may be used to refer to two diseases, each with a different bacterial agent.

Clostridial Infections

Deep, penetrating animal bites always present the hazard of **clostridial infections.** The ***Clostridium*** genus contains gram-positive, endospore-forming bacilli, which are obligate anaerobes normally residing in the human or animal intestinal tract. The mouths of animals may be transiently contaminated with endospores of clostridial species picked up from soil or with vegetative bacilli derived from fecal sources. These organisms are dangerous contaminants of bite wounds because, if they establish a foothold in human tissues and multiply there anaerobically, they produce highly toxic substances having a systemic effect. Tetanus and gas gangrene, two serious clostridial diseases acquired through infection of injured soft tissues, are commonly incurred from accidental "street wounds" and are discussed in detail under Diseases Acquired through Infection of Wounds. They also must be considered as a possibility when animal bite wounds are inflicted. The management and treatment of such wounds should therefore include immunoprophylaxis and/or booster vaccination for tetanus and débridement with aeration as indicated to prevent development of gas gangrene.

Pasteurellosis

Organisms of the genus *Pasteurella* are widely distributed among animals and birds. They may cause epidemic and septicemic diseases in animals, and occasionally infect humans, chiefly through animal bites. Of the four species in this genus, one, *Pasteurella multocida,* is the major human pathogen causing **pasteurellosis** (the term *multocida* literally means "many killing").

The pasteurellae are small, nonmotile, gram-negative rods that are facultatively anaerobic. Virulent strains of *P. multocida* are encapsulated. They are often present as normal flora in many domestic animals. The nasopharynx of cats and the tonsils of dogs are commonly colonized with *P. multocida,* but they are found also in cattle, sheep, and fowl. Colonized animals are usually asymptomatic, but under stressful situations *P. multocida* may cause outbreaks similar to cholera in birds, hemorrhagic septicemia in cattle, and primary or secondary pneumonias.

Human disease caused by *P. multocida* (rarely by other pasteurellae) is commonly transmitted by animal bites or scratches. Such wounds may develop frank infection, with redness, pain, swelling, and a serosanguinous drainage.

Systemic complications such as infected joints and osteomyelitis develop in about 40% of patients. Regional lymphadenitis followed by septicemia may occur, especially in persons with impairments of the mononuclear phagocyte system, such as cirrhosis of the liver or rheumatoid arthritis. Secondary pulmonary infections with *P. multocida* are sometimes seen in patients with underlying chronic lung disease (carcinoma, emphysema). Upper respiratory infections may also be complicated by this organism. Local extensions or bacteremia may lead, more rarely, to such problems as meningitis, brain abscess, or endocarditis.

Unlike many gram-negative rods, the pasteurellae are susceptible to penicillin which is the drug of choice, but tetracyclines are also effective. Specific chemotherapy should be initiated at the time of an animal bite to prevent *Pasteurella* infection. The wound should be carefully cleansed and débrided and suturing avoided if possible. Efforts to develop effective vaccines to control veterinary disease have not been successful thus far.

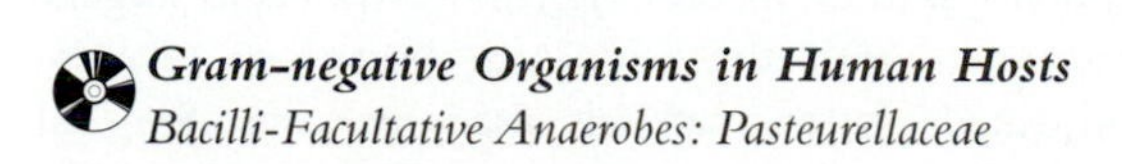
Gram-negative Organisms in Human Hosts
Bacilli-Facultative Anaerobes: Pasteurellaceae

Rat Bite Fever

Two types of bacterial infection are associated with rat bites. One type of **rat bite fever** seen in the United States is caused by an unusual gram-negative bacillus that is pleomorphic, forming long filaments. This organism, *Streptobacillus moniliformis,* is of interest because of its ability to enter spontaneously into a stage of growth in which it lacks a cell wall. In this respect it resembles the mycoplasmas described in chapter 12.

The bacillus frequently infects rodents, which transmit it to humans by biting. After an incubation period of 3 to 10 days, the organism progresses from the primary lesion to regional lymph nodes and then to the bloodstream, producing lymphadenitis, petechial skin rash, joint involvement (arthritis), and attacks of recurring fever. Diagnosis is made on the basis of a history of rat bite and by bacteriologic demonstration of the organism in pus from the skin lesion or lymph nodes, in joint fluid, and in blood.

Penicillin or tetracycline drugs are effective in treatment. Control depends on eradication of rats, especially from buildings where people live or work. This disease can also be transferred through unpasteurized or contaminated milk. The source of infection in milk is not known, but a few epidemics have been related to this food. One of these occurred in the town of Haverhill, Massachusetts, and the milkborne disease came to be known as *Haverhill fever.*

Another type of rat bite fever is caused by a bacterium called *Spirillum minor.* This is a gram-negative, motile organism, spiraled in two or three short, rigid coils. It produces a clinical disease similar to that caused by *Streptobacillus,* except that the joints are seldom involved. This form of rat bite fever occurs more frequently in the Far East (where it is called *sodoku*) than in the United States. The organism does not grow on ordinary laboratory culture media but can sometimes be demonstrated in the blood of infected patients or by inoculating blood into mice or guinea pigs. Penicillin is the drug of choice for *Spirillum minor* infection. Prevention of this disease depends on the control or eradication of rats.

Treatment and Control

The treatment of bite wounds begins with prompt surgical cleansing and débridement to clear away contaminating organisms, aerate tissues, and ensure adequate blood supply so that healing can proceed normally. If the wound is extensive or deep, antimicrobial agents may be given prophylactically and tetanus antiserum and/or toxoid administered. The question of rabies prophylaxis must also be considered (see table 23.1).

The control of infected wounds centers on efforts to establish drainage and to prevent cellulitis or lymphangitis from getting out of hand. Surgical incision and drainage may be indicated, or continuous hot soaks may be applied to encourage spontaneous suppuration. The affected part is kept elevated, if possible, to prevent venous or lymphatic stasis. Specific chemotherapy, if warranted, is chosen on the basis of antimicrobial susceptibility tests of the isolated causative organism(s).

Diseases Acquired Through Infection of Wounds

Tetanus (Lockjaw)

The Clinical Disease

Tetanus, or **lockjaw,** is an acute disease caused by the neurotoxin of *Clostridium tetani*. This organism is not invasive and remains localized in the area where its endospores were introduced into the body. If anaerobic conditions prevail in the tissues at the site of entry, the endospores germinate and the vegetative bacterial cells multiply, elaborating toxin.

Local infection is often insignificant, producing little tissue damage, but exotoxin is absorbed from the area and extends along peripheral motor nerve trunks to the spinal cord. Within the cord, the toxin causes increased reflex excitability, whereas its action on peripheral nerves interferes with normal transmission of nerve impulses to muscles. Severe muscle spasms occur first at the area of infection and toxin production, then along the route of nerve trunk involvement. The muscles of the jaw and neck contract convulsively and remain "locked" so that the mouth cannot be opened, and swallowing is very difficult.

Other voluntary muscles also become involved, and widespread muscle spasms may lead to opisthotonos (backward flexion of the head and feet, with arching of the body and boardlike rigidity of the abdomen). Respiratory paralysis is the most serious complication and may be fatal. The toxin does not inflict permanent damage, and patients who recover do so completely but slowly. Tetanus fatalities are related to the time lapse between the dissemination of toxin and the institution of antitoxin and other therapy as well as the age and general condition of the patient. The average mortality rate is about 35%.

Laboratory Diagnosis

I. Identification of the Organism

(Clostridium tetani).

A. *Microscopic Characteristics* The tetanus bacillus is a gram-positive, anaerobic, endospore-forming bacillus. The endospore is characteristically wider than the diameter of the bacillus and forms at one end so that the rod has a drumstick appearance. Vegetative tetanus bacilli have an even diameter and are motile.

B. *Culture Characteristics* *C. tetani* is a strict anaerobe. It grows on blood agar plates incubated at 37°C in an anaerobic jar or an incubator from which air has been evacuated and replaced by an inert gas. A mixture of nitrogen with 10% CO_2 is commonly used to provide necessary atmospheric conditions. The organism can also be cultivated in fluid media having a low oxidation-reduction potential. Because many different anaerobic and facultative bacteria grow well in these media, they cannot be relied on to isolate *C. tetani* or other anaerobes in pure culture.

On the surface of blood agar, *C. tetani* colonies have a filamentous periphery marking an area of swarming growth, and they are usually hemolytic. The clostridial species are distinguished from one another by the patterns of their carbohydrate fermentations and their proteolytic activities. Some can digest gelatin and milk proteins, whereas others have only a weak action on proteins, or none at all. The tetanus bacillus does not ferment carbohydrates and is weakly proteolytic.

C. *Toxin Production* Definitive identification of *C. tetani* is also obtained by demonstrating its capacity for toxin production, the pathogenic effect of the toxin for laboratory animals and the specific neutralization of this effect by tetanus antitoxin.

II. Identification of Patient's Serum Antibodies

Serologic procedures have no diagnostic value. Patients who develop tetanus as a result of infection do so because they have no protective antitoxin in their bloodstream.

III. Specimens for Diagnosis

Diagnosis of tetanus is generally based on the clinical picture because the highly anaerobic organism is difficult to isolate in culture. The site of injury and infection may be insignificant, the wound may be closed, and the area of clostridial growth may be inapparent. When feasible, pus or tissue from the wound should be submitted for smear and anaerobic culture, but neither the diagnosis nor the institution of therapy can safely wait for laboratory findings, nor can negative findings be considered conclusive.

Epidemiology

I. Communicability of Infection

A. *Reservoirs, Sources, and Transmission Routes* The normal habitat of the tetanus bacillus is the intestinal tract of humans and animals. Endospores of the organism are widely disseminated in nature and are commonly present in the dust of streets as well as in soil. Tetanus endospores are generally introduced into the body through wounds of various kinds, not all of which may appear important at the time. Puncture or crushing wounds; injuries caused by bullets, knives, or clubs; obstetric or abortion wounds; an unhealed umbilical stump (tetanus neonatorum); burns; and surgical wounds all may offer an entry point for the organisms. Endospores are introduced by the object that inflicts the injury (a rusty nail, a thrown or falling object, a knitting needle used to attempt abortion, an unsterilized surgical knife), or infections result from subsequent exposure of the wound to an environmental source. Wounded soldiers or automobile accident cases may lie on the ground or be exposed to sources arising from attempts to treat wounds under field conditions. Surgical wounds sometimes are contaminated by inadequately sterilized instruments, sutures, or dressings, and orthopedic surgery is confronted with the occasional hazard of tetanus-contaminated plaster materials used for casts. Surgical wounds are also threatened by subsequent exposure to endogenous fecal sources of infection during the patient's postoperative course, but tetanus is not likely to occur in this situation unless the wound provides the necessary anaerobic conditions for growth of this noninvasive organism. Good surgical technique and patient care play a large role in preventing this type of infection.

B. *Incubation Period* The nature, extent, and location of the wound, as well as the numbers of infecting organisms, influence the incubation period. If the tetanus bacilli can grow immediately and multiply rapidly, production of toxin may begin within a few days (2 to 6, with an average of 4), but incubation sometimes requires 2 or 3 weeks or even longer.

C. *Communicable Period* Tetanus is not directly communicable from person to person.

D. *Immunity* Specific protection is provided by antibodies directed against tetanus toxin. Immunization with tetanus toxoid provides the most solid protection. Passive immunization may be necessary when the threat of infection arises in nonimmune persons. Susceptibility to tetanus infection is universal among the nonimmunized.

II. Control of Active Infection

A. *Transmission-based Precautions* Standard precautions as applied to all hospitalized patients (see chap. 11).

B. *Other Precautions* None.

C. *Treatment* *Tetanus antitoxin* can neutralize toxin before it becomes attached to nerve tissue but not afterward. For this reason, antitoxin is administered immediately when nonimmune persons sustain traumatic injuries and there is no history of recent active immunization. Human tetanus immune globulin is given in a dose of 250 to 500 units and should be protective for about a month. When human antiserum is not available, horse or other animal antitoxins are used, but care must be taken with these preparations to avoid anaphylactic hypersensitivity reactions (see chap. 6).

Antimicrobial agents are given to kill vegetative tetanus bacilli and to control simultaneous mixed infections. Penicillin is the drug of choice for prohibiting further growth and toxin production.

Surgical débridement of a wound with obviously active infection is essential to remove dead tissue and provide aeration.

Tetanus toxoid is given to stimulate active immunity when a potentially dangerous wound is sustained and if the person has not had a booster dose within the past 5 years. If the patient has not been immunized or has received fewer than two doses of toxoid, toxoid injections are given along with antitoxin, but they must be injected at different anatomic sites.

D. *Control of Contacts* The chief problem in obstetric, pediatric, and surgical situations is to locate the possible source of infection so that further cases may be prevented. For this reason case reports are mandatory in most states and countries. If the source is associated with inadequate surgical or sterilization techniques, these can be identified and controlled before they lead to serious outbreaks in hospital or midwife practice. Sporadic cases acquired from environmental sources of infection require only individual control, but prevention of such cases can be achieved by adequate programs of immunization.

III. Prevention by Immunization

Active immunization with tetanus toxoid provides certain and durable protection and should be initiated during infancy or early childhood. The preferred program is the immunization of infants beginning at 2 months of age with triple vaccine containing diphtheria and tetanus toxoids and pertussis vaccine (DTP) (see table 10.1). Booster doses at regular intervals reinforce immunity during childhood and adolescence.

For adults, boosters at 10-year intervals maintain a safe level of immunity, but if a serious injury is sustained, a single injection of toxoid is given immediately if 5 years has elapsed since the last booster. The need for passive protection with antitoxin is obviated for people with a history of regularly scheduled toxoid immunization (see fig. 23.5).

GAS GANGRENE

The Clinical Disease

Clostridial **gas gangrene** is a fulminating infection that can arise in necrotic tissue and spread rapidly. When the organisms are introduced into tissues in which conditions permit anaerobic multiplication, they use proteins and carbohydrates freed from dead or dying cells, producing gas as an end product. As multiplication proceeds, clostridia secrete enzymes such as lecithinase, collagenase, and hyaluronidase (see chap. 5), which are destructive to adjacent normal tissues, and also to red blood cells so that the area of necrosis is extended (colorplate 12e). Bacterial growth accelerates and advances under these conditions, eventually resulting in massive hemolysis and toxemia.

Gas production in the tissues produces crepitation, one of the diagnostic criteria of this destructive infection. Muscle tissue is characteristically involved (*clostridial myositis*) and is rapidly invaded because of its rich content of carbohydrate. The organisms may also advance through connective and subcutaneous tissues, producing progressive anaerobic cellulitis. Acute toxemia may result in shock and rapid death, depending on the speed and efficacy of therapeutic intervention, the location of infection, and the general condition of the patient.

Clostridial species capable of tissue invasion are often present in the human intestinal tract. They sometimes induce gangrenous infections of the intestinal wall and the peritoneum in patients with bowel obstructions. Risk of clostridial infection from endogenous sources is constantly present in abdominal surgery or postoperatively if segments of bowel become ischemic (lacking in adequate blood supply) or nonviable for any reason.

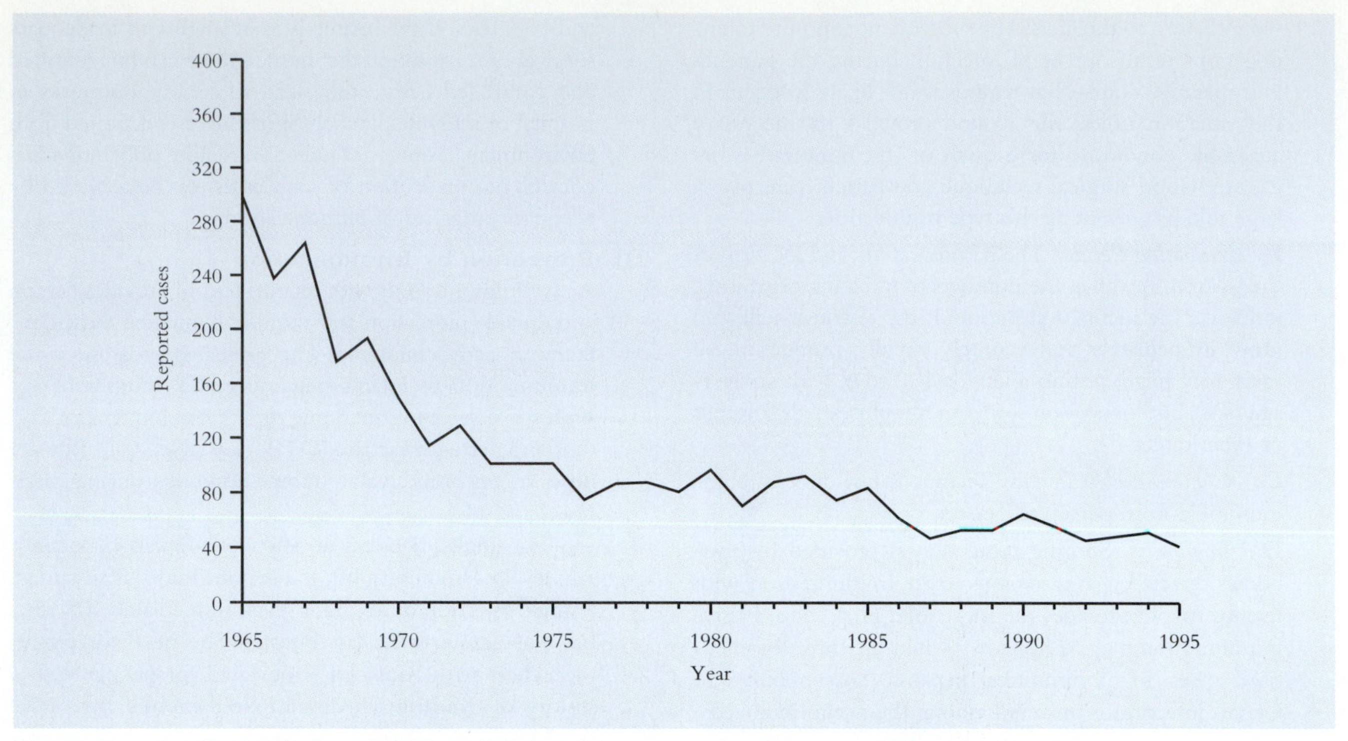

FIGURE 23.5 Tetanus cases in the United States from 1965 to 1995. The decline of tetanus is related to the early immunization of infants and the maintenance of lifelong immunity especially in high risk individuals. Tetanus toxoid was first available in 1933.

Centers for Disease Control and Prevention *MMWR Summary of Notifiable Diseases. United States, 1995* (October 25, 1996, Vol. 44: 53).

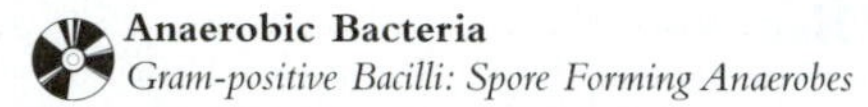

Anaerobic Bacteria
Gram-positive Bacilli: Spore Forming Anaerobes

Microbial Pathogenesis of Infectious Diseases
Toxins: Exotoxins

Clostridial species may occur in the genital flora in about 5% of normal women. Organisms derived from genital or intestinal sources may induce severe uterine infections as a result of septic instrumental abortion.

Some strains of *Clostridium perfringens* can be responsible for acute food poisoning if they are introduced into food and have an opportunity to multiply there before it is eaten. Meat provides the best medium for clostridial growth, particularly if it has been previously cooked in a stew or meat pie. Contaminated leftover meat dishes not thoroughly reheated before eating are the usual source of clostridial enteritis (see chap. 16).

LABORATORY DIAGNOSIS

I. Identification of the Organism

(Clostridium perfringens, Clostridium novyi, Clostridium septicum, Clostridium histolyticum.)

These organisms are species most commonly associated with gas gangrene, which is often a mixed infection. More than one clostridial species may be present, as well as streptococci or staphylococci and gram-negative enteric bacteria.

It must be emphasized that a diagnosis of gas gangrene is primarily made on clinical grounds. The mere finding of clostridia in wounds does not necessarily incriminate them as the cause of infection. Endospores of these organisms may be found as transient contaminants of surface tissues in wounds. Here they would be unable to vegetate and multiply in the absence of necrotic tissue and anaerobic conditions. On the other hand, their isolation from exudates of gangrenous, crepitant lesions confirms the bacterial nature of the clinical disease.

Microscopic and Culture Characteristics Like the tetanus bacillus, these members of the *Clostridium* genus are anaerobic, endospore-forming bacilli. They differ from *C. tetani* in that their endospores usually are placed in subterminal positions. However, endospores are almost never produced in infected tissue, and they are not seen microscopically in clinical material. These bacteria are isolated and identified by methods described in the section on laboratory diagnosis of tetanus. Unlike the fastidious tetanus bacillus, *C. perfringens* is more aerotolerant and readily grown. Most strains actively attack carbohydrates and are strongly proteolytic.

II. Identification of Patient's Serum Antibodies

Serologic diagnosis is of no value in clostridial infections (except botulism).

III. Specimens for Diagnosis

Exudates and tissue from infected wounds should be submitted in anaerobic transport containers for smears and culture. Collect this material with care from the active depths of the wound so that misleading surface contaminants may be avoided.

Epidemiology

I. Communicability of Infection

A. *Reservoirs, Sources, and Transmission Routes* See the corresponding discussion in the section on tetanus.

B. *Incubation Period* This is usually very short, with infection arising in 1 to 3 days following wound contamination.

C. *Communicable Period* Clostridial infections are ordinarily not transmissible from person to person. In the hospital surgical service, however, the infected patient may be a rich source of the organism in the operating room or on the surgical ward. Every precaution must be taken to prevent transfer of clostridial infection to other surgical patients by direct or indirect routes.

D. *Immunity* Susceptibility to clostridial wound infection is general. Antitoxin immunity may arise as a result of infection, but its durability and protective value are questionable. Passively administered antitoxin may be used in treatment of active infection but, again, is of questionable value.

II. Control of Active Infection

A. *Transmission-based Precautions* Patients with gas gangrene should be placed on standard precautions (see chap. 11) to prevent transfer of pathogenic clostridia and other infecting organisms to other surgical cases. There is no danger of transmission except to persons with surgical or traumatic wounds that offer suitable oxygen-free conditions for clostridial multiplication.

B. *Other Precautions* Attendant medical staff must take suitable precautions to prevent transmission of infection from an active case to other patients. Gowns and gloves should be worn by those who treat the wound or attend the patient closely, and hand washing is an urgent need. Individualized instruments and equipment should be used and sterilized by autoclaving. Boiling and chemical disinfection do not kill resistant clostridial endospores easily. For this reason, concurrent and terminal disinfection pose a problem. The most satisfactory program is one that aims at destroying the richest sources of the organism (dressings incinerated, bed linens autoclaved before being laundered, unprotected mattresses and pillows sterilized in an ethylene oxide chamber) and removing environmental contamination by thorough disinfection-cleaning of the patient's unit. Good disinfection will kill vegetative bacilli, and careful cleaning will remove endospores mechanically.

In any case, clostridial endospores are constant environmental contaminants, and large numbers of vegetative cells are found in the intestinal tract of many patients.

C. *Treatment* The most urgently important treatment is surgical removal of necrotic tissue and aeration of the wound. Antimicrobial therapy is instituted promptly, penicillin being the drug of choice.

OTHER ANAEROBIC INFECTIONS

Tetanus and gas gangrene are among the most dramatic and serious infections caused by anaerobic organisms, but they are uncommon in comparison with the processes associated with other **anaerobic infections.** The nonendospore-forming anaerobes make up the largest portion of the normal flora colonizing skin and mucous membranes of the upper respiratory, gastrointestinal, and urogenital tracts. In these sites, they outnumber aerobic and facultative organisms from 3 to 1,000 times.

Although they are ordinarily of low virulence, anaerobes may be involved in infections whenever tissues adjacent to sites that they colonize are damaged, with subsequent lowering of the oxidation-reduction potential. For example, tooth abscesses, pneumonia from aspiration of upper respiratory flora, peritonitis from ruptured appendices, and septic abortions all involve predominantly anaerobic organisms. Usually, these infections are polymicrobial; that is, more than one microorganism is isolated from the infection site. Sometimes only anaerobes are present, but often the mixture contains both anaerobic and facultative organisms.

Table 23.2 presents the common anaerobic organisms, their usual human habitat, and the infections with which they are associated. As a general rule, anaerobes residing in the upper respiratory tract cause infections above the diaphragm, whereas intestinal and genitourinary anaerobes cause infections below the diaphragm. However, any species can be found at any site. The anaerobes that most frequently cause infection are *Bacteroides fragilis,* a few related *Bacteroides* species, and *Peptostreptococcus* species (anaerobic streptococci).

Actinomyces species, especially *Actinomyces israelii,* are responsible for a specific infection known as *actinomycosis.* In this infection, soft tissue and bone are destroyed by a chronic suppurative or granulomatous process. Multiple-draining sinus tracts may be formed that drain off pus to adjacent areas or to the surface. The drainage contains visible "sulfur granules," which are actually yellow-tinted, tangled masses of the organisms (fig. 23.6). Classically, the disease affects the mouth and jaw, thorax, or abdomen, but any part of the body may be involved.

Specimens for Diagnosis

Proper specimen collection is essential for accurate diagnosis of anaerobic infections. Because the same organisms are found at the infection site and as part of the adjacent normal flora, only specimens obtained in a way such as to avoid contamination with colonizing bacteria are appropriate for sending to the laboratory. These include normally sterile

TABLE 23.2 ANAEROBES COMMONLY ISOLATED FROM HUMAN CLINICAL SPECIMENS

Organism	Usual Body Site	Infectious Processes or Site of Anaerobic Infection
Gram-Positive Cocci		
Peptostreptococcus spp.	Mouth, intestine, genital tract	Bacteremia, brain abscess, eye, head and neck, lung abdomen, female genital tract, bone and joint
Gram-Negative Cocci		
Veillonella spp.	Mouth, intestine	Oral, bite wound, head and neck, soft tissue
Gram-Positive Bacilli		
Clostridium spp.	Intestine	Bacteremia, abdomen, female genital tract
Eubacterium spp.	Mouth, skin	Abscesses, wounds, periodontal disease
Actinomyces spp.	Mouth	Actinomycosis, brain abscess, periodontal pockets, eye, abdomen
Lactobacillus spp.	Mouth, intestine, genital tract	Seldom clinically significant
Propionibacterium spp.	Skin	Bacteremia, eye
Bifidobacterium spp.	Intestine, skin, genital tract	Seldom clinically significant
Gram-Negative Bacilli		
Bacteroides fragilis	Intestine	Bacteremia, brain abscess, abdomen, female genital tract, bone and joint
Bacteroides spp.	Intestine, genital tract	Bacteremia, brain abscess, head and neck, abdomen
Fusobacterium spp.	Mouth, intestine, genital tract	Brain abscess, head and neck, lung, abdomen
Porphyromonas spp.	Mouth, genital tract	Oral, head and neck, femalegenital tract
Prevotella spp.	Mouth	Head and neck, lung

Anaerobic Bacteria
Normal Flora: Anaerobic Normal Flora Sites. Gram-positive Bacilli: Non-Spore Forming Anaerobes. Gram-positive Cocci: Taxonomy; Habitat; Diseases. Gram-negative Bacilli: Bacteroides; Fusobacteria; Porphyromonas; Prevotella. Gram-negative Cocci: Veillonella.

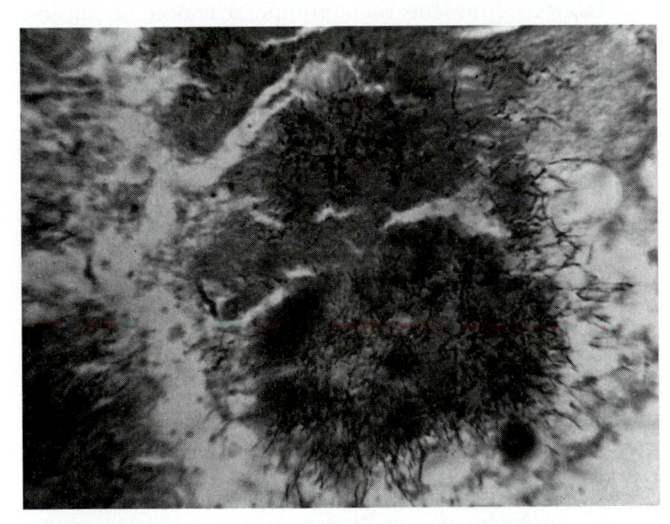

FIGURE 23.6 Actinomycotic sulfur granule. Note the tangled filamentous, branching bacilli (×1,000).

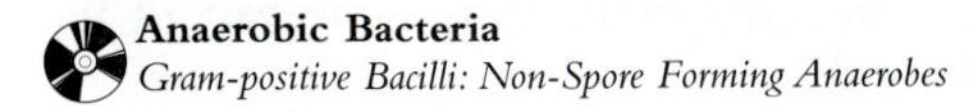
Anaerobic Bacteria
Gram-positive Bacilli: Non-Spore Forming Anaerobes

fluids such as blood, pleural, and joint fluid, or purulent material aspirated from deep abscesses or through decontaminated surfaces. Sputum, urine, and skin or urogenital swabs are all subject to contamination with normal flora anaerobes when they are collected and so are not appropriate for anaerobic culture.

Once they are properly collected, specimens for anaerobic culture must be delivered to the laboratory in oxygen-free transport containers. Inoculated culture media are incubated in oxygen-free anaerobe jars or incubators in addition to routine incubators. Identification to genus and species is made primarily on the basis of biochemical reactions. Commercial kits are available to simplify this process. As yet, no standard method has been devised for antimicrobial susceptibility testing of anaerobes. Most clinicians rely on published reports of susceptibility tests performed in reference laboratories.

Control and Treatment

Because most anaerobic infections are derived from the patient's endogenous flora, the threat of transmission is negligible. Trauma, surgery, and malignancies predispose to infection; therefore, physicians must be alert to the possibility that anaerobes are involved in infections resulting from these conditions. (Some of the best early studies on anaerobic infections were done at Cook County Hospital in Chicago where a large number of abdominal gunshot and knife-wound infections were encountered.) Surgical débridement and drainage of abscesses bring oxygenated blood to the area and are used in conjunction with antimicrobial agents such as metronidazole, chloramphenicol, and imipenem. In the past few years, several anaerobe species, including *B. fragilis,* have become more resistant to antimicrobial agents commonly used to treat anaerobic infections.

Bacterial Infections of "Street Wounds"

Soft tissues subjected to traumatic injury may be infected with any of a variety of bacteria derived from the environment or the patient's skin. An organism frequently associated with traumatic injuries and burns is *Pseudomonas aeruginosa*, a gram-negative, aerobic bacillus that may be found in the normal intestinal tract as well as the environment. This organism induces suppurative lesions in soft tissues and is characterized by its production of a blue-green pigment (pyocyanin) that imparts its color and sickly sweet odor to pus.

Traumatic wounds are most frequently infected with a mixed flora, containing both gram-positive cocci and gram-negative bacilli. Deep wounds often support anaerobic bacteria, such as species of *Bacteroides* or peptostreptococci, as well as clostridia. These organisms should always be sought by probing deep into the wound for material to be submitted to the laboratory for anaerobic culture. Aerobic organisms recovered from the surfaces of such wounds may be mere contaminants that are not necessarily multiplying actively or injuring the tissue.

Burn Wound Infections

Burns represent one of the most difficult problems in patient management because the nature of the injury predisposes exposed surface tissues to infection. **Burn wound infections** are the chief cause of death in patients who have survived the first three or four days of massive injury, surface fluid loss, pulmonary edema, or shock following a severe burn injury. Devitalized surface tissue offers suitable conditions for the growth of many opportunistic organisms, notably streptococci, staphylococci, gram-negative enteric bacteria, and anaerobes.

The organisms most frequently encountered in burn wound infections are *Pseudomonas aeruginosa* and *Staphylococcus aureus.* These infections present very difficult problems because the organisms are resistant to many clinically useful antimicrobial agents.

When efforts to control burn infections fail, septicemia is the usual consequence and the cause of death. The wound itself becomes overwhelmed with colonizing bacteria, and systemic invasion may induce fatal shock. In recent years many efforts have been made to develop methods that would prevent the complication of "burn wound sepsis," including the provision of closed hospital areas in which strict aseptic control can be maintained against airborne infection. The most important factors, however, are removal of all necrotic tissue and rapid wound closure, often by skin grafting from nonburned areas.

Questions

1. What is the reservoir for rabies? How is the disease transmitted?
2. When is active immunization indicated for rabies?
3. Why do dog and human bites often produce serious infection?
4. Why do animal bites present the hazard of clostridial infection?
5. What characteristics of *Clostridium tetani* contribute to its pathogenicity?
6. How is tetanus prevented? What is the treatment for a tetanus infection?
7. What precautions are necessary when caring for a patient with gas gangrene? Why must these precautions be taken?
8. How do low-virulence anaerobes become pathogenic?
9. Why are burns highly susceptible to infection?
10. What precautions are necessary in the care of a burn patient?

Bibliography

Balows, A., W. J. Hausler, Jr., M. Ohashi, and A. Turano, eds. 1988. *Laboratory diagnosis of infectious diseases principles and practice,* Bacterial, mycotic and parasitic diseases. vol. 1. New York: Springer-Verlag.

Benenson, A. S. 1995. *Control of communicable diseases in man.* 16th ed. Washington, D.C.: American Public Health Association.

Centers for Disease Control and Prevention. 1995. Human rabies–Washington 1995. *Morbid. Mortal. Weekly Report* 44 (34): 625.

Centers for Disease Control and Prevention. 1995. Mass treatment of humans exposed to rabies–New Hampshire, 1994. *Morbid. Mortal. Weekly Report* 44 (26): 484.

Culpepper, M. M., and P. G. Adams. 1988. Nursing in the civil war. *Am. J. Nurs.* 88 (7): 981–984.

Kaplan, M. M., and H. Koprowski. 1980. Rabies. *Sci. Am.* 242 (1): 120–134.

Lennette, E. H., P. Halonen, and F. A. Murphy, eds. 1988. *Laboratory diagnosis of infectious diseases principles and practice,* Viral, rickettsial and chlamydial diseases, vol. 2. New York: Churchill Livingstone.

Mandell, G. L., J. E. Bennett and R. Dolin. 1995. *Principles and practice of infectious diseases,* 3d ed. New York: Churchill Livingstone.

Margileth, A. M. 1985. Cat-scratch fever. *Hosp. Med.* 21 (5): 57–67.

Overturf, O. 1989. Bacterial and rickettsial zoonosis associated with tick and flea bites. *TOP Emerg. Med.* 10 (4): 67–79.

Roueche, B. 1991. Three sick babies. In *The medical detectives.* New York: Plume.

———. The west branch study. In *The medical detectives.* New York: Plume.

———. The incurable wound. In *The medical detectives.* New York: Plume.

SUMMARY TABLE 23.3 INFECTIOUS DISEASES RESULTING FROM ACCIDENTAL INJURY

Clinical Disease	Causative Organism	Other Possible Entry Routes	Incubation Period	Communicable Period
I. Infectious Diseases Acquired through Animal Bites				
Rabies	Rabies virus	Respiratory	3–12 weeks, sometimes very prolonged	Virus is present in biting animals for several days before their symptoms appear and throughout the course of their illness; bats may shed virus for weeks
Cat scratch disease	*Bartonella henselae*		1 or 2 weeks from initial injury, may be shorter	Has not been clarified, but infection is not directly transmissible from person to person
Pasteurellosis	*Pasteurella multocida*		Few days	Not transmissible from person to person
Rat bite fever	*Streptobacillus moniliformis* *Spirillum minor*		Few days	Same as above
II. Diseases Acquired through Infection of Wounds				
Tetanus	*Clostridium tetani*		2–6 days, average 4; sometimes 2 or 3 weeks or longer	Not communicable from person to person
Gas gangrene	*Clostridium perfringens* *Clostridium* spp.		1–3 days	In the hospital surgical service, infected patient may be a rich source of organisms in operating room or on surgical ward
Other anaerobic infections	See table 23.2		Variable	Not communicable person to person
III. Burn Wound Infections				
Local infection, superficial or deep; sepsis	*Pseudomonas aeruginosa* *Staphylococcus aureus* Streptococci Gram-negative enteric bacilli Anaerobes		1 to several days	Infected patient may be a source of infection for others with open wounds

Sanford, J. P. 1990. Humans and animals: Increasing contacts, increasing infections. *Hosp. Pract.* 25 (2): 123–140.

Smith, J. S. 1996. New aspects of rabies with emphasis on epidemiology, diagnosis, and prevention of the disease in the United States. *Clin. Microbiol. Rev.* 9 (2): 166–176.

Technique with multidose vials leads to pseudomonad infections. 1990. *Hosp. Infect. Control* 17 (6): 73–74.

Thomas, D. J. 1989. Neonatal nosocomial infections. *Crit. Care Nurs. Q.* 11 (4): 83–94.

Specimens Required	Laboratory Diagnosis	Immunization	Treatment	Transmission-Based Precautions
Saliva (from submaxillary gland) Brain of biting animal if available	Fluorescent antibody technique for direct examination; cell culture and animal inoculation	Inactivated virus (administered to persons at high risk after exposure is confirmed or reasonably certain)	Active vaccination and immune serum	Contact isolation; articles contaminated by saliva may be infectious; protect skin, clothing from saliva; avoid injury to hands from patient's teeth
Pus or biopsy from suppurating lymph nodes	Smear and culture; silver stains	None	Ciprofloxacin Trimethoprim-sulfameth-oxazole	Standard precautions; concurrent disinfection of pus from suppurating lesion or lymph node
Acute and convalescent sera	Rise in antibody titer			
Skin test (not routine)	Positive response			
Pus from wound, lymph nodes	Smear and culture	None	Penicillin Tetracyclines	Same as above
Blood	Culture			
Same as above	Culture Animal inoculation	None	Penicillin Tetracyclines	Standard precautions
Pus or tissue from wound if feasible	Smear and anaerobic culture	Toxoid for active immunization; antitoxin for passive immunization after injury has occurred	Antitoxin Penicillin	Standard precautions
Exudates and tissue from wound	Smears and anaerobic culture	None	Penicillin Metronidazole Clindamycin	Standard precautions; protect other patients with surgical wounds; gowns and gloves when changing dressings; autoclave all equipment; incinerate dressings; *hand washing*
Carefully collected exudates and tissue	Same as above	None	As indicated by published antimicrobial susceptibility studies	Standard precautions
Exudates and tissue from wounds	Smears and culture	None	Antimicrobial agents as indicated by laboratory studies of organisms Wound débridement Skin grafts	Contact precautions if wounds draining, otherwise standard precautions Private room or special burn center; *hand washing*
Blood	Culture			

GLOSSARY

A

abscess A localized collection of pus surrounded by inflamed tissue, induced by microbial infection.

Acanthamoeba A genus of ameba that produces corneal infections in those whose contact lenses or lens solutions are contaminated by this protozoan.

acid-fast A term referring to bacteria, especially mycobacteria, that after staining with basic fuchsin cannot be decolorized by acid alcohol.

acid-fast stain *See* **Ziehl-Neelsen stain.**

acquired immunodeficiency syndrome (AIDS) A currently pandemic disease syndrome that results from infection with the human immunodeficiency virus (HIV) and that is transmitted sexually, congenitally, neonatally, or via contaminated blood and blood products. Characterized by an initial loss of normal cell-mediated immune responses due to viral inactivation of T cells, followed by abnormally increased susceptibility to opportunistic infections and heightened risk of some cancers.

actively acquired immunity Specific immunity acquired through natural exposure to a particular microbial antigen (infection or infectious disease) or through artificial introduction of the agent into the body (vaccination).

acute infections Microbial diseases characterized by a relatively rapid onset. Symptoms persist until host defenses and/or antimicrobial treatment become effective, after which they usually subside and disappear. Some persist as chronic infection.

acyclovir An antimicrobial agent used in the treatment of some viral diseases, including those caused by the herpes simplex viruses, cytomegalovirus, and Epstein-Barr virus. Known in some cases to interfere with DNA synthesis and replication.

adenine A purine found in the nucleotides incorporated into nucleic acids.

adenoviruses Primary DNA viral agents of respiratory disease in humans of all ages and of gastroenteritis in children and adults, spread directly from person to person.

adherence The attachment of invading microbes to a specific type of cell or substrate by means of specialized structures (e.g., pili) or by reactions of microbial surface components with specific receptors on the host cell.

aeration In community water purification, a process that increases the oxygen content of water and speeds oxidative metabolism of organic material by bacteria; also displaces other dissolved gases that contribute unpleasant odors and taste to polluted water.

aerobe An organism that grows only in the presence of free molecular oxygen.

aerobic respiration A metabolic process in which energy is released as a result of the oxidation of organic compounds (e.g., carbohydrates, with free oxygen acting as an electron acceptor).

AFB Acid-fast bacilli.

African sleeping sickness A trypanosomal disease prevalent in parts of Africa where the tsetse fly vector abounds.

agammaglobulinemia An immunodeficiency disease resulting from a congenital defect that results in the complete failure of immunoglobulin production.

agglutination The clumping that occurs in vitro when antibodies combine with antigens on the surface of particles such as microbial cells, red blood cells, or latex particles.

agglutinins The antibodies that cause clumping of particles whose surfaces are coated with specific antigen.

AIDS *See* **acquired immunodeficiency syndrome.**

AIDS-related complex (ARC) An early combination of symptoms (e.g., weight loss, persistent diarrhea, fatigue, and low-grade opportunistic infections) caused by the HIV infection that may lead to full-blown AIDS.

allergen An antigen (immunogen) that elicits an immediate type of hypersensitivity reaction.

allergic contact dermatitis (ACD) A type of delayed hypersensitivity induced by many simple chemical allergens (e.g., soaps) or plant substances (e.g., poison ivy). The allergens react with sensitized T cells to produce minor skin rashes or enlarged blisterlike lesions that ooze and become crusted and extremely itchy.

allergy A type of immediate hypersensitivity that occurs when antibodies combine with allergens to produce conditions such as asthma, hay fever, or allergic dermatitis.

alpha-hemolysis A zone of incomplete clearing, often accompanied by a greenish discoloration, around bacterial colonies growing on blood agar culture media; caused by incomplete lysis of erythrocytes.

alveolar macrophages Phagocytic cells in the alveoli of the lungs; part of the mononuclear phagocyte system (MNPS).

amantadine An antimicrobial agent used in the prophylactic treatment of influenza A virus infection.

ameba (*pl.* amebae) A protozoan belonging to the phylum Rhizopoda.

ameba meningitis A meningoencephalitis caused by the protozoan *Naegleria fowleri,* found in fresh-water lakes or ponds where people swim.

amebiasis An intestinal or invasive disease caused by the protozoan *Entamoeba histolytica* that has two developmental stages: trophozoite and cyst. Also called amebic dysentery.

amebula (*pl.* amebulae) A small trophozoite emerging with others from the cysts of *Entamoeba histolytica.*

American trypanosomiasis A form of trypanosomal disease of Central and South America; often referred to as Chagas' disease. Transmitted by the bite of the reduviid bug.

aminoglycosides A group of antimicrobial agents containing amino sugars that bind to microbial ribosomes and inhibit protein synthesis.

amphotericin B An antimicrobial agent used to treat systemic fungal diseases. Acts by binding to sterols in fungal cell membranes, increasing membrane permeability and resulting in leakage of cell constituents.

amplification techniques Methods for greatly increasing nucleic acid sequences of microbes in culture or in patient specimens. Used primarily to facilitate their detection with nucleic acid probes when only small numbers of organisms are available for study.

anaerobe An organism that grows only in the absence of free molecular oxygen.

anaerobic respiration A metabolic process in which energy is released by the breakdown of organic compounds in the absence of free oxygen; an inorganic molecule other than oxygen acts as an electron acceptor.

anaphylaxis An immediate, sometimes fatal, hypersensitivity reaction that follows contact by injection or ingestion of a specific allergen into an individual sensitized to that antigen through previous exposure.

Anopheles A genus of mosquito that carries and transmits the agent of malaria.

antibiotic An antibacterial agent; now commonly referred to as an antimicrobial agent.

antibody An immunoglobulin (Ig) produced in response to the presence and activity of an antigen and having the ability to combine specifically with the antigen that stimulated its production.

anticodon A sequence of three nucleotides in the tRNA molecule that is complementary to the three nucleotides in a codon; these sequences function together in the process of translation.

antigen A foreign substance, usually of high molecular weight (e.g., a protein or polysaccharide), that induces an immune response when introduced into the body. Also called *immunogen.*

antigenic determinants Various chemical groups on the antigen molecule responsible for stimulating antibody formation and for the specificity of its reactions.

antigenic drift The gradual changes in the antigenic proteins of microorganisms caused by genetic mutations coding for altered amino acid sequences.

antigenic shift Alteration in the surface proteins of microorganisms that produce new subtypes unrecognized by host immune mechanisms. In influenza A virus, caused by reassortment of viral RNA from a human strain with RNA from an animal strain.

antimetabolites Antimicrobial agents that interfere with growth by acting as analogues, or substitutes, for compounds normally used by bacteria in synthetic or enzyme functions.

antimicrobial agent A substance naturally produced by microorganisms (especially many fungi), or a synthetic derivative of it, that is inhibitory or lethal for other microorganisms (particularly bacteria).

antimicrobial susceptibility testing Techniques for assessing the susceptibility or resistance of microorganisms (chiefly bacteria) to clinically useful antimicrobial agents.

antisepsis The chemical disinfection of skin or other living tissue.

antiserum (*pl.* antisera or antiserums) Serum containing antibodies.

antistreptolysins Specific antibodies that react with streptococcal hemolysins.

antitoxin An antibody to a microbial toxin with which it combines, either in vivo or in vitro, specifically to neutralize it.

Apicomplexa A phylum of single-celled protozoa having a complex reproductive cycle that includes both asexual and sexual stages; cells produced sexually mature as sporozoites.

arbovirus encephalitides A group of encephalitic diseases caused by arboviruses and transmitted among humans, horses, birds, and some other animals by mosquitoes and, sometimes, by ticks.

ARC *See* **AIDS-related complex.**

archaebacteria A group of bacteria included with the procaryotes but possessing unique characteristics. Their metabolic processes indicate that they may have evolved very early in the formation of the earth.

arthropod An insect of the invertebrate phylum Arthropoda having jointed feet.

ascariasis An intestinal nematode disease caused by the adult form of the roundworm *Ascaris lumbricoides.*

Ascomycota Fungal kingdom containing the true yeasts, some mildews, and molds. Characteristic sexual spores, or ascospores, are formed within a sac called an ascus.

ascospore *See* **Ascomycota.**

asepsis A method or methods for keeping an environment or object free from pathogenic microorganisms.

aseptic meningitis A nonbacterial, nonpurulent meningitis often associated with viruses such as mumps and enteroviruses.

aspergillosis An opportunistic mycotic infection caused by *Aspergillus fumigatus* and other species of *Aspergillus* that produces granulomatous lesions on the lungs or other tissues.

assimilation A metabolic process in the nitrogen cycle; plants convert nitrites into the amine groups (NH_2) necessary for the synthesis of amino acids and proteins.

astroviruses Small, star-shaped viruses involved in a small percentage of cases of gastroenteritis among young children.

athlete's foot A fungal infection of tissues of the feet. Also called *tinea pedis.*

atopic A term referring to allergies, probably hereditary in origin, produced by natural exposure to allergens (rather than by inoculation).

atopic allergies Immediate types of hypersensitivity (e.g., asthma, hay fever, hives, and other skin rashes) occurring through the inhalation, ingestion, or physical contact with an allergen; based on antibody-antigen reactions. Differ from anaphylaxis which requires injection of the allergen.

autoclave *See* **steam sterilizer.**

autoimmune disease An immune response of the body to its own cells and tissues and involving failure of the immune system to recognize "self."

autotroph An organism that is relatively independent of other forms of life, being capable of using CO_2 as its only or principle source of carbon.

AZT *See* **zidovudine.**

B

babesiosis A protozoan infection caused by species of *Babesia* and transmitted among humans and animals by a tick vector.

bacillary dysentery *See* **shigellosis.**

Bacillus A genus of gram-positive, aerobic endospore-forming bacilli, mostly nonpathogenic. *B. anthracis* produces anthrax and *B. cereus* has been associated with outbreaks of food poisoning.

bacillus (*pl.* bacilli) A bacterium shaped like a small stick or rod.

bacteremia The presence of bacteria in the bloodstream.

bacteria (*sing.* bacterium) True procaryotic, unicellular microorganisms with the simplest type of cell organization.

bacterial food poisoning A type of disease, usually gastrointestinal, induced by ingesting food contaminated by bacteria or their toxins.

bacterial vaginosis (BV) A common vaginal infection presumed to be sexually transmitted, but of unknown etiology; *Gardnerella vaginalis,* a gram-variable bacillus, and *Mobiluncus* species, gram-negative curved or straight bacilli, have been incriminated.

bactericide Any physical or chemical agent that kills bacteria.

bacteriophage A virus that parasitizes bacteria.

bacteriostasis The arrest of bacterial growth; multiplication is halted, but bacteria are not killed and may resume growth when the bacteriostatic agent is removed.

balantidiasis An intestinal disease caused by the protozoan *Balantidium coli* (phylum Ciliophora). The organism has two developmental stages: a motile, ciliated trophozoite and a nonmotile cyst.

Basidiomycota Fungal phylum containing mushrooms, toadstools, plant rusts, and smuts. Characteristic sexual spores, called basidiospores, are produced in a club-shaped structure called a basidium.

basidiospore *See* **Basidiomycota.**

basophils White blood cells with blue staining granules that produce histamine, a blood vessel dilator released in allergic reactions, and heparin, an anticoagulant that prevents blood clotting.

B cell A type of lymphocyte, derived from stem cells, that matures in bone marrow and later migrates to lymph nodes and spleen. B cells have antigen-specific receptors on their surfaces. On contact with a foreign antigen they proliferate into clones of new identical cells. Each B cell can mature into a plasma cell that synthesizes and secretes antibodies (immunoglobulins or Igs) specific for the antigen involved. Igs circulate in the blood and lymph, providing humoral immunity. Also called *B lymphocyte.*

BCG ("bacille de Calmette et Guerin") A bacterial vaccine used to provide increased resistance to tuberculosis in persons at particular risk of acquiring the disease (named after the French workers who first developed it).

beef tapeworm *See* **taeniasis.**

beta-hemolysis A zone of clearing around bacterial colonies growing on plated blood agar culture media; caused by complete lysis of red blood cells.

beta-lactamase A microbial enzyme that breaks open the beta-lactam rings of some penicillins and cephalosporins; also called *penicillinase* or *cephalosporinase.*

beta-lactam ring A ring structure within the penicillin and cephalosporin molecules that may be broken by microbial enzymes called beta-lactamases, rendering these antimicrobial agents inactive.

biogenesis The development of life from pre-existing forms of life; self-reproduction.

biosphere That portion of the planet earth where life exists and is maintained by the constant cycling of energy and chemical supplies among diverse living organisms.

bladder worm The larval form of the tapeworm *Taenia* species in the tissues of infected animals. In humans, tissue invasion by *Taenia solium* produces cysticercosis.

blastocystosis An intestinal disease caused by the protozoan *Blastocystis hominis,* an organism of uncertain taxonomy now tentatively placed in the phylum Rhizopoda along with amebae.

blastomycosis A systemic mycotic disease caused by *Blastomyces dermatitidis* and characterized by both pulmonary and subcutaneous lesions.

blastospore A budding cell which is the asexual spore of yeasts.

blocking antibody An antibody molecule with only one combining site for antigen, so that it is incomplete compared with those antibodies that have two or more combining sites. Prevents formation of antigen-antibody complexes.

Bordetella pertussis The bacterial agent of whooping cough.

botulism A type of food poisoning resulting from ingesting the exotoxin of *Clostridium botulinum* in improperly canned or preserved foods.

bronchopneumonia A type of pneumonia involving the bronchial tree rather than the alveoli; more diffuse and less localized than lobar pneumonia.

Brucella A genus of small, fastidious, gram-negative coccobacilli containing three pathogenic species, *B. abortus, B. melitensis,* and *B. suis,* which cause disease in cattle and other domestic animals as well as humans.

brucellosis A systemic disease of animals and humans caused by species of the bacterial genus *Brucella.*

C

caliciviruses Viral agents of gastroenteritis having cup-shaped depressions on their surfaces (*calyx* = cup or chalice) and transmitted from person to person or in contaminated shellfish and drinking water.

Campylobacter A genus of slender, curved, gram-negative bacilli associated with various diseases of animals and with diarrheal infections of humans.

Candida A genus of medically important yeasts often found on human mucosal surfaces; may cause superficial or more serious infections of deep tissues.

candidiasis Usually an opportunistic local or systemic mycotic infection caused by *Candida albicans* and other species of *Candida,* often involving skin and mucosa where the organism is a member of the normal flora.

capsid The protein coat surrounding the core of DNA or RNA in a virus.

capsule A viscous polysaccharide or protein coat surrounding the cell wall of some microorganisms. Also called a *slime layer.*

carrier A person infected with a transmissible microorganism who shows few or no symptoms but is a potential source of infection for others. Carriers form an important epidemiologic link in the chain of infection.

catalyst A substance that promotes, or accelerates, a chemical reaction without being permanently changed in the process. Enzymes are important examples.

cat scratch disease An infection often associated with a scratch wound inflicted by a cat (or other animal) involving swelling of lymph nodes, recurring fever, and general malaise. The agent has been identified as *Bartonella henselae.*

CD4 A glycoprotein receptor on the surface of some T cells (CD4+ lymphocytes) and other body cells to which HIV attaches and through which it enters, multiplies, and eventually kills the cell. Accessory factors are also required for attachment.

cell-mediated immunity (CMI) Immunity resulting from the physical reaction of the T cells (lymphocytes) of the immune system with foreign (or infected) cells to destroy them.

cell membrane A thin membrane surrounding the cytoplasm and beneath the cell wall of microorganisms that mediates the transport of materials into the cell.

cell wall The rigid, multilayered structure that lies outside the microbial cell membrane, giving it shape and protecting it from rupture. It is chemically complex and contains many antigenic determinants that are distinctive for individual genera or species.

cephalosporin A class of antimicrobial agents, including cephalothin, cefoxitin, and cefotaxime, which inhibit cell-wall synthesis in gram-positive and some gram-negative bacteria.

cephalosporinase A bacterial enzyme (beta-lactamase) that inactivates some cephalosporins by breaking down their beta-lactam ring.

cercaria (*pl.* cercariae) The final free-swimming larval form in the life cycle of trematodes.

cestodes A group of platyhelminths (flattened, segmented worms) known as the tapeworms, each of whose segments contains both male and female sex organs.

chancre The primary superficial lesion of syphilis that appears at the site where the organism entered.

chancroid A sexually transmitted disease caused by the gram-negative bacillus *Haemophilus ducreyi.*

chemoprophylaxis Administration of antimicrobial agents to prevent disease in persons who have been exposed to, or are at great risk of, dangerous infection.

chemosynthetic autotrophs Organisms in a few genera of bacteria that use CO_2 as their principal carbon source, but obtain energy from the chemical oxidation of inorganic compounds rather than from light.

chemosynthetic heterotrophs Organisms that depend on the chemical energy released from a variety of oxidation-reduction reactions of organic compounds; human beings, animals, protozoa, fungi, and most bacteria belong to this group.

chemotaxis A chemical stimulus that attracts phagocytes, such as neutrophils, to a site where a foreign substance, especially a living microorganism, is present.

chemotherapy The treatment of infectious disease with antimicrobial agents.

chickenpox *See* **varicella-zoster virus.**

Chlamydia (*pl.* chlamydiae) A genus of small, pathogenic coccoid bacteria that depend on infected host cells for energy required in their metabolism; includes *C. pneumoniae, C. psittaci,* and *C. trachomatis.*

chlamydial (nongonococcal) urethritis A sexually transmitted disease commonly caused by the organism *Chlamydia trachomatis.*

chloramphenicol A broad spectrum antimicrobial agent that acts against both gram-positive and gram-negative organisms as well as rickettsiae; microbial ribosomal protein synthesis is affected.

chlorination The final step in community water purification; the process involves the use of gaseous chlorine or one of its compounds (e.g., chlorinated lime or sodium hypochlorite) to ensure disinfection of filtered water.

chlorophyll The green, light-sensitive pigment in plants that enables them to use light energy to manufacture carbohydrate from water and atmospheric CO_2.

chloroquine An antimicrobial agent used to treat or prevent malaria, and to treat *E. histolytica* infection; binds to nucleoproteins of the parasite and interferes with protein synthesis.

cholera An acute diarrheal disease caused by an enterotoxin produced by the gram-negative bacillus *Vibrio cholerae.*

chromosome One of the bodies of nuclear material in all living cells; contains most or all of the DNA constituting the genes of each cell.

chronic infections Microbial diseases produced by organisms that can survive host defenses and maintain some level of damaging activity over a relatively long time.

-cide A noun suffix meaning "killer" or "killing" (e.g., germicide). The adjectival suffix is *-cidal.*

Ciliophora A phylum of protozoa made up of cells whose outer surface is covered by fine, minute hairs called cilia and that possess both a macro- and micronucleus.

clonal selection theory The theory that clones of lymphocytes (B and T cells) bear distinct antibody molecules or other receptors on their cell membranes and that this is genetically determined rather than dependent on prior contact with any antigen.

clone An aggregate of asexually produced cells or organisms arising from one parent.

clonorchiasis An invasive disease of bile ducts and liver caused by the trematode liver fluke *Clonorchis sinensis.*

Clostridium A genus of gram-positive, anaerobic, endospore-forming bacilli associated with serious human infections such as gas gangrene, food poisoning, pseudomembranous colitis, botulism, and tetanus.

CMV *See* **cytomegalovirus infection.**

coagulation Precipitation of microorganisms and organic material out of water by the addition of chemicals (e.g., aluminum sulfate) to create colloidal aggregates (floc) that settle out of solution. Used in the purification of community water supplies. In clinical terms, coagulation refers to blood clot formation.

coccidioidomycosis A systemic mycotic disease caused by *Coccidioides immitis,* endemic in dry areas such as the southwestern deserts of the United States.

coccus (*pl.* cocci) A bacterium that is spherical in shape.

codon A combination of three specific nucleotides in mRNA that codes for the amino acid sequence in the synthesis of each cell protein during the process of translation.

coliform A term referring to gram-negative, nonsporing, fermentative bacilli that ferment lactose with gas formation. Because they are common inhabitants of the colon of humans and animals, they are used as an index of fecal pollution in the bacteriologic analysis of water.

collagenase A microbial enzyme that disintegrates the proteins of collagen tissue in the host.

colonization The process by which microbes adhere to and multiply on a nutrient surface such as a mucosal surface or a culture medium.

colostrum An antibody-rich, milky fluid secreted by the mammary glands a few days before and after birth, prior to the production of milk.

combined immunodeficiency The congenital absence of bone-marrow stem cells resulting in a dual deficiency of B and T cells. This condition is usually fatal within the first year of life unless bone marrow, carrying competent stem cells, can be transplanted.

commensalism A type of mutualistic or symbiotic relationship between two organisms in which one benefits while the other is unaffected.

communicable disease A disease caused by an infectious agent or its toxic products and which is transmissible from an animate or inanimate reservoir, directly or indirectly, to a susceptible host.

communicable period The period during which an infectious agent may be transferred from an infected person to another person, animal, or arthropod, or from an infected animal to humans.

complement A complex of proteins normally present in plasma that contributes to the action of antibodies in bringing about the destruction of microbial or other foreign cells.

condyloma acuminata *See* **genital warts; human papillomaviruses (HPV).**

congenital syphilis Syphilis acquired by a fetus from an infected mother when the spirochetes cross the placental barrier.

conjugation In bacteria, the process of direct transfer of DNA from one cell (the donor) to another (the recipient), requiring specific interaction between the two. The process is mediated by an F plasmid in the donor cell, which possesses a sex pilus through which the DNA is transferred. The donor cell is designated F^+; the recipient is F^-.

conjugative plasmid A plasmid that carries the genes for sex pili. During conjugation it transfers copies of itself to other bacteria.

conjunctivitis Microbial infection of the conjunctiva that may be caused by bacteria, including *Chlamydia trachomatis* and some viruses.

convalescent period The phase of recovery from acute infectious disease during which fever declines, antibody production reaches a peak, and improvement begins.

coronaviruses A group of viruses that cause upper respiratory infections and possibly infant gastroenteritis.

coxsackieviruses A group of enteroviruses that may cause a variety of diseases (e.g., aseptic meningitis, encephalitis, herpangina, pleurodynia, respiratory infections, and neonatal myocarditis).

C-reactive protein A globulin, normally present in plasma in trace amounts, that increases during the inflammatory response.

cryptococcosis A systemic mycotic disease caused by *Cryptococcus neoformans.* Commonly responsible for the slow development of chronic meningitis but may also involve subcutaneous tissues, lymph nodes, and lungs; an opportunistic infection in some AIDS patients.

Cryptococcus A medically important genus of yeasts classed with the basidiomycetes.

cryptosporidiosis An invasive protozoan disease caused by *Cryptosporidium* species (usually *C. parvum* in humans). Widespread among young domestic animals, but self-limited in healthy humans. In AIDS or other immunocompromised patients it may become chronic and eventually fatal.

cyanobacteria A large group of photosynthetic bacteria formerly known as the blue-green algae.

cyst A microbial cell enclosed within a wall; formed by some species of protozoa. May be dormant, resting cells, resistant to adverse environmental conditions, or reproductive forms that are part of a cell's life cycle. *See* **encystment; excystment.**

cysticercosis An invasive cestode disease caused by the bladder worm, the larval form of the tapeworm *Taenia solium.*

cytokines Chemical substances released by white blood cells in response to immunogenic stimuli. These substances act as mediators that influence other cells.

cytomegalovirus (CMV) infection A common, usually asymptomatic, herpes virus infection of young adults that may cause serious congenital or neonatal disease. Especially severe in immunocompromised patients, including those with AIDS.

cytoplasm The colloidal material, or protoplasm, filling a cell.

cytosine A pyrimidine found in the nucleotides incorporated into nucleic acids.

cytotoxic Having a toxic or destructive effect on cells.

D

dark-field microscopy The examination of unstained microorganisms using a special condenser on the light microscope to focus a cone of light on the specimen. Objects in the specimen are brightly illuminated, standing out from a nearly black background.

deamination A metabolic process in the nitrogen cycle; certain bacteria break down dead organic protein, converting amine groups into ammonia.

definitive host The human or animal host infected by the adult stage of a parasitic worm. Also called a *final host.*

delayed hypersensitivity A condition mediated by sensitized T cells; operative in microbial allergies (allergies of infection) and allergic contact dermatitis.

dendritic cells Specialized branching cells in lymph nodes that participate, along with tissue macrophages, in the initial uptake and processing of immunogens.

denitrification A metabolic process in the nitrogen cycle; certain bacteria transform nitrates and nitrites into free molecular nitrogen.

deoxyribonuclease A bacterial enzyme that contributes to virulence by helping to break down host cell DNA.

deoxyribonucleic acid *See* **DNA.**

dermatophyte A member of a group of fungi causing superficial infections known as dermatophytoses (ringworm).

Deuteromycota A phylum within the kingdom *Fungi;* contains many different genera in which a recognizable sexual stage of reproduction has not been found, so classification remains uncertain; type of asexual spore production is an important basis for species identification. Group contains most of the agents of mycotic disease in humans.

diapedesis The migration of phagocytes through the walls of capillaries into adjacent tissue.

didanosine A drug for the treatment of AIDS, used especially in patients who do not respond to treatment with AZT (zidovudine).

DiGeorge's syndrome Congenital absence of the thymus. Children born without a thymus cannot produce T cells and so lack cell-mediated immunity (CMI); condition creates increased susceptibility to severe microbial infections.

dimorphic fungi Agents of mycotic disease displaying two types of morphology: one in the host's tissues, and another in the environment; usually temperature dependent.

diphtheria An acute bacterial disease caused by toxin-producing strains of *Corynebacterium diphtheriae.*

diphyllobothriasis An intestinal cestode disease caused by the adult form of the fish tapeworm *Diphyllobothrium latum.*

dipicolinic acid (DPA) A substance present in bacterial endospores that may contribute to their heat resistance.

diplococcus (*pl.* diplococci) A spherical bacterium that occurs in pairs.

diploid Having two sets of chromosomes as in a eucaryotic cell.

disinfection The destruction of pathogenic microorganisms (except bacterial endospores) on inanimate objects using chemical or physical agents.

disseminated gonococcal infection (DGI) A systemic extension of gonorrhea following entry of the organism into the bloodstream; involves one or more joints, the skin, and may result in meningitis, endocarditis, myocarditis, or pericarditis.

DNA Deoxyribonucleic acid. A polynucleotide constituting the genetic material of all living cells; composed of deoxyribonucleotides linked by phosphate bridges.

dominant gene An altered gene in one set of chromosomes of a diploid cell whose effect is expressed in the progeny of that cell, while its matched, unaltered partner gene in the other chromosome set is masked. *See also* **recessive gene.**

double helix The molecular structure of DNA; complementary double strands of nucleotides recur in a specific order.

DPT Diphtheria-pertussis-tetanus vaccine. Contains three antigens to actively immunize children and adults against the agents of diphtheria, whooping cough, and tetanus.

drug allergies Allergic reactions to antimicrobial agents or to other drugs. Patient becomes sensitized following first administration; allergic reaction takes place when given again. Also called *drug hypersensitivities.*

dwarf tapeworm *See* **hymenolepiasis.**

dysentery An intestinal disease characterized by severe diarrhea with passage of blood and mucus; usually caused by infection with gram-negative enteric bacilli (e.g., species of *Shigella*). *See* **shigellosis.**

E

EAgg *See Escherichia coli*

Ebola virus An agent of viral hemorrhagic fever (VHF) recently recognized in outbreaks of severe disease in Africa. The disease is characterized initially by fever, diarrhea, and vomiting, which may either resolve or progress to multiorgan dysfunction and disseminated intravascular coagulation.

EBV *See* **Epstein-Barr virus.**

echinococcosis An invasive cestode disease caused by the larval form of the tapeworm *Echinococcus granulosus.* Also called *hydatid cyst disease.*

echoviruses A group of enteroviruses frequently found in the human intestinal tract; sometimes associated with diseases similar to those caused by coxsackieviruses.

ecology Science of the biologic interrelationships between organisms and between organisms and their natural environments.

EHEC *See* ***Escherichia coli.***

ehrlichiosis An acute febrile infectious disease caused by organisms of the genus *Ehrlichia.* Most species infect animals, but two, *E. sennetsu* (mostly seen in Japan) and *E. chaffeensis* (recognized in the United States), are associated with human infections characterized by white blood cell invasion. The latter species is known to be transmitted through tick bites.

EIA *See* **enzyme immunoassay.**

EIEC *See* ***Escherichia coli.***

Elek test A method for testing strains of *Corynebacterium diphtheriae* for toxin production using the immunodiffusion technique.

elementary body (EB) The infectious stage in the life cycle of chlamydiae. The EB enters the host cell and develops within a membranous vacuole into the reticulate body, the multiplying stage. Some reticulate bodies transform to EBs, which are released from the cell and can infect new host cells.

emetine An antimicrobial agent sometimes used in the treatment of *E. histolytica* infection; causes nuclear degeneration, interfering with trophozoite multiplication.

enanthem A lesion of the mucous membranes seen in some microbial infections; characteristics are often diagnostic of a particular disease (e.g., Koplik spots in measles).

encephalitis Inflammation of the brain, often associated with microbial infection.

encystment The process of cyst formation in protozoa.

endemic Descriptive of a disease that is persistent in a population, with a low but relatively uniform frequency of occurrence (e.g., the common cold).

endocarditis *See* **infective endocarditis.**

endogenous infection Infection caused by an organism growing within a host's body. Organism may be pathogenic or a member of the normal flora; gains access to tissues beyond its original or usual site where it is more capable of causing damage.

endoplasm *See* **cytoplasm; protoplasm.**

endospore A highly resistant, dehydrated, thick-walled structure in some gram-positive bacilli. Forms as the cell's vital constituents condense, permitting the cell to survive unfavorable environmental conditions. *See* **germination.**

endotoxin The lipopolysaccharide in the outer membrane of gram-negative bacterial cell walls. The lipid portion is responsible for most of the symptoms suffered by hosts infected by gram-negative organisms.

Entamoeba histolytica *See* **amebiasis.**

enteric bacilli *See* ***Enterobacteriaceae.***

enteric infection A microbial disease of the intestinal tract.

enteritis Inflammation of the intestines usually caused by microbial infection or by microbial toxins.

enteritis necroticans A rare but acute type of food poisoning caused by *Clostridium perfringens.*

Enterobacteriaceae A family of bacteria normally present in the intestinal tract of humans and animals. Organisms are gram-negative, straight rods, motile or nonmotile, facultatively anaerobic, and have simple nutritional requirements.

enterobiasis A common intestinal nematode disease caused by the roundworm *Enterobius vermicularis* (pinworm).

enterococci Cocci in chains that are part of the normal flora of the intestinal tract of humans and many animals; often associated with urinary tract or wound infections and with endocarditis.

enterotoxin A toxic microbial product affecting the cells of the intestinal mucosa; usually causes nausea, vomiting, and diarrhea.

enterovirus Any of several viruses that characteristically enter the human body by way of the alimentary tract (e.g., polioviruses, coxsackieviruses, and echoviruses); classed as picornaviruses.

enzyme Organic catalysts that promote the uptake and use of raw materials by living organisms, the synthesis of materials needed for cell energy and function, and the degradation of unneeded substances that are then returned to the environment.

enzyme immunoassay (EIA) A test to detect a positive serologic reaction. In the test system, either the antigen or antibody is linked to an enzyme. A chromogenic substrate specific for the enzyme breaks down to a colored end product if the antibody and antigen have combined; this indicates a positive test. Also called *enzyme-linked immunosorbent assay (ELISA).*

eosinophils White blood cells with red-staining granules that appear to play a role in allergies; their numbers characteristically increase in allergic reactions.

EPEC *See* ***Escherichia coli.***

epidemic Descriptive of a disease that suddenly increases in large numbers of people simultaneously or in rapid succession; also, the increase in incidence of a disease higher than expected for a given population, even though the number of people involved may be small.

epidemic diarrhea A contagious disease of newborns, especially in hospital nurseries, caused by enteropathogenic strains of *Escherichia coli* (EPEC).

epidemiology The science that deals with the determinants, incidence, distribution, and control of health and disease in a population.

Epidermophyton A fungal genus that causes superficial infections of the skin (dermatomycoses).

Epstein-Barr virus (EBV) A herpes virus that causes infectious mononucleosis; also associated with nasopharyngeal carcinomas and a lymphoma known as Burkitt's tumor.

erysipelas A highly contagious skin infection caused by strains of group A beta-hemolytic streptococci.

erythema infectiosum A usually mild exanthematous disease of children; a bright red facial rash with "slapped cheek" appearance is characteristic. Also called *fifth disease.*

erythrocyte A red blood cell.

erythromycin An antimicrobial agent that affects microbial ribosome protein synthesis; useful in the treatment of pneumonias caused by *Mycoplasma* or *Legionella* species.

Escherichia coli A species of gram-negative, nonsporulating, motile, fermentative bacillli normally found in the intestinal tract of animals and humans, but notable also as the cause of certain enteric infections in humans. Various strains are known to be *enterotoxigenic* (ETEC), *enteroinvasive* (EIEC), *enteropathogenic* (EPEC), *enterohemorrhagic* (EHEC), or *enteroaggregative* (EAggEC), depending on their ability to produce toxic substances or to invade the intestinal mucosa.

ETEC *See* ***Escherichia coli.***

ethylene oxide A gaseous agent used to sterilize heat-sensitive objects that would be destroyed in a hot-air oven or a steam sterilizer.

eucaryotic A term describing a cell whose nucleus is separated from the cell protoplasm by a membrane. Compare with procaryotic.

exanthem An eruption or rash of the skin present in many microbial infections; characteristics are often diagnostic of a particular disease (e.g., measles, smallpox, scarlet fever).

excystment The process in which one or more active cells emerge from a cyst, as in the developmental stages of some protozoa.

exogenous infection Infection acquired from sources and reservoirs outside the host's body.

exotoxin Heat-sensitive protein product produced by certain microorganisms during their normal metabolic growth; usually excreted into the environment in which the organisms are growing. Highly toxic for humans (e.g., *C. botulinum* toxin).

F

FA *See* **fluorescent antibodies.**

facultative anaerobe A microorganism that does not require oxygen for growth but that grows better in its presence.

fascioliasis An invasive trematode disease caused by the liver fluke *Fasciola hepatica.*

fasciolopsiasis An invasive trematode disease caused by the intestinal fluke *Fasciolopsis buski.*

febrile reactions Clinically, reactions induced by fever and its results (e.g., elevated pulse and respiration rates, restlessness, irritability, headache, or diffuse pains).

fermentation A metabolic process in which energy is released under anaerobic conditions; organic molecules act as electron donors and/or acceptors.

F factor The fertility factor in some bacterial cells. *See* **F plasmid.**

fibrin A protein formed from fibrinogen during blood clotting; facilitates phagocytosis when antibody (opsonin) is unavailable, serving as a trap for microorganisms.

fibrinolysin An enzyme capable of dissolving fibrin clots.

fifth disease *See* **erythema infectiosum.**

flagellates A term for protozoans of the phylum Zoomastigina, which produce flagella.

flagellum (*pl.* flagella) A fine, threadlike appendage that provides motility to many unicellular microorganisms.

flocculation In water purification, the addition of chemicals (e.g., aluminum sulfate) causing colloidal aggregates of contaminants to form and settle out of solution.

flucytosine An antimicrobial agent used in the treatment of *Cryptococcus* and *Candida* infections; acts as an antimetabolite and also inhibits DNA synthesis.

flukes *See* **trematodes.**

fluorescence microscopy The examination of microorganisms under an ultraviolet light source after staining with fluorescent dyes or fluorescent antibodies; the stain combines with specific substances within or on the microbial cells to make them visible under the microscope.

fluorescent antibodies (FA) Antibodies that are "labeled" by treatment with a fluorescent dye.

foodborne disease A type of disease developing after the ingestion of chemically or microbially contaminated food; may be manifest as a gastroenteritis or a systemic disease, depending on the agent. Also called *food poisoning.*

F plasmid A plasmid present in some bacterial cells which functions in the process of bacterial conjugation; codes for its own replication. Also called *F factor.*

F prime factor An F factor in the bacterial cell that carries chromosomal genes but lies outside of the chromosome. *See* also **Hfr.**

frameshift A mutation in a bacterial cell; one or more bases in the DNA are added or deleted, causing the codons in that segment of the DNA to change. Proteins may be altered or nonfunctional as a result.

FTA-ABS Fluorescent treponemal antibody-absorbed. Serologic test for the presence of specific treponemal antibody in the serum of patients with syphilis.

fungus (*pl.* fungi) A true eucaryotic, unicellular or multicellular microorganism usually having a walled thallus. Yeasts are unicellular fungi; molds are multicellular fungi with complex cells but primitive differentiation of tissues.

furuncle An abscess localized in the hair follicles, typically of the axillae, back of the neck, buttocks, or eyelashes, usually caused by *Staphylococcus aureus.*

G

gamete A mature male or female sex cell carrying half the number (haploid) of chromosomes characteristic of a given species and capable of fusion with a gamete of the opposite sex to produce a new individual with the full (diploid) number of chromosomes.

gamma globulins Immunoglobulins (protein constituents of plasma) synthesized by lymphocytes in the development of immunity. *See* also **antibody.**

ganciclovir An antimicrobial agent used in the treatment of cytomegalovirus infections; interferes with DNA synthesis.

gastroenteritis Inflammation of the linings of the stomach and intestines; sometimes caused by chemical irritants in food, but more often by ingested pathogenic microorganisms and/or their endotoxins (e.g., species of *Salmonella* and *Staphylococcus,* and viruses).

gene splicing *See* **recombinant DNA technology.**

genital herpes A sexually transmitted infection caused by herpes simplex virus type 2.

genital warts A sexually transmitted disease caused by human papillomaviruses (HPV). Also called *condyloma acuminata.*

German measles *See* **rubella.**

germination The process by which bacterial endospores differentiate back into metabolically active, dividing, vegetative cells.

Ghon complex A healed tuberculosis lesion that shows the calcification and fibrosis indicative of arrested or healed primary tuberculosis.

giardiasis An intestinal disease caused by the protozoan *Giardia lamblia* (phylum Zoomastigina).

globulins Protein constituents of plasma classified in three major groups: alpha, beta, and gamma.

glomerulonephritis An inflammatory disease of the renal glomeruli often appearing as a sequel of streptococcal diseases; thought to result from deposition of circulating streptococcal antigen-antibody (immune) complexes on basement membranes of glomerular capillaries.

glycocalyx A polysaccharide slime produced by several strains of bacteria that allows them to adhere to surfaces. An important clinical example is glycocalyx-producing strains of *S. epidermidis,* which can adhere to the smooth surface of catheter tips.

gonococcus A common term for *Neisseria gonorrhoeae.*

gonorrhea A sexually transmitted disease caused by the gram-negative diplococcus *Neisseria gonorrhoeae.*

graft rejection The result of the operation of T cells that distinguish between "self" and "nonself."

Gram stain A differential staining technique that places bacteria into gram-positive or gram-negative groups depending on their ability to retain crystal violet when decolorized with an organic solvent such as alcohol.

granulocytes Polymorphonuclear white blood cells containing granules with different staining characteristics: neutrophils (neutral-staining), eosinophils (acidophilic), and basophils (basophilic).

granuloma An organization of body cells resembling a tumor but designed to restrain and eliminate an infectious microorganism that has invaded tissue.

granuloma inguinale A sexually transmitted disease caused by *Calymmatobacterium granulomatis,* a gram-negative, intracellular bacillus.

griseofulvin An antimicrobial agent used in the treatment of ringworm; acts by interrupting fungal cell mitosis, causing defective DNA production.

growth curve A graphic representation of the rise and decline of microbial growth in a stationary liquid culture.

guanine A purine found in the nucleotides incorporated into nucleic acids.

Guillain-Barré syndrome (GBS) A relatively rare central nervous system disease involving generalized paralysis; cause is unknown but is often linked with influenza infection or vaccination, and possibly other virus infections.

H

Haemophilus A genus of small, gram-negative bacilli containing two important species: *Haemophilus influenzae,* an agent of upper and lower respiratory infections and of meningitis in infants; and *Haemophilus ducreyi,* the agent of chancroid, an STD.

halophilic Literally, "salt-loving." Refers to microorganisms that grow best in media containing increased amounts of sodium chloride.

Hansen disease (HD) A chronic, progressive, disfiguring skin and nerve disease caused by *Mycobacterium leprae.* Also called *leprosy.*

Hantavirus A member of the family of Bunyaviruses that has been identified as the agent of an acute pulmonary syndrome.

H antigen An antigenic protein present on the flagella of motile, gram-negative organisms (e.g., *Salmonella*) and useful in serotyping.

haploid Having only one set of chromosomes or half the number of chromosomes characteristic of a parent or somatic cell.

hapten *See* **incomplete antigen.**

HbOC *Haemophilus influenzae* type b conjugate vaccine.

helical Having a form of a helix, or spiral. Some viruses have a helical nucleocapsid.

Helicobacter A genus of curved, gram-negative bacilli once classified as *Campylobacter. H. pylori* is a major cause of human gastritis and peptic and duodenal ulcer. *H fenelliae* and *H. cinaedi* are causes of proctitis, enteritis, and systemic infections in homosexual men.

helminth A parasitic worm that may belong to one of two major groups: *nemathelminths* (roundworms) or *platyhelminths* (flatworms).

helper T cells T lymphocytes that join with B cells to destroy antigens having more than one kind of determinant; for example, when an incomplete antigen (hapten) is linked to a carrier molecule, B cells react with the hapten and T cells with the carrier to stimulate the antibody response.

hemagglutination (HA) The clumping of red blood cells by antibodies, or by surface protein antigens on myxoviruses (e.g., influenza, measles, and mumps).

hemagglutination-inhibition (HI) A serologic method for preventing hemagglutination by coating a myxovirus with specific antibody to prevent attachment to red blood cells; used to identify a particular virus if combination with its antibody prevents hemagglutination.

hemagglutinin An antibody that causes hemagglutination; also, an important surface protein antigen on the influenza virus and on other myxoviruses that agglutinates red blood cells. Viral hemagglutinins enable the virus to attach to host cells to initiate infection.

hemoflagellates Flagellated protozoans that are transmitted to humans by biting insects, enter the bloodstream, and live in soft tissues (e.g., *Trypanosoma* and *Leishmania* species).

hemolytic uremic syndrome A disease, especially of children, that follows intestinal infection with enterohemorrhagic strains of *Escherichia coli* (EHEC); characterized by acute renal failure, decreased platelets, and hemolytic anemia.

hepatitis An inflammation of the liver often related to virus infection or to chemically toxic substances.

hepatitis viruses A group of viruses, designated as A, B, C, D, and E, involved as major agents of acute liver disease throughout the world. A and E are transmitted through the alimentary route; the others are transmitted parenterally, primarily by blood transfusion.

herpangina An acute but brief illness caused by a coxsackievirus and characterized by vesicular lesions in the pharynx, fever, vomiting, and abdominal pain.

herpes simplex virus (HSV) A virus associated with a wide spectrum of infections (e.g., labial herpes, genital herpes, keratoconjunctivitis, eczema, meningoencephalitis, and neonatal infections acquired from maternal genital lesions).

heterophile antibodies Antibodies that are hemagglutinins for sheep red blood cells and develop in the serum of many patients with infectious mononucleosis.

heterotroph An organism whose principal carbon source is organic material preformed by other living things.

Hfr High frequency of recombination. A bacterial cell possessing an F plasmid that is integrated into the main chromosomal DNA. This situation is reversible (i.e., F can leave the integrated (Hfr) state and become autonomous (F^+), lying separately in the cell). F sometimes takes chromosomal genes with it when it separates. Autonomous F with one or more chromosomal genes attached is called an F prime factor (F').

histoplasmosis A systemic mycotic disease caused by *Histoplasma capsulatum;* characterized by pulmonary infection which may disseminate to lymph nodes, spleen, and liver.

HIV Human immunodeficiency virus. A retrovirus associated with the onset of AIDS.

holozoic A term describing a type of heterotroph capable of ingesting solid food. Amebae and many other protozoa are holozoic, having specialized structures that permit them to ingest and excrete particulate matter.

host Any organism (human, animal, plant, or microbial) that harbors a parasitic microorganism and provides conditions necessary for the parasite's survival.

human papillomaviruses (HPV) A group of viruses that cause wartlike lesions. When transmitted by certain types of the virus through sexual intercourse, these lesions appear on the genitalia of men and women. In some women, the presence of HPV is associated with the development of cervical cancer.

humoral immunity Immunity mediated by antibodies (immunoglobulins) circulating in the blood.

hyaluronic acid A mucoid substance found in human connective tissue and in the capsule of group A streptococci; one of the factors contributing to the virulence of the streptococci.

hyaluronidase A microbial enzyme that breaks down hyaluronic acid, the gel-like matrix of connective tissue; also called *spreading factor.*

hybridomas Hybrid cells formed by fusing antibody-producing plasma (B) cells with rapidly multiplying myeloma (bone marrow tumor) cells in vitro. Each hybridoma quickly produces clones of cells secreting a specific, monoclonal antibody.

hydatid cyst A larval stage in the life cycle of the tapeworm *Echinococcus granulosus* in which fluid-filled sacs are formed in the tissues of the animal or human host.

hymenolepiasis An intestinal cestode disease produced by the adult form of the tapeworms *Hymenolepis nana* (dwarf tapeworm) and *Hymenolepis diminuta* (rat tapeworm).

hypergammaglobulinemia An immunodeficiency disease resulting from an overabundance of immunoglobulin production by B cells. This condition is found in diseases such as rheumatoid arthritis and systemic lupus erythematosus.

hypersensitivity A condition in which the body's immune responses to an antigen are excessive; resulting damage to tissues may range from minor to lethal. These responses may be mediated by antibodies (immediate hypersensitivity) or by T cells (delayed hypersensitivity).

hypha (*pl.* hyphae) An individual filament, or branch, within the mycelium of a mold.

hypogammaglobulinemia An immunodeficiency disease resulting from decreased synthesis of immunoglobulins. *See* **agammaglobulinemia.**

I

icosahedron A solid form having 20 equilateral triangular faces and 12 corners. Some viruses have a nucleocapsid that is icosahedral in shape.

IgA A class of immunoglobulins (*A*=alpha) found in external secretions of mucosal surfaces (e.g., tears, saliva, bile, urine, and colostrum). "Secretory" antibodies (sIgA) are important in resistance to infection of the surface tissues of the body.

IgD A class of immunoglobulins (*D*=delta) found in the cell membranes of human B lymphocytes where it is thought to function as an antigen receptor in the initial stages of the immune response.

IgE A class of immunoglobulins (*E*=epsilon) that plays an important role in hypersensitivity of the immediate (anaphylactic) type. IgE attaches to mast cells and basophils, triggering the release of histamine with damaging effects to surrounding tissues; may be expressed as anaphylactic shock, hay fever, urticaria (hives), or allergic dermatitis.

IgG A class of immunoglobulins (*G*=gamma) comprising the principal antibodies of the late phase of the immune response; appear in serum following exposure to the antigens that initially stimulate IgM production.

IgM A class of immunoglobulins (*M*=mu) comprising the major component of the primary antibody reaction. Appears first in a given immune response; gives the host some advantage over an invading microorganism while it attempts to gain a foothold and spread.

immediate hypersensitivity A condition mediated by antibodies attached to tissue cells, such as mast cells or basophils, at the sites where they combine with antigen. The antigen-antibody combination causes the release of physiologically active substances, including histamine; results are dilation of blood vessels, smooth muscle spasm, and increased capillary permeability with edema. Various symptoms of this condition are collectively known as allergy; the most serious is anaphylaxis.

immune complex diseases Diseases caused by antigen-antibody combinations (immune complexes) that fix complement; injury results from small Ig-antigen deposits in blood vessel walls; examples include serum sickness, rheumatoid arthritis, lupus, and glomerulonephritis.

immune system The complex of functions of lymphoid tissue, lymphocytes, and macrophages that enable the body to recognize an invading foreign agent as not being "self" and to take specific action to neutralize, destroy, and eliminate it.

immunity A state of resistance to infection; may be natural (based on racial, species, or individual factors) or specific (based on the functions and mechanisms of the immune system).

immunoblot assay An immunologic method used especially for identifying HIV antibody in the serum of patients infected with AIDS.

immunodeficiency diseases Diseases caused by disorders of the immune system; may be primary (congenital), iatrogenic (induced by immunosuppressive drugs), or caused by malignancies of the lymphatic system or by infection with HIV. Examples include hypergammaglobulinemias, hypogammaglobulinemias, and AIDS.

immunodiffusion A serologic technique in which a soluble antigen and its antibody (precipitin) are allowed to diffuse through a semisolid gel. If they are specific for each other, visible precipitation lines form where they meet and combine in the gel; if they do not combine, no precipitate is seen.

immunogen An antigen that can induce an immune response as well as bind specifically to the components of the immune response (antibodies or lymphocytes).

immunoglobulin (Ig) *See* **antibody; gamma globulin.**

immunoprophylaxis The prevention of infectious disease by active or passive immunization.

immunotherapy The treatment of infectious disease with specific antisera.

impetigo A skin infection, most common in children, caused by strains of group A beta-hemolytic streptococci; characteristically appears as thick, raised, crusted lesions. Also called *streptococcal pyoderma.*

inclusion conjunctivitis An acute conjunctivitis of the newborn resulting from maternal sexually transmitted infection with *Chlamydia trachomatis.*

incomplete antibody *See* **blocking antibody.**

incomplete antigen A molecule that is not immunogenic itself but becomes so after linkage with a macromolecule (e.g., a protein or polysaccharide).

incubation period The initial stage of microbial disease during which an invading microorganism attempts to establish itself, eliciting host defense mechanisms. No outward signs of infection appear until this period comes to an end, and the pathogen begins to create damaging effects. Period varies in length but averages 1 to 2 weeks.

induction A process in bacteriophage reproduction in which a prophage hitherto integrated into the host cell DNA and replicating with it (lysogeny) becomes "awakened." The viral DNA is then fully expressed with resultant virion production and lysis of the host cell.

infant pneumonitis A respiratory disease of infants caused by *Chlamydia trachomatis;* acquired during birth from the infected maternal genital tract and becomes manifest at 4 to 16 weeks of age.

infection A state in which the host harbors microorganisms that survive and multiply in or on body tissues; includes commensalism as well as infectious disease but does not necessarily imply the latter. *See* also **commensalism; mutualism; symbiosis; infectious disease.**

infectious disease The damaging result of infection of a host by a microorganism whose injurious properties break through the host's defenses and cause illness.

infectious drug resistance The spread of resistance to antimicrobial agents in bacterial populations through the transfer of R factors (plasmids).

infectious mononucleosis An acute, infectious, viral disease caused by a herpes virus called the Epstein-Barr virus (EBV). Because the major route of transmission is apparently through intimate oral contact, the infection has been called the "kissing disease."

infective endocarditis (IE) A subacute or acute bacterial infection of the endocardium usually involving localization of normal flora (e.g., streptococci, staphylococci, gram-negative bacilli) at some site of previous injury to the endocardium, usually the heart valves.

inflammatory response The process by which circulating white blood cells react protectively to injury in order to limit and control it locally; characterized by redness, heat, swelling, and pain.

influenza An acute viral disease of the respiratory tract caused by three major types of influenza virus (A, B, and C) distinguished by their nucleocapsid antigens; may occur as individual cases or in epidemics and pandemics. Also called *flu.*

Insecta A class of arthropod including flies, mosquitoes, fleas, lice, and true bugs, many of which are important vectors of infection.

interferon (IFN) One of a group of proteins produced by connective tissue cells and leukocytes or, potentially, by any infected body cell; production may be elicited by virus infection and other stimulants of lymphocyte division. In RNA virus infection, interferon is released from infected cells; it then attaches to the surface of uninfected cells, causing them to synthesize a new protein that blocks new infection.

intermediate host A host in which larval stages of development of a parasitic worm or protozoan develop.

intestinal flagellates Flagellated protozoans that live in the human bowel or on other mucosal surfaces (e.g., *Trichomonas* and *Giardia* species).

invasiveness The ability of a microorganism to grow and multiply in host tissues, to protect itself from host defenses, and to penetrate and spread in the body of the host to which it has gained access.

ionizing radiation Radiation of very short wave length or high energy (X rays, gamma and beta rays, and high-speed electrons); lethal for many types of cells (tissue cells, bacteria, viruses, and others).

IPV Inactivated poliovirus vaccine. Also called *Salk vaccine.*

isolation The separation of infected humans or animals in places and under conditions that prevent direct or indirect transmission of the infectious agent; separation is maintained for the period of communicability.

isoniazid An antimicrobial agent that inhibits cell-wall synthesis; used primarily for treating tuberculosis.

J

Jarisch-Herxheimer reaction A febrile response to specific therapy of secondary or later stages of syphilis; may be caused by the release of endotoxinlike substances from spirochetes killed by the antimicrobial agent.

jumping genes *See* **transposons.**

K

kala-azar A severe infectious disease caused by a protozoan hemoflagellate of the genus *Leishmania,* transmitted to humans by the bite of sandflies.

Kaposi sarcoma A type of cancer frequently seen in AIDS patients that is characterized by pigmented nodules usually on the head and neck. May also occur at scattered skin sites and affect viscera such as the gastrointestinal tract and oral cavity.

ketoconazole An antimicrobial agent used to treat systemic fungi and ringworm; acts by altering cell membranes, increasing cell permeability.

killer T cells Specialized T cells having cytotoxic effects on cells they recognize as foreign, or "nonself."

Koch's postulates A series of procedures used as an aid in proving that a particular microorganism is the cause of a specified disease.

Koplik spots Enanthematous lesions of the mouth and throat characteristic of measles; appear as bright red areas with whitish centers.

Kupffer cells Macrophages functioning in the liver; part of the mononuclear phagocyte system (MNPS).

L

Lancefield grouping A serologic system for classifying streptococci according to the nature of their carbohydrate antigens. Important strains belong to groups A, B, C, or D; some strains remain ungrouped.

larva (*pl.* larvae) An immature stage of development of an animal parasite.

latent infection Infection in which a pathogenic microorganism persists in host tissues but produces no reaction. Organism may emerge to produce an acute episode of infection at a time when host defenses are lowered.

latex agglutination A serologic test in which particles coated with antibody molecules are used to detect soluble antigens present in body fluids or culture broths. The combination of antigen with the antibody-coated latex particles results in visible clumping of the particles, which is a positive test.

lecithinase A microbial enzyme destructive for lecithin, one of the components of many human cells, including red blood cells.

Legionnaires disease A pulmonary infection caused by bacteria of the genus *Legionella* (e.g., *L. pneumophila*).

Leishmania A genus of protozoan hemoflagellates carried by insects; transmitted through insect bites to humans and other animals. Diseases are known, collectively, as leishmaniasis.

lentivirus A subfamily of the retroviruses in which HIV is classified.

lepromas Granulomatous nodules in the skin, mucous membranes, and some visceral organs, characteristic of the lepromatous form of Hansen disease.

lepromin A skin-testing antigen used to distinguish between cases of tuberculoid (usually positive) and lepromatous (negative) cases of Hansen disease.

leprosy *See* **Hansen disease.**

leukocidin A microbial product capable of destroying leukocytes (white blood cells).

leukocytes Cellular components of blood known as white blood cells, including granulocytes, monocytes, and lymphocytes.

leukocytosis An increase in the number of leukocytes, especially polymorphonuclear cells, in the circulating blood; particularly characteristic of bacterial infection.

L-forms Variants of bacteria that have lost their capacity to produce cell walls either by mutation or by treatment with agents (e.g., penicillin) that inhibit cell-wall synthesis; may revert back to their normal cell-wall producing forms. Also called *cell-wall deficient bacteria.*

life cycle The patterned development of parasites, especially protozoans and helminths, involving their sexual and asexual stages of development in one host or another, or in the environment.

lipopolysaccharide (LPS) A compound of lipid and polysaccharide in the outer membrane of the gram-negative cell wall; constitutes the endotoxin of gram-negative bacteria.

lipoteichoic acid A virulence factor in the wall of group A streptococci enabling them to bind to the surface of human epithelial cells.

Listeria monocytogenes *See* **listeriosis.**

listeriosis Probably a foodborne disease caused by the gram-positive bacillus *Listeria monocytogenes;* most noted for its damaging or lethal effects on newborns (who acquire it congenitally through the placenta) and adults with immunosuppressive or other debilitating diseases.

lobar pneumonia A type of pneumonia (usually bacterial) in which the alveolar spaces of an involved lobe of the lung fill with exudate containing red and white blood cells and the lobe becomes consolidated.

Lyme disease A chronic, systemic disease caused by the spirochete *Borrelia burgdorferi,* transmitted to humans from an animal reservoir by a tick.

lymphatic system A network of tissues scattered through the body in which lymphocytes are concentrated in nodes connected by vessels that drain fluid (lymph) from body tissues into lymphatic vessels.

lymphocytes Cells of the blood (white blood cells) and lymphatic system that play a major role in immune and hypersensitivity mechanisms. *See* **B cells; T cells.**

lymphogranuloma venereum A sexually transmitted disease caused by specific serotypes of *Chlamydia trachomatis.*

lymphoid organs Sites where lymphocytes are formed or accumulate. The bone marrow and thymus are primary lymphoid organs; lymph nodes and spleen are secondary.

lymphokines Chemical factors released by sensitized lymphocytes, especially T cells, that contribute to the destruction of the sensitizing antigen.

lysis The disintegration of a cell, through rupture of its wall or membrane, and the dissolution of its contents.

lysogenic conversion A process in which bacterial host cells carrying a prophage may acquire new properties (e.g., toxin production). Toxins are products of the prophage rather than of the bacterial cell; diphtheria exotoxin and scarlet fever pyrogenic exotoxin are examples.

lysogeny A state of infection of a bacterial cell (lysogen) by a prophage that reproduces along with the host rather than taking over and destroying it.

lysosome A small membranous sac, or organelle, within phagocytic cells containing various hydrolytic enzymes (e.g., lysozyme); these enzymes are important in the intracellular digestion of particles taken up by the phagocyte.

lysozyme An antimicrobial enzyme present in phagocytes and in some body secretions (e.g., mucus, tears, and saliva); breaks down the peptidoglycan present in some bacterial cell walls.

M

macrophages Large, mononuclear phagocytic cells derived from monocytes present in blood, lymph, and other tissues.

major histocompatibility complex A group of molecules found on tissues and phagocytic cells that play an important role in recognition of foreign antigens by T cells, especially those responsible for organ rejection.

malaria A serious infection of human red blood cells caused by various species of the protozoan genus *Plasmodium* and transmitted to humans by the bite of mosquito vectors.

Mantoux test A skin test for tuberculosis requiring intracutaneous injection of purified antigen (PPD) of *M. tuberculosis.*

mastigote An intermediate developmental form of the hemoflagellated protozoans *Leishmania* and *Trypanosoma.* Various stages referred to as "amastigote" or "promastigote" in *Leishmania,* and "trypomastigote" or "epimastigote" in *Trypanosoma,* may be found throughout the life cycles of the parasite.

measles A viral infection characteristically beginning as a severe cold but later emerging as an exanthematous disease producing a generalized skin rash with fever and bronchitis. Mucosal lesions in the mouth and throat (Koplik's spots) are often clinically diagnostic. The virus is a paramyxovirus with hemagglutinating properties. Also called *rubeola.*

mebendazole An antimicrobial agent used in the treatment of pathogenic roundworms and some tapeworms; acts by inhibiting uptake of glucose, depleting glycogen stores, and interfering with absorptive and secretory function. Trade name Vermox.

meninges The membranes covering brain surfaces.

meningitis An acute inflammation (septic or aseptic) of the membranes of the brain and spinal cord; may be due to nonspecific injury, but is commonly caused by microbial infection (bacteria, viruses, fungi, or some protozoa may be the agents).

meningococcal meningitis A potentially epidemic form of bacterial meningitis caused by *Neisseria meningitidis.*

meningoencephalitis A disease involving both the meninges of the brain and brain tissue itself.

metronidazole An antimicrobial agent that interferes with nucleic acid replication or metabolism; useful in the treatment of some protozoan and anaerobic gram-negative bacillary infections (e.g., those caused by *Giardia* and gastrointestinal strains of *Bacteroides*). Trade name Flagyl.

miconazole An antimicrobial agent used in the treatment of systemic and ringworm fungi; acts by altering cell membranes and by interfering with intracellular enzymes.

microglia Macrophages functioning in the central nervous system; part of the mononuclear phagocyte system (MNPS).

microsporidiosis A gastrointestinal disease most commonly occurring in severely immunocompromised persons, caused by obligate intracellular protozoa of the phylum *Microspora,* order *Microsporidia.*

Microsporum A fungal genus that causes superficial infections of the hair and skin.

missense A change in the DNA resulting from transition or from transversion mutations, in which the codon for one amino acid is changed to the codon for a different amino acid.

mitochondria (*sing.* mitochondrion) Specialized membranous bodies, or organelles, in the cytoplasm of eucaryotic cells where respiratory functions occur; provide the cell with energy required for vital processes.

MMR Measles-mumps-rubella vaccine.

mold *See* **fungus.**

Monera A kingdom that includes the "true bacteria" (eubacteria).

monoclonal antibodies Highly specific antibodies produced by a population of genetically identical plasma cells (clones); produced by a hybrid cell culture line derived from the fusion of a cancer cell with plasma cells activated to produce a specific antibody. *See* also **hybridomas.**

monocytes Mononuclear white blood cells that are the precursors of macrophages. Like neutrophils, they are chemotactically attracted out of the bloodstream to sites of infection in tissues where they become mature macrophages.

mononuclear phagocyte system (MNPS) A collective term referring to the mononuclear phagocytic cells that are widely distributed in the body in organs or bloodstream. The macrophages of the MNPS function at many sites (spleen, liver, lungs, lymph nodes) to remove foreign particles, living or dead, through phagocytosis, and clear the body of unwanted debris.

M protein An important virulence factor of group A streptococci lying beneath and extending into the capsule; prevents phagocytosis of the organism and is antigenic, conferring type specificity on group A strains.

mucociliary escalator A layer of ciliated epithelial cells and mucus lining most of the upper and lower respiratory tract; traps microorganisms or other foreign particles carried into the lower airways and sweeps them back upward into the throat by the beating action of the cilia.

mumps A viral infection of the parotid glands that may later localize in other salivary glands, testes or ovaries, the thyroid gland, and occasionally the central nervous system; one of the paramyxoviruses that agglutinates red blood cells, and can also lyse them.

mutagen A chemical or physical agent that induces mutations in cellular genes.

mutation An inheritable change in the characteristics of a cell due to gene alteration.

mutualism A type of interrelationship between two organisms in which both gain benefit and are dependent upon each other for survival. *See* also **symbiosis.**

mycelium The mat of branching, filamentous tubes, or hyphae, that make up the body of a mold. May be vegetative (that part extending downward and into the surrounding environment) or aerial (that part extending above the surface on which it is growing).

mycetoma A subcutaneous mycotic disease caused by numerous soil fungi; characterized by nodule and abscess formation, usually on the foot, with extension through sinus tracts to muscles and bones. Also called *maduromycosis.*

***Mycobacterium avium* complex** Mycobacterial agents associated with respiratory infections and increasingly seen as the cause of disseminated infections in AIDS patients.

Mycobacterium bovis The agent of tuberculosis in cattle; may be transmitted to humans.

Mycobacterium leprae The causative agent of Hansen disease (leprosy).

Mycobacterium tuberculosis The causative agent of tuberculosis. Commonly called "tubercle bacilli" because organisms are found in tubercles formed during active infection. Also called AFB because of their acid-fast staining reaction.

Mycoplasma A genus of bacteria lacking the capacity to form cell walls but like true bacteria in other respects. One species causes primary atypical pneumonia.

mycosis (*pl.* mycoses) A superficial or systemic disease caused by a fungus. Also called *mycotic disease.*

mycotic Referring to a fungal infection or disease.

myocarditis Inflammation of the myocardium sometimes caused by microbial infection, notably coxsackievirus infection in newborn infants.

myxoviruses RNA viruses such as influenza whose surface proteins (hemagglutinins) provide an affinity for mucin or mucoprotein. Since red blood cells have a surface mucoprotein to which virus hemagglutinin attaches, these viruses bring about hemagglutination.

N

Naegleria A genus of ameba that can cause human brain or eye infections. Central nervous system disease caused by this protozoan is almost invariably fatal. Penetrates through the nasal passages of swimmers, especially divers, who come in contact with infected water.

necrotizing fasciitis An invasive infection of deep subcutaneous tissue caused by group A streptococci (referred to in this case as "flesh-eating bacteria"). Systemic signs may be referred to as streptococcal toxic-shock syndrome.

Neisseria A genus of gram-negative diplococci containing two important pathogenic species, *Neisseria gonorrhoeae* and *Neisseria meningitidis,* as well as a number of nonpathogenic species.

Nemathelminthes A class of helminths commonly known as roundworms or nematodes, many of which are parasitic.

nematode *See* ***Nemathelminth.***

neuraminidase A surface protein enzyme and antigen of influenza virus that breaks down respiratory mucus, allowing virus hemagglutinins to attach to cells more readily; may also help to bring about release of virus from the host cell surface.

neutralization The inactivation of a toxin or virus by combination with its specific antibody which neutralizes it, rendering it harmless to its host; may occur in vivo as well as in vitro.

neutrophils Polymorphonuclear granulocytes (often called "polys" or PMNs); the most numerous and most phagocytic of the white blood cells.

niclosamide An antimicrobial agent used in the treatment of some tapeworms and flukes; inhibits mitochondrial oxidative phosphorylation in worms.

nitrification An aerobic metabolic process in the nitrogen cycle in which certain bacteria convert ammonia to nitrates and nitrites.

nitrogen cycle The regeneration of nitrogen in nature. Certain bacteria help to transform inorganic, gaseous nitrogen into ammonia, nitrites, and nitrates; nitrates may be converted into amino groups or transformed back into gaseous nitrogen as the cycle continues.

nitrogen-fixation A metabolic process in which atmospheric, free nitrogen is converted to ammonia by "nitrogen-fixing" bacteria.

Nocardia A genus of gram-positive, acid-fast, filamentous bacteria containing several pathogenic species; may cause the chronic pulmonary disease in humans referred to as nocardiosis.

nonsense A change in the DNA resulting from transition or from transversion mutations, in which the codon for an amino acid is changed to one of the stop signal codons.

nonspecific resistance The general, innate defense mechanisms (other than specific immunity) by which the body resists infection or the invasion of foreign objects or substances. The most important of these include the protective barriers of skin and mucous membranes and the phagocytic cells of the mononuclear phagocyte system (MNPS).

normal flora The bacteria and other microorganisms always associated with the superficial tissues of the human or other animal body; may be "resident" or "transient" as local conditions permit.

Norwalk viruses A group of viruses (named after the town of Norwalk, Ohio) involved in epidemic outbreaks of gastroenteritis and primarily affecting older children and adults. A group of related viruses, called Norwalk-like viruses, has been similarly incriminated in such epidemics.

nosocomial infection An infection developing in a patient in a hospital or other health care facility; includes infections that were not present at the time of admission, lingering infections acquired during a previous admission, and such infections among staff.

nucleic acid probes Short sequences of nucleic acids used for specifically identifying microbes in clinical specimens or in culture by recognition of their complementary, unique DNA or RNA sequences.

nucleic acids DNA and RNA; formed from combinations of nucleotides. DNA comprises the genetic material of cells; RNA is important in protein synthesis.

nucleocapsid The nucleic acid core and capsid of a virus.

nucleoside A compound consisting of a sugar (ribose in RNA, deoxyribose in DNA) combined with a purine or pyrimidine base.

nucleoside analogues A class of drugs active against the human immunodeficiency virus that act by inhibiting the viral reverse transcriptase enzyme, thereby preventing the conversion of viral RNA ro DNA; includes drugs such as AZT (zidovudine), DDI (didanosine), 3TC (lamivudine), and DDC (zalcitabine).

nucleotide A compound made up of three kinds of molecules: a sugar (ribose in RNA, deoxyribose in DNA), a phosphate group, and a nitrogen base. Each nucleotide consists of one sugar molecule, the attached nitrogen base, and the adjacent phosphate molecule.

nucleus An organelle within eucaryotic cells, enclosed by a double membrane, that contains the cell's chromosomes and controls reproduction and protein synthesis. In procaryotic cells this nuclear material is not enclosed within a membrane but lies free in the cytoplasm.

nystatin An antimicrobial agent used in the treatment of *Candida* infections; acts by binding to sterols in fungal cell membranes, altering permeability.

O

O antigen An antigenic protein on the outer membrane of some gram-negative bacteria (e.g., *Salmonella*); part of the lipopolysaccharide of the cell walls. Useful in serotyping these bacteria.

obligate intracellular parasite A microorganism that must live within a host cell upon which it depends for metabolic and/or reproductive support; often causes harm to or the death of the host cell.

oocyst A cyst formed around the zygote of *Plasmodium* or *Toxoplasma* in hosts harboring the sexual stage of such protozoans; also the infective form of coccidian parasites.

ophthalmia neonatorum Gonococcal infection of the eyes of a human newborn; infection occurs during passage through the birth canal of a mother with active gonorrhea.

opportunistic infection Infection caused by a microorganism that is generally harmless, but finds an opportunity to cause infection in a person whose resistance is lowered by underlying disease or an immunodeficiency.

opsonic effect The enhancement of phagocytosis of microorganisms, other foreign particles, or the body's own dead cells by antibodies (opsonins) coating their surfaces, making them more easily ingested.

OPV Oral poliovirus vaccine. Also called *Sabin vaccine.*

organelle A small organ or structure within the body of a cell that performs specialized functions (e.g., a mitochondrion).

osteomyelitis An infection in bone tissue; often associated with staphylococcal invasion.

oxidation-reduction Reactions involving the transfer of electrons or hydrogen atoms from one compound to another.

P

pandemic Descriptive of an increase in the occurrence of a disease within large populations, often with a wide geographic, even global, distribution.

paracoccidioidomycosis A systemic mycotic disease caused by *Paracoccidiodes brasiliensis;* endemic in Latin America, particularly Brazil. Also called *South American blastomycosis.*

paragonimiasis An invasive trematode disease caused by the lung fluke *Paragonimus westermani.* Infection is acquired by ingestion of crustaceans such as crayfish or crabs containing encysted larvae. The adult worm lodges in pulmonary tissue causing local damage.

parainfluenza viruses A group of viruses responsible for mild or acute upper and lower respiratory infections.

paramyxoviruses RNA viruses such as measles and mumps; larger than the related myxoviruses (e.g., influenza) but like myxoviruses, possess hemagglutinins.

parasite An organism that depends on its host for its existence; lives on or within the tissues of the host and obtains nutrient at the expense of the host.

parenteral A term referring to a direct route of access to body tissues, by-passing mucosal surfaces. Biting insects, hypodermic needles, animal bites, accidental or surgical wounds may all provide microorganisms with a route of entry into deep tissues.

paresis An incomplete paralysis; may be a symptom of CNS involvement in the tertiary stage of untreated syphilis.

paronychia Inflammation of the soft tissues around the nails; often caused by *Staphylococcus aureus* or fungal agents.

passive artificial immunization The production of temporary immunity to a hazardous infection by injecting a serum containing antibodies into an exposed, nonimmune individual. Serum is obtained from humans or animals that have been actively immunized by infection or vaccination; cellular immunity can be transferred by injecting sensitized lymphocytes.

passively acquired immunity Specific protection from an infectious disease acquired through either natural or artificial means. Natural immunity involves the placental transfer of maternal antibodies (serum IgG) to the blood of the developing fetus in utero; artificial immunity is provided through injection into nonimmune individuals of serum containing antibodies.

pasteurization The process of heating milk or other fluids to destroy nonsporing pathogenic bacteria under specific conditions of time and temperature; process does not spoil the flavor or quality of the beverage.

pathogen A microorganism capable of causing disease.

pathogenesis The manner in which a disease originates and develops.

pathogenicity An expression of the ability of a microorganism to cause disease.

penicillin One of a class of antimicrobial agents useful in treating infections caused by gram-positive bacteria; agents inhibit cell-wall formation by acting on peptidoglycan synthesis.

penicillinase A bacterial enzyme (beta-lactamase) that inactivates penicillin and some of its derivatives by breaking down their beta-lactam ring.

peptidoglycan A constituent of bacterial cell walls that provides their strength and rigidity; a complex of polysaccharides and peptide chains occurring in greater amounts in gram-positive than in gram-negative cell walls.

pertussis *See* **whooping cough.**

Peyer's patches Deep, lymphoid tissue of the intestinal wall; the first site of *Salmonella typhi* localization following ingestion in contaminated food or water.

pH An expression of the acid-alkaline balance of a solution or medium given as a mathematical value from 1 to 14. A middle value of 7 indicates neutrality; values below indicate acidity and those above signify alkalinity.

phagocytosis The ingestion of foreign particles, including microorganisms, by holozoic cells. In human beings and other animals, such cells include the white blood cells and macrophages.

phagolysosome The membranous sac within a phagocyte that results from the fusion of a phagosome with a lysosome.

phagosome A membranous vacuole within phagocytes. *See* also **phagolysosome.**

phase-contrast microscopy The examination of unstained microorganisms using a special condenser and objectives on the light microscope. Microorganisms stand out sharply and internal structures in large cells are distinguishable because their density varies from the cytoplasm to affect light waves differently.

photosynthesis The process by which green plants trap light energy and convert it to chemical energy for use in manufacturing carbohydrate from water and atmospheric CO_2.

photosynthetic autotrophs Organisms such as plants, algae, and some bacteria that use CO_2 from the atmosphere and light energy to manufacture organic materials.

photosynthetic heterotrophs A small group of bacteria that rely on light to provide energy for metabolizing organic compounds.

picornaviruses The smallest of the RNA viruses (*pico* = very small; *rna* indicates the type of nucleic acid they possess); includes enteroviruses and rhinoviruses.

pinworm *See* **enterobiasis.**

piperazine An antimicrobial agent sometimes used in the treatment of *Ascaris* and *Enterobius* infections; paralyzes the worm by inhibiting nerve transmissions at neuromuscular junctions.

plague A virulent disease caused by the gram-negative bacillus *Yersinia pestis.* Transmitted to humans from a rodent reservoir by a flea, it results in swollen lymph nodes, called "buboes," and in this clinical form it is called "bubonic" plague. Systemic spread to the lungs results in "pneumonic" plague, which is directly transmissible from one person to another. In epidemic form in the past it was known as "the black death" because of the dark, necrotic lesions appearing in lymph nodes and skin.

plasma The fluid portion of blood consisting of a colloidal solution of proteins in water; also contains carbohydrates, lipids, vitamins, hormones, electrolytes, and dissolved gases.

plasma cell A specialized lymphocyte (B cell) whose chief function is to synthesize and release antibody into the circulating blood.

plasmid A DNA molecule with extra genetic information enabling it to exist and duplicate itself independently of the chromosome or to be associated with it.

Plasmodium A parasitic genus of protozoans that are the agents of malaria.

Platyhelminthes A class of helminths, commonly known as flatworms, containing two major parasitic groups: the cestodes (tapeworms) and the trematodes (flukes).

pleomorphic Having variable forms (e.g., bacteria that may vary in shape, length, width, or other characteristics).

pleurodynia A coxsackievirus disease characterized by fever and acute muscle pain in the chest wall.

pneumococci Members of the genus *Streptococcus;* characteristically gram-positive diplococci. May be members of the normal flora of the upper respiratory tract, or associated with pneumonia.

pneumocystosis An acute and often fatal pulmonary or systemic disease in immunocompromised patients caused by *Pneumocystis carinii,* an organism of uncertain classification but thought to be a fungus.

poliomyelitis A contagious virus disease caused by polioviruses; may occur as a nonparalytic illness with mild gastrointestinal disturbance, an aseptic meningitis, or a paralytic disease of varying severity.

polymerase An enzyme that catalyzes the formation of DNA or RNA from preexisting strands; acts as a template, or model.

polymerase chain reaction (PCR) A method for amplifying DNA sequences before using nucleic acid probes to identify microorganisms.

polymorphonuclear Having multilobed nuclei; descriptive of one type of white blood cells.

pork tapeworm *See* **taeniasis; cysticercosis.**

potable Suitable for drinking, used in reference to water.

PPD Purified protein derivative. A skin-testing antigen derived from tubercle bacilli and used in the Mantoux test for tuberculosis.

praziquantel An antimicrobial agent used in the treatment of some fluke and tapeworm infections; mode of action is unknown, but results in paralysis of worm suckers causing their dislodgement from tissues.

precipitation In vitro reaction that occurs when antibodies combine with soluble free antigens to form a visible precipitate.

precipitins Antibodies that cause precipitation of soluble antigens in vitro and in vivo. Soluble free antigen combines with antibody to form complexes that become visible as a fine precipitate settling out of solution; may be responsible for immune complex disease when they form in body tissues.

primaquine An antimicrobial agent used in the treatment of malaria; interferes with DNA function of the parasite.

primary atypical pneumonia A clinical syndrome implying an acute lower respiratory infection of primary origin; may be caused by any of a large group of heterogeneous viruses or by the bacterium *Mycoplasma pneumoniae.*

prions Infectious particles responsible for certain "slow" diseases (such as "mad cow disease") in animals and humans; they are smaller than any known viroid, consisting only of a protein component and no detectable nucleic acid.

procaryotic Having a nucleus not confined within a membrane but lying free in the cell protoplasm; bacteria (including the rickettsiae and chlamydiae) are procaryotic organisms.

prodomal period of infection A brief time (usually 1 to 2 days) prior to the development of acute illness during which early symptoms appear.

proglottid An individual segment of a tapeworm; arise from an area below the scolex (head), developing and maturing in a series. Mature segments that become filled with eggs are said to be gravid.

properdin A serum protein that may participate with complement in destroying microorganisms and does not require the presence or action of antibodies to do so.

prophage Bacteriophage DNA that is integrated into DNA of the host cell and replicates with it; the viral DNA is not expressed (i.e., is not translated into viral proteins) but remains latent, and the host cell is not destroyed.

protease inhibitors A class of drugs active against human immunodeficiency virus that act by inhibiting viral aspartyl protease, an enzyme that is vital to the final stages of viral replication. Examples are saquinavir, indinavir, and ritonavir. The drugs result in synthesis of noninfectious virus particles and are highly effective when given in combination with reverse transcriptase inhibitors.

Protista A kingdom of unicellular eucaryotic organisms; may occur singly or in colonies that lack tissue organization. Members are called protists, and include the algae and protozoa. Also called *Protoctista.*

protoplasm A colloidal mixture of proteins, lipids, and nucleic acids forming the matrix of living cells.

protoplast The structure produced when a gram-positive bacterium is treated with agents (e.g., lysozyme or penicillin) that prevent cell-wall formation.

protozoa True, eucaryotic, unicellular microorganisms having a wide range of cellular activities; may be free-living or parasitic.

pseudomembrane A tough, membranous lesion formed on mucosal tissues of the upper respiratory tract in cases of diphtheria; also a lesion of the mucosal lining of the lower intestinal tract caused by *Clostridium difficile* infection.

pseudomembranous colitis An infection of the lower bowel caused by *Clostridium difficile* involving the formation of plaques, or pseudomembranes, on the lining of the colon.

pseudopod Literally, "false foot." The protrusion of the cytoplasm of an ameba which, when withdrawn and reextruded, gives the cell a gliding motion; provides the means of food ingestion by reaching out to surround and engulf nutrient particles.

psittacosis A respiratory and systemic disease transmitted by psittacine birds (e.g., parrots and parakeets), pigeons, and farmyard fowl; caused by the organism *Chlamydia psittaci.* Also called *ornithosis, parrot fever.*

purine A nitrogen-containing molecule with a structure of joined rings; adenine and guanine, components of nucleic acids, are purines.

pus An exudate composed of leukocytes and tissue debris, usually containing the infecting microorganisms that elicited the response.

pustular acne vulgaris A form of acne in which staphylococci play an aggravating role in creating infected pimples.

pyelonephritis A microbial infection of the kidney.

pyogenic A term descriptive of microbial agents producing purulent exudates (pus).

pyrimethamine An antimicrobial agent used in the treatment of toxoplasmosis and malaria; acts as a folic acid inhibitor.

pyrimidine A nitrogen-containing molecule with a structure of one ring. Three important pyrimidines—thymine, cytosine, and uracil—are components of nucleic acids.

pyrogen A fever-producing substance or microorganism.

Q

Q fever A bacterial interstitial pneumonia caused by *Coxiella burnetii.* Natural reservoirs are domestic animals, ticks, and other arthropod vectors; humans are incidental victims.

quarantine When *complete,* the limitation of freedom of any healthy person or domestic animal exposed to a communicable disease for a time equal to the longest incubation period of the disease; exercised to prevent effective contact with others not so exposed. When *modified,* the selective, partial limitation of freedom of persons or domestic animals; based on differences in susceptibility or danger of disease transmission (e.g., exclusion of children from school, or restriction of military personnel to base or quarters).

quellung reaction A serologic test in which capsular antibodies cause an apparent swelling (quellung) of the capsules of intact bacteria. Soluble capsule antigen precipitates when combined with antibody; the surface tension and density of the capsule increases to make the capsule more visible and give it a swollen appearance.

quinacrine An antimicrobial agent used in the treatment of giardiasis, tapeworm infections, and malaria; mode of action is unknown. Trade name Atabrine.

R

rabies An acute viral disease of the central nervous system, usually fatal, transmitted to humans through the bites of infected animals (dogs, bats, foxes, and other wild animals). Characterized by respiratory paralysis, spasms of the throat muscles, fear of swallowing (hydrophobia), peripheral vascular collapse, and eventual coma before death.

rash A characteristic of the acute phase of some infectious diseases; appearance and distribution on the body is often helpful in diagnosing such infections as measles, scarlet fever, chickenpox, or smallpox.

rat tapeworm *See* **hymenolepiasis.**

reagin A nontreponemal antibody present in the serum of patients with syphilis; can be detected by serologic methods such as the VDRL and RPR tests.

recessive gene An unaltered gene in one set of chromosomes in a diploid cell whose effect in the progeny of that cell is masked by the expression of an altered, matching partner gene in the other chromosome set. *See* also **dominant gene.**

recombinant DNA technology The biochemical techniques used to insert DNA molecules of different origins into a cell and to engineer its genetic output according to the potential of the inserted genes.

reduviid bug The insect vector of American trypanosomiasis.

reservoir of infection A local environment or host that supports the survival and multiplication of pathogenic microorganisms.

resistance transfer factor (RTF) *See* **R factor.**

respiratory syncytial virus A major cause of lower respiratory tract infection in infants and young children; also causes acute upper respiratory disease in older children and adults. May be hospital-acquired (nosocomial).

restriction endonucleases Enzymes that can make cuts in DNA molecules at specific base pair sequences, enabling genes to be spliced into either plasmid or chromosomal DNA; employed in recombinant DNA technology.

reticulate body (RB) A form in the life cycle of chlamydiae that develops from the elementary body. In this stage the organism grows and divides within a vacuole in the infected host cell. Some RBs transform into elementary bodies that when freed, can infect another cell.

retroviruses RNA viruses possessing reverse transcriptase, an enzyme that transcribes the RNA genome into a DNA copy that can be integrated into the host's chromosome and persist in latent form. This is a reverse of the normal flow of genetic information from DNA to RNA (*See* **transcription**). HIV is classified as a member of a subfamily of this group (*See* **lentivirus**).

reverse transcriptase *See* **retroviruses.**

reverse transcriptase inhibitors Classes of drugs that inhibit the activity of the retroviral reverse transcriptase enzyme, especially the human immunodeficiency virus. Two types are currently available, nucleoside analogues (*See* **nucleoside analogues**) and nonnucleoside analogues such as nevirapine and delaviridine.

Reye syndrome An acute encephalitis with liver dysfunction that occurs most often in children treated with aspirin during influenza or chickenpox infection; cause is unknown.

R factor A plasmid carrying one or more genes for resistance to antimicrobial agents; can be transferred from one bacterial cell to another of the same or a different species (or genus), resulting in a rapid spread of drug resistance in bacterial populations.

rheumatic fever A sequel of streptococcal diseases and believed to be a consequence of humoral and cellular responses to streptococcal antigens; characterized by fever, carditis, and migratory inflammation of the joints.

rhinoviruses A large group of viruses associated with common colds; classed as picornaviruses.

Rhizopoda A phylum of protozoa comprised of simple ameboid forms that move by bulging and retracting their protoplasm in any direction.

Rhizopus nigricans A black bread mold; a member of the fungus phylum of Zygomycota.

ribavirin An antimicrobial agent that is active in vitro against many RNA and DNA viruses; interferes with nucleic acid synthesis, inhibiting viral replication. In the U.S., available only for aerosol therapy of infants with severe respiratory syncytial virus infections.

ribonucleic acid *See* **RNA.**

ribosome An organelle that is the site of protein synthesis in all living cells; makeup is approximately two-thirds RNA and one-third protein.

Rickettsia (*pl.* rickettsiae) A genus of very short, rod-shaped bacteria that multiply only in living cells; associated with serious human diseases (e.g., typhus fever, Rocky Mountain spotted fever, and others).

rifampin An antimicrobial agent that interferes with nucleic acid replication or metabolism; used in the treatment of tuberculosis and leprosy.

ringworm A superficial mycotic infection caused by fungal agents. *See* **tinea.**

RNA Ribonucleic acid, a polynucleotide composed of ribonucleotides linked by phosphate bridges; functions primarily to direct protein synthesis in living cells, under the control of DNA.

rose spots The characteristic rash of typhoid fever and sometimes of paratyphoid fever.

rotaviruses A group of RNA viruses shaped like a wheel because of their double-layered capsid (*rota* = wheel); responsible worldwide for acute diarrheal illness in infants and young children as well as in young domestic animals.

roundworm *See* ***Nemathelminthes.***

RPR Rapid Plasma Reagin; a serologic test for syphilis.

rubella One of the frequent but mild viral exanthems of children or young adults; may cause serious and permanent damage in infants born of infected mothers because the virus crosses the placenta and damages embryonic tissue.

rubeola *See* **measles.**

S

Sabin vaccine *See* **OPV.**

Salk vaccine *See* **IPV.**

Salmonella A genus of gram-negative, nonsporulating, fermentative bacilli containing many species widespread in nature; most are pathogenic for humans, animals, and birds.

salmonellosis An infection caused by species of the genus *Salmonella;* usually manifest as a gastroenteritis and caused by ingestion of food contaminated with the organisms. May develop into systemic disease (e.g., typhoid or paratyphoid fever, or *Salmonella* bacteremia).

sanitization The removal of pathogenic microorganisms from inanimate objects by mechanical and chemical cleaning; does not imply sterilization or complete disinfection.

saprophyte An organism that lives on dead or decaying organic matter, obtaining nutrients from the products of such decay.

saprophytic A term describing a heterotroph that lives on dead organic material which it must absorb in soluble form. Many bacteria and fungi are saprophytic because their rigid cell walls are capable of exchanging only soluble materials in their environment. Compare with **holozoic.**

scalded skin syndrome A disseminated toxic disease of children caused by strains of staphylococci that produce an epidermolytic toxin; characterized by large, superficial blisters that peel off in sheets from the underlying epidermal tissue.

scanning electron microscope An electron microscope that employs an electron beam for scanning a specimen coated with a thin metallic film; electrons are scattered to form an image of the specimen surface on a viewing screen.

scarlet fever A systemic disease characterized by a skin rash and fever; results when a streptococcal strain causing sore throat produces streptococcal pyrogenic (erythrogenic) exotoxin to which the patient is not immune.

scolex (*pl.* scolices) The head of a tapeworm possessing muscular suckers for attaching to the intestinal lining of the host. Some species also have a circlet of hooks for attachment.

sedimentation An initial process in community water purification during which water is held in a large basin until large particulate matter settles out.

sepsis A toxic condition resulting from the spread of microorganisms or their products from a focus of infection.

septicemia Sepsis or infection of the blood.

septic meningitis An acute meningitis characterized by an outpouring of purulent exudate over membranes of the brain in response to bacterial infection.

serologic reactions The interactions of antigens and antibodies occurring in vitro.

serology The in vitro study of serum or other body fluids to determine their antibody or antigen content.

serotyping A serologic technique for differentiating strains of microorganisms whose cells or products differ in their antigenic composition.

serum The fluid portion of the blood that remains after clotting.

serum sickness A type of immediate hypersensitivity, or an immune complex disease, that may occur after a large amount of serum is administered for passive immunization.

sexduction The transfer of F prime factor by conjugation; results in the formation of two F prime donor cells.

sexually transmitted disease (STD) Any microbial infection transmitted through sexual activity.

Shigella A genus of gram-negative, nonsporulating, nonmotile, facultatively anaerobic bacilli resembling other enteric bacteria but which are uniformly pathogenic for humans; spread by fecal-oral contact or fecally contaminated objects, water, or food.

shigellosis An acute dysentery caused by bacilli belonging to the genus *Shigella:* notably *S. dysenteriae, S. flexneri, S. boydii,* and *S. sonnei.*

shingles *See* **zoster.**

shunt-associated meningitis Infection of the meninges caused by contamination of indwelling intracranial shunts used to draw off fluid from the brain after surgery or trauma; usually caused by normal flora of the skin (e.g., *Staphylococcus epidermidis* or diphtheroids).

sIgA *See* **IgA.**

simian immunodeficiency virus (SIV) A viral agent similar to HIV found in African green monkeys; possibly derived from a common precursor.

smallpox A severe, acute infectious viral disease that was often fatal but has now been eradicated through effective vaccination programs; acquired through the respiratory tract and characterized by an intense vesicular, pustular rash on all parts of the body. Also called *variola major.*

South American blastomycosis *See* **paracoccidioidomycosis.**

specific immune response The immunologic mechanisms by which the body protects itself from foreign agents (antigens) and acts to destroy them. These mechanisms involve the activity of lymphocytes, some of which (B cells) produce antibodies that react specifically with the soluble antigen that invoked their production. Others (T cells) enter into physical contact with particulate antigens (e.g., foreign cells) to destroy them.

specificity In an immunologic context, the reciprocal structure of antigen and antibody combining sites permitting them to "dovetail," like a key in a lock, when they interreact and bind together.

specific resistance to infection Immunity based on immunologic mechanisms in which lymphocytes respond to specific foreign agents (antigens) and act in various ways to eliminate them.

spheroplast The structure that results when a gram-negative bacterium has been treated with agents, such as lysozyme, that remove the peptidoglycan but not the lipopolysaccharide layer from its cell wall, leaving only a partial cell wall.

spirillum (*pl.* spirilla) A spiral-shaped bacterium having one or more fixed curves in a rigid body.

spirochete A bacterium curved like a spirillum, but much longer, very flexible, and having many tight or loose coils. It may undulate and/or rotate rapidly around its long axis.

spontaneous generation The discounted theory that new life can be created, or can arise, from nonliving matter.

sporadic disease A disease that occurs occasionally and at random intervals in a population.

sporangiospore An asexual spore of a mold, found within a sporangium typically filled with such spores. When the sporangium ruptures, each spore released may germinate to produce new growth of hyphae and mycelium.

sporangium A round, saclike structure appearing at the tip of specialized hyphae; extends aerially from a mold and contains many asexual spores.

spore In fungi, a reproductive cell.

sporotrichosis A subcutaneous mycotic disease caused by *Sporothrix schenkii;* the fungus characteristically enters through the skin of an extremity and produces subcutaneous lesions along the path of lymphatic drainage.

sporozoite A stage in the life cycle of protozoans such as *Plasmodium.* In *Plasmodium* species, sporozoites are produced in the insect host and injected into the next victim of the insect's bite.

sporulation The process of endospore formation in certain bacterial species.

spreading factor *See* **hyaluronidase.**

Staphylococcus A bacterial genus of gram-positive cocci morphologically arranged in grapelike clusters; associated with many human and animal infections or with toxic diseases.

-stasis A noun suffix meaning "halted" or "arrested" (e.g., bacteriostasis). The adjectival suffix is *-static.*

STD Sexually transmitted disease.

steam sterilizer An apparatus for sterilizing objects by steam under pressure; an autoclave.

stem cells Cells in the bone marrow that are precursors of the mature red and white blood cells of the circulating blood. Lymphoid stem cells differentiate into mature lymphocytes.

sterilization In microbiology, the complete destruction of all forms of microbial life using physical or chemical agents. Objects so treated are said to be sterile; the term *sterility* implies "free of any microorganisms."

streptococcal pyoderma *See* **impetigo.**

streptococcal pyrogenic exotoxin A substance produced by some strains of beta-hemolytic streptococci; responsible for the skin rash of scarlet fever. Strains producing it are infected by a specific bacteriophage. Also called *erythrogenic toxin.*

streptococcal sore throat A severe type of pharyngitis generally caused by group A streptococci.

Streptococcus A bacterial genus of gram-positive cocci morphologically arranged in long or short chains and associated with many human and animal infections; the Lancefield serologic method places them in groups A–V.

streptokinase An enzyme produced by virulent streptococci; dissolves clots formed by the body to entrap the organisms.

streptolysin O (SLO) A type of hemolysin produced by group A streptococci. Inactivated by oxygen, and thus is designated "O"; causes beta-hemolysis when blood agar plate cultures of the organism are incubated anaerobically.

streptolysin S (SLS) A type of hemolysin produced by group A streptococci growing aerobically on blood agar plates; can kill leukocytes that have phagocytized the streptococcal cell.

strobila Entire tapeworm; length varies with the type of worm and ranges from three or four proglottids to several hundred.

subacute infection An infection less severe than an acute microbial disease, but having a longer duration; may or may not become chronic.

subclinical A term referring to inapparent infection.

subcutaneous Literally, "under the skin"; may refer to a type of infection or to a method of injection.

substrate A substance on which an enzyme acts specifically.

sudden infant death syndrome (SIDS) The largely unexplained sudden death of human infants up to 3 months of age; sometimes attributable to botulism acquired from spores of *Clostridium botulinum* ingested with honey.

sulfonamides A group of antimicrobial agents that act as antimetabolites; used in the treatment of some urinary tract infections.

sulfones A group of antimicrobial agents that act as antimetabolites; used in the treatment of Hansen disease (leprosy).

suppressor T cells T lymphocytes that suppress the immune response and which may be responsible for preventing autoimmune disease (i.e., immune reactions against "self" antigens).

suppuration The formation or discharge of pus.

surgical asepsis The use of techniques, during and after surgery, designed to exclude all microorganisms.

symbiosis Literally, "living together"; an interdependency between different kinds of organisms that results in mutual support.

syphilis A sexually transmitted disease caused by the spirochetal agent *Treponema pallidum.*

T

tabes dorsalis A symptom of CNS involvement in the tertiary stage of untreated syphilis, manifest by a shuffling walk. Also called *locomotor ataxia.*

taeniasis An intestinal cestode disease caused by the adult forms of the tapeworms *Taenia saginata* (beef tapeworm) or *Taenia solium* (pork tapeworm). The larval form of *T. solium* (bladder worm) has the capacity to produce the invasive disease cysticercosis.

taxonomy The science of the classification of living things according to their presumed structure and function.

T cell A type of lymphocyte derived from stem cells; matures under the influence of the thymus and later migrates to lymph nodes, spleen, bone marrow, and the bloodstream. On contact with certain antigens, T cells become immunologically active; interactions with antigens provide cell-mediated immunity (CMI). Also called *T lymphocyte.*

TD Tetanus-diphtheria vaccine.

teichoic acids Constituents of the cell wall of gram-positive bacteria made up of repeating units of sugar alcohols (glycerol or ribitol); may play a role in regulating cell-wall synthesis in gram-positive bacteria.

temperate bacteriophage A bacteriophage that reproduces in the host cell by either the lytic or the lysogenic cycle.

tetracyclines A group of antimicrobial agents produced by *Streptomyces* or semisynthetically; broad spectrum drugs that act by inhibiting microbial ribosomal protein synthesis. Examples are chlortetracycline and oxytetracyclines.

thallophyte A simple plant having no roots, stems, or leaves (e.g., a fungus).

thallus A type of body characteristic of thallophytes (i.e., lacking roots, stems, and leaves); found in many fungi.

thiabendazole An antimicrobial agent used in the treatment of most roundworm infections; mode of action is unknown.

thrush A mycotic infection of the mucous membranes of the mouth caused by *Candida albicans.*

thymine A pyrimidine found in the nucleotides incorporated into nucleic acids.

tinea A fungal infection (ringworm) of superficial areas of the body (e.g., tinea capitis [scalp], tinea corporis [body], tinea pedis [feet], and tinea unguium [nails]) caused by three genera of fungi: *Epidermophyton, Microsporum,* or *Trichophyton.*

titer The reciprocal of the highest dilution of an antiserum that reacts visibly with antigen in a given serologic test.

toxic shock syndrome (TSS) A systemic toxic disease associated with virulent toxin-producing strains of *Staphylococcus aureus.*

The toxin known as (TSST-1), has been associated with the use of tampons in menstruating women but can also cause the syndrome in men, women, and children of all ages in whom predisposing factors may play a role. The symptoms of TSS may also occur in necrotizing fasciitis caused by group A streptococci, probably through the action of streptococcal pyrogenic exotoxin.

toxigenic Capable of producing a toxin or toxins.

toxigenicity The ability of a microorganism to produce toxic substances that contribute to its virulence.

toxocariasis An invasive nematode disease caused by the roundworms *Toxocara canis* or *Toxocara cati;* infection in young children is caused by the larvae of animal ascarids that normally infect dogs and cats. Also called *visceral larva migrans* (*VLM*).

toxoids Nontoxic but antigenic substances derived from toxins; used as vaccines to provide active immunity against diseases caused by microbial toxins (e.g., diphtheria, tetanus).

toxoplasmosis An invasive parasitic disease caused by the intracellular protozoan *Toxoplasma gondii,* classified in the phylum Apicomplexa. An orally acquired disease often unremarkable and unrecognized in adult, healthy humans. Congenital toxoplasmosis, transmitted to the infant through the placenta from an infected mother, may be severe or fatal. In AIDS patients, cerebral toxoplasmosis has become a leading cause of death.

trachoma A serious communicable disease of the conjunctiva and cornea caused by *Chlamydia trachomatis;* may lead to corneal scarring and blindness.

transcription The first step in the synthesis of cell proteins; an enzyme mediates the production of a single-stranded RNA copy from a complementary template strand of DNA. The copy is known as messenger RNA (mRNA); the genetic message in DNA is transcribed into mRNA. *See* also **translation**.

transduction A process involving the transfer of DNA from one bacterial cell to another by means of a bacterial virus (bacteriophage) that carries the DNA from donor to recipient.

transformation A process in which naked DNA released by a bacterial cell into the surrounding medium is taken up by another bacterial cell. Naked DNA is naturally released after bacteria die; it also may be experimentally isolated in the laboratory.

transition mutations Substitutions at a single site in the DNA where one purine base is changed to the other or one pyrimidine base is changed to the other.

translation The second step in the synthesis of cell proteins in which mRNA molecules attach to ribosomes. The genetic message directs the synthesis of protein components (amino acids and polypeptides); RNA molecules, called transfer RNA (tRNA) translate the message by directing both the sequence of placement of each amino acid at the proper site on the ribosome and the joining of amino acids into polypeptides. During translation, mRNA, tRNA, and ribosomes interact functionally to synthesize each protein molecule. See also **transcription.**

transmission electron microscope A microscope that passes a beam of electrons through a specimen, refocusing the scattered electrons with magnetic lenses to form an image on a fluorescent screen.

transposition The movement of a piece of DNA within a bacterial cell from plasmid to chromosome, chromosome to plasmid, or plasmid to plasmid.

transposons DNA segments that move within a bacterial cell from one plasmid to another, or from plasmid to chromosome, or from chromosome to plasmid. Process is called transposition; DNA segments are called jumping genes.

transversion mutations Changes that occur when a purine base is replaced by either of the pyrimidine bases or when a pyrimidine base is replaced by either of the purine bases.

traveler's diarrhea A gastroenteritis often experienced by travelers to areas where food or water is microbially contaminated; usually caused by enterotoxigenic *Escherichia coli* (ETEC). Also called *turista*.

trematodes A group of nonsegmented platyhelminths known as flukes. Includes two groups: those that are flattened, leaf-shaped, and hermaphroditic (intestinal, liver, and lung flukes); and those that are elongated, round, and bisexual (blood flukes).

Treponema pallidum The spirochetal agent of syphilis.

trichinosis An invasive nematode disease caused by the larval form of the roundworm *Trichinella spiralis;* this roundworm develops both adult and larval forms within the same animal or human host.

Trichomonas A genus of protozoan intestinal flagellates commonly found in the human intestine as well as on the genital mucosa of both men and women. *Trichomonas vaginalis* is frequently a cause of troublesome, sexually transmitted genital irritations called trichomoniasis.

trichomoniasis *See* ***Trichomonas.***

Trichophyton A fungal genus that causes superficial infections of the hair, skin, or nails.

trichuriasis An intestinal nematode disease caused by the adult form of the roundworm *Trichuris trichiura* (whipworm).

trimethoprim An antimicrobial agent that acts as an antimetabolite; used with other drugs in the treatment of leprosy, and in combination with sulfamethoxazole to treat many bacterial infections.

trophozoite An asexual stage of an actively metabolizing and motile protozoan (e.g., amebae, malarial parasites, *Toxoplasma*).

Trypanosoma A genus of protozoan hemoflagellates carried by insects and transmitted through their bites to humans or other animals; diseases caused by this organism are known collectively as trypanosomiasis.

trypanosomiasis An infection caused by trypanosomes injected by insect vectors into human or animal hosts; organisms live and multiply in the blood or lymph of the infected host. Examples are African sleeping sickness and American trypanosomiasis, Chagas' disease.

tsetse fly The insect vector of African trypanosomiasis (African sleeping sickness).

TSS *See* **toxic shock syndrome.**

TSST-1 *See* **toxic shock syndrome.**

tubercle A type of lesion commonly seen in tuberculosis; tissue cells are organized in a definite pattern designed to wall off the infecting organism.

tuberculoid (neural) Hansen disease A mutilating form of the disease in which lesions form around peripheral nerves; lesions cause sensory damage, anesthesia, and atrophy of tissues, particularly in the extremities.

tuberculosis A chronic infectious disease caused by *Mycobacterium tuberculosis;* characterized by tubercle formation and tissue necrosis as a result of inflammation and host hypersensitivity to the bacterial antigens. Commonly affects the lungs but may occur in other body organs.

tumor immunity The propensity of the cellular immune system, mediated by T cells, to destroy cancer cells.

tumor necrosis factor A cytokine that enhances the inflammatory response, activates inflammatory leukocytes to kill microbes, stimulates the production of cytokines, and induces acute phase reactants.

turista *See* **traveler's diarrhea.**

typhoid fever A severe systemic disease of humans caused by *Salmonella typhi,* ingested in contaminated food or water. The organism localizes first in intestinal lymphoid tissue (Peyer's patches) and is then widely disseminated to other tissues, especially the liver and gallbladder.

U

ultraviolet radiation Wavelengths in the ultraviolet spectrum from 250 to 280 nm used as a method for destroying microorganisms in the air and on inanimate surfaces. These wavelengths are the most destructive for microorganisms because they are absorbed by nucleic acids. UV penetrates poorly and is therefore of limited effectiveness.

unicellular Comprised of one cell.

uracil A pyrimidine found in the nucleotides incorporated into RNA.

V

vaccines Injectable preparations of microorganisms, or their toxic products, that have either been killed or weakened (attenuated) to eliminate their pathogenic properties without inactivating their antigens. Vaccines are used to provide the active immunity that prevents infectious disease.

vancomycin One of the antimicrobial agents that inhibits cell-wall synthesis and is useful in treating bacterial infections.

varicella-zoster virus The viral agent of chickenpox acquired through the respiratory tract. Chickenpox is a mild but highly contagious disease of childhood, characterized by fever, malaise, and an itching vesicular skin eruption.

VDRL A serologic test for the reagin that occurs in the serum of patients with syphilis.

vector An agent or living organism (often an insect) that can transmit infectious organisms from one host to another.

vegetative In microbiology, a term referring to a growing cell, or one having the capacity for growth and multiplication.

venereal disease A term formerly used to refer to a sexually transmitted disease (STD).

Vibrio A genus of small, gram-negative, aerobic, nonsporulating bacilli, curved in the shape of a comma, with a single polar flagellum. Two species are of particular importance in human disease: *V. cholerae,* the agent of cholera; and *V. parahaemolyticus,* another agent of a less severe diarrhea.

viral encephalitis An infection of brain tissue involving cerebral dysfunction as well as meningitis; caused by such agents as mumps, enteroviruses, herpesviruses, and arboviruses.

viral hemorrhagic fever (VHF) An uncommon but serious human disease associated with several viruses, e.g., Ebola virus, Marburg virus, Lassa virus, and Crimean-Congo virus. Recent outbreaks of Ebola VHF in Africa have had mortality rates ranging from 50 to 90%.

viral meningitis *See* **aseptic meningitis.**

viremia The presence of a virus in the bloodstream.

virion A complete virus particle.

viroids The smallest known agents of infectious diseases consisting only of short strands of RNA and lacking a protein coat.

virucide Any physical or chemical agent that kills viruses.

virulence A term describing those properties of infecting microorganisms that enable them to cause disease of an infected host; properties include primarily invasiveness and toxigenicity. *See* also **pathogenicity.**

virulent bacteriophage A bacteriophage that reproduces only by the lytic cycle, causing the death of the bacterial host cell.

virus An infectious, submicroscopic agent without independent metabolic or reproductive functions; capable of living only within viable cells. Simple, acellular structure varies but is basically a single type of nucleic acid (DNA or RNA) wrapped in a protein coat, with or without a lipid envelope.

virustasis The arrest or inhibition of viral growth.

visceral larva migrans (VLM) *See* **toxocariasis.**

W

wandering cells Macrophages of the MNPS that move about with ameboid activity, responding to chemotactic stimuli; include the histiocytes of connective tissue and the monocytes attracted into the tissues from the bloodstream.

whipworm *See* **trichuriasis.**

whooping cough An acute respiratory disease, especially of children, caused by the bacterial species *Bordetella pertussis;* characterized by a paroxysmal cough.

Widal test A serologic method for detecting agglutinating antibodies in the serum of patients who have systemic *Salmonella* infection (e.g., typhoid or paratyphoid fever).

Y

yeast *See* **fungus.**

Yersinia A genus of gram-negative bacilli that contains two notable pathogenic species: *Y. enterocolitica,* the agent of a severe enterocolitis in humans, and *Y. pestis,* the agent of plague.

Z

zidovudine An antimicrobial agent active against retroviruses, including human immunodeficiency virus (HIV); inhibits reverse transcriptase enzyme in retroviruses. Also called *AZT.*

Ziehl-Neelsen stain A differential staining technique that distinguishes between bacteria that can retain a red dye after washing with an acid-alcohol solution (acid-fast), and those that cannot (non-acid-fast). Also called *acid-fast stain.*

Zoomastigina A phylum of protozoa comprised of cells possessing long, whiplike flagella whose lashing provides rapid motility. Also called flagellates.

zoonoses Infectious diseases of animals, many of which can be transmitted to humans.

zoster A recurrent infection with varicella-zoster virus, the same virus that causes chickenpox; virus persists in latent form in nerve ganglia, appearing sporadically in adults as a painful inflammation of posterior nerve roots and ganglia. The path of the affected sensory nerves is marked by crops of vesicular eruptions of the skin of the area. Also called *shingles.*

zygomycosis A type of opportunistic mycotic infection caused by species of *Mucor, Rhizopus, Absidia;* characterized by massive invasion of blood vessels (with formation of emboli), pulmonary lesions, involvement of orbital and nasal sinuses, gastrointestinal ulceration, and central nervous system invasion. Also called *mucormycosis, phycomycosis.*

Zygomycota A phylum of the kingdom of fungi characterized by sexual reproduction in which a zygote is formed from the fusion of a pair of gametes. Reproduce asexually by means of sporangiospores borne within a sporangium; an outstanding example is the common black bread mold *Rhizopus nigricans.*

zygote A cell formed from the union of two sex cells, or gametes, and containing the diploid number of chromosomes.

Index

Page numbers in **boldface** refer to illustrations; those in **boldface** followed by **t** refer to tables. **CP** followed by a **boldface number and letter** refers to a colorplate figure.

A

B

C

D

E

F

G

H

I

J

K

L

M

N

O

P

Q

R

S

T

U

V